ENCYCLOPEDIA OF STATISTICAL SCIENCES

VOLUME 9

**Strata Chart
to Zyskind–Martin Models**

Cumulative Index, Volumes 1–9

ENCYCLOPEDIA OF STATISTICAL SCIENCES

VOLUME 9

STRATA CHART
to ZYSKIND–MARTIN MODELS
CUMULATIVE INDEX, VOLUMES 1–9

A WILEY-INTERSCIENCE PUBLICATION

John Wiley & Sons

NEW YORK · CHICHESTER · BRISBANE · TORONTO · SINGAPORE

Library of Congress Cataloging in Publication Data:

Encyclopedia of statistical sciences.

"A Wiley-Interscience publication."
Includes bibliographies.
Contents: v. 1. A to Circular probable error—
v. 3. Faà di Bruno's formula to Hypothesis testing—
[etc.]—v. 9. Strata chart to
Zyskind–Martin models.
Cumulative Index, Vols. 1–9.
1. Mathematical statistics—Dictionaries.
2. Statistics—Dictionaries. I. Kotz, Samuel.
II. Johnson, Norman Lloyd. III. Read, Campbell B.

QA276.14.E5 1982   519.5′03′21   81-10353
ISBN 0-471-85474-3 (v. 9)

Printed in the United States of America

10 9 8 7 6 5 4 3

CONTRIBUTORS

J. R. Abernathy, *University of North Carolina, Chapel Hill, North Carolina.* Vital Statistics

S. N. Afriat, *University of Ottawa, Ottawa, Canada.* Test Approach to Index Numbers, Fisher's

H. Ahrens, *Akademie der Wissenschaften der DDR, Berlin, German Democratic Republic.* Unbalancedness of Designs, Measures of

F. B. Alt, *University of Maryland, College Park, Maryland.* Taguchi Method for Off-line Quality Control

T. Amemiya, *Stanford University, Stanford, California.* Two-stage Least Squares

P. K. Andersen, *Statistical Research Unit, Copenhagen, Denmark.* Survival Analysis

O. D. Anderson, *Temple University, Philadelphia, Pennsylvania.* Times Series Analysis and Forecasting (TSA & F) Society

C. E. Antle, *Pennsylvania State University, University Park, Pennsylvania.* Weibull Distribution

S. F. Arnold, *Pennsylvania State University, University Park, Pennsylvania.* Sufficient Statistics; Union–Intersection Principle; Wishart Distribution

A. C. Atkinson, *Imperial College, London, England.* Transformations

L. J. Bain, *University of Missouri, Rolla, Missouri.* Weibull Distribution

K. S. Banerjee, *University of Maryland, Baltimore County, Catonsville, Maryland.* Weighing Designs

R. E. Barlow, *University of California, Berkeley, California.* System Reliability

G. A. Barnard, *Brightlingsea, Essex, England.* Two-by-two (2 × 2) Tables

V. Barnett, *University of Sheffield, Sheffield, England.* Teaching Statistics

N. R. Bartlett, *Uncle Ben's of Australia, Wodonga, Victoria, Australia.* Testing, Destructive

W. Bell, *U.S. Bureau of the Census, Washington, D.C.* Time Series, Nonstationary

R. J. Beran, *University of California, Berkeley, California.* Weak Convergence, Statistical Applications

S. Berg, *University of Lund, Lund, Sweden.* Urn Models; Voting Paradox

B. L. Bergman, *Linköping University, Linköping, Sweden.* Total Time on Test Transform

H. Bernhardt, *Humboldt Universität, Berlin, German Democratic Republic.* Von Mises, Richard Martin Elder

J. B. Birch, *Virginia Polytechnic Institute and State University, Blacksburg, Virginia.* Untilting

R. Bohrer, *University of Illinois, Urbana, Illinois.* Studentized Maximal Distributions, Centered

L. Bondesson, *The Swedish University of Agricultural Sciences, Umeå, Sweden.* T_1- and T_2-Classes of Distributions

R. A. Bradley, *University of Georgia, Athens, Georgia.* Trend-free Block Designs; Wilcoxon, Frank

K. R. W. Brewer, *Bureau of Agricultural Eco-*

nomics, *Lyneham, ACT, Australia.* Stratified Designs; *Unequal Probability Sampling*

B. M. Brown, *University of Tasmania, Hobart, Tasmania, Australia.* Tukey's Median Polish

K. G. Brown, *Chapel Hill, North Carolina.* Sub-balanced Data; Williams' Test (of Trend)

E. Carlstein, *University of North Carolina, Chapel Hill, North Carolina.* Typical Values

P. L. Chapman, *Colorado State University, Fort Collins, Colorado.* Whittemore's Collapsibility

W. P. Cleveland, *Federal Reserve Board, Washington, D.C.* X-11 Method

D. R. Cox, *Imperial College, London, England.* Transformations

N. A. C. Cressie, *Iowa State University, Ames, Iowa.* Variogram

D. E. Critchlow, *Purdue University, West Lafayette, Indiana.* Ulam's Metric

J. Cuzick, *Imperial Cancer Research Fund, London, England.* Trend Tests

C. Dagum, *University of Ottawa, Ottawa, Canada.* Trend

E. B. Dagum, *Statistics Canada, Ottawa, Canada.* Trend

H. A. David, *Iowa State University, Ames, Iowa.* Studentized Range

J. A. Deddens, *University of Cincinnati, Cincinnati, Ohio.* Survival Analysis, Grouped Data

M. H. DeGroot, *Carnegie-Mellon University, Pittsburgh, Pennsylvania.* Well-calibrated Forecasts

M. S. Dueker, *University of Connecticut, Stamford, Connecticut.* Supersaturated Designs

R. G. Easterling, *Sandia National Laboratories, Albuquerque, New Mexico.* Technometrics

A. R. Eckler, *Morristown, New Jersey.* Target Coverage

A. S. C. Ehrenberg, *London Business School, London, England.* Television Viewing, Statistical Aspects

M. Engelhardt, *University of Missouri, Rolla, Missouri.* Weibull Process

P. C. Fishburn, *AT&T Bell Laboratories,* Murray Hill, New Jersey. Utility Theory

J. L. Fleiss, *Columbia University, New York, New York.* Stuart–Maxwell Test

D. A. S. Fraser, *York University, North York, Ontario, Canada.* Structural Inference; Structural Models; Structural Prediction; Structural Probability

G. H. Freeman, *University of Warwick, Coventry, England.* Systematic Designs

M. Frisén, *University of Göteborg, Göteborg, Sweden.* Unimodal Regression

K. R. Gabriel, *University of Rochester, Rochester, New York.* Weather Modification I

J. Galambos, *Temple University, Philadelphia, Pennsylvania.* Truncation Methods in Probability Theory; Two-series Theorem

D. J. Gans, *Bristol-Myers Products, Hillside, New Jersey.* Trimmed and Winsorized Means, Tests for; Tukey's Quick Test

S. I. Gass, *University of Maryland, College Park, Maryland.* Traveling Salesman Problem

J. D. Gibbons, *University of Alabama, University, Alabama.* Truncation, Coefficient of; Wilcoxon-type Scale Tests

D. B. Gillings, *Quintiles, Inc., Chapel Hill, North Carolina.* Tests, One-sided versus Two-sided

V. P. Godambe, *University of Waterloo, Waterloo, Ontario, Canada.* Uninformativeness of a Likelihood Function

I. J. Good, *Virginia Polytechnic Institute and State University, Blacksburg, Virginia.* Surprise Index

W. S. Griffith, *University of Kentucky, Lexington, Kentucky.* Wear Processes

R. M. Groves, *University of Michigan, Ann Arbor, Michigan.* Telephone Surveys, Computer Assisted; Variance Interviewer

I. Guttman, *University of Toronto, Toronto, Canada.* Tolerance Regions, Statistical

P. Hall, *Australian National University, Canberra, ACT, Australia.* Vacancy

W. J. Hall, Jr., *University of Rochester, Rochester, New York.* Unlikelihood

E. J. Hannan, *Australian National University, Canberra, ACT, Australia.* Wiener, Norbert

B. Harris, *University of Wisconsin, Madison, Wisconsin.* Tetrachoric Correlation Coefficient

R. E. Hausman, *AT & T, Basking Ridge, New Jersey.* Stratified Multistage Sampling

N. W. Henry, *Virginia Commonwealth University, Richmond, Virginia.* Traceline

T. P. Hettmansperger, *Pennsylvania State University, University Park, Pennsylvania.* Walsh Averages; Weighted Least Squares Rank Estimator

B. M. Hill, *University of Michigan, Ann Arbor, Michigan.* Tail Probabilities

W. G. S. Hines, *University of Guelph, Guelph, Ontario, Canada.* T-Square Sampling; Wandering-Quarter Sampling

M. Hollander, *Florida State University, Tallahassee, Florida.* Testing for Symmetry; Wilcoxon, Frank; Wilcoxon-type Tests for Ordered Alternatives in Randomized Blocks

H. Holling, *Osnabrück, Federal Republic of Germany.* Suppressor Variables

J. R. M. Hosking, *IBM, Thomas J. Watson Research Center, Yorktown Heights, New York.* Van Montfort–Otten Test

H. K. Hsieh, *University of Massachusetts, Amherst, Massachusetts.* Taha Test; Thoman–Bain Test

R. Hultquist, *Pennsylvania State University, University Park, Pennsylvania.* Variance Components

H.-L. H. Hwang, *Bell Communications Research, Red Bank, New Jersey.* t Designs

J. P. Imhof, *Université de Génève, Geneva, Switzerland.* Wiener Measure

A. K. Jain, *Bell Communications, Holmdel, New Jersey.* Stratified Multistage Sampling

N. T. Jazairi, *York University, North York, Ontario, Canada.* Terms of Trade; Test Approach to Index Numbers, Fisher's

J. N. R. Jeffers, *Institute of Terrestrial Ecology, Grange-Over-Sands, Cumbria, England.* Systems Analysis in Ecology

M. E. Johnson, *Los Alamos National Laboratories, Los Alamos, New Mexico.* Tests Based on Empirical Probability Measures

I. Johnstone, *Stanford University, Stanford, California.* Wald's Decision Theory

K. Kafadar, *Hewlett Packard Corporation, Palo Alto, California.* Twicing

C. J. Kahane, *National Highway Traffic Safety Administration, Washington, D.C.* Vehicle Safety, Statistics in

G. Kalton, *University of Michigan, Ann Arbor, Michigan.* Survey Sampling; Systematic Sampling

T. Kariya, *Hitotsubashi University, Kunitachi, Tokyo, Japan.* Zellner Estimator

N. Keiding, *Statistical Research Unit, Copenhagen, Denmark.* Truncation, Nonparametric Estimation Under

J. S. Kim, *GTE Laboratories, Waltham, Massachusetts.* Total Positivity

B. Klefsjö, *University of Lulea, Lulea, Sweden.* Total Time on Test Transform

M. Knott, *London School of Economics, London, England.* Von Mises Expansions

G. G. Koch, *University of North Carolina, Chapel Hill, North Carolina.* Survival Analysis, Grouped Data; Tests, One-sided versus Two-sided

H. S. Konijn, *Tel Aviv University, Tel Aviv, Israel.* Stratifiers, Selection of; Symmetry Tests, Pure and Combined

K. Krickenberg, *Université René Descartes, Paris, France.* Zeitschrift für Wahrscheinlichkeitstheorie und Verwandte Gebiete

P. M. Kroonenberg, *University of Leiden, Leiden, The Netherlands.* Three-mode Analysis

N. Laird, *Harvard School of Public Health, Boston, Massachusetts.* Sundberg Formulas

J. L. G. Lanke, *University of Lund, Lund, Sweden.* Unicluster Designs

Y. J. Lee, *National Institutes of Health, Bethesda, Maryland.* Trend in Count Data, Test for

E. L. Lehmann, *University of California, Berkeley, California.* Unbiasedness

G. Leunbach, *Danish Institute for Educational Research, Copenhagen, Denmark.* Sufficient Estimation and Parameter-free Inference

J. Q. Longyear, *Wayne State University, Detroit, Michigan.* Strength and Other Properties of an Array

T. Lwin, *CSIRO, Melbourne, Victoria,*

Australia. Testing, Destructive

D. B. MacKay, *Indiana University, Bloomington, Indiana.* Thurstone's Theory of Comparative Judgment

B. F. J. Manly, *University of Otago, Dunedin, New Zealand.* Van Valen's Test

C. R. Mann, *Mann Associates, Inc., Washington, D.C.* Utilization Analysis

B. H. Margolin, *University of North Carolina, Chapel Hill, North Carolina.* Trend in Proportions, Test for

J. S. Marron, *University of North Carolina, Chapel Hill, North Carolina.* Window Width

P. McCullagh, *Imperial College, London, England.* Tensors

P. W. Mielke, Jr., *Colorado State University, Fort Collins, Colorado.* Valand's (Mantel and Valand's) Nonparametric MANOVA Technique; Whittemore's Collapsibility

R. Mojena, *University of Rhode Island, Kingston, Rhode Island.* Ward's Clustering Algorithm

R. Morton, *CSIRO, Canberra ACT, Australia.* Ultrastructural Relationships

R. J. Muirhead, *University of Michigan, Ann Arbor, Michigan.* Zonal Polynomials

M. W. Muller, *National Institute for Personnel Research, South Africa.* Unfolding

H. Neudecker, *University of Amsterdam, Amsterdam, The Netherlands.* Varimax Method

H. Niederhausen, *Florida Atlantic University, Boca Raton, Florida.* Takacs' Goodness-of-fit Distribution

P. L. Odell, *Richardson, Texas.* Variance, Upper Bounds

J. K. Ord, *Pennsylvania State University, University Park, Pennsylvania.* Time Series

G. P. Patil, *Pennsylvania State University, University Park, Pennsylvania.* Sum-symmetric Power Series Distributions; Weighted Distributions

S. C. Pearce, *University of Kent, Canterbury, Kent, England.* Strip Plots; Total Balance

W. Philipp, *University of Illinois, Urbana, Illinois.* Uniform Distributions Modulo 1

F. C. Powell, *University of Cambridge, Cambridge, England.* Westenberg Test of Dispersion

P. Prescott, *The University of Southampton, England.* Student's *t*-Tests

F. Proschan, *Florida State University, Tallahassee, Florida.* Total Positivity

F. L. Ramsey, *Oregon State University, Corvallis, Oregon.* Transect Methods

R. H. Randles, *Florida State University, Gainesville, Florida.* Theil Test and Estimator of Slope; Triangle Tests; Tukey's Confidence Interval for Location; Weighted Symmetry; Wilcoxon Signed Rank Test

B. L. S. P. Rao, *Indian Statistical Institute, New Delhi, India.* Sudakov's Lemma

C. R. Rao, *University of Pittsburgh, Pittsburgh, Pennsylvania.* Weighted Distributions

M. V. Ratnaparkhi, *Wright State University, Dayton, Ohio.* Sum-symmetric Power Series Distributions

C. B. Read, *Southern Methodist University, Dallas, Texas.* Studentization; Studentized Extreme Deviates; Successive Differences; *t*-Distribution, Tukey's Test for Nonadditivity; Uniform (or Rectangular) Distributions; Wald's Equation; Weighted Least Squares; Zipf's Law

E. Regazzini, *Università di Milano, Milan Italy.* Subjective Probabilities

B. Reiser, *University of Haifa, Haifa, Israel.* Stress-strength Models

H. T. Reynolds, *University of Delaware, Newark, Delaware.* Test Factor Stratification

H. Riedwyl, *University of Berne, Berne, Switzerland.* *V*-Statistics

A. J. Roth, *Ciba-Geigy Pharmaceuticals, Summit, New Jersey.* Welch Tests

D. Ruppert, *University of North Carolina, Chapel Hill, North Carolina.* Trimming and Winsorization

L. M. Ryan, *Harvard School of Public Health, Boston, Massachusetts.* Weighted Normal Plots

J. Sanchez, *Instituto de Matemática, Cibernética y Computación, Havana, Cuba.* Unbalancedness of Designs, Measures of

J. G. Saw, *University of Florida, Gainesville, Florida.* Ultraspherical Polynomials

H. T. Schreuder, *U.S. Forest Service, Fort Collins, Colorado.* Tin's Ratio Estimators

J. W. Seaman, Jr., *University of Southwestern Louisiana, Lafayette, Louisiana.* Variance, Upper Bounds

A. R. Sen, *University of Calgary, Calgary, Alberta, Canada.* Wildlife Sampling

P. K. Sen, *University of North Carolina, Chapel Hill, North Carolina.* Subhypothesis Testing; Time-sequential Inference; Weighted Empirical Processes:-Genesis and Applications

E. Seneta, *University of Sydney, Sydney, NSW, Australia.* Vestnik Statistiki; Yanson, Yulii Eduardovich

R. J. Serfling, *Johns Hopkins University, Baltimore, Maryland.* U-Statistics

M. Shaked, *University of Arizona, Tucson, Arizona.* Variance Dilation

R. Shealy, *University of Illinois, Urbana, Illinois.* Unidimensionality, Tests of

G. Shorack, *University of Washington, Seattle, Washington.* Variance Comparisons Based in Permutation Theory

A. F. Siegel, *University of Washington, Seattle, Washington.* Zero Degrees of Freedom

M. P. Singh, *Statistics, Canada, Ottawa, Canada.* Survey Methodology

R. D. Snee, *E. I. du Pont de Nemours & Co., Inc., Wilmington, Delaware.* Window Plot

H. Solomon, *Stanford University, Stanford, California.* Variance, Sample

J. D. Spurrier, *University of South Carolina, Columbia, South Carolina.* Working–Hotelling–Scheffé Confidence Bands

R. G. Stanton, *University of Manitoba, Winnipeg, Manitoba, Canada.* Witt Designs

M. A. Stephens, *Simon Fraser University, Burnaby, British Columbia, Canada.* Uniformity, Tests of; Variance, Sample; Watson's U^2

W. Stout, *University of Illinois, Urbana, Illinois.* Unidimensionality, Tests of

H. Strecker, *Universität Tübingen, Tübingen, Federal Republic of Germany.* Variate Difference Method

P. H. Swain, *Purdue University, West Lafay-*ette, Indiana. Swain–Fu Distance

L. Takács, *Case Western Reserve University, Cleveland, Ohio.* Takács Process

J. C. Tanner, *Road Research Laboratory, Crowthorne, Berkshire, England.* Traffic Flow Problems

M. S. Taqqu, *Cornell University, Ithaca, New York.* Weak Stationarity

M. E. Terry, *AT & T Bell Laboratories, Murray Hill, New Jersey.* Terry–Hoeffding Test

M. E. Thompson, *University of Waterloo, Waterloo, Ontario, Canada.* Superpopulation Models

G. Tintner*, *Technische Universität Wien, Vienna, Austria.* Variate Difference Method

D. M. Titterington, *University of Glasgow, Glasgow, Scotland.* Subsurvival Functions

R. L. Trader, *University of Maryland, College Park, Maryland.* Super-Bayesian

M. Vaeth, *Aarhus University, Aarhus, Denmark.* Survival Analysis

K. Wakimoto, *Okayama University, Tsushima, Okayama, Japan.* Sun Charts

T. Wansbeek, *University of Groningen, Groningen, The Netherlands.* Vec Operator

J. H. Ware, *Harvard School of Public Health, Boston, Massachusetts.* Tracking

G. S. Watson, *Princeton University, Princeton, New Jersey.* Wheeler and Watson's Test

G. H. Weiss, *National Institutes of Health, Bethesda, Maryland.* Tauberian Theorems; Wald's Identity—Applications

L. Weiss, *Cornell University, Ithaca, New York.* Wald, Abraham; Weiss Test of Independence (of Variables); Weiss-type Estimators of Shape Parameters; Wolfowitz, Jacob

M. Westberg, *University of Göteborg, Göteborg, Sweden.* Test, Westberg Adaptive Combination of

---

*Deceased

H. White, *University of California, La Jolla, California.* White Tests of Misspecification

P. Whittle, *University of Cambridge, Cambridge, England.* Whittle Equivalence Theorem

R. A. Wijsman, *University of Illinois, Urbana, Illinois.* Wijsman's Representation Theorem

R. R. Wilcox, *University of Southern California, Los Angeles, California.* Strong True-score Theory

J. C. Wilkin, *Social Security Administration, Baltimore, Maryland.* Whittaker–Henderson Graduation Formulas

E. Willekens, *Katholieke Universiteit, Leuven,* *Belgium.* Subexponential Distributions

O. D. Williams, *University of North Carolina, Chapel Hill, North Carolina.* Target Population

A. Wörgötter, *Technische Universität Wien, Vienna, Austria.* Variate Difference Method

G. Wörgötter, *Technische Universität Wien, Vienna, Austria.* Variate Difference Method

C.-M. Yeh, *State University of New York, Buffalo, New York.* Trend-free Block Designs

M. Zelen, *Harvard School of Public Health, Boston, Massachusetts.* Weighted Distributions

ACKNOWLEDGMENT

The Editors would like to take this opportunity to acknowledge the impact and contributions of many institutions and individuals. For better or worse, they have affected the process of implementation of this ambitious decade-long project (much more ambitious than we realized originally).

Our first and foremost thanks are due to the five members of the Advisory Board and to some 300 contributors from all over the world whose names appear in the front matter of the individual volumes. These devoted individuals, who agreed to write on the subject matter of their statistical expertise and to boil down the knowledge of decades into a small compass, represent the backbone of this project. They reflect the vigorous idealism and dedication prevalent throughout the international statistical community.

Perceptive readers may notice that a few (fortunately, very few) names of internationally renowned statisticians are absent from the lists of contributors. In many cases, their expertise was indeed solicited, but they chose, for no doubt valid reasons, not to accept our invitations.

On the other hand, it is with deep gratitude, mixed with sadness, that the Editors acknowledge the interest in the early stages of this project rendered by the late Jerzy Neyman, who energetically and enthusiastically communicated to us his suggestions about the structure of the ESS and the choice of particular entries and appropriate authors.

After his much to be lamented demise, E. L. Lehmann continued as of 1983 to serve, on his own initiative, as an unofficial advisor. His erudition, good judgment, and experience substantially improved the quality of the last four volumes of the series. Special thanks are also due to Jean D. Gibbons and P. K. Sen for their interest and advice far beyond the call of duty. We are indebted to Dean R. P. Lamone, College of Business and Management, University of Maryland, B. L. Golden, Chairperson of the Department of Management Science and Statistics at the University, and to the Chairpersons of the Departments of Statistics at Pennsylvania State University, the University of North Carolina, and Southern Methodist University—W. L. Harkness; G. Simon, W. L. Smith, S. Cambanis; D. B. Owen and W. Schucany—for providing the facilities and environment that facilitated the implementation of numerous tasks and chores associated with the ESS.

As to the "home front," our special thanks are due to C. B. Read, who joined us in 1980 as an associate editor and soon proved to be an indispensable full partner. Dr. Read became executive editor in 1986. He has shown remarkable devotion to the cause of the statistical sciences by expertly, patiently, and carefully steering the project to its conclusion.

We are particularly grateful to June Maxwell, who cheerfully and efficiently handled the daily tasks of administering this complicated project, through thick and thin, from its inception. Librarians at various universities, government agencies, and research institutions throughout the world (too numerous to mention individually) also provided invaluable assistance.

The Editors, of course, bear the responsibility for the inevitable deficiencies in coordination, misstatements, and omissions of some important topics and possibly unbalanced emphasis on others. It should however be noted that nonscientific but practical considerations have restricted our ability to carry out certain essential, last minute alterations. We have also been constrained in the scope of a projected supplementary volume, which hopefully will be available to our readers in the very near future.

While working on this project for over a decade, we have had an opportunity to read and reflect on the history, the unprecedented growth, and current status of the statistical sciences. Although our strong belief in the overall unity of these sciences and the pervasiveness of statistical methodology in a variety of diverse fields remains unshaken, we could not help noticing the process of natural fragmentation of these sciences into well-defined separate, though interconnected, disciplines such as data collection, data presentation, data summarizing, data analysis, design of experiments, prediction and decision making, model building, and model verification, etc. It would perhaps be appropriate for editors of statistical journals and authors of monographs and books to be guided by this newly emerging configuration in their decisions and activities which shape the future direction of statistical sciences.

The readers who are interested in more details on the making of this encyclopedia are referred to our recent paper describing the development of the idea and execution of the actual work in the *Journal of Official Statistics* **3**(1), 93–99 (1987).

The Editors are grateful for the positive response to the earlier volumes and hope that this final volume will also meet with a favorable reception.

SAMUEL KOTZ
NORMAN L. JOHNSON

College Park, Maryland
Chapel Hill, North Carolina
January 1988

ENCYCLOPEDIA OF STATISTICAL SCIENCES

VOLUME 9

**Strata Chart
to Zyskind–Martin Models**

Cumulative Index, Volumes 1–9

S continued

STRATA CHART

This is also known as a *band chart* or a *zone chart*. It can be regarded as a compound (or component) set of bar charts*. In standard multiple bar charts, each bar is separate. In strata charts the *width* of each stratum represents the size of the relevant item. For example, Fig. 1 (from ref. 1) represents exports of coffee, sugar, cocoa, etc., from Brazil during the years 1955–1965. In such charts, the shading (or coloring) of the strata so as to distinguish one from another is of importance.

Reference

[1] Edwards, B. (1972). *Statistics for Business Students*. Collins, London, p. 254.

(GRAPHICAL REPRESENTATION OF DATA
  PIE CHART)

STRATEGY *See* DECISION THEORY; GAME THEORY

STRATIFICATION *See* OPTIMUM STRATIFICATION; STRATIFIED DESIGNS; STRATIFIED MULTISTAGE SAMPLING; STRATIFIERS (SELECTION OF)

STRATIFIED DESIGNS

INTRODUCTION

Stratification is a technique widely employed in *finite population sampling**. It is used when the population units can easily be divided into groups such that the members of each group have some property or set of properties in common, relevant to the investigation at hand, which is not shared by the members of the other groups. If the sample selection within each group is carried out independently of the selection in every other group, the groups are called *strata*; any sample design that involves dividing the population into strata is a *stratified design*.

The usual purpose of stratification is to increase the efficiency of the sample design —either by decreasing the *mean square errors** of the sample estimates while keeping the *sample size** or the cost of the survey constant, or alternatively by decreasing the sample size or the survey cost for the same mean square errors. Sometimes, however, the strata themselves are of primary interest.

The following are typical examples of stratified populations:

1. A population of school students stratified by school class and by sex. A typical stratum would be "class 5H, boys."

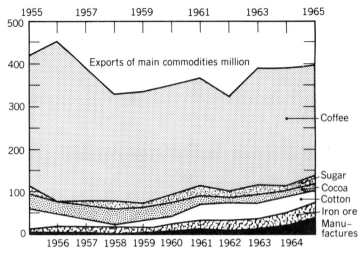

Figure 1 Brazilian exports. Source: *Barclays Bank Review*.

2. A population of individual taxpayers stratified by state, sex, and range of reported income. The income ranges are arbitrary. One possibility is (a) up to $4,999; (b) $5,000–$9,999; (c) $10,000–$19,999; (d) $20,000–$49,999; (e) $50,000 and over. In that case, a typical stratum would be "South Australia, females, $10,000–$19,999."

3. A population of households in a country, treating each region as a separate stratum. (Note that it would not be necessary to have a list of every population unit in order to divide the populations into such strata or, indeed, to carry out selection. *See* AREA SAMPLING*.) The "regions" may have boundaries defined by some outside body or may be chosen as a matter of convenience.

4. A population of retail establishments stratified by state, by description (grocer, butcher, etc.), and by range of annual sales. Here again there is some flexibility in the definitions of the "descriptions," while the ranges of annual sales chosen are entirely arbitrary.

Some stratification criteria are sharply defined (school class, sex, state), some admit of a degree of subjective judgment (region, description of retail establishment), while a third group (age, range of reported income, range of annual sales), being quantitative in

character, leads to arbitrary stratum boundaries. Some choices of boundaries are better than others from the point of view of achieving an efficient sample design. The boundaries corresponding to the most efficient sample design for a particular item of interest provide *optimum stratification**.

THEORETICAL ADVANTAGES OF STRATIFICATION

In this section, based on Evans [3], the effects of using a stratified in place of an unstratified design are presented, first in general and then for a simple example. Because the effects of stratification can be derived much more simply for sampling with replacement than for sampling without replacement, we consider first the case where the population total is to be estimated from a simple random sample selected with replacement.

Capital letters will be used to denote population values and lower-case letters to denote sample values. Let N be the number of units in the population, n be the number of units in a sample, Y_i be the value for the ith population unit of an item whose total is being estimated, the population total itself being denoted by $Y = \sum_{i=1}^{N} Y_i$, and the population mean by $\bar{Y} = N^{-1}Y$. Then the simplest estimator of Y is the *expansion* or *number-raising* estimator, defined by

$$y_u' = n^{-1}N \sum_{i=1}^{n} y_i = N\bar{y}.$$

This estimator is unbiased and its variance is

$$\sigma_{y_u'}^2 = n^{-1}N^2\sigma^2,$$

where $\sigma^2 = N^{-1}\sum_{i=1}^{N}(Y_i - \bar{Y})^2$ is the population variance.

Now suppose that the population is divided into strata and samples of size n_h ($h = 1, 2, \ldots L$), are selected independently from the units in each stratum with equal probabilities and with replacement. The stratified form of the number-raising estimator of Y is

$$y_s' = \sum_{h=1}^{L} y_h' = \sum_{h=1}^{L} n_h^{-1}N_h \sum_{i=1}^{n_h} y_{hi}.$$

This also is unbiased and its variance is

$$\sigma_{y_s'}^2 = \sum_{h=1}^{L} \sigma_{y_h'}^2 = \sum_{h=1}^{L} n_h^{-1}N_h^2\sigma_h^2,$$

where

$$\sigma_h^2 = N_h^{-1} \sum_{i=1}^{N_h} \left(Y_{hi} - \bar{Y}_h\right)^2.$$

Three special cases are of particular interest: The first is *proportional sampling**, where $n_h \propto N_h$; the second is *Neyman allocation**, where the n_h are chosen to minimize the variance of y_s' given a fixed sample size n; the third is *optimum allocation*, a variant of Neyman allocation used where some strata are more expensive to sample from than others and the n_h are chosen to minimize the variance of y_s' given a fixed survey cost.

Consider first the case of proportional sampling. The difference between the variances of y_u' and y_s' can be shown to be

$$\sigma_{y_u'}^2 - \sigma_{y_s'}^2 = n^{-1}N^2 \sum_{h=1}^{L} P_h(\bar{Y}_h - \bar{Y})^2,$$

where $P_h = N_h/N$ is the proportion of population units in the hth stratum. This difference is nonnegative, being proportional to the weighted variance of the $\bar{Y}_h$ with weights P_h. Consequently, given this type of sampling and estimation, it is impossible to lose efficiency as a result of stratification, and the greatest efficiency is achieved when the

stratum means are as different from each other as possible.

In the case of *Neyman allocation*, the method of undetermined multipliers can be used to show that the n_h must be proportional to $N_h\sigma_h$ and the further reduction in variance as compared with proportional sampling is

$$n^{-1}N^2 \left[\sum_{h=1}^{L} P_h\sigma_h^2 - \left(\sum_{h=1}^{L} P_h\sigma_h\right)^2 \right].$$

The expression in square brackets is the weighted variance of the σ_h with weights P_h. Thus this form of optimum allocation has two reductions in variance compared with simple random sampling, one term proportional to the weighted variance of the stratum means and the other proportional to the weighted variance of the stratum standard deviations. The weights in each case are the proportions of population units in the strata.

In the more general case of optimum allocation, where the cost of sampling a unit in the hth stratum is proportional to C_h, the method of undetermined multipliers can be used to show that the n_h should be chosen to be proportional to $N_h\sigma_h C_h^{-1/2}$. The further reduction in variance as compared with proportional sampling is then

$$n^{-1}N^2 \left(\sum_{h=1}^{L} P_h\sigma_h^2 - \sum_{h=1}^{L} P_h\sigma_h C_h^{1/2} \right.$$
$$\left. \times \sum_{h=1}^{L} P_h\sigma_h C_h^{-1/2} \right).$$

The expression in parentheses is now the weighted covariance between the $\sigma_h C_h^{1/2}$ and the $\sigma_h C_h^{-1/2}$, again with weights P_h. Thus this more general form of optimum allocation also has two reductions in variance compared with simple random sampling*, one term being proportional to the same weighted variance of the stratum means, but the other now being proportional to a similarly weighted covariance, between the stratum standard deviation multiplied by the square root of the cost of sampling and the same stratum standard deviation divided by the square root of the cost of sampling. Since this comparison retains the same value of n throughout, the reduction in variance will be greatest when the costs of sampling are equal from stratum to stratum and optimum allocation coincides with Neyman allocation.

These results will now be applied to the population of 64 cities given by Cochran [2, pp. 93–94]. The 16 largest cities are in stratum 1 and the remaining 48 in stratum 2. The observed variable Y_i is the 1930 population of the ith city, in thousands.

The population total is $Y = 19,568$ and the population variance is $\sigma^2 = 51,629$. If a simple random sample of 24 is selected with replacement (in Cochran's book the sample is selected without replacement) the variance of the number-raising estimator is $\sigma^2_{y_u'} = 8811,344$.

For the individual strata, the population variances are

$$\sigma_1^2 = 50,477.374, \quad \sigma_2^2 = 5464.651.$$

In proportional allocation we have $n_1 = 6$ and $n_2 = 18$. The variance of the stratified number-raising estimator is

$$\sigma^2_{y_s'} = 2153,717 + 699,475$$
$$= 2853,192.$$

The difference between $\sigma^2_{y_u'}$ and $\sigma^2_{y_s'}$ is 5958,152. This figure may also be obtained as follows. The population means in stratum 1, stratum 2 and the population as a whole are

$$\overline{Y}_1 = 629.375, \qquad \overline{Y}_2 = 197.875$$

and

$$\overline{Y} = 305.75.$$

Then

$$24^{-1} \times 64^2$$
$$\times \left[64^{-1} \times 16 \times (629.375 - 305.75)^2 \right.$$
$$\left. + 64^{-1} \times 48 \times (197.875 - 305.75)^2 \right]$$
$$= 170.66667 \times (0.25 \times 104,733.1406$$
$$+ 0.75 \times 11,637.0156)$$
$$= 5958,152.$$

For Neyman allocation the sample is allocated proportionally to $N_h \sigma_h$. Now

$$N_1 \sigma_1 = 16 \times 50,477.734^{0.5}$$
$$= 16 \times 244.672505 = 3594.76$$

and

$$N_2 \sigma_2 = 48 \times 5464.651^{0.5}$$
$$= 48 \times 73.923279 = 3548.32$$

These are nearly equal, so the Neyman allocation is $n_1 = n_2 = 12$. The variance of the stratified number-raising estimator is then

$$\sigma^2_{y_s'} = 1076,858 + 1049,213$$
$$= 2126,071.$$

The reduction in the value of $\sigma^2_{y_s'}$ in moving from proportional to Neyman allocation is $2853,192 - 2126,071 = 727,121$. Because the values of n_1 and n_2 were rounded to the nearest unit, this is not exactly the same as is given by the preceding expression, i.e.,

$$24^{-1} \times 64^2 \left[0.25 \times 50,477.734 \right.$$
$$+ 0.75 \times 5464.651$$
$$- (0.25 \times 224.672505$$
$$\left. + 0.75 \times 73.923279)^2 \right]$$
$$= 170.66667(12,619.434 + 4098.488$$
$$- 111.610585^2)$$
$$= 170.66667(16,717.922 - 12,456.923)$$
$$= 170.66667 \times 4260.999$$
$$= 727,211.$$

The latter is, however, close to 727,121, which is the actual reduction.

The latter is also the maximum extent of the reduction in variance as long as the sample size remains constant at 24. If the cost of sampling varies from stratum to stratum, optimum sampling will differ from Neyman allocation sampling, but if the sample is kept at 24 units, the departure from Neyman allocation must increase the variance at the same time that it reduces the cost. Suppose for instance that it cost four times as much to sample one unit from the large-city stratum 1 as from stratum 2 ($C_1 = 4C_2$). Then the optimum values of n_h will be proportional to $N_h \sigma_h C_h^{-1/2}$. Now

$$N_1 \sigma_1 C_1^{-1/2} = 3594.76 \times 0.5 C_2^{-1/2}$$
$$= 1797.38 \times C_2^{-1/2},$$

while

$$N_2 \sigma_2 C_2^{-1/2} = 3548.32 \times C_2^{-1/2}.$$

The optimum allocation is then $n_1 = 8$ and $n_2 = 16$, which at $4 \times 8 + 16 = 48$ stratum-2 city equivalents is cheaper than $n_1 = n_2 = 12$, which costs $4 \times 12 + 12 = 60$ stratum-2 city equivalents. But the variance for $n_1 = 8$ and $n_2 = 16$ exceeds the variance for $n_1 = n_2 = 12$.

Somewhat similar but more complex and less readily interpretable results can be obtained for sampling without replacement. In extreme cases where each stratum has the same—or nearly the same—mean, stratified random sampling without replacement can be less efficient than simple random sampling without replacement.

If the number of strata to be formed is held constant but the positions of the stratum boundaries are allowed to vary, optimum positions may be calculated for them. This topic is treated in OPTIMUM STRATIFICATION.

Optimum allocation and optimum stratification can be carried out only for one variable at a time, and different variables will, in general, give different sample numbers and stratum boundaries. Typically, a sample survey is used to estimate several means or totals. However, there is usually one important variable which will be less accurately measured than the others, almost regardless of the sample design, and this is the obvious choice for optimization. The multiparametric case is considered in OPTIMUM STRATIFICATION.

STRATIFICATION AND RATIO ESTIMATION

Stratification may be used either as an alternative to ratio estimators* or in combination with them. Both are ways of using relevant supplementary information—stratification in the selection of the sample and ratios in the estimation process. Where this supplementary information is purely qualitative or descriptive, stratification is the only possibility. Where it is quantitative, either technique may be used, or both together. In the limit, as the number of strata is allowed to increase indefinitely, stratification approximates to unequal probability sampling (see Brewer [1]).

Where stratified sampling is used in conjunction with ratio estimators, there is a choice between separate ratio estimation (also known as stratum-by-stratum ratio estimation) and combined ratio estimation (also known as across-stratum ratio estimation).

The separate ratio estimator is the sum of the ratio estimators defined for each stratum separately. If the population values of the supplementary or benchmark variable are denoted by X_{hi}, the sample values by x_{hi}, and the hth stratum total by X_h, then the separate ratio estimator may be written

$$Y_s'' = \sum_{h=1}^{L} (y_h'/x_h') x_h',$$

where y_h' is the unbiased number-raising estimator of Y_h and x_h' is the same estimator of X_h.

The combined ratio estimator is the ratio of the sum of the number-raising estimators for each stratum to the corresponding sum of the same estimators for the benchmark variable, multiplied by the population total of the benchmark variable; that is,

$$y_c'' = \left(\sum_{h=1}^{L} y_h' \middle/ \sum_{h=1}^{L} x_h' \right) X.$$

The choice between the separate and combined ratio estimators depends on the relative importance of the variance and of the squared bias in the mean squared error of the separate ratio estimator. The variance of the separate ratio estimator is the sum of the individual stratum variances and the squared bias is the square of the sum of the individual stratum biases. The variance of the combined ratio estimator is generally a little greater than that of the separate ratio estimator, but its bias is smaller (since it depends on a larger sample than that found in each individual stratum). Thus one would tend to prefer the separate ratio estimator if the ratio estimator biases in the individual strata are negligible, but the combined ratio estimator may be preferred if they are appreciable. A rule of thumb is sometimes used, that the sample size in each individual stratum must be at least 6, or that some degree of combined ratio estimation should otherwise be adopted (not necessarily over all strata at once).

Cochran [2] compares separate and combined ratio estimation for two small artificial populations. The example population on page 177 has three four-unit strata, each with a very different ratio of Y_h to X_h. Two units are selected from each stratum. Every one of the 216 possible samples is enumerated. The combined ratio estimator has high variance 262.8 and low squared bias 6.5, while the separate ratio estimator has low variance 35.9 and high squared bias 24.1 (Several other unbiased and low-bias ratio estimators are compared in the same example.) The population on page 187 has two four-unit strata, but these have very similar ratios of Y_h to X_h. Again, two units are selected from each stratum and each of the 36 possible samples is enumerated. This time the combined ratio estimator has both the smaller variance (40.6 as opposed to 46.4) and the smaller squared bias (0.004 as opposed to 0.179). This second example shows that when the ratios of Y_h to X_h are nearly constant the separate ratio estimator cannot always be relied upon to have the smaller variance.

The choices of sample allocation (proportional or optimum) and of stratum boundaries (arbitrary or optimum) are subject to the same considerations with ratio estimation as with number-raising estimation, except that the relevant population variances are naturally those appropriate to ratio estimation. In situations where stratification on a quantitative variable and ratio estimation are both appropriate, it may be preferable to use unequal probability sampling in place of stratification. The abolition of stratum boundaries based on a quantitative variable allows further refinement in qualitative stratification, e.g., by industry or by type of establishment. The simultaneous optimization of the estimator and of the selection probabilities in this situation is considered by Brewer [1].

THE USE OF STRATIFIED DESIGNS IN PRACTICE

Sudman [4] distinguishes four situations in which stratified sampling may be used:

1. The strata themselves are of primary interest.
2. Variances differ between the strata.
3. Costs differ by strata.
4. Prior information differs by strata.

If the strata themselves are of primary interest—if for instance the separate unemployment rates for persons aged 15–19 and for persons aged 20 and over are both important —the statistician must consider whether the users need equal accuracies in the two estimates (as measured by their coefficients of variation*) or whether they are prepared to accept a higher coefficient of variation for the smaller group, in this case the teenagers. If the sample fractions are small and the population coefficients of variation roughly equal, then equal accuracies imply roughly equal sample sizes. Proportional sampling leads to the estimator for the larger population (here, those aged 20 and over) being the more accurate. Sudman suggests a number of possible compromises. If only the individual strata are relevant, if the loss function is quadratic and the loss for each stratum is proportional to its size, then the optimal sample numbers are proportional to the square roots of the population numbers.

Where variances and costs differ between strata, the optimum allocation formulae given previously are relevant. This happens chiefly with economic populations, such as farms or businesses. In human populations the population variances and costs seldom differ greatly from stratum to stratum, and it is usually better in these circumstances to retain the simplicity of proportional allocation.

The last case, in which prior information differs from stratum to stratum, arises only in Bayesian* analysis. Sudman gives an example of optimum sampling for nonresponse* in a human population where, given very high sampling costs and some prior information, it may be decided not to attempt to sample at all from a " very difficult" stratum. He points out, that either explicitly or implicitly, most researchers employ some prior beliefs about omitted potential respondents. Once this " very difficult" stratum is omitted, proportional sampling is nearly as efficient as optimum sampling, and simpler to use.

CONCLUSION

There are three basic elements in stratified designs. Criteria for stratification must be chosen, strata formed, and sample numbers allocated between them. If ratio estimation is employed, there is a further choice between separate and combined ratios.

Except where the strata themselves are of primary importance, the aim of stratification is to decrease the mean square errors of the estimates. This is done by forming strata within which the units are as similar to each other as possible, while each stratum differs from every other stratum as much as possible.

Again, unless the strata themselves are of primary importance, human populations should be proportionately sampled, but economic populations, by virtue of their great differences in unit size, should be as nearly as possible optimally sampled.

References

[1] Brewer, K. R. W. (1979). *J. Amer. Statist. Ass.*, **74**, 911–915. (Demonstrates relationship between Neyman allocation and optimum unequal probability sampling.)

[2] Cochran, W. G. (1977). *Sampling Techniques*, 3rd ed. Wiley, New York. (Updated version of a classic work first published in 1953.)

[3] Evans, E. D. (1951). *J. Amer. Statist. Ass.*, **46**, 95–104. (Demonstrates theoretical advantages of stratification.)

[4] Sudman, S. (1976). *Applied Sampling*. Academic Press, New York, Chap. 6.

Bibliography

Stratification is an ubiquitous technique. Accounts can be found in all major sampling textbooks, some of which appear in the following list.

Armitage, P. (1974). *Biometrika*, **34**, 273–280. (Compares stratified and unstratified random sampling.)

Barnett, V. (1974). *Elements of Sampling Theory*, Hodder and Stoughton, London, England. (A concise modern treatment at an intermediate mathematical level.)

Bryant, E. C., Hartley, H. O. and Jessen, R. J. (1960). *J. Amer. Statist. Ass.*, **55**, 105–124. (Two-way stratification.)

Chatterjee, S. (1967). *Skand. Aktuarietidskr.*, **50**, 40–44. (Optimum stratification.)

Chatterjee, S. (1968). *J. Amer. Statist. Ass.*, **63**, 530–534. (Multivariate stratification.)

Chatterjee, S. (1972). *Skand. Aktuarietidskr.*, **55**, 73–80. (Multivariate stratification.)

Cochran, W. G., (1946). *Ann. Math. Statist.*, **17**, 164–177. (Compares systematic and stratified random sampling.)

Cochran, W. G. (1961). *Bull, Int. Statist. Inst.*, **38**(2), 345–358. (Compares methods for determining stratum boundaries.)

Cornell, F. G. (1949). *J. Amer. Statist. Ass.*, **42**, 523–532. (Small example.)

Dalenius, T. (1957). *Sampling in Sweden*. Almqvist and Wicksell, Stockholm. (Includes comprehensive discussion on optimum stratification.)

Dalenius, T. and Gurney, M. (1951). *Skand. Aktuarietidskr*, **34**, 133–148. (Optimum stratification.)

Dalenius, T. and Hodges, J. L., Jr. (1959). *J. Amer. Statist. Ass.*, **54**, 88–101. (Optimum stratification.)

Ericson, W. A. (1965). *J. Amer. Statist. Ass.*, **60**, 750–771. (Bayesian analysis of stratification in single stage sampling.)

Ericson, W. A. (1968). *J. Amer. Statist. Ass.*, **63**, 964–983. (Bayesian analysis of stratification in multistage sampling.)

Fuller, W. A. (1970). *J. R. Statist. Soc. B*, **32**, 209–226. (Sampling with random stratum boundaries.)

Hagood, M. J. and Bernert, E. H. (1945). *J. Amer. Statist. Ass.*, **40**, 330–341. (Component indexes as a basis for stratification.)

Hansen, M. H., Hurwitz, W. N. and Madow, W. G. (1953). *Sample Survey Methods and Theory*, 2 vols., Wiley, New York. (Encyclopaedic.)

Hartley, H. O., Rao, J. N. K. and Kiefer, G. (1969). *J. Amer. Statist. Ass.*, **64,** 841–851. (Variance estimation with one unit per stratum.)

Hess, I., Sethi, V. K. and Balakrishnan, T. R. (1966). *J. Amer. Statist. Ass.*, **61,** 74–90. (Practical investigation.)

Huddleston, H. F., Claypool, P. L. and Hocking, R. R. (1970). *Appl. Statist.*, **19,** 273–278. (Optimum allocation using convex programming.)

Keyfitz, N. (1957). *J. Amer. Statist. Ass.*, **52,** 503–510. (Variance estimation with two units per stratum.)

Kish, L. (1965). *Survey Sampling.* Wiley, New York, (Practically oriented.)

Kokan, A. R. (1963). *J. R. Statist. Soc. A*, **126,** 557–565. (Optimum allocation in multivariate surveys.)

Mahalanobis, P. C. (1946). *Philos. Trans. R. Soc. London B*, **231,** 329–451. (Stratified designs in large-scale sample surveys.)

Mahalanobis, P. C. (1946). *J. R. Statist. Soc.*, **109,** 326–370. (Samples used by the Indian Statistical Institute.)

Moser, C. A. and Kalton, G. (1971). *Survey Methods in Social Investigation*, 2nd ed. Heinemann Educational Books, London, England. (Includes 15 pages on stratification from the viewpoint of the social scientist.)

Murthy, M. N. (1967). *Sampling Theory and Methods.* Statistical Publication Society, Calcutta, India. (Includes a particularly useful bibliography for sampling articles published up to that date.)

Neyman, J. (1934). *J. R. Statist. Soc.*, **97,** 558–606. (The classical paper which established randomization theory as the only acceptable basis for sampling inference for over three decades. Includes the derivation of Neyman allocation.)

Nordbotten, S. (1956). *Skand. Aktuarietidskr.*, **39,** 1–6. (Allocation to strata using linear programming.)

Raj, D. (1968. *Sampling Theory.* McGraw-Hill, New York.

Rao, J. N. K. (1973). *Biometrika*, **60,** 125–133. (Double sampling, analytical surveys.)

Sethi, V. K. (1963). *Aust. J. Statist.*, **5,** 20–33. (Uses normal and chi-square distributions to investigate optimum stratum boundaries.)

Slonim, M. J. (1960). *Sampling in a Nutshell.* Simon and Schuster, New York. (A slightly humorous nontechnical approach.)

Stephan, F. F. (1941). *J. Marketing*, **6,** 38–46. (Expository.)

Sukhatme, P. V. and Sukhatme, B. V. (1970). *Sampling Theory of Surveys with Applications.* Food and Agricultural Organization, Rome, Italy.

Tschuprow, A. A. (1923). *Metron*, **2,** 461–493, 646–683. (Includes an anticipation of the Neyman allocation formulae.)

U. S. Bureau of the Census (1963). The Current Population Survey—A Report on Methodology. *Tech. Paper No. 7*, U. S. Government Printing Office, Washington, D. C.

(AREA SAMPLING
FINITE POPULATION SAMPLING
MULTISTRATIFIED SAMPLING
NEYMAN ALLOCATION
OPTIMUM STRATIFICATION
PROBABILITY PROPORTIONAL TO SIZE
 (PPS) SAMPLING
PROPORTIONAL SAMPLING
RATIO ESTIMATORS
STRATIFIERS, SELECTION OF)

K. R. W. Brewer

STRATIFIED MULTISTAGE SAMPLING

Stratified multistage sampling is an efficient sampling method which combines the techniques of stratified sampling (*see* STRATIFIED DESIGNS) and multistage sampling. Stratified multistage sampling is based on grouping units into subpopulations called *strata* and then using a hierarchical structure of units within each stratum. For contrast, simple random sampling*, the simplest sampling scheme, is a method of selecting units from the population in *one step* such that all units in the population have an equal chance of selection and the selections are independent. Of course, as with all sampling schemes, the purpose of choosing the sample is still to make inferences about the population (*see* SURVEY SAMPLING).

In this article, stratified multistage sampling is introduced in three parts: (i) two-stage sampling without stratification, (ii) generalization to more than two stages without stratification, and (iii) stratification of the population before multistage sampling. The first two parts describe two-stage/multistage sampling, while the last part combines the techniques of multistage sampling and stratification.

TWO-STAGE SAMPLING

Suppose a radio station wants to conduct a sample survey to estimate the proportion of persons in city A who listen to this station. However, nobody has a list of all persons in this city. How can one select a random sample of persons? A random sample of persons in city A can be selected in two stages by using a list of dwellings or households. First, a random sample of households is selected from the household list. Second, for each selected household, a list of persons is prepared and a random sample of persons is chosen. The sample of persons chosen by this process is a *two-stage sample* and can be used for the radio station study. Note that two-stage sampling is different from two-phase sampling, in which a sample of units is selected in the first phase and then a subset of these sampled units is selected in the second phase (*see* MULTIPHASE SAMPLING and DOUBLE SAMPLING). In two-stage sampling, the sampling units in the second-stage are different from the sampling units in the first stage.

If the above-mentioned two-stage sample included *all* persons in each selected household, it would be a cluster sample*. Since a two-stage sample usually includes only a subset of second-stage units, it is also referred to as a *subsample*. First-stage units are also referred to as *primary units** and second-stage units are referred to as *secondary units* or *subunits* (see Cochran [2], Hansen et al. [3], Jessen [7], and Stuart [10]). For surveys of large populations that have a hierarchical structure, two-stage sampling is often more convenient and economical than one-stage simple random sampling.

Let N be the number of first-stage units, M_i the number of second-stage units in the ith first-stage unit, and y_{ij} the value of the jth second-stage unit in the ith first-stage unit. Suppose that n first-stage units are selected in the sample and m_i second-stage units are chosen in the ith first-stage unit. The sampling units may be chosen with or without replacement; we assume that sampling is without replacement and that sec-

ond-stage units are chosen with equal probabilities within a particular first-stage unit, while the first-stage units are chosen with equal probabilities or with probabilities Z_i (usually an estimate of size of first-stage units). Let $\bar{Y}_i$ and $\bar{y}_i$ denote the population and sample means for the ith first-stage unit, respectively.

First, consider the case of equal probability selection for the first-stage units. One estimator of the population total Y is $\hat{Y} = (N/n)\sum_{i=1}^{n} M_i \bar{y}_i$. This estimator becomes self-weighting (i.e., all second-stage units in the sample get equal weight in the computation of $\hat{Y}$) when the m_i are proportional to the M_i. This estimator is unbiased* (i.e., in repeated sampling, the average of $\hat{Y}$ tends to Y). An unbiased sample estimate of the variance of this estimator is

$$v(\hat{Y}) = \frac{N(N-n)}{n(n-1)} \sum_{i=1}^{n} \left(M_i \bar{y}_i - \frac{\hat{Y}}{N} \right)^2 + \frac{N}{n} \sum_{i=1}^{n} \frac{M_i(M_i - m_i)}{m_i} s_{2i}^2,$$

where s_{2i}^2, the sample variance for second-stage units in the ith first-stage unit, is given by

$$s_{2i}^2 = \frac{1}{m_i - 1} \sum_{j=1}^{m_i} (y_{ij} - \bar{y}_i)^2.$$

Another estimator of Y in the case of equal probability selection is the ratio estimator*

$$\hat{Y}_R = M_0 \sum_{i=1}^{n} M_i \bar{y}_i \bigg/ \sum_{i=1}^{n} M_i,$$

where $M_0 = \sum_{i=1}^{N} M_i$. A sample estimate of the variance of this estimator is

$$v(\hat{Y}_R) = \frac{N(N-n)}{n(n-1)} \sum_{i=1}^{N} M_i^2 \left(\bar{y}_i - \frac{\hat{Y}_R}{M_0} \right)^2 + \frac{N}{n} \sum_{i=1}^{N} \frac{M_i(M_i - m_i)}{m_i} s_{2i}^2.$$

For a more detailed discussion of the preceding estimators, see Cochran [2] or Hansen et al. [3].

Next, consider the case of unequal probabilities of selection for first-stage units. Suppose these units are chosen with probabilities proportional to Z_i (ppz). As before, the second-stage units are chosen with equal probabilities within first-stage units. An unbiased estimator of the population total Y is

$$\hat{Y}_{ppz} = \frac{1}{n} \sum_{i=1}^{n} \frac{M_i \bar{y}_i}{Z_i}.$$

An unbiased sample estimator of the variance of $\hat{Y}_{ppz}$ is

$$\nu\left(\hat{Y}_{ppz}\right) = \frac{1}{n(n-1)} \sum_{i=1}^{n} \left\{\frac{M_i \bar{y}_i}{Z_i} - \hat{Y}_{ppz}\right\}^2.$$

If $Z_i = M_i/M_0$, the selection of first-stage units is with probability proportional to size* (pps). In this special case, the preceding estimates simplify to the expressions

$$\hat{Y}_{pps} = \frac{M_0}{n} \sum_{i=1}^{n} \left(\sum_{j=1}^{m_i} \frac{y_{ij}}{m_i}\right),$$

$$\nu\left(\hat{Y}_{pps}\right) = \frac{1}{n(n-1)} \sum_{i=1}^{n} \left(M_0 \bar{y}_i - \hat{Y}_{pps}\right)^2. \tag{1}$$

An examination of (1) reveals that $\hat{Y}_{pps}$ becomes self-weighting when $m_i = m$ (i.e., all subsamples are of the same size). In this case, the estimator of Y is

$$\hat{Y}_{pps} = \frac{M_0}{mn} \sum_{i=1}^{n} \sum_{j=1}^{m} y_{ij}.$$

How does the variance of the estimator of Y from two-stage sampling compare with those from one-stage simple random sampling (srs) and cluster sampling for the same total number of second-stage units? Assuming that the intraclass correlation* between second-stage units is positive, which is usually the case, the relationship is

$$\nu\left(\hat{Y} \text{ srs}\right) \leqslant \nu\left(\hat{Y} \text{ two-stage}\right) \leqslant \nu\left(\hat{Y} \text{ cluster}\right).$$

As discussed by Sukhatme and Sukhatme [11], the increase in variance for two-stage sampling over simple random sampling is approximately proportional to the intraclass correlation. The reduction in variance for two-stage sampling over cluster sampling is inversely proportional to the sampling fraction of second-stage units.

Murthy [9] has described a two-stage survey for the estimation of total area under paddy. In this survey, the first-stage units (villages) were chosen with probabilities proportional to size and the second-stage units (plots) were also chosen with probabilities proportional to size. He also compared the cost of this two-stage sampling of plots with that of a corresponding one-stage sampling of plots.

MULTISTAGE SAMPLING

Many large-scale surveys involve three or more stages of sampling units. The procedure for selecting the units at different stages and the estimation of population parameters for two-stage sampling can be generalized to more than two stages of sampling. Of course, as the number of stages increases, the number of possible methods of sample selection also increases. For example, there are four methods for selecting first-stage and second-stage units in a three-stage sample, since first-stage and second-stage units may each be chosen either with equal probabilities or with probabilities proportional to a measure of size.

Here, we consider the simplest case of three-stage sampling, in which first-stage units are of equal sizes as are second-stage units. Let N be the number of first-stage units, M the number of second-stage units in each first-stage unit, K the number of third-stage units in each second-stage unit, and y_{ijr} the value of the rth third-stage unit in the jth second-stage unit in the ith first-stage unit. Suppose that units are selected by simple random sampling at each stage as follows: (i) n first-stage units, (ii) m second-stage units from each first-stage unit in the sample, and (iii) k third-stage units each second-stage unit in the sample.

An unbiased estimator of the population total Y is

$$\hat{Y} = \frac{N}{n} \sum_{i=1}^{n} \left\{ \frac{M}{m} \sum_{j=1}^{m} \left(\frac{K}{k} \sum_{r=1}^{k} y_{ijr} \right) \right\}$$

$$= \frac{NMK}{nmk} \sum_{i=1}^{n} \sum_{j=1}^{m} \sum_{r=1}^{k} y_{ijr}.$$

An estimator of the population mean at the third-stage unit level is

$$\bar{\bar{\bar{y}}} = \frac{\hat{Y}}{NMK} = \frac{1}{nmk} \sum_{i=1}^{n} \sum_{j=1}^{m} \sum_{r=1}^{k} y_{ijr}.$$

Finally, an unbiased sample estimate of the variance of $\bar{\bar{\bar{y}}}$ is

$$v\left(\bar{\bar{\bar{y}}}\right) = \frac{1 - f_1}{n} s_1^2 + \frac{f_1(1 - f_2)}{nm} s_2^2$$

$$+ \frac{f_1 f_2 (1 - f_3)}{nmk} s_3^2,$$

where $f_1 = n/N$, $f_2 = m/M$, and $f_3 = k/K$ are the sampling fractions at the three stages and s_1^2, s_2^2, *and* s_3^2 are the sample variances associated with the three stages

$$s_3^2 = \frac{\sum_{i=1}^{n} \sum_{j=1}^{m} \sum_{r=1}^{k} \left(y_{ijr} - \bar{y}_{ij}\right)^2}{nm(k - 1)},$$

$$s_2^2 = \frac{\sum_{i=1}^{n} \sum_{j=1}^{m} \left(\bar{y}_{ij} - \bar{\bar{y}}_i\right)^2}{n(m - 1)},$$

$$s_1^2 = \frac{\sum_{i=1}^{n} \left(\bar{\bar{y}}_i - \bar{\bar{\bar{y}}}\right)^2}{n - 1}.$$

The number of bars on top of y indicate the number of stages over which the mean has been computed. Note that the variance associated with each stage is simply the mean squared difference between the values (means) at that stage and the mean at the next higher stage. The procedures and estimators discussed here for two-stage and three-stage sampling can be extended to more than three stages.

STRATIFICATION OF FIRST-STAGE UNITS

So far, we have discussed multistage sampling in the context of a single homogeneous population. In many large scale surveys, it is efficient to divide the population into several homogeneous subpopulations of first-stage units (strata) before selecting a multistage sample from each stratum. In stratified sampling, the sampling procedure used in one stratum may be different from those used in other strata. For convenience, stratified two-stage sampling is discussed here, but this procedure can be easily generalized to stratified multistage sampling.

Let the population be divided into L strata with N_k first-stage units in stratum h, M_{hi} second-stage units in the ith first-stage unit in stratum h, and y_{hij} the value of the jth second-stage unit in the ith first-stage unit in stratum h. The population mean for second-stage units in stratum h is

$$\bar{\bar{Y}}_h = \frac{1}{M_{h0}} \sum_{i=1}^{N_h} \sum_{j=1}^{M_{hi}} y_{hij},$$

where $M_{h0} = \sum_{i=1}^{N_h} M_{hi}$. By combining strata means, the overall population mean for second-stage units is

$$\bar{\bar{Y}} = \sum_{h=1}^{L} M_{h0} \bar{\bar{Y}}_h \bigg/ \sum_{h=1}^{L} M_{h0} = \sum_{h=1}^{L} W_h \bar{\bar{Y}}_h,$$

where $W_h = M_{h0} / \sum_{h=1}^{L} M_{h0}$.

Let n_h be the number of first-stage units selected in the sample from stratum h and m_{hi} the number of second-stage units selected from the ith first-stage unit in the sample from stratum h. Suppose that the units are selected by simple random sampling at each stage within each stratum. An unbiased estimator of $\bar{\bar{Y}}$ is

$$\bar{\bar{y}} = \sum_{h=1}^{L} W_h \bar{\bar{y}}_h,$$

where

$$\bar{\bar{y}}_h = \frac{1}{M_{h0}} \frac{N_h}{n_h} \sum_{i=1}^{n_h} \left(\frac{M_{hi}}{m_{hi}} \sum_{j=1}^{m_{hi}} y_{hij} \right).$$

An unbiased sample estimate of the variance of $\bar{\bar{y}}$ is

$$
\nu(\bar{\bar{y}}) = \sum_{h=1}^{L} W_h^2 \left\{ \frac{1 - f_{1h}}{n_h} s_{1h}^2 \right.
$$
$$
\left. + \frac{f_{1h}}{n_h^2} \sum_{i=1}^{n_h} \left(\frac{M_{hi}}{\bar{M}_h} \right)^2 \frac{1 - f_{2hi}}{m_{hi}} s_{2hi}^2 \right\},
$$

where s_{1h}^2 and s_{2hi}^2 are the sample variances of first-stage units and second-stage units, respectively, and $\bar{M}_h = M_{h0}/N_h$.

Here, we have not addressed the following two major aspects of stratification: (i) use of stratification variables to define strata and (ii) allocation of the total sample to the different strata. Kpedekpo [8] has discussed the determination of optimum stratification* points and the choice of the number of strata. The problem of optimum allocation in stratified sampling has been discussed by Causey [1] and Sukhatme and Tang [12]. Jarque [6] has discussed the problem of optimum stratification in multivariate sampling (*see* MULTISTRATIFIED SAMPLING). Hausman and Jain [4] have discussed a multivariate methodology for analyzing data from stratified multi-stage sampling. Finally, if a fixed stratified multistage sample is used for periodic estimation, it may be necessary to make an adjustment for changes in strata composition [5].

An excellent example of a stratified multistage sampling survey is the Current Population Survey (CPS; *see* BUREAU OF LABOR STATISTICS) conducted monthly by the Bureau of the Census [13]. The CPS has been designed to obtain estimates of employment, unemployment, income, and other characteristics of the U.S. population. The primary sampling units (PSUs) consisting of either large individual counties or groups of smaller counties in all 50 states and the District of Columbia are grouped into strata. In each stratum, a sample of household addresses is chosen in three stages: (i) selection of one PSU from the stratum, (ii) selection of census enumeration districts in each selected PSU, and (iii) selection of household addresses in each selected census enumeration district.

References

[1] Causey, B. D. (1983). *SIAM J. Sci. Statist. Comput.*, **4**, 322–329.

[2] Cochran, W. G. (1977). *Sampling Techniques*, 3rd ed. Wiley, New York. (A comprehensive text on sampling theory.)

[3] Hansen, M. H., Hurwitz, W. N., and Madow, W. G. (1953). *Sample Survey Methods and Theory*, 2 vols. Wiley, New York. (A comprehensive text with the most detailed discussion of sample survey methods.)

[4] Hausman, R. E. and Jain, A. K. (1982). *Proc. Section on Survey Research Methods*, Amer. Statist. Ass., 111–116.

[5] Jain, A. K. (1982). *Ann. Inst. Statist. Math. Tokyo*, **34**, 59–71.

[6] Jarque, C. M. (1981). *Appl. Statist.*, **30**, 163–169.

[7] Jessen, R. J. (1978). *Statistical Survey Techniques*. Wiley, New York. (An easy to read text.)

[8] Kpedekpo, G. M. K. (1973). *Metrika*, **20**, 54–64.

[9] Murthy, M. N. (1977). *Sampling Theory and Methods*. Statistical Publishing Society, Calcutta, India.

[10] Stuart, A. (1962). *Basic Ideas of Scientific Sampling*. Hafner, New York. (An easy to read text.)

[11] Sukhatme, P. V. and Sukhatme, B. V. (1970). *Sampling Theory of Surveys with Applications*. Iowa State University Press, Ames, IA. (A good discussion of both theory and applications.)

[12] Sukhatme, B. V. and Tang, K. T. (1975). *J. Amer. Statist. Ass.*, **70**, 175–179.

[13] U.S. Bureau of the Census (1963). The Current Population Survey: A Report on Methodology. *Tech. Paper No. 7*, U.S. Government Printing Office, Washington, D.C.

(CLUSTER SAMPLING
DOUBLE SAMPLING
MULTIPHASE SAMPLING
MULTISTRATIFIED SAMPLING
OPTIMUM STRATIFICATION
STRATIFIED DESIGNS
SURVEY SAMPLING)

ARIDAMAN K. JAIN
ROBERT E. HAUSMAN

STRATIFIED SAMPLING See MULTISTRATIFIED SAMPLING; NEYMAN ALLOCATION; OPTIMUM STRATIFICATION; PROBABILITY PROPORTIONAL TO SIZE (pps) SAMPLING; PROPORTIONAL ALLOCATION; STRATIFIED DESIGNS; STRATIFIED MULTISTAGE SAMPLING

STRATIFIERS, SELECTION OF

Stratifiers or *stratifying variables* are variables whose joint values are used to classify a population into several classes, called *strata*, from each one of which a sample is drawn independently. Both numerical and nonnumerical variables may be used for creating strata. Thus in sampling retail stores for the purpose of estimating total sales of each of several food items in a given area, one may use as stratifying variables: some measure of size of the store (or of its food department), type of store (department store, general food store, specialized food store), location, etc. (The strata need not constitute a complete cross-classification of the population by the classifiers. In the example just given, different size classes may well be used for the different types of stores, and in certain locations a different type of classification may be appropriate.)

In any given investigation there are often many potential stratifiers, so that a choice may be necessary. The following considerations are relevant to this choice:

1. For the most important properties of the units under investigation, the strata should be more homogeneous than the population as a whole. Thus in the preceding example, the variances of the sales of most of the food items among the stores in any one of the strata should be much smaller than the variances of the sales of these items among all the stores in the population. Clearly, strata defined by size and type of store may be expected to have this property and a further classification by location could lead to further substantial reduction of the variance. In general, stratification reduces the variance of estimators of an overall average or total by a fraction equal to somewhat less than the square of the correlation ratio* $\eta_{y|x}^2 = \text{var } E[Y|X]/\text{var}(Y)$ (where X is the joint stratifier) if sampling of the strata is proportional to stratum size. Disproportional sampling can lead to a further reduction; see, e.g.,

Cochran [1] and Konijn [2]. (This relation between the variance of estimators and $\eta_{y|x}^2$ is not emphasized in the literature in this subject.)

2. The organization of the investigation may favor separate sampling of certain subpopulations. In the preceding example, the surveying organization may have separate branches in each of a number of locations, each having their own staff for recruiting, training, and supervising the necessary personnel. It is then natural to use location as a stratifier, even if this does not lead to increased homogeneity. If there are separate, easily available lists that between them cover all or most of the population, it will often be worthwhile to include among the stratifiers the ones whose values define the different lists. Moreover, in addition to measures for the population as a whole, we may require separate measures for certain parts of the population (in our example these may be certain size and/or location classes). Hence we may need a lower bound on the number of individuals selected from any one subpopulation.

3. Prior information, variances, and costs of obtaining information among different parts of the population can be used to improve the efficiency of the design. The larger retail stores may have the desired information readily available, while for the smaller stores the investigators may have to obtain much of the information themselves. This tends to make it advantageous to include disproportionally few small stores in the sample.

4. The sampling distribution of the usual estimator of the total or other desired function from an unstratified sample may not be well approximated by a standard family of limit distributions (such as Gaussian), but may be reasonably well approximated by such a family if data from a suitable stratifier are used, so that approximate confidence intervals

may be computed; see Konijn [2, Sec. III-14.1]. Thus in a distribution of household incomes, some indicator related to the value of dwellings—used as a stratifier—may lead to an estimator of average income that has a sample distribution that is close to normal, whereas without using this stratifier this may not be so.

5. A potential stratifier can often be used to modify the estimating formula rather than to create strata. Such an alternative is particularly relevant if the investigation is multipurpose, since estimating formulae can be chosen to be quite different for the different quantities estimated.

6. The use of certain stratifiers may involve substantially extra costs, but poststratification may cut down on these extra costs, since only the sampled units need to be classified.

7. In cluster sampling* it is often efficient to stratify by size of cluster. An alternative is to select the clusters with probabilities proportional to some indicator of their size.

8. One method of adjusting for nonresponse bias involves poststratification in which values of the stratifier are different levels of *probability* of nonresponse*, assigned on the basis of some response model. In a particular case the model may incorporate knowledge of the skill, sex, and age of the interviewer and/or of the interviewee (Särndal and Swensson [3]).

References

[1] Cochran, W. G. (1977). *Sampling Techniques*, 3rd ed. Wiley, New York.

[2] Konijn, H. S. (1973). *Statistical Theory of Sample Survey Design and Analysis*. North-Holland, Amsterdam, Netherlands.

[3] Särndal, C. E. and Swensson, B. (1985). Incorporating nonresponse modelling in a general randomization theory approach. *Bull. Intern. Statist. Inst.*, **51**(3), 15.2:1–15.

(CLUSTER SAMPLING
OPTIMUM STRATIFICATION
PROBABILITY PROPORTIONAL TO SIZE
  (pps) SAMPLING
STRATIFIED DESIGNS
SURVEY SAMPLING)

H. S. KONIJN

STRAW POLL *See* ELECTION PROJECTIONS

STRENGTH OF A SAMPLING PLAN

A sampling plan* is said to be of *strength* $(p_1, \alpha; p_2, \beta)$ if the probability of rejecting a lot with proportion p of nonconforming items $Q(p)$ satisfies the equations

$$Q(p_1) = \alpha \quad \text{(producer's risk*)},$$

$$Q(p_2) = 1 - \beta$$
$$(\beta \text{ is the consumer's risk*}).$$

(QUALITY CONTROL, STATISTICAL SAMPLING PLANS)

STRENGTH OF A TEST

This is usually measured by a pair of numbers (α, β), the first of which (α) is the significance level [probability of rejecting the hypothesis tested (H_0) when it is valid] and the second (β) is the probability of rejecting H_0 when some other hypothesis (H) is valid (the power* of the test with respect to H). Statements of strength should be qualified by stating the alternative hypothesis* (H) to which β refers.

(HYPOTHESIS TESTING)

STRENGTH AND OTHER PROPERTIES OF AN ARRAY

A matrix **A** is an *array of strength s* if for every subset S of s rows of **A**, the number of columns of **A** having exactly i nonzero entries among the rows of S depends only on i, for each $i \leqslant s$. An array $\mathbf{A}$ of strength s is *balanced* if also for every i, each of the $\binom{s}{i}$ possible columns with i nonzero entries among the rows of S appears equally often.

The arrays with strength and balance include the balanced incomplete block designs*, the orthogonal arrays*, the sets of mutually orthogonal Latin squares*, and the group codes, as well as being particularly useful for factorial designs*.

If the number of columns in an array $\mathbf{A}$ of strength s is large compared to s, then $\mathbf{A}$ may always be reduced (by omitting some of the columns) to a balanced array of strength s.

If $\mathbf{A}$ is an array of strength s, then for each $i \leqslant s$ the number of columns of $\mathbf{A}$ having nonzero entries among the rows of S is some integer $\sigma(i)$. *The weights of* $\mathbf{A}$ *are* $(\sigma(0), \sigma(1), \ldots, \sigma(s))$. If $\mathbf{A}$ is also balanced, then each $\sigma(i) = c(s, i)\tau(i)$. Many authors use $(\tau(0), \tau(1), \ldots, \tau(s))$ for the weights of balanced arrays of strength s.

If $\mathbf{A}$ is an array of strength s with weights $(\sigma(0), \sigma(1), \ldots, \sigma(s))$, t all-zero columns, and v nonzero columns, then each all-zero column (each nonzero column) of $\mathbf{A}$ contributes exactly t to $\sigma(0)$ [exactly v to $\sigma(s)$], so the array $\mathbf{A}^* = \mathbf{A}$ with all of its completely zero columns and all of its nonzero columns deleted is an array of strength s with weights $(\sigma(0) - t, \sigma(1), \ldots, \sigma(s-1), \sigma(s) - v)$. If $\mathbf{A}$ is balanced, then so is $\mathbf{A}^*$, which is called the *trim array* of $\mathbf{A}$. If $\mathbf{A}$ had no such columns to begin with, then $\mathbf{A}$ is called a *trim array*.

Bibliography

Bose, R. C. and Bush, K. A. (1952). *Ann Math. Statist.*, **23**, 508–524.

Chopra, D. V. (1979). *Gujarat Statist. Rev.*, **6**, 1–8.

Srivastava, D. V. (1972). *J. Comb. Theory A*, **13**, 198–206.

Srivastava, D. V. and Chopra, D. V. (1973). Balanced arrays and orthogonal arrays. *A Survey of Combinatorial Theory*. North-Holland, Amsterdam, pp. 411–428.

(BLOCKS, BALANCED INCOMPLETE DESIGN OF EXPERIMENTS)

JUDITH Q. LONGYEAR

STRESS   See MULTIDIMENSIONAL SCALING

STRESS-STRENGTH MODELS

Stress-strength or interference theory models (in engineering terminology) arise mainly in mechanical reliability* theory (see, e.g., Haugen [6]). They can be classified as either dynamic or static.

DYNAMIC MODELS

These are models of systems subject to stresses occurring randomly in time. The stressors are governed by a counting process $\{N(t): t > 0\}$. Usually two groups of stresses can be distinguished: (i) *environmental* stresses such as humidity, pressure, temperature, etc., and (ii) *operating* stresses such as friction, load, vibrations, etc. It is assumed that, at a specified time t, the ith stress has an associated random value $X_i(t)$ and that the stresses vary randomly and independently in time. A typical stress-strength model (see Ebrahimi [3, 4]) for a system with M different stresses combining additively would have lifetime given by

$$T_1 = \inf\left\{ s: \sum_{i=1}^{M} X_i(s) > Y(s)\right\},$$

where $Y(t)$ denotes the *strength* of the system at time t. An alternative (nonadditive) model is

$$T_2 = \inf\{ s: \max(X_1(s), \ldots, X_m(s)) \\ > Y(s)\}.$$

Further details and references are available in Esary et al. [5].

STATIC MODELS

In many situations it can be assumed that neither stress nor strength depend on time. Interest is focused on modeling the reliability. Such models depend on the physical setup. The simplest static model is given by

$$R = \Pr(Y > X),$$

where Y and X are independent random variables denoting strength and stress, respectively. For reviews of these models and their associated statistical inference problems, both parametric and nonparametric, see Basu [1] and Bhattacharyya and Johnson [2] and references therein. Details on the special but widely used case where X and Y are normally distributed are given in Reiser and Guttman [7].

References

[1] Basu, A. P. (1977). Proc. 22nd Conf. on the Design of Experiments in Army Research Development and Testing. *ARO Report No. 77-2*, U.S. Army Research Office, Durham, N.C., pp. 97–110.

[2] Bhattacharyya, G. K. and Johnson, R. A. (1981), Proc. 26th Conf. on the Design of Experiments in Army Research Development and Testing. *ARO Report No. 81-2*, U.S. Army Research Office, Durham, N.C., pp. 531–548.

[3] Ebrahimi, N. (1985a). *J. Appl. Prob.*, **22**, 467–472.

[4] Ebrahimi, N. (1985b). *Statist. Prob. Lett.*, **3**, 295–297.

[5] Esary, J. D., Marshall, A. W., and Proschan F. (1973). *Ann. Prob.*, **11**, 627–649.

[6] Haugen, E. B. (1980). *Probabilistic Mechanical Design*. Wiley, New York.

[7] Reiser, B. and Guttman, I. (1986). *Technometrics*, **28**, 253–257.

(CATASTROPHE THEORY
CUMULATIVE DAMAGE MODELS
DAMAGE MODELS
FATIGUE MODELS
HAZARD RATE AND OTHER
   CLASSIFICATIONS OF DISTRIBUTIONS
RELIABILITY (various entries)
RISK THEORY
SHOCK MODELS
STATISTICAL CATASTROPHE THEORY)

BENJAMIN REISER

STRICT COHERENCE, PRINCIPLE OF

This is a principle in subjective probability theory that prescribes the avoidance of bets that must lead either to loss or to zero gain whatever be the "state of nature." See, for example, Fellner [1] for details and for the origins of this term.

Reference

[1] Fellner, W. J. (1965). *Probability and Profit*. Irwin, Homewood, Ill.

(COHERENCE
GAMBLING, STATISTICS IN)

STRIP PLOTS

Strip-plot designs, also called *criss-cross* designs, were proposed by Cochran and Cox [1] for experiments in which two factors both require plots to be large in size and awkward in shape. They are, therefore, a development of split plots*. In its simplest form, the design is made up of a series of rectangles, each of which is divided into rows for the application of one factor and into columns for the application of the other, the units (or plots) being formed by the intersection of the two. Given one factor with three levels, A, B, and C, and a second with four, a, b, c, and d, two rectangles might look like this:

Ba	*Ca*	*Aa*		*Cb*	*Bb*	*Ab*
Bd	*Cd*	*Ad*		*Cc*	*Bc*	*Ac*
Bb	*Cb*	*Ab*		*Ca*	*Ba*	*Aa*
Bc	*Cc*	*Ac*		*Cd*	*Bd*	*Ad*

The rows and columns often represent strips in space, but one could represent experimental units and the other, occasions. (It is, of course, assumed that there are no residual effects from one occasion to the next.) There is no objection to factors, as already defined, being composite, e.g., the design would not be changed if the four levels, a, b, c, and d, could also be written 1, x, y, and xy, being themselves factorial in structure.

Such designs should be distinguished on the one hand from change-over designs* and on the other from lattice designs*. In a change-over design the units are given a fresh set of treatments in each period, as can happen with a strip-plot design, but residual effects are taken into account by arranging for all sequences of two treatments to arise in a balanced manner. A strip-plot design, however, is necessarily factorial, one factor being applied to the units and the other to periods, so that all units in a rectangle change together, residual effects being ignored. The relationship to lattice squares (*see* LATTICE DESIGNS), however, is much closer. In a strip-plot, design main effects* are confounded between rows or columns; with lattice squares, the confounding* is of balanced sets of contrasts*. Latin squares* are quite different because they give no confounding, each treatment occurring once in each row and each column.

The analysis of variance* for a strip-plot design assigns each effect its own interaction* with rectangles as error. Thus, if factor A has a levels and factor B has b and if there are p rectangles, the analysis has the form in Table 1.

Sums of squared deviations are readily found because all effects are othogonal. A numerical example appears in SPLIT PLOTS.

Usually both factors are assigned strips at random within each rectangle. Sometimes, however, one of them is difficult to randomize. In that case, the analysis for its main effect is vitiated but the other two are valid.

It is sometimes claimed for strip-plot designs that they give a precise study of the interaction, because its error, i.e., the interaction of blocks, Factor A and Factor B is likely to be less than those for the two main effects. In general the argument holds, but it cannot be relied upon absolutely.

Table 1

Source	Degrees of Freedom
Factor A	$a - 1$
Error (A)	$(a - 1)(p - 1)$
Factor B	$b - 1$
Error (B)	$(b - 1)(p - 1)$
Interaction (I)	$(a - 1)(b - 1)$
Error (I)	$(a - 1)(b - 1)(p - 1)$
Total	$abp - 1$

Reference

[1] Cochran, W. G. and Cox, G. M. (1950). *Experimental Designs*. Wiley, New York.

(CHANGEOVER DESIGNS
DESIGN OF EXPERIMENTS
FACTORIAL EXPERIMENTS
LATTICE DESIGNS
SPLIT PLOT DESIGNS)

S. C. PEARCE

STRIP SAMPLING *See* STATISTICS IN ANIMAL SCIENCE

STRONG LAWS OF LARGE NUMBERS *See* LAWS OF LARGE NUMBERS

STRONGLY ERGODIC CLASS *See* MARKOV PROCESSES

STRONG TRUE-SCORE THEORY

In mental test theory a general goal is to use observed test scores to make inferences about an unknown parameter θ that represents an examinee's ability in a certain area such as arithmetic reasoning, vocabulary, spatial ability, etc. The parameter θ is frequently called an examinee's *true score*. There are several types of true scores [3]; differences among them are not discussed. True-score models are just probability models that yield methods for estimating θ or making inferences about the characteristics of a test. The term *strong true-score theory* was introduced by Lord [2] to make a distinction between "weak" theories that cannot be contradicted by data and "strong" theories where assumptions are made about the distribution of observed test scores. Strictly speaking, latent trait* models (also known as *item response theories*) fall within this definition, but the term *strong true-score model* is usually reserved for models based on the binomial or some related probability distribution, apparently because the main focus of

Lord's paper was a model based on the binomial probability function.

Consider a single examinee responding to n dichotomously scored items. As just indicated, the best known strong true-score model assumes that the probability of x correct responses is given by

$$f(x|\theta) = \binom{n}{x}\theta^x(1-\theta)^{n-x}. \qquad (1)$$

In addition to specifying a probability function for an examinee's observed score x, strong true-score models typically specify a particular family of distributions for θ over the population of examinees. When (1) is assumed, the family of beta* densities is commonly used, where $g(\theta)$, the probability density function of θ, is given by

$$g(\theta) = \frac{\Gamma(r+s)}{\Gamma(r)\Gamma(s)}\theta^{r-1}(1-\theta)^{s-1},$$

$$0 < \theta < 1 \qquad (2)$$

where $r, s > 0$ are unknown parameters. Estimates of the parameters r and s are easily obtained with the method of moments* [7] and maximum likelihood* estimates are available from ref. 1. Basically the beta-binomial* model falls within the realm of empirical Bayesian* techniques, as do most strong true-score models. The beta-binomial model frequently gives a good fit to data and it provides a solution to many measurement problems [7]. Included are methods of equating tests and methods of estimating test accuracy and reliability.

Several objections have been raised against the beta-binomial model, but from ref. 7 the only objection that seems to have practical importance is that the model ignores guessing. Here a correct guess refers to the event of a correct response to a randomly sampled item that the examinee does not know. For a strong true-score model where a correct guess is defined in terms of randomly sampled examinees (and where items are fixed), see ref. 15.

Suppose every item has t alternatives and for a specific examinee, let ζ be the probability of knowing a randomly sampled item. Morrison and Brockway [4] assumed random

guessing, in which case

$$\theta = \zeta + t^{-1}(1 - \zeta)$$

and the density of θ is

$$\frac{t}{t-1}g\left(\frac{t\theta - 1}{t - 1}\right), \qquad t^{-1} \leqslant \theta \leqslant 1.$$

Unfortunately it appears that the random guessing assumption is unsatisfactory. The only model that has given good results is one proposed by Wilcox [9, 10] based on an answer-until-correct scoring procedure and the assumption that an examinee's guessing ability is a monotonic function of θ. By an answer-until-correct scoring procedure is meant that an examinee chooses responses to a multiple-choice test item until the correct alternative is chosen. These tests are usually administered by having an examinee erase a shield on especially designed answer sheets. Under the shield is a letter indicating whether the correct answer was chosen. If not, another shield is erased and the process continues until the correct alternative is selected.

Let ζ_i be the probability that an examinee can eliminate i distractors from a randomly sampled item, $i = 0, 1, \ldots, t - 1$. It is assumed that when an examinee does not know, there is at least one distractor that cannot be eliminated through partial information and so $\zeta_{t-1} = \zeta$. It is also assumed that an examinee eliminates as many distractors as possible and then guesses at random from among the alternatives that remain. For empirical evidence in support of this last assumption, see ref. 14. If θ_i is the probability of a correct answer on the ith try of a randomly sampled item, then

$$\theta_i = \sum_{j=0}^{t-i} \zeta_j/(t - j),$$

and so the ζ_j's can be estimated. If x_i is the number of items requiring i attempts, it is assumed that the x_i's have a multinomial* probability function, that θ_1 has a beta density with parameters r and s, and that

$$E\left(\frac{\theta_2}{1 - \theta_1}\bigg|\theta_1\right) = c\int_0^{\theta_1}h(u)\,du + t^{-1}, \quad (3)$$

where c is an unknown parameter and $h(u)$ is also a beta density but with parameters a and b. The model implies that

$$\theta_1 \leqslant \theta_2 \leqslant \cdots \leqslant \theta_t, \qquad (4)$$

and so the lower limit for the integral in (3) should be t^{-1}, but this modification has not yet been applied to real data. Equation (3) is based on the assumption that the more items an examinee knows, the higher the probability will be that an examinee will give a correct guess to an item that is not known. The parameters a, b, and c are currently estimated using what is basically the method of moments; for the lengthy details, see ref. 9.

As a final note, there are now extensions of strong true-score models based on closed sequential sampling* techniques that might be useful in computerized testing. By closed sequential sampling is meant that items are randomly sampled and administered until some criterion is met; the criterion used will depend on the purpose of the test.

Consider, for example, a criterion-referenced test where the goal is to determine whether $\theta \geqslant \theta_0$, where θ_0 is a known constant. Suppose $\theta \geqslant \theta_0$ is decided if and only if $x \geqslant c$, where c (a positive integer) is some known passing score. Given that $\theta \geqslant \theta_0$ (or that $\theta < \theta_0$), the probability of a correct decision is available immediately (given θ) if the binomial model is assumed. For related results, see ref. 5.

Suppose instead that items are randomly sampled until an examinee gets c items correct (or $m = n - c + 1$ items wrong). Let $x(y)$ be the number of correct (incorrect) responses when the sampling of items terminates. The joint probability function of x and y is

$$f(x, y|\theta) = L\frac{(x + y - 1)!}{x!y!}\theta^x(1 - \theta)^y,$$

where $x = c$ and $0 \leqslant y \leqslant m - 1$ (or where $y = m$ and $0 \leqslant x \leqslant c - 1$) and where $L = m$ if $y = m$, and otherwise $L = c$. Wilcox [8] showed that the probability of a correct decision under the closed sequential procedure is exactly the same as it is under the bi-nomial model, but the expected number of items is always less. For results on estimating θ under the closed sequential procedure, see ref 16. For extensions to the multivariate case, including an application to answer-until-correct tests, see refs. 12 and 13.

References

[1] Griffiths, D. A., (1973). *Biometrika*, **29**, 637–648.

[2] Lord, F. M., (1965). *Psychometrika*, **30**, 239–270.

[3] Lord, F. M. and Novick, M. R. (1968). *Statistical Theories of Mental Test Scores*. Addison-Wesley, Reading, Mass. (The current classic on mental test theory.)

[4] Morrison, D. G. and Brockway, G. (1979). *Psychometrika*, **44**, 427–442.

[5] Wilcox, R. R. (1979). *Psychometrika*, **44**, 55–68. (This paper considers the problem of determining whether an examinee has a true score greater than or less than θ_0, an unknown parameter that characterizes a control group.)

[6] Wilcox, R. R. (1980). *Appl. Psychol. Meas.*, **4**, 425–446.

[7] Wilcox, R. R. (1981a). *J. Educ. Statist.*, **6**, 3–32. (A review of the beta-binomial model with an emphasis on mental test theory.)

[8] Wilcox, R. R. (1981b). *Brit. J. Math. Statist. Psychol.*, **34**, 238–242.

[9] Wilcox, R. R. (1982a). *Brit. J. Math. Statist. Psychol.*, **35**, 57–70. (The only item sampling model that has given satisfactory results when dealing with guessing.)

[10] Wilcox, R. R. (1982b). *J. Educ. Meas.*, **19**, 67–74.

[11] Wilcox, R. R. (1982c). *J. Exper. Educ.*, **50**, 219–222.

[12] Wilcox, R. R. (1982d). *Brit. J. Math. Statist. Psychol.*, **35**, 193–207.

[13] Wilcox, R. R. (1982e). *Educ. Psychol. Meas.*, **42**, 789–794.

[14] Wilcox, R. R. (1983a). *Appl. Psychol. Meas.*, **8**, 239–240.

[15] Wilcox, R. R. (1983b). *Psychometrika*, **48**, 211–222.

[16] Wilcox, R. R. (1983c). *Educ. Psychol. Meas.*, **43**, 43–51.

(EDUCATIONAL STATISTICS
GROUP TESTING
PSYCHOLOGICAL TESTING THEORY)

RAND R. WILCOX

STRUCTURAL DISTRIBUTION *See* STRUCTURAL INFERENCE

STRUCTURAL EQUATION MODELS

See the section Simultaneous Equation Models in the entry ECONOMETRICS. For a recent reference, see Judge et al. [1, pp. 561–695] and the references listed therein. *See also* FIX-POINT METHOD.

Reference

[1] Judge, G. G., Griffith, W. E., Hill, R. C., Luetkepohl, H., and Lee, T.-C. (1985). *The Theory and Practice of Econometrics*, 2nd ed. Wiley, New York.

STRUCTURAL ESTIMATION *See* STRUCTURAL INFERENCE

STRUCTURAL INFERENCE

Structural inference is statistical inference* in the presence of model structure *beyond* that provided by the ordinary statistical model, which is taken as a class of probability measures or density functions. A central part of structural inference, representing the origins of the topic, is concerned with the statistical analysis of structural models*, with special emphasis on the logical and deductive aspects that devolve from the structure.

Model structure beyond that provided by the ordinary model as given by the class of measures or densities can include metric and continuity properties for the observations and parameter spaces, particular component variables specified as *givens* as opposed to constructs, separate model components corresponding to physically distinct elements in the investigation, and other *structural* elements in various directions away from the minimum space-algebra-measures model. Also, in a sort of reverse way, the omission of elements from the ordinary model can be of structural interest—the risks of going

negative on structure relative to the ordinary model, at its technical minimum; for example, see ref. 5.

AN EXAMPLE

Consider a simple example introduced in STRUCTURAL MODELS. A measurement process known to be bias-free has measurement error* that is Student-t on 8 degrees of freedom with scaling factor 2.7. For two independent measurements y_1 and y_2 on a quantity θ we have the structural model $y_1 = \theta + e_1$, $y_2 = \theta + e_2$, where (e_1, e_2) is a sample from the error distribution $(2.7)^{-2} \Pi_1^2 f(e_i/2.7 : 8)$, where $f(\cdot : \lambda)$ is a Student density on λ degrees of freedom. The data recorded there are $(123.1, 126.5)$.

From the equations for measuring θ we have by subtraction $y_2 - y_1 = e_2 - e_1$, which for the data, gives $e_2 - e_1 = 126.5 - 123.1 = 3.4$. Thus a full one-dimensional aspect of the errors in the application can be calculated directly from the data.

The discussion of modelling criteria in ref. 10 (Chap. 1) focuses on what is required for the modelling of unknowns in an investigation. For the present example, the difference $d = e_2 - e_1 = 3.4$ is known for the actual measurement errors, but the location of the errors as given by e_1, say, is unknown. An analogous case more central to routine probability theory* is that of a bridge player who sees his own and his partner's hands and asks concerning the split of the remaining 26 cards between the opponents. The applied-probability model is the conditional model given the observed hands [6, p. 10; 10, Sec. 4.1]. For the measurement process, the same model criteria predicate the conditional distribution of (e_1, e_2) or equivalently e_1, given $e_2 - e_1 = 3.4$. The corresponding density function is $g(e_1) = k^{-1} f(e_i/2.7 : 8) f((e_1 + 3.4)/2.7 : 8)$, where the norming constant k is obtained by computer or even graphical numerical integration* and the related equation is $y_1 = \theta + e$; the observed data value is $y_1 = 123.1$. For the corresponding response distribution of (y_1, y_2) or of y_1 conditional on the observed $y_2 - y_1 = e_2 - e_1 = 3.4$, we have the density

$$g(y_1 - \theta) = k^{-1} f((y_1 - \theta)/2.7 : 8)$$
$$\times f((y_1 + 3.4 - \theta)/2.7 : 8),$$

where k is as already described; the data value is of course $y_1 = 123.1$.

This conditioning from $\mathbb{R}^2$ for (e_1, e_2) or (y_1, y_2) to $\mathbb{R}$ for, say, e_1 or y_1, is viewed as *necessary* in the framework of the modelling criteria [10, Chap. 1]. "Necessary" was correspondingly used as part of the title and discussion in ref. 7. This and its analog more generally provide a substantial *reduction* in going from an initial model to a *reduced model**.

With the reduced model obtained by conditioning on the observed characteristic $e_2 - e_1 = 3.4$, the routes for statistical inference are largely well determined. We have $y_1 = \theta + e_1$ together with the density $g(e_1)$ already derived, or in ordinary model form we have $g(y_1 - \theta)$ using the same function $g(\cdot)$. The data comprise the single value $y_1 = 123.1$.

For estimation, the choice of estimate depends in essence only on the type of estimate wanted. The unbiased estimate would be $y_1 - \hat{d} = 123.1 - \hat{d}$, where $\hat{d}$ is the bias in the error distribution $g(\cdot)$: $\hat{d} = \int e_1 g(e_1)\, de$. The model estimate as available would be $y_1 - \hat{d} = 123.1 - \hat{d}$, where $\hat{d}$ is the mode (as available) of $g(\cdot)$. Similarly, median or mean-square error estimates are directly available in a corresponding manner.

For testing the hypothesis $\theta = \theta_0 = 110$, say, we note that under the hypothesis an additional characteristic of the error can be calculated, $e_1 = y_1 - \theta_0 = 123.1 - 110 = 13.1$. This error can then be compared with the error distribution $g(\cdot)$ and an observed level of significance (OLS) calculated in terms of a tail area of $g(\cdot)$ outward from the value 13.1,

$$\text{OLS} = \int_{13.1}^{\infty} g(e_1)\, de_1 \bigg/ \int_{\hat{d}}^{\infty} g(e_1)\, de_1,$$

using, say, the conical test methods in ref. 12.

For forming a confidence region for θ, the distribution $g(\cdot)$ of $e_1 = y_1 - \theta$ can provide a central $1 - \alpha$ interval, say 95%, of values (e_1', e_1''). The corresponding confidence interval is $(123.1 - e_1'', 123.1 - e_1')$.

For a detailed numerical analysis of a more complicated example see ref. 7.

LOCATION MODELS

Consider the statistical analysis of the structural version of the location model. The example in the preceding section illustrated the case $n = 2$.

The structural location model M for a sample of n has the form

$$y_1 = \theta + e_1, \ldots, y_n = \theta + e_n,$$

where $\mathbf{e}$ is distributed as $\Pi f_\lambda(e_i)$. The primary (structural) parameter is θ and the shape parameter for the error distribution is λ. We examine the analysis of this model M in the presence of response data $y_1^0, \ldots, y_n^0$.

In accord with the model, the data $y_1^0, \ldots, y_n^0$ have come from realized values for the errors $e_1, \ldots, e_n$. This permits us to calculate significant characteristics of the realized error values:

$$e_2 - e_1 = y_2^0 - y_1^0, \ldots, e_n - e_1 = y_n^0 - y_1^0.$$

All the differences among the e-values are observable as the corresponding differences among the y^0 values. In fact there is only one freedom left for the vector $(e_1, \ldots, e_n)$ of errors, namely the location of the n values on the real line. In the preceding section, we used e_1 to measure or record this location and obtained the residual equation $y_1^0 = \theta + e_1$. We could, in fact, have used any of a wide range of possibilities to describe the location of the sample, and the results in terms of where the sample is located and in terms of inferences for θ would be the same. For the present discussion we choose the average $\bar{e}$ to describe location; it has computational advantages and *the results are independent of the particular choice*.

Note then that we have *observed* the differences $e_1 - \bar{e}, \ldots, e_n - \bar{e}$: Their values are, respectively, $y_1^0 - \bar{y}^0, \ldots, y_n^0 - \bar{y}^0$, recorded, say, as $d_1^0, \ldots, d_n^0$. The single unobserved characteristic of the error is $\bar{e}$. In accord with discussion in the preceding section [6, p.10; 10, Sec. 4.1], we describe $\bar{e}$ by means of the conditional distribution given $(e_1 - \bar{e}, \ldots, e_n - \bar{e}) = \mathbf{d}^0$. This reduction is based on the modelling criteria mentioned in the preceding section, which can be summarized briefly thus: *a model presents the information concerning the unknowns of the application*. For a different kind of use of these criteria, see ref. 11. This separation into observable and unobservable is a central logical aspect that arises in the analysis of structural models [6, 10].

The marginal density $h_\lambda(\mathbf{d})$, for $\mathbf{d}$ which takes values in $\mathscr{L}^\perp(\mathbf{1})$ (the orthogonal complement of the one vector), is available by one-dimensional computer or graphical integration:

$$h_\lambda(\mathbf{d}) = \int_{-\infty}^{\infty} \Pi f_\lambda(\bar{e} + d_1, \ldots, \bar{e} + d_n) \sqrt{n} \, d\bar{e}.$$

For statistical inference concerning λ, we have an observed value $\mathbf{d}^0$ from the distribution $h_\lambda(\mathbf{d})$ on $\mathscr{L}^\perp(\mathbf{1})$. This is an ordinary statistical model situation. The complication typically is that $h_\lambda(\mathbf{d})$ is available only numerically, not in closed functional form; a few exceptions exist and seem to be related to the normal* and exponential* models. As a consequence, in most cases, the only readily available procedure involves the examination of the observed likelihood function $L(\lambda) = ch_\lambda(\mathbf{d}^0)$; this can be plotted against λ. Note that this is a by-product of the determination of $h_\lambda(\mathbf{d}^0)$, which is the norming constant for the conditional density of $\bar{e}$ or $\bar{y}$.

The conditional distribution of $\bar{e}$ is

$$g_\lambda(\bar{e}) = h^{-1}(\mathbf{d}^0) \Pi f_\lambda(\bar{e} + d_i^0) \sqrt{n} \, ;$$

the related equation is $\bar{y} = \theta + \bar{e}$. The corresponding response distribution of $(y_1, \ldots, y_n)$ or of $\bar{y}$, given the observed differences, has density

$$g_\lambda(\bar{y} - \theta) = h_\lambda^{-1}(\mathbf{d}^0) \Pi f_\lambda(\bar{y} - \theta + d_i^0) \sqrt{n}$$

with, of course, the same norming constant. For statistical inference concerning θ, we have the observed $\bar{y}^0$, the equation $\bar{y} = \theta + \bar{e}$, and the distribution $g_\lambda(\bar{e})$, or equivalently $\bar{y}^0$ and the distribution $g_\lambda(\bar{y} - \theta)$.

The various routes for inference follow the pattern of the example in the preceding section. For estimation, we have $\hat{\theta} = \bar{y}^0 - \hat{d}$, where $\hat{d}$ is the appropriate mean, median, mode, or other characteristic of the distribution $g_\lambda(\cdot)$. This is of course conditional on a chosen or selected value or values for λ, perhaps selected on the basis of the likelihood* analysis just discussed.

For testing the hypothesis $\theta = \theta_0$, we compare the observed value $\bar{y}^0 - \theta_0$ for $\bar{e}$ with the distribution $g_\lambda(\cdot)$, perhaps calculating a tail area as an observed level of significance.

For a confidence interval, we calculate a $1 - \alpha$ interval for $g_\lambda(\cdot)$, say $(\bar{e}', \bar{e}'')$, and then obtain the confidence interval $(\bar{y}^0 - \bar{e}'', \bar{y}^0 - \bar{e}')$. This same interval can be derived, if desired, from the pivotal quantity $\bar{y} - \theta$ examined conditionally, given the differences $\mathbf{d}^0$.

More generally, location models follow closely the preceding pattern of analysis. Typically, say for a three-dimensional parameter, there would correspond a three-dimensional error distribution. The functional form of this distribution would be available immediately from the basic error distribution, but the norming constant would require some type of three-dimensional integration. This of course fits in with the calculation of any error probability, which would also require a three-dimensional integration.

LOCATION-SCALE MODELS*

Consider a sample of size n from a response y whose error pattern is given by a distribution $f_\lambda(\cdot)$ and manifests itself in the response at the location μ with scaling σ; the parameter λ can allow for some uncertainty concerning the shape of the distribution. The structural location-scale model has the form

$$y_1 = \mu + \sigma e_1, \ldots, y_n = \mu + \sigma e_n,$$

where $\mathbf{e}$ is distributed as $\Pi f_\lambda(e_i)$. The primary parameters are μ for location and σ for scaling. Note that the standardizations used for the error density $f_\lambda(\cdot)$ are interdependent with the definitions adopted for μ and σ as the response location and scaling parameters. Various details of this are discussed in refs. 6–8 and 10.

For observed data $y_1^0, \ldots, y_n^0$, the *relative* spacings for the underlying errors are observable. This can be expressed in various ways, for example,

$$\left(\frac{e_1 - \bar{e}}{s_e}, \ldots, \frac{e_n - \bar{e}}{s_e} \right)$$
$$= \left(\frac{y_1^0 - \bar{y}^0}{s_{y^0}}, \ldots, \frac{y_n^0 - \bar{y}^0}{s_{y^0}} \right) = \mathbf{d}^0.$$

The length of this vector is $(n - 1)^{1/2}$ if based on, say, the use of the standard deviation s_e and the average $\bar{e}$.

The distribution for the observable $\mathbf{d}$ is obtained by two-dimensional integration:

$$h_\lambda(\mathbf{d}) = \int_{-\infty}^{\infty} \int_0^{\infty} \Pi f_\lambda(\bar{e} + s_e d_i)$$
$$\times \sqrt{n}\, s_e^{n-2} (n - 1)^{(n-1)/2}\, ds_e\, d\bar{e}.$$

This is the model that describes the observed $\mathbf{d}^0$. Unfortunately, the model $h_\lambda(\mathbf{d})$ is rarely obtained in closed form after the integrations. If not in closed form, then only likelihood-function $(L(\lambda) = ch_\lambda(\mathbf{d}^0))$ analysis concerning λ seems to be easily available.

For a given λ value, the distribution

$$g_\lambda(\bar{e}, s_e) = h_\lambda^{-1}(\mathbf{d}^0) \Pi f_\lambda(\bar{e} + s_e d_i^0)$$
$$\times \sqrt{n}\, s_e^{n-2} (n - 1)^{(n-1)/2}$$

for $\bar{e}$ and s_e and the equations

$$\bar{y} = \mu + \sigma e,$$
$$s_y = \sigma s_e,$$

together with the observed $\bar{y}^0$ and s_{y^0} provide the inference base for μ and σ.

The equations factor in a unique manner as

$$\frac{\bar{y} - \mu}{s_y} = \frac{\bar{e}}{s_e} = \frac{t}{\sqrt{n}}, \qquad \frac{s_y}{\sigma} = s_e.$$

In fact, any other separation of μ and σ provides equivalent individual equations.

The marginal distribution of $t/\sqrt{n}$ provides inference concerning μ, essentially in the pattern in the preceding section. Similarly the marginal distribution of s_e provides inference concerning σ, again essentially in the pattern of the preceding section (use logarithms).

As part of the preceding, it can be noted that any other location-scale measures, say $\tilde{y}$ the median and R the range, would lead to the same statistical inferences. For example the change from $\bar{e}$ to $\tilde{e}$ would correspond to a location-scale change and the same change would be exactly reflected in the change of distribution from that for $\bar{e}$ to that for $\tilde{e}$.

For more leisurely discussion of this inference process together with data see refs. 7, 8, and 10.

More general location and scale models have analogous analyses. An implementation limitation arises with respect to the size of numerical integrations* that are feasible on the available computer systems; however, an alternative that accomplishes most of the preceding results with modest calculations is now in development [12].

GENERAL MODEL

A general structural model takes the following form. The response Y is a transformation θ of a realized error E. The transformations θ belong to a transformation group G and the error E is a realized value from a distribution D_λ with parameter λ.

For an application, this model should provide a reasonable approximation to the available information concerning the system under investigation; for example, in terms of the distribution for error and in terms of the group-closure property, say, holding for some reasonable neighbourhood on the E or θ spaces. For a general theoretical examination of the need for the closure property *see* STRUCTURAL MODELS and refs. 1–3, 11, 12, and 15 in that article.

Consider the analysis of the general model $Y = \theta E$, $E \sim D_\lambda$ with data Y_0, for which the realized E is an element of $\{\theta^{-1}Y_0 : \theta \in G\} = \{gY_0 : g \in G\} = GY_0$, that is, $GE = GY_0$ and the observable characteristic of the error is GE which takes the value GY_0. Thus for λ we have the marginal model for GE and the observed value GY_0. In most cases this will lead at least to an accessible likelihood function analysis.

For inference based on a selected λ value, we have necessarily (relative to the modelling criteria [10]) the conditional model for E given $GE = GY_0$ and the equation $Y = \theta E$ together with data Y_0. Alternately, we have the conditional model for $Y = \theta E$, given $GY = GE = GY_0$, together with data Y_0.

With regular models and an identifiable parameter θ, the variable E, conditionally given $GE = GY_0$, can be represented by a group element $[E]$ relative to a reference point $D(E) = D(Y_0)$ chosen on the orbit set $GE = GY_0$: $E = [E]D(Y_0)$. The conditional model then describes $[E]$ for fixed $D(E) = D(Y_0)$ or describes $[Y] = \theta[E]$ for fixed $D(Y_0)$.

For a discussion of various continuous variable examples of this general type of model see ref. 10.

STRUCTURAL PROBABILITY

Consider the simple example in the introductory section An Example. The analysis of the model and data gives the distribution

$$g(e_1) = k^{-1}f(e_1/2.7 : 8)$$
$$\times f((e_1 + 3.4)/2.7 : 8)$$

for the error e_1 and the equation $y_1 = \theta + e_1$, together with the data value $y_1 = 123.1$.

A basic question is whether the observed value $y_1 = 123.1$ together with the equation $y_1 = \theta + e_1$, where θ is unknown, gives any grounds for altering the probability description $g(e_1)$ for the unknown e_1. Certainly the description $g(e_1)$ for an unobserved realized value is standard in applied probability*; recall the hands-of-bridge example in the

earlier discussion of the present example. The added item here is then that the observed y_1 has come via the equation $y_1 = \theta + e_1$, with the θ value unknown. Arguments have been given [6], [10] that the probability description remains valid. Some disagreements with this result, not, however, with the arguments themselves, have been expressed in terms of betting assessments of probabilities and answered in ref. 9.

The *structural probability distribution* describing information concerning θ is obtained by using $g(e_1)$ to describe the e_1 in the equation $123.1 = \theta + e_1$ and then "inverting" to obtain a distribution for θ:

$$k^{-1}f((123.1 - \theta)/2.7 : 8)$$
$$\times f((123.1 - \theta + 3.4)/2.7 : 8)$$

If a 95% interval (e_1', e_1'') is chosen for e_1, then a 95% structural interval $(123.1 - e_1'', 123.1 - e_1'')$ is obtained for θ. As this is standard pivotal inversion, the interval is also a 95% confidence interval.

Cases where concern has been expressed are those where the 95% interval for e_1 is varied in its location in a way dependent on the observed y_1. Of course, unusual results occur with confidence intervals if the same dependence is admitted. The concerns that have been expressed for the structural interval correspond to cases where structure on the θ values is involved: the discussion in ref. 9 showed that a different structure on the θ values could switch the coverages the other way.

These questions concerning validity when the regions for the error are allowed to depend on the response values seem now largely to rest with complicated groups, involving nonamenability. Such groups indeed pose difficulties in other areas of statistics.

Consider the general location model

$$y_1 = \theta + e_1, \ldots, y_n = \theta + e_n, \mathbf{e} \sim \Pi f(e_i)$$

with data $y_1^0, \ldots, y_n^0$. The reduced model is $\bar{y} = \theta + \bar{e}$, $\bar{e} \sim g(\bar{e})$ with data $\bar{y}^0$;

$$g(\bar{e}) = h^{-1}\Pi f(\bar{e} + d_i).$$

The *structural distribution* for θ is obtained from $\bar{y}^0 = \theta + \bar{e}$ using the distribution $g(\bar{e})$

for $\bar{e}$,

$$h^{-1}\Pi f\left(\bar{y}^0 - \theta + d_i\right) = h^{-1}\Pi f\left(y_i^0 - \theta\right);$$

this is the normalized likelihood function. The basic question is whether $\bar{y}^0$ in the equation $\bar{y} = \theta + e_1$ gives any grounds to alter the probability description $g(\bar{e})$. This has been addressed in refs. 6 and 10; arguments support the probability description; betting arguments suggest some caution, but these are in a context of a probability distribution for θ, which favors the centre of the line and does not cover, for example, diverging sequences of θ-values.

For the general model $Y = \theta E$, $E \sim D$ with data Y_0, we obtained the reduced model $Y = \theta E$, where E is distributed as E conditional on $GE = GY_0$. The structural distribution is obtained by transferring this distribution for E to θ via the equation $Y_0 = \theta E$. For the regular models, this gives $\theta = [Y_0][E]^{-1}$, where $[E]$ has the conditional distribution given $D(E) = D(Y_0)$.

In most cases the structural distribution coincides with the Bayesian posterior based on a flat prior, flat with respect to a right invariance prior*; see refs. 6 and 10.

A structural prediction distribution is a probability distribution presenting information concerning a future observation or data array. It provides probabilities that future data fall in intervals or regions of interest; it can also provide point estimates or interval estimates that have desirable properties in relation to the prediction distribution.

The structural prediction distribution is obtained by combining the model for the future observation with the structural distribution for the parameter and integrating out the parameter.

For example, with the simple location model in the section Location Models, suppose that a future observation w has density $k(w : \theta)$. The joint distribution for w and θ is $g_\lambda(\bar{y} - \theta)k(w : \theta)$ and the marginal distribution for w is then $\int g_\lambda(\bar{y} - \theta)k(w : \theta)\, d\theta$; this is called the *prediction distribution*.

For the general structural model $Y = \theta E$, $E \sim D_\lambda$, we obtained the reduced model Y = θE, where E has the conditional distribution given $GE = GY_0$. Suppose the equation $Y_0 = \theta E$ can be solved as $\theta = \mathrm{Str}(Y_0, E)$ which in the regular case becomes $\theta = [Y_0][E]^{-1}$; the conditional distribution for E then gives the structural distribution for θ. Consider a future observation W with density $k(W : \theta)$. The prediction density for W is then obtained by averaging $k(W : \theta)$ with respect to the distribution given by $\mathrm{Str}(Y_0, E)$ for θ. In the regular case, where in addition $W = \theta E_2$, we obtain its conditional prediction distribution as given by $[Y_0][E]^{-1}E_2$.

COMPARISONS

As noted in the Introduction, structural inference is concerned with statistical inference in the presence of model structure beyond that of the ordinary statistical model, with special attention to the analysis of error (*structured* and *structural*) models. In particular, it focuses on deductive steps that proceed from such structure. In these terms it is not a method of inference in the ordinary sense, as, say, confidence methods, significance tests, and UMV unbiased estimation are. Indeed, these various methods are all natural terminal methods to follow after the deductive steps.

As the preceding indicates, structural inference does not directly address statistical inference in a general context; rather it addresses these aspects that follow from special elements of structure. However, in the general context the structural approach has produced a direct argument for the likelihood as being all that the model provides concerning a particular data point [6, p. 186ff; 8, p. 310ff; 10 p. 70ff]; this can be viewed as stronger than the support that derives from sufficiency. That structural inference does not directly address statistical inference in general is not to suggest a lack of immediate concern for such, but rather that there is a place for determining and evaluating the special consequences of structure in statistical inference.

For making comparisons among methods of statistical inference, there is a need to isolate the ingredients for statistical inference: *The Given*, the elements that are in correspondence (to a reasonable approximation) with aspects of the system under investigation—*the model and the data*; *The Assumptions*, the various additional ingredients or *additives* introduced as part of the "inference process"—choice of principle, choice of ancillary*, choice of pivotal quantity*, choice of estimation methods, etc. These latter ingredients are not taken to be in direct correspondence with the investigation, although a metamodelling process might embrace some, as in correspondence with a larger physical context.

Fiducial inference*, alongside *likelihood* and *sufficiency*, was one of Fisher's major contributions to statistics in the 1930s and 1940s. That it did not develop as a concept as well as the others may only be that it was attempting to go farther, perhaps prematurely. It was directed exclusively toward posterior distributions* and has resisted acceptable formalization with ordinary models; it did not follow from the *given*. Structural analysis originated with the use of error models to provide a framework for fiducial-type analyses.

General error models, called *structured models*, were introduced [14] and validity was examined for various conditional probabilities. Bunke [3] called such models *functional models* and Dawid and Stone [4] used them to reexamine certain fiducial distributions*; neither noted that the nominal conditional probabilities can be invalid, as discussed in [14]. For this, the structured model would be taken as part of the given, but questions concerning the identification of the error complement were not examined.

Pivotal inference* takes the ordinary model as *the given* and then adds pivotal quantities* and ancillary statistics*. The pivotal quantities are the ordinary means for generating tests and confidence intervals and ancillary statistics provide a conditioning concept from Fisher. If the pivotal quantities are taken as part of the given, then the enlarged model is an inverse-function reexpression of structured models [2]. If the pivotal quantities are additives, then questions concerning the arbitrariness of the choice arise. As an exploratory procedure for examining models and data, they are part of confidence theory analyses.

As an example illustrating the arbitrary choice involved with pivotals, consider (x_1, x_2) with a bivariate normal distribution having means 0, variances 1, and correlation ρ. A first pivotal pair is

$$x_1 \sim N(0,1),$$
$$(x_2 - \rho x_1)/(1 - \rho^2)^{1/2} \sim N(0,1);$$

an alternate pair is

$$x_2 \sim N(0,1),$$
$$(x_1 - \rho x_2)/(1 - \rho^2)^{1/2} \sim N(0,1).$$

The 95% central confidence interval from the first is

$$\frac{x_1 x_2 \pm 1.96(1.96^2 + x_1^2 - x_2^2)^{1/2}}{x_1^2 + 1.96^2}$$

and from the second is obtained by interchanging x_1 and x_2; these intervals can be quite different, say, nonoverlapping, or one empty.

For some recent applications of structural techniques in general statistical inference, see refs. 12 and 13.

References

[1] Brenner, D., Fraser, D. A. S., and Monette, G. (1981). *Statist. Hefte*, **22**, 231–234.

[2] Brenner, D., Fraser, D. A. S., and Monette, G. (1983). *Statist. Hefte*, **24**, 7–20.

[3] Bunke, H. (1975). *Math. Operationsforschung Statist.*, **6**, 667–676.

[4] Dawid, P. and Stone, M. (1982). *Ann. Statist.*, **10**, 1054–1074.

[5] Feuerverger, A. and Fraser, D. A. S. (1980). *Canad. J. Statist.*, **8**, 41–45.

[6] Fraser, D. A. S. (1968). *The Structure of Inference*. DAI, University of Toronto Textbook Store, Toronto, Ontario.

[7] Fraser, D. A. S. (1976a). *J. Amer. Statist. Ass.*, **71**, 99–110.

[8]   Fraser, D. A. S. (1976b). *Probability and Statistics: Theory and Applications*. DAI, University of Toronto Textbook Store, Toronto, Ontario.

[9]   Fraser, D. A. S. (1977). *Ann. Statist.*, **5**, 892–898.

[10]   Fraser, D. A. S. (1979). *Inference and Linear Models*. DAI, University of Toronto Textbook Store, Toronto, Ontario.

[11]   Fraser, D. A. S. (1982). In *Foundations of Statistical Inference*, V. P. Godambe and D. A. Sprott, eds. Holt, Rinehart and Winston, Toronto, Ontario, pp. 32–55.

[12]   Fraser, D. A. S. and Massam, H. (1985). Conical tests: Observed levels and confidence regions of significance. *Statist. Hefte*, **26**, 1–18.

[13]   Fraser, D. A. S. and Massam, H. (1984). Second order inference for generalized least squares. *Canad. J. Statist.*, **15**, 21–30.

[14]   Fraser, D. A. S., Monette, G., and Ng, K.-W. (1983). Marginalization, likelihood and structured models, In *Multivariate Analysis VI*, P. R. Krishnaiah, ed. North-Holland, Amsterdam, Netherlands.

(FIDUCIAL INFERENCE
INFERENCE, STATISTICAL
PIVOTAL INFERENCE
REDUCED MODEL
STRUCTURAL MODELS
STRUCTURAL PREDICTION
STRUCTURAL PROBABILITY)

D. A. S. FRASER

STRUCTURAL MODELS

A structural model is an *error model* wherein the error variable is taken formally to be the source for variation in the response variable. As such, a structural model provides directly a description of the error or variation source in a process or system under investigation, and in addition provides a parameter-dependent function or transformation that presents the response variable in terms of the error variable.

The structural model gives a more detailed description of a process or system, more detailed, that is, than the ordinary statistical model, which provides a set of possible distributions for the response of the process or system. This more detailed statistical model predicates certain inference methods of analysis; *see* STRUCTURAL INFERENCE. A generalization of the structural model, called a *structured model*, omits a closure property of structural models, a property that may be essential for the validity of some of the inference methods based on error variables. (*See* STRUCTURAL INFERENCE and ref. 1 therein.)

After a simple example of a structural model, we describe briefly the background of the structural/error models and then outline some examples indicating directions in which the model can be developed. For the *analysis* of structural models with data *see* STRUCTURAL INFERENCE.

AN EXAMPLE

Consider a measurement process that is bias-free and has error that is closely approximated by the Student-t distribution* on 8 degrees of freedom and with scaling factor 2.7. For two independent measurements y_1, y_2 of a physical quantity θ, we can contemplate two different modellings.

The ordinary statistical model presents (y_1, y_2) as a sample of 2 from the Student(8) distribution, scaled by 2.7 and relocated to θ; that is, with density $(2.7)^{-2}\Pi_1^2 f((y_i - \theta)/2.7:8)$ on $\mathbb{R}^2$, where $f(\cdot : \lambda)$ is the Student density on λ degrees of freedom. The formally presented model has $F = \{(2.7)^{-2}\Pi_1^2 f((y_i - \theta)/2.7:8) : \theta \in \mathbb{R}\}$ as the class of response distributions on $\mathbb{R}^2$. Note that with $\theta \in \mathbb{R}$, this is an *infinite* class of distributions for the response. Also note that for the model with observed data, the statistical problem (*see* INFERENCE, STATISTICAL-II) is to organize, summarize, and present the available information concern-ing θ.

By contrast, the structural model derives from the explicit modelling of the error or measurement process. For this let e_1, e_2 designate the errors in the operation of the measuring instrument; these have the *known* distribution $(2.7)^{-2}\Pi_1^2 f(e_i/2.7 : 8)$ on $\mathbb{R}^2$. The use of the instrument for the measurement of θ then gives the response variable y_1, y_2 as tranformations or presentations $y_1 = \theta + e_1$, $y_2 = \theta + e_2$ of the error variables e_1, e_2. This model has a single (known) distribution for the error and two equations for the physical measurement of θ: $M = ((2.7)^{-2}\Pi_1^2 f(e_i/2.7 : 8)$; $y_1 = \theta + e_1$, $y_2 = \theta + e_2)$. For this model coupled with data, the statistical problem is to organize, summarize and present the available information concerning the true θ.

With the first statistical model we have an infinite class of probability distributions; with the second statistical model we have a single probability distribution, the distribution that describes the measurement instrument or process. In the second case we are in the position of being able to use the actual probabilities of the process or system. The advantage of the more detailed structural model then shows in the logical implications that arise as part of statistical inference. These implications and advantages of the explicit modelling of error were developed in a series of papers commencing with [9, 10].

For the example, suppose that the data are $y_1 = 123.1$, $y_2 = 126.5$. The ingredients for statistical inference with the structural model consist then of the preceding model M and the data (123.1, 126.5). We examine the implications and inferences for this model-data combination as an initial simple example in STRUCTURAL INFERENCE.

BACKGROUND

References to error and the use of symbols for it may be found in the literature of probability and statistics from the time of Gauss and Laplace to that of the common use of $\mathbf{y} = \mathbf{X}\beta + \mathbf{e}$ for the linear model. This familiar use of error variables provides a compact means of referring to the class of possible response distributions and does not commonly connote a substantive presentation of the randomness in the system under investigation. The formality of an explicit error space with error distribution (a *probability space**) is recent in statistics (see refs. 9 and 10). In particular, the implication from explicit error to conditional error distributions was initiated in a case of "automatic conditioning."

The development of structural models arose as part of an examination [8, 10] of inference procedures in R. A. Fisher*'s statistical papers, with attention in large part focused on the fiducial approach to statistics. The structural model provided a framework in which certain of the fiducial examples could be reformulated in such a way that the procedures acquired a logical or deductive validity as opposed to their apparently somewhat arbitrary nature in the original context; they also opened up a range of problems to largely unequivocal solution by standard inference methods. The structural model can thus lead to *structural probability** (the developed alternative to *fiducial probability**), but it also leads to a full spectrum of terminal inference procedures (*see* INFERENCE, STATISTICAL-II and STRUCTURAL INFERENCE); this was indicated in ref. 10 and elaborated on later in ref. 15.

Structural models were generalized in ref. 11 by the removal of the closure (group) property for the set of functions that presents the response in terms of the error. It was noted there and formalized [1] that this could remove justification for (indeed invalidate) the central inference step to a conditional error distribution and thus lead to possible erroneous results.

COMPARISONS

We briefly compare the structural model with the ordinary statistical model and use the preceding simple example for illustration.

We first note that *given* the structural model, we can deduce the ordinary model by presenting only the induced distributions for the response variable (y_1, y_2). On the other hand, given the ordinary model, we are in a position only to acknowledge the structural model as a possible description; it cannot be deduced. This second direction amounts to seeking a variable that has a fixed distribution, that is, a *pivotal* variable. Such a constructed pivotal variable might or might not correspond to an objective error variable in the system under investigation. Such a question can only be decided by actual reference to the process or system. If such an error variable were in fact identified, then modelling criteria argue that it should be an objective element in the model, thus requiring in fact the use of the structural model or an isomorphic equivalent. In short, to go from the ordinary model to the structural model, we need additional information from the measurement process.

Thus the structural model makes a stronger statement or description of the physical situation. Where such a stronger statement can be made, basic modelling criteria [15, Chap. 1] say it *must* be made, unless other outside information or theory develops to show that it is inappropriate or irrelevant. Thus, in harmony with using all the substantiated information, we would use the structural model in cases where error was identifiable. For discussion of modelling criteria, see ref. 15, Chap. 1.

We are of course left with a relevant and significant question. Does the difference in model matter for the subsequent inference processes, that is, does the additional information in the structural model matter?

An answer to this is partly involved with the flexibility of notation in statistics, itself perhaps a reaction to the strong formality of decision theory*. The method of analysis determined by the structural model with data (*structural inference*) is one of a series of methods that exists or can be offered for the ordinary model with data. One can say then that "the same result can be obtained from the ordinary model with data." This is true, in the sense of the flexibility just mentioned. More substantively, however, it is not true, as the particular method of inference can

only be obtained from the ordinary model with data by adding material that determines the particular choice of method, or by adding a principle or criterion that does the same. Such additional material could be the specification of the error variable, but that then amounts to specifying the structural model. Without such an objective reference, the additional material amounts to an arbitrary element, an *additive* in the sense discussed in [3]. Why make the special arbitrary choice? To agree with that from the structural model? But then, properly, the starting point should involve the modelling of the error, the *structural model* in the present notation.

GENERAL MODELS

The structural model in the simple example involves an error variable that describes the randomness* inherent in the operation of the measuring instrument.

For a general version of this, consider a physical system where the randomness source has been identified and can be described by an error variable E with distribution D on a space S. Let $Y = T(\theta)E$ be the transformation that presents the response variable Y as a function of the error E, where $T(\theta)$ depends on a parameter θ in Ω; for simplicity write $Y = \theta E$. The model can then be written $M = (E \sim D, Y = \theta E, \theta \in \Omega)$. Such a model is called a *structured model* [11]; more recently, Bunke [5] has used the term *functional model*, and with the model written in terms of the inverse transformation $T^{-1}(\theta) = \theta^{-1}$, Barnard has used the term *pivotal model*. A more general version involves replacing the single distribution D by a class of distributions $\{D_\lambda: \lambda \in \Lambda\}$; see refs. 10 and 15.

An important aspect of the example mentioned at the beginning is the automatic conditioning that occurs with data (*see* STRUCTURAL INFERENCE). This conditioning was first noted in ref. 9 and was central to the development in refs. 10 and 15. The validity of such conditioning requires the transformations θ to be closed under composition and inverse, that is, to form a transformation group [11, 1]. In such cases the model is called a *structural model* [11]: $M = (E \sim D, Y = \theta E, \theta \in \Omega)$, where the collection of transformations θ is a transformation group on the sample space.

In many contexts where a structural model is appropriate, the available information may not allow a reasonable approximation by a single error distribution but yet allow approximation by one in a class of distributions $f_\lambda(E): \lambda \in \Lambda\}$, where λ can be referred to as the shape parameter, shape of the error distribution. This generalization is also called a *structural model*: $(E \sim D_\lambda, Y = \theta E, \lambda \in \Lambda, \theta \in \Omega)$, where again $\{\theta\} = \Omega$ is a transformation group on the sample space. For a general discussion with examples, see ref. 15.

An explicit and clearly identifiable error variable may be found in many applications. The dilution and bioassay* contexts with randomness proposed by Fisher [6] lead to an extraordinary but very clearcut example of explicit error [17, 15}.

An alternative and more general approach to support and justify structural models may be found in ref. 15, based on ref. 1. This involves the assessment of a response variable by means of a family of scans [2, 4]. The ability of the scans to identify an error distribution form is tied to the same closure or group property that is needed to validate the conditioning process; some risks associated with omitting the group property may be found in ref. 12.

LOCATION MODELS

We first survey various location models that extend and generalize the simple example discussed earlier.

Consider the measurement of a real quantity θ and suppose that the measurement error* has density $f_\lambda(e)$ with shape parameter λ in Λ. In a more general context, in accord with the concluding paragraph in the

section General Models, we could consider a *response* location θ and an identified response error with error density $f_\lambda(e)$. In either context, the structural model for a response sample $y_1, \ldots, y_n$ is given by $y_1 = \theta + e_1, \ldots, y_n = \theta + e_n$; $\mathbf{e} \sim \Pi f_\lambda(e_i)$; $\lambda \in \Lambda, \theta \in \mathbb{R}$. The corresponding ordinary model is $\Pi f_\lambda(y_i - \theta)$, $(\theta, \lambda) \in \mathbb{R} \times \Lambda$. The analysis of the structural model is straightforward following the general patter in ref. 15; for the details of a more restricted analysis, see ref. 10. If λ embraces scale change, a location-scale model (to follow) would be required.

A generalized linear model* extension of the preceding would take the following form: The response vector $\mathbf{y}$ has location $\mathbf{X}\beta$ and the response error $\mathbf{e}$ has density $\Pi f_\lambda(e_i)$. The structural model is

$$\mathbf{y} = \mathbf{X}\beta + \mathbf{e}; \quad \mathbf{e} \sim \Pi f_\lambda(e_i);$$

$$\lambda \in \Lambda, \beta \in \mathbb{R}^r.$$

The corresponding ordinary model is

$$f_\lambda(\mathbf{y} - \mathbf{X}\beta), \quad (\beta, \Lambda) \in \mathbb{R}^r \times \Lambda,$$

with $f_\lambda(\mathbf{e})$ designating $\Pi f_\lambda(e_i)$. Again the structural analysis is straightforward in the pattern in ref. 15; the details for a more restricted analysis are indicated by problems in ref. 7, Chap 3.

The preceding can be modified to cover location on the plane $\mathbb{R}^2$ or in $\mathbb{R}^n$. Consider a response $\mathbf{y}$ with location θ and response error $f_\lambda(\mathbf{e})$. Then for a sample $Y = (\mathbf{y}_1, \ldots, \mathbf{y}_n)$, the structural model is given by:

$$\mathbf{y}_1 = \theta + \mathbf{e}_1, \ldots, \mathbf{y}_n = \theta + \mathbf{e}_n;$$

$$\mathbf{E} = (e_1, \ldots, e_n) \sim \Pi f_\lambda(\mathbf{e}_i),$$

$$\lambda \in \Lambda, \theta \in \mathbb{R}^r.$$

Location can also arise on the circle, or on a sphere in three or higher dimensions. Consider a response $Y = (y_1, Y_2)'$ giving direction on the plane and recorded with $y_1^2 + y_2^2 = 1$. Let $(\cos \theta, \sin \theta)$ be the general direction of the response with error $\mathbf{e} = (e_1, e_2)' = (\cos a, \sin a)'$ having a density function $f_\kappa(a)$ on the unit circle. The structural model for a sample $\mathbf{Y} = (\mathbf{y}_1, \ldots, \mathbf{y}_n)$ is given as

$$\mathbf{y}_1 = \mathbf{\Theta}\mathbf{e}_1, \ldots, \mathbf{y}_n = \mathbf{\Theta}\mathbf{e}_n;$$

$$\mathbf{E} = (\mathbf{e}_1, \ldots, \mathbf{e}_n) \sim \Pi f_\kappa(\mathbf{e}_i);$$

$$\kappa \in K; \; \theta \in [0, 2\pi),$$

where

$$\mathbf{\Theta} = \begin{pmatrix} \cos\theta, & \sin\theta \\ -\sin\theta, & \cos\theta \end{pmatrix}$$

is the 2×2 rotation matrix. For the analysis, see Fraser [15, Chap. 9]. This can be generalized to incorporate both scale and shape in the structural component [15, Sec. 9.3].

The circle-sphere example can be modified by having a mean $\boldsymbol{\theta}$ on a circle of radius ρ_0, say, with center at the origin, and having a rotationally symmetric error distribution $f(e_1, e_2)$ for departure from $\boldsymbol{\theta}$; thus, for example, $\mathbf{y} = [\mathbf{\Theta}](\mathbf{e} + \rho_0(1, 0)')$, where $[\mathbf{\Theta}]$ is the 2×2 rotation matrix that rotates through an angle, say θ, from $(1, 0)'$. This model generalized to $\mathscr{R}^n$ provides the basis for second order inference with nonlinear least squares.

Consider a response y with general location μ and error scaling σ, and suppose the error has density $f_\lambda(e)$ known up to a shape parameter λ. Then for a sample $y_1, \ldots, y_n$, the structural model is

$$y_1 = \mu + \sigma e_1, \ldots, y_n = \mu + \sigma e_n;$$

$$\mathbf{e} \sim \Pi f_\lambda(e_i), \qquad \lambda \in \Lambda, \mu \in \mathbb{R}, \sigma \in \mathbb{R}^+.$$

For detailed discussion see refs. 15, Chap. 2, and 13, Chap. 11; further discussion together with adaptive methods may be found in ref. 14.

A generalized-linear-model extension of the preceding yields the structural model version of the common regression* or linear model. The response vector $\mathbf{y}$ has location $\mathbf{X}\boldsymbol{\beta}$ and error scaling σ; the error $\mathbf{e}$ has density function $f_\lambda(\mathbf{e})$, which typically would have the sample form $\Pi f_\lambda(e_i)$. The structural model then takes the form

$$\mathbf{y} = \mathbf{X}\boldsymbol{\beta} + \sigma\mathbf{e};$$

$$\mathbf{e} \sim f_\lambda(\mathbf{e}), \qquad \lambda \in \Lambda, \boldsymbol{\beta} \in \mathbb{R}^r, \sigma \in \mathbb{R}^+.$$

For the analysis, see refs. 14, Chap. 6, and

13, Chap. 11. The corresponding ordinary model is

$$\sigma^{-n} f_\lambda\big(\sigma^{-1}(\mathbf{y} - \mathbf{X}\boldsymbol{\beta})\big),$$

$$(\boldsymbol{\beta}, \sigma, \lambda) \in \mathbb{R}^n \times \mathbb{R}^+ \times \Lambda.$$

For multivariate models, scaling can enter in a variety of ways and to various depths. For example, in the bivariate case we could consider (a) coordinate-by-coordinate scaling giving equations $y_1 = \mu_1 + \sigma_1 e_1$, $y_2 = \mu_2 + \sigma_2 e_2$ or (b) scaling with successive error linear dependence giving equations $y_1 = \mu_1 + \sigma_1 e_1$, $y_2 = \mu_2 + \gamma_{21} e_1 + \sigma_{(2)} e_2$ or (c) full positive linear reexpression giving equations $y_1 = \mu_1 + \gamma_{11} e_1 + \gamma_{12} e_2$, $y_2 = \mu_2 + \gamma_{21} e_1 + \gamma_{22} e_2$ with $|\gamma_{ij}| > 0$; or indeed various modifications of these. For various multivariate and multivariate regression models, see refs. 15 and 7.

If the error distribution in the preceding multivariate models has rotational symmetries, then the ordinary model can be the same for two different structural models. On the assumption that a model as used is supported by the context being described (say by objective error or scanned error), then the preceding causes no problem: The appropriate structural model would be used, and the ordinary model would be seen to omit essential information on error properties.

References

[1] Brenner, D. and Fraser, D. A. S. (1979). *Statist. Hefte*, **20**, 1–22.

[2] Brenner, D. and Fraser, D. A. S. (1980). *Statist. Hefte*, **21**, 296–304.

[3] Brenner, D., Fraser, D. A. S. and Monette, G. (1981). *Statist. Hefte*, **22**, 231–234.

[4] Brenner, D., Fraser, D. A. S., Evans, M., Massam, H., and Rost, E. (1984). *Statist. Hefte*, **25**, 61–68.

[5] Bunke, H. (1975). *Math. Operationsforschung Statist.*, **6**, 667–676.

[6] Fisher, R. A. (1935). *J. R. Statist. Soc.*, **98**, 39–54.

[7] Fraser, D. A. S. (1958). *The Structure of Inference*. DAI, University of Toronto Textbook Store, Toronto, Ontario.

[8] Fraser, D. A. S. (1961). *Biometrika*, **48**, 261–280.

[9] Fraser, D. A. S. (1966). *Biometrika*, **53**, 1–9.

[10] Fraser, D. A. S. (1968). *The Structure of Inference*, 2nd ed. DAI, University of Toronto Textbook Store, Toronto, Ontario.

[11] Fraser, D. A. S. (1972). In *Foundations of Statistical Inference* (V. P. Godambe and D. A. Sprott, eds.), Holt, Rinehart, and Winston, Toronto, Ontario.

[12] Fraser, D. A. S. (1973). *J. Amer. Statist. Ass.*, **68**, 101–104.

[13] Fraser, D. A. S. (1976a). *Probability and Statistics; Theory and Applications*. DAI, University of Toronto Textbook Store, Toronto, Ontario

[14] Fraser, D. A. S. (1976b). *J. Amer. Statist. Ass.*, **71**, 99–110, 112–113.

[15] Fraser, D. A. S. (1979). *Inference and Linear Models*. DAI, University of Toronto Textbook Store, Toronto, Ontario.

[16] Fraser, D. A. S. and Massam, H. (1984). Second order inference for nonlinear least squares. Unpublished.

[17] Fraser, D. A. S. and Prentice, R. L. (1971). *Ann. Math. Statist.*, **42**, 141–146.

(FIDUCIAL INFERENCE
INFERENCE, STATISTICAL-I, II
STRUCTURAL INFERENCE
STRUCTURAL PREDICTION
STRUCTURAL PROBABILITY)

D. A. S. FRASER

STRUCTURAL PARAMETER *See* STRUCTURAL INFERENCE

STRUCTURAL PREDICTION

A *structural prediction distribution* is a probability distribution describing a future observation or a future data array. It provides probabilities that future data fall in intervals or regions of interest; it can also be used to obtain point estimates or interval estimates that have desirable properties in relation to the prediction distribution itself.

The structural prediction distribution is obtained by combining the model distribution for the future observation with the structural probability* distribution for the parameter of the model; the structural distribution in turn is based on a structural model* together with current data. For brief discussion and an example, *see* STRUCTURAL INFERENCE.

D. A. S. FRASER

STRUCTURAL PROBABILITY

A structural probability distribution is a posterior distribution* that is obtained from a structural model* together with data. The posterior distribution gives a description for the unknown parameter value of the model and thus, for any interval or set, produces a corresponding *structural probability* for the parameter lying in the interval.

For some discussion and examples *see* STRUCTURAL INFERENCE, which is the analysis of structural models, and for related discussion, *see* INFERENCE, STATISTICAL-II.

D. A. S. FRASER

STRUCTURAL REGRESSION *See* STRUCTURAL INFERENCE

STRUCTURAL ZEROS

It not uncommonly happens that some cells in a contingency table* are empty (the observed frequencies are zeros). Empty cells may occur because of paucity of data or because certain combinations of factor levels are impossible from logical considerations. Empty cells of the latter kind are called *structural zero cells*. Methods of analyzing data allowing for structural zeros are described in Bishop et al. [1] and Haberman [2, Chap. 7].

References

[1] Bishop, Y. M., Fienberg, S., and Holland, P. W. (1975). *Discrete Multivariate Analysis: Theory and Methods*. MIT Press, Cambridge, Mass., (Chap. 5.)

[2] Haberman, S. (1979). *Analysis of Qualitative Data*, Vol. 2. Academic, New York.

STRUCTURE FUNCTION *See* COHERENT STRUCTURE THEORY

STRUCTURED MATRICES *See* PATTERNED COVARIANCES

Table 1   Data from a Study of Two Matched Samples[a]

Category for Sample 1	Category for Sample 2				Total
	1	2	$\cdots$	k	
1	n_{11}	n_{12}	$\cdots$	n_{1k}	$n_{1.}$
2	n_{21}	n_{22}	$\cdots$	n_{2k}	$n_{2.}$
$\vdots$	$\vdots$	$\vdots$	$\cdots$	$\vdots$	$\vdots$
k	n_{k1}	n_{k2}	$\cdots$	n_{kk}	$n_{k.}$
Total	$n_{.1}$	$n_{.2}$	$\cdots$	$n_{.k}$	$n_{..}$

[a] Each cell entry is a number of *pairs* of observations.

STUART–MAXWELL TEST

When the observations in two samples are paired, with each observation classified into one of k mutually exclusive and exhaustive categories, the resulting data may be arrayed as in Table 1. Stuart [9] (and, independently, Maxwell [4] and Pike et al. [7]) developed a test statistic for testing the hypothesis of marginal homogeneity (*see also* MARGINAL SYMMETRY). If the $n_{..}$ pairs of observations have a multinomial distribution*, with underlying probability P_{ij} that an observation from sample 1 falls in category i and the paired observation from sample 2 falls in category j ($i, j = 1, \ldots, k$), the hypothesis of marginal homogeneity is that $P_{i.} = \Sigma_j P_{ij}$ is equal to $P_{.i} = \Sigma_j P_{ji}$ for all i.

Define $\mathbf{d} = (d_1, \ldots, d_k)'$, where $d_i = n_{i.} - n_{.i}$. Under the hypothesis, $\mathbf{d}$ has a mean vector of $\mathbf{0}$ and an estimated covariance matrix $\mathbf{V} = \|v_{ij}\|$, where $v_{ii} = n_{i.} + n_{.i} - 2n_{ii}$ and, for $i \neq j$, $v_{ij} = -2\bar{n}_{ij}$, say, where $\bar{n}_{ij} = (n_{ij} + n_{ji})/2$. Because $\Sigma d_i = 0$, $\mathbf{V}$ is singular.

Let $\mathbf{A}$ be any $(k-1) \times k$ matrix of rank $k-1$. As shown by Stuart [9], the test statistic

$$X^2 = (\mathbf{Ad})'(\mathbf{AVA'})^{-1}(\mathbf{Ad}) \qquad (1)$$

has, asymptotically, a chi-square distribution* with $k-1$ degrees of freedom when the hypothesis of marginal homogeneity is true. The hypothesis is rejected if $X^2 > \chi^2_{k-1, \alpha}$, where $\chi^2_{\nu, p}$ denotes the $100(1-p)$ percentile of the chi-square distribution* with ν degrees of freedom. Stuart also showed that the value of X^2 does not depend on the

particular $(k-1) \times k$ matrix $\mathbf{A}$ used, so long as it has rank $k-1$. It is simplest to take $\mathbf{A} = \|a_{ij}\|$ with $a_{ii} = 1$ for $i = 1, \ldots, k-1$ and $a_{ij} = 0$ otherwise.

Simple formulas for X^2 are available when $k = 2$ and $k = 3$. For $k = 2$,

$$X^2 = \frac{(n_{12} - n_{21})^2}{n_{12} + n_{21}}, \qquad (2)$$

identical to the McNemar statistic* [5] without the continuity correction. For $k = 3$, Fleiss and Everitt [3] showed that

$$X^2 = \frac{\bar{n}_{23}d_1^2 + \bar{n}_{13}d_2^2 + \bar{n}_{12}d_3^2}{2(\bar{n}_{12}\bar{n}_{13} + \bar{n}_{12}\bar{n}_{23} + \bar{n}_{13}\bar{n}_{23})}. \qquad (3)$$

If X^2 is found to be significant, comparisons of various kinds between the two marginal distributions will usually be made. The following are the three most important kinds of comparisons in practice.

(a) The two marginal distributions are to be compared with respect to the simple dichotomy, category i versus all other categories combined.

(b) The two marginal distributions are to be compared with respect to the dichotomy of a single composite category formed by combining categories $i_1, \ldots, i_m$ into one, versus all remaining categories combined.

(c) When the k categories are ordered, there is usually interest in whether one of the two marginal cumulative distributions is shifted to the left of or to the right of the other.

Associated with each of these comparisons is a contrast of the form $\sum c_i(P_{i.} - P_{.i})$, where $c_1, \ldots, c_k$ are k constants, at least two of which are unequal. For comparisons of the kind described in (a), $c_i = 1$ and all other c_j's are equal to 0. For comparisons of the kind described in (b), $c_{i_1} = \cdots = c_{i_m} = 1$ and the remaining $k - m$ c_j's are equal to 0. The simplest set of constants sensitive to the shifts described in (c) consists of the equally spaced integers $c_i = k + 1 - 2i, i = 1, \ldots, k$ (see Agresti [1]).

The hypothesis that $\sum c_i(P_{i.} - P_{.i}) = 0$ may be tested as follows. Define the vector $\mathbf{c} = (c_1, \ldots, c_k)'$. Under the hypothesis, $C = \mathbf{c'd}$ has an expectation of zero and an estimated variance of $\mathrm{Var}_0(C) = \mathbf{c'Vc}$. The ratio $C^2/\mathrm{Var}_0(C)$ may be compared to one of two different critical values depending on the number of contrasts being tested (see Miller [6]). If many contrasts are of interest, the Scheffé criterion leads to a critical value of $\chi^2_{k-1, \alpha}$. If the number of contrasts specified beforehand is moderate, the Bonferroni* criterion leads to a critical value of $\chi^2_{1, \alpha/K}$, where K is the number of prespecified contrasts (see MULTIPLE COMPARISONS).

The data in Table 2 have become classic in the literature of the analysis of categorical data* (see, e.g., Bishop et al. [2, p. 284]). The vector of differences is $\mathbf{d} = (69, 34, -51, -52)'$, with an estimated covariance matrix of

$$\mathbf{V} = \begin{Vmatrix} 843 & -500 & -241 & -102 \\ -500 & 1454 & -794 & -160 \\ -241 & -794 & 1419 & -384 \\ -102 & -160 & -384 & 646 \end{Vmatrix}.$$

For any 3×4 matrix $\mathbf{A}$ of rank 3, the value

of the test statistic in (1) is $X^2 = 11.96$ with 3 degrees of freedom. Because $\chi^2_{3, 0.05} = 7.81$, the distribution of right eyes across the four grades of unaided vision is significantly different at the 5% level from the distribution of left eyes.

The four categories are ordered, so the kind of comparison described in (c) is indicated. With the vector of constants given by $\mathbf{c} = (3, 1, -1, -3)'$, $C = \mathbf{c'd} = 448$ and $\mathrm{Var}_0(C) = \mathbf{c'Vc} = 16,800$. The ratio $448^2/16,800 = 11.95$ is significant according to the Scheffé criterion, so the conclusion may be drawn that unaided vision in the right eye tends to be *better than* (not just different from) unaided vision in the left eye.

For further discussion, see Read [8].

References

[1]   Agresti, A. (1983). *Biometrics*, **39**, 505–510.
[2]   Bishop, Y. M. M., Fienberg, S. E., and Holland, P. W. (1975). *Discrete Multivariate Analysis*. MIT Press, Cambridge, Mass.
[3]   Fleiss, J. L. and Everitt, B. S. (1971). *Brit. J. Math. Statist. Psychol.*, **24**, 117–123.
[4]   Maxwell, A. E. (1970). *Brit. J. Psychiatry*, **116**, 651–655.
[5]   McNemar, Q. (1947). *Psychometrika*, **12**, 153–157.
[6]   Miller, R. G. (1981). *Simultaneous Statistical Inference*, 2nd ed. Springer, New York.
[7]   Pike, M. C., Casagrande, J., and Smith, P. G. (1975). *Brit. J. Prevent. Soc. Med.*, **29**, 196–201.
[8]   Read, C. B. (1978). *Psychometrika*, **43**, 409–420.
[9]   Stuart, A. (1955). *Biometrika*, **42**, 412–416.

(BOWKER'S TEST FOR SYMMETRY
CATEGORICAL DATA
CONTINGENCY TABLES
MARGINAL SYMMETRY

Table 2   Unaided Distance Vision of Females Aged 30–39 Employed in Royal Ordnance Factories[a]

Right Eye Grade	Left Eye Grade				
	Highest	Second	Third	Lowest	Total
Highest	1520	266	124	66	1976
Second	234	1512	432	78	2256
Third	117	362	1772	205	2456
Lowest	36	82	179	492	789
Total	1907	2222	2507	841	7477

[a] Data from Stuart [9].

McNEMAR STATISTIC
PAIRED COMPARISONS
QUASI-SYMMETRY)

JOSEPH L. FLEISS

"STUDENT"   *See* GOSSET, WILLIAM SEALY

STUDENTIZATION

This term harks back to the research of William Sealy Gosset* ("Student") in studying the distribution of $\sqrt{n}(\bar{X} - \mu)/S$ [7], where μ and $\bar{X}$ are the population and sample means for a random sample of size n from a normal population and S is the sample standard deviation (SD), replacing the unknown SD σ in the standard measure $\sqrt{n}(\bar{X} - \mu)/\sigma$. The basic property of studentization is that the scale-dependent numerator is divided by an estimator S of the scale parameter σ and renders the resulting ratio distributionally free of σ. As a consequence, hypothesis testing* problems that are analyzed through the use of studentized test statistics remain invariant under scale transformations (*see* INVARIANCE CONCEPTS IN STATISTICS; SIMILAR REGIONS AND TESTS).

Gosset himself may have introduced the word "studentization" into the language. In a letter to Egon Pearson* dated January 29, 1932, he wrote [6]:

> I have been meaning to write to you for some time re the proposals for the use of range and sub-range which I made in my last letter to you. Of course there is a serious crab which I had at one time recognized and then forgotten in that the thing would have to be "Studentised": the only measure of the S.D. is provided by a limited number of degrees of freedom.

Studentization took on broader meanings as time passed by. The earliest restricted it to normally distributed samples and required the scale estimator of σ in the denominator to be independent of the numerator; this meant, in general, that the scale estimator had to be obtained from a different sample than the numerator.

The key results and related bibliography of this early work and of research through the 1960s are given in David [3]. Let $X_{(1)} \leq \cdots \leq X_{(n)}$ be the order statistics* from a random sample of size n from an $N(\mu, \sigma^2)$ population. For constants $a_1, \ldots, a_n$, David defined the class of studentized statistics as having the form

$$U = \sum_{i=1}^{n} a_i X_{(i)}/S_\nu, \qquad (1)$$

where $\nu S_\nu^2/\sigma^2$ has a chi-squared distribution with ν degrees of freedom. He referred to *external studentization* when S_ν is replaced by S, based on the same sample. Among the statistics whose properties were studied were:

(a)   the studentized range*, $(X_{(n)} - X_{(1)})/S_\nu$ or $(X_{(n)} - X_{(1)})/S$;

(b)   the studentized extreme deviates*

$$\max_{i=1,\ldots,n} |X_i - \bar{X}|/S_\nu$$

and

$$\max_{i=1,\ldots,n} |X_i - \bar{X}|/S,$$

of interest in the detection of outliers* (other forms are listed by David [3, Chap. 8]);

(c)   the noncentral studentized maximal distributions*, i.e., of $\max(X_i)/S_\nu$ and $\max|X_i|/S_\nu$, respectively, $i = 1, \ldots, n$;

(d)   studentized residuals* in regression.

Pooled studentization arises when σ^2 is estimated by

$$\left[\nu S_\nu^2 + (n-1)S^2\right]/(\nu + n - 1).$$

Externally studentized variables tended to be studied first, because the independence of numerator and denominator made their distributional properties easier to obtain. Interest switched to the more naturally occurring *internal studentization* with the development of computers, but also because U and S are independent whenever $\sum_{i=1}^{n} a_i = 0$ [3, Chap. 5]. If

$$U' = \sum_{i=1}^{n} b_i X_i / S,$$

then U', $\overline{X}$, and S are independent whenever $\sum_{i=1}^{n} b_i = 0$. These results derive from the independence of complete sufficient statistics and statistics with distributions free of σ, or of μ and σ (*see* BASU THEOREMS), and lead to such properties as

$$E(U^k) = E\left[\left(\sum_{i=1}^{n} a_i X_{(i)}\right)^k\right] \Big/ E(S^k),$$

$$k = 1, 2, \ldots,$$

for moments of the studentized range and studentized extreme deviate, for example.

In the 1970s, "studentization" was broadened further in meaning to include parent distributions other than the normal, particularly the family of gamma distributions* and others, like the Weibull*, derived from it. Investigations of many of the "new" statistics covered by the term had been continuing already for some time, so that the change was largely one of semantics. However, it allowed some unification of general results. Thus Bondesson [1] proved that the set of studentized differences $(X_1 - \overline{X})/S, \ldots,$ $(X_n - \overline{X})/S$ is independent of $\overline{X}$ if and only if the X_i's are normal or linearly transformed gamma variables. More important, Margolin [4] showed that for a large class of problems involving distributions related to the normal, multinormal, and gamma distributions, the distribution of internally studentized statistics can be derived from that of the corresponding unstudentized statistics by means of an inverse Laplace transform (*see* INTEGRAL TRANSFORMS). He applied this technique to random divisions of an interval, to studentized residuals, and to the latent roots* of a Wishart* matrix.

The *Current Index to Statistics** between 1976 and 1985 lists papers devoted to diverse studentized statistics. Articles can be found pertaining to M-estimators*, U-statistics*,

the bivariate range in correlated samples, the sample median, the two-sample Wilcoxon statistic, and jackknifed L-estimates, inter alia. Butler [2] extended the concept to multivariate problems, in which vectors $\hat{e}_i$ of residuals in multivariate regression become studentized as quadratic forms $\hat{e}_i' S^{-1} \hat{e}_i$, where S estimates the underlying covariance matrix. The most general use of the term appears to be that of Murota and Takeuchi [5]; they describe as studentized the empirical characteristic function* (ECF)

$$n^{-1} \sum_{j=1}^{n} \exp(itX_j)$$

of a sample $X_1, \ldots, X_n$ when the X_j's are replaced by the statistics $X_1/S, \ldots, X_n/S$. One could argue that the ECF is not studentized by this, but only that it becomes a function of studentized statistics.

References

[1] Bondesson, L. (1976). *Ann. Statist.*, **4**, 668–672.

[2] Butler, R. W. (1983). *J. R. Statist. Soc. B*, **45**, 120–132.

[3] David, H. A. (1981). *Order Statistics*, 2nd ed. Wiley, New York, Chaps. 5 and 8.

[4] Margolin, B. H. (1977). *Biometrika*, **64**, 573–582.

[5] Murota, K. and Takeuchi, K. (1981). *Biometrika*, **68**, 55–65. (The studentized ECF is used to test for the shape of distribution.)

[6] Pearson, E. S. (1939). *Biometrika*, **30**, 210–250. [See pp. 245–246. Reprinted (1970) in *Studies in the History of Probability and Statistics*, Vol. 1, E. S. Pearson and M. G. Kendall, eds. Hafner, New York.]

[7] "Student" (1908). *Biometrika*, **6**, 1–25.

(INVARIANCE CONCEPTS IN STATISTICS
NONCENTRAL STUDENTIZED MAXIMAL
    DISTRIBUTIONS
STUDENTIZED EMPIRICAL
    CHARACTERISTIC FUNCTION
STUDENTIZED EXTREME DEVIATES
STUDENTIZED RANGE
STUDENTIZED RESIDUAL
STUDENT'S *t*-TESTS)

CAMPBELL B. READ

STUDENTIZED EMPIRICAL CHARACTERISTIC FUNCTION

The empirical characteristic function (ECF), based on a set of n independent identically distributed (i.i.d.) random variables X_1, $X_2, \ldots, X_n$ is

$$c_n(t) = n^{-1} \sum_{j=1}^{n} \exp(itX_j),$$

where t is a real number.

The *studentized* ECF is defined [1] as

$$\tilde{c}_n(t) = n^{-1} \sum_{j=1}^{n} \exp(itX_j/S),$$

where

$$S^2 = (n-1)^{-1} \sum_{j=1}^{n} \left(X_j - \bar{X}\right)^2,$$

$$\bar{X} = n^{-1} \sum_{j=1}^{n} X_j.$$

The function $\left|\tilde{c}_n(t)\right|^2$ is invariant under shifts and scale changes.

Using this property Murota and Takeuchi [1] investigate the application of the studentized ECF in tests of shape of a distribution —in particular, tests of normality*, giving tables of estimated 5, 10, 50, 90, and 95% points of the statistic $\left|\tilde{c}_n(t)\right|^2$ for $n = 10(5)35$, 50, and $t = 0.5(0.5)2.0$.

Reference

[1] Murota, K. and Takeuchi, K. (1981) *Biometrika*, **68**, 55–65.

(CHARACTERISTIC FUNCTIONS
DEPARTURE FROM NORMALITY, TESTS
  FOR
STUDENTIZATION)

STUDENTIZED EXTREME DEVIATES

Let $X_{(1)} \leqslant \cdots \leqslant X_{(n)}$ be order statistics* from a random sample of size n from a normal population,

$$\bar{X} = n^{-1} \sum_{i=1}^{n} X_i$$

and

$$S^2 = (n-1)^{-1} \sum_{i=1}^{n} \left(X_i - \bar{X}\right)^2.$$

If an independent root-mean-square estimator S_ν^2 of σ^2 is available (so that $\nu S_\nu^2/\sigma^2$ has a chi-squared distribution* with ν degrees of freedom), let

$$S'^2 = \left[(n-1)S^2 + \nu S_\nu^2\right]/(\nu + n - 1)$$

be the pooled estimator of σ^2. The studentized extreme deviates (SEDs), of interest in testing normal data for outliers*, are

(i) $(X_{(n)} - \bar{X})/S'$ or $(\bar{X} - X_{(1)})/S'$ and

(ii) $\max_{i=1,\ldots,n} \left|X_{(i)} - \bar{X}\right|/S'$, sometimes called *studentized maximum absolute deviates*.

The case $\nu = 0$ corresponds to *internal studentization*, $\nu \geqslant 1$ to *pooled studentization*, and the case in which $S'^2 \equiv S_\nu^2$ to *external studentization* (David [2, Chaps. 5 and 8]).

(i)   Tables of upper percent points were developed by Nair [7], David [1], and Pillai [9] for external SEDs; these appear as Table 26 of Pearson and Hartley [8] for $n = 3(1)10, 12, \nu = 5(1)9, \alpha = 0.05, 0.01$; and $\nu = 10(1)20, 24, 30, 40, 60, 120, \infty, \alpha = 0.10, 0.025, 0.005, 0.001$. Grubbs and Beck [6] compiled tables of upper percent points of internal SEDs for the same α values, but for $n = 3(1)147$. Quesenberry and David [10] tabulated upper 5 and 1% points of pooled and interval SEDs for $n = 3(1)10, 12, 15, 20$ and $\nu = 0(1)10, 12, 14, 20, 24, 30, 40, 50$; these appear as Table 26a of Pearson and Hartley [8].

(ii) Tables of bounds on the upper 5 and 1% points of pooled and interval SEDs for $n = 3(1)10, 12, 15, 20$, and $\nu = 0(1)10, 12, 14, 20, 24, 30, 40, 50$ also appear in Quesenberry and David [10]; when $0 \leqslant \nu \leqslant 5$, (upper bound − lower bound) $\leqslant 0.001$ and when $n \geqslant 5$ for all ν, (upper bound − lower bound) $\leqslant 0.006$. Pearson and Hartley [8, Table 26b] tabulate exact percent points from these bounds.

An excellent summary of the performance of SEDs in testing for outliers, and of other competing test statistics, is David [2, Sec. 8.4]. The internal SEDs perform best when only one outlier is present, but they are less satisfactory in the presence of two or more outliers from a $N(\mu + \lambda\sigma, \sigma^2)$ population, due to the masking* effect. See also Dixon [4], David and Paulson [3] and Ferguson [5].

However, Rosner [12] developed a stepwise test for several outliers, in which the data point furthest from the sample mean is deleted, and the procedure is repeated k times, say, deleting the data point furthest from the mean of each successive subset. A test for a single outlier is performed at each stage. For k = 2, the SED (ii) performs best, by a slight margin, against three other statistics; sample estimates of the 5, 1, and 0.5% points of the SED are given for $n = 10(1)20(5)50$. Charts extending the 5% points to $k = 1(1)10$ outliers are given by Chhikara and Feiveson [1].

References

[1] Chhikara, R. S. and Feiveson, A. L. (1980). *Commun. Statist. B*, **9**, 155–166.

[2] David, H. A. (1956). *Biometrika*, **43**, 449–451.

[3] David, H. A. (1981). *Order Statistics*, 2nd ed. Wiley, New York.

[4] David, H. A. and Paulson, A. S. (1965). *Biometrika*, **52**, 429–436.

[5] Dixon, W. J. (1950). *Ann. Math. Statist.*, **21**, 488–506.

[6] Ferguson, T. S. (1967). In *Proc. Fourth Berkeley Symp. Prob. Statist.*, Vol 1. University of California Press, Berkeley, Calif. pp. 253–287.

[7] Grubbs, F. E. and Beck, G. (1972). *Technometrics*, **14**, 847–854.

[8] Nair, K. R. (1948). *Biometrika*, **35**, 118–144.

[9] Pearson, E. S. and Hartley, H. O. (1966). *Biometrika Tables for Statisticians*, 3rd ed., Vol. 1.

Cambridge University Press, Cambridge, England. (In addition to Table 26, 26a, and 26b, see the discussion on pp. 49–52.)

[10] Pillai, K. C. S. (1959). *Biometrika*, **46**, 473–474.

[11] Quesenberry, C. P. and David, H. A. (1961). *Biometrika*, **48**, 379–390.

[12] Rosner, B. (1975). *Technometrics*, **19**, 307–312.

(MEAN SLIPPAGE TESTS
ORDER STATISTICS
OUTLIERS
STUDENTIZATION)

CAMPBELL B. READ

STUDENTIZED MAXIMAL DISTRIBUTIONS, CENTERED

[*Editors' note.* This entry should be read in conjunction with NONCENTRAL STUDENTIZED MAXIMAL DISTRIBUTIONS. Observations X_j are independent normal random variables with mean vector μ and common variance σ^2, and are independent of S, where $\nu S^2/\sigma^2 \sim \chi_\nu^2$.]

CENTERED STUDENTIZED MAXIMAL DISTRIBUTIONS AND OUTLIER DETECTION

To detect whether any of the observations X_j, $j = 1(1)k$, is significantly larger than the others, a natural test statistic is the *centered studentized maximum* (CSM),

$$\text{CSM} = \max\left(X_j - \overline{X} \right)/S,$$

where $\overline{X}$ is the average of the observations and S is an estimator of standard deviation. For detecting outliers* that are significantly either large or small, the natural analog of CSM is the *centered studentized maximum modulus* (CSMM),

$$\text{CSMM} = \max\left| X_j - \overline{X} \right| / S.$$

The (central) distributions of CSM and CSMM have been derived and tabulated for the following two cases, in each of which the X_j are a random sample of normal variates.

Case 1. For the case where S is the sample standard deviation of the X_j, with $k - 1$ degrees of freedom, CSM tabulation was done first in ref. 5. In this case, note that S is not independent of the X_j. Tables for CSM and CSMM are given in ref. 2 and generalized to cases wherein the sample estimator of σ is augmented by an independent estimator, based on ν degrees of freedom, in ref. 6.

Case 2. For the case where the X_j are independent of S, where $\nu S^2/\sigma^2$ has distribution χ_ν^2, the CSM distribution was first derived by Nair [4]. Tables for CSM are given in ref. 4 and extended and corrected in ref. 1, while tables for CSMM are given in ref. 3. This case occurs, for example, when the X_j are cell means in a one-way layout with n observations per cell; the denominator is then $S/\sqrt{n}$, where S^2 is the mean square for error.

Inequalities can make these tables of somewhat wider use. For example, if the X_j in Case 2 have common correlation $\rho > 0$, then the $X_j - \bar{X}$ have common correlation $-1/(k - 1)$, independent of ρ, and variance $(1 - \rho)(k - 1)/(nk)$, less than when $\rho = 0$. Hence, the Case 2 tables [1, 3, 4] can be applied conservatively.

References

[1] David, H. A. (1956). *Biometrika*, **43**, 449–451. [Tables of CSM(2) for $k = 3(1)10, 12$. $\nu = 10(1)20, 24, 30, 40, 60, 120, \infty$; and $\alpha = 0.1, 0.05, 0.025, 0.01$.]

[2] Grubbs, F. (1950). *Ann. Math. Statist.*, **21**, 27–58. [Tables of CSM(1) for $k = 3(1)25$ and $\alpha = 0.1, 0.05, 0.025, 0.01$.]

[3] Halperin, M., Greenhouse, S. W., Cornfield, J., and Zalokar, J. (1955). *J. Amer. Statist. Ass.*, **50**, 185–195. [Tables of CSMM(2) for $k = 3(1)10, 15, 20, 30, 40, 60$, $\nu = 3(1)10, 15, 20, 30, 40, 60, 120, \infty$, and $\alpha = 0.05, 0.01$.]

[4] Nair, K. R. (1948). *Biometrika*, **35**, 118–144.

[5] Pearson, E. S. and Chandrasekar, C. (1936). *Biometrika*, **28**, 308–320. [Tables of CSM(1) for $k = 3(1)19$ and $\alpha = 0.1, 0.05, 0.025, 0.01$.]

[6] Quesenberry, C. P. and David, D. A. (1961). *Biometrika*, **48**, 379–390. [Tables of CSM for $k = 3(1)10, 12, 15, 20$, $\nu = 0(1)10, 12, 15, 20, 24, 30, 40, 50$, and $\alpha = 0.05, 0.01$. Tables of CSMM for $k = 3(1)10, 12, 15, 20$, $\nu = 0(1)10, 12, 15, 20, 24, 30, 40, 50$, and $\alpha = 0.05, 0.01$.]

(NONCENTRAL STUDENTIZED MAXIMAL DISTRIBUTIONS
OUTLIERS
STUDENTIZATION)

ROBERT BOHRER

STUDENTIZED RANGE

The term *studentized range* is commonly used in two rather different senses. Let $X_1, \ldots, X_n$ be a random sample from a normal $N(\mu, \sigma^2)$ population and let $X_{(1)} \leqslant \cdots \leqslant X_{(n)}$ be the corresponding order statistics*. Then $W = X_{(n)} - X_{(1)}$ is the sample range. The dependence of the distribution of W on the generally unknown σ can be removed by studentization*, i.e., dividing W by an estimator of σ. Let $S^2 = \Sigma(X_i - \bar{X})^2/(n - 1)$ and let S_ν^2 be such that $\nu S_\nu^2/\sigma^2$ has a χ^2 distribution with ν degrees of freedom (DF), *independent* of W. To recognize the distinction between W/S and W/S_ν, both commonly referred to as studentized ranges, we call (David [3])

$Q = W/S$ the internally studentized range

and

$Q_\nu = W/S_\nu$ the externally studentized range.

It should be noted that this distinction and many of the methods to follow apply equally to the class of studentized statistics obtained on replacing W by T, where T is any statistic such that T/σ is location and scale free (*see* STUDENTIZATION and STUDENTIZED EXTREME DEVIATES).

EXTERNALLY STUDENTIZED RANGE

There are a number of possible approaches to obtaining the distribution of externally studentized statistics. In the case of Q_ν, these are

(a) **Fitting a Curve by Moments.** Note that for $0 \leqslant k < \nu$,

$$E[Q_\nu^k] = E[S_\nu^{-k}] E[W^k]$$

$$= \frac{\nu^{k/2} \Gamma[\frac{1}{2}(\nu - k)] E[(W/\sigma)^k]}{2^{k/2} \Gamma(\frac{1}{2}\nu)}.$$

(b) **Quadrature.** Use

$$\Pr[Q_\nu < q] = \int_0^\infty \Pr\{W < s_\nu q\} f(s_\nu) \, ds_\nu$$

$$= \frac{2(\frac{1}{2}\nu)^{\nu/2}}{\Gamma(\frac{1}{2}\nu)} \int_0^\infty s^{\nu-1} \exp\left(-\frac{1}{2}\nu s^2\right)$$

$$\times \Pr\{W < sq\} \, ds.$$

(c) **Hartley's Method of Studentization.** Hartley [16] obtained the expansion

$$\Pr[Q_\nu < q] = \Pr\{W < q\} + a_1 \nu^{-1}$$

$$+ a_2 \nu^{-2} + \cdots,$$

where a_1, a_2 are functions of n and q which have been tabulated by Pearson and Hartley [26] for $n \leqslant 20$ and $\nu \geqslant 10$.

It will be seen that (a) provides a useful approximation for studentized statistics for which the lower moments of the numerator are known. Since good tables of the cumulative distribution function (CDF) of W are available (Hartley [15]; Harter [14]), method (b) is preferable to (a) in the case of Q_ν and has (essentially) been used in the definitive tables [14] to give the CDF and percentage points of Q_ν. Hartley's three-term representation (c) is useful for ν large enough (e.g., $\nu \geqslant 20$) and has been combined with (b) to provide percentage points of Q_ν (Pearson and Hartley [27]). A computing algorithm for upper tail CDF and percentage points of Q_ν is given by Lund and Lund [19].

INTERNALLY STUDENTIZED RANGE

Study of the distribution of Q is facilitated by the independence of Q and S in normal

samples. This result follows immediately from the fact that Q, having a distribution free of μ and σ, is independent of the complete sufficient statistic $(\overline{X}, S)$ (*see* BASU THEOREMS and Basu [1]). Similar results clearly hold for other internally studentized statistics. In analogy to the previous section, we now have

(a)

$$E[Q^k] = \frac{E[W^k]}{E[S^k]}$$

$$= \left[\frac{1}{2}(n - 1)\right]^{k/2}$$

$$\times \frac{\Gamma[\frac{1}{2}(n - 1)]}{\Gamma[\frac{1}{2}(n - 1 + k)]} E\left[\left(\frac{W}{\sigma}\right)^k\right],$$

(b)

$$\Pr\{W < w\} = \int_0^\infty \Pr\left\{Q < \frac{w}{s}\right\} f(s) \, ds.$$

By fitting Pearson-type curves, method (a) has been used in the main to provide approximate percentage points of Q [27, Table 29c]. Sufficiently extreme upper α significance points q_α of Q were computed exactly from the following relation [4], valid for $q_\alpha \geqslant \{\frac{3}{2}(n - 1)\}^{1/2}$:

$$q_\alpha^2 = 2(n - 1) t_{\nu;\alpha'}^2 / (\nu + t_{\nu;\alpha'}^2),$$

where $t_{\nu;\alpha'}$ is the upper $\alpha' = \alpha/\{n(n - 1)\}$ significance point of Student's t with $\nu = n - 2$ DF. Further exact results are given by Currie [2]. The integral equation* (b) for the CDF is due to Hartley and Gentle [18], who also investigate methods of solution.

A general approach to the distribution of internally studentized statistics via Laplace transform inversion is provided by Margolin [20].

APPLICATIONS

Important applications of Q_ν occur in the analysis of variance* for balanced classifications. For example. let

$$X_{ij} = \mu_i + B_j + Z_{ij},$$

$$i = 1, \ldots, n, \, j = 1, \ldots, m,$$

where the μ_i are fixed "treatment" effects and the random "block" effects B_j and error variates Z_{ij} are mutually independent having, respectively, normal $N(0, \sigma_B^2)$ and $N(0, \sigma_Z^2)$ distributions. Then, on the null hypothesis $\mu_1 = \mu_2 = \cdots = \mu_n$,

$$\sqrt{m} \ \text{range}(\bar{X}_{i.})/S_\nu$$

is distributed as Q_ν with $\nu = (n - 1)(m - 1)$ DF, where

$$\nu S_\nu^2 = \sum_{i=1}^{n} \sum_{j=1}^{m} \left(X_{ij} - \bar{X}_{i.} - \bar{X}_{.j} + \bar{X}_{..} \right)^2.$$

Thus referring the test statistic to tables of percentage points $q_{n,\nu;\alpha}$ of Q_ν gives a test similar to the corresponding F-ratio test. The latter test has no strong optimality properties; a comparison of respective power functions is given in David et al. [5]. Moreover, the studentized range test is particularly convenient as a multiple comparison* procedure (Tukey [32]) since, if we refer any number of studentized pairs $\sqrt{m}(\bar{x}_{i.} - \bar{x}_{i'})/s_\nu$ to $q_{n,\nu;\alpha}$, we will have at most an overall error rate of α if H_0 is true. This is so because the range is the largest of all the pairwise differences. Simultaneous confidence intervals for the $\frac{1}{2}n(n - 1)$ differences $\mu_i - \mu_{i'}$ follow readily (cf. ORDER STATISTICS). The studentized range test is also the starting point of several multiple range test procedures (e.g., Duncan [7]). A graphical approach is provided by Feder [8]. For a general account of simultaneous inference*, see Miller [21].

The internally studentized range Q has been proposed as a test of normality (David et al. [4]). In empirical sampling studies, Shapiro et al. [28] have found that Q "has particularly good properties against symmetric, especially short-tailed (e.g., the uniform) distributions but seems to have virtually no power with respect to asymmetry."

Even before any appropriate theory was developed, Q was used by Snedecor [29] and several other applied statisticians as a check on calculations of S. Unusual values of Q pointed to discrepancies. While Snedecor had approximate underlying normality in mind,

it should be noted (Thomson [31]) that Q is bounded for *any* parent distribution:

Upper bound of Q: $[2(n - 1)]^{1/2}$,
Lower bound of Q: $2[(n - 1)/n]^{1/2}$, for n even; $2[n/(n + 1)]^{1/2}$, for n odd.

Gross errors in the computation of S can therefore be detected at once, whatever the distribution.

Snedecor [29] also pointed out that, for a normal sample, W can be predicted from S. A precise result follows again from the independence of Q and S. The regression of Q on S is linear through the origin [6],

$$E[W|S = s] = (E[W]/E[S])s,$$

and $\text{var}(W|S = s) \propto s^2$. A well-known related result is (e.g., Hartley [17]) that $\text{corr}(W, S) = \sqrt{E_W}$, where E_W is the efficiency of W as an estimator of σ, i.e.,

$$E_W = \text{var}(S/E[S])/\text{var}(W/E[W]).$$

RELATED STATISTICS

1. Closely related to the externally studentized range are the *studentized maximum modulus* (see NONCENTRAL STUDENTIZED MAXIMAL DISTRIBUTIONS),

$$M_\nu^* = \max_{i=1,\ldots,n} |Y_i|/S_\nu,$$

where the Y_i are independent $N(0, \sigma^2)$ variates, and the *studentized augmented range* [9],

$$Q_\nu' = \max(Q_\nu, M_\nu).$$

Tables of upper percentage points of Q_ν' are close to those of Q_ν (Stoline [30]). Applications of Q_ν' in multiple comparison procedures for unbalanced designs are given in, e.g., ref. 30 and Felzenbaum et al. [9].

2. Let (X_i, Y_i), $i = 1, \ldots, n$, be a random sample from a bivariate normal* $N(\mu_X, \mu_Y, \sigma_X^2, \sigma_Y^2, \rho)$ distribution. A two-dimensional generalization of the internally studentized range is the *studentized*

bivariate range $R/_2 S$, where

$$R^2 = \max_{i,j} \left[(X_i - X_j)^2 + (Y_i - Y_j)^2 \right]$$

and

$$_2 S^2 = \frac{1}{2(n-1)} \times \left[\sum_{i=1}^{n} (X_i - \overline{X})^2 + \sum_{i=1}^{n} (Y_i - \overline{Y})^2 \right].$$

$R/_2 S$ and related statistics are of special interest in ballistics and are described by Grubbs [13]. It is commonly assumed that $\sigma_X^2 = \sigma_Y^2$ and $\rho = 0$ (circular normality); some percentage points of $R/_2 S$ have been obtained in this case by Gentle et al. [12]. See also Pagurova et al. [22a]. The general situation is considered by Patil and Liu [23].

HISTORICAL NOTES

The beginning of interest in the (externally) studentized range is attributed by Pearson [25] to a letter he received from "Student" (W. S. Gosset*) in 1932. Referring to the comparison of *selected* differences in variety means in a field experiment, Gosset writes: "Of course, there is a serious crab … in that the thing [i.e., ranges] would have to be 'Studentised'." Newman [22] published approximate tables of upper percentage points of Q_ν by using quadrature on Pearson's [24] Pearson-type approximation to the distribution of range in normal samples. When the exact distribution of range became available (Hartley [15]), Pearson and Hartley [26] improved these percentage points and prepared tables from which the probability integral of Q_ν could be obtained for ν sufficiently large.

The internally studentized range was first studied theoretically in 1954 [4]. In his treatment of this statistic, Snedecor [29] disregarded the distinction between internal and external studentization. The basic result on the independence of Q and S in normal samples can also be established by arguments used in related contexts by Fisher [10] and more explicitly by Geary [11].

References

[1] Basu, D. (1955). *Sankhyā*, **15**, 377–380.

[2] Currie, I. D. (1980). *Scand. J. Statist.*, **7**, 150–154.

[3] David, H. A. (1970, 1981). *Order Statistics*. Wiley, New York.

[4] David, H. A., Hartley, H. O., and Pearson, E. S. (1954). *Biometrika*, **41**, 482–493.

[5] David, H. A., Lachenbruch, P. A., and Brandis, H. P. (1972). *Biometrika*, **59**, 161–168.

[6] David, H. A. and Perez, C. A. (1960). *Biometrika*, **47**, 297–306.

[7] Duncan, D. B. (1955). *Biometrics*, **11**, 1–42.

[8] Feder, P. I. (1975). *Technometrics*, **17**, 181–188.

[9] Felzenbaum, A., Hart, S., and Hochberg, Y. (1983). *Ann. Statist.*, **11**, 121–128.

[10] Fisher, R. A. (1930). *Proc. Roy. Soc. London A*, **130**, 16–28.

[11] Geary, R. C. (1933). *Biometrika*, **25**, 184–186.

[12] Gentle, J. E., Kodell, R. L., and Smith, P. L. (1975). *Technometrics*, **17**, 501–506.

[13] Grubbs, F. E. (1964). *Statistical Measures of Accuracy for Riflemen and Missile Engineers*. Edwards Brothers, Ann Arbor, Mich.

[14] Harter, H. L. (1970a,b). *Order Statistics and Their Uses in Testing and Estimation*, Vols. 1 and 2. U.S. Government Printing Office, Washington, D.C.

[15] Hartley, H. O. (1942). *Biometrika*, **32**, 334–348.

[16] Hartley, H. O. (1944). *Biometrika*, **33**, 173–180.

[17] Hartley, H. O. (1955). *Comm. Pure Appl. Math.*, **8**, 47–72.

[18] Hartley, H. O. and Gentle, J. E. (1975). In *A Survey of Statistical Design and Linear Models*, J. N. Srivastava, ed. North-Holland, Amsterdam, Netherlands, pp. 197–207.

[19] Lund, R. E. and Lund, J. R. (1983). *Appl. Statist.*, **32**, 204–210.

[20] Margolin, B. H. (1977). *Biometrika*, **64**, 573–582.

[21] Miller, R. G., Jr. (1981). *Simultaneous Statistical Inference*, 2nd ed. Springer, New York.

[22] Newman, D. (1939). *Biometrika*, **31**, 20–30.

[22a] Pagurova, V. I., Rodionov, K. D. and Rodionova, M. V. (1981). *Theor. Prob. Appl.*, **26**, 366–371.

[23] Patil, S. A. and Liu, A. H. (1981). *Sankhyā B*, **43**, 172–186.

[24] Pearson, E. S. (1932). *Biometrika*, **24**, 404–417.

[25] Pearson, E. S. (1938). *Biometrika*, **30**, 210–250.

[26] Pearson, E. S. and Hartley, H. O. (1943). *Biometrika*, **33**, 89–99.

[27] Pearson, E. S. and Hartley, H. O. (1954, 1970). *Biometrika Tables for Statisticians*, Vol. 1. Cambridge University Press, London.

[28] Shapiro, S. S., Wilk, M. B., and Chen, H. J. (1968). *J. Amer. Statist. Ass.*, **63**, 1343–1372.

[29] Snedecor, G. W. (1937). *Statistical Methods*. Collegiate Press, Ames, Iowa.

[30] Stoline, M. R. (1978). *J. Amer. Statist. Ass.*, **73**, 656–660.

[31] Thomson, G. W. (1955). *Biometrika*, **42**, 268–269.

[32] Tukey, J. W. (1953). The problem of multiple comparisons. Unpublished memorandum.

Acknowledgment

This work was supported by the U.S. Army Research Office.

(DEPARTURES FROM NORMALITY, TESTS FOR
MULTIPLE COMPARISONS
NONCENTRAL STUDENTIZED MAXIMAL
  DISTRIBUTIONS
ORDER STATISTICS
RANGES
STUDENTIZATION
STUDENTIZED EXTREME DEVIATES
STUDENTIZED MAXIMAL
  DISTRIBUTIONS, CENTERED)

H. A. DAVID

STUDENTIZED RESIDUAL

This is an estimated residual* divided by an estimate of its standard deviation. For general linear models* with normal homoscedastic and independent residual variation, the estimators of the residuals are normally distributed and the estimator of their standard deviation is usually a known multiple of the square root of the residual mean square (*see* ANALYSIS OF VARIANCE). This is not, in general, independent of the estimated residuals. If it were (for example, if it were based on another set of data), each studentized residual would have a *t*-distribution* if the model were valid.

(ANALYSIS OF COVARIANCE
ANALYSIS OF VARIANCE
STUDENTIZATION)

STUDENT'S *t*-TESTS

Student's *t*-tests are used to make inferences from small samples and usually involve applying significance tests* to hypotheses concerning population means.

"Student" was the pseudonym of W. S. Gosset* (1876–1937), whose need to quantify the results of small scale experiments motivated him to develop and tabulate the probability integral of the ratio $z = x/s$, where x is the distance of the sample mean from the population mean and s is the sample standard deviation* (defined with the sample size as the divisor as was the practice at that time); see ref. 8.

Gosset corresponded about the distribution of z with R. A Fisher* (1890–1962), who in 1912, in a letter now lost, gave a rigorous geometric derivation of the distribution of z which was eventually published in 1923 and later [5] given in terms of $t = z(n-1)^{1/2}$. In further correspondence during the period 1915–1924, summarised in ref. 4, other applications of the z-distribution were discussed and it is evident that it was Fisher who realised that a unified treatment of tests of significance of a mean, of the difference between two means, and of simple and partial coefficients of correlation and regression could be achieved more readily in terms of $t = z\nu^{1/2}$, where ν is the number of degrees of freedom* associated with the sum of squares used in defining z.

This unification resulted in Student's *t*-distribution* being appropriate for testing hypotheses in an abundance of practical situations. Amongst the most common areas of application are the comparison of means and the examination of regression coefficients* in the development of linear models using least squares*.

THE *t*-STATISTIC AND ITS DISTRIBUTION

If Y is a normally distributed variable with expectation zero and standard deviation σ_Y and if s_Y^2 is an estimate of σ_Y^2, independent of Y, such that $\nu s_Y^2/\sigma_Y^2$ has a chi-squared

Table 1   Critical Values for One- and Two-Sided *t*-Tests Using 5 and 1% Significance Levels.[a]

Degrees of Freedom ν	One-Sided Critical Values		Two-Sided Critical Values	
	5%	1%	5%	1%
5	2.015	3.365	2.571	4.032
10	1.812	2.764	2.228	3.169
15	1.753	2.602	2.131	2.947
20	1.725	2.528	2.086	2.845
30	1.697	2.457	2.042	2.750
60	1.671	2.390	2.000	2.660
∞	1.645	2.326	1.960	2.576

[a] This table is compiled with permission from Table III of Fisher and Yates, *Statistical Tables for Biological, Agricultural and Medical Research*, published by Oliver and Boyd Limited, Edinburgh, and from Table 12 of *Biometrika Tables for Statisticians*, Vol 1, published by the Biometrika Trust.

distribution* with ν degrees of freedom, then the ratio $t = Y/s_Y$ is distributed as a Student's *t*-variable with density function

$$f(t) = \frac{\Gamma\{\frac{1}{2}(\nu + 1)\}}{\Gamma(\frac{1}{2}\nu)(\nu\pi)^{1/2}}\left[1 + \frac{t^2}{\nu}\right]^{-(\nu+1)/2},$$

$$-\infty < t < \infty,$$

which depends only on the degrees of freedom ν.

Percentage points of the corresponding distribution function of t may be found in most statistical texts and are included in all sets of standard statistical tables; see for example Pearson and Hartley [11] or Fisher and Yates [7]. Critical values suitable for one-sided or two-sided tests of hypotheses may be found from such tables and are usually provided in addition to the distribution function. Table 1 gives a few critical values, from which it may be seen that the values for the *t*-test approach those of the corresponding normal test as the number of degrees of freedom increases. *See also* *t*-DISTRIBUTION.

THE SIGNIFICANCE OF A MEAN OF A SINGLE SAMPLE

One of the simplest applications of a Student's *t*-test involves using a sample of

independent observations $x_1, \ldots, x_n$ from a normal population with mean μ and variance σ^2 to test the null hypothesis* H_0: $\mu = \mu_0$ against the alternative hypothesis* H_1: $\mu = \mu_1$. These are composite hypotheses* since σ^2 is not specified. The statistic $u = \Sigma(x_i - \mu_0)^2$ is complete and sufficient for σ^2 under H_0, but not otherwise. These properties imply that a most powerful similar test of H_0 against H_1 may be found by examining the regions of constant u; *see* SIMILAR REGIONS AND TESTS and NEYMAN STRUCTURE, TESTS WITH.

Since

$$u = \Sigma(x_i - \bar{x})^2 + n(\bar{x} - \mu_0)^2$$

$$= \Sigma(x_i - \bar{x})^2\left(1 + \frac{n(\bar{x} - \mu_0)^2}{\Sigma(x_i - \bar{x})^2}\right),$$

where $\bar{x} = n^{-1}\Sigma x_i$, the best critical region for constant u consists of small values of $\Sigma(x_i - \bar{x})^2$, or equivalently of large values of $n(\bar{x} - \mu_0)^2\{\Sigma(x_i - \bar{x})^2\}^{-1}$; see ref. 10, p. 197 for further details.

If Y and s_Y^2 are defined as $Y = \bar{x} - \mu_0$ and $s_Y^2 = n^{-1}s^2$, where $s^2 = (n - 1)^{-1}\Sigma(x_i - \bar{x})^2$ is the sample variance, then $t = Y/s_Y$ has a Student's *t*-distribution with $n - 1$ degrees of freedom and the best critical region previously described consists of large values of $t^2(n - 1)^{-1}$. Since t has a distribu-

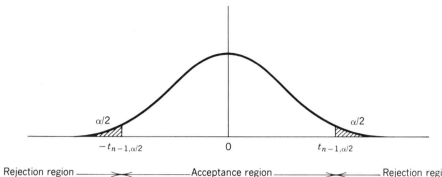

Figure 1   Distribution of $t = (\bar{x} - \mu_0)\sqrt{n}/s$ when the null hypothesis is true, showing the acceptance and rejection regions for a two-sided alternative hypothesis and significance level α.

tion not depending on σ^2, it is independent of the complete statistic u.

One-sided tail regions of this *t*-distribution with $n - 1$ degrees of freedom provide the critical regions for the most powerful tests of H_0 against H_1 for either $\mu_1 < \mu_0$ or $\mu_1 > \mu_0$. The corresponding two-sided critical values with significance level α (*see* SIGNIFICANCE TESTS) to test H_0 against H_1: $\mu \neq \mu_0$ are $\pm t_{n-1, \alpha/2}$, where

$$\Pr\left(|t| > t_{n-1, \alpha/2}\right) = \alpha.$$

The test procedure now involves dividing the range of values for t into two regions, an acceptance region $|t| \leqslant t_{n-1, \alpha/2}$ supporting the null hypothesis and a rejection region $|t| > t_{n-1, \alpha/2}$ supporting the alternative hypothesis.

Formally the test procedure in this case would be: Accept H_0: $\mu = \mu_0$ if $t = (\bar{x} - \mu_0)\sqrt{n}/s$ lies in the acceptance region $|t| \leqslant t_{n-1, \alpha/2}$; otherwise reject H_0. The corresponding regions for this two-sided test are shown in Fig. 1.

As an illustration of this test procedure consider the following experiment with maize in which the yield per plot is measured in kilograms of dry matter. At a standard plant density, the average yield per plot is 25 kg. In order to investigate the effect of plant density on yield, 10 plots were planted at a different plant density with the following results: 24.1, 26.8, 30.4, 25.5, 28.2, 24.5, 27.6, 28.4, 25.3, 26.5. Test whether there has been a change in mean yield.

Here the null hypothesis is H_0: $\mu = \mu_0 = 25$, while the alternative hypothesis is H_1: $\mu \neq 25$. The critical values for 9 degrees of freedom and a 5% level of significance are ± 2.262. The sample mean is $\bar{x} = 26.73$ and the standard deviation is $s = 1.963$. The *t*-ratio is

$$t = (\bar{x} - 25)\sqrt{10}/s$$

$$= (26.73 - 25)\sqrt{10}/1.963 = 2.787.$$

Since 2.787 is greater than 2.262, the value of t falls in the rejection region and H_0 is rejected at the 5% level of significance. The conclusion is that there is sufficient evidence to reject the null hypothesis that the mean yield is 25 kg. The problem now becomes one of estimation of the new level of yield (*see* CONFIDENCE INTERVALS AND REGIONS).

A multivariate generalisation of this *t*-test used in testing hypotheses on multivariate means was proposed by Hotelling* [9] (*see* HOTELLING'S T^2). The test statistic T^2 reduces to t^2 in the univariate case.

THE SIGNIFICANCE OF THE DIFFERENCE BETWEEN TWO SAMPLE MEANS

Probably the most common application of a Student's *t*-test involves the comparison of two normal populations with means μ and $\mu + \delta$ and common variance σ^2. The hypotheses are H_0: $\delta = 0$ against H_1: $\delta = \delta_1$, which are to be tested using two independent samples of size n_1 and n_2 denoted by x_{1i}, $i = 1, \ldots, n_1$, and x_{2i}, $i = 1, \ldots, n_2$, with means $\bar{x}_1$ and $\bar{x}_2$ and combined sample size $n = n_1 + n_2$. As in the one-sample case, these hypotheses are composite since μ and σ^2 are not specified. However, the overall mean $\bar{x}$ and sum of squares S^2 defined by

$$n\bar{x} = n_1\bar{x}_1 + n_2\bar{x}_2,$$

$$S^2 = \sum_{j=1}^{2} \sum_{i=1}^{n_j} (x_{ji} - \bar{x})^2,$$

form a pair of statistics $(\bar{x}, S^2)$ which is complete and sufficient for (μ, σ^2) when H_0 is true. This implies that a most powerful similar test for H_0 against H_1 will have a best critical region consisting of large values of $(\bar{x}_1 - \bar{x}_2)^2/S^2$, which is equivalent to large values of

$$\frac{(\bar{x}_1 - \bar{x}_2)^2}{(n_1 - 1)s_1^2 + (n_2 - 1)s_2^2},$$

where s_1^2 and s_2^2 are the two sample variances (for details see ref. 10, p. 197).

If $Y = \bar{x}_1 - \bar{x}_2$ and $s_Y^2 = s^2(n_1^{-1} + n_2^{-1})$, where s^2 is the pooled estimate of σ^2 given by

$$s^2 = \frac{(n_1 - 1)s_1^2 + (n_2 - 1)s_2^2}{n_1 + n_2 - 2},$$

then $t = Y/s_Y$ has a t-distribution with $n_1 + n_2 - 2$ degrees of freedom and the best critical region consists of large values of $t^2(n_1^{-1} + n_2^{-1})(n_1 + n_2 - 2)^{-1}$ or equivalently large values of t^2. Therefore, tail regions of Student's t-distribution with $n_1 + n_2 - 2$ degrees of freedom provide the appropriate critical regions to test H_0 against H_1.

To illustrate this procedure, suppose that two groups of patients are treated with an active drug or a placebo in a clinical trial to investigate problems of premature labour. One of the variables of interest is the final length of pregnancy and it is required to compare the mean duration of pregnancy for the two groups of patients. The details of the two samples are summarised in the table:

	Active	Placebo
Mean duration of pregnancy (in weeks)	37.7	36.7
Standard deviation	1.42	1.98
Number of patients	25	23

Before testing H_0: $\delta = 0$ against H_1: $\delta = \delta_1 > 0$ with these data, the assumption of equal variances for the two groups should be examined using the ratio s_1^2/s_2^2 of the two sample variances (*see* F-TESTS). This ratio is not significant, so it is appropriate to compare the means using the two-sample t-test. The pooled estimate of σ^2 is

$$s^2 = \left\{24(1.42)^2 + 22(1.98)^2\right\}/46 = 2.927$$

and the appropriate test statistic is

$$t = \frac{\bar{x}_1 - \bar{x}_2}{\sqrt{\left\{s^2(1/n_1 + 1/n_2)\right\}}}$$

$$= \frac{37.7 - 36.7}{\sqrt{\left\{s^2(1/25 + 1/23)\right\}}} = 2.023.$$

The corresponding one-sided critical value obtained from the t-distribution with 46 degrees of freedom and significance level 5% is 1.68. Since 2.023 is greater than 1.68, H_0 is rejected at the 5% level with the conclusion that the mean duration of pregnancy is greater for those patients given the active drug.

COMPARISON OF THE MEANS OF MATCHED PAIRS

Data collected as pairs of observations (*see* MATCHED PAIRS), in which one observation receives one treatment while the other receives a different treatment, may be analysed using a matched-pairs t-test*. Comparison of the treatments is based on a single sample t-test applied to the differences between the responses within each pair of observations.

SIGNIFICANCE OF A REGRESSION COEFFICIENT

Student's *t*-tests may also be applied in regression * problems. In particular, in a simple linear regression analysis, the data, consisting of *n* pairs of observations (x_i, y_i) for $i = 1, \ldots, n$, may be modelled using

$$y_i = \alpha + \beta x_i + \epsilon_i,$$

where the ϵ_i's are random errors assumed to be independently normally distributed with means zero and variances σ^2. The least squares estimates of α and β are

$$a = \bar{y} - b\bar{x}$$

and

$$b = \sum_{i=1}^{n} (x_i - \bar{x})(y_i - \bar{y}) \bigg/ \sum_{i=1}^{n} (x_i - \bar{x})^2.$$

The estimate *b* is normally distributed with mean β and variance $\sigma^2 \{ \Sigma (x_i - \bar{x})^2 \}^{-1}$. If σ^2 is estimated by the residual mean square s^2 (*see* LINEAR REGRESSION), then

$$t = (b - \beta) \left\{ \sum_{i=1}^{n} (x_i - \bar{x})^2 \right\}^{1/2} \bigg/ s$$

has a *t*-distribution with $n - 2$ degrees of freedom.

As before, considerations of completeness and sufficiency imply that the most powerful similar test of H_0: $\beta = 0$ against H_1: $\beta = \beta_1$ has a critical region based on large values of t^2. This test is equivalent to the corresponding *F*-test* in the analysis of variance*, since, if $\beta = 0$,

$$t^2 = b^2 \sum (x_i - \bar{x})^2 / s^2$$

$$= \frac{\text{regression mean square}}{\text{residual mean square}} = F,$$

where *F* has an *F*-distribution* with 1 and $n - 2$ degrees of freedom. It is also equivalent to using the sample correlation* coefficient

$$r = \frac{\sum_{i=1}^{n} (x_i - \bar{x})(y_i - \bar{y})}{\left\{ \sum_{i=1}^{n} (x_i - \bar{x})^2 \sum_{i=1}^{n} (y_i - \bar{y})^2 \right\}^{1/2}}$$

to test the hypothesis that the population correlation coefficient $\rho = 0$ against H_1: $\rho = \rho_1$; the relationship in this case is

$$t^2 = (n - 2) r^2 (1 - r^2)^{-1}.$$

COMPARISON OF TWO REGRESSION COEFFICIENTS

The results of the previous section may be applied to the problem of comparing two regression coefficients, β_1 and β_2 say, for two simple linear regression models fitted to independent samples of n_1 and n_2 pairs of observations (x_{1i}, y_{1i}), $i = 1, \ldots, n_1$, and (x_{2i}, y_{2i}), $i = 1, \ldots, n_2$.

If b_1 and b_2 are the least squares estimates of the two regression slopes, then they will be normally distributed with means β_1 and β_2 and variances

$$\sigma^2 \left\{ \sum_{i=1}^{n_1} (x_{1i} - \bar{x}_1)^2 \right\}^{-1}$$

and

$$\sigma^2 \left\{ \sum_{i=1}^{n_2} (x_{2i} - \bar{x}_2)^2 \right\}^{-1},$$

respectively, provided that there is a common error variance σ^2. An appropriate estimate s^2 of σ^2 is obtained by pooling the sums of squares of the residuals from the two fitted regression lines. The degrees of freedom for this pooled estimate s^2 will be $\nu = n_1 - n_2 - 4$ and the ratio

$$t = \frac{(b_1 - b_2) - (\beta_1 - \beta_2)}{\left\{ s^2 \left[\left\{ \sum_{i=1}^{n_1} (x_{1i} - \bar{x}_1)^2 \right\}^{-1} + \left\{ \sum_{i=1}^{n_2} (x_{2i} - \bar{x}_2)^2 \right\}^{-1} \right] \right\}^{1/2}}$$

will have a Student's *t*-distribution with ν degrees of freedom.

FURTHER APPLICATIONS

Tests based on Student's *t* are used in other applications also, but in many of these areas the test results require careful interpretation, either in terms of the meaning of a significant result or in terms of the true level of significance associated with the test.

For example, in a multiple linear regression analysis, in which several regression coefficients are fitted by least squares, each estimated coefficient, when divided by its standard error, may be tested using a *t*-test provided the normality assumptions for the errors are justified. However, it is not in general valid to draw conclusions about one coefficient independently of the other terms in the model. Further details may be found in MULTIPLE LINEAR REGRESSION; REGRESSION COEFFICIENTS; REGRESSION, POLYNOMIAL; STEPWISE REGRESSION; PARTIAL CORRELATION.

In designed experiments, an analysis of variance may indicate that a number of mean values contain significant differences. Repeated application of Student's *t*-tests to pairs of means is inappropriate because of the difficulties of determining the critical values corresponding to a specified overall significance level. It is preferable to use one of several available multiple comparison* procedures which attempt to make suitable adjustments for this.

Other tests related to Student's *t*-tests include

1. *The Fisher–Behrens test* for the equality of population means when the variances need to be estimated separately. For historical details see Behrens [2], Sukhatme [13], and Fisher [6]; *see also* BEHRENS–FISHER PROBLEM.

2. *Sequential t-tests* applied to observations taken in time and where, following each observation or group of observations, the decisions are either to accept or reject the null hypothesis or to continue sampling. Details are given in Wald [15]; *see also* SEQUENTIAL ANALYSIS.

3. *Tests for the detection of outliers** using maximum Studentized residuals, where appropriate critical values are obtained by applying Bonferroni bounds to Student's *t*-distribution; see Barnett and Lewis [1] and STUDENTIZED MAXIMAL DISTRIBUTIONS.

4. *Tests with indifference zones* considered by Brown and Sackrowitz [3] for which there is a region of values of μ near μ_0 where the decision to accept H_0 is not a significant error; when there is such an indifference zone, the usual one-sided Student's *t*-test is inadmissible*. For this problem an alternative test is suggested which dominates the *t*-test. However, the difference in power is very small unless n is small or the indifference zone is large compared with the standard deviation. A modified test is also presented when the alternative hypothesis is two-sided, but in this case it is shown that not only are the powers similar but that the two-sided Student's *t*-test is admissible.

POWER, INVARIANCE, AND ROBUSTNESS OF STUDENT'S *t*-TESTS

Student's *t*-tests depend on the ratio $t = Y/s_Y$ having a *t*-distribution with ν degrees of freedom when the null hypothesis H_0 is true. The probability of rejecting H_0 when it is true is α, the significance level, but the probability of rejecting H_0 when it is false depends on δ, the mean value of Y under some specific alternative hypothesis. This probability is the power of the test relative to that particular alternative, and it may be determined by integrating the noncentral *t*-distribution*, with noncentrality parameter δ/σ_Y and degrees of freedom ν, over the appropriate rejection region for the test. Considered as a function of δ/σ_Y, this integral provides the power function for the test. Charts of this power function are available (see Pearson and Hartley [11]) that may be used to determine the sample size required to ensure that any specific difference δ would be detected with a given probability.

Provided the normality assumptions are valid, the problem of testing H_0: $\mu = 0$ against the alternative $\mu \neq 0$ remains invariant under the group of scale transformations, and t^2 is maximal invariant in the sample space. This together with the principle of sufficiency* implies that the usual two-sided t-test is uniformly most powerful invariant for testing $\mu = 0$ against $\mu \neq 0$ (*see* INVARIANCE CONCEPTS IN STATISTICS and NEYMAN STRUCTURE, TESTS WITH).

However, if these assumptions are not valid, alternative test statistics could be more powerful than the t-test. Student's t is particularly vulnerable to long-tailed nonnormality and a variety of substitute t-statistics have been proposed to guard against this situation. Tukey and McLaughlin [14] consider a one-sample trimmed-t statistic involving the ratio of a trimmed mean to a Winsorized sum of squared deviations. The resulting statistic is distributed approximately as a t-variable with reduced degrees of freedom. Yuen and Dixon [16] extend the trimmed t-statistic to the two-sample situation and compare the behaviour of Student's t and the trimmed t. A simple alternative involving a trimmed mean and a linear estimator of scale is compared with Student's t for a range of nonnormal situations by Prescott [12]. Most of these alternatives are almost as powerful as t when the populations are normal and are more powerful when the populations have long-tailed distributions; *see* TRIMMING AND WINSORIZATION.

The Wilcoxon test based on rank sums provides a useful alternative for the comparison of two population means provided the sample sizes are not too small. For normal distributions it has asymptotic efficiency 0.955.

References

[1]   Barnett, V. and Lewis, T. (1984). *Outliers in Statistical Data*, 2nd ed. Wiley, New York.

[2]   Behrens, W. V. (1929) *Landw. Jb.* **68**, 807–37.

[3]   Brown, L. D. and Sackrowitz, H. (1984). *Ann. Statist.*, **12**, 451–469.

[4]   Eisenhart, C. (1979). *Amer. Statist.*, **33**, 6–10.

[5]   Fisher, R. A. (1925). *Metron*, **5**, 90–104.

[6]   Fisher, R. A. (1939). *Ann. Eugenics*, **9**, 174–180.

[7]   Fisher, R. A. and Yates, F. (1953). *Statistical Tables for Biological, Agricultural and Medical Research*. Oliver and Boyd, London.

[8]   Gosset, W. S. (1908). *Biometrika*, **6**, 1–25.

[9]   Hotelling, H., (1931). *Ann. Math. Statist.*, **2**, 360–378.

[10] Kendall, M. G. and Stuart, A. (1961). *The Advanced Theory of Statistics*, Vol. 2. Griffin, London.

[11] Pearson, E. S. and Hartley, H. O. (1966). *Biometrika Tables for Statisticians*, 3rd ed., Vol I. Cambridge University Press, Cambridge, England.

[12] Prescott, P. (1975). *Appl. Statist.*, **24**, 210–217.

[13] Sukhatme, P. V. (1938). *Sankhyā*, **4**, 39–48.

[14] Tukey, J. W. and McLaughlin, D. H. (1963). *Sankhyā A*, **25**, 331–352.

[15] Wald, A. (1947). *Sequential Analysis*, Wiley, New York.

[16] Yuen, K. K., and Dixon, W. J. (1973). *Biometrika*, **60**, 369–374.

(ANALYSIS OF VARIANCE
BEHRENS–FISHER PROBLEM
BONFERRONI INEQUALITIES AND
   INTERVALS
CONFIDENCE INTERVALS AND REGIONS
F-TESTS
HOTELLINGS T^2
HYPOTHESIS TESTING
INVARIANCE CONCEPTS IN STATISTICS
LINEAR REGRESSION
MATCHED PAIRS t-TESTS
MULTIPLE COMPARISONS
MULTIPLE LINEAR REGRESSION
NEYMAN STRUCTURE, TESTS WITH
NONCENTRAL t-DISTRIBUTION
PARTIAL CORRELATION
REGRESSION, POLYNOMIAL
ROBUST REGRESSION
SEQUENTIAL ANALYSIS
SIGNIFICANCE TESTS
SIMILAR REGIONS AND TESTS
STEPWISE REGRESSION
STUDENTIZED MAXIMAL DISTRIBUTIONS
t-DISTRIBUTION
TRIMMING AND WINSORIZATION
WELCH TESTS)

P. Prescott

STURGES' RULE

Sturges [1] proposed the formula

$$1 + \log n = 1 + 2.303 \log_{10} n$$

as an (approximate) rule for determining the number of groups to be used for a data set containing n data points. For example, with $n = 57$, the rule gives $1 + (2.303 \times 1.756) \doteq 5$ groups.

Reference

[1] Sturges, H. A. (1926). *J. Amer. Statist. Ass.*, **21**, 65–66.

(GROUPED DATA)

STUTTERING POISSON DISTRIBUTIONS

The original stuttering Poisson distribution was defined by Galliher et al. [2] as a special case of the distribution of $X = \sum_{j=1}^{m} X_j$, where X_j, $j = 1, \ldots, m$, are independent and

$$\Pr\left[X_j = jk \right] = e^{-\lambda_j}\lambda_j^k/k!,$$
$$k = 0, 1, \ldots, \infty. \qquad (1)$$

In particular

$$\Pr[X = 0] = \exp\left(- \sum_{j=1}^{m} \lambda_j \right)$$

and recursively,

$$\Pr[X = x + 1] = \frac{1}{x+1} \sum_{i=1}^{x(m)} (i + 1)\lambda_{i+1}$$
$$\times \Pr[X = x - i],$$

where $x(m) = \min(x, m)$.

The probability generating function* of this distribution is

$$\prod_{j=1}^{m} \exp\left[-\lambda_j(1 - t^j) \right].$$

and—as Suzuki [3] points out—this is not a usual compounding mixture* of several Poisson distributions*. (Compare with Neyman's type A* and Thomas' distributions*.)

Galliher et al. [2] considered the special two-parameter case where

$$\lambda_j = (1 - p)p^{j-1}\lambda, \qquad j = 1, 2 \ldots.$$

The general case (1) was first introduced by Adelson [1], who called it a "quasicompound" Poisson distribution. More recently, however (see, e.g., Suzuki [3]), the term "stuttering Poisson" has been applied to Adelson's class of distributions.

References

[1] Adelson, R. M. (1966). *Operat. Res. Quart.*, **17**, 73–75.

[2] Galliher, H. P., Morse, P. M., and Simond, M. (1959). *Operat. Res.*, **7**, 362–384.

[3] Suzuki, G. (1980). *Ann. Inst. Statist. Math.*, **32**, Part A, 143–159.

(MIXTURE DISTRIBUTIONS
NEYMAN'S TYPE A, B AND C
   DISTRIBUTIONS
POISSON DISTRIBUTION
THOMAS DISTRIBUTION)

SUB-BALANCED DATA

An analysis of variance* model, a general formulation of which is given in [1], is often used to ascribe differences in experimental outcomes to qualitative factors of an experiment. If one is interested in hypothesis testing* or confidence* regions on either the fixed effects or the variance components*, a common way to proceed is to first partition the total sum of squares into an analysis of variance (ANOVA). Most methods of inference for an analysis of variance are based on terms assumed to be independent and, suitably scaled, distributed as chi-squared variables. When these properties are satisfied for all terms in an analysis of variance over all possible values of the unknown parameters, it is sometimes called a *proper* analysis of variance (p-ANOVA) [1].

The usual procedure of partitioning a total sum of squares by adjusting sequentially for the factors in the model produces an ANOVA with the chi-squared and independence properties when the data are balanced (or equivalently, balanced in both the fixed and random effects, as will be defined). However, without balanced data, this procedure will not always produce a p-ANOVA when there is one. A p-ANOVA is a preferable alternative in some cases, so it is of interest to know when such a partition of the sum of squares is possible. The minimal condition that an experimental design must satisfy is that the random effects are sub-balanced, a condition easily checked in practice so that a p-ANOVA can be constructed when possible. Note that the minimal condition for a p-ANOVA is a statement about random effects only and not the fixed effects. It is trivially satisfied when all effects are fixed ($k = 1$ in the model to follow), regardless of whether they are unbalanced or not, and is satisfied if the random effects are balanced. Further conditions may be required for a p-ANOVA to contain terms that may be needed for inference on specific parameters of interest, particularly involving the fixed effects [1, 2].

Before technically defining "balanced" and "sub-balanced," we consider a simple example that will help to illustrate the practical difference. Suppose tests are conducted to ascertain the content level of a chemical used in pesticides in a brand of commercial flour and that batches of the flour are randomly divided into samples and sent to more than one laboratory for testing. "Batches" and "laboratories" might both be treated as random effects. A balanced design will produce a p-ANOVA, but that may be unnecessarily stringent. A balanced design would require that the same number of samples are analyzed from each batch, with the same number of tests on each sample, and that each laboratory receives a sample from an equal number of batches and conducts the same number of tests on each sample. However, there may be reasons for testing some batches more thoroughly than others or for conducting more tests at some laboratories than others. Both of these objectives can be achieved within a sub-balanced design, within limits. To achieve a sub-balanced design, one might divide the laboratories into mutually exclusive groups of arbitrary size (including one) and then assign each batch to a single group of laboratories. Laboratories within a group would each conduct the same number of tests on a single sample from each batch, but that number could vary across groups of laboratories.

It may be useful to explain the meaning of sub-balanced data in relation to balanced data, even though the meaning of "balance" as used in the statistical literature is highly contextual [3]. A definition of balanced data as used here is given in Scheffé [4, p. 224]. Alternatively, we may define two effects (fixed or random) as *balanced* if their incidence matrix* [the (k, l)th term of which contains the number of observations in which the kth element (level) of the first effect occurs with the lth element of the second effect] satisfies the following three requirements:

1. All nonzero elements are identical.
2. Row totals are equal and column totals are equal.
3. All rows (or all columns) are either pairwise orthogonal or identical.

Hence, two effects are balanced if, ignoring the other effects, the design corresponds either to what is usually called a balanced two-way layout or to a balanced nested classification. Two effects are *sub-balanced* if, in their incidence matrix,

1. All nonzero elements in a given row are identical.
2. Any given pair of rows is either orthogonal or identical.

A design in which all pairs of effects, fixed and random, are balanced (sub-balanced) may be called a *balanced (sub-balanced) design*, or be said to have *balanced (sub-balanced) data*. If the condition is restricted to random effects, we may say that the design is *balanced (sub-balanced) in the random effects*, which is all that is required for an ANOVA with the independence and chi-squared properties already described to be possible.

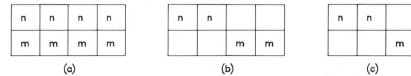

Figure 1 Three possible configurations of an incidence matrix in the example with two batches of flour (rows) and four laboratories (columns).

Figure 1 shows three design configurations for only two batches and four laboratories that will serve to illustrate further the distinction between balanced and sub-balanced data. Batches are represented by rows and laboratories by columns, with the incidence denoted by m, n, or blank (for zero). In Fig. 1(a), where each laboratory performs n tests on the first batch and m tests on the second, there is no difference between balance and sub-balance and either obtain if and only if $n = m$. Figure 1(b) is a suitable design for random effects, even though batches and laboratories would be confounded if they were treated as fixed effects. In that figure, where two laboratories perform n tests on the first batch and the remaining two laboratories conduct m tests on the second batch, the effects are sub-balanced for arbitrary n and m and are balanced if and only if $n = m$. In Fig. 1(c), which might apply to the situation of Fig. 1(b) with the data from one laboratory missing, the effects are sub-balanced for any values of n and m, but never balanced. In general, two effects are sub-balanced if, by reordering rows and columns, a block-diagonal matrix can be formed in which the elements within each block are identical. When this is possible but there is only a single block or all blocks are identical with a single row (or a single column), then the effects are balanced.

where Y is a random n-vector; $X_1, \ldots, X_t$, $U_1, \ldots, U_k$ are given matrices containing only 0s and 1s, with a single 1 in each row of each U_i matrix, $U_k = I$; $\beta_1, \ldots, \beta_t$ are unknown vector parameters; $\xi_1, \ldots, \xi_k$ are unobservable vectors with $\xi_i \sim N(0, I\sigma_i^2)$, $i = 1, \ldots, k$; and the random elements on the right side of (1) are assumed to be mutually independent. The covariance matrix of Y is $\Sigma = \sum_{i=1}^{k} V_i \sigma_i^2$, where $V_i = U_i U_i'$. Letting $\sigma^2 = (\sigma_1^2, \ldots, \sigma_k^2)'$, the parameter space of σ^2 is defined to be $\Omega = \{\sigma_i^2 : \sigma_i^2 \geqslant 0, i = 1, \ldots, k - 1, \sigma_k^2 > 0\}$. Thus Σ is positive definite for all points in the parameter space. The variables $\xi_1, \ldots, \xi_{k-1}$ are the random effects, ξ_k is the error term, and the parameters $\beta_1, \ldots, \beta_t$ are fixed effects. Special cases include the fixed effects model ($k = 1$) and the random effects or variance components model ($t = 1$ and X_1 is a column of ones).

A design satisfying (1) is sub-balanced in the random effects when the incidence matrix of random effects ξ_i and ξ_j, given by $U_i'U_j$, satisfies the two requirements of the preceding definition, for all $i, j = 1, \ldots, k - 1$. A necessary and sufficient condition for a partition of $Y'Y$ into independent terms with the chi-squared property as previously described (a p-ANOVA) is that V_i, V_j commute ($V_iV_j = V_jV_i$) for all i, j or equivalently, that the design is sub-balanced in the random effects [1].

DESCRIPTION OF THE MIXED MODEL ASSUMED

A general formulation of the mixed model of the analysis of variance is described by

$$Y = X_1\beta_1 + \cdots + X_t\beta_t + U_1\xi_1 + \cdots U_k\xi_k,$$
(1)

References

[1] Brown, K. G. (1983). *J. Amer. Statist. Ass.*, **78**, 162–167.

[2] Brown, K. G. (1984). *Ann Statist.*, **12**, 1488–1499.

[3] Preece, D. A. (1982). *Utilitas Mathematica*, **21C**, 85–186.

[4] Scheffé, H. (1959). *The Analysis of Variance*. Wiley, New York.

(BALANCING IN EXPERIMENTAL DESIGN
FIXED-, RANDOM-, AND MIXED-EFFECTS
   MODELS
GENERAL BALANCE)

KENNETH G. BROWN

SUBDIVIDED-SURFACE CHART *See*
STRATA CHART

SUBEXPONENTIAL DISTRIBUTIONS

Let F be a cumulative distribution function
(CDF) on the positive half line $(0, \infty)$ such
that $F(x) < 2$ for every $x \in \mathbb{R}$. Then F is
called *subexponential* $(F \in \mathscr{S})$ iff

$$\lim_{x \to \infty} \frac{1 - F^{*2}(x)}{1 - F(x)} = 2. \qquad (1)$$

[Here * denotes the usual (Stieltjes) convolution* product.] The class $\mathscr{S}$ was originated
independently by Chistyakov [1] and Chover
et al. [2]. Chistyakov noticed that as a direct
implication of the definition of (1), F also
satisfies

$$\lim_{x \to \infty} \frac{1 - F(x - y)}{1 - F(x)} = 1$$

$$\text{for every } y \in \mathbb{R}. \qquad (2)$$

Distributions satisfying (2) are called *long-
tailed* and the class of such CDFs is denoted
by $\mathscr{L}$. It is not hard to show that, if $F \in \mathscr{L}$,

$$e^{-\epsilon x} = o(1 - F(x)) \qquad (x \to \infty)$$

$$\text{for every } \epsilon > 0,$$

meaning that the tail of F decreases to zero
more slowly than any negative power of the
exponential function. This justifies the name
subexponential but also shows the inconsistency in terminology, since only CDFs in
$\mathscr{S}$ are called subexponential, while it is
known that $\mathscr{S}$ is a proper subclass of $\mathscr{L}$. To
give a probabilistic interpretation of the definition in (1), let X_1, X_2 be two independent
random variables with CDF F. Then (1) is

equivalent to

$$\Pr[X_1 + X_2 > x]$$
$$\sim \Pr[\text{Maximum}(X_1, X_2) > x]$$
$$\text{as } x \to \infty,$$

which is in turn equivalent to

$$\Pr[\text{Maximum}(X_1, X_2)$$
$$\leqslant x | X_1 + X_2 > x] \to 0 \quad \text{as } x \to \infty.$$

This relation illustrates that a large sum of
two subexponential observations is very
likely to come from one very large and one
relatively small observation, which makes the
class $\mathscr{S}$ ideal to model distributions with fat
(heavy) tails.

An important implication of the definition
in (1) is that subexponential distributions
also satisfy (1) with the number 2 replaced
by any positive integer $m \geqslant 2$. Even more
general if $(p_n)_n$ is a discrete probability
measure on Z^+ such that $\sum_{n=0}^{\infty} p_n z^n$ is analytic at $z = 1$, and if $G = \sum_{n=0}^{\infty} p_n F^{*n}$, then
the following statements are equivalent:

(i) $F \in \mathscr{S}$

(ii) $G \in \mathscr{S}$ and $1 - G \neq o(1 - F)$

(iii) $\displaystyle\lim_{x \to \infty} \frac{1 - G(x)}{1 - F(x)} = \sum_{n=1}^{\infty} n p_n.$

This theorem shows that $\mathscr{S}$ is the class that
characterizes this special type of limit behaviour of compound distributions*. Since
such distributions frequently arise in applied
stochastic models such as risk theory*,
queueing theory*, random walks*, branching
theory, etc. ...., applications of the theorem
above may be found in various domains of
stochastic processes*. A nice survey paper
listing both applications and theoretical
properties of the class $\mathscr{S}$ and providing the
appropriate references is Embrechts [3].

References

[1] Chistyakov, V. P. (1964). A theorem on sums of
    independent positive random variables and its ap-
    plication to branching processes. *Theory Prob.
    Appl.*, **9**, 640–648.

[2] Chover, J., Ney, P., and Wainger, S. (1973). Func-
    tions of probability measures. *J. d Anal. Math.*,
    26, 255–302.

[3] Embrechts, P. (1982). Subexponential distribution
    functions and their applications, a review. *Proc.
    7th Brasov Conf. Probability Theory.*

Bibliography

Cline, D. B. H. (1987). Convolutions of distributions with exponential and subexponential tails. *J. Austral. Math. Soc.*, **44** (to appear). (Extends and unifies previously known results by Embrechts and Goldie.)

Embrechts, P. and Goldie, C. M. (1980). On closure and factorization properties of subexponential and related distributions, *J. Austral. Math. Soc.* (*Series A*), **29**, 243–256. (Describes basic properties of the class $\mathscr{S}$.)

Omey, E. and Willekens, E. (1987). Second order behaviour of distributions subordinate to a distribution with finite mean, *Stoch. Models*, **3** (to appear). (Higher order theory for the class $\mathscr{S}$, with applications.)

(HEAVY TAILED DISTRIBUTIONS
TAIL PROBABILITIES)

ERIC WILLEKENS

SUBGRADUATION   *See* GRADUATION

SUBHYPOTHESIS TESTING

A simple hypothesis* specifies completely the statistical model, while, due to the presence of nuisance parameter(s)*, a composite hypothesis* fails to do so. In univariate as well as multivariate linear models (and in some other statistical models), the set of parameters of interest may be partitioned into two subsets: Plausible hypotheses relate only to the first subset, while the second one contains the nuisance parameters, irrespective of such hypotheses. For example, in the classical (normal theory) two-factor fixed-effects analysis of variance model, the first subset contains the main effects* and interactions*, and the error variance constitutes the second subset. We are generally interested in testing for the main effects and interactions. In this setup, if an hypothesis relates only to the main effects, treating the interactions as nuisance parameters (or vice versa), we have a subhypothesis testing model. In general, for a linear model, an hypothesis relating to a subset of the regression parameters without assuming that the complementary subset is redundant (nor assuming the error variance to be known) is termed a *subhypothesis*. If, in the same setup, an hypothesis relates to the *entire* set of regression parameters (error variance unknown), though we still have a composite hypothesis, nevertheless, it is not termed a subhypothesis.

In a subhypothesis testing model, we have thus a hierarchy of two subsets Θ_1 and Θ_2 of the parameters, such that Θ_2 is nuisance (irrespective of any plausible hypothesis on Θ_1), and Θ_1 in turn is partitioned into two subsets Θ_{11} and Θ_{12}, where the hypothesis relates to Θ_{11}, treating Θ_{12} as also nuisance. Different subhypotheses may relate to different partitioning of Θ_1, but Θ_2 remains the same, In a parametric model, Θ_2 is generally a vector of (scale) parameters, while in nonparametric and semiparametric models (see Puri and Sen [1] and Sen [2]) Θ_2 may generally stand for a suitable class of distributions for the error variables associated with the model; in all cases, Θ_1 refers to the parameter of interest.

Though a subhypothesis is a particular case of a composite hypothesis, in general, the hierarchy of the parameter space may allow one to construct suitable tests more conveniently and to characterize their optimality under the same set of regularity conditions (as pertaining to the overall test for Θ_1). In nonparametric or semiparametric problems, there is, however, a basic difference between a full hypothesis and a subhypothesis testing problem. In a full hypothesis testing model, generally, one has a genuinely distribution-free* test (in the univariate case) or a permutationally (conditionally) distribution-free test (in the multivariate case), generated by invariance* under suitable groups of transformations on the original observations. However, in a subhypothesis testing model, due to the nuisance parameter Θ_{12}, such invariance structure may no longer hold and, hence, exact distribution-free tests may not exist. Nevertheless, asymptotically distribution-free tests can be constructed under parallel conditions and these are generally robust (see Sen and Puri [3] and Sen [2]). For nonlinear (parametric) models too, exact tests for subhypotheses may not exist, but asymptotic ones do.

References

[1] Puri, M. L. and Sen, P. K. (1985). *Nonparametric Methods in General Linear Models.* Wiley, New York.

[2] Sen, P. K. (1984). *J. Statist. Plann. Infer.*, **10**, 31–42.

[3] Sen, P. K. and Puri, M. L. (1977). *Z. Wahrsch. verw. Geb.*, **39**, 175–186.

(COMPOSITE HYPOTHESIS
HYPOTHESIS TESTING
NUISANCE PARAMETERS
SIMPLE HYPOTHESIS)

P. K. SEN

SUBJECTIVE PROBABILITIES

The subjectivistic approach considers probability as a measure of the degree of belief* of a given subject in the occurrence of an event or, more generally, in the veracity of a given assertion. This point of view has been shared by some of the greatest pioneers of the mathematical theory of probability such as J. Bernoulli* and P. S. de Laplace*, but in the first part of the twentieth century, increasing attention was paid to definitions of subjective probability (hereafter designated S.P.) from which one could deduce the whole calculus of probability. The contributions in this field by Ramsey [18], de Finetti* [5, 6, 7], and Savage [19] constitute a relevant chapter in the foundations of probability*. Their treatments of S.P. are based on *suitable conditions of coherence** (or *admissibility*) when dealing with the rational behaviour of a person in the face of uncertainty. Ramsey and, chiefly, Savage consider S.P. an element of the more general problem of the numerical representation for the coherent ordering of acts, and they develop a theory of decision making which establishes both the properties of probability and utility* (*see also* DECISION THEORY). Even though de Finetti does not reject the decisional approach, he prefers to separate probability from utility. His position can be seen as a particular case of that of Ramsey and Savage by supposing that the decision maker has a linear utility or by assuming that the monetary sums that are at stake are small enough to justify a linear approximation to the actual decision-maker utility function [9, Sec. 3.2].

Other authors, supporting views which agree only in part with the subjectivistic one, are Keynes [15] and Koopman [16]. According to Keynes [15], probability is the *rational* degree of belief in a proposition on the basis of others. *Rational* means that probability is seen as a logical relation between one set of propositions and another. Such a view has been defended by Carnap [2]. In most cases, it amounts to presenting as valid the selection of just one of the coherent probability distributions on the grounds of conditions, such as considerations of symmetry that, in actual fact, are the result of a subjective opinion and not of a premise of an objective assessment.

We will first explain de Finetti's theory, in view of its simplicity and lucidity, and will evaluate the fundamental contribution of Savage* to qualitative probability. Then we will describe Savage's decisional approach and, finally, a comparison between the subjectivistic and objectivistic approach will be presented.

EVENTS AND RANDOM QUANTITIES

The domain of application of probability is given, roughly speaking, by the class of all random entities implicated in order to state precisely the problems which arise under uncertainty. In a subjectivistic framework, *random (for a given person)* means that the entity in question is unequivocally identified and that its true realization is unknown to him. In the present article, we will confine ourselves to considering *random events* and *random quantities* (hereafter r.e. and r.q., respectively) according to the following meanings (*see also* AXIOMS OF PROBABILITY):

A r.e. is a proposition, admitting at most two values, true and false, which is random for a given person since his state of information does not permit him to deduce its logical value.

A r.q. is a number belonging to a specified set of real numbers, called the set of the possible values of the r.q.. The number is random for a given person because his state of information does not permit him to deduce its true value.

We will indicate by inf X (sup X, respectively) the greatest lower bound (g.l.b.) [or least upper bound (l.u.b.), respectively] of the set of possible values of the r.q. X. When $-\infty < \inf X \le \sup X < \infty$, we say that X is *bounded*. The event that admits only the value "true" is called the *sure event* (in symbols, Ω), whereas the event that admits only the value "false" is called the *impossible event* (in symbols, ϕ). Each event can be assimilated to a r.q. in the following way: Given a r.e. E, consider the r.q. that assumes the value 1 when E is true and the value 0 when E is false. Such a r.q. is called the *indicator* of E; we will adopt de Finetti's suggestion that the same symbol that designates an event also designates its indicator.

PROBABILITY AND PREVISION IN DE FINETTI'S QUANTITATIVE APPROACH

Firstly we define the prevision* of a bounded r.q. and subsequently the probability of a r.e., which is seen as the prevision of its indicator.

According to de Finetti's betting scheme [9, Chap. 3], a person who wants to summarize his degree of belief in the different values of a r.q. X by a number $\bar{x}$, is supposed to be obliged to accept any bet on X with gain $c(\bar{x} - X)$, where X is the actual value of the r.q. considered and c is an

arbitrary real number chosen by an opponent.

Since c may be positive or negative, there is no advantage for our fictitious bank in deviating from the value $\bar{x}$ that makes the bets with gains $c(\bar{x} - X)$ and $-c(\bar{x} - X)$ indifferent.

The condition of coherence assumed by de Finetti presupposes essentially that $\bar{x}$ has to be a value such that there is no choice of c for which the realizations of the gain are all uniformly negative; that is, given $c \in \mathbb{R}$, there is no $\gamma > 0$ for which $c(\bar{x} - X) < -\gamma$ for all values of X.

In view of the arbitrariness of c, this condition is equivalent to

$$\inf Y \le 0 \le \sup Y \ \forall c \in \mathbb{R},$$
$$\text{where } Y = c(\bar{x} - X).$$

When $\bar{x}$ satisfies this condition, it is called the *prevision* of X (in symbols, $P[X]$).

It is possible to give another criterion for the evaluation of $P[X]$, based on a penalty proportional to the square of the deviation between X and $\bar{x}$ (*see* PREVISION). The two criteria are equivalent as shown by de Finetti [9, Sec. 3.3.7].

De Finetti maintains that coherence* is all that theory can prescribe. The actual choice of a coherent prevision is the task of whatever person is concerned.

The concept of coherence may be extended to a class $\mathcal{X}$ of bounded r.q.s in the following way. We say that the real function P on $\mathcal{X}$ is a *prevision* on $\mathcal{X}$ if, for any finite subclass $\{X_{ji(1)}, \ldots, X_{j(n)}\}$ of $\mathcal{X}$ and any choice of $(c_1, \ldots, c_n) \in \mathbb{R}^n$, $n = 1, 2, \ldots$, the gain

$$Y = \sum_{k=1}^{n} c_k \{ P[X_{j(k)}] - X_{j(k)} \}$$

is such that $\inf Y \le 0 \le \sup Y$.

If E is a r.e., then the prevision $P[E]$ of its indicator is called the *probability* of E.

The problem we ought now to consider is that of making sure of the existence of a prevision on any class of bounded r.q.s. We will consider this question after recalling some elementary conditions under which a real-valued function represents a prevision.

The proof of these conditions can be found in refs. 9, Chap. 3, and 10, Sec. 15 of the Appendix.

Given $\mathscr{X} = \{X\}$, P is a prevision on $\mathscr{X}$ if and only if

$$\inf X \leqslant P[X] \leqslant \sup X.$$

This fact establishes that the family of the classes of bounded r.q.s that are able to support a prevision is nonempty. The following are necessary conditions for the coherence of P on any class $\mathscr{X}$ of bounded r.q.s:

(p$_1$) $X \in \mathscr{X} \Rightarrow \inf X \leqslant P[X] \leqslant \sup X$;

(p$_2$) $X_i \in \mathscr{X}$, $\lambda_i \in \mathbb{R}$ $(i = 1, \ldots, n)$,

$$\sum_{i=1}^{n} \lambda_i X_i \in \mathscr{X}$$

$$\Rightarrow P\left[\sum_{i=1}^{n} \lambda_i X_i\right] = \sum_{i=1}^{n} \lambda_i P[X_i].$$

Now let $\mathscr{X}_1$ be a set of bounded r.q.s and P_1 be a prevision on it. If there exists a prevision P_2 on $\mathscr{X}_2 \supset \mathscr{X}_1$ such that $P_1[X] = P_2[X]$ for all X in $\mathscr{X}_1$, then P_2 is said *to extend* P_1 to $\mathscr{X}_2$ and P_2 is an *extension* of P_1 to $\mathscr{X}_2$. Assuming the axiom of choice, by induction (transfinite if $\mathscr{X}_2$ is nondenumerably infinite) de Finetti [8, Sec. 9; 10, Sec. 15 of the Appendix] has demonstrated that an extension of a coherent prevision does always exist (notice that it need not be unique). Then, for any class of bounded r.q.s, one can assess a prevision on such a class and so the previous concept of coherence is well defined. Consequently, coherent probabilities do not suffer from the restrictions that affect the assessment of a probability measure* according to Kolmogorov such as, for instance, the impossibility of defining a probability measure without atoms on the power set of a nondenumerably infinite set (a result obtained by Ulam [21]). This positive feature of de Finetti's condition of coherence is es-

sentially due to the fact that it does not imply σ-additivity of probability that, on the contrary, is axiomatically assumed in Kolmogorov's setting of probability theory (*see also* AXIOMS OF PROBABILITY). It is worth investigating these aspects further.

In order to compare de Finetti's theory with others, in particular with Kolmogorov's, it is useful to suppose that $\mathscr{X}$ has the structure of a linear space, including the r.q. identically equal to 1, with respect to the usual operations of addition of r.q.s and multiplication of a r.q. by a real number. When $\mathscr{X}$ has such a structure, conditions (p$_1$) and (p$_2$) become sufficient in order for a real function P on $\mathscr{X}$ to be a prevision. Similarly, when $\mathscr{A}$ is an algebra of events, a real-valued function P on $\mathscr{A}$ is a coherent probability if and only if

(a$_1$) $E \in \mathscr{A} \Rightarrow P[E] \geqslant 0$;

(a$_2$) $P[\Omega] = 1$;

(a$_3$) $E_1, E_2 \in \mathscr{A}$ with $E_1 \cap E_2 = \phi \Rightarrow$

$$P[E_1 \cup E_2] = P[E_1] + P[E_2]$$

(finite additivity condition).

For the proof of these results, see refs. 6, Sec. 9, and 10, Sec. 15 of the Appendix. The last statement clearly shows that a S.P. law according to de Finetti need not satisfy the σ-additivity condition and that any σ-additive probability is coherent.

Another important feature of de Finetti's framework is that coherence is preserved in a passage to the limit. In fact, if $\{P_n; n \geqslant 1\}$ is a sequence of coherent probabilities on a class of events $\mathscr{F}$ and if $\mathscr{F}_l \neq \phi$ is the subset of the elements of $\mathscr{F}$ for which the limit P of $\{P_n; n \geqslant 1\}$ exists, then P is also a coherent probability on $\mathscr{F}_l$. On the contrary, even if $\{P_n; n \geqslant 1\}$ is a sequence of σ-additive probability laws on $\mathscr{F}$, P need not be σ-additive on $\mathscr{F}_l$. This result clearly also holds for a sequence of previsions defined on a class of bounded r.q.s.

MORE ABOUT PREVISION

In de Finetti's quantitative approach, prevision such as probability is defined via a scheme by which it is possible to directly elicit its value. Usually a r.q. is defined as a real function on the set Ω of all the possible (mutually exclusive and exhaustive) outcomes of a random experiment: $X = X(\omega)$, $\omega \in \Omega$. If a coherent probability P is defined on an algebra $\mathscr{A}$ of subsets of Ω, does the evaluation of the prevision of X depend on P? Generally speaking, the answer is yes, but without further conditions, it may happen that $P[X]$ is not uniquely determined by the knowledge of P on $\mathscr{A}$ [9, Chap. 6]. The knowledge of P on $\mathscr{A}$ uniquely determines $P[X]$ whenever

$$X = X(\omega) = \sum_{i=1}^{n} x_i E_i,$$

where $\{E_1, \ldots, E_n\} \subset \mathscr{A}$ is a partition of Ω. In such a case, from $(\mathbf{p_2})$ one deduces

$$P[X] = \sum_{i=1}^{n} x_i P[E_i].$$

Furthermore, if X is a bounded r.q. such that, for every $\epsilon > 0$, there exists a finite partition of Ω, $\{E_1, E_2, \ldots, E_n\} \subset \mathscr{A}$, such that

$$\sum_{i=1}^{n} \left\{ \sup_{\omega \in E_i} X(\omega) - \inf_{\omega \in E_i} X(\omega) \right\} P[E_i] < \epsilon,$$

then one obtains

$$P[X] = \int_{\Omega} X(\omega) P(d\omega), \qquad (1)$$

where the integral is of the Stieltjes type in the framework of finitely additive probability spaces [1, Sec. 4.5].

It may be that $P[X]$ is not uniquely determined by P on $\mathscr{A}$ as in the following case. Put $\Omega = (0,1]$ and let $\mathscr{A}$ consist of all finite disjoint unions of right-semiclosed intervals $(a, b] \subset \Omega$. Define $X = X(\omega)$ as the r.q. that assumes the value 0 when ω is rational in Ω and the value 1 when ω is irrational in Ω. So, given any finite partition

$\{E_1, E_2, \ldots, E_n\}$ of Ω in $\mathscr{A}$, one has

$$\sum_{i=1}^{n} \left\{ \sup_{\omega \in E_i} X(\omega) - \inf_{\omega \in E_i} X(\omega) \right\} P[E_i] = 1.$$

This shows that, whatever P on $\mathscr{A}$ may be, a coherent prevision of X exists by virtue of the extension theorem quoted in the previous section, but it is not deducible from (1).

In almost all treatments of probability, one deals exclusively with a particular meaning of the concept of prevision: that of expected value*. It is defined for those $X = X(\omega)$ such that $\{\omega \in \Omega: \; X(\omega) \leqslant x\} \in \mathscr{A}$ for each $x \in \mathbb{R}$ and $\mathscr{A}$ is a σ-algebra (such a r.q. is called a *random variable*). Furthermore, the concept of expected value is introduced, after defining probability measures, by an integral of the Lebesgue–Stieltjes type in the following manner. Let X be a random variable defined on Ω. Arrange the values of X into classes such that the kth of them, $A_{k,h}$, contains the values between kh (excluded) and $(k + 1)h$ (included) for $h > 0$ and $k = 0, \pm 1, \pm 2, \ldots$. Then the expected value of X, $E[X]$, is defined by

$$E[X] = \lim_{h \to 0} \sum_{k=-\infty}^{\infty} (k + 1) h P[A_{k,h}], \quad (2)$$

provided that this limit exists. In such a case, $E[X]$ represents a coherent evaluation for the prevision of X. Clearly, when X is a bounded random variable, its expected value exists and it has the same value as the Stieltjes integral (1). But there are cases in which this integral exists even though the r.q. X is not a random variable with respect to a given class of subsets of Ω. Here is an example where $\Omega = (0, 1]$ and $\mathscr{A}$ consists of all finite disjoint unions of right-semiclosed subintervals of Ω. For any

$$A = \bigcup_{i=1}^{n} (a_i, b_i]$$

belonging to $\mathscr{A}$, let

$$P[A] = \sum_{i=1}^{n} (b_i - a_i)$$

and let us consider the r.q. $X = \sin(1/\omega)$ with $\omega \in \Omega$. Since, for $x \in (-1, 1)$, the set

$\{\omega \in \Omega : \sin(1/\omega) \leqslant x\}$ is a denumerably infinite disjoint union of closed subintervals of Ω not belonging to $\mathscr{A}$, $P[A_{k,h}]$ is not assigned and formula (2) cannot be applied. As regards the prevision of X, let us fix $\epsilon \in (0,1)$ and consider a partition of $(\epsilon, 1]$ determined by

$$\epsilon = a_0 < a_1 < \cdots < a_{n-1} < a_n = 1.$$

Hence, by virtue of $(\mathbf{p}_1)$,

$$-\epsilon + \sum_{i=1}^{n} (a_i - a_{i-1})\inf\{\sin(1/x): \\ x \in (a_{i-1}, a_i]\} \leqslant P[X]$$

$$\leqslant \sum_{i=1}^{n} (a_i - a_{i-1})\sup\{\sin(1/x): \\ x \in (a_{i-1}, a_i]\} + \epsilon.$$

Keeping ϵ fixed and indefinitely increasing the number of divisions of $(\epsilon, 1]$, one obtains

$$-\epsilon + \int_{\epsilon}^{1} \sin(1/x)\, dx \leqslant P[X] \\ \leqslant \int_{\epsilon}^{1} \sin(1/x)\, dx + \epsilon;$$

that is,

$$P[X] = \int_{0}^{1} \sin(1/x)\, dx,$$

where the integral is of the Riemann type. If P is extended from $\mathscr{A}$ to the minimal σ-algebra $\mathscr{B}$, including $\mathscr{A}$ in such a way as to preserve σ-additivity, then $X = \sin(1/\omega)$ becomes a random variable with respect to $\mathscr{B}$ and its expected value coincides with $P[X]$.

For further developments on this topic and the related problem of the prevision of unbounded r.q.s, see ref. 9, Chaps. 3 and 6.

CONDITIONAL PROBABILITY IN DE FINETTI'S QUANTITATIVE APPROACH

As regards the concept of conditional probability*, de Finetti [7, pp. 13–16; 9, Chap. 4] states that the *conditional event* $E|H$ is a logical entity which is true if $E \cap H$ is true, false if E is false and H is true, void if H is false. The probability $P[E|H]$ of $E|H$ can be defined by using the same criterion based on bets, assuming that the bet is annulled if H turns out to be false and strengthening the condition of coherence by hypothesizing that the evaluations conditional on H must turn out to be coherent under the hypothesis that H is true [10, Sec. 16 of the Appendix]. As a consequence of these qualitative conditions, an interesting characterization of coherent conditional probability laws can be obtained:

Let $\mathscr{E}$ and $\mathscr{H}$ be two algebras of events such that $\mathscr{H} \subset \mathscr{E}$; let $P[\cdot|\cdot\cdot]$ be a real function defined on $\mathscr{E} \times \mathscr{H}^0$ where $\mathscr{H}^0 = \mathscr{H} \setminus \{\phi\}$. Then $P[\cdot|\cdot\cdot]$ is a (coherent) conditional probability on $\mathscr{E} \times \mathscr{H}^0$ if and only if:

(**c**$_1$) $P[\cdot|H]$ is a probability according to $(\mathbf{a}_1)$–$(\mathbf{a}_3)$ on $\mathscr{E}$ for each $H \in \mathscr{H}^0$;

(**c**$_2$) $P[H|H] = 1$ for all $H \in \mathscr{H}^0$;

(**c**$_3$) $P[A \cap B|C] = P[A|B \cap C]|P[B|C]$ whenever $C, B \cap C \in \mathscr{H}^0$ and $A, B \in \mathscr{E}$.

In this framework, conditional probability is defined also when, provided $H \neq \phi$, one has $P[H] = 0$. Furthermore a probability law satisfying $(\mathbf{c}_1)$–$(\mathbf{c}_3)$ need not be disintegrable [11] or conglomerable [4; 8, Secs. 26, 30, 31; 9, Chap. 4].

QUALITATIVE SUBJECTIVE PROBABILITIES

Another approach to S.P. considers a system of axioms of a purely qualitative nature concerning the comparison between events such as: an event is not less probable (for a specified person) than another. S.P. so defined is called *qualitative* (or *comparative*) probability. The first precise formulation of such a system of axioms was given by de Finetti [6]. Many authors have contributed, successively, to the theory of qualitative S.P. Their works can be found via the references quoted in refs. 12, Chaps. 2 and 17, and via Wakker [23] and Suppes and Zanotti [20].

According to this approach, one can suppose that there exists an algebra $\mathscr{A}$ of events on which an individual defines a binary relation such that, given $A, B \in \mathscr{A}, A \geqslant B$ means that, for him or her, event A is not less probable than the event B. $A > B$ stands for not $-(B \geqslant A)$ and $A \sim B$ for $(A \geqslant B$ and $B \geqslant A)$. De Finetti's axioms of qualitative S.P. are usually formulated in the following fashion:

(q_1) If $A \geqslant B$ and $B \geqslant C$, then $A \geqslant C$ (transitivity).
(q_2) Either $A \geqslant B$ or $B \geqslant A$ (comparability).
(q_3) $A \geqslant \phi$.
(q_4) $\Omega > \phi$.
(q_5) If $A \cap C = B \cap C = \phi$, then $A \geqslant B$ if and only if $A \cup C \geqslant B \cup C$.

One calls *qualitative probability structure* (Q.P.S.) the couple $(\mathscr{A}, \geqslant)$ if $\geqslant$ satisfies (q_1)–(q_5) for all A, B, and C in $\mathscr{A}$.

In fact, axiom (q_3) represents a weaker version of that proposed by de Finetti, according to which $A > \phi$ for every logically possible event A. Even if this assumption is preferable from an intuitive point of view, we will accept the weaker version in view of the solution of the problem that we are about to present. This problem regards the statement of the existence of a *quantitative* S.P., denoted by P, satisfying (a_1)–(a_3) and such that

$$P[A] \geqslant P[B] \quad \text{if } A \geqslant B.$$

If such a P exists, then we say that it *almost agrees* with $\geqslant$. We say that P *agrees* with $\geqslant$ when

$$P[A] \geqslant P[B] \quad \text{if and only if } A \geqslant B.$$

In order to solve the preceding problem, de Finetti [6, Sec. 10] claimed, for every $n = 1, 2, \ldots,$ the existence of a partition of Ω in n events of $\mathscr{A}$ that are considered to be equally probable. Roughly speaking, this allows a subject to represent his qualitative S.P. via the ratio between the number of favourable cases and the number of possi-

bles cases. De Finetti's hypothesis was weakened by Savage [19, p. 34] who claimed that

(S) For every $n = 1, 2, \ldots,$ there exists a partition of Ω in $\mathscr{A}$ such that $A \geqslant B$ $(A, B \in \mathscr{A})$ whenever A and B are unions of $(r + 1)$ and r elements $(r = 1, \ldots, n - 1)$ of the partition, respectively.

Savage showed [19, Theorem 2, p. 34] that:

If the Q.P.S. $(\mathscr{A}, \geqslant)$ satisfies **(S)**, then there exists a unique probability P which almost agrees with $\geqslant$.

Further results hold, concerning also strictly agreeing probabilities, assuming hypotheses alternative to **(S)** such as $\geqslant$ is fine or/and tight. These results, essentially due to Savage [19, pp. 37–39], have been generalized by many other authors (see Wakker [23]).

Suppes and Zanotti [20] have proposed a satisfactory approach to qualitative conditional probability. Their axioms are strong enough to prove the existence and the uniqueness of a quantitative conditional probability that strictly agrees with the given qualitative one.

SAVAGE'S APPROACH VIA DECISION THEORY

Slightly differing from de Finetti, who assumes that the meaning of the binary relation $\geqslant$ between events is evident to a person, Savage starts from a system of axioms about rational preference ordering of acts in the face of uncertainty (*see* DECISION THEORY and GAME THEORY). Consequently, probability appears as a necessary ingredient of the numerical representation of the preference ordering of acts as well as the utility* function. Savage's axioms, first formulated and discussed in Chaps. 3 and 5 of ref. 19 (see also ref. 12, Chap. 8), constitute an extension to acts of (q_1)–(q_5). In order to describe these axioms, some preliminary definitions are necessary. Savage defines *the world* as "the object about which the person is concerned" and *a state* of the world as a "description of the world, leaving no relevant aspect undescribed" [19, p. 9]. Since every action produces some consequence, a set $\mathscr{C}$ of consequences has to be defined. Denoting by Ω the set of the states, an act is seen as a function f on Ω to $\mathscr{C}$, i.e., $f(s)$ is the consequence that occurs if f is implemented and $s \in \Omega$ turns out to be true. On the set $\mathscr{D}$ of available acts, a binary relation $\gtrdot$ is defined; for $f, g \in \mathscr{D}$, $f \gtrdot g$ means that g *is not preferred to* f. When $f \gtrdot g$ and $g \gtrdot f$ hold simultaneously, then f and g are said to be *equivalent* (in symbols: $f \doteq g$). A concept of *conditional preference* is also introduced in this way: $f \gtrdot g$ given $A \subset \Omega$, means that $f' \gtrdot g'$ whenever $f = f'$ and $g = g'$ on A and $f' = g'$ on $\Omega \setminus A$. $A \subset \Omega$ is said to be *null* if $f \doteq g$ whenever $f = g$ on $\Omega \setminus A$. Given that $f = x$ and $g = y$ on Ω—where $x, y \in \mathscr{C}$—we write $x \gtrdot y$ if and only if $f \gtrdot g$. Suppose now that A and B are subsets of Ω. Then we write $B \geqslant A$ (see the previous section) if and only if when $x > x'$ in $\mathscr{C}$ one has $f_B \gtrdot f_A$, where $f_A(s) = x$ for $s \in A$, $f_A(s) = x'$ for $s \in \Omega \setminus A$, $f_B(s) = x$ for $s \in B$, and $f_B(s) = x'$ for $s \in \Omega \setminus B$. Finally, we write $x \gtrdot f$ given $B \subset \Omega$, with $x \in \mathscr{C}$ and $f \in \mathscr{D}$, if and only if $h \gtrdot f$ given B, when $h(s) = x$ for every s.

At this point, Savage's axioms may be formulated as follows:

(S₁) $\gtrdot$ is a simple order in the sense that, for all $f, g, h \in \mathscr{D}$, either $f \gtrdot g$ or $g \gtrdot f$ and if $f \gtrdot h$ and $h \gtrdot g$, then $f \gtrdot g$.

(S₂) For every $f, g \in \mathscr{D}$ and $A \subset \Omega$, $f \gtrdot g$ given A or $g \gtrdot f$ given A.

(S₃) If $f(s) = x$ and $f'(s) = x'$ for every $s \in A \subset \Omega$, where A is not null, then $f' \gtrdot f$ given A if and only if $x' \gtrdot x$.

(S₄) For every $A, B \subset \Omega$, $B \geqslant A$ or $A \geqslant B$.

(S₅) $x' \gtrdot x$ for some $x', x \in \mathscr{C}$, where $x' \gtrdot x$ stands for not $x \gtrdot x'$.

(S₆) If $h \gtrdot g$ and x is any consequence, then there exists a partition $\pi =$ $\{A_1, A_2, \ldots, A_n\}$ of Ω such that, for $A_i \in \pi$ and

$$h_i(s) = \begin{cases} h(s) & \text{if } s \notin A_i, \\ x & \text{if } s \in A_i, \end{cases}$$

$$g_i(s) = \begin{cases} g(s) & \text{if } s \notin A_i, \\ x & \text{if } s \in A_i, \end{cases}$$

one obtains $h \gtrdot g_i$ and $h_i \gtrdot g$, respectively.

(S₇) If $g(s) \gtrdot f$ given B for every $s \in B$, then $g \gtrdot f$ given B.

Axiom **(S₁)** states that all pairs of decisions are comparable, while Axiom **(S₂)** asserts the same thing for conditional decisions. **(S₃)** is a version of the *sure-thing principle*. **(S₄)** claims that all pairs of events are comparable according to the order induced between them by $\geqslant$. Axiom **(S₅)** excludes the trivial case in which all decisions are equivalent. **(S₆)** is fundamental in order to establish the existence of a quantitative probability strictly agreeing with $\geqslant$. **(S₇)** is a necessary condition of dominance in order to deduce a quantitative representation of $\geqslant$ when $\mathscr{C}$ is infinite.

Savage shows that, under **(S₁)**–**(S₇)** there is a unique quantitative probability P on the set of all subsets of Ω which strictly agrees with the relation $\geqslant$. Furthermore there is a real-valued function u on $\mathscr{C}$ for which

$$f \gtrdot g \Leftrightarrow \int_\Omega u[f(s)] P(ds)$$

$$\geqslant \int_\Omega u[g(s)] P(ds).$$

u is the decision-maker subjective utility function. Such a function turns out to be bounded and unique up to a positive linear transformation.

Savage's axiomatic approach to preference and other theories dealing with the same topic are well explained in two valuable books by Fishburn [13, 14].

SUBJECTIVISTIC AND OBJECTIVISTIC INTERPRETATIONS

We conclude with some remarks concerning the comparison between the subjectivistic point of view and the objectivistic one which characterizes, in a different way, the classical approach (based on the notion of equally probable cases) and the frequentistic approach (*see* FREQUENCY INTERPRETATION IN PROBABILITY AND STATISTICAL INFERENCE).

The classical definition of probability can easily be justified from the subjectivistic point of view by stating that when a person judges the n events of an n-partition of Ω as being equally probable, then coherence [via (a_3)] implies that an event that is the union of m elements of the partition envisaged has probability m/n, i.e., the ratio of the number of favorable cases to the number of possible cases.

The analysis of the frequentist point of view is more complex. De Finetti proposes [7, Chaps. II–VI] to separate the analysis into two phases and explains their subjectivistic foundations. The first phase deals with the relations between evaluations of probabilities and the prevision of future frequencies; the second concerns the relationship between the observation of past frequencies and the prevision of future frequencies.

As regards the first phase, let us consider a sequence of events $E_1, E_2, \dots$ relative to a sequence of trials and suppose that, under the hypothesis H_N stating a certain result of the first N events, a person considers equally probable the events $E_{N+1}, E_{N+2}, \dots$. Then, denoting by $\bar{f}_{H_N}$ the prevision of the random relative frequency of occurrence of the n events $E_{N+1}, \dots, E_{N+n}$, conditional on H_N, the well known properties of a prevision yield

$$p_{H_N} = \bar{f}_{H_N},$$

where p_{H_N} indicates the probability of each $E_{N+1}, E_{N+2}, \dots$ conditional on H_N. Hence, by estimating $\bar{f}_{H_N}$ via the observation of past frequencies, one obtains an evaluation of p_{H_N}. But when is it permissible to esti-

mate $\bar{f}_{H_N}$ in such a manner? This is the problem of the second phase. De Finetti's answer is: When the events considered are supposed to be elements of a stochastic process* whose probability law, conditional on large samples, admits, as prevision of the future frequency, a value approximately equal to the frequency observed in these samples. Since the choice of the probability law governing the stochastic process is subjective, the prediction of a future frequency based on the observation of those past is naturally subjective. De Finetti shows that the procedure is perfectly admissible when the process is exchangeable (*see* EXCHANGEABILITY), that is, when only information about the number of successes and failures is relevant, irrespective of just which trials are successes or failures. This hypothesis is fundamental from a statistical point of view because it states precisely the concept of homogeneous trials of the same phenomenon.

In order to develop the ideas just sketched, consider an infinite sequence of exchangeable events $E_1, \dots, E_n, \dots$ and denote by $f_N(E)$ the number of those $E_1, \dots, E_N$ which occur. Then a fundamental theorem by de Finetti [7, pp. 32–33] asserts that the sequence $\{ f_N(E)/N \}_{N \geqslant 1}$ converges almost surely to a random quantity $\tilde{\theta}$ whose CDF we will indicate by $F_{\tilde{\theta}}(\cdot)$, and that for $t = 0, 1, \dots, N$ and $N = 1, 2, \dots,$

$$\Pr[f_N(E) = t]$$
$$= \binom{N}{t} \int_0^1 \theta^t (1 - \theta)^{N-t} \, dF_{\tilde{\theta}}(\theta) \quad (3)$$

[*see* EXCHANGEABILITY, eq. (7)].

Consequently, denoting by $f_{N,n}(E)$ the number of those $E_{N+1}, \dots, E_{N+n}$ which occur, one obtains

$$\Pr[f_{N,n}(E) = k | f_N(E) = t]$$
$$= \binom{n}{k} \int_0^1 \theta^k (1 - \theta)^{n-k} \, dF_{\tilde{\theta}}(\theta | t, N),$$

where

$$F_{\tilde{\theta}}(\theta | t, n) = \frac{\displaystyle\int_0^\theta \theta^t (1 - \theta)^{N-t} \, dF_{\tilde{\theta}}(\theta)}{\displaystyle\int_0^1 \theta^t (1 - \theta)^{N-t} \, dF_{\tilde{\theta}}(\theta)}.$$

Hence, the prevision $\bar{f}_{H_N}$ of $f_{N,n}(E)/n$ is given by

$$\bar{f}_{H_N} = \int_0^1 \theta \, dF_{\tilde{\theta}}(\theta \mid t, N).$$

Now, if $x = t/N$ and

$$F_{\tilde{\theta}}(x + \epsilon) - F_{\tilde{\theta}}(x - \epsilon) > c\epsilon^\rho$$

for small $\epsilon > 0$ and c, $\rho > 0$, then, from a result of von Mises [22, pp. 341–342] one deduces

$$\bar{f}_{H_N} \sim x \quad \text{as} \quad N \to \infty,$$

which is the desired result.

So the notion of exchangeability introduced by de Finetti in 1928 [3] proves to be fundamental for an explanation of the subjectivistic foundations of the classical frequentist view under conditions of homogeneity of the trials. Furthermore, de Finetti's representation theorem (3) for exchangeable events gives a completely satisfactory explanation of the scheme, usually adopted in Bayesian inference*, according to which the events are assumed to be independent and equally distributed, conditional on the value of a stochastic parameter $\tilde{\theta}$. In view of de Finetti's strong law of large numbers for exchangeable processes, such a parameter represents the random relative frequency of occurrence in the long run.

References

[1]   Bhaskara Rao, K. P. S. and Bhaskara Rao, M. (1983). *Theory of Changes. A Study of Finitely Additive Measures*. Academic, London, England. (A systematic and detailed study of finitely additive measures.)

[2]   Carnap, R. (1962). *Logical Foundations of Probability*, 2nd ed. University of Chicago Press, Chicago, IL. (1st. ed., 1950. A fundamental treatise on the logical viewpoint of probability.)

[3]   de Finetti, B. (1929). *Atti del Congresso Internazionale dei Matematici, Bologna, October 3–10, 1928*, **6**, 179–190. Zanichelli, Bologna, Italy. (Introduces the concept of exchangeable sequences of events and enunciates the relative representation theorem.)

[4]   de Finetti, B. (1930). *Rend. R. Ist. Lombardo (Milano)*, **63**, 414–418. (Includes the discovery of nonconglomerability.)

[5]   de Finetti, B. (1931a). *Probabilismo. Saggio Critico sulla Teoria delle Probabilità e sul Valore della Scienza*. Libreria Editrice F. Perrella, Naples, Italy. (Deals with the philosophical bases of subjective probability.)

[6]   de Finetti, B. (1931b). *Fund. Math.*, **17**, 298–329. (The first systematic treatment of de Finetti's condition of coherence and of qualitative subjective probability.)

[7]   de Finetti, B. (1937). *Ann. Inst. Henri Poincaré*, **7**, 1–68. (The classical reference for de Finetti's theory of probability and for the inductive reasoning according to this theory.)

[8]   de Finetti, B. (1949). *Ann. Triestini, Ser. 2*, **19**, 29–81. [English translation in de Finetti, B. (1972). *Probability, Induction and Statistics*. Wiley, London, England, Chap. 5. A penetrating comparison between de Finetti's theory and Kolmogorov's axiomatic approach.]

[9]   de Finetti, B. (1974). *Theory of Probability. A Critical Introductory Treatment*, Vol. 1. Wiley, London, England.

[10]   de Finetti, B. (1975). *Theory of Probability. A Critical Introductory Treatment*, Vol. 2. Wiley, London, England. (Refs. [9] and [10] constitute the most penetrating and serious attempt to present the theory of probability from a subjectivistic point of view.)

[11]   Dubins, L. E. (1975). *Ann. Prob.*, **3**, 89–99. (The concepts of conglomerability and disintegrability are analysed. A fundamental extension of a finitely additive conditional probability is given.)

[12]   Fine, T. L. (1973). *Theories of Probability*. Academic, New York. (This book critically surveys several theories of probability and constitutes a good general reference.)

[13]   Fishburn, P. C. (1970). *Utility Theory of Decision Making*. Wiley, New York.

[14]   Fishburn, P. C. (1982). *The Foundations of Expected Utility*. Reidel, Dordrecht, The Netherlands. (Refs. [13] and [14] treat decision and preference extensively and, consequently, the decisional approach to subjective probability.)

[15]   Keynes, J. M. (1921). *A Treatise on Probability*, 1st ed. Macmillan, London, England. (2nd ed., 1929. A development of the theory of logical probability.)

[16]   Koopman, B. O. (1940). *Bull. Amer. Math. Soc.*, **46**, 763–774. (Comparative probability is analysed from a logical viewpoint.)

[17]   Kyburg, H. E., Jr. and Smokler, H. E., eds. (1980). *Studies in Subjective Probability*. Krieger, Huntington, NY. (An anthology consisting of works by Ramsey [19], de Finetti [7], Koopman [16], Good, Savage, de Finetti, and Jeffreys which also includes an excellent introduction and a list

of articles and books relevant to the subject of subjective probability.)

[18] Ramsey, F. P. (1926). Truth and probability. In *Studies in Subjective Probability*, H. E. Kyburg, Jr. and H. E. Smokler, eds. Krieger, Huntington, NY, pp. 23–52. (Penetrating development of a subjectivistic view of probability and utility.)

[19] Savage, L. J. (1954). *The Foundations of Statistics*. Wiley, New York (2nd ed. 1972 by Dover. The classical reference for Savage's fundamental approach to probability, utility, and inductive reasoning).

[20] Suppes, P. and Zanotti, M. (1982). *Z. Wahrsch. verw. Geb.* **60**, 163–169.

[21] Ulam, S. (1930). *Fund. Math.*, **16**, 140–150.

[22] von Mises, R. (1946). *Mathematical Theory of Probability and Statistics*. Academic, New York. (A unified presentation of the theory of probability and statistics from the frequentistic point of view.)

[23] Wakker, P. (1981). *Ann. Statist.*, **9**, 658–662.

(AXIOMS OF PROBABILITY
BAYESIAN INFERENCE
BELIEF FUNCTIONS
CHANCE
DE FINETTI, BRUNO
DEGREE OF CONFIRMATION
DEGREES OF BELIEF
EXCHANGEABILITY
FOUNDATIONS OF PROBABILITY
INFERENCE, STATISTICAL-I, -II
LOGIC OF STATISTICAL REASONING
PREVISION
PROBABILITY MEASURE
RAMSEY'S PRIOR
SCIENTIFIC METHOD AND STATISTICS
UTILITY THEORY)

EUGENIO REGAZZINI

SUBMEAN

The *mathematical expectation** of a function $G(X)$ of a random variable X is defined as

$$E[G(X)] = \int_{-\infty}^{\infty} G(x)\, dF(x),$$

where $F(x)$ is the cumulative distribution function* of X. The submean $M(a, b)$ of X over the interval $(a, b]$ is defined as the mathematical expectation of the truncated distribution* obtained by removing values

$(X \leqslant a)$ and $(X > b)$, and is given by

$$M(a, b) = \int_{b}^{a} x\, dF(x) \Big/ \int_{a}^{b} dF(x).$$

For example, if the distribution of X is normal with expected value μ and standard deviation σ, then

$$M(\mu + \sigma, \infty)$$

$$= \mu + \frac{(\sqrt{2\pi})^{-1} \int_{1}^{\infty} u e^{-u^2/2}\, du}{(\sqrt{2\pi})^{-1} \int_{1}^{\infty} e^{-u^2/2}\, du} \cdot \sigma$$

$$= \mu + \frac{(\sqrt{2\pi})^{-1} e^{-1/2}}{1 - \Phi(1)} \cdot \sigma$$

$$= \mu + 1.525\sigma.$$

Generally, if $\mu = 0$ and $\sigma = 1$,

$$M(t, \infty) = (\sqrt{2\pi})^{-1} e^{-t^2/2} / \{1 - \Phi(t)\}.$$

(MILLS' RATIO)

SUBNORMAL DISPERSION *See* LEXIS, WILHELM

SUBSURVIVAL FUNCTION

Suppose, for each $i = 1, \ldots, n$, independently, (X_{i1}, X_{i2}) are two independent random variables. For each $j = 1, 2$, X_{ij} are identically distributed, with survival function $S_j(t)$, so that $S_j(t) = P(X_{ij} > t)$, for any i. Define, for each i, $X_i = \min(X_{i1}, X_{i2})$ and

$$\delta_i = \begin{cases} 1 & \text{if } X_i = X_{i1}, \\ 2 & \text{if } X_i = X_{i2}. \end{cases}$$

(It is assumed that $X_{i1} \neq X_{i2}$, with probability 1.)

Then the subsurvival functions $S_1^*(t)$ and $S_2^*(t)$ are defined by

$$S_j^*(t) = P(X_i > t, \delta_i = j), \qquad j = 1, 2.$$

The corresponding *empirical subsurvival functions* are defined, for $j = 1, 2$, by

$$\hat{S}_j^*(t) = n^{-1} \sum_{i=1}^{n} I[X_i > t, \delta_i = j],$$

where $I[\cdot, \cdot]$ denotes the indicator function.

For $j = 1$ or 2, the survival function $S_j(\cdot)$ for X_{1j}, say, can be expressed as a functional form in $S_1^*(\cdot)$ and $S_2^*(\cdot)$ ([2]; *see also* KAPLAN–MEIER ESTIMATOR). An equivalent relationship based on the empirical sub-survival functions provides an estimate $\hat{S}_j(\cdot)$ of $S_j(\cdot)$. If $\{X_{i1}, i = 1, \ldots, n\}$ denotes failure times and $\{X_{i2}, i = 1, \ldots, n\}$ denotes censoring times, then $\hat{S}_1(\cdot)$ turns out to be the Kaplan–Meier estimator.

In general, the (X_{i1}, X_{i2}) can be regarded as failure times corresponding to two competing risks*. The definition of subsurvival function can be extended to the case of r competing risks [1], [3, p. 419]). If $X_i = \min(X_{i1}, \ldots, X_{ir})$, if $\delta_i = j$ if $X_i = X_{ij}$, and if A is some subset of $\{1, \ldots, r\}$, then the subsurvival function $S_A^*(\cdot)$ is defined by

$$S_A^*(t) = P(X_i > t, \delta_i \in A).$$

References

[1] Peterson, A. V. (1975). *Tech. Rep. 13*, Dept. of Statistics, Stanford University, Stanford, CA.

[2] Peterson, A. V. (1977). *J. Amer. Statist. Ass.*, **72**, 854–858.

[3] Prakasa Rao, B. L. S. (1983). *Nonparametric Functional Estimation*. Academic, New York.

(CENSORED DATA
COMPETING RISKS
KAPLAN–MEIER ESTIMATOR
SURVIVAL ANALYSIS)

D. M. TITTERINGTON

SUCCESS RUNS *See* RUNS

SUCCESSION RULE, LAPLACE'S *See* LAPLACE'S LAW OF SUCCESSION

SUCCESSIVE DIFFERENCES

Let $x_1, x_2, \ldots, x_n$ be a set of data, assumed to be obtained from independent random variables $X_1, X_2, \ldots, X_n$, having common variance σ^2. The subscripts indicate the order in which the data are taken. Then $x_2 - x_1, x_3 - x_2, \ldots, x_n - x_{n-1}$ are the *succes-*

sive differences, or in the terminology of quality control* engineers, *running ranges of two*. Let $s^2 = (n-1)^{-1}\sum_{i=1}^{n}(x_i - \bar{x})^2$, the sample variance, where $\bar{x} = (\sum_{i=1}^{n}x_i)/n$. If the data come from a normally distributed population with fixed mean μ, s^2 is the most efficient unbiased estimator of σ^2. However, if there is a trend* in the data, there is a positive bias in estimating σ^2 by s^2. Problems of trend and nonrandomness first attracted attention in ballistics and weapons testing, where wind variation along with heat and wear affect the dispersion of distances traveled by projectiles. Estimation of σ by successive difference methods goes back to Vallier [15] in 1894, but it was the research of von Neumann, Hart, and others on ballistics problems at Princeton University in the Second World War (*see* MILITARY STATISTICS) that provided effective techniques both for estimation of σ^2 and of σ while eliminating trend effects and for testing data for the presence of trend (see refs. 4, 5, 6, 9, 16, and 17). Direct research on successive differences ceased, at least for the time being, in 1967 (*see*, however, TREND and VARIATE DIFFERENCE METHOD, which makes use of related quantities in time series*).

ESTIMATION

Estimators of σ^2 and of σ that are designed to eliminate trend include the following:

(i) The *mean successive difference*

$$d = (n-1)^{-1}\sum_{i=1}^{n-1}|x_{i+1} - x_i|. \quad (1)$$

Under normality assumptions in the absence of trend (Kamat [7])

$$E(d) = 2\sigma/\sqrt{\pi},$$

$$\mathrm{Var}(d) = \left\{\left(\frac{8}{3} + \frac{4\sqrt{3} - 12}{\pi}\right)\frac{1}{n-1}\right.$$

$$\left. - \left(\frac{2}{3} + \frac{4\sqrt{3} - 8}{\pi}\right)\frac{1}{(n-1)^2}\right\}\sigma^2$$

If $E(X_i) = \theta_i$ and if $(\theta_{i+1} - \theta_i)/\sigma$ is small $(i = 1, \ldots, n - 1)$, then [7]

$$E(d) \simeq \frac{2\sigma}{\sqrt{\pi}} \left\{ 1 + \frac{\sum_{i=1}^{n-1}(\theta_{i+1} - \theta_i)^2}{4(n-1)\sigma^2} \right\}.$$

If $X_i \sim N(\mu, \sigma^2)$ for each i, then $z = d\sqrt{\pi}/2$ is an unbiased estimator of σ and (Hoel [6]) $\sqrt{n}(z - \sigma)$ is asymptotically $N(0, (2\pi/3) + \sqrt{3} - 3)$ as $n \to \infty$. The asymptotic relative efficiency* (ARE) of z relative to the estimator based on s is 0.605.

(ii) The *mean square successive difference*

$$\delta^2 = (n-1)^{-1} \sum_{i=1}^{n-1} (x_{i+1} - x_i)^2 \quad (2)$$

and the *mean half-square successive difference*

$$q = \tfrac{1}{2}\delta^2, \quad (3)$$

which is unbiased. Von Neumann et al. [17] fitted Pearson curves (*see* PEARSON DISTRIBUTIONS) to the distribution of δ^2 when $X_i \sim N(\mu, \sigma^2)$ for each i. They showed that then

$$\mathrm{Var}(\delta^2) = 4\sigma^4 \left[(3n - 4)/(n-1)^2 \right],$$

derived third and fourth moments of δ^2, and showed that the ARE of q relative to s^2 is $\tfrac{2}{3}$. For data from other distributions and in the absence of trend, Moore [9] showed that

$$\mu_2(\delta^2) = (n-1)^{-2}$$
$$\times \left[2(2n - 3)\mu_4(X) + 2\mu_2^2(X) \right]$$

and that the RE of q relative to s^2 is

$$\frac{2}{n} \frac{(n-1)^2\beta_2(X) - (n-1)(n-3)}{(2n-3)\beta_2(X) + 1},$$

$$\to 1 - \frac{1}{\beta_2(X)}$$

as $n \to \infty$, where $\mu_r(X)$ is the rth central moment of X, where X has the distribution of the data and where $\beta_2(X) = \mu_4(X)/\mu_2^2(X)$. Geisser [1] gives the exact density of δ^2. Harper [3] tabulates upper and lower 0.5, 1.0, 2.5, 5.0, 10, and 25 percent points of the distribution of q/σ^2 for $n - 1 = 2(1)30(2)60(5)100$ under normality when no

trend is present. He shows that

$$(n-1)\left[(q/\sigma^2) - 1 \right]/\sqrt{3n - 5}$$

is asymptotically $N(0,1)$, but provides some improved approximations. Kamat [7] showed that under normality assumptions, when $E(X_i) = \theta_i$,

$$E(\delta^2) = 2\sigma^2 + \sum_{i=1}^{n-1} (\theta_{i+1} - \theta_i)^2/(n-1).$$

(iii) The *second variable mean difference*

$$d_2 = (n-2)^{-1} \sum_{i=1}^{n-2} |x_{i+2} - 2x_{i+1} + x_i| \quad (4)$$

and the *second variable mean square difference*

$$\delta_2^2 = (n-2)^{-1} \sum_{i=1}^{n-2} (x_{i+2} - 2x_{i+1} + x_i)^2, \quad (5)$$

which are estimators of σ and of σ^2, respectively. Each term consists of the second backward difference*

$$\Delta^2 x_{i+2} = \Delta x_{i+2} - \Delta x_{i+1}$$
$$= (x_{i+2} - x_{i+1}) - (x_{i+1} - x_i).$$

These estimators may be more successful at eliminating trend than d and δ^2, respectively, but are also less efficient; see refs. 8 and 13. Kamat [8] gives moments when $X_i \sim N(\mu, \sigma^2)$: If $\sigma'^2 = 6\sigma^2$,

$$E(d_2) = \sigma'\sqrt{2/\pi},$$

$$\mathrm{Var}(d_2) = \frac{2}{\pi}\sigma'^2 \left\{ \frac{1.062321}{n - 2} - \frac{0.519368}{(n-2)^2} \right\};$$

$$E(\delta_2^2) = \sigma'^2,$$

$$\mathrm{Var}(\delta_2^2) = \frac{35(n-2) - 18}{9(n-2)^2}\sigma'^4.$$

Rao [13] tabulates the efficiency of $\delta_2^2/6$ and of $\delta^2/2$ relative to s^2 for general underlying distributions, $n = 5(5)25, \infty$; $\beta_2(X) = 1(1)6, 10, \infty$.

Generalizations of d_2 and δ_2^2 based on higher order backward differences $\Delta^r x_i = \Delta^{r-1} x_i - \Delta^{r-1} x_{i-1}$ are defined by unbiased estimators

$$d_r = \sum_{i=r+1}^{n} |\Delta^r x_i| \Big/ \left[(n-4)\sqrt{2\binom{2r}{r}/\pi} \right],$$

$$\delta_r^2 = \sum_{i=r+1}^{n} (\Delta^r x_i)^2 \Big/ \left[(n-r)\binom{2r}{r} \right]$$

(6)

of σ and σ^2, respectively, under normality; see Guest [2], who gives expressions for the first two moments, and Morse and Grubbs [10].

Sathe and Kamat [14] approximate the distributions of d, δ^2, d_2, and δ_2^2 for $n \geqslant 5$ by $(\chi_\nu^2/c)^{1/\lambda}$, where χ_ν^2 represents chi-square with ν degrees of freedom; values of λ can be kept constant as n increases. They provide tables of values of ν and of $\log_{10} c$.

TESTING FOR TREND

A hypothesis of no trend can be tested under normality assumptions by

(a) the *mean successive difference ratio* $W = d/s$;

(b) the *mean square successive difference ratio* $\eta = \delta^2/s^2$.

W and η are each independent of s and s^2 (*see* BASU THEOREMS), so that

$$\mu_r(W) = \mu_r(d)/\mu_r(s),$$

$$\mu_r(\eta) = \mu_r(\delta^2)/\mu_r(s^2).$$

Thus

$$E(W) = \left[\Gamma\left(\frac{n-1}{2}\right) \Big/ \Gamma\left(\frac{n}{2}\right) \right]$$
$$\times \sqrt{\frac{2}{(n-1)\pi}},$$

$$E(\eta) = 2,$$

$$\mathrm{Var}(\eta) = 4\frac{n-2}{n^2-1}.$$

Kamat [7] gives approximate 0.5, 1.0, 2.5, and 5.0 percent points of the null distributions of W/s for $n = 10(5)30(10)50$. The null distribution of η was obtained by von

Neumann [16]. Hart and von Neumann [5] tabulate cumulative probabilities; Hart [4] gives lower 0.1, 1.0, and 5.0 percent points of η for $n = 4(1)60$. In defining W and η, early writers divided $\Sigma(x_i - \bar{x})^2$ by n rather than by $n-1$ in the expression for s^2, which affects their computations. More recent tables of lower percent points of η are those of Owen [12, pp. 149–150] (whose f is the sample size n), to four decimal places, and Nelson [11], who gives lower 1.0, 5.0, and 10.0 lower percent points to three decimal places, for $n = 10(1)30(2)50(5)100(10)$ $200(50)500, 600, 800, 1000$. Since the null distribution of η is symmetric about the mean value 2, the upper percent point is equal to 4 minus the corresponding lower percent point.

References

[1] Geisser, S. (1956). *Ann. Math. Statist.*, **27**, 819–824.

[2] Guest, P. G. (1951). *J. R. Statist. Soc. B*, **8**, 233–237.

[3] Harper, W. M. (1967). *Biometrika*, **54**, 419–433.

[4] Hart, B. I. (1942). *Ann. Math. Statist.*, **13**, 445–447.

[5] Hart, B. I. and von Neumann, J. (1942). *Ann. Math. Statist.*, **13**, 207–214.

[6] Hoel, P. G. (1946). *Ann. Math. Statist.*, **17**, 475–482.

[7] Kamat, A. R. (1953). *Biometrika*, **40**, 116–127.

[8] Kamat, A. R. (1954). *Biometrika*, **41**, 1–11.

[9] Moore, P. G. (1955). *J. Amer. Statist. Ass.*, **50**, 434–456.

[10] Morse, A. P. and Grubbs, F. E. (1947). *Ann. Math. Statist.*, **18**, 194–214.

[11] Nelson, L. S. (1980). *J. Qual. Tech.*, **12**, 174–175.

[12] Owen, D. B. (1962). *Handbook of Statistical Tables*. Addison-Wesley, Reading, MA.

[13] Rao, J. N. K. (1959). *J. Amer. Statist. Ass.*, **54**, 801–806.

[14] Sathe, Y. R. and Kamat, A. R. (1957). *Biometrika*, **44**, 349–359.

[15] Vallier, E. (1894). *Balistique Experimentale*. Berger-Levrault, Paris, France.

[16] von Neumann, J. (1941). *Ann. Math. Statist.*, **12**, 367–395.

[17] von Neumann, J., Kent, R. H., Bellinson, H. R., and Hart, B. I. (1941). *Ann. Math. Statist.*, **12**, 153–162.

(FINITE DIFFERENCES, CALCULUS OF
MILITARY STATISTICS
TREND
VARIATE DIFFERENCE METHOD)

CAMPBELL B. READ

SUDAKOV'S LEMMA

Use of sequential sampling* in statistical inference problems when the observations are independent and identically distributed is well known. It is important, both from a theoretical and practical point of view, to find out whether one can extend classical results of sequential analysis* to stochastic processes*. One of the important tools in the study of sequential plans for estimation of parameters of stochastic processes is Sudakov's lemma.

Let $\{\mathbf{X}(t), t \in T\}$ be a stochastic process defined on a probability space $(\Omega, \mathscr{F}, P_\theta)$, where $\theta \in \Theta$ and $T = [0, \infty)$. Suppose the process is stochastically continuous, i.e.,

$$\mathbf{X}(t) \xrightarrow{\mathrm{P}} \mathbf{X}(t_0) \quad \text{as } t \to t_0 \text{ for every } t_0 \geqslant 0.$$

Let $\mathscr{X}$ be the space of possible values of the random vector $\mathbf{X}(t)$ and $\mathscr{F}_t$ be the sub σ-algebra of $\mathscr{F}$ generated by the random vectors $\mathbf{X}(s)$, $0 \leqslant s \leqslant t$. Denote the restriction of P_θ to $\mathscr{F}_t$ by P_θ^t. Suppose P_θ^t is absolutely continuous with respect to a certain measure $P_{\theta_0}^t$ for any $t < \infty$ and for every $\theta \in \Theta$. Let τ be a stopping time taking values in T adapted to the family of σ-algebras $\{\mathscr{F}_t, t \geqslant 0\}$. Define

$$\mathbf{Z}(\omega) = (\tau(\omega), \mathbf{X}(\tau(\omega), \omega)),$$

mapping Ω into $T \times \mathscr{X}$. For any Borel set $C \subset T \times \mathscr{X}$, define $Q_\theta(C) = P_\theta(\mathbf{Z}^{-1}C)$. Under some further assumptions to be stated, Sudakov's lemma allows one to conclude that Q_θ is absolutely continuous with respect to the measure Q_{θ_0} and the Radon–Nikodym* derivative of Q_θ with respect to Q_{θ_0} can be explicitly computed.

Suppose the density of $\mathbf{X}(t)$ for any given t exists with respect to a σ-finite measure ν. Let $p_{\theta_0}(t, x; \theta)$ denote the ratio of the density of $\mathbf{X}(t)$ when θ is the true parameter to the density of $\mathbf{X}(t)$ for a fixed value $\theta = \theta_0$ in Θ.

Sudakov's lemma. *Suppose the process $\mathbf{X}(t)$ is right-continuous for almost every $\omega \in \Omega$ with respect to P_{θ_0}. Further assume that the function $p_{\theta_0}(t, x; \theta)$ is jointly continuous in (t, x) on $T \times \mathscr{X}$ for any fixed θ. If $\mathbf{X}(t)$ is a sufficient statistic for θ, given $\{\mathbf{X}(s), 0 \leqslant s \leqslant t\}$, then the measure Q_θ is absolutely continuous with respect to the measure Q_{θ_0} and, in fact,*

$$dQ_\theta/dQ_{\theta_0} = p_{\theta_0}(t, x; \theta).$$

This lemma was proved by Sudakov [5]. For a detailed proof and applications, see Kagan et al. [2] or Basawa and Prakasa Rao [1]. The usefulness of Sudakov's lemma is in giving explicit conditions under which different measures generated by stochastic processes stopped at a stopping time are absolutely continuous, and in giving sufficient conditions for the explicit determination of the corresponding Radon–Nikodym derivative. As a consequence of the latter determination, one can obtain a sequential version of the Cramér–Rao* inequality to determine optimal unbiased sequential plans (see Magiera [3]) for some special classes of processes.

We shall now describe an application of Sudakov's lemma to the so-called exponential classes of processes with independent increments. Let $\mathscr{X} = R^m$, $\theta \in \Theta \subset R^k$. We say that the process

$$\{\mathbf{X}(t), t \geqslant 0\},$$

$$\mathbf{X}(t) = (X_1(t), \dots, X_m(t))',$$

belongs to the *exponential class* if $\mathbf{X}(t)$ is a right-continuous homogeneous process with independent increments with $P(\mathbf{X}(0) = \mathbf{0}) = 1$ and $EX_i^2(t) < \infty$ $(i = 1, 2, \dots, m; t \geqslant 0)$, and the random vector $\mathbf{X}(t)$ has density

$$f(\mathbf{x}, t, \theta)$$

$$= g(\mathbf{x}, t) \exp\left\{ \sum_{i=1}^m a_i(\theta) x_i + b(\theta) t \right\}$$

with respect to a σ-finite measure ν, where $x = (x_1, \ldots, x_m)'$ and g is a nonnegative function. This class includes Bernoulli process*, negative binomial* process, gamma* process, Poisson process*, and Wiener process when $m = 1$. Let τ be a finite stopping time and define the measure Q_θ as before. Then (Winkler and Franz [6])

$$P_\theta\{(\tau, \mathbf{X}(\tau)) \in S\}$$

$$= \int_{\Omega_S} \exp\left\{ \sum_{i=1}^m \alpha_i(\theta) X_i(\tau) + \beta(\theta)\tau \right\} dP_{\theta_0},$$

where $\Omega_S = \{\omega: (\tau, \mathbf{X}(\tau)) \in S\}$, $S \subset T \times \mathcal{X}$, $\alpha_i(\theta) = a_i(\theta) - a_i(\theta_0)$, $\beta(\theta) = b(\theta) - b(\theta_0)$. Therefore,

$$dQ_\theta/dQ_{\theta_0} = \exp\left\{ \sum_{i=1}^m \alpha_i(\theta) X_i + \beta(\theta)t \right\}.$$

This gives an explicit formula for the Radon–Nikodym derivative in Sudakov's lemma for the exponential class of processes. For a slightly more general version of Sudakov's lemma and its application to sequential estimation*, see Rozanski [4].

References

[1] Basawa, I. V. and Prakasa Rao, B. L. S. (1980). *Statistical Inference for Stochastic Processes.* Academic, London, England.

[2] Kagan, A. M., Linnik, Yu. V. and Rao, C. R. (1973). *Characterization Problems in Mathematical Statistics.* Wiley, New York.

[3] Magiera, R. (1974). *Zastos. Mat.*, **14**, 227–235.

[4] Rozanski, R. (1980). *Zastos. Mat.*, **17**, 73–86.

[5] Sudakov, V. N. (1969). *Zap. Naučn. Sem. Leningrad. Otdel. Mat. Inst. Şteklov*, **12**, 157–164. (In Russian.)

[6] Winkler, W. and Franz, J. (1979). *Scand. J. Statist.*, **6**, 129–139.

(ABSOLUTE CONTINUITY
MEASURE THEORY IN PROBABILITY AND
  STATISTICS
POISSON PROCESSES
RADON–NIKODYM THEOREM
SEQUENTIAL ANALYSIS
STOCHASTIC PROCESSES)

B. L. S. Prakasa Rao

SUFFICIENCY   *See* SUFFICIENT STATISTICS

SUFFICIENT ESTIMATION AND PARAMETER-FREE INFERENCE

SUFFICIENT ESTIMATION

Suppose that n variables $x_1, \ldots, x_n$ are given, with n distribution functions $f_1(x_1, \xi), \ldots, f_n(x_n, \xi)$, which depend on a parameter ξ of one or several dimensions. [The distribution functions may be identical, $f_1(x, \xi) = \cdots = f_n(x, \xi)$, or they may be different; they may for instance depend on a set of other parameters.]

An estimator $t = u(x_1, \ldots, x_n)$ has a distribution function $h(t, \xi)$ derived from the f-functions. Then t (which must be of the same dimension as ξ) is a *sufficient estimator* of ξ if

$$f_1(x_1, \xi) \times \cdots \times f_n(x_n, \xi)$$

$$= g(x_1, \ldots, x_n) \times h(t, \xi),$$

where the function g does not depend on ξ (but perhaps on other parameters).

This means that the analysis may be separated into two parts. In one part, we estimate ξ from t in a suitable way, given the function h. In the other part we study the conditional distribution of $x_1, \ldots, x_n$, given t, that is, the function g.

All information on ξ is contained in the first part and hypotheses about ξ are tested here; hence, the name sufficient. But the distribution of t may depend on other parameters and, if so, part one must be done after they have been estimated in part two. If there are no other parameters, part two of the analysis may fall under the heading of parameter-free inference.

The concept is connected with the name of R. A. Fisher* [2].

For a more mathematical discussion, *see* SUFFICIENT STATISTICS.

Example 1. n variables all have the same Gaussian (normal) distribution. The parameter ξ has two dimensions,

$$\xi_1 = \mu \quad \text{(mean)}$$

and

$$\xi_2 = \sigma \quad \text{(standard deviation).}$$

A set of sufficient estimators is

$$t_1 = (x_1 + \cdots + x_n)/n = m$$

and

$$t_2 = \left\{ (x_1 - m)^2 + \cdots \right.$$
$$\left. + (x_n - m)^2 \right\}/(n - 1).$$

m alone is a sufficient estimator of μ alone, but if the standard deviation is not assumed to be known, the estimator also depends on this other parameter.

Example 2. n variables are given as numbers of successes in experiments with $r_1, \ldots, r_n$ trials. Each distribution is binomial and we indicate them as (r_i, θ). θ is the only parameter.

The sufficient estimator is $t = x_1 + \cdots + x_n$, with a binomial distribution (R, θ), where $R = r_1 + \cdots + r_n$. The conditional distribution g is a multiple hypergeometric distribution*.

PARAMETER-FREE INFERENCE

The term is used for methods of hypothesis testing* in models with a more or less systematic parameter structure, where derived distributions exist which are independent of all parameters. Such derived distributions may be conditional on a sufficient estimator (as before) or they can have different origins. In all cases, they are closely linked with the mathematical presentation of the model and are not transferable to other models with similar numerical properties. In the case of conditional distributions, the conditional variable is often called ancillary*. In most cases it can be classified as mathematical

ancillarity, as opposed to experimental ancillarity (*see* the closing paragraph of ANCILLARY STATISTICS).

The concept must not be confused with nonparametric methods. The latter make use of those characteristics of probability distributions which are independent of the mathematical structure of the distribution, including the parameter structure. Such characteristics include the median and other percentiles*, and rank order* tests.

Within linear models based on the Gaussian distribution, the use of conditional distributions is not needed for the elimination of parameters. μ is a location parameter. In fact the formula for t_2 in Example 1 shows how μ is eliminated by simply using the difference between two values (a single observation and the average of all). If we were to introduce the distribution of $x_1, \ldots, x_n$ conditional on m, it would prove to be the same, independent of the value of m, and, therefore, only an unnecessary complication.

Similarly σ is a scale parameter and is eliminated by forming the ratio between two sums of squares. This idea was first expressed by "Student" (W. S. Gosset*) [3] in 1908 when he introduced what was later to be known as the t-test, using $m - \mu$ divided by the square root of the sum of squares as a test statistic of a hypothesized μ value. This is the principle behind the analysis of variance*.

Conditional distributions are the way to obtain parameter-free probability formulas in discrete distributions such as the binomial. The classical example is the hypergeometric distribution* (see Example 2) as a testing method for the hypothesis that the probability of success is the same in two experiments. We shall illustrate the method by a somewhat more complex example.

Example 3. As in Example 2, the variables are the numbers of successes in n experiments. The probability of success is assumed to increase by equal steps from one experiment to the next.

The concept of "equal steps" is defined by the formula

$$\theta_i/(1 - \theta_i) = \gamma^i \sigma,$$

which assures that θ is between 0 and 1 for all positive values of the parameters. With other definitions of the concept of equal steps, artificial boundaries on the parameters would have to be introduced.

Sufficient estimators for the parameters are

$$t_1 = x_1 + x_2 + \cdots + x_n$$

and

$$t_2 = 1x_1 + 2x_2 + \cdots + nx_n.$$

A computer program for estimation is contained in the GLIM* system (Baker and Nelder [1]).

Of interest here are the probabilities of all possible outcomes conditional on these estimators. The possible outcomes for given $r_1, \ldots, r_n$ and t_1 and t_2 are found by enumeration, and their probabilities are proportional to the products of the binomial coefficients entering into each single probability. It is here illustrated by an example with small r values, which makes the enumeration easy. The r-values are 8, 4, 7, and 5 and $t_1 = 9$ and $t_2 = 28$. Table 1 gives all possible outcomes together with the binomial coefficients needed to compute probabilities. The conditional probabilities are found by dividing each product by the sum which, in the example, is 11005.

To utilize these probabilities for testing purposes, we need a "measure of extremeness" with respect to the model. In the Gaussian models already mentioned, these measures are sums of squares of deviations between observed and expected values. Chi-square tests* actually answer the question: What is the probability of the observed result and all results more extreme? F-tests*, which are obtained by dividing one sum of squares into another, have the same purpose, with the numerator as the measure of extremeness.

In discrete distributions, it is often possible to find a sum of squares of deviations. It may also be possible to find a comparable value by the likelihood ratio* method (in Gaussian models this usually gives exactly the same value; in binomial models the two values will often be approximately equal).

There is, however, one method that will nearly always produce a reasonable measure of extremeness, that is, to sort the possible outcomes by magnitude of probability. Table 2 shows the possible outcomes ordered by decreasing extremeness, with cumulative probabilities. The last entries are the cases that show the most regular progression of the percentages. This trait is also found in examples with larger numbers.

As the example goes, a set of observed numbers of successes of 0-1-6-2 will lead to the conclusion that this outcome, together with all that are more extreme, have a total probability of 0.0255. At a certain significance level (e.g., 0.05), this means that the

Table 1 Feasible Outcomes in Example 3

x_1 x_2 x_3 x_4	$\binom{r_1}{x_1}$	$\binom{r_2}{x_2}$	$\binom{r_3}{x_3}$	$\binom{r_4}{x_4}$	Product
2 0 2 5	28	1	21	1	588
1 2 1 5	8	6	7	1	336
0 4 0 5	1	1	1	1	1
1 1 3 4	8	4	35	5	5600
0 3 2 4	1	4	21	5	420
1 0 5 3	8	1	21	10	1680
0 2 4 3	1	6	35	10	2100
0 1 6 2	1	4	7	10	280
					11005

Table 2 Probability Distribution in Example 3

x_1	x_2	x_3	x_4	Prob.	Cumul. Prob.
0	4	0	5	0.0001	0.0001
0	1	6	2	0.0254	0.0255
1	2	1	5	0.0305	0.0561
0	3	2	4	0.0382	0.0942
2	0	2	5	0.0534	0.1477
1	0	5	3	0.1527	0.3003
0	2	4	3	0.1908	0.4911
1	1	3	4	0.5089	1.0000

assumption of a progression of probability by equal steps is rejected.

With larger numbers of trials and with more groups, the enumeration of possible outcomes becomes impractical, but in this model and other models based on binomial distributions, the exact distribution can be approximated by chi-square formulas.

In models of other mathematical types this is not always the case, but this problem has not been explored systematically.

References

[1] Baker, R. J. and Nelder, J. A. (1978). *General Linear Interactive Modelling*. Oxford University Press, London, England.

[2] Fisher, R. A. (1922). *Philos. Trans. Roy. Soc.* A, **222**, 309–368.

[3] Student (1908). *Biometrika*, **6**, 1–25.

(ANCILLARY STATISTICS
HYPOTHESIS TESTING
SUFFICIENT STATISTICS)

GUSTAV LEUNBACH

SUFFICIENT PARTITION *See* CONDITIONAL INFERENCE

SUFFICIENT STATISTICS

In an experimental situation, a *statistic* is a function of the observations only (so that it does not depend on any unknown constants.) A *sufficient statistic* is a statistic such that the conditional distribution of the original observations given that statistic also does not depend on any unknown constants. The concept of sufficiency was first defined in Fisher [8]. Fisher [9] said that the sufficient statistic "is equivalent, for all subsequent purposes of estimation, to the original data from which it was derived." Therefore, good estimators should depend on the original observations only through a sufficient statistic. Later this principle was extended to say

that any inference (estimators, tests, confidence intervals, etc.) should be based only on the sufficient statistic. This is often called the *sufficiency principle* and is accepted by most statisticians.

To better understand the basis for this principle, consider the experiment of tossing a coin twice, independently, with unknown probability θ of heads. Suppose that we want to estimate θ on the basis of the outcomes of these tosses. Let X_i be 1 if the ith toss is heads and 0 if the ith toss is tails, and let $T = X_1 + X_2$. Then the conditional distribution of the original observations $\mathbf{X} = (X_1, X_2)$ given the statistic T is as follows: If $T = 0$, then $\mathbf{X} = (0, 0)$ with probability 1; if $T = 1$, then $\mathbf{X} = (0, 1)$ or $(1, 0)$, each with probability $\frac{1}{2}$; and if $T = 2$, then $\mathbf{X} = (1, 1)$ with probability 1. Note that this conditional distribution does not depend on the unknown constant θ, so that T is sufficient. Now, let $d(\mathbf{X})$ be an estimator based on the original observations $\mathbf{X}$. We next find a procedure based on T which should be just as good as $d(\mathbf{X})$. We observe T and use a computer to generate new random variables $\mathbf{X}^* = (X_1^*, X_2^*)$ from the conditional distribution of $\mathbf{X}$ given T. Then the (unconditional) joint distribution of $\mathbf{X}^*$ is the same as the joint distribution of $\mathbf{X}$ and, therefore, the rule d applied to $\mathbf{X}^*$ should be just as good as the rule d applied to $\mathbf{X}$.

Note that the discussion in the preceding example did not depend in any important way on the fact that we were flipping coins or estimating an unknown constant. More generally, consider an experiment in which we observe $\mathbf{X} = (X_1, \ldots, X_n)$ and let $\mathbf{T}(\mathbf{X})$ be a sufficient statistic. Let $d(\mathbf{X})$ be any procedure based on $\mathbf{X}$. We generate a new procedure based on $\mathbf{T}$ in the following way. We first observe $\mathbf{T}$, then generate $\mathbf{X}^*$ from the conditional distribution of $\mathbf{X}$ given $\mathbf{T}$ (which we can do because that distribution is completely specified). The joint distribution of $\mathbf{X}^*$ is the same as that of $\mathbf{X}$, so that $d(\mathbf{X}^*)$ should be just as good as $d(\mathbf{X})$.

In the next section, we give the definition of sufficient and minimal sufficient statistics and give the basic criteria used to find them.

Then we state some elementary results which all show, in different ways, that nothing is lost if we reduce to a sufficient statistic. In the following sections, we discuss complete sufficient statistics and their applications, and a Bayesian interpretation for sufficient statistics. We also give a brief introduction to sufficiency for sequential models, in which there may be some loss when we reduce by sufficiency. Finally we discuss the relationship between sufficiency, conditionality, and the likelihood principle*.

BASIC DEFINITIONS AND RESULTS

A statistical model consists of a random n-dimensional vector $\mathbf{X} \in \chi$ of observations having joint distribution function $F(\mathbf{x}; \boldsymbol{\theta})$ depending on the unknown p-dimensional vector of constants $\boldsymbol{\theta} \in \Theta$. We assume that F is known for each $\boldsymbol{\theta}$. We call χ the *sample space* and Θ the *parameter space*. A *statistic* $\mathbf{T}(\mathbf{X})$ is a (possibly vector-valued) function of $\mathbf{X}$ which does not depend on $\boldsymbol{\theta}$. A *sufficient statistic* is a statistic $\mathbf{T}$ such that the conditional distribution of $\mathbf{X}$ given $\mathbf{T}$ does not depend on $\boldsymbol{\theta}$. Let $\mathbf{T}$ have distribution function $G(\mathbf{t}; \boldsymbol{\theta})$. Then the model in which we observe $\mathbf{T}$ having distribution $G(\mathbf{t}; \boldsymbol{\theta})$ is also a statistical model. We call this the *reduced model* and the model in which we observe $\mathbf{X}$ having distribution $F(\mathbf{x}; \boldsymbol{\theta})$ is called the *original model*. Many of the results in this article show that if a procedure $d(\mathbf{T})$ has some property (such as maximum likelihood estimator) for the reduced model, it has that property for the original model.

Throughout this article we assume that $\mathbf{X}$ is either a discrete or continuous random vector with joint density function $f(\mathbf{x}; \boldsymbol{\theta})$ and that $\mathbf{T}$ has joint density function $g(\mathbf{t}; \boldsymbol{\theta})$. (Actually, the results can be extended directly to any model for which $\mathbf{X}$ and $\mathbf{T}$ have joint density functions with respect to some dominating σ-finite measure.) Even in the case of continuous random vectors, it is necessary to use measure theory* to state the results precisely. This notation is necessary

because conditional distributions and joint density functions are really only defined up to sets of measure 0. However, in this article, we shall be somewhat imprecise and ignore such sets of probability 0. For a more rigorous approach to this material, see Halmos and Savage [11], Lehmann [13, pp. 47–50], Zacks [18, pp. 29–99], or Bahadur [2].

For most common models, it is quite difficult to find the conditional distribution of the original observations $\mathbf{X}$ given a statistic $\mathbf{T}$. In the continuous case, in fact, such conditional distributions are typically singular and have no density functions. Therefore, it is important to have some other way to determine whether a statistic is sufficient. The following criterion is the primary such result. It was proved in various degrees of rigor by Fisher [8], Neyman [16], and Halmos and Savage [11].

Factorization criterion

1. Let $\mathbf{T}(\mathbf{X})$ be a statistic. $\mathbf{T}(\mathbf{X})$ is sufficient if and only if $f(\mathbf{x}; \boldsymbol{\theta}) = k(\mathbf{x})h(\mathbf{T}(\mathbf{x}); \boldsymbol{\theta})$ for some functions k and h.
2. Let $\mathbf{T}$ have density $g(\mathbf{t}; \boldsymbol{\theta})$. Then $\mathbf{T}$ is sufficient if and only if $f(\mathbf{x}; \boldsymbol{\theta}) = k(\mathbf{x})g(\mathbf{T}(\mathbf{x}); \boldsymbol{\theta})$ for some function $k(\mathbf{x})$.

The first version of this criterion is useful for proving that a statistic is sufficient, while the second is for proving results about sufficiency (e.g., proving that the maximum likelihood* estimator for the reduced model is also the maximum likelihood estimator for the original model). It is a straightforward corollary of this criterion that an invertible function of a sufficient statistic is also sufficient. This should not be surprising, since if h is invertible, then someone who knows $\mathbf{T}^* = h(\mathbf{T})$ also knows $\mathbf{T}$.

Example. Let $X_1, \ldots, X_n$ be independent, $X_i \sim N(\mu, \sigma^2)$. Then $\mathbf{X} = (X_1, \ldots, X_n)'$, $\boldsymbol{\theta} = (\mu, \sigma^2)'$. Let

$$\mathbf{T}(\mathbf{X}) = (T_1, T_2)' = \left(\Sigma X_i^2, \Sigma X_i \right)'.$$

Then

$$f(\mathbf{x};\boldsymbol{\theta}) = (2\pi)^{-n/2}\sigma^{-n}$$

$$\times \exp\left(-\frac{T_1}{2\sigma^2} + \frac{T_2\mu}{\sigma^2} - \frac{n\mu^2}{2\sigma^2}\right)$$

and, therefore, **T** is sufficient. Now, let

$$\overline{X} = T_2/n,$$

$$S^2 = \frac{T_1 - T_2^2/n}{n-1}$$

be the sample mean and the sample variance. Then $(\overline{X}, S^2)$ is an invertible function of **T** and is also a sufficient statistic.

In any model, it is trivial to find a sufficient statistic, since the original observation vector **X** is always sufficient. What we would like is to find the "smallest" sufficient statistic. We say that a sufficient statistic **T** is *minimal sufficient* if it is a function of any other sufficient statistic. A minimal sufficient statistic always exists and any invertible function of a minimal sufficient statistic is also minimal sufficient.

We now present a method (due to Lehmann and Scheffé [14]) for finding a minimal sufficient statistic. Define an equivalence relation on the sample space by $\mathbf{x}_1 \equiv \mathbf{x}_2$ if

$$f(\mathbf{x}_1;\boldsymbol{\theta}) = k(\mathbf{x}_1,\mathbf{x}_2)f(\mathbf{x}_2;\boldsymbol{\theta})$$

for all $\boldsymbol{\theta}$, where k does not depend on $\boldsymbol{\theta}$. This equivalence relation divides the sample space into equivalence classes. Let $\mathbf{T}(\mathbf{x})$ be any function which is constant on equivalence classes and different on different equivalence classes. Then $\mathbf{T}(\mathbf{X})$ is a minimal sufficient statistic. This result can be used to prove the following:

Exponential criterion I

Let

$$f(\mathbf{x};\boldsymbol{\theta}) = k(\mathbf{x})p(\boldsymbol{\theta})\exp(\mathbf{c}(\boldsymbol{\theta}))'\mathbf{T}(\mathbf{x}).$$

If the components of $\mathbf{c}(\boldsymbol{\theta})$ are linearly inde-

pendent over Θ, then $\mathbf{T}(\mathbf{X})$ is minimal sufficient.

Example. In the previous example, we see that $f(x;\theta)$ has the desired form with

$$k(x) = (2\pi)^{-n/2},$$

$$p(\boldsymbol{\theta}) = \sigma^{-n}\exp\{-n\mu^2/(2\sigma^2)\},$$

$$\mathbf{c}(\boldsymbol{\theta}) = \left(-(2\sigma^2)^{-1}, \mu/\sigma^2\right)',$$

$$\mathbf{T}(\mathbf{X}) = \left(\sum X_i^2, \sum X_i\right)'.$$

Therefore, $(\sum X_i^2, \sum X_i)$ is minimal sufficient. Since $(\overline{X}, S^2)$ is an invertible function of **T**, $(\overline{X}, S^2)$ is also minimal sufficient.

In this example, as in many others, the dimension of the minimal sufficient statistic is the same as the dimension of the parameter. However, this need not always happen. For example, let $X_1, \ldots, X_n$ be independently normally distributed with mean μ and variance μ^2. Then there is only one parameter, but the minimal sufficient statistic is still $(\overline{X}, S^2)$, which has dimension 2.

We now present two more situations in which minimal sufficient statistics are easily derived. Both these results follow directly from the Lehmann–Scheffé algorithm previously given. First, suppose that Θ consists of only two points θ_0 and θ_1. Let $T(\mathbf{X}) = f(\mathbf{X};\theta_0)/f(\mathbf{X};\theta_1)$ be the Neyman–Pearson test statistic for testing that $\theta = \theta_0$ against $\theta = \theta_1$. Then $T(\mathbf{X})$ is minimal sufficient. Second, suppose that $\Theta \subset R^1$ is an interval and that $f(\mathbf{x};\theta)$ has monotone likelihood ratio* in $T(\mathbf{x})$. Then $T(\mathbf{X})$ is a minimal sufficient statistic.

We have already noted that the minimal sufficient statistic is not unique, since any invertible function of it is also minimal sufficient. We now present an alternative approach to sufficiency which removes this nonuniqueness. For any statistic $\mathbf{T}(\mathbf{X})$, define an equivalence relation on the sample space by $\mathbf{x}_1 \equiv \mathbf{x}_2$ if $\mathbf{T}(\mathbf{x}_1) = \mathbf{T}(\mathbf{x}_2)$. The partition associated with **T** is the set of equivalence classes for this relation. The *minimal* *sufficient partition* is defined as the partition associated with a minimal sufficient statistic. Note that if T_1 is an invertible function of T_2, then T_1 and T_2 have the same partition. This implies that the minimal sufficient partition is unique. Note also, that this partition is the one used earlier in the Lehmann–Scheffé approach to finding a minimal sufficient statistic. That approach makes it clear that the partition is what is important. We can define the minimal sufficient statistic on the sets of the partition arbitrarily.

SOME ELEMENTARY RESULTS

In this section we consider some elementary results about sufficient statistics; all lend support to the statement that no important information is lost when we reduce to a sufficient statistic. We continue to assume the $T(X)$ is a sufficient statistic and refer to the model in which we observe X having density $f(x; \theta)$ as the original model and the model in which we observe T having density $g(t; \theta)$ as the reduced model. We do not assume that T is minimal sufficient.

Consider first estimating the possibly vector-valued function $\tau(\theta)$. A *loss function* $L(a, \theta)$ is a function which measures our loss when we estimate τ by a and θ is the true value of the parameter. The *risk function* of an estimator $d(X)$ is

$$R(d; \theta) = E_\theta L(d(X); \theta),$$

the expected loss from d. For example, if $\tau \in R^1$, we often use $L(a; \theta) = (a - \tau(\theta))^2$ for the loss function squared error and have the mean squared error* $E_\theta(d(X) - \tau(\theta))^2$ for the risk function. We now give some results relating sufficiency and estimation*. For any estimator $d(X)$, define $d^*(T) = Ed(X)|T$. Then d^* does not depend on θ, so that d^* is also an estimator.

(a) $Ed(X) = Ed^*(T)$ so that d^* is unbiased if d is.

(b)   (Rao–Blackwell*.) If $L(a, \theta)$ is a convex function of a, then $R(d, \theta) \geqslant R(d^*, \theta)$.

(c)   If $\hat{\tau}$ is a maximum likelihood* estimator (MLE) of τ for the reduced model, then $\hat{\tau}$ is a MLE of τ for the original model.

Part (c) implies that if the MLE is unique, then it is a function of the sufficient statistic. Part (b) implies that for any rule based on X there is a rule based on T which has no greater risk. In this case, we say that the rules based on T form an *essentially complete class*. In fact, it can be shown that if L is strictly convex in a, then d^* has smaller risk than d unless d is already a function of T [i.e., unless $d(X) = h(T(X))$]. In this case, we say that the rules based on T form a *complete class*. The function $L(a, \theta) = (a - \tau(\theta))^2$ is strictly convex, so these results are applicable to mean square error.

Now consider testing that $\theta \in N$ against $\theta \in \Theta - N$. Let $\phi(X)$ be any critical function and let $\phi^*(T) = E\phi(X)|T$. Then ϕ^* is also a critical function (i.e., $0 \leqslant \phi^* \leqslant 1$; ϕ^* does not depend on θ).

(d) ϕ and ϕ^* have the same size and power function*. If ϕ is unbiased then so is ϕ^*.

(e)   The likelihood ratio test* (LRT) statistic is the same for the reduced model as for the original model.

Result (e) implies that the LRT depends only on the sufficient statistic, while result (d) implies that the rules based on T form an essentially complete class. Note, however, that ϕ^* may be a randomized rule, even though ϕ is nonrandomized.

Sufficiency is not the only method used to reduce models. Another method is reduction by invariance*. However, when reducing by invariance some information is lost. Ferguson [7, p. 157] therefore suggests reducing first by sufficiency before considering invariance. Two questions which arise from this approach are the following:

1. Will the reduced problem be invariant?
2. Could there be an invariant rule for the original model which is better than any invariant rule for the reduced model?

The following results from Arnold [1] answer these questions.

(f) If an invariant model is reduced to a minimal sufficient statistic, then the reduced model is invariant.

(g) If the reduced model is invariant under an affine group and $\mathbf{d}(\mathbf{X})$ is an invariant rule for the original model, then $\mathbf{d}^*(\mathbf{T}) = E\mathbf{d}(\mathbf{X})|\mathbf{T}$ is an invariant rule for the reduced model.

Results (b), (d), and (g) imply that for any invariant rule for the original model, there is an invariant rule for the reduced model which is "just as good." For more detailed results about the relationship between sufficiency and invariance for testing problems, see Hall et al. [10].

We close this section with a result about Fisher information* which shows that, in this sense, no information is lost when we reduce to a sufficient statistic.

(h) Let $\mathbf{T}(\mathbf{X})$ be a statistic and let $\theta \in R^1$. Then the Fisher information in $\mathbf{T}$ is less than or equal to the Fisher information in $\mathbf{X}$, with equality if and only if $\mathbf{T}$ is sufficient.

COMPLETE SUFFICIENT STATISTICS

A statistic $\mathbf{T}$ has a complete family of distributions if $Eh(\mathbf{T}) = 0$ for all θ implies that $P(h(\mathbf{T}) = 0) = 1$ for all θ. A *complete sufficient statistic* is a sufficient statistic with a complete family of distributions. Any invertible function of a complete sufficient statistic is also a complete sufficient statistic. A complete sufficient statistic is minimal suffi-

cient, but a minimal sufficient statistic need not be complete.

If $\mathbf{T}$ is a complete sufficient statistic and $\mathbf{d}_1(\mathbf{T})$ and $\mathbf{d}_2(\mathbf{T})$ are both unbiased estimators of a function $\tau(\theta)$, then $\mathbf{d}_1(\mathbf{T}) = \mathbf{d}_2(\mathbf{T})$ with probability 1. Therefore, there is at most one unbiased estimator of any function $\tau(\theta)$ based on $\mathbf{T}$. This fact, together with the Rao–Blackwell theorem already given, implies the following result.

(a) (Lehmann–Scheffé.) Let $\mathbf{T}$ be a complete sufficient statistic and let $\mathbf{d}(\mathbf{t})$ be an unbiased estimator of $\tau(\theta)$. Then $\mathbf{d}(\mathbf{T})$ is the minimum variance unbiased estimator* of $\tau(\theta)$. For any convex loss function, $\mathbf{d}(\mathbf{T})$ minimizes the risk among all unbiased estimators.

This result implies that the estimator $\mathbf{d}(\mathbf{T})$ is the only unbiased estimator that we should consider using.

Now, consider testing that $\theta \in N$ against $\theta \in \Theta - N$. Let B be the boundary between the null set N and the alternative set $\Theta - N$, and let $\mathbf{T}$ be a complete sufficient statistic for the model in which we observe $\mathbf{X}$ having density $f(\mathbf{x}; \theta)$, $\theta \in B$.

(b) Let $\phi(X)$ be an unbiased size α test. If ϕ has a continuous power function*, then $E_\theta \phi(\mathbf{X})|\mathbf{T} = \alpha$ for all $\theta \in B$.

This result is often useful in finding uniformly most powerful (UMP) unbiased size α tests, when all tests have continuous power functions and there exists a complete sufficient statistic $\mathbf{T}$ for the boundary. In that case, we consider the problem conditionally on $\mathbf{T}$. We find a UMP unbiased size α test $\phi_{\mathbf{T}}^*(\mathbf{X})$ for the conditional problem. Result (b) implies that the test $\phi(\mathbf{X}) = \phi_{\mathbf{T}(\mathbf{X})}^*(\mathbf{X})$ is UMP unbiased size α for the unconditional problem. Tests that are conditionally size α are called *tests with Neyman structure**. For more details and examples of this method for deriving optimal tests, see Lehmann [13, pp. 134–212] or Ferguson [7, pp. 224–235]; *see also* UNBIASEDNESS.

It is often difficult to show a statistic is complete from the definition of completeness. We now give a criterion which is often useful. For a proof, see Lehmann [13, pp. 132].

Exponential criterion II

Let

$$f(\mathbf{x};\boldsymbol{\theta}) = k(\mathbf{x})p(\boldsymbol{\theta})\exp(\mathbf{c}(\boldsymbol{\theta}))'\mathbf{T}(\mathbf{X}).$$

If $\mathbf{c}(\Theta) = \{\mathbf{c}(\boldsymbol{\theta});\ \boldsymbol{\theta} \in \Theta\}$ contains an open set, then $\mathbf{T}(\mathbf{X})$ is a complete sufficient statistic.

Example. We return to the example in which we observe X_i independent, $X_i \sim N(\mu,\sigma^2)$. Then

$$\mathbf{T}(\mathbf{X}) = \left(\Sigma X_i^2, \Sigma X_i\right)',$$

$$\mathbf{c}(\boldsymbol{\theta}) = \left(-1/(2\sigma^2), \mu/\sigma^2\right)'.$$

The image of $\mathbf{c}$ is the set of all vectors whose first coordinate is negative, which contains an open set, and hence $\mathbf{T}$ is a complete sufficient statistic. Since $(\overline{X}, S^2)$ is an invertible function of $\mathbf{T}$, $(\overline{X}, S^2)$ is also complete and sufficient.

BAYESIAN SUFFICIENCY

In the previous discussion, we have assumed that the parameter vector $\boldsymbol{\theta}$ is an unknown constant. A Bayesian statistician takes a different approach and assumes that $\boldsymbol{\theta}$ is an unknown random vector having a known density $\pi(\boldsymbol{\theta})$, the *prior distribution** of $\boldsymbol{\theta}$. We can assume without loss of generality that $\pi(\boldsymbol{\theta}) > 0$ for all $\boldsymbol{\theta}$. [If $\pi(\boldsymbol{\theta}) = 0$ for some $\boldsymbol{\theta}$, a Bayesian could throw that point out of the parameter space.] A Bayesian would also know $f(\mathbf{x};\boldsymbol{\theta})$, the conditional distribution of $\mathbf{X}$ given $\boldsymbol{\theta}$. He could then compute the conditional distribution of $\boldsymbol{\theta}$ given $\mathbf{X}$, the *posterior distribution** of $\boldsymbol{\theta}$. He then bases his procedures only on this posterior distribution. The following results establish that the Bayes

procedures are the same whether computed from the original or the reduced model.

(a) If $\mathbf{T}$ is a sufficient statistic, then the conditional distribution of $\boldsymbol{\theta}$ given $\mathbf{T}$ is the same as the conditional distribution of $\boldsymbol{\theta}$ given $\mathbf{X}$.

(b) Let $\mathbf{S}(\mathbf{X})$ be a statistic. The conditional distribution of $\boldsymbol{\theta}$ given $\mathbf{X}$ depends on $\mathbf{X}$ only through $\mathbf{S}(\mathbf{X})$ if and only if $\mathbf{S}(\mathbf{X})$ is a sufficient statistic.

Kolmogoroff [12] suggested that a sensible definition of sufficiency for a Bayesian is that a statistic $\mathbf{T}(\mathbf{X})$ is sufficient if the posterior distribution of Θ given $\mathbf{X}$ depends on $\mathbf{X}$ only through $\mathbf{T}(\mathbf{X})$. Result (b) implies that the Bayesian definition of sufficiency is the same as the frequentist definition given earlier, at least under the regularity conditions we are assuming. (For a nonregular example where the definitions are different, see Blackwell and Ramanoorthi [6].)

SEQUENTIAL SUFFICIENCY

We now consider a sequential decision problem in which there are

1. a sequence of observable random variables $X_1, X_2, \ldots$, where $X_i \in \chi_i$, $(X_1, \ldots, X_j)$ has joint density $f_j(x_1, \ldots, x_j;\boldsymbol{\theta})$, $\boldsymbol{\theta} \in \Theta$;

2. a cost function $c_j(\boldsymbol{\theta}, x_1, \ldots, x_j)$ such that $c_0(\boldsymbol{\theta}) \geqslant 0$, and c_j increases monotonically to ∞ as $j \to \infty$;

3. a space of possible actions $\mathbf{a}$ and a loss function $L(\mathbf{a};\boldsymbol{\theta})$ which is a convex function of $\mathbf{a}$.

Let $\mathbf{T}_1(X_1), \mathbf{T}_2(X_1, X_2), \ldots$, be a sequence of statistics such that $\mathbf{T}_j$ depends only on $(X_1, \ldots, X_j)$. Then $\{\mathbf{T}_j\}$ is a *sufficient sequence of statistics* if for all j, $\mathbf{T}_j$ is sufficient for the fixed sample model in which we observe $(X_1, \ldots, X_j)$ having density $f_j(x_1, \ldots, x_j;\boldsymbol{\theta})$, $\boldsymbol{\theta} \in \Theta$. We might suppose that for any stopping* rule and terminal decision rule based on the original observations $\{X_j\}$, there would be a stopping rule and a terminal decision rule based on the sufficient sequence $\{T_j\}$ which is just as good. Unfortunately, this need not happen. For one thing, the cost function may not be completely determined by $\{T_j\}$. (In this case, it does not seem that $\{T_j\}$ is really "sufficient" for the given sequential problem.) However, there are even examples with cost $c_j = j$, in which the rules based on a sufficient sequence do not form an essentially complete class. A sufficient sequence $\{T_j\}$ is called *transitive* if for all j and all bounded functions f_j,

$$E\Big(f_j\big(X_1,\ldots,X_j\big)|(T_j,T_{j+1})\Big)$$

$$= E\big(f_j\big(X_1,\ldots,X_j\big)|T_j\big).$$

The following summarizes the basic results in sequential sufficiency. We assume that c_j depends only on T_j and θ.

(a) Let $\{T_j\}$ be a sufficient sequence. For any stopping rule ϕ and any terminal decision rule δ based on the original observations $\{X_j\}$, there is a terminal decision rule δ^* based on $\{T_j\}$ such that (ϕ, δ^*) is as good as (ϕ, δ).

(b) If $\{T_j\}$ is a transitive sufficient sequence, ϕ is any stopping rule and δ any terminal decision rule based on the original observations $\{X_j\}$, then there exists a stopping rule ϕ^* and a terminal decision rule δ^* based only on $\{T_j\}$ such that (ϕ^*, δ^*) is as good as (ϕ, δ).

Result **(a)** implies that the class of rules in which the terminal decision rule is based on the sufficient sequence is an essentially complete class of rules. Result **(b)** implies that the class of rules in which both the stopping rule and terminal decision rule are based on the sufficient sequence is essentially complete as long as the sufficient sequence is transitive. See Ferguson [7, pp. 329–340] for proofs of these results. See Bahadur [2] for a more detailed discussion of sequential sufficiency.

SUFFICIENCY, CONDITIONALITY, AND THE LIKELIHOOD PRINCIPLE

The likelihood* function $L_x(\theta)$ is defined by $L_x(\theta) = f(x;\theta)$ and is thought of as a function of θ for each fixed x. The *likelihood principle** says that a statistical procedure should depend only on $L_x(\theta)$, where x is the vector of actual observations. Further, it says that if two different experiments (with the same parameter θ) lead to the same or proportional likelihoods (for the actual outcomes observed), then the same conclusion should be drawn from both experiments.

In particular, any conclusions should only depend on the outcome observed and not on any other outcome we might have observed. This implies that the notions of unbiasedness* (either in estimation or testing), invariance, risk functions, power functions, confidence coefficients, admissibility*, and minimaxity all violate the likelihood principle. Maximum likelihood estimators do satisfy the likelihood principle, but size α likelihood ratio tests do not (since size does not).

Any Bayes procedures which are based on the posterior distribution do satisfy the likelihood principle, and the principle is often used as part of the argument in favor of a Bayesian approach to statistics (see Berger and Wolpert [3], pp. 124–136).

As stated in the introduction, the *sufficiency principle* says that any inference should be based only on a sufficient statistic. The *conditionality principle* says that if an experiment involving θ is randomly chosen from a collection of experiments (independently of θ), then any experiment not chosen is irrelevant to the statistical analysis. That is, the analysis should be performed conditionally on the experiment chosen (*see* CONDITIONAL INFERENCE). (The experiment chosen is an ancillary for this problem, so that this principle is a version of conditioning on an ancillary statistic*.)

Birnbaum [4] showed for discrete distributions that the sufficiency principle and the conditionality principle together are equivalent to the likelihood principle. Berger and

Wolpert [3] extended this result to general distributions. Therefore, if we believe the sufficiency principle and the conditionality principle, we have to accept the likelihood principle and should therefore eliminate much of statistics as it is now practiced.

FURTHER COMMENTS

We have stated that most statisticians accept the principle that statistical analysis should depend only on a sufficient statistic. However, if we really believed this principle (and our models), we would not do residual analysis in multiple regression models. The difficulty is that we rarely believe that our models are exactly correct for the experiment. However, the sufficiency or insufficiency of a statistic is determined only by the exact model. "Nearby" models may have sufficient statistics which are quite different. Therefore, in practice, the sufficiency principle is often tempered by our lack of confidence in our models.

Most textbooks on mathematical statistics have discussions of sufficient statistics. Lindgren [15] has a particularly nice discussion, with no measure theory, of sufficiency, minimal sufficiency, complete sufficiency, and the minimal sufficient partition. Ferguson [7] treats sufficiency, completeness, and sequential sufficiency and proves complete class theorems for rules based on sufficient statistics, again with no measure theory. Lehmann [13] and Zacks [18] both present measure-theoretic treatments of properties of sufficient statistics.

References

[1] Arnold, S. F. (1985). Sufficiency and invariance. *Lett. Statist.*, **3**, 275–281.

[2] Bahadur, R. R. (1954). Sufficiency and statistical decision functions. *Ann. Math. Statist.*, **25**, 423–462.

[3] Berger, J. O. and Wolpert, R. L. (1984). *The Likelihood Principle*. Lecture Notes No. 6. Inst. Math. Statist., Hayward, Calif.

[4] Birnbaum, A. (1961). On the foundations of statistical inference: binary experiments. *Ann. Math. Statist.*, **32**, 414–435.

[5] Blackwell, D. (1947). Conditional expectation and unbiased sequential estimation. *Ann. Math. Statist.*, **18**, 105–110.

[6] Blackwell, D. and Ramanoorthi, R. V. (1982). A Bayes but not classically sufficient statistic. *Ann. Math. Statist.*, **10**, 1025–1026.

[7] Ferguson, T. S. (1967). *Mathematical Statistics, A Decision Theoretic Approach*. Academic, New York.

[8] Fisher, R. A. (1922). On the mathematical foundation of theoretical statistics. *Philos. Trans. R. Soc. (Lond.) Ser. A*, **222**, 309–368.

[9] Fisher, R. A. (1925). Theory of statistical estimation, *Proc. Camb. Philos. Soc.*, **22**, 700–725.

[10] Hall, W. J., Wijsman, R. A., and Ghosh, J. K. (1965). The relationship between sufficiency and invariance with applications in sequential analysis. *Ann. Math. Statist.*, **36**, 575–614.

[11] Halmos, P. R. and Savage, L. J. (1949). Applications of the Radon–Nikodym theorem to the theory of sufficient statistics. *Ann. Math. Statist.*, **20**, 225–241.

[12] Kolmogoroff, A. N. (1942). Definitions of center of dispersion and measure of accuracy from a finite number of observations. *Izv. Akad. Nauk. SSSR Ser. Mat.*, **6**, 3–32. (In Russian.)

[13] Lehmann, E. L. (1959). *Testing Statistical Hypotheses*. Wiley, New York.

[14] Lehmann, E. L. and Scheffé, H. (1950, 1955). Completeness, similar regions and unbiased estimation, *Sankhyā*, **10**, 305–340 and **15**, 219–236.

[15] Lindgren, B. W. (1976). *Statistical Theory*, 3rd Ed., MacMillan, New York.

[16] Neyman, J. (1935). Sur un teorema concernte le cosidette statistiche sufficienti, *Giorn. Ist. Ital. Att.*, **6**, 320–334.

[17] Rao, C. R. (1945). Information and accuracy attainable in estimation of statistical parameters. *Bull. Calcutta Math. Soc.*, **37**, 81–91.

[18] Zacks, S. (1971). *The Theory of Statistical Inference*. Wiley, New York.

(ANCILLARY STATISTICS
BAYESIAN INFERENCE
COMPLETENESS
CONDITIONAL INFERENCE
ESTIMATION, POINT
INFERENCE, STATISTICAL: I
INFERENCE, STATISTICAL: II
INVARIANCE CONCEPTS IN STATISTICS
LEHMANN–SCHEFFÉ THEOREM
LIKELIHOOD PRINCIPLE
MINIMUM VARIANCE UNBIASED
    ESTIMATION
NEYMAN STRUCTURE, TESTS WITH
POSTERIOR DISTRIBUTIONS

RAO–BLACKWELL THEOREM
SUFFICIENT ESTIMATION AND
   PARAMETER-FREE INFERENCE)

STEVEN F. ARNOLD

SUKHATME SCALE TEST   *See*   SCALE
TESTS

SUMCOR   *See*   GENERALIZED CANONICAL
VARIABLES

SUMMATION [*n*]

The symbol [*n*] is used as an operator to
denote a summation of *n* terms.
   If *n* is odd,

$$[n]u_x = u_{x-(n-1)/2} + u_{x-(n-3)/2}$$
$$+ \cdots + u_{x+(n-3)/2}$$
$$+ u_{x+(n-1)/2};$$

if *n* is even,

$$[n]u_x = u_{x-n/2} + u_{x-n/2+1}$$
$$+ \cdots + u_{x+n/2-1} + u_{x+n/2}.$$

The symbol is used to provide compact representation of certain graduation* formulae.
Repeated application of the operator is denoted by juxtaposition. Thus,

$$[3][5]u_x = [3](u_{x-2} + u_{x-1} + u_x$$
$$+ u_{x+1} + u_{x+2})$$
$$= (u_{x-3} + u_{x-2} + u_{x-1})$$
$$+ (u_{x-2} + u_{x-1} + u_x)$$
$$+ \cdots + (u_{x+1} + u_{x+2} + u_{x+3})$$
$$= u_{x-3} + 2u_{x-2} + 3(u_{x-1} + u_x$$
$$+ u_{x+1}) + 2u_{x+2} + u_{x+3}$$
$$= [5][3]u_x.$$

Symbolically (for *n* even or odd),

$$[n] \equiv (E^{n/2} - E^{-n/2})/(E^{1/2} - E^{-1/2}),$$

where *E* is the shift operator. In terms of
central differences*,

$$[n] \equiv n + \frac{1}{24}(n^2 - 1)\delta^2$$
$$+ \frac{1}{1680}(n^2 - 1)(n^2 - 9)\delta^4$$
$$+ \text{terms in } \delta^6 \text{ and higher orders,}$$

when δ is the central operator ($\delta u_x = u_{x+1/2} - u_{x-1/2}$).

(FINITE DIFFERENCES, CALCULUS OF
GRADUATION
MOVING AVERAGES)

SUM-QUOTA SAMPLING

This is a term coined by Kremers [1] to
describe a system of sampling that is relevant when total allowable cost is specified in
advance, but actual cost of sampling varies
from individual to individual.
   Sampling proceeds sequentially, obtaining
observed values of random variables
$X_1, X_2, \ldots, X_n, \ldots$ until the total cost C_n ($= c_1 + c_2 + \cdots + c_n$) first exceeds the specified amount Q (the "quota"). Construction
of efficient unbiased estimators of population parameters and other statistical inference procedures are described in ref. 1.

Reference

[1]   Kremers, W. K. (1985). Ph,D. thesis, Cornell University, Ithaca, NY.

(QUOTA SAMPLING
SEQUENTIAL SAMPLING)

SUM-SYMMETRIC POWER SERIES DISTRIBUTIONS

The sum-symmetric power series distributions (SSPSD) constitute a subclass of multivariate power series distributions. The multinomial* (M), negative multinomial (NM),
multiple Poisson* (MP), multivariate logarithmic series* distributions (MLS), and their
origin-truncated versions belong to this subclass. The SSPSD class was introduced by
Patil [5] for the distributions occurring in
direct and inverse sampling* with replacement from populations with multiple char-
acters. See also Barndorff-Nielsen [1] and
Patil et al. [6], [7]. The estimation results for
SSPSDs have been found of some use in

diversity* related research. See Smith et al. [8].

DEFINITIONS AND PROPERTIES

Let T be a subset of the s-fold Cartesian product of the set I of nonnegative integers. Define $f(\boldsymbol{\theta}) = \Sigma_\mathbf{x} a(\mathbf{x})\Pi_1^s \theta_i^{x_i}$, where the summation extends for $\mathbf{x} = (x_1, x_2, \ldots, x_s)$ over T and $a(\mathbf{x}) > 0$, with $\boldsymbol{\theta} \in \Omega$ the parameter space such that $\theta_i > 0$, $i = 1, 2, \ldots, s$, and $f(\boldsymbol{\theta})$ is finite and differentiable. Then an s-dimensional random vector $\mathbf{X}$ with probability function (pf)

$$p(\mathbf{x}) = p(\mathbf{x}; \boldsymbol{\theta}) = a(\mathbf{x})\prod_1^s \theta_i^{x_i}/f(\boldsymbol{\theta}),$$

$$x \in T \quad (1)$$

is said to have the *multivariate power series distribution* (MPSD) with *range* T and *series function* (sf) $f(\boldsymbol{\theta})$. Further, $\boldsymbol{\theta}$ is the series parameter, which takes values in the region of convergence of power series of $f(\boldsymbol{\theta})$ in powers of θ_i's. The $a(\mathbf{x})$ is the *coefficient function*.

Now, in the MPSD given by (1), let the sf $f(\boldsymbol{\theta})$ be sum-symmetric in that $f(\boldsymbol{\theta}) = g(\theta_1 + \theta_2 + \cdots + \theta_s)$. Further, let the sf $g(\cdot)$ be such that $g(\theta_1 + \theta_2 + \cdots + \theta_s) = \Sigma_{z=0}^\infty a(z)(\Sigma_1^s \theta_i)^z$, where $z = \Sigma_1^s x_i$; then the pf of $\mathbf{X}$ is of the form

$$p(\mathbf{x}; \boldsymbol{\theta}) = \frac{(\Sigma_1^s x_i)!}{\Pi_1^s x_i!}\frac{a(\Sigma_1^s x_i)\Pi_1^s \theta_i^{x_i}}{g(\Sigma_1^s \theta_i)}. \quad (2)$$

The random vector $\mathbf{X}$ having the pf given by (2) is said to follow the *SSPSD with parameters* θ_i, $i = 1, 2, \ldots, s$, and the sf $g(\theta_1 + \theta_2 + \cdots + \theta_s)$.

Property 1. If $\mathbf{X} \sim \text{SSPSD}(\boldsymbol{\theta}, g(\Sigma_1^s \theta_i))$, then

(i) $\Sigma_1^s X_i \sim \text{PSD}(\Sigma_1^s \theta_i, g(\Sigma_1^s \theta_i))$.
(ii) $(\mathbf{X}|\Sigma_1^s x_i) \sim$ singular multinomial $(\theta_i^*, i = 1, 2, \ldots, s, \Sigma_1^s x_i)$, where $\theta_i^* = \theta_i/\Sigma_1^s \theta_j$ for $i = 1, 2, \ldots, s$.

Moreover, the class of SSPSDs consists of the power series mixtures* on the index

parameter n of the multinomial distributions.

(iii) $(X_1, X_2, \ldots, X_r)|(X_{r+1} = x_{r+1}, \ldots, X_s = x_s) \sim \text{SSPSD}((\theta_1, \theta_2, \ldots, \theta_r), h(\theta_1 + \theta_2 + \cdots + \theta_r))$, where $h(\theta_1 + \theta_2 + \cdots + \theta_r)$ depends in form on $g(\theta_1 + \theta_2 + \cdots + \theta_s)$. Further, it depends in value on the sum $x_{r+1} + x_{r+2} + \cdots + x_s$ (and not on the individual values), but is independent of θ_{r+1}, $\theta_{r+2}, \ldots, \theta_s$.

Property 2. The means, variances, covariances, and the moments of the SSPSD take the following forms, where $X = \Sigma X_i$:

(i) $E[X_i] = \mu_i = (\theta_i/\theta)\mu = (\theta_i/\theta)E[X]$, so that μ_i is proportional to θ_i, $i = 1, 2, \ldots, s$.
(ii) $\text{var}(X_i) = (\theta_i/\theta)[\mu + (\theta_i/\theta)(\sigma^2 - \mu)]$; $\sigma^2 = \text{var}(X)$.
(iii) Covariance $(X_i, X_j) = \sigma_{ij} = (\theta_i\theta_j/\theta^2)(\sigma^2 - \mu)$.
(iv) $E[X_1^{(r_1)}X_2^{(r_2)} \cdots X_s^{(r_s)}] = \mu_{[r]} = [\Pi(\theta_i/\theta)^{r_i}]\mu_{(r)}$, where $\mu_{(r)} = E[X^{(r)}]$ and $X^{(r)} = X(X - 1) \cdots (X - r + 1)$ is the rth descending factorial power* of X.

Property 3. The multinomial, negative multinomial, multivariate logarithmic series distributions, and their origin-truncated versions are special cases of the SSPSD as follows:

(i) $g(\Sigma \theta_i) = (1 + \Sigma_1^s \theta_i)^n$, n a positive integer, for multinomial.
(ii) $g(\Sigma \theta_i) = (1 - \Sigma_1^s \theta_i)^{-k}$, $k > 0$, for negative multinomial.
(iii) $g(\Sigma \theta_i) = \exp(\Sigma_1^s \theta_i)$, for multivariate Poisson.
(iv) $g(\Sigma \theta_i) = -\log(1 - \Sigma_1^s \theta_i)$, for multivariate log series.

In each of the special cases (i)–(iii), if the sf is $g(\Sigma \theta_i) - 1$, we get the origin-truncated versions of the respective distributions.

Theorem 1. Let $\mathbf{X} \sim \text{SSPSD}(\boldsymbol{\theta}, g(\Sigma_1^s \theta_i))$. Let $p_r^{(i)}$ denote the probability $P(X_i = r)$, where X_i is the ith component of $\mathbf{X}$. If $p_0^{(i)} p_2^{(i)} / [p_1^{(i)}]^2$ is a constant function of $\boldsymbol{\theta}$, then $p_r^{(i)} p_{r+2}^{(i)} / [p_{r+1}^{(i)}]^2$ is a constant function of $\Sigma_1^s \theta_i$ for $r = 0, 1, 2, \ldots$, and the SSPSD is either the multinomial, or negative multinomial, or the multiple Poisson.

Theorem 2. Let $\mathbf{X} \sim \text{SSPSD}(\boldsymbol{\theta}, g(\Sigma_1^s \theta_i))$. Define $S = \Sigma_1^s X_i$. If for the distributions of S and any S_i the sfs are identical in form and if $\boldsymbol{\theta}$ occurs in the sf of the distribution of X_i only through its power series parameter, then the SSPSD is either the multinomial or the negative multinomial or the multiple Poisson.

The characterization results for certain members of the bivariate SSPSD and its weighted* version are discussed in Mahfoud and Patil [4].

Definition 1. Let (X_1, X_2) be the nonnegative r.v.s having the pdf $p(x_1, x_2)$. Let $w(x_1, x_2) > 0$ be the weight function such that $E[W(X_1, X_2)] < \infty$. Then the distribution of (X_1^w, X_2^w), the weighted version of (X_1, X_2), is given by

$$p^w(x_1, x_2) = \frac{w(x_1, x_2) p(x_1, x_2)}{E[W(X_1, X_2)]}.$$

Theorem 3. Let $(X_1, X_2) \sim \text{SSPSD}(\theta_1, \theta_2, g(\theta_1 + \theta_2))$. Let $w(x_1, x_2) = x_1$ be the weight function. Then $\eta^2(X_1 | X_1 + X_2) = \eta^2(X_1^w | X_1^w + X_2^w)$ if and only if (X_1, X_2) is distributed as double Poisson, bivariate multinomial, or bivariate negative multinomial, where $\eta^2(U|V)$ denotes the correlation ratio of U on V.

ESTIMATION

The SSPSD being a special case of the MPSD, the estimators of $\boldsymbol{\theta}$ can be obtained using the methods described for MPSD (*see*

MPSD). The properties of the minimum variance unbiased (MVU) estimators of $\boldsymbol{\theta}$ are discussed in Joshi and Patil [2], [3].

Let $\mathbf{X} \sim \text{MPSD}(\boldsymbol{\theta}, f(\boldsymbol{\theta}))$ with $\mathbf{x} \in I_s$, where I is the set of nonnegative integers and I_s denotes the s-fold Cartesian product of I with itself. Let the $s \times m$ matrix $X = [X_{ij}]$, $i = 1, 2, \ldots, s$, $j = 1, 2, \ldots, m$ be a random sample of size m from the distribution of $\mathbf{X}$. Define $\mathbf{y}' = (y_1, y_2, \ldots, y_s)$, where $y_i = \Sigma_{j=1}^m X_{ij}$, $i = 1, 2, \ldots, s$.

Theorem 4. $\mathbf{X}$ has an SSPSD if the MVU estimator of θ_i is proportional to y_i, $i = 1, 2, \ldots, s$.

Theorem 5. The probability function $p = p(\mathbf{k}; \boldsymbol{\theta})$ of the SSPSD given by (2) is MVU estimable for every sample size m and the MVU estimator is given by

$$\tilde{p} = \tilde{P} \Pi \binom{y_i}{k_i} \bigg/ \binom{y}{k},$$

where ($\sim$) stands for the MVU estimator of the parametric function under it and where P is the pf of the component sum $X = \Sigma X_i$ of the SSPSD.

Theorem 6. The only parametric functions $\beta(\boldsymbol{\pi})$ that are MVU estimable on the basis of a random sample of size m from an SSPSD (2) are polynomials in $\pi_1, \pi_2, \ldots, \pi_s$ of degree not exceeding $\min\{m[s]\}$, the minimum of the m-fold sum of the range s of $y = \Sigma y_i$.

Theorem 7. For the SSPSD given by (2), if $\alpha(\boldsymbol{\theta})$ and $\beta(\boldsymbol{\pi})$, polynomials in $\pi_1, \pi_2, \ldots, \pi_s$, are MVU estimable on the basis of a random sample of size m, then their product is also MVU estimable for the same sample size m and the MVU estimator of the product is the product of the individual MVU estimators.

See MULTIVARIATE POWER SERIES DISTRIBUTIONS and WEIGHTED DISTRIBUTIONS for specific results and related details.

References

[1] Barndorff-Nielsen, O. (1977). *Information and Exponential Families*. Wiley, New York, pp. 205–208.

[2] Joshi, S. W. and Patil, G. P. (1971). *Sankhyā A*, **33**, 175–184.

[3] Joshi, S. W. and Patil, G. P. (1972). *Sankhyā A*, **34**, 377–386. [Also, in *Theory Prob. Appl. Moscow*, **19**, (1974).]

[4] Mahfoud, M. and Patil, G. P. (1982). *Statistics and Probability. Essays in Honor of C. R. Rao*. North-Holland, Amsterdam, Netherlands, pp. 479–492.

[5] Patil, G. P. (1968). *Sankhyā B*, **30**, 355–366.

[6] Patil, G. P., Boswell, M. T., Joshi, S. W., and Ratnaparkhi, M. V. (1984). In *Dictionary and Classified Bibliography of Statistical Distributions in Scientific Work*, Vol. 1: International Co-operative Publishing House, Fairland, Md.

[7] Patil, G. P., Boswell, M. T., Ratnaparkhi, M. V., and Roux, J. J. J. (1984). In *Dictionary and Classified Bibliography of Statistical Distributions in Scientific Work*, Vol. 3: International Co-operative Publishing House, Fairland, Md.

[8] Smith, W., Grassle, J. F. and Kravitz, D. (1979). In *Ecological Diversity in Theory and Practice*, Statistical Ecology Series, Vol. 6, J. F. Grassle, G. P. Patil, W. Smith, and C. Taillie, eds. International Co-operative Publishing House, Fairland, Md., pp. 177–191.

(MODIFIED POWER SERIES
    DISTRIBUTIONS
MULTIVARIATE POWER SERIES
    DISTRIBUTIONS
MULTINOMIAL DISTRIBUTIONS
MULTIVARIATE DISTRIBUTIONS
MULTIVARIATE LOGARITHMIC SERIES
    DISTRIBUTION
POWER SERIES DISTRIBUTIONS
WEIGHTED DISTRIBUTIONS)

G. P. PATIL
M. V. RATNAPARKHI

SUN CHART

HOW TO CONSTRUCT A SUN CHART

Let us denote a permutation of the consecutive positive integers from 1 to k by $\mathbf{p} = (p(1), p(2), \ldots, p(k))$ and denote k-variate data of size n by

$$x_\alpha(\mathbf{p}) = (x_{p(1)\alpha}, x_{p(2)\alpha}, \ldots, x_{p(k)\alpha}),$$
$$\alpha = 1, 2, \ldots, n,$$

where $x_{p(r)\alpha}$ $(r = 1, 2, \ldots, k)$ is a nonnegative real value and k is larger than 2.

The procedure for constructing a sun chart follows the steps shown below:

Step 1. Consider a coordinate on the rectangular-coordinate graph corresponding to the data $x_\alpha(\mathbf{p})$ such as

$$\left(\frac{1}{Mk} \sum_{r=1}^{k} x_{p(r)\alpha} \cos\left(\frac{\pi r}{k+1} \right), \right.$$
$$\left. \frac{1}{Mk} \sum_{r=1}^{k} x_{p(r)\alpha} \sin\left(\frac{\pi r}{k+1} \right) \right)$$
$$= (X_{\mathbf{p}\alpha}, Y_{\mathbf{p}\alpha}),$$

say, where

$$M = \max_{\substack{1 \le r \le k \\ 1 \le \alpha \le n}} x_{p(r)\alpha}.$$

Denote the distances from the point of the mean value in k and two dimensions by

$$d_{k\alpha} = \sqrt{\sum_{r=1}^{k} \left(x_{p(r)\alpha} - \bar{x}_{p(r)} \right)^2}$$

and

$$d_{2\alpha}(\mathbf{p}) = \sqrt{\left(X_{\mathbf{p}\alpha} - \bar{X} \right)^2 + \left(Y_{\mathbf{p}\alpha} - \bar{Y} \right)^2},$$
$$\alpha = 1, 2, \ldots, n,$$

respectively, where

$$x_{p(r)} = \frac{1}{n} \sum_{\alpha=1}^{n} x_{p(r)\alpha}, \qquad r = 1, 2, \ldots, k,$$

$$\bar{X} = \frac{1}{n} \sum_{\alpha=1}^{n} X_{\mathbf{p}\alpha}, \qquad \bar{Y} = \frac{1}{n} \sum_{\alpha=1}^{n} Y_{\mathbf{p}\alpha}.$$

Step 2. Consider all the permutations of size $k!$ such as

$$\mathbf{p}_1 = (1, 2, \ldots, k-1, k),$$
$$\mathbf{p}_2 = (1, 2, \ldots, k, k-1),$$
$$\vdots$$
$$\mathbf{p}_{k!} = (k, k-1, \ldots, 2, 1).$$

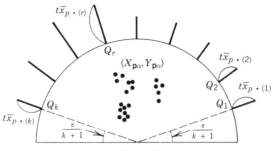

Figure 1

Construct the following measure of the similarity between the k- and two-dimensional configurations (*see* MEASURES OF SIMILARITY, DISSIMILARITY AND DISTANCE):

$$D(k, \mathbf{p}) = \frac{1}{n} \sum_{\alpha=1}^{n} \left(\frac{d_{2\alpha}(\mathbf{p})}{d_{k\alpha}} - \bar{d} \right)^2,$$

where

$$\bar{d} = \frac{1}{n} \sum_{\alpha=1}^{n} \frac{d_{2\alpha}(\mathbf{p})}{d_{k\alpha}}.$$

Step 3. Denote $\max\{D(k, \mathbf{p}_1), D(k, \mathbf{p}_2), \ldots, D(k, \mathbf{p}_k)\}$ by $D(k, \mathbf{p}^*)$, where $\mathbf{p}^* = (p^*(1), p^*(2), \ldots, p^*(k))$.

Step 4. Draw a bar chart* with the length of $t\bar{x}_{p^*(r)}$ on the extension of the line from

Table 1   Record of four subjects

No.	Jap. (x_1)	Soc. (x_2)	Math. (x_3)	Sci. (x_4)
①	10	9	8	9
②	5	5	6	7
③	2	3	2	3
④	9	10	9	9
⑤	6	6	6	7
⑥	3	3	1	1
⑦	4	4	3	8
⑧	3	6	0	2
⑨	6	5	3	6
⑩	8	9	10	10
Mean value	5.6	6.0	4.8	6.2

the origin O to the point $(\cos[\pi r/(k + 1)], \sin[\pi r/(k + 1)])$, $r = 1, 2, \ldots, k$, as in Fig. 1, where $\bar{x}_{p^*(r)} = (1/n)\sum_{\alpha=1}^{n} x_{p^*(r)\alpha}$, $r = 1, 2, \ldots, k$, and t is a scale parameter which is a positive real-valued constant. Then a chart consisting of the n points $(X_{p^*(1)\alpha}, Y_{p^*(1)\alpha}), \ldots, (X_{p^*(k)\alpha}, Y_{p^*(k)\alpha})$ in the unit semicircle together with the bar charts of size k will be named a "sun chart."

Example. Suppose that the records in four subjects $(x_{1\alpha}, x_{2\alpha}, x_{3\alpha}, x_{4\alpha})$, $\alpha = 1, 2, \ldots, 10$, of 10 junior high school pupils are as given in Table 1. Here, $x_{j\alpha}$ satisfies $0 \leqslant x_{j\alpha} \leqslant 10$, $j = 1, 2, 3, 4$, $\alpha = 1, 2, \ldots, 10$, and the four subjects are Japanese, social studies, mathematics, and science.

By putting $k = 4$ and $M = 10$, we get the values of $D(4, \mathbf{p})$ for $\mathbf{p}_1, \mathbf{p}_2, \ldots, \mathbf{p}_{24}$ as shown in Table 2. From Table 2 we find $\mathbf{p}^* = (2, 1, 3, 4)$ and we get the coordinates in Table 3 to draw a sun chart for $\mathbf{p}^*$. Then we obtain a sun chart as shown in Fig. 2 by putting $t = 0.05$ in Step 4, where symbol $ⓐ$ in the chart shows the ordinal number of each pupil.

The following facts are observed from Fig. 2.

1. All the pupils are clustered into three groups denoting high, middle, and low averages with respect to the four subjects.

2. By looking at the bar charts and the location of a point $(X_{\mathbf{p}7}, Y_{\mathbf{p}7})$, we may see that the No. ⑦ pupil has low records especially in Math., Jap., and

Table 2

p	$D(4, \mathbf{p})$
$\mathbf{p}_1 = (1, 2, 3, 4)$	0.88853
$\mathbf{p}_2 = (1, 2, 4, 3)$	0.95068
$\mathbf{p}_3 = (1, 4, 2, 3)$	0.96617
$\mathbf{p}_4 = (4, 1, 2, 3)$	0.92393
$\mathbf{p}_5 = (1, 3, 2, 4)$	0.91883
$\mathbf{p}_6 = (1, 3, 4, 2)$	0.96336
$\mathbf{p}_7 = (1, 4, 3, 2)$	0.95040
$\mathbf{p}_8 = (4, 1, 3, 2)$	0.89483
$\mathbf{p}_9 = (3, 1, 2, 4)$	0.93313
$\mathbf{p}_{10} = (3, 1, 4, 2)$	0.97309
$\mathbf{p}_{11} = (3, 4, 1, 2)$	0.94972
$\mathbf{p}_{12} = (4, 3, 1, 2)$	0.88415
$\mathbf{p}_{13} = (2, 1, 3, 4)$	0.88415
$\mathbf{p}_{14} = (2, 1, 4, 3)$	0.94972
$\mathbf{p}_{15} = (2, 4, 1, 3)$	0.97309
$\mathbf{p}_{16} = (4, 2, 1, 3)$	0.93313
$\mathbf{p}_{17} = (2, 3, 1, 4)$	0.89483
$\mathbf{p}_{18} = (2, 3, 4, 1)$	0.95040
$\mathbf{p}_{19} = (2, 4, 3, 1)$	0.96336
$\mathbf{p}_{20} = (4, 2, 3, 1)$	0.91883
$\mathbf{p}_{21} = (3, 2, 1, 4)$	0.92393
$\mathbf{p}_{22} = (3, 2, 4, 1)$	0.96617
$\mathbf{p}_{23} = (3, 4, 2, 1)$	0.95068
$\mathbf{p}_{24} = (4, 3, 2, 1)$	0.88853

Soc. in comparison with Sci., and belongs to the group denoting middle average.

COMPARISON WITH PRINCIPAL COMPONENT SCORES CHART

Let us denote the first and second principal component* scores for Table 1 by $(z_{1\alpha}, z_{2\alpha})$, $\alpha = 1, 2, \ldots, 10$. We show the values of $(z_{1\alpha}, z_{2\alpha})$, $\alpha = 1, 2, \ldots, 10$, in Table 4 and plot the points $(z_{1\alpha}, z_{2\alpha})$, $\alpha = 1, 2, \ldots, 10$, in Fig. 3. Figure 3 shows the chart of principal component scores (CPCS).

Comparing Figs. 2 and 3, we may see:

1. The patterns of scattered points in the sun chart and CPCS are similar in appearance.

2. The sun chart has merits in comparison with CPCS, in grasping the correspondence between the location of scattered

Table 3

α	$X_{\mathbf{p}\alpha}$	$Y_{\mathbf{p}\alpha}$
1	-0.005	0.707
2	-0.048	0.456
3	-0.001	0.195
4	0.020	0.712
5	-0.020	0.494
6	0.055	0.166
7	-0.072	0.347
8	0.102	0.198
9	0.006	0.381
10	-0.036	0.717

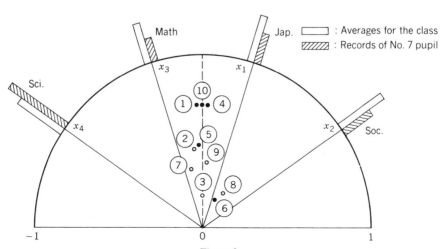

Figure 2

Table 4

α	$z_{1\alpha}$	$z_{2\alpha}$
1	2.309	−0.367
2	−0.007	0.567
3	−2.161	0.236
4	2.459	−0.463
5	0.375	0.216
6	−2.428	−0.350
7	−0.672	0.921
8	−1.842	−0.983
9	−0.412	0.053
10	2.379	0.169

points $(X_{\mathbf{p}\alpha}, Y_{\mathbf{p}\alpha})$, $\alpha = 1, 2, \ldots, 10$, and the four variates x_1, x_2, x_3, x_4 with the bar charts. See Wakimoto [1] for further details.

Reference

[1] Wakimoto, K. (1980). *Ann. Inst. Statist. Math. Tokyo*, **32B**, 303–310.

(BAR CHARTS
COMPONENT ANALYSIS
GRAPHICAL REPRESENTATION OF DATA)

K. WAKIMOTO

SUNDBERG FORMULAS

Exponential families* of distributions play an important role in statistical inference, especially when we deal with maximum likelihood estimation*. The special form of the probability function associated with an exponential families data vector allows one to easily identify the sufficient statistics* and to express the likelihood equations in terms of sufficient statistics and their first moments. In addition, the Fisher information* matrix is the variance–covariance matrix of the sufficient statistics. These facts are central in deriving the properties of maximum likelihood estimates based on exponential families data and in characterizing their asymptotic distribution.

Many important statistical applications involve distributions that do not fit directly into the regular exponential family class, for example censored, grouped*, or otherwise incomplete data*, variance components*, finite mixtures, and other random and mixed effects models, to name a few. These distributions and others, can be represented as models for incomplete data from exponential family distributions. We will show how the maximum likelihood theory for regular exponential families extends naturally when

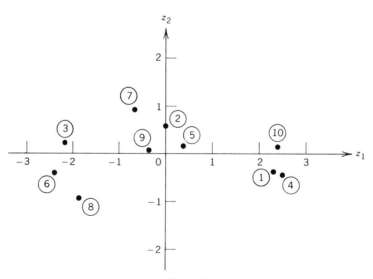

Figure 3

sampling from incomplete exponential families. In particular, with incomplete data*, the likelihood equations and Fisher information matrix take on special forms that can usefully be regarded as generalizations of the corresponding complete data equations. The maximum likelihood theory for these incomplete data samples was first published in Sundberg [8], hence the term Sundberg formulas. He ascribes the original results to lecture notes of P. Martin-Löf.

Following the notation in EXPONENTIAL FAMILIES, we say that a data vector **x** has an *exponential families representation* if its probability function can be written as

$$p(\mathbf{x}; \boldsymbol{\omega}) = a(\boldsymbol{\omega})\exp\{\boldsymbol{\theta}(\boldsymbol{\omega})^T\mathbf{t}(\mathbf{x})\}, \quad (1)$$

defined over the sample space χ. The vector $\boldsymbol{\theta} = \boldsymbol{\theta}(\boldsymbol{\omega})$ is the canonical parameter and $\mathbf{t} = \mathbf{t}(\mathbf{x})$ is the vector of sufficient statistics. Since $\boldsymbol{\theta}$ is one-to-one with $\boldsymbol{\omega}$, we generally write $p(\mathbf{x}; \boldsymbol{\theta})$, $a(\boldsymbol{\theta})$, etc. We let k be the smallest dimension for $\mathbf{t}$ and $\boldsymbol{\theta}$ such that (1) holds. It determines the order of the family and the dimension of the minimal sufficient statistic. The density is regular if the parameter space Θ is some open subset of R^k containing all values of $\boldsymbol{\theta}$ such that

$$a(\boldsymbol{\theta})^{-1} = \int_{\chi}\exp\{\boldsymbol{\theta}^T\mathbf{t}\}\,d\mathbf{x} < \infty. \quad (2)$$

With regular families, the maximum likelihood estimate of $\boldsymbol{\theta}$, say $\hat{\boldsymbol{\theta}}$, is uniquely determined by the solution of

$$E_{\boldsymbol{\theta}}(\mathbf{t}) = \mathbf{t} \quad (3)$$

and its Fisher information matrix is given by

$$\mathbf{I}_x = \text{var}_{\boldsymbol{\theta}}(\mathbf{t}). \quad (4)$$

Equations (3) and (4) are derived by differentiating both sides of (2) with respect to $\boldsymbol{\theta}$, to show that

$$\partial\log a(\boldsymbol{\theta})/\partial\boldsymbol{\theta} = -E_{\boldsymbol{\theta}}(\mathbf{t}) \quad (5)$$

and

$$\partial^2\log a(\boldsymbol{\theta})/\partial\boldsymbol{\theta}\partial\boldsymbol{\theta}^T = -\text{var}_{\boldsymbol{\theta}}(\mathbf{t}). \quad (6)$$

Equations (1)–(6) summarize some important features of complete data vectors from

exponential families, which are relevant to computing maximum likelihood estimates and their asymptotic variances. See, for instance, refs. 8 and 5 or EXPONENTIAL FAMILIES. We now derive the parallel formulas for incomplete data vectors.

With incomplete data, we do not observe **x** directly, but rather a many-to-one mapping $\mathbf{x} \rightarrow \mathbf{y} = \mathbf{y}(\mathbf{x})$. In this case, the probability function $g(\mathbf{y}; \boldsymbol{\omega})$ of **y** can be obtained by integrating (1) over $\chi_{\mathbf{y}}$, which is the region in the original sample space χ where **x** is known to be, once we observe $\mathbf{y} = \mathbf{y}(\mathbf{x})$.

Example 1. Censored Exponential Samples. Suppose n items with exponentially distributed lifetimes are put to test. Let ω denote the mean lifetime. If all items are observed to time of failure, the complete data **x** is the vector of n observed lifetimes. It has probability function

$$p(\mathbf{x}; \omega) = \omega^{-n}e^{-\Sigma x_i/\omega},$$
$$x_i \geq 0, \quad i = 1, \dots, n.$$

This follows the form given in (1), where the canonical parameter is the hazard rate, $\theta = \omega^{-1}$, $a(\theta) = \theta^n$, and $t = -x_+ \equiv -\Sigma x_i$. Since $E_{\theta}(t) = -n\omega$, the maximum likelihood estimate for ω is $\hat{\omega} = -t/n$; the corresponding estimate for θ is $\hat{\theta} = -n/t$, with asymptotic variance

$$\text{var}(\hat{\theta}) = [\text{var}_{\theta}(t)]^{-1} = (n\omega^2)^{-1}.$$

Incomplete data arise if the experiment is terminated at time T and some items have not yet failed at T, i.e., are right censored at T. The observed data **y** now consist of the n vectors (z_i, δ_i), where $z_i = x_i$ if $\delta_i = 1$ and $z_i = T \leq x_i$ if $\delta_i = 0$. The probability function of **y** is

$$g(\mathbf{y}; \omega) = \omega^{-\delta_+}e^{-z_+/\omega}.$$

It is easily seen that $g(\mathbf{y}; \omega)$ can be expressed as the $(n - \delta_+)$-fold integral of $g(\mathbf{x}; \omega)$, i.e., as

$$\int_T^{\infty}\cdots\int_T^{\infty}\omega^{-n}e^{-x_+/\omega}\,dx_1\cdots dx_k,$$

where we assume, without loss of generality,

that the first k have $\delta_i = 0$ and the remainder have $\delta_i = 1$.

By definition, the general form of $g(\mathbf{y}; \boldsymbol{\theta})$ with incomplete data can be expressed as

$$g(\mathbf{y}; \boldsymbol{\theta}) = \int_{X_y} p(\mathbf{x}; \boldsymbol{\theta}) \, d\mathbf{x},$$

$$= a(\boldsymbol{\theta})/a(\boldsymbol{\theta};\mathbf{y}), \qquad (7)$$

where

$$a(\boldsymbol{\theta};\mathbf{y})^{-1} = \int_{X_y} \exp\{\boldsymbol{\theta}^T \mathbf{t}\} \, d\mathbf{x}. \qquad (8)$$

Note the similarity between (2), which defines $a(\boldsymbol{\theta})$, and (8), which defines $a(\boldsymbol{\theta}; \mathbf{y})$. From (1) and (7), it follows directly that the conditional distribution of $\mathbf{x}$ given $\mathbf{y}$ has probability function

$$k(\mathbf{x}; \boldsymbol{\theta}, \mathbf{y}) = p(\mathbf{x}; \boldsymbol{\theta})/g(\mathbf{y}; \boldsymbol{\theta})$$

$$= a(\boldsymbol{\theta}; \mathbf{y})\exp\{\boldsymbol{\theta}^T \mathbf{t}\}.$$

This shows that the conditional distribution of $\mathbf{x}$ given the observed data $\mathbf{y}$ has the exponential families representation with the same canonical parameters and sufficient statistics; however, the sampling and parameter spaces will, in general, be different. This fact plays an important role in characterizing the derivatives of $\log g(\mathbf{y}; \boldsymbol{\theta})$.

Using (7), we have

$\partial \log g(\mathbf{y}; \boldsymbol{\theta})/\partial \boldsymbol{\theta}$

$\qquad = \partial \log a(\boldsymbol{\theta})/\partial \boldsymbol{\theta} - \partial \log a(\boldsymbol{\theta}; \mathbf{y})/\partial \boldsymbol{\theta}.$

It follows from (5) and its generalization for $a(\boldsymbol{\theta}; \mathbf{y})$ that

$$\partial \log g(\mathbf{y}; \boldsymbol{\theta})/\partial \boldsymbol{\theta} = E_{\boldsymbol{\theta}}(\mathbf{t}) - E_{\boldsymbol{\theta};\mathbf{y}}(\mathbf{t}),$$

where $E_{\boldsymbol{\theta};\mathbf{y}}$ denotes expectation over $\mathbf{x}$ given $\mathbf{y}$. Thus, the likelihood equations based on incomplete data $\mathbf{y}$ require us to solve

$$E_{\boldsymbol{\theta}}(\mathbf{t}) = E_{\boldsymbol{\theta};\mathbf{y}}(\mathbf{t}). \qquad (9)$$

Contrasting this with the complete data likelihood (3) based on $\mathbf{x}$, we see that $\mathbf{t}$ (which is not completely determined by the observed data $\mathbf{y}$) has been replaced by its conditional expectation given $\mathbf{y}$.

The term Sundberg formulas refers to (7)–(9), which determine the special form of

the density of incomplete exponential families data and the corresponding likelihood equations, as well as higher order derivatives of $\log g(\mathbf{y}; \boldsymbol{\theta})$.

Differentiating $\log g(\mathbf{y}; \boldsymbol{\theta})$ twice with respect to $\boldsymbol{\theta}$, the observed information in $\mathbf{y}$, say $\mathbf{O_y}$, can be expressed as

$$\mathbf{O_y} = \mathrm{var}_{\boldsymbol{\theta}}(\mathbf{t}) - \mathrm{var}_{\boldsymbol{\theta};\mathbf{y}}(\mathbf{t}) \equiv \mathbf{I_x} - \mathbf{I_{x;y}},$$

where by definition,

$$\mathbf{I_{x;y}} = \mathrm{var}_{\boldsymbol{\theta};\mathbf{y}}(\mathbf{t})$$

is the Fisher information* in the conditional distribution of $\mathbf{x}$ given $\mathbf{y}$. In contrast to regular complete data exponential families, the second derivative matrix, in general, depends upon observed data. It may not be negative definite everywhere in Θ and the likelihood equations are not guaranteed to have a unique solution.

The Fisher information $\mathbf{I_y}$ in $\mathbf{y}$ is obtained by taking expectations of $\mathbf{O_y}$:

$$\mathbf{I_y} = \mathrm{var}\{E_{\boldsymbol{\theta};\mathbf{y}}(\mathbf{t})\},$$

or alternatively

$$\mathbf{I_y} = \mathbf{I_x} - E\{\mathbf{I_{x;y}}\}.$$

$\mathbf{I_x} - \mathbf{I_y}$ is denoted the expected "lost" information and $[\mathbf{I} - \mathbf{I_y}\mathbf{I_x}^{-1}]$ is the expected "fraction of information lost" by observing $\mathbf{y}$ rather than $\mathbf{x}$.

Example 1 (continued).   We now continue Example 1 to see how censoring modifies the likelihood equations and Fisher information. Recall that the complete data sufficient statistic is $t = -\Sigma x_i$, $E_{\boldsymbol{\theta}}(t) = -n\omega$, and $\mathrm{var}_{\boldsymbol{\theta}}(t) = n\omega^2$. To calculate expectations conditional on $\mathbf{y}$, notice that if $\delta_i = 1$, then $z_i = x_i$, so $E_{\boldsymbol{\theta};y}(x_i) = z_i$ and $\mathrm{var}_{\boldsymbol{\theta};y}(x_i) = 0$. If censoring occurs ($\delta_i = 0$), it is easily seen that x_i given $x_i \geqslant T$ is again exponential, now with mean $\omega + T$ and variance ω^2. Using (9) to derive the likelihood equations based on $\mathbf{y}$ we have

$$n\omega^* = E_{\boldsymbol{\theta};y}(x_+)$$

$$= \sum [\delta_i z_i + (1 - \delta_i)(T + \omega^*)]$$

$$= z_+ + (n - \delta_+)\omega^*$$

or

$$\omega^* = z_+/\delta_+.$$

The observed information is

$$\text{var}_\theta(t) - \text{var}_{\theta;y}(t) = n\omega^2 - (n - \delta_+)\omega^2$$
$$= \delta_+\omega^2.$$

It depends upon data (δ_+), but is always greater than 0 if $\delta_+ > 0$; thus ω^* is uniquely defined. Taking the expectation of O_y, we have

$$I_y = \omega^2 E_\theta(\delta_+) = n\omega^2(1 - e^{-T/\omega}).$$

This shows that the expected fraction of lost information is $e^{-T/\omega}$. If T is large relative to ω, most items will fail by T and little information is lost by censoring.

The censored exponentials case is special, in that its distribution and associated maximum likelihood theory can be derived directly from first principles, without resorting to Sundberg's formulas. In addition, it is easily seen that the likelihood equations have a unique closed form solution. In general, however, incomplete data likelihood equations may be quite complex and require iterative solution. Here the special form of (9) for the likelihood equations may simplify their derivation considerably.

In addition, (9) also suggests an iterative algorithm for computing the maximum likelihood estimate of θ, useful when no closed form solution exists. Start with some θ^0 and use it to calculate $t^0 = E_{\theta^0;y}(t)$; solve $t^0 = E_\theta(t)$ to get θ^1; set $\theta^0 = \theta^1$ and repeat until convergence. This algorithm was labeled EM by Dempster et al. [1], who also present much of the general theory underlying the algorithm and give many special examples. The name EM comes from the two explicit steps, expectation and maximization, at each iteration of the algorithm. Starting with θ^P, each iteration requires:

E Step.   Set

$$t^P = E_{\theta;y}(t)\big|_{\theta=\theta^P}$$

and

M Step.   Solve

$$E_\theta(t) = t^P$$

to find θ^{P+1}.

Clearly the usefulness of the EM algorithm in any particular incomplete data problem depends upon how easy it is to implement the E and M steps, and what alternative algorithms are available. Notice that the M step always has a unique solution; the EM will be most easily implemented if simple closed form expressions can be found for both steps. *See also* MISSING INFORMATION PRINCIPLE.

In the next example, the observed data likelihood equations are complicated to derive and no closed form solution exists. Yet it is easy to see how to apply Sundberg's formulas and the EM algorithm.

Example 2. A Mixture of Normals. For simplicity, we consider a two point mixture with equal known variances. Extension to complex mixtures is very straightforward. The data consist of a sample from a mixture of two normals:

$$g(y; \omega) = \prod_{i=1}^{n} \bigg[pe^{-(y_i-\mu_1)^2/2}$$

$$+ (1 - p)e^{-(y_i-\mu_2)^2/2}\bigg]\bigg/\sqrt{2\pi}.$$

We desire the maximum likelihood estimate for $\omega = (p, \mu_1, \mu_2)$. This can be represented as incomplete data by taking x to be a sample of independently distributed observations as follows. Let $x_i = (z_i, \delta_i)$ with $z_i \sim N(\mu_1, 1)$ if $\delta_i = 1$ and $z_i \sim N(\mu_2, 1)$ if $\delta_i = 0$; let δ_i be Bernoulli with parameter p. Then $y = z$; δ is not observed. To derive the likelihood equations and implement the EM, it is not necessary to determine $p(x; \omega)$ or θ; only to formulate x as having a regular exponential family density and to determine the sufficient statistics. With mixtures, we formulate complete data to be independent samples from known populations. The observed data lack the information as to which population an observation comes from. In this case the complete data are merely two independent normal samples. Thus

$$t = \bigg(\delta_+, \sum_i \delta_i z_i, \sum(1 - \delta_i)z_i\bigg)$$

is sufficient for $\omega = (p, \mu_1, \mu_2)$.

It is now easier to use (9) to derive the likelihood equations rather than differentiate $p(\mathbf{y}; \omega)$ directly. Here

$$E_{\theta}(\mathbf{t}) = (np, np\mu_1, n(1 - p)\mu_2).$$

Given $\mathbf{y} = \mathbf{z}$, each z_i is fixed at y_i, but δ_i is again Bernoulli with probability $p(y_i)$, given by Bayes rule:

$$p(y_i; \omega) = \frac{p\exp(f_1)}{p\exp(f_1) + (1 - p)\exp(f_2)},$$

where $f_j = -\frac{1}{2}(y_i - \mu_j)^2$, $j = 1, 2$. Thus

$$E_{\theta;y}(\mathbf{t}) = \Big(\sum p(y_i), \sum p(y_i)y_i,$$
$$\sum (1 - p(y_i))y_i\Big).$$

The likelihood equations require an iterative solution, and the EM can be implemented very simply: Given $\omega^P = (p^P, \mu_1^P, \mu_2^P)$:

E Step.

$$t_1^P = \sum p(y_i; \omega^P),$$
$$t_2^P = \sum p(y_i; \omega^P)y_i,$$
$$t_3^P = \sum y_i - t_2^P;$$

M Step.

$$p^{P+1} = t_1^P/n,$$
$$\mu_1^{P+1} = t_2^P/t_1^P,$$
$$\mu_2^{P+1} = t_3^P/t_1^P.$$

At convergence we have ω^* as well as an estimate for each i of the probability that the observation arises from the first component ($p(y_i; \omega^*)$).

Each iteration of EM increases the likelihood [1]; however, in some cases, local maxima of the likelihood may exist. Sundberg [9] and Wu [10] give conditions for convergence of the EM to a relative maximum of the likelihood. The rate of convergence of the algorithm is geometric. In large samples its convergence factor is approximately equal to the largest eigenvalue of $\mathbf{I} - \mathbf{O}_y\mathbf{I}_x^{-1}$, where the parameters in $\mathbf{O}_y$ and $\mathbf{I}_x$ are evaluated at the point of convergence. Sundberg [9] derives the convergence factor for a number

of examples, including censored, mixed, and grouped normals. Louis [6] shows how to exploit the Sundberg formulas to calculate the observed information when using the EM and suggests a method for improving convergence. He gives the mixture of two normals as one example.

The next example, a variance component* model, is another where the EM plays a useful role; it also illustrates potential problems with convergence to boundary points.

Example 3: A Variance Component* Model. The observed data consist of a vector $\mathbf{y}$ of independent observations, where each y_i is $N(0, D_i + \sigma^2)$ for some known D_i. If the D_i are not all equal, computing the MLE of σ^2 requires iteration. To formulate this as an incomplete data problem, let $x_i = (\mu_i, e_i)$, $y_i = \mu_i + e_i$, and μ_i and e_i be independent normals with zero means and $\text{var}(u_i) = D_i$, $\text{var}(e_i) = \sigma^2$. The complete data sufficient statistic for σ^2 is $t = \sum e_i^2$. If t were observed, $\hat{\sigma}^2 = t/n$.

Given y_i, each e_i is $N(w_iy_i, D_iw_i)$, where

$$w_i = \sigma^2(D_i + \sigma^2)^{-1}.$$

Thus, the likelihood equations require

$$n\sigma^2 = E_{\theta;y}(t) = \sum \{D_iw_i + (w_iy_i)^2\}.$$

The E step at the pth iteration of the EM calculates

$$t^P = \sum \{(w_i^Py_i)^2 + w_i^PD_i\}$$

for

$$w_i^P = (\sigma^P)^2(D_i + \sigma^{P2})^{-1}$$

and the M step sets

$$(\sigma^{(P+1)})^2 = t^P/n.$$

The observed and expected information matrices are calculated using the higher order moments of normals:

$$I_x = \text{var}_{\theta}(t) = 2\sigma^4n,$$
$$I_y = \text{var}\, E_{\theta;y}(t) = 2\sigma^4\sum w_i^2,$$
$$f = 1 - I_y/I_x = 1 - \sum w_i^2/n.$$

If each D_i is small relative to σ^2, $w_i \doteq 1$, and

f is small. In this case each $u_i \doteq 0$ and little has been lost by adding u_i to e_i. Conversely, if each D_i is large relative to σ^2, there will be very little information in $\mathbf{y}$ about σ^2, and $f \doteq 1$.

Convergence of the EM depends upon the "observed fraction lost," which is

$$\sum \left[D_i^2 w_i^2 + 2 D_i \sigma^2 y_i^2 w_i^3 \right] / (n\sigma^4).$$

Notice that if there is a local maximum at the boundary, $\hat{\sigma}^2 = 0$, and the observed fraction lost is 1. In this case, convergence of the EM is not guaranteed; practical experience suggests that large numbers of iterations are required to calculate "zero" variance components. If the local maximum occurs for a relatively large value of σ^2, convergence should be rapid.

Using the EM algorithm to calculate the variance components in the mixed model analysis of variance is equivalent to Henderson's algorithm [3]. From this example, it is clear that the algorithm easily extends to a more general setting. Dempster et al. [2] and Laird and Ware [4] give other variance component applications; Rubin and Thayer [7] discuss factor analysis*.

Sundberg's formulas for incomplete data and the EM algorithm can be generalized to deal with both distributions outside the exponential families class (see Dempster et al. [1]) and with nonparametric estimation. In the latter case, the extension refers to the self-consistency principle and the self-consistency algorithm. Details and references can be found in SELF-CONSISTENCY.

References

[1] Dempster, A. P., Laird, N. M., and Rubin, D. B. (1977). *J. R. Statist. Soc. B*, **39**, 1–38, with discussion. (Introduces the general form of the EM algorithm and gives many examples.)

References [2]–[4] discuss application of the EM to variance component models.

[2] Dempster, A. P., Rubin, D. B., and Tsutakawa, R. K. (1981). *J. Amer. Statist. Ass.*, **76**, 341–353.
[3] Laird, N. M. (1982). *J. Statist. Comput. Simul.*, **14**, 295–303.
[4] Laird, N. M. and Ware, J. H. (1982). *Biometrics*, **38**, 963–974.
[5] Lehmann, E. H. (1959). *Testing Statistical Hypotheses*. Wiley, New York, (Basic reference for exponential family distributions.)
[6] Louis, T. A. (1982). *J. R. Statist. Soc. B*, **44**, 226–233. (Shows how to calculate the observed information using EM and suggests a method for speeding convergence.)
[7] Rubin, D. B. and Thayer, D. T. (1982). *Psychometrika*, **47**, 69–76. (Applies EM to factor analysis models.)
[8] Sundberg, R. (1974). *Scand. J. Statist.*, **1**, 49–58. (Develops maximum likelihood theory for incomplete data from an exponential family distribution.)
[9] Sundberg, R. (1976). *Commun. Statist. B*, **5**, 55–64. (Presents the EM algorithm for exponential families.)
[10] Wu, C. F. J. (1983). *Ann. Statist.*, **11**, 95–103. (Discusses convergence properties of the EM algorithm.)

Acknowledgment

This work was supported by Grant GM29745 from the National Institutes of Health.

(EXPONENTIAL FAMILIES
INCOMPLETE DATA
MAXIMUM LIKELIHOOD ESTIMATION
MISSING INFORMATION PRINCIPLE
SELF-CONSISTENCY)

NAN LAIRD

SUNFLOWERS

A modification in scatter plots* to reduce the problem of overlap, suggested by Cleveland and McGill [1]. The "sunflower" symbols use a code to indicate the number of individuals within a square.

A dot means one observation.

A dot and two lines $\left(\vcenter{\hbox{$\vdots$}} \right)$ means two observations.

A dot and three lines $\left(\vcenter{\hbox{$\vdots$}} \right)$ means three observations, and so on.

Reference

[1] Cleveland, W. S. and McGill, R. (1984). *J. Amer. Statist. Ass.*, **79**, 807–822.

(GRAPHICAL REPRESENTATION OF DATA
SNOWFLAKES
TRIPLE SCATTER PLOT)

SUPERADDITIVE AND SUBADDITIVE ORDERING

Given two CDFs F and G, F is said to be superadditive (subadditive) with respect to G if

$$G^{-1}F(x+y) \geqslant (\leqslant)G^{-1}F(x) + G^{-1}F(y)$$

for all x and y in the support* of F. Symbolically

$$F \overset{su}{\prec} G \left(F \underset{su}{\prec} G \right).$$

Superadditive ordering neither implies nor is implied by ordering by dispersion*—that is, the ordering $F \overset{disp}{\prec} G$, defined by

$$G^{-1}(\beta) - G^{-1}(\alpha) \geqslant F^{-1}(\beta) - F^{-1}(\alpha)$$

for all $0 < \alpha < \beta < 1$ [or, equivalently, $G^{-1}F(x) - x$ is a nondecreasing function of x]. However, if

$$F \overset{su}{\prec} G \quad and \quad F \overset{st}{\prec} G$$

(where $\overset{st}{\prec}$ denotes stochastic ordering), then also

$$F \overset{disp}{\prec} G.$$

If F and G are absolutely continuous*, with $F(0) = G(0) = 0$, and their PDFs f and g satisfy $f(0) > 0$ and $g(0) > 0$, then

$$F \overset{su}{\prec} G \quad implies \quad F \overset{disp}{\prec} G.$$

(See Ahmed et al. [1].)

Reference

[1] Ahmed, A. N., Alzaid, A., Bartosziewicz, J., and Kocher, S. C. (1986). *Adv. Appl. Prob.*, **18**, 1019–1022.

(DEPENDENCE, CONCEPTS OF
ORDERING DISTRIBUTIONS BY
   DISPERSION
ORDERING OF DISTRIBUTIONS, PARTIAL
ORDERING, STARSHAPED
STOCHASTIC ORDERING)

SUPER-BAYESIAN

The term "super-Bayesian" refers to a Bayesian expert or investigator who is consulted to help a group of Bayesian decision makers with different priors to reach a joint decision. In ref. 5, where the term originates, it is also assumed that the Bayesians have different utility* functions, making interutility comparisons difficult. The super-Bayesian is assumed to be disinterested in the decision to be made and supplies no utility function of his own. The group decision is taken to be the Nash solution (which may be a randomized solution); such a solution does not require that comparisons among utility functions be made; *see* NASH EQUILIBRIUM. Related consideration of multi-Bayesian decision making is given in ref. 11.

The idea of combining judgments (that are expressed in terms of probability distributions) from multiple experts to achieve a consensus distribution that can be used for making inferences or decisions is an important one. For example, an individual may need to combine different meteorological forecasts in order to plan a day's activities; a patient may need to reconcile differing medical prognoses in order to make an informed decision regarding treatment. Perhaps the earliest consideration of the consensus problem is given in ref. 3, where the parimutuel method is employed to form a consensus distribution. Early mathematical and empirical comparisons of methods for combining probability distributions from several experts into a single distribution are given in ref. 12. An "opinion pool," or linear combination of expert distributions, is discussed in refs. 2, 7, and 10; *see also* POOLS, OPINION. A detailed framework for evaluating the information provided to a decision maker by an expert and for incorporating judgments by multiple experts is presented in refs. 8 and 9.

Since experts' assessments of probabilities may violate the laws of probability, i.e., they may be incoherent (*see* COHERENCE), it is important to examine methods that allow a decision maker to reconcile the incoherent probability assessments. Two models for doing this are discussed in ref. 6. In combining judgments from multiple experts, possible sources of dependence must be considered. In ref. 4, the importance of considering the relationship between the decision maker's and an expert's opinion is investigated. The possibility of stochastic dependence in information from multiple sources is modelled explicitly in refs. 1 and 13.

References

[1] Agnew, C. E. (1985). *J. Amer. Statist. Ass.*, **80**, 343–347.

[2] DeGroot, M. H. (1974). *J. Amer. Statist. Ass.*, **69**, 118–121.

[3] Eisenberg, E. and Gale, D. (1959). *Ann. Math. Statist.*, **30**, 165–168.

[4] French, S. (1980). *J. R. Statist. Soc. A*, **143**, 43–48.

[5] Garisch, I., deWaal, D. J., and Groenewald, P. C. N. (1984). *S. Afr. Statist. J.*, **18**, 111–122.

[6] Lindley, D. V., Tversky, A., and Brown, R. V. (1979). *J. R. Statist. Soc. A*, **142**, 146–180. (One of the most important contributions on reconciliation of probability assessments. Internal reconciliation and the use of an external observer to achieve reconciliation are both considered. Provides a good starting point for the reader interested in combining expert judgments that may be incoherent.)

[7] McConway, K. J. (1981). *J. Amer. Statist. Ass.*, **76**, 410–414.

[8] Morris, P. A. (1974). *Manag. Sci.*, **20**, 1233–1241.

[9] Morris, P. A. (1977). *Manag. Sci.*, **23**, 679–693. (These two papers present an analytic framework for using experts in decision situations that is totally consistent with the subjective interpreta-

tion of probability. Both single and multiple experts are considered.)

[10] Stone, M. (1961). *Ann. Math. Statist.*, **32**, 1339–1345.

[11] Weerahandi, S. and Zidek, J. V. (1983). *Ann. Statist.*, **11**, 1032–1046.

[12] Winkler, R. L. (1968). *Manag. Sci.*, **15**, B61–B75.

[13] Winkler, R. L. (1981). *Manag. Sci.*, **27**, 479–488.

Bibliography

Genest, C. and Zidek, J. V. (1986). *Statist. Sci.*, **1**, 114–148. (A detailed discussion of the super-Bayesian approach, with an extensive annotated bibliography.)

(BAYESIAN INFERENCE
COHERENCE
DECISION THEORY
NASH EQUILIBRIUM
POOLS, OPINION
PREVISION
UTILITY THEORY)

R. L. TRADER

SUPEREFFICIENCY   *See* HODGES SUPER-EFFICIENCY

SUPERLATIVE INDEX NUMBER   *See* DIVISIA INDICES

SUPERNORMAL DISPERSION   *See* LEXIS, WILHELM

SUPERPOPULATION MODELS

DEFINITION AND EXAMPLES

In finite population sampling*, a superpopulation model is essentially a probability model for the population characteristics values. Some examples follow, in which the characteristic is denoted by Y:

1. **Hospital Discharges** (Herson [14]).   For a population of short stay hospitals, the *regression model*

$$Y_i = \beta x_i + e_i$$

has been suggested, where

$$x_i = \text{number of beds in hospital } i,$$

$$Y_i = \text{number of discharges from}$$
$$\text{hospital } i \text{ in a given month,}$$

and $e_1, e_2, \ldots, e_N$ are independent random variables, e_i having mean zero and variance $\sigma^2 x_i$.

2. **Repeated Surveys** (Scott and Smith [26]). In a monthly economic survey, each household may remain in the sample for several successive months. A model that is conceptually useful in planning such surveys is

$$Y_{ij} = \mu_i + e_{ij},$$

where Y_{ij} is the characteristic value (e.g., expenditure on food) for household i and month j, and the e_{ij} have mean 0, variance σ^2, and covariance

$$E[e_{ij}e_{ik}] = \sigma^2 \rho(k - j).$$

3. **Auditing** (Andrews and Godfrey [1]). In the estimation of the total dollar value of a specific account balance in a financial statement consisting of a finite number of component items, a model like the following might be appropriate:

$$Y_i = Z_i \theta x_i + (1 - Z_i) x_i,$$

where

$$Y_i = \text{audit value of item } i,$$

$$x_i = \text{book value of item } i,$$

and Z_i is a Bernoulli random variable taking value 0 or 1.

4. **Small Area Statistics** (Holt et al. [15]). It is sometimes necessary to use a sample survey to provide estimates for subpopulation totals over an area which is so small that it contains only one or two sampled points. This task is facilitated if it can be assumed, for example, that

$$Y_{ij} = \beta + e_{ij},$$

where

$$Y_{ij} = \text{characteristic value for household}$$
$$i \text{ in small area } j,$$

and the e_{ij} are uncorrelated with mean 0 and variance σ^2. Since β is independent of the area j, information about the given area can be "borrowed" from sample points in neighbouring areas.

In the approach to sampling theory given in most current textbooks, estimation procedures have been based on the distributions of sample quantities induced by a probability sampling design. However, increasing attention has been paid to the use of superpopulation models, which have appeared in the literature primarily in two contexts. The first is the selection of sampling designs and estimators, using criteria which also involve the design distribution. The second is the development of inference procedures not depending explicitly on a sampling design. These areas will be described in greater detail. (The citations by no means exhaust the literature.)

FORMAL DESCRIPTION

For a description of recent research some basic sampling notation is required.

Consider a finite population $P = \{1, \ldots, i, \ldots, N\}$ of labelled individuals. Let y_i denote the value of a variate y (or *characteristic*) associated with individual i. A *sample* is a subset s of individuals that is somehow selected from the population P for examination. The problem of finite population sampling is to make inferences, given a *sample observation*

$$\tau_s = \{(i, y_i) : i \in s\},$$

concerning some function ϕ of the *population vector*

$$\mathbf{y} = (y_1, y_2, \ldots, y_N),$$

such as the *population total*

$$T(\mathbf{y}) = \sum_{i=1}^{N} y_i.$$

A *superpopulation assumption* is that, with respect to the characteristic y, the population at hand is generated at random from an infinite hypothetical population (the super-population) of finite populations of size N; formally, $\mathbf{y}$ is assumed to be a realization of a random vector

$$\mathbf{Y} = (Y_1, Y_2, \ldots, Y_N).$$

A *superpopulation* (SP) *model* specifies a family $C = \{\xi\}$ of possible joint distributions ξ envisaged for $Y_1, \ldots, Y_N$. Some simple examples:

M1. $Y_1, \ldots, Y_N$ are independent and identically distributed, and $N(\mu, \sigma^2)$ with μ and σ^2 unknown (here C is a two-parameter family of joint distributions).

M2. $Y_1/\alpha_1, \ldots, Y_N/\alpha_N$ are exchangeably (symmetrically) distributed, $\alpha_1, \ldots, \alpha_N$ being known positive numbers (*see* EXCHANGEABILITY).

M3. (A regression model) $Y_i = \beta x_i + e_i$, $E_\xi[e_i] = 0$, $\mathrm{Var}_\xi(e_i) = \sigma^2 x_i$, and $E_\xi[e_i e_j] = 0$ for $i \neq j$, $x_1, \ldots, x_N$ known, β and σ^2 unknown.

M4. $Y_1, \ldots, Y_N$ is a sequence of variates forming a stationary time series*.

M3 has already been mentioned in connection with hospital discharges.

SP MODELS IN SELECTION OF DESIGNS AND ESTIMATORS

Minimizing Expected Variance

Let p be a *probability sampling design* [essentially a probability function $p(s)$ on the collection of subsets s of P] and let $e(\tau_s)$ be an *estimator* of the population function $\phi(\mathbf{y})$. Then $e(\tau_s)$ is said to be *p-unbiased* if

$$E_p[e] = \sum_{s \in S} p(s) e(\tau_s) = \phi(\mathbf{y})$$

for every possible value of the vector $\mathbf{y}$. For example, if

$$\pi_i = \sum_{s:\, i \in s} p(s)$$

is the inclusion probability for the individual i,

$$e^0(\tau_s) = \sum_{i \in s} (y_i/\pi_i)$$

is *p-unbiased* for $T(\mathbf{y})$, since

$$E_p[e^0(\tau_s)] = \sum_{s \in S} p(s) \sum_{i \in s} (y_i/\pi_i)$$

$$= \sum_{i=1}^{N} (y_i/\pi_i) \sum_{s:\, i \in s} p(s) = T(\mathbf{y}).$$

If e is *p-unbiased*, its variance, which is a function of $\mathbf{y}$, is given by

$$\mathrm{Var}_p(e) = \sum_{s \in S} p(s)[e(\tau_s) - \phi(\mathbf{y})]^2$$

$$= E_p[(e - \phi)^2].$$

The nonexistence theorems of Godambe [11] and Godambe and Joshi [12] and their generalizations imply that "optimal" estimators minimizing $\mathrm{Var}_p(e)$ for all $\mathbf{y}$ do not exist in general. However, as indicated in the same papers, for certain important superpopulation models $C = \{\xi\}$ and functions $\phi(\mathbf{y})$, it is possible to find sampling designs p and/or *p-unbiased* estimators e which minimize the expected variance

$$E_\xi\left[E_p\left[(e - \phi(\mathbf{Y}))^2\right]\right] \qquad (1)$$

for all $\xi \in C$. Hence such models can be used to justify the selection of particular estimator–design pairs.

Some Optimality Results

The first result of this type may well have been that of Cochran [13], who showed that if $N\bar{y}_s$ ($N \times$ the sample mean) is to be used to estimate $T(\mathbf{y})$ and if model **M4** with convex autocorrelation function obtains, then systematic sampling* produces smaller average variance (1) than simple or stratified random sampling with proportional allocation*. This result was generalized by Madow [17], Gautschi [10], and further by Hájek [13]. If $(Y_1/\alpha_1, \ldots, Y_N/\alpha_N)$ forms a stationary series with convex autocorrelation function, Hájek's elegant theorem proves the optimality of a generalized systematic sampling design among all designs making

$$\frac{A}{n} \sum_{i \in s} (y_i / \alpha_i) \qquad (2)$$

with $A = \sum_{i=1}^{N} \alpha_i$ unbiased for $T(\mathbf{y})$.

There are many theorems in the literature dealing with models akin to **M2** (Cassel et al. [4, p. 87]). More recent developments are discussed by Thompson [29]. One version of the basic result is that the estimator (2) and a design with inclusion probabilities π_i proportional to α_i are jointly optimal for estimating $T(\mathbf{y})$. The proof can be extended to the case of models under which the distribution of $(Y_1 / \alpha_1, \ldots, Y_N / \alpha_N)$ is not necessarily exchangeable, but invariant under a *subgroup* of the group of all permutations of the population *labels* 1, ..., i, ..., N. The introduction of "random permutation" models akin to **M2** has also been used as a device for proving minimax* properties of estimator–design pairs (see Cassel et al. [4, Sec. 3.6]).

SP MODELS IN INFERENCE ABOUT A POPULATION FUNCTION $\phi(\mathbf{y})$

In the presence of a model such as **M1–M4**, $\phi(\mathbf{y})$ is a realization of the random variate $\phi(\mathbf{Y})$, and in principle, any such function is estimated once the unseen coordinates of $\mathbf{Y}$ have been *predicted* from the seen values in τ_s. The problem of predicting $\mathbf{Y}$ from τ_s using the model $C = \{\xi\}$ is part of the mainstream of statistical theory, and has been solved in many instances. The solutions generally have nothing to do with the design used in selecting the sample.

An example is provided by the *subjective Bayesian* approach to finite population inference (Ericson [9] and Scott and Smith [26]). Here a single prior distribution* is placed upon the parameters of a family of joint distributions for $Y_1, \ldots, Y_N$, or directly upon the vector $\mathbf{y}$. The inference about $\phi(\mathbf{y})$ is the posterior distribution* of $\phi(\mathbf{Y})$ given the observation τ_s, and this posterior distri-

bution is mathematically independent of the sampling design.

It is clear that the Bayesian approach is a special case of the use of superpopulation models, in which C is a singleton family. Because the vector $\mathbf{y}$ in a sense parametrizes the distribution of τ_s, other superpopulation models are sometimes called (classes of) priors in the literature.

In another mode of inference, in the case of model **M3** we might ask that for all $\xi \in C$,

$$E_\xi [e - T(\mathbf{Y})] = 0,$$

and that e be homogeneous linear in y_i, $i \in s$. Then the estimator e which minimizes the *predictive mean squared error*

$$E_\xi \left[(e - T(\mathbf{Y}))^2 \right] \quad (s \text{ fixed})$$

is the ratio estimator* for $T(\mathbf{y})$, namely

$$e_R = \left(\sum_{i=1}^{N} x_i \right) \left(\sum_{i \in s} Y_i \right) \Big/ \left(\sum_{i \in s} x_i \right),$$

regardless of how s is chosen. Predictive interval estimates for $T(\mathbf{y})$ can be based on e_R and sample estimates of $E_\xi [(e_R - T(\mathbf{Y}))^2]$. (See Brewer [2] and Royall [19, 20].)

Such methods as proposed by Royall have been termed the "classical" *predictive approach* by Cassel et al. [4]. The Bayesian approach previously described and the fiducial* approach as applied by Kalbfleisch and Sprott [16] to models such as **M1** are also predictive in nature, producing predictive distributions for functions $\phi(\mathbf{Y})$ based on the model and not on the sampling design.

Proponents of predictive methods are arguing in effect that the availability of a useful SP model eliminates the need to bring the sampling design into the inference process, either in the choice of estimator (as in the section Minimizing Expected Variance) or in the construction of design-based confidence intervals as in traditional survey practice. (Carried to extremes, the argument would deny the importance of randomizing to obtain the sample in the first place, although few predictivists would adopt this position.) Brewer and Sarndal [3] have attempted to classify various approaches to sampling inference according to the roles assigned to model and design, and the discussion following their paper is illuminating. (See also Smith [28] and Royall [21].) It is suggested by Thompson [29] that, in a sense, even traditional sampling inference may be regarded as predictive.

The approximate nature of tractable SP models makes it important to ensure that inferences based partially or completely upon them are robust to model departures. Most recent work on inference involving models has been concerned with robustness in one sense or another. For example, Royall and Herson [25] show that properly "balanced" (purposive) samples (*see* REPRESENTATIVE SAMPLING) can make the optimality of the ratio estimator robust against polynomial departures from $E_\xi[Y_i] = \beta x_i$ in **M3**. Scott et al. [27] generalize this result. Royall and Eberhardt [24] and Royall and Cumberland [22, 23] develop and assess estimators of uncertainty for ratio and more general regression estimators of $T(\mathbf{y})$ which are robust against departures from the assumed variance function for **Y**. Some further references are given in the Bibliography.

PRACTICAL CONSIDERATIONS

It should be noted that in many problems of survey practice there is quite general agreement that the adoption of some sort of superpopulation model is desirable or even necessary. One obvious example: A sample of persons has been selected systematically from a list and it is intended that the results should be analyzed by methods traditional for simple random sampling*. This procedure is justifiable if **M2** with $\alpha_1 = \cdots = \alpha_N$ (or some submodel) is adopted (Madow and Madow [18]). Such examples, where the appropriate randomization has not been carried out for some reason, are legion.

Even when a randomized design has been used, models must be introduced explicitly or implicitly to deal with nonsampling errors resulting from nonresponse, inadequate frames, response errors, etc. Some useful references are given by Cochran [7] and Cassel et al. [5].

In geostatistics* (David [8]), models of spatially distributed characteristics are useful because the quantities $\phi(\mathbf{y})$ to be estimated often pertain to small areas represented by only a few sample points. The use of models in population surveys was mentioned in the first section of this article.

References

[1] Andrews, R. W. and Godfrey, J. T. (1979). Superpopulation models on auditing populations. *ASA Proceedings of Business and Economics Statistics Section*, American Statistical Association, Washington, D.C., pp. 397–400.

[2] Brewer, K. R. W. (1963). Ratio estimation in finite populations: some results deducible from the assumption of an underlying stochastic process. *Austral. J. Statist.*, **5**, 93–105.

[3] Brewer, K. R. W. and Sarndal, C. E. (1981). Six approaches to enumerative survey sampling. To appear in *Proceedings of Symposium on Incomplete Data*. U.S. Department of Health, Education and Welfare, Washington, D.C.

[4] Cassel, C. M., Sarndal, C. E., and Wretman, J. H. (1977). *Foundations of Inference in Survey Sampling*. Wiley, New York. (Survey of literature on the predictive approach up to 1975. A general reference; refs. 6, 10, 13, 17, and 29 and annotated entries in the Bibliography supplement the bibliography in this book.)

[5] Cassel, C. M., Sarndal, C. E., and Wretman, J. H. (1981). Some uses of statistical models in connection with the non-response problem. *Proceedings of Symposium on Incomplete Data*. U.S. Department of Health, Education and Welfare, Washington, D.C.

[6] Cochran, W. G. (1946). Relative accuracy of systematic and stratified random samples for a certain class of populations. *Ann. Math. Statist.*, **17**, 164–177.

[7] Cochran, W. G. (1977). *Sampling Techniques*, 3rd ed. Wiley, New York.

[8] David, M. A. (1978). Sampling and estimation problems for three dimensional spatial stationary and nonstationary stochastic processes as encountered in the mineral industry. *J. Statist. Plann. Inf.*, **2**, 211–244.

[9] Ericson, W. A. (1969). Subjective Bayesian models in sampling finite populations (with discussion). *J. R. Statist. Soc. B*, **31**, 195–233.

[10] Gautschi, W. (1957). Some remarks on systematic sampling. *Ann. Math. Statist.*, **28**, 385–394.

[11] Godambe, V. P. (1955). A unified theory of sampling from finite populations. *J. R. Statist. Soc. Series B*, **17**, 269–278.

[12] Godambe, V. P. and Joshi, V. M. (1965). Admissibility and Bayes estimation in sampling from finite populations-I. *Ann. Math. Statist.*, **36**, 1707–1722.

[13] Hájek, J. (1959). Optimum strategy and other problems in probability sampling. *Casopis Pro. Pest. Mat.*, **84**, 387–423.

[14] Herson, J. (1976). An investigation of relative efficiency of least squares prediction to conventional probability sampling plans. *J. Amer. Statist. Ass.*, **71**, 700–703.

[15] Holt, D., Smith, T. M. F. and Tomberlin, T. J. (1979). A model-based approach to estimation for small sub-groups of a population. *J. Amer. Statist. Ass.*, **74**, 405–410.

[16] Kalbfleisch, J. D. and Sprott, D. A. (1969). Applications of likelihood and fiducial probability to sampling finite populations. In *New Developments in Survey Sampling*, N. L. Johnson and H. Smith, eds. Wiley Interscience, New York.

[17] Madow, W. G. (1949). On the theory of systematic sampling II. *Ann. Math. Statist.*, **20**, 333–354.

[18] Madow, W. G. and Madow, L. H. (1944). On the theory of systematic sampling I. *Ann. Math. Statist.*, **15**, 1–24.

[19] Royall, R. M. (1970). On finite population sampling theory under certain linear regression models. *Biometrika*, **57**, 377–387.

[20] Royall, R. M. (1971). Linear regression models in finite population sampling theory. In *Foundations of Statistical Inference*. V. P. Godambe and D. A. Sprott, eds. Holt, Rinehart and Winston of Canada, Toronto.

[21] Royall, R. M. (1976). Current advances in sampling theory: implications for human observational studies. *Amer. J. Epid.*, **104**, 463–474.

[22] Royall, R. M. and Cumberland, W. G. (1978). Variance estimation in finite population sampling. *J. Amer. Statist. Ass.*, **73**, 351–358.

[23] Royall, R. M. and Cumberland, W. G. (1981). An empirical study of the ratio estimator and estimators of its variance. *J. Amer. Statist. Ass.*, **76**, 66–88.

[24] Royall, R. M. and Eberhardt, K. R. (1975). Variance estimates for the ratio estimator. *Sankhyā*, *Ser. C*, **37**, 43–52.

[25] Royall, R. M. and Herson, J. (1973). Robust estimation in finite populations, I and II. *J. Amer. Statist. Ass.*, **68**, 880–889, 890–893.

[26] Scott, A. J. and Smith, T. M. F. (1969). Estimation in multistage surveys. *J. Amer. Statist. Ass.*, **64**, 830–840.

[27] Scott, A. J., Brewer, K. R. W. and Ho, E. W. H. (1978). Finite population sampling and robust estimation. *J. Amer. Statist. Ass.*, **73**, 350–361.

[28] Smith, T. M. F. (1976). The foundations of survey sampling: a review. *J. R. Statist. Soc. A*, **139**, 183–204.

[29] Thompson, M. E. (1983). The likelihood principle and randomization in sampling theory. Unpublished.

[30] Thompson, M. E. (1987). Model and design correspondence in finite population sampling. Unpublished.

Bibliography

Bellhouse, D. R., Thompson, M. E. and Godambe, V. P. (1977). Two-stage sampling with exchangeable prior distributions. *Biometrika*, **64**, 97–103. (Supplements the bibliography in ref. 4.)

Brewer, K. R. W. (1979). A class of robust sampling designs for large-scale surveys. *J. Amer. Statist. Ass.*, **74**, 911–915. (Emphasizes robustness.)

Godambe, V. P. (1966). A new approach to sampling from finite populations I, II. *J. R. Statist. Soc. B*, **28**, 310–328. (Discussion of the relationship of the predictive approach to traditional sampling estimation.)

Godambe, V. P. (1982). Estimation in survey-sampling: robustness and optimality. *J. Amer. Statist. Ass.*, **77**, 393–406. (Emphasizes robustness.)

Godambe, V. P. and Thompson, M. E. (1973). Estimation in sampling theory with exchangeable prior distributions. *Ann. Statist.*, **1**, 1212–1221. (Supplements the bibliography in ref. 4.)

Hartley, H. O. and Sielken, R. L. (1975). A "superpopulation viewpoint" for finite population sampling. *Biometrics*, **31**, 411–412. (Discussion of the relationship of the predictive approach to traditional sampling estimation.)

Joshi, V. M. (1979). The best strategy for estimating the mean of a finite population. *Ann. Statist.*, **7**, 531–536. (Supplements the bibliography in ref. 4.)

Laplace, P. S. (1812). *Théorie Analytique des Probabilités*, 391–394. (An early use of predictive finite population inference.)

Liu, T. P. (1979). A general completeness theorem in sampling theory (abstract). *Bulletin IMS*, **8**, 281. (Supplements the bibliography in ref. 4.)

Rao, J. N. K. and Bellhouse, D. R. (1978). Optimal estimators of a finite population mean under certain random permutation models. *J. Statist. Plann. Inf.*, **2**, 125–142. (Supplements the bibliography in ref. 4.)

Sugden, R. A. (1979). Inference on symmetric functions of exchangeable populations. *J. R. Statist. Soc. Series B*, **41**, 269–273. (Supplements the bibliography in ref. 4.)

(FINITE POPULATIONS, SAMPLING FROM)

M. E. THOMPSON

SUPERSATURATED DESIGNS

Many experiments in physical, chemical, or industrial research are designed to determine what effect changing one or more controllable variables (factors) will have on the response variable (yield or quality of the product, for example). When the number of factors being considered for study is large and the number of trials that can be run is small, the experimenter has the choice of either selecting a few factors for study or running all the factors in a supersaturated design, a design which has a number of trials less than the number of factors being studied.

Before supersaturated designs were devised, Plackett and Burman [3] described a set of orthogonal experimental designs that they called "optimum multifactorial experiments," chosen to require the minimum number of trials needed to get estimates of all main effects* but no interactions* and to get these results with minimum variance; *see* PLACKETT AND BURMAN DESIGNS.

In 1962, Booth and Cox [2] developed a set of designs with fewer trials than factors (supersaturated designs) that come as close as possible to the orthogonal designs* described by Plackett and Burman. They devised a set of conditions for a quality that they called *near orthogonality*:

(i) All columns consist of $n/2$ values at the high level ($+$) and $n/2$ values at the low level ($-$).

(ii) For all possible designs, choose the one with the minimum value of the maximum absolute value of the dot products of all pairs of columns.

(iii) Of two designs with the same value for **(ii)**, choose the one in which the number of pairs of columns achieving the maximum value is a minimum.

Using these conditions, Booth and Cox used a computer to search systematically for supersaturated designs that are nearly orthogonal. They listed seven such designs: 16 factors in 12 trials, 20 factors in 12 trials, 24 factors in both 12 trials and 18 trials, 30 factors in 18 trials and 24 trials, and 36 factors in 18 trials.

To use these designs efficiently, the experimenter must be aware of the confounding* patterns, so that when several factors show an effect, an intelligent choice can be made as to whether these are actually different effects or just different expressions of the same effect. Since some of the factors are more likely to have an effect than others, it would be best if these likely factors could be kept unconfounded with each other, even though it is impossible to keep them from confounding the other factors. As an example of a Booth and Cox supersaturated design, the design for 24 factors in 12 trials is shown in Table 1. The 7 factors, A, B, C, D, E, F, and G are all mutually orthogonal. All other factors are more or less partially confounded with some of these factors and with each other.

A completely different type of saturated design* was introduced by Watson [3] in 1961 when he developed the concept of group screening experimental designs. In this design, the k factors are grouped into g groups. Each group is then tested as a single factor in a systematic design. If a group factor is found to have no effect on the response, all original factors in the group are considered to have no effect. If a group factor is found to have an effect, all original factors in the group must be studied further, either individually for a final decision or in smaller groups, which will require further testing.

The number of factors that should be placed in each group depends on an a priori estimate of the probability (p) that each

Table 1

Trial Number	A	B	C	D	E	F	G	H	I	J	K	L	M	N	O	P	Q	R	S	T	U	V	W	X
1	+	−	+	−	+	+	+	+	−	−	−	+	−	−	+	−	−	+	−	−	+	−	−	−
2	−	+	+	+	+	−	−	−	−	−	+	−	−	−	−	−	−	−	−	+	+	−	−	−
3	−	+	+	−	−	+	+	−	+	−	+	−	+	+	−	−	+	−	−	−	+	−	−	−
4	−	−	−	−	+	+	−	−	−	−	+	−	+	−	−	+	−	+	+	−	−	+	−	+
5	+	+	+	−	+	−	−	−	+	+	+	+	−	−	−	+	+	+	+	+	−	+	−	−
6	+	−	+	+	−	+	−	−	−	+	−	−	+	−	+	+	+	−	−	−	+	−	+	+
7	+	−	−	+	+	−	+	−	+	−	−	−	−	+	−	−	−	−	+	−	+	−	+	−
8	−	−	+	+	−	−	+	+	−	+	+	+	−	+	−	+	−	−	−	−	+	+	+	+
9	+	+	−	−	−	−	+	+	+	−	−	+	+	−	+	+	−	−	+	−	−	−	+	−
10	+	+	−	+	−	+	−	+	−	−	+	−	−	+	+	−	+	+	−	+	−	−	−	+
11	−	+	−	+	+	+	+	+	+	+	−	+	+	+	+	−	+	+	+	+	+	+	+	+
12	−	−	−	−	−	−	−	+	+	+	+	+	+	+	+	+	+	+	+	+	+	+	+	+

factor will have an effect on the response variable. Those factors that are likely to have an effect ($p > 0.3$) should be tested individually or, at most, in pairs. Those factors that may have an effect ($0.05 \leqslant p \leqslant 0.3$) can be put in groups of 3–5 in size. Very unlikely factors ($p < 0.01$) can be put in groups of 10 or more. In this way large numbers of factors can be tested using few groups and therefore few trials.

As an example, consider an experiment of 20 factors: 6 of the factors are assumed to be likely to have an effect, $p = 0.3$ and factors are put into 3 groups of 2 each; 4 of the factors may have an effect, $p = 0.09$ for each, and they are all put into 1 group; the

Table 2

Trial Number	Group Factors				
	A	B	C	D	E
1	−	−	−	+	+
2	+	−	−	−	−
3	−	+	−	−	+
4	+	+	−	+	−
5	−	−	+	+	−
6	+	−	+	−	+
7	−	+	+	−	−
8	+	+	+	+	+

Table 3

Group Factor	A	B	C	D	E
Probability that a factor will have an effect	0.3	0.3	0.3	0.09	0.01
Number of factors in the group	2	2	2	4	10
Probability that at least one factor in the group has an effect	0.51	0.51	0.51	0.31	0.096
Probability of no effect	0.49	0.49	0.49	0.69	0.904

10 remaining factors are unlikely to have an effect, $p = 0.01$, and they are also all placed into 1 group. This gives 5 group factors which can be run in a fractional factorial* using 8 trials, as shown in Table 2. All factors in each group are run at the same level in each trial.

The probabilities that each of these group factors will have an effect and therefore will need further trials are given in Table 3. Using these probabilities, the expected total number of trials is 16.6.

References

[1] Booth, K. H. V. and Cox, D. R. (1962). Some systematic supersaturated designs. *Technometrics*, **4**, 489–495.

[2]   Plackett, R. L. and Burman, J. P. (1946). The design of optimum multifactorial experiments. *Biometrika*, **33**, 305–325.

[3]   Watson, G. S. (1961). A study of the group screening method. *Technometrics*, 3, 371–388.

Bibliography

Anderson, V. L. and McLean, R. A. (1974). *Design of Experiments*. Dekker, New York.

Patel, M. S. (1962). Group-screening with more than two stages. *Technometrics*, **4**, 209–217.

(CONFOUNDING
DESIGN OF EXPERIMENTS
FRACTIONAL FACTORIAL DESIGNS
MAIN EFFECTS
PLACKETT AND BURMAN DESIGNS
SATURATED DESIGNS)

MARILYNN S. DUEKER

SUPPLEMENTED BALANCE   *See* GENERAL BALANCE

SUPPORT

(i)   A point on the real line is said to be *in the support* of the cumulative distribution function* F of a random variable X if, for all $\epsilon > 0$, the open interval $(x - \epsilon, x + \epsilon)$ has positive probability. The set S of all such points is the *support* of F (or of X) and is a closed set [1].

When X is absolutely continuous, the nonuniqueness of its probability density function* (PDF) creates a fuzziness in defining the PDF, which can be changed on any countable set of points and given arbitrary values on that set without affecting F. In contrast, the support S of X is unique. On S the PDF is positive almost surely and off S it is zero almost surely.

Examples

1.   Let X have a geometric distribution*. The support of X is the set of integers $\{1, 2, 3, \ldots\}$.

2.   Let X have a standard beta distribution*. The support of X is the closed interval $[0, 1]$.

3.   It is possible for a discrete distribution to have a closed interval as its support. Let $r_1, r_2, \ldots, r_k, \ldots$ be an ordering of the rationals in the unit interval $[0, 1]$, $[0, 1)$, $(0, 1]$, or $(0, 1)$. The distribution giving probability 2^{-k} to r_k is a discrete distribution whose support is $[0, 1]$ (Ferguson [2, Sec. 2.3]).

The last example is measure-theoretic, involving the topology of the real line. The rationals are countable but also dense, so that, for any x in $[0, 1]$, there is no open neighborhood of x that does not contain at least one rational.

If two probability measures are equivalent (namely each is absolutely continuous* with respect to the other), then they have the same support. The converse, however, is not true in general. See Lehmann [3, p. 19] for more details.

(ii)   Let B_n be the field of Borel sets in n-dimensional Euclidean space, i.e., generated by half-open n-dimensional rectangles, and let μ be a measure on B_n. A point $\mathbf{x}$ is said to be *in the support* of μ if and only if every open neighborhood of $\mathbf{x}$ has strictly positive measure. The set of such points is the *support* of μ. It is a closed set [1]. The support of a probability measure on B_1 is the same as that of its distribution function F, already defined.

References

[1]   Chung, K. L. (1974). *A Course in Probability Theory*, 2nd ed. Academic, New York, pp. 10 and 31.

[2]   Ferguson, T. S. (1967). *Mathematical Statistics: A Decision Theoretic Approach*. Academic, New York.

[3]   Lehmann, E. L. (1983). *Theory of Point Estimation*. Wiley, New York.

(MEASURE THEORY IN PROBABILITY
AND STATISTICS)

SUPPORTING HYPERPLANE THEO-REM *See* GEOMETRY IN STATISTICS; CONVEXITY

SUPPRESSION

The term suppression was apparently coined by Horst [4] in 1941 and is widely used in statistical applications of multiple linear regression* to psychology* and education* (see, e.g., Cohen and Cohen [1] and Conger [2]).

The original meaning of suppression relates to the concept of suppressor variable* which is defined as a second independent (regressor) variable X_2 that is uncorrelated with the "dependent" variable Y, but is such that it substantially increases the observed multiple correlation* coefficient. In this case, X_2 makes up for its lack of correlation with Y by correlation with that part of X_1 which is orthogonal to Y, thus "suppressing" some of the variance in X_1 that is irrelevant to Y.

In current literature, the term suppression refers more broadly to the "bizarre" behavior of fitted regression coefficients exhibited by the phenomenon that the fitted coefficient of, say, X_1 in the regression of Y on X_1 may differ by orders of magnitude or be of opposite sign as compared with the fitted coefficient of X_1 in bivariate regression on X_1 and X_2—see discussions by Currie and Korabinksi [3] and Lewis and Escobar [5], using the intuitively appealing geometry of bivariate regression.

The concept of suppression is closely related to that of *enhancement* [3, 5], which occurs when the squared multiple correlation* coefficient exceeds the sum of squared simple correlations with Y.

References

[1]   Cohen, J. and Cohen, P. (1975). *Applied Multiple Regression/Correlation Analysis for the Behavioral Sciences*. L. Erlbaum, N.J.

[2]   Conger, A. (1974). *Educ. Psychol. Meas.*, **34**, 35–46.

[3]   Currie, I. and Korabinski, A. (1984). *The Statistician*, **33**, 283–293.

[4]   Horst, P. (1941). *Soc. Sci. Res. Bull.* (*New York*), **40**.

[5]   Lewis, W. J. and Escobar, L. A. (1986). *Statistician*, **35**, 17–26.

(GEOMETRY IN STATISTICS
MULTIPLE CORRELATION
MULTIPLE LINEAR REGRESSION)

SUPPRESSOR VARIABLES

Suppressor variables are a subset of predictor variables in multiple linear regression models, given the linear function $Y = b_0 + b_1 X_1 + \cdots + b_p X_p + e$ and least-squares* estimation of the unknown parameters $b_0, b_1, \ldots, b_p$ (for further explanations, *see* LINEAR REGRESSION). Furthermore, we assume positive validities (predictor–criterion correlations) by an adequate polarization of the predictor variables, i.e., $r_{y,i} > 0$, $i = 1, \ldots, p$. The regression weights (coefficients) b_i, $i = 1, \ldots, p$, refer to effects of the orthogonal part of X_i with respect to the other independent variables $X_1, X_2, \ldots, X_{i-1}, X_{i+1}, \ldots, X_p$, denoted by $X_{i|1,2,\ldots,i-1,i+1,\ldots,p}$. These coefficients in general are not standardized (*see also* PARTIAL REGRESSION).

A standardized measure for the association of Y and $X_{i,1,2,\ldots,i-1,i+1,\ldots,p}$, however, is the *semipartial correlation coefficient*

$$r_{Y(i|1,2,\ldots,i-1,i+1,\ldots,p)}$$
$$= \frac{s_{Yi|1,2,\ldots,i-1,i+1,\ldots,p}}{s_Y s_{i|1,2,\ldots,i-1,i+1,\ldots,p}},$$

with $s_{Yi|1,2,\ldots,i-1,i+1,\ldots,p}$ and $s_{i|1,2,\ldots,i-1,i+1,\ldots,p}$ as conditional covariances and variances.

Horst [5] introduced the term "suppressor variables" as independent variables in multiple regression, which have zero or negligible correlations with the criterion but appreciable correlations with the other independent variables. He could prove that, in general, suppressor variables have negative regression coefficients. The following (hypothetical) example may illustrate such a suppressor variable (X_2) in a regression model with two predictors (to which the concept of suppressor variables has been limited by most authors dealing with this subject). Let Y be a rating of job performance of programmers by supervisors, X_1 a test for reasoning, and X_2 a measure of test anxiety. Then the following correlations seem to be reasonable: $r_{Y_1} = 0.60$, $r_{Y_2} = 0.00$, and $r_{12} = 0.50$. A multiple regression leads to the regression coefficients (given Y, X_1, X_2 as standardized variables) $b_1 = 0.80$ and $b_2 = -0.40$ and the multiple correlation $R = 0.69$. The negative regression weight of the suppressor variable (test anxiety) has often been interpreted as subtracting irrelevant variance from the predictor variable (reasoning). Darlington [2] proved, for regression models with two predictor variables, that a predictor variable (X_2) will be a suppressor variable, i.e., gets a negative regression weight, if the inequality $r_{Y_2} - r_{Y_1} r_{12} < 0$ holds, given the validities $r_{Y_2} > 0$ and $r_{Y_1} > 0$.

Conger [1] defined a suppressor variable "to be a variable which increases the predictive validity of another variable (or set of variables) by its inclusion in a regression equation. This variable is a suppressor only for those variables whose regression weights are increased" (Conger, [1, p. 36ff]). According to Conger an increased prediction for an independent variable X_1 is given if the regression weight β_i surpasses the corresponding validity r_{Yi}, i.e., $|\beta_i| > |r_{Yi}|$, given the same sign for both coefficients.

In the case of regression models with two predictor variables, Conger identified three types of suppressor variables:

(1) "classical" suppressor variables, as defined by Horst.

(2) "negative" suppressor variables, as defined by Darlington.

(3) "reciprocal" suppressor variables, a "new" type of suppressor variables, specified by positive validities and $r_{12} < 0.0$. In this case both regression weights surpass the corresponding validities.

Velicer [6] criticized that semipartial regression coefficients are not standardized as validities and these measures cannot be compared. He proposed to speak of suppression as present when the squared semipartial correlation of at least one predictor (X_1) is greater than the squared simple predictor–criterion correlation

$$r^2_{Y(i|1,2,\ldots,i-1,i+1,\ldots,p)} > r^2_{Yi}.$$

Unlike Conger, Velicer does not take account of the direction of influence, which cannot be neglected for many problems. Considering this argument, Holling [3] proposed the following definition: Suppression occurs when the semipartial predictor–criterion correlation of at least one predictor (X_i) is numerically greater than the corresponding simple predictor–criterion correlation, i.e., $|r_{Y(i|1,2,\ldots,i-1,i+1,\ldots,p)}| > |r_{Yi}|$, given that both indices have the same sign.

Especially in the case of regression models with more than two predictors, it seems more appropriate to speak of suppressor structures without identifying single variables as suppressor variables, because in most cases several predictors together lead to suppression.

Holling [3] proved the following equivalent condition for suppression according to his definition:

$$|r_{Y^i(i|1,2,\ldots,i-1,i+1,\ldots,p)}| > |r_{Yi}|,$$

given equal signs for both coefficients with

$$Y^i = b_0 + b_1 X_3 + \cdots + b_p X_p.$$

This condition refers only to linear combinations of some variables $X_1, X_2, \ldots, X_p$ and allows a generalization of the concept of suppression to given linear combinations of random variables and especially to all statistical procedures within the general linear model (Holling [3]).

Conger's reported classification of suppressor variables in regression models with two predictor variables has often been the foundation of systems for such regression models. Because of the underlying criticized suppressor concept those classifications have to be revised. Holling [4], using his definition of suppression, developed the following exhaustive system of three mutually exclusive categories of regression models with two predictor variables. Given none, one, respectively, two independent variables, whose semipartial correlation with the criterion is numerically greater than the corresponding validity (given the same signs), we have a "simple predictor structure," "simple suppressor structure," respectively, "reciprocal suppressor structure."

References

[1] Conger, A. J. (1974). *Educ. Psychol. Meas.*, **34**, 35–46.

[2] Darlington, R. B. (1968). *Psychol. Bull.*, **69**, 161–182.

[3] Holling, H. (1983). *Educ. Psychol. Meas.*, **40**, 1–9.

[4] Holling, H. (1986). *Biom. J.*, **23**, 783–790.

[5] Horst, P. (1941). The role of prediction variables which are independent of the criterion. In: The Prediction of Personal Adjustment, P. Horst, Ed. *Soc. Sci. Res. Bull.*, **48**, 431–436.

[6] Velicer, W. F. (1978). *Educ. Psychol. Meas.*, **38**, 953–958.

(CORRELATION COEFFICIENT
REGRESSION (various entries))

HEINZ HOLLING

SURPRISE INDEX

INTRODUCTION

Perhaps the main function of a feeling of surprise is to make us reconsider the validity of our previous assumptions [8, p. 1131]. It can provoke us to change our subjective (personal) probabilities* of various hypotheses and often to generate hypotheses that we had not previously entertained. These comments apply in ordinary life, in statistics, and even in mathematics. The topic of surprise is also of interest to some economists, psychologists, and philosophers [3, 26]. Measures of surprise might also be of value (i) for constructing artificial music and other arts, where, to avoid monotony, some surprise is necessary, but not so much as to destroy the unity of the work; (ii) in a theory of humor where sudden changes in the frame of reference occur.

The first reasonable measure of surprise in terms of probability, apart from tail-area probabilities, was apparently proposed by Weaver [27, 28]. Meanwhile, the economist Shackle [24, 25] had proposed that the concept of potential surprise was fundamental in business decisions. This article surveys these ideas and later developments.

WEAVER'S SURPRISE INDEX

Although surprise is subjective, Weaver [27, 28] suggested that an objective index of surprise could be defined that would measure the extent to which you (the subject) *ought* to be surprised. He emphasized that it would be an error to assume that an event of small probability should cause surprise: It has to be small compared with the probabilities of alternative outcomes. He considered an experiment having a discrete set of possible outcomes having probabilities $p_1, p_2, p_3, \ldots$. Then, if the ith of these outcomes occurs, his index of surprise is defined as $E_j[p_j]/p_i$, where E_j denotes an expectation over the random variable j. This index is equal to ρ/p_i where $\rho = \sum p_j^2$ is Gini's index of homogeneity or the "repeat rate" in Turing's later and self-explanatory terminology.

For the evaluation of Weaver's surprise index for the Poisson and binomial distributions*, see refs. 22 and 13, p. 562. For continuous distributions, one works with probability densities instead of probabilities, but the surprise index is then invariant only under linear transformations of the independent variable.

Any measure of surprise has to depend on the assumptions H that we have before the observations are made, and a precise measure requires that $P(E|H)$ should be precise for each possible outcome E. In other words, either H is a simple statistical hypothesis or else we need to be "sharp Bayesians" to have a sharp measure of surprise.

GENERALIZATIONS OF WEAVER'S INDEX: ENTROPY*

Weaver's index is multiplicative if two independent experiments are combined into one, and there is a single-parameter generalization having the same property [7, 8]. This is

$$\lambda_c = \left(\Sigma p_j^{c+1} \right)^{1/c} / p_i, \qquad c > 0, \qquad (1)$$

where $c = 1$ gives Weaver's index, while the limit as $c \to 0$ gives

$$\lambda_0 = p_i^{-1} \Pi p_j^{p_j} = p_i^{-1} \exp\left\{ E_j\left[\log p_j \right] \right\}. \quad (2)$$

The further generalization

$$p_i^{-1} \phi^{-1} \left\{ E_j\left[\phi(p_j) \right] \right\},$$

where ϕ is a monotonic increasing function, is not multiplicative if ϕ is not a power or logarithm.

An additive index of surprise is $\Lambda_c = \log \lambda_c$, equal to the amount of information [6, p. 75] in the ith event minus the entropy [14]. The expression $\Lambda_c + \log p_i$ is sometimes called *Rényi's generalized entropy* because of ref. 23, which, however, did not mention surprise indexes because he was unaware of refs. 7 and 8.

Because $E[\Lambda_0] = 0$, and because of its close relationship to entropy and information, it seems that Λ_0 is the most natural of these additive surprise indexes, and λ_0 the most natural multiplicative one. Negative values of Λ_0 correspond, so to speak, to nondescript outcomes or "antisurprise" [15]. Bartlett [2] discussed Λ_0 in relation to "the significance of odd bits of information," but without explicit reference to surprise indexes.

For the k-dimensional multivariate normal distribution $\mathcal{N}(\boldsymbol{\mu}, \mathbf{C})$, we have [8]

$$\Lambda_c = \tfrac{1}{2}(\mathbf{x} - \boldsymbol{\mu})'\mathbf{C}^{-1}(\mathbf{x} - \boldsymbol{\mu})$$
$$\quad - (k/2c)\log(c + 1), \qquad c > 0, \quad (3)$$
$$\Lambda_0 = \tfrac{1}{2}(\mathbf{x} - \boldsymbol{\mu})'\mathbf{C}^{-1}(\mathbf{x} - \boldsymbol{\mu}) - \tfrac{1}{2}k$$
$$\quad = \tfrac{1}{2}(D^2 - k), \qquad (4)$$

where $\mathbf{x}$ is the observed value of the random vector and D^2 is the Mahalanobis (squared) distance* between $\mathbf{x}$ and $\boldsymbol{\mu}$. Note that D^2 has a chi-squared distribution* with k degrees of freedom.

DEPENDENCE ON THE CATEGORIZATION

Although the probabilities of all hands of 13 cards in the game of bridge have equal probabilities, some hands, such as all 13 spades, are of special human interest and for that reason would be surprising. Many other hands are also interesting to various degrees. As emphasized in ref. 8, the surprise indexes defined so far depend very much on the way that you have categorized the outcomes. This fact largely undermines the objectivity (impersonal character) of the surprise indexes in many circumstances. Previous information and hypotheses will also change the degree of surprise because they change your subjective probabilities. For example, an all-spades hand is much less surprising if, before the cards were dealt, you had noticed the other players exploding with irrepressible mirth.

SHACKLE'S POTENTIAL SURPRISE

Unaware of Weaver's note, Shackle [24, 25] used the concept of potential surprise, instead of degrees of belief*, to attack the question of how people, especially entrepreneurs, make decisions. He considered that the "interestingness" of an imagined outcome was a function of its desirability and of its potential surprise, and that people, when deciding on an action, usually concentrate on two "focus outcomes" of maximum interestingness, one desirable and the other undesirable. A feeling of surprise is an emotion, whereas a judgment of subjective probability is more intellectual, so perhaps decisions, especially emotional ones, are often made somewhat along Shackelian lines.

Good [7] argued that, since surprise could be given various meanings in terms of subjective probability*, it should be possible to use judgments of probabilities to sharpen judgments of potential surprise and vice versa. In this way, your entire body of judgments might be improved. Krelle [20] argued that there is a one-to-one relationship between degree of potential surprise and subjective probability.

A WEAKNESS OF ADDITIVE SURPRISE INDEXES: TAIL-AREA PROBABILITIES

The example of the multivariate normal reveals a weakness in all these surprise indexes, namely that they can more easily exceed a given value when the dimensionality is increased. This weakness is a simple consequence of the additivity property alone. It may be better to treat a surprise index, as defined so far, as a statistic whose tail-area probability P, or better P^{-1}, is used as the revised index of surprise. This would be a kind of surprise-Fisher compromise. It would be consistent with the treatment of "the significance of odd bits of information" by Bartlett [2].

It would be possible to use, as a measure of surprise, P^{-1}, where P is the tail-area probability of any statistic $\mathscr{S}$ used for testing our prior beliefs. To take $\mathscr{S}$ as Λ_0 is equivalent to defining a surprise index as the sum of the probabilities of all events whose probabilities do not exceed $P(E|H)$. This definition still depends, for discrete variables, on how the events are categorized, and, for continuous variables, is not invariant under nonlinear transformations of the independent variable(s). The possibility of attaining invariance will be discussed later in this article.

A PRINCIPLE OF LEAST SURPRISE

Good [9, 16] suggested, but did not strongly advocate, the possibility, after an observa-

tion is made, of selecting a hypothesis H (or estimating a parameter, which is logically the same thing), by a principle of least surprise, and that, if this is done, the prior (initial) probability $P(H)$ should also be taken into account. A special case of this suggestion, without allowing for $P(H)$, was proposed by Barndorff-Nielsen [1], who was unaware of ref. 9.

If the index Λ_0 is used, the expressions to be minimized are, respectively,

$$-\log P(E_i|H) + \sum_j P(E_j)\log P(E_j|H)$$

$$(5)$$

and

$$-\log P(E_i \,\&\, H) + \sum_j P(E_j)\log P(E_j|H),$$

$$(6)$$

depending on whether $P(H)$ is or is not taken into account. Here $E_1, E_2, \ldots$ denote the possible outcomes of an observation and E_i is the one that occurred. Non-Bayesians might prefer to minimize (5) rather than (6), though the two procedures are equivalent if all the hypotheses under consideration have equal prior probabilities.

Fortunately, simple hypotheses often have higher prior probabilities* than complicated ones, so that the capacity of surprise leads to the discovery of new truths [8, p. 1131]. As often happens in the application of significance tests*, a surprising outcome can cause us to look for new models or hypotheses. This is true whether the assumed model is non-Bayesian or Bayesian.

AN INVARIANT INDEX RELATED TO COMPLEXITY

The concept of surprise is closely connected with those of complexity and coincidences, but the surprise indexes mentioned before do not explicitly allow for complexity. To understand the connections, let us consider as an example the true mathematical assertion E, that

$$\left|163 - \left[\pi^{-1}\log_e(640320^3 + 744)\right]^2\right|$$

$$< 10^{-32}. \tag{7}$$

For this example, it seems difficult or impossible to apply the surprise indexes mentioned before.

To decide whether we should be surprised by (7), define a proposition F as the logical disjunction of all propositions $E_{a,b}$ of the form that the difference between $[\pi^{-1}\log_e(a^3 + b)]^2$ and the closest integer to it is less than 10^{-32}, where a and b are positive integers and $a < 10^6$, $b < 10^3$. Given a naive (but by no means stupid) state of mathematical knowledge, H_0, the probability of $E_{640320,744}$ is 2×10^{-32}. (This is a "dynamic probability," whereas the logical probability is 1; *see* AXIOMS OF PROBABILITY.) But (7) is much less surprising than if two genuine independent randomizing devices, when started, both produced the same sequence of 32 decimal digits. This is largely because we must "pay" for the complexity of F. [For attempts to measure complexity, see, for example, refs. 19, 17, (pp. 155 and 235), and 4.] To estimate an upper bound to a measure of the complexity, we may generously allow one "decimal unit of surprise" for each of π, -1, $\log_e$, a, 3, $+$, b, 2, 10, 6, 10, and 3 and at most four more for the remaining syntactic structure of the statement, and thus count the complexity of F as no more than 16 decimal units. We are still left with at least another 15.7 decimal units of surprise. So (7) is much too surprising to be a coincidence and it must have a non-number-crunching "explanation," whether or not any one knows it. The explanation is in fact known: It involves the theory of the elliptic modular function [29, p. 461; 10]. When a mathematician confidently conjectures a theorem, it is perhaps because he believes that the evidence would be too surprising if the theorem were known to be false.

This example suggests that a reasonable index of surprise, conditional on previous assumptions H_0, is

$$S(E|H_0) = -\log_{10}P(E|H_0) - \chi(E|H_0), \tag{8}$$

where $\chi(E|H_0)$ is an additive measure, in decimal units, of the complexity of the part of E that goes beyond what is *known* to follow from H_0. The formula (8) differs from that for Λ_0 in that the entropy term in Λ_0 is replaced by a complexity term. Note that, should E contain real parameters measured to unnecessarily many decimal places, the value of $S(E|H_0)$ would be unchanged. This is a necessary invariance property for any satisfactory index of surprise. Further, if H_0 were replaced by a full explanation of E, then both terms of (8) would vanish, that is, there would be no surprise, as is appropriate.

Now consider the following statistical example. Pearson [21] collected 12448 two-by-two contingency tables*, each with sample size 35, and with independent row and column classifications. Later, Fisher [5] found the average chi-squared for these tables was 1.00001. He rightly regarded this as especially surprising because he knew that the theoretical expectation was 35/34. Pearson had not been concerned with the values of chi-squared. There are several relevant possible hypotheses, so we omit details, but, at first sight at least, a measure of surprise is in the region of four or five decimal units. A hypothesis of least surprise, if its utterly negligible initial probability is not taken into account, is that Pearson thought the expected value of chi-squared was 1 and that he cheated, without even mentioning chi-squared in his paper! Never would anyone go to so much trouble with so little purpose. This example illustrates that, when we are convinced that the surprise in an event can be removed only by making an entirely unreasonable hypothesis, then the event is called a *pure coincidence* [11, p. 169; 17, p. 146]. The example shows that a principle of least surprise that does not allow for prior probabilities of hypotheses is a seriously incomplete recipe for selecting a hypothesis.

References

[1]   Barndorff-Nielsen, O. (1976). *J. R. Statist. Soc. B*, **38**, 103–131 (with discussion). (Suggests a principle of least surprise, without allowing for prior probabilities of hypotheses. Compare ref. 9.)

[2]   Bartlett, M. S. (1952). *Biometrika*, **39**, 328–337. (Uses Λ_0 to measure the significance of odd bits of information as defined in ref. 9, without reference to surprise.)

[3]   Carter, C. F., Meredith, G. P., and Shackle, G. L. S., eds. (1954), *Uncertainty and Business Decisions*, University Press, Liverpool, England. (Proceedings of a conference centered on Shackle's "potential surprise.")

[4]   Cover, T. M. (1973). *Ann. Statist.*, **1**, 862–871. (Suggests that the complexity of H could be measured by the number of hypotheses at least as "simple" as H. The logarithm of this number is perhaps preferable.)

[5]   Fisher, R. A. (1926). *Eugenics Rev.*, **18**, 32–33. (Found that the average chi-squared of 12448 two-by-two tables was remarkably close to 1. See ref. 21.)

[6]   Good, I. J. (1950). *Probability and the Weighing of Evidence*, Griffin, London/Hafner, New York, (Defines the amount of information in a proposition as minus the logarithm of its probability.)

[7]   Good, I. J. (1953). In *Uncertainty and Business Decisions*. University Press, Liverpool, England, pp. 19–34. Partly reprinted in ref. 17. (Suggested that Shackle's "potential surprise" could be expressed in terms of subjective probability. Generalizes Weaver's surprise index.)

[8]   Good, I. J. (1956). *Ann. Math. Statist.* **27**, 1130–1135; **28**, 1055. (Discusses surprise in general terms and computes surprise indexes for the multivariate normal distribution.)

[9]   Good, I. J. (1971a). In *Foundations of Statistical Inference*, V. P. Godambe and D. A. Sprott, eds. Holt, Rinehart and Winston of Canada, Toronto, p. 368. (Suggests a principle of least surprise for hypothesis selection, allowing for the prior probabilities.)

[10]   Good, I. J. (1971b). *Pi Mu Epsilon J.*, **5**, 314–315. (Asks what approximate integer is the most surprising.)

[11]   Good, I. J. (1981). In *Philosophy of Economics*, J. C. Pitt, ed. Reidel, Dordrecht, The Netherlands, pp. 149–174; reprinted in [17]. (Mentions surprise and coincidences in relation to hypothesis testing.)

[12]   Good, I. J. (1982a). *J. Amer. Statist. Ass.*, **77**, 342–344. (Emphasizes that if an observation is surprising given both the null and the non-null hypothesis, then a search for other hypotheses is sensible.)

[13]   Good, I. J. (1982b). *J. Amer. Statist. Ass.*, **77**, 561–563. (Relates surprise indexes to measures of diversity. Mentions that Weaver's surprise index for binomials is expressible in terms of Legendre polynomials.)

[14]   Good, I. J. (1983a). *Behav. Brain Sci.*, **6**, 70. (Points out that the basic formula in the article under discussion is that for Λ_0.)

[15]   Good, I. J. (1983b). *J. Statist. Comp. Simul.*, **17**, 69–71. (The maximum possible "antisurprise" is computed for multinomials and a paradox arises analogous to "the least uninteresting integer.")

[16]   Good, I. J. (1983c). *J. Statist. Comp. Simul.*, **18**, 215–218. (Relates refs. 1 and 9.)

[17]   Good, I. J. (1983d). *Good Thinking: The Foundations of Probability and its Applications*. University of Minnesota Press, Minneapolis, MN. (Contains several mentions of surprise; see its indexes.)

[18]   Good, I. J. (1984). *J. Statist. Comp. Simul.*, **20**, 294–299. (Suggests that, given some observation, the maximum explicativity is a better criterion than least surprise, when choosing a hypothesis. Dynamic probability must be used if the hypotheses are unconstrained.)

[19]   Kolmogorov, A. N. (1983). In *Probability Theory and Mathematical Statistics: Proc. USSR–Japan Symp., 1982*, K. Ito and J. V. Prokhorov, eds. Springer-Verlag, Berlin, Germany, pp. 1–6. (Discusses a measure of complexity of sequences, with references.)

[20]   Krelle, W. (1957). *Econometrica*, **25**, 618–619. (Claims that Shackle's "potential surprise" is expressible in terms of subjective probability. Compare ref. 7.)

[21]   Pearson, E. S. (1925). *Biometrika*, **17**, 388–442. (Discusses his collection of 12448 two-by-two contingency tables that were later used in ref. 5.)

[22]   Redheffer, R. M. (1951). *Ann. Math. Statist.*, **22**, 128–130. (Computes λ_1 for the binomial and Poisson distributions.)

[23]   Rényi, A. (1961). *Proc. Fourth Berkeley Symposium Math. Statist. Prob.*, Vol. 1, University of California Press, Berkeley, Calif., pp. 547–561. (Discusses λ_c in relation to information theory with no mention of surprise.)

[24]   Shackle, G. L. S. (1949). *Expectation in Economics*. Cambridge University Press, London, England. (Claims that potential surprise is psychologically more basic than subjective probability and admits no relationship between them.)

[25]   Shackle, G. L. S. (1954). In *Uncertainty and Business Decisions*. University Press, Liverpool,

England, pp. 90–97. (Summarizes aspects of ref. 24.)

[26] Slovic, P. and Fischhoff, B. (1977). *J. Exp. Psychol.*, **3**, 544–551. (Discusses experiments on the psychology of surprise.)

[27] Weaver, W. (1948). *Sci. Monthly*, **67**, 390–392. (Source paper for λ_1, but somewhat inconsistent.)

[28] Weaver, W. (1963). *Lady Luck: the Theory of Probability*. Doubleday, New York. (A convenient source for the material of ref. 27.)

[29] Weber, H. (1908). *Lehrbuch der Algebra*, Vol. 3. Reprinted by Chelsea Publishing Co., New York. (For the background of the surprising approximate integer.)

(BAYESIAN INFERENCE
COMBINATION OF DATA
DEGREES OF BELIEF
FIDUCIAL INFERENCE
INFERENCE, STATISTICAL: I, II
PRIOR PROBABILITIES
SUBJECTIVE PROBABILITIES
TWO-BY-TWO TABLES)

I. J. GOOD

SURROGATE RESPONSE *See* MEASUREMENT ERROR

SURVEILLANCE

In quality control* this term means monitoring or observation to identify whether an item or activity conforms to specified requirements. It may involve observations over a period of time to measure degradation or degradation associated with shelf life.

For a theory of replacement taking this kind of deterioration into account, *see* RELEVATION.

SURVEY METHODOLOGY

Survey Methodology is a statistical journal published biannually by Statistics Canada*, the Canadian federal statistical agency. As stated in the editorial policy, the journal publishes articles dealing with various aspects of statistical development relevant to a statistical agency such as:

design issues in the context of practical constraints

use of different data sources and collection techniques

total survey error

survey evaluation

research in survey methodology

time series analysis

seasonal adjustment

demographic studies

data integration

estimation and data analysis methods

general survey systems development

Emphasis is placed on the development and evaluation of specific methodologies as applied to data collection* or data themselves. All papers submitted for publication are refereed. The editorial address is: Editor, *Survey Methodology*, Methodology Branch, Statistics Canada, Ottawa, Ontario, Canada K1A 0T6.

A great deal of applied research is conducted at various statistical agencies throughout the world to meet the ever increasing demands for data. The benefits of exchanges of experience and ideas among researchers are obvious. *Survey Methodology* is intended to serve as a medium for communication among applied researchers at different statistical agencies and universities.

The journal was established in 1975, as an in-house journal primarily intended "to provide a forum in a Canadian context for publication of articles on the practical applications of the many aspects of survey methodology." Its basic objectives and policy remained unchanged for about 10 years. During this period, however, the journal gradually grew to the point that the pressing demands and interests could not be met within the restrictive framework established at its inception.

The year 1984 was a turning point. Several major changes were introduced, such as broadening the scope of the editorial policy, expansion of the Editorial Board, improvement of the appearance and quality of the printing, and the introduction of a price for the journal. A management board was also established. The management function had previously been amalgamated within the Editorial Board, which had consisted of a few people from Statistics Canada. The Editorial Board, which includes the Management Board members, was greatly expanded and now consists of 19 persons representing all areas covered by the journal. They are: R. Platek (Chairman and founder), M. P. Singh (Editor and co-founder), and, in alphabetical order, J. Armstrong, K. G. Basavarajappa, D. R. Bellhouse, L. Biggeri, E. B. Dagum, W. A. Fuller, J. F. Gentleman, G. J. C. Hole, T. M. Jeays, G. Kalton, H. Lee, C. Patrick, J. N. K. Rao, C. E. Särndal, F. Scheuren, V. Tremblay, and K. M. Wolter.

One of the unique features of the journal is that, having started in English, it became fully bilingual in Canada's two official languages in 1981 starting with Volume 7, Number 2. The French name of the journal is *Techniques d'Enquête*.

Statistics Canada is the sole sponsor of the journal. However, closer collaboration with statistical associations and societies has been and is being pursued. The journal is made available to members of these organizations at a reduced price. Agreements have been reached with the International Association of Survey Statisticians and the Statistical Society of Canada. A similar agreement with the American Statistical Association* is being pursued.

From time to time special editions of *Survey Methodology* contain proceedings of symposia on specific topics. Special editions include the December 1980 issue (Volume 6, Number 2) on "Survey Research for the 1980's," the June 1984 issue (Volume 10, Number 1) on "Analysis of Survey Data—Issues and Methods," and the June 1986 issue (Volume 12, Number 1) on "Missing Data in Surveys." The following articles, selected from past issues, illustrate the types included in the journal:

"Measurement of Response Errors in Censuses and Sample Surveys" by G. J. Brackstone, J. F. Gosselin, and B. E. Garton (Vol. 1.2).

"Non-Response and Imputation" by R. Platek and G. B. Gray (Vol. 4.2).

"Data, Statistics, Information—Some Issues of the Canadian Social Statistics Scene" by I. P. Fellegi (Vol. 5.2).

"On the Variance of Asymptotically Normal Estimators From Complex Surveys" by D. A. Binder (Vol. 7.2).

"Estimating Monthly Gross Flows in Labour Force Participation" by S. E. Fienberg and E. A. Stasny (Vol. 9.1).

"Evaluation of Composite Estimation for the Canadian Labour Force Survey" by S. Kumar and H. Lee (Vol. 9.2).

"On Analytical Statistics from Complex Samples" by L. Kish (Vol. 10.1).

"Least Squares and Related Analyses for Complex Survey Designs" by W. A. Fuller (Vol. 10.1).

"Post '81 Censal Redesign of the Canadian Labour Force Survey" by M. P. Singh, J. D. Drew, and G. H. Choudhry (Vol. 10.2).

"Conditional Inference in Survey Sampling" by J. N. K. Rao (Vol. 11.1).

"An Empirical Study of Some Regression Estimators for Small Domains" by M. A. Hidiroglou and C. E. Särndal (Vol. 11.1).

"Seasonal Adjustment of Labour Force Series during Recession and Non-Recession Periods" by E. B. Dagum and M. Morry (Vol. 11.2).

"Estimating Population by Age and Sex for Census Divisions and Census Metropolitan Areas" by R. B. P. Verma, K. G. Basavarajappa, and R. K. Bender (Vol. 11.2).

"Handling Missing Survey Data" by G. Kalton and D. Kasprzyk (Vol. 12.1).

"Basic Ideas of Multiple Imputation for Nonresponse" by D. B. Rubin (Vol. 12.1).

M. P. SINGH

SURVEY SAMPLING

HISTORICAL PERSPECTIVE

Nowadays sample surveys are in widespread use by government agencies, market researchers, opinion pollsters, social researchers, and many others, yet survey research has a relatively short history. The development of the survey method has taken place in the past 100 years, with the major expansion occurring since the 1930s. A critical factor enabling survey research to attain its current prominence has been the development of efficient sampling methods for the economic collection of survey data.

At several meetings of the International Statistical Institute* (ISI) around the turn of this century, statisticians debated the scientific validity of any form of sampling for surveying finite populations*. The Norwegian Kaier argued for the use of a form of sampling that he called the representative method, but he faced strong criticism from those who were convinced that only a complete coverage of the population would suffice. Kaier* and his supporters won the day in 1903 when the ISI adopted a resolution recommending the use of the representative method (*see* REPRESENTATIVE SAMPLING). Since that time, the principles of sampling and statistical inference for survey samples have been refined and a range of sampling techniques has been developed. A major advance came in 1934 with the classic paper by Neyman [9], in which he provided a theory of inference based on confidence intervals*, demonstrated the weakness of purposive compared with probability sampling, introduced probability sampling for clusters, and obtained the optimum allocation formula for stratified sampling (*see* OPTIMUM STRATIFICATION). This paper was followed in the 1940s by a number of papers on the main probability sampling methods by statisticians such as Cochran, Deming, Hansen, Hurwitz, Madow, Mahalanobis, and Yates. By the end of that decade, the subject

of survey sampling was sufficiently well consolidated for the appearance of the first set of specialist textbooks. The years since that time have seen many theoretical and practical extensions within the basic framework of the earlier research, and in addition the development of theoretical research on the foundations of statistical inference* in survey sampling. Recent reviews of the developments in survey sampling are provided by Hansen et al. [3], Kruskal and Mosteller (until 1940) [8], O'Muircheartaigh and Wong [10], and Smith [12].

FEATURES OF SURVEY SAMPLING

The subject of survey sampling is concerned with the process of selecting members of the population to be included in a survey and also, since the choice of estimator depends on the sample design, with the estimation process. A sample design needs to be developed to meet the survey objectives, and should be an integral part of the overall survey design. Thus, for instance, a survey aiming to compare the employment experiences of college and high school graduates would call for a different sample design from one aiming to describe the employment experiences of the total labor force. A survey of the U.S. population in which the data are to be collected by face-to-face interviews would call for a different sample design from one in which the data are to be collected by telephone interviews.

Cost considerations feature prominently in determining an efficient sample design. If, say, the survey data are to be collected cheaply by mail questionnaire from one stratum of the population and more expensively by face-to-face interviews from another stratum, then, other things being equal, it is efficient to sample the former stratum at a higher rate than the latter. If the interviews in a face-to-face interview survey take only five minutes to complete, so that most of the interviewers' time is spent on travelling and making contact with respondents, it will be efficient to cluster the sampled members close to one another in order to reduce the travel component of the costs. If, however, the interview is a lengthy one, a more widespread sample is desirable. To design an efficient sample, a cost model of the relevant aspects of the survey process is required. The aim of sample design can then be specified as either minimizing the total cost of the survey while achieving desired levels of variance for the survey estimates or, alternatively, as minimizing the variance of the estimates for a given total cost.

A feature that distinguishes survey sampling from the sampling in the mainstream statistical literature is that the population from which the survey sample is drawn is finite in size. It may be—and often is—very large, such as the approximately 163 million adults in the U.S. population, but it is not the infinite population assumed in standard statistical theory. A closely associated feature is that a great deal of information on the members of the population is often available. For instance, the members of the U.S. population can be classified by their areas of residence: the region of the country, whether they live in urban or rural areas, the economic status of the area, etc. Different ways of using such supplementary information have led to the development of a variety of survey sampling procedures. Supplementary information may be used either at the selection stage (such as for ensuring the desired sample sizes for different segments of the sample—stratification—or for clustering the sampled elements for efficient data collection), or at the estimation stage (e.g., for post-stratification, or for ratio or regression estimation), or at both stages.

PROBABILITY VERSUS NONPROBABILITY SAMPLING

A basic distinction in sampling methods is that between probability and nonprobability sampling. With probability sampling, each population element has a known and nonzero probability of being selected. The selection probabilities arise from the use of a randomized procedure, such as random number tables. Nonprobability sampling is any form of sampling that fails to meet the conditions for probability sampling.

Probability sampling requires the existence of a sampling frame* from which the sample can be drawn. In the simplest form, the frame is a list of the population elements, but more generally it is a means of identifying the elements. In area sampling*, for instance, each element is associated with a particular area. Samples of areas are chosen, and either all or samples of elements in the selected areas are included in the survey. Nonprobability sampling does not need a sampling frame.

The major advantage of probability sampling is that statistical theory can be employed to derive the properties of the sample estimators. Bias in sample selection is avoided and estimators with no, or little, bias can be employed in the analysis. Confidence intervals for the population parameters can be constructed from the sample data. No such theoretical development is possible for nonprobability sampling. Instead, the user of results from a nonprobability sample has to rely on a subjective evaluation. While apparently successful past experience with a nonprobability sampling procedure may give some confidence, there can be no guarantee that the procedure continues to operate satisfactorily.

Despite the theoretical weakness of nonprobability sampling, various forms of it are widely used, primarily for reasons of cost and convenience. Three common forms of nonprobability sampling are:

1. **Haphazard, Convenience or Accidental Sampling.** With this form of sampling, the sampled elements are chosen for convenience or haphazardly, with the purpose of making inferences about some general population. Examples include: a sample of volunteers; street corner interviews; and pull-out questionnaires in a magazine.

2.   Judgment or Purposive Sampling or Expert Choice. In this form of sampling, the elements are carefully selected to provide a "representative" sample. Studies have demonstrated that selection bias can arise with expert choice (Cochran and Watson [2]; Yates [15]), but nevertheless the method may well be appropriate for very small samples when the researcher has a good deal of information about the population elements. Thus, for instance, if costs dictated that a sample had to be confined to one town, it would be preferable to choose the town purposively rather than rely on a random choice.

3.   Quota Sampling*. In this method of sampling, which is mainly used in market research, interviewers are assigned quotas of respondents of different types to interview. For example, an interviewer may be required to interview seven men under 35 years old, five men 35 and older, six employed women, and eight unemployed women. The quotas are usually chosen to be in proportion to the estimated population numbers for the various types, often based on past census data. The areas within which the interviews are to be taken may be chosen by a probability design and the interviewers may be assigned routes to follow to collect the interviews. However, the ultimate choice of respondents is not made by a probability mechanism.

Since no theoretical treatment is possible for nonprobability sampling methods, no general results can be obtained. Each application of a nonprobability sampling method must be evaluated individually. In view of this, the remainder of this article will be confined to probability sampling methods.

PROBABILITY SAMPLING METHODS

A range of probability sampling methods has been developed to serve various purposes. In practice, several methods are often used together to give rise to a *complex sample design*. This section starts with a brief review of the more common sampling methods and then illustrates how they may be combined. Readers requiring more details about the individual methods are referred to the related entries.

The basic probability sampling methods are:

1.   Simple Random Sampling*. With this method every possible set of n different elements from the N population elements is equally likely to be the sample. The sample may be drawn by allotting a different number to each population element and then drawing numbers from a table of random numbers until n different elements are selected. The sample is selected without replacement, so that an element cannot be selected more than once. In common with a number of other sample designs, simple random sampling is an equal probability selection method (*epsem*).

2.   Systematic Sampling*. This method selects every k th element on the population list, beginning from a random start between 1 and k. Systematic sampling is also epsem. If the list is randomly ordered, it is equivalent to simple random sampling. Since the method is simple to apply and easy to check, it is widely used in practice.

3.   Stratified Sampling*. In stratified sampling, supplementary information is used to divide the population into groups, or strata, and then separate samples are selected within each stratum. Using this procedure, the sample sizes to be taken from each stratum are specified in advance to suit the survey objectives. Often the same sampling fraction is used in all strata, producing an epsem design: This is known as proportionate stratification. A uniform sampling fraction is not essential, however, and in many cases disproportionate stratification is used.

When the strata are sampled at varying rates, weighting adjustments are needed in the analysis in compensation (see the following text). Stratification is widely used to improve the precision of survey estimators. When results are needed separately for some strata, disproportionate allocation may be used to provide samples of sufficient sizes from these strata to produce estimators of the desired levels of precision. Stratification features in nearly all survey sample designs.

4. **Cluster Sampling***. As with stratified sampling, the population is made up of a set of groups of elements. However, with cluster sampling only a sample of the groups (clusters) is selected. Strictly defined, cluster sampling involves the inclusion of all the elements in the selected clusters in the sample, but the term is also widely used more loosely to cover situations in which only samples of elements from the selected clusters are included (see Multistage Sampling that follows). The elements within clusters usually tend to be somewhat homogeneous in the survey variables. In consequence, a cluster sample usually produces less precise estimators than a simple random sample of the same size. The main use for cluster sampling is when it gives rise to economies in data collection*, thus permitting a larger sample size to be obtained than with an unclustered sample. The use of a cluster sample is then appropriate when its estimators are more precise than those from an unclustered sample for the same total budget. Cluster sampling is also sometimes used when the sampling frame* lists clusters rather than elements, as for instance when the frame lists dwellings and the elements are persons.

5. **Multistage Sampling***. When the population clusters are large, a sample of complete clusters would be restricted to a small number of clusters, and if the clusters are somewhat internally homogeneous in the survey variables, the re-

sulting estimators would be imprecise. A way to obtain a more widespread sample is to take a larger sample of clusters and to take only a sample of elements within the selected clusters: This design is a two-stage sample. The approach can be readily extended to more stages: At the first stage, the initial clusters are selected, at the second stage subclusters are selected within selected clusters, and so on until at the last stage elements are selected within the final stage clusters. The first stage clusters are commonly termed *primary sampling units** (PSUs). *See also* STRATIFIED MULTISTAGE SAMPLING.

6. **Probability Proportional to Size Sampling***. The clusters used in multistage sampling are as a rule naturally occurring population groupings, such as geographical districts, hospitals and wards, and schools and classes. As such, they vary in size—that is, in the numbers of elements they contain—often to a very marked extent. An efficient way to obtain an epsem sample of elements from a multistage design with unequal sized clusters is to sample the clusters at each stage with probabilities proportional to their sizes (PPS). Consider, for example, a two-stage design in which n PSUs are selected at the first stage with PPS and m elements are selected with equal probability from each selected PSU. Assuming for simplicity that $nN_i < N$, where N_i is the number of elements in PSU i and $N = \Sigma N_i$, the overall selection probability for an element in PSU i is given by the selection equation

$$(nN_i/N)(m/N_i) = mn/N,$$

a constant. This PPS procedure thus produces a fixed total sample size (mn) with the same sample size (m) taken from each selected PSU and an overall equal probability for all elements.

In practice, the true cluster sizes are seldom known, but reasonable estimates of them are often available. These estimated sizes, or measures of size, may be used for sampling with probabilities proportional to estimated sizes, which is sometimes abbreviated to PPES. The selection equation for an epsem sample of elements with a two-stage PPES design is

$$(nM_i/M)(m/M_i) = nm/M,$$

where M_i denotes the estimated size of PSU i and $M = \Sigma M_i$. With this design, the application of the second stage sampling fraction of (m/M_i) to the N_i elements actually in selected PSU i yields a sample of (mN_i/M_i) elements from that PSU. This sample size varies from one PSU to another, and hence the total sample size is also not fixed. However, providing the M_i and N_i are close, the variability in sample size will be small and tolerable. Sampling with PPES is widely used to handle unequal sized clusters in multistage designs.

The preceding methods are widely used in sample design. In the case of a small compact population for which a list of elements is available, a simple single stage design without clustering is generally appropriate. The design might well consist of dividing the population into strata and then taking a systematic sample of elements within each stratum. Either proportionate or disproportionate stratification might be used, depending on the survey objectives, the variances of the survey variables within the strata, and cost factors.

In the case of a face-to-face interview survey of a widespread population, some clustering of interviews is almost certainly needed. The sample design for a national face-to-face interview survey is typically a complex one involving several stages of sampling with PPES selection and stratification at each stage. In the United States such a sample of persons might consist of the following stages:

(1) Selecting a highly stratified sample of standard metropolitan statistical areas, counties, or groups of counties with probability proportional to some measure of size (e.g., the number of occupied housing units they contain).

(2) Within the selected first stage units, selecting a PPES stratified sample of blocks in urban areas where census block statistics are available and census enumeration districts elsewhere.

(3) Dividing the larger selected second stage units into clearly defined segments, assigning approximate measures of size to the segments, and taking a PPES sample of them (with each of the smaller second stage units being treated as a single segment).

(4) Listing the housing units in the selected segments and taking a systematic sample from the list.

(5) Listing eligible persons within the selected housing units and selecting one person at random from each housing unit for the sample.

For descriptions of U.S. national area sample designs, see U.S. Bureau of the Census [13] for the Current Population Survey, Hess [4], and Kish [6].

The sampling methods previously outlined are those widely used in practice, but the list is not exhaustive. Other methods include:

7. **Two-Phase Sampling.** In two-phase or double sampling* a large sample is selected at the first phase and a subsample of the first phase sample is selected at the second phase. When information needed for sample design or for estimation is not available for the entire population, it is sometimes possible to collect it cheaply for a large first phase sample. The information may then be used in the design of the second phase sample (e.g., for stratification) or in the estimation from the second phase sample (e.g., in ratio* or regression estimation). *See* also DOUBLE SAMPLING and MULTIPHASE SAMPLING.

8. **Replicated Sampling.** In replicated or interpenetrating* sampling the overall

sample is composed of a set of subsamples, each of the identical sample design. Replicated sampling is used to investigate variable nonsampling errors, such as interviewer variance or coder variance. It is also used to provide easily computed standard error estimates for complex estimates and for complex subsample designs.

9. **Panel Designs.** When sampling on two or more successive occasions, some overlap in the samples is generally advantageous for measuring change. Sometimes complete overlap of the samples is undesirable because sampled units may refuse to remain in the panel or they may become conditioned by their panel membership. In such circumstances, a partial overlap may be appropriate. *See* PANEL DATA *and* LONGITUDINAL DATA ANALYSIS.

PRACTICAL CONSIDERATIONS

Two important practical considerations with probability sampling are the sampling frame and nonresponse*. The way in which the sampling frame is constructed exerts a strong influence on the choice of sample design. Thus, for instance, the division of the frame into groups facilitates the use of these groups for clusters or for strata. Ideally the frame should comprise a listing of each population element once and once only, and should contain no other listings. In practice, this ideal is seldom realized: Some elements may be missing from the list, some may be listed as clusters rather than individually, some may appear more than once on the frame, and the frame may contain listings that do not relate to population elements. See Kish [6] for a discussion of these four frame problems and ways in which attempts may be made to handle them. Often the most serious problem is noncoverage, that is, an incomplete frame with some missing elements: For such elements, the probability sampling requirement that all elements have a nonzero selection probability fails to hold.

The theory of probability sampling assumes that the survey responses are obtained for all the sampled elements. In practice, this assumption is almost always invalidated by some degree of nonresponse. Total nonresponse occurs when a sampled element fails to provide any survey responses, generally because the selected respondent either refuses to participate or is not at home when contacts are attempted. There has been considerable concern that the rate of total nonresponse has been increasing in recent years. Weighting adjustments are sometimes used in an attempt to counteract the possible bias in survey estimates caused by differential response rates among different population groups. Item nonresponse, which occurs when a responding unit provides acceptable responses to some but not all of the survey questions, may arise because the respondent does not know the answers to certain questions, because he or she refuses to answer certain questions, or because some answers are deleted on the grounds that they are inconsistent with other responses. Item nonresponse is sometimes handled by imputation*, that is, by assigning values for the missing responses.

ANALYSIS OF SURVEY DATA

The analysis of survey data depends on the sample design employed in two main ways. In the first place, the sample estimators themselves depend on the selection probabilities of the elements and, second, the variances of the sample estimates depend on the joint inclusion probabilities of pairs of elements.

When some form of equal probability sampling is used, sample estimators may generally be formed in the standard way. For instance, the simple sample mean $\bar{y} = \sum y_i/n$ is an unbiased estimator of the population mean $\bar{Y}$ for any equal probability sampling scheme with a fixed sample size n

($\bar{y}$ may also be used to estimate $\bar{Y}$ if n is not fixed, providing it is under reasonable control). If, however, an unequal probability sampling scheme is employed, weights made inversely proportional to the selection probabilities are needed in the analysis to redress the balance of the sample. If the ith sampled element has selection probability p_i, then the population mean $\bar{Y}$ is estimated by $\bar{y}_w = \Sigma w_i y_i / \Sigma w$; where $w_i \propto 1/p_i$. In addition to adjusting for unequal selection probabilities, weights are often used to adjust for differential response rates among different subgroups, and for post-stratification (see the following text).

The variances of estimators given by standard statistical theory apply only with simple random sampling with replacement (a form of simple random sampling in which the elements are allowed to appear in the sample more than once). They produce close approximations to the variances of estimators with simple random sampling without replacement when the sampling fraction is small (as is usually the case), but they can be misleading when applied with other designs. The variances of estimators with simple random sampling provide a useful benchmark against which to gauge the variances of the estimators with other designs. The ratio of the variance of an estimator with a given design to the variance of the estimator with a simple random sample of the same size is the *designed effect**. Design effects for estimators from complex stratified multistage designs are usually greater than 1. The use of formulae from standard statistical theory for estimating the standard errors of estimators obtained from a complex sample design thus tend to overstate the precision of the estimators. Approaches used for computing appropriate standard error estimates for complex sample designs include the Taylor series linearization or delta method (*see* STATISTICAL DIFFERENTIALS, METHOD OF), balanced repeated replications, and jackknife* methods (Kalton [5], Kish and Frankel [7], and Rust [11]). A number of computer programs for calculating sampling errors from com-

plex designs are available. Wolter [14] provides a detailed account of procedures for variance estimation with complex designs and describes some of the computer software available for the purpose.

Another feature of estimation in sample surveys is that supplementary information is sometimes used in the analysis to improve the precision of the estimators. One use of this information is for poststratification, or stratification after selection. With this procedure the sample elements are assigned weights to make the sample's weighted distribution for one or more variables conform to the known population distribution of those variables. Routinely employed, poststratification weighting adjustments apply to all the survey estimators. Another use of supplementary information is for adjusting specific estimators. A general form of linear adjustment using the supplementary variable x available for the total population in estimating the population mean of variable y is

$$\bar{y}_r = \bar{y} + k(\bar{X} - \bar{x}),$$

where $\bar{y}$ and $\bar{x}$ are the sample means of variables y and x, $\bar{X}$ is the population mean of variable x, and k may be chosen in various ways. If k is a constant, $\bar{y}_r$ is termed a *difference estimator*; if k is the sample regression coefficient for the regression of y on x, $\bar{y}_r$ is a *regression estimator*; if $k = \bar{y}/\bar{x}$, $\bar{y}_r$ is a *ratio estimator*; and if $k = -\bar{y}/\bar{X}$, $\bar{y}_r$ is a *product estimator*. Extensions of these estimators that use several auxiliary variables are also available. The use of supplementary information to improve the precision of sample estimators is discussed in Cochran [1] and all the other major survey sampling texts.

NONSAMPLING ERRORS

Sampling errors are not the only source of error in the survey process. Nonresponse and noncoverage can cause biases in survey estimators, and both variable errors and biases can occur at the data collection and processing stages. Response errors, which arise when the information collected is inaccurate, are often a major cause of concern. An efficient survey design is one that minimizes the total survey error. Often an efficient design involves striking an economic balance between sampling and nonsampling errors. It may, for instance, be necessary to choose between an expensive data collection procedure with a small sample and a less expensive procedure with a large sample. While the former may have little response error, it will have a large sampling error. On the other hand, with the latter procedure, the reverse may hold. For these reasons, the sampling process cannot be divorced from the rest of the survey design. The subject of survey sampling, therefore, covers the development of models for all types of survey error, and most sampling texts devote a chapter to the subject.

References

[1] Cochran, W. G. (1977). *Sampling Techniques*, 3rd ed. Wiley, New York.

[2] Cochran, W. G. and Watson, D. J. (1936). *Empire J. Exp. Agric.*, **4**, 69–76.

[3] Hansen, M. H., Dalenius, T., and Tepping, B. J. (1985). In *A Celebration of Statistics. The ISI Centenary Volume*, A. C. Atkinson and S. E. Fienberg, eds. Springer-Verlag, New York, pp. 327–354.

[4] Hess, I. (1985). *Sampling for Social Research Surveys, 1947-1980*. Institute for Social Research, Ann Arbor, Mich.

[5] Kalton, G. (1977). *Bull. Int. Statist. Inst.*, **47**, 495–514.

[6] Kish, L. (1965). *Survey Sampling*. Wiley, New York.

[7] Kish, L. and Frankel, M. R. (1974). *J. R. Statist. Soc. B*, **36**, 1–37.

[8] Kruskal, W. and Mosteller, F. (1980). *Int. Statist. Rev.*, **48**, 169–195.

[9] Neyman, J. (1934). *J. R. Statist. Soc.*, **97**, 558–625.

[10] O'Muircheartaigh, C. and Wong, S. T. (1981). *Bull. Int. Statist. Inst.*, **49**, 465–493.

[11] Rust, K. F. (1985). *J. Official Statist.*, Sweden., **1**, 381–397.

[12] Smith, T. M. F. (1976). *J. R. Statist. Soc. A*, **139**, 183–204.

[13] U.S. Bureau of the Census (1978). *The Current Population Survey: Design and Methodology*. U.S. Government Printing Office, Washington, D.C.

[14] Wolter, K. M. (1985). *Introduction to Variance Estimation*. Springer-Verlag, New York.

[15] Yates, F. (1934/35). *Ann. Eugenics*, London, **6**, 202–213.

Bibliography

Barnett, V. (1974). *Elements of Sampling Theory*. English Universities Press, London, England. (A short introductory text on sampling theory.)

Cassel, C-M., Sarndal, C-E., and Wretman, J. H. (1977). *Foundations of Inference in Survey Sampling*. Wiley, New York. (An advanced theoretical text on foundations of survey sampling inference.)

Cochran, W. G. (1977). *Sampling Techniques*, 3rd ed. Wiley, New York. (A widely used text on sample survey theory.)

Deming, W. E. (1950). *Some Theory of Sampling*. Dover, New York. (A text on sampling theory and practice.)

Deming, W. E. (1960). *Sample Design in Business Research*. Wiley, New York. (A text on sampling practice and theory, with an emphasis on the use of replicated sampling.)

Hájek, J. (1981). *Sampling from a Finite Population*. Dekker, New York. (An advanced monograph on sampling theory.)

Hansen, M. H., Hurwitz, W. N., and Madow, W. G. (1953). *Sample Survey Methods and Theory: Methods and Applications*, Vol. I; *Theory* Vol. II. Wiley, New York. (A comprehensive text on sampling theory and practice.)

Jessen, R. J. (1978). *Statistical Survey Techniques*. Wiley, New York. (An intermediate level text on sampling.)

Johnson, N. L. and Smith, H., eds. (1969). *New Developments in Survey Sampling*. Wiley, New York. (A collection of papers presented at a symposium on the foundations of survey sampling.)

Kalton, G. (1983). *Introduction to Survey Sampling*. Sage, Beverly Hills, Calif. (A short nonmathematical introduction to sampling practice.)

Kendall, M. G., Stuart, A., and Ord, K. J. (1983). *The Advanced Theory of Statistics: Design and Analysis, and Time-Series*, Vol. 3, 4th ed. Macmillan, New York. (Chapters 39 and 40 provide an advanced treatment of sampling theory.)

Kish, L. (1965). *Survey Sampling*. Wiley, New York. (A comprehensive text on sampling practice.)

Konijn, H. S. (1973). *Statistical Theory of Sample Survey Design and Analysis*. American Elsevier, New York. (An advanced text on sampling theory.)

Krewski, D., Platek, R., and Rao, J. N. K., eds. (1981). *Current Topics in Survey Sampling*. Academic, New

York. (A collection of papers from a symposium on survey sampling.)

Levy, P. S. and Lemeshow, S. (1980). *Sampling for Health Professionals*. Lifetime Learning Publications, Belmont, Calif. (An intermediate level text on sampling methods.)

Mendenhall, W., Ott, L., and Scheaffer, R. L. (1971). *Elementary Survey Sampling*. Wadsworth, Belmont, Calif. (An elementary text for students with limited mathematical backgrounds.)

Moser, C. A. and Kalton G. (1971). *Survey Methods in Social Investigation*, 2nd ed. Heinemann, London, England. (A text on survey methods, with a full nonmathematical introduction to sampling methods in Chapts. 4–8.)

Murthy, M. N. (1967). *Sampling Theory and Methods*. Statistical Publishing Society, Calcutta, India. (An advanced text on sampling theory and practice, with an extensive bibliography.)

Namboodiri, N. K., ed. (1978). *Survey Sampling and Measurement*. Academic, New York. (A collection of papers presented at a symposium.)

Raj, D. (1968). *Sampling Theory*. McGraw-Hill, New York. (An advanced text on sampling theory.)

Raj, D. (1972). *The Design of Sample Surveys*. McGraw-Hill, New York. (The first half is an intermediate text on sampling practice. The second half reviews surveys in different fields of application.)

Rossi, P. H., Wright, J. D., and Anderson, A. B., eds. (1983). *Handbook of Survey Research*. Academic, New York. (A handbook on social survey methods, with sizeable chapters on sampling theory by M. R. Frankel and on applied sampling by S. Sudman.)

Sampford, M. R. (1962). *An Introduction to Sampling Theory*. Oliver and Boyd, London, England. (An intermediate level introduction with agricultural examples.)

Satin, A. and Shastry, W. (1983). *Survey Sampling: A Non-mathematical Guide*. Statistics Canada, Ottawa, Canada. (A short and simple introduction to survey sampling.)

Smith, T. M. F. (1976). *Statistical Sampling for Accountants*. Haymarket Publishing, London, England. (An intermediate text on sampling methods in accounting.)

Som, R. K. (1973). *A Manual of Sampling Techniques*. Heinemann, London, England. (An intermediate level text.)

Stuart, A. (1984). *The Ideas of Sampling*, rev. ed. Griffin, London, England. (A short text illustrating the basic ideas of survey sampling nonmathematically with a small numerical example.)

Sudman, S. (1976). *Applied Sampling*. Academic, New York. (An intermediate text on sampling practice.)

Sukhatme, P. V., Sukhatme, B. V., Sukhatme, S., and Asok, C. (1984). *Sampling Theory of Surveys with Applications*, 3rd ed. Iowa State University Press, Ames, IA. (An advanced text on sampling theory.)

Williams, W. H. (1978). *A Sampler on Sampling*. Wiley, New York. (An intermediate level introduction to sampling methods.)

Yates, F. (1981). *Sampling Methods for Censuses and Surveys*, 4th ed. Griffin, London, England. (An advanced text on sampling practice.)

(AREA SAMPLING
CLUSTER SAMPLING
DOUBLE SAMPLING
FINITE POPULATIONS, SAMPLING FROM
INTERPENETRATING SUBSAMPLES
LONGITUDINAL DATA ANALYSIS
MULTIPHASE SAMPLING
NEYMAN ALLOCATION
NONRESPONSE (IN SAMPLE SURVEYS)
OPTIMUM STRATIFICATION
PANEL DATA
PRIMARY SAMPLING UNITS
PROBABILITY PROPORTIONAL TO SIZE
    SAMPLING
PROPORTIONAL ALLOCATION
PUBLIC OPINION POLLS
QUOTA SAMPLING
REJECTIVE SAMPLING
REPRESENTATIVE SAMPLING
SAMPLING FRAME
SIMPLE RANDOM SAMPLING
STRATIFIED DESIGNS
STRATIFIED MULTISTAGE SAMPLING
SYSTEMATIC SAMPLING)

GRAHAM KALTON

SURVIVAL ANALYSIS

SURVIVAL DATA

Survival analysis is concerned with statistical models and methods for analysing data representing life times, waiting times, or more generally times to the occurrence of some specified event. Such data, denoted as *survival data*, can arise in various scientific fields including medicine, engineering, and demography*. In a clinical trial* the object of the study is the comparison of survival times with different treatments in some chronic disease; in an engineering reliability* experiment, a number of items could be put on test simultaneously and observed until failure (*see also* LIFE TESTING); a demographer could be interested in studying the length of time that a group of workers stay in a particular job. Thus, survival data are basically nothing but realisations of nonnegative random variables. What distinguishes survival analysis from other fields of statistics is, however, the almost inevitable presence of (right) censoring*, in that limitations in time and other restrictions on data collection* prevent the experimenter from observing the event in question for every individual or item under study. Rather, for some individuals only partial information will be available about their survival times, namely, that they exceed some observed censoring times, whereas the survival times themselves are not observed or cannot be observed. It depends on the design of the study which model for the type of censoring will be adequate in a given context (*see* CENSORED DATA). Most methods for survival analysis are based on the following two assumptions [12, p. 120]:

(1) Given all that has happened up to time t, the failure mechanisms for different individuals act independently over the interval $[t, t + dt)$.

(2) For an individual alive and *uncensored* at t, the conditional probability of failing in $[t, t + dt)$ given all that has happened up to time t coincides with the conditional probability of failing in $[t, t + dt)$ given survival up to time t.

[For a more rigorous formulation of assumptions (1) and (2), see ref. 9, p. 27.] An implication of assumption (2) is the exclusion of censoring mechanisms withdrawing individuals from risk when they appear to have particularly high or low risk of failure. A simple censoring model, often applicable in biomedical contexts and fulfilling both assumptions, is random censorship in which survival times and censoring times are stochastically independent. The object of a survival analysis to draw inferences about

the distribution of the survival times T. This distribution can be characterised by the survival distribution function (SDF),

$$S(t) = \Pr[T > t], \qquad t \geqslant 0$$

or equivalently (provided that S is differentiable) by the hazard rate* function (HRF)

$$\lambda(t) = -d(\log S(t))/dt = f(t)/S(t),$$
$$t \geqslant 0, \qquad (1)$$

where $f(t)$ is the probability density function* (PDF) or the cumulative hazard rate function (CHRF)

$$\Lambda(t) = \int_0^t \lambda(u)\, du, \qquad t \geqslant 0; \qquad (2)$$

see also RELIABILITY, PROBABILISTIC. Statistical models for continuous survival data can be specified via any of these quantities. In applications, one may have some qualitative information about the way in which the *instantaneous* risk of the event changes over time and thus the models are often most conveniently specified via the HRF.

Throughout it will be assumed that the available data are of the form $t_1, \ldots, t_d$, $t_{d+1}^*, \ldots, t_n^*$, where t_i, $i = 1, \ldots, d$ are survival times and t_i^*, $i = d + 1, \ldots, n$, are censoring times, the censoring mechanism fulfilling the assumptions (1) and (2). We define $\tau_i = t_i$ for $i = 1, \ldots, d$ and $\tau_i = t_i^*$ for $i = d + 1, \ldots, n$. For methods where $S(t)$ is discontinuous, corresponding to survival distributions with discrete components, see ref. 9. Methods for dealing with grouped survival data do exist (*see* SURVIVAL ANALYSIS, GROUPED DATA), the classical actuarial life table* being the main example. Such methods are applicable in situations where the sample size n is so large that it is not feasible to record the exact survival and censoring times but only to which of a number of prespecified time intervals they belong. They also apply in situations where individuals are only observed at prespecified follow-up times so that it is only known in which interval between successive follow-up times that the event under study has occurred. A somewhat extreme example would be to analyse only the status alive/dead after a single time interval. This is the situation in the analysis of quantal response* data in bioassay*.

NONPARAMETRIC SURVIVAL MODELS

Presence of censored observations implies that classical nonparametric methods based on ranks are not directly applicable. In particular, standard graphical procedures such as an empirical CDF or a histogram* cannot be used with censored data. To present the required modifications, it is convenient to consider the quantities

$$N(t) = \#\{i = 1, \ldots, d: t_i \leqslant t\}$$

and

$$Y(t) = \#R(t),$$

where

$$R(t) = \{i = 1, \ldots, n: \tau_i \geqslant t\}.$$

Thus $N(t)$ is the number of failures before or at t and $Y(t)$ is the number of individuals *at risk* at $t -$ (that is, just before t).

The SDF can be estimated by the Kaplan–Meier* or product limit estimate [13]

$$\hat{S}(t) = \prod_{t_i \leqslant t} \left[1 - \frac{\Delta N(t_i)}{Y(t_i)} \right],$$

where $\Delta N(t) = N(t) - N(t -)$ is the number of failures at t. Similarly, the CHRF can be estimated by the Nelson–Aalen estimate [1, 16]

$$\hat{\Lambda}(t) = \sum_{t_i \leqslant t} \frac{\Delta N(t_i)}{Y(t_i)}.$$

Conditions [including (**1**) and (**2**)] can be found [1, 4, 9] under which $\hat{S}(t)$ and $\hat{\Lambda}(t)$ behave asymptotically ($n \to \infty$) as normal processes, and approximate standard errors can be calculated as

$$\hat{\sigma}(\hat{\Lambda}(t))$$
$$= \left[\sum_{t_i \leqslant t} \frac{\Delta N(t_i)}{Y(t_i)(Y(t_i) - \Delta N(t_i) + 1)} \right]^{1/2},$$
$$\hat{\sigma}(\hat{S}(t)) = \hat{S}(t) \, \hat{\sigma}(\hat{\Lambda}(t)).$$

Estimates $\hat{\lambda}(t)$ of the HRF can be obtained, e.g., by smoothing $\hat{\Lambda}(t)$ using some kernel function [19]. Also, approximate standard errors of $\hat{\lambda}(t)$ can be obtained in this way.

Sometimes it is of interest to compare an observed survival distribution with an SDF $\exp(-\Lambda^*(t))$ that is known, for example, from the life tables for some reference population. Conditions can be found in ref. 2 under which a test statistic of the form

$$\sum_{i=1}^{d} K(t_i) \frac{\Delta N(t_i)}{Y(t_i)} - \int_0^\infty K(u) \, d\Lambda^*(u)$$

has an asymptotic standard normal distribution (as $n \to \infty$) when normalized by

$$\left[\sum_{i=1}^{d} K^2(t_i) \frac{\Delta N(t_i)}{Y^2(t_i)} \right]^{1/2}$$

or by

$$\left[\int_0^\infty K^2(u) I(Y(u) > 0) \frac{d\Lambda^*(u)}{Y(u)} \right]^{1/2}.$$

Here $K(t)$ is a stochastic "weight" process; special choices of K yield various test statistics discussed in the survival data literature [2]. $I(\cdot)$ is the indicator function.

Nonparametric comparison of the survival distributions in $k \geqslant 2$ groups of individuals can be based on test statistics of the form

$$Z_j = \sum_{i=1}^{d} \left[K(t_i) Y_j(t_i) \right.$$
$$\left. \times \left\{ \frac{\Delta N_j(t_i)}{Y_j(t_i)} - \frac{\Delta N(t_i)}{Y(t_i)} \right\} \right],$$
$$j = 1, \ldots, k.$$

Here $\Delta N(t)$ and $Y(t)$ are as before, the quantities $\Delta N_j(t)$ and $Y_j(t)$ being defined similarly for each group j, $j = 1, \ldots, k$, and $K(t)$ is again a stochastic process, KY_j giving weights to the differences between the jumps in the Nelson–Aalen estimates in each group j and those in the combined estimate based on the k samples. Defining the $k \times k$ matrix $\mathbf{V} = (V_{jl})$ by

$$V_{jl} = \sum_{i=1}^{d} \left[K^2(t_i) \frac{Y_j(t_i)}{Y(t_i)} \right.$$
$$\left. \times \left\{ \delta_{jl} - \frac{Y_l(t_i)}{Y(t_i)} \right\} \Delta N(t_i) \right],$$

where δ_{jl} is the Kronecker delta*, conditions can be found [2] under which $\mathbf{Z'V^-Z}$ has an asymptotic chi-squared distribution with $k - 1$ degrees of freedom (as $n \to \infty$). Here $\mathbf{Z} = (Z_1, \ldots, Z_k)'$ and $\mathbf{V^-}$ is a generalised inverse of $\mathbf{V}$.

Special choices of K yield various well known test statistics; in particular $K(t) = I(Y(t) > 0)$ corresponds to the logrank test*, in which case Z_j, reduces to the difference between the observed $O_j = N_j(\infty)$ and the "expected" number of failures in group j:

$$E_j = \sum_{i=1}^{d} \left[\frac{Y_j(t_i)}{Y(t_i)} \Delta N(t_i) \right].$$

In the case $k = 2$, an equivalent test statistic is

$$Z = \sum_{i=1}^{d} \left[K(t_i) \left\{ \frac{\Delta N_2(t_i)}{Y_2(t_i)} - \frac{\Delta N_1(t_i)}{Y_1(t_i)} \right\} \right],$$

and under suitable conditions [1, 9] $ZV^{-1/2}$, where

$$V = \sum_{i=1}^{d} \left\{ K^2(t_i) \frac{\Delta N_1(t_i) + \Delta N_2(t_i)}{Y_1(t_i) Y_2(t_i)} \right\},$$

has an asymptotic standard normal distribution (as $n \to \infty$). In this two-sample case, the choice $K = Y_1 Y_2 / Y$ yields the logrank test.

PARAMETRIC SURVIVAL MODELS

In the analysis of survival data, the exponential*, Weibull*, gamma*, and lognormal* distributions play a central role. The basic properties of these four distributions will be briefly reviewed, the emphasis being on the aspects of the distributions that are important in survival analysis. For further reading, see the book by Lawless [15] and the references in the bibliography. *See also* HAZARD RATE AND OTHER CLASSIFICATIONS OF DISTRIBUTIONS.

The simplest lifetime distribution, the exponential distribution, is characterized by a constant HRF $\lambda(t) = \lambda$ for all $t \geqslant 0$, and the SDF has the form $S(t) = \exp(-\lambda t)$.

Although the assumption of a constant hazard rate is very restrictive, the exponential distribution was the first survival model to become widely used, partly due to its computationally very attractive features. However, theoretical investigations initiated by Zelen and Dannemiller [23] have revealed that these methods are very sensitive to departures from the exponential model and, consequently, should be applied with caution. The appropriateness of the exponential distribution for a given set of survival data may be checked by plotting $\hat{\Lambda}(t)$ or equivalently $-\log \hat{S}(t)$ versus t. Such a plot should approximate a straight line through the origin.

The Weibull distribution is probably the most widely used parametric survival model in technical as well as biomedical applications. The Weibull model provides a fairly flexible class of distributions and includes the exponential distribution as a special case. The HRF and the SDF have the form

$$\lambda(t) = \lambda \rho (\lambda t)^{\rho - 1}, \qquad t \geqslant 0,$$

and

$$S(t) = \exp(-(\lambda t)^{\rho}), \qquad t \geqslant 0, \quad (3)$$

where $\lambda > 0$ is an inverse scale parameter and $\rho > 0$ is a shape parameter. The HRF is monotonically increasing if $\rho > 1$, monotone decreasing if $0 < \rho < 1$, and constant if $\rho = 1$. The Weibull distribution appears as one of the asymptotic distributions of the smallest extreme (*see* EXTREME VALUE DISTRIBUTIONS) and this fact motivates its use in certain applications. From (3) it follows that

$$\log \Lambda(t) = \rho \log \lambda + \rho \log t,$$

and the appropriateness of the Weibull model can therefore be checked by plotting $\log \hat{\Lambda}(t)$ versus $\log t$.

The gamma model provides an alternative two-parameter family of distributions that includes the exponential distribution as a special case. The gamma distribution has a PDF of the form

$$f(t) = \lambda (\lambda t)^{\alpha - 1} \exp(-\lambda t) / \Gamma(\alpha),$$
$$t \geqslant 0,$$

where $\lambda > 0$ is an inverse scale parameter and $\alpha > 0$ is a shape parameter. The SDF can be expressed as

$$S(t) = 1 - I(\alpha, \lambda t),$$

where $I(\alpha, x)$ is the incomplete gamma function

$$I(\alpha, x) = \Gamma(\alpha)^{-1} \int_0^x u^{\alpha - 1} e^{-u} \, du.$$

Closed form expressions are only available for integer values of α and the gamma model is therefore less attractive than the Weibull model. The HRF is monotone increasing if $\alpha > 1$ with $\lambda(0) = 0$ and $\lim_{t \to \infty} \lambda(t) = \lambda$ and monotone decreasing if $0 < \alpha < 1$ with $\lim_{t \to 0} \lambda(t) = +\infty$ and $\lim_{t \to \infty} \lambda(t) = \lambda$. For $\alpha = 1$, the HRF $\lambda(t)$ is constant ($= \lambda$).

The lognormal distribution distribution has also been widely used in survival analysis. As for the gamma model, no closed form expressions are available for $S(t)$ and $\lambda(t)$. The distribution is most easily specified through $\log T$ having a normal distribution with mean μ and variance σ^2, say, when T is a lognormal variate. With this parametrization the lognormal SDF is

$$S(t) = 1 - \Phi\left(\frac{\log t - \mu}{\sigma} \right),$$

where Φ is the CDF of the standard normal distribution. The lognormal HRF has the value 0 at $t = 0$, increases to a maximum, and then decreases with a limiting value of 0 as t tends to infinity. This behaviour may be unattractive in certain applications.

The generalized gamma model introduced by Stacy [20] provides a flexible three-parameter family of distributions that includes the preceding four models as special cases. Other useful parametric models are distributions for which $\lambda(t)$ or $\log \lambda(t)$ is a polynomial of low order, the inverse Gaussian distribution (see Chhikara and Folks [6]), the log-logistic model (see, ref. 12, Sec. 2.2.6), and the piecewise exponential model.

All the parametric models may be modified to allow for an initial failure-free period by introduction of a threshold parameter or "guarantee" time Δ by replacing T by $T' = T - \Delta$.

INFERENCE FOR PARAMETRIC MODELS

Maximum likelihood estimation* and large-sample* likelihood* methods are the inference procedures generally used, as the presence of censoring makes exact distributional results extremely complicated in most situations. In the case of no censoring or type II censoring (see CENSORED DATA), several alternative procedures, including methods giving exact distributional results for the parameter estimates, are available for the exponential and the Weibull models. A review of these methods is given by Lawless [15].

For the class of censoring schemes satisfying (1) and (2) the likelihood function becomes (apart from a constant of proportionality)

$$L(\theta) = \prod_{i=1}^d f(t_i; \theta) \prod_{i=d+1}^n S(t_i^*; \theta), \quad (4)$$

where θ is the vector of unknown parameters to be estimated. From (4) it is apparent that survival models admitting a closed form expression for the SDF are computationally attractive. In the exponential model ($\theta = \lambda$), the maximum likelihood estimate becomes $\hat{\lambda} = d/(\Sigma t_i + \Sigma t_i^*)$, but in general an iterative procedure must be applied to find the maximum likelihood estimates. If the model includes a threshold parameter, additional problems arise. In most cases the variances of the parameter estimates have to be estimated from the observed information matrix (see FISHER INFORMATION)

$$-\left(\left(\frac{\partial^2 \log L(\theta)}{\partial \theta_i \, \partial \theta_j} \Bigg|_{\theta = \hat{\theta}} \right) \right),$$

since calculation of the expected information requires detailed knowledge of the distribution of the censoring times.

The theoretical justification of the asymptotic normality* of the maximum likelihood estimate and the limiting χ^2 distribution of the likelihood ratio statistic with censored data has been established in many special cases. A unified approach to the asymptotic theory of maximum likelihood estimation for parametric survival models with censoring satisfying (**1**) and (**2**) has been given by Borgan [5].

REGRESSION MODELS

Evaluation of the influence (in this context often denoted the "prognostic value"), of one or several concomitant variables* measured on each individual is often an important part of survival analysis. Various types of regression models* have been developed to deal with these types of problems, the two most important models being Cox's proportional hazards regression model and the accelerated failure time regression model. In both models the concomitant variables enter the regression analysis as independent variables, while the survival time is treated as the dependent variable.

In *Cox's semiparametric regression model* [7], the HRF for a subject with covariates $\mathbf{x} = (x_1, \ldots, x_p)$ is given by

$$\lambda(t, \mathbf{x}) = \lambda_0(t)\exp(\boldsymbol{\beta}'\mathbf{x}), \qquad (5)$$

where $\boldsymbol{\beta}$ is a p-dimensional vector of unknown regression coefficients reflecting the effects of $\mathbf{x}$ on survival and $\lambda_0(t)$ is an unspecified function of time. The basic features of the model (5) are the assumption of proportional hazards, that is, $\lambda(t, \mathbf{x}_1)/\lambda(t, \mathbf{x}_2)$ does not depend on t, and the assumption of log linearity in $\mathbf{x}$ of the HRF. The statistical analysis of Cox's regression model* is based on the partial likelihood* function that in the case of no ties between the survival times is given by

$$L(\boldsymbol{\beta}) = \prod_{i=1}^{d} \left(\frac{\exp(\boldsymbol{\beta}'\mathbf{x}_i)}{\sum_{j \in R(t_i)} \exp(\boldsymbol{\beta}'\mathbf{x}_j)} \right). \qquad (6)$$

Here $\mathbf{x}_i$ is the covariate vector of the individual with survival time t_i and $R(t_i)$ is the risk set at t_i. Estimates of the parameters $\boldsymbol{\beta}$ are obtained by maximizing $L(\boldsymbol{\beta})$ and the usual type of large-sample likelihood methods also apply to partial likelihoods when the censoring satisfies the assumptions (**1**) and (**2**); see, e.g., refs. 3 and 21. The CHRF $\Lambda_0(t)$ corresponding to $\lambda_0(t)$ can be estimated by

$$\hat{\Lambda}_0(t) = \sum_{t_i \leqslant t} \left(\sum_{j \in R(t_i)} \exp(\hat{\boldsymbol{\beta}}'\mathbf{x}_j) \right)^{-1}$$

in the case of no tied survival times.

Cox's regression model can be generalized to allow for time-dependent covariates. Comprehensive reviews of the model are given by Kalbfleisch and Prentice [12] and Oakes [17].

Parametric proportional hazards models are obtained by replacing the arbitrary function $\lambda_0(t)$ by a function belonging to some parametrized family and then statistical inference is usually based on a likelihood function analogous to (4). The simplest examples arise when $\lambda_0(t)$ is replaced by a constant (exponential regression) or by a power function of time (Weibull regression).

The *accelerated failure time regression model* is conveniently introduced via the logarithm of the survival time T. The model specifies a linear relationship

$$\log T = \boldsymbol{\gamma}'\mathbf{x} + \sigma W, \qquad (7)$$

where σ is a scale parameter and W is a random variable giving the error. Various choices of the error distribution lead to regression versions of the parametric survival models already discussed. Specifically, if W has an extreme value distribution* (Gumbel distribution), a Weibull regression model is obtained, the exponential regression being a special case corresponding to $\sigma = 1$.

A lognormal regression model is obtained if W is a standard normal variate. Parametric statistical inference for the model (7) based on a likelihood function analogous to (4) has been described by Kalbfleisch and Prentice [12, Chap. 3] and Lawless [15, Chap. 6]. Nonparametric analysis of the model (7) has been developed by Prentice [18]; see also ref. 14.

The accelerated failure time regression model can alternatively be formulated via the HRF, viz.,

$$\lambda(t, \mathbf{x}) = \lambda_u[t \exp(-\boldsymbol{\gamma}'\mathbf{x})] \exp(-\boldsymbol{\gamma}'\mathbf{x}),$$

where $\lambda_u(\cdot)$ denotes the HRF of the random variable $U = \exp(\sigma W)$. The only such models that are also proportional hazards models are the exponential and the Weibull regression models.

SOME EXTENSIONS

The models mentioned so far can be thought of as describing transitions from one state "alive" to another state "dead," the transition rate* being the HRF $\lambda(t)$. Many of the methods described can be extended to situations where more than one type of transition can occur, the simplest being the competing

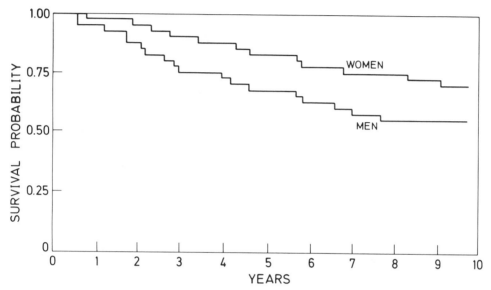

Figure 1   Kaplan–Meier plot for grouping according to sex.

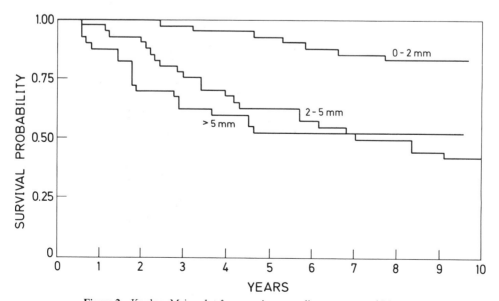

Figure 2   Kaplan–Meier plot for grouping according to tumour thickness.

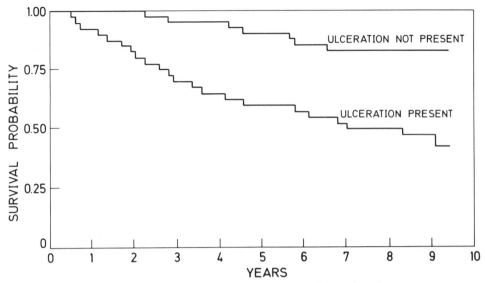

Figure 3 Kaplan–Meier plot for grouping according to ulceration.

risks* model where the state "dead" is split into, say, a states: "dead from cause 1,"...,"dead from cause a." In this context, statistical models for cause-specific HRFs can be analysed [12, Sec. 7.2]. Similarly, one can analyse hazard rate models for recurrent events (see, e.g., MULTIVARIATE COX REGRESSION MODEL).

In the survival data models described, heterogeneities between individuals have been taken into account only in the form of observable covariates $\mathbf{x}_i$. As an alternative, an unobservable random individual "frailty*" V_i can be introduced yielding models where the HRF for individual i is given by

$$\lambda_i(t) = v_i \lambda(t), \qquad (8)$$

when $V_i = v_i$ and $\lambda(t)$ is the HRF for an individual with unit frailty [11, 22]. Models of the form (8) are analogous to variance components* models and they are in several ways qualitatively different from ordinary life-table models. In particular, the population HRF $\bar{\lambda}(t)$ at age t,

$$\bar{\lambda}(t) = \lambda(t) E(V|T \geqslant t),$$

depends on the mean frailty $E(V|T \geqslant t)$ among survivors up to t, which decreases with t. Thus, the HRF for a heterogeneous population is decreasing even if the individual hazards are constant.

Table 1   Single Factor Logrank Analyses

Factor	No. of Patients n_j	No. of Deaths O_j	"Expected" No. of Deaths E_j	Logrank Test Statistic (D.F.)
Sex				
Females	126	28	37.1	6.5
Males	79	29	19.9	(1)
Tumour thickness				
0–2 mm	109	13	33.7	31.6
2–5 mm	64	30	16.4	(2)
> 5 mm	32	14	6.9	
Ulceration				
Not present	115	16	35.8	29.6
Present	90	41	21.2	(1)

Statistical methods for analysing models like (8) are discussed in refs. 10 and 11.

AN EXAMPLE

In the period 1964–1973, 205 patients with malignant melanoma had radical surgery

Table 2   Regression Analyses.

Variable x	Scoring	Cox Regression Model $\hat{\beta}$ [S.E. $(\hat{\beta})$]	Exponential Regression Model $\hat{\gamma}$ [S.E. $(\hat{\gamma})$]
1: Sex	0 = male		
	1 = female	−0.38 (0.27)	−0.36 (0.27)
2: Tumour thickness (mm)	log (tumour thickness) −0.7	0.58 (0.18)	0.53 (0.18)
3: Ulceration	0 = not present		
	1 = present	0.94 (0.32)	0.96 (0.33)

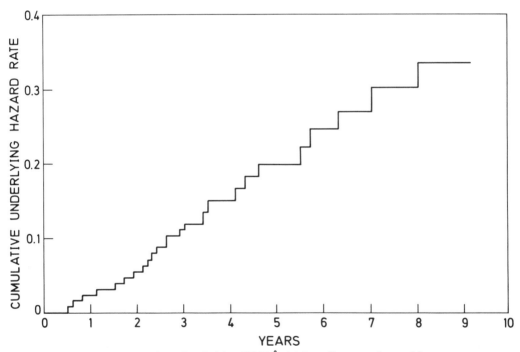

Figure 4   Estimated underlying CHRF $\hat{\Lambda}_0(t)$ from Cox regression model.

performed at the Department of Plastic Surgery, University Hospital of Odense, Denmark [8]. At the end of the follow-up period (1 January 1978), 57 of the patients had died from the disease. Among the variables registered were sex, tumour thickness, and ulceration. Figures 1–3 show the Kaplan–Meier estimates of the survival functions in groups defined by these variables. Table 1 shows the results of three logrank analyses comparing the survival distributions in the groups. It is seen that, when considered separately, each of the variables has a significant influence on survival.

The three variables were included in a Cox regression model:

$$\lambda(t) = \lambda_0(t)\exp(\beta_1 x_1 + \beta_2 x_2 + \beta_3 x_3);$$

see Table 2. Comparing $\hat{\beta}_1$ with its estimated standard error, it is seen that sex is insignificant when ulceration and tumour thickness are included in the model. The estimated correlation between the regression coefficients corresponding to sex ($\hat{\beta}_1$) and tumour thickness ($\hat{\beta}_2$) is -0.39, reflecting the fact that males tend to have thicker tumours than females. Figure 4 shows the estimated underlying CHRF $\hat{\Lambda}_0(t)$ and indicates that an exponential regression model might be appropriate, $\hat{\Lambda}_0(t)$ looking roughly linear. Table 2 also shows the estimates from a model

$$\lambda(t) = \lambda_0 \exp(\gamma_1 x_1 + \gamma_2 x_2 + \gamma_3 x_3),$$

and it is seen that the $\hat{\gamma}$'s are in good agreement with the $\hat{\beta}$'s from the Cox regression model. The estimated value of λ_0 is $\hat{\lambda}_0 = 0.031$ (yr^{-1}).

References

[1]   Aalen, O. O. (1978). *Ann. Statist.*, **6**, 701–726.

[2]   Andersen, P. K., Borgan, Ø., Gill, R. D., and Keiding, N. (1982). *Int. Statist. Rev.*, **50**, 219–258.

[3]   Andersen, P. K. and Gill, R. D. (1982). *Ann. Statist.*, **10**, 1100–1120.

[4]   Breslow, N. E. and Crowley, J. J. (1974). *Ann. Statist.*, **2**, 437–453.

[5]   Borgan, Ø. (1984). *Scand. J. Statist.*, **11**, 1–16; correction, **11**, 275.

[6]   Chhikara, R. S. and Folks, J. L. (1977). *Techometrics*, **19**, 461–468.

[7]   Cox, D. R. (1972). *J. R. Statist. Soc. B*, **34**, 187–220.

[8]   Drzewiecki, K. T. and Andersen, P. K. (1982). *Cancer*, **49**, 2414–2419.

[9]   Gill, R. D. (1980). *Censoring and Stochastic Integrals. Mathematical Centre Tracts* **124**. Mathematisch Centrum, Amsterdam, Netherlands.

[10] Heckman, J. J. and Singer, B. (1982). In *Multidimensional Mathematical Demography*, Land and Rogers, eds. Academic, New York, pp. 567–599.

[11] Hougaard, P. (1984). *Biometrika*, **71**, 75–83.

[12] Kalbfleisch, J. D. and Prentice, R. L. (1980). *The Statistical Analysis of Failure Time Data*. Wiley, New York.

[13] Kaplan, E. L. and Meier, P. (1958). *J. Amer. Statist. Ass.*, **53**, 457–481.

[14] Koul, H., Susarla, V., and Van Ryzin, J. (1981). *Ann. Statist.*, **9**, 1276–1288.

[15] Lawless, J. F. (1982). *Statistical Models and Methods for Lifetime Data*. Wiley, New York.

[16] Nelson, W. (1969). *J. Qual. Tech.*, **1**, 27–32.

[17] Oakes, D. (1981). *Int. Statist. Rev.*, **49**, 265–284.

[18] Prentice, R. L. (1978). *Biometrika*, **65**, 167–179.

[19] Ramlau-Hansen, H. (1983). *Ann. Statist.*, **11**, 453–466.

[20] Stacy, E. W. (1962). *Ann. Math. Stat.*, **33**, 1187–1192.

[21] Tsiatis, A. A. (1981). *Ann. Statist.*, **9**, 93–108.

[22] Vaupel, J. W., Manton, K. G., and Stallard, E. (1979). *Demography*, **16**, 439–454.

[23] Zelen, M. and Dannemiller, M. (1961). *Technometrics*, **3**, 29–49.

Bibliography

Andersen, P. K., Borgan, Ø., Gill, R. D., and Keiding, N. (1982). *Int. Statist. Rev.*, **50**, 219–258. (Review paper on nonparametric one- and *k*-sample tests for survival data.)

Cox, D. R. and Oakes, D. (1984). *Analysis of Survival Data*. Chapman and Hall, London and New York.

Elandt-Johnson, R. C. and Johnson, N. L. (1980). *Survival Models and Data Analysis*. Wiley, New York. (Gives a broad description of several aspects of survival data models, in particular various versions of the life table.)

Gross, A. J. and Clark, V. A. (1975). *Survival Distributions: Reliability Applications in the Biomedical Sciences*. Wiley, New York. (The presentation focuses on parametric survival models and maximum likelihood inference for these models.)

Kalbfleisch, J. D. and Prentice, R. L. (1980). *The Statistical Analysis of Failure Time Data*. Wiley, New York. (A comprehensive account on regression models for survival data with medical applications. Describes parametric, semiparametric, as well as nonparametric inference procedures. Several data sets and computer programs are included.)

Lawless, J. F. (1982). *Statistical Models and Methods for Lifetime Data*. Wiley, New York. (The book gives a thorough unified treatment of the statistical models and methods used in the analysis of life time data with particular emphasis on parametric inference.)

Lee, E. T. (1980). *Statistical Methods for Survival Data Analysis*. Lifetime Learning Publications, Belmont, Calif. (A comprehensive introduction to and review of the field. Several examples with data, illustrations, and computer programs makes the book particularly suitable for applied statisticians.)

Mann, N. R., Schafer, R. E., and Singpurwalla, N. D. (1974). *Methods for Statistical Analysis of Reliability and Lifetime Data*. Wiley, New York. (Mainly statistical methods relevant for reliability applications. Contains much material on different inference procedures in the Weibull model.)

Miller, R. G. (1981). *Survival Analysis*. Wiley, New York. (A set of lecture notes that in a condensed form covers a lot of material.)

Nelson, W. (1982). *Applied Life Data Analysis*. Wiley, New York. (Probably the main reference for technical applications.)

Oakes, D. (1981). *Int. Statist. Rev.*, **49**, 265–284. (Review paper on solved and unsolved problems in connection with the proportional hazards regression models.)

Peto, R., Pike, M. C., Armitage, P., Breslow, N. E., Cox, D. R., Howard, V., Mantel, N., McPherson, K., Peto, J., and Smith, P. G. (1976, 1977). *Br. J. Cancer*, **34**, 585–612; **35**, 1–39. (A nontechnical, but thorough account on the use of Kaplan–Meier plots and logrank tests in the analysis of survival data especially from clinical trials.)

Acknowledgment

This research was supported by the Danish Medical Research Council and the Danish Social Science Research Council.

(CENSORED DATA
CLINICAL TRIALS
COX'S REGRESSION MODEL
DEMOGRAPHY
FOLLOW UP
FRAILTY
HAZARD RATE CLASSIFICATION OF
    DISTRIBUTIONS
KAPLAN–MEIER ESTIMATOR
LIFE TESTING
PARTIAL LIKELIHOOD
RELIABILITY, PROBABILISTIC
SURVIVAL ANALYSIS, GROUPED DATA)

PER KRAGH ANDERSEN
MICHAEL VÆTH

SURVIVAL ANALYSIS, GROUPED DATA

There are many types of investigations in which data for life times (or perhaps waiting times until certain events) are expressed as classifications into a set of mutually exclusive intervals rather than as specific values. Some examples that illustrate such grouped survival data are as follows:

E1. An experiment is undertaken to compare treatments for survival of bacteria infected mice. The mice are inspected every 6 h for the event of death; see Bowdre et al. [2].

E2. Patients for whom a health disorder was recently treated successfully are evaluated for recurrence by diagnostic procedures at a specific set of follow-up times; see Johnson and Koch [14] and Example 3 of POISSON REGRESSION.

E3. Data from a large study of graft survival after kidney transplant operations are summarized for a cross-classification of donor relationship and match grade in life tables* with a specific set of intervals; see Laird and Olivier [18].

E4. A large study of female patients with cancer of the cervix uteri has its data summarized in a life table that encompasses deaths due to cancer of the cervix uteri and deaths due to other causes for a specific set of intervals; see Chiang [4] and Example 12.1 in Elandt-Johnson and Johnson [8].

Grouped survival data arise in studies like **(E1)** and **(E2)** because of the periodic monitoring of subjects in their research designs. This method of observation only provides information on whether the event of interest occurred between two follow-up assessments rather than the exact life time. However, for many situations, it is the only feasible method with respect to ethical or resource considerations. Some additional examples for which periodic monitoring is used are: the detection of health outcomes (e.g., abnormal electrocardiogram or failure of dental restorations), the emergence of certain psychological characteristics (e.g., memory skills), or the inspection of equipment for maintenance purposes.

For studies like **(E3)** and **(E4)**, the grouped survival data occur as a consequence of the structure from life tables. Relatedly, when sample sizes are large, a life table can arise from the aggregation* for the level of measurement which is used for actual life times (e.g., times of death expressed in months instead of days). In other words, the concept of grouped survival data applies to situations where a continuous monitoring process yields many ties due to roundoff error. Another consideration is that survival times may be determined in different ways for two or more potential causes of failure (e.g., in days for death and in 3-month intervals for recurrence), and so a life table provides a convenient format for summarizing this information in a common way. The discussion here has emphasized how life tables result from primary survival data for subjects, but often they serve as the only available data source for secondary analyses.

As is often the case for studies of life times, grouped survival information can involve censored data*. For example, in a clinical trial* (E2), a subject might be lost to follow-up* or withdrawn from risk at its termination date before experiencing recurrence. There also can be vectors of explanatory variables for subjects with grouped survival data; see (E3). These might refer to treatment group, demographic characteristics, or medical history status. They can be either continuous or categorical. Continuous explanatory variables are often grouped into a finite number of categories for situations with grouped survival data. Finally, in some studies, there might be several competing risks* as potential causes of failure; see (E4).

DATA STRUCTURE

Since grouped survival data are based on classifications of the follow-up status of subjects with respect to the interval of occurrence of some event, two general formats for their display are contingency tables* and life tables. For the contingency table, there are s rows and $(rc + 1)$ columns. The s rows correspond to s distinct samples of subjects and the $(rc + 1)$ columns encompass the potential outcomes of final observation during one of r time intervals due to one of c causes or the maintenance of survival of all causes through the entire follow-up period. The entries in the contingency table are the

frequencies $\{n_{ijk}\}$, where $i = 1, 2, \ldots, s$ indexes the set of samples, $j = 1, 2, \ldots, r$ indexes the successive time intervals $(y_{j-1}, y_j]$ with $y_0 = 0$, and $k = 1, 2, \ldots, c$ indexes the causes for the termination of observation (e.g., death due to cause of interest, death due to some other cause, lost to follow-up, etc.). Also, the combination $j = r$, $k = 0$ denotes survival of all causes for all time intervals. The size of the ith sample is

$$n_i = \sum_{j=1}^{r} \sum_{k=1}^{c} n_{ijk} + n_{ir0}.$$

Alternatively, for each of the s samples, a life table can be constructed from the corresponding row of the contingency table. It has r rows from the respective time intervals and $(c + 1)$ columns for the c causes and the outcome of survival of all causes during the time interval of the row; thus, the entries in its first c columns are the $\{n_{ijk}\}$ and those in its last column are the numbers of subjects

$$n_{ij0} = \sum_{j'=j+1}^{r} \sum_{k=1}^{c} n_{ij'k} + n_{ir0} \qquad (1)$$

who survive all causes through the end of the jth interval (i.e., final observation occurs after the jth interval).

Most applications involve only two causes for termination of observation. One $(k = 1)$ corresponds to the occurrence of a failure event such as death or recurrence of a disorder, and the other $(k = 2)$ corresponds to a withdrawal from risk (i.e., censoring) event such as lost to follow-up or violations of research design specifications (e.g., usage of supplementary treatments). Also, it is usually assumed that the withdrawal events are unrelated to the failure events; *see* LIFE TESTING and SURVIVAL ANALYSIS. Subsequent discussion is primarily concerned with this specific type of situation. Further consideration of the more general framework with c causes is given in Chiang [5], Elandt-Johnson and Johnson [8], Gail [10], Johnson and Koch [14], Larson [19], and COMPETING RISKS.

STATISTICAL METHODS

The main strategies for analyzing grouped survival data are nonparametric methods and model fitting methods based on maximum likelihood* and weighted least squares*. When only minimal assumptions (e.g., random sampling, data integrity) apply, a nonparametric approach provides useful ways for constructing descriptive estimates and hypothesis tests concerning the observed experience of the respective groups. For example, alternative estimates include the following functions of the frequencies $\{n_{ijk}\}$ in life table format for $c = 2$ causes:

(i) The ratio for the number of failures per unit time

$$f_i = \sum_{j=1}^{r} n_{ij1} \bigg/ \sum_{j=1}^{r} N_{ij}. \qquad (2)$$

Here $N_{ij} = (n_{ij0} + 0.5n_{ij1} + 0.5n_{ij2})(y_j - y_{j-1})$ denotes the total exposure time for which subjects in the ith sample have risk of failure during the jth interval under the approximation that failures and withdrawals occur, on average, at the midpoints of their intervals.

(ii) The actuarial survival rates through the end of the jth interval

$$G_{ij} = \prod_{j'=1}^{j} g_{ij'}, \qquad (3)$$

where

$$g_{ij'} = \frac{n_{ij'0} + 0.5n_{ij'2}}{n_{ij'0} + n_{ij'1} + 0.5n_{ij'2}}.$$

(iii) The actuarial hazard function (or failure rate) for the jth interval

$$h_{ij} = \frac{2(1 - g_{ij})}{(y_j - y_{j-1})(1 + g_{ij})}. \qquad (4)$$

The significance of the association of survival with groups can be evaluated through censored data rank tests. In this regard, the log-rank* test is widely used. For the case of $s = 2$ samples, it can be computed as the Mantel–Haenszel statistic* for the set of (2×2) tables

	Failure	Not failure	Total
Sample 1	n_{1j1}	$n_{1j0} + n_{1j2}$	$n_{1, j-1, 0}$
Sample 2	n_{2j1}	$n_{2j0} + n_{2j2}$	$n_{2, j-1, 0}$
Total	n_{+j1}	$n_{+j0} + n_{+j2}$	$n_{+, j-1, 0}$

$$(5)$$

corresponding to the r intervals. A more specific expression is

$$Q_C = \left\{ \sum_{j=1}^{r} (n_{1j1} - m_{1j1}) \right\}^2 \bigg/ \left\{ \sum_{j=1}^{r} v_j \right\}, \qquad (6)$$

where $m_{1j1} = \{n_{1, j-1, 0}n_{+j1}/n_{+, j-1, 0}\}$ is the expected value of n_{1j1} and

$$v_j = \frac{m_{1j1}n_{2, j-1, 0}(n_{+j0} + n_{+j2})}{n_{+, j-1, 0}(n_{+, j-1, 0} - 1)} \qquad (7)$$

is its variance under the hypothesis of no association between survival and sample. When the sample sizes $\{n_i\}$ are moderately large (e.g., ≥ 20 with ≥ 10 failures), the log-rank statistic Q_C approximately has the chi-square distribution with 1 degree of freedom. The log-rank statistic for the comparison of s samples is the extension of (6) to a quadratic form with $(n_{1j1} - m_{1j1})$ replaced by the vector for a set of $(s - 1)$ samples and with v_j replaced by the corresponding covariance matrix; this statistic approximately has the chi-square distribution with $(s - 1)$ degrees of freedom. When the samples are cross-classified with strata based on one or more explanatory variables, the computation of Q_C becomes based on the combined set of (2×2) tables in (5) for all intervals within all strata. For additional discussion of log-rank tests, see CHI-SQUARE TESTS and LOG-RANK SCORES, STATISTICS, AND TESTS; for more general consideration of censored data rank tests, see Peto and Peto [20], Prentice [21], and Prentice and Marek [23].

For sufficiently large sample sizes, functions of the life table frequencies $\{n_{ijk}\}$ such as **(i)**, **(ii)**, or **(iii)** approximately have multivariate normal distributions. Also, consistent estimates for the corresponding covariance matrix can be based on linear Taylor series methods (*see* STATISTICAL DIFFERENTIALS). On this basis, weighted least squares procedures can be used to fit linear models which describe the variation among these functions; the estimated parameters for such models approximately have multivariate normal distributions, and Wald statistics* for goodness of fit approximately have chi-square distributions. Additional discussion of this type of methodology is given in CHI-SQUARE TESTS. Applications are described for the actuarial survival rates (ii) in Koch et al. [16], for the actuarial hazard function (iii) in Gehan and Siddiqui [11], and for functions from a competing risks framework in Johnson and Koch [14].

Maximum likelihood* (ML) methods enable the fitting of models with an assumed structure for the underlying distribution of the grouped survival data. A general specification of such models is

$$\pi_{ijk} = \psi_{ijk}(\boldsymbol{\theta}, \mathbf{x}_{ij}), \qquad (8)$$

where the $\{\pi_{ijk}\}$ denote the probabilities of the (j, k)th outcome for a randomly obtained observation from the ith population, $\mathbf{x}_{ij}$ is a $(u \times 1)$ vector of explanatory variables for the ith sample during the jth time interval, and $\boldsymbol{\theta}$ is a vector of unknown parameters. The ML estimates $\hat{\boldsymbol{\theta}}$ are obtained by substituting (8) into the product multinomial likelihood

$$\phi(\{n_{ijk}\} | \{\pi_{ijk}\})$$

$$= \prod_{i=1}^{s} n_i! \frac{\pi_{ir0}^{n_{ir0}}}{n_{ir0}!} \left[\prod_{j=1}^{r} \prod_{k=1}^{2} \frac{\pi_{ijk}^{n_{ijk}}}{n_{ijk}!} \right] \quad (9)$$

for the contingency table format of the data and then maximizing the resulting function; this can be done by solving the equations

$$\frac{\partial}{\partial \boldsymbol{\theta}} \left\{ \log_e \left[\phi(\{n_{ijk}\} | \boldsymbol{\theta}) \right] \right\} = \mathbf{0} \quad (10)$$

for $\hat{\boldsymbol{\theta}}$ by iterative methods.

A framework for which the determination of ML estimates is relatively straightforward involves the assumption that the time until the failure event has a piecewise exponential distribution; i.e., for each of the s samples,

there are independent exponential distributions with hazard parameters $\{\lambda_{ij}\}$ for the respective time intervals. Some additional assumptions which are usually made for this model are

(a) Withdrawal events occur uniformly in the respective intervals and their censoring process is unrelated to failure events in the noninformative sense of Lagakos [17].

(b) The within-interval probabilities of failure $(\pi_{ij1}/\pi_{i, j-1,0})$ are small, where

$$\pi_{i, j-1,0} = \sum_{j'=j}^{r} \sum_{k=1}^{2} \pi_{ij'k} + \pi_{ir0}$$

denotes the probability of surviving all causes through the end of the $(j - 1)$th interval.

From the conditions specified here, it follows that maximizing (9) is approximately equivalent to maximizing the piecewise exponential likelihood function

$$\phi_{PE} = \prod_{i=1}^{s} \prod_{j=1}^{r} \lambda_{ij}^{n_{ij1}} \left\{ \exp(-\lambda_{ij} N_{ij}) \right\}. \quad (11)$$

For the likelihood (11), the relationship of the failure event to the explanatory variables is specified through models for the $\{\lambda_{ij}\}$. A useful model for many applications has the log-linear specification

$$\lambda_{ij} = \exp(\mathbf{x}'_{ij} \boldsymbol{\theta}). \quad (12)$$

A convenient feature of (12) is that the ML estimates $\hat{\boldsymbol{\theta}}$ for its parameters can be readily obtained by Poisson regression* computing procedures for fitting log-linear models to sets of observed counts or contingency tables; these include both Newton–Raphson* procedures and iterative proportional fitting*. Additional discussion of statistical methods for piecewise exponential* models is given in Aitkin and Clayton [1], Frome [9], Holford [12, 13], Laird and Olivier [18], Whitehead [24], and POISSON REGRESSION.

The simplification of the log-linear model (12) to

$$\lambda_{ij} = \exp(\eta_j + \mathbf{x}'_i \boldsymbol{\beta}) \quad (13)$$

has *proportional hazards* structure; here $\exp(\eta_j)$ denotes the constant value of the hazard function within the jth interval for a reference population with $\mathbf{x} = \mathbf{0}$, and $\boldsymbol{\beta}$ is the vector of parameters for the relationship of the hazard function for the ith population with its explanatory variables $\mathbf{x}_i$. The more general formulation in Cox [6] of the proportional hazards model is

$$h(y, \mathbf{x}) = h_0(y)\{\exp(\mathbf{x}'\boldsymbol{\beta})\}, \quad (14)$$

where y denotes continuous time and $h_0(y)$ is the hazard function for the reference population. When there are no ties, the maximizing of the partial likelihood* of Cox [7] with respect to $\boldsymbol{\beta}$ is computationally straightforward, and the resulting estimator $\hat{\boldsymbol{\beta}}$ has the usual ML properties; *see* COX'S REGRESSION MODEL. However, for grouped survival data and other situations with many ties, modified strategies are necessary. One approach is to work with a piecewise exponential model like (13) and use Poisson regression computing procedures. Additional discussion of methods for dealing with ties in analyses with the proportional hazards model is given in Breslow [3], Kalbfleisch and Prentice [15], and Prentice and Gloeckler [22].

EXAMPLES

The data in Table 1 are from a study of Bowdre et al. [2] to compare treatments for mice infected with *Vibrio vulnificus* bacteria. The survival status of each mouse was assessed at 6, 12, 18, 24, 30, 36, 48, 60, 72, and 96 h. The numbers of deaths per hour $\{f_i\}$ from (2) are

$$f_1 = 0.058 \quad \text{for carbenicillin,}$$

$$f_2 = 0.023 \quad \text{for cefotaxime.}$$

The log-rank (Mantel–Haenszel) statistic $Q_C = 4.52$ from (6) indicates that the survival experience of the cefotaxime group is significantly better.

An example is given in POISSON REGRESSION for surgically treated duodenal ulcer

Table 1   Survival Status of Mice

Interval for	Number of Deaths	
Death (in hours)	Carbenicillin	Cefotaxime
0–6	1	1
6–12	3	1
12–18	5	1
18–24	1	0
24–30	1	2
30–36	0	0
36–48	0	2
48–60	1	1
60–72	0	0
72–96	0	1
Alive at 96	0	1
Total	12	10

patients who were evaluated at 6, 24, and 60 months for recurrence. The data are summarized in life table format. A piecewise exponential model is used and its parameters are estimated by maximum likelihood. Another example for the piecewise exponential model is the graft survival data (E3) that is discussed in Laird and Olivier [18]. Weighted least squares* methods are illustrated in refs. 11, 14, and 16; applications include the fitting of Weibull* and other probability distributions to life tables.

References

[1] Aitkin, M. and Clayton, D. (1980). *J. R. Statist. Soc. C*, **29**, 156–163.

[2] Bowdre, J. H., Hull, J. H., and Cocchetto, D. M. (1983). *J. Pharm. Exp. Therapeutics*, **225**, 595–598.

[3] Breslow, N. E. (1974). *Biometrics*, **30**, 89–99.

[4] Chiang, C. L. (1961). *Biometrics*, **17**, 57–78.

[5] Chiang, C. L. (1968). *Introduction to Stochastic Processes in Biostatistics*. Wiley, New York.

[6] Cox, D. R. (1972). *J. R. Statist. Soc B*, **34**, 187–220.

[7] Cox, D. R. (1975). *Biometrika*, **62**, 269–276.

[8] Elandt-Johnson, R. C. and Johnson, N. L. (1980). *Survival Models and Data Analysis*. Wiley, New York.

[9] Frome, E. L. (1983). *Biometrics*, **39**, 665–674.

[10] Gail, M. H. (1975). *Biometrics*, **31**, 209–222.

[11] Gehan, E. A. and Siddiqui, M. M. (1973). *J. Amer. Statist. Ass.*, **68**, 848–856.

[12] Holford, T. R. (1976). *Biometrics*, **32**, 587–597.

[13] Holford, T. R. (1980). *Biometrics*, **36**, 299–306.

[14] Johnson, W. D. and Koch, G. G. (1978). *Int. Statist. Rev.*, **46**, 21–51.

[15] Kalbfleisch, J. D. and Prentice, R. L. (1980). *Statistical Analysis of Failure Time Data*. Wiley, New York.

[16] Koch, G. G., Johnson, W. D., and Tolley, H. D. (1972). *J. Amer. Statist. Ass.*, **67**, 783–796.

[17] Lagakos, S. W. (1979). *Biometrics*, **35**, 139–156.

[18] Laird, N. and Olivier, D. (1981). *J. Amer. Statist. Ass.*, **76**, 231–240.

[19] Larson, M. G. (1984). *Biometrics*, **40**, 459–469.

[20] Peto, R. and Peto, J. (1972). *J. R. Statist. Soc. A*, **135**, 185–206.

[21] Prentice, R. L. (1978). *Biometrika*, **65**, 167–179.

[22] Prentice, R. L. and Gloeckler, L. A. (1978). *Biometrics*, **34**, 57–67.

[23] Prentice, R. L. and Marek, P. (1979). *Biometrics*, **35**, 861–867.

[24] Whitehead, J. (1980). *Appl. Statist.*, **29**, 268–275.

Acknowledgments

The authors would like to thank J. H. Bowdre, J. H. Hull, and D. M. Cocchetto for permission to use the data in Table 1. They also would like to express appreciation to Amy Goulson and Ann Thomas for editorial assistance. This research was partially supported by the U.S. Bureau of the Census through Joint Statistical Agreement JSA 84-5.

(ACTUARIAL STATISTICS—LIFE
CENSORED DATA
CHI-SQUARE TESTS
CLINICAL TRIALS
COMPETING RISKS
CONTINGENCY TABLES
COX'S REGRESSION MODEL
GROUPED DATA
LIFE TABLES
LIFE TESTING
LOG-RANK SCORES, STATISTICS, AND
  TESTS
MANTEL–HAENSZEL STATISTIC
POISSON REGRESSION
SURVIVAL ANALYSIS)

James A. Deddens
Gary G. Koch

SUSARLA–VAN RYZIN ESTIMATOR

This is a nonparametric estimator of the expected value of a continuous life distribution* with survival function (SF) $S(x)$, based on censored survival data $Z_1, Z_2, \ldots, Z_n; \delta_1, \delta_2, \ldots, \delta_n$, where

$$Z_i = \min(X_i, Y_i),$$

$$\delta_i = \begin{cases} 1, & X_i \leqslant Y_i, \\ 0, & X_i > Y_i, \end{cases} \quad i = 1, 2, \ldots, n,$$

with all X's and Y's mutually independent with SFs $S(x)$ and $G(y)$, respectively. (In the context of survival data, X_i can be regarded as representing time of death and Y_i as time of censoring for the ith individual.) The Susarla–van Ryzin estimator is

$$\hat{\mu} = n^{-1} \sum_{i=1}^{n} \delta_i Z_i \{\hat{G}(Z_i)\}^{-1},$$

where

$$\hat{G}(t) = \begin{cases} \prod_{Z_i \leqslant t} \left[\{N(Z_i) - 1\} \{N(Z_i)\}^{-1} \right]^{1-\delta_i}, \\ \qquad \text{if } t \leqslant \max(Z_1, \ldots, Z_n), \\ 0, \qquad \text{if } t > \max(Z_1, \ldots, Z_n), \end{cases}$$

$$(1)$$

and $N(t)$ denotes the number of individuals surviving up to time t. [So that $N(Z_i)$ is the *risk set* at time Z_i—see KAPLAN–MEIER ESTIMATOR.]

The limiting distribution (as $n \to \infty$) of $(\hat{\mu} - \mu)\sqrt{n}$ is normal with expected value zero and variance

$$-\int_0^\infty \left\{ \int_t^\infty S(x)\, dx \right\}^2$$
$$\times \{S(t)\}^{-2} \{G(t)\}^{-1}\, dS(t).$$

The estimator was proposed in ref. 2. It coincides with Buckley and James' estimator [1] for parameters in a linear regression* model in the special case of the expected value (see, e.g., Susarla et al. [3]).

References

[1] Buckley, J. and James, I. (1979). *Biometrika*, **66**, 427–436.

[2]   Susarla, V. and Van Ryzin, J. (1980). *Ann. Statist.*, **8**, 1001–1016.

[3]   Susarla, V., Tsai, W. Y., and Van Ryzin, J. (1984). *Biometrika*, **71**, 624–625.

(CENSORED DATA
KAPLAN–MEIER ESTIMATOR
SURVIVAL ANALYSIS)

SUSPECT OBSERVATIONS   *See* OUT-LIERS

SWAIN–FU DISTANCE

The Swain–Fu distance is a measure of "distance" or difference between two distributions. It is relatively simple to calculate, depending only on the first and second moments of the distributions [2].

The Swain–Fu distance can be interpreted geometrically as follows (see Fig. 1). Let P_1 and P_2 be n-dimensional distributions, and let μ_i, and Σ_i be the mean vector and covariance matrix, respectively, associated with distribution P_i ($i = 1, 2$). Let D_1 be the distance along the direction ($\mu_2 - \mu_1$) from μ_1 to the surface of the ellipsoid of concentration [1] for the distribution P_1, and let D_2 be the corresponding distance for P_2. The Swain–Fu distance $\mathscr{D}_{12}$ between P_1 and P_2

is defined to be

$$\mathscr{D}_{12} = \frac{|\mu_1 - \mu_2|}{D_1 + D_2}.$$

The distance D_i ($i = 1, 2$) previously described geometrically can be calculated by the formula [3]

$$D_i = \left\{ \frac{|\mu_1 - \mu_2|^2 (n + 2)}{\mathrm{tr}\left[\Sigma_i^{-1}(\mu_1 - \mu_2)(\mu_1 - \mu_2)'\right]} \right\}^{1/2}.$$

The ellipsoid of concentration for a distribution P is the ellipsoid over which a uniform distribution has the same first and second moments as the distribution P. In a sense it provides a generic characterization of the location and dispersion of the distribution, without accounting for its skewness or higher order effects. Wacker [3] showed that the Swain–Fu distance $\mathscr{D}$ and the divergence J are related by

$$\mathscr{D} \le \frac{1}{2}\sqrt{\frac{J}{n + 2}}.$$

References

[1]   Cramér, H. (1946). *Mathematical Methods of Statistics*, Princeton Univ. Press, Princeton, N.J.

[2]   Swain, P. H. (1970). Nonparametric and Linguistic Approaches to Pattern Recognition Ph.D. Thesis, Purdue University, W. Lafayette, IN.

[3]   Wacker, A. G. (1971). The Minimum Distance Approach to Classification, Ph.D. Thesis, Purdue University, W. Lafayette, IN.

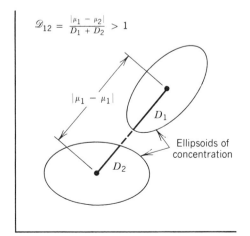
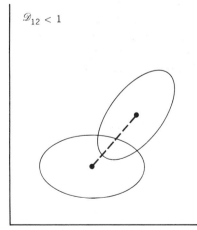

Figure 1   Two-dimensional illustration of the Swain–Fu distance.

(DIVERSITY INDICES
MAHALANOBIS D^2
MEASURES OF SIMILARITY,
   DISSIMILARITY, AND DISTANCE)

P. H. SWAIN

SWAMPING   *See* MASKING AND SWAMP-ING

SWING

In cooperative simple *n*-person games*, a coalition containing player *i* is called a *swing* with regard to this player if the coalition is winning but becomes nonwinning when *i* is removed. Evidently a winning coalition may provide a swing for more than one player or no player at all.

(GAME THEORY
POWER INDEX OF A GAME
SIMPLE *n*-PERSON GAME)

SWITCH-BACK DESIGNS   *See* REVERSAL DESIGNS

SWITCHING RULES

In quality control literature, switching rules are guidelines within a sampling scheme for shifting from one sampling plan to another, as experience indicates, on the basis of the demonstrated quality history. These rules are an essential part of many acceptance inspection* schemes.

(QUALITY CONTROL, STATISTICAL
SAMPLING PLANS)

SYLVESTER MATRIX

A Sylvester matrix is a Hadamard matrix* of order 2^s, its rows forming a set of 2^s point Walsh functions*. These matrices occur in coding theory*, signal processing*, and sampling.

SYMBOLIC CALCULUS   *See* FINITE DIF-FERENCES, CALCULUS OF

SYMBOLIC SCATTALLY

This is a method of plotting two integer variables with many duplications of pairs of values. It is both a scattergram* and a tally*. The type of tally used (as described in ref. 1 is shown in Fig. 1. As an example, the data in Table 1 would plot as shown in Fig. 2. If still a third variable is to be represented, different symbols are used for the points to indicate the different values of this third variable. This is a *symbolic scattally*. An ex-

Table 1

x	0	0	0	0	0	1	1	2	2	0	1	0	3	0
y	1	1	1	0	0	2	2	1	2	2	2	1	2	1

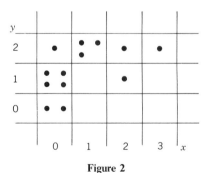

Figure 2

Figure 1

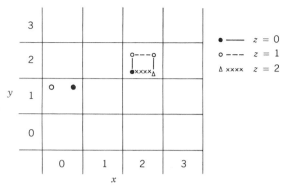

Figure 3

Table 2

x	0	2	2	2	2	2	0	2	2	2	2
y	1	2	2	2	2	2	1	1	2	2	2
z	1	1	1	0	2	0	0	1	1	1	2

ample is shown in Fig. 3, from data in Table 2.

Reference

[1]   Capobianco, M. F. (1985). Private communication.

(GRAPHICAL REPRESENTATION OF
   DATA)

SYMMETRIC DIFFERENCE

IN SET THEORY

The symmetric difference of two sets A and B is the set $A \triangle B$ composed of those elements which belong to *just one* of the two sets. In symbols,

$$A \triangle B = (A \cap \overline{B}) \cup (\overline{A} \cap \overline{B}).$$

The $\triangle$ operator is both commutative and associative.

IN FINITE DIFFERENCE CALCULUS

The *first symmetric difference* of a function $f(x)$ at point x_1, x_2 is

$$(x_1, x_2) = \{ f(x_2) - f(x_1) \}/(x_2 - x_1).$$

The *second symmetric difference* at points x_1, x_2, x_3 is

$$(x_1, x_2, x_3)$$
$$= \{ (x_2, x_3) - (x_1, x_2) \}/(x_3 - x_1)$$
$$= \{ (x_1, x_3) - (x_1, x_2) \}/(x_3 - x_2)$$
$$= \{ (x_2, x_3) - (x_1, x_3) \}/(x_2 - x_1).$$

Note that

$$(x_1, x_2, x_3) = (x_2, x_1, x_3) = (x_2, x_3, x_1)$$
$$= (x_3, x_2, x_1) = (x_3, x_1, x_2)$$
$$= (x_1, x_3, x_2);$$

that is, (x_1, x_2, x_3) is a *symmetric* function of x_1, x_2, and x_3.

Symmetric differences are also called *divided differences*.

(FINITE DIFFERENCES, CALCULUS OF
UNION OF SETS)

SYMMETRIC FUNCTIONS

In general, a function $g(x_1, \ldots, x_k)$ of k variables is *symmetric* if its value is unchanged by any reordering of the variables. For example, with $k = 3$, $g(x_1, x_2, x_3)$ is symmetric if and only if

$$g(x_1, x_2, x_3) = g(x_2, x_1, x_3)$$
$$= g(x_2, x_3, x_1)$$
$$= g(x_1, x_3, x_2)$$
$$= g(x_3, x_1, x_2)$$
$$= g(x_3, x_2, x_1).$$

In statistical applications we are mostly concerned with *monomial symmetric functions* (msfs) which are sums of terms

$$\prod_{j=1}^{m} \left\{ \prod_{g=1}^{\lambda_j} x_{n_{j-1}+g}^{p_j} \right\},$$

(where $n_j = \sum_{i=1}^{j} \lambda_i$, $n_0 = 0$, $\sum_{j=1}^{m} \lambda_j \leqslant k$, with $p_1 > p_2 > \cdots > p_k > 0$ for positive integers $p_1, \ldots, p_k$) taken over all possible selections of distinct subsets of sizes λ_1, $\lambda_2, \ldots, \lambda_m$ from the k x's, yielding

$$k! \left[\left(k - \sum_{j=1}^{m} \lambda_j \right)! \prod_{j=1}^{m} \lambda_j! \right]^{-1}$$

terms in all. This msf corresponds to the partition

$$\left(p_1^{\lambda_1} p_2^{\lambda_2} \cdots p_m^{\lambda_m} \right)$$

of the number $\sum_{i=1}^{m} \lambda_i p_i$, called the *weight* of the partition, and is conventionally represented by the same symbols. The related *augmented* msfs

$$\left[p_1^{\lambda_1} p_2^{\lambda_2} \cdots p_m^{\lambda_m} \right] = \left\{ \sum_{j=1}^{m} \lambda_j! \right\}$$
$$\times \left(p_1^{\lambda_1} p_2^{\lambda_2} \cdots p_m^{\lambda_m} \right)$$

are often more convenient to use.

In statistical theory, msfs are especially useful in calculations relating to moments* and cumulants*; *see* POLYKAYS. Of special interest are:

(a) The *power sums* $s_r = (r) = \sum_{i=1}^{k} x_i^r$.

(b) The *unitary functions* $a_r = (1^r) = \sum_{i_1 < \cdots < i_r}^{k-r+1} \cdots \sum^{k} \prod_{g=1}^{r} x_{i_g}$.

(c) The *homogeneous product sums* $h_r = \sum (p_1^{\lambda_1} \cdots p_k^{\lambda_k})$, where the summation is over all partitions of weight r. For example,

$$h_3 = (3) + (21) + (1^3).$$

The generating functions* of these functions are, respectively,

$$S(t) = \sum_{r=1}^{\infty} r^{-1} s_r t^r = \sum_{j=1}^{k} \log(1 - tx_j),$$

$$A(t) = \sum_{r=0}^{k} a_r t^r = \prod_{j=1}^{k} (1 + tx_j),$$

$$H(t) = \sum_{r=1}^{\infty} h_r t^r = \left\{ \prod_{j=1}^{k} (1 - tx_j) \right\}^{-1}$$
$$= \exp(-S(t)).$$

Relationships among the functions include:

$$a_r = (r!)^{-1} \times \begin{vmatrix} s_1 & 1 & 0 & \cdots & 0 \\ s_2 & s_1 & 2 & \cdots & 0 \\ \vdots & \vdots & \vdots & & \vdots \\ s_r & s_{r-1} & s_{r-2} & \cdots & r \end{vmatrix}$$

and

$$h_r = (-1)^r \sum_{g=1}^{r} (-1)^g g! \sum_{(\lambda)_g} \prod_{j=1}^{g} \left(\frac{a_{p_j}^{\lambda_j}}{\lambda_j!} \right),$$

where summation over $(\lambda)_g$ means summation over all g-part partitions of weight r.

David et al. [1] give tables expressing msfs, power sums, unitary functions, and homogeneous product sums in terms of each other for all pairs (except power sums and homogeneous product sums) for weights up to and including 12. They also give useful accounts of applications of these functions.

Bipartite symmetric functions are related to polynomials in two sets of variables $(x_1, \ldots, x_k)$ and $(y_1, \ldots, y_k)$. In particular the bipartite power sums are

$$s_{uv} = \sum_{i=1}^{k} x_i^u y_i^v \qquad (s_{u0} = s_u)$$

and the bipartite unitary functions are

$$a_{uv} = ((10)^u, (01)^v)$$
$$= \sum \cdots \sum \left(\prod_{j=1}^{u} x_{i_j} \right) \left(\prod_{g=1}^{v} y_{i_{u+g}} \right),$$

summation being over all sets of suffixes

$\{i_j\}$ with

$$i_1 < i_2 < \cdots < i_u;$$
$$i_{u+1} < i_{u+2} < \cdots < i_{u+v}.$$

A basic text is MacMahon [3]. Some more mathematical properties of symmetric functions and uses are described in ref. 2.

References

[1] David, F. N., Kendall, M. G., and Barton, D. E. (1966). *Symmetric Functions and Allied Tables*. Cambridge University Press, London, England.

[2] Macdonald, I. G. (1979). *Symmetric Functions and Hall Polynomials*. Clarendon Press, Oxford, England.

[3] MacMahon, P. A. (1960). *Combinatory Analysis*, 2 vols. Chelsea, New York. (Originally published in 1915, 1916).

(CUMULANTS
FAÀ DI BRUNO'S FORMULA
GENERATING FUNCTIONS
POLYKAYS
POWER PRODUCT SUMS)

SYMMMETRIC MEANS *See* ANGLE BRACKETS

SYMMETRY TESTS, PURE AND COMBINED

This article discusses distribution-free* tests for the hypothesis H_0 of symmetry about zero of independent, but not necessarily identically distributed random variables, based on (1) the sign test* statistic B, (2) some other test statistic V^+ for symmetry about zero, or (3) some combination of these. Let the random variables whose values turn out to take on nonzero values be denoted by $Z_1, \ldots, Z_n$; all tests will be conducted under the condition that exactly n of the random variables fulfill this condition. For any n numbers $b_1, \ldots, b_n$, let $\rho_n(b_i)$ denote the midrank of b_i among the b's. Let $I(E)$ be 1 when the event E occurs and 0 otherwise, and let $\zeta_c(b_i) = I(b_i > c)$. The entry TESTING FOR SYMMETRY mentions for V^+ the signed-rank statistic T^+, which is the Wilcoxon two-sample test statistic T applied to the positive Z_i and the absolute values of the negative Z_i, respectively, as can be seen if T^+ is written in the form

$$T^+ = \Sigma \zeta_0(Z_i)\rho_n(|Z_i|).$$

In general, V^+ will here be confined to statistics based on two-sample rank tests V. This includes modifications of T^+ in which all ranks $\rho_n(|Z_i|)$ are replaced by rank scores $\phi(\rho_n(|Z_i|)/(n+1))$, where the score function ϕ on $(0,1)$ is nonnegative and nondecreasing [and for T^+, $\phi(u) = (n+1)u$]. Some important examples are:

(i) $$\phi(u) = \Phi_+^{-1}(u),$$

with Φ_+ the distribution function of the absolute value of a standard normal variate —the analogous V was introduced by van der Waerden.

(ii) $$\phi(u) = (n+1)\min(u, 1-\gamma)$$

for some γ strictly between 0 and 1. This operator is called *Winsorization**.

(iii) $$\phi(u) = \zeta_c(u)$$

with $c = n/[2(n+1)]$, but with $\rho_n(b_i)$ the max rank of b_i. The analogous V was introduced by Westenberg and is also referred to as a *median test statistic*. (This definition, to also cover the case of ties, has been given by Hemelrijk [3, 4], though in a different form; it equals the number of the positive Z's greater than *or equal* to the median $\tilde{M}$ of all the $|Z|$'s.)

All such tests are distribution-free*, but, when the Z_i do not have a continuous distribution, the tests are only so when conducted conditionally, given the distribution of the ties. In many cases that makes the test rather clumsy to apply (a recursion formula for the conditional distribution of

$$2V^+ - \tfrac{1}{2}\sum_i \phi\big(\rho_n(|Z_i|)\big)/(n+1)$$

is given in [12]). For (iii), however, the application is very easy, since the conditional probability that the statistic take on the value w, given that n_1 of the Z's are positive, n_2

are negative, and that the number of $|Z|$'s greater than or equal to $\tilde{M}$ equals r, is simply

$$\binom{r}{w}\binom{n_1 + n_2 - r}{n_1 - w}\Big/\binom{n_1 + n_2}{n_1}.$$

The combined tests discussed in the literature are mostly based on a linear combination of B and V^+. Ruist [10] calls a test based on $B(V^+) = kB + V^+$, with k any number exceeding the maximum possible value of V^+, a *modified sign test*, and for $V^+ = T^+$ a *Wilcoxon sign test*. For such a test, if, for some integer d, H_0 is rejected by the sign test at level $d/2^n$ but not at level $(d + 1)/2^n$, a randomized sign test at a level α between those two would reject by means of an auxiliary experiment, whereas a modified sign test would also take into account the value of V^+ in order to come closer to the desired level of significance. In general, an important advantage of a combined test is that near α it makes available more levels of significance than the pure tests. (Randomized tests* are never used in practice, but theoretical investigations, such as those to be mentioned, always assume randomized tests in order to facilitate power comparisons between tests.) The statistic M^+ based on the Mann–Whitney statistic M, is $M^+ = T^+ - \frac{1}{2}B(B + 1)$, a nonlinear function of T^+ and B.

We shall discuss a.o. the following composite alternative to H_0 (the notation is not standard):

H_{s+}: each Z_i has a distribution which is

symmetric about a positive point;

a subscript q indicates for each component i by how much the value of the distribution function of Z_i at 0 (or the midvalue if there is a jump there) falls short of $\frac{1}{2}$ for each i for some q_i in $(0, \frac{1}{2})$; the superscript ω relates to the values of ω_{jk}, the conditional probability that the absolute value of the kth negative observation falls short of the value of the jth positive observation, *minus* $\frac{1}{2}$. When all the q_i or ω_{jk} are equal to some common value, we denote it by q or ω, respectively. Let u denote unimodality* and $\bar{u}$ strong unimodality.

To keep the discussion short, we assume equal distributions of the Z's under the alternative hypothesis (for a treatment of the unequal case, see Konijn [7] and Sen [11].) Since discreteness of the distributions can cause some complications (for an example, see ref. 3, pp. 39–40), we also shall assume continuity. Moreover, we shall only consider one-sided tests that reject for large values of the statistic, and corresponding alternatives.

By far the most important application of symmetry tests is to the first case mentioned in the entry TESTING FOR SYMMETRY, in which, for each pair of subjects, one subject is treated (giving observation Y_i) and the other serves as control (giving observation X_i); $Z_i = Y_i - X_i$, and we are interested in whether or not the treatment has an effect. If the effect of the treatment is to add a non-random quantity θ_i to the treated subject in the ith pair, Z_i is symmetrically distributed about θ_i.

Under the preceding conditions and if the distribution of $Z_i - \theta_i$ is not flat at $\frac{1}{2} - q_i$, the alternative $H_{s+}(\theta)$ for $\theta_i > 0$ equals H_{sq} for some q_i in $(0, \frac{1}{2})$ determined by θ_i for each i. [Some discussion of the case $\theta_i = \theta(X_i)$ is found on p. 68 in ref. 8.] A more general class of alternatives $H_>$ states that the Y_i tend to be larger than the corresponding X_i or the class $H_>^Z$ for which the positive Z's tend to be larger than the absolute values of the negative Z's. (Y tends to be larger than X, or is *stochastically larger* *than* X, if for all t, $\Pr[Y > t] \geqslant \Pr[X > t]$, and if, for a set of t to which the distributions attach positive probability, the inequality is sharp.) Doksum and Thompson [1] considered subclasses of the latter class of alternatives and examined certain linear combinations of B and V^+.

Hoeffding [6] showed that for each H_{q+} the (randomized) sign test maximizes the minimum power; this is an important property, since in addition the power of the sign test increases with q. It was also proved for H_{sq+} by Ruist [10]. Later Fraser [2] showed that the sign test is uniformly most powerful in the union of H_{q+}, and Ruist [10] tried to show that the latter result also holds for H_{sq+}. However, among the tests based on

$B(V^+)$ admitted by Ruist in the comparison were only those in which the V^+ were "the most common" rank statistics. The situation is the same for H_{suq+}, but for H_{suq+}^{ω} there exist n and α for which a modified sign test is uniformly more powerful than the comparable sign test. For $H_{s\bar{u}q+}^{\omega}$ he found an entire class of modified sign tests that are uniformly more powerful than the corresponding sign test (or in exceptional cases as powerful as this test). One of these is the modified Wilcoxon test.

In choosing among distribution-free tests, one sometimes singles out a particular parametric family of distributions (and distributions close to it in some sense) as much more relevant than any other families; or one wants the outcome of the test to be not too much influenced by "wild" data. The choice of score function usually has to do with these considerations, e.g., one would choose (i) if the distinguished family is normal and (ii) (with a suitable choice of γ) when there are believed to be outliers*. One can also allow the data to affect the choice (including score functions which are linear combinations of several standard ones), giving so-called adaptive* statistics; see refs. 5 and 9, where one also finds data on power functions, which may help in deciding how many subjects to observe.

Frequently one cannot find a test which is optimum in some desired sense, or even nearly so; but one would wish the test to be at least consistent and unbiased for all the most likely alternatives. B/n has mean $\frac{1}{2} + q$, so the sign test is unbiased and consistent for H_{q+}. Since a distribution may well have $q = 0$ without being symmetric (or even being symmetric but not about 0), the sign test is not consistent for alternatives which include such a distribution.

$$(T^+ - [(n+1)/2]B)/\{n(n-1)\}$$

has mean

$$\omega\left(\tfrac{1}{2} + q\right)\left(\tfrac{1}{2} - q\right) = \Pr[Z_1 > -Z_2 > 0]$$
$$- \tfrac{1}{2}\Pr[Z_1 > 0]\Pr[-Z_2 > 0]$$

and so generates a test which is unbiased and consistent for H_+^{ω}. It is interesting to

note that under the null hypothesis this statistic is uncorrelated with B, which can be shown using Example 6 of the Appendix in ref. 8 or the last expression in (6.23) of ref. 12, when corrected as indicated in the references; see ref. 1 for a certain asymptotic minimax property of this statistic in a subclass of H^Z with zero medians.

Theorem VI of ref. 12 showed that the test based on T^+ (or on $T^+ - B$) is unbiased and consistent for the union of H_{sq+}. This holds even for $H_{>}^Z$, since $T^+ - B$ has mean

$$\tfrac{1}{2}n(n-1)\int\{1 - Q(-Z)\}\,dQ(Z)$$

$$< \tfrac{1}{2}n(n-1)\int Q(Z)\,dQ(Z)$$

$$= \tfrac{1}{4}n(n-1),$$

where Q is the common distribution of the Z's. From the relation of M^+ to T^+ it follows that the M^+-test is not unbiased and not consistent for H_{s+}. The median test as previously defined is consistent for H_{s+}, since n^{-1} times the statistic then converges in probability to $1 - Q(\gamma)$, provided Q is not flat at γ. Here γ is the solution of $Q(\gamma) - Q(-\gamma) = \tfrac{1}{2}$. (The consistency condition in ref. 3 is much too strong.)

The case in which X_i and Y_i are independent has been discussed in Konijn [7]. The T^+-test is then unbiased and consistent for $H_{>}$ [using the result of Problem 46 (iii) of ref. 8, Chap. 4].

References

[1] Doksum, K. and Thompson, R. (1971). Power bounds and asymptotic minimax results for one-sample rank tests. *Ann. Math. Statist.*, **42**, 12–34.

[2] Fraser, D. A. S. (1953). Non-parametric theory: scale and location parameters. *Canad. J. Math.*, **6**, 46–68.

[3] Hemelrijk, J. (1950a). Symmetrietoetsen. Ph.D dissertation, Mathematisch Centrum, Amsterdam, Netherlands. [Printed by Excelsior Foto-offset.]

[4] Hemelrijk, J. (1950b). A family of parameter-free tests for symmetry with respect to a given point. *Indag. Math.*, **12**, 340–350, 419–431. K. Nederl. Akad. Wetensch., Proc. (Sci.), **53**, 945–955, 1186–1198.

[5] Hettmansperger, T. P. (1984). *Statistical Inference Based on Ranks*. Wiley, New York.

[6] Hoeffding, W. (1951). "Optimum" nonparametric tests. *Proc. Second Berkeley Symp. Math. Statist. and Prob.*, 83–92.

[7] Konijn, H. S. (1957). Some nonparametric tests for treatment effects in paired replications. *J. Indian Soc. Agric. Statist.*, **9**, 145–167. [Among the more disturbing misprints: (9.5) should read $\frac{1}{4}n$; in line 5 from the bottom on p. 163 the bracket should close before the $+$; on p. 155, δ is the upper α-point of Φ. Moreover, on p. 149, "in $\Omega - \Omega_0$" should read "for each θ," and the remarks on Ruist should be those of the present paper.]

[8] Lehmann, E. L. (1975). *Nonparametrics. Statistical Methods Based on Ranks*. Holden-Day, San Francisco. [The last summand of (A.76) has to be multiplied by 2.]

[9] Randles, R. H. and Wolfe, D. A. (1979). *Introduction to the Theory of Nonparametric Statistics*. Wiley, New York.

[10] Ruist, E. (1954). Comparison of tests for nonparametric hypotheses. *Arkiv Mat.*, **3**, 133–163. [The author communicated to me that Table 1 and the reference to it on p. 160 are in error. Of course, ψ on p. 158 should be ψ_2.]

[11] Sen, P. K. (1968). On a further robustness property of the test and estimator based on Wilcoxon's signed rank statistic *Ann. Math. Statist.*, **39**, 282–285.

[12] van Eeden, C. and Benard, A. (1957). A general class of distributionfree tests for symmetry containing the tests of Wilcoxon and Fisher. *Indag. Math.*, **19**, 381–408. *K. Nederl. Akad. Wetensch., Proc.* (*Sci.*), **60**, 381–408. [(6.13) and all subsequent general expressions for covariance and correlation should be multiplied by 2.]

(ADAPTIVE METHODS
DISTRIBUTION-FREE METHODS
MANN–WHITNEY–WILCOXON
   STATISTICS
PAIRED COMPARISONS
RANK TESTS
SIGN TESTS
TRIMMING AND WINSORIZATION
WILCOXON SIGNED-RANK TEST)

H. S. KONIJN

SYNERGISM

A term used to denote mutual reinforcing action of two stimuli. More precisely, synergism is said to exist when the combined effect of two stimuli is greater than the sum of their separate effects. It can be regarded as an interaction* of positive sign between the two effects.

Interaction of the opposite sign, when the combined effect is less than the sum of the separate effects, can be called *antagonism*, though some workers prefer to reserve this term for the situation when the combined effect is less than *either* of the two separate effects.

(INTERACTION)

SYNTHETIC ESTIMATORS

These estimators were introduced by the U.S. National Center for Health Statistics* [4] in 1968, and have been studied by Gonzalez and Waksberg [1] and Laake [2, 3], among others. They are used to estimate characteristics of finite populations* within subdomains of the complete domain of study (for example, counties within a state or age groups within a school population).

The central idea is to form strata ("poststrata") on the basis of observed sample values (attempting to minimize interstrata variation); then calculate sample means for each poststratum—subdomain combination and form weighted means of these sample means. It is necessary to have available values for the weights, e.g., from census results.

Laake [2] calculated estimates of mean square error* (MSE) for synthetic estimators for a Norwegian Labor Force Survey. The results indicated that no reduction in MSE was effected by introducing additional information beyond the numbers of individuals in the subdomains and poststrata. Further details are given in ref. 3.

References

[1] Gonzalez, M. E. and Waksberg, J. (1973). Estimation of the Error of Synthetic Estimators. Presented at *Internat. Assoc. Survey Statist.*, Vienna, Austria.

[2] Laake, P. (1978). *Scand. J. Statist.*, **5**, 57–60.

[3] Laake, P. (1979). *J. Amer. Statist. Ass.*, **74**, 355–356.

[4] U.S. National Center for Health Statistics (1968). *Synthetic State Estimation of Probability*, U.S. Public Health Service, Washington, D.C.

(FINITE POPULATIONS, SAMPLING FROM NEYMAN ALLOCATION SURVEY SAMPLING)

SYSTEMATIC DESIGNS

Systematic and chance elements are both present in the design of any trial. The choice as to when and where to do the trial at all is a systematic one, even if the particular points in time and space are arbitrary. Also, it is usually, though not always, true that there is not a totally random choice of experimental units, e.g., field plots, patients, machines. However, systematic designs in experimentation are defined here as those in which experimental treatments are allocated to the units in a systematic pattern. This was the usual practice until the 1920s, the same pattern being commonly followed in every repetition of the treatments, even if the allocation of the treatments in the first repeat was haphazard. Thereafter, the concept of *randomisation** in experimental design, originally due to Fisher* [13], took over, and most subsequent statistically designed experiments had a considerable element of randomness in them.

There is no doubt that randomisation is usually justified, because the random allocation of treatments to experimental units permits the calculation of unbiased estimates of treatment means and variances with a minimum of distributional assumptions. Nevertheless, not every statistician was immediately convinced of the advantages of randomisation, and the argument rumbled on into the 1930s, with Fisher advocating randomisation and others, especially Gosset* ("Student"), in favour of systematic designs [16]. For practical reasons also, less statistically inclined experimenters remained unhappy about randomisation, arguing that difficulties in the actual random allocation made randomised experiments less accurate than they should be. The new orthodoxy of randomisation made systematic designs a statistical heresy, but more recently some statisticians have proposed the conscious use of systematic designs in special circumstances, and these are reviewed here.

There has always been a use for the systematic arrangement of a grid of control or standard plots, even in trials where new test materials are randomised. The standard plots serve as a series of reference points against which to assess the new materials, and this is very valuable, especially in chemical determinations where there may be a drift with time, or in field trials on land with a marked fertility trend. With truly independent experimental units, as in some industrial experimentation, randomisation is not very important [29]. Indeed, Youden [32] suggested that in such circumstances randomisation was unnecessary, and sometimes even positively harmful. Where an element of randomness is needed, he advocated constrained randomisation, a halfway house between full randomisation and systematic designs, as being preferable for reducing the variability associated with the position of the experimental units. The whole position was summarized by Cox [7], who distinguished three types of situation. In the first, randomisation is used to remove subjective errors by a process of concealment, and failure to randomise would be a serious error. In the second, a very common practical situation, randomisation is a safeguard against the unexpected, and failure to randomise is not disastrous. Finally, however, there are times when randomisation is a bad thing, either because of practical difficulties or because the variation is in the form of a trend.

TREND-FREE AND BALANCED DESIGNS

Amongst the earliest of the more modern uses of systematic designs were those of Cox [5, 6], who was concerned with trend-free designs either for agricultural trials with plots laid out in a line or for industrial experiments with the experimental units equally spaced in time. He recommended [6] the use of systematic designs where there was likely to be a gain in accuracy or convenience, provided that this gain was not outweighed by doubts about the assumptions underlying the systematic design; practical knowledge of the experimental situation is necessary to assess these points. Most of Cox's examples have only a few treatments, two to four, but one large example on the drawing and spinning of wool is a single replicate of a 2^3 factorial with one treatment on main plots and the other two on subplots.

Balanced designs (*see* BALANCING IN EXPERIMENTAL DESIGN) in which the treatments are arranged in blocks in a one-dimensional sequence, but where pairs of treatments occur equally often next to each other, were described by Williams [31], who gave two series of designs for up to 10 treatments, depending on whether or not a treatment could occur next to itself. He also found a smaller number of designs in which each treatment occurred equally often next but one to every other treatment. These designs refer to unordered pairs of neighbours, that is, an occurrence of (12) is regarded as the same as that of (21). The problem for ordered pairs of neighbours, where (12) and (21) are different, is more difficult, but Dyke and Shelley [10] found many solutions for nine blocks of four treatments, with a dummy plot at each end. The uses of such designs for investigating interactions* between neighbouring plots in trials for the control of barley mildew were described by Jenkyn et al. [18]. It is also sometimes possible to have two-dimensional designs balanced for trend [23, 8]. Further, there has recently been an interest in two dimensional designs balanced for the effects of nearest neighbours* [15, 30].

TRIALS WITH QUANTITATIVE LEVELS

Perhaps one of the most important practical uses of systematic designs is in field trials where the treatment levels are quantitative, particularly where the amount of land available is severely limited. Then, it makes good sense to have only gradual changes of the levels of a treatment such as spacing or the amount of an applied chemical; guard rows or discard areas are not needed between the plots within a replicate, but are still needed at the edges of a complete replicate. Systematic designs are especially useful in two types of trial, *preliminary* and *scientific*. In a preliminary trial, where little is known about likely responses, a systematic design permits the exploratory use of many levels of a factor in a small space. By contrast, in a more scientific trial the aim may be the precise assessment of the response to several levels of a quantitative factor: Alternatively, the responses to levels of one such factor may need to be compared at different levels of a second factor, either quantitative or qualitative [12]. In either situation, the small area of each replicate of a systematic design will permit accurate estimates without the use of an enormous area of land.

A technique permitting a gradual change in dosage of a chemical that is sprayed onto field plots is essential for a systematic design with such materials. A variable dosage sprayer was first constructed for use primarily with herbicides [22], and the principle was rapidly adopted for field trials of pesticides by the agricultural chemicals industry. In all such trials the dose of applied chemical varies logarithmically with distance along the row of plots. Experimental designs using the log-dose principle were considered in detail by Thompson and Wheatley [28] for trials comparing different insecticides as well as different doses of the same insecticide. A 16-fold dose range was achieved in rows of 22 spaced cauliflower plants and a similar range in a 7 m. row of carrots. These trials also had a systematic grid of untreated control plots.

For spacing experiments, Nelder [19] pointed out that randomised designs have the disadvantage of either requiring a constant number of plants per plot, when it is difficult to fit different sized plots into a block, or of having all plots the same size, when closely spaced plots may contain an unnecessarily large number of plants. Another factor to be considered in spacing trials is *rectangularity*, that is, the ratio of distances between and within rows of plants. Nelder constructed several types of design for spacing trials; in some of these, spacing and rectangularity both varied, while in others one remained constant and the other changed. These arrangements, and a variant specifically designed so that all plantings are in rows [3], have been much used by agriculturists. The commonest in practice seems to have been Nelder's design Ia, consisting of a series of concentric circles with radii increasing in geometric progression and rows of individual plants radiating from the centre like spokes on a wheel. If the angle between neighbouring spokes is 5° and the ratio of successive radii is 1.1, then the rectangularity is near unity, so that every plant is nearly equidistant from its four nearest neighbours. A set of 20 such circles means that the distance between the outermost pair of circles is more than 6 times that between the innermost pair. Sometimes it is not possible to fit complete circles into the available land, and segments of circles are used, usually not all oriented in the same direction [14].

Systematic designs have also been used for fertiliser trials. Cleaver et al. [4] described a set of designs where the levels of fertiliser treatments change gradually and systematically from one end of a block to another. These designs can be used for many levels of one fertiliser, possibly up to nine or more; an alternative use is for two fertilisers applied orthogonally such that the level of fertiliser A increases progressively from one column of plots to the next and the level of B increases from one row to the next.

RANDOMISATION IN SYSTEMATIC TRIALS

The use of systematic designs does not entirely preclude randomisation. Dyke and Shelley [10] recommend choosing a design for a particular trial at random from one of the many possible systematic layouts satisfying their requirements. In the trials of Cleaver et al. [4] the blocks of plots are replicated and an element of randomness introduced by arranging that the direction of increase in the different replicates is chosen at random: If there are two fertilisers, a random choice is also possible within each replicate as to whether A varies with columns and B with rows or vice versa. In log-dose insecticide trials, the principle is commonly adopted of having the dose of all chemicals in a block increasing in the same sense but choosing the direction of increase randomly in each block; the positions of different insecticides within a block are also random. A form of restricted randomisation has been suggested for trials with varieties of apple trees [27]: All the different varieties are present once each in a block, but the removal of alternate trees in alternate rows leaves an arrangement in which the remaining trees consist of full sets of all the varieties in blocks twice the size of the original blocks.

There may also be a place for balance and restricted randomisation in clinical trials* in which patients are available to the experimenter sequentially. Efron [11] introduced what he called a *biased coin design*, where the probability of a subject being allocated to the treatment that has previously had fewer subjects is between 0.5 and 1.0. Balance is possible over several prognostic factors, as shown by Pocock and Simon, together or separately [24–26]. An improved procedure was described by Begg and Iglewicz [2], who stress the advantages of the method, including flexibility of design. However, a strong body of opinion, illustrated by Peto et al. [21], holds that complete randomisation is desirable in clinical trials; the merits and demerits of some systematic allocation of patients to treatments are thus as yet unresolved.

ANALYSIS OF SYSTEMATIC DESIGNS

A feature common to many of the trials using systematic designs is that their primary purpose is either, at one extreme, just to find out what variation occurs over a wide range of levels of the factor under test or, at the other, to provide data for estimating parameters in a previously developed model. Cleaver et al. [4] demonstrated that the pattern of increase of yield with fertiliser was usually similar in different blocks, and also that there was little variation according to whether plants were in a part of a plot adjacent to a higher or lower fertiliser level; the results of their systematic trials provided data for predictive models of fertiliser requirements, e.g., ref. 17 and subsequent papers.

Unconventional methods of analysis have also been used in designs balanced for nearest neighbours, where the method of adjusting plot values by covariance on neighbouring plot values due to Papadakis [20] has recently been reexamined [1]; suggestions have been put forward that the Papadakis* method has flaws and that improved procedures are possible [30], but at the time of writing the topic is still under active investigation. Jenkyn et al. [18] used conventional analysis of variance*, treating their experiments as if they were in randomised blocks*, but also analysis of covariance*, with a Fourier series as covariate, to allow for interactions between neighbours. Alternative methods of analysis for these data have been suggested, based on recent nearest neighbour models [2a, 9, 16a, 30].

CONCLUSION

In all the types of trial quoted here and in many others, the use of systematic designs has permitted considerably more information to be obtained from a given amount of resources than would be possible using conventional randomised designs. The main advantages found are as follows: (i) less land is needed where quantitative treatments change only gradually; (ii) designs balanced for nearest neighbours help to overcome positional effects; (iii) a deterministic element in sequential experiments prevents extreme lack of balance; (iv) in some industrial trials it may be impossible to have some levels of a factor occurring after others. For all these reasons systematic designs have a place in modern experimentation, even if the more conventional forms of statistical analysis such as analysis of variance are not possible.

References

[1] Bartlett, M. S. (1978). *J. R. Statist. Soc. B*, **40**, 147–174. (The paper that reexamined the methods of nearest neighbour analysis.)

[2] Begg, C. B. and Iglewicz, B. (1980). *Biometrics*, **36**, 81–90.

[2a] Besag, J. and Kempton, R. (1986). *Biometrics*, **42**, 231–251.

[3] Bleasdale, J. K. A. (1967). *Expl. Agric.*, **3**, 73–85.

[4] Cleaver, T. J., Greenwood, D. J., and Wood, J. T. (1970). *J. Hort. Sci.*, **45**, 457–469. (The introduction of systematic designs for fertiliser trials.)

[5] Cox, D. R. (1951). *Biometrika*, **38**, 312–323.

[6] Cox, D. R. (1952). *J. R. Statist. Soc. B*, **14**, 211–219. (The case for systematic designs in particular situations firmly stated.)

[7] Cox, D. R. (1961). *J. R. Statist. Soc. A*, **124**, 44–48.

[8] Cox, D. R. (1979). *J. R. Statist. Soc. B*, **41**, 388–389.

[9] Draper, N. R. and Guttman, I. (1980). *Appl. Statist.*, **29**, 128–134.

[10] Dyke, G. V. and Shelley, C. F. (1976). *J. Agric. Sci., Camb.*, **87**, 303–305.

[11] Efron, B. (1971). *Biometrika*, **58**, 403–417. (The planned introduction of a systematic element into sequential clinical trials.)

[12] Finch, S., Skinner, G., and Freeman, G. H. (1976). *Ann. Appl. Biol.*, **83**, 191–197.

[13] Fisher, R. A. (1926). *J. Min. Agric. G. Br.*, **33**, 503–513. (The first firm statement of the benefits of randomisation, though the idea had been mentioned in the first edition of *Statistical Methods for Research Workers* a year earlier.)

[14] Freeman, G. H. (1964). *Biometrics*, **20**, 200–203.

[15] Freeman, G. H. (1979). *J. R. Statist. Soc. B*, **41**, 88–95.

[16] Gosset, W. S. (1936). *Supp. J. R. Statist. Soc.*, **3**, 115–136. (The final battle in the "randomised *v* systematic design" war of the 1930s.)

[16a] Green, P. J., Jennison, C., and Seheult, A. (1985). *J. R. Statist. Soc. B*, **47**, 299–315.

[17] Greenwood, D. J., Cleaver, T. J., and Turner, M. K. (1974). *Proc. Fertil. Soc.*, **145**, 1–30. (A paper making use of data from a systematic trial for model building.)

[18] Jenkyn, J. F., Bainbridge, A., Dyke, G. V., and Todd, A. D. (1979). *Ann. Appl. Biol.*, **92**, 11–28.

[19] Nelder, J. A. (1962). *Biometrics*, **18**, 283–307. (The introduction of systematic designs for spacing trials.)

[20] Papadakis, J. S. (1937). *Bull. Inst. Amel. Plantes à Salonique*, **23**. (The origin of the method of adjusting plot values by their nearest neighbours.)

[21] Peto, R., Pike, M. C., Armitage, P., Breslow, N. E., Cox, D. R., Howard, S. V., Mantel, N., McPherson, K., Peto, J., and Smith, P. G. (1976). *Br. J. Cancer*, **34**, 585–612. (Firm defence of complete randomisation in clinical trials.)

[22] Pfeiffer, R., Brunskill, R. T., and Hartley, G. S. (1955). *Nature, Lond.*, **176**, 472–473.

[23] Phillips, J. P. N. (1968). *Appl. Statist.*, **17**, 162–170.

[24] Pocock, S. J. (1979). *Biometrics*, **35**, 183–197.

[25] Pocock, S. J. and Simon, R. (1975). *Biometrics*, **31**, 103–115. (Suggested improvement to Efron's procedure.)

[26] Simon, R. (1979). *Biometrics*, **35**, 503–512.

[27] Taylor, J. (1949). *J. Agric. Sci. Camb.*, **39**, 303–308. (Example of restricted randomisation in field trials.)

[28] Thompson, A. R. and Wheatley, G. A. (1977). *Pestic. Sci.*, **8**, 418–427. (Full description of systematic log dose insecticide trials.)

[29] Tippett, L. H. C. (1935). *Suppl. J. R. Statist. Soc.*, **1**, 27–62. (Statement of the advantages of systematic designs for industrial experiments.)

[30] Wilkinson, G. N., Eckert, S. R., Hancock, T. W., and Mayo, O. (1983). *J. R. Statist. Soc. B*, **45**, 151–211. (Introduction of new models for nearest neighbour analysis and of systematic designs for use in the field.)

[31] Williams, R. M. (1952). *Biometrika*, **39**, 151–167. (Introduction of designs balanced for serial trends in one dimension.)

[32] Youden, W. J. (1972). *Technometrics*, **14**, 13–22.

(AGRICULTURE STATISTICS IN
BALANCING IN EXPERIMENTAL DESIGN
CLINICAL TRIALS
DESIGN OF EXPERIMENTS
NEAREST NEIGHBOR METHODS
PAPADAKIS METHOD
RANDOMISATION
TREND-FREE BLOCK DESIGNS)

G. H. Freeman

SYSTEMATIC SAMPLING

Systematic sampling is a widely used simple selection procedure. Suppose that a sample of n elements is required from a population list of N elements and assume for now that $k = N/n$ is an integer. A systematic sample is obtained by taking every kth element on the list; k is the *sampling interval*. As a rule, the first sampled element is determined by the selection of a random number between 1 and k, say r. The selected sample then comprises the rth, $[r + k]$th, $[r + 2k]$th, …, and $[r + (n - 1)k]$th elements on the list. The use of a random start gives every population element the same selection probability $1/k$. The joint selection probability for the ith and jth population elements is $1/k$ if $i = j + mk$, where m is an integer, and 0 otherwise.

In practice the sampling interval k is seldom an integer. Noninteger values of k may be handled in several ways. One is to round k to the nearest integer and apply the preceding procedure. The resulting sample size will differ from the initial choice, but in many cases this will be acceptable. A second way is to round k down to an integer, to select a random start throughout the whole population, and then select $(n - 1)$ additional elements by successively adding the rounded-down k to the random start; the list is treated as circular with the last element being followed by the first. A third way is to randomly remove a number (t) of elements from the population so that $(N - t)/n$ is an integer, after which the procedure previously described can be applied. A fourth way is to employ a fractional sampling interval k, choosing a fractional random number, and successively adding k to it; the resulting numbers are then rounded down to identify the selected elements.

Systematic sampling is equivalent to the selection of a single cluster in cluster sampling*. Each of the k possible random starts defines a population cluster, with the jth cluster comprising the population elements $j, [j + k], [j + 2k], …, [j + (n - 1)k]$. The random start chooses one of the population clusters to be the sample. Let Y_{ij} denote the value of the variable Y for element i in cluster j, let $\overline{Y}_j$ denote the mean for the elements in cluster j, let $\bar{y} = \Sigma y_i/n$ be the sample mean (i.e., the mean of the selected cluster), and assume that $k = N/n$ is an integer. Then $\bar{y}$ is an unbiased estimator of the population mean $\overline{Y} = \Sigma\Sigma Y_{ij}/N = \Sigma\overline{Y}_j/k$ and its variance is

$$V(\bar{y}) = \Sigma(\overline{Y}_j - \overline{Y})^2/k.$$

Since systematic sampling selects only one cluster and replication of a sampling process is needed for variance estimation, $V(\bar{y})$ cannot be estimated from the sample without invoking some assumption about the formation of the clusters or equivalently about the order of the list. A variety of alternative variance estimators have been proposed based on different assumptions about the order of the list. Wolter [11] reports on a theoretical and empirical comparison of eight of these variance estimators.

One frequently used assumption for variance estimation is that the list is randomly ordered with respect to the survey variables. Then systematic sampling is equivalent to simple random sampling*, and $V(\bar{y})$ may be estimated by the simple random sampling formula $(N - n)s^2/(Nn)$, where $s^2 = \Sigma(y_i - \bar{y})^2/(n - 1)$.

Often the list is ordered in groups, as for instance when a firm's employees are listed in the departments (groups) in which they work. Systematic sampling from such a list ensures that each group is represented in the sample in approximately the same proportion as in the total population. Assuming a random ordering of elements within groups, the sample design closely resembles proportionate stratified sampling (*see* PROPORTIONAL ALLOCATION), the groups being the strata. $V(\bar{y})$ may then be estimated using the formula for proportionate stratified sampling. With proportionate stratification the desired stratum sample sizes are generally fractional and hence have to be rounded to the nearest integer; when the sample sizes are very small, rounding may cause distor-

tions. Since systematic sampling avoids the need for this rounding, it is often used when a detailed stratification is required: The elements are listed by strata, with careful attention to the ordering of the strata, and then a systematic sample is taken throughout the list, yielding an "implicit stratification."

The methods of variance estimation already described require assumptions about the order of the population list, because a systematic sample selects only a single cluster. The need for such assumptions can be avoided by selecting several clusters, that is, by taking several random starts. Thus, for instance, instead of a single systematic sample with an interval of k, c systematic samples could be selected with intervals of kc, starting with c different random starts from 1 to kc. Assuming that the population size is a multiple of kc, the variance of the overall sample mean from the c samples ($\bar{y}$) may then be estimated by

$$[1 - (1/k)]\Sigma(\bar{y}_\gamma - \bar{y})^2/\{c(c - 1)\},$$

where $\bar{y}_\gamma$ is the mean of the γth subsample.

A number of theoretical and empirical studies have been conducted to examine the efficiency of systematic sampling in specific situations. A simple theoretical model for the order of the list has the Y values following a linear trend. Under this model, the sample mean from a systematic sample is more precise than the mean from a simple random sample of the same size, but less precise than the mean from a proportionate stratified sample in which one element is selected from each of n equal-sized strata (the first k population elements comprising the first stratum, the next k elements the next stratum, etc.).

Several approaches have been proposed to improve the precision of systematic sampling in the case of a linear trend. One is to change the weights of the sampled elements that are lowest and highest in the order of the population list (Yates [12]; Bellhouse and Rao [1]). A second is to take a centrally located sample: Instead of starting with a random number between 1 and k, the middle element of the sampling interval is taken. The sample mean of a centrally located systematic sample has a lower mean square error than that of a random start systematic sample when the population follows a monotonic trend (Madow [7]); however, a centrally located systematic sample is not a probability sample. A third approach, termed *balanced systematic sampling* (Murthy [8]), takes two balanced starts in the sampling interval $2k$ (with n even), the first being a random number (r) from 1 to k and the second being ($2k - r + 1$). Another variant is to take one-half of the systematic sample working forward through the list and the other half working backward through the list from the end, using the same random start for both halves (Singh et al. [9]).

Systematic sampling would fare badly if the survey variable followed a periodic variation in the population list and the sampling interval coincided with a multiple of the periodicity. Consider, for instance, a list of married couples, with husbands listed before their wives. An even sampling interval would produce a sample of either all husbands or all wives. An odd sampling interval, however, would produce equal—or almost equal—proportions of husbands and wives. When periodicities are present, a sampling interval that is a multiple of the periodicity should generally be avoided. In practice, regular periodic cycles are rarely encountered.

Systematic sampling has been extended into two (and more) dimensions. This extension may be applied, for instance, in sampling geographical areas as in agricultural surveys. Consider a square field of $n^2 k^2$ square unit areas of which n^2 are required for the sample. The field could be divided into n^2 subsquares of dimensions ($k \times k$), with one unit area to be selected from each. The choice of two random numbers r and r' between 1 and k would fix the coordinates of the selected unit square in the top left-hand subsquare of the field. The remaining selections could then be determined by successively adding k to r and r'. This procedure, which results in the sampled unit squares having the same location in each subsquare, produces an *aligned sample*. A modification is to randomly choose different horizontal coordinates for the first row and different vertical coordinates for the first column, leading to an unaligned sample. Further extensions of this approach lead to lattice sampling.

The preceding discussion relates to applications of systematic sampling for sampling units with equal probability. It is also widely used for sampling units with unequal probability, such as sampling with probability proportional to size*(PPS). The procedure is best described by means of an example. Suppose that three units are to be selected from the following six units with probabilities proportional to their measures of size M_i:

Unit (i):	1	2	3	4	5	6	Total
M_i	10	3	14	12	9	18	66
Cumulative M_i	10	13	27	39	48	66	

The cumulative totals of the size measures are calculated as in the last row; using these totals, unit 1 is associated with the numbers 1–10, unit 2 with the numbers 11–13, etc. Dividing the overall total (66) by the number of units to be selected (3) gives the sampling interval of 22. A random start between 1 and 22, say 12, is chosen; adding 22 to the random start gives 34 and adding 22 again gives 56. The three selections are thus units 2, 4, and 6. Provided that the sizes of all the units are smaller than the sampling interval, no unit can be selected more than once. This systematic procedure provides a simple means of selecting units with unequal probabilities without replacement.

Systematic sampling features as a component of many sample designs. It is described in survey sampling* texts such as Cochran [2], Hansen, Hurwitz, and Madow [3], Kish [5], Konijn [6], Murthy [8], Sukhatme et al. [10], and Yates [13]. A useful recent review is by Iachan [4].

References

[1] Bellhouse, D. R. and Rao, J. N. K. (1975). *Biometrika*, **62**, 694–697.

[2]   Cochran, W. G. (1977). *Sampling Techniques*, 3rd ed. Wiley, New York.

[3]   Hansen, M. H., Hurwitz, W. N., and Madow, W. G. (1953). *Sample Survey Methods and Theory*, Vols. 1 and 2. Wiley, New York.

[4]   Iachan, R. (1982). *Int. Statist. Rev.*, **50**, 293–303.

[5]   Kish, L. (1965). *Survey Sampling*. Wiley, New York.

[6]   Konijn, H. S. (1973). *Statistical Theory of Sample Survey Design and Analysis*. North-Holland, Amsterdam, Netherlands.

[7]   Madow, W. G. (1953). *Ann. Math. Statist.*, **24**, 101–106.

[8]   Murthy, M. N. (1967). *Sampling Theory and Methods*. Statistical Publishing Society, Calcutta, India.

[9]   Singh, D., Jindal, K. K., and Garg, J. N. (1968). *Biometrika*, **55**, 541–546.

[10]   Sukhatme, P. V., Sukhatme, B. V., Sukhatme, S., and Asok, C. (1984). *Sampling Theory of Surveys with Applications*, 3rd ed. Iowa State University Press, Ames, IA.

[11]   Wolter, K. M. (1984). *J. Amer. Statist. Ass.*, **79**, 781–790.

[12]   Yates, F. (1948). *Philos. Trans. R. Soc. Lond. A*, **241**, 345–377.

[13]   Yates, F. (1981). *Sampling Methods for Censuses and Surveys*, 4th ed. Griffin, London, England.

(AREA SAMPLING
CLUSTER SAMPLING
PROBABILITY PROPORTIONAL TO SIZE
    (PPS) SAMPLING
PROPORTIONAL ALLOCATION
STRATIFIED DESIGNS
SURVEY SAMPLING)

GRAHAM KALTON

SYSTEM RELIABILITY

System reliability analysis problems arise in many practical engineering areas. Some of these include communication networks, electrical power systems, water transmission systems, nuclear power reactors, and transportation systems. The purpose of a system reliability analysis is to acquire *information* about a system of interest relative to *making decisions* based on considerations of availability, reliability*, and safety as well as any inherent engineering* risks. The philosophy and guidelines for a system analysis have been discussed in several excellent introductory chapters by Haasl in a *Fault Tree Handbook* [6]. Broadly speaking, there are two important aspects to a system analysis: (1) an INDUCTIVE ANALYSIS stage and (2) a DEDUCTIVE ANALYSIS stage.

In the INDUCTIVE ANALYSIS stage we gather and organize available information on the system. We define the system, describe its functional purpose, and determine its critical components. At this stage, we ask the question: WHAT can happen to the system as a result of a component failure or a human error? We hypothesize and guess possible system failure scenarios as well as system success modes. A *preliminary hazard analysis* is often performed at the system level. A *failure and effects analysis* is conducted at the component level.

The DEDUCTIVE ANALYSIS aspect of a system reliability analysis answers the question: HOW can a system fail (or succeed) or be unavailable? A logic network (or fault tree* if we are failure oriented) is often the best device for deducing how a major system failure event could possibly occur. However, its construction depends on a thorough understanding of the system and the results of the system inductive analysis. A block diagram or a network is a useful device for representing a successfully functioning system. Since the network is close to a system functional representation, it cannot capture abstract system failure and human error events as well as the logic network representation. However, from the point of view of probability analysis, the network representation seems to be easier to analyze.

CALCULATION OF SYSTEM RELIABILITY

There are at least five general methods for calculating system reliability. Because of the generality of these methods, the computational running time is, in the worst case, exponential in the number of system components. For that reason, recent research has concentrated on finding *efficient* (i.e., polynomial running time) algorithms for systems of special structure.

The five general methods are:

(1) State space enumeration and Markov chains.
(2) The inclusion–exclusion principle* based on the set of system success (or failure) events.
(3) The sum-of-disjoint-products method, also based on these same events.
(4) Pivotal decomposition based on network or logic network representations.
(5) Bayes' theorem* applied to dependent probability nodes in a logic network (or fault tree).

State enumeration consists of listing the system success and failure states. Continuous time finite state Markov chains (*see* MARKOV PROCESSES) are often used to model system operation relative to component failures and repairs, preventative maintenance, etc. [2]. Although often used, this method is intractable for very large systems. To implement it, a directed graph is constructed whose nodes correspond to system states. The logical and statistical dependence between states is indicated by arrows. Associated with each node is a conditional probability function. In the Markov chain approach, the probability function for a specified node is conditional on the states of those nodes which have arrows immediately directed into the specified node. Based on this graph, differential equations are calculated and solved (if possible) to obtain the time dependent state transition probabilities. Besides the computational difficulty, this method can be criticized because of the Markov (conditional independence) assumption, which may be hard to justify in practice.

Let E_i, $i = 1, \ldots, p$, correspond to system success events; then

$$R = \sum_i P(E_i) - \sum_{i<j} P(E_i E_j)$$

$$+ \cdots + (-1)^{p-1} P(E_1 E_2 \cdots E_p)$$

is the inclusion–exclusion formula for system reliability. There are $2^p - 1$ terms. Part of its inefficiency is due to there being many cancelling terms in general, irrespective of the probability measure. Likewise we can compute

$$R = P(E_1) + P(\overline{E}_1 E_2)$$

$$+ \cdots + P(\overline{E}_1 \overline{E}_2 \ldots \overline{E}_{p-1} E_p),$$

where the product events are disjoint by construction. Unfortunately this general method is no better than the inclusion–exclusion method.

NETWORK RELIABILITY

Recent research has resulted in algorithms that allow exact reliability calculation for very large networks. Reliability is measured relative to network connectivity. Let G be a graph* with e edges and n nodes of which k ($1 \leqslant k \leqslant n$) nodes are distinguished. Edge i may fail with probability p_i and edge failures are independent. The problem is to calculate the probability that the k distinguished nodes (in the distinguished node set K) can communicate. The network is directed if all edges are directed. By replacing undirected edges by two edges in antiparallel that are completely dependent probabilistically, any undirected network problem can be treated as a directed network problem. The main theoretical formula is the *topological formula* for directed networks, i.e.,

$$R_K(G) = \sum_i (-1)^{e_i - n_i + 1} P(G_{ai}), \quad (1)$$

where $R_K(G)$ is the probability that all K nodes can communicate and $P(G_{ai})$ is the probability that all edges and nodes in G_{ai} work. G_{ai} is an acyclic K-subgraph with e_i edges and n_i nodes. (A K-subgraph is a union of K-trees; a K-tree is a tree all of whose pendant nodes belong to K.) An algorithm for finding all G_{ai} is given in ref. [4]. This is the best algorithm for this class of problems, based on the inclusion–exclusion principle.

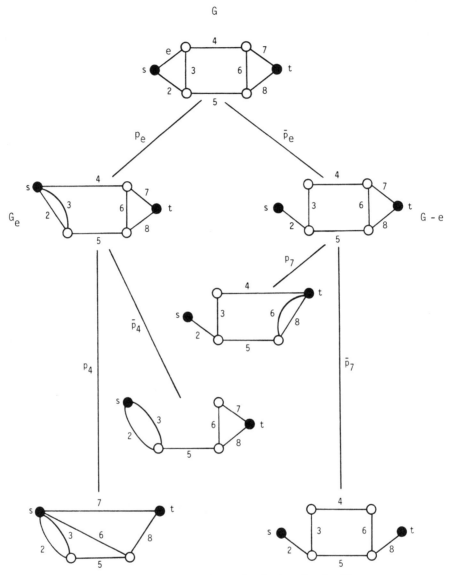

Figure 1 Binary computational tree using the factoring algorithm.

Figure 1 illustrates the pivoting algorithm relative to an undirected network with two distinguishable nodes. The first pivot is on edge 1, then on edges 4 and 7, respectively. The networks corresponding to the leaves of the binary computational tree are series-parallel reducible so that no further pivoting is required. The algorithm requires that all possible series and parallel probability reductions be performed at each stage before pivoting. By choosing pivot edges so that the resulting subgraphs have no irrelevant edges, an optimal algorithm is obtained within the class of algorithms based on pivoting and series-parallel probability reductions [3].

References

[1] Agrawal, A. and Barlow, R. E. (1984). *Operat. Res.*, **32**, 478–492.

[2]   Barlow, R. E. and Proschan, F. (1965). *Mathematical Theory of Reliability*. Wiley, New York.

[3]   Satyanarayana, A. and Chang, M. (1983). *Networks*, **13**, 107–120.

[4]   Satyanarayana, A. and Prabhakar, A. (1978). *IEEE Trans. Rel.*, **R-27**, 82–100.

[5]   Skwirzynski, J. K., ed. (1983). *Electronic Systems Effectiveness and Life Cycle Costing*. Springer, Berlin, Germany, pp. 3–24.

[6]   Vesely, W. E., Goldberg, F. F., Roberts, N. H., and Haasl, D. F. (1981). *Fault Tree Handbook*. Office of Nuclear Regulatory Research, NUREG-0492, Washington, D.C.

(FAULT TREE ANALYSIS
FLOWGRAPH ANALYSIS
HAZARD PLOTTING
INCLUSION-EXCLUSION METHOD
NETWORK ANALYSIS
NUCLEAR MATERIAL SAFEGUARDS
RELIABILITY, PROBABILISTIC)

RICHARD E. BARLOW

SYSTEMS ANALYSIS IN ECOLOGY

The origins of systems analysis lie in engineering* and cybernetics, originally concerned principally with the analysis of electronic communication systems. Gradually, however, systems analysis has been applied to a growing range of practical problems and, in so doing, has changed much in its concept and content. A simple definition is that *systems analysis* is the orderly and logical organization of data and information into models, followed by the rigorous testing and exploration of these models necessary for their validation and improvement. The purpose of such an analysis may be the explanation of physical, chemical, or biological processes, or it may be the provision of guidance on the management or manipulation of those processes.

Thus, systems analysis is not a mathematical technique. It is a broad research strategy that certainly uses mathematical techniques and concepts, but in a systematic scientific approach to the solution of complex problems. As such, it provides a framework of thought designed to help decision-makers to choose a particular course of action or to predict the outcome of one or more courses of action that intuitively seem desirable. In particularly favourable cases, the course of

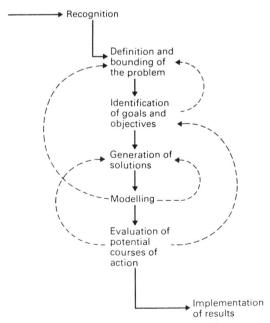

Figure 1   Diagram of the phases of systems analysis.

action that is indicated by the systems analysis will be the "best" choice in some defined way. A more detailed description of the various phases of systems analysis is given by Jeffers [1] as summarized in Fig. 1.

Because systems analysis is a framework of thought rather than a detailed prescription, not all of the steps listed above need to be included in every investigation. Alternatively, the order in which the steps are undertaken may be varied, or it may be necessary to iterate through them in various patterns. The relevance of the objectives of the analysis, or the importance of excluded factors, may have to be re-examined periodically. The most useful models will mimic reality with sufficient precision to serve a broad spectrum of decisions and decision-makers.

The development of systems analysis followed shortly after the definition of the *ecosystem* by Tansley [2] as "the system including not only the organism complex, but also the whole complex of physical factors forming what we call the environment of the bio—the habitat factors in the widest sense." As a result, there has been some confusion between ecosystem ecology and systems ecology in dealing with the structure and function of levels of biological organization beyond that of the individual and species.

Ecologists in the early and middle 1950s began studies of energetics in field ecosystems in an effort to model the flow of energy between trophic levels. These early studies were followed in the 1960s by the prodigious efforts to apply systems analysis to biological and ecological research during the International Biological Programme (IBP), and no history of ecosystem ecology or systems ecology is complete without consideration of the IBP. Nevertheless, the attempts to use systems analysis in this context were not particularly successful, principally because ecologists tried to apply the methods of systems analysis to data which they had already collected, instead of using systems analysis itself to define what kinds of data were actually required to solve the problems set by the IBP.

A further attempt was made to introduce systems analysis in the solution of large-scale problems in UNESCO's Man and the Biosphere Programme (MAB). The original intention was that systems and modelling approaches would bring about a greater integration and coordination of scientific activity, with mathematical models playing an important role in the understanding and optimal management of natural resources and, particularly, of renewable resources. However, it very quickly became apparent that very few of the developing countries had the expertise to engage in systems analysis, while the rapid development of systems analysis techniques and applications, especially at the International Institute of Applied Systems Analysis (IIASA) at Laxenburg near Vienna, made it difficult for scientists to keep up with the state of the art, even in developed countries. By 1981, it had become necessary to review the role of systems analysis in MAB and to propose a more limited deployment of the available expertise. It was stressed, nevertheless, that systems analysis and modelling of the sustained and profitable utilization of renewable resources would provide a better theoretical basis for future decisions than many of the current empirical and experimental approaches.

Today, systems analysis remains an integral part of the training and research of ecologists, demanding a level of numeracy* and quantitative logic which was hardly to be found in the traditional biologist or ecologist. The almost parallel development of the computer, and especially the microprocessor, has brought within the reach of almost all scientists powerful mathematical tools which, within a framework of logic and hypothesis testing, can greatly extend their comprehension and control of complex systems. The mathematical models used in ecology as part of that systems analysis range from differential and difference equations*, through a wide range of statistical formulations such as the analysis of variance* in multiple regression* and cluster analysis, to the topological models of catastrophe theory* and fractal* geometry. However, it should be emphasized that this use of models does not spring from any compulsion to use mathematics for its own sake. It is the inherent complexity of ecological relationships, the characteristic variability of living organisms, and the apparently unpredictable effects of the deliberate modification of ecosystems by man, that requires the ecologist to impose an orderly and logical organization of his research that goes beyond the sequential application of simple tests, although the "appeal to nature" invoked by the experimental method remains at the heart of that organization.

References

[1] Jeffers, J. N. R. (1978). *An Introduction to Systems Analysis: With Ecological Applications*. Arnold, London, England.

[2] Tansley, A. G. (1935). The use and abuse of vegetational concepts and terms. *Ecology*, **16**, 284–307.

Bibliography

Häfele, W. (1981). *Energy in a Finite World: A Global Systems Analysis*. Ballinger, Cambridge, Massachusetts. (Presents a report by the Energy Systems Program Group of the International Institute for Applied Systems Analysis on one of the most intensive systems analyses ever completed.)

McIntosh, R. P. (1985). *The Background of Ecology: Concept and Theory*. Cambridge University Press, Cambridge, England. (A useful review of the development of ecology as a science, containing a critical examination of the role of systems analysis in ecological research. Strongly recommended.)

Odum, H. T. (1983). *Systems Ecology: An Introduction*. Wiley, New York. (One of the principal textbooks for the school of systems analysis that puts the greatest emphasis on energy transfers in ecological systems.)

(ECOLOGICAL STATISTICS
STATISTICS IN ANIMAL SCIENCE
STATISTICS IN FORESTRY)

J. N. R. JEFFERS

SYSTEMS, GAMBLING   *See* GAMBLING, STATISTICS IN

SYSTEMS OF FREQUENCY CURVES
See FREQUENCY CURVES, SYSTEMS OF

SYSTEMS OF FREQUENCY SURFACES
See FREQUENCY SURFACES, SYSTEMS OF

SZROETER'S TEST OF HOMOSCEDASTICITY

A somewhat neglected test of homoscedasticity against ordered alternatives in a linear regression model*. In the model

$$Y_t = \mathbf{X}'_t\boldsymbol{\beta} + Z_t \qquad (t = 1, 2, \ldots T)$$

(with obvious notation), where the Z_t's are mutually independent normal variables, with $E[Z_t] = 0$ and $\mathrm{Var}(Z_t) = \sigma_t^2$, it is desired to test the null hypothesis $\sigma_1 = \sigma_2 = \cdots = \sigma_T$ against the alternatives $\sigma_1 \leqslant \sigma_2 \leqslant \cdots \leqslant \sigma_T$ (with at least one strict inequality).

If the observed residuals* from a fitted regression $y = X'\hat{\beta}$ (where $\hat{\beta}$ is any consistent* estimator of β) are denoted by

$$E_t = Y_t - \mathbf{X}'_t\hat{\beta},$$

then Szroeter's [7] test statistic is

$$Q = T(\tilde{H} - \bar{h}) \Big/ \left\{ 2 \sum_{t=1}^{T} \left(h_t - \bar{h} \right)^2 \right\}^{1/2},$$

where

$$\tilde{H} = \Sigma_{t=1}^{T} W_t h_t$$

with $W_t = E_t^2 / (\Sigma_{j=1}^{T} E_j)^2$, $\bar{h} = T^{-1}\Sigma_{t=1}^{T} h_t$, and $h_1 \leqslant h_2 \leqslant \cdots \leqslant h_T$ being a "suitably chosen" set of T fixed numbers

Under the null hypothesis, Q has asymptotically (as $T \to \infty$) a unit normal distribution. If one of the alternative hypotheses is valid, Q tends to be larger than zero.

If we choose $h_t = t$ $(t = 1, 2, \ldots, T)$, then

$$Q = \left(\frac{\sigma T}{T^2 - 1} \right)^{1/2} \left(\tilde{H} - \frac{T+1}{2} \right)$$

with

$$\tilde{H} = \frac{\Sigma_{t=1}^{T} t E_t^2}{\Sigma_{t=1}^{T} E_t^2}.$$

Usually, ordinary least squares* estimators of β are used for $\hat{\beta}$.

Griffiths and Surekha [3] have investigated the power of the test, using simulation*, and compared it with the powers of competitors such as the Goldfield–Quandt* test, tests based on Lagrange multipliers*, tests developed by Silvey [5] and Rao [4], and modified by Godfrey [2], and Breusch and Pagan's test [1]. A measure of heteroscedasticity* suggested by Surekha [6] was used to provide a basis for comparison. Griffiths and Surekha's [3] results were quite favorable to Szroeter's test.

References

[1] Breusch, T. S. and Pagan, A. R. (1979), *Econometrica*, **47**, 1284–1294.

[2] Godfrey, L. G. (1978). *J. Econometrics*, **8**, 227–236.

[3] Griffiths, W. E. and Surekha, K. (1986). *J. Econometrics*, **31**, 219–231.

[4] Rao, C. R. (1948). *Proc. Camb. Philos. Soc.*, **44**, 50–57.

[5] Silvey, S. D. (1959). *Ann. Math. Statist.*, **30**, 389–407.

[6] Surekha, K. (1980). Contributions to Bayesian Analysis in Heteroscedastic Model, Ph.D. thesis, University of New England, Armidale, Australia.

[7] Szroeter, J. (1978). *Econometrica*, **46**, 1311–1327.

(HETEROSCEDASTICITY
LINEAR REGRESSION)

T

T_1- AND T_2-CLASSES OF DISTRIBUTIONS

In 1977 Thorin [14, 15] introduced a class of infinitely divisible* distributions on $[0, \infty)$, called generalized Γ-convolutions, but here called T_1-distributions, which made it possible for him to prove that in particular the lognormal* distribution is infinitely divisible. Though based on complex analysis and mathematically complicated, his technique turned out to be very powerful and now it is known that many of the most common distributions are in T_1, or in its extension $\bar{T}_1$ to distributions on $(-\infty, \infty)$, and thus infinitely divisible. (This may indicate that infinite divisibility is a less important concept than the early prominent probabilists of this century were led to believe.) The T_2-class is an extension of T_1, containing all mixtures of exponential distributions* and being, like T_1, closed with respect to (weak) limits, convolution*, and convolution roots (if $F = F_n^{n*}$, then F_n is an nth *convolution root* of F).

THE T_1-CLASS

A probability distribution on $[0, \infty)$ with Laplace (–Stieltjes) transform* $\varphi(s)$ is said to be a T_1-distribution if

$$\frac{\varphi^*(s)}{\varphi(s)} = -a - \int \frac{1}{t+s} U(dt),$$

where $a \geq 0$ and $U(dt)$ is a nonnegative measure on $(0, \infty)$. Here a and $U(dt)$ are unique and a is the left extremity of the distribution. The T_1-class is obtained by taking limits of finite convolutions of gamma distributions. For a gamma distribution*, $U(dt)$ is concentrated at one point.

A nondegenerate T_1-distribution with $a = 0$ and $\beta = \int U(dt) < \infty$ is a scale mixture of gamma distributions with shape parameter β and has thus a density $f(x)$. Moreover [6], $\beta = \sup\{\alpha; \lim_{x \downarrow 0} f(x)/x^{\alpha-1} = 0\}$. More surprising is that a scale mixture* of gamma distributions belongs to T_1 if the scale mixing distribution does; see ref. 6. In fact, if X and Y are independent random variables, then

$$X \sim T_1, Y \sim \text{gamma}$$
$$\Rightarrow X \cdot Y \sim T_1 \text{ and } X/Y \sim T_1$$

where $\sim$ stands for "has its distribution in." A consequence is that, for $X \sim T_1$,

$$f_n(x) = \frac{(-1)^n}{n!} \left(\frac{n}{x}\right)^{n+1} \varphi_X^{(n)}\left(\frac{n}{x}\right),$$
$$x > 0, \ n = 1, 2, \ldots,$$

is a sequence of T_1-densities converging (weakly) to the density of X.

157

Another property of T_1 is the following:

$$X \sim T_1, \quad Y \sim N(0,1) \Rightarrow \sqrt{X} \cdot Y \sim \overline{T}_1,$$

where $\overline{T}_1$ is the class of extended generalized Γ-convolutions, i.e., the class of limits of finite convolutions of gamma distributions on $(0, \infty)$ as well as on $(-\infty, 0)$; see refs. 16, 5, and 1. Nondegenerate $\overline{T}_1$-distributions are absolutely continuous and unimodal* since $\overline{T}_1$ is a subclass of the L-class*. Stable distributions* belong to $\overline{T}_1$ and positive stable distributions to T_1. More important is that large classes of densities of distributions in T_1 and $\overline{T}_1$ have been found. Distributions with densities of the form

$$f(x) = Cx^{\beta - 1} \prod_{i=1}^{N} (1 + c_i x)^{-\gamma_i}, \quad x > 0,$$

where the parameters are positive, are in T_1; see ref. 3. The class of these distributions and their limits is denoted by B. Of course, the gamma distributions* are in B. The B-class is closed with respect to multiplication of positive powers of densities (with proper normalization) but it has also other attractive properties:

(i) $X \sim B \Rightarrow X^q \sim B$ if $|q| \geq 1$.

(ii) $X \sim B$, $Y \sim$ gamma (X, Y independent) $\Rightarrow X \cdot Y \sim B$, and $X/Y \sim B$.

(iii) $X_1, \ldots, X_n \sim$ gamma ($X_1, \ldots, X_n$ independent) $\Rightarrow X_1^{q_1} \cdot \cdots \cdot X_n^{q_n} \sim B$ if $|q_j| \geq 1$, $j = 1, \ldots, n$.

The F-, lognormal, and many other common distributions on $(0, \infty)$ are contained in B and hence in T_1 and L.

If $f(x)$ is a symmetric density on $(-\infty, \infty)$ and $f(\sqrt{s})$, $s \geq 0$, is proportional to the Laplace transform of a B-distribution, then $f(x)$ is a $\overline{T}_1$-density; see ref 5. The infinite divisibility of the t-distribution*, first established by Grosswald [7] though implicit in ref. 8, follows immediately from this result. Another consequence is the infinite divisibility of the hyperbolic distribution*.

The probabilistic interpretation of the B-class is still unclear. The class would be more understandable if, for c, $\gamma \geq 0$, $C(1 + cx)^{-\gamma} f(x)$ is a T_1-density whenever

$f(x)$ is. This has only been verified to be true when γ is an integer (unpublished).

Discrete T_1-distributions have been studied in ref. 4.

THE T_2-CLASS

A distribution on $[0, \infty)$ with Laplace transform $\varphi(s)$ is said to be a T_2-distribution if

$$\frac{\varphi'(s)}{\varphi(s)} = -a - \int \frac{1}{(t + s)^2} Q(dt),$$

where $a \geq 0$ and $Q(dt)$ is a nonnegative measure on $(0, \infty)$. The name is motivated by the exponent 2, but other names have been used as well.

The T_2-distributions were introduced in ref. 5 to obtain a class of infinitely divisible distributions containing mixtures of exponential distributions* and being closed under convolution. Mixtures of exponential distributions are infinitely divisible (the Goldie–Steutel theorem) and are those T_2-distributions for which $a = 0$ and $Q(dt)$ is absolutely continuous* with a density bounded by 1. The T_1-distributions are those T_2-distributions for which $Q(dt)$ is absolutely continuous with a nondecreasing density.

The Laplace transform $\varphi(s)$ of an infinitely divisible distribution on $[0, \infty)$ satisfies the relation

$$\varphi'(s)/\varphi(s) = -a - \int e^{-sy} y N(dy),$$

where $N(dy)$ is a nonnegative measure on $(0, \infty)$; the Lévy measure. The distribution is in T_2 if and only if $N(dy)$ has a completely monotone density $n(y)$ [i.e., the derivatives of $n(y)$ have alternating signs]. The distribution is in T_1 if and only if $yn(y)$ is completely monotone.

The T_2-distributions (with $a = 0$) appear naturally in the following ways:

(i)   As the class of marginal distributions for shotnoise processes* of the form $X(t) = \sum_{\tau_k \leq t} g(t - \tau_k) V_k$, $t \in R$, where g is a nonnegative function on $[0, \infty)$, $\{\tau_k\}_{-\infty}^{+\infty}$ points in a Poisson point process on $(-\infty, \infty)$, and the

V_k's are independent random variables with a common exponential distribution. (For a similar representation of the T_1-distributions, see SHOT-NOISE PROCESSES.)

(ii) As the smallest class of distributions that is closed with respect to convolution and weak limits and that contains all first passage time distributions for birth–death and diffusion processes. (Here the T_2-distributions may have probability mass $\leqslant 1$.)

First passage time distributions for diffusion processes* as T_2-distributions have been studied by Kent [11, 12]. In fact, the T_2-class appeared originally in ref. 9, pp. 214–217, in this context. For birth–death* and diffusion processes, every first passage time distribution is representable (though not uniquely) as a PF_∞-distribution (i.e., a denumerable convolution of exponential distributions) convolved with a mixture of exponential distributions as shown by Keilson [10]. If the class of these latter distributions is denoted by K, then $T_1 \subset K \subset T_2$. Every distribution in K is a limit of first passage time distributions for a sequence of birth–death processes and hence so is every distribution in T_1, cf. ref. 5.

For a stationary autoregressive process of order 1,

$$X_n = cX_{n-1} + \epsilon_n, \qquad n = 1, 2, \ldots,$$

where $X_n \sim T_1$, the innovations ϵ_n have their common distribution in T_2. Another similar relation between the T_1- and T_2-distributions has been found by Berg and Forst [2]. Multivariate T_2-distributions have been studied by Kent [13].

References

[1] Barndorff-Nielsen, O., Kent, J., and Sørensen, M. (1982). *Int. Statist. Rev.*, **50**, 145–159.

[2] Berg, C. and Forst, G. (1982). *Scand. Actu. J.*, **65**, 171–175.

[3] Bondesson, L. (1979a). *Ann. Prob.*, **7**, 965–979. (Very technical content.)

[4] Bondesson, L. (1979b). *Scand. Actu. J.*, **62**, 125–166.

[5] Bondesson, L. (1981). *Z. Wahrsch. verw. Geb.*, **57**, 39–71. Correction and Addendum, **59**, 277. (Contains many references.)

[6] Bondesson, L. (1985). *Scand. Actu. J.*, **68**, 197–209.

[7] Grosswald, E. (1976). *Z. Wahrsch. verw. Geb.*, **36**, 103–109.

[8] Hammersley, J. (1961). *Proc. Fourth Berkeley Symp.*, *III*, J. Neyman, ed. Univ. of California Press, pp. 17–78.

[9] Itô, K. and McKean, H. P. (1965). *Diffusion Processes and their Sample Paths*. Springer, Berlin, Germany.

[10] Keilson, J. (1981). *Statist. Neerland.*, **35**, 49–55.

[11] Kent, J. (1980). *Z. Wahrsch. verw. Geb.*, **52**, 309–319.

[12] Kent, J. (1982). *Ann. Prob.*, **10**, 207–219.

[13] Kent, J. (1983). In *Probability*, *Statistics and Analysis*, J. F. C. Kingman and G. E. H. Reuter, eds. Cambridge University Press, London, England, pp. 161–179.

[14] Thorin, O. (1977a). *Scand. Actu. J.*, **60**, 31–40. (The first paper on T_1-distributions; stable and Pareto distributions are shown to be in T_1.)

[15] Thorin, O. (1977b). *Scand. Actu. J.*, **60**, 121–148.

[16] Thorin, O. (1978). *Scand. Actu. J.*, **61**, 141–149.

(BIRTH AND DEATH PROCESSES
DIFFUSION PROCESSES
INFINITE DIVISIBILITY
L-CLASS LAWS
SHOT-NOISE PROCESSES)

L. BONDESSON

t DESIGNS

Designs for arranging v treatments in b blocks, each of size $k < v$ are called *proper incomplete block designs.** A *t* design is a proper incomplete block design with the additional property that every subset of t treatments appears together in the same number λ_t of blocks. Traditionally, t-(v, k, λ_t) denotes such a *t* design.

t designs with $\lambda_t = 1$, known as Steiner systems, were first introduced by Woolhouse [24]. The appeal was purely combinatorial. *Balanced incomplete block* (BIB) *designs* (*see* BLOCKS, BALANCED INCOMPLETE), which are *t* designs with $t = 2$, were first utilized by

Yates [25] in statistical experiments when equal precision was desired on all treatment comparisons. *t* designs with $t = 3$ are called *doubly BIB designs* by Calvin [1], who also pointed out the usefulness of such designs for experiments in which the treatment effects are correlated. Hedayat and John [5] showed that under the homoscedastic additive linear model, a BIB design is resistant* (that is, retains its variance balance) upon removal of any single treatment if and only if it is a 3 design. Most [17] generalized the preceding result and showed that a BIB design is resistant upon removal of any *n* treatments if and only if it is an $(n + 2)$ design. Moreover, the symmetries of *t* designs provide many statistical optimalities as can be seen in refs. 15 and 16. For more applications and a detailed survey of the literature on *t* designs, see refs. 7, 8, and 14, which also provide an extensive bibliography on the literature of *t* designs.

Example. Let $v = 8$, $k = 4$, and label the 8 treatments as $1, 2, \ldots, 8$. Then the following 14 columns (blocks)

$$
\begin{array}{cccccccccccccc}
1 & 1 & 1 & 1 & 1 & 1 & 1 & 2 & 2 & 2 & 2 & 3 & 3 & 4 \\
2 & 2 & 2 & 3 & 3 & 4 & 5 & 3 & 3 & 4 & 6 & 4 & 5 & 5 \\
3 & 4 & 5 & 4 & 7 & 6 & 6 & 4 & 5 & 5 & 7 & 6 & 6 & 7 \\
6 & 8 & 7 & 5 & 8 & 7 & 8 & 7 & 8 & 6 & 8 & 8 & 7 & 8 \\
\end{array}
$$

form a 2-(8, 4, 3) design. It is also a 3-(8, 4, 1) design. Under the homoscedastic additive linear model, this design is variance balanced even when all of the experimental units assigned to one of the treatments have been lost.

It is worth noting that a BIB design with $v = 2k$ together with its complement form a 3 design no matter how the BIB design is constructed [21].

NECESSARY CONDITIONS

A t-(v, k, λ_t) design in which every subset of k treatments occurs as a block exactly m ($\geqslant 1$) times is called a *trivial t design*. Necessarily, a *t* design with $t = k$ or $k = v - 1$ is trivial. In search of nontrivial t-(v, k, λ_t) de-

signs with $t < k$ and $k \leqslant v - 2$, there are necessary constraints. By observing that every t-(v, k, λ_t) design is an s-(v, k, λ_s) design with

$$
\lambda_s = \lambda_t \binom{v - s}{t - s} \bigg/ \binom{k - s}{t - s}
$$

for all $s < t$, a set of necessary conditions for the existence of a nontrivial *t* design is that

$$
\lambda_t \binom{v - s}{t - s} \equiv 0 \left(\bmod \binom{k - s}{t - s} \right),
$$
$$
s = 0, \ldots, t - 1. \quad (1)
$$

The conditions in (1) are also sufficient when λ_t is sufficiently large [23].

BOUNDS ON THE NUMBER OF BLOCKS *b*

To avoid a large number of blocks and hence an inordinately large and costly number of experimental units involved in a *t* design, an important consideration to practitioners as well as theoreticians is the number of blocks needed to form a *t* design. A list of lower bounds on the number of blocks *b* of a *t* design is:

1. When $v \geqslant k + 2$, $b \geqslant (t - 1)(v - t + 2)$ [18, Chap. 7].
2. When $v \geqslant k + t - 1$, $b \geqslant 2^{t-2}(v - t + 2)$ [2].
3. When $t = 2s$ and $v \geqslant k + s$, $b \geqslant \binom{v}{s}$ [20].
4. When $t = 2s + 1$ and $v \geqslant k + s + 1$, $b \geqslant 2 \binom{v - 1}{s}$ [20].
5. When $t = 2s + 1$ and $v \geqslant k + s$, $b \geqslant v - s \left\{ k \binom{v}{s} \right\}^{-1}$ [13].

Further discussion on the bounds on *b* can be found in ref. 7.

TABLES OF *t* DESIGNS

The problem of the existence and construction of *t* designs has been of considerable interest to both statisticians and algebraists since the 1930s. The following tables should be beneficial.

1. Fisher and Yates tables [3]. The original tables listed all parameter combinations and their solutions for 2 designs in the range $v, b \leqslant 100$ and $r, k \leqslant 10$. This list was extended by Rao [19] to include parameter combinations with $r, k \leqslant 15$, and further by Sprott [22] to include the cases $r, k \leqslant 20$. Such tables and the methods of their construction are available in various textbooks such as refs. 4 and 18.

2. Kageyama and Hedayat [14] provide 8 tables, including: a table of parameter combinations that satisfy the necessary conditions, but the corresponding *t* designs do not exist; a table of parameter combinations that satisfy the necessary conditions while the existence or nonexistence of the corresponding *t* designs is not established in the published literature; a table of small size *t* designs, $t \geqslant 3$; etc.

THE TRADE-OFF METHOD

While the primary interest of using a *t* design only relies on every subset of *t* treatments appearing together in the same number of blocks, there is no need to restrict the blocks of a *t* design to be all distinct. From the point of view of applications, some treatment combinations may be more desirable than others. Since many tables of small size *t* designs are available, the experimenter can then choose a *t* design from the existent tables and replace some blocks in the design by an equal number of blocks to obtain another *t* design which perhaps will be more economical and fits the individual situation better. This is the idea of *trade-off* [6]. In order that the resultant design retain the structure of being a *t* design, the collection of blocks (T) to be removed from a given design and the collection of blocks ($\overline{T}$) to be

added into the design must satisfy the following property:

Every subset of *t* treatments appears together the same number of times in blocks of T as it appears together in blocks of $\overline{T}$.

Such pairs of collections T and $\overline{T}$ are said to form a *trade on the given design*. For example, let the design be

$$
\begin{array}{cccccccccccccc}
1 & 1 & 1 & 1 & 1 & 1 & 2 & 2 & 2 & 2 & 3 & 3 & 4 & 4 \\
2 & 2 & 3 & 3 & 5 & 6 & 3 & 3 & 4 & 5 & 4 & 6 & 5 & 5 \\
4 & 6 & 4 & 5 & 7 & 7 & 5 & 7 & 7 & 6 & 6 & 7 & 6 & 7
\end{array}
$$

This is a 2-$(7, 3, 2)$ design with 14 distinct blocks. Take

T				$\overline{T}$			
1	1	2	4	3	2	1	1
2	3	3	5	5	4	4	2
4	5	7	7	7	7	5	3

Then T and $\overline{T}$ form a trade on the preceding design. If the blocks in T in the preceding design are replaced by the blocks in $\overline{T}$, a new design that contains 13 distinct blocks with the block (247) repeated twice is obtained. This new design is again a 2-$(7, 3, 2)$ design.

For more details about trades and their applications see refs. 9–12.

References

[1] Calvin, L. D. (1954). *Biometrics*, **10**, 61–88.

[2] Dey, A. and Saha, G. M. (1974). *Ann. Inst. Statist. Math.*, **26**, 171–173.

[3] Fisher, R. A., and Yates, F. (1963). *Statistical Tables for Biological Agricultural and Medical Research*. Oliver and Boyd, London, England.

[4] Hall, M., Jr. (1967). *Combinatorial Theory*. Blaisdell, Waltham, Mass.

[5] Hedayat, A. and John, P. W. M. (1974). *Ann. Statist.*, **2**, 148–158.

[6] Hedayat, A. and Li, S. Y.-R. (1979). *Ann. Statist.*, **7**, 1277–1287.

[7] Hedayat, A. and Kageyama, S. (1980). *J. Statist. Plann. Inf.*, **4**, 173–212.

[8] Hedayat, A. and Kageyama, S. (198x). *J. Statist. Plann. Inf.* To appear.

[9] Hedayat, A. and Hwang, H. L. (1983). *Commun. Statist. Simul. Comp.*, **12**, 109–125.

[10] Hwang, H. L. (1982). On (k, t) trades and the construction of BIB designs with repeated blocks. Ph.D. dissertation, University of Illinois at Chicago, Chicago, Ill.

[11] Hwang, H. L. (1983a). A characterization of the trades on *t*-designs. *IMS Bull.*, **67**, 85.

[12] Hwang, H. L. (1983b). Further investigation of (k, t) trades. *IMS Bull.*, **70**, 226.

[13] Kageyama, S. (1975). *Ann. Inst. Statist. Math.*, **27**, 529–530.

[14] Kageyama, S. and Hedayat, A. (1983). *J. Statist. Plann. Inf.*, **7**, 257–287.

[15] Kiefer, J. (1958). *Ann. Math. Statist.*, **29**, 675–699.

[16] Kiefer, J. (1975). In *A Survey of Statistical Designs and Linear Models*. North-Holland, Amsterdam, Netherlands, pp. 333–353.

[17] Most, B. M. (1975). *Ann. Statist.*, **3**, 1149–1162.

[18] Raghavarao, D. (1971). *Construction and Combinatorial Problems in Design of Experiments*. Wiley, New York.

[19] Rao, C. R. (1961). *Sankhyā*, **23**, 117–127.

[20] Ray-Chaudhuri, D. K. and Wilson, R. M. (1975). *Osaka J. Math.*, **12**, 737–744.

[21] Sprott, D. A. (1955). *Ann. Math. Statist.*, **26**, 752–758.

[22] Sprott, D. A. (1962). *Sankhyā*, **24**, 203–204.

[23] Wilson, R. M. (1973). *Utilitas Math.*, **4**, 207–215.

[24] Woolhouse, W. S. B. (1844). Prize question 1733. *Lady's and Gentleman's Diary*, London.

[25] Yates, F. (1936). *Ann. Eugenics (Lond.)*, **7**, 121–140.

(BLOCKS, BALANCED INCOMPLETE
DESIGN OF EXPERIMENTS
INCOMPLETE BLOCK DESIGNS)

HUEY-LUEN HWANG

t DISTRIBUTION

The *t* distribution is that of the ratio of a standard normal variable to an independent root-mean-square chi-squared variable, written succinctly as

$$t_\nu = Z \big/ \sqrt{\chi_\nu^2 / \nu}, \qquad (1)$$

where ν denotes the *degrees of freedom**. The genesis of *t* was the investigation by Gosset* ("Student") [18] into the distribution of $\sqrt{n}\,(\overline{X} - \mu)/S$, where $\overline{X}$ and S^2 are the sample mean and sample variance of a random sample of size n from a normal $N(\mu, \sigma^2)$ population; this has a t_{n-1} distribution. A key feature here is that the distribution is free of the unknown scale parameter σ and depends only on its degrees of freedom $\nu = n - 1$ (*see also* STUDENTIZATION). Many test statistics have a *t* distribution under certain null hypotheses of interest (*see* STUDENT *t*-TESTS). Related confidence intervals and regions* can be constructed (*see also* PIVOTAL QUANTITIES).

The distribution of t_ν is bell-shaped, symmetric about zero, and infinitely divisible [4, 6]. The probability density function (PDF) is

$$\left[\sqrt{\nu}\, B\big(\tfrac{1}{2}, \tfrac{1}{2}\nu\big)\right]^{-1}\!\big(1 + t^2/\nu\big)^{-\frac{1}{2}(\nu + 1)},$$
$$|t| < \infty, \qquad (2)$$

which converges to the standard normal PDF $(\sqrt{2\pi})^{-1}e^{-t^2/2}$ as $\nu \to \infty$. The cumulative distribution function* (CDF) is

$$\Pr(t_\nu \leqslant t) = 1 - \tfrac{1}{2}I_{\nu/(\nu + t^2)}\big(\tfrac{1}{2}\nu, \tfrac{1}{2}\big), \quad (3)$$

where $I(\cdot, \cdot)$ is the incomplete beta function ratio*. Owen [13] gives an exact expression suitable for computation, as follows: For ν odd,

$$\Pr(t_\nu \leqslant t) = \frac{1}{2} + \tan^{-1}\!\left(\frac{t}{\sqrt{\nu}}\right)$$
$$+ \frac{t\sqrt{\nu}}{\nu + t^2} \sum_{j=0}^{(\nu-3)/2} b_j B^j,$$
$$b_0 = 1, \qquad b_j = \{2j/(2j + 1)\}b_{j-1},$$
$$j = 1, \ldots, (\nu - 3)/2,$$
$$B = \big(1 + t^2/\nu\big)^{-1}; \qquad (4a)$$

for ν even,

$$\Pr(t_\nu \leqslant t) = \frac{1}{2} + \frac{t}{2\sqrt{\nu + t^2}} \sum_{j=0}^{(\nu-2)/2} c_j B^j,$$
$$c_0 = 1, \; c_j = \{(2j - 1)/(2j)\}c_{j-1},$$
$$j = 1, \ldots, (\nu - 2)/2. \qquad (4b)$$

The case $\nu = 1$ yields the Cauchy distribution*.

There are other exact expressions and several approximations to the CDF [7, 9, 14], of which Johnson and Kotz [7] give a comprehensive survey up to 1970. A more recent normalizing transformation [1],

$$z = \pm \left\{ \frac{8\nu + 1}{8\nu + 9} \right\}$$

$$\times \left[\left(\nu + \frac{19}{12} \right) \log \left\{ 1 + \frac{t^2}{\nu + 12} \right\} \right]^{1/2}, \quad (5)$$

is uniformly accurate to $O(\nu^{-2})$ as well as being concise. Soms [17] derived bounds on a quantity analogous to Mills' ratio*. For $\nu \leqslant 6$, however, the expressions (4) for the CDF are as concise as most approximations and have the advantage of being exact.

If $\nu > r$ and r is odd, the rth moment of t_ν is zero (if $\nu \leqslant r$, the rth moment is undefined). If $\nu > r$ and r is even, the rth moment is

$$\mu_r = \nu^{r/2} \frac{1.3 \cdots (r-1)}{(\nu - r)(\nu - r + 2) \cdots (\nu - 2)} \quad (6)$$

(if $\nu \leqslant r$, the rth moment is infinite).

The mean of t_ν is thus zero ($\nu > 1$) and the variance is $\nu/(\nu - 2)$, for $\nu > 2$. The kurtosis* μ_4/μ_2^2 is $3 + 6/(\nu - 4)$, $\nu > 4$.

Approximations for computation of the quantiles and percentage points of t have been surveyed in several sources [7, 14, 15]. These are compared for accuracy with existing tables and with the Cornish–Fisher expansion* [5, 7]. Bukač and Burstein [3] give polynomial approximations at the commonest α levels with relative errors less than 0.00005. Two sources make special provision for percentage points of t between 0.01 and 0.00001 (these are required when using the Bonferroni method in multiple comparisons*); Moses [11] gives charts which provide near two-decimal accuracy, and Koehler [8] provides the approximation

$$\left(t_{\nu, \alpha/2} \right)^{-1} = -0.0953 - (0.631)/(\nu + 1)$$

$$+ 0.81 \left[-\ln(2\alpha - \alpha^2) \right]^{-1/2}$$

$$+ 0.076 \left[\left(\sqrt{2\pi} - \tfrac{1}{2} \right) \alpha \sqrt{\nu} \right]^{1/\nu}, \quad (7)$$

for which the relative error does not exceed 1.4% if $\nu \geqslant 8$ and $0.00001 \leqslant \alpha \leqslant 0.2$.

Johnson and Kotz [7] give a detailed discussion of the coverage and accuracy up to 1970 of tables of the CDF, PDF, and percent points of t, including extreme tail percentiles, and of two nomographs*. The earliest tables were compiled by "Student" himself [18], and the most extensive comprise the collection by Smirnov [16]. Tables published by the Japanese Standards Association [19] include percent points to three decimal places for $0.0005 \leqslant \alpha \leqslant 0.25$ and $\nu = 1(1)50, 60, 80, 120, 240, \infty$; also ordinates and tail probabilities, both for $t = 0(0.1)5.0$ and $\nu = 1(1)10, 12, 15, 20$. Mardia and Zemroch's tables [10] give upper percent points to five significant figures for fractional degrees of freedom $\nu = 0.1(0.1)3.0$ $(0.2)7.0(0.5)11(1)40, 60, 120, \infty$, and $0.0001 \leqslant \alpha \leqslant 0.5$. A nomograph by Nelson [12] provides upper percent points of t for $\nu \geqslant 4$, $0.0001 \leqslant \alpha \leqslant 0.10$.

The t distribution is related to certain others. t_ν^2 has an F distribution* with 1 and ν degrees of freedom. If the numerator in (1) has a normal $N(\delta, 1)$ rather than a $N(0, 1)$ distribution, then t_ν has a noncentral t distribution*. For discussion of multivariate versions, see the list of related entries.

Limited study has been done on t-type statistics based on nonnormal distributions. Bowman et al. [2] examine the effect of normal mixtures and Pearson Type I distributions* on skewness and kurtosis, and list references on earlier studies. Zolatarev [20] gives the limiting distribution as $n \to \infty$ of nonnormal t statistics of the form

$$t_n = \sum X_i \Big/ \left[\sum \left(X_i - \overline{X} \right)^2 \right]^{1/2},$$

where $X_1, \ldots, X_n$ constitute a random sample from a parent distribution belonging to the domain of attraction of a symmetric stable law* with characteristic function* of the form $\exp(-\lambda |t|^\alpha)$, $\lambda > 0$, $0 < \alpha < 2$.

References

[1] Bailey, B. J. R. (1980). *Appl. Statist.*, **29**, 304–306.

[2] Bowman, K. O., Beauchamp, J. J., and Shenton, L. R. (1977). *Int. Statist. Rev.*, **45**, 233–242.

[3] Bukač, J. and Burstein, H. (1980). *Commun. Statist. B*, **9**, 665–672.

[4] Epstein, B. (1977). *Sankhyā B*, **39**, 103–120.

[5] Fisher, R. A. and Cornish, E. A. (1960). *Technometrics*, **2**, 205–226.

[6] Ismail, M. E. H. (1977). *Ann. Prob.*, **5**, 582–585.

[7] Johnson, N. L. and Kotz, S. (1970). *Distributions in Statistics: Continuous Univariate Distributions*, 2. Wiley, New York. (See Chapter 27.)

[8] Koehler, K. J. (1983). *Technometrics*, **25**, 103–105.

[9] Ling, R. F. (1978). *J. Amer. Statist. Ass.*, **73**, 274–283.

[10] Mardia, K. V. and Zemroch, P. J. (1978). *Tables of the F- and Related Distributions with Algorithms*. Academic, New York.

[11] Moses, L. E. (1978). *Commun. Statist. B*, **7**, 479–490.

[12] Nelson, L. S. (1975). *J. Qual. Tech.*, **7**, 200–201.

[13] Owen, D. B. (1968). *Technometrics*, **10**, 445–478.

[14] Patel, J. K. and Read, C. B. (1982). *Handbook of the Normal Distribution*. Dekker, New York, (See Section 7.10.)

[15] Prescott, P. (1974). *Biometrika*, **61**, 177–180.

[16] Smirnov, N. V. (1961). *Tables for the Distribution and Density Functions of t-Distribution ("Student's" Distribution)*. Pergamon, Oxford, England.

[17] Soms, A. P. (1980). *J. Amer. Statist. Ass.*, **75**, 438–440.

[18] "Student" (1908). *Biometrika*, **6**, 1–25.

[19] Yamauti, Z., ed. (1972). *Statistical Tables and Formulas with Computer Applications*. Japanese Standards Association, Tokyo. (Tables B3.)

[20] Zolatarev, V. M. (1985). *Problemy Ustoĭchivosti Stokhastich. Modelei*. Nauka, Moscow, pp. 57–63.

(GOSSET, WILLIAM SEALY
MATRIC-*t* DISTRIBUTION
MULTIPLE COMPARISONS
MULTIVARIATE *t* DISTRIBUTION
NONCENTRAL *t* DISTRIBUTION
PIVOTAL QUANTITIES
STUDENT'S *t* TESTS)

CAMPBELL B. READ

T-SQUARE SAMPLING

A practical method of sampling spatial distributions of hidden or unmapped (point) items, proposed by Besag and Gleaves [1]. The measurements produced provide infor-

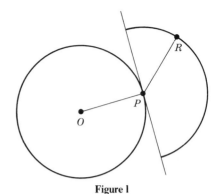

Figure 1

mation both about the probability mechanism generating the locations of the items [for example, a spatial (Poisson) process*] and about the density of the items. Unlike quadrat sampling*, which requires prior decisions about the approximate density of the items, the distances used in T-square sampling are determined by the population sampled.

T-square sampling is a variant of nearest neighbor sampling. In each sampling scheme, sampling sites $O_1, O_2, \ldots, O_n$, are chosen randomly in the region of interest, for example by choosing exact coordinates on a map of the region. Each sampling site is then visited and the location of the nearest item to that sampling site is then found, say at P_i for sampling site O_i. The sampling schemes differ in that, while nearest neighbor sampling involves searching for the location of the nearest item to P_i, T-square sampling involves searching only that half-plane on the far side from O_i of a line through P_i perpendicular to O_iP_i. This restriction, which can be implemented by requiring that the compass bearings from all searched areas be within 90° of the compass bearing from P_i to O_i, simplifies the distributional properties of the bivariate distance measurements. A variant of T-square sampling, wandering quarter sampling*, can be used to obtain multivariate distance measurements with similar simple distributional properties.

Reference

[1] Besag, J. E. and Gleaves, J. T. (1973). *Bull. Int. Statist. Inst.*, **45**(1), 153–158.

Bibliography

Diggle, P. J. (1983). *Statistical Analysis of Spatial Point Processes*. Academic, London, England.

Ripley, B. D. (1981). *Spatial Statistics*. Wiley, New York.

(COEFFICIENT OF VARIATION
SPATIAL PROCESSES
SPATIAL SAMPLING)

W. G. S. HINES

t TEST, STUDENT'S   *See* STUDENT'S *t*-TESTS

TABLES   *See* NUMERACY

TABULATION   *See* NUMERACY

TAGGING   *See* CAPTURE–RECAPTURE METHODS

TAGUCHI METHOD FOR OFF-LINE QUALITY CONTROL

Off-line quality control* methods are the measures taken at the product and process design stages to improve product quality. G. Taguchi has developed a systematic approach to off-line quality control that has been used to a moderate extent in Japan and has attracted attention in a number of other countries, including the United States. What follows is a summary of the features of the Taguchi method with emphasis on the statistical properties.

Taguchi defines the quality of a product to be "the loss imparted by the product to the society from the time the product is shipped" (Phadke [5]). The emphasis is on losses caused by deviation of the product's functional characteristic (Y) from a desired target value (τ). Taguchi believes that the behavior of loss can be approximated in some instances by a quadratic function

$$L = k(Y - \tau)^2 \qquad (1)$$

where k is a coefficient. Thus, the mere satisfaction of specified tolerances (such as occurs with a pass/fail interpretation of loss) is less desirable than the attainment of optimal conditions. Furthermore, since the expected loss is proportional to the variance of the functional characteristic when $E[Y] = \tau$, there should be continuous efforts to identify and reduce the variation in the product's functional characteristic.

Factors affecting the performance of Y can be categorized as *controllable factors* and *noise factors*. The latter include *outer noise* (such as the variation in operational environment), *inner noise* (such as the deterioration in the product or process), and *between product noise* (such as that due to manufacturing imperfections). Rather than attempt to find and control the noise factors, Taguchi advocates a three-stage design procedure for off-line quality control: (a) system design, (b) parameter design, and (c) tolerance design. In the first stage, a system is designed to fulfill a specific function. The second step, parameter design, attempts to find levels of the controllable factors such that the effect of the noise factors on the functional characteristic is minimized. In tolerance design, it may be necessary to specify narrower tolerances for some of the factors. This final stage is considered only if the reduction in variation achieved at the parameter design stage is insufficient. Thus, parameter design is the key stage.

To illustrate these concepts, consider the example (see Table 1) presented by Byrne and S. Taguchi [1] in which an objective is to maximize the pull-off force of nylon tubing inserted into an elastomeric connector for use in automotive engine components. The four controllable factors are: interference between the tubing and the connector (A), wall thickness of the connector (B), insertion depth of the tubing into the connector (C), and percent adhesive (D). Three levels were chosen for each of the controllable factors. The three noise factors are conditioning time (E), conditioning temperature (F), and conditioning relative humidity (G). Each noise factor was set at two levels (values representative of what the product would experience in the engine).

Table 1   Data from Byrne and Taguchi [1].

									Outer Array (L_8)
			8	7	6	5	4	3	2	1	Run No.

Inner Array (L_9)												

		8	7	6	5	4	3	2	1	Run No.
		2	2	2	2	1	1	1	1	E
		2	2	1	1	2	2	1	1	F
		1	1	2	2	2	2	1	1	Ex F
		2	1	2	1	2	1	2	1	G
		1	2	1	2	2	1	2	1	Ex G
		1	2	2	1	1	2	2	1	Fx G
		2	1	1	2	1	2	2	1	e

Run No.	A	B	C	D								S/N ratio
1	1	1	1	1	19.1	20.0	19.6	19.6	19.9	16.9	9.5	15.6	24.025
2	1	2	2	2	21.9	24.2	19.8	19.7	19.6	19.4	16.2	15.0	25.522
3	1	3	3	3	20.4	23.3	18.2	22.6	15.6	19.1	16.7	16.3	25.335
4	2	1	2	3	24.7	23.2	18.9	21.0	18.6	18.9	17.4	18.3	25.904
5	2	2	3	1	25.3	27.5	21.4	25.6	25.1	19.4	18.6	19.7	26.908
6	2	3	1	2	24.7	22.5	19.6	14.7	19.8	20.0	16.3	16.2	25.326
7	3	1	3	2	21.6	24.3	18.6	16.8	23.6	18.4	19.1	16.4	25.711
8	3	2	1	3	24.4	23.2	19.6	17.8	16.8	15.1	15.6	14.2	24.832
9	3	3	2	1	28.6	22.6	22.7	23.1	17.3	19.3	19.9	16.1	26.152

The design that Taguchi recommends is actually the direct product of two designs. A so-called *inner array* is constructed to study the effects of the controllable factors themselves, and for each cell of this inner array, a design for the noise factors is run, called the *outer array*. The mean and variance of the functional characteristic over the outer array are computed for each cell of the inner array and then analyzed with respect to the controllable factors. Taguchi himself recommends that orthogonal arrays* be used for both the inner and outer arrays, but this choice has been criticized on the grounds that it ignores the possibility of important interaction effects between the controllable factors. (See Hunter [2].) Also, it is at least questionable whether a highly systematic arrangement for the outer array such as a fractional factorial design* can be expected to reflect accurately the true behavior of the noise factors, which behavior is by definition unsystematic and possibly random.

In order to facilitate the study of the variation in the response (noise) as well as the mean response (signal) for each row of the innter array, Taguchi introduces a *signal-to-noise index*. Kackar [3] provides an overview of these indices and demonstrates how they are related to the expected loss function. For the Byrne and Taguchi example, the appropriate index is

$$S/N = -10 \log \left[\sum_{i=1}^{n_2} y_i^{-2} \Big/ n_2 \right];$$

the values are shown in Table 1. Here, $n_2 = 8$.

Although a formal analysis of variance* can be conducted to determine which factors are significant, Taguchi advocates graphical methods also. For each controllable factor, this would include plots of the average S/N ratios for each level, as well as plots of the mean response. Such an analysis for the Byrne and Taguchi example yielded the following choice of levels for the controllable factors: A_2 (medium interference), B_2 (medium wall thickness), C_3 (deep insertion), and D_1 (low percent adhesive). However, cost considerations resulted in using B_1 (thin wall thickness) in place of B_2.

Taguchi has been enthusiastically praised for emphasizing the use of designed experiments not only to set a product's characteristics at target values but also to reduce variation around these targets, and also for putting forward a specific strategy for effecting these goals. On the other hand, certain important statistical details of Taguchi's proposed strategy—the choice of experimental design, the method of analysis, etc.—have been criticized as simplistic: the fact that the orthogonal arrays that Taguchi recommends as designs make it impossible to sort out possibly important interaction effects is one example. It should be noted, however, that the straightforward, systematic nature of Taguchi's program makes it relatively easy to introduce to experimenters and hence easy to implement. And an experimental design in the factory is worth two in the book!

A full discussion of the pros and cons of off-line quality control can be found in the article by Kackar [3] and the ensuing discussions by Box, Easterling, Freund, Lucas, and Pignatiello and Ramberg, as well as the article by Hunter [2].

After the operating conditions are determined using off-line quality control, it is necessary to follow this up with those quality control activities needed during manufacturing. Taguchi refers to these as on-line activities which include (a) diagnosis and adjustment of process, (b) forecasting and correction, (c) measurement (inspection) and disposition, and (d) afterservice by the sales department. For additional detail, refer to Taguchi [7]. G. Taguchi [6] has also developed accumulation analysis, a technique for testing independence in ordered categorical data. This procedure is reviewed by Nair [4].

References

[1] Byrne, D. M. and Taguchi, S. (1986). *Ann. Qual. Conf. Trans.* American Society for Quality Control, pp. 168–177.

[2] Hunter, J. S. (1985). *J. Qual. Tech.*, **17**, 210–221.

[3] Kackar, R. N. (1985). *J. Qual. Tech.*, **17**, 176–183; Discussion, 190–209.

[4] Nair, V.N. (1986). *Technometrics*, **28**, 283–294; Discussion, 295–311.

[5] Phadke, M. S. (1982). *Proc. Sect. Statist. Educ.*, *Amer. Statist. Ass.*, pp. 11–20.

[6] Taguchi, G. (1974). *Saishin Igaku*, **29**, 806–813.

[7] Taguchi, G. (1976). *An Introduction to Quality Control*, Central Japan Quality Control Association, Nagoya, Japan.

Bibliography

Barker, T. B. (1986). *Quality Prog.*, **19**, 32–42.

Box, G. E. P. and Draper, N. R. (1987). *Empirical Model Building and Response Surfaces*. Wiley, New York.

Kackar, R. N. (1986). *Quality Prog.*, **19**, 21–29.

Phadke, M. S., Kackar, R. N., Speeney, D. V., and Grieco, M. J. (1983). *Bell System Tech. J.*, **62**, 1273–1310.

Plackett, R. L. and Burman, J. P. (1946). *Biometrika*, **33**, 305–325.

Taguchi, G. and Wu, Y. I. (1980). *Introduction to Off-line Quality Control*, Central Japan Quality Control Association, Nagoya, Japan.

(CONTROL CHARTS
DESIGN OF EXPERIMENTS
FRACTIONAL FACTORIAL DESIGNS
ORTHOGONAL ARRAYS
QUALITY CONTROL, STATISTICAL)

F. B. Alt

TAHA TEST

Taha [17–20] suggests a rank test for comparing two distributions. The test has considerable power in comparison to either the Mann–Whitney–Wilcoxon* [9, 26], or Mood's dispersion test* [13], for one-sided asymmetrical distributions.

Let $x_1, x_2, \ldots, x_m$ and $y_1, y_2, \ldots, y_n$ be two independent samples of sizes m, n ($m + n = N$) taken from populations with continuous cumulative distribution functions (CDFs), say $F(x)$ and $G(x)$, respectively. The problem is to test the null hypothesis H_0: $F(x) = G(x)$ against the alternative H_1: $G(x) = F(\theta x)$, $\theta < 1$. The alternative implies that G is stochastically larger than F. Let the two samples be rearranged in one sequence

$$Z_1 < Z_2 < \cdots < Z_N.$$

Table 1   Values t_p such that $P_{H_0}[L \leqslant t_p] = p$.

			p	
m	n	0.90	0.95	0.975
3	3	70	77	—
4	4	147	158	174
5	5	265	285	298
6	6	436	466	487
7	7	666	708	742
8	8	966	1024	1072

Let $R(y_i)$ be the rank of the ith y in the sequence of the combined sample. The test statistic suggested by Taha [17] is the sum of squared ranks

$$L = \sum_{i=1}^{n} R^2(y_i).$$

The alternative H_1 is to be accepted if the observed value of L is too great. Taha [17] provides critical values for the test based on L for $m, n = 3(1)8$. Some of them are given in Table 1. Conover and Iman [1] give more extensive tables for the distribution of L. They also give a formula to approximate the distribution for larger sample sizes.

The mean and variance of L under H_0 are, respectively,

$$E(L) = n(N + 1)(2N + 1)/6$$

and

$$V(L) = nm(N + 1)$$
$$\times (2N + 1)(8N + 11)/180.$$

Taha [17] also gives the skewness and the kurtosis of L under H_0, and shows that the normal approximation to the null distribution of the standardized L is satisfactory, at least, for $m = n = 8$. The mean and variance of L under H_0, for data with ties, have been given by Mielke [10].

Example. We consider two sets of air conditioner failure data (in hours) given by Proschan [15]. For Plane 7909, the failure times are $90, 100, 160, 346, 407$ and for Plane 8045, the failure times are $102, 307, 321, 378,$ and 432. Assume that the samples are taken from CDFs $F(x)$ and $G(x)$, respectively. We want to test H_0 against H_1. The ranks associated with the second sample are $3, 5, 6, 8,$ and 10. The computed L statistic is

$$L = 3^2 + 5^2 + 6^2 + 8^2 + 10^2 = 234,$$

which is less than 265, the 90th percentile of L (Table 1) under H_0. The test does not reject H_0 at the 10% level.

Taha [17] also studies the asymptotic relative efficiencies (AREs) (Noether [14]; Gibbons [5, Chap. 14]) of the Taha test with respect to the Mann–Whitney–Wilcoxon test, the Mood test, and the standard F test* for dispersions, against the gamma distribution.* The results are 1.076, 2.5, and 1.318, respectively, if both F and G are single parameter exponential CDFs. James [8], however, shows that the ARE of the Taha test with respect to the Savage test* [16] is about 0.87 for the exponential distribution. Further considerations of the ARE are given by Duran and Mielke [3].

Test statistics similar to L are suggested by many authors, including Tamura [24], Taha [19], Mielke [12], Duran [2], Fligner and Killeen [4], and Conover and Iman [1]. Mielke [11] studies the asymptotic behavior of tests based on powers of ranks. Tests based on polynomials of ranks are investigated by Taha and El Nadi [23], Taha [20, 21], and Habib [6]. Tests for comparing three or more populations, based on squared ranks, have been discussed by Taha [18] and Tsai et al. [25].

Tests based on powers of ranks are especially useful in weather modification* experiments (see Mielke [12] and Hanson et al. [7]). Taha [22] gives a review of such tests and indicates that Rochetti (by personal communication) has found that the Taha test is robust in the sense of Hampel, but the Mann–Whitney–Wilcoxon test and the Savage test are not.

References

[1]   Conover, W. J. and Iman, R. L. (1978). *Commun. Statist. B*, **7**, 491–514. (Asymptotic normality and efficiency of some nonparametric statistics.)

[2] Duran, B. S. (1976). *Commun. Statist. A*, **5**, 1287–1312. (A survey of nonparametric tests for scale.)

[3] Duran, B. S. and Mielke, P. W., Jr. (1968). *J. Amer. Statist. Ass.*, **63**, 338–344. (Robustness of sum of squared ranks test.)

[4] Fligner, M. A. and Killeen, T. J. (1976). *J. Amer. Statist. Ass.*, **71**, 210–213.

[5] Gibbons, J. D. (1984). *Nonparametric Statistical Inference.* 2nd edn., Dekker, New York. (Good introduction to the theory of nonparametric procedures, with an understandable chapter on the asymptotic relative efficiency including examples.)

[6] Habib, M. G. (1975). Study of some functions of ranks. M. Sc. Thesis, Ain Shams University, Cairo, Egypt.

[7] Hanson, M. A., Barker, L. E., Bach, C. L., and Cooley, E. A. (1979). *Commun. Statist. A.* **8**, 1129–1147. (A bibliography of weather modification experiments.)

[8] James, B. R. (1967). *Proc. Fifth Berkeley Symp. Math. Statist. Prob.*, University of California Press, Berkeley, CA, pp. 389–393.

[9] Mann, H. B. and Whitney, D. R. (1947). *Ann. Math. Statist.*, **18**, 50–60. (Proposes the sum of ranks statistic and proves the asymptotic normality of the null distribution.)

[10] Mielke, P. W., Jr. (1967). *Technometrics*, **9**, 312–314.

[11] Mielke, P. W., Jr. (1972). *J. Amer. Statist. Ass.*, **67**, 850–854. (Asymptotic behavior of two-sample tests based on powers of ranks for detecting scale and location alternatives.)

[12] Mielke, P. W., Jr. (1974). *Technometrics*, **16**, 13–16. (Squared rank test appropriate to weather modification crossover design.)

[13] Mood, A. (1954). *Ann. Math. Statist.*, **25**, 514–522.

[14] Noether, G. E. (1955). *Ann. Math. Statist.*, **26**, 64–68. (Develops Pitman's notion of asymptotic relative efficiency.)

[15] Proschan, F. (1963). *Technometrics*, **5**, 375–384. (Theoretical explanation of observed decreasing failure rate, with an analysis of airplane air conditioner failure data.)

[16] Savage, I. R. (1956). *Ann. Math. Statist.*, **27**, 590–616. (Locally most powerful rank tests for Lehmann alternatives.)

[17] Taha, M. A. H. (1964). *Publ. Inst. Statist.*, **27**, 169–179. (Rank test appropriate for asymmetrical one-sided distribution.)

[18] Taha, M. A. H. (1970). *Proc. Sixth Conference on Statist. and Computational Sciences, Cairo*, pp. 24–26. (A rank test for scale parameters for several samples from asymmetrical one-sided distributions.)

[19] Taha, M. A. H. (1973a). *Egyptian Statist. J.*, **17**, 27. (A comparative study of two rank tests.)

[20] Taha, M. A. H. (1973b). *Bull. Inter. Statist. Inst.*, **45**(2), 485–496. (Polynomial approximation to locally most powerful rank tests.)

[21] Taha, M. A. H. (1977). *Bull. Inter. Statist. Inst.*, **47**(4), 575–580. (On the locally most powerful rank test.)

[22] Taha, M. A. H. (1983). Powers of rank test statistics. Presented at The 1st Saudi Symposium on Statistics, Riyad, Saudi Arabia. (A review of tests based on powers of ranks.)

[23] Taha, M. A. H. and El Nadi (1973). *Proc. of the 9th Conference on Statistics and Computational Sciences, Cairo, Egypt*, Vol. 1, p. 12. (Simple function of ranks.)

[24] Tamura, T. (1963). *Ann. Math. Statist.*, **34**, 1101–1103. (Rank tests based on powers of ranks.)

[25] Tsai, W. S., Duran, B. S., and Lewis, T. O. (1975). *J. Amer. Statist. Ass.*, **70**, 791–796. (Small sample behavior of some multisample nonparametric tests for scale.)

[26] Wilcoxon, F. (1945). *Biometrics*, **1**, 80–83. (One of the many original sources of the sum of ranks test.)

Acknowledgment

The author is grateful to Professor Mohamed A. H. Taha for kindly providing him with the unpublished manuscript [22].

(DISTRIBUTION-FREE METHODS
MANN–WHITNEY–WILCOXON STATISTIC
MOOD'S DISPERSION TEST
RANK TESTS
SCALE TESTS)

H. K. HSIEH

TAIL ORDERING

This concept was introduced by Doksum [3].

Given two (absolutely) continuous cumulative distribution functions G and F on $[0, \infty]$, G is said to be *tail-ordered* with respect to F if $F^{-1}[G(x)] - x$ is a nondecreasing function of x. (Compare with star ordering as defined in (18) of ORDERING, STARSHAPED.) Deshpande and Kochar [2] showed that ordering by dispersion* and tail ordering are mutually equivalent. Kochar [5] obtained a relation between hazard (failure) rates* and tail ordering. A typical result is: If either F or G is an increasing hazard* failure rate (IHR) distribution, then if G is tail-ordered with respect to F, the *hazard rate* $\lambda_F(x)$ of F does not exceed the hazard rate $\lambda_G(x)$ of G for all $x > 0$. This is used to construct a distribution-free* test for testing $\lambda_F(x) = \lambda_G(x)$ against $\lambda_F(x) \leqslant \lambda_G(x)$ for all x.

Recently, Bogai and Kochar [1] investigated the relationship between tail ordering and TP_2* ordering. See also Keilson and Sumita [4].

References

[1]   Bagai, I. and Kochar, S. C. (1986). *Commun. Statist. Theor. Math.*, **15**, 1377–1388.

[2]   Deshapande, J. V. and Kochar, S. C. (1983). *Adv. Appl. Prob.*, **15**, 686–687.

[3]   Doksum, D. (1969). *Ann. Math. Statist.*, **40**, 1167–1176.

[4]   Keilson, J. and Sumita, U. (1982). *Canad. J. Statist.*, **10**, 181–191.

[5]   Kochar, S. C. (1985). *Bull. Inst. Math. Statist.*, **14** (78), 47. (Abstract 85t-14.)

(ORDERING DISTRIBUTIONS BY DISPERSION
ORDERING OF DISTRIBUTIONS, PARTIAL ORDERING, STARSHAPED STOCHASTIC ORDERING)

TAIL PROBABILITIES

We shall discuss two aspects of the tails of distributions as pertaining to statistical inference* and decision theory*. The first aspect (A) concerns inference about the parameters of the tails of a distribution when the global form of the distribution is unknown, but the mathematical form of one or both tails of the distribution is known. The second aspect (B) concerns the robust Bayesian approach to inference using algebraic-tailed prior distributions* for some of the parameters of the distribution.

ASPECT A

Many empirical phenomena, although only partially, are nonetheless quite usefully characterized by the mathematical form of the tails of the distribution. Consider a random variable X with probability distribution function $F(x)$ and $G(x) = 1 - F(x)$, so that $G(x)$ is the probability that $X > x$. Often the mathematical form of $G(x)$ is known, at least approximately, for large x, and takes a particularly simple form, while the distribution in its entirety is quite complicated and is less well known. Suppose that one is primarily interested in establishing the form of the upper tail distribution $G(x)$ and in estimating the parameters that determine the upper tail, rather than in estimating the entire distribution. This type of problem does not fall into either of the two conventional modes of statistical inference: parametric and nonparametric, respectively. It is not fully parametric, since it is only the tail of the distribution that is parametrized, nor is it fully nonparametric, since we are exploiting a parametric model for the part of the distribution that is of special interest to us, say the upper tail, and assuming little or nothing about the distribution elsewhere. Not surprisingly, appropriate methods for statistical inference about the tails of a distribution are quite different in character than those of conventional parametric and nonparametric statistics. Thus, for inference about the upper tail of a distribution, it is natural to condition upon the upper r order statistics* of the data, for some r that can be determined in a variety of ways, including both data-analytic and decision-theoretic, and to base inference primarily upon these order statistics. Before discussing such methods, let us consider a few empirical phenomena for which such methods are appropriate. Consider the proportion of units (say light-bulbs) in a specified population of such units, whose lifetimes exceed x, where x is "large" and by "large" we mean relative to the "true" distribution of such lifetimes, i.e., corresponding to one of the upper percentiles of the population. Such an upper tail is often found to be of exponential* form

$$G(x) = \exp(-\alpha x), \qquad \alpha > 0.$$

In fact, it is often assumed that the distribution is exponential in its entirety, although this is a much stronger assumption than is usually either realistic or requisite for the purpose of estimating α. Next consider a population of incomes of individuals, or of city sizes. These are two well known examples of Zipf's law*, which generalizes the Pareto law, as is discussed in Zipf [18], Johnson and Kotz [17], and Ijiri and Simon [16]. For these populations, the form of the upper tail is algebraic rather than exponential, i.e.,

$$G(x) = Cx^{-\alpha}, \quad \text{where } C > 0, \alpha > 0.$$

Here again it would ordinarily be appropriate only to assume that such a relationship holds approximately in the upper tail, rather than globally. Another familiar example, which includes the exponential distribution as a special case, is the Weibull distribution*, with

$$G(x) = \exp\left[-(\beta x)^s\right], \qquad s > 0, \beta > 0.$$

This distribution is widely used in reliability theory* [1]. We now proceed to show how the parameters of such distributions can be estimated using only the assumption that the form of the distribution is appropriate in the upper tail.

For convenience, we discuss the upper tails of a population consisting of positive values, i.e., the upper tail of the distribution of a positive random variable, although the methods are quite general. Results for the lower tail can be obtained by taking reciprocals. The study can be broken into three distinct parts. The first concerns estimation or inference about the parameters of the upper tail when the mathematical form of the tail is known. The second part concerns the construction of probability models that yield the form of observed data, for example, models for the exponential and Zipf laws, and the third part concerns testing for the appropriateness of a specific model.

We shall begin with statistical inference about the parameters of the upper tail, assuming that the form of the tail distribution is known. Consider, for example, the distribution of city sizes in the United States in a given year. The empirical distribution of such data is highly complicated, but it has been demonstrated that the upper tail is approximately of the form $Cx^{-\alpha}$. For inference about α one wishes to use this knowledge, but one does not want to assume that such a relationship holds for all x. In Hill [14] a simple general approach was suggested for drawing inferences about C and α in both a conditional maximum-likelihood* sense, and a Bayesian* sense. Suppose that $Y(i)$ denotes the order statistics of the sample in descending order, so that $Y(1) > Y(2) > \cdots > Y(n)$. Suppose also that $G(x)$ is of the form $Cx^{-\alpha}$ for some positive C and α, and for all x larger than some D (possibly unknown). If

$$Y(r + 1) > D > Y(r + 2),$$

then [14] (see also David [5, p. 18]) the $E(i) = i\nu(i)$ are independent exponential random variables, with parameter α, where $\nu(i) = \ln[Y(i)/Y(i + 1)]$ for $i = 1, \ldots, r$. Hence the conditional maximum-likelihood estimate of α, given that the largest $r + 1$ order statistics exceed D, is $r/(\Sigma E(i))$, and corresponding Bayes modifications are easily obtained.

To implement this approach one must either know D or use data-analytic Bayes methods to determine the choice of r, as described in Hill [14]. In this approach only the $r + 1$ upper order statistics are used for inference about α, rather than all the data, and in this way one avoids the complications due to the global form of the distribution being unknown. See DuMouchel [7] for a comparison between such conditional inference and inference based upon a global stable*distributional model.

Finally, this method is not only simple but general, in that if the form of the upper tail is known, no matter what it may be, then a known transformation can always be employed to reduce the problem to the previous case of an algebraic tailed distribution. For example, suppose that the upper tail is of the Weibull form already described. Then the transformation $Z = \exp[Y^s]$ reduces the problem to the previous form, with $\alpha = \beta^s$.

Now let us turn to the question of modelling the tail of a distribution. One can begin with exponential tailed distributions for positive random variables. The exponential distribution* follows from extremely simple and natural assumptions [8, p. 220]. However, many real world populations are not of this form, and in fact have the type of algebraic tail previously discussed. The connection is quite simple and direct. Mixtures of geometric distributions* (or more generally, mixtures of negative binomial distributions*) tend to have the long tailed behavior that we have discussed. In Hill [11, 12], Hill and Woodroofe [15], and Chen [3, 4], the way in which such mixtures arise is demonstrated. See also Johnson and Kotz [17, pp. 128, 350].

The model proposed by Hill, which is an urn model*, is as follows. Suppose that a large number N of objects must be distributed to M nonempty cells, where $M < N$, and let $L(i)$ be the number of items in cell i. Conditional upon M and N, it is assumed that the vector $L = [L(1), \ldots, L(M)]$ has a symmetrical Dirichlet-multinomial distribution, including as special cases the Bose–Einstein and Maxwell–Boltzmann distributions (*see* FERMI–DIRAC STATISTICS). An equivalent derivation of Zipf's law can be obtained by starting with a sequence of independent identically distributed negative binomial random variables and conditioning upon their sum [3, 4]. Next it is assumed that M itself is a random variable, with a distribution that depends upon N, and that as N grows large the distribution of M/N, given N, converges to a distribution F. If this limiting distribution F is degenerate, so that M/N converges to a constant k, the expected proportion of $L(i)$ equal to s is of negative binomial* form. On the other hand [11, 12, 15], if M/N has a nondegenerate limiting distribution, then the resulting distribution is a mixture of negative binomial*

distributions, and if $F(x)$ is of algebraic form for x near 0, this yields the algebraic long-tailed distributions previously discussed in both weak and strong forms, and with parameter α a known function of the parameter of $F(x)$ as x approaches 0.

Next consider testing for a specific form of tail distribution. Hill [14] proposes two types of test based upon the fact that if the data are in the upper tail, i.e., exceed D, then the distribution of the $iV(i)$ is the distribution of independent exponential random variables, and both standard tests and data-analytic Bayes procedures can be used to determine the adequacy of the assumed form of tail distribution. See Haeusler and Teugels [9], Hall and Welsh [10], and Csörgő et al. [4a] for recent results on the behavior of this test.

ASPECT B

Another topic of interest concerns the form of the tail of a prior distribution for certain parameters. Consider inference about the mean of a normal population. Many have suggested that one should use a conjugate prior distribution for such a problem (*see* CONJUGATE FAMILIES OF DISTRIBUTIONS). Others, including Dickey [6], Hill [13], and Berger [2], have argued that such a prior is not realistic, and lacks robustness* properties. Specifically, Dickey argues for the use of multivariate t prior densities, with the location parameter vector distributed independently of the error variance, on the grounds that the use of the natural conjugate prior distribution incorporates an unrealistic dependency between these parameters. Hill proves a theorem concerning the limiting posterior distribution* of the location parameter vector for extreme data as the norm of the data vector $\mathbf{X}$ goes to infinity, which suggest that it is algebraic tailed prior densities (of which the multivariate t density is an important special case) that both yield desirable posterior robustness properties and correspond to the usual view of how to interpret such extreme data. Berger then demonstrates, from a decision-theoretic risk function point of view, that such algebraic tailed prior densities are preferable to the natural-conjugate prior densities when one takes into account the various types of robustness that seem desirable; for example, the risk function that arises in using a normal prior distribution for a normal mean can be strikingly bad as compared to that obtained by using a Cauchy* or other algebraic tailed prior distribution.

Robust Bayesian inference is insensitive, in an appropriate sense, to the precise form of the prior distribution of the parameters; in a predictive setting it is appropriately insensitive to the specified joint distribution of the past and future observations. It is thus intimately related to the nature of the tails of such distributions.

References

[1] Barlow, R. E. and Proschan, F. (1975). *Statistical Theory of Reliability and Life Testing Probability Models*, Holt, Rinehart and Winston, New York.

[2] Berger, J. (1984). The robust Bayesian viewpoint. In *Robustness of Bayesian Analyses*, J. Kadane, ed. North-Holland, Amsterdam, Netherlands.

[3] Chen, W. (1978). *On Zipf's Law*. Ph.D. Dissertation, University of Michigan, Ann Arbor, MI.

[4] Chen, W. (1980). *J. Appl. Prob.*, **17**, 611–622.

[4a] Csörgő, S., Deheuvels, P., and Mason, D. (1985). *Ann. Statist.*, **13**, 1050–1077.

[5] David, H. A. (1981). *Order Statistics*, 2nd. edn. Wiley, New York.

[6] Dickey, J. (1974). Bayesian alternatives to the *F* test. In *Studies in Bayesian Econometrics and Statistics in Honor of Leonard J. Savage*, S. Fienberg and A. Zellner, eds. North-Holland, Amsterdam, Netherlands.

[7] DuMouchel, W. (1983). *Ann. Statist.*, **11**, 1019–1031.

[8] Feller, W. (1950). *An Introduction to Probability Theory and Its Applications*. Wiley, New York.

[9] Haeusler, E. and Teugels, J. L. (1985). *Ann. Statist.*, **13**, 743–756.

[10] Hall, P. and Welsh, A. H. (1984). *Ann. Statist.*, **12**, 1079–1084.

[11] Hill, B. M. (1970). *J. Amer. Statist. Ass.*, **65**, 1220–1232.

[12] Hill, B. M. (1974). *J. Amer. Statist. Ass.*, **69**, 1017–1026.

[13] Hill, B. M. (1974). On coherence, inadmissibility, and inference about many parameters in the theory of least squares. In *Studies in Bayesian Econometrics and Statistics in Honor of Leonard J. Savage*, S. Fienberg and A. Zellner, eds. North-Holland, Amsterdam, Netherlands.

[14] Hill, B. M. (1975). *Ann. Statist.*, **3**, 1163–1174.

[15] Hill, B. M. and Woodroofe, M. (1975). *J. Amer. Statist. Ass.*, **70**, 212–219.

[16] Ijiri, Y. and Simon, H. A. (1977). *Skew Distributions and the Sizes of Business Firms*. North-Holland, Amsterdam, Netherlands.

[17] Johnson, N. L. and Kotz, S. (1977). *Urn Models and Their Applications*. Wiley, New York.

[18] Zipf, G. K. (1949). *Human Behavior and the Principle of Least Effort*. Addison-Wesley, Cambridge, Mass.

Bibliography

Daniels, H. E. (1987). *Intern. Statist. Rev.*, **55**, 37–48. (Discussion of saddlepoint approximations for the tail probability of a sample mean.)

(BAYESIAN INFERENCE
CONJUGATE FAMILIES OF
  DISTRIBUTIONS
HEAVY-TAILED DISTRIBUTIONS
ORDER STATISTICS
PRIOR DISTRIBUTIONS
ROBUST ESTIMATION
TAIL PROBABILITIES: THEOREMS
ZIPF'S LAW)

BRUCE M. HILL

TAIL PROBABILITIES: THEOREMS

The following theorems, relating to tail probabilities, have applications in statistical inference. In all cases they relate to independent and identically distributed random variables $X_1, X_2, \ldots$.

1. Hsu–Robbins Theorem [2, 3]. $E[X_i] = 0$ and $E[X_i^2] < \infty$ if and only if for every $\epsilon > 0$,

$$\sum_{n=1}^{\infty} \Pr[|S_n| > n\epsilon] < \infty,$$

where $S_n = \sum_{i=1}^{n} X_i$.

2. Katz Theorem [1, 4]. $E[X_i] = 0$ and $E[|X_i|^r] < \infty$ if and only if for every $\epsilon > 0$,

$$\sum_{n=1}^{\infty} n^{r-2} \Pr[|S_n| > \epsilon n] < \infty.$$

3. Spitzer Theorem [5]. $E[X_i] = 0$ and $E[|X^i|] < \infty$ only if for every $\epsilon > 0$,

$$\sum_{n=1}^{\infty} n^{-1} \Pr[|S_n| > \epsilon n] < \infty.$$

Note that the Katz theorem, with $r = 2$, gives the Hsu–Robbins and with $r = 1$, gives the Spitzer theorem. For further details, and generalizations, see Stout [6].

References

[1] Baum, L. E. and Katz, M. (1965). *Trans. Amer. Math. Soc.*, **120**, 108–123.

[2] Erdös, P. (1949), *Ann. Math. Statist.*, **20**, 286–291.

[3] Hsu, P. L. and Robbins, H. E. (1947). *Proc. Nat. Acad. Sci. USA*, **33**, 25–31.

[4] Katz, M. (1963). *Ann. Math. Statist.*, **34**, 312–318.

[5] Spitzer, F. (1956). *Trans. Amer. Math. Soc.*, **82**, 323–339.

[6] Stout, W. F. (1974). *Almost Sure Convergence.* Academic, New York.

(LIMIT THEOREMS
TAIL PROBABILITIES)

TAKÁCS' GOODNESS-OF-FIT DISTRIBUTION

Takács' distribution is a generalization of the Kolmogorov–Smirnov distribution. In the one-sample case, let $X_1, \ldots, X_n$ be i.i.d.

random variables with empirical distribution function (EDF) $F_n(x)$, and let $F(x)$ be a continuous cumulative distribution function (CDF). If the (one-sided) Kolmogorov–Smirnov statistic* $D^+ = \max\{F_n(x) - F(x)\}$ exceeds a critical value $d_0 = d_0(\alpha, n)$, the null hypothesis* H_0: "$F(x)$ is the parent CDF of the sample" is rejected at level α (*see* KOLMOGOROV–SMIRNOV-TYPE TESTS OF FIT). In other words, H_0 is rejected if $F_n(x)$ crosses the boundary (acceptance band) $F(x) - d_0$ at least once, where a crossing occurs if

$$F(X_{i:n}) - d_0 \leqslant F_n(X_{i:n})$$
$$< F(X_{i+1:n}) - d_0$$

for some $i = 0, \ldots, n$, where $F(X_{0:n}) = 0$, $F_n(X_{i:n}) = i/n$, and $F(X_{n+1:n}) = 1$ for $X_0 = -\infty$ and $X_{n+1} = +\infty$.

Similarly, one can use Takács' distribution to calculate a critical value $d_s = d_s(\alpha, n)$ and reject H_0 if $F_n(x)$ crosses $F(x) - d_s$ more than s times.

Example. In Example 1 of KOLMOGOROV–SMIRNOV-TYPE TESTS OF FIT the case $n = 20$ and $D^+ = 0.286$ is considered. If this discrepancy would have been exceeded at least *twice* in the sample, the test would be significant at about half the level of D^+ or less. The significance probabilities can be easily calculated from (3) below. Some critical values are given in Table 1.

ONE-SAMPLE DISTRIBUTION

Already in 1939, Smirnov [11] had derived the asymptotic distribution of the number of

Table 1   Critical Values $d_s(\alpha, n)$

	$n = 10$			$n = 20$			$n = 50$		
$\alpha \setminus s$	0	1	2	0	1	2	0	1	2
0.01	0.4566	0.3992	0.3415	0.3287	0.3011	0.2735	0.2106	0.2000	0.1894
0.05	0.3687	0.3133	0.2575	0.2647	0.2378	0.2109	0.1696	0.1591	0.1486
0.10	0.3226	0.2679	0.2135	0.2316	0.2050	0.1784	0.1484	0.1380	0.1276

crossings. Darling [1] improved Smirnov's result, showing that for $c = 1$, $b = 0$, and $B = n$,

$$\Pr[cF(X_{i:n}) \leqslant i + a/n < cF(X_{i+1:n}) \text{ for}$$

more than s of the subscripts

$$i = b, \ldots, B | H_0]$$

$$= \left\{ 1 - \frac{s + 4a + s(s + 2a)^2/n}{6n} \right\}$$

$$\times \exp\left\{ - \frac{(s + 2a)^2}{2n} \right\} + O(n^{-1}). \quad (1)$$

Without proof, Darling [1] also gave the exact distribution (1) for $c = 1$, $b = 0$, and $B = n$. Under the additional assumption that a is an integer, the exact formula was derived by Nef [5].

In 1971, Takács [13] proved that for any positive c and $b = \max\{0, -a\}$, $B = \min\{n, cn - a\}$, the probability (1) equals

$$\frac{n!}{(cn)^n} \sum_{i=b+s}^{B} \frac{(cn - a - i)^{n-j}}{(n - j)!}$$

$$\times \sum_{i=s}^{j-b} \frac{(a + j - i)^{j-i}}{(j - i)!(i - s)!}$$

$$\times \left[i^{i-s-2}(s^2 + s - i) \right], \quad (2)$$

where the expression in brackets [ ] equals 1 if $i = s = 0$. If $a \geqslant 0$, (2) reduces to

$$\frac{n!(a + s)}{(cn)^n} \sum_{j=s}^{B} \frac{(cn - a - j)^{n-j}}{(n - j)!}$$

$$\times \frac{(j + a)^{j-s-1}}{(j - s)!} \quad (3)$$

GENERALIZATIONS

If crossings of $F_n(x)$ with the tails of $F(x)$ are not to be considered, the bounds b and B in (1) may be chosen between $\max\{0, -a\}$ and $\min\{n, cn - a\}$; formula (2) still remains valid [7]. The same is true for (3), if $b = 0$.

Note that (1) does not change if a is replaced by $(c - 1)n - a$, b by $n - B$, and B by $n - b$. Furthermore, (1) can be written

as

$$\Pr[U_{i:n} \leqslant (i + a)/(cn) < U_{i+1:n} \text{ for more}$$

than s of the subscripts $i = b, \ldots, B]$,

where $U_{1:n}, \ldots, U_{n:n}$ are the uniform $(0, 1)$ order statistics. For power calculations it becomes necessary to find probabilities of the form

$$\Pr[U_{i:n} \leqslant a_i \leqslant U_{i+1:n} \text{ for more than } s \text{ of}$$

the subscripts $i = b, \ldots, B]$, (4)

where $a_b \leqslant \cdots \leqslant a_B$ is a given sequence, depending on $F(x)$ and the alternative. An algorithm for computing (4) is described in ref. 6, p. 171.

MATCHINGS

A match occurs in the kth interval if

$$(k - 1)/n < F(X_{k:n}) \leqslant k/n.$$

H_0 can be rejected if the number of matches is less than a critical value $r_{n,\alpha} \sim (-21n \times (1 - \alpha))^{1/2}$ (for large n). Siddiqui [10] proved the consistency of this test under any continuous alternative CDF G which coincides with F only on a set with Lebesgue measure 0. For more details about the matching test see refs. 2, 8, and 10.

To find the power of the matching test, one has to calculate probabilities of the form

$$\Pr[a_{i-1} < U_{i:n} \leqslant a_i \text{ for at least } r_{n,\alpha} \text{ of the}$$

subscripts $i = 1, \ldots, n]$

$$= \Pr[U_{i:n} \leqslant a_i < U_{i+1:n} \text{ for more than}$$

$r_{n,\alpha}$ of the subscripts $i = 0, \ldots, n]$. (5)

If $a_0 > 0$ or $a_n < 1$ the phrase "for more than" in (5) has to be replaced by "for at least".

(5) can be calculated from (2) if $a_i = i + a/n$. For $a = 0$ we get the exact distribution of the matching test

$$\Pr[\text{less than } s \text{ matches}]$$

$$= 1 - (1 - 1/n) \cdots \{1 - (s - 1)/n\},$$

$$1 \leqslant s \leqslant n + 1.$$

TWO-SAMPLE TESTS

Let $Y_1, \ldots, Y_m$ be a second sample of i.i.d. random variables with EDF G_m. Intersections between F_n and G_m can be described by considering the associated lattice path (Gnedenko path). This path goes upward one unit in its ith step, if the ith largest value in the combined sample is a Y value. The path goes one unit to the right otherwise. Let $0 \leqslant a_b \leqslant \cdots \leqslant a_B \leqslant m$. If both samples have the same underlying continuous CDF,

Pr[the associated path reaches at least s of

the points (i, a_i) from below, $i = b, \ldots, B$]

$$(6)$$

can be written as a double sum [7, eq. (11)] if for some integers a and p, $p > 0$, we have $a_i = pi + a + 1$ and $\max(0, -a/p) \leqslant b < B \leqslant \min(n, (m - a - 1)/p)$. The double sum reduces to

$$p^s \binom{m+n}{n}^{-1} \sum_{j=n-B}^{n-s} \binom{j + jp + m - pn - a - 1}{j}$$

$$\times \binom{(n-j)(p+1) + a + 1}{n - s - j}$$

$$\times \frac{s(p+1) + a + 1}{(n-j)(p+1) + a + 1} \qquad (7)$$

if $a \geqslant 0$ and $b = 0$. Takács [14] derived (7) for the important case $m = pn$. For arbitrary $p > 0$, Smirnov [12] obtained the same limiting distribution as in the one-sample case, defining x and y by $a = y\sqrt{np(p+1)}$ and $s = x\sqrt{np/(p+1)}$). Mihalevič [3] considered the special case $n = m$, $p = 1$, and $a \geqslant 0$, where (7) simplifies to

$$\binom{2n}{n + s + a + 1} \bigg/ \binom{2n}{n}.$$

In general, (6) can only be calculated recursively. Using Sheffer polynomials* for the backward difference* operator ∇, a matrix form of such a recursion is described in [6, eq. (1.1)].

Instead of only reaching the points (i, a_i) from below, the path can cross through such points (from below or from the right). For

the distribution of crossings and related problems see refs. 4, 6, and 9.

If $a_i = pi + a + 1$ and $m = pn$, (6) is the probability that

$$G_m(Y_{i-1:m}) \leqslant F_n(Y_{i:m}) + a/m < G_m(Y_{i:m})$$

for at least s of the subscripts $i = b, \ldots, B$, which can be seen as the probability of at least s crossing of G_m over $F_n + a/m$. A match occurs at the ith place if

$$X_{i:n} < Y_{[im/(n+1):m]} < X_{i+1:n}.$$

Such a match can be defined by any $m \geqslant n$, but if $m = pn$, the matching distribution can be calculated from (7) as

Pr[at least s matches]

$$= p^{s+1} \binom{mp + m}{m - s - 1} \binom{mp + m}{m}^{-1}$$

[14, eq. (17)]. For more details about the case $m = n$ see ref. 8.

References

[1] Darling, D. A. (1960). *Theory Prob. Appl.* **5**, 356–361.

[2] Khidr, A. M. (1981). *Indian J. Pure Appl. Math.*, **12**, 1402–1407.

[3] Mihalevič, V. S. (1952). *Dokl. Akad. Nauk SSSR*, **85**, 485–488. [Transl. (1961). *Soviet Math. Dokl.*, **1**, 63–67.]

[4] Mohanty, S. G. (1968). *Studia Sci. Math. Hung.*, **3**, 225–241.

[5] Nef, W. (1964). *Z. Wahrsch. verw. Geb.*, **3**, 154–162.

[6] Niederhausen, H. (1982). *Congr. Numerantium*, **36**, 161–173.

[7] Niederhausen, H. (1983). *Ann. Statist.* **11**, 600–606.

[8] Rao, J. S. and Tiwari, R. C. (1983). *Statist. Prob. Lett.*, **1**, 129–135.

[9] Sen, K. and Saran, J. (1983). *J. Statist. Plann. Inf.*, **7**, 371–385.

[10] Siddiqui, M. M. (1982). *J. Statist. Plann. Inf.*, **6**, 227–233.

[11] Smirnov, N. V. (1939a). *Mat. Sbornik*, **6**, 3–26.

[12] Smirnov, N. V. (1939b). *Bull. Math Univ. Moscow Ser. A*, **2**, 3–14.

[13] Takács, L. (1971a). *J. Appl. Prob.*, **8**, 321–330.

[14] Takács, L. (1971b). *Ann. Math. Statist.*, **42**, 1157–1166.

H. NIEDERHAUSEN

TAKÁCS PROCESS

This process is the *virtual waiting time process* for a single-server queueing model in which in the time interval $(0, \infty)$ customers arrive singly at a counter in accordance with a point process and are served singly by one server in order of arrival. Denote by $\tau_1, \tau_2, \ldots, \tau_n, \ldots$ the successive arrival times and by $\chi_1, \chi_2, \ldots, \chi_n, \ldots$ the corresponding service times. Let us imagine that the server uses a reading (or chemical) timer which has a clock mechanism and each time a customer arrives, the server sets the hand forward by the future service time of the customer. Since this clock runs as long as the server is busy, it shows at any given time t the time that a customer would wait if he arrived at that time. The time on the timer at time t, denoted by $\eta(t)$, is the virtual *waiting time* at time t; it can also be interpreted as the *total workload* of the server at time t, that is, the time needed to complete the serving of all those customers who arrive

before time t. If at time $t = 0$ the server is already busy, then $\eta(0)$ is defined as the initial workload of the server. Otherwise $\eta(0) = 0$.

If the arrival times or the service times or both are random variables, then $\eta(t)$ is a random variable for every t and the family of random variables $\{\eta(t), 0 \leqslant t < \infty\}$ forms a stochastic process*, the so-called *virtual waiting time process*. The virtual waiting time increases by χ_n at time τ_n and otherwise decreases linearly with slope one. If the virtual waiting time reaches 0, it remains 0 until the next customer arrives; see Fig. 1. The *actual waiting time* of the customer arriving at time τ_n is $\eta_n = \eta(\tau_n - 0)$ for $n = 1, 2, \ldots$.

The process $\{\eta(t), 0 \leqslant t < \infty\}$ also appears naturally in the theories of storage*, dams* or reservoirs, and insurance risk. In these cases we can interpret $\eta(t)$ as the total quantity of gasoline in a refinery, the content of a dam or reservoir, and the risk reserve in an insurance company at a given time t. See Takács [11, 12, 15].

A queueing model with a wide variety of applications is the following: Customers arrive according to a Poisson process* of density λ. The service times are independent random variables each having the same distribution function $H(x)$. The arrival times and service times are also independent and they are independent of the initial workload of the server. This is the so-called M/G/1

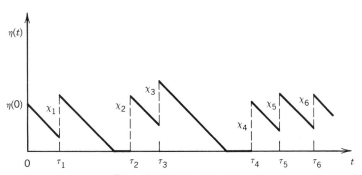

Figure 1   Virtual waiting time.

queue. We also let

$$a = \int_0^\infty x \, dH(x) \qquad (1)$$

and

$$\psi(s) = \int_0^\infty e^{-sx} \, dH(x) \qquad (2)$$

for $\mathrm{Re}(s) \geqslant 0$. In this model the virtual waiting time process $\{\eta(t), 0 \leqslant t < \infty\}$ is a Markov process*. See Takács [4–7]. The distribution function $P\{\eta(t) \leqslant x\} = W(t, x)$ satisfies Kolmogorov's equation

$$\frac{\partial W(t, x)}{\partial t}$$

$$= \frac{\partial W(t, x)}{\partial x}$$

$$- \lambda \left[W(t, x) - \int_0^x H(x - y) \, d_y W(t, y) \right] \quad (3)$$

for almost all (t, x), $t \geqslant 0$, $x \geqslant 0$. The initial condition $P\{\eta(0) \leqslant x\} = W(0, x)$ and (3) uniquely determine $W(t, x)$ for all $t \geqslant 0$ and $x \geqslant 0$. An explicit solution of (3) is given by Takács [7, 8, 10, and 13]. The Laplace–Stieltjes transform

$$\Omega(t, s) = E\{e^{-s\eta(t)}\} \qquad (4)$$

exists for $\mathrm{Re}(s) \geqslant 0$ and satisfies

$$\frac{\partial \Omega(t, s)}{\partial t} = [s - \lambda + \lambda\psi(s)]\Omega(t, s)$$

$$- sW(t, 0) \qquad (5)$$

for $t > 0$. By (5) it can be proved that if $\lambda a < 1$, then the limit distribution

$$\lim_{t \to \infty} P\{\eta(t) \leqslant x\} = W^*(x) \qquad (6)$$

exists, is independent of the initial distribution, and

$$\Omega(s) = \int_0^\infty e^{-sx} \, dW^*(x) = \frac{1 - \lambda a}{1 - \lambda a\psi^*(s)} \qquad (7)$$

for $\mathrm{Re}(s) \geqslant 0$, where

$$\psi^*(s) = \frac{1 - \psi(s)}{as} \qquad (8)$$

$$= \int_0^\infty e^{-sx} \, dH^*(x), \qquad (9)$$

and

$$H^*(x) = \frac{1}{a} \int_0^x [1 - H(y)] \, dy \qquad (10)$$

for $x \geqslant 0$.

In 1930, Pollaczek [3] and, in 1932, Khintchine [1] had already proved that if $\lambda a < 1$, then the limit distribution

$$\lim_{n \to \infty} P\{\eta_n \leqslant x\} = W(x) \qquad (11)$$

exists and its Laplace–Stieltjes transform is given by (7). Accordingly, $W^*(x) = W(x)$ for an M/G/1 queue. See also Takács [16, 17].

The situation is different for a GI/G/1 queue. For such a queue the interarrival times $\tau_n - \tau_{n-1}$ ($n = 1, 2, \ldots$; $\tau_0 = 0$) and the service times χ_n ($n = 1, 2, \ldots$) are independent sequences of independent and identically distributed random variables and all these variables are independent of the initial workload of the server. The following notations are used:

$$P\{\tau_n - \tau_{n-1} \leqslant x\} = F(x),$$

$$P\{\chi_n \leqslant x\} = H(x),$$

$$a = \int_0^\infty x \, dH(x), \quad b = \int_0^\infty x \, dF(x).$$

If $a < b$ and if $F(x)$ is not a lattice distribution* function, then the limit distribution function $\lim_{t \to \infty} P\{\eta(t) \leqslant x\} = W^*(x)$ exists, is independent of the distribution of $\eta(0)$, and

$$W^*(x) = \left(1 - \frac{a}{b}\right) + \frac{a}{b} W(x) * H^*(x) \qquad (12)$$

for $x \geqslant 0$, where $H^*(x)$ is defined by (10) and $W(x) = \lim_{n \to \infty} P\{\eta_n \leqslant x\}$ is the limit distribution function of the actual waiting time of the nth customer (Takács [9]). The distribution function $W(x)$ can be determined by a theorem of Lindley [2]. The distribution function of $\eta(t)$ for $t > 0$ can be obtained by the results of Takács [14, 18, and 19].

References

[1]   Khintchine, A. Ya. (1932). *Mat. Sbornik*, **39**(4), 73–84.

[2]   Lindley, D. V. (1952). *Proc. Camb. Philos. Soc.*, **48**, 277–289.

[3]   Pollaczek, F. (1930). *Math. Z.*, **32**, 64–100, 729–750.

[4]   Takács, L. (1954). *Magy. Tud. Akad. III. Oszt. Közl.*, **4**, 543–570.

[5]   Takács, L. (1955). *Acta Math. Acad. Sci. Hung.*, **6**, 101–129.

[6]   Takács, L. (1961). *Proc. Fourth Berkely Symp. Math. Statist. Prob.*, **2**, Univ. California Press, Berkeley, CA, pp. 535–567.

[7]   Takács, L. (1962a), *Introduction to the Theory of Queues*. Oxford University Press, New York.

[8]   Takács, L. (1962b). *Ann. Math Statist.*, **33**, 1340–1348.

[9]   Takács, L. (1963a). *Sankhyā A*, **25**, 91–100.

[10] Takács, L. (1963b). *Operat. Res.*, **11**, 261–264.

[11] Takács, L. (1964). *J. Appl. Prob.*, **1**, 69–76.

[12] Takács, L. (1966). *J. Math. Mech.*, **15**, 101–112.

[13] Takács, L. (1967). *Combinatorial Methods in the Theory of Stochastic Processes*. Wiley, New York.

[14] Takács, L. (1970a). *Ann. Inst. Statist. Math.*, **22**, 339–348.

[15] Takács, L. (1970b). *Skand. Aktuar.*, **53**, 64–75.

[16] Takács, L. (1970c). *Adv. Appl. Prob.*, **2**, 344–354.

[17] Takács, L. (1975). *Scand. Actu. J.*, 65–72.

[18] Takács, L. (1976). *Adv. Appl. Prob.*, **8**, 548–583.

[19] Takács, L. (1978). *Adv. in Math.*, *Suppl. Studies*, **2**, 45–93.

(DAM THEORY
QUEUEING THEORY
STOCHASTIC PROCESSES)

LAJOS TAKÁCS

TALWAR–GENTLE SCALE TEST

Talwar and Gentle [4] proposed a modified Mann–Whitney–Wilcoxon* test, based on a Levene* type transformation of the sample data.

We suppose that $X_1, \ldots, X_m$ and $Y_1, \ldots, Y_n$ represent values from random samples of sizes m, n from populations I and II, respectively. Suppose that the population distributions in I and II belong to the same location-scale family* but with location and scale parameters μ_1, σ_1 and μ_2, σ_2, respectively. Let

$$X_i' = |X_i - \overline{X}|, \qquad Y_j' = |Y_j - \overline{Y}|,$$

where

$$\overline{X} = m^{-1} \sum_{i=1}^{m} X_i, \qquad \overline{Y} = n^{-1} \sum_{j=1}^{n} Y_j.$$

Then the Mann–Whitney statistic is

$$U_{m, n} = \sum_{i=1}^{m} \sum_{j=1}^{n} D_{ij},$$

where

$$D_{ij} = \begin{cases} 1 & \text{if } X_i' > Y_j', \\ 0 & \text{if } X_i' < Y_j', \end{cases}$$

$$(i = 1, \ldots, m; \; j = 1, \ldots, n).$$

Large or small values of $U_{m, n}$ are regarded as significant for rejection of the null hypothesis $\sigma_1 = \sigma_2$. Simulation* studies indicate that tests based on $U_{m, n}$ are more robust than the two-sample t test proposed by Brown and Forsythe [1].

Talwar and Gentle [4] recommend using 25% trimmed means* in place of $\overline{X}$ and $\overline{Y}$.

Conover et al. [2] study the performance of the test when $\overline{X}, \overline{Y}$ are replaced by median $(X_1, \ldots, X_n)$ and median $(Y_1, \ldots, Y_n)$.

References

[1]   Brown, M. B. and Forsythe, A. B. (1974). *J. Amer. Statist. Ass.*, **69**, 364–367.

[2]   Conover, W. J., Johnson, M. E., and Johnson, M. M. (1981). *Technometrics*, **23**, 351–361.

[3]   Levene, H. (1960). Robust tests for equality of variances. In *Contributions to Probability and Statistics*, I. Olkin, ed. Stanford University Press, Palo Alto, Calif.

[4]   Talwar, P. P. and Gentle, J. E. (1977). *Commun. Statist. A*, **6**, 363–369.

(BARTLETT'S TEST OF HOMOGENEITY OF
   VARIANCES
LEVENE'S ROBUST TEST OF
   HOMOGENEITY OF VARIANCES
MANN–WHITNEY–WILCOXON STATISTIC
SCALE TESTS)

TAPER *See* VARIANCE, INNOVATION

TARGET COVERAGE

When a planar target is attacked by an artillery shell, a bomb dropped from an airplane, or an intercontinental ballistic missile, random errors in the weapon impact point are introduced by physical effects such as variations in the propulsive force or atmospheric conditions in the vicinity of the target. A family of mathematical models developed since World War II relates the expected fraction of the target destroyed to:

1 **The Characteristics of the Target.** Usually assumed to have value distributed according to a *"cookie-cutter"* function (uniform inside a circle of radius K) or a *bivariate Gaussian* function centered at the origin,

$$\exp\{-(x^2 + y^2)/(2\sigma_T^2)\}.$$

2 **The Characteristics of the Attacking Weapon.** Usually assumed to destroy all target points within a radius D of impact [the "cookie-cutter" damage function, with probability $d(r) = 1$ for $r \leqslant D$ and $d(r) = 0$ otherwise] or the diffused Gaussian damage function which destroys a target point at distance r from the impact point with probability $d(r) = \exp\{-r^2/(2b^2)\}$.

If the radius of the target (K or σ_T) is small with respect to the radius of effective action of the weapon (D or b), the target can be approximated by a point, and the expected fraction destroyed is replaced with the probability of target destruction.

The probability density function of the weapon impact point with respect to the target is usually assumed to be bivariate Gaussian centered on the target. Sometimes, different variances σ_x^2 and σ_y^2 are assumed for the components of weapon impact error parallel and perpendicular to the weapon trajectory projected on the target plane. A

few models allow for an offset aiming point (x_o, y_o); these are chiefly used to evaluate collateral damage to targets in the vicinity of a more important one, or to assess the effect of spreading out the impact points of a salvo when the target is large with respect to the radius of effective action of each weapon.

The various models of target damage can all be summarized by the expected fraction of target damage $E[f]$:

$$E[f] = \int_{-\infty}^{\infty}\int_{-\infty}^{\infty}\int_{-\infty}^{\infty}\int_{-\infty}^{\infty} T(x_t, y_t)$$
$$\times d(x - x_t, y - y_t)$$
$$\times p(x, y)\, dx_t\, dy_t\, dx\, dy,$$

where $T(x_t, y_t)$ is the distribution of target value, $d(x - x_t, y - y_t)$ is the damage function, and $p(x, y)$ is the probability density function of the weapon impact point. For the Gaussian model of target value, damage function, and impact point distribution,

$$E[f] = $$

$$\frac{b^2 \exp\left\{-\frac{1}{2}\left(\dfrac{x_o^2}{b^2 + \sigma_x^2 + \sigma_T^2} + \dfrac{y_o^2}{b^2 + \sigma_y^2 + \sigma_T^2}\right)\right\}}{(b^2 + \sigma_x^2 + \sigma_T^2)^{1/2}(b^2 + \sigma_y^2 + \sigma_T^2)^{1/2}},$$

where (x_o, y_o) is the offset aiming point of the weapon relative to the target (centered at the origin), σ_x^2 and σ_y^2 the variances of the weapon impact point, σ_T^2 the variance of target value, and b^2 the damage function constant introduced previously. For the cookie-cutter target with radius K, $x_o = y_o = 0$, and $\sigma_x^2 = \sigma_y^2 = \sigma^2$, the corresponding formula is

$$E[f] = 2(b^2/K^2)$$
$$\times \left[1 - \exp\{-\tfrac{1}{2}K^2/(\sigma^2 + b^2)\}\right].$$

For the special case of a point target, the probability of target destruction P is given by the preceding formula with $\sigma_T^2 = 0$. The probability of target destruction can, for this target only, also be evaluated in closed form assuming $x_o = y_o$ and a cookie-cutter damage function of radius D and $\sigma^2 = \sigma_x^2 = \sigma_y^2$:

$$P = 1 - \exp\{-D^2/(2\sigma^2)\}.$$

The value $D/\sigma = 1.1774$ results in a probability of destruction equal to 0.5; this is known as the *circular probable error**.

Tables are available for calculating the corresponding probabilities for unequal variances σ_x^2 and σ_y^2 and offset aiming points. (See ref. 5 for a list of tables and their parameter ranges.) However, little is known about how to aim a single weapon at a configuration of two or more point targets to maximize the probability that at least one target is destroyed. (Marsaglia [8] shows, for example, that if two equal-valued points are sufficiently close, one should aim midway between their locations; otherwise, one should pick one target and aim at a calculable offset point in the direction of the other target.)

If the probability of destruction P of a point target can be calculated for a single weapon, it is easy to determine the probability of destruction $P(n)$ for a salvo of n identical weapons independently arriving at the target with the same probability density function of impact points:

$$P(n) = 1 - (1 - P)^n.$$

However, this formula is only an upper bound to the expected fraction destroyed of a cookie-cutter or Gaussian target attacked by n independently targeted weapons. Normally, each weapon in a salvo is assumed to have the same offset aiming point (x_o, y_o) but independent Gaussian errors around this point with variances σ_x^2 and σ_y^2.

Grubbs [7] introduces a method of calculating the expected damage for both the Gaussian and cookie-cutter targets, but his series approximation converges slowly. Schroeter [11] presents formulas for the variance of the damage by a salvo of weapons, and presents graph to estimate both the expected damage and the variance for a restricted range of parameter values. Schroeter [10] generalizes the salvo model to include multiple salvos with offset aiming points (x_i, y_i) drawn independently and at random from a bivariate Gaussian probability density function. The expected damage is a function of seven parameters and can only

be evaluated by numerical methods using a digital computer. However, he provides several graphs to estimate the expected damage for a restricted range of parameter values.

For targets that are large with respect to the weapon impact point distribution, two strategies are available for increasing the expected fraction of target destroyed by a weapon salvo: multiple aiming points and an increase in the weapon impact point variances σ_x^2 and σ_y^2 (analogous to adjusting the choke on a gun using birdshot as ammunition). Little can be done analytically with either model. However, to maximize the expected fraction of target destroyed, a rule of thumb based on parametric studies suggests [4, 5] that the weapon impact point variance $\sigma^2[\sigma_x^2 = \sigma_y^2 = \sigma^2]$ should be set according to the formula

$$[\sigma/\sigma_T]^2 = 0.34\sqrt{n}\,[D/\sigma_T],$$

where n is the salvo size, the damage function is a cookie-cutter of radius D, and σ_T^2 is the variance of the value of a Gaussian target. Walsh [12] considers the far more difficult analytic problem of optimizing the probability density function of weapon impacts, not merely adjusting the variance of a Gaussian attack. He derives an upper bound to the expected fraction of target damage based on an impossible-to-realize weapon impact distribution, but which can be used as a yardstick against which to judge impact distributions of actual weapons. If L is the total lethality of a single weapon, defined by $L = \int \int d(x, y)\, dx\, dy$, and A is the area of a uniform-valued target, Walsh's upper bound becomes $E[f] = 1 - (1 - L/A)^n$, the same form as the probability of destruction of a point target by a salvo of independent weapons. Curran et al. [2] derive an exact expression for expected target damage to replace Walsh's upper bound; this can be evaluated only by numerical methods.

Galiano and Everett [6] and Duncan [3] show that, when the number n of weapons in a salvo approaches infinity and the cookie-cutter radius D of each weapon approaches zero such that $nD^2 = c$, there is a bivariate distribution of weapon impact points on a Gaussian target which maximizes the expected fraction of target damage:

$$E[f]_{max} = 1 - \left\{1 + \left(c/\sigma_T^2\right)^{1/2}\right\}$$

$$\times \exp\left\{-\left(c/\sigma_T^2\right)^{1/2}\right\},$$

often referred to in the literature as the *square-root law* of target damage. The effectiveness of physically realizable distribution functions of weapon impact points can be compared with this ideal. For finite n, Curran et al. [2] point out that the square-root law of target damage is a lower bound, although usually quite good (no more than 3% error for $n > 7$ and $0.6 < D/\sigma_T < 4.0$).

If a salvo is directed against a target one weapon at a time, the possibility exists of correcting an offset aiming point by observing where successive weapons land with respect to the target. Techniques involving shoot-adjust-shoot have been used by artillerymen for a long time. However, these techniques are restrictive in two ways; they make corrections based on the sign of the error, not its actual value, and corrections are limited to a set of equally spaced aiming points, not continuous. Barr and Piper [1] consider the artillery problem; Nadler and Eilbott [9] consider the more general model in which the actual values of past impact points are observable and a correction of any magnitude can be made. Among all aiming-point correction procedures which are linear in the past impact-point observations, the following procedure maximizes the probability of target destruction. Consider a one-dimension attack against a point target in which the weapon has a random Gaussian error with variance σ^2 about its aiming point, and which destroys a target within a distance D of its impact. Letting x_i denote the impact point of the ith weapon with respect to the target for $i = 0, 1, 2 \ldots$, the corrections are $0, -x_0, -(1/2)x_1, -(1/3)x_2, \ldots$. If these corrections are independently applied to the x and y components of the observed impacts points (x_i, y_i), the probability that the target is destroyed with n cookie-cutter weapons is

$$P = 1 - \exp\left\{-\tfrac{1}{2}D^2\sigma^{-2}\sum_{i=1}^{n} i/(i+1)\right\},$$

which can be evaluated with tables of the digamma function*. The summation is approximated by $n - \ln(2n + 3)/3$ for large n. This probability is less than $1 - \exp(-nD^2/(2\sigma^2))$, the probability that a salvo of n cookie-cutter weapons with no offset aiming point will destroy a target; in other words, if the aiming point offset is small with respect to σ and n is not too large, it is better *not* to apply the corrections given.

For further information, see the review paper by Eckler [4] and the monograph by Eckler and Burr [5, Chap. 2].

References

[1] Barr, D. R. and Piper, L. D. (1972). *Operat. Res.*, **20**, 1033–1043.

[2] Curran, R. T., Jacquette, S. C., and Politzer, J. L. (1979). *Naval Res. Logist. Quart.*, **26**, 545–550.

[3] Duncan, R. L. (1964). *SIAM Rev.*, **6**, 111–114.

[4] Eckler, A. R. (1969). *Technometrics*, **11**, 561–589. (This review paper contains approximately 60 references.)

[5] Eckler, A. R. and Burr, S. A. (1972). *Mathematical Models of Target Coverage and Missile Allocation*. Military Operations Research Society, pp. 15–76. (This is the second chapter of a monograph which discusses optimal strategies for allocating offensive and defensive weapons to one or more targets, under various assumptions about the degree of knowledge possessed by one side about the capabilities of the other.)

[6] Galiano, R. J. and Everett, H. (1967). Defense Models IV: Some Mathematical Relations for Probability of Kill; Family of Damage Functions for Multiple Weapon Attacks. Paper 6, Lambda Corporation, Arlington, Va.

[7] Grubbs, F. E. (1968). *Operat. Res.*, **16**, 1021–1026.

[8] Marsaglia, G. (1965). *Operat. Res.*, **13**, 18–27.

[9] Nadler, J. and Eilbott, J. (1971). *Operat. Res.*, **19**, 685–697.

[10] Schroeter, G. (1980). *Operat. Res.*, **28**, 1299–1315.

[11] Schroeter, G. (1982). *Naval Res. Logist. Quart.*, **29**, 97–111.

[12] Walsh, J. E. (1956). *Operat. Res.*, **4**, 204–212.

(BIVARIATE NORMAL DISTRIBUTION
MILITARY STATISTICS
MULTINORMAL DISTRIBUTIONS)

A. R. ECKLER

TARGET POPULATION

The word *population* has several meanings, with each area of statistical application providing the potential for different interpretations. The phrase *target population* [1–4] typically refers to a human population, specifically that group of individuals to which one would like to apply the results of a particular research investigation. The phrase *reference population* [2] has a similar meaning.

The target population may not be equivalent to the group studied since operational or practical limitations can make the full target population, or a carefully constructed representative sample, inaccessible. Thus *target population* is used in contrast to *survey population* [3], *sample population* [1], *study population* [4], and *actual population* [4]. These latter terms typically refer to the population that is operationally available. For this population it is often possible to select a representative sample, since by definition it takes into account practical or operational limitations. Statistical inferences from such a sample are, by necessity, restricted to this *actual population*. Making the same inferences about the target population, while ignoring the potential impact of the operational or practical limitations that led to the creation of the actual population, can be problematic, since these limitations can be the source of serious biases.

As an example, consider an urban district for which an assessment of teenage health attitudes is desired. The approach is a telephone survey with a representative sample selected from a master list (*see also* SAMPLING FRAME) of telephone subscribers. In this case, the *target population* could be all teenage citizens of the district as of a specified time. The *actual* or *survey population*

would be those teenagers accessible through a telephone on the master list. In this example, the actual population is a subset of the target population. The differences between the target and actual populations could vary enormously according to the country and city within which this urban district was located.

Another example of the use of the phrase "target population" occurs in the context of clinical trials*. A given trial, for example, may have as participants men aged 35–59 years who have had an initial heart attack. The purpose could be to compare the efficiencies of two drug regimens for delaying or presenting the second attack. For this situation, the target population would be all men meeting the same general conditions as those actually included in the study. This usage is less formal, since there typically are no provisions for selecting the clinical trial participants, by some random or other representative process, from the broader target population. In this case, inferences from the group studies to the target population are scientific judgements based on aspects other than statistical inference.

References

[1] Cochran, W. G. (1963). *Sample Techniques*, 2nd ed. Wiley, New York, p. 6.

[2] Elandt-Johnson, R. C. and Johnson, N. L. (1980). *Survival Models and Data Analysis*. Wiley, New York, p. 17.

[3] Kish, L. (1965). *Survey Sampling*. Wiley, New York, pp. 6–7.

[4] Kleinbaum, D. G., Kupper, L. L., and Morgenstern, H. (1982). *Epidemiologic Research: Principles and Quantitative Methods*. Lifetime Learning Publications, Hayward, Calif., pp. 20–21.

[5] Fisher, R. A. (1973). *Statistical Methods and Scientific Inference*, 3rd ed. Hafner, New York, p. 81.

(CLINICAL TRIALS
INFERENCE, STATISTICAL-I, II
SAMPLING FRAME
SURVEY SAMPLING)

O. D. WILLIAMS

TAUBERIAN THEOREMS

A *Tauberian theorem* is one that infers the asymptotic behavior of the coefficients of a power series from the singular behavior of the function represented near its radius of convergence [10, 15]. The converse of a Tauberian theorem is known as an *Abelian theorem*. Similar theorems are available for function pairs related by a variety of integral transforms* [18, 19]. Such theorems relate the behavior of a function and its transform in a way exemplified by the following exposition for the Laplace transform: Let $F(t)$ be a nondecreasing function, having the property that the Laplace transform

$$f(s) = \int_0^\infty e^{-st} \, dF(t) \qquad (1)$$

converges for $\mathrm{Re}(s) > 0$. The Tauberian theorem specifies the behavior of $F(t)$ for large t in terms of that of $f(s)$ in a neighborhood of $s = 0$ or it can relate the behavior of $F(t)$ for small t to that of $f(s)$ for large $|s|$. For the purpose of more exactly specifying the Tauberian theorem, we need to define the notion of a slowly varying function [8]. The function $L(t)$ will be slowly varying at $t = \infty$ if, for every $c > 0$, it satisfies

$$\lim_{t \to \infty} [L(ct)/L(t)] = 1.$$

A good example of such a function is $L(t) = \log(t)$. The same definition applies for $t = 0$ when the limit $t \to 0$ replaces $t \to \infty$ in the preceding definition.

The Tauberian theorem states that if, for some nonnegative number γ,

$$f(s) \approx s^{-\gamma} L(1/s) \qquad (2)$$

as $s \to 0$, then as $t \to \infty$, $F(t)$ has the asymptotic behavior

$$F(t) \approx [t^\gamma L(t)]/[\Gamma(\gamma + 1)]. \qquad (3)$$

The same theorem holds if the limits are interchanged, i.e., $s \to \infty$ and $t \to 0$ replaces those given previously. If the derivative $F'(t)$ also exists, its value can be calculated by differentiating eq. (3).

Since $f(s)$ in eq. (1) is defined as a Stieltjes integral, it implies a corresponding theorem

for generating functions*. Specifically, if

$$f(z) = \sum_{n=0}^{\infty} f_n z^n \qquad (4)$$

is a power series with a radius of convergence equal to 1 and if, as $z \to 1$,

$$f(z) \approx A(1 - z)^{-\gamma} L[1/(1 - z)], \qquad (5)$$

then

$$\sum_{j=0}^{n} f_j \approx \frac{An^\gamma}{\Gamma(\gamma + 1)} L(n) \qquad (6)$$

as $n \to \infty$. If the f_j's are monotonic, then one can infer the behavior of f_n from (6) by taking the first difference.

Abelian theorems, the earliest of which dates back to 1828 [1], allow one to specify the analytic behavior of $f(z)$ near its radius of convergence in terms of the asymptotic behavior of the f_n. The earliest Tauberian theorem appeared in 1897 [15]. Tauberian theorems are intrinsically harder to prove than Abelian theorems because the simple converse is not generally true, but further conditions are required on the f_n, typified by Tauber's original condition $f_n = O(n^{-1})$. Much weaker conditions also suffice, as was proved in a long series of investigations by Hardy and Littlewood, some of which is summarized in Hardy [10]. The most frequently cited proofs of major Tauberian theorems rely on the work of Karamata [12], but the proof cited in Feller's text [8] is simple and elegant, and suffices for most probabilistic applications.

Tauberian theorems are important in probability because of the central role played by generating functions for discrete random variables and characteristic functions* for continuous variables. Feller appears to have been the first to systematically apply Tauberian theorems to derive results in renewal theory* [6, 7]. Since then they have been used in a variety of contexts in probability. While Tauberian methods do not always yield the most general results for a given problem, they generally are easy to apply in practice because they require very little information about the function whose coefficients are being sought.

As an example of the use of Tauberian theorems for generating functions, consider the calculation of the asymptotic behavior of the expected number of distinct sites visited by an n-step random walk* in one dimension. Let $p(j)$ be the probability that a single step of the random walk is equal to j, where j can take on values $0, \pm 1, \pm 2, \ldots$. Further define the generating function

$$\lambda(\theta) = \sum_{j=-\infty}^{\infty} p(j)\exp(ij\theta). \quad (7)$$

One can derive, by a probabilistic argument, the generating function of the expected number of distinct sites visited in an n-step random walk [16]:

$$S(z) \equiv \sum_{n=0}^{\infty} S_n z^n = z\left[(1-z)^2 P(z)\right]^{-1}, \quad (8)$$

where $P(z)$ is the integral

$$P(z) = \frac{1}{2\pi}\int_{-\pi}^{\pi} \frac{d\theta}{1 - z\lambda(\theta)}. \quad (9)$$

While it is feasible to derive expressions for S_n for small n from $S(z)$ by differentiation, it is clearly impractical to do so for large n. Since $\lambda(0) = 1$ and $|\lambda(z)| < 1$ for $|z| < 1$, it follows that $P(z)$ will be singular at $z = 1$, and for no smaller value of $|z|$. If the mean step size is equal to zero and the variance

$$\sigma^2 = \sum_{j=-\infty}^{\infty} j^2 p(j) \quad (10)$$

is finite, then the singular behavior of $P(z)$ is [16]

$$P(z) \approx \left[2\sigma^2(1-z)^3\right]^{-1/2} \quad (11)$$

in the neighborhood of $z = 1$. Application of the Tauberian theorem of (4)–(6) together with the obvious monotonicity of the S_n implies that for large n,

$$S_n \approx \sigma(8n/\pi)^{1/2}. \quad (12)$$

Analogous results can be derived for $p(j)$ which correspond to stable laws* [9] for which $\sigma^2 = \infty$ in (10), as well as for higher dimensions (*see* RANDOM WALKS) [16].

Applications of Tauberian theorems abound in probability theory*. Many of the elementary asymptotic results in renewal theory are derivable using these techniques [6, 7]. Feller has given an extremely elegant derivation of the Tauberian theorem in [4]–[6] and applied the general methodology to characterize properties of the distribution function of stable laws defined on $(0, \infty)$ [8]. As mentioned, there are applications in renewal theory [6, 7, 13], queueing theory* [4, 17], and branching theory [11]. Several investigators have made use of Tauberian theorems in studies of random walks [3, 5, 14, 16] and the central limit theorem [2]. While Tauberian theorems are not probabilistic in content, they have nevertheless proved to be a convenient computational tool in many theoretical and applied investigations in probability.

References

[1] Abel, N. H. (1828). *J. Reine Angew. Math.*, **3**, 79–82.

[2] Bingham, N. H. (1981). *Ann. Prob.*, **9**, 221–231.

[3] Darling, D. A. and Kac, M. (1957). *Trans. Amer. Math. Soc.*, **84**, 444–458. (Tauberian theorems are applied to find asymptotic forms for the time of occupation of an arbitrary set by a diffusion process or random walk. An elegant demonstration of the power of Tauberian techniques.)

[4] De-Meyer, A. and Teugels, J. L. (1980). *J. Appl. Prob.*, **17**, 802–813.

[5] Diaconis, P. and Stein, C. (1978). *Ann. Prob.*, **6**, 483–490.

[6] Feller, W. (1941). *Ann. Math. Statist.*, **12**, 243–267.

[7] Feller, W. (1949). *Trans. Amer. Math. Soc.*, **67**, 98–119.

[8] Feller, W. (1971). *An Introduction to Probability Theory and Its Applications*, Vol. 2, 2nd ed. Wiley, New York. (Contains one of the slickest and simplest proofs of the Tauberian theorems mentioned in the text as well as several applications. Noteworthy related material is to be found on properties of slowly varying functions.)

[9] Gillis, J. and Weiss, G. H. (1970). *J. Math. Phys.*, **11**, 1308–1313.

[10] Hardy, G. H. (1949). *Divergent Series*. Oxford University Press, Oxford, England. (Much of the early work developing different forms of Tauberian theorems for power series was done by Hardy and Littlewood. A chapter in this mono-

graph gives many of these results as well as the flavor of British work in this area.)

[11] Jakymiv, A. L. (1981). *Math. Sbornik*, **115**, 463–477.

[12] Karamata, J. (1931). *J. Reine Angew. Math.*, **164**, 27–39.

[13] Leung, L. T. (1976). *Ann. Prob.*, **4**, 628–643.

[14] Spitzer, F. (1960). *Trans. Amer. Math. Soc.*, **94**, 150–169.

[15] Tauber, A. (1897). *Monat. Math. Phys.*, **8**, 273–277. (The original paper on the subject with the simplest theorem for power series.)

[16] Weiss, G. H. and Rubin, R. J. (1983). *Adv. Chem. Phys.*, **52**, 363–504. (Summarizes some applications of Tauberian theorems to random walks in physics and chemistry.)

[17] Whitt, W. (1972). *Z. Wahrsch. verw. Geb.*, **22**, 251–267.

[18] Widder, D. (1941). *The Laplace Transform*. Princeton University Press, Princeton, NJ. (Gives a comprehensive discussion of Tauberian theorems for Laplace and Stieltjes transforms and some material related to Tauberian theorems for Fourier transforms.)

[19] Wiener, N. (1959). *The Fourier Integral and Certain of its Applications*. Reprint, Dover, New York. (Develops a series of Tauberian theorems for Fourier transform as well as applications to the prime number theorem.)

(BRANCHING PROCESSES
CHARACTERISTIC FUNCTIONS
GENERATING FUNCTIONS
INTEGRAL TRANSFORMS
LIMIT THEOREM, CENTRAL
QUEUEING THEORY
RANDOM WALKS
RENEWAL THEORY)

GEORGE H. WEISS

TAUSWERTHE METHOD (FOR GENERATION OF PSEUDORANDOM VARIABLES)

This method (Tauswerthe [2]) is based on the sequence $\{a_k\}$ of 0's and 1's produced by a linear recurrence relation

$$a_k = c_p a_{k-p} + c_{p-1} a_{k-p+1} + \cdots + c_1 a_{k-1}$$

$$(\mathrm{mod}\, 2) \qquad (1)$$

based on a set $(c_1, \ldots, c_p)$ of 0's and 1's with

$c_p = 1$. The corresponding random integers (to base 2) are formed by successive L-tuples of the a's, namely

$$Y_1 \equiv a_1 a_2 \cdots a_L \quad (\mathrm{base}\, 2),$$
$$Y_2 \equiv a_{L+1} a_{L+2} \cdots a_{2L} \quad (\mathrm{base}\, 2),$$

and generally

$$Y_j \equiv a_{(j-1)L+1} a_{(j-1)L+2} \cdots a_{jL} \quad (\mathrm{base}\, 2).$$

This is said to form a *L-wise decimation* of $\{a_k\}$. If L is prime relative to $2^p - 1$, the decimation is said to be *proper*. A generator using this method is called a *Tauswerthe generator*.

Toothill et al. [3] showed that if nL is prime relative to $2^p - 1$, then the sequence $\{U_j\}$ with

$$U_j = 2^{-L} Y_j$$

has n-dimensional uniformity.

Tauswerthe generators are examples of "feedback shift register" methods. For further details see refs. 3 and 4 and especially ref 1.

References

[1] Kennedy, W. J. and Gentle, J. E. (1980). *Statistical Computing*. Dekker, New York.

[2] Tauswerthe, R. C. (1965). *Math. Comp.*, **19**, 201–209.

[3] Toothill, J. P. P., Robinson, W. D., and Adams, A. G. (1971). *J. Amer. Soc. Comp. Mach.*, **18**, 381–399.

[4] Toothill, J. P. P., Robinson, W. D., and Eagle, D. J. (1973). *J. Amer. Soc. Comp. Mach.*, **20**, 469–48.

(GENERATION OF RANDOM VARIABLES
PSEUDORANDOM NUMBER
GENERATORS)

TAXI PROBLEM

Schrödinger (see ref. 1, p. 127) considered the following problem: Given observed values of n mutually independent random variables $X_1, \ldots, X_n$, each having the discrete uniform distribution*

$$\Pr[X = x] = N^{-1}, \qquad x = 1, 2, \ldots, N,$$

construct an estimate of N.

This problem can be formulated in terms of estimating the number (N) of taxis in a city, from observations of the numbers $X_1, \ldots, X_n$ of n taxis, assuming (a) taxis are numbered $1, 2, \ldots, N$, and (b) each taxi is equally likely to be observed on each occasion. This description corresponds to sampling with replacement* (i.e., it is possible for the same taxi to be observed more than once); a related problem corresponds to sampling without replacement.

For the continuous analog of the problem —that of estimating the upper limit (θ) of a continuous rectangular (uniform*) distribution with probability density function

$$f_x(x) = \theta^{-1}, \qquad 0 < x < \theta,$$

the maximum likelihood estimator is M, the minimum mean squared error* estimator is $(n + 2)M/(n + 1)$, and the closest (in Pitman* sense) estimator is $2^{1/n}M$, where $M = \max(x_1, \ldots, x_n)$. (See Geary [1] and Johnson [2].)

Analogous results hold, approximately, for the discrete case (see, e.g., Rohatgi [3, p. 300]).

References

[1] Geary, R. C. (1944). *Biometrika*, **33**, 123–238.
[2] Johnson, N. L. (1950). *Biometrika*, **37**, 281–287.
[3] Rohatgi, V. K. (1984). *Statistical Inference*. Wiley, New York.

(ESTIMATION, POINT)

TAYLOR EXPANSION

This expansion is named after Brook Taylor (1685–1731), although it was known to James Gregory (1638–1675) and John Bernoulli (1667–1748). It is the expansion of a function $f(x)$ as a polynomial in x together with estimates on the remainder (R_n):

$$f(x) = f(a) + f'(a)(x - a)$$
$$+ f''(a)(x - a)^2/2!$$
$$+ f'''(a)(x - a)^3/3! + \cdots$$
$$+ f^{(n-1)}(a)(x - a)^{n-1}/(n - 1)!$$
$$+ R_n.$$

The remainder can be expressed in several different forms. *Lagrange's form*

$$R_n = \frac{h^n}{n!} f^{[n]}(a + \theta h),$$

(where θ is some number between 0 and 1 and $h = x - a$) and *Cauchy's form*

$$R_n = \frac{h^n(1 - \theta)^{n-1}}{(n - 1)!} f^{[n]}(a + \theta h)$$

are often used in application. Another (the so-called integral) form for the remainder is

$$R_n = \frac{1}{(n - 1)!} \int_a^x (x - t)^{n-1} f^{(n)}(t)\, dt.$$

For yet further formulas for R_n, see ref. 1. For a function $f(x_1, x_2, \ldots, x_k)$ of k variables, the *multivariate Taylor expansion* is

$$f(\mathbf{x}) = f(\mathbf{a}) + \sum_{j=1}^{k} (x_j - a_j) \frac{\partial f}{\partial x_j}\bigg|_{\mathbf{x}=\mathbf{a}}$$

$$+ \frac{1}{2!} \left\{ \sum_{j=1}^{k} (x_j - a_j) \frac{\partial}{\partial x_j} \right\}^2 f \bigg|_{\mathbf{x}=\mathbf{a}}$$

$$+ \frac{1}{3!} \left\{ \sum_{j=1}^{k} (x_j - a_j) \frac{\partial}{\partial x_j} \right\}^3 f \bigg|_{\mathbf{x}=\mathbf{a}}$$

$$+ \cdots,$$

where $\mathbf{x} = (x_1, \ldots, x_k)$, $\mathbf{a} = (a_1, \ldots, a_k)$, and the operators $\partial/\partial x_j$ are multiplied formally so that, for example, the third term in the expression is

$$\frac{1}{2} \left\{ \sum_{j=1}^{k} (x_j - a_j)^2 \frac{\partial^2 f}{\partial x_j^2}\bigg|_{\mathbf{x}=\mathbf{a}} \right.$$

$$\left. + 2 \sum_{i<j}^{k-1} \sum^{k} (x_i - a_i)(x_j - a_j) \frac{\partial^2 f}{\partial x_i\, \partial x_j}\bigg|_{\mathbf{x}=\mathbf{a}} \right).$$

Reference

[1] Iyanaga, S. and Kawada, Y., eds. (1977). *Encyclopedic Dictionary of Mathematics*, Vol. 2. MIT Press, Cambridge, Mass.

(CORNISH–FISHER AND EDGEWORTH
   EXPANSIONS
GENERATING FUNCTIONS
STATISTICAL DIFFERENTIALS, METHOD
   OF)

TEACHING STATISTICS

The journal *Teaching Statistics* publishes material on the teaching of statistics at the school level. It is an international journal for teachers of pupils ages 9–19. It seeks to help teachers of applied subjects such as biology, economics, geography, social science, the sciences, etc., to employ statistical ideas and methods to illuminate the work in their special subjects and to incorporate statistical methods in their teaching. It is also concerned with assisting in the teaching of statistics as part of a mathematics course. The material in the journal emphasises classroom teaching practice. *Teaching Statistics* was first published in 1979 and currently produces three issues a year constituting a single volume of about 100 pages. The present Editor is David Green, assisted by a Managing Editor (Vic Barnett), an Editorial Board, and an Advisory Panel with broad international representation. The current (1984) editorial address is The Editor, Teaching Statistics, Department of Probability and Statistics, University of Sheffield, Sheffield S3 7RH, UK. The subscription rate for Volume 9 (1987) is £5.75 for UK subscribers and £7.25 for overseas subscribers.

Teaching Statistics is published under the joint sponsorship of the International Statistical Institute*, The Institute of Statisticians*, The Royal Statistical Society*, and the Applied Probability Trust (which also publishes the *Journal of Applied Probability*, *Advances in Applied Probability*, and a school-journal called *Mathematical Spectrum*). The journal operates under a charitable trust, The Teaching Statistics Trust, which is concerned with the promotion of statistical education*.

Teaching Statistics publishes short articles providing specific illustration of the ways in which statistics can be usefully introduced into, or taught within, the school curriculum across the whole range of different subjects. It also contains regular book reviews, letters, news and notes, and editorial comment. Many of the articles are of UK origin but there is a firm journal policy to provide material of international interest. This is reflected by articles [in Volume 5 (1983)] from Canada, Israel, New Zealand, Puerto Rico, South Africa, Sweden, the United Kingdom, and the United States. A German edition of the journal, entitled *Stochastik in der Schule*, which arose from a mutual interest in statistical education between the Departments of Statistics in Sheffield University and Dortmund University, West Germany, was launched in 1979. The initiative for the German edition was taken by Friedhelm Eicker of Dortmund University, who remains (1984) one of the present editors. (*Stochastik in der Schule* is predominantly a German translation of *Teaching Statistics* but is, to an increasing extent, introducing revisions of the material of the English edition to suit the German market and some original material from German sources.)

Volume 5 (1983) of *Teaching Statistics* contained such articles as:

"Bivariate normal model: a classroom project" by M. W. Maxfield and B. C. Lyon.

"IN STEP With the Microcomputer" by C. W. Anderson and V. Barnett.

"How to convince a student that an estimator is a random variable" by K. Vännman.

"Interpretation of statistical results—a real world example" by H. Tamura.

"A review of publications on the teaching of probability and statistics" by H. Sahai.

"A practical study of the capture/recapture method of estimating population size" by B. A. C. Dudley.

"Continuous expectation: just as expected" by P. De Roos.

All published articles are refereed and the journal seeks to present material of classroom relevance firmly rooted in practical problems. Authors are encouraged to produce articles which are light and readable as well as informative. Average times for refereeing and from acceptance to publication are about 2 months and 9 months, respectively. The Institute of Statisticians offers each year a C. Oswald George Prize of £50 for an article of outstanding interest and merit published in *Teaching Statistics*.

VIC BARNETT

TCHEBYSHEFF *See* CHEBYSHEV

TECHNOMETRICS

Technometrics was established in 1959 as a joint publication of the American Statistical Association* and the American Society for Quality Control*. Its purpose then, as now, was the development and use of statistical methods in the physical, chemical, and engineering sciences. Thus, the journal's emphasis is on applications—on the publication of statistical methods that can immediately be used on real problems in these sciences—and on a style of presentation appropriate for a broad readership of statisticians, scientists, and engineers. The journal seeks to publish papers describing new statistical techniques, papers illustrating innovative application of known statistical methods, expository papers on particular statistical methods, and papers dealing with the philosophy and problems of applying statistical methods, when such papers are consistent with the journal's objectives. Book reviews, queries, and letters to the editor are also published. Issues are published quarterly in February, May, August, and November. The journal has been self-supported since its origin, receiving all its income solely from subscriptions. Circulation in 1983 was about 5500 copies.

The Founding Editor of *Technometrics* was J. S. Hunter, who served from 1959 through 1963. He was followed by Fred C.

Leone (1964–1968), Harry Smith, Jr. (1969–1971), Donald A. Gardiner (1972–1974), William H. Lawton (1975–1977), John W. Wilkinson (1978–1980), Robert G. Easterling (1981–1983), and Jerald F. Lawless (1984–1986) and William Q. Meeker (1987–). The journal is managed by a Management Committee that consists of three members selected by each sponsoring society plus a jointly selected chairman.

Technometrics was created, primarily, to meet the needs of industrial scientists and statisticians. Hence its cosponsorship, and indeed its initial impetus, came from the American Society for Quality Control. The process leading to the journal's creation has been described by J. S. Hunter in "The Birth of a Journal," February 1983, pp. 1–7. This article was the first of several papers especially invited to mark *Technometrics'* 25th year of publication (1983). Subsequent invited review papers include "Outlier …s" by R. J. Beckman and R. D. Cook (May 1983), "Developments in Linear Regression Methodology: 1959–1982" by R. R. Hocking (August 1983), and "Statistical Methods in Reliability" by J. F. Lawless (November 1983). The February 1984 issue featured "The Importance of Practice in Statistics" by George E. P. Box, who played a vital role in the creation of *Technometrics*. There appeared in May 1984 a review paper by D. M. Steinberg and W. G. Hunter on experimental design. The titles of these invited papers effectively capsulize the journal's emphasis over its 25 years of publication

The contents of Vol. 25, No. 4, November 1983, in addition to the paper by Lawless, included:

"Transformations and Influential Cases in Regression" by R. D. Cook and P. C. Wang.

"Two-Level Multifactor Designs for Detecting the Presence of Interactions" by M. D. Morris and T. J. Mitchell.

"Selecting Check Points for Testing Lack of Fit in Response Surface Models" by J. T. Shelton, A. I. Khuri, and J. A. Cornell.

"A New Algorithm for Extreme Vertices Designs for Linear Mixture Models" by A. K. Nigam, S. C. Gupta, and S. Gupta.
"The Derivative of $|X'X|$ and Its Uses" by D. Bates.
"A Two-Sample Test of Equal Gamma Distribution Scale Parameters With Unknown Common Shape Parameter" by W. Shiue and L. J. Bain.

The review papers mentioned document and illustrate the important role *Technometrics* has played in developing and disseminating statistical methods and ideas. Present industrial concern about quality and productivity promise to keep *Technometrics* in the forefront of developing statistical methods for use in the physical, chemical, and engineering sciences.

ROBERT G. EASTERLING

TELEPHONE SURVEYS, COMPUTER ASSISTED

Computer assisted telephone surveys (also labelled CATI, computer assisted surveys, computer based interviewing) use computer terminals (usually cathode ray tube terminals) to display survey questions to an interviewer, who reads them over the telephone to a respondent and then enters into the terminal the responses obtained to the questions. CATI systems often check the numeric answers to determine if they lie within a valid range. They also route the interviewer automatically to the succeeding questions that are proper for the particular respondent. Some CATI systems also manage the sample cases and assign cases to interviewers, permit the monitoring of interviewers' terminals by supervisors, alter the question wording* for particular respondents, randomize the order of questions or of response categories (permitting estimation of order effects on survey statistics), and randomly assign cases to interviewers (permitting estimation of interviewer effects on survey statistics).

Most CATI systems use minicomputers (250,000–4,000,000 bytes of core memory) to control 10–60 terminals used by interviewers conducting the same or different surveys simultaneously. They are often used in centralized telephone interviewing facilities, where interviewers work in carrels that contain a telephone and a terminal. Some larger CATI systems are installed on different computers in regional offices of survey organizations and are linked to one another in a network of machines that share the software for the CATI system and pass completed data records to a central machine. Using newer, more powerful microcomputers in intelligent terminals, some CATI systems employ a central computer as a storage device rather than as the controller over the interviews of each interviewing station. Thus, some systems have much of the software for the questionnaire stored in the interviewer's display unit rather than in the central machine. Such a design for a CATI system permits the use of a much smaller central machine, and allows each terminal to continue with an interview even if the central machine is overburdened or down. The design is part of an inevitable movement to independence of the interviewer's terminal from the central computer and facilitates the use of computer assistance in personal interview surveys as well as telephone surveys.

Although CATI systems differ in the set of functions they offer, the crucial performance features of any CATI system include the speed at which the next question is displayed on the screen after the interviewer enters a response, the speed with which completed interview data can be retrieved from the system for analysis after the last interview is taken, and the efficiency with which the system schedules callbacks and administers the sample.

CATI can be viewed as a technology that combines several discrete steps of survey activities. The development of a printed questionnaire, assignment of the sample to interviewers, creation of a numeric machine readable data set, and post survey cleaning of the data are all performed at the point of data collection* under CATI. It is the characteristic of reducing the number of separate steps in collecting and processing survey data that leads many to suspect that CATI surveys can collect data at a lower cost than non-CATI surveys.

Of interest to survey statisticians when considering the use of CATI for surveys is the relative size of survey costs and survey errors that are present with CATI relative to other methods. Those relevant to a comparison of CATI and non-CATI telephone surveys are

1. With CATI there are no additional costs of printing questionnaires, key entry, or post survey data processing prior to analysis.

2. Since editing that locates invalid numeric codes for responses is done by CATI at the time of data collection, some of these errors can be corrected by the respondent. No such capability is typically enjoyed by non-CATI surveys.

3. Errors by the interviewer in following instructions regarding the flow of the questionnaire are eliminated by CATI because the machine determines the order of display of questions.

4. The selections of cases to be called next and the administration of the sample can be controlled by CATI; thus some of the clerical duties of non-CATI surveys are eliminated.

5. After the last interview is completed on the survey, all numeric data are available for statistical analysis (some systems permit intermediate analysis of subsamples during the interviewing itself).

6. Since CATI permits randomization of sample case assignment to interviewer, randomized ordering of questions, and monitoring of interviews, it can produce a survey with more fully measured non-sampling errors than is feasible with a non-CATI survey.

The relative costs and errors of CATI and non-CATI surveys are dependent on how each of the modes is implemented; CATI systems are too new at this writing to claim

assured preference for their use over non-CATI survey methods. It is clear, however, that the *potential* gains of computer assistance in the data collection phase of surveys are large and inevitable.

Bibliography

Groves, R. M. and Mathiowetz, N. A. (1984). Computer-assisted telephone interviewing: Effects on interviewers and respondents. *Public Opinion Qtly.*, **48**, 356–369.

Groves, R. M. and Nicholls, II, W. L. (1986). The status of computer-assisted telephone interviewing: Part II—Data quality issues. *J. Official Statist.* (*Sweden*), **2**, 117–134.

Harlow, B. L., Rosenthal, J. F. and Ziegler, R. G. (1985). A comparison of computer-assisted and hardcopy telephone interviewing. *Amer. J. Epidemiology*, **122**, 335–340.

Nicholls, II, W. L. and Groves, R. M. (1986). The status of computer-assisted telephone interviewing: Part I—Introduction and cost and timeliness of survey data. *J. Official Statist.* (*Sweden*), **2**, 93–115.

(COMPUTERS AND STATISTICS
SURVEY SAMPLING)

ROBERT M. GROVES

TELESCOPING

This is the name given to a major type of memory error that causes distortion in the results of sample surveys* involving behavioral questions.

It consists of misremembering the date an event occurred, in such a way as to bring it into the time period being discussed. Sudman [2], based on U.S. experience in the period 1960–1980, concluded that reports of purchases of grocery products within the last seven days were likely to include overestimates of the order of 50%. He suggested that the proportion of overestimates due to telescoping can be approximated by the formula $t^{-1} \log(b_2 t)$, where t is the length of time over which recall is required, and b_2 varies from 2.5–3.5, for large expenditure on durable goods, to 4–5 for smaller expenditures. For long intervals (large t) overestimates due to telescoping are relatively small, but still exist.

See also Neter and Waksberg [1] for a detailed discussion of an experimental study, relating to telescoping of expenditure data, conducted by the U.S. Bureau of the Census* in the early 1960s.

References

[1] Neter, J. and Waksberg, J. (1964). *J. Amer. Statist. Ass.*, **59**, 18–55.

[2] Sudman, S. (1983). In *Statistical Methods and the Improvement of Data Quality*, T. Wright, ed. Academic, New York, pp. 85–115.

(NONSAMPLING ERRORS IN SURVEYS
QUESTION-WORDING EFFECTS IN
  SURVEYS
SURVEY SAMPLING)

TELEVISION VIEWING, STATISTICAL ASPECTS

Television viewing has become people's largest leisure activity in developed countries. The amount of viewing per week follows an approximately normal distribution with a mean of about 25 and a standard deviation of about 10 hr. in the U.S.A., where it adds up to over 200 billion hr. of viewing a year.

The cost of making watchable TV programs is large, about $100,000 an hour in many cases, over $1 million for *Dallas*, but with audience ratings up to 20%—e.g., 40 million Americans all watching the same thing—the cost per viewer is a few cents. The U.S. is unique in that almost all TV programming is funded as a cross-subsidy from advertising (some $20 billion a year) and therefore appears free. In other western countries there is usually mixed funding, advertising and a license fee or grants out of taxes (e.g., Barwise and Ehrenberg [1]).

AUDIENCE MEASUREMENT

Television viewing leaves no physical transaction record and audiences have, therefore, to be measured using sampling methods (e.g., Twyman [10].) The samples used are mostly large enough for sampling errors to be less important than noncooperation bias and measurement errors*. Technicalities are changing and can be obtained from rating firms (e.g., Nielsen, Arbitron, AGB, etc. in the U.S.).

Television viewing is such an extensive, yet mostly low, involvement activity that ordinary questionnaire techniques (e.g., "How often did you see the Thursday NBC Late Night News/*Dallas*/or whatever in the last four weeks?") can lead to large exaggeration. Coincidental interviewing—i.e., well-timed calls asking "What are you watching *now?*"—is more accurate but too expensive for regular use. Careful "aided recall" interviewing could work well when there were few channels, but for national data and in the largest cities, meters are attached to television sets in a panel of households, to measure minute by minute whether the set is on and to which channel. Some 10–20% of the time, however, no one may be in the room.

For data on individuals' viewing and also for smaller areas, one-week diaries have been widely used in the U.S. for people to record their viewing quarter hourly. In Britain and elsewhere, diaries have been run in conjunction with set meters on an ongoing panel basis. Some of these diary methods are now being replaced by so-called "people meters," using electronic push-button devices, to measure increasing numbers of channels and possibly also VCR recording and replay. These various procedures have been extensively investigated and mostly have been found to give rather consistent results when well operated.

All the measurement procedures for individuals tend to reflect whether a person is in the viewing room with the set on. A substantial proportion (say 40% by and large) of such people are, however, not actively or exclusively involved in viewing but are known to be "nipping" (momentarily out of the room) or "napping" (asleep), or reading, talking, or doing something else as well. Measurement problems of this kind are reasonably well understood, but not routinely measured nor always taken into account (e.g., TAA [9]).

TV audiences are measured partly for programming purposes, but more as the only basis for buying and selling advertising airtime. Nowadays, added complications are "zapping" and "zipping," i.e., fast switching between channels using remote control and running through commercials on "fast forward" when playing back time-shift VCR recordings. Many hundreds of millions of measurements are collected over a year in the U.S.A.

VIEWING PATTERNS

Most of the resulting data are used merely to present "the ratings," either the percentage of households with sets on to a given channel or the percent of all individuals or of demographic subgroups viewing in a given quarter hour. Total viewing levels tend to be steady from day to day, with the total audience largest at about 9 P.M. each evening, but the shares of the different TV channels vary with their programming. This is the point of major week-by-week interest.

Being collected on a panel basis for a week or more, the data are, however, also a source for analysing people's viewing patterns (e.g., Nielsen [8], Goodhardt et al. [6]). Some of these are surprising.

The audience size or rating of regular programs tends to be steady from week to week (or day to day for "stripped" series), but there is much stochastic movement under the surface. Only about half (50% or so) the program's viewers this week view it again next week; it is less than half for low-rating programs. This is mainly because about a third of this week's viewers do not watch TV at all at the same time next week. If they do, they then mostly tend to watch the same program as before. The repeat-viewing levels generally do not vary by type of program (Barwise [1], Ehrenberg and Wakshlag [4], and Goodhardt et al. [6]).

On television, therefore, "nobody reads the whole book through." For a popular series over two or three months, few people watch all episodes or nearly all. The number of episodes seen generally follows a beta-binomial distribution* to a very close degree of approximation (e.g., Goodhardt et al. [6], Greene [7]). This is shown in Table 1 for *Coronation Street* in the U.K., the longest running TV soap opera. Predictably, only some 10% of the population (or 16% of those seeing at least one episode), saw all or all but one of the eight episodes.

The overlap between the audiences to different programs also follows a simple pattern, the so-called duplication of viewing law. This states that the percentage $r_{s/t}$ of the audience of program t who also watch program s is a constant times the rating r_s of the latter program: $r_{s/t} = kr_s$. The constant k, known as the *duplication coefficient*, is higher if the two programs are on the same channel (typically roughly 1.5 for U.S. prime-time network programs) than if they are on different channels (typically about 1, i.e., viewing of s and t is then uncorrelated). There are certain subpatterns like the "lead-in" effect, i.e., high duplication between consecutive programs on the same channel. This is, however, only of short duration. People's

Table 1   Percent Viewing 0 to 8 Successive Episodes of *Coronation Street* (Observed and Beta Binomial Distribution)

London Men 1971	Episodes Seen								
	0	1	2	3	4	5	6	7	8
Obs. (%)	41	13	8	7	8	6	6	6	5
BBD[a] (%)	37	15	10	8	7	6	6	5	5

[a] Fitted by mean and average repeat rate.

availability to view (e.g., during the weekday or late in the evening) also affects the value of k.

Surprisingly perhaps, k does not vary by program type, as is illustrated in Table 2 for duplication between two channels, where $k \doteq 1$; irrespective of the programs broadcast (not even identified here), the duplication percent $r_{s/t}$ is close to the Tuesday ratings r_s; the averages in the last line but one are *very* close. Such duplication patterns have been empirically studied in many hundred thousands of cases over the last 20 years. There is also well-established stochastic theory (e.g., Goodhardt [5], Goodhardt et al. [6]).

The extract in Table 3 further illustrates the absence of large program type effects and also shows how people choose to watch a variety of programs in the U.K. (Ehrenberg [3]). It shows that people of different kinds spent much the same proportion of their week's viewing time on the different types of program.

Table 2   Audience Overlap: The Duplication Law (Two US Networks, 6 Days Apart)

Viewers of ABC on Wednesday at	New York Housewives, 1974				
	Percent who also viewed CBS on the following Tuesday at				
	8*	7:30	9:30	11	11:30 pm
4 pm[a]	34	16	10	7	4
4:30 pm	22	14	11	2	3
6 pm	26	11	13	3	3
7 pm	29	12	14	6	4
7:30 pm	29	12	10	5	2
8 pm	29	8	12	7	3
10 pm	28	13	15	6	5
11 pm	24	10	15	5	3
11:30 pm	23	9	14	7	2
Average	27	12	12	5	3
% of pop., viewing on Tuesday	26	14	11	7	3

[a] $\frac{1}{4}$-hour start-times for different programs; * columns arranged in order of rating.

Table 3   Viewing of Different Program Types in Britain

Britain 1985	% of time spent viewing							
	Light Entertainment	Light Drama	Films	Sport	Drama Arts etc.	Information	News	Other Incl. Childrens'
Socio-economic status								
Higher (%)	18	20	7	9	9	20	11	6
Lower (%)	17	22	9	9	7	20	10	6
Viewers of								
Dynasty (%)	17	22	7	8	8	20	12	6
Panorama (%)	17	20	11	10	8	17	10	7

THE EFFECTS OF TELEVISION

Much concern tends to be expressed over the large amounts of television most of us watch. How does it affect child development, violence, sex, the making of the President, materialism, and so on.

Large numbers of studies have been carried out and large numbers of books and articles written. The main result is that nothing seems to have been proved—one interpretation is that TV is a low involvement medium and largely washes over us (e.g., Barwise and Ehrenberg [2]). On this view, even TV advertising either merely makes us aware of new products or services (the vast majority of which we do not buy) or is defensive, the advertiser seeking to keep existing customers.

References

[1] Barwise, T. P. (1986). Repeat-viewing of prime-time TV series. *J. Advert. Res.*, **26**, 9–14.

[2] Barwise, T. P. and Ehrenberg, A. S. C. (1988). *Television and Its Audience*. Sage, London, England.

[3] Ehrenberg, A. S. C. (1985). *Advertisers or Viewers Paying?* Admap Publications, London, England.

[4] Ehrenberg, A. S. C., and Wakshlag, J. (1987). Repeat-viewing with people meters, *J. Advert. Res.*, **27**, 9–13.

[5] Goodhardt, G. J. (1966). The constant in duplicated television viewing, *Nature*, **212**, 1616.

[6] Goodhardt, G. J., Ehrenberg, A. S. C., and Collins, M. A. (1975, 1986). *The Television Audience: Patterns of Viewing*. Saxon House, London. 2nd ed. Gower Press, Farnborough and Brookfield, Vt.

[7] Greene, J. D. (1982). *Consumer Behavior Models for Non-Statisticians*. Praeger, New York.

[8] Nielsen (1985). *The Television Audience 1985*. A. C. Nielsen Company, Northbrook, IL.

[9] TAA (1982). *The Audience Rates Television*. Television Audience Assessment, Cambridge, Mass.

[10] Twyman, W. A. (1986). Television media research, in *Consumer Market Research Handbook*, 3rd ed. R. Worcester and J. Downhan, eds. North-Holland/ESOMAR, Amsterdam, Netherlands.

(PUBLIC OPINION POLLS
SURVEY SAMPLING)

A. S. C. EHRENBERG

TEMPERED DISTRIBUTIONS   *See* DIRAC DELTA FUNCTION

TENSORS

DEFINITION

A *contravariant tensor* of order d, written $\omega^{i_1 i_2 \cdots i_d}$, is a d-dimensional array of real-valued functions defined relative to a given coordinate system $x = x^1, \ldots, x^p$ in R^p. The essential feature of a tensor is that the value of ω in any other coordinate system $y = y^1, \ldots, y^p$ is given by

$$\bar{\omega}^{r_1 r_2 \cdots r_d} = a^{r_1}_{i_1} a^{r_2}_{i_2} \cdots a^{r_d}_{i_d} \omega^{i_1 i_2 \cdots i_d}, \quad (1)$$

where $a^r_i = \partial y^r / \partial x^i$ and the summation convention applies to the repeated indices $i_1, \ldots, i_d$ [i.e., the right-hand side of (1) is summed with respect to the i's]. The transformation law for a *covariant tensor* of order d, written $\omega_{i_1 i_2 \cdots i_d}$, is

$$\bar{\omega}_{r_1 r_2 \cdots r_d} = b^{i_1}_{r_1} b^{i_2}_{r_2} \cdots b^{i_d}_{r_d} \omega_{i_1 i_2 \cdots i_d}, \quad (2)$$

where $b^i_r = \partial x^i / \partial y^r$, the matrix inverse of a^r_i, is assumed to have full rank. It is possible also to define hybrid or mixed tensors that have both subscripts and superscripts. The transformation law for such tensors involves a combination of (1) and (2) in the obvious way.

These are the definitions commonly found in the literature on differential geometry, where x and y are typically alternative coordinate systems on some p-dimensional surface embedded in R^n and ω is a combination of partial derivatives. Geometrical interpretation demands invariance under nonlinear but invertible transformation of coordinates. Other definitions are sometimes used in which only linear or affine transformation is considered (Greub [1]). In the physics literature, the term "Cartesian tensor" is sometimes used in this context (Jeffreys [2]).

The primary advantage of working with tensors, as opposed to arbitrary arrays of functions, is that the calculations are essentially independent of the coordinate system used. This feature is particularly appealing in statistical calculations connected with log likelihood* functions, where, in principle, the choice of parametrization should not affect the conclusions reached.

PROPERTIES

(i) **Multiplication.** If ω_{ij} and ω^{ijkl} are two tensors, by notation covariant of order 2 and contravariant of order 4, respectively, then $\gamma_{ij}^{klmn} = \omega_{ij}\omega^{klmn}$ is a mixed tensor of covariant order 2 and contravariant order 4. Similarly for the product of two covariant or two contravariant tensors.

(ii) **Contraction.** If γ_{ij}^{klmn} is a tensor, then by summing over pairs of indices, we may form new tensors such as $\gamma_i^{klm} = \gamma_{ij}^{jklm}$ and $\gamma^{kl} = \gamma_{ij}^{ijkl}$.

(iii) **Scalars.** If ω_{ij} and ω^{ijkl} are two tensors, then $\gamma = \omega^{ijkl}\omega_{ij}\omega_{kl}$ is a tensor of order zero or an invariant. It has the same value in all coordinate systems.

(iv) **Matrix Inversion.** If ω_{ij} is a covariant tensor, not necessarily of full rank, and ω^{ij} is any generalized inverse, then ω^{ij} is a contravariant tensor. Also, $\omega_{ij}\omega^{ij} = \text{rank}(\omega_{ij})$ is an invariant and is independent of the choice of generalized inverse.

Note that γ_{ij}^{jklm} is not, in general, the same as γ_{ij}^{klmj} or γ_{ij}^{iklm} unless, by some added convention, arrays are assumed to be symmetrical under index permutation. Without such a convention, there are three distinct invariants that can be formed by combining the tensors ω_{ij} and ω^{ijkl} as described in (iii). It is highly desirable for statistical work to arrange matters so that all arrays are symmetrical under index permutation. Of course, subscripts must not be interchanged with superscripts.

LOG LIKELIHOOD DERIVATIVES

Tensor Properties

To develop a standard example, let $l(\theta; Y) = \log f_Y(Y; \theta)$ be the log likelihood function expressed in terms of the p-dimensional parameter $\theta = \theta^1, \ldots, \theta^p$. We may write the arrays of partial derivatives as $U_i = \partial l/\partial \theta^i$, $U_{ij} = \partial^2 l/\partial \theta^i \partial \theta^j$, and so on. If $\phi = \phi^1, \ldots, \phi^p$ is an alternative parametrization, the log likelihood derivatives with respect to the components of ϕ may be written

$$U_r^* = U_i\theta_r^i, \qquad U_{rs}^* = U_{ij}\theta_r^i\theta_s^j + U_i\theta_{rs}^i, \quad (3)$$

where $\theta_r^i = \partial\theta^i/\partial\phi^r$ and $\theta_{rs}^i = \partial^2\theta^i/\partial\phi^r\partial\phi^s$. The main emphasis here is that, given the relationship between ϕ and θ, it is possible to determine U_r^* from U_i alone, but it is not possible to determine U_{rs}^* from U_{ij} alone. We say that U_i is a tensor but that U_{ij} is not. On the other hand, $\kappa_{ij} = E\{U_{ij}; \theta\}$ is a tensor because U_i has zero expectation at the true parameter point. In addition, if $\hat{\theta}$ is the maximum likelihood estimate of θ, then the observed Fisher information* $-U_{ij}(\hat{\theta}; Y)$ is a tensor because $U_i(\hat{\theta}; Y) = 0$. Finally, the residual second derivative

$$V_{ij} = U_{ij} - \beta_{ij}^k U_k,$$

where β_{ij}^k is the least squares regression coefficient* of U_{ij} on U_k, is a tensor. The regression coefficient β_{ij}^k is not a tensor.

Invariant Statistics

The maximized log likelihood ratio* statistic $\Lambda = 2l(\hat{\theta}; Y) - 2l(\theta; Y)$ is an invariant random variable in the sense that its value is independent of the parametrization used. The usual first-order approximation in terms of log likelihood derivatives involves the quadratic score statistic*

$$Q = U_i U_j \kappa^{i,j},$$

where $\kappa^{i,j}$ is the matrix inverse or generalized inverse* of $\kappa_{i,j} = \mathrm{cov}(U_i, U_j)$. By virtue of its tensor construction, Q is an invariant random variable whose expectation is $\kappa_{i,j}\kappa^{i,j} = \mathrm{rank}(\kappa_{i,j})$. Other invariants that arise in the second-order approximation of Λ include

$$U_r U_s V_{tu}\kappa^{r,t}\kappa^{s,u}, \qquad \nu_{rs,tu}\kappa^{r,s}\kappa^{t,u}$$

$$\text{and} \quad \nu_{rs,tu}\kappa^{r,t}\kappa^{s,u},$$

where $\nu_{rs,tu} = \mathrm{cov}(V_{rs}, V_{tu})$. For example, in the usual normal-theory nonlinear regression model with known variance $\sigma^2 = 1$, the mean of Λ is

$$p\left\{1 + \kappa^{r,s}\kappa^{t,u}(2\nu_{rt,su} - \nu_{rs,tu})/(4p)\right\}$$
$$+ O(n^{-2}),$$

where n is the sample size. The coefficient of p is the *Bartlett factor* and has an interpretation in terms of curvature of the model surface in R^n.

CUMULANTS* AND GENERALIZED CUMULANTS

Definition. Let $X = X^1, \ldots, X^p$ be a p-dimensional random variable whose moment generating function $M_X(\xi) = E\{\exp(\xi_i X^i)\}$ is assumed to have the expansion

$$M_X(\xi) = 1 + \xi_i\kappa^i + \xi_i\xi_j\kappa^{ij}/2!$$
$$+ \xi_i\xi_j\xi_k\kappa^{ijk}/3! + \cdots.$$

The cumulants $\kappa^i, \kappa^{i,j}, \kappa^{i,j,k}, \ldots$ are defined as the coefficients in the series expansion of $K_X(\xi) = \log M_X(\xi)$,

$$K_X(\xi) = \xi_i\kappa^i + \xi_i\xi_j\kappa^{i,j}/2!$$
$$+ \xi_i\xi_j\xi_k\kappa^{i,j,k}/3! + \cdots.$$

We find, for example, that

$$\kappa^{ij} = \kappa^{i,j} + \kappa^i\kappa^j,$$
$$\kappa^{ijk} = \kappa^{i,j,k} + \kappa^i\kappa^{j,k} + \kappa^j\kappa^{i,k}$$
$$+ \kappa^k\kappa^{i,j} + \kappa^i\kappa^j\kappa^k.$$

More generally, the sum on the right extends over all partitions of the set of indices on the left. If $\mathscr{D}^* = \{(i,j,k,\ldots)\}$, we may write

formally

$$\kappa(\mathscr{D}^*) = \sum_{\mathscr{D}} \kappa(D_1) \cdots \kappa(D_\nu), \quad (4)$$

where the sum extends over all ν and over all partitions $\mathscr{D} = \{D_1, D_2, \ldots, D_\nu\}$ of the set of indices into ν nonempty blocks.

Tensor Properties

If Y is an affine transformation of X given by $Y^r = a^r + a_i^r X^i$, then the cumulants of Y are

$$a^r + a_i^r\kappa^i, \qquad a_i^r a_j^s\kappa^{i,j}, \qquad a_i^r a_j^s a_k^t\kappa^{i,j,k},$$

and so on. Thus the cumulants of order 2 or more behave like contravariant tensors but only under affine transformation of X. However, a_i^r need not have full rank.

Generalized Cumulants

Let

$$\kappa^{i,jk} = \mathrm{cov}(X^i, X^j X^k),$$
$$\kappa^{ij,kl} = \mathrm{cov}(X^i X^j, X^k X^l),$$

and

$$\kappa^{i,j,kl} = \mathrm{cum}(X^i, X^j, X^k X^l),$$

the third-order mixed cumulant of the three variables X^i, X^j, and $X^k X^l$. This notation should not be confused with that for the joint cumulants of log likelihood derivatives. More generally, if $\mathscr{D}^* = \{D_1^*, \ldots, D_a^*\}$ is a partition of the indices into a nonempty blocks, $\kappa(\mathscr{D}^*)$ is the joint cumulant of the a variables

$$Y^1 = \prod_{j \in D_1^*} X^j, \qquad Y^2 = \prod_{j \in D_2^*} X^j,$$
$$\ldots, Y^a = \prod_{j \in D_a^*} X^j.$$

The fundamental identity giving the generalized cumulant $\kappa(\mathscr{D}^*)$ in terms of products of ordinary cumulants is

$$\kappa(\mathscr{D}^*) = \sum_{\mathscr{D} \vee \mathscr{D}^* = 1} \kappa(D_1) \cdots \kappa(D_\nu), \quad (5)$$

where the sum is over all $\mathscr{D} = \{D_1, \ldots, D_\nu\}$ such that $\mathscr{D}$ and $\mathscr{D}^*$ are not both subpartitions of any partition other than the full unpartitioned set. We say that the sum is over all partitions *complementary* to $\mathscr{D}^*$. Alternatively, if $\mathscr{D}$ and $\mathscr{D}^*$ are represented as two graphs on the same vertices, labelled $1, \ldots, p$, the sum in (5) extends over all $\mathscr{D}$ such that the edge sum graph $\mathscr{D} \oplus \mathscr{D}^*$ is connected (McCullagh [3]). Tables of connected pairs up to order 8 are given by McCullagh and Wilks [4].

By way of illustration, the third-order cumulant of the three variables X^i, X^j and the product $X^k X^l$ is

$$\kappa^{i,j,kl} = \kappa^{i,j,k,l} + \kappa^k \kappa^{i,j,l}[2] + \kappa^{i,k}\kappa^{j,l}[2],$$

where the four partitions with two blocks are

$$\{(k),(i,j,l)\}, \qquad \{(l),(i,j,k)\},$$
$$\{(i,k),(j,l)\}, \qquad \{(i,l),(j,k)\};$$
$$\kappa^k \kappa^{i,j,l}[2] = \kappa^k \kappa^{i,j,l} + \kappa^l \kappa^{i,j,k}$$

and

$$\kappa^{i,k}\kappa^{j,l}[2] = \kappa^{i,k}\kappa^{j,l} + \kappa^{i,l}\kappa^{j,k}.$$

Note that $\{(i,j),(k,l)\}$ is not included in this list even though it is not a subpartition of $\{(i),(j),(k,l)\}$.

The converse of (5) giving generalized cumulants in term of moments is

$$\kappa(\mathscr{D}^*) = \sum_{\mathscr{D} \geqslant \mathscr{D}^*} (-1)^{\nu-1}(\nu - 1)!$$
$$\times \mu(D_1) \cdots \mu(D_\nu),$$

where the sum extends over all $\mathscr{D}$ such that $\mathscr{D}^*$ is a subpartition of $\mathscr{D}$.

The fundamental identity (5) plays an important role in computing the joint cumulants of an arbitrary polynomial function of X (McCullagh [3]).

HERMITE TENSORS

One definition of ordinary Hermite tensors $h^i(x;\lambda)$, $h^{ij}(x;\lambda)$, ... uses the generating function*

$$\exp\left\{\xi_i(x^i - \lambda^i) - \xi_i\xi_j\lambda^{i,j}/2!\right\},$$

where, for example, $h^{ijk}(x;\lambda)$ is the coefficient of $\xi_i\xi_j\xi_k/3!$ in the Taylor expansion*. Alternatively, on taking the logarithm of the

generating function, we may define

$$h^i(x;\lambda) = x^i - \lambda^i,$$
$$h^{i,j}(x;\lambda) = -\lambda^{i,j}, \qquad (6)$$
$$h^{i,j,k}(x;\lambda) = \cdots = 0,$$

analogous to the relationship between moments and cumulants. Generalized Hermite tensors $h^{i,jk}$, $h^{ij,kl}$ and so on, are defined using (5) with the cumulants replaced by expressions (6). For example,

$$h^{ij,kl} = h^{i,k}h^{j,l}[2] + h^i h^k h^{j,l}[4]$$
$$= -(x^i - \lambda^i)(x^k - \lambda^k)\lambda^{j,l}[4]$$
$$+ \lambda^{i,k}\lambda^{j,l}[2],$$

where, for example,

$$h^i h^k h^{j,l}[4] = h^i h^k h^{j,l} + h^i h^l h^{j,l}$$
$$+ h^j h^k h^{i,l} + h^j h^l h^{i,j},$$

is an array of polynomials each of degree 2. As can be seen from (6), the generalized Hermite tensors are formally identical to the generalized cumulants of the normal distribution.

The subscripted versions of the Hermite tensors are obtained by multiplying by $\lambda_{i,j}$, the matrix inverse of $\lambda^{i,j}$, as often as necessary. For example, $h_i(x;\lambda) = \lambda_{i,j}h^j(x;\lambda)$, $h_{i,j}(x;\lambda) = \lambda_{i,k}\lambda_{j,l}h^{k,l}(x;\lambda) = -\lambda_{i,j}$, provided that $\lambda^{i,j}$ has full rank or that the Moore–Penrose inverse* is used.

The more usual definition of Hermite tensors is based on derivatives of the normal density. If $\phi(x;\lambda)$ is the normal density with mean vector λ^i and covariance matrix $\lambda^{i,j}$, then the partial derivatives with respect to the components of x are $-h_i\phi(x;\lambda)$, $h_{ij}\phi(x;\lambda)$, $-h_{ijk}\phi(x;\lambda)$, and so on with signs alternating. In the univariate case, these reduce to the Hermite polynomials suitably scaled and centered (*see* CHEBYSHEV–HERMITE POLYNOMIALS).

The Hermite tensors are mutually orthogonal in the sense that $\int h^i h^{jk}\phi\,dx = 0$, $\int h^{ij}h^{klm}\phi\,dx = 0$, and so on with integration over R^p. In addition, $\int h^i h^j\phi\,dx = \lambda^{i,j}$, $\int h^{ij}h^{kl}\phi\,dx = \lambda^{i,j}\lambda^{j,l}[2!]$, and so on.

Hermite tensors arise naturally in multivariate Edgeworth expansions. Generalized Hermite tensors arise in the Edgeworth approximation for conditional cumulants (McCullagh, [3], Sec. 6]).

References

[1] Greub, W. H. (1967). *Multilinear Algebra*. Springer, New York.

[2] Jeffreys, H. (1952). *Cartesian Tensors*. Cambridge University Press, New York.

[3] McCullagh, P. (1984). Tensor notation and cumulants of polynomials. *Biometrika*, **71**, 461–476.

[4] McCullagh, P. and Wilks, A. R. (1985). Complementary set partitions. TM 11214-950328-07, AT & T Bell Labs, Murray Hill, N.J.

(CORNISH–FISHER AND EDGEWORTH EXPANSIONS
CUMULANTS
GENERALIZED INVERSE
PROBABILITY SPACES, METRICS AND DISTANCES ON
SCORE STATISTICS)

PETER MCCULLAGH

TEORIYA VEROYATNOSTEI I EE PRIMENENIYA See THEORY OF PROBABILITY AND ITS APPLICATIONS

TEORIYA VEROYATNOSTEI I MATEMATISCHESKAYA STATISTIKA See THEORY OF PROBABILITY AND MATHEMATICAL STATISTICS

TERÄSVIRTA-MELLIN MODEL SELECTION TEST See LINEAR MODELS, SELECTION OF (*Supplement*)

TERMINAL DECISION See SEQUENTIAL ANALYSIS

TERMS OF TRADE

The "terms of trade" are a broad class of what are basically ratios of two ratio estimators*. For example, one of the two ratios, say R_1, might be the ratio of the prices farmers receive in the current period relative to a past period, and the other, say R_2, is the corresponding ratio of the prices they pay, then the quotient of R_1 and R_2, say $R = (R_1/R_2)$ is a terms of trade measure or index for farmers. If $R > 1$, then the farmers' terms of trade have "improved" or are "favourable," and if $R < 1$, they have "deteriorated" or are "unfavourable." Terms of trade concepts and corresponding measures can be applied to parties engaged in any kind of exchange process, but theories, estimation methods, and applications are almost entirely confined to international trade.

BRIEF HISTORY

The early history of terms of trade can be traced back to the 1600s when some of the English mercantilist writers [21] argued that it was in the interest of England if her export prices were higher than her import prices, which is in modern terminology a terms of trade argument. However, economic theories of terms of trade were the contribution of English classical economists in the 1900s, who formulated the theory of international trade within the framework of three terms of trade questions:

Which goods are traded between nations and why?

What are the relative prices of exports and imports?

Is trade gainful for the trading nations taken together and separately, and if so how are the gains divided among them?

But modern terms of trade concepts and measurement problems were defined and discussed early in this century, and attempts at empirical measures of some of these concepts were made in the first half of this century [10, 15, 17].

CONCEPTS AND MEASURES

There are several concepts and measures of terms of trade and gains from trade, which may be found with some overlap in refs. 9, 12, 18, and 21. In this article, we shall concentrate on those most frequently discussed and used:

1. The *net barter* terms of trade is defined as $T = (P_x/P_m)$, where P_x is an index number* of export prices in the current period relative to the base period and P_m is the corresponding index number of import prices.

2. The *gross barter* terms of trade is defined as $T^* = (Q_x/Q_m)$, where Q_x and Q_m are quantity index numbers of exports and imports, respectively, in the current period relative to the base period.

3. The *double factoral* terms of trade is defined as $t^{**} = (P_x/P_m)(Z_x/Z_m)$, where P_x and P_m are as previously defined and Z_x and Z_m are the corresponding productivity index numbers in the exporting and importing countries, respectively. (If we set $Z_m = 1$, then we obtain another version of T^{**} which is called the single factoral terms of trade.)

4. The *gains from trade* is defined as $G = (V_x/P_m) - (V_x/P_x)$, where P_m and P_x are as previously defined for the periods $t = 0, 1, \ldots, N$, with P_x and P_m equal to 100 in period 1, and V_x is the corresponding value of exports, expressed at current prices in, say, £ millions, for the $(N + 1)$ periods. Sometimes G is expressed as an index number of relative change between the current period and the base period; in that case V_x is replaced by $(P_x Q_x)$, as previously defined, and now the gains from trade defined as an index number is $G^* = Q_x(T - 1)$, where T is as previously defined. (A version of G^*, called the *income* terms of trade, is simply Q_x multiplied by T.)

THE PROBLEM OF MEASUREMENT

In all the measures already defined, the indexes to be constructed are P_x, P_m, Q_x, Z_x, and Z_m. We postpone discussion of Z_x and Z_m, and deal first with P_x, P_m, Q_x, and Q_m. These are price and quantity index numbers (*see* INDEX NUMBERS), *except* that the commodities to which they refer are very different, as we shall see, from those used, for example, in the construction of the consumer price index*. It is in this exception that almost all the difficulties lie in estimating the terms of trade and gains from trade. For simplicity we discuss only P_x and Q_x, since the method and problems of constructing P_m and Q_m are similar.

Let $A = \{a_1, a_2, \ldots, a_n\}$ be a set of n commodities exported in the base period 0 and in the current period 1, and let P_{ti} and q_{ti} be the price and quantity of a_i for $t = 0, 1$ and $i = 1, 2, \ldots, n$. Then P_x is the export price index in period 1 relative to period 0 and Q_x is the corresponding export quantity index. If a_i is indeed a commodity, in the sense that it is a single homogeneous item such as sirloin steak, and all the a_i are homogeneous in this sense and remain the same items in the two periods, then given P_{ti} and q_{ti} for each a_i, P_x and Q_x could be obtained by using a formula such as Laspeyres, Paasche*, or Fisher's ideal index*; these are the formulae most commonly used in practice, especially by national and international statistical agencies. Many other forms have been used in individual studies [8, 11].

But the formula is not the main issue in measuring the terms of trade. There are two approaches to estimating P_x and Q_x, regardless of which formula is used. These two approaches are distinguished on the basis of how to define a_i and obtain p_{ti} and q_{ti}, and they are referred to as the *unit value approach* and the *survey price approach* [19], which we now discuss.

THE UNIT VALUE APPROACH

The unit value approach [1, 2, 4, 16] is the dominant approach in practice. Basically, the product is not a commodity, but a category (a group of commodities) which contains say m unknown commodities, for which the total value and some measure of quantity (e.g., "shipping tons") are recorded in the external trade statistics of the country in question. Since the prices of the commodities are unknown, they are replaced by the unit value which is obtained by dividing the total value of that category by its quantity. After that, the construction of terms of trade indexes proceeds as usual. But P_x is based on unit values, not on prices, and therefore it is likely to be very inaccurate. For a unit value index of (say) export prices to be equal to the corresponding index of export prices, the following three conditions must hold:

1. The variance of the m base period prices in each category is zero.
2. The variance of the m quantity ratios between the current period and the base period in each category is zero.
3. The correlation between the m base period prices and the corresponding m quantity ratios in each category is zero.

These conditions are discussed in detail in refs. 2, 14, and 19, among others.

THE SURVEY PRICE APPROACH

This is a very recent approach to terms of trade measurement; at present it is applied only in the United States [20] and the Federal Republic of Germany [6]. To estimate P_x and Q_x, a sample of exporters is selected (in the United States the sample is random and in the Federal Republic of Germany it is nonrandom) and asked at regular time intervals to provide information on the prices and quantities of the products they had exported during the past periods (e.g., months). The exporters are also asked to specify these products in great detail so that they can be interpreted as homogeneous commodities. If the survey price approach is based on a good sampling plan* (in the present context, a major practical problem), it is the best method of estimating P_x and Q_x. Even within this ideal approach, there remain problems intrinsic to international trade which make

the construction of accurate terms of trade measures difficult.

FACTORAL TERMS OF TRADE

For the double factoral terms of trade $T^{**} = (P_x/P_m)(Z_x/Z_m)$, we need to estimate Z_x and Z_m, in addition to P_x and P_m. To begin with, Z_x and Z_m may be indexes of either partial productivity or total factor productivity, whose methods of estimation in principle are exactly those of productivity measurement* in general. However, in the present context, there is little experience, let alone a documented and accepted methodology, of how to estimate Z_x and Z_m. The few empirical studies which have attempted to produce a proxy estimate of T^{**} have used various ad hoc procedures [3, 5, 12, 13, 18]. A hypothetical example may be useful. Suppose (P_x/P_m) is equal to 1.05. Then there is a 5% improvement in T. But suppose further that Z_x of the exporting country is equal to 1.10, and Z_m of the country which produces the imports is equal to 1.02. Then $(Z_x/Z_m) = 1.08$, and hence $T^{**} = 1.13$, which is interpreted to mean that for the home country the exporting of one unit of the productive resources now buys 13% more of the productive resources of the foreign country relative to the base period.

USE AND INTERPRETATION

A typical example of the use of some terms of trade measures is shown in Table 1. In this example, P_x, P_m, Q_x, and Q_m were calculated by the Laspeyres formula. Those indexes provide the basis for studying the price competitiveness and the trend in the prices and quantities of the United Kingdom exports and imports. Thus T shows that the U.K. terms of trade in 1983 relative to 1980 have deteriorated by 1.6 percentage points; the relative increase in the import prices was more than that in the export prices. This does not necessarily mean that the British exports have become more price competitive, because T^* shows that in 1983 relative to 1980 the increase in the volume of the British exports compared with that of their imports has deteriorated by 4.9 percentage points. However, neither T nor T^* can mean much by themselves in terms of who is winning and who is losing from trade and by how much.

This is how G becomes an important measure of the terms of trade effect. G measures the additional volume of imports valued at current prices which can be purchased from the proceeds of the actual exports. If we write G in index number form $[Q_x(T - 1)]$, we see that the gain from trade in the current period varies directly with the change in T and in Q_x, relative to the base period. In Table 1, the gain from trade in 1981 was positive, but in 1982 and 1983 it was negative. In general if $G > 0$, then trade is gainful *since* the base period, and if $G < 0$, then there is a loss from trade in the current period, *since* the base period in which $G = 0$. Therefore, if $G < 0$, this does not mean that there was no gain from trade; it means the gain is lower than what it was in the base period. Further details may be found in refs. 7, 9, 12, 18, and 21, among others.

References

[1] Allen, R. G. D. (1953). In *International Trade Statistics*, R. G. D. Allen and J. E. Ely, eds. Wiley, New York, pp. 186–211. (Elementary; numerical examples.)

[2] Allen, R. G. D. (1975). *Index Numbers in Theory and Practice*. Macmillan, London, England.

[3] Appelyard, D. R. (1974). *Indian Econ. J.*, **22**, 36–49. (Application.)

[4] Brenna, S. (1983). In *Price and Quantity Measurement in External Trade*, Series M, No. 76. United Nations, New York, pp. 42–108.

[5] Devons, E. (1954). *Manchester School Econ. Social Stud.*, **22**, 158–275. (Application.)

[6] Gucks, S. (1983). In *Price and Quantity Measurement in External Trade*, Series M, No. 76. United Nations, New York, pp. 1–39.

[7] Greenfield, C. C. (1984). *The Statistician*, **33**, 371–79.

[8] Hansen, B. and Lucas, E. F. (1984). *Rev. Income Wealth*, **30**, 25–38.

[9] Hibbert, J. (1975). *Econ. Trends*, **25**, xxviii–xxxv. (Concepts.)

[10] Hilgerdt, F. (1944). *The Network of World Trade*. League of Nations, Geneva, Switzerland. (Of historical interest.)

[11] Jazairi, N. T. (1971). *Bull. Oxford Econ. Statist.*, **33**, 181–195.

[12] Kindleberger, C. P. (1956). *The Terms of Trade: A European Case Study*. M.I.T. Press, Cambridge, Mass.

[13] Lipsey, R. E. (1963). *Price and Quantity Trends in the Foreign Trade of the United States*. Princeton University Press, Princeton, N.J.

[14] Maizels, A. (1957). *J. R. Statist. Soc. A*, **120**, 215–219.

[15] Mitchell, W. C. (1913). *Business Cycles*. Burt Franklin, New York. (Of historical interest.)

[16] Sellwood, R. (1975). *Econ. Trends*, **258**, 95–104. (U.K. practice.)

[17] Silverman, A. G. (1930). *Rev. Econ. Statist.*, **12**, 139–148.

Table 1   United Kingdom Terms of Trade and Gains From Trade 1980–1983.

	Measure	1980	1981	1982	1983
(1)	P_x	100	108.8	116.7	126.6
(2)	P_m	100	108.1	117.9	128.6
(3)	Q_x	100	99.2	101.5	102.3
(4)	Q_m	100	96.1	100.7	107.6
(5)	V_x (£ million)	63,209	67,848	73,167	79,653
(6)	$T = (P_x/P_m)$	100	100.7	99.0	98.4
(7)	$T^* = (Q_x/Q_m)$	100	103.2	100.8	95.1
(8)	$G(\text{£ million}) = \left[V_x\left(\dfrac{1}{P_m} - \dfrac{1}{P_x} \right) \right]$		$+404$	-638	-978

Source: United Kingdom *Monthly Digest of Statistics*, No. 463, July 1984, H.M.S.O., London, England. [This source contains (1)–(6); (7)–(8) were calculated from (1)–(5).]

[18] Spraos, J. (1983). *Inequalizing Trade*. Clarendon, Oxford, England. (Terms of trade of developing countries.)

[19] United Nations (1981). *Strategies for Price and Quantity Measurement in External Trade*, Series M, No. 69. United Nations, New York. (Elementary.)

[20] United States Bureau of Labor Statistics (1982). *BLS Handbook of Methods*, Vol. 1. U. S. Bureau of Labor Statistics, Washington, D.C.

[21] Viner, J. (1937). *Studies in the Theory of International Trade*. Allen and Unwin, London, England.

(ECONOMETRICS
FISHER'S IDEAL INDEX NUMBER
FOREIGN TRADE STATISTICS,
   INTERNATIONAL
INDEX NUMBERS
PAASCHE–LASPEYRES INDEX NUMBERS
PRODUCTIVITY MEASUREMENT
TEST APPROACH TO INDEX NUMBERS,
   FISHER'S)

NURI T. JAZAIRI

TERRY–HOEFFDING TEST

This is a statistical procedure for testing whether two CDFs $F_1(\cdot)$ and $F_2(\cdot)$ are identical, given random samples containing n_1 and n_2 observations, respectively, from these CDFs. It is more sensitive than the better known Mann and Whitney U test [2] or the Wilcoxon test* [3], with respect to alternatives in which one population distribution is stochastically greater (or less) than the other [i.e., $F_1(x) < (>)F_2(x)$ for all x].

The test is carried out by ranking the $n_1 + n_2$ observations into a single sequence, then selecting either sample, transforming the ranks into "scores" using Table XX of Fisher and Yates [1], and then summing these scores as the linear rank statistic G_1 (*see* NORMAL SCORES TEST).

The exact distribution and critical values of G_1 for small N ($= n_1 + n_2$) are given in ref. 1. To obtain critical values for samples with $N > 10$ or to find the probability point associated with a given value of statistic G_1, one may use either the statistic $g(G_1)$ (to be

defined), whose null distribution is approximated by that of Student's t or the statistic, or $u(G_1)$ (also to be defined), for which the null distribution is approximated by the standard normal.

We define

$$h(G_1) = G_1 \Big/ \big[n_1 n_2 N^{-1} S_{2,N} \big]^{1/2},$$

where $S_{2,N}$ is the sum of squares of expected values of order statistics in a random sample of size N from a unit normal distribution. Then

$$g(G_1) = \frac{h(G_1)(N-2)^{1/2}}{\big[1 - \{h(G_1)\}^2 \big]^{1/2}}.$$

As $m, n \to \infty$ with n_1/N bounded away from 0 or 1, the distribution of $g(G_1)$ is approximated by that of Student's t. Similarly, the limiting distribution of

$$u(G_1) = G_1(N-1)^{1/2}/(n_1 n_2)^{1/2}$$

is the standard normal distribution. This normalized function of the test statistic can be used to combine or "pool" results of several experiments.

References

[1] Fisher, R. A. and Yates, F. (1983). *Statistical Tables for Biological, Agricultural and Medical Research*. Oliver and Boyd, London, England.

[2] Terry, M. E. (1952). *Ann. Math. Statist.*, **23**, 346–366.

[3] Wilcoxon, F. (1945). *Biometrics Bull.*, **1**, 80–83.

Bibliography

Hoeffding, W. (1951). Optimum nonparametric tests. In *Proc. 2nd Berkeley Symp., Math. Statist. Prob.* University of California Press, Berkeley, Calif.

Mann, H. B. and Whitney, D. R. (1947). *Ann. Math. Statist.*, **18**, 50–60.

(NORMAL SCORES TEST)

M. E. TERRY

TEST APPROACH TO INDEX NUMBERS, FISHER'S

Index number formulae are ratio estimators* that may be required to satisfy certain desired properties. For example, the value of the index must be independent of the ordering of commodities, and when all prices change from one period to another at the same proportion, the change in the value of the index must be equal to the common factor of proportionality. Properties of this type are referred to in the index number literature as "tests" or "axioms." Fisher [8] emphasized and employed some of these tests in judging the relative merits of various index number formulae. Frisch [10] marked the distinction between Fisher's approach to index numbers, where there are tests of formulae, and the direct "economic–theoretic" approach in terms of utility [1, 2]. Alternative approaches, especially those based on regression methods, may also be found in the literature [15]. The internal inconsistency of Fisher's tests, and their suitability as a procedure to derive index number formulae continue to be debated [2, 3, 6, 11, 18].

THE GENESIS OF FISHER'S TESTS

Index number tests are sets of arithmetic properties used to characterize price and quantity index formulae (*see* INDEX NUMBERS). The self-evident justification for most of these tests is that they hold for the price and quantity ratios of individual commodities, and hence it was thought (see for example, [8, pp. 64 and 75]) that ideally they should also hold for the corresponding ratio of the prices and quantities of an *aggregate* or composite of commodities. In some cases, however, these tests are justified on the basis of economic arguments concerning the definition of cost of living and welfare indices and functional relations assumed in economic theory between commodity prices and quantities [1, 17, 20, 22].

However, a fundamental question in aggregation theory is the consistency of the macrosystem with the microsystem through an aggregation function. Viewed from this perspective, Fisher's test approach requires a particular type of consistency called "characteristic consistency" [14] in the sense that,

say, the composite price index retains the characteristics of the individual price ratios under the aggregation function, which could be Laspeyres*, Paasche*, or some other index number formula.

Alternatively, demand data provide vectors of prices and quantities demanded, and a price index is given by a formula P_{rs} having reference to two periods r and s, the base and current periods. Conventionally, and with Fisher, this is an algebraical formula involving the demand data just for these two periods alone. He considered a large assortment of formulae for this purpose, and proposed his "tests" in order to discriminate among them [1, 2].

THE PRINCIPAL TESTS

The original tests discussed by Fisher and others [7–9, 23] have since multiplied and many of them are often defined differently by different authors [5, 6, 16, 19, 20]. New and old tests have also been used recently in the characterization of particular index number formulae such as Fisher's ideal formula itself [3, 5, 11]. In this article we shall concentrate on the principal tests, especially those found in refs. 7–9, and in much of the literature.

Let $A = \{a_1, a_2, \ldots, a_n\}$ be a set of n commodities; p_{it} and q_{it} are the price and quantity of a_i observed at time $t = 0, 1, \ldots, T$ for $i = 1, 2, \ldots, n$. The price and quantity indices in period s relative to period r are P_{rs} and Q_{rs}, and the value index is

$$V_{rs} = \left(\sum_{i=1}^{n} p_{is} q_{is} \bigg/ \sum_{i=1}^{n} p_{ir} q_{ir} \right).$$

Also let P_r and P_s be the "price levels" in periods r and s; p_r, p_s, q_r and q_s are the price and quantity vectors of the n commodities in the same two periods, and the vector products are such as $p_r q_s = p_r \cdot q_s = \sum_{i=1}^{n} p_{ir} q_{is}$. The better-known tests are the following:

(i)   The Commodity Reversal Tests.   This requires that the values of P_{rs} and Q_{rs}

are independent of the ordering (permutations) of the n commodities.

(ii) The Identity Test. This requires that $P_{tt} = 1$ and $Q_{tt} = 1$; that is, when the prices or quantities in period t are compared with themselves, the change in the value of the index should be zero.

(iii) The Commensurability Test. P_{rs} and Q_{rs} are invariant under a dimensional change in the units of measurement of the commodity prices and quantities, respectively.

(iv) The Determinateness Test. If the price (quantity) of an individual commodity becomes zero, then P_{rs} (Q_{rs}) does not become zero, indeterminate, or infinite.

(v) The Proportionality Test. $P_{rs} = \lambda$ if $P_{is} = \lambda p_{ir}$ and $Q_{rs} = \lambda$ if $q_{is} = \lambda q_{ir}$ for all i and a scalar $\lambda > 0$.

(vi) The Base Test. For any three arbitrary periods, say k, r, and s, the price index $P_{rs} = P_{ks}/P_{kr}$ and the quantity index $Q_{rs} = Q_{ks}/Q_{kr}$; that is, a comparison between the prices or quantities in periods s and r via a third period k shall be independent of this third period k.

(vii) The Circular or Transitivity Test. $P_{rs}P_{st} = P_{rs}$ and $Q_{rs}Q_{st} = Q_{rt}$ for any $r \neq s \neq t$ and $r, s, t = 0, 1, \ldots, T$.

(viii) The Time or Point Reversal Test. $P_{rs}P_{st} = P_{rs}$ and $Q_{rs}Q_{st} = 1$ for $r \neq s$ and $r, s = 0, 1, \ldots, T$; that is, the index formula for two periods r and s is symmetric with respect to the comparison base.

(ix) The Factor Reversal Test. P_{rs} and Q_{rs} are two indices of the *same* form, except for "interchanging the prices and quantities" [8, p. 72], which multiply out to the corresponding value index V_{rs}.

Tests **(i)**–**(v)** are usually passed by all standard index number formulae, although a popular version of the logarithmic indices [15] may fail the proportionality test. Test **(vii)** implies test **(vi)**, but the converse is not true, and both these tests, especially the transitivity test, are highly desirable in international comparisons (*see* PURCHASING POWER PARITY). Test **(viii)** requires symmetry between the current and the base year when they are interchanged, and it is also implied by test **(vii)**, but the converse is not true. Test **(ix)** requires a consistent matching of the price index and the quantity index on the hypotheses that V_{rs} can be factored into P_{rs} and Q_{rs}. A version of **(ix)** called the *product test* or the *weak factor reversal test* requires only that P_{rs} and Q_{rs} multiply out to V_{rs}, without necessarily being both of the same form; thus, for example, P_{rs} could be based on Lespeyres formula and Q_{rs} on Paasche, or vice versa. Fisher [8, p. xiii] considered tests **(viii)** and **(ix)** as the "two legs on which index numbers can be made to walk." Lespeyres and Paasche formulae fail these two tests, but Fisher's ideal index passes them. (*See* FISHER'S IDEAL INDEX NUMBER, and PAASCHE–LASPEYRES INDEX NUMBERS.)

INCONSISTENCY OF FISHER'S TESTS AND ECONOMIC EXTENSIONS

An outstanding concern with these tests is the impossibility of satisfying them, by any formula! Frisch [9] was the first to raise this kind of issue. Then it was in regard to the factor reversal test, and Eichhorn [5] found his argument wrong. Wald [21] offered another argument, as did Swamy [19] who was convinced Wald's argument was wrong. Eichhorn and Voeller [6] propose several new theorems which show relations, especially inconsistencies within a collection of properties that include Fisher's tests. They made a similar enquiry about consistency in a functional equation point of view of Fisher's "equation of exchange."

In the economic–theoretic approach, even the conventional idea of what constitutes a formula for an index, algebraic and taking the data in isolated pairs, which is at the start of Fisher's approach, can be put aside. Instead, one is led to consider numbers (for details see ref. 2) P_r, Q_r such that $p_r q_r = P_r Q_r$ and $p_r q_s \geqslant P_r Q_s$, and hence also $p_r q_s / p_s q_s \geqslant P_r / P_s$ or $p_r q_s / p_r q_r \geqslant Q_s / Q_r$. By taking any solution of the system and setting $P_{rs} = P_r / P_s$, $Q_{rs} = Q_r / Q_s$, we have true price and quantity indices, with an indeterminacy that just reflects the limitation of the data and is progressively diminished as data are added. The Fisher ratio tests are certainly satisfied, since the indices are actually constructed as ratios. The other tests, with some appropriate reformulation, can be brought to bear also. Indices so constructed are those that are "true," or identical with those that are "exact," in the sense of Byushgens [4], for some homogeneous utility that fits the data. The condition for a solution of the inequalities is a strengthening of the consistency condition of Houthakker [13], resulting from the imposition of homogeneity. Applied to two periods alone, it reduces to the familiarly considered relation between Paasche and Laspeyres indices, termed the *index number theorem* by Hicks [12], which is that the one should not exceed the other, so uncovering the essential significance of this relation.

References

[1] Afriat, S. N. (1977). *The Price Index*. Cambridge University Press.

[2] Afriat, S. N. (1987). *Logic of Choice and Economic Theory*. Clarendon Press, Oxford. (Technical, economic theory, extensive analysis.)

[3] Balk, B. M. (1985). *Statist. Hefte*, **26**, 59–63.

[4] Byushgens, S. S. (1925). *Mat. Sbornik*, **32**, 625–631; *Recueil Math.* [*Moscow*], **32** (translation).

[5] Eichhorn, W. (1976). *Econometrica*, **44**, 247–256. (Proof of inconsistency.)

[6] Eichhorn, W. and Voeller, J. (1977). *Theory of the Price Index: Fisher's Test Approach and Generalizations*. Springer, Berlin. (Intermediate, general survey.)

[7] Fisher, I. (1911). *The Purchasing Power of Money*. Macmillan, New York.

[8] Fisher, I. (1982). *The Making of Index Numbers*. Houghton Mifflin, Boston.

[9] Frisch, R. (1930). *J. Amer. Statist. Ass.*, **25**, 397–406. (First proof of inconsistency.)

[10] Frisch, R. (1936). *Econometrica*, **4**, 1–38. (Survey article.)

[11] Funke, H. and Voeller, J. (1978). *Theory and Applications of Economic Indices*, W. Eichhorn, R. Henn, O. Opitz, and R. W. Shephard, eds. Physica-Verlag, Berlin.

[12] Hicks, J. R. (1956). *A Revision of Demand Theory*. Clarendon Press, Oxford. (Economic theory.)

[13] Houthakker, H. S. (1950). *Economica*, **17**, 159–174. (Economic theory.)

[14] Ijiri, Y. (1971). *J. Amer. Statist. Ass.*, **66**, 766–782. (Aggregation theory, survey article.)

[15] Jazairi, N. T. (1983). *Bull. Int. Statist. Inst.*, **50**, 122–147.

[16] Pfouts, R. W. (1966). *Rev. Int. Statist. Inst.*, **34**, 174–185.

[17] Samuelson, P. A. and Swamy, S. (1974). *Amer. Econ. Rev.*, **64**, 566–593.

[18] Silver, M. S. (1984). *Statistician* (*London*), **33**, 229–237. (Elementary survey.)

[19] Swamy, S. (1965). *Econometrica*, **33**, 619–623. (Proof of inconsistency.)

[20] Vartia, Y. O. (1976). *Relative Changes and Index Numbers*. Research Institute of the Finnish Economy, Helsinki.

[21] Wald, A. (1937). *Z. Nationalökon.*, **8**, 179–219.

[22] Wald, A. (1939). *Econometrica*, **7**, 319–335.

[23] Walsh, C. M. (1901). *The Measurement of General Exchange Value*. Macmillan, New York. (Of historical interest.)

(FISHER'S IDEAL INDEX NUMBER
INDEX NUMBERS
PAASCHE-LASPEYRES INDEX
  NUMBERS)

SYDNEY N. AFRIAT
NURI T. JAZAIRI

TEST FACTOR STRATIFICATION

Test factor stratification emerged in the early 1950s as a way to analyze multivariate categorical data*. (See especially Lazarsfeld [3], Hyman [2], and Rosenberg [4, 5].) Relying on the technology of the day, it could easily be carried out with IBM card sorters and mechanical calculators, while providing a systematic and objective method for analyzing the flood of survey data that were becoming available to the behavioral sciences. Although the technique has largely been superseded by log-linear modeling and other procedures, it still has great value in its own right.

Test factor stratification begins with an "original" relationship between two variables, usually presented in the form of a two-way cross-classification table. The goal is to understand or "elaborate" this relationship by introducing additional variables called *test factors*. Elaborating the original association requires that one stratify the data according to one or more control variables, the test factors, thereby producing a multidimensional cross-classification. If the factor, which can be multivariate, has K levels, the data are divided into K contingency tables*. One then examines the relationships within each of these subtables to see what effect the test factor(s) has had.

The proponents of this method looked for three general kinds of models:

(i) Spuriousness. The two main variables are not directly related but only appear to be so, due to their association with a common test factor (this model is sometimes called *conditional independence and association*).

(ii) Developmental Sequence. The independent variable "causes" the test factor which in turn "causes" the dependent variable (again, the terms "conditional independence and association" are frequently applied in this context): *see also* CAUSALITY *and* PATH ANALYSIS.

(iii) Specification. The nature and magnitude of the association varies according to categories of the third variable (this idea is often called *interaction**).

By assuming a certain time order or causal priority among the variables and by comparing the partial relationships with the original one, it is possible to test these interpretations.

A simple example may further clarify the advantages and disadvantages of test factor stratification. Hyman [2] discusses data originally analyzed by Gosnell [1], who wanted to see whether a propaganda campaign would be effective in stimulating Chicagoans to register to vote. The original relationship, presented in Table 1, suggests that the experiment was moderately successful because people who received the material were more likely to register than the control group. Gosnell did not rest with this interpretation, but sought to discover the conditions under which the information campaign would be most successful. Thus, he introduced a test factor—level of political interest—that was measured by participation in the previous seven elections. The contingency tables appear in Table 2. Judging from the percentages, the propaganda had little effect among highly interested citizens but was quite successful on the less interested ones. In this sense, then, the relationship has been "specified": the association is virtually nil at one level of the test factor but is quite pronounced at another.

Hyman and others relied mainly on percents and frequencies to arrive at their conclusions. There was no formal statistical theory used to carry out tests of significance*, and since the models were not parametrized, there was no estimation*. Therefore, the determination of what the data "said" rested largely on judgment and common sense. These qualities are of course, essential for any technique, however sophisticated, but test factor analysis seemed somewhat indeterminate and subjective. One had to peruse a series of contingency tables which could become numerous as the categories of the test factor increased. Because percents are the main standard of comparison, the analyst must make certain that a sufficient number of cases are present in each partial table, a requirement that is often difficult to meet, even in large surveys.

In looking for spuriousness, as an example, the control variable was supposed to "reduce" the original relationship. Yet, what

**Table 1   The Effects of Propaganda on Registration (%).
Figures in Parentheses are Numbers of Cases.**

	Experimental Group (Propaganda) (%)	Control Group (No propaganda) (%)
Registered	74	64
Did Not Register	26	36
	100	100
	(2612)	(2204)

Source: Gosnell [1] (quoted from Hyman [2]).

**Table 2   Effects of Propaganda on Registration According to Political Participation (%).
Figures in parentheses are numbers of cases.**

	High Participation		Low Participation	
	Exp. Group	Control Group	Exp. Group	Control Group
Reg.	89	82	61	44
Not Reg.	11	18	39	56
	100	100	100	100
	(1229)	(1114)	(1383)	(1090)

Source: Gosnell [1] (quoted from Hyman [2]).

constitutes a reduction? Did the association have to disappear entirely? In how many of the partial tables? Multivariate methods such as log-linear analysis have largely replaced this informal procedure, partly because they provide more satisfactory answers to questions of this sort.

Nonetheless, test factor stratification is still a useful tool, particularly in the social sciences where categorical data* abound. For one thing, it provides a quick and simple way to undertake a preliminary analysis of the data. Indeed, some of the more advanced techniques really build upon this framework. Second, like any multivariate method, it requires the investigator to make explicit the assumptions about causal priorities and hence encourages serious theoretical thinking. Hyman's work is a masterpiece in this regard. Finally, this is an especially good way to introduce novices to categorical data analysis. After all, many of the main ideas of, say log-linear analysis, appear in simplified form in this procedure. Thus, far

from relegating it to the dustbin, it is well worth keeping in one's statistical toolbox.

References

[1] Gosnell, H. (1927). *Getting Out the Vote: An Experiment in the Stimulation of Voting.* University of Chicago Press, Chicago, Ill.

[2] Hyman, H. (1955). *Survey Design and Analysis.* The Free Press, New York. (This classic in the analysis of nominal data remains useful today.)

[3] Lazarsfeld, P. (1955). Interpretation of statistical relationships as a research operation. In *The Language of Social Research.* F. Lazarsfeld and M. Rosenberg, eds. The Free Press, New York.

[4] Rosenberg, M. (1962). *Social Forces,* **41**, 53–61.

[5] Rosenberg, M. (1968). *The Logic of Survey Analysis.* Basic Books, New York.

(CATEGORICAL DATA
CAUSALITY
CONTINGENCY TABLES
MULTIDIMENSIONAL CONTINGENCY
    TABLES

NOMINAL DATA
PUBLIC OPINION POLLS
SURVEY SAMPLING)

H. T. REYNOLDS

TESTIMATOR

This is a term introduced by Sclove et al. [1] to describe estimators based on inferences derived from preliminary test(s). It has some similarity to a Bayes estimator, but in the latter case the source of the prior knowledge is not so precisely specified.

A simple example is in pooling* two (or more) estimators of variances where the decision to pool depends on the result of a test of the equality of the two (or more) corresponding population variances. Sclove et al. [1] apply the method to estimating the mean of a multinormal distribution.

See Waikar et al. [2] for a more detailed discussion and bibliography.

References

[1] Sclove, S. L., Morris, C., and Radhakrishnan, R. (1972). *Ann. Math. Statist.*, **43**, 1481–1490.
[2] Waikar, V. B., Schuurmann, F. J., and Raghunathan, T. E. (1984). *Commun. Statist.—Theory Meth.*, **13**, 1901–1913.

(BAYESIAN INFERENCE)

TESTING, DESTRUCTIVE

GENERAL PRINCIPLES

Destructive testing is generally necessary in situations where inadequate performance or failure of materials leads to high penalties for replacement or recovery. In some instances failure of critical components can cause very serious losses. Strength and durability (i.e., nonfailure under stress or force or use) are usually the desired qualities. Properties such as time, force, or wear required to achieve a certain state of collapse or destruction are measured to assess the desired qualities of units of materials or components or systems.

Where the units to be tested are relatively inexpensive or readily replaced, destructive tests can be used directly by employing a sampling scheme for selection of the individual units for testing. The main questions here concern the choice of a suitable sampling rate (the proportion to be tested) and a sampling rule (the mode by which individuals are chosen) which will allow proper inferences about the reliability of the whole batch or population of units. Much literature exists on applicable sampling theory and practice (*see* CENSORED DATA *and* LIFE TESTING), which are based on knowledge of the probability distributions of the properties being measured.

PROBABILITY MODELS

In general, the measured properties such as time, force or stress have long-tailed or skewed distributions, reflecting a small proportion of individuals that "withstand" higher levels of stress or survive for a relatively long time when compared with the vast majority of seemingly identical individuals. Examples of such distributions are the standard exponential*, lognormal*, Weibull*, gamma*, extreme value*, similar continuous univariate distributions with a shifted origin, truncated versions of such distributions, and mixtures of similar continuous univariate distributions (*see also* HEAVY-TAILED DISTRIBUTIONS, Johnson and Kotz [7], and Mann et al. [9]). The strength properties of timber, for instance, have a distribution that is very closely approximated by a lognormal or by a mixture of lognormal distributions. The requirement of long-tailed distributions in inferential problems implies the use of summary statistics (such as geometric mean* and harmonic mean*) different from the usual sample mean or median.

STATISTICAL INFERENCE

An arithmetic average of observations from a long-tailed distribution will also have a skewed distribution. Such a summary statistic can make the reporting of results and drawing conclusions from a sample a perilous task. It is much more appropriate to use statistics that have symmetric distributions for these purposes. For example, geometric means* are appropriate for lognormal data. However, reporting results using geometric means or other more suitable summary statistics is not as straightforward as the usual arithmetic mean. When the experimenter is required to use a distribution with a certain unknown displaced parameter, the usual maximum likelihood* technique could have problems. Adequate methods of inference for this case have been given by Cheng and Amin [3].

In some destructive testing, it is readily possible to obtain two sample quantiles* of strength property (such as breaking load of single threads of yarn) instead of individual measurements of the strength. Problems of estimating the parameters of the distribution of strength property by using sample estimates of two or more quantiles arose in this connection. Such a problem has been considered by Stout and Stern [11].

The preceding discussion is confined mainly to sampling inspection by variables. There exists extensive literature on sampling inspection (*see also* INSPECTION SAMPLING) of attributes; estimation of the probability that a defective item fails in a destructive test was considered in this context by Nelson et al. [10]. The destructive nature of these tests is such that when a test item fails, the test equipment used for it also is lost. Optimal sampling plans* for using equipment (a limited resource) to obtain an estimate of the probability of defective failure at a required precision level is discussed.

Study of strength properties by destructive sampling is closely linked to the inspection of times in life testing experiments. Sequential sampling* designs have been considered by Bergman and Turnbull [2] in such experiments where testing is destructive. Here, although the theory considers an underlying variable for the life time, the attribute sampling plans were the main concern where an item was observed to have failed or not at the time of inspection.

DECISION MAKING AND COST CONSIDERATION

The ultimate purpose of sampling for destructive tests often requires assessment of plans in terms of total cost, and making decisions having least possible costs. Ladany [8] studied least cost acceptance sampling* plans for lots where the strength of the material is controlled at a fixed level. The plans considered are for lot-to-lot inspection by attributes. When the cost of destructive units is very high, such plans may not be economical. In such a case, a skip-lot* plan was proposed by Hsu [6], who also considered a cost analysis similar to that of lot-to-lot plans. Minimization of the total cost was also considered by Hillier [5] for continuous sampling plans* in the context of destructive testing.

STATISTICAL RELATIONSHIPS

Testing until destruction may be the ultimate operation but it can be very impractical. Nondestructive tests are much more desirable and can be used once models and relationships have been established between the results of nondestructive testing and destructive testing. Multivariate probability models are now required to describe the relationships. Furthermore, nonstandard sampling schemes are required to estimate these models and/or the relationships implied by them. For example, Bartlett and Lwin [1] and Evans et al. [4] use multivariate lognormal models to describe the strength properties of timber. Special experimental designs are needed to obtain information on critical parameters such as the correlation* coefficient between the two variables repre-

coefficient between the two variables representing results from two different destructive tests, e.g., the modulus of rupture under tension and bending loads, respectively; bending test with a mild load was a preliminary requirement to eliminate really weak pieces of timber in the lot. Relationships between the measurement of elasticity (nondestructive) and that of modulus of rupture under tension load (destructive and time consuming) under the specified bending test are required for long term use.

Once an appropriate model is established for the variables representing the strength properties and a valid sampling scheme is employed to obtain relevant data, it becomes possible to estimate a working relationship between destructive and non-destructive variables. This would enable the experimenter to perform only nondestructive tests and a small scale destructive test in the future to predict the likely result of a full scale destructive test. In the timber example, for instance, by loading to a modest level of bending load (thereby only destroying a few weak pieces of timber), it is possible to establish a high likelihood of the surviving timber being within certain specifications. The added advantage of using bending load only in the future is the reduction in time of using bending tests as compared to tension tests.

References

[1] Bartlett, N. R. and Lwin, T. (1984). *J. R. Statist. Soc. C*, **33**, 65–73.

[2] Bergman, S. W. and Turnbull, B. W. (1983). *Biometrika* **70**, 305–314.

[3] Cheng, R. C. H. and Amin, N. A. K. (1983). *J. R. Statist. Soc. B*, **45**, 394–403.

[4] Evans, J. W., Johnson, R. A., and Green, D. W. (1984). *Technometrics*, **26**, 285–290.

[5] Hillier, F. S. (1964). *J. Amer. Statist. Ass.*, **59**, 376–402.

[6] Hsu, J. I. S. (1977). *IEEE Trans. Rel.*, **26**, 70–72.

[7] Johnson, N. L. and Kotz, S. (1970a, b). *Distributions in Statistics. Continuous Univariate Distributions*—1, 2 Wiley, New York.

[8] Ladany, S. P. (1975). *J. Qual. Tech.*, **7**, 123–126.

[9] Mann, N. R., Schafer, R. E., and Singpurwalla, N. D. (1974). *Methods for Statistical Analysis of Reliability and Life Data*. Wiley, New York.

[10] Nelson, A. C., Williams, J. S., and Fletcher, N. T. (1963). *Technometrics*, **5**, 459–468. (Estimation.)

[11] Stout, H. P. and Stern, F. (1961). *J. R. Statist. Soc. B*, **23**, 434–443.

(HEAVY-TAILED DISTRIBUTIONS
INSPECTION SAMPLING
QUALITY CONTROL, STATISTICS IN
SAMPLING PLANS)

N. R. Bartlett

T. Lwin

TESTING FOR SYMMETRY

INTRODUCTION

Let $Z_1, Z_2, \ldots, Z_n$ be a random sample from the continuous distribution function $F(z) = P(Z \leqslant z)$. We are interested in tests of

$$H_0: P(Z \leqslant z) = P(-Z \leqslant z) \quad \text{for all } z,$$

namely, F is symmetric about (its median) 0, against the alternative that F is slanted toward positive values.

An equivalent way to write H_0 is

$$H_0: F(z) = 1 - F(-z) \quad \text{for all } z.$$

The testing problem can arise in a number of situations including the following:

(i) Consider a paired-sample problem where we have n independent pairs $(X_1, Y_1), \ldots, (X_n, Y_n)$ of subjects and, within each pair, one subject is randomly assigned the "treatment" while the other subject serves as a "control." The hypothesis of "no treatment effect" says that the joint distribution of the response (X, Y) is exchangeable*, i.e.,

$$H_0^*: P(X \leqslant x; Y \leqslant y)$$
$$= P(X \leqslant y; Y \leqslant x),$$
$$\text{for all} (x, y).$$

H_0^* is often called the hypothesis of *bivariate symmetry*. It implies that the distribution of $Z = Y - X$ is the same as that of $-Z =$

$X - Y$ (we write this as $Z \stackrel{d}{=} -Z$ where $\stackrel{d}{=}$ means "has the same distribution as ") or equivalently, the distribution of Z is symmetric about 0. Thus one way to test H_0^* is to use the Z's and test H_0. If one suspects that the treatment will tend to increase a subject's response, one considers the alternative that the distribution of Z is slanted toward positive values, Of course, whereas H_0^* implies H_0, H_0 does not imply H_0^*; there are joint distributions $P(X \leqslant x; Y \leqslant y)$ ($= G(x, y)$, say) for which the induced distribution of Z is symmetric about 0 but G is not exchangeable. (For tests of H_0^* which are not based solely on the Z's, see, for example, Sen [23], Bell and Haller [3], Hollander [14], Kepner and Randles [18], and references therein.)

(ii) A single sample $Z_1, \ldots, Z_n$ is obtained. Here the Z's are not differences but, for example, can represent independent measurements of some physical constant. Thus, suppose the parameter of interest is the ratio of the mass of the earth to that of the moon. It is hypothesized that the ratio is 81.3035 and that the distribution of measurements is symmetric about this specified value. Under that hypothesis,

$$Z' = Z - 81.3035$$

is symmetric about 0, and thus the hypothesis of interest can be tested by testing that the Z''s are symmetric about 0. More generally, if you hypothesize that a Z distribution is symmetric about a *known* value θ_0, this is equivalent to the hypothesis that the distribution of $Z - \theta_0$ is symmetric about 0.

In a parametric setting, where F is assumed to be a normal distribution, H_0 can be tested using the *t-test* based on the statistic $t = \overline{Z} / \widehat{SD}$, where $\overline{Z} = \sum_{i=1}^n Z_i / n$ and $\widehat{SD} = s / n^{1/2}$ with $s^2 = \sum_{i=1}^n (Z_i - \overline{Z})^2 / (n - 1)$.

In a nonparametric setting with no parametric assumptions about F, popular and useful tests of H_0 include the *sign test* based on the number of positive Z's; the Wilcoxon *signed-rank test** based on the statistic

$$T^+ = \sum_{i \leqslant j}^n I\{ Z_i + Z_j > 0 \},$$

where $I\{A\} = 1$ if A occurs, 0 otherwise; the Smirnov tests (Smirnov [24]; Butler [7]) based on the statistics

$$\sup_{-\infty < z \leqslant 0} Q_n(z), \qquad \sup_{-\infty < z \leqslant 0} |Q_n(z)|,$$

where

$$Q_n(z) = (n/2)^{1/2} \{ F_n(z) + F_n(-z) - 1 \}$$

and F_n is the empirical distribution function (*see* EDF STATISTICS) of the Z's, and the Cramér–von Mises-type statistic $V^2 = \int_{-\infty}^0 Q_n^2(z) \, dF_n(z)$ (Orlov [20]; also see Rothman and Woodroofe [22], Koziol [19], and the references therein). These tests are discussed elsewhere; in particular *see* BUTLER–SMIRNOV TEST and KOLMOGOROV–SMIRNOV-TYPE TESTS OF SYMMETRY. We turn to a situation for which procedures are not as well known, namely, where the center of symmetry is *not known*.

TESTS FOR SYMMETRY ABOUT AN UNKNOWN MEDIAN

Let $Z_1, \ldots, Z_n$ be a random sample from the continuous distribution $F(z - \theta)$, where $F(0) = \frac{1}{2}$. The null hypothesis asserts that the distribution of Z is symmetric about the unknown median θ, equivalently, that $Z - \theta \stackrel{d}{=} \theta - Z$ or $Z - 2\theta \stackrel{d}{=} -Z$. The null hypothesis is of interest because many classical procedures (both parametric and nonparametric) assume symmetry and it is useful to test the validity of the assumption. Investigating the presence or absence of symmetry is also important to the statistician in terms of deciding what parameter to estimate. When F is symmetric about θ, the natural measure of location for F is the point of symmetry θ. Bickel and Lehmann [5] argue that whereas there are many reasonable estimators of the center of symmetry of such a distribution, in the asymmetric case, after focussing on a location parameter $\theta(F)$ (say), there is typically only one natural estimator of $\theta(F)$ and that is $\theta(F_n)$, where F_n is the empirical distribution function. In the spirit of deciding what parameter should be estimated in the asymmetric case, Doksum [9] first considers the defining property $Z - 2\theta \stackrel{d}{=} -Z$ enjoyed by the point of symmetry θ in the symmetric model. By allowing parameters to be function-valued, he shows that when F is continuous there is essentially only one reasonable function that satisfies the property, namely

$$\theta_F(z) = \tfrac{1}{2}\big[z - \bar{F}^{-1}(F(z))\big], \quad z \, \varepsilon \, S(F),$$

where $S(F) = \{z : \ 0 < F(z) < 1\}$ is the support* of F and $\bar{F}(z) = 1 - F(-z)$, $-z \, \varepsilon \, S(F)$. When F is continuous, $Z - 2\theta_F(Z) \stackrel{d}{=} -Z$. Doksum advocates $\theta_F(\cdot)$ as a measure of how much F deviates from symmetry about its median; F is symmetric if and only if $\theta_F(\cdot)$ is constant. He estimates $\theta_F(z)$ by the *empirical symmetry function*

$$\hat{\theta}_n(z) = \tfrac{1}{2}\big[z - \bar{F}_n^{-1}(F_n(z))\big],$$

where F_n and $\bar{F}_n$ denote the empirical distribution functions of $Z_1, \ldots, Z_n$ and $-Z_1, \ldots, -Z_n$, respectively.

The problem of testing symmetry versus asymmetry when θ is unknown is receiving much attention in the current literature. The goal is to design tests which (1) have good power and asymptotic efficiency properties for interesting alternatives to symmetry and (2) are either asymptotically distribution-free* under H_0 or (at least) have a type-I error probability that stays relatively constant for F's satisfying H_0.

Some standard (classical) tests of symmetry versus asymmetry include:

a. Test Based on Moment Ratio*. Consider the statistic

$$b_1 = m_3(m_2)^{-3/2},$$

where m_k is the kth sample moment about the sample mean $\bar{Z}$. An asymptotically distribution-free test of symmetry uses a standardized form of b_1, namely $n^{1/2} b_1 \hat{\sigma}^{-1}$, where $\hat{\sigma}^2$ is a consistent estimator of the asymptotic variance of $n^{1/2} b_1$. See Gupta [13] for additional details.

Of course, when the mean μ of Z is known, one can base a test on a modified version of b_1, where m_k is redefined to be the kth sample moment about μ.

b. Test Based on the Difference Between the Sample Mean and the Sample Median. In a symmetric population, the median, mode, and mean (if it exists) coincide. Thus it is natural to take the difference between the median and mean (or the mode and mean) as a measure of skewness of the distribution. Hotelling and Solomons [16] (see also Kendall and Stuart [17, p. 95]) proposed the population parameter $(\mu - \theta)/\sigma$ as a measure of skewness (μ is the mean, θ the median, and σ the standard deviation), with corresponding sample statistic

$$D = n\big(\bar{Z} - \tilde{\theta}\big) \bigg/ \bigg\{ n^{-1} \sum_{i=1}^n \big(Z_i - \bar{Z}\big)^2 \bigg\}^{1/2},$$

where $\tilde{\theta}$ is the sample median (*see* MEAN, MEDIAN, MODE AND SKEWNESS). Using an expression given by Gastwirth [12] for the asymptotic variance (τ^2, say) of $n^{-1/2}D$, a consistent estimate of τ^2 can be given to provide an asymptotically distribution-free test of symmetry. Gastwirth shows that if F has a density f which is unimodal and symmetric about 0, then $\tfrac{1}{4} \leqslant \tau^2 \leqslant 1$, but setting the level of the D test conservatively (by using $\tau^2 = 1$) can lead to a true level significantly below the nominal level.

Some nonparametric tests of symmetry about an unknown median θ include:

c. Modified Sign Test. The ordinary sign statistic can be written as $B = \sum_{i=1}^n I\{Z_i > 0\}$, when testing for symmetry about 0, and $B_{\theta_0} = \sum_{i=1}^n I\{Z_i > \theta_0\}$, when testing for symmetry about a *known* median $\theta = \theta_0$. When θ is unknown, Gastwirth [12] reasoned that θ can be estimated by the sample mean $\bar{Z}$, and his modified sign statistic is $B^* = \sum_{i=1}^n I\{Z_i > \bar{Z}\}$. Unlike B, B^* is not distribution-free under the null hypothesis. Under H_0, $n^{-1/2}(B^* - n/2)$ is asymptotically normal with asymptotic mean 0, but its asymptotic variance depends on F and can be quite different from the asymptotic variance of B. To obtain an asymptotically distribution-free test based on B^*, one can use

Gastwirth's [12] expression for the asymptotic variance of B^* to devise consistent estimates of that asymptotic variance; if (instead of consistently estimating the asymptotic variance of B^*) one naively refers B^* to the distribution of B, the results are unsatisfactory in that the true level of the test may be quite different than the nominal value.

d. Triples Test.   Randles et al. [21] proposed a test which considers observation triples (Z_i, Z_j, Z_k) and scores each triple via the function

$$\psi(Z_i, Z_j, Z_k) = \tfrac{1}{3}\{\operatorname{sgn}(Z_i + Z_j - 2Z_k)$$
$$+ \operatorname{sgn}(Z_i + Z_k - 2Z_j)$$
$$+ \operatorname{sgn}(Z_j + Z_k - 2Z_i)\},$$

where $\operatorname{sgn}(a) = -1$, 0, or 1 according as $a <$, $=$, or > 0. (Z_i, Z_j, Z_k) forms a *right triple* when $\psi(Z_i, Z_j, Z_k) = \tfrac{1}{3}$. This occurs when the middle observation is closer to the smallest observation than it is to the largest observation (such a case "looks" skewed to the right). Similarly, a triple "looks" skewed to the left when the middle observation is closer to the largest observation than it is to the smallest observation. In such a case, (Z_i, Z_j, Z_k) forms a *left triple*, and then $\psi(Z_i, Z_j, Z_k) = -\tfrac{1}{3}$. The event $\psi(Z_i, Z_j, Z_k) = 0$ is neither "left" nor "right" and occurs with probability zero when F is continuous. The Randles et al. test is based on the statistic

$$R = \binom{n}{3}^{-1} \sum_{i<j<k} \psi(Z_i, Z_j, Z_k).$$

R is not distribution-free under the null hypothesis (the null asymptotic variance depends on F), but Randles et al. give a consistent estimate of the asymptotic variance of $n^{1/2}R$ which enables them to provide an asymptotically distribution-free test of symmetry versus asymmetry based on R.

e. Modified Signed-Rank Test.   Gupta [13] proposed a test of symmetry versus asymmetry based on the number of positive deviations from the sample median $\tilde{\theta}$ that exceed the absolute values of the negative deviations from $\tilde{\theta}$. Gupta's statistic J can be written as

$$J = \binom{n}{2}^{-1} \sum_{i<j} \big[I\{Z_i + Z_j > 2\tilde{\theta}\}$$
$$- I\{Z_i > \tilde{\theta}; Z_j > \tilde{\theta}\}\big].$$

J is not distribution-free under the null hypothesis but Gupta (see also Hollander and Wolfe [15]) provides a consistent estimate of the asymptotic variance of $n^{1/2}J$ which is used to form an asymptotically distribution-free test. Related modified Wilcoxon tests are proposed by Antille et al. [2] and Bhattacharya et al. [4].

All the test statistics mentioned here must be interpreted with caution in that they measure different notions of asymmetry. Gastwirth's statistic estimates $P(Z > \mu)$, where $\mu = E[Z]$, and rejects when the estimate is too far from its value under the null hypothesis of symmetry $\tfrac{1}{2}$. But there are asymmetrical populations for which $P(Z > \mu) = \tfrac{1}{2}$ and Gastwirth's modified sign test will not be able to detect such populations. The Randles et al. statistic estimates

$$\eta = P(Z_1 + Z_2 - 2Z_3 > 0)$$
$$- P(Z_1 + Z_2 - 2Z_3 < 0);$$

the test will not detect asymmetrical populations for which $\eta = 0$. Similarly, Gupta's test will not detect asymmetrical populations for which $E[J] \doteq \tfrac{1}{4}$ (the latter being its asymptotic expectation under the null hypothesis). The statistic b_1^2 estimates the population parameter $\beta_1^2 = \mu_3^2/\mu_2^3$, where $\mu_r = E[(Z - E[Z])^r]$. β_1 is known as the *coefficient of skewness* and is a measure of asymmetry. If the distribution of Z is symmetric, $\beta_1 = 0$ (since $\mu_3 = 0$); however, there exist asymmetrical distributions for which $\mu_3 = 0$, and the test based on b_1 will not detect such populations. Similarly, the test based on D will not detect asymmetrical populations for which the mean and median coincide.

The Randles et al. article [21] contains a Monte Carlo* study comparing various tests of symmetry versus asymmetry. Randles et al. show that the tests based on R and b_1

(with suitable choices of the consistent estimators of the asymptotic variances) hold their α levels well, but that Gupta's test is extremely conservative. Also see Antille et al. [2], Boos [6], and Bhattacharya et al. [4] for further comparisons of power efficiency and robustness of level for various competing procedures. Other tests available for the case when θ is unknown may be found in these references, and in Doksum [9], Antille and Kersting [1], Doksum et al. [10], Finch [11], and Davis and Quade [8].

References

[1] Antille, A. and Kersting, G. (1977). *Z. Wahrsch. verw. Geb.*, **39**, 235–255. (Technical paper; proposes a trimmed sign statistic based on the differences of symmetrically located intervals of order statistics and also studies modified sign and Wilcoxon statistics based on estimating the center of symmetry by the sample mean or sample median.)

[2] Antille, A., Kersting, G., and Zucchini, W. (1982). *J. Amer. Statist. Ass.*, **77**, 639–646. (Technical paper; proposes and investigates trimmed Wilcoxon tests and trimmed tests based on gaps.)

[3] Bell, C. B. and Haller, H. S. (1969). *Ann. Math. Statist.*, **40**, 259–269. (Technical paper; considers various formulations of symmetry and proposes various tests for bivariate symmetry.)

[4] Bhattacharya, P. K., Gastwirth, J. L., and Wright, A. L. (1982). *Biometrika*, **69**, 377–382. (Technical paper; proposes modifications of the Wilcoxon test for symmetry about the sample median, including tests also proposed in Ref. 2.)

[5] Bickel, P. J. and Lehmann, E. L. (1975). *Ann. Statist.*, **3**, 1045–1069. (Technical paper; studies desirable criteria for measures of location. Compares different measures in terms of asymptotic efficiencies of their estimators.)

[6] Boos, D. B. (1982). *J. Amer. Statist. Ass.*, **77**, 647–651. [Technical paper; proposes a new test for symmetry based on minimizing a Cramér–von Mises distance (that measures deviation from symmetry) and dividing the minimum distance by a constant times Gini's mean difference.]

[7] Butler, C. C. (1969). *Ann. Math. Statist.*, **40**, 2209–2210. (Technical paper; independently proposes and obtains the distributions of the Smirnov statistics for testing symmetry.)

[8] Davis, C. E. and Quade, D. (1978). *Commun. Statist. A*, **7**, 413–418. (Technical paper; proposes various U-statistics as measures of and tests for symmetry against skewness alternatives.)

[9] Doksum, K. A. (1975). *Scand. J. Statist.*, **2**, 11–22. (Technical paper; introduces the empirical symmetry function for estimating how much a distribution deviates from symmetry about its median.

[10] Doksum, K. A., Fenstad, G., and Aaberge, R. (1977). *Biometrika*, **64**, 473–487. (Technical paper; studies plots for checking symmetry and deviations from symmetry and proposes tests of symmetry for the case where the point of symmetry is known and also for the case where the point of symmetry is unknown.)

[11] Finch, S. J. (1977). *J. Amer. Statist. Ass.*, **72**, 387–392. (Technical paper; introduces a test of symmetry based on the gaps of the order statistics.)

[12] Gastwirth, J. L. (1971). *J. Amer. Statist. Ass.*, **66**, 821–823. (Technical paper; considers the effect of estimating the center of symmetry by the sample mean on the sign test for symmetry).

[13] Gupta, M. K. (1967). *Ann. Math. Statist.*, **38**, 849–866. (Technical paper; proposes an asymptotically distribution-free test of symmetry based on J.)

[14] Hollander, M. (1971). *Biometrika*, **58**, 203–212. (Technical paper; introduces an omnibus test of bivariate symmetry.)

[15] Hollander, M. and Wolfe, D. A. (1973). *Nonparametric Statistical Methods*. Wiley, New York. (General reference for nonparametric methods.)

[16] Hotelling, H. and Solomons, L. M. (1932). *Ann. Math. Statist.*, **3**, 141–142. [Technical paper; proposes the difference (suitably standardized) between the sample mean and the sample median as a natural test of asymmetry.]

[17] Kendall, M. G. and Stuart, A. (1977). *The Advanced Theory of Statistics*, Vol. 1, 4th ed. Macmillan, New York. (General reference for statistical theory).

[18] Kepner, J. L. and Randles, R. H. (1982). *J. Amer. Statist. Ass.*, **77**, 475–482. (Technical paper; proposes a conditional test that uses the null hypothesis of bivariate symmetry to detect unequal marginal scales in a bivariate population.)

[19] Koziol, J. A. (1980). *J. Amer. Statist. Ass.*, **75**, 161–167. (Technical paper; decomposes the Orlov statistic into orthogonal components, suggests the first component as an alternative test statistic, and shows that the components are asymptotically equivalent to a class of linear rank statistics.)

[20] Orlov, A. I. (1972). *Theory Prob. Appl.*, **17**, 357–361. (Technical paper; proposes the Cramér–von Mises-type statistic for testing symmetry.)

[21] Randles, R. H., Fligner, M. A., Policello, G. E., and Wolfe, D. A. (1980). *J. Amer. Statist. Ass.*, **75**, 168–172. (Technical paper; proposes an asymptotically distribution-free test of symmetry based on R.)

[22] Rothman, E. D. and Woodroofe, M. (1972). *Ann. Math. Statist.*, **43**, 2035–2038. (Technical paper; proposes a variant of the Orlov statistic.)

[23] Sen, P. K. (1967). *Sankhyā A*, **29**, 351–372. (Technical paper; uses a conditional approach to derive distribution-free bivariate symmetry tests.)

[24] Smirnov, N. V. (1947). *Dokl. Akad. Nauk. SSR*, **56**, 11–14. (Technical paper; introduces the Smirnov tests of symmetry.)

Acknowledgment

Research supported by the Air Force Office of Scientific Research, AFSC, USAF, under Grant AFOSR 82-K-0007 to Florida State University and by Public Health Service Grant 5R01GM21215-07 to Stanford University.

(DISTRIBUTION-FREE METHODS
EXCHANGEABILITY
SIGN TEST
STUDENT'S t-TESTS
SYMMETRY TESTS, PURE AND COMBINED
U-STATISTICS
WILCOXON SIGNED-RANK TEST)

MYLES HOLLANDER

TESTS AND *P*-VALUES, COMBINING

Tippett [1] suggested the use of the criterion $P_{[1]} = \min(P_1, P_2, \ldots, P_k)$ for combining *P*-values* from *k* mutually independent significance tests*.

If all null hypotheses are valid, then the event

$$P_{[1]} < 1 - (1 - \alpha)^{1/k}$$

has probability α.

Wilkinson [2] suggested using the *r*th smallest *P*-value $P_{[r]}$ in place of $P_{[1]}$. If all *k* null hypotheses are valid, then $P_{[r]}$ has a standard beta distribution* with parameters $r, k - r + 1$. This is used to establish critical regions of the form $P_{[r]} < \omega_{k,r,\alpha}$ of size α.

References

[1] Tippett, L. M. C. (1931). *The Methods of Statistics*. Williams and Norgate, London, England.

[2] Wilkinson, B. (1951). *Psychol. Bull.*, **48**, 156–158.

(COMBINATION OF DATA
P-VALUES
TESTS, WESTBERG ADAPTIVE
  COMBINATION OF)

TESTS BASED ON EMPIRICAL PROBABILITY MEASURES

Standard goodness-of-fit* tests involve comparison of an hypothesized model represented by a distribution function F and an empirical model—the sample distribution function F_n, which is computed from the data. The extent to which F and F_n agree corresponds to the amount of faith we might place in F as a reasonable representation of the underlying process. The Kolmogorov–Smirnov* statistic specifically requires the calculation of

$$\sup_x |F_n(x) - F(x)| \qquad (1)$$

to carry out the comparison. Foutz [7] noted that statistics such as (1) are restrictive in the sense that the hypothesized model and the data are compared only with respect to Borel sets of the form $(-\infty, x]$. He argues that the statistic

$$F = \sup_{B \in \text{Borel sets}} |P_n(B) - P(B)| \qquad (2)$$

be evaluated where P is a measure representing the hypothesized model (corresponding to F) and P_n is a particular empirical probability measure based on the data. To compute (2), Foutz defines the empirical measure as

$$P_n(B) = n^{-1} \sum_{i=1}^{n+1} P(B \cap B_i)/P(B_i),$$

where B_i are statistically equivalent blocks based on the data [1]. This empirical probability measure is convenient, to say the least, since F can then be computed directly as

$$F = \max_{k=1,2,\ldots,n-1} \left[\frac{k}{n} - (D_1 + D_2 + \cdots + D_k) \right],$$

$$(3)$$

where $D_1 < D_2 < \cdots < D_n$ are the ordered

values of $P(B_1), \ldots, P(B_n)$. Hence, the seemingly hopeless task of optimizing over the Borel sets in (2) is reduced to the straightforward calculation (3).

To use F in practice requires knowledge of its null distribution. Assuming P is the true probability measure governing the process, F has the same distribution as $\max[k/n - (U_{(1)} + U_{(2)} + \cdots + U_{(k)})]$, where the $U_{(i)}$'s represent the n ordered interval widths formed from $n - 1$ independent uniform* $(0, 1)$ variates. It is remarkable that this is the null distribution of F regardless of the hypothesized measure P and the dimension of X. For small n ($\leqslant 5$), the null distribution can be derived explicitly [6, 7]. For intermediate n, the critical values of F have been estimated by Monte Carlo [6]. Finally, for large n, $n^{1/2}(F - e^{-1})$ is asymptotically normal with mean zero and variance $2e^{-1} - 5e^{-2}$ [7]. This result can be proved using the machinery of Brownian bridges [5]; *see* PROCESSES, EMPIRICAL.

Foutz and Birch [2, 3, 8] have specialized F to the univariate two-sample problem, obtaining a test superior to the Kolmogorov–Smirnov, Mann–Whitney–Wilcoxon*, Kuiper, and Cramér–von Mises tests for selected normal mixture alternatives. A multivariate version of this test has also been developed [9].

An open area of investigation is the performance of F if the hypothesized measure P requires the estimation of unknown parameters [4].

References

[1] Anderson, T. W. (1966). *Proc. Int. Symp. Multivariate Anal.*, P. R. Krishnaiah, ed. Academic, New York, pp. 5–27.

[2] Birch, J. B. and Foutz, R. V. (1984). *Commun. Statist.—Simul. Comp.*, **13**, 397–405.

[3] Birch, J. B. and Foutz, R. V. (1985). *Commun. Statist.—Simul. Comp.*, **14**, 397–405.

[4] Booker, J. M., Johnson, M. E., and Beckman, R. J. (1984). *Proc. Statist. Comp. Sec. Amer. Statist. Ass.*, pp. 208–213.

[5] Csörgő, M. (1981). Analytical Methods in Probability Theory. *Lect. Notes. Math.*, **861**, Springer, Berlin, pp. 25–34.

[6] Franke, R. and Jayachandran, T. (1984). *J. Statist. Comp. Simul.*, **20**, 101–114.

[7] Foutz, R. V. (1980). *Ann. Statist.*, **8**, 989–1001.

[8] Foutz, R. V. and Birch, J. B. (1982). *Commun. Statist. A*, **11**, 1839–1853.

[9] Kim, K.-K. (1984). Nonparametric multivariate two-sample tests based on empirical probability measures. Ph.D. dissertation, Virginia Polytechnical Institute and State University, Blacksburg, VA.

(EDF STATISTICS
GOODNESS-OF-FIT
KOLMOGOROV–SMIRNOV-TYPE TESTS OF FIT
PROBABILITY SPACES, METRICS AND DISTANCES ON)

MARK E. JOHNSON

TESTS FOR CENSORING

Censoring* is omission of predetermined order statistics*. It is to be distinguished from truncation* (omission of values in certain ranges). *See* CENSORING and TRUNCATION, but note that in CENSORED DATA these definitions are interchanged.

Given a set of ordered sample values $X_1 \leqslant X_2 \leqslant \cdots \leqslant X_r$, how can the hypothesis that it represents a complete random sample be tested, against the alternative that it is the remainder of such a sample after removal of certain order statistics? To facilitate discussion, the following notation is useful. The most general form of censoring (of a complete random sample of size n) leaving r observed values would be to remove all but the $(s_0 + 1)$th, $(s_0 + s_1 + 1)$th, ... and $(s_0 + s_1 + \cdots + s_{r+1} + 1)$th order statistics. There are thus s_0 missing values less than or equal to X_1; s_1 in the interval $[X_1, X_2]$; $\cdots$; s_{r-1} in $[X_{r-1}, X_r]$, and $s_r = n - r - s_0 - \cdots - s_{r-1}$ greater than or equal to x_r. The hypothesis that such censoring has occurred could be denoted by $H_{s_0, s_1, \ldots, s_r}$; the hypothesis of no censoring would be $H_{0,0,\ldots,0}$. Such elaborate schemes are, however, of little practical relevance. Almost always it is a question of testing for censoring of extreme values, taking $s_1 = s_2 = \cdots =$

$s_{r-1} = 0$ and testing the hypothesis $H_{0,0}$ (no censoring) against alternatives H_{s_0,s_r} ($s_0 + s_r > 0$). Particular attention is given to the cases $s_0 = 0$ (censoring from above, or "to the right"), $s_r = 0$ (censoring for below, or "to the left"), and $s_0 = s_r > 0$ (symmetrical censoring), with observed values which are mutually independent and correspond to absolutely continuous* random variables with a common density function (PDF) $f(x)$.

In order to make inferences about $H_{0,0}$ versus H_{s_0,s_r}, it is necessary to have some further information about the common distributions, apart from its continuity. If $f(x)$ is known, there is a uniformly most powerful test of $H_{0,0}$ against alternatives $H_{s_0,\theta s_0}$ (that is, with $s_r/s_0 = \theta$, fixed). It has a critical region of the form

$$Y_1(1 - Y_r)^\theta > \text{constant},$$

where $Y_j = \int_{-\infty}^{X_j} f(x)\, dx$. (See refs. 1 and 3.)

If $\theta = \infty$ (censoring from right) the region is $1 - Y_r >$ constant. (See refs. 1 and 3.) If the ratio s_r/s_0 is not known, a "general purpose" test, with critical region

$$Y_1 + (1 - Y_r) > \text{constant}$$

is suggested by an application of Roy's union–intersection principle* [2].

Even when the PDF is not known exactly, a test can be constructed if a complete random sample from the same distribution is available, in addition to the possibly censored sample [3].

Indirect censoring, wherein there is censoring on an unobserved variable that has a (known) joint distribution with the observed variable(s), can be tested in a similar way. The critical region is now defined in terms of the *conditional expected value* of $Y_1(1 - Y_r)^\theta$, given the value(s) of the observed variable(s) ([5], [6]).

References

[1] Johnson, N. L. (1966). Tests of sample censoring. *Proc. 20th Tech. Conf. Amer. Soc. Qual. Control,* pp. 699–703.

[2] Johnson, N. L. (1970). A general purpose test of censoring of extreme sample values. In *S. N. Roy Memorial Volume.* University of North Carolina Press, Chapel Hill, N.C., pp. 377–384.

[3] Johnson, N. L. (1971). *Austral. J. Statist.,* **12**, 1–6.

[4] Johnson, N. L. (1979). In *Robustness in Statistics,* R. L. Launer, ed. Academic, New York, pp. 127–146. (A general survey with several further references.)

[5] Johnson, N. L. (1980). *J. Multivariate Anal.,* **10**, 351–362.

[6] Johnson, N. L. and Dawson, J. E. (1985). In *Biostatistics: Statistics in Public Health and Environmental Sciences (Essays in Honor of B. G. Greenberg),* P. K. Sen, ed. Elsevier, New York, pp. 345–356.

(CENSORED DATA
CENSORING
PROGRESSIVE CENSORING)

TESTS OF NORMALITY See DEPARTURES FROM NORMALITY, TESTS FOR

TESTS OF RANDOMNESS See RANDOMNESS, TESTS FOR

TESTS, ONE-SIDED VERSUS TWO-SIDED

The choice between a one-sided test or a two-sided test for a univariate hypothesis depends on the objective of statistical analysis prior to its implementation. The underlying issue is whether the alternative* against which the (null) hypothesis is to be assessed is one-sided or two-sided. The alternative is often one-sided in a clinical trial* to determine whether active treatment is better than placebo; a two-sided alternative is usually of interest in a clinical trial to determine which of two active treatments is better.

The principal advantage of a one-sided test is greater power* for the contradiction of the null hypothesis when the corresponding one-sided alternative applies. Conversely, for alternatives on the opposite side, its lack of sensitivity represents a disadvantage. Thus, if alternatives on both sides of a null hypothesis are considered to be of inferential interest, a two-sided test is necessary. However, where the identification of one direction of alternatives is actually the objective of an investigation, the cost of the broader scope of a two-sided test is the larger sample size it requires to have the same power for this direction as its one-sided counterpart. The benefit provided by the increase in sample size is power for alternatives in the opposite direction. If this purpose for increased sample size is not justifiable on economic, ethical, or other grounds, then a one-sided test for a correspondingly smaller sample size becomes preferable. Thus, both one-sided tests and two-sided tests are useful methods, and the choice between them requires careful judgement.

The statistical issues can be clarified further by considering the example of the hypothesis of equality of two population means μ_1 and μ_2. The null hypothesis has the specification

$$H_0: \mu_1 - \mu_2 = \delta = 0. \qquad (1)$$

Suppose $\bar{y}_1$ and $\bar{y}_2$ are sample means based on large sample sizes n_1 and n_2 (e.g., $n_i \geq 40$) from the two populations; also suppose the population variances σ_1^2 and σ_2^2 are essentially known through their consistent estimation by the sample variances s_1^2 and s_2^2. Then the statistic

$$z = d/\sigma_d, \qquad (2)$$

where $d = (\bar{y}_1 - \bar{y}_2)$ and $\sigma_d = \{(s_1^2/n_1) + (s_2^2/n_2)\}^{1/2}$, approximately has the standard normal distribution with expected value 0 and variance 1.

A two-sided test for the hypothesis H_0 in (1) has the two-sided rejection region

$$R_2(\alpha_L, \alpha_U) = \left\{ \begin{array}{l} \text{any observed } z \text{ such that} \\ z \leq z_{\alpha_L} \text{ or } z \geq z_{1-\alpha_U} \end{array} \right\},$$
$$(3)$$

where z_{α_L} and $z_{1-\alpha_U}$ are the $100\alpha_L$ and $100(1 - \alpha_U)$ percentiles of the standard normal distribution and $(\alpha_L + \alpha_U) = \alpha$ is the specified significance level* (or Type I error). For most applications, (3) is symmetric with $\alpha_L = \alpha_U = (\alpha/2)$, and $z_{\alpha_L} = z_{\alpha/2} = -z_{1-(\alpha/2)} = -z_{1-\alpha_U}$; this structure is assumed henceforth for two-sided tests of H_0 in (1).

The one-sided test for assessing H_0 in (1) relative to the alternative

$$H_\delta: \mu_1 - \mu_2 = \delta > 0 \qquad (4)$$

of a larger mean for population 1 than population 2 has the one-sided rejection region

$$R_U(\alpha) = R_2(0, \alpha)$$

$$= \left\{ \begin{array}{l} \text{any observed } z \\ \text{such that } z \geq z_{1-\alpha} \end{array} \right\}; \qquad (5)$$

similarly, if the alternative (4) specified $\delta < 0$, the one-sided rejection region would be $R_L(\alpha) = R_2(\alpha, 0)$. Thus, the symmetric two-sided test based on $R_2(\alpha/2, \alpha/2)$ is equivalent to simultaneous usage of the two one-sided tests based on $R_L(\alpha/2)$ and $R_U(\alpha/2)$.

The power of the one-sided test (5) with respect to H_δ in (4) is

$$\psi_U(\delta|\alpha) = \Pr\{ R_U(\alpha) | H_\delta \}$$
$$= 1 - \Phi\{ z_{1-\alpha} - (\delta/\sigma_d) \}, \qquad (6)$$

where $\Phi( )$ is the cumulative distribution function of the standard normal distribution. The power of the two-sided test (3) for this situation is

$$\psi_2(\delta|\alpha) = \Pr\{ R_2(\alpha/2, \alpha/2) | H_\delta \}$$
$$= \left[1 - \Phi\{ z_{1-(\alpha/2)} - (\delta/\sigma_d) \} \right.$$
$$\left. + \Phi\{ z_{\alpha/2} - (\delta/\sigma_d) \} \right]. \qquad (7)$$

When $\delta > 0$, $\psi_U(\delta|\alpha) > \psi_2(\delta|\alpha)$, and the one-sided test is more powerful. However, when $\delta < 0$, $\psi_2(\delta|\alpha) > \alpha/2 > \psi_U(\delta|\alpha)$, and so the one-sided test's power is not only much poorer, but is also essentially negligible. Also, in the very rare situations where rejection is indicated, it is for the wrong reason [i.e., H_0 is contradicted by large z in $R_U(\alpha)$ when actually $\delta < 0$].

When one direction of alternatives such as (4) is of primary interest, the two-sided test, which achieves the same power ψ for specific α and δ as its one-sided counterpart, requires sample sizes that are $\lambda(\alpha, \psi)$ times larger [where $\lambda(\alpha, \psi) \geq 1$]. For usual significance levels $0.01 \leq \alpha \leq 0.05$ and power $\psi \geq 0.50$, the two-sided test multiplier $\lambda(\alpha, \psi)$ of the one-sided test sample sizes n_1 and n_2 is

Table 1   Multiplier of One-Sided Test Sample Sizes for Two-Sided Test to Have the Same Power.

Power	α 0.01	0.02	0.05
0.50	1.23	1.28	1.42
0.60	1.20	1.25	1.36
0.70	1.18	1.22	1.31
0.80	1.16	1.20	1.27
0.90	1.14	1.17	1.23

given by

$$\lambda(\alpha, \psi) = \left\{ \frac{z_{1-(\alpha/2)} + z_\psi}{z_{1-\alpha} + z_\psi} \right\}^2. \quad (8)$$

In Table 1, values of $\lambda(\alpha, \psi)$ are reported for $\alpha = 0.01, 0.02, 0.05$ and $\psi = 0.50, 0.60, 0.70, 0.80, 0.90$. For the typical application of power $\psi = 0.80$ and significance level $\alpha = 0.05$, the sample size required for a two-sided test is 27% greater than for its one-sided counterpart. Also, the multipliers $\lambda(\alpha, \psi)$ can be seen to decrease as either α decreases or ψ increases.

Some further insight about one-sided and two-sided tests can be gained from their relationship to confidence intervals*. The one-sided test based on $R_U(\alpha)$ in (5) corresponds to the one-sided lower bound confidence interval

$$\delta \geq d - z_{1-\alpha}\sigma_d = d_{L, \alpha}. \quad (9)$$

If $d_{L, \alpha} > 0$, then H_0 is contradicted relative to the alternative H_δ in (4); if $d_{L, \alpha} \leq 0$, then there is not sufficient evidence to support H_δ. In this latter context, δ may be near 0 or less than 0; but the distinction between these interpretations is not an inferential objective of a one-sided confidence interval or hypothesis test. For the purpose of the more refined assessment of whether δ is greater than 0, near 0, or less than 0, a two-sided test is needed; its corresponding confidence interval is

$$d_{L, \alpha/2} \leq \delta \leq d_{U, \alpha/2}, \quad (10)$$

where $d_{L, \alpha/2} = \{d - z_{1-(\alpha/2)}\sigma_d\}$ and $d_{U, \alpha/2} = \{d + z_{1-(\alpha/2)}\sigma_d\}$. If $d_{L, \alpha/2} > 0$,

then H_0 is contradicted with respect to $\delta > 0$; if $d_{U, \alpha/2} < 0$, then H_0 is contradicted with respect to $\delta < 0$; and if $d_{L, \alpha/2} \leq 0 \leq d_{U, \alpha/2}$, then H_0 is not contradicted and δ is interpreted as being near 0 in the sense of the confidence limits $(d_{L, \alpha/2}, d_{U, \alpha/2})$. When support for $\delta > 0$ is the objective of an investigation, the cost for the two-sided confidence interval's or test's additional capability for distinguishing between $\delta < 0$ or δ near 0 is either reduced power for the same sample size or increased sample size for the same power.

A third way to specify one-sided and two-sided tests is through one-sided and two-sided p-values*; the one-sided p-value for assessing the one-sided alternative H_δ in (4) through z in (2) is

$$p_U(z) = 1 - \Phi(z); \quad (11)$$

if $p_U(z) \leq \alpha$, then z is interpreted as contradicting H_0 on the basis of the small probability $\leq \alpha$ for repeated sampling under H_0 to yield values $\geq z$. For symmetric two-sided tests of H_0 in (1), the two-sided p-value* is

$$p_2(z) = 2\{1 - \Phi(|z|)\}; \quad (12)$$

if $p_2(z) \leq \alpha$, then H_0 is contradicted. The definition of two-sided p-values for asymmetric situations is more complicated; it involves considerations of extreme outcomes for a test statistic in both directions from H_0. For summary purposes, the rejection region, confidence interval, and p-value specifications of a one-sided test are equivalent in the sense of yielding the same conclusion for H_0; this statement also applies to symmetric two-sided tests.

A concern for any one-sided test is the interpretation of values of the test statistic which would have contradicted the hypothesis if a two-sided test were used. From the inferential structure which underlies one-sided tests, such outcomes are judged to be random events compatible with the hypothesis, no matter how extreme they are. However, their nature can be a posteriori described as "exploratory information supplemental to the defined (one-sided) objective" of an investigation. This perspective enables suggestive statements to be made about opposite direction findings; their strengthening to inferential conclusions would require confirmation by one or more additional investigations.

Another issue sometimes raised is that one-sided tests *seem to make it easier to contradict a hypothesis and thereby to have a weaker interpretation than would have applied to two-sided tests.* However, when the null hypothesis H_0 is true, the probability of its contradiction is the significance level α regardless of whether a one-sided test or a two-sided test is used. It is *easier* for the one-sided test to contradict H_0 when its one-sided alternative applies, but this occurs because the one-sided test is *more powerful* for such alternatives.

Some additional practical comments worthy of attention are as follows:

(i)   Among the commonly used statistical tests for comparing two population means, z and t-tests lead to one-sided or two-sided tests in a natural manner such as (3) and (5) due to the symmetry about zero of their standardized distributions. Chi-square and F-tests* for such comparisons involve squared quantities and so lead to two-sided tests. One-sided counterparts for chi-square and F-test p-values are usually computed indirectly using $p_1 = (p_2/2)$ if the difference is in the same direction as the alternative hypothesis, and $p_1 = 1 - (p_2/2)$ if the difference is in the opposite direction, where p_1 and p_2 are one-sided and two-sided p-values, respectively.

(ii)   Fisher's exact test* for independence in a 2×2 contingency table* leads naturally to either a one-sided or two-sided test since the discrete event probabilities for it pertain to one or the other side of the underlying permutation distribution. However, this test is often asymmetric and then a one-sided p-value (less than 0.5) cannot be doubled to give the corresponding two-sided p-value.

(iii)   Fisher's method of combining c independent tests (see Fisher [2] and Folks [3]) is analogous to a one-sided test when its power is directed at a one-sided alternative.

For this test, the one-sided p-value is the probability of larger values of

$$Q_F = -2 \sum_{k=1}^{c} \log p_k$$

with respect to the χ^2 distribution with $2c$ degrees of freedom where the $\{p_k\}$ are one-sided p-values in the direction of the one-sided alternative of interest for the c respective tests. The p-value corresponding to the opposite side is obtained by the same type of computation with $\{p_k\}$ replaced by their complements $\{1 - p_k\}$.

(iv)   A practical advantage of one-sided p-values is their descriptive usefulness for summarizing the results of hypothesis tests; such p-values contain more information than their two-sided counterparts because the one-sided version identifies the direction of any group difference as well as providing the criterion for evaluating whether the hypothesis of no difference is contradicted. This additional descriptive feature eliminates the need for identifying the direction of difference as would be necessary in summary tables of two-sided p-values.

Additional discussion of one-sided and two-sided tests is given in many textbooks dealing with statistical methodology, e.g., see Armitage [1], Hogg and Craig [4], Mendenhall et al. [5]. Also, *see* HYPOTHESIS TESTING.

References

[1]   Armitage, P. (1971). *Statistical Methods in Medical Research*. Wiley, New York.

[2]   Fisher, R. A. (1932). *Statistical Methods for Research Workers*, 4th ed. Oliver and Boyd, Edinburgh, Scotland.

[3]   Folks, J. L. (1984). Combination of independent tests. In *Handbook of Statistics: Nonparametric Methods*, Vol. 4, P. R. Krishnaiah and P. K. Sen, eds. North-Holland, Amsterdam, Netherlands, pp. 113–121.

[4]   Hogg, R. V. and Craig, A. T. (1978). *Introduction to Mathematical Statistics*, 4th ed. Macmillan, New York.

[5]   Mendenhall, W., Scheaffer, R. L., and Wackerly, D. D. (1981), *Mathematical Statistics with Applications*, 2nd ed. Duxbury Press, Boston, Mass.

Acknowledgments

This research was supported in part by the U. S. Bureau of the Census through Joint Statistical Agreement JSA-

84-5. The authors would like to express appreciation to Ann Thomas for editorial assistance.

(CONFIDENCE INTERVALS AND REGIONS
EXPLORATORY DATA ANALYSIS
FISHER'S EXACT TEST
HYPOTHESIS TESTING
INFERENCE, STATISTICAL-I, II
POWER
P-VALUES
SIGNIFICANCE TESTS, HISTORY AND
    LOGIC)

GARY G. KOCH
DENNIS B. GILLINGS

TESTS, STATISTICAL See HYPOTHESIS TESTING (*See also tests for specific cases*)

TESTS, TIPPETT'S AND WILKINSON'S COMBINATIONS OF See TESTS AND P-VALUES, COMBINING

TESTS, WESTBERG ADAPTIVE COMBINATION OF

INTRODUCTION

In statistical work one often encounters the problem of how to combine results from statistical tests. Suppose, for example, that k experiments have been performed to detect a certain effect, the magnitude of which may be measured by a different parameter θ_i in each experiment ($i = 1, \ldots, k$). The ith experiment can be used to test the null hypothesis H_{0i}: $\theta_i = 0$. Call the attained significance level (P-value*) P_i. If the test statistics have continuous distributions, then when H_{0i} is true, P_i is uniformly distributed over the interval [0, 1]. Testing the combined null hypothesis

$$H_0: H_{0i} \text{ is true for all } i$$

against

$$H_1: H_{0i} \text{ is false for at least one } i$$

presents a problem in combination of tests.

Many procedures have been proposed for combining the P-values arising from several independent tests in order to test whether all null hypotheses are true. Two commonly used methods of combining independent significance levels $P_1, \ldots, P_k$ are Fisher's procedure [1], based on the product of the P-values, and Tippett's procedure [5], based on the minimum P-value.

In studies by Frisen [2] and Westberg [6], Fisher's method was compared with Tippett's according to the power. These studies show that neither of the methods is generally more powerful than the other.

Where a high power is desired when just one of the hypotheses H_{0i} is false and the deviation from H_0 is large, Tippett's method is preferable. Where it is more important to detect alternatives for which many of the hypotheses H_{0i} might be false to a comparable degree, Fisher's method is likely to be preferable. One way of indicating if there are few or many hypotheses that are false is to use plots of the P-values themselves in order to evaluate the number of valid hypotheses, a method proposed by Schweder and Spjøtvoll [4]. Another kind of indication of the number of false hypotheses is provided in an adaptive way, by deciding on an appropriate test statistic. The adaptive method to be described tends to be similar to Fisher's procedure when many or all of the hypotheses H_{0i} are false, and tends to be similar to Tippett's when only one or very few of the hypotheses are strongly violated.

The procedure is stepwise. It should not be confused with the multiple test procedure proposed by Holm [3], which is applicable when it is desired to know *which* of the hypotheses are false.

THE ADAPTIVE METHOD

Denote the k ordered P-values by

$$P_{(1)} \geqslant P_{(2)} \geqslant \cdots \geqslant P_{(k-1)} \geqslant P_{(k)}.$$

Choose constants a_i ($i = 1, \ldots, k$) such that

$$1 > a_1 > a_2 > \cdots > a_k > 0.$$

The test is based on the statistic

$$Z(n) = \prod_{i=k-n+1}^{k} \left(P_{(i)} / a_{k-n+1} \right), \qquad n > 0,$$

where the random variable n is the greatest integer such that $P_{k-n+1} < a_{k-n+1}$. If there is no such integer (i.e., $n = 0$), take $Z(0) = 1$. When $n = k$, the test statistic is identical to Fisher's, and when $n = 1$, the test statistic is identical to Tippett's, but the procedures are not the same, because n is stochastic.

Conditional on n, $-2 \ln Z(n)$ is distributed as chi-squared* with $2n$ degrees of freedom. Hence the critical region $-2 \ln Z(n) > \chi^2_{2n, \alpha'}$ will provide a test with significance level $\{1 - \Pr[n = 0]\} \alpha'$. ($\chi^2_{2n, \alpha'}$ is the upper $100\alpha'\%$ point of the chi-squared distribution* with $2n$ degrees of freedom.)

It is suggested that the a_i's should be such that

$$a_i = a_1 \{1 - (i - 1)/k\}.$$

This can be shown to lead to

$$\Pr[n = 0] = 1 - a_1,$$

$$\Pr[n = 1] = a_1(1 - a_1)(1 - a_1/k)^{k-2},$$

$$\Pr[n = k] = a_1^k.$$

Westberg [7] suggests choosing a_1 to make $\Pr[n = 1] = \Pr[n = k]$, and her Table 1 provides appropriate values for a_1.

The critical region of the test is then

$$-2 \ln Z(n) > \chi^2_{2n, \alpha/a_1}.$$

Example. The attained significance levels of three tests are $P_{(1)} = 0.60$, $P_{(2)} = 0.20$, and $P_{(3)} = 0.10$, so $k = 3$ and $a_1 = 0.581$. Since $P_{(1)} = 0.60 > a_1 = 0.581$, $P_{(1)}$ will not be included in the test statistic. But $P_{(2)} = 0.20 < a_2 = 0.387$ and so P_2 and P_3 are included in the test statistic, which is then

$$Z(2) = \prod_{i=2}^{3} (P_{(i)}/a_2),$$

with

$$-2 \ln Z(2) = -2\{\ln(0.20/0.387)$$
$$+ \ln(0.10/0.387)\} = 4.027.$$

If the desired significance level is $\alpha = 0.05$, then $\alpha' = 0.05/0.581 = 0.086$. Since 4.027 is less than $\chi^2_{4, 0.086} = 8.16$, H_0 is not rejected.

POWER

In the cases examined, the new method is always better than the worst of Fisher's and

Tippett's, and sometimes even better than either. This latter case arises when there are substantial departures from several (but not a high proportion) of the H_{0i}'s. Detailed numerical comparisons can be found in Westberg [7].

References

[1] Fisher, R. A. (1932). *Statistical Methods for Research Workers*, 4th ed. Oliver and Boyd, Edinburgh, Scotland.

[2] Frisen, M. (1974). *Stochastic Deviation from Elliptical Shape*. Almqvist and Wiksell, Stockholm, Sweden.

[3] Holm, S. (1979). A simple sequentially rejective multiple test procedure. *Scand. J. Statist*, **6**, 65–70.

[4] Schweder, T. and Spjøtvoll, E. (1982). Plots of *P*-values to evaluate many tests simultaneously. *Biometrika*, **69**, 493–502.

[5] Tippett, L. H. C. (1931). *The Methods of Statistics*. Williams & Norgate, London, England.

[6] Westberg, M. (1985a). Combining independent statistical tests. *The Statistician*, *London*, **34**, 287–296.

[7] Westberg, M. (1985b). An adaptive method of combining independent statistical tests. *Research Report 1985:6*, Statistiska Institutionen, Göteborgs Universitet, Sweden.

(ADAPTIVE METHODS
COMBINATION OF DATA
P-VALUES
TESTS AND *P*-VALUES, COMBINING)

MARGARETA WESTBERG

TETRACHORIC CORRELATION COEFFICIENT

Consider a fourfold (2×2 contingency) table with frequencies given by

	A	A^c	
B	a	b	$a + b$
B^c	c	d	$c + d$
	$a + c$	$b + d$	N

Assume that this table has been obtained from a random sample $(x_1, y_1), (x_2, y_2)$,

$\ldots,(x_N, y_N)$ from a bivariate normal* distribution, where $A = \{(x, y); x < x_0\}$, $B = \{(x, y); y < y_0\}$ and a is the number of pairs (x_i, y_i), $i = 1, 2, \ldots, N$ with $x_i < x_0$, $y_i < y_0$.

Let $\phi(x, y; \rho)$ denote the standard bivariate normal density with correlation coefficient ρ and let $\phi(x)$ and $\phi(y)$ be the corresponding marginal density functions. Let $\Phi(x, y; \rho)$, $\Phi(x)$, and $\Phi(y)$ be the corresponding cumulative distribution functions. With no loss of generality, assume that $a + c \geqslant b + d$ and $a + b \geqslant c + d$. Let $h = (x_0 - \mu_x)/\sigma_x$ and $k = (y_0 - \mu_y)/\sigma_y$. One can estimate h and k by

$$\hat{h} = \Phi^{-1}\left(\frac{a + c}{N}\right), \quad \hat{k} = \Phi^{-1}\left(\frac{a + b}{N}\right).$$

Then we can estimate ρ by solving

$$\frac{d}{N} = \int_{\hat{h}}^{\infty} \int_{\hat{k}}^{\infty} \frac{1}{2\pi(1 - \hat{\rho}^2)^{1/2}} \exp\left[\frac{-1}{2(1 - \hat{\rho}^2)}\right.$$
$$\left. \times \left(x^2 - 2\hat{\rho}xy + y^2\right)\right] dy\, dx$$
$$= \overline{\Phi}(\hat{h}, \hat{k}, \hat{\rho})$$

for $\hat{\rho}$ given $\hat{h}$ and $\hat{k}$. The solution is unique (Slepian [22]). The unique solution $\hat{\rho}$ is usually denoted by r_t and called the *tetrachoric correlation coefficient*. It was introduced by Pearson [19]. A closely related concept is the polychoric correlation coefficient (*see* POLYCHORIC AND POLYSERIAL CORRELATION), which is employed when the two variables x and y are divided into k_1 and k_2 ordered categories, respectively, and at least one of k_1, k_2 exceeds 2. The original technique for solving the equation is to expand $\Phi(h, k, \rho)$ in the tetrachoric series and employ tables compiled by Everitt [8] to locate the solution numerically. This procedure is generally regarded as tedious; many publications have been devoted to approximations and computing aids. In particular, see refs. 4, 5, 7, 13, 15, 17, 18, 24, and 27.

Pearson [19] proposed a number of approximations, one of which,

$$Q_1 = \frac{ad - bc}{ad + bc},$$

is also known as Yule's [25, 26] coefficient of association. On the basis of numerical experimentation, Pearson recommended three as being reasonably accurate. They are

$$Q_3 = \sin\frac{\pi}{2} \frac{\sqrt{ad} - \sqrt{bc}}{\sqrt{ad} + \sqrt{bc}},$$

$$Q_4 = \sin\frac{\pi}{2}\left\{1 + \frac{2bcN}{(ad - bc)(b + c)}\right\}^{-1},$$

and

$$Q_5 = \sin\frac{\pi}{2}(1 + \kappa^2)^{-1/2},$$

where

$$\kappa^2 = 4abcdN^2\Big/\left[(ad - bc)^2(a + d)(b + c)\right].$$

Castellani [3] and Dingby [6] also studied various approximations to r_t.

To facilitate the use of r_t in statistical inference, Pearson [20] gave a series expansion for, and considered the problem of obtaining large sample approximations to, the probable error of r_t. Many others [1, 2, 9, 11, 12, 14, 27] investigated the behavior of the standard error of r_t. In particular, Hamdan [12] showed that r_t is the maximum likelihood estimator of ρ for a fourfold table. Using the asymptotic theory for maximum likelihood estimators, he obtained an asymptotic approximation to the standard error of r_t. Tallis [23] also considered the maximum likelihood estimation* of ρ from fourfold tables. His results differ from those of Hamdan, since he simultaneously estimated ρ, h, and k.

Yule [25, 26] proposed analyzing fourfold tables by quantities now know as measures of association*. He proposed

$$Q_1 = \frac{ad - bc}{ad + bc}$$

and

$$Q_2 = \frac{\sqrt{ad} - \sqrt{bc}}{\sqrt{ad} + \sqrt{bc}}.$$

Q_1 is *Yule's coefficient of association* and Q_2 the *coefficient of colligation* (*see* ASSOCIATION, MEASURES OF). These papers generated a highly acrimonious debate, the details of which can be gleaned from Pearson and

Heron [21] and Heron [16]. Yule held the opinion that discrete quantities such as Q_1 and Q_2 are appropriate for the analysis of fourfold tables. Pearson and Heron maintained that even in seemingly discrete cases, the normal distribution or perhaps some other continuous model should be employed as the underlying model for the data. In Yule's analysis of the effectiveness of smallpox inoculations [10], the two dichotomies are inoculated versus not-inoculated and surviving versus dead. Pearson had maintained that the inoculation classification was really continuous and reflected the degree of immunity, which depends on genetic factors, the length of elapsed time since the last inoculation, and so forth. Likewise, survival is continuous, since this reflects resistance, virulence of the attack of the disease, and so on. Another example cited by Pearson uses color of horse compared with color of sire. The classifications in both instances are into light and dark, which Pearson also regarded as a dichotomization of continuous variables. One can view this debate as a precursor of debates on the relative merits of parametric methods versus nonparametric methods.

Much of the use of the tetrachoric correlation coefficient as a descriptive and inferential measure has been made by psychologists, educational psychologists, and psychometricians. This is readily evidenced by the list of references.

References

[1] Brown, M. B. (1977). *Appl. Statist.*, **26**, 343–351.

[2] Brown, M. B. and Benedetti, J. K. (1977). *Psychometrika*, **42**, 347–356.

[3] Castellani, N. L., Jr. (1966). *Psychometrika*, **31**, 67–73.

[4] Chesire, L., Saffir, M., and Thurstone, L. L. (1933). *Computing Diagrams for the Tetrachoric Correlation Coefficient*. University of Chicago, Chicago, Ill.

[5] Davidoff, M. D. and Goheen, H. W. (1953). *Psychometrika*, **18**, 115–121.

[6] Dingby, P. G. N. (1983). *Biometrics*, **39**, 753–757.

[7] Dingvi, D. R. (1979). *Psychometrika*, **44**, 169–172.

[8] Everitt, P. F. (1910). *Biometrika*, **7**, 437–452.

[9] Goheen, H. W. and Kavruck, S. (1948). *Psychometrika*, **13**, 279–280.

[10] Greenwood, M., Jr. and Yule, G. U. (1915). *Proc. Roy. Soc. Medicine*, **8**(II), 113–194.

[11] Guilford, J. P. and Lyons, T. C. (1942). *Psychometrika* **7**, 243–249.

[12] Hamdan, M. A. (1970). *Biometrika*, **57**, 212–215.

[13] Hamilton, M. (1948). *Psychometrika*, **13**, 259–269.

[14] Hayes, S. P., Jr. (1943). *Psychometrika*, **8**, 193–203.

[15] Hayes, S. P., Jr. (1946). *Psychometrika*, **11**, 163–172.

[16] Heron, D. (1911). *Biometrika*, **8**, 101–123.

[17] Jenkins, W. L. (1955). *Psychometrika*, **20**, 253–258.

[18] Kirk, D. B. (1973). *Psychometrika*, **38**, 259–268.

[19] Pearson, K. (1901). *Philos. Trans. R. Soc. A*, **195**, 1–47. (A paper of historical significance on the analysis of fourfold tables.)

[20] Pearson, K. (1913). *Biometrika*, **9**, 22–27.

[21] Pearson, K. and Heron, D. (1913). *Biometrika*, **9**, 159–315. (This paper details the dispute between K. Pearson and G. Udny Yule on the analysis of contingency tables.)

[22] Slepian, D. (1982). *Bell System Tech. J.*, **41**, 463–501.

[23] Tallis, G. M. (1962). *Biometrics*, **18**, 342–353.

[24] Welsh, G. S. (1955). *Psychometrika*, **20**, 83–85.

[25] Yule, G. U. (1900). *Philos. Trans. R. Soc. A*, **194**, 257–319. (A historical paper providing the origins of the analysis of fourfold tables using measures of association.)

[26] Yule, G. U. (1912). *J. R. Statist. Soc.*, **75**, 579–652. (This paper gives Yule's side of the dispute with Pearson regarding the use of association measures as opposed to correlation coefficients.)

[27] Zalinski, J., Abrahams, N. M., and Alf, E., Jr. (1979). *Educ. Psych. Meas.*, **39**, 267–275.

(ASSOCIATION, MEASURES OF
CONTINGENCY TABLES
POLYCHORIC AND POLYSERIAL
   CORRELATION
TWO-BY-TWO TABLES)

BERNARD HARRIS

TETRAD (TETRAD DIFFERENCE)

The *tetrad* (or *tetrad difference*) for the ordered set of four variables X_1, X_2, X_3, X_4

is

$$\tau_{1234} = \rho_{12}\rho_{34} - \rho_{13}\rho_{24},$$

where ρ_{ij} is the coefficient of correlation* between X_i and X_j. Note that $\tau_{1234} = \tau_{2143} = \tau_{2134} = \tau_{1243}$ and $\tau_{1324} = -\tau_{1234}$ while τ_{1423} ($= \rho_{14}\rho_{23} - \rho_{12}\rho_{34}$) has no simple relationship to the previous τ's.

The term is sometimes used in a broader sense with the ρ's representing frequencies in four adjacent cells of a contingency table*.

(TETRACHORIC CORRELATION
  COEFFICIENT)

TETRANOMIAL DISTRIBUTION

A multinomial distribution* with four cells, also called a *quadrinomial* distribution. Although the distribution is conveniently expressed as

$$\Pr\left[\bigcap_{j=1}^{4} (N_j = n_j)\right] = \frac{N!}{\prod_{j=1}^{4} n_j!} \prod_{j=1}^{4} p_j^{n_j}$$

$$\left(\sum_{j=1}^{4} n_j = N; \sum_{j=1}^{4} p_j = 1\right),$$

it is really a *trivariate* distribution because of the condition on the n's.

(MULTINOMIAL DISTRIBUTIONS)

THE PROFESSIONAL STATISTICIAN
See THE STATISTICIAN

THEIL TEST AND ESTIMATOR OF SLOPE

In a series of papers, Theil [11] proposed an estimator of the slope for a simple linear regression* setting, and inference procedures based on that estimator. Let $x_1 < x_2 < \cdots < x_n$ denote n values of an independent variable. Corresponding to each x_i, a dependent variable Y_i is also observed, which is related to x_i via

$$Y_i = \alpha + \beta x_i + E_i.$$

The parameters α and β are unknown, while $E_1, \ldots, E_n$ are assumed to be independent errors, each with the same continuous distribution.

To test H_0: $\beta = \beta_0$, we let B denote the number of pairs $1 \leqslant i < j \leqslant n$ for which

$$(Y_j - \beta_0 x_j) - (Y_i - \beta_0 x_i) > 0.$$

The test statistic is then

$$K = 2B - n.$$

For detecting H_a: $\beta > \beta_0$ ($\beta < \beta_0$ or $\beta \neq \beta_0$), we reject H_0 for $K \geqslant k_\alpha$ ($K \leqslant -k_\alpha$ or $|K| \geqslant k_{\alpha/2}$), where k_α denotes the upper αth quantile* of the null distribution of K. Under H_0, $K/\binom{n}{2}$ has the same null distribution as Kendall's tau* and thus tables are readily available. See, for example, Table A.21 in Hollander and Wolfe [3]. When n is large the test is based on the fact that

$$\frac{K}{\{n(n-1)(2n+5)/18\}^{1/2}}$$

has an approximate standard normal distribution under H_0.

The confidence interval for β that corresponds to the two-sided version of this test, is constructed by forming the slope of the line connecting (x_i, Y_i) to (x_j, Y_j), namely,

$$\beta_{ij} = (Y_i - Y_j)/(x_i - x_j)$$

for each of the $N \equiv \binom{n}{2}$ pairs of $1 \leqslant i < j < n$. Let $\beta_{(1)} < \beta_{(2)} < \cdots < \beta_{(N)}$ denote the ordered values of these slopes. A $100(1 - \alpha)\%$ confidence interval for β takes the form

$$[\beta_{(m)}, \beta_{(N+1-m)}],$$

where m is the integer satisfying $2m = N - k$ with

$$P[-k \leqslant K \leqslant k | H_0] = 1 - \alpha.$$

The value k may be obtained from the tabulated null distribution of K or approximated by

$$(z_{\alpha/2})[n(n-1)(2n+5)/18]^{1/2}$$

when n is large, where $z_{\alpha/2}$ denotes the upper $\alpha/2$ quantile of a standard normal distribution. The point estimate of β is

$$\hat{\beta} = \underset{1 \leqslant i < j \leqslant n}{\text{median}} (\beta_{ij}).$$

Table 1

x_i	1	2	3	4	10	12	18
y_i	9	15	19	20	45	55	78

Table 2

1	2.5	2.88	3.67	3.71	3.75	3.88
3.93	3.94	4	4	4	4	4.06
4.14	4.18	4.25	4.75	5	5	6

Example. Data used by Sen [9] relate the time (x_i) to the distance travelled (y_i) of an object; see Table 1. The $N = \binom{n}{2}$ slopes computed from the pairs of points (arranged in ascending order) are shown in Table 2. Therefore, the point estimate of β is $\hat{\beta} = 4$. From the null distribution of K, we see that

$$P[-11 \leqslant K \leqslant 11] = 0.93.$$

Thus, $m = 5$ and a 93% confidence interval for β is [3.71, 4.25].

Sen [9] described the asymptotic properties of the test and estimators including asymptotic efficiencies. He also discussed the important extension to settings in which the independent variable values are not unique. Beran [1] proved an optimal property for the hypothesis test proposed by Theil. Jaeckel [5] investigated point estimators which are weighted medians of the β_{ij} slopes. Additional properties of these weighted median estimators and the corresponding rank tests were investigated by Scholz [8] and Sievers [10].

Other variations on the Theil procedures include inferences on the intercept α (Maritz [6]), truncation* on Y values (Bhattacharya et al. [2]), and estimators requiring the computation of fewer pairwise slopes (Theil [11], Hussain and Sprent [4], and Markowski [7]).

References

[1] Beran, R. J. (1971). *Ann. Math. Statist.*, **42**, 157–168. (Shows an optimal property of Theil test.)

[2] Bhattacharya, P. K., Chernoff, H., and Yang, S. S. (1983). *Ann. Statist.*, **11**, 505–514. (Extends the procedures to cases in which the Y values are truncated and cannot be observed if they exceed y_0.)

[3] Hollander, M. and Wolfe, D. A. (1973). *Nonparametric Statistical Methods*. Wiley, New York. (Introductory text containing descriptions and tables for the Theil procedures, as well as references to related work.)

[4] Hussain, S. S. and Sprent, P. (1983). **146**, 182–191. (Compares the performance of Theil estimator to reduced computation versions.)

[5] Jaeckel, L. A. (1972). *Ann. Math. Statist.*, **43**, 1449–1458. (Investigates weighted medians of the pairwise slopes.)

[6] Maritz, J. S. (1979). *Aust. J. Statist.*, **21**, 30–35. (Performs inferences on the intercept in the fashion of Theil's on the slope.)

[7] Markowski, E. P. (1984). *Biometrika*, **71**, 51–56. (Studies a reduced computation version which uses the pairwise slopes over pairs $1 \leqslant i < j \leqslant n$ for which $j - i \geqslant k_n > 1$.)

[8] Scholz, F.-W. (1978). *Ann. Statist.*, **6**, 603–609. (Studies the asymptotic properties of weighted medians of pairwise slopes.)

[9] Sen, P. K. (1968). *J. Amer. Statist. Ass.*, **63**, 1379–1389. (Develops the asymptotic properties of the Theil procedures and extends them to cases with nonunique x values.)

[10] Sievers, G. L. (1978). *J. Amer. Statist. Ass.*, **73**, 628–631. (Develops weighted rank statistics for testing and the corresponding estimators.)

[11] Theil, H. (1950). *Proc. Kon. Ned. Akad. Wet. A*, **53**, 386–392, 521–525, 1397–1412. (Tests for slope confidence intervals and point estimates for regression parameters. Includes multiple independent variable and reduced computation estimators.)

(KENDALL'S TAU
REGRESSION (various entries))

RONALD H. RANDLES

THEORY OF PROBABILITY AND ITS APPLICATIONS

This is an English translation of the Russian journal *Teoriya Veroyatnostei i ee Primeneniya* (Academy of Sciences of USSR, Moscow). It is published by the Society of Industrial and Applied Mathematics (SIAM), 1400 Architects Building, 117 South 17th Street, Philadelphia, PA 19103-5052, and Winterstoke Road, Bristol BS3 2NT, England.

Translations appear about 9 months to a year after original publication. Volume 31 of the Russian journal appeared in 1986.

THEORY OF PROBABILITY AND MATHEMATICAL STATISTICS

This is an English translation of the Russian journal *Teoriya Veroyatnostei i Matematicheskaya Statistika* (Kiev University, USSR). It is published by the American Mathematical Society, P. O. Box 1571, Annex Station, Providence, RI 02901-9930.

Translations appear about a year after original publication. Number 34 of the Russian journal appeared in 1986.

THIELE'S INTERPOLATION FORMULA
See RECIPROCAL DIFFERENCES

THIRD KIND, ERROR OF THE   *See*
CONSULTING, STATISTICAL

THOMAN–BAIN TESTS

Let $x_1, x_2, \ldots, x_m$ and $y_1, y_2, \ldots, y_n$ be two independent samples taken from Weibull* populations with cumulative distribution functions (CDFs)

$$F_1(x) = 1 - \exp\left[-(x/b_1)^{c_1}\right] \quad (1)$$

Table 1   Percentage Points t_p such that $P(T_1 \leqslant t_p \mid H_0) = p$, $m = n$

$n \backslash p$	0.90	0.95	0.98
5	2.152	2.725	3.550
10	1.655	1.897	2.213
15	1.485	1.654	1.870
20	1.396	1.534	1.708
30	1.304	1.409	1.541
40	1.255	1.342	1.453
50	1.224	1.299	1.396
60	1.203	1.268	1.355
80	1.174	1.227	1.301
100	1.155	1.199	1.266

and

$$F_2(x) = 1 - \exp\left[-(x/b_2)^{c_2}\right], \quad (2)$$

respectively. Two hypothesis testing* problems are of interest: (i) testing H_0: $c_1 = c_2$ against H_1: $c_1 < c_2$ and (ii) testing H_0': $b_1 = b_2$ and $c_1 = c_2$ against H_1': $b_1 < b_2$ and $c_1 = c_2$. Thoman and Bain [8] consider tests based on maximum likelihood* estimators (MLEs) of parameters from each sample. The MLEs $\hat{b}_1$ and $\hat{c}_1$ of b_1 and c_1, respectively, based on the first sample, are solutions to the equations:

$$m/\hat{c}_1 - m\left(\sum_{i=1}^{m} x_i^{\hat{c}_1}\ln x_i\right)\Big/\left(\sum_{i=1}^{m} x_i^{\hat{c}_1}\right)$$

$$+ \sum_{i=1}^{m} \ln x_i = 0, \quad (3)$$

$$\hat{b}_1 = \left(\sum_{i=1}^{m} x_i^{\hat{c}_1}/m\right)^{1/\hat{c}_1}. \quad (4)$$

Equation (3) can be solved iteratively using Newton–Raphson's method*. An estimate of c_1 from a Weibull probability plot* (see, e.g., Mann et al. [4]) can be selected as an initial value for solving (3). Here $\hat{c}_1$ is the positive root of eq. (3). Similarly, one obtains the MLEs $\hat{b}_2$ and $\hat{c}_2$ of b_2 and c_2, respectively, based on the second sample.

Thoman and Bain [8] suggest using the statistic $T_1 = \hat{c}_2/\hat{c}_1$ to test H_0, rejecting H_0 if T_1 is too great. The distribution of T_1 under H_0 is parameter free, but is intractable. Using Monte Carlo methods*, Thoman and Bain give percentage points of T_1 under H_0, for $m = n = 5(1)20$, 22(2)80, 90, 100, and 120. Some of these are reproduced in Table 1.

For the second problem, they suggest using the statistic

$$T_2 = \left[(\hat{c}_1 + \hat{c}_2)/2\right]\left(\ln \hat{b}_2 - \ln \hat{b}_1\right),$$

rejecting H_0' if T_2 is too great. The distribution of T_2 under H_0' is also parameter free, but its exact form is also intractable. Again using Monte Carlo methods, Thoman and Bain give percentage points of T_2 under H_0', for $m = n = 5(1)20$, 22(2)80, 90, 100, and 120; some are reproduced in Table 2.

Table 2   **Percentage Points** t_p **such that** $P(T_2 \leqslant t_p \mid H_0') = p$, $m = n$

$n \backslash p$	0.90	0.95	0.98
5	1.226	1.670	2.242
10	0.704	0.918	1.195
15	0.544	0.704	0.904
20	0.459	0.593	0.755
30	0.366	0.472	0.595
40	0.314	0.404	0.507
50	0.279	0.360	0.450
60	0.254	0.328	0.408
80	0.219	0.284	0.352
100	0.196	0.255	0.315

Table 3

Tester I	93.4	98.7	116.6	117.8	132.7
	136.6	140.3	158.0	164.8	183.9
Tester II	152.7	172.0	172.5	173.3	193.0
	204.7	216.5	234.9	262.6	422.6

Example.   Consider data in Table 3 of Mc-Cool [5] (see, also [1]) on fatigue lives (in hours) for 10 bearings in each of two testers. Assuming these to be independent samples from Weibull CDFs (1) and (2), respectively, we obtain $\hat{c}_1 = 5.38$, $\hat{b}_1 = 145.57$ for tester I and $\hat{c}_2 = 2.94$, $\hat{b}_2 = 246.41$ for tester II. To test $c_1 = c_2$ against $c_1 > c_2$, we compute $\hat{c}_1/\hat{c}_2 = 5.38/2.94 = 1.83$, which is less than 1.897, the 95th percentage point of T_1 under H_0 for $m = n = 10$ (Table 1). The test does not reject the hypothesis $c_1 = c_2$ at the 5% level. To test H_0' against H_1', we compute $T_2 = [(5.38 + 2.94)/2](\ln 246.41 - \ln 145.57) = 2.19$, which is greater than 1.195, the 98th percentage point of T_2 under H_0' for $m = n = 10$ (Table 2). The test rejects H_0' at the 2% level.

These test procedures can also be applied to Type II censored data*. See McCool [5, 6] for percentage points. McCool [6], Lawless and Mann [3], and others have suggested approximating the distributions of $\hat{b}_1/b_1$ by chi-square, which is applied to approximate T_1 by an F distribution*. See also Lawless [2, pp. 180–182]. For problem (ii), Schafer

and Sheffield [7] suggest a statistic similar to T_2, but the average of $\hat{c}_1$ and $\hat{c}_2$ in the expression of T_2 is replaced by the MLE of the common shape parameter c ($= c_1 = c_2$) obtained from the combined sample. This leads to an improvement, in terms of power, over the test based on T_2. Likelihood ratio tests* for the equality of three or more Weibull shape parameters (or scale parameters) are developed, for example, in Lawless [2, pp. 182–188].

References

[1] Engelhardt, M. and Bain, L. J. (1979). *Technometrics*, **21**, 233–237. (Prediction limits, tests for comparing the shape and scale parameters of Weibull distributions, based on order statistics.)

[2] Lawless, J. F. (1982). *Statistical Models and Methods for Lifetime Data*. Wiley, New York. (Broad coverage, including a lot of references of recent works.)

[3] Lawless, J. F. and Mann, N. R. (1976). *Commun. Statist. A*, **5**, 389–405. (Tests for homogeneity for extreme value scale parameters.)

[4] Mann, N. R., Schafer, R. E., and Singpurwalla, N. D. (1974). *Methods for Statistical Analysis of Reliability and Life Data*. Wiley, New York. (An excellent practical reference book.)

[5] McCool, J. I. (1974). Inferential techniques for Weibull distributions. *Aerospace Research Laboratories Report*, ARL TR *74-0180*, Wright-Patterson AFB, Ohio.

[6] McCool, J. I. (1975). Inferential techniques for Weibull populations II. *Aerospace Research Laboratories Report ARL TR 75-0233*, Wright–Patterson AFB, Ohio.

[7] Schafer, R. E. and Sheffield, T. S. (1976). *Technometrics*, **18**, 231–235.

[8] Thoman, D. R. and Bain, L. J. (1969). *Technometrics*, **11**, 805–815.

(WEIBULL DISTRIBUTION)

H. K. Hsieh

THOMAS DISTRIBUTION

Thomas [2] constructed a model for the distribution of the number of plants of a given species in randomly placed quadrats. The probability of x plants in any given quadrat is given by

$$P(X = 0) = e^{-\lambda},$$

$$P(X = x) = \frac{e^{-\lambda}}{x!} \sum_{k=1}^{x} \binom{x}{k} \lambda^k (k\phi)^{x-k-k\phi}$$

$$(x \geq 1).$$

This is also called a *double Poisson distribution*. It is both a *compound* (mixture) and a generalized distribution (see, e.g., Johnson and Kotz [1]) and is similar to the Neyman Type A distribution.

References

[1] Johnson, N. L. and Kotz, S. (1969). *Distributions in Statistics; Discrete Distributions*. Wiley, New York.

[2] Thomas, M. (1949). *Biometrika*, **36**, 18–25.

(NEYMAN'S TYPE A, B, AND C
  DISTRIBUTIONS
QUADRAT SAMPLING)

THOMAS–FIERING MODEL *See* HY-
DROLOGY, STOCHASTIC

THOMPSON'S CRITERION

A criterion for detection of outliers* in random samples from normal populations was proposed by Thompson [2]. It is based on the values of the sample standardized deviations

$$D_i = (X_i - \overline{X})/S \qquad (i = 1, 2, \ldots, n),$$

where $X_1, X_2, \ldots, X_n$ represent the sample values,

$$\overline{X} = n^{-1}\Sigma_{i=1}^{n} X_i,$$

and

$$S^2 = (n-1)^{-1}\Sigma_{i=1}^{n}(X_i - \overline{X})^2.$$

Observations for which the absolute value of this quantity exceeds a specified value are regarded as outliers.

The sampling distribution of D_i is the same as that of $N^{-1/2}(N-1) \times$ [correlation

coefficient* from random sample of size n of two independent normal populations] or, equivalently,

$$\frac{D_i \sqrt{N(N-2)}}{\sqrt{\{(N-1)^2 - ND_i^2\}}}$$

has a Student's t distribution* with $(N-2)$ degrees of freedom.

Pearson and Chandra Sekar [1] point out that the use of this criterion can be especially misleading when more than one outlier is present. For further details see OUTLIERS.

References

[1] Pearson, E. S. and Chandra Sekar, C. (1935). *Biometrika*, **28**, 308–320.

[2] Thompson, W. R. (1935). *Ann. Math. Statist.*, **6**, 214–219.

(CHAUVENET'S CRETERION
OUTLIERS
PEIRCE'S CRITERION)

THREE-EIGHTHS RULE

This is the following quadrature* formula, using values of the integrand at four equally spaced values of the variable:

$$\int_{a}^{a+3h} f(x)\, dx$$

$$\doteq \frac{3h}{8}[f(a) + f(a + 3h)$$

$$+ 3\{f(a + h) + f(a + 2h)\}].$$

The formula is exact if $f(x)$ is a polynomial of degree 3 or less. If the fourth derivative, $f^{(4)}(x)$, of $f(x)$ exists and is continuous, the remainder term is

$$-1.6 \times 10^{-4}(3h)^5 f^{(4)}(\xi)$$

$$= -0.03888h^5 f^{(4)}(\xi)$$

for some ξ, $a \leq \xi \leq a + 3h$.

The name *three-eighths rule* is also given to the formula

$$\int_a^{a+3h} f(x)\, dx$$

$$\doteq \frac{3h}{8}\left[3\left\{f\left(a+\frac{1}{2}h\right)+f\left(a+\frac{5}{2}h\right)\right\}\right.$$

$$\left.+2f\left(a+\frac{3}{2}h\right)\right].$$

The remainder term is

$$1.4\times 10^{-4}(3h)^5 f^{(4)}(\xi) = 0.03402 h^5 f^{(4)}(\xi).$$

(NUMERICAL INTEGRATION
SIMPSON'S RULE
TRAPEZOIDAL RULE
WEDDLE'S RULE)

THREE-MODE ANALYSIS

Three-mode analysis refers to a collection of multivariate techniques to analyse data which can be classified in three ways, for instance by subjects, variables, and experimental conditions. Extensions of principal component analysis* and factor analysis* to three-mode data fall into this class. Three-mode analysis is closely linked to multidimensional scaling* of three-way data, not discussed here (see refs. 2, 6, and 8).

OVERVIEW

Three-mode analysis had its origin in psychometrics (*see* STATISTICS IN PSYCHOLOGY). It has been developed largely by Ledyard Tucker at the University of Illinois, culminating in his papers on its mathematical foundations [16] and on its relation to multidimensional scaling* [17]. Later contributors are Bentler and Lee [1], Harshman [5, 6], Kroonenberg and De Leeuw [7–9], McDonald [13], and Sands and Young [14]. Three-mode analysis arose out of a need within various fields in psychology to analyse both multivariate data collected on different occasions and multistimulus-multitrait data

Table 1   *m* Variables Observed on a Sample of *n* Individuals at *p* Occasions

Individuals	Variables			
	V_1	V_2	$\cdots$	V_m
Occasion O_1				
I_1	x_{111}	x_{121}	$\cdots$	x_{1m1}
I_2	x_{211}	x_{221}	$\cdots$	x_{2m1}
I_n	x_{n11}	x_{n21}	$\cdots$	x_{nm1}
Occasion O_p				
I_1	x_{11p}	x_{12p}	$\cdots$	x_{1mp}
I_2	x_{21p}	x_{22p}	$\cdots$	x_{2mp}
I_n	x_{n1p}	x_{n2p}	$\cdots$	x_{nmp}

collected from several individuals. The latter necessitated methods to analyse simultaneously the structure of the stimuli, attributes, and of differences between individuals, as well as their interrelationships or interactions.

Three-Mode Data: Models and Analysis

The essential nature of three-mode data is illustrated in Table 1, with m variables observed on each of n individuals on p occasions. A data point x_{ijk} indicates the value given to variable j by individual i on occasion k. The values may be arranged in many ways, for instance, as p matrices $\mathbf{X}_k$, as a three-way array $\mathbf{X}$ of dimensions $n \times m \times p$, or, neglecting the three-way structure, as an array of order $(n \times p) \times m$. For stochastic models one generally refers to the column vector of observations $\mathbf{x} = (x_{11}, x_{21}, \ldots, x_{mp})'$.

Two main types of models have been developed for three-mode data, both of which can be traced back to Tucker [15]. One extends principal component models to three modes, and all modes including that of the observational units are treated nonstochastically. Therefore, it belongs primarily to the realm of data analysis, rather than statistics; the data are treated as coming from a population rather than a sample. The other type of model is a direct extension of factor analysis, which falls within the framework of the theory of covariance structures. In such models, the variables at each occasion are treated stochastically, and it is not the raw data, but the covariance matrix that is modeled.

The general aim of three-mode analysis is to fit a model to the data with a low-dimensional representation so that the basic underlying structure can be more readily discerned and interpreted. This is achieved by computing (principal) components or factors for two or three modes and, depending on the model, a core matrix. The *core matrix* is a three-way array **G**, which contains information on the relations between the components of the modes (see the following text). The parameters are estimated by minimizing a loss function using generalized least-squares methods (factor analysis models) or alternating least-squares methods (component models).

COMPONENT MODELS

Component models are descriptive, exploratory, and nonstochastic. The disregard for stochastic variation leads to a formulation couched entirely in terms of linear algebra. The models proposed are presented as decompositions of data matrices or as lower-rank approximations to such decompositions.

In the two-mode case one generally seeks a lower-rank approximation ($\nu \leqslant m$) to the decomposition using a least-squares loss function

$$\|\mathbf{X} - \mathbf{AB}'\|^2, \qquad (1)$$

where $\| \cdot \|$ indicates the Euclidean norm, **X** is the data matrix (n persons $\times$ m variables), **A** (n persons $\times$ ν components) representing the weights (or "scores") of the subjects on the components, and **B** (m variables $\times$ ν components) representing the weights (or "loadings") of the variables on the components. Instead of (1), the model equation can be defined as

$$x_{ij} = \sum_{\alpha=1}^{\nu} a_{i\alpha} b_{j\alpha} + e_{ij}$$
$$(i = 1, \ldots, n; \; j = 1, \ldots, m), \quad (2)$$

where the e_{ij} are the residuals or deviations from the model.

Harshman's Model [5]

A direct extension to three-mode data can be made from (2) when the same variables have been observed on the same persons on several occasions, by requiring that the structure in (2) remains the same over all occasions, but that the importance or weights of entire components may be different on each occasion of measurement:

$$x_{ijk} = \sum_{\alpha=1}^{\nu} a_{i\alpha} b_{j\alpha} c_{k\alpha} + e_{ijk}, \qquad (3)$$

where $c_{k\alpha}$ represents the weights of occasion k for component α. The definition of the model implies that it is not the relative positions of variables (or subjects) on a component which are assumed to vary, but only the relative importance of the component itself for each occasion. In practice, this basic assumption of "parallel profiles" must be true or acceptable, given the data and research design, before the model can be applied.

The roles of the persons, variables, and occasions in the model are symmetric, and the preceding explanation could have started with variables as well. One attractive feature of the model is that under mild regularity conditions on the data, it is identified up to permutations of the order of the components and multiplicative factors per component. In particular, the orientation of the components, which are not necessarily orthogonal, cannot be changed without changing the fit of the model, i.e., the minimum value of the loss function associated with (3). This uniqueness-of-orientation property is not present in regular principal component analysis and exploratory factor analysis.

Within model (3) all modes have weights on the same components, and one single interpretation of these components suffices, usually based on the relative weights of the variables on those components. The subject and occasion weights are interpreted accordingly.

Tucker's Model [16]

A different extension of principal component analysis leads to a less restricted three-mode model, which at the same time requires a more complex interpretation. The starting point will first be explained for two-mode data. In (1) the data matrix was decomposed into weights for persons and for variables on the same components. It is now assumed that variables and persons have their own set of components, and furthermore, that it is possible to estimate the strength of the relation between the components of each set. This leads to a loss function of the form

$$\|\mathbf{X} - \mathbf{AGB}'\|^2, \qquad (4)$$

in which $\mathbf{A}$ (n persons $\times$ ν components) is the set of components for the persons, $\mathbf{B}$ (m variables $\times$ ν components) the set of components for the variables, and $\mathbf{G}$ the matrix which contains the measures for indicating the direction and the strength of the relations between the components of each set. The simplest solution of (4), which occurs when $\mathbf{A}$ and $\mathbf{B}$ are restricted to be column-wise orthonormal, is essentially unique up to permutation of the columns of $\mathbf{A}$ and $\mathbf{B}$. $\mathbf{A}$ is the eigenvector matrix of $\mathbf{XX}'$, $\mathbf{B}$ the eigenvector matrix of $\mathbf{X}'\mathbf{X}$, and $\mathbf{G}$ a diagonal matrix with the square roots of the eigenvalues of $\mathbf{X}'\mathbf{X}$ and $\mathbf{XX}'$, also called *singular values* of $\mathbf{X}$. The corresponding model equation of (4) is

$$x_{ij} = \sum_{\alpha=1}^{\nu} a_{i\alpha}b_{j\alpha}g_{\alpha\alpha} + e_{ij}. \qquad (5)$$

The diagonality of $\mathbf{G}$ implies that the αth component of $\mathbf{A}$ is exclusively linked to the αth component of $\mathbf{B}$, i.e., there is no relation between the αth component of $\mathbf{A}$ and the α'th component of $\mathbf{B}$. Thus the αth components of $\mathbf{A}$ and $\mathbf{B}$ may be given the same

interpretation and may be equated. Even though for two-mode data the two conceptualizations of components (2) and (5) are thus equivalent, for three-mode data the two conceptualizations lead to different models.

In the case of three-mode data, three separate sets of components may be defined, not only for subjects ($\mathbf{A}$) and variables ($\mathbf{B}$) as before, but also for occasions ($\mathbf{C}$). Since each of the sets of components partitions the variation in the data set, the components of the sets must be related. Therefore, the three-mode model must also specify the extent of these relations, which leads to the model equation

$$x_{ijk} = \sum_{\alpha=1}^{\nu} \sum_{\beta=1}^{\mu} \sum_{\gamma=1}^{\pi} a_{i\alpha}b_{j\beta}c_{k\gamma}g_{\alpha\beta\gamma} + e_{ijk}, \qquad (6)$$

in which $g_{\alpha\beta\gamma}$ contains the information on the relation between the αth component of the subjects, the βth component of the variables, and the γth component of the occasions. The $g_{\alpha\beta\gamma}$ may be collected in a three-way array $\mathbf{G}$, called the *core matrix* because it is assumed to contain the basic relationships in the data. When $\nu = \mu = \pi$ and $g_{\alpha\beta\gamma} = 0$ unless $\alpha = \beta = \gamma$, (6) reduces to (3), provided the $g_{\alpha\alpha\alpha}$ are absorbed in some way in the $a_{i\alpha}$, $b_{j\alpha}$, and/or $c_{k\alpha}$.

The loss function corresponding to (6) has the form

$$\|\mathbf{X} - \mathbf{AG}(\mathbf{B}' \otimes \mathbf{C}')\|^2, \qquad (7)$$

where the three-way data array $\mathbf{X}$ is an $n \times mp$ matrix, the three-way core matrix $\mathbf{G}$ is $\nu \times \mu\pi$, and ($\mathbf{B} \otimes \mathbf{C}$) refers to the right Kronecker product of matrices ($\mathbf{B} \otimes \mathbf{C}$) = $[b_{j\beta}\mathbf{C}]$.

To explore the relation between components from the various sets, we return again to the two-mode case. The squared $g_{\alpha\alpha}$'s of $\mathbf{G}$ in (5) express the amount of variance accounted for by the αth components of $\mathbf{A}$ and $\mathbf{B}$. If $\mathbf{A}$, $\mathbf{B}$, and $\mathbf{C}$ are columnwise orthonormal, exactly the same interpretation holds for $g_{\alpha\beta\gamma}^2$, i.e., $g_{\alpha\beta\gamma}^2$ is the amount of variance (or sums of squares) jointly accounted for by the αth, βth, and γth component of **A**, **B**, and **C**, respectively.

Model (6), too, just as (3), is symmetric in its components. In practice this symmetry is more apparent than real, as one of the modes is generally the source of the data (e.g., persons), while the others are not. Various asymmetric models have been proposed (see refs. 2, 6, and 8).

FACTOR ANALYSIS MODELS (BENTLER AND LEE [1])

Component models are deterministic, while factor analysis models are probabilistic; it is assumed that m random variables are each observed at p occasions, $\mathbf{x} = (x_{11}, x_{21}, \ldots, x_{mp})$. A rather general three-mode model for $\mathbf{x}$ can be formulated as

$$x_{jk} = \mu_{jk} + \sum_{\alpha=1}^{\nu} \sum_{\beta=1}^{\mu} \sum_{\gamma=1}^{\pi} b_{j\beta} c_{k\gamma} g_{\alpha\beta\gamma} y_{\alpha} + \epsilon_{jk},$$
(8)

with x_{jk} the j, kth element of the vector $\mathbf{x}$ of observations, μ_{jk} the corresponding elements of the vector $\boldsymbol{\mu}$ of population means, and ϵ_{jk} the j, kth element of the random vector $\boldsymbol{\epsilon}$ of residual variates representing sources of variation affecting only x_{jk}. In the model part of (8), $b_{j\beta}$ and $c_{k\gamma}$ are again the weights for the components of variables and occasions, respectively, the $g_{\alpha\beta\gamma}$ are the elements of the core matrix, and y_{α} is the αth element of $\mathbf{y} = (y_1, \ldots, y_{\nu})$, the random vector of subject components (or "scores"). The matrix formulation is

$$\mathbf{x} = \boldsymbol{\mu} + (\mathbf{B} \otimes \mathbf{C})\mathbf{G}\mathbf{y} + \boldsymbol{\epsilon},$$
(9)

in which $\mathbf{G}$ is the core matrix of order ($\mu\pi \times \nu$). The estimation of the parameters in (9) is solved via the covariance matrix of $\mathbf{x} - \boldsymbol{\mu}$,

$$\boldsymbol{\Sigma}_x = (\mathbf{B}' \otimes \mathbf{C}')\mathbf{G}'\boldsymbol{\Sigma}_y\mathbf{G}(\mathbf{B} \otimes \mathbf{C}) + \boldsymbol{\Sigma}_{\epsilon}, \quad (10)$$

in which $\boldsymbol{\Sigma}_y$ is the covariance matrix of y and $\boldsymbol{\Sigma}_{\epsilon}$ is the covariance matrix of ϵ, assumed to be diagonal. In going from (9) to (10), use has been made of the (standard) assumptions about the statistical independence of **y** and $\boldsymbol{\epsilon}$ and the mutual independence of the residual variates.

APPLICATIONS

Most published applications have come from psychology, using model (6) or some variant; recently, more applications of model (3) have appeared, due to the wider availability of adequate computer programs. Applications of (8) and its variants are virtually restricted to those in the original papers. We summarize two studies which have benefitted from a three-mode analysis.

Tongue Shapes (Harshman et al. [5])

Systematic patterns of the relationships between tongue shapes for producing different vowels have been observed both within and between speakers. A problem however, is, to represent the curved surface of the tongue so that these relationships can be easily identified and measured. In this study tracings were made of X-rays taken during the pronunciation of 10 English vowels by five speakers. The positions of the tongue in these 50 vowels were quantified in terms of 13 superimposed grid lines. Model (3) was used to analyse the 10(vowels) × 5(speakers) × 13(tongue positions) three-way array. The analysis showed that the data could be described by two components common to all modes. One component generates a forward movement of the root of the tongue accompanied by an upward movement of the front of the tongue (position mode). Movement from front to back vowels involves decreasing amounts of this component (vowel mode). Different speakers use the two components in different degrees, which may be associated with their individual anatomy (speaker mode).

Job Classification (Cornelius et al. [3])

When developing a performance appraisal system in an organization, one must decide how many different rating instruments are needed to provide useful administrative data. Essentially this is a question of identifying homogeneous groups of employees to be combined for evaluation on a single form. The amount of time spent on 153 worker-oriented job elements was averaged over individuals (US Coast Guard enlisted personnel) per job–rank combination. Model (6) was used to analyse the 153(job elements) $\times$ 18(jobs) $\times$ 5(ranks) three-way array employing the original non-least-squares methods of Tucker [16]. Two components were derived for the ranks model grouping the lower and higher ranks, respectively. Five components were deemed necessary for the jobs mode, separating such groups as aviation, service and clerical, electronics, engineering, and deck and watch. Seven components were retained for the job-elements model identifying groups of activities like machine-tending, managing, cooking, machine repair, clerical and contact with others, boating, and air crew tasks. The core matrix, which contains the interactions of the job element, rank, and job components, revealed that in different job groups lower-rank officers spend different amounts of time on the various job-element groups, while this was less so for higher-rank officers. Consequently, five different evaluation forms were developed for lower-rank officers, and only one for higher-rank officers.

Other Applications

The two examples differ in an important aspect other than subject matter. The first data set consists of profile data, i.e., persons (speakers) have scores on variables (tongue positions) under several conditions (vowels). In the second data set, the x_{ijk} are scores on one dependent variable, "time spent on a job element," and the data correspond to a $153 \times 18 \times 5$ fully-crossed factorial design* and could, in principle, have been tackled by analysis of variance*, given suitable research questions. Many other types of data have been analysed by three-mode models, such as multivariate time series, three-way similarity data, multiple correlation* or covariance matrices, multivariate growth curves*, three-way interactions from log-linear models or analyses of variance, multitrait-multimethod matrices*, etc.

FURTHER READING

The basic treatises on the theory of three-mode analysis are refs. 1, 4, 8, 9, 14, 16, and 17. Relatively elementary expositions can be found in chapters of refs. 10–12 and 15. Discussions of relationships with multidimensional scaling are contained in refs. 2 and 8, and in chapters of ref. 10. An annotated bibliography which includes both theoretical and virtually all published applications up to 1983 is ref. 7. The most comprehensive book is ref. 8; the widest perspective on multimode analysis is provided in ref. 10.

References

[1] Bentler, P. M. and Lee, S.-Y. (1979). *Brit. J. Math. Statist. Psychol.*, **32**, 87–104.

[2] Carroll, J. D. and Wish, M. (1974). In *Contemporary Developments in Mathematical Psychology*, Vol. 2, D. H. Krantz, R. C. Atkinson, R. D. Luce, and P. Suppes, eds. Freeman, San Francisco, pp. 57–105.

[3] Cornelius III, E. T., Hakel, M. D., and Sackett, P. R. (1979). *Personnel Psychol.*, **32**, 283–297.

[4] Harshman, R. A. (1970). *UCLA Working Papers in Phonetics*, **16**, 1–84. (Xerox University Microfilms, Ann Arbor, MI, report no. 10,085.)

[5] Harshman, R. A., Ladefoged, P., and Goldstein, L. (1977). *J. Acoust. Soc. Amer.*, **62**, 693–707.

[6] Harshman, R. A. and Lundy, M. E. (1984). In *Research Methods for Multi-Mode Data Analysis*, H. G. Law, C. W. Snyder, Jr., J. A. Hattie, and R. P. McDonald, eds. Praeger, New York, pp. 122–284.

[7] Kroonenberg, P. M. (1983a). *Brit. J. Math. Statist. Psychol.*, **36**, 81–113.

[8] Kroonenberg, P. M. (1983b). *Three-Mode Principal Component Analysis: Theory and Applications*. DSWO Press, Leiden, The Netherlands.

[9] Kroonenberg, P. M. and De Leeuw, J. (1980). *Psychometrika*, **45**, 69–97.

[10] Law, H. G., Snyder, Jr., C. W., Hattie, J. A., and McDonald, R. P., eds. (1984). *Research Methods for Multi-Mode Data Analysis*. Praeger, New York.

[11] Levin, J. (1965). *Psychol. Bull.*, **64**, 442–452.

[12] Lohmöller, J.-B. (1979). *Arch. Psychologie*, **131**, 137–166. (In Germany.)

[13] McDonald, R. P. (1984). In *Research Methods for Multi-Mode Data Analysis*, H. G. Law, C. W. Snyder, Jr., J. A. Hattie, and R. P. McDonald, eds. Praeger, New York, pp. 285–307.

[14] Sands, R. and Young, F. W. (1980). *Psychometrika*, **45**, 39–67.

[15] Tucker, L. R. (1965). In *Proceedings of the 1964 Invitational Conference on Testing Problems*. Educational Testing Service, Princeton, N.J. [Also in *Testing Problems in Perspective*, A. Anastasi, ed. American Council on Education, Washington, D.C. (1966).]

[16] Tucker, L. R. (1966). *Psychometrika*, **31**, 279–311.

[17] Tucker, L. R. (1972). *Psychometrika*, **37**, 3–27.

(COMPONENT ANALYSIS
FACTOR ANALYSIS
MULTIDIMENSIONAL SCALING
MULTIVARIATE ANALYSIS
PSYCHOLOGICAL SCALING
PSYCHOLOGICAL TESTING THEORY
STATISTICS IN PSYCHOLOGY)

PIETER M. KROONENBERG

THREE *R* (3*R*)   *See* RUNNING MEDIAN

THREE-SERIES THEOREM   *See* KOLMOGOROV'S THREE-SERIES THEOREM; STRONG LAW OF LARGE NUMBERS

THREE-SIGMA LIMITS   *See* CONTROL CHARTS

THREE-SIGMA (3σ) RULE

The 3σ *rule* states that for "most commonly encountered" random variables X, the inequality

$$\Pr\left[|X - E[X]| \geqslant 3\sigma(X)\right] \leqslant 0.05 \quad (1)$$

[where $\sigma(X)$ is the standard deviation of X] holds. Vysočanskiĭ and Petunin [2] have shown that this rule holds for *all* unimodal distributions. More precisely, for such distributions

$$\Pr\left[|X - E[X]| \geqslant 3\sigma(X)\right] \leqslant 4/81 \approx 0.049. \quad (2)$$

In fact, the Camp-Meidell inequality*

$$\Pr\left[|X - E[X]| \geqslant t\sigma(X)\right] \leqslant \tfrac{4}{9}t^{-2},$$

which is known to be true for *symmetric* unimodal distributions, is valid for *arbitrary* unimodal distributions, provided that

$$t \geqslant \sqrt{8/3} \approx 1.633.$$

A one-sided 3σ rule is valid for *all* unimodal distributions with finite variance [3]. Specifically, for any $\epsilon > 0$,

$$\max_{x \in v_\sigma} \Pr\left[X \geqslant E[X] + \epsilon\right]$$

$$= \begin{cases} \dfrac{3\sigma^2 - \epsilon^2}{3(\sigma^2 + \epsilon^2)} & \text{for } 3\epsilon^2 < 5\sigma^2, \\[2ex] \dfrac{4\sigma^2}{9(\sigma^2 + \epsilon^2)} & \text{for } 3\epsilon^2 > 5\sigma^2, \end{cases} \quad (3)$$

where v_σ is the set of all random variables having finite variance σ^2.

A simplified proof of (2) and (3) appears in ref. 1.

References

[1] Dharmadhikari, S. W. and Joag-Dev, K. (1985). *Theor. Prob. Appl.*, **30**, 867–871.

[2] Vysočanskiĭ, D. F. and Petunin, Yu. I. (1980). *Theor. Prob. Math. Statist.*, **21**, 25–36.

[3] Vysočanskiĭ, D. F. and Petunin, Yu. I. (1984). *Teor. Veroyat. Mat. Stat.*, **31**, 26–31.

(BERNSTEIN'S INEQUALITY
CAMP–MEIDELL INEQUALITY
CHEBYSHEV'S INEQUALITY
PROBABILITY INEQUALITIES FOR
  RANDOM VARIABLES
UNIMODALITY
ZELEN'S INEQUALITIES)

THRESHOLD PARAMETER

For continuous distributions, this term denotes a parameter separating regions wherein a probability density function (PDF) is positive from those where it is zero. For example, in the beta distribution* with PDF

$$f_x(x) = \{B(\theta, \phi)\}^{-1} \frac{(x-a)^{\theta-1}}{(b-a)}$$

$$\times \frac{(b-x)^{\phi-1}}{(b-a)} \cdot \frac{1}{b-a}$$

$$(a \leqslant x \leqslant b; \theta, \phi > 0),$$

the parameters a and b are threshold parameters and

$$B(\theta, \phi) = \int_0^1 y^{\theta-1}(1-y)^{\phi-1}\, dy$$

is a beta function*. For the lognormal distribution* of X, defined by

$$Y = \gamma + \delta \log(X - \xi) \qquad (X > \xi),$$

having a unit normal distribution, ξ is the threshold parameter.

(SUPPORT)

THROWBACK

A device for modifying printed differences in a table to produce improvement in interpolated values. Everett's central difference* formula to fourth central differences (accurate to fifth differences) is

$$f(x) \doteqdot xf(1) + (1-x)f(0)$$

$$-\epsilon_2(x)\delta^2 f(1) - \epsilon_2(1-x)\delta^2 f(0)$$

$$+\epsilon_4(x)\delta^4 f(1) + \epsilon_4(1-x)\delta^4 f(0),$$

where δ denotes central difference $[\delta f(x) = f(x + \frac{1}{2}) - f(x - \frac{1}{2})$, etc.] and

$$\epsilon_{2k}(x) = \{(2k+1)!\}^{-1} x(1-x^2)$$

$$\times (4-x^2)\cdots(k^2-x^2).$$

This can be written

$$f(x) \doteqdot xf(1) + (1-x)f(0)$$

$$-\epsilon_2(x)\left(\delta^2 - \frac{\epsilon_4(x)}{\epsilon_2(x)}\delta^4\right)f(1)$$

$$-\epsilon_2(1-x)\left(\delta^2 - \frac{\epsilon_4(1-x)}{\epsilon_2(1-x)}\delta^4\right)f(0).$$

Since $[\epsilon_4(x)]/[\epsilon_2(x)]$, i.e. $(4-x^2)/20$, lies

between 0.15 and 0.2 for x between 0 and 1, it follows that the formula

$$f(x) \doteqdot xf(1) + (1-x)f(0) - \epsilon_2(x)\delta'^2 f(1)$$

$$-\epsilon_2(1-x)\delta'^2 f(0),$$

with $\delta'^2 = \delta^2 - \theta\delta^4$ (modified second differences) for some "average" value θ between 0.15 and 0.2, will not differ much from the full fourth difference formula. Choice of θ is arbitrary; conventionally one assumes a uniform "distribution" for x, leading to

$$\theta = \frac{1}{20}\int_0^1(4 - x^2)\, dx = \frac{1}{5} - \frac{1}{60} = 0.183.$$

Some tables have modified differences $\delta'^2 = \delta^2 - 0.183\delta^4$ (or $\delta'^2 = \delta^2 - 0.184\delta^4$) printed. Higher-order differences can be modified in a similar way, but it is necessary to include further adjustments to allow for the effect of using modified lower-order differences.

Further details can be found, for example, in ref. 1.

Reference

[1] Fox, L. (1956). *Mathematical Tables* 2. National Physical Laboratory, H. M. Stationery Office, London, England.

(EVERETT'S CENTRAL DIFFERENCE
  FORMULA
FINITE DIFFERENCES, CALCULUS OF
INTERPOLATION)

THURSTONE'S THEORY OF COMPARATIVE JUDGMENT

In the late nineteenth and early twentieth century, psychophysicists were working on empirical discrimination experiments. The distinction between a physical continuum and a sensory continuum had been made and research efforts were aimed at understanding the magnitudes required on physical and sensory continua for subjects to perceive differences in stimuli. Stimuli of psychophysical interest included such objects as weight, pitch, color, and brightness. *See* PSYCHOPHYSICS, STATISTICAL METHODS IN.

In 1860, Fechner [3] proposed a logarithmic relationship between magnitudes on the physical and sensory dimensions. This proposal was based on earlier work by Weber [16] which suggested that the probability p_{ij} of a stimulus object, o_i, dominating another, o_j, was a function of the ratio of the physical magnitude of o_i over o_j. Fechner's work led to the principle by Fullerton and Cattell [4] that equally often noticed differences are equal. Letting μ_i be the magnitude of o_i on a sensory dimension, this principle can be restated as follows:

$$p_{ij} = p_{mn} \Rightarrow f(\mu_i - \mu_j) = f(\mu_m - \mu_n).$$

At Columbia in the early twentieth century, Thorndike [12] and others had been using pair comparison experiments to test hypotheses in discrimination. They hypothesized that the difference in values of pairs of stimuli along a psychological continuum $[(\mu_i - \mu_j) - (\mu_m - \mu_n)]$ was a function of $[\Phi^{-1}(p_{ij}) - \Phi^{-1}(p_{mn})]$, where Φ^{-1} is the inverse normal CDF. A basic assumption was that the random variables were independent and identically distributed.

Thurstone's [13] comparative judgment model, first published in 1927, was also based on pair comparisons. It assumed that when a subject is asked to express a dominance judgment about two stimuli, the outcome is determined by a comparison of the values for two random variables. If X_i is the random variable associated with stimulus object o_i and X_j is the random variable associated with stimulus object o_j, then the subject is said to judge o_i as having more of the attribute of interest than o_j if $X_i > X_j$.

Thurstone proposed several versions or cases of the comparative judgment model and additional extensions have been offered by others. The case I model, his most complete expression of the comparative judgment model, was as follows:

$$\mu_i - \mu_j = z_{ij}\left(\sigma_i^2 + \sigma_j^2 - 2r_{ij}\sigma_i\sigma_j\right)^{1/2},$$

where z_{ij} is the normal deviate corresponding to the proportion of times stimulus object o_i is judged to dominate o_j and r_{ij} is the correlation* between X_i and X_j.

Two other Thurstone versions, case III, where the covariances $r_{ij}\sigma_i\sigma_j$ are assumed equal, and case V, where the X_i are assumed to be independent identically distributed random variables, are commonly referenced.

Thurstone's model showed that equally often noticed differences need not be equal, and provided a mathematical foundation for the work being done at Columbia on pair comparison experiments. By allowing $\sigma_i \neq \sigma_j$, p_{ij} could equal p_{mn} yet $(\mu_i - \mu_j)$ need not equal $(\mu_m - \mu_n)$. By providing a mathematical foundation that did not depend upon the existence of a physical continuum, the domain of research in mathematical psychology was considerably expanded. Besides allowing what Thurstone later described as "more interesting" variables, a greater understanding of psychological measurement was obtained.

Though different, the Fechner and Thurstone models are in the same tradition. Both assume that the probability of confusing two stimuli is a function of their psychological separation or distance. From an estimation point of view, this is the same as saying that psychological scales (estimates of μ_i) are a function of subjects' inabilities to discriminate among stimuli. (Probabilities of 0 or 1 are, in fact, not allowed.) Critics of the indirect scaling method advocated by Thurstone suggest that this is comparable to saving the noise and throwing out the information.

The original comparative judgment model has, in turn, been generalized and altered by others. One object of change has been the form of the PDF. Thurstone was not dogmatic about the use of the normal distribution and considered the choice of an appropriate distribution an experimental issue. Of the various alternatives proposed, the two perhaps most worthy of noting are the lognormal* and double exponential distributions. If the lognormal is used, then $\sigma_i = k\mu_i$. This property is related to Weber's observation (it is not the same since the right-hand side deals with a psychological, not a physical stimulus) and has empirical support for a number of stimuli. For the double exponential distribution, Yellott [17] has shown that the case V model is equivalent to another popular psychological model, Luce's [6] choice axiom, under several experimental conditions (see LUCE'S CHOICE AXIOM AND GENERALIZATIONS).

The spirit of Thurstone's model has also been used in the development and evaluation of models in other areas of mathematical psychology. A good example would be in multidimensional scaling* where the object is to recover the multidimensional coordinates of a set of stimuli from estimates of the distances among the stimuli (see MULTIDIMENSIONAL SCALING). Hefner [5] showed that the comparative judgment model can be extended to a multidimensional space. If the projections of the stimuli on each dimension are normally distributed with the standard deviation of each stimulus i on dimension k (σ_{ik}) equal for all dimensions ($\sigma_{ik} = \sigma_i$), then the PDF of the distances is related to the noncentral chi-square distribution*.

APPLICATIONS

Thurstone's theory postulates a psychological process that represents stimulus objects by random variables defined over a unidimensional space. The dimension is an attribute of interest defined by the experimenter. Attributes may have corresponding physical dimensions—weight, brightness, loudness—or they may have no corresponding physical dimension—social values, aesthetics, utility. If utility is the attribute of interest, subjects are called upon to make affective dominance judgments or choices among the stimuli. In other situations, nonaffective dominance judgments are expressed.

A wide variety of stimuli have been evaluated with Thurstone's model. Due no doubt to its simplicity, the great majority of applications have involved the case V model, but interest in more complex cases is increasing. Stimuli evaluated in some recent studies include politicians [11], symptoms of sleeplessness [9], and public transportation alternatives [7].

One of the early applications by Thurstone [14] involved social values. His experiment considered 19 offenses (stimuli), ranging alphabetically from abortion to vagrancy. Subjects were shown all 171 pairs ($i < j$) of offenses and asked to judge which offense in each pair was more serious. Tied judgments were not allowed. (A recent extension of the case V model that permits ties is ref. 1.) For each pair, the number of times n_{ij} offense i was judged more serious than offense j was tabulated. Frequencies n_{ij} were then divided by the number of subjects N ($N = n_{ij} + n_{ji}$) to give the proportions p_{ij} that were then transformed to normal deviates z_{ij}. A case V model with identical standard deviations was assumed and the expected values of the 19 random variables $x_i = (1/N)\Sigma_j z_{ij}$ were estimated. The expected values were interpreted as psychological measures of the seriousness of the offenses.

A host of experimental procedures can be used to collect data for the comparative judgment model. Most popular has been the method of pair comparisons (see PAIRED COMPARISONS). Pair comparisons have the advantage of not forcing transitivity on the judgments. Since the case V model has the property of strong stochastic transitivity

$$(p_{ij} \geqslant 0.5, p_{jk} \geqslant 0.5 \Rightarrow p_{ik} \geqslant \max[p_{ij}, p_{jk}])$$

and the case III and case I models possess the property of weak stochastic transitivity

$$(p_{ij} \geqslant 0.5, p_{jk} \geqslant 0.5 \Rightarrow p_{ik} \geqslant 0.5),$$

the models are falsifiable [10].

Pair comparison judgments may, for large numbers of stimuli, require too many judgments. Most of the alternatives to pair comparison experiments require substantially fewer judgments. One alternative, for n stimuli, is to collect n rating scale judgments per subject. Also referred to as the method of successive intervals or categorical judgments, this approach treats the boundaries between the alternative rating categories as stimuli and estimates scale values for the actual stimuli and the boundaries (*see* PSYCHOLOGICAL SCALING).

ESTIMATION

For the comparative judgment model, parameter estimates must be derived from the observed frequency n_{ij} or the proportion of times p_{ij} that o_i dominates o_j. A variety of statistical and nonstatistical criteria have been proposed for obtaining estimates of the parameters of the case III and case V models. Most commonly used is the least-squares* solution which seeks parameter estimates that minimize the quantity $Q = \sum_{i<j}(z_{ij} - \hat{z}_{ij})^2$ where $z = \Phi^{-1}(p_{ij})$ and Φ is the CDF of the normal. For the case V model, $\hat{z}_{ij} = x_i - x_j$ and for case III, $\hat{z}_{ij} = (x_i - x_j)/(s_i^2 + s_j^2)^{1/2}$, where x_i is the estimate of μ_i and s_i is the estimate of σ_i.

Maximum likelihood* estimates may be found by maximizing the likelihood function

$$L = \prod_{i<j}\binom{N}{n_{ij}}\hat{p}_{ij}^{n_{ij}}(1 - \hat{p}_{ij})^{N-n_{ij}}.$$

For the case V model, $\hat{p}_{ij} = \Phi(x_i - x_j)$ and for case III, $\hat{p}_{ij} = \Phi((x_i - x_j)/(s_i^2 + s_j^2)^{1/2})$.

Minimum chi-square* criteria can also be used. The criterion most often associated with the comparative judgment model is the minimum normit chi square [2], which minimizes the sum of the squared differences between transformations of the proportions, specifically, the quantity

$$Q = \sum_{i<j} w_{ij}(z_{ij} - \hat{z}_{ij})^2,$$

where $w_{ij} = [n_{ij}(\phi^2(\hat{z}_{ij}))]/[\hat{p}_{ij}(1 - \hat{p}_{ij})]$ and ϕ is the PDF of the normal. Depending upon how p_{ij} is defined, this criterion may be used with the case V or case III model.

In general, the case III model is the most complex form of the comparative judgment model to be estimated. The case I model has too many parameters to estimate. Recently, though, an approximation of case I, called a *factorial model* by Takane [11], has been proposed which avoids the overparameterization problem and uses ML estimation.

Least squares is the most popular estimation procedure. For the case V model, the LS solution is $x_i = (1/N)\sum_j z_{ij}$. However, the LS solution is not as statistically efficient as the ML solution. Unfortunately, there is no algebraic solution for the ML estimates and iterative numerical optimization procedures must be used. An algebraic solution for the case V minimum normit criterion is given in ref. 2. The case III model requires numerical methods for all statistical criteria. Estimation of initial values for numerical solution methods is covered in ref. 8.

Scale value estimates x_i are unique up to a linear transformation and standard deviation estimates s_i are assumed to be unique up to a proportionality transformation. The indeterminacy of the case V model is usually solved by assuming a constant value for the s_i and fixing one of the x_i or requiring their sum to equal zero. For case III, one of the s_i is usually fixed. The ability of alternative estimation criteria to recover the parameters of case V and case III models from pair comparison data is reported in ref. 8. Estimation for experimental methods other than pair comparisons is given in refs. 2 and 15.

GOODNESS OF FIT

When ML estimates are used, likelihood ratios can be employed to test the fit of the alternative models. Since (n_{ij}/N) is a maximum likelihood estimate of the population probability ρ_{ij} when the restrictive conditions of the case V or case III model are removed, it is also possible to estimate the likelihood for this general situation.

When LS estimates are used, goodness of fit* can be measured by comparing the observed p_{ij} with the $\hat{p}_{ij}$ predicted by the Thurstone model. An intuitive measure would be to just take the average absolute difference of the two values. A chi-square test of the agreement between observed and estimated proportions is also possible. In either case, a common problem is that the null hypothesis is that the model being evaluated is correct. Thus, failure to reject the null hypothesis and, therefore, acceptance of the model is abetted by poor experimental procedures.

References

[1] Batchelder, W. H. (1979). *J. Math. Psychol.* **19**, 39–60.

[2] Bock, R. D. and Jones, L. V. (1968). *The Measurement and Prediction of Judgment and Choice.* Holden-Day, San Francisco, CA. (Contains an extensive description of how the comparative judgment model is used in psychophysical settings. Numerous examples are included.)

[3] Fechner, G. T. (1860). *Elemente der Psychophysik.* Breitkopf and Hartel, Leipzig, Germany. (An English translation, published in 1966, is available as *Elements of Psychophysics*, Holt, Rinehart and Winston, New York.)

[4] Fullerton, G. S. and Cattell, J. (1892). *On the Perception of Small Differences.* Publications of the University of Pennsylvania Series in Philosophy No. 2, Philadelphia, PA.

[5] Hefner, R. A. (1958). Extensions of the law of comparative judgment to discriminable and multidimensional stimuli. Ph.D. dissertation, University of Michigan, Ann Arbor, MI.

[6] Luce, R. D. (1959). *Individual Choice Behavior.* Wiley, New York.

[7] MacKay, D. B. and Chaiy, S. (1982a). In *Patronage Behavior and Retail Management*, W. Darden and R. Lusch, eds. Elsevier, New York, pp. 71–85.

[8] MacKay, D. B. and Chaiy, S. (1982b). *Psychometrika*, **47**, 353–359.

[9] McKenna, S. P., Hund, S. M., and McEwen, J. (1981). *Int. J. Epidemiology*, **10**, 93–97.

[10] Suppes, P. and Zinnes, J. L. (1963). In *Handbook of Mathematical Psychology*, R. D. Luce, R. R. Bush, and E. Galanter, eds. Wiley, New York, pp. 1–76. (The most complete statement of the measurement theory behind mathematical psychology.)

[11] Takane, Y. (1980). *Japanese Psychol. Res.*, **22**, 188–196.

[12] Thorndike, E. L. (1910). *Teachers College Record*, **11**, 1–93.

[13] Thurstone, L. L. (1927a). *Amer. J. Psychol.* **38**, 368–389. (A classic article, thought by Thurstone to be his best.)

[14] Thurstone, L. L. (1927b). *J. Abnormal and Social Psychol.*, **21**, 384–400.

[15] Torgerson, W. S. (1958). *Theory and Methods of Scaling.* Wiley, New York. (An excellent introduction to Thurstonian models.)

[16] Weber, E. H. (1846). In *Handwörterbuch der Physiologie*, Vol. 3, R. Wagner ed. Vieweg und Sohn, Braunschweig, Germany, 481–588.

[17] Yellott, J. I. (1977). *J. Math. Psychol.*, **15**, 109–142.

Bibliography

Guilford, J. P. (1954). *Psychometric Methods.* McGraw-Hill, New York. (An older text with a good description of the classical methods in psychophysics and mathematical psychology.)

Luce, R. D. (1977). *Psychometrika*, **42**, 461–489. (A modern psychophysical insight upon selected issues in Thurstonian scaling, presented on the 50th anniversary of Thurstone's basic papers on comparative judgment at the annual meeting of the Psychometric Society.)

Thurstone, L. L. (1959). *The Measurement of Values.* University of Chicago Press, Chicago, IL. (A collection of Thurstone's major articles.)

(LUCE'S CHOICE AXIOM AND
  GENERALIZATIONS
PAIRED COMPARISONS
PSYCHOLOGY, STATISTICS IN
PSYCHOPHYSICS, STATISTICAL METHODS
  IN)

DAVID B. MACKAY

TILT *See* UNTILTING

TIME-REVERSAL TEST *See* INDEX NUMBERS; TEST APPROACH TO INDEX NUMBERS, FISHER'S

TIME-SEQUENTIAL INFERENCE

In classical sequential analysis*, an estimator (of a parameter of interest) or a test of significance (of a plausible statistical hypothesis) is based on a random sample whose *size* is not predetermined but is governed by some well defined stopping rule* and hence is a positive integer-valued random variable. The average sample number (ASN*) plays a basic role in the study of the performance characteristics of sequential inference procedures. In problems of reliability*, life testing*, survival analysis*, and in other longitudinal* (follow-up) studies, the sample observations typically relate to the times to failure or occurrences of some characteristic events (taking place sequentially over time), along with other plausible concomitant variables.

As a result, a more general situation may arise where the stopping rule may rest on a more intricate picture (generally termed the *history process*) depending on the flow of outcomes over time (viz., the sample order statistics*, total time on test*, and/or other related features of the experiment). In this context, the actual sample size may not be random, although the stochastic duration of the follow-up study (or some other time events) is an essential feature capable of yielding a well defined stopping rule on which a (time wise) sequential inference procedure may be devised along the lines of the traditional sequential procedures. In the literature, these are known as *time-sequential inference* procedures (see, for example, Sen [11, 15]).

Traditionally, such procedures arose from realizations of some (continuous or discrete) time-parameter stochastic processes, such as the Poisson and Wiener processes*. Whereas these processes are characterized by independent and mostly homogeneous increments, in a general time-sequential inference problem, neither this independence nor the homogeneity of the increments may be taken for granted. Further, in a time-sequential procedure time and cost constraints on the follow-up scheme often dictate the formulation of the stopping rule, and the traditional role of the ASN is replaced by a somewhat different entity (viz., expected total time on test, etc.). Time-sequential inference problems thus need to be treated in a somewhat different manner from classical sequential problems.

We discuss briefly traditional procedures related to the Poisson and Wiener processes.

The basic methodology is due to Dvoretzky et al. [5, 6]. Suppose that one observes continuously a process $\{X(t), t \geqslant 0\}$ (beginning at $t = 0$) and that we have two specified density functions for $X(t)$, namely, $f_i(x, t)$, $i = 1, 2$. We wish to decide as soon as possible whether f_1 or f_2 is the true density for the process under consideration. If T denotes the stopping time, the objective is to minimize $E[T|f_i]$, $i = 1, 2$, subject to the usual constraint that the error probabilities (of the first and second kind) do not exceed some specified α_1 and α_2, respectively, where $\alpha_1 + \alpha_2 < 1$. For every $t \geqslant 0$, let

$$Z(t) = \log\{ f_2(X(t), t)/f_1(X(t), t)\},$$
$$t > 0 \; (Z(0) = 0), \tag{1}$$

be the usual log-likelihood ratio, and based on the process $Z = \{ Z(t), t \geqslant 0\}$, one seeks to draw sequentially a conclusion on the acceptability of f_1 or f_2. For both Poisson and drifted Wiener processes, Z has stationary homogeneous increments, and the classical Wald sequential probability ratio test (SPRT) applies [19]. Actually, an exact determination of the optimal sequential procedure is possible in the continuous time-parameter case, and this provides good approximations for the discrete case when the unit of time is small. For both the Poisson and Wiener processes, the history process is given by $\{ X(s), s \leqslant t\}$, for $t \geqslant 0$.

Consider next the estimation problem. Assume that the density of $f_\theta(x, t)$ of $X(t)$ depends on a parameter θ (possibly vector-valued), so that for any estimator $\hat{\theta}(t)$, based on $\{ X(s), s \leqslant t\}$ and on a properly defined stopping time $T \; (> 0)$, one may consider the *risk function*

$$\rho(T, \hat{\theta}) = E_\theta\big[c(T) + L\big(\hat{\theta}(T), \theta\big)\big], \tag{2}$$

where $c(t), t > 0$ is a given *cost function* and $L(\hat{\theta}, \theta)$ is a *loss function* depicting the loss due to estimating θ by $\hat{\theta}$. Typically, $c(t) = ct$ for some positive $c \; (< \infty)$, and $L(\cdot)$ is the absolute or square error loss. The problem is to choose the stopping time T so that the risk in (2) is a minimum; this is the *minimum risk point estimation problem*, and sequential methods generally work out well for standard situations like the Poisson and Wiener processes. One may also consider the problem of finding a bounded width confidence interval* for θ; for Poisson and Wiener processes, sequential procedures have been worked out.

We consider next a typical time-sequential procedure arising in life testing (Epstein and Sobel [7]), n items being drawn at random from an exponential distribution with the density function $\theta^{-1}\exp(-x/\theta)$, $x \geqslant 0$, and all placed simultaneously on a life test. In the replacement case, failed items are immediately replaced by new items, so that the sample size remains equal to n, while in the nonreplacement case, at the kth failure point, there remain $n - k$ active units whose life times are right-censored at that point, for $k = 1, \ldots, n$. Usually, in this case, for a prefixed r $(1 < r < n)$, the experiment is terminated at the rth failure, if it has not been done earlier by the adapted time-sequential rule.

Suppose that one wants to test H_0: $\theta = \theta_0$ against H_1: $\theta = \theta_1$ $(< \theta_0)$ with prescribed Type I and II errors α_1 and α_2, respectively. Let $X_{(1)}, \ldots, X_{(r)}$ denote the ordered failure time points and let $X_{(0)} = 0$. Then at a time point $t(\geqslant 0)$, the total time on test $V_n(t)$ is equal to nt (in the replacement case) and to

$$\sum_{i \leqslant r(t)} X_{(i)} + (n - r(t))(t - X_{(r(t))})$$

(in the nonreplacement case), where $r(t)$ stands for the number of failures in $[0, t]$, $t \geqslant 0$. In the replacement case, through the classical Poisson process scheme, the Wald SPRT works out well. In the nonreplacement case, close approximations for the OC and ASN functions are available in Epstein and Sobel [7].

Consider next estimation problems relating to θ. In the replacement scheme, because of the reduction to a standard Poisson process, standard theory for the latter applies to the former as well. For the nonreplacement

scheme, a more elaborate analysis is necessary. An exact treatment seems to be prohibitive when n and r are large. However, when the cost per unit of time is small, an asymptotic solution can be obtained; see Sen [12], where the time-sequential point estimation problem for the mean of an exponential distribution has been treated in detail. An extension of this theory (incorporating random censoring) is due to Gardiner et al. [8].

The simple exponential life testing model has been extended to a variety of reliability models in Barlow and Proschan [2]; their emphasis is mostly on the underlying nonsequential probability models. It is quite reasonable to use time-sequential procedures for these models too. As an illustration, consider a single-unit system supported by a repair facility and a single spare. The system fails when the unit currently operating fails before the repair of the latest failed unit is completed. The *availability* of the system is defined by $\theta = E[O]/\{E[O] + E[D]\}$, where O is the time until the system failed and D is the system downtime. If the lifetime (X) and repair time (Y) are both exponentially distributed with parameters μ_1 and μ_2, respectively, then

$$\theta = (1 + \rho)/(1 + \rho + \rho^2),$$

where $\rho = \mu_2/\mu_1$. However, in the negation of the independence of X and Y or any departure from their exponentiality, θ may not be a simple function of ρ or of the two means μ_1 and μ_2. Also, $O = X_1 + \cdots + X_n$ and $D = Y_N - X_N$, where N $(= \min[k: X_k < Y_k])$ is a positive integer-valued random variable. Thus, we have an extended renewal* process and again, Wiener process results can be adapted to provide good approximations to the performance characteristics of time-sequential tests and estimates; for details, see Sen and Bhattacharjee [17] and Sen [16].

The problem of (time-) sequential life testing in a more general setting has received considerable attention. First, the simple exponential model may not be tenable in the majority of the cases. Second, due to possible incorporation of concomitant variables, one needs to consider a model depicting the covariates in a suitable manner; the conventional linear regression* model may not work out well in this respect. For this reason, Cox [4], suggested a proportional hazard model (PHM), where the hazard function is of arbitrary form (*see* COX'S REGRESSION MODEL), but the covariates enter into the hazard in a multiplicative factor. For such a PHM, time-sequential tests have been worked out in Sen [13, 14].

In a nonparametric setup, the formulation in terms of the hazard functions is not that necessary, nor is the assumption of proportionality of the hazards that crucial. A typical nonparametric time-sequential procedure has been described in PROGRESSIVE CENSORING SCHEMES: see also Chatterjee and Sen [3], Majumdar and Sen [10], and Sinha and Sen [8]. The approach, unified in Sen [14, 15], rests on a common theme; consider a basic stochastic process related to the sequence of test statistics or estimators cropping up in a time-sequential scheme, provide a suitable Wiener (or Bessel) process approximation to the distribution theory of such a process, and then use the standard results on sequential inference for Wiener (or Bessel) processes to provide good approximations for the case at hand.

The partial likelihood* approach of Cox [4] for the proportional hazard model has been extended for certain classes of counting processes as well. Though most of these developments are nonsequential in nature (Anderson and Gill [1]), the fact that the events in a counting process occur sequentially in time may prompt one to consider time-sequential inference procedures for such a general counting process as well. By incorporating random intensity functions, some doubly stochastic Poisson process representations for some counting processes arising in the context of spike trains (related to discharge activities of neurons) have been worked out by Habib and Sen [9]. Time-sequential inference procedures may also be generally adapted in this context through the Wiener process approximations for such nonhomogeneous Poisson processes.

References

[1] Anderson, P. K. and Gill, R. D. (1982). *Ann. Statist.*, **10**, 1100–1120.

[2] Barlow, R. E. and Proschan, F. (1975). *Statistical Theory of Reliability and Life Testing: Probability Models*. Holt, Rinehart and Winston, New York.

[3] Chatterjee, S. K. and Sen, P, K. (1973). *Calcutta Statist. Ass. Bull.*, **22**, 13–50.

[4] Cox, D. R. (1972). *J. R. Statist. Soc. B*, **34**, 187–220.

[5] Dvoretzky, A., Kiefer, J., and Wolfowitz, J. (1953a). *Ann. Math. Statist.*, **24**, 254–264.

[6] Dvoretzky, A., Kiefer, J., and Wolfowitz, J. (1953b). *Ann. Math. Statist.*, **24**, 403–415.

[7] Epstein, B. and Sobel, M. (1955). *Ann. Math. Statist.*, **26**, 82–93.

[8] Gardiner, J. C., Susarla, V., and Van Ryzin, J. (1986). *Ann. Statist.* **14**, 607–618.

[9] Habib, M. K. and Sen, P. K. (1985). In *Biostatistics: Statistics in Biomedical, Public Health and Environmental Sciences, The B. G. Greenberg Volume*, P. K. Sen, ed. North-Holland, Amsterdam, The Netherlands, pp. 481–509.

[10] Majumdar, H. and Sen, P.K. (1978). *J. Multivariate Anal.*, **8**, 73–95.

[11] Sen, P. K. (1978). *Commun. Statist. A*, **7**, 307–311.

[12] Sen, P. K. (1980). *Commun. Statist. A*, **9**, 27–38.

[13] Sen, P. K. (1981a). *Ann. Statist.*, **9**, 109–121.

[14] Sen, P. K. (1981b). *Sequential Nonparametrics: Invariance Principles and Statistical Inference*. Wiley, New York.

[15] Sen, P. K. (1985a). *Theory and Application of Sequential Nonparametrics*. SIAM, Philadelphia, PA.

[16] Sen, P. K. (1985b). In *Reliability and Quality Control*, A. P. Basu, ed. North-Holland, Amsterdam, The Netherlands, pp. 297–310.

[17] Sen, P. K. and Bhattacharjee, M. C. (1985). In *Reliability and Quality Control*, A. P. Basu, ed. North Holland, Amsterdam, The Netherlands, pp. 280–296

[18] Sinha, A. N. and Sen, P. K. (1982). *Sankhyā B*, **44**, 1–18.

[19] Wald, A. (1947). *Sequential Analysis*. Wiley, New York.

(AVERAGE SAMPLE NUMBER (ASN)
BROWNIAN MOTION
COX'S REGRESSION MODEL

P. K. SEN

TIME SERIES

A time series is created when the status of an observational unit is recorded through time. This definition is sufficiently broad to include most of our everyday existence, and further qualification is necessary.

The process under observation may develop continuously in time and may be so recorded as, for example, the trace of an electrocardiograph (ECG). More commonly, the process is observed only at discrete points in time (height of an individual, crop yield) whether or not that process is developing continuously in time. Further, continuous records like an ECG trace may be digitized to produce discrete-time records since analysis using digital computers is quicker than with analog devices. Therefore, we shall concentrate mainly upon *discrete*-time series defined at points in time $\{ t_i, \ i = \ \cdots, -1, 0, 1, \cdots \}$ with corresponding random variables $Y(t_i)$, while recognizing that the continuous-time representation may be convenient for theoretical purposes. (*See* STOCHASTIC PROCESSES for further discussion of continuous-time models.)

The random variables may be continuous or discrete, although we shall not consider point processes for which the variable of interest is the time of occurrence. The times between successive observations will be taken to be equal, since most of the developments in time series make this assumption. Unequal spacings are considered briefly at the end of our discussion. In practice, modest irregularities, such as the number of (trading) days in a month, can often be dealt with by a suitable adjustment. For example, we might divide a monthly total by the number of days in that month. A further practical matter which is often ignored is that data may be aggregated over time to produce monthly or annual figures. Such data are generally treated as though recorded at the midpoint of the time interval in question. For further discussion *see* AGGREGATION.

Our basic definition of a time series is thus that we observe a sequence of random variables $Y(t)$ or Y_t equally spaced in time at $t = \ \cdots, -1, 0, 1, \cdots$, after a suitable rescaling of the time axis if necessary.

The traditional approach to time series modeling recognizes four components of a time series: trend*, or long-term movements in the series (T); seasonal fluctuations, which are of known periodicity (S); cyclic variations of a nonseasonal variety, whose periodicity is unknown (C); and R, the random error component.

The models developed within this framework are typically of the forms

$$Y = T + C + S + R,$$

$$Y = TCSR,$$

$$Y = TCS + R.$$

When the trend is the component of primary concern, it may be modeled directly in the form

$$Y_t = f(t, \boldsymbol{\theta}) + \epsilon_t,$$

where $f(t, \boldsymbol{\theta})$ denotes the deterministic component of the series and ϵ_t is the error term (or nondeterministic component). Alternatively, trend might be removed by a combination of transformations and differencing to produce a stationary series (see following text). For fuller details, *see* TREND.

When all four components are relevant, a variety of moving averages* and filtering operations may be applied to the data, usually to estimate the seasonal component and thus to produce a deseasonalized series. See

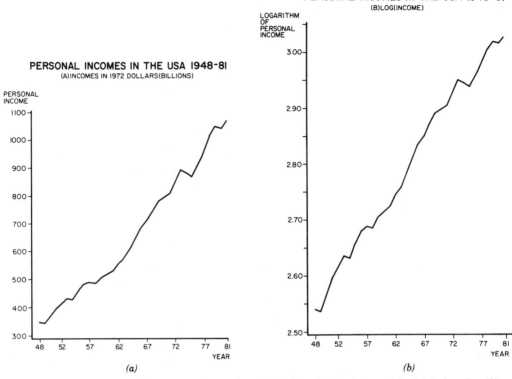

Figure 1 Annual U.S. personal incomes in 1972 dollars for 1948–1981 (*a*) Original scale; (*b*) logarithmic scale. [Source: U.S. Survey of Current Business.]

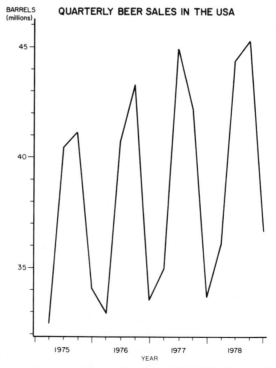

Figure 2 Quarterly beer sales in the Unites States for 1975–1978. [Source: U.S. Survey of Current Business.]

246

ref. 28 (Chaps. 3–5) for a discussion of such methods. The best known of these is the Census X-11 procedure developed by Shiskin and his co-workers.

For many years, it appeared that this strand of research was quite separate from the rest of time series. However, recent work [10] has shown that the X-11 method* may be closely approximated by an autoregressive integrated moving average* (ARIMA) model, and this recognition has led to improvements in seasonal adjustment procedures.

For further details of these and related models, *see* SEASONALITY.

The study of cycles* typically uses sine waves to describe strict periodicities. The perturbation of such waves leads naturally to the consideration of spectral analysis*.

Data analysis for time series may be performed either in the *time domain* or the *frequency domain*. Time domain methods represent the random variable $Y(t)$ in terms of its past history and uncorrelated error terms, the class of linear autoregressive moving average (ARMA*) schemes being the main foundation upon which such analyses are based. The structure of such processes may be described in terms of the autocorrelations.

In the frequency domain, the stationary process is described by a set of cosine waves which vary in angular frequency ω and amplitude. The wavelength, or distance between successive peaks, is $2\pi/\omega$. The spectral density is the plot of the squared amplitude against ω, which indicates the power (or "variance explained") at frequency ω.

A fundamental result in time series is that the autocorrelation function and the spectral density function form a Fourier transform pair so that the information contained in one function is formally equivalent to that contained in the other. Nevertheless, given a finite data record, one approach may do better. In the physical sciences, the study of wave-like phenomena often leads to the use of frequency domain analysis, whereas in the social and economic sciences, the regression type structure of the time domain is usually preferred.

EXAMPLES OF TIME SERIES

Figure 1(*a*) shows a plot of U.S. personal incomes (annual total) for 1948–1981, measured in 1972 dollars. Real growth in the economy produces a compound growth pat-

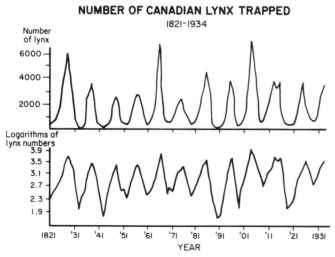

Figure 3   Numbers of lynx trapped in Mackenzie river district of NW Canada, 1821–1934. [Source: Elton and Nicholson (1942). *J. Animal Ecol.* **11**, 215–244.]

tern over the period. Transforming to logarithms, as in Fig. 1(*b*), produces a more or less linear trend, around which certain cyclical variations are apparent.

Figure 2 shows quarterly beer sales in the United States for 1975–1978. The seasonal pattern is evident, although there is a slight upward trend during the period.

Figure 3 shows one of the most famous and most analyzed data sets in the time series literature. The lynx data, taken originally from the records of the Hudson's Bay Company, show a pronounced 10-year cycle. Interestingly, mathematical ecologists (notably R. M. May) have recently developed population models which display such stable limit cycles.

A HISTORICAL PERSPECTIVE

Although it is possible to uncover plots of time series dating back to the eleventh century, the first recognizable plot appears to be owing to Playfair [36] in 1821; see ref. 40 for a more complete discussion. During the nineteenth century, the actuarial profession developed several methods for smoothing data using moving averages [29, pp. 452–461]. However, these techniques were applied to life tables* and not to time series as such. The work of Buys–Ballot led ultimately to the development by Schuster, in 1898, of the periodogram* and this ushered in a period of searching for hidden periodicities and seeking appropriate explanations. It was not until the 1940s that the periodogram was shown to be an inconsistent estimator for the spectrum, and modern spectral analysis* was born.

In 1927, the development of autoregressive series by Yule [46] represented a major breakthrough in time domain modeling. This, combined with the ergodic theorem* of Birkhoff [6] and Khinchin [30] and the demonstration by Wold [43] that the nondeterministic part of any stationary series may be represented as a moving average, paved the way for modern time domain analysis using autoregressive moving average

(ARMA) processes. In the frequency domain, the work of Wiener* [41, 42] was of paramount importance, although its full impact was not realized until the publication of his 1949 monograph [42]; see Brillinger [9] for a fuller discussion of these developments.

Other major contributions, which failed to be given due recognition at the time, were the work of Abbe* in 1863 on the distribution of the serial correlation* coefficient (see ref. 27) and Bachelier's use of the random walk* to describe the movements of stock market prices. Also, Lauritzen [31] points out that in 1880 T. N. Thiele anticipated many of the ideas of Kalman filtering*.

The more recent history of time series methodology may be summarized under one of several major headings:

1. Single series modeling in the time domain following the seminal work of Box and Jenkins [7]; *see* BOX–JENKINS MODELS.

2. Spectral* and cross-spectral analysis with the development of optimal smoothing criteria.

3. Transfer functions* and intervention models* to enhance the regression approach to time series modeling; see ref 8.

4. The development of filtering models (*see* KALMAN FILTERING) which enables the parameter estimates and forecasts to be sequentially updated in an efficient fashion. Harvey [22] develops an elegant algorithm for ARIMA models using this approach.

5. Multiple time series* modeling using both ARIMA and state-space approaches.

6. Seasonal models and seasonal adjustment procedures (*see* SEASONALITY).

7. Comparison and validation of forecasting* techniques.

For more comprehensive reviews of recent developments, see Cox [11], Newbold [34], and Durbin [15].

STATIONARITY

The time series $Y(t)$ is said to be *weakly* (or wide sense or covariance or second order) stationary if

$$E[Y(t)] = \mu, \quad \text{all } t, \qquad (1)$$

$$\text{Var}[Y(t)] = \sigma^2, \quad \text{all } t, \qquad (2)$$

and

$$\text{cov}[Y(t), Y(t-s)] = \gamma(s) = \sigma^2\rho(s),$$
$$\text{all } t \text{ and } s. \quad (3)$$

The series is *strictly* stationary if the joint distribution of $\{Y(t_1), \ldots, Y(t_k)\}$ is the same as that of $\{Y(t_1 + h), \ldots, Y(t_k + h)\}$ for all $\{t_i\}$, k, and h. If the joint distributions are multivariate normal* and the process is weakly stationary, it follows that the process is strictly stationary.

The ergodic theorem of Birkhoff and Khinchin tells us that when the process is weakly stationary and

$$\lim_{n \to \infty} \left[n^{-1}\Sigma_{s=1}^n \rho(s) \right] = 0,$$

the parameters μ, σ^2, and $\rho(s)$ may be estimated consistently by the corresponding sample values. This result provides the basis for statistical inference in time series analysis, because it is usually possible to observe only a single realization through time.

Central limit theorems* may be proved for time series provided that the dependence through time is not "too persistent"; see, for example, refs. 17 (Chap. 6) and 2 (Chap. 7).

If an observed time series appears to be nonstationary, in that any of (1)–(3) fail to hold, the series must be differenced and/or transformed in order to induce stationarity; see TIME SERIES, NONSTATIONARY and VARIATE DIFFERENCE METHOD. Hereafter, we shall assume that $Y(t)$ is weakly stationary unless stated to the contrary.

THE FREQUENCY DOMAIN

If we think of the time series $Y(t)$ as being made up of a weighted average of sine waves, we may represent the series in the *frequency domain* by

$$Y(t) = \int_0^{\pi} [A(\omega)\cos \omega t + B(\omega)\sin \omega t] \, d\omega,$$

where $A(\omega)$ and $B(\omega)$ are orthogonal random processes and ω represents the frequency (number of cycles/unit time) of the sine wave. It follows (*see* SPECTRAL ANALYSIS) that the proportion of the total variance of Y attributable to waves in the frequency band $(0, \omega_0]$ is

$$F(\omega_0) = \int_0^{\omega_o} f(\omega) \, d\omega, \qquad (4)$$

where

$$f(\omega) = \pi^{-1}[1 + 2\Sigma\rho(s)\cos \omega s] \qquad (5)$$

and the $\rho(s)$ are given in (3); $f(\omega)$ is the *spectral density function* of the time series. Further,

$$\rho(s) = \int_0^{\pi} \cos \omega s f(\omega) \, d\omega, \qquad (6)$$

as $\rho(s)$ and $f(\omega)$ form a Fourier transform pair. That is, the information contained in the autocorrelation function is mathematically equivalent to that given by the spectral density function. Therefore, the choice of analysis rests upon the quality of the estimators and the objectives of the study; *see* PERIODOGRAM and SPECTRAL ANALYSIS for further details and an example.

When several time series are studied simultaneously, we may use cross-spectral analysis to examine the interrelationships between the series at different frequencies (*see* MULTIPLE TIME SERIES and SPECTRAL ANALYSIS).

THE TIME DOMAIN

When the time series is weakly stationary, it may be described by the mean, variance, and correlation properties given in (1)–(3). The plot of $\rho(s)$ against $s = 1, 2, \ldots$ is known as the autocorrelation function (ACF).

The natural class of linear models for a time series is the autoregressive integrated moving average* (ARIMA) class of the form

$$Y_t = \phi_1 Y_{t-1} + \cdots + \phi_p Y_{t-p} + a_t$$

$$- \phi_1 a_{t-1} - \cdots - \phi_q a_{t-q} \quad (7)$$

or

$$\phi(B)Y_t = \theta(B)a_t, \quad (8)$$

where $BY_t = Y_{t-1}$ and

$$\phi(B) = 1 - \phi_1 B - \cdots - \phi_p B^p, \quad (9)$$

$$\theta(B) = 1 - \theta_1 B - \cdots - \theta_q B^q. \quad (10)$$

Y_t is assumed to be stationary and may correspond to $\nabla^d Z_t$, where $\nabla Z_t = Z_t - Z_{t-1}$, and d differences are required to induce stationarity. The random errors a_t are independent and identically distributed with mean zero and variance σ^2. The a_t are known as a "white-noise" process (*see* NOISE). The autocorrelations $\rho(s)$ may be expressed in terms of the ϕ's and θ's.

The use of ARIMA schemes was developed into a coherent framework for time series modeling by the seminal work of Box and Jenkins [7]. They suggest the paradigm:

$$\downarrow \qquad \qquad \uparrow$$
$$\text{model identification} \quad |$$
$$\downarrow \qquad \qquad |$$
$$\text{model estimation} \quad | \quad \text{iterate if necessary}$$
$$\downarrow \qquad \qquad |$$
$$\text{diagnostic checking} \quad |$$
$$\downarrow$$
$$\text{generation of forecasts}$$

For further discussion and examples *see* BOX–JENKINS MODELS.

A major tool for model identification is the sample autocorrelation function (ACF), based on the sample autocorrelations or serial correlations*

$$r_s = \frac{\displaystyle\sum_{t=s+1}^{n} (y_t - \bar{y})(y_{t-s} - \bar{y})}{\displaystyle\sum_{t=1}^{n} (y_t - \bar{y})^2},$$

$$s = 1, 2, \ldots \quad (11)$$

[several variants of (11) exist].

The sampling distributions of the r_s have been the subject of extensive research; see ref. 29 (Chap. 48). Recent developments include improved approximations developed by Durbin [13] using Edgeworth series* expansions. Further results on the mean and variance of r_s for general ARMA schemes have been obtained by several authors; see, for example ref. 1. Other criteria for model selection include the sample partial autocorrelation function [*see* AUTOREGRESSIVE MOVING AVERAGE (ARMA) MODELS] and the sample inverse autocorrelation function, defined as the ACF of the "inverted" process

$$\theta(B)z_t = \phi(B)a_t; \quad (12)$$

for details of the evaluation of these functions see ref. 29 (p. 521). For a review of several other model identification procedures, see ref. 12.

Once the model has been identified, estimation proceeds either by nonlinear least squares* or by maximum likelihood* if normality is assumed. Several algorithms have been developed over the years [cf. ref. 29, (pp. 623–36)]; that of Ansley [3] is now regarded as perhaps the most accurate. Model goodness-of-fit* may be examined using the ACF of the residuals; see, for example ref. 29 (pp. 623 and 640–41).

For a discussion of forecasting methods, using both the ARIMA models and a variety of other approaches, *see* FORECASTING and PREDICTION AND FORECASTING.

Nonlinear Models

Although the bulk of the time series literature deals with linear models such as (4), some progress has been made using nonlinear models. Granger and Andersen [19] consider bilinear models like

$$Y_t = \phi_1 Y_{t-1} + a_t - \theta_1 a_{t-1} - \beta_{11} a_{t-1} Y_{t-1}. \quad (13)$$

These schemes are suggested by proportional growth models such as the bilinear process

$$(Y_t - Y_{t-1})/Y_{t-1} = a_t - \theta_1 a_{t-1}. \quad (14)$$

References 20, 23, and 37 provide further discussion of these models.

STATE-SPACE MODELS

Rather than specify an ARIMA model directly, we may consider a time series to be made up of several components, such as mean level and trend, and allow these components to change over time. Thus, allowing for possible measurement (or transient) errors, we may specify a state-space model* in stages as, for example:

observation equation

$$Y_t = \mu_t + \epsilon_t; \qquad (15)$$

state equations

$$\mu_t = \gamma_1 \mu_{t-1} + \beta_t + \delta_{1t}, \qquad (16)$$

$$\beta_t = \gamma_2 \beta_{t-1} + \delta_{2t}, \qquad (17)$$

where the ϵ_t and δ_{jt} ($j = 1, 2$) are mutually independent white-noise* processes with zero means and variances σ_ϵ^2, σ_1^2, and σ_2^2; γ_1 and γ_2 are parameters. For a more general discussion *see* LINEAR SYSTEMS, STATISTICAL THEORY.

Substituting for β_t and μ_t in turn, we find that (15)–(17) reduce to an ARIMA ($k, 2 - k, 2$) scheme when k of the γ_j are less than 1 in absolute value (rather than $\gamma_j = 1$). In general, linear state-space models may be represented as ARIMA schemes, although the latter are only representable in state-space form under certain conditions on the parameters [cf. ref. 29 (pp. 528–530)].

The state-space models are also known as *Kalman filtering** models. Forecasting using (15)–(17) involves the systematic updating of estimates from one time period to the next, which may be performed using Bayes theorem*; the method is then known as *Bayesian forecasting* (Harrison and Stevens [21]).

In addition to providing an efficient algorithm for updating estimates and forecasts, filtering methods provide a unity between the forecasting and time series literature which has sometimes been lacking

in the past. Gardner et al. [18] provide a computer algorithm for the exact maximum likelihood fitting of ARMA models using the Kalman filter; Harvey and Pierse [23] extend this to allow for missing values.

TRANSFER FUNCTIONS AND INTERVENTIONS

The random variable Y_t may be considered as the output from a linear system which contains one or more input variables. For simplicity, we consider one such variable, X_t. The ARMA model discussed earlier may be extended to include the effect of X as

$$Y_t = \frac{\omega(B)}{\delta(B)} X_{t-b} + \frac{\theta(B)}{\phi(B)} a_t, \qquad (18)$$

where $\omega(\cdot)$ and $\delta(\cdot)$ are again polynomials in B, and b represents the delay before the input has an effect upon the output; a_t represents white noise as before. Equation (18) represents a transfer function model*, first proposed in this form by Box and Jenkins [7, Chaps. 10 and 11]. Transfer functions include, as special cases, other approaches to time series model building such as distributed lag models (*see* LAG MODELS, DISTRIBUTED) and regression with autocorrelated errors (*see* LINEAR REGRESSION).

The development of a time series may be interrupted by an event, such as resetting a machine or a policy change, which may have either a temporary or a permanent effect upon the level of the series. Such events are known as *interventions*; appropriate statistical methods have been developed by Box and Tiao [8]. *See* INTERVENTION MODEL ANALYSIS for further details.

In place of a single response, we may have a vector of random variables at each point in time. Vector ARIMA models are discussed in MULTIPLE TIME SERIES.

OTHER TOPICS

At the outset, we assumed that the data were recorded at equally spaced points in time.

With the exception of an early paper by Quenouille [37], irregularly spaced data have received little attention. However, Jones [26] has developed continuous time autoregressive models for unequally spaced time series. The recent volume edited by Parzen [35] contains several papers of interest on this topic.

Missing data is another topic for which few results were available until recently. Jones [25] and Ansley and Cohen [4] use the state-space formulation to develop exact likelihood estimation procedures when data are missing; Harvey and Pierse [23] give a Kalman filter method which allows interpolation of missing observations.

LITERATURE

A comprehensive bibliography of the time series literature up to 1965 was provided by Wold [44]. The review papers by Cox [11], Newbold [34], and Durbin [15] provide more recent coverage.

The major journals in statistics continue to publish substantial numbers of papers on time series, and the popularity of the subject is further demonstrated by the introduction of four new journals devoted primarily to this area: *Journal of Time Series Analysis**, *Journal of Business and Economic Statistics*, *Journal of Forecasting*, and *International Journal of Forecasting*. The related areas of econometrics*, engineering, and operations research* continue to contribute substantially to both the theory and practice of the subject.

Among the specialist texts listed in the bibliography, the edited volumes [8, 14, and 16] give the most complete account of recent developments in the subject.

COMPUTER PROGRAMS

The availability of standard main-frame computer packages for time series analysis may be summarized as follows (*see also* STATISTICAL SOFTWARE):

Univariate ARMA	BMDP, SAS, SPSS, MINITAB
Intervention analysis⎱ Transfer functions⎰	BMDP, SAS
Spectral analysis	BMDP, SAS
Census X-11	SAS
State space	SAS

A large number of time series and forecasting programs are available for microcomputers; see Beaumont et al. [5]. Of particular interest is the AUTOBOX package of Reilly [39], which not only provides the usual univariate ARMA, intervention and transfer function analyses, but also provides for fully automated model identification in these areas.

FUTURE DEVELOPMENTS

In the past 20 years, the analytical framework for linear, univariate series with normally distributed errors has been substantially completed; nevertheless, several important areas of activity remain. These include the development of second order approximations to sampling distributions [13, 16], improved methods of model identification [12], and models for series that are irregularly spaced or have missing values [23, 35]. Statisticians have been rather slow to recognize the benefits of the sequential updating algorithm developed as Kalman filtering [22, 23], but the conceptual and computational benefits make this an attractive area for further development as well as providing stronger links between the time series and forecasting literatures; see Durbin [14].

The other major trends are the relaxation of one or more of the categories: linear, univariate, and normally distributed. The class of bilinear models has already been mentioned, and other nonlinear model formulations are becoming more popular; see, for example, Young and Ord [45]. The further advances in multiple time series modeling and econometrics show a drawing together of these research traditions and a recognition that both a systems approach and careful specification of the error structure are important for successful modeling (Hendry and Richard [24]). In a series of papers, Martin [33] has developed robust estimators for both the time and frequency domains and the effects of additive outliers clearly require further investigation.

References

[1] Anderson, O. D. and DeGooijer, J. G. (1983). *Sankhyā B*, **45**, 249–256.

[2] Anderson, T. W. (1971). *The Statistical Analysis of Time Series*. Wiley, New York.

[3] Ansley, C. F. (1979). *Biometrika*, **66**, 59–66.

[4] Ansley, C. F. and Kohn, R. (1983). *Biometrika*, **70**, 275–278.

[5] Beaumont, C., Mahmoud, E., and McGee, V. E. (1985). *J. Forecasting*, **4**, 305–311.

[6] Birkhoff, G. D. (1931). *Proc. National Academy Sciences USA*, **17**, 656–660.

[7] Box, G. E. P. and Jenkins, G. M. (1970). *Time Series Analysis, Forecasting and Control*. Holden-Day, San Francisco, CA. (Rev. ed. 1976.)

[8] Box, G. E. P. and Tiao, G. C. (1975). *J. Amer. Statist. Ass.*, **70**, 70–79.

[9] Brillinger, D. R. (1976). In *On the History of Statistics and Probability*, D. B. Owen, ed. Marcel Dekker, New York, pp. 267–280.

[10] Cleveland, W. P. and Tiao, G. C. (1976). *J. Amer. Statist. Ass.*, **71**, 581–587.

[11] Cox, D. R. (1981). *Scand. J. Statist.*, **8**, 93–115.

[12] DeGooijer, J. G., Abraham, B., Gould, A., and Robinson, L. (1985). *Int. Statist. Rev.*, **53**, 301–329.

[13] Durbin, J. (1980). *Biometrika*, **67**, 311–333 and 335–349.

[14] Durbin, J. (1983). In *Time Series Analysis: Theory and Practice*, Vol. 3, O. D. Anderson, ed. North-Holland, Amsterdam, The Netherlands.

[15] Durbin, J. (1984). *J. R. Statist. Soc. Ser. A*, **147**, 161–173.

[16] Durbin, J. (1985). In *Essays in Time Series and Allied Processes*, J. Gani and M. B. Priestely, eds. Applied Probability Trust, Sheffield, England.

[17] Fuller, W. A. (1976). *Introduction to Statistical Time Series*. Wiley, New York.

[18] Gardner, G., Harvey, A. C., and Phillips, G. D. A. (1980). *Appl. Statist.*, **29**, 311–322.

[19] Granger, C. W. J. and Andersen, A. P. (1987). *An Introduction to Bilinear Time Series Models*. Vandenhoeck and Ruprecht, Göttingen, Germany.

[20] Haggan, V., Heravi, S. M., and Priestley, M. (1984). *J. Time Series Anal.*, **5**, 69–102.

[21] Harrison, P. J. and Stevens, C. F. (1976). *J. R. Statist. Soc. Ser. B*, **38**, 205–247.

[22] Harvey, A. C. (1984). *J. Forecasting*, **3**, 245–283.

[23] Harvey, A. C. and Pierse, C. R. (1984). *J. Amer. Statist. Ass.*, **79**, 125–131.

[24] Hendry, D. F. and Richard, J. F. (1983). *Int. Statist. Rev.*, **51**, 111–164.

[25] Jones, R. H. (1980). *Technometrics*, **22**, 389–395.

[26] Jones, R. H. (1981). In *Applied Time Series Analysis*, Vol. II, D. F. Findley, ed. Academic, New York, pp. 651–682.

[27] Kendall, M. G. (1971). *Biometrika*, **58**, 369–373.

[28] Kendall, M. G. (1980). *Time-Series*. Griffin, High Wycombe, England, and Macmillan, New York.

[29] Kendall, M. G., Stuart, A., and Ord, J. K. (1983). *The Advanced Theory of Statistics*, Vol. 3, 4th ed. Griffin, High Wycombe, England, and Macmillan, New York.

[30] Khinchin, A. Y. (1932). *Math. Ann.*, **107**, 485–488.

[31] Lauritzen, S. L. (1981). *Int. Statist. Rev.*, **49**, 319–331.

[32] Li, W. K. (1984). *J. Time Series Anal.*, **5**, 173–181.

[33] Martin, R. D. (1981). In *Applied Time Series*, Vol. 2, D. F. Findley, ed. Academic, New York, pp. 683–759.

[34] Newbold, P. (1981). *Int. Statist. Rev.*, **49**, 53–66. (A review of recent developments with extensive bibliography.)

[35] Parzen, E., ed. (1984). *Time Series Analysis of Irregularly Observed Data*. Lecture Notes in Statistics, Vol. 25. Springer, New York.

[36] Playfair, W. (1821). *A Letter on our Agricultural Distress*. London, England.

[37] Quenouille, M. H. (1958). *Metrika*, **1**, 21–27.

[38] Rao, T. S. and Gabr, M. M. (1984). *An Introduction to Bispectral Analysis and Bilinear Time Series Models*. Springer, New York.

[39] Reilly, D. (1984). *Proc. Bus. Econ. Statist. Amer. Statist. Ass.*, 539–542.

[40] Royston, E. (1956). *Biometrika*, **43**, 241–247.

[41] Wiener, N. (1930). *Acta Math.*, **55**, 117–258.

[42] Wiener, N. (1949). *The Extrapolation, Interpretation, and Smoothing of Time Series*. Wiley, New York.

[43] Wold, H. O. (1938). *A Study in the Analysis of Stationary Time Series*. Almquist and Wiksell, Uppsala, Sweden.

[44] Wold, H. O. (1965). *Bibliography on Time Series and Stochastic Processes*. Oliver and Boyd, Edinburgh, Scotland.

[45] Young, P. and Ord, J. K. (1985). *Tech. Forecasting and Social Change*, **28**, 263–274.

[46] Yule, G. U. (1927). *Philos. Trans. R. Soc., London Ser. A*, **226**, 267–313.

Bibliography

Bibliographies for cross-referenced entries should also be consulted for particular topics.

Abraham, B. and Ledolter, J. (1983). *Statistical Methods for Forecasting*. Wiley, New York. (Coverage of both regression and ARIMA models and their inter-relationships.)

Anderson, B. D. O. and Moore, J. B. (1979). *Optimal Filtering*. Prentice-Hall, Englewood Cliffs, NJ. (Kalman filtering from the engineering standpoint.)

Anderson, O. D. (1975). *Time Series Analysis and Forecasting: The Box–Jenkins Approach*. Butterworth, London, England. (Introduction to univariate ARIMA models.)

Anderson, T. W. (1971). *The Statistical Analysis of Time Series*. Wiley, New York. (Theoretical; extended coverage of simultaneous tests of hypotheses and distributions of serial correlations.)

Bloomfield, P. (1976). *Fourier Analysis of Time Series: An Introduction*. Wiley, New York. (Introduction to frequency domain methods and their application.)

Box, G. E. P. and Jenkins, G. M. (1970). *Time Series Analysis, Forecasting and Control*. Holden-Day, San Francisco, CA. (Rev. ed., 1976.) (Classic exposition of the Box–Jenkins approach.)

Brillinger, D. R. (1975). *Time Series Data Analysis and Theory*. Holt, Rinehart and Winston, New York. (Expanded edition, 1981.) (Covers both theory and practice, primarily in the frequency domain.)

Brillinger, D. R. and Krishnaiah, P. R. (1983). *Time Series in the Frequency Domain*. North-Holland, Amsterdam, The Netherlands. (Multi-author review of recent research in the frequency domain.)

Chatfield, C. (1984). *The Analysis of Time Series*, 3rd ed. Chapman and Hall, London, England. (Introduction to both time and frequency domain analyses.)

Cryer, J. D. (1985). *Time Series Analysis with MINITAB*. Duxbury, Boston. (Introduction to univariate ARIMA models, integrated use of MINITAB computer system.)

Fuller, W. A. (1976). *Introduction to Statistical Time Series*. Wiley, New York. (Theoretical; includes extended coverage of asymptotic distribution theory.)

Granger, C. W. J. and Hatanaka, M. (1964). *Spectral Analysis of Economic Time Series*. Princeton University Press, Princeton, NJ. (Frequency domain analysis particularly oriented to economic applications.)

Granger, C. W. J. and Newbold, P. (1977). *Forecasting Economic Time Series*. Academic, New York. (Introduction with an emphasis on economic applications.)

Gregson, R. A. M. (1983). *Time Series in Psychology*, Erlbaum, Hillsdale, NJ. (Introduction to use of time series in psychology.)

Grenander, U. and Rosenblatt, M. (1984). *Statistical Analysis of Stationary Time Series*, 2nd ed. Chelsea, New York. (An updated and expanded version on the theory of time series analysis.)

Griliches, Z. V. and Intriligator, M. D., eds. (1981, 1984). *Handbook of Econometrics*, Vols. I and II. North-Holland, Amsterdam, The Netherlands. (Multi-author survey of recent developments in econometrics.)

Hannan, E. J. (1970). *Multiple Time Series*. Wiley, New York. (Rigorous development of theory for vector processes.)

Hannan, E. J., Krishnaiah, P. R., and Rao, H. H. (1985). *Time Series in the Time Domain*. North-Holland, Amsterdam, The Netherlands. (Multi-author survey of recent research in time domain.)

Harvey, A. C. (1981). *Time Series Models*. Philip Allan, Oxford, England. (Introduction to time and frequency domain analysis, with particular emphasis on likelihood estimation.)

Jenkins, G. M. (1979). *Practical Experiences with Modelling and Forecasting Time Series*. GJP Publications, St. Helier, Jersey, Channel Islands. (Case studies using ARIMA and transer-function models.)

Jenkins, G. M. and Watts, D. G. (1968). *Spectral Analysis and Its Applications*. Holden-Day, San Francisco, CA. (Intermediate introduction to frequency domain analysis.)

Kendall, M. G. (1980). *Time-Series*. Griffin, High Wycombe, and Macmillan, New York. (Introduction including both classical decomposition methods and time and frequency domain analyses.)

Kendall, M. G., Stuart, A., and Ord, J. K. (1983). *The Advanced Theory of Statistics*, Vol. 3, 4th ed. Griffin, High Wycombe, England, and Macmillan, New York. (Theoretical; coverage of both time and frequency domains.)

Koopmans, L. M. (1974). *The Spectral Analysis of Time Series*. Academic, New York. (Theoretical; frequency domain analysis.)

Nelson, C. R. (1973). *Applied Time Series Analysis for Managerial Forecasting*. Holden-Day, San Francisco, CA. (Introduction to univariate ARIMA models.)

Nerlove, M., Grether, D. M., and Carvallo, J. L. (1979). *Analysis of Economic Time Series*. Academic, New York. (Covers time and frequency domain analysis, with particular emphasis on distributed lags and seasonal adjustment.)

Pankratz, A. (1983). *Forecasting with Univariate Box–Jenkins Models: Concepts and Cases*. Wiley, New York. (Introduction to univariate ARIMA models with many case studies.)

Priestley, M. B. (1981). *Spectral Analysis and Time Series*, 2 vols. Academic, London, England. (Theoretical; primary emphasis on the frequency domain.)

Quenouille, M. H. (1957). *The Analysis of Multiple Time*

Series. Griffin, London, England. (A classic on multiple time series.)

Roberts, H. V. (1984). *Time Series and Forecasting with IDA*. McGraw-Hill, New York. (Introduction to regression and time series using the interactive IDA computer language.)

Taylor, S. (1986). *Modeling Financial Time Series*. Wiley, Chichester, England and New York. (An introduction, emphasizing financial forecasting and tests of efficient markets.)

Vandaele, W. (1983). *Applied Time Series and Box–Jenkins Models*. Academic, New York. (Introduction to time domain methods including transfer functions and intervention analysis.)

Young, P. (1986). *Recursive Estimation and Time Series Analysis: An Introduction*. Springer, Berlin, W. Germany, and New York. (Emphasizes the role of recursive estimation in time series analysis, with special reference to control engineering.)

Zurbenko, I. G. (1986). *The Spectral Analysis of Time Series*. North-Holland, Amsterdam, Netherlands. (Theoretical discussion of frequency domain methods.)

(AGGREGATION
AUTOREGRESSIVE-INTEGRATED
   MOVING AVERAGE (ARIMA) MODELS
AUTOREGRESSIVE-MOVING AVERAGE
   (ARMA) MODELS
BOX–JENKINS MODELS
CYCLES
FORECASTING
INTERVENTION MODEL ANALYSIS
KALMAN FILTERING
LAG MODELS, DISTRIBUTED
LINEAR SYSTEMS, STATISTICAL THEORY
MOVING AVERAGES
MULTIPLE TIME SERIES
PERIODOGRAM
PREDICTION AND FORECASTING
SEASONALITY
SPECTRAL ANALYSIS
STATE-SPACE MODELS
STOCHASTIC PROCESSES
TIME SERIES, NONSTATIONARY
TRANSFER FUNCTION
TREND
VARIATE DIFFERENCE METHOD)

J. KEITH ORD

TIME SERIES ANALYSIS AND FORECASTING (TSA & F) SOCIETY

A *time series* is a set of observations ordered in time; for instance, the monthly unemployment figures, an EEG trace (of brain activity), or the hourly concentration readings from a chemical process.

Time series analysis with its vital applications, such as to forecasting*, signal extraction, or optimal control, is arguably now one of the most important branches of statistical practice, and, to help circulate information, a number of projects have been initiated.

These include the running of sequences of instructional courses and international conferences, the holding of seminars and the publication of several organs for communication—a monthly broadsheet, a quarterly newsletter, and an academic journal. A world-wide network of experts has been set up to coordinate the efforts of the associated special interest group; the most recent step was the formation of a TSA & F Society.

Time series methods are attracting the steadily increasing interest of quantitative workers from all around the world, whose main occupation may range from forestalling cardiac arrest to finding oil. This broad relevance to many fields was reflected by the numerous delegates, from various countries, who attended the first TSA & F Conference at Cambridge University, England, in July 1978. However, practitioners are highly dispersed (and often relatively isolated); this has given rise, in the past, to much duplicated effort and ignorance of TSA & F advances in unfamiliar areas. To help combat such difficulties, the various activities outlined in this article have been started.

INSTRUCTIONAL COURSES AND SEMINARS

Courses usually last one working week, although a shortened two-day version is also available. They have always concentrated on univariate (single series) analysis, as it is believed that effective treatment of this simplest case is a prerequisite for all further work, and people tend mistakenly to become involved with multivariate methods before they are really equipped and ready to do so. Sixteen (predominantly time domain) introductory courses have been held in London, Nottingham, Oslo, Santander (Spain), Tampere (Finland), and Tucuman (Argentina). Such courses are intended as an introduction to the theory, techniques, and application of modern time series analysis.

Over the last ten years, TSA & F Research Seminars have also been arranged in Amsterdam (2), Brussels (2), Buenos Aires, Caracas, Dublin, Graz (Austria), London (2), Montreal, Olso, Ottawa, Prague, Rio de Janeiro, Southampton, Valencia (3), West Berlin (2), and Wisla (Poland).

INTERNATIONAL CONFERENCES

It is intended that a regular sequence of international time series meetings (ITSM) should be held around the world. These are aimed at disseminating recent developments in the art, science, and practice of time series analysis, and are seen as a means of bringing practitioners together from diverse parent disciplines and geographical locations.

To date, 15 major events have taken place: at Cambridge (2), Nottingham (3), and on Guernsey (all in the British Isles); elsewhere in Europe—Dublin and Valencia (2); Cincinnati and Houston (2) in the United States; and Burlington and Toronto (2) in Canada. The Proceedings have been published and speakers at these conferences have included many known experts.

TSA & F NEWS

This started as a quarterly publication (in January 1979) and now has a circulation of 3000 copies, airmail to 71 countries. Each issue contains around 25 pages of typescript, covering useful news about TSA & F activities—past, present, or future. There are sections for correspondence, reports on meetings, notices of courses and conferences, a calendar of events, details and reviews of publications, comparative costs and effectiveness of computer packages, news about people and research in progress, listing of jobs vacant and wanted, a personal column,

visits and exchanges, working papers, a teachers' corner, a practitioners' forum, and advertisements. All suggestions concerning content and suitable contributed items are always welcome.

In June 1980, a monthly broadsheet (the *TSA & F Flyer*) was introduced to supplement the newsletter and speed the spread of more pressing news. It currently has a circulation of 3500 copies going to 72 countries, and appears about 10 times a year.

JOURNAL OF TIME SERIES ANALYSIS

This academic journal started production in October 1980 and is published and distributed by Tieto Ltd., Bank House, Clevedon, Avon BS21 7HH, England, under the sponsorship of the Bernoulli Society*.

TSA & F Network

An inner core of 140 especially motivated subscribers to the newsletter provides a world-wide network for information collection and distribution, which spans much of the globe and a great many parent subject disciplines.

Representatives are needed in all countries to collect and distribute information nationally. At present, people agree to post notices of TSA & F Activities on display boards or circulate copies to colleagues and friends and to generally promote the projects locally. They also pass back details about events occurring in their own areas.

The network forms the basis for the TSA & F Society.

TSA & F SOCIETY

The magnitude of response to the ongoing TSA & F projects suggested the formation of a nonprofit-making professional society as a natural development. In countries where there are several network members, individuals are chosen to liaise with national bodies concerned with allied topics. It is envisaged that Directors should always be selected from network representatives. As far as the network goes, the policy is that anyone with enthusiasm and drive will always be very welcome.

Membership of the Society consists of Honorary Fellows, Fellows, Members, Associates, and Corporate or Institutional Members.

OTHER GOALS

Evidently a vast amount of work still has to be done in organising and coordinating the areas of activity discussed. We anticipate shortly investigating the possibility of at least one major fresh proposal—that of a TSA & F Institute. Whereas the Society will be mainly oriented towards the needs of people practising TSA & F, the Institute would be concerned primarily with the art and science of the subject, and is likely to devote itself to research and teaching.

It is also believed that, eventually, many of the activities described in this article will be taken over by national and international bodies, now that sufficient interest from individuals has been established. Whether or not the Society stays autonomous, or becomes a section of some adopting parent organisation, remains to be seen.

TSA & F Publications

Apart from the *News*, *Flyer*, and *Journal*, there are also the TSA & F Conference Proceedings in the North-Holland Series: *Forecasting* (1979), *Time Series* (1980), *Analysing Time Series* (1980), *Forecasting Public Utilities* (1980), *Time Series Analysis* (1981), *Applied Time Series Analysis* (1982), and *Time Series Analysis: Theory and Practice* [Vols. 1 and 2 (1982); 3 and 4 (1983); 5–7 (1984)].

O. D. ANDERSON

TIME SERIES, NONSTATIONARY

A time series* is *strictly stationary** if the joint distribution of any finite set of random variables from the series depends only on their relative positions, more explicitly, a time series z_t is strictly stationary if

$$\Pr[z_{t_1} \leqslant z_1, \ldots, z_{t_m} \leqslant z_m]$$
$$= \Pr[z_{t_1+k} \leqslant z_1, \ldots, z_{t_m+k} \leqslant z_m]$$

for any set of time points $t_1, \ldots, t_m$ and any k. The conditions for z_t to be *weakly stationary* (or second-order stationary or wide-sense stationary) are that the mean, variance, and autocovariances (assumed to exist) do not depend on t:

$$E[z_t] = \mu_t = \mu_z \quad \text{for all } t, \quad (1)$$
$$\text{var}(z_t) = \sigma_t^2 = \sigma_z^2 \quad \text{for all } t, \quad (2)$$
$$\text{cov}(z_t, z_{t+k}) = \gamma_k \quad \text{for all } t. \quad (3)$$

Conditions (2) and (3) imply that the autocorrelations do not depend on t:

$$\text{corr}(z_t, z_{t+k}) = \rho_k \quad \text{for all } t. \quad (4)$$

Strict stationarity and existence of second moments implies weak stationarity*. The reverse does not hold in general, though it does hold for Gaussian time series. Here we will deal only with first and second moments, so we shall use stationary to mean weakly stationary. Also, we will consider only discrete time series where t takes on equally spaced values labelled $0, \pm 1, \pm 2, \ldots$.

A *nonstationary time series* fails to satisfy one or more of the conditions for stationarity. The conditions (1)–(4) are quite restrictive and can be violated by trend*, seasonality*, periods of volatility, structural changes, varying measurement errors*, or effects due to unusual events or efforts made to control the series. Thus, nonstationary time series are the rule and stationary time series the exception. Since many classical results in time series assume stationarity, very often nonstationary series are dealt with by trying to remove the nonstationarity in z_t and relate z_t to a stationary series.

REMOVING NONSTATIONARY MEANS AND VARIANCES

If $\mu_t = E(z_t)$ depends on t, we can try to find a parametric model for μ_t, estimate the parameters of the model, and take $z_t - \hat{\mu}_t$ to remove this source of nonstationarity. A popular approach is to parameterize μ_t as a linear function of some set of observed variables $\mathbf{X}_t = (X_{1t}, \ldots, X_{mt})'$ and parameters $\boldsymbol{\beta} = (\beta_1, \ldots, \beta_m)'$, leading to

$$z_t = \beta_1 X_{1t} + \cdots + \beta_m X_{mt} + N_t. \quad (5)$$

More generally, we could have $\mu_t = f(\mathbf{X}_t; \boldsymbol{\beta})$, where f is nonlinear. The X_{it} may be nonstochastic variables, such as polynomial or trigonometric functions of time, or seasonal indicator variables (*see* SEASONALITY), or they may be other (stochastic) time series, including lagged time series (*see* TRANSFER FUNCTION). Also quite useful are indicator variables for the occurrence of an unusual event thought to affect the time series— use of such variables is called *intervention analysis* [5].

Model (5) differs from the standard regression* model in that the noise* or error series, N_t, will typically be autocorrelated and may itself be nonstationary. This generally makes ordinary least-squares* estimation of $\boldsymbol{\beta}$ inefficient, and the usual regression standard errors for the estimates can be severely biased. Given a time series model for N_t such as the ARIMA models discussed later, $\boldsymbol{\beta}$ and the time series parameters can be estimated jointly by maximum likelihood* assuming normality, and inferences can be made (see ref. 29 for linear f and ref. 12 for nonlinear f). Alternatively, a spectral approach can be used instead of modeling N_t [16]. Actually, for certain specific types of μ_t (including polynomials, trigonometric functions, and seasonal means), ordinary least-squares estimates of $\boldsymbol{\beta}$ are asymptotically

efficient if N_t is stationary with continuous spectral density [30, Sec. 7.7].

If σ_t^2 depends on t, we can try to parametrize and estimate it and model $(z_t - \hat{\mu}_t)/\hat{\sigma}_t$. In time series, it is often more appropriate to divide $z_t - \mu_t$ by the innovation variance* of N_t (*see* TIME SERIES), say, $\sigma_{a,t}^2 = \text{var}(a_t)$. Expressions used for σ_t^2 (or $\sigma_{a,t}^2$) tend to be simple, such as letting it take on one value over one set of time points and another value over another set. More complex parametrizations of $\sigma_{a,t}^2$ in autoregressive models are given in refs. 9 and 38.

Another technique for eliminating nonstationary variances is to apply a variance stabilizing transformation. Power or Box–Cox* transformations* z_t^λ, of which the logarithmic transformation is a limiting case, are one possibility.

Example. For an example we use the monthly average temperature in Washington, D.C. from January 1961 to December 1980 (taken from ref. 39). Figure 1(a) shows the last 10 years of the data. A suitable mean function for this series is

$$\mu_t = \beta_0 + \beta_1 \cos(2\pi t/12) + \beta_2 \sin(2\pi t/12)$$
$$+ \beta_3 \cos(4\pi t/12) + \beta_4 \sin(4\pi t/12).$$

The β_i's were estimated by least squares*; the resulting $\hat{\mu}_t$ is plotted in Fig. 1(b). The sample standard deviations about zero ($\hat{\sigma}_t$) of the residuals for each month are:

Jan.	Feb.	March	April	May	June
4.2	3.9	3.5	2.6	2.8	2.0

July	Aug.	Sept.	Oct.	Nov.	Dec.
1.5	2.0	2.8	2.9	3.0	3.7

Notice that $\hat{\sigma}_t$ is higher in the winter months. The series $(z_t - \hat{\mu}_t)/\hat{\sigma}_t$ is plotted in Fig. 1(c).

EXTENSIONS

Model (5) can be generalized by replacing $\boldsymbol{\beta}$ by $\boldsymbol{\beta}_t$ and letting this vary over time, for example letting it change values at certain time points. For standard regression models

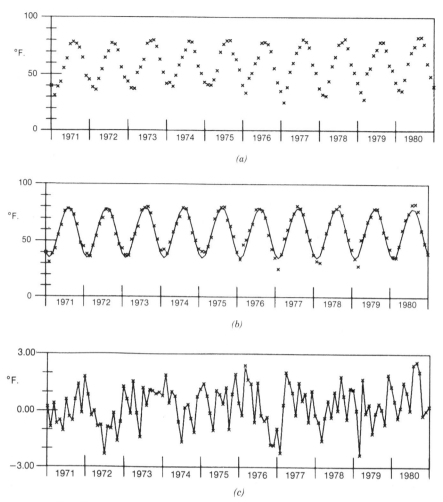

Figure 1 (*a*) Monthly average temperature, Washington, D.C. (January 1961 to December 1980). (*b*) Monthly average temperature and fitted mean function ($\hat{\mu}_t$). (*c*) Standardized monthly average temperature ($z_t - \hat{\mu}_t$)/$\hat{\sigma}_t$.

the problem of detecting shifts in β at unknown time points has been considered in refs. 2, 6, 31, and 32. Tests for shifts in variance (σ_t^2) have been applied, especially to time series of stock market price changes or rates of return; see refs. 1, 21, 22, and 41.

HOMOGENEOUS AND EXPLOSIVE NONSTATIONARITY

Another approach to transforming a nonstationary time series z_t to stationarity is to apply a (finite) linear filter*. This may be expressed as

$$\left(1 - \alpha_1 B - \cdots - \alpha_p B^p\right) z_t = w_t, \quad (6)$$

where B is the backward-shift operator $B^j z_t = z_{t-j}$, $\alpha(B) = 1 - \alpha_1 B - \cdots - \alpha_p B^p$ is the (finite) linear filter, and w_t is a stationary time series. For simplicity assume $E[z_t] = 0$ (or that we have already taken $z_t - E[z_t]$) and consider only finite ($p < \infty$) one-sided filters. ($\alpha(B)$ contains no negative powers of B and thus only shifts t backwards.)

If the zeroes of $\alpha(B)$, viewed as a polynomial in B, lie outside the unit circle, then (6) can be inverted to $z_t = \alpha(B)^{-1} w_t$ and z_t

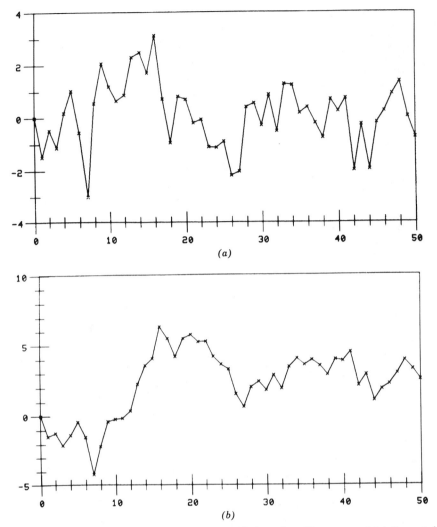

Figure 2   50 observations (a) from $(1 - 0.5B)z_t = a_t$, (b) from $(1 - B)z_t = a_t$, and (c) (truncated at -200) from $(1 - 1.15B)z_t = a_t$.

will be stationary. Model (6) becomes a nonstationary model if we let $\alpha(B)$ have zeroes on or inside the unit circle. For example, consider the three first-order autoregressive (AR)* models

$$(1 - 0.5B)z_t = a_t, \qquad (7)$$

$$(1 - B)z_t = a_t, \qquad (8)$$

$$(1 - 1.15B)z_t = a_t, \qquad (9)$$

where a_t is white noise, for which the zero of $\alpha(B)$ is outside, on, and inside the unit circle, respectively. Figures 2(a), (b), and (c) show 50 observations generated from these models with the a_t's independent normal

$(0, 1)$ random variables. The nonstationary models (8) and (9) lead to time series smoother in appearance than the stationary model (7), and the time series from (9) eventually takes off toward $-\infty$. Models where $\alpha(B)$ has zeroes on the unit circle are *homogeneously nonstationary* and models where $\alpha(B)$ has zeroes inside the unit circle are *explosively nonstationary.*

DIFFERENCING

Explosive models are unrealistic in many situations; however, homogeneously nonsta-

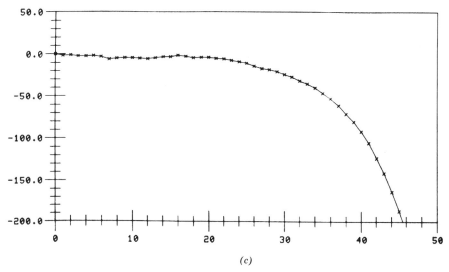

(c)

Figure 2 (*Continued*).

tionary models such as (8) are quite useful. In fact, many nonstationary series appear to be made possibly stationary by *differencing*: taking $\nabla z_t = z_t - z_{t-1}$, where $\nabla = 1 - B$ is the backward difference* operator. More generally, we can take $\nabla^d z_t = \nabla[\nabla^{d-1} z_t]$ for $d > 0$, although d should be small, and frequently first differencing will suffice. Very useful for seasonal series is the seasonal difference of period s,

$$\nabla_s z_t = (1 - B^s) z_t = z_t - z_{t-s}.$$

Figures 3(a), (b), and (c) show P_t, the annual mid-year U.S. population from 1930–1982 (taken from ref. 7), and ∇P_t and $\nabla^2 P_t$. P_t is certainly nonstationary, with a strong upward trend, and ∇P_t exhibits wandering behavior that may reflect the need to take $\nabla^2 P_t$.

There is a connection between differencing and removing nonstationary means, since taking $\nabla^d z_t$ removes a polynomial μ_t of degree $d - 1$, and seasonal differencing removes a periodic mean such that $\mu_t = \mu_{t-s}$. Early efforts at dealing with nonstationary series often involved fitting a polynomial for μ_t. However, nonstationarity is typically better handled through differencing. This is illustrated for the population series in Fig. 4(a) and 4(b), which show the result of fitting the quadratic $\mu_t = a + bt + ct^2$ to P_t.

While the fit appears good [Fig. 4(a)], the residuals from the fit [Fig. 4(b)] still appear nonstationary and are certainly not independent. Differencing a series for which fitting a polynomial or periodic μ_t is really appropriate (known as overdifferencing) can be discovered and corrected in the subsequent analysis [3].

STARTING VALUES AND NONSTATIONARITY

For the stationary AR(1) model,

$$z_t = \alpha z_{t-1} + a_t, \qquad |\alpha| < 1,$$

it can be shown that $z_t = a_t + \alpha a_{t-1} + \alpha^2 a_{t-2} + \cdots$. For $|\alpha| \geqslant 1$, this does not converge. However, given a *starting value*, z_0 say, z_t for $t > 0$ can be generated recursively from $z_t = \alpha z_{t-1} + a_t$, so that

$$z_t = \alpha^t z_0 + a_t + \alpha a_{t-1} + \cdots + \alpha^{t-1} a_1. \tag{10}$$

For the random walk* model (8) set $\alpha = 1$ in (10); if the starting value is $z_0 = 0$ (fixed), then (10) becomes $z_t = \sum_{j=1}^{t} a_j$ and

$$\mathrm{var}(z_t) = t\sigma_a^2, \tag{11}$$

$$\mathrm{cov}(z_t, z_s) = \min(s, t)\sigma_a^2. \tag{12}$$

Note that (11) and (12) violate conditions (2) and (3) for stationarity.

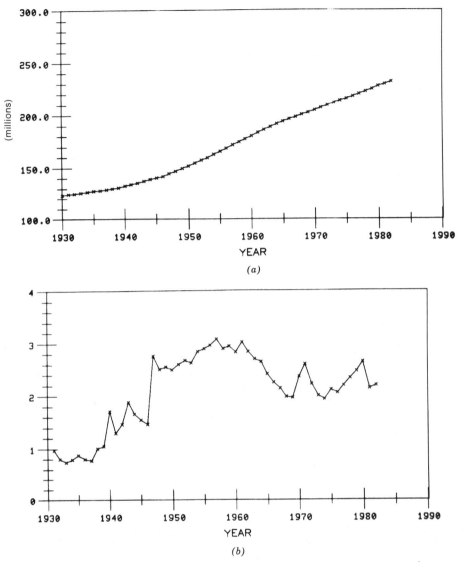

Figure 3 (*a*) P_t, the U.S. population at mid-year (in millions) 1930–1982. (*b*)∇P_t. (*c*)$\nabla^2 P_t$.

In general, if z_t is stationary it can be represented in terms of present and past innovations a_{t-j} $j \geq 0$ (the Wold decomposition*). But if we need to take $(1 - \alpha_1 B - \cdots - \alpha_d B^d)z_t$ to get stationarity, such as taking $\nabla^d z_t$, then z_t will require d starting values. The requirement of starting values is a fundamental difference between stationary series and this type of nonstationary series.

In (13) we assume

$$E[z_t] = \mu_t = 0$$

for simplicity, though we could replace z_t by $z_t - \mu_t$.

ARIMA MODELS

A useful model for nonstationary series is the autoregressive integrated moving average* [ARIMA(p, d, q)] model

$$\left(1 - \phi_1 B - \cdots - \phi_p B^p\right)\nabla^d z_t$$
$$= \left(1 - \theta_1 B - \cdots - \theta_q B^q\right)a_t, \quad (13)$$

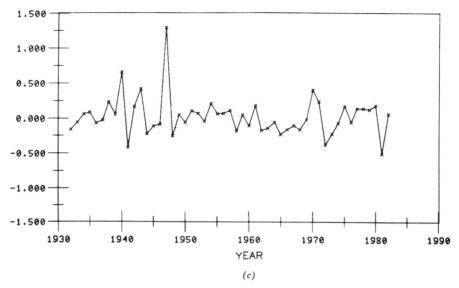

(c)

Figure 3 (*Continued*).

or

$$\alpha(B)z_t = \theta(B)a_t,$$

where $\alpha(B) = 1 - \alpha_1 B - \cdots - \alpha_{p+d}B^{p+d}$ $= \phi(B)\nabla^d$ and $\phi(B) = 1 - \phi_1 B - \cdots - \phi_p B^p$ has zeroes outside the unit circle. If $d = 0$, then z_t is stationary and model (13) is called simply an autoregressive moving average* (ARMA) model (I stands for "integrated" or "summed", the inverse of differencing). For seasonal data of period s, we can use multiplicative seasonal ARIMA models (see ARIMA models [4, Chap. 9]). Here we concentrate on the nonseasonal model (13).

SAMPLE AUTOCORRELATIONS—IDENTIFYING THE DEGREE OF DIFFERENCING

For a stationary ARMA* model, the autocorrelations (*see* SERIAL CORRELATION) $\rho_k = \mathrm{corr}(z_t, z_{t+k})$ die out exponentially with k; for large n, the same can be expected of the sample autocorrelations r_k from data $z_1, \ldots, z_n$. For the ARIMA($p, 1, q$) model, for large n the r_k's tend to decrease only slowly with k, their magnitude and pattern of decrease depending on the AR and MA parameters. For example, for the $(0, 1, 1)$

model, the decrease will be approximately linear, with the r_k's near 1 for θ_1 near 0, but considerably less than 1 for θ_1 near 1 [4, 18, 40].

These results suggest that if the sample autocorrelations of z_t do not die out, then one should difference z_t until the r_k's of the differenced series die out. In this way we can try to identify d. An analogous approach can be used for seasonal series by examining $r_s, r_{2s}, r_{3s}, \ldots$. Figure 5(a), (b), and (c) show the r_k's for the U.S. population series p_t and for ∇p_t and $\nabla^2 P_t$. For P_t, it appears we need $d = 2$.

The asymptotic distribution of $n(r_k - 1)$ for the ($p, 1, q$) model is obtained in ref. 18. It is nonnormal, in contrast to the asymptotic normality* of $n^{1/2}(r_k - \rho_k)$ in the stationary case.

ESTIMATION OF UNIT AND EXPLOSIVE ROOTS

For the stationary invertible ARMA model, the maximum likelihood* (or least-squares*) parameter estimates are consistent, and $n^{1/2}(\hat{\phi}_i - \phi_i)$ and $n^{1/2}(\hat{\theta}_i - \theta_i)$ are asymptotically normal with mean zero. Work in the nonstationary case has concentrated on

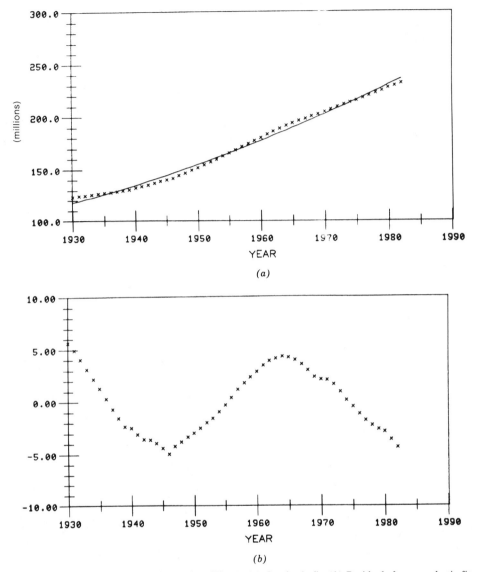

Figure 4 U.S. population at mid-year (in millions). (*a*) Quadratic fit. (*b*) Residuals from quadratic fit.

purely autoregressive models. A detailed review of the literature in this area is ref. 11.

For the AR(p) model

$$\left(1 - \alpha_1 B - \cdots - \alpha_p B^p\right)z_t = a_t,$$

the least-squares estimates $\hat{\alpha}_1, \ldots, \hat{\alpha}_p$ obtained by regressing z_t on $z_{t-1}, \ldots, z_{t-p}$ (taking $z_1, \ldots, z_p$ as fixed) are consistent estimates of $\alpha_1, \ldots, \alpha_p$ regardless of the location of the zeros of $\alpha(B)$ [26]. For the ARIMA(p, d, q) model (13) with $q > 0$, if

we estimate an AR(r) model with $r \geq d$ by least squares, then the unit roots of $\alpha(B) = \phi(B)\nabla^d$ [i.e., the $\nabla^d = (1 - B)^d$ part] are estimated consistently, but the stationary roots in $\phi(B)$ are not, due to the MA part [36]. Consistent iterative least-squares estimates of the AR parameters in ARIMA(p, d, q) are given in ref. 37.

For the stationary AR(1) model ($|\alpha_1| < 1$), $n^{1/2}(\hat{\alpha}_1 - \alpha_1)$ is asymptotically normal. For $\alpha_1 = \pm 1$, $n(\hat{\alpha}_1 - \alpha_1)$ has a skewed nonnormal asymptotic distribution, tabled in ref. 10. For $|\alpha_1| > 1$, $|\alpha_1|^n(\hat{\alpha}_1 - \alpha_1)$ has an asymptotic Cauchy distribution* if the a_t's are normal (and another distribution otherwise). The usual regression t statistic $(\Sigma_2^n z_{t-1}^2)^{1/2}(\hat{\alpha}_1 - \alpha_1)/\hat{\sigma}_a$ is asymptotically normal $(0,1)$ for $|\alpha_1| > 1$ if the a_t's are normal, and has a nonnormal asymptotic distribution for $\alpha_1 = \pm 1$, tabled in ref. 10. These results can be extended to unit and explosive root estimators in more complicated models such as AR(p), ARIMA(p, d, q), seasonal AR models, and AR models with regression terms (see ref. 11).

In practice, unit AR roots are thus estimated more accurately than stationary roots, and explosive roots more accurately than unit roots. (For the series following models (7)–(9), $\hat{\alpha}_1$ is 0.438, 0.947, and 1.1502 respectively.) Also, the asymptotic normal

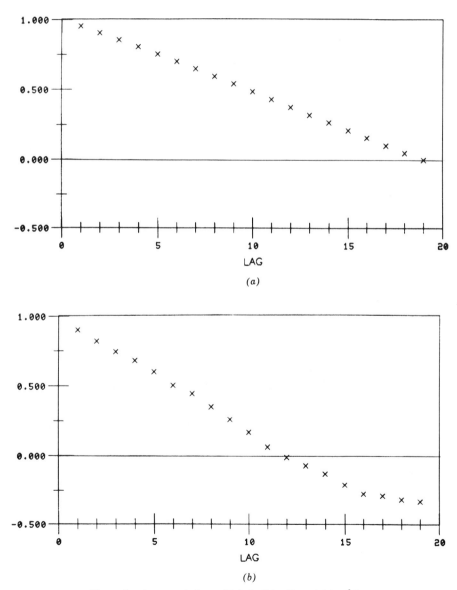

Figure 5   Autocorrelations of (a) P_t, (b) ∇P_t, and (c) $\nabla^2 P_t$.

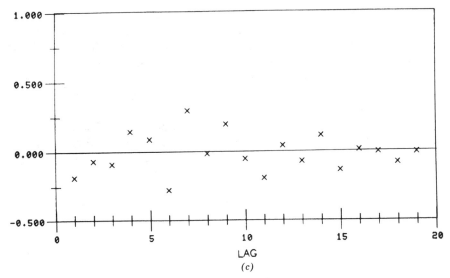

Figure 5 (*Continued*).

tion of $n^{1/2}(\hat{\alpha}_1 - \alpha_1)$ may be a bad approximation for α_1 near, but not equal to 1, so it may be better to use the asymptotic distribution for $\alpha_1 = 1$ in this case.

One way to decide if differencing is needed is to estimate an AR model and look for unit roots. Estimating an AR(2) model for the population series gives

$$P_t = 245 + 1.9033 P_{t-1} - 0.9034 P_{t-2} + a_t$$

or, approximately,

$$\nabla P_t = 245 + \hat{\alpha}_1 \nabla P_{t-1} + a_t \quad (14)$$

with $\hat{\alpha}_1 = 0.90$. Testing $\alpha_1 = 1$ in (14) yields a nonsignificant result at the 10% level, so we might use either (14) or $\nabla^2 P_t = a_t$, a model that says that the second differences of the U.S. mid-year population behave like a sequence of uncorrelated random shocks.

FORECASTING*

For the ARIMA(p, d, q) model (13), $\hat{z}_n(l)$, the minimum mean squared error forecast of z_{n+l} given $z_n, z_{n-1}, \ldots$, satisfies the homogeneous difference equation

$$\hat{z}_n(l) = \alpha_1 \hat{z}_n(l-1) + \cdots \quad (15)$$
$$+ \alpha_{p+d} \hat{z}_n(l - p - d), \quad l > q,$$

with starting values $\hat{z}_n(q), \ldots, \hat{z}_n(q - p - d + 1)$ [$\hat{z}_n(l) = z_{n+l}$ for $l \leqslant 0$]. The forecast error $z_{n+l} - \hat{z}_n(l)$ has variance

$$V(l) = \left(1 + \psi_1^2 + \cdots + \psi_{l-1}^2\right)\sigma_a^2, \quad (16)$$

where $\psi_j = \alpha_1 \psi_{j-1} + \cdots + \alpha_{p+d}\psi_{j-p-d} - \theta_j$ ($\psi_i = 0$ for $i < 0$ and $\theta_j = 0$ for $j > q$). In practice estimates of parameters are substituted in (15) and (16). [See refs. 4 (Chap. 5) and 30 (Sec. 10.2) for discussions of forecasting with ARIMA models; *see also* PREDICTION AND FORECASTING.]

Properties of $\hat{z}_n(l)$ and $V(l)$ can be deduced from properties of solutions to difference equations*. To summarize results as $l \to \infty$:

$d = 0$ (stationary):

$$\hat{z}_n(l) \to 0 \quad (\text{or} \to \mu_z \text{ if } \mu_z \neq 0),$$

$$V(l) \to \left(\sum_0^\infty \psi_j^2\right)\sigma_a^2 = \sigma_z^2;$$

$d > 0$ (nonstationary):

$\hat{z}_n(l)$ dominated by polynomial in l of degree $d - 1$,

$$V(l) \to \infty.$$

For $d = 1$, $\hat{z}_n(l)$ approaches a constant, but for $d > 1$, $\hat{z}_n(l) \to \pm\infty$. Since $V(l) \to \infty$ for $d > 0$, for forecast intervals for series

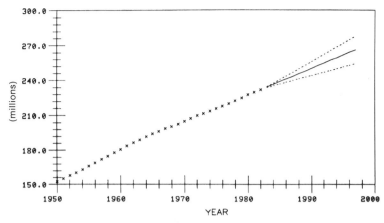

Figure 6   U.S. population at mid-year (in millions) forecasting from 1982.

requiring differencing can become quite wide as l increases.

A useful extension to the ARIMA (p, d, q) model is

$$\phi(B)\nabla^d z_t = \theta_0 + \theta(B)a_t.$$

In this case $\phi(B)\nabla^d \hat{z}_n(l) = \theta_0$ for $l > q$, a nonhomogeneous difference equation. $\hat{z}_n(l)$ will eventually be dominated by a polynomial of degree d in l, with $\theta_0/d!$ the coefficient of l^d. Figure 6 shows 20 forecasts and approximate 95% prediction limits $\hat{z}_n(l) \pm 2V(l)^{1/2}$ for the population series using the model (14); $\hat{z}_n(l)$ is approximately linear in l.

VARIATIONS AND EXTENSIONS

Proposed variations and extensions of ARIMA models include regression models with coefficients following ARIMA models [8, 17, 34], ARIMA models with time-varying parameters [24, 27, 33], periodic models for seasonal series, where the parameters and possibly the ARIMA orders vary seasonally over time [23, 28, 35], and unobserved components models, where z_t is the sum of two or more time series each following an ARIMA model or variant (*see* SMOOTHNESS PRIORS [15, 25]). The computations for some of these models are conveniently handled with the Kalman filter*. Finally, taking $\nabla^d z_t$ has been extended to noninteger d to yield "fractional differencing" or "long memory" models [14, 20], which are nonstationary for $d \geqslant \frac{1}{2}$.

NONSTATIONARY SPECTRAL ANALYSIS

Some nonstationary series appear to be changing character only slowly over time. The changing spectral content of such series can be examined by estimating the spectrum (*see* SPECTRAL ANALYSIS) separately for moving segments of the data over which the series might be assumed approximately stationary. However, it is not immediately clear what this approach estimates, since the classical definition of the spectrum (spectral density),

$$f(\omega) = (2\pi)^{-1} \sum_{-\infty}^{\infty} \gamma_k e^{-ik\omega}, \quad (17)$$

depends fundamentally on stationarity [so $\text{cov}(z_t, z_{t+k}) = \gamma_k$ does not depend on t].

Hatanaka and Suzuki [19] define the *pseudo-spectrum* for a (zero mean) series over time points $t = 1, \ldots, n$ as

$$p(\omega) = (2\pi n)^{-1} E\left[\left|\sum_{t=1}^{n} z_t e^{-i\omega t}\right|^2\right],$$

noting $\lim_{n \to \infty} p(\omega)$ gives the usual spectrum if z_t is stationary. $p(\omega)$ and the pseudo-autocovariance function $n^{-1}\sum_{t=1}^{n-k}E(z_t z_{t+k})$ $(k \geq 0)$ are a Fourier transform pair (see INTEGRAL TRANSFORMS). Cross pseudo-spectra are also defined. Ignoring the effect of spectral windows, $p(\omega)$ is the expectation of the usual spectral estimates using $z_1, \ldots, z_n$, providing an interpretation for what these are estimating in the nonstationary case.

Several approaches have been suggested to defining a time-varying spectrum [30]. Attempting to generalize (17) directly with time-varying autocovariances can lead to difficulties such as negative values for the "spectrum." Instead, Priestley [30] defines an *evolutionary spectrum* by generalizing the spectral decomposition of a stationary series to

$$z_t = \int_{-\pi}^{\pi} A_t(\omega)\, dZ(\omega), \quad (18)$$

$$E\big[|dZ(\omega)|^2\big] = h(\omega)\, d\omega.$$

Here $Z(\omega)$ is a continuous stochastic process* on $\omega \in [-\pi, \pi]$ with orthogonal increments and $h(\omega)$ is the spectrum of the stationary series $\int_{-\pi}^{\pi} e^{it\omega} Z(d\omega)$. The evolutionary spectral density is

$$f_t(\omega) = |A_t(\omega)|^2 h(\omega)$$

[for normalization, $A_0(\omega) = 1$], assuming that the Fourier transform $(2\pi)^{-1}$ $\sum_{t=-\infty}^{\infty} A_t(\omega) e^{-it\lambda}$ of the sequence $\{A_t(\omega)\}$ in t is concentrated about $\lambda = 0$ for each ω, which implies $A_t(\omega)$ varies slowly over time. There is latitude in defining $A_t(\omega)$ and hence $f_t(\omega)$; see ref. 30.

Granger and Hatanaka [13] observe that the usual spectral estimators applied to z_t satisfying (18) approximately estimate $n^{-1}\sum_{t=1}^{n} f_t(\omega)$. To estimate $f_t(\omega)$ they take $|U_t|^2$, where $U_t = \sum_j g_j z_{t-j} e^{-i\omega(t-j)}$ with $\{g_j\}$ the weights for a low-pass filter—this is called *complex demodulation*. Using this, and its cross-spectral analogue, they investigate changing spectral content and changing relationships for several economic time series. Priestley [30] suggests smoothing the $|U_t|^2$'s to estimate $f_t(\omega)$ by $\sum_l w_l |U_{t-l}|^2$.

Another approach to nonstationary spectral analysis is to find a nonstationary model for z_t and take what would correspond to the spectrum if the model were stationary. For example, if z_t follows the ARIMA model (13), we might say the spectrum is

$$f(\omega) = \sigma_a^2 \frac{|\theta(e^{-i\omega})|^2}{|\phi(e^{-i\omega})|^2} |1 - e^{-i\omega}|^{2d}, \quad (19)$$

although this goes to $+\infty$ as $\omega \to 0$. At least $|1 - e^{-i\omega}|^{2d} f(\omega)$ gives the spectrum of the stationary series $\nabla^d z_t$, and (19) has a connection with the pseudo-spectrum [19]. One might also proceed this way with time-varying models, using $\theta_t(e^{-i\omega})$ and $\phi_t(e^{-i\omega})$ in (19). Kitagawa [25] takes this approach with time-varying autoregressive models.

References

[1] Ali, M. M. and Giacotto, C. (1982). *J. Amer. Statist. Ass.*, **77**, 19–28. (Testing for changing variances over time.)

[2] *Annals of Economics and Social Measurement* (1973). **2**(4). (Entire issue on time-varying parameters in regression.)

[3] Box, G. E. P. and Abraham, B. (1978). *Appl. Statist.*, **27**, 120–130. (Discusses relation between forecasting with deterministic functions and forecasting with ARIMA models.)

[4] Box, G. E. P. and Jenkins, G. M. (1970). *Time Series Analysis: Forecasting and Control.* Holden-Day, San Francisco, CA. (Applied time series text with detailed discussion of ARIMA modelling.)

[5] Box, G. E. P. and Tiao, G. C. (1975). *J. Amer. Statist. Ass.*, **70**, 70–79. (Intervention analysis.)

[6] Brown, R. L., Durbin, J., and Evans, J. M. (1975). *J. R. Statist. Soc. B* (with discussion), **37**, 149–192. (Testing for changes in regression coefficients over time.)

[7] Bureau of the Census (1965, 1974, 1982). *Current Population Reports*, Series P-25, Nos. 311, 519, 922. Government Printing Office, Washington, D.C.

[8] Cooley, T. F. and Prescott, E. C. (1976). *Econometrica*, **44**, 167–184. (Regression models with ARIMA coefficients.)

[9] Engle, R. F. (1982). *Econometrica*, **50**, 987–1007. [ARCH (autoregressive conditional heteroscedasticity) models, modeling changing innovation variances. Under certain conditions, the unconditional variances do not change over time, so these models can be stationary.]

[10] Fuller, W. A. (1976). *Introduction to Statistical Time Series*. Wiley, New York. (Time series text covering estimation of nonstationary autoregressive models.)

[11] Fuller, W. A. (1985). *Handbook of Statistics: Time Series in the Time Domain*, Vol. 5, E. J. Hannan, P. R. Krishnaiah, and M. M. Rao, eds. North-Holland, Amsterdam, The Netherlands, pp. 1–23. (Review of literature on estimation and inference in nonstationary autoregressive models.)

[12] Gallant, A. R. and Goebel, J. J. (1976). *J. Amer. Statist. Ass.*, **71**, 961–967. (Nonlinear regression in time series.)

[13] Granger, C. W. J. and Hatanaka, M. (1964). *Spectral Analysis of Economic Time Series*. Princeton University Press, Princeton, NJ. (Text on stationary and nonstationary spectral analysis, with examples.)

[14] Granger, C. W. J. and Joyeux, R. (1980). *J. Time Series Anal.*, **1**, 15–29. (Fractional differencing models.)

[15] Grether, D. M., Nerlove, M., and Carvalho, J. L. (1979). *Analysis of Economic Time Series: A Synthesis*. Academic, New York. (Unobserved components models.)

[16] Hannan, E. J. (1970). *Multiple Time Series*. Wiley, New York. (Text written at a high mathematical level.)

[17] Harrison, P. J. and Stevens, C. F. (1976). *J. R. Statist. Soc. B* (with discussion), **38**, 205–247. (Regression models with ARIMA coefficients.)

[18] Hasza, D. P. (1980). *J. Amer. Statist. Ass.*, **75**, 349–352. (Behavior of sample autocorrelations from nonstationary ARIMA models.)

[19] Hatanaka, M. and Suzuki, M. (1967). *Essays in Mathematical Economics in Honor of Oskar Morgenstern*, M. Shubik, ed. Princeton University Press, Princeton, NJ, pp. 443–466. (Nonstationary spectral analysis.)

[20] Hosking, J. R. M. (1981). *Biometrika*, **68**, 165–176. (Fractional differencing models.)

[21] Hsu, D. A. (1977). *Appl. Statist.*, **26**, 279–284. (Testing for changing variances over time.)

[22] Hsu, D. A. (1979). *J. Amer. Statist. Ass.*, **74**, 31–40. (Testing for changing variances over time.)

[23] Jones, R. H. and Brelsford, W. M. (1967). *Biometrika*, **54**, 403–408. (Periodic models for seasonal series.)

[24] Kitagawa, G. (1983). *J. Sound Vibration*, **89**, 433–445. (Time-varying autoregressive models used for nonstationary spectral analysis. ARIMA models with time-varying parameters.)

[25] Kitagawa, G. and Gersch, W. (1984). *J. Amer. Statist. Ass.*, **79**, 378–389. (Unobserved components models.)

[26] Lai, T. L. and Wei, C. Z. (1983), *J. Multivariate Anal.*, **13**, 1–23. (Estimation and inference in nonstationary ARIMA models.)

[27] Melard, G. and Kiehm, J. L. (1981). *Time Series Analysis*, O. D. Anderson and M. R. Perryman, eds. North-Holland, Amsterdam, The Netherlands, pp. 355–363. (ARIMA models with time-varying parameters.)

[28] Pagano, M. (1978). *Ann. Statist.*, **6**, 1310–1317. (Periodic models for seasonal series.)

[29] Pierce, D. A. (1971). *Biometrika*, **58**, 299–312. (Linear regression in time series.)

[30] Priestley, M. B. (1981). *Spectral Analysis and Time Series*, Vols. 1 and 2. Academic, London, England. [Fairly mathematical time series text with discussion of nonstationary spectral analysis (Vol. 2, Chap. 11).]

[31] Quandt, R. E. (1958). *J. Amer. Statist. Ass.*, **53**, 873–880. (Testing for changes in regression coefficients over time.)

[32] Quandt, R. E. (1960). *J. Amer. Statist. Ass.*, **55**, 324–330. (Testing for changes in regression coefficients over time.)

[33] Subba Rao, T. (1970). *J. R. Statist. Soc. B*, **32**, 312–322. (ARIMA models with time-varying parameters.)

[34] Swamy, P. A. V. B. and Tinsley, P. A. (1980). *J. Econometrics*, **12**, 103–142. (Regression models with ARIMA coefficients.)

[35] Tiao, G. C. and Grupe, M. R. (1980). *Biometrika*, **67**, 365–373. (Periodic models for seasonal series.)

[36] Tiao, G. C. and Tsay, R. S. (1983). *Ann. Statist.*, **11**, 856–871. (Estimation in nonstationary ARIMA models.)

[37] Tsay, R. S. and Tiao, G. C. (1984). *J. Amer. Statist. Ass.*, **79**, 84–96. (Estimation in nonstationary ARIMA models.)

[38] Tyssedal, J. S. and Tjostheim, D. (1982). *J. Time Series Anal.*, **3**, 209–217. (Parametrizing changing variances in autoregressive models.)

[39] U.S. Department of Commerce (1980)). *Local Climatological Data, 1980: Washington, D.C., National Airport*. National Climatic Center, Asheville, N.C.

[40] Wichern, D. W. (1973). *Biometrika*, **60**, 235–239. (Behavior of sample autocorrelations from nonstationary ARIMA models.)

[41] Wichern, D. W., Miller, R. B., and Hsu, D. A. (1976). *Appl. Statist.*, **25**, 248–256. (Testing for changing variances over time.)

(AUTOREGRESSIVE-INTEGRATED
MOVING AVERAGE (ARIMA) MODELS
FORECASTING
GRADUATION
MOVING AVERAGES
PERIODOGRAM ANALYSIS

WILLIAM BELL

TIN'S RATIO ESTIMATORS

These are estimators of the ratio $R = E[Y]/E[X]$ of expected values of random variables Y and X, based on a simple random sample of n values (x_i, y_i) from a population of size N $(i = 1, \ldots, n)$. Tin [11] proposed a modified ratio estimator*, as follows. Let

$$\bar{x} = \left(\sum_{i=1}^{n} x_i \right) \Big/ n$$

and

$$\bar{y} = \left(\sum_{i=1}^{n} y_i \right) \Big/ n,$$

so that the simple ratio estimator is defined by

$$r = \bar{y}/\bar{x}.$$

Tin's estimator of R is given by

$$t_{T_1} = r \left[1 + \theta_1 \left(\frac{S_{xy}}{\bar{x}\bar{y}} - \frac{S_x^2}{\bar{x}^2} \right) \right],$$

where

$$\theta_1 = n^{-1} - N^{-1},$$

$$S_x^2 = (n - 1)^{-1} \sum_{i=1}^{n} (x_i - \bar{x})^2,$$

$$S_{xy} = (n - 1)^{-1} \sum_{i=1}^{n} (\bar{x}_i - \bar{x})(y_i - \bar{y}).$$

He also proposed a modification, obtained by subtracting an estimate of the bias of r from t_{T_1}, which yields

$$t_{T_2} = r \left[1 + \theta_1 \left(\frac{S_{xy}}{\bar{x}\bar{y}} - \frac{S_x^2}{\bar{x}^2} \right) \left(1 - 3\theta_1 \frac{S_x^2}{\bar{x}^2} \right) \right].$$

He further suggested two ratio estimators

based on extensions of Quenouille's [5] bias reduction method (*see* JACKKNIFE METHODS). The first one was

$$t_{T_3} = \frac{g}{g - 1} r - \frac{1}{g(g - 1)} \sum_{i=1}^{g} r_i',$$

where g $(2 \leq g \leq n)$ is the number of equal parts into which the sample of n is divided and r_i' is the ratio of sample means of y and x from the ith part of the sample where $i = 1, 2, 3, \ldots, g$. According to Pascual [4] and Sastry [10], H. O. Hartley had proposed this estimator earlier, with $g = n$. The second one was

$$t_{T_4} = \tfrac{8}{3} r - (r_1 + r_2)$$
$$+ \tfrac{1}{12}(r_{11} + r_{12} + r_{21} + r_{22}),$$

where r_1 and r_2 are the ratio estimators from the first and second halves $(g = 2)$ and r_{11}, r_{12}, r_{21}, and r_{22} are analogous ratio estimators obtained from dividing the first and second halves of the sample into two equal parts $(g = 4)$.

These estimators all have biases of order n^{-2}.

Tin compared t_{T_1} with r, with a special case of Quenouille's estimator* $T_{Q_2} = 2r - \bar{r}_2$, where $\bar{r}_2 = \sum_{i=1}^{2} r_j/2$, and with Beale's ratio estimator

$$t_B = r \left(1 + \theta_1 \frac{S_{xy}}{\bar{x}\bar{y}} \right) \Big/ \left(1 + \theta_1 \frac{S_x^2}{\bar{x}^2} \right).$$

It is assumed that $|(\bar{x} - E[x])/E[x]| \ll 1$ for n large enough.

Analytical comparisons were made for finite populations, infinite populations in which X and Y have a bivariate normal distribution*, and for infinite populations in which $Y = \alpha + \beta X + e$, where the e's are uncorrelated with zero mean and constant variance δ for fixed X, and X has a gamma distribution* with parameter h. He concluded that generally t_{T_1} is the most efficient, followed by t_B and t_{Q_1} in that order, and that t_{T_2} and t_{T_4} are less attractive than the alternatives considered.

See MICKEY'S UNBIASED RATIO AND REGRESSION ESTIMATORS and QUENOUILLE'S ESTIMATOR for discussions of comparisons of

Tin's with other ratio estimators by Hutchinson [3], Rao and Beegle [6], Rao and Webster [7], deGraft-Johnson and Sedransk [2], and Rao [8]. Sahoo [9] studied the estimator

$$r_S = r\left\{1 - \theta\left(\frac{S_{xy}}{\overline{xy}} - \frac{S_x^2}{\overline{x}^2}\right)\right\}^{-1}.$$

In summary, it appears that the Tin estimator t_{T_1} is one of the more efficient ratio estimators available, but it is somewhat more difficult to compute than alternatives and has no readily computable variance estimate.

References

[1] Beale, E. M. L. (1962). *Ind. Organ.* **31**, 27–28.

[2] deGraft-Johnson, K. T. and Sedransk, J. (1974). *Ann. Inst. Statist. Math.*, **26**, 339–350.

[3] Hutchinson, M. C. (1971). *Biometrika*, **58**, 2,313–321.

[4] Pascual, J. N. (1961). *J. Amer. Statist. Ass.*, **56**, 70–87.

[5] Quenouille, M. H. (1956). *Biometrika*, **43**, 353–360.

[6] Rao, J. N. K. and Beegle, L. D. (1967). *Sankhyā B*, **29**, 47–56.

[7] Rao, J. N. K. and Webster, J. T. (1966). *Biometrika*, **53**, 571–577.

[8] Rao, P. S. R. S. (1981). *J. Amer. Statist. Ass.*, **76**, 434–442.

[9] Sahoo, L. N. (1985). *J. Statist. Res.*, **17**, 1–6.

[10] Sastry, K. V. R. (1965). *Indian Soc. Agric. Statist.*, **17**, 19–29.

[11] Tin, M. (1965). *J. Amer. Statist. Ass.*, **60**, 294–307.

(MICKEY'S UNBIASED RATIO AND
  REGRESSION ESTIMATORS
PASCUAL'S ESTIMATOR
QUENOUILLE'S ESTIMATOR
RATIO ESTIMATORS)

HANS T. SCHREUDER

TOBIT *See* ECONOMETRICS; QUANTAL RESPONSE ANALYSIS

TOEPLITZ LEMMA

A form of this lemma, which is used in the study of strong convergence of sums of independent random variables* is as follows:

Let $\{a_n\}$ be a sequence of nonnegative real numbers and

$$b_n = \sum_{j=1}^{n} a_j,$$

where $b_1 > 0$ (and so $b_n > 0$ for all n) and $b_n \to \infty$ as $n \to \infty$.

Also let $\{x_n\}$ be a sequence of real numbers with $x_n \to x$ as $n \to \infty$. Then

$$b_n^{-1} \sum_{j=1}^{n} a_j x_j \to x \quad \text{as } n \to \infty.$$

(In particular if $a_n = 1$ for all n, $\overline{x} = n^{-1}\sum_{j=1}^{n} x_j \to x$ as $n \to \infty$.)

See, and compare with, KRONECKER LEMMA.

Reference

[1] Toeplitz, O. (1911). *Prace Mat.-Fiz.*, **22**, 113–119.

(LAWS OF LARGE NUMBERS)

TOEPLITZ MATRICES

A *Toeplitz matrix* is a square matrix with (i) all values in the principal diagonal equal; (ii) all values in each subdiagonal equal. Thus if $\mathbf{A} = (a_{ij})_{m \times m}$ is a Toeplitz matrix, then $a_{ij} = a_{i-j}$, and in particular,

$$a_{ii} = a_0 \qquad (i = 1, \ldots, m).$$

If $\mathbf{A}$ is a symmetrical Toeplitz matrix, then $a_{i-j} = a_{j-i}$.

If $X_1, X_2, \ldots, X_m$ are in a time series* with lag correlations $\rho(X_i, X_j) = \rho_{|i-j|}$, then their correlation matrix* is a Toeplitz matrix with

$$a_{ii} = 1, \qquad a_{ij} = \rho_{|i-j|}.$$

More generally, the covariance matrix of a stationary stochastic process* is a Toeplitz matrix.

Application of Toeplitz matrices in graduation* is described in refs. 1–4. In this connection, *banded* Toeplitz matrices, for

which
$$a_{i-j} = 0 \quad \text{for } i - j > r$$
or
$$j - i > s \quad (r, s \geq 0, r + s \leq m)$$
are of special importance. If $r + s < m$, the matrix is *strictly banded*.

A strictly banded matrix which is the inverse of a Toeplitz matrix is a *Trench matrix* (Trench [4]; Greville and Trench [3]).

References

[1] Greville, T. N. E. (1980). *J. Approx. Theory*, **33**, 43–58.

[2] Greville, T. N. E. (1981). *Scand. Actu. J.*, **64**, 39–55, 65–81.

[3] Greville, T. N. E. and Trench, W. F. (1979). *Linear Algebra Appl.*, **27**, 199–209.

[4] Trench, W. F. (1967). *SIAM J. Appl. Math.*, **15**, 1502–1510.

(GRADUATION
MATRICES (various entries))

TOLERANCE DISTRIBUTION *See* BIO-ASSAY

TOLERANCE INTERVALS *See* NON-PARAMETRIC TOLERANCE LIMITS; TOLERANCE REGIONS, STATISTICAL

TOLERANCE LIMITS

(i) In quality control*, limits defining the boundaries of acceptable quality for an individual unit of a manufacturing or service operation. The term *specification limits* is frequently used interchangeably, but it is desirable to use this term for categorizing *stated requirements* rather than for *evaluation*.

(ii) The limits of a tolerance interval* in estimating a distribution.

TOLERANCE REGIONS, STATISTICAL

INTRODUCTION

Historically, the subject of statistical tolerance regions arose and developed in response to engineers' concern with *ordinary tolerance regions*. For example, in a mass-production process, industry-wide specifications might dictate that any component that measures less than W_1 or greater than W_2 be considered as unsatisfactory. That is to say, a certain variability is tolerated and, indeed, the interval $[W_1, W_2]$ may arise from design considerations and/or cost break-even points, etc.

In fact, the manufacturer might well like to know how successfully the production process is performing in the sense that he may wish information on the probability

$$\Pr[W_1 \leqslant Y \leqslant W_2], \qquad (1)$$

where the random variable Y is the measurement of interest. As the reader can see from (1), tolerance intervals (and in general, tolerance regions) are intimately connected with the problem of *prediction*, for (1) asks: With what probability do you *predict* that Y will fall in $[W_1, W_2]$?

As a check on the tolerance interval $[W_1, W_2]$, we may wish to take sample data $\mathbf{Y} = (Y_1, \ldots, Y_n)$ and construct an interval S, with

$$S(Y_1, \ldots, Y_n)$$

$$= [L_1(Y_1, \ldots, Y_n), L_2(Y_1, \ldots, Y_n)], \quad (2)$$

where L_1 and L_2 are chosen according to different critera. S is a *statistical tolerance interval* and can be compared with the (ordinary) tolerance interval $[W_1, W_2]$.

DEFINITIONS

Suppose Y is a random variable or vector with distribution P_Y^θ, where $\theta \in \Omega$ and where outcomes y of Y belong to the sample space $\mathcal{Y}$ of Y. Suppose a sample $(Y_1, \ldots, Y_n)$ of n independent observations on Y is to be taken.

Definition 1. A *statistical tolerance region* $S(Y_1, \ldots, Y_n)$ is a (set) statistic defined over $\mathcal{Y}^n = \mathcal{Y}_1 \times \cdots \times \mathcal{Y}_n$ which takes values in the sample space $\mathcal{Y}$ of the random variable Y. ($\mathcal{Y}_i = \mathcal{Y}$ here.)

There are, of course, many types of statistical tolerance regions that may be constructed, and we now define them.

Definition 2. $S(Y_1, \ldots, Y_n)$ is a *β-content* tolerance region at *confidence level* γ if

$$\Pr\left[P_Y^\theta\left[S(Y_1, \ldots, Y_n)\right] \geq \beta\right] \geq \gamma. \quad (3)$$

In words, a *β-content* tolerance region $S(Y_1, \ldots, Y_n)$ contains at least $100\beta\%$ of the population or process being sampled with γ-confidence level (i.e., probability γ).

Let us return to the example in the Introduction. Suppose it is known to the manufacturer that unless 90% of his production is acceptable, he will lose money. Hence, he might well want to construct a tolerance interval S (we are dealing with one characteristic, for example, Y = thickness in millimeters) in such a way that the probability of the content of S, where the content is

$$P_Y^\theta[S] = P_Y^\theta\{[L_1, L_2]\} = \int_{L_1}^{L_2} f(y|\theta) \, dy,$$

$$(4)$$

being greater than $\beta = 0.90$, is at least $\gamma = 0.95$, say. Here $f(y|\theta)$ is the probability density function of Y. That is, we wish to determine $L_1(Y_1, \ldots, Y_n)$ and $L_2(Y_1, \ldots, Y_n)$ such that

$$\Pr\left[\int_{L_1}^{L_2} f(y|\theta) \, dy \geq 0.90\right] \geq 0.95. \quad (5)$$

Note that $[L_1, L_2] \in \mathcal{Y} = R^1 = [0, \infty)$, the sample space of Y.

In general, we refer to the content of a tolerance region

$$\int_S f(y|\theta) \, dy = C_\theta(S) = P_Y^\theta[S(Y_1, \ldots, Y_n)]$$

$$(6)$$

as the *coverage* of the region S. The coverage of S is, of course, a random variable. In certain applications (e.g., reliability of com-

ponents), S is one-sided, specifically

$$S(Y_1, \ldots, Y_n) = [L_1(Y_1, \ldots, Y_n), \infty], \quad (7)$$

and is used as a check on the (ordinary) tolerance interval $[W_1, \infty)$, etc.

Definition 3. $S(Y_1, \ldots, Y_n)$ is a *β-expectation* region if

$$E\{P_Y^\theta[S(Y_1, \ldots, Y_n)]\} = \beta. \quad (8)$$

Condition (8) simply requires that the coverage $P_Y^\theta(S)$ of S has expectation (mean value) β.

Note that Definitions 1–3 are quite general and may be used in a *classical sampling* approach or in a *Bayesian** approach to the problem of tolerance region construction. (3) and (8) are evaluated in the former using the sampling distribution of $C(S)$ (discussed in the next section) and for the latter using the posterior distribution* of $C(S)$ (discussed later).

THE CLASSICAL SAMPLING THEORY APPROACH

Distribution-Free Tolerance Regions

The first type of tolerance region construction proceeded using the classical sampling theory approach and involved a *distribution-free** region to be defined and discussed in the following text. Important results in this area are due to Wilks [26], Wald [23] and Tukey [22], amongst others, which we now describe succinctly, without proofs. (*See* Guttman [12] and NONPARAMETRIC CONFIDENCE INTERVALS for further discussion.)

Consider first sampling from a univariate population with distribution function

$$F_\theta(y) = \int_{-\infty}^{y} f_\theta(t) \, dt, \quad (9)$$

where θ indexes the continuous probability density functions defined on $R^1 = \mathcal{Y}$. Let a sample of n independent observations be taken and denote the order statistics* by $(Y_{(1)}, \ldots, Y_{(n)})$. Now the $(n + 1)$ intervals

$$(-\infty, Y_{(1)}], (Y_{(1)}, Y_{(2)}],$$

$$\ldots, (Y_{(n-1)}, Y_{(n)}], (Y_{(n)}, \infty) \quad (10)$$

are termed *blocks*. Their probability contents

$$F_\theta(Y_{(j)}) - F_\theta(Y_{(j-1)}), \quad j = 1, \ldots, n+1,$$

$$(11)$$

where $Y_{(0)} = -\infty$, $Y_{(n+1)} = \infty$, are the *coverages* of the blocks $(Y_{(j-1)}, Y_{(j)}]$; see (6). Let $T_i = F(Y_i)$; the T_i's are uniformly and independently distributed on $(0, 1)$ (*see* PROBABILITY INTEGRAL TRANSFORMATION), so that the ordered T's, $T_{(1)}, \ldots, T_{(n)}$, where $T_{(j)} = F(Y_{(j)})$, have joint density function

$$g(t_{(1)}, \ldots, t_{(n)})$$

$$= \begin{cases} n! & \text{if } 0 < t_{(1)} < \cdots < t_{(n)} < 1, \\ 0 & \text{otherwise.} \end{cases}$$

$$(12)$$

Note that

$$U_j = T_{(j)} - T_{(j-1)} \tag{13}$$

is the coverage of the jth block $(Y_{(j-1)}, Y_{(j)}]$ defined in (11); the density function of $U_1, \ldots, U_n$, given by

$$h(u_1, \ldots, u_n) = \begin{cases} n! & \text{if } 0 < u_i < 1, \text{ and} \\ & \sum_{i=1}^{n} u_i < 1, \\ 0 & \text{otherwise,} \end{cases}$$

$$(14)$$

is completely symmetrical in its arguments. Hence the coverage of *any* particular block will have the same properties as the coverage of any other block, for example,

$$E[U_i] = E[U_j] = 1/(n+1). \tag{15}$$

It is for this reason that the blocks $(Y_{(j-1)}, Y_{(j)}]$ corresponding to the coverages $U_1, \ldots, U_n$, $U_{n+1} = 1 - U_1 - \cdots - U_n$ are referred to as *statistically equivalent blocks*, a term due to Tukey [22], who proved the following theorem.

Theorem 1. The sum, T, of any k coverages has the beta distribution* function $I_t(k, n-k+1)$, where, in general,

$$I_t(p, q) = \int_0^t \frac{\Gamma(p+q)}{\Gamma(p)\Gamma(q)} v^{p-1}(1-v)^{q-1} \, dv.$$

$$(16)$$

Further, if C is the coverage of k blocks, then

$$E[C] = k/(n+1). \tag{17}$$

The proof may be found in Guttman [12]. This theorem holds for any distribution being sampled, so long as it is continuous, that is, belongs to the family (9). Thus, for example, if S is chosen to be

$$S(Y_1, \ldots, Y_n) = (Y_{(r)}, Y_{(n-r+1)}],$$

$$r < (n+1)/2, \tag{18}$$

then the coverage of this interval has the beta distribution with $(p, q) = (n - 2r + 1, 2r)$, so that the distribution of the coverage does not depend on the distribution of the Y's, so long as that distribution is continuous. We call such tolerance intervals *distribution-free*. Formally, we have (Y's may be vector-valued, etc.):

Definition 4. $S(Y_1, \ldots, Y_n)$ is a *distribution-free* tolerance region for $\{F_\theta(y) | \theta \in \Omega\}$ if the induced probability distribution of the coverage of S [see (6)] is independent of the parameter $\theta \in \Omega$.

Example 1. Table 1 gives the ordered urinary excretion rates (z) of cortisone found in 27 patients with cases of Cushing's syndrome, after certain treatment. (Data taken from Aitchison and Dunsmore [1].) It is standard practice to work with $y = \log_{10} z$.

Suppose the experimenter is unwilling to assume a particular distributional form for the distribution of the log rates—clearly the distribution is continuous in some interval. He (or she) might then want to construct a distribution-free interval which estimates where the central 90% of the distribution of these rates is located.

To this end, suppose he first wishes to construct a distribution-free interval S of 90% expectation. Because $n + 1 = 28$, the arithmetic is such that he might choose $r = 1$, so that $k = 26$, with

$$S = (Y_{(1)}, Y_{(27)}] = (-1.10, -0.22]. \tag{19}$$

Table 1

Patient No.	13	18	11	6	3	7	21	24	2
z = Cortisone rate	0.08	0.10	0.13	0.14	0.15	0.16	0.16	0.16	0.18
$y = \log_{10} z$	-1.10	-1.00	-0.89	-0.85	-0.82	-0.80	-0.80	-0.80	-0.74
Patient No.	8	26	25	10	14	17	19	23	9
z = Cortisone rate	0.18	0.18	0.19	0.20	0.22	0.24	0.24	0.26	0.32
$y = \log_{10} z$	-0.74	-0.74	-0.72	-0.70	-0.66	-0.62	-0.62	-0.59	-0.49
Patient No.	4	12	27	15	1	16	20	22	5
z = Cortisone rate	0.33	0.33	0.35	0.36	0.38	0.39	0.42	0.48	0.60
$y = \log_{10} z$	-0.48	-0.48	-0.46	-0.44	-0.42	-0.41	-0.38	-0.32	-0.22

From Theorem 1, the expectation of the coverage of S is

$$E[C(S)] = E\left[\int_{Y_{(1)}}^{Y_{(27)}} f(y|\theta)\, dy\right]$$

$$= 26/28 = 0.928; \qquad (20)$$

that is, $(-1.10, -0.22)$ is an estimate of where the central 92.8% of the distribution lies and/or we would predict that a future log rate Y falls in S, with probability 0.928. Or the experimenter may wish to have a statistical tolerance interval that is distribution-free* and of content 0.90, at a certain confidence level γ. It turns out that (19) is of content $\beta = 0.90$ with confidence level $\gamma = 0.784$, found using Theorem 1. Tables to facilitate the choice of content β at confidence level γ for various n have been computed by Somerville [20]; the preceding numerics were found using them—see also Guttman [12, pp. 22–23]. We will return to the data of Table 1 subsequently.

The intriguing and interesting point about Theorem 1 is that for suitably defined blocks and their coverages, it holds for two- or more dimensional random variables and regions, provided only that sampling is from a population with continuous probability density function. We illustrate with two dimensions; the extension to more than two will be obvious.

Suppose then, that sampling of (Y_1, Y_2) is from

$$F_\theta(y_1, y_2) = \int_{-\infty}^{y_1} \int_{-\infty}^{y_2} f_\theta(t_1, t_2)\, dt_2\, dt_1,$$

$$(21)$$

where f_θ is continuous. Let $h_s(y_1, y_2)$, $s = 1, \ldots, n$, be a set of n *ordering functions* defined in $\mathcal{Y} = R^2$ [it could be that $h_s(y_1, y_2) = h(y_1, y_2)$, all s] and having a continuous probability density function. Let

$$V_i^{(s)} = h_s(Y_{i1}, Y_{i2}), \qquad (22)$$

where the (Y_{i1}, Y_{i2}), $i = 1, \ldots, n$, are the n independent observations taken on (Y_1, Y_2). For each s we obtain a set of V_i's which can be ordered, for example $V_{(1)}^{(1)}, \ldots, V_{(n)}^{(1)}$, etc.

Now define a set in R^2, say

$$B^{(1)} = \left\{ (Y_1, Y_2) | h_1(Y_1, Y_2) < V_{(1)}^{(1)} \right\}, \quad (23)$$

and denote the complement of $B^{(1)}$ in R^2 by $\bar{B}^{(1)}$. The curve in R^2 defined by $h_1(Y_1, Y_2) = V_{(1)}^{(1)}$ is called a *cut*. We further define as first coverage, the content of $B^{(1)}$, labelled U_1^*, that is,

$$U_1^* = \iint_{B^{(1)}} f_\theta(t_1, t_2) \, dt_1 \, dt_2. \quad (24)$$

Now *delete from further consideration that observation* (Y_{i1}, Y_{i2}) *satisfying* $h_1(Y_{i1}, Y_{i2}) = V_{(1)}$. Continuing with the remaining $(n - 1)$ observations, we use the curve $V_{(1)}^{(2)} = h_2(Y_1, Y_2)$, which cuts the set $\bar{B}^{(1)}$ into two sets $B^{(2)}$ and $\bar{B}^{(2)}$, where $B^{(2)} \subset B^{(1)}$ is such that $h_2(X, Y) < V_{(1)}^{(2)}$. In this way, we can define sets $B^{(1)}, \ldots, B^{(n)}$ such that

$$B^{(2)} \subset \bar{B}^{(1)}, \ldots, B^{(n)} \subset \bar{B}^{(n-1)} \quad (25)$$

and a residual set $B^{(n+1)}$ such that

$$B^{(1)} \cup B^{(2)} \cup \cdots \cup B^{(n)} \cup B^{(n+1)} = R^2. \quad (26)$$

We define coverages $U_1^*, \ldots, U_n^*$ such that

$$U_j^* = \int_{B^{(j)}} \cdots \int f_\theta(y_1, y_2) \, dy_1 \, dy_2. \quad (27)$$

Tukey [22] proved the following:

Theorem 2. The coverages U_i^*, $i = 1, \ldots, n$, for the blocks $B^{(i)}$ have the same distribution as the coverages U_i for the blocks

of Theorem 1, namely

$$h(u_1^*, \ldots, u_n^*) =$$

$$\begin{cases} n! \, du_1^* \cdots du_n^*, & \text{if } 0 < u_i^* < 1, \sum_1^n u_i^* < 1, \\ 0, & \text{otherwise.} \end{cases}$$

$$(28)$$

Hence we are at the same place as in (14), so that *Theorem 1 now holds* for any of the k coverages U_i^* of Theorem 2.

Another particularly interesting point is that if, at any stage, ordering functions are used which depend on the observed values of the previous ordering functions, then the coverages of any n of the $(n + 1)$ blocks will still have the distribution specified by (28) (see Fraser [7] and Kemperman [14]). It affords great flexibility in constructing distribution-free tolerance regions and allows us to arrive at a region with predetermined shape. We illustrate all this with the following example.

Example 2. Table 2 gives the urinary excretion rates (in units of mg/24 h) for cortisol (z_1) and cortisone (z_2) of 27 patients with Cushing's disease. (The data of Table 2 are actually the ordered z_2's.) Since it is standard practice to work with log rates, Table 2

Table 2   Data from 27 patients with Cushing's disease

Patient No.	(z_1, z_2)	(y_1, y_2)	Patient No.	(z_1, z_2)	(y_1, y_2)
1	0.41, 0.38	−0.39, −0.42	15	0.48, 0.36	−0.32, −0.44
2	0.16, 0.18	−0.80, −0.74	16	0.80, 0.39	−0.10, −0.41
3	0.26, 0.15	−0.59, −0.82	17	0.40, 0.24	−0.40, −0.62
4	0.34, 0.33	−0.47, −0.48	18	0.22, 0.10	−0.66, −1.00
5	1.12, 0.60	0.05, −0.22	19	0.24, 0.24	−0.62, −0.62
6	0.15, 0.14	−0.82, −0.85	20	0.56, 0.42	−0.25, −0.38
7	0.20, 0.16	−0.70, −0.80	21	0.40, 0.16	−0.40, −0.80
8	0.26, 0.18	−0.59, −0.74	22	0.88, 0.48	−0.06, −0.32
9	0.56, 0.32	−0.25, −0.49	23	0.44, 0.26	−0.36, −0.59
10	0.26, 0.20	−0.59, −0.70	24	0.24, 0.16	−0.62, −0.80
11	0.16, 0.13	−0.80, −0.89	25	0.27, 0.19	−0.57, −0.72
12	0.56, 0.33	−0.25, −0.48	26	0.18, 0.18	−0.74, −0.74
13	0.33, 0.08	−0.48, −1.10	27	0.60, 0.35	−0.22, −0.46
14	0.26, 0.22	−0.59, −0.66			

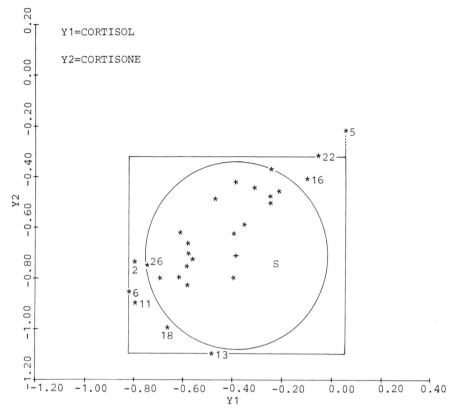

Figure 1 A circular 0.679 expectation tolerance region S (based on the data of Table 2) composed of 19 blocks. The numbers shown beside the 9 points (Y_{i1}, Y_{i2}) that do not lie in S are the corresponding patient numbers i.

also gives $y_i = \log_{10} z_i$. (Data from Aitchison and Dunsmore [1].)

The $n = 27$ observations (y_{i1}, y_{i2}) are plotted in Fig. 1. Consider the ordering functions

$$V^{(1)} = h_1(Y_1, Y_2) = Y_1,$$
$$V^{(2)} = h_2(Y_1, Y_2) = Y_2,$$
$$V^{(3)} = h_3(Y_1, Y_2) = -Y_1,$$
$$V^{(4)} = h_4(Y_1, Y_2) = -Y_2. \tag{29}$$

Recall that we delete from further consideration the observation that is such that $h_s(Y_{i1}, Y_{i2}) = V_1^{(s)}$ as s goes from $1, \ldots, 4$, etc. Then for the data of Table 2,

$$v_{(1)}^{(1)} = -0.82,$$
$$v_{(1)}^{(2)} = -1.10,$$
$$v_{(1)}^{(3)} = -0.05,$$
$$v_{(1)}^{(4)} = 0.32. \tag{30}$$

This means for example, that $B^{(1)}$ is the set (see Fig. 1)

$$B^{(1)} = \{(y_1, y_2)|y_1 < -0.082\}, \tag{31}$$

while $B^{(4)}$ is the set (contained in $\overline{B}^{(3)}$) such that

$$B^{(4)} = \{(y_1, y_2)|y_2 < -0.32\}; \tag{32}$$

these correspond to the cuts

$$y_1 = 0.082 \quad \text{and} \quad y_2 = -0.32, \quad \text{etc.} \tag{33}$$

Having used the four ordering functions (29), suppose we now use the ordering functions for the remaining $N - 4$ points (located interior to the rectangle found using the obvious four cuts; see Fig. 1):

$$h_{j+4} = h(Y_1, Y_2)$$
$$= -\left\{(Y_1 - Y_{1c})^2 + (Y_2 - Y_{2c})^2\right\}^{1/2}, \tag{34}$$

where (Y_{1c}, Y_{2c}) is the center of the rectangle previously described, given by

$$Y_{1c} = \left(V_{(1)}^{(1)} - V_{(1)}^{(3)} \right)/2,$$
$$Y_{2c} = \left(V_{(1)}^{(2)} - V_{(1)}^{(4)} \right)/2, \quad (35)$$

which, for the data of Table 2, are observed to be

$$y_{1c} = (-0.82 + 0.05)/2 = -0.385,$$
$$y_{2c} = (-1.10 - 0.32)/2 = -0.71. \quad (36)$$

Clearly h of (36) depends on h_i, $i = 1, 2, 3$ and 4, and clearly the h_i and h generate $(n + 1)$ blocks $B^{(j)}$, with the last 24 ($= n - 3$) generated using (34) with (35), namely,

$$h(y_1, y_2) = -\left\{ (y_1 + 0.385)^2 \right.$$
$$\left. + (y_2 + 0.710)^2 \right\}^{1/2}. \quad (37)$$

If we wish our tolerance region to be circular, we may use s inner circular blocks; using the theorems of Fraser [7] and Kemperman [14], the coverage of these s blocks has distribution $I_t(s, n - s + 1)$. Figure 1 shows a circular region composed of the blocks $B^{(10)}, \ldots, B^{(28)}$. This is a region of expectation $100\beta = 100s/(n + 1) = 67.9\%$.

Put another way, we would predict a Cushing syndrome patient to have (Y_1, Y_2) = (log cortisol, log cortisone) falling in this circular region with probability 0.679. Also, we can view this region as being of content $\beta = 0.679$ at confidence level $\gamma = 0.472$, as determined from Somerville's tables, referred to previously. We will return to this data subsequently.

Returning to β-expectation regions, the reader will have noticed that we continually have made the intuitive interpretation of predicting a future observation to fall in the region of β-expectation with probability β. This can be made precise using a lemma due to Paulson [18].

Paulson's Lemma. If on the basis of a given sample on $\mathbf{Y}$, a t-dimensional random variable, a t-dimensional confidence (prediction) region S of *level* β is found for a future observation, say $\mathbf{Y}^{(f)}$, and if C denotes the probability that $\mathbf{Y}^{(f)}$ will fall in S, then

$$E[C] = \beta,$$

that is, the probability that $\mathbf{Y}^{(f)} \in S$ is the expectation of the coverage of S.

For discussion and proof, see Paulson [18] and Guttman [12]. This lemma explains why $(Y_{(r)}, Y_{(n-r+1)}]$, which is of expectation $\beta = (n - 2r + 1)/(n + 1)$, can be used as a prediction interval of level β for a future Y or, in Example 2, why the circular region of expectation $100\beta\% = 67.9\%$ can be used as a prediction region for a future (Y_1, Y_2), etc. In fact this lemma is very general and applies to other types (other than distribution free) of regions which are of β-expectation, to be discussed in the following text.

Other properties of distribution-free tolerance regions are discussed in Fraser and Guttman [8]. One that should be mentioned here is that there *do not* exist distribution-free tolerance regions $S(Y_1, \ldots, Y_n)$, symmetric in the Y_i's, if sampling is from discrete distributions.

Parametric Tolerance Regions of β-Expectation

In this section, we discuss tolerance regions $S(Y_1, \ldots, Y_n)$ (the Y_i may be vector-valued) which possess the property (8) of β-expectation. Indeed, we have seen how to construct distribution-free tolerance regions which are, for example, of $s/(n + 1)$ expectation, and succinctly, this is accomplished by using s out of the $(n + 1)$ statistically equivalent blocks, where the blocks are determined by certain ordering functions and related cuts, etc. The distribution being sampled was assumed continuous, but otherwise unknown.

Now when we are in the position of being able to assume that P_Y^θ, the distribution being sampled, is of a certain functional form, this added information leads to regions different from those provided by the distribution-free case. We now show how to utilize this type of information along with Paulson's lemma to construct β-expectation regions.

Suppose then, that the distribution being sampled is one of the class of normal probability measures, $\{P^\theta_Y | \theta \in \Omega\}$, where

$$\theta = (\mu, \sigma^2),$$

$$\Omega = \left\{ (\mu, \sigma^2) \,\middle|\, \begin{array}{c} -\infty < \mu < \infty \\ 0 < \sigma^2 \end{array} \right\},$$

$$P^\theta_Y(y) = \int_{-\infty}^y (2\pi\sigma^2)^{-1/2} \qquad (38)$$

$$\times \exp\left[-(t - \mu)^2/(2\sigma^2) \right] dt.$$

Suppose $Y_1, \ldots, Y_n$ is a random sample of n independent observations from one of the distributions in this class, and denote the sufficient statistic* for θ by

$$(\overline{Y}, V^2) = \left(n^{-1} \sum_1^n Y_i, \right.$$

$$\left. (n - 1)^{-1} \sum_{i=1}^n (Y_i - \overline{Y})^2 \right). \qquad (39)$$

Suppose we wish to construct a tolerance interval with ability to pick up the center $100\beta\%$ of the normal distribution $N(\mu, \sigma^2)$ being sampled. We appeal to Paulson's lemma. Suppose $Y^{(f)}$ is a further independent observation from $N(\mu, \sigma^2)$, that is, $(Y_1, \ldots, Y_n, Y^{(f)})$ are all independent, and let us construct an interval which is a prediction interval of level β for $Y^{(f)}$, that is, because of the preceding considerations, we wish to find

$$(\overline{Y} - KV, \overline{Y} + KV] \qquad (40)$$

such that

$$\beta = \Pr(\overline{Y} - KV < Y^{(f)} \le \overline{Y} + KV). \qquad (41)$$

But (41) is clearly equivalent to

$$\beta = \Pr\left[-K/\sqrt{(n + 1)/n} \right.$$

$$\left. < W \le K/\sqrt{(n + 1)/n} \right], \qquad (42)$$

where

$$W = (Y^{(f)} - \overline{Y})/\left[\sqrt{(n + 1)/n}\, V \right] = t_{n-1}, \qquad (43)$$

with t_{n-1} denoting a Student t variable with $n - 1$ degrees of freedom. Hence, if

$$K = \sqrt{(n + 1)/n}\, t_{n-1;(1-\beta)/2}, \qquad (44)$$

then (42) is satisfied, so that (40) is a prediction (confidence) interval for $Y^{(f)}$ at level β. (In general, $t_{m;\gamma}$ denotes the point exceeded with probability γ using the Student t distribution* with m degrees of freedom.) Hence from Paulson's lemma,

$$\left(\overline{Y} - \sqrt{(n + 1)/n}\, Vt_{n-1;(1-\beta)/2}, \right.$$

$$\left. \overline{Y} + \sqrt{(n + 1)/n}\, Vt_{n-1;(1-\beta)/2} \right] \qquad (45)$$

is a region of β-expectation. This result was first obtained by Wilks [26]. The β-expectation tolerance region may be found using techniques of hypothesis testing*, and hence has certain optimum properties; see the summary in Guttman [12, pp. 34–39, etc.].

Example 3. We return to the data of Table 1. For comparison with the distribution-free interval (19) of 0.928 expectation, suppose we may assume normality and that we wish to construct an interval of the type (45) of 0.928 expectation. For the data of Table 1, we find

$$\bar{y} = -0.6404, \qquad v = 0.2141, \qquad n = 27,$$

$$\beta = 0.928, \qquad t_{26;0.036} = 1.902, \qquad (46)$$

so that the 0.928-expectation tolerance interval, based on normality, is observed to be

$$(-1.0551, -0.2257]. \qquad (47)$$

The length of the interval (47) is, to two decimal places, 0.83, while the length of the distribution-free interval (19) is 0.88; recall that both intervals are of 0.928 expectation. This is a rather typical result, for obvious reasons.

In his important paper, Paulson [18] also discussed sampling from the bivariate normal distribution*. We will return to his solution after discussing the methodology for sampling from the k-variate normal distribution $N(\mathbf{\mu}, \mathbf{\Sigma})$, $\mathbf{\Sigma}$ a positive-definite variance–covariance matrix of order $(k \times k)$, etc. Let $(\mathbf{Y}, \ldots, \mathbf{Y}_n)$ be a sample of n independent observations from this process, and let

$$\mathbf{y} = n^{-1} \sum_{i=1}^{n} \mathbf{Y}_i,$$

$$\mathbf{V} = (n-1)^{-1} \sum_{i=1}^{n} (\mathbf{Y}_i - \overline{\mathbf{Y}})(\mathbf{Y}_i - \overline{\mathbf{Y}})' \quad (48)$$

$$= (v_{ij})$$

denote the sufficient statistics* for (μ, Σ). Further, let

$$U^2 = (\mathbf{Y}^{(f)} - \overline{\mathbf{Y}})'\mathbf{V}^{-1}(\mathbf{Y}^{(f)} - \overline{\mathbf{Y}}), \quad (49)$$

where $\mathbf{Y}^{(f)}$ denotes a further independent observation from $N(\mu, \Sigma)$, etc. Suppose we use as a prediction interval for $\mathbf{Y}^{(f)}$, based on the data $(\mathbf{Y}_1, \ldots, \mathbf{Y}_n)$, the (ellipsoidal) region

$$S(\mathbf{Y}_1, \ldots, \mathbf{Y}_n) = \left\{ \mathbf{Y}^{(f)} | (\mathbf{Y}^{(f)} - \overline{\mathbf{Y}})'\mathbf{V}^{-1} \right.$$
$$\left. \times (\mathbf{Y}^{(f)} - \overline{\mathbf{Y}}) \le C_\beta \right\}, \quad (50)$$

and that we wish this region to be of the level β. Then

$$U^2 = [(n+1)/n]T^2$$
$$= [(n+1)/n][(n-1)k/(n-k)]$$
$$\times F_{k, n-k}, \quad (51)$$

where T^2 is a Hotelling-T^{2*} statistic (*see* MULTIVARIATE ANALYSIS), so that C_β of (50) is given by

$$C_\beta = [(n+1)/n]$$
$$[(n-1)k/(n-k)]F_{k, n-k; 1-\beta}, \quad (52)$$

where, in general, $F_{m_1, m_2; \gamma}$ denotes the point exceeded with probability γ of F-distribution* with (m_1, m_2) degrees of freedom. Tables of C_β for various values of n and β, for $k = 2$, 3, and 4, are available in Guttman [12]. For $k = 1$ and 2, we are led to the solutions first given by Paulson [18]. For the case $k = 1$, and hence dealing with the interval (45), tables of C_β are also available in Guttman [12], for various values of n and β. Also (50) may be found using techniques of hypothesis testing as described in Guttman [12, pp. 44–49, etc.].

Example 4.   We return to the data of Table 2. It turns out that $(k = 2)$

$$\overline{\mathbf{y}} = (-0.4650, -0.6404)',$$

$$\mathbf{V} = \begin{bmatrix} 0.05560272 & 0.04030955 \\ 0.04030955 & 0.04581825 \end{bmatrix}. \quad (53)$$

If we choose $\beta = 0.90$, then the observed elliptical region (50) becomes

$$S = \left\{ (Y_1^{(f)}, Y_2^{(f)}) | U^2 \le C_{0.90} \right\}, \quad (54)$$

where

$$U^2 = 49.6535 \left(Y_1^{(f)} + 0.4650 \right)^2$$

$$+ 60.2570 \left(Y_2^{(f)} + 0.6404 \right)^2$$

$$- 2(43.6837) \left(Y_1^{(f)} + 0.4650 \right)$$

$$\times \left(Y_2^{(f)} + 0.6404 \right), \quad (55)$$

with

$$C_{0.90} = 5.454, \quad (56)$$

using the tables referenced already. Figure 2 gives a plot of the data of Table 2 and the region (50), based on normality, for $\beta = 0.90$ and $\beta = 0.679$, the latter for comparison with Fig. 1. The shapes of the regions in Figs. 1 and 2 (for $\beta = 0.679$) are different, of course, because here we are using the assumption of normality.

The case where the distribution being sampled is the single exponential is discussed in detail in Guttman [12], as well as the case of the double exponential with mean known; needed tables are also given there.

Tolerance Regions for β Content

In this section, we discuss the construction of tolerance regions S that are of β content at level γ, that is, satisfying Definition 2. We have already met such regions, specifically, those which are distribution free (see for instance, Example 1). The ingredients there are: Y (possibly vector-valued) has continuous distribution function P_Y^θ; by means of ordering functions, s out of $(n+1)$ blocks are chosen as the statistical tolerance region S. Then the coverage $C(S)$ of S, where

$$C(S) = \int_S (f(y)|\theta) \, dy = P_Y^\theta(S), \quad (57)$$

has the standard beta distribution* $I(s, n-$

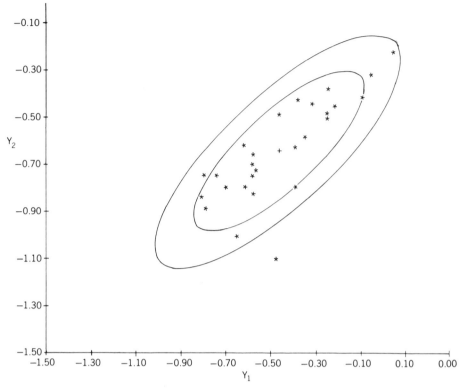

Figure 2 β-expectation tolerance regions based on the assumption of bivariate normality for the data of Table 2. [Outer ellipse for $\beta = 0.90$, inner ellipse for $\beta = 0.679$; see (50), etc.]

$s + 1$). Thus

$$\Pr[C \geq \beta] = \int_{\beta}^{1} \frac{\Gamma(n + 1)}{\Gamma(s)\Gamma(n - s + 1)}$$

$$\times u^{s-1}(1 - u)^{n-s} du, \quad (58)$$

and various tables exist, such as those referenced in Distribution-Free Tolerance Regions, to aid in the choice of n for given (s,β) to make (58) have value γ, etc.

However, in many instances, as indicated in Parametric Tolerance Regions, the experimenter is willing to assume a given functional form for the distribution being sampled. This changes the nature of the regions so found.

When sampling is from the univariate normal distribution, it is often of interest to estimate where the central $100\beta\%$ of the distribution lies, that is, to estimate the set

$$A_c = (\mu - \sigma z_{(1 - \beta)/2}, \mu + \sigma z_{(1 - \beta)/2}), \quad (59)$$

where, in general z_δ is the point exceeded with probability δ when using the standard

normal $N(0, 1)$ distribution. If $(Y_1, \ldots, Y_n)$ is a sample of n independent observations from $N(\mu, \sigma^2)$ and we wish to construct $S(Y_1, \ldots, Y_n)$ which is β-content at level γ and which will serve as estimator of A_c, then a reasonable candidate is

$$S(Y_1, \ldots, Y_n) = (\overline{Y} - KV, \overline{Y} + KV] \quad (60)$$

with

$$(\overline{Y}, V^2) = \left(n^{-1} \sum_{i=1}^{n} Y_i, \right.$$

$$\left. (n - 1)^{-1} \sum_{i=1}^{n} (Y_i - \overline{Y})^2\right),$$

and where K is chosen so that this tolerance interval is β content at confidence level γ. For given K, the interval S has coverage C, where

$$C = \Phi\left(\frac{\overline{Y} - \mu}{\sigma} + K\frac{V}{\sigma}\right)$$

$$- \Phi\left(\frac{\overline{X} - \mu}{\sigma} - K\frac{V}{\sigma}\right) \quad (61)$$

and $\Phi(z)$ is the cumulative distribution function of the $Z = N(0, 1)$ random variable. From the results of Parametric Tolerance Regions,

$$E[C] = \beta', \tag{62}$$

where β' is the root of the equation

$$t_{n-1;(1-\beta')/2} = K/(1 + n^{-1})^{1/2}. \tag{63}$$

In general, the distribution of C is very complicated and the choice of K which will make (60) of β content at level γ is difficult. However, Wald and Wolfowitz [24] gave an approximation for K which is extremely good, even for values of n as low as 2, provided β and γ are both greater than 0.95. It is as follows: Set

$$K = k_1 k_2, \tag{64}$$

where $k_1 = k_1(n; \beta)$ satisfies

$$\Phi(n^{-1/2} + k_1) - \Phi(n^{-1/2} - k_1) = \beta \tag{65}$$

and $k_2 = k_2(n - 1; \gamma)$ is such that

$$k_2 = \left[(n - 1)/\chi^2_{n-1; \gamma}\right]^{1/2}; \tag{66}$$

here $\chi^2_{m; \delta}$ is the point exceeded with probability δ when using the chi-square distribution* with m degrees of freedom.

Theorem 3 (Ellison [5]). If $(Y_1, \ldots, Y_n)$ are n independent observations from the $N(\mu, \sigma^2)$ distribution and if we wish to construct a β-content tolerance interval at confidence level γ of the form (60) with $K = k_1 k_2$, where the k_i are defined in (65) and (66), then S has coverage C given by (61) with $K = k_1 k_2$, which satisfies

$$\text{Pr}_{\mu, \sigma^2}[C \geq \beta] = \gamma' \tag{67}$$

with $|\gamma - \gamma'| = O(n^{-1})$. Further, n^{-1} is the exact rate of convergence, except when $\gamma = \frac{1}{2}$.

To facilitate construction of such intervals, Bowker [2] has given a table of K based on (64) for various γ, β and extensive values of n. Weissberg and Beatty [25] tabulate $k_1 = k_1(n; \beta)$ and $k_2 = k_2(f; \gamma)$

separately, where f is the degrees of freedom associated with the estimator of σ^2, assumed independent of the estimator of $\mu = E[Y]$. For the case discussed so far, $f = n - 1$, but the Beatty and Weissberg tables can be utilized for more complicated situations, such as linear regression* in p independent variables, that is, $E[Y] = x'\theta$, so that (in standard notation)

$$V^2 = (\mathbf{Y} - \mathbf{X}\hat{\theta})'(\mathbf{Y} - \mathbf{X}\hat{\theta})/(n - p)$$

has $f = n - p$ degrees of freedom associated with it, etc.

As to the coverage of S, the following theorem gives an approximation to its distribution.

Theorem 4 (Guttman [12, pp. 61–62]). If the coverage C of a tolerance interval S of the form (60) has mean and variance μ_c and σ_c^2, then to terms of order $1/n$,

$$\mu_c = [2\Phi(K) - 1] - K\Phi(K)(2n)^{-1},$$
$$\sigma_c^2 = [2K^2\phi^2(K)]/n \tag{68}$$
$$= [K^2\exp\{-K^2\}]/(\pi n).$$

Since C lies in $(0, 1)$, a solution found to be very good for $n \geq 100$ is to approximate the distribution of C by a standard beta distribution with

$$p = [\mu_c^2(1 - \mu_c) - \mu_c\sigma_c^2]/\sigma_c^2,$$
$$q = [\mu_c(1 - \mu_c)^2 - (1 - \mu_c)\sigma_c^2]/\sigma_c^2, \tag{69}$$

where μ_c and σ_c^2 are given in (68). The approximation is good in the sense that the percentage error of K determined by Theorem 4 with (69) in relation to the K given by Bowker [2] is small; see Guttman [11].

When sampling from the univariate normal, interest often focuses on one of the tails of the distribution, for example, in estimating where the left-hand tail A_L of β content of the $N(\mu, \sigma^2)$ distribution being sampled lies, with

$$A_L = (-\infty, \mu + \sigma z_{1-\beta}]. \tag{70}$$

If $(Y_1, \ldots, Y_n)$ is a sample of n independent observations from this distribution, then an interval of the form

$$S(Y_1, \ldots, Y_n) = (-\infty, \overline{Y} + K'V] \quad (71)$$

seems sensible. [In order to make the tolerance interval S of β expectation, K' is set equal to $(1 + n^{-1})^{1/2} t_{n-1; 1-\beta}$, as seen from (44).] Now the coverage C of (71) is

$$C = \Phi\left(\frac{\overline{Y} - \mu}{\sigma} + K'\frac{V}{\sigma}\right) \quad (72)$$

and if we wish to select $K' = K'(n; \gamma, \beta)$ so that S of (71) is of β content at level γ, we need the following theorem:

Theorem 5 (Guttman [12]). Sampling from a $N(\mu, \sigma^2)$ population and using the preceding notation, the coverage C defined by (72) of the interval S given by (71) is such that

$$\Pr[C \geq \beta] = \Pr\left[T_{n-1}^*\left(\sqrt{n}\, z_{1-\beta}\right) \leq \sqrt{n}\, K'\right], \quad (73)$$

where $T_f^*(\delta)$ is a noncentral t^* variable with f degrees of freedom and noncentrality parameter δ.

Tables of K' are given in Owen [15–17] and reproduced in Guttman [12]. The cases of sampling from the exponential and k-variate normal, which proceed in the same fashion as previously, are discussed in Guttman [12].

THE BAYESIAN APPROACH

Tolerance Regions of β Expectation

For the experimenter/statistician who approaches statistical inference from the Bayesian* route, the coverage of a tolerance interval S is a function of the parameters θ, once having seen the data $(Y_1, \ldots, Y_n) = (y_1, \ldots, y_n)$, so that, given the data,

$$C(S) = \int_S f(y|\theta)\, dy = P_Y^\theta[S(y_1, \ldots, y_n)] \quad (74)$$

depends on θ. Here the posterior distribu-

tion* $p(\theta|y_1, \ldots, y_n)$ of θ is given by

$$p(\theta|y_1, \ldots, y_n) = Kl(\theta|y_1, \ldots, y_n)p(\theta); \quad (75)$$

l denotes the likelihood* of θ based on the data $(y_1, \ldots, y_n)$, and $p(\theta)$ is the prior distribution* of θ, with the normalizing constant K such that

$$K^{-1} = \int_{\theta \in \Omega} p(\theta|y_1, \ldots, y_n)p(\theta)\, d\theta. \quad (76)$$

(Note that θ and y_j's may be vector-valued.) Thus, to a Bayesian, Definition 3 implies that $C(S) = C(S; \theta)$ has expectation β, where the expectation is taken with respect to the posterior distribution of θ given by (76), so that we require S to be such that

$$E[C(S; \theta)|y_1, \ldots, y_n] = \beta. \quad (77)$$

The consequences of (77) are very interesting. To begin with, it may be rewritten as

$$\beta = \int_\Omega \int_S f(y|\theta)p(\theta|y_1, \ldots, y_n)\, dy\, d\theta, \quad (78)$$

where $f(y|\theta)$ is the distribution being sampled. Assuming that the conditions of Fubini's theorem hold, we may invert the order of integration, so that

$$\beta = \int_S \int_\Omega f(y|\theta)p(\theta|y_1, \ldots, y_n)\, d\theta\, dy$$
$$= \int_S h(y|y_1, \ldots, y_n)\, dy, \quad (79)$$

where

$$h(y|y_1, \ldots, y_n)$$
$$= \int_\Omega f(y|\theta)p(\theta|y_1, \ldots, y_n)\, d\theta \quad (80)$$

is the conditional distribution of Y, given the data $y_1, \ldots, y_n$, where Y is an additional observation from $f(y|\theta)$, independent of $Y_1, \ldots, Y_n$. The density $h(y|y_1, \ldots, y_n)$ has been called the *predictive* or *future distribution* of Y. It is argued elsewhere (e.g., Guttman [10] and Raiffa and Schlaifer [19]) that for prediction purposes, the density h is all that is necessary. For example, the modes of h or $E(Y|y_1, \ldots, y_n)$ are often used as predictors of a future Y, etc.

Additionally, (79) says that a tolerance region S is a β-expectation tolerance region if it is a β-confidence region for Y, where Y has the predictive distribution given by (80). This is the analogue of Paulson's lemma. Thus, for given $f(y|\theta)$, we need only find $h(y|y_1, \ldots, y_n)$ and a region S such that

$$P(Y \in S|y_1, \ldots, y_n)$$
$$= \int_S h(y|y_1, \ldots, y_n)\, dy = \beta. \quad (81)$$

The preceding then says that S will be a β-expectation region; in other words, find a *predictive* region S of level β and then S will be of expectation β.

To illustrate, consider first the case where sampling is from the (univariate) exponential* distribution given by

$$f(y|\sigma) = \sigma^{-1}\exp(-y/\sigma), \qquad y > 0, \sigma > 0, \quad (82)$$

a distribution that arises in many life-testing* and reliability* situations. The likelihood function based on n independent observations $(y_1, \ldots, y_n)$ is

$$l(\sigma|y_1, \ldots, y_n) \propto \sigma^{-n}\exp(-t/\sigma) \quad (83)$$

with $t = \sum_{i=1}^n y_i$. A prior $p(\sigma)$ that is often used in this situation for σ is the *conjugate prior* (*see* CONJUGATE FAMILIES OF DISTRIBUTIONS) with parameters (n_0, t_0), such that

$$p(\sigma) \propto \sigma^{-(n_0+1)}\exp\{-t_0/\sigma\}. \quad (84)$$

The posterior distribution of σ given $(y_1, \ldots, y_n)$ is

$$p(\sigma|y_1, \ldots, y_n) = \frac{(t + t_0)^{n+n_0}}{\Gamma(n + n_0)}$$
$$\times \sigma^{-(n+n_0+1)}\exp\left[-\frac{(t_0 + t)}{\sigma}\right], \quad (85)$$

so that $\sigma = 2(t + t_0)/\chi^2_{2(n+n_0)}$, a posteriori. Using (80), we find that the predictive distribution of y, given $y_1, \ldots, y_n$, is

$$h(y|y_1, \ldots, y_n)$$
$$= (n + n_0)[1 + y/(t + t_0)]^{-(n+n_0+1)}. \quad (86)$$

Suppose now that we are interested in the right hand $100\beta\%$ of the exponential process

being sampled and that we wish to find an interval of the form $S = [a(y_1, \ldots, y_n), \infty)$, such that

$$\int_a^\infty h(y|y_1, \ldots, y_n)\, dy$$
$$= (n + n_0)$$
$$\times \int_a^\infty [1 + y/(t + t_0)]^{-(n+n_0+1)}\, dy$$
$$= \beta. \quad (87)$$

This implies that

$$a = (t + t_0)d_\beta,$$
$$d_\beta = [(1/\beta)^{1/(n+n_0)} - 1]. \quad (88)$$

Hence the interval

$$S(y_1, \ldots, y_n) = [(t + t_0)d_\beta, \infty] \quad (89)$$

is of β-expectation. That is, in predicting that a (future) Y will fall in S as given by (89) with probability β, we know that the coverage of $C(S)$ has (posterior) expectation β.

If we allow the parameter (n_0, t_0) to tend to zero in such a way that $p(\sigma)$ of (84) becomes more and more diffuse, then $p(\sigma)$ tends to $p^{(1)}(\sigma)$, where

$$p^{(1)}(\sigma) \propto 1/\sigma. \quad (90)$$

This is the *in-ignorance* prior. Its use in the steps outlined previously results in a predictive $h^{(1)}$ given by

$$h^{(1)}(y|y_1, \ldots, y_n) = n[1 + y/t]^{-(n+1)}, \quad (91)$$

so that the β-expectation tolerance interval takes the form

$$S^{(1)} = [td_\beta^{(1)}, \infty),$$
$$d_\beta^{(1)} = [(1/\beta)^{1/n} - 1]. \quad (92)$$

This corresponds to the well-known sampling theory result. Tables of d_β and $d_\beta^{(1)}$ are readily available (Guttman [12]).

We turn now to sampling from the k-variate normal $N(\boldsymbol{\mu}, \boldsymbol{\Sigma})$, where $\boldsymbol{\Sigma}$ is $(k \times k)$, symmetric, and positive definite. The conjugate prior, often used in this situation, is given by

$$p\left(\mu, \Sigma^{-1}\right) \propto |\Sigma^{-1}|^{(n_0-k-1)/2}$$

$$\times \exp\left\{-\tfrac{1}{2}\mathrm{tr}\, \Sigma^{-1}\left[(n_0-1)\mathbf{V}_0\right.\right.$$

$$\left.\left.+n_0(\mu-\bar{\mathbf{y}}_0)(\mu-\bar{\mathbf{y}}_0)'\right]\right\}, \quad (93)$$

where $\bar{\mathbf{y}}_0$ is a $(k \times 1)$ vector of (known) constants, $\mathbf{V}_0$ is a $(k \times k)$ symmetric positive definite matrix of known constants, and $\mathrm{tr}\,\mathbf{A}$ denotes the trace* of the matrix $\mathbf{A}$. For $k = 1$, Σ^{-1} is $1/\sigma^2$, and after suitably transforming (93), reduces to the well-known conjugate prior for the univariate normal situation,

$$p\left(\mu, \sigma^2\right) \propto \left(\sigma^2\right)^{-(n_0/2)-1}$$

$$\times \exp\left\{-\left[(n_0-1)v_0^2\right.\right.$$

$$\left.\left.+n_0(\mu-\bar{y}_0)^2\right]/(2\sigma^2)\right\}. \quad (94)$$

For general k, as (93) becomes more and more diffuse, it approaches the noninformative prior

$$p\left(\mu, \Sigma^{-1}\right) \propto |\Sigma^{-1}|^{-(k+1)/2}, \quad (95)$$

advocated by Geisser [9]. If $k = 1$, after transformation,

$$p\left(\mu, \sigma^2\right) \propto 1/\sigma^2. \quad (96)$$

If we carry through the steps necessary to calculate the predictive distribution (80), with $\theta = (\mu, \Sigma)$, based on the data $\mathbf{y}_1, \ldots, \mathbf{y}_n$, and the prior (93), etc., then (see Guttman [12]) the predictive distribution h is given by

$$h(\mathbf{y}|\mathbf{y}_1, \ldots, \mathbf{y}_n)$$

$$= \frac{(n+n_0)^{k/2}\Gamma((n+n_0)/2)|\mathbf{Q}^{-1}|^{1/2}}{(n+n_0+1)^{k/2}\pi^{k/2}\Gamma((n+n_0-k)/2)}$$

$$\times \left\{1 + \frac{n+n_0}{n+n_0+1}(\mathbf{y}-\bar{\bar{\mathbf{y}}})'\mathbf{Q}^{-1}(\mathbf{y}-\bar{\bar{\mathbf{y}}})\right\}^{-(n+n_0)/2};$$

$$(97)$$

here, denoting the sufficient statistics as $(\bar{\mathbf{y}}, \mathbf{V})$ with $\bar{\mathbf{y}} = n^{-1}\sum_{\alpha=1}^{n}\mathbf{y}_\alpha$ and $(n-1)\mathbf{V} = \sum_{\alpha=1}^{n}(\mathbf{y}_\alpha-\bar{\mathbf{y}})(\mathbf{y}_\alpha-\bar{\mathbf{y}})'$, we have

$$\bar{\bar{\mathbf{y}}} = (n_0+n)^{-1}(n_0\bar{\mathbf{y}}_0+n\bar{\mathbf{y}}),$$

$$\mathbf{Q} = (n_0-1)\mathbf{V}_0 + (n-1)\mathbf{V}$$

$$+ \frac{n_0 n}{n+n_0}(\bar{\mathbf{y}}-\bar{\mathbf{y}}_0)(\bar{\mathbf{y}}-\bar{\mathbf{y}}_0)'. \quad (98)$$

The result (97) connects $\mathbf{Y}$, given $\mathbf{y}_1, \ldots, \mathbf{y}_n$, to the k-variate t distribution with degrees of freedom $(n + n_0 - k)$ and to the quadratic form* $\mathbf{Q}$. From properties of the multivariate-t distribution*, it is known that [given $(\mathbf{y}_1, \ldots, \mathbf{y}_n)$]

$$\frac{n_0+n}{n+n_0+1}(\mathbf{Y}-\bar{\bar{\mathbf{y}}})'\mathbf{Q}^{-1}(\mathbf{Y}-\bar{\bar{\mathbf{y}}})$$

$$= \frac{k}{n+n_0-k}F_{k,n+n_0-k}. \quad (99)$$

This result will be put to work in the following text.

Suppose we are interested in the central $100\beta\%$ of the normal distribution $N(\mu, \Sigma)$ being sampled, that is, we wish to estimate where the set A_c^k lies, where

$$A_c^k = \left\{\mathbf{y}|(\mathbf{y}-\mu)'\Sigma^{-1}(\mathbf{y}-\mu)\right.$$

$$\left.\leq \chi_{k;1-\beta}^2\right\}. \quad (100)$$

A sensible choice for S, based on $\mathbf{y}_1, \ldots, \mathbf{y}_n$, is

$$S(\mathbf{y}_1, \ldots, \mathbf{y}_n)$$

$$= \left\{\mathbf{y}\left|\frac{n_0+n}{n_0+n+1}(\mathbf{y}-\bar{\bar{\mathbf{y}}})'\right.\right.$$

$$\times \left[\frac{\mathbf{Q}}{n_0+n-k}\right]^{-1}(\mathbf{y}-\bar{\bar{\mathbf{y}}})$$

$$\left.\leq kF_{k,n+n_0-k;1-\beta}\right\}. \quad (101)$$

Using (99), S as defined by (101) is a $100\beta\%$ confidence region for the additional observation $\mathbf{Y}$, where, conditional on $(\mathbf{y}_1, \ldots, \mathbf{y}_n)$, $\mathbf{Y}$ has the distribution (97). Hence, (101) is of β-expectation. If the prior (95) is used, the appropriate β-expectation region is given by

$$S(\mathbf{y}_1, \ldots, \mathbf{y}_n) = \left\{\mathbf{y}\left|\frac{n}{n+1}(\mathbf{y}-\bar{\mathbf{y}})'\right.\right.$$

$$\times \left[\frac{(n-1)\mathbf{V}}{n-k}\right]^{-1}(\mathbf{y}-\bar{\mathbf{y}})$$

$$\left.\leq kF_{k,n-k;1-\beta}\right\}. \quad (102)$$

in agreement with the standard classical sampling result.

β-CONTENT TOLERANCE REGIONS

Finally, we briefly describe some results for the construction of tolerance regions that satisfy (3), that is, are of β-content at confidence level γ. To illustrate, consider sampling from the exponential distribution (82), with a (conjugate) prior of the form (84). Then, based on a sample of size n, the posterior distribution of σ is given by (85), which is to say, a posteriori,

$$\sigma = 2(t + t_0)/\chi^2_{2(n+n_0)}. \qquad (103)$$

Suppose also, that we are interested in the $100\beta\%$ right-hand tail of the distribution (82), that is, in the unknown set

$$[\sigma \ln(1/\beta), \infty). \qquad (104)$$

This being so, we would choose S to be of the form

$$S(y_1, \ldots, y_n) = [a(y_1, \ldots, y_n), \infty). \qquad (105)$$

The question is how to choose a so that S of (105) is of β-content at level γ, (β, γ) being fixed known numbers. Now the coverage of (96) is

$$C(S) = \int_a^\infty \sigma^{-1} \exp(-y/\sigma)\, dy$$
$$= \exp(-a/\sigma). \qquad (106)$$

Hence, if S is to be of β-content at (posterior) confidence level γ, we have

$$\gamma = \Pr[C(S) \geq \beta | y_1, \ldots, y_n]$$
$$= \Pr[\sigma \geq a/[\ln(1/\beta)] | y_1, \ldots, y_n]. \qquad (107)$$

But a posteriori σ has the distribution (103), so that (107) may be written

$$\gamma = \Pr\left[\chi^2_{2(n+n_0)} \leq 2(t + t_0)[\ln(1/\beta)]/a\right] \qquad (108)$$

or

$$a = a(y_1, \ldots, y_n)$$
$$= 2(t + t_0)[\ln(1/\beta)]/\chi^2_{2(n+n_0); 1-\gamma}. \qquad (109)$$

If (n_0, t_0) tends to zero in such a way that

$p(\sigma)$ of (84) tends to the "noninformative" prior $p^{(1)}(\sigma)$ of (96), then $a(y_1, \ldots, y_n)$ of (109) tends to $a^{(1)}$, where

$$a^{(1)} = 2t[\ln(1/\beta)]/\chi^2_{2n; 1-\gamma}, \qquad (110)$$

corresponding to the sampling theory result.

In summary, if the prior (84) applies, then, using (109), the tolerance interval of β content at confidence level γ is

$$S(y_1, \ldots, y_n) = [a, \infty), \qquad (111)$$

where $a = a(y_1, \ldots, y_n)$ is given by (109); but if the prior (90) is applicable, the interval

$$S(y_1, \ldots, y_n) = [a^{(1)}, \infty) \qquad (112)$$

is of β content with confidence level γ, where $a^{(1)}$ is given by (110). The interpretation of (111) and (112) is that in light of the data and prior information that we have about σ, the confidence or degrees of belief* that we now have in $C(S)$ exceeding β is γ.

Results for other cases, such as the univariate normal are also known (see for example, Guttman [12, pp. 140–143]).

References

[1] Aitchison, J. and Dunsmore, I. R. (1975). *Statistical Prediction Analysis*. Cambridge University Press, Cambridge, England. (Advanced level.)

[2] Bowker, A. H. (1947). *Techniques of Statistical Analysis*, Chap. 2. McGraw-Hill, New York. (Intermediate level.)

[3] Breth, M. (1979). *Biometrika*, **66**, 641–644 [A (nonparametric) Bayesian approach when the (continuous) distribution function is not specified —very interesting paper, but at an advanced level.]

[4] Chatterjee, S. K. and Patra, N. K. (1980). *Calcutta Statist. Ass. Bull.*, **29**, 73–93. (Advanced level—gives theoretical development of β-content tolerance sets when sampling on multivariate random variables for large samples.)

[5] Ellison, B. E. (1964). *Ann. Math. Statist.*, **35**, 762–772. (Advanced level.)

[6] Evans, M. (1980). *Canad. J. Statist.*, **8**, 79–85. [Advanced level—discussion of construction of tolerance regions that cover an event (set) of interest at confidence γ; example is in terms of the usual normal model with set of interest chosen to be the usual ellipsoidal contour of content β.]

[7] Fraser, D. A. S. (1953). *Ann. Math. Statist.*, **24**, 44–55. (Advanced level.)

[8]   Fraser, D. A. S. and Guttman, I. (1956). *Ann. Math. Statist.*, **27**, 162–179. (Advanced level.)

[9]   Geisser, S. (1965). *J. Amer. Statist. Ass.*, **60**, 602–607.

[10] Guttman, I. (1963). *J. R. Statist. Soc. Ser. B.*, **25**, 368–376. (Advanced level.)

[11] Guttman, I. (1970a). *Ann. Math. Statist.*, **41**, 376–400. (Advanced level.)

[12] Guttman, I. (1970b). *Statistical Tolerance Regions*, Number 26. Statistical Monographs, Charles Griffin and Company, London, England. (Advanced level; contains exposition of both classical and Bayesian approaches.)

[13] Hall, I. J. and Sheldon, D. D. (1979). *J. Qual. Technol.*, **11**, 13–19 (Very readable; gives an alternative approach for constructing β-content level γ regions under bivariate normality.)

[14] Kemperman, J. H. B. (1956). *Ann. Math. Statist.*, **27**, 180–186. (Advanced level.)

[15] Owen, D. B. (1958). *Tables of Factors for One-Sided Tolerance Limits for a Normal Distribution*. Monograph No. SCR-13, Sandia Corporation, Albuquerque, NM. (Intermediate level.)

[16] Owen, D. B. (1962). *Handbook of Statistical Tables*. Addison-Wesley, Reading, Mass. (Elementary level; very useful in many ways.)

[17] Owen, D. B (1963). *Factors for One-Sided Tolerance Limits and for Variable Sampling Plans*. Monograph No. SCR-607, (19th ed.) Sandia Corporation, Albuquerque, NM. (Elementary level; directly useful if setting one-sided limits, as the title suggests.)

[18] Paulson, E. (1943). *Ann. Math. Statist.*, **14**, 90–93. (Advanced level.)

[19] Raiffa, H. and Schlaifer, R. (1967). *Applied Statistical Decision Theory*, Harvard University Press, Cambridge, Mass. (Advanced level.)

[20] Somerville, P. N. (1958). *Ann. Math. Statist.*, **29**, 559–601. (Elementary level.)

[21] Tietjen, G. L. and Johnson, M. E. (1979). *Technometrics*, **21**, 107–110. (Tolerance limits for a future sample standard deviation are found, after a discussion of the reliability problem that requires this—intermediate level.)

[22] Tukey, J. W. (1947). *Ann. Math. Statist.*, **18**, 529–539. (Advanced level.)

[23] Wald, A. (1943). *Ann. Math. Statist.*, **14**, 45–55. (Intermediate level—an important paper, historically.)

[24] Wald, A. and Wolfowitz, J. (1946). *Ann. Math. Statist.*, **17**, 208–215. (Advanced level.)

[25] Weissberg, A. and Beatty, G. H. (1960). Tables of tolerance limit factors for normal distributions. *Technometrics*, **2**, 483–500. (Intermediate level—very useful tables.)

[26] Wilks, S. S. (1941). *Ann. Math. Statist.*, **12**, 91–96. (Intermediate level—historically important, as this is the first paper that dealt with nonparametric tolerance limits, as well as the setting of tolerance limits when sampling from the univariate normal.)

(CONFIDENCE INTERVALS AND REGIONS
NONPARAMETRIC TOLERANCE
  INTERVALS
ORDER STATISTICS
QUALITY CONTROL, STATISTICAL)

IRWIN GUTTMAN

TONG ESTIMATORS   *See* MARKOV DECISION PROCESSES

TORNQUIST INDEX   *See* INDEX NUMBERS; LOG-CHANGE INDEX NUMBERS

TORNQUIST–THEIL APPROXIMATION   *See* DIVISIA INDICES

TOTAL BALANCE

A design for a comparative experiment is said to be *totally balanced* if all contrasts* are estimated with equal effectiveness. It has two forms. In *total variance balance*, all standardized contrasts are estimated with the same variance. (An obvious example is a design in balanced incomplete blocks*.) In *total efficiency balance*, blocks have been introduced in such a way that all contrasts suffer the same proportionate increase in variance, i.e., all have the same efficiency factor.

Both properties depend upon the weighted concurrences. If Treatment A occurs a times in a block and Treatment B occurs b times, there being k plots in all, the block contributes ab/k to the weighted concurrence of A and B. For example, let a design have two blocks, one with three plots and the other with four, and let the first contain treatments A, B, and C, once each, while the other has a second plot of C; Table 1 sets out the calculations.

For a design to have *total variance balance* the weighted concurrences must sum to the same value w for all contrasts. For example,

Table 1

| Block | Occurrences | | | | Weighted concurrences | | |
	A	B	C	Size	A and B	A and C	B and C
I	1	1	1	3	$(1 \times 1)/3$	$(1 \times 1)/3$	$(1 \times 1)/3$
II	1	1	2	4	$(1 \times 1)/4$	$(1 \times 2)/4$	$(1 \times 2)/4$
					$7/12$	$5/6$	$5/6$

Table 2

| Block | Treatments | Size | Weighted concurrences | | |
			A and B	A and C	B and C
I	ABCC	4	$(1 \times 1)/4$	$(1 \times 2)/4$	$(1 \times 2)/4$
II	ABCC	4	$(1 \times 1)/4$	$(1 \times 2)/4$	$(1 \times 2)/4$
III	ABBC	4	$(1 \times 2)/4$	$(1 \times 1)/4$	$(2 \times 1)/4$
IV	ABBC	4	$(1 \times 2)/4$	$(1 \times 1)/4$	$(2 \times 1)/4$
V	AB	2	$(1 \times 1)/2$	$(1 \times 0)/2$	$(1 \times 0)/2$
VI	AC	2	$(1 \times 0)/2$	$(1 \times 1)/2$	$(1 \times 0)/2$
			2	2	2

suppose someone were called upon to design an experiment with three treatments ($v = 3$) and there were available four blocks with four plots and a supply of others with only two. A possible solution is given in Table 2. In this instance, all pairs of treatments have the same weighted concurrences ($w = 2$), so it is as if all treatments had six ($vw = 6$) replicates in an orthogonal design* (Rao [7], Pearce [1], [3]).

In *total efficiency balance*, the weighted concurrence of the ith and j^{th} treatment always equals $\alpha r_i r_j$, where α is a constant and r_i is the actual replication of the ith treatment. In that case, all contrasts have an efficiency factor of $n\alpha$, where n is the num-

ber of plots in the experiment (Puri and Nigam [5], [6], Pearce [3]). For example, let an experiment be designed as in Table 3. Since replications are, respectively, 6, 3, and 3, that makes $\alpha = 2/27$ for all contrasts. Also $n = 12$, so all comparisons should have an efficiency factor of $8/9$. In fact, the variance of the contrast between the parameters for A and B is $9\sigma^2/16$. If there had been no blocks, it would have been $(1/6 + 1/3)\sigma_0^2 = \sigma_0^2/2$, where σ_0^2 is the variance from a completely randomized design. The ratio of the coefficients, $1/2$ and $9/16$, is indeed $8/9$. Similarly, for the contrast of B and C, the variances are respectively $3\sigma^2/4$ and $2\sigma_0^2/3$, the efficiency factor being again $8/9$.

Table 3

| Block | Treatments | Size | Weighted concurrences | | |
			A and B	A and C	B and C
I	AAB	3	$(2 \times 1)/3$	$(2 \times 0)/3$	$(1 \times 0)/3$
II	AAC	3	$(2 \times 0)/3$	$(2 \times 1)/3$	$(0 \times 1)/3$
III	ABC	3	$(1 \times 1)/3$	$(1 \times 1)/3$	$(1 \times 1)/3$
IV	ABC	3	$(1 \times 1)/3$	$(1 \times 1)/3$	$(1 \times 1)/3$
			$4/3$	$4/3$	$2/3$

Given total variance balance, it is possible to partition the treatment sum of squares in any way desired (Pearce, [2], [3]). (Those who think that the contrasts of interest should be specified at the design stage and all other contrasts subordinated to them will be unimpressed.) Total efficiency balance has less obvious practical advantages. It permits freedom of partition among the basic contrasts (Pearce et al. [4]), but that is not often needed.

References

[1] Pearce, S. C. (1964). *Biometrics*, **20**, 699–706.

[2] Pearce, S. C. (1982). *Utilitas Math.*, **21B**, 123–139.

[3] Pearce, S. C. (1983). *The Agricultural Field Experiment: a Statistical Examination of Theory and Practice.* Wiley, Chichester, England, Sects. 5.1 and 5.2.

[4] Pearce, S. C., Caliński, T., and Marshall, T. F. de C. (1974). *Biometrika*, **61**, 449–460.

[5] Puri, P. D. and Nigam, A. K. (1975a). *J. R. Statist. Soc. B*, **37**, 457–458.

[6] Puri, P. D. and Nigam, A. K. (1975b). *Sankhyā, Ser. B*, **37**, 457–460.

[7] Rao, V. R. (1958). *Ann. Math. Statist.*, **29**, 290–294.

(BLOCKS, BALANCED INCOMPLETE DESIGN OF EXPERIMENTS GENERAL BALANCE OPTIMUM DESIGN OF EXPERIMENTS)

S.C. PEARCE

TOTALLY DIAGONAL LATIN SQUARES
See LATIN SQUARES

TOTAL POSITIVITY

INTRODUCTION

The theory of total positivity has been extensively applied in several domains of mathematics, statistics, economics, and mechanics. In statistics, totally positive functions are fundamental in permitting characterizations of best statistical procedures for decision problems. The scope and power of this concept extend to ascertaining optimal policy for inventory* and system supply problems, to clarifying the structure of stochastic processes* with continuous path functions, to evaluating the reliability of coherent systems, and to understanding notions of statistical dependency.

In recent years Samuel Karlin has made brilliant contributions in developing the intrinsic relevance and significance of the concept of total positivity to probability and to statistical theory. In 1968, he wrote a classical book [13] devoted to this vast subject; it presents a comprehensive, detailed treatment of the analytic structure of totally positive functions and conveys the breadth of the great variety of fields of it applications. This book, together with Karlin's other fundamental papers, inspired many new developments and discoveries in many areas of statistical applications. Frydman and Singer [8] obtained a complete solution to the embedding problem for the class of continuous-time Markov chains (*see* MARKOV PROCESSES); the class of transition matrices for the finite state time-inhomogeneous birth and death processes* coincides with the class of nonsingular totally positive stochastic matrices*. Keilson and Kester [21] employed total positivity to characterize a class of stochastically monotone Markov chains with the property that the expectation of unimodal functions of the chain is itself unimodal in the initial state. To help unify the area of stochastic comparisons, Hollander et al. [9] introduced the concept of *functions decreasing in transposition* [DT]. In the bivariate case, a function $f(\lambda_1, \lambda_2; x_1, x_2)$ is said to have the DT property if

(a) $f(\lambda_1, \lambda_2; x_1, x_2) = f(\lambda_2, \lambda_1; x_2, x_1)$

and

(b) $\qquad \lambda_1 < \lambda_2, \qquad x_1 < x_2$

imply that

$$f(\lambda_1, \lambda_2; x_1, x_2) \geqslant f(\lambda_1, \lambda_2; x_2, x_1);$$

i.e., transposing from the natural order (x_1, x_2) to (x_2, x_1) decreases the value of the function. In their paper, total positivity is essential in showing that $P_\lambda(\mathbf{R} = \mathbf{r})$, the probability of rank order λ, is a DT function.

Karlin and Rinott [18, 19] extended the theory to multivariate cases. Multivariate total positivity properties are instrumental in refs. 18 and 19 for the results which are applied to obtain positive dependence of random vector components and related probability inequalities (*see also* DEPENDENCE, CONCEPTS OF).

For an excellent global view of the theory, as well as for references, see Karlin [13].

DEFINITION AND BASIC PROPERTIES

Definition. A function $f(x, y)$ of two real variables ranging over linearly ordered one-dimensional sets X and Y, respectively, is said to be *totally positive of order k* (TP$_k$) if for all $x_z < x_2 < \cdots < x_m$, $y_1 < y_2 \cdots < y_m$ (x_i in X; y_i in Y), and all $1 \leqslant m \leqslant k$,

$$f\begin{pmatrix} x_1, x_2, \ldots, x_m \\ y_1, y_2, \ldots, y_m \end{pmatrix}$$

$$= \begin{vmatrix} f(x_1, y_1) & f(x_1, y_2) & \cdots & f(x_1, y_m) \\ f(x_2, y_1) & f(x_2, y_2) & \cdots & f(x_2, y_m) \\ \vdots & \vdots & \ddots & \vdots \\ f(x_m, y_1) & f(x_m, y_2) & \cdots & f(x_m, y_m) \end{vmatrix}$$

$$\geqslant 0.$$

Typically, X and Y are either intervals of the real line or a countable set of discrete values on the real line, such as the set of all integers or the set of nonnegative integers. When X or Y is a set of integers, the term "sequence" rather than "function" is used. If $f(x, y)$ is TP$_k$ for all positive integers $k = 1, 2, \ldots$, then $f(x, y)$ is said to be *totally positive of order ∞*, written TP$_\infty$ or TP.

A related weaker property is that of sign regularity. A function $f(x, y)$ is *sign regular of order k* (SR$_k$) if for every $x_1 < x_2 < \cdots < x_m$, $y_1 < y_2 < \cdots < y_m$, and $1 \leqslant m \leqslant k$, the sign of

$$f\begin{pmatrix} x_1, x_2, \ldots, x_m \\ y_1, y_2, \ldots, y_m \end{pmatrix}$$

depends on m alone.

Many well known families of density functions (both continuous and discrete) are totally positive; TP$_2$ is the order of TP-ness which has found greatest application. In the context of statistics, the TP$_2$ property is the monotone likelihood ratio* property. Higher order TP-ness is hardly used in application except for the occasional use of TP$_3$.

Some examples of functions that possess the TP property are:

(i) $f(x, y) = e^{xy}$ is TP in $x, y \in (-\infty, \infty)$, so that $f(x, y) = x^y$ is TP in $x \in (0, \infty)$ and $y \in (-\infty, \infty)$.

(ii) $f(k, t) = e^{-\lambda t}[(\lambda t)^k / k!]$ is TP in $t \in (0, \infty)$ and $k = \{0, 1, 2, \ldots\}$.

(iii) $f(x, y) = \begin{cases} 1 & \text{if } a \leqslant x \leqslant y \leqslant b, \\ 0 & \text{if } a \leqslant y \leqslant x \leqslant b. \end{cases}$

PF$_k$ as Special Case of Interest

The concepts of TP$_1$ and TP$_2$ densities are familiar; every density is TP$_1$, while the TP$_2$ densities are those having a monotone likelihood ratio.

A further important specialization occurs if a TP$_k$ function may be written as a function $f(x, y) = f(x - y)$ of the difference of x and y, where x and y traverse the entire real line; $f(u)$ is then said to be a *Pólya frequency function of order k* (PF$_k$). Note that a Pólya frequency function is not necessarily a probability frequency function in that $\int_{-\infty}^{\infty} f(u)\, du$ need not be 1 nor even finite.

The class of PF$_2$ functions is particularly important and has rich applications to decision theory* [10–12, 18], reliability theory* [3], and the stochastic theory of inventory control* models [1, 16]. *See also* PÓLYA TYPE 2 FREQUENCY (PF$_2$) DISTRIBUTIONS.

Every PF$_2$ function is of the form $e^{-\psi(x)}$, where $\psi(x)$ is convex. On the other hand, there exists no such simple representation for PF$_k$, $k \geqslant 3$. Probability densities which are PF$_2$ abound.

Probability densities which decrease to zero at an algebraic rate in the tails are not PF$_2$. For example,

(i) Weibull with shape parameter less than 1:

$$f(x) = \alpha\lambda(\lambda x)^{\alpha-1}\exp[-(\lambda x)^{\alpha}],$$

$$x \geqslant 0, \quad \lambda > 0, \quad 0 < \alpha < 1,$$

and (ii) Cauchy:

$$f(x) = (1 + x^2)/\pi, \quad -\infty < x < \infty$$

are not PF_2.

Intriguing results in the structure theory of PF_k functions can be found in Karlin and Proschan [16], Karlin et al. [17], and Barlow and Marshall [2].

Variation Diminishing Property

An important feature of totally positive functions of finite or infinite order is their variation diminishing property; if $F(x, y)$ is TP_k and $g(y)$ changes sign at most $j \leqslant k - 1$ times, then $h(x) = \int f(x, y)g(y)\,dy$ changes sign at most j times; moreover, if $h(x)$ actually changes sign j times, then it must change sign in the same order as $g(y)$. It is this distinctive property which makes TP so useful. The variation diminishing property is essentially equivalent to the determinantal inequalities (1). Greater generality in stating this property is possible. The interested reader is referred to Karlin [13, Chap. 1]. A more direct approach to the theory is taken by Brown et al. [5], giving appropriate definitions and criteria for checking directly whether a family of densities possesses the variation diminishing property.

Composition and Preservation Properties

Many of the structural properties of TP_k functions are deducible from the following indispensible basic identity.

Basic Composition Formula. Let $h(x, t) = \int f(x, y)g(y, t)\,d\sigma(y)$ converge absolutely, where $d\sigma(y)$ is a sigma-finite measure. Then

$$h\binom{x}{t} = \int \cdots \int_{A_h} f\binom{x}{y}g\binom{y}{t}\prod_{i=1}^{n}d\sigma(y_i),$$

$$(2)$$

where $\mathbf{x} = (x_1, \ldots, x_n)$, $\mathbf{y} = (y_1, \ldots, y_n)$, $\mathbf{t} = (t_1, \ldots, t_n)$ and $y_1 < y_2 < \cdots < y_n$ in A_n.

A direct consequence of the composition formula is: If $f(x, y)$ is TP_m and $g(y, t)$ is TP_n, then

$$h(x, t) = \int f(x, y)g(y, t)\,d\sigma(y)$$

(the convolution of f and g) is $TP_{\min(m,n)}$. In many statistical applications, this consequence is exploited, principally in the case when f and g are Pólya frequency densities. That is, if $f(x)$ is PF_m and $g(x)$ is PF_n, then $h(x) = \int f(x - t)g(t)\,dt$ is $PF_{\min(m,n)}$. A key result follows.

Theorem 1. Let $f_1, f_2, \ldots$ be density functions of nonnegative random variables with each f_i in the class PF_k. Then $g(n, x) = f_1 * f_2 * \cdots * f_n(x)$ (* indicates convolution) is TP_k in the variables n and x, where n ranges over $1, 2, \ldots$ and x traverses the positive real line.

The case when the random variables are not restricted to be nonnegative is discussed in Karlin and Proschan [16]. These composition and preservation properties allow us to generate other totally positive functions, making it easy to enlarge the TP or PF classes and to determine whether the TP property holds.

Unimodality and Smoothness Properties

A function totally positive or more generally sign regular is endowed with certain structural properties pertaining to unimodality* and smoothing properties. From the definition of PF_2 we can derive

$$\begin{vmatrix} f(x_1 - y) & -f'(x_1 - y) \\ f(x_2 - y) & -f'(x_2 - y) \end{vmatrix} \geqslant 0 \quad (3)$$

for $x_1 < x_2$ and y arbitrary. In the event that $f'(u_0) = 0$, (3) implies that $f'(u) \geqslant 0$ for $u \leqslant u_0$ and $f(u) \leqslant 0$ for $u > u_0$. This implies that $f(u)$ is PF_2, $f(u)$ is unimodal. In particular, every PF_2 density is a unimodal density. The unimodality result is valid in case f is a PF_2 sequence.

We now describe a smoothing property possessed by the transformation under which convexity in $g(x)$ is carried over into convexity in $h(x)$, viz.,

$$h(n) = \int f^{(n)}(x)g(x)\, dx \quad \text{for } n = 1, 2, \ldots,$$
$$\tag{4}$$

where $f^{(n)}(x)$ is the n-fold convolution* of f. To make this notion precise, assume $f(x)$ is PF_3 and $g(x)$ is convex. Let $u = \int x f(x)\, dx$. Note that for arbitrary real constants a_0 and a_1,

$$\int \{ g(x) - [(a_0/u)x + a_1] \} f^{(n)}(x)\, dx$$

$$= h(n) - (a_0 n + a_1). \tag{5}$$

Since $g(x)$ is convex, $g(x) - [(a_0/u)x + a_1]$ has at most two changes of sign and if two changes of sign actually occur, they occur in the order $+ - +$ as x traverses the real axis from $-\infty$ to $+\infty$. Since f is PF_3, $f^{(n)}(x)$ is TP_3 in the variables n and x by Theorem 1. The variation diminishing property implies that $h(n) - (a_0 n + a_1)$ will have at most two changes of sign. Moreover, if $h(n) - (a_0 n + a_1)$ has exactly two changes of sign, these will occur in the same order as those of $g(x) - [(a_0/u)x + a_1]$, namely $+ - +$. Since a_0, a_1 are arbitrary, $h(n)$ is a convex function of n. Similar results apply for concavity.

APPLICATIONS TO STATISTICAL DECISION THEORY*

Historically this is perhaps the first area of statistics benefiting from the application of TP, due to the great papers of Karlin [10–12]. Consider testing a null hypothesis against an alternative, i.e., a two-action statistical decision problem. There exist two loss functions L_1 and L_2 on the parameter space, where $L_i(\theta)$ is the loss incurred if action i is taken when θ is the true parameter value. The set in which $L_1(\theta) < (>)L_2(\theta)$ is the set in which action 1 (action 2) is preferred when θ is the true state of nature. The two actions are indifferent at all other points. We assume that $L_1(\theta) - L_2(\theta)$ changes sign exactly n times at $\theta_1, \theta_2, \ldots, \theta_n$.

Let ϕ be a randomized decision procedure which is the probability of taking action 2 (accepting the alternative hypothesis) if x is the observed value of the random variable X. Let $\mathscr{C}_n$ be the class of all monotone randomized decision procedures defined by

$$\phi(x) = \begin{cases} 1 & \text{for } x_{2i} < x < x_{2i+1}, \\ & \quad i = 0, 1, \ldots, [n/2], \\ \lambda_j & \text{for } x = x_j, 0 \leqslant \lambda_j \leqslant 1, \quad (6) \\ & \quad j = 1, 2, \ldots, n, \\ 0 & \text{elsewhere}, \end{cases}$$

where $[a]$ denotes the greatest integer less than or equal to a and $x_0 = -\infty$.

Using the variation diminishing property, Karlin [11] obtained the main results:

Theorem 2. Let $f(x, \theta)$ be a strictly TP_{n+1} density and

$$\rho(\theta, \phi)$$

$$= \int [(1 - \phi(x))L_1(\theta) + \phi(x)L_2(\theta)]$$

$$\times f(x, \theta)\, d\mu(x).$$

Then for any randomized decision procedure ϕ not in $\mathscr{C}_n$ there exists a unique ϕ^0 such that $f(\theta, \phi^0) \leqslant f(\theta, \phi)$ with inequality everywhere except for $\theta = \theta_1, \theta_2, \ldots, \theta_n$.

Theorem 3. If ϕ and ψ are two procedures in $\mathscr{C}_n$ and f is strictly TP_{n+1}, then $\int [\phi(x) - \psi(x)] f(x, \theta)\, d\mu(x)$ has less than n zeros counting multiplicities.

Assume $f(x, \theta)$ is strictly TP_2. For a one-sided testing problem, existence of a uniformly most powerful level α test can be easily established by Theorems 2 and 3.

For further discussion and other applications see Karlin [10–12] and Karlin and Rubin [20].

APPLICATIONS IN PROBABILITY AND STOCHASTIC PROCESSES*

Let $P(t, x, E)$ be the transition probability function of a homogeneous strong Markov process* whose state space is an interval on the real line and that possesses a realization in which almost all sample paths are continuous. Karlin and McGregor [14] established the intimate relationship between the general theory of TP functions and the theory of diffusion processes*. Their main result shows the transition probability function $P(t, x, E)$ to be totally positive in variables x and E. That is, if $x_1 < x_2 < \cdots < x_n$ and $E_1 < E_2 < \cdots < E_n$ ($E_i < E_j$ denotes that $x < y$ for every $x \in E_i$ and $y \in E_j$), then $\det \|P(t, x_i, E_j)\| \geqslant 0$ for every $t > 0$ and integer n. This relation introduces the concept of a TP set function $f(x, E) = P(t, x, E)$, where t is fixed, x ranges over a subset of the real line, and E is a member of a given sigma field of sets on the line.

If the state space of the process is countably discrete, then continuity of the path functions means that in every transition of the process the particle changes "position," moving to one of its neighboring states. Thus, discrete state continuous path processes coincide with the so-called birth–death processes* (Karlin and McGregor [15]) which are stationary Markov processes whose transition probability matrix

$$P_{ij}(t) = \Pr(x(t) = j | x(0) = i)$$

is totally positive in the values i and j for every $t > 0$.

Two concrete illustrations of transition probability functions that arise from suitable diffusion processes are [14]:

(i) Let $L_n^\alpha(x)$ be the usual Laguerre polynomial*, normalized so that $L_n^\alpha(0) = \binom{n+\alpha}{n}$, and let $P(t)$ be the infinite matrix with elements

$$P_{mn}(t) = \int_0^\infty e^{-xt} L_n^\alpha(x) L_m^\alpha(x) x^\alpha e^{-x} dx.$$

Then $P(t)$ is strictly TP for each fixed $t > 0$ and $\alpha > -1$. This is an example of a transition probability matrix for a birth–death process.

(ii) The Wiener process on the real line is a strong Markov process with continuous path functions (see BROWNIAN MOTION). The

direct product of n copies of this process is the n-dimensional Wiener process, known to be a strong Markov process. Therefore, the transition probability function

$$P(t, x, E)$$

$$= \{1/(\sqrt{4\pi t})\} \int_E \exp[-(x-y)^2/(4t)] dy$$

is totally positive for $t > 0$.

APPLICATIONS IN INVENTORY PROBLEMS

Suppose that the probability density $f(x)$ of demand for each period is a PF$_3$. The policy followed is to maintain the stock size at a fixed level S which will be suitably chosen so as to minimize appropriate expected costs or is determined by a fixed capacity restriction. At the end of each period, an order is placed to replenish the stock consumed during that period so that a constant stock level is maintained on the books. Delivery takes place n periods later. The expected cost for a stationary period as a function of the lag is

$$L(n) = \int_0^S h(S - y) f^{(n)}(y) \, dy$$
$$+ \int_S^\infty \rho(y - S) f^{(n)}(y) \, dy, \quad (7)$$

where S is fixed, h represents the storage cost function, and ρ is the penalty cost function.

Let h and ρ be convex increasing functions with $h(0) = \rho(0) = 0$. Then we may write

$$L(n) = \int r(y) f^{(n)}(y) \, dy,$$

where

$$r(y) = \begin{cases} h(S - y) & \text{for } 0 \leqslant y \leqslant S, \\ \rho(y - S) & \text{for } S < y. \end{cases} \quad (8)$$

Now $r(y)$ is a convex function. Hence by the convexity preserving property of (4), $L(n)$ is a convex function. Thus, if the length of lag increases, the marginal expected loss increases.

Interesting applications of total positivity are found in system supply problems. Suppose we wish to determine the initial spare-parts kit for a complex system which provides maximum assurance against system shutdown due to shortage of essential components during a period of length t under a budget for spares c_0. We assume that a failed component is instantly replaced by a spare, if available. Only spares originally provided may be used for replacement, i.e., no resupply of spares can occur during the period. The system contains d_i operating components of type i, $i = 1, 2, \ldots, k$. The length of life of the jth operating component of the ith type is assumed to be an independent random variable with PF_k density f_{ij}, $j = 1, 2, \ldots, d_i$. The unit cost of a component of type i is c_i.

Our problem is to find n_i, the number of spares initially provided of the ith type, such that $\prod_{i=1}^{k} P_i(n_i)$ is maximized subject to $\sum_{i=1}^{k} n_i c_i \leq c_0$ and $n_i = 0, 1, 2, \ldots$, for $i = 1, 2, \ldots, k$, where $P_i(m)$ is the probability of experiencing at most m failures of type i.

In Black and Proschan [4], a detailed discussion of methods is given for computing the solution when each $\ln P_i(m)$ is concave in m, or equivalently, when each $P_i(n - m)$ is a TP_2 sequence in n and m. To show $P_i(n - m)$ is a TP_2 sequence in n and m, we note:

1. $c_{ij}(n)$, the probability of requiring n replacements of operating component i, j, is a PF_2 sequence in n for each fixed i and j.

2. $\rho_i(n)$, the probability of requiring n replacements of type i, is a PF_2 sequence in n for each i, since

$$\rho_i(n) = c_{i1} * c_{i2} * \cdots * c_{id_i}(n).$$

3. $P_i(n - m)$ is a TP_2 sequence in n and m for each i, since

$$\text{(a)} \ P_i(n) = \sum_{m=-\infty}^{\infty} \rho(n - m) q(m),$$

where

$$q(m) = \begin{cases} 1 & \text{for } m = 0, 1, 2, \ldots, \\ 0 & \text{otherwise,} \end{cases}$$

(b) $q(m)$ is a PF_∞ sequence, and (c) the convolution of PF_k is PF_k.

A procedure for computing the optimal spare parts kit in terms of $P_i(m)$ is given in ref. 4: For arbitrary $r > 0$, and for those i such that $\ln P_i(1) - \ln P_i(0) < rc_i$, define $n_i^*(r) = 0$; for the remaining i, define $n_i^*(r)$ as $1 +$ [largest n such that $\ln P_i(n + 1) - \ln P_i(n) \geq rc_i$]. Compute $c[n^*(r)] = \sum_{i=1}^{k} c_i n_i^*(r)$. Then n^* is optimal when c_0 is one of the values assumed by $c[n^*(r)]$ as r varies over $(0, \infty)$.

APPLICATIONS IN RELIABILITY AND LIFE TESTING

A life distribution F is said to have *increasing* (*decreasing*) *failure rate*, denoted by IFR (DFR), if $\log[1 - F(t)] \equiv \log \bar{F}(t)$ is concave [convex on $[0, \infty)$]. If F has a density f, then the *failure rate at time* t is defined by $r(t) = f(t)/\bar{F}(t)$ for $F(t) < 1$. Distributions with monotone failure rate are of considerable practical interest and such distributions constitute a very large class (*see* HAZARD RATE AND OTHER CLASSIFICATIONS OF DISTRIBUTIONS and RELIABILITY, PROBABILISTIC).

The monotonicity properties of the failure rate function $r(t)$ are intimately connected with the theory of total positivity. The statement that a distribution F has an increasing failure rate is equivalent to the statement that $\bar{F}(x - y)$ is TP_2 in x and y or $\bar{F}(x)$ is PF_2.

The concept of TP yields fruitful applications in shock models*. We say that a distribution F has *increasing failure rate average* (IFRA) if $(1/t)[-\log \bar{F}(t)]$ is increasing in $t > 0$ or, equivalently, $[\bar{F}(t)]^{1/t}$ is decreasing in $t > 0$. An IFRA distribution provides a natural description of coherent system life when system components are independent IFR. The IFRA distribution also arises naturally when shocks occur randomly according to a Poisson process* with intensity λ. The *i*th shock causes a random amount X_i of damage, where $X_1, X_2, \ldots$ are independently distributed with common distribution F.

A device fails when the total accumulated damage exceeds a specified capacity or threshold x. Let $\bar{H}(t)$ denote the probability that the device survives $[0, t]$. Then

$$\bar{H}(t) = \begin{cases} \sum_{k=0}^{\infty} \bar{P}_k e^{-\lambda t}(\lambda t)^k / k! \\ \qquad\qquad \text{for } 0 \leqslant t < \infty, \\ 1 \qquad\qquad \text{for } t < 0. \end{cases} \tag{9}$$

Note that $e^{-\lambda t}[(\lambda t)^k / k!]$ represents the Poisson* probability that the device experiences exactly k shocks in $[0, t]$, while $\bar{P}_k = F^{(k)}(x)$ represents the probability that the total damage accumulated over the k shocks does not exceed the threshold x, with $1 = \bar{P}_0 \geqslant \bar{P}_1 \geqslant \bar{P}_2 \geqslant \ldots$.

As key tools in deriving the main result in shock models, the methods of total positivity (in particular, the variation diminishing property of TP functions) are employed. If $\bar{P}_k^{1/k}$ is decreasing in k, $\bar{P}_k - \xi$, $0 \leqslant \xi \leqslant 1$, has at most one sign change, from $+$ to $-$ if one occurs. It follows from the variation diminishing property that $[\bar{H}(t)]^{1/t}$ is decreasing in t, i.e., H is IFRA.

The following implications are readily checked:

PF_2 density $\rightarrow$ IFR distribution

$\rightarrow$ IFRA distribution*.

For further discussion and illustrations of the usefulness of total positivity in reliability practise, see Barlow and Proschan [3].

MULTIVARIATE TOTAL POSITIVITY

The following natural generalization of TP$_2$ was introduced and studied by Karlin and Rinott [18].

Definition. Consider a function $f(\mathbf{x})$ defined on $\mathscr{X} = \mathscr{X}_1 \times \mathscr{X}_2 \times \cdots \times \mathscr{X}_k$, where each $\mathscr{X}_i$ is totally ordered. We say that $F(\mathbf{x})$ is *multivariate totally positive of order* 2 or MTP$_2$ if

$$f(\mathbf{x} \vee \mathbf{y}) f(\mathbf{x} \wedge \mathbf{y}) \geqslant f(\mathbf{x}) f(\mathbf{y})$$
$$\text{for every } \mathbf{x}, \mathbf{y} \in \mathscr{X}, \tag{10}$$

where

$$\mathbf{x} \vee \mathbf{y} = (\max(x_1, y_1), \max(x_2, y_2)$$
$$, \ldots, \max(x_n, y_n)),$$
$$\mathbf{x} \wedge \mathbf{y} = (\min(x_1, y_1), \min(x_2, y_2)$$
$$, \ldots, \min(x_n, y_n)).$$

In order to verify (10) it suffices to show that $f(\mathbf{x}) > 0$ is TP$_2$ in every pair of variables when the remaining variables are held fixed.

Multivariate normal distributions constitute an important class of MTP$_2$ probability densities. Let $\mathbf{X}$ follow the density

$$f(\mathbf{x}) = (2\pi)^{-n/2} |\Sigma|^{-1/2}$$
$$\times \exp\left[-\tfrac{1}{2}(\mathbf{x} - \mu)' \mathbf{B}(\mathbf{x} - \mu)\right],$$

where $\Sigma^{-1} = \mathbf{B} = \|b_{ij}\|_{i,j=1}^n$. This density is TP$_2$ in each pair of arguments and hence MTP$_2$ if and only if $b_{ij} \leqslant 0$ for all $i \neq j$.

In many situations, the random variables of interest are not independent. To appropriately model these situations, Esary et al. [6] introduced the concept of association of random variables. Random variables $X_1, X_2, \ldots, X_n$ are said to be *associated* if $\text{Cov}(f(\mathbf{X}), g(\mathbf{X})) \geqslant 0$ for all pairs of increasing functions f and g.

If $\mathbf{X} = (X_1, X_2, \ldots, X_n)$ has a joint MTP$_2$ density, then [18]

$$E[\phi(\mathbf{X})\psi(\mathbf{X})] \geqslant E[\phi(\mathbf{X})] E[\psi(\mathbf{X})],$$

provided the functions ϕ and ψ are simultaneously monotone increasing (or decreasing). Equivalently, $\text{Cov}(\phi(\mathbf{X}), \psi(\mathbf{X})) \geqslant 0$. Thus an MTP$_2$ random vector $\mathbf{X}$ consists of associated random variables.

The union of independent sets of associated random variables produces an enlarged set of associated random variables. Clearly, increasing functions of associated random variables are again associated. Hence, if $\mathbf{X}$ and $\mathbf{Y}$ are independent random variables each with associated components, then the components of $\mathbf{Z} = \mathbf{X} + \mathbf{Y}$ are associated. In particular, if **X** and **Y** both have MTP$_2$ densities, then association of $(Z_1, Z_2, \ldots, Z_n)$ is retained. However, **Z** need not have a joint MTP$_2$ density.

A key to many of the results on positive dependence and probabilistic inequalities for the multinormal*, multivariate t^*, and Wishart* distributions obtained by Karlin and Rinott [19] is the degree of MTP$_2$ property inherent in these distributions. Their main theorem delineates a necessary and sufficient condition that the density of $|\mathbf{X}| = (|X_1|, |X_2|, \ldots, |X_n|)$, where $\mathbf{X} = (X_1, X_2, \ldots, X_n)$ is governed by $N(\mathbf{0}, \Sigma)$, be MTP$_2$ is that there exists a diagonal matrix **D** with diagonal elements ± 1 such that the off-diagonal elements of $-\mathbf{D}\Sigma^{-1}\mathbf{D}$ are all nonnegative.

For an illustration of the power of this theorem, consider $|\mathbf{X}|$ possessing a joint MTP$_2$ density where $\mathbf{X} \sim N(\mathbf{O}), \Sigma)$. Define $S_i = \Sigma_{\nu-1}^{p} X_{i\nu}^2$, $i = 1, 2, \ldots, n$, where $\mathbf{X}_\nu = (X_{1\nu}, X_{2\nu}, \ldots, X_{n\nu})$, $\nu = 1, 2, \ldots, p$, are independent and identically distributed (i.i.d.) random vectors satisfying the condition of the theorem. The random variables $S_1, S_2, \ldots, S_n$ are associated and have the distribution of the diagonal elements of a random positive definite $n \times n$ matrix **S**, where **S** follows the Wishart distribution* $W_n(p, \Sigma)$ with p degrees of freedom and parameter Σ. It is established in ref. 19 that

$$\Pr[S_1 \geqslant c_1, S_2 \geqslant c_2, \ldots, S_n \geqslant c_n]$$

$$\geqslant \prod_{i=1}^{n} \Pr[S_i \geqslant c_i]$$

for any positive c_i. For other applications and ramifications of MTP$_2$, see Karlin and Rinott [18, 19]. Fahmy et al. [7] exploited the concept of MTP$_2$ to obtain interesting results on assessing the effect of the sample on the posterior distribution in the Bayesian* context.

References

[1] Arrow, K. J., Karlin, S., and Scarf, H. E. (1958). *Studies in the Mathematical Theory of Inventory and Production*, Stanford University Press, Stanford, CA.

[2] Barlow, R. E. and Marshall, A. W. (1964). *Ann. Math. Statist.*, **35**, 1234–1274.

[3] Barlow, R. E. and Proschan, F. (1981). *Statistical Theory of Reliability and Life Testing: Probability Models*. To Begin With, Silver Spring, MD. (A clear, systematic and detailed treatment of applications of TP in reliability and life-testing theory is presented in Chaps. 3–5.)

[4] Black, G. and Proschan, F. (1959). *Operat. Res.*, **7**, 581–588.

[5] Brown, L. D., Johnstone, I. M., and MacGibbon, K. B. (1981). *J. Amer. Statist. Ass.*, **76**, 824–832. (A largely expository article which gives a more direct account of the variation diminishing property. Their approach avoids the extensive mathematical preliminary and isolates the more important statistical property.)

[6] Esary, J. D., Proschan, F., and Walkup, D. W. (1967). *Ann. Math. Statist.*, **38**, 1466–1474.

[7] Fahmy, S., Pereira, C., Proschan, F., and Shaked, M. (1982). *Commun. Statist. A*, **11**, 1757–1768.

[8] Frydman, H. and Singer, B. (1979). *Math. Proc. Camb. Philos. Soc.*, **86**, 339–344.

[9] Hollander, M., Proschan, F., and Sethuraman, J. (1977). *Ann. Statist.*, **5**, 722–733.

[10] Karlin, S. (1956). *Proc. Third Berkeley Symposium on Probability and Statistics*, Vol. 1, University of California Press, Berkeley, CA, pp. 115–129.

[11] Karlin, S. (1957a). *Ann. Math. Statist.*, **28**, 281–308.

[12] Karlin, S. (1957b). *Ann. Math. Statist.*, **28**, 839–860.

[13] Karlin, S. (1968). *Total Positivity*. Stanford University Press, Stanford, CA. (A comprehensive discussion of the theory, together with a discussion of the key references, is presented.)

[14] Karlin, S. and McGregor, J. L. (1959a). *Pacific J. Math.*, **9**, 1141–1164.

[15] Karlin, S. and McGregor, J. L. (1959b). *Pacific J. Math.*, **9**, 1109–1140.

[16] Karlin, S. and Proschan, F. (1960). *Ann. Math. Statist.*, **31**, 721–736.

[17] Karlin, S., Proschan, F., and Barlow, R. E. (1961). *Pacific J. Math.*, **11**, 1012–1033.

[18] Karlin, S. and Rinott, Y. (1980). *J. Multivariate Anal.*, **10**, 467–498. (This paper provides basic information on multivariate total positivity theory.)

[19] Karlin, S. and Rinott, Y. (1981). *Ann. Statist.*, **9**, 1035–1049.

[20] Karlin, S. and Rubin, H. (1956). *Ann. Math. Statist.*, **27**, 272–299.

[21] Keilson, J. and Kester, A. (1978). *Stoch. Processes*, **7**, 179–190.

JEE SOO KIM
FRANK PROSCHAN

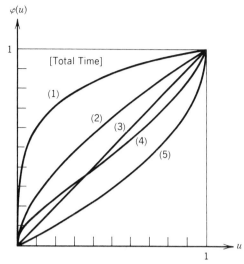

Figure 1 Scaled TTT transforms from five different life distributions: 1. normal with $\mu = 1$, $\sigma = 0.3$ (increasing failure rate); 2. gamma distribution with shape parameter 2.0 (increasing failure rate); 3. exponential distribution (constant failure rate); 4. lognormal (i.e., $\log_e Y$ normal) with $\mu = 0$, $\sigma = 1$; 5. Pareto distribution with $S(t) = (1 + t)^{-2}$, $t \geq 0$ (decreasing failure rate).

TOTAL PROBABILITY THEOREM

The events $E_1, E_2, \ldots, E_n$ are called a *partition of the sample space* Ω if $E_i \cap E_j = \varnothing$ for all $i \neq j$ and $\bigcup_{i=1}^n E_i = \Omega$.

If $A \subset \Omega$ is any event, then $E_1, E_2, \ldots, E_n$ also partition A, since $A = \bigcup_{i=1}^n (A \cap E_i)$ with $(A \cap E_i) \cap (A \cap E_j) = \varnothing$ for all $i \neq j$. Consequently, $P(A) = \sum_{i=1}^n P(A \cap E_i)$. This result is referred to as the *theorem of total probability*. It is used in the derivation of Bayes' theorem* among other applications.

TOTAL TIME ON TEST TRANSFORM

Let $0 = x_{0:n} \leqslant x_{1:n} \leqslant \cdots \leqslant x_{n:n}$ denote an ordered sample from a life distribution F [i.e., a distribution function with $F(0-) = 0$] with survival function $S = 1 - F$ and finite mean $\mu = \int_0^\infty S(t)\, dt$. The *total time on test* (TTT) statistics $T_i = \sum_{j=1}^i (n - j + 1) \times (x_{j:n} - x_{j-1:n})$, $i = 1, 2, \ldots, n$, were introduced by Epstein and Sobel [16] in connection with inference problems concerning the exponential distribution*. By plotting and connecting the points $(i/n, T_i/T_n)$, $i = 0, 1, \ldots, n$, $T_0 = 0$, by straight line segments we obtain a curve called the TTT plot. This plotting technique was first suggested by Barlow and Campo [4].

By using $G^{-1}(u) = \inf\{t : G(t) > u\}$ for a life distribution G, we can write (see ref. 2)

$$T_i = \int_0^{F_n^{-1}(i/n)} (1 - F_n(t))\, dt;$$

here F_n is the empirical distribution function* determined from the ordered sample.

Then (see ref. 21) with probability 1,

$$T_i \to \int_0^{F^{-1}(u)} S(t)\, dt, \qquad 0 \leqslant u \leqslant 1,$$

when $n \to \infty$ and $i/n \to u$, if F is strictly increasing.

The function

$$H^{-1}(u) = \int_0^{F^{-1}(u)} S(t)\, dt, \qquad 0 \leqslant u \leqslant 1,$$

is the *TTT transform* of F; the mean of F is given by $\mu = H^{-1}(1)$.

The scale invariant transform

$$\varphi(u) = \frac{1}{\mu} H^{-1}(u) = \frac{1}{\mu} \int_0^{F^{-1}(u)} S(t)\, dt,$$
$$0 \leqslant u \leqslant 1,$$

is the *scaled TTT transform* of F, introduced by Barlow and Campo [4]; examples are illustrated in Fig. 1.

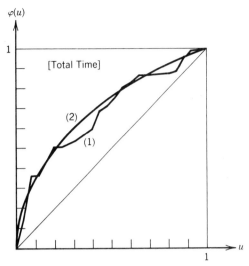

$\varphi(u)$

[Total Time]

(2)

(1)

u

1

Figure 2 The TTT plot (1) is based on a simulated sample with $n = 25$ from a Weibull distribution with shape parameter $\beta = 2.0$. The corresponding scaled TTT transform is illustrated by the curve (2).

The TTT plot will approach the graph of the scaled TTT transform of F as n, the number of observations, increases to infinity; cf. Fig. 2. When the ordered sample is from an exponential distribution*, T_i/T_n, $i = 1, 2, \ldots, n$, behaves like an ordered sample from a uniform distribution* on $[0, 1]$. This means that, as $n \to \infty$, the TTT plot approaches the diagonal of the unit square, which is the scaled TTT transform of the exponential distribution; cf. Fig. 1. For a detailed discussion of the convergence, see, e.g., ref. 15.

The scaled TTT transform and the TTT plot have proven to be very useful tools in different reliability* applications (see, e.g., ref. 11), some of which will be outlined.

MODEL IDENTIFICATION

In ref. 4 the scaled TTT transform and the TTT plot were used for model identification purposes. Since the TTT plot under rather general conditions will approach the TTT transform of F when n, the number of observations, increases to infinity, it is possible to make model identifications by comparing

the TTT plot with transparencies of scaled TTT transforms corresponding to different life distributions; see Fig. 2. The independence of scale and its simplicity make this technique a useful complement to other methods used in practice. Generalizations to censored or truncated samples are indicated in refs. 4 and 11.

AGING PROPERTIES

Different aging properties can be characterized by using the TTT transform. For instance, a life distribution is IFR (increasing failure rate) if and only if its scaled TTT transform is concave; see refs. 3, 4 and 21. This fact is illustrated by the transforms in Figs. 1 and 2. Another useful characterization is that F is NBUE (new better than used in expectation) if and only if $\varphi(u) \geqslant u$, $0 \leqslant u \leqslant 1$: see ref. 7. More about relations between $\varphi(u)$ and different aging properties of F can be found in refs. 3, 4, and 18; *see also* RELIABILITY, PROBABILISTIC.

Since the TTT plot converges to the corresponding scaled TTT transform when $n \to \infty$, the TTT plot based on a sample from an IFR life distribution should be close to a concave curve. The plot in Fig. 2 indicates that the underlying life distribution is IFR. In an analogous manner, it is possible to get subjective tests against other aging properties. For objective methods for testing exponentiality against different forms of aging, some of which are based on the TTT transform, see, e.g. refs. 1, 17 and 19.

AGE REPLACEMENT

Suppose that a certain type of unit is needed in a production process. The unit is replaced at age T at a cost c and at failure at a cost $(c + K)$. The average long run cost per unit time is then given by

$$C(T) = \{ c + KF(T) \} \Big/ \int_0^T S(t)\, dt.$$

The optimal age T^* minimizing $C(T)$ is, if it

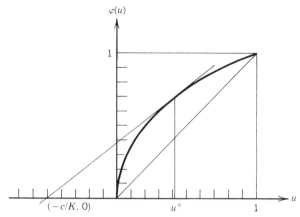

Figure 3 Graphical determination of the optimal replacement age when $F(t) = 1 - \exp(-t^2/100)$, $t \geqslant 0$, and $K = 2c$. The slope of the line through $(-c/K, 0)$ and $(u, \varphi(u))$ is maximized when $u = u^* = 0.42$. We get $T^* = 7.4$ by solving $F(T^*) = u^*$.

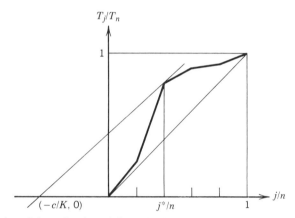

Figure 4 Illustration of the estimation of the optimal replacement age. The slope of the line through $(-c/K, 0)$ and $(j/n, T_j/T_n)$ is maximized for $j = j^*$. The estimate of T^* is then $x_{j^*:n}$.

exists (see ref. 7), $T^* = F^{-1}(u^*)$, where u^* maximizes $\varphi(u)/(u + c/K)$); cf. Fig. 3.

If we do not know the life distribution, but have an ordered sample of observations, we can estimate T^*, as is illustrated in Fig. 4, by using the TTT plot in a way analogous to the one based on the scaled TTT transform. These graphical procedures can also be used to find criteria for a unique optimal replacement age to exist (see ref. 9) and for sensitivity analysis with respect to the costs or times involved (see ref. 8). Furthermore, the idea can be used when the costs are discounted (see ref. 10) and in many other replacement models (see refs. 9 and 13).

GENERALIZATIONS

The TTT-plotting technique has great potential also in areas of reliability not covered here. For instance, ref. 10 discusses how to use the TTT concept when analyzing burn-in problems.

An interesting generalization of the TTT-plot intended for analyzing dependent failure data is presented in [2].

Finally it is worth to note the close relationship between the TTT-transform and the Lorenz curve, which is widely used in economics to illustrate income distributions (see [14] and [20]).

References

[1] Aarset, M. (1987). *IEEE Trans. Rel.*, **R-36**, 106–108.

[2] Akersten, P. A. (1987). "The double TTT-plot—a tool for the study of non-constant failure intensities." In *Proc. Reliability '87, Birmingham, England*, 2B/3/ 1–8.

[3] Barlow, R. E. (1979). *Naval Res. Logist. Quart.*, **26**, 393–402.

[4] Barlow, R. E. and Campo, R. (1975). In *Reliability and Fault Tree Analysis*, R. E. Barlow, J. Fussell, and N. D. Singpurwalla, eds. SIAM, Philadelphia, PA, pp. 451–481. (Introduces the scaled TTT transform and the TTT plot.)

[5] Bergman, B. (1977a). In 1977 *Annual Reliability and Maintainability Symposium*, pp. 467–471.

[6] Bergman, B. (1977b). *Scand. J. Statist.*, **4**, 171–177.

[7] Bergman, B. (1979). *Scand. J. Statist.*, **6**, 161–168.

[8] Bergman, B. (1980). *Microelectronics and Rel.*, **20**, 895–896.

[9] Bergman, B. and Klefsjö, B. (1982). *IEEE Trans. Rel.*, **R-31**, 478–481.

[10] Bergman, B. and Klefsjö, B. (1983). *Naval Res. Logist. Quart.*, **30**, 631–639.

[11] Bergman, B. and Klefsjö, B. (1984). *Oper. Res.*, **31**, 596–606. (An expository paper with many references.)

[12] Bergman, B. and Klefsjö, B. (1985). *QRE Int.*, **1**, 125–130.

[13] Bergman, B. and Klefsjö, B. (1987). *Eur. J. Oper. Res.*, **28**, 302–307.

[14] Chandra, M. and Singpurwalla, N. D. (1981). *Math. Oper. Res.*, **6**, 113–121.

[15] Csörgő, M., Csörgő, S. and Horváth, L. (1986). *An Asymptotic Theory for Reliability and Concentration Processes*, Lecture Notes in Statistics, **33**, Springer, Heidelberg, W. Germany. (Unified asymptotic theory for empirical time on test, Lorenz and concentration processes.)

[16] Epstein, L. and Sobel, M. (1953). *J. Amer. Statist. Ass.*, **48**, 486–502.

[17] Hollander, M. and Proschan, F. (1982). Nonparametric concepts and methods in reliability. In *Handbook of Statistics*, Vol. 4, P. R. Krishnaiah and P. K. Sen, eds. Elsevier, Amsterdam, The Netherlands, pp. 613–655.

[18] Klefsjö, B. (1982). *Scand. J. Statist.*, **9**, 37–41.

[19] Klefsjö, B. (1983). *Commun., Statist. A*, **12**, 907–927.

[20] Klefsjö, B. (1984). *Naval Res. Logist. Quart.*, **31**, 301–308.

[21] Langberg, N. A., Leone, R. B., and Proschan, F. (1980). *Ann. Prob.*, **8**, 1163–1170.

Acknowledgment

The research of Bengt Klefsjö was supported by Swedish Natural Science Council Postdoctoral Fellowship F-PD 1564-100.

(EXPONENTIAL DISTRIBUTION
FAULT TREE ANALYSIS
LIFE TESTING
RELIABILITY, PROBABILISTIC
RENEWAL THEORY
SURVIVAL ANALYSIS)

BO BERGMAN
BENGT KLEFSJÖ

TOTAL VARIATION, DISTANCE OF

Given two cumulative distribution functions $F_1(\cdot)$, $F_2(\cdot)$, the metric defined by

$$\sigma(F_1, F_2) = \tfrac{1}{2} \int_{-\infty}^{\infty} d\left\{ |F_1(x) - F_2(x)|^2 \right\}$$

is called their distance *of total variation*.

(HELLINGER DISTANCE
LEVY DISTANCE (METRIC)
PROBABILITY SPACES, METRICS AND
DISTANCES ON)

TOURNAMENTS *See* ROUND ROBIN; SPORTS, STATISTICS IN

TRABAJOS DE ESTADÍSTICA

This journal commenced publication in 1950. From Vol. 14 (1963) to Vol. 36 (1985), inclusive, the title was *Trabajos de Estadística y de Investigación Operativa*. In 1986 this was split into two journals—*Trabajos de Estadística* (starting again with Vol. 1) and *Trabajos de Investigación Operativa**.

Trabajos de Estadística accepts papers in Spanish, English, and French only. There is one volume per year, published in three parts, containing in total about 350 pages. Subscription price is 3300 Spanish pesetes (US $30). The Editor is Francisco Javier Girón González-Torre, Dept. Estadística, Facultad de Ciencias, 29071 Málaga, Spain.

The journal is published by the Confederación Española de Centros de Investigación Matemática y Estadística.

TRABAJOS DE INVESTIGACIÓN OPERATIVA

This journal commenced publication in 1986, as a consequence of splitting *Trabajos de Estadística y de Investigación Operativa* into two parts (*see* TRABAJOS DE ESTADÍSTICA).

Trabajos de Investigación Operativa accepts papers in Spanish, English, and French only. There is one volume per year, published in three parts, containing in total about 350 pages. Subscription price is 3300 Spanish pesotes (US $30). The Editor is Marco Antonio López Cerdá, Dept. Matemáticas, Facultad de Ciencia, 03071, Alicante, Spain.

The journal is published by the Confederación Española de Centros de Investigación Matemática y Estadística.

TRACE (1)

The trace of a $n \times n$ matrix **A** denoted by $\mathrm{tr}(\mathbf{A})$ is, by definition, the sum of its n diagonal elements.

The following four of its basic properties are often useful in statistical multivariate analysis*:

1. $\mathrm{tr}(\mathbf{AB}) = \mathrm{tr}(\mathbf{BA})$, where **A** and **B** are any two matrices for which the products **AB** and **BA** are defined.

2. For any sequence of $n \times n$ matrices $\mathbf{A}_1, \mathbf{A}_2, \ldots, \mathbf{A}_n$,

$$\mathrm{tr}\left(\sum_{i=1}^{n} \mathbf{A}_i \right) = \sum_{i=1}^{n} \mathrm{tr}(\mathbf{A}_i).$$

3. If $\lambda_1, \lambda_2, \ldots, \lambda_n$ are the eigenvalues* of the matrix **A**, then

$$\mathrm{tr}(\mathbf{A}^s) = \sum_{i=1}^{n} \lambda_i \ (s = \cdots -1, 0, 1, \ldots).$$

4. If **P** is an orthogonal matrix ($\mathbf{PP}' = \mathbf{I}$), then

$$\mathrm{tr}(\mathbf{PAP}') = \mathrm{tr}(\mathbf{A}).$$

(MULTINORMAL DISTRIBUTION
PILLAI'S TRACE
QUADRATIC FORMS)

TRACE (2)

A term used in exploratory data analysis* to denote any form of "line" (straight, broken, or curved). It is a statistical analog of the mathematical term "curves" (which includes straight lines as special cases).

Reference

[1] Tukey, J. W. (1977). *Exploratory Data Analysis*. Addison-Wesley, Reading, MA.

TRACEABILITY

This means "ability to trace"—for example in going back from installation to source. The term is used in quality control* in (at least) three different senses [1].

 (i) **Calibration:** Ability to trace the calibration of measuring equipment to a certain standard.

 (ii) **Data:** Ability to trace the operational, computational, and recording steps of measurement or evaluation of an item, process, or source.

(iii) **Distribution:** Ability to trace the history, application, or location of an item (or items similar to a particular item) by means of records identifying it (or them).

Reference

[1] Standards Committee of American Society for Quality Control (ASQC) (1978). *ASQC Standard A3.*

TRACE CORRELATION *See* GLAHN AND HOOPER CORRELATION COEFFICIENTS

TRACELINE (TRACE LINE; TRACE-LINE)

The term *trace line* was coined by Paul F. Lazarsfeld during the 1940s as part of the early development of latent structure analysis*. It appears to have been used for the first time in Chaps. 10 and 11 of *Measurement and Prediction*, Volume IV of the monumental series *Studies in Social Psychology in World War II* [7]. Samuel Stouffer, in his influential introduction to the volume, puts quotation marks around the phrase, and Lazarsfeld italicizes it in his first usage, so it is clear that there is not supposed to be a previously accepted meaning.

A trace line is a function that gives, for each value x of a latent trait, the probability of responding positively to a particular item that serves as an indicator of the trait. Thus, proper usage would be "trace line of an item." Mokken [5, p. 117] used the term *trace function* to emphasize the point that the probability of positive response depends not only on a respondent's position on the latent continuum, but also on properties of the item or question. It should be emphasized that even in the earliest usage there was no restriction to linear functions: Trace lines need not be straight lines!

The term was closely identified with Lazarsfeld in the 1950s: for example, Marschak [4], writing on how the concept of probability was used in the social sciences, refers to "Lazarsfeld's 'tracelines'." The same function played a central role in the development of test theory, under a different name. As Lord explained [2]:

> Lazarsfeld calls such a curve the "trace line" of the item; here we will follow Tucker's terminology [in a 1946 article in *Psychometrika*] and call it the *item characteristic curve*.

The two traditions were still distinct in 1968 when Lazarsfeld and Lord published their comprehensive books *Latent Structure Analysis* [1] and *Statistical Theories of Mental Test Scores* [3].

In the past decade, latent trait theory has become more and more identified with the name of George Rasch, whose monograph [6] appears to be independent of both latent structure and classical test theory. In recent writing on item response theory, the terms *item response function* and *Rasch model* are used instead of "traceline" or "item characteristic curve," perhaps in keeping with modern preferences for algebraic over geometric terminology.

References

[1] Lazarsfeld, P. F. and Henry, N. W. (1968). *Latent Structure Analysis*. Houghton Mifflin, Boston, MA.

[2] Lord, F. M. (1953). The relation of test score to the trait underlying the test. *Educ. Psychol. Meas.*, **XIII**.

[3] Lord, F. M. and Novick, M. R. (1968). *Statistical Theories of Mental Test Scores*. Addison-Wesley, Reading, MA.

[4] Marschak, J. (1954). Probability in the social sciences. In *Mathematical Thinking in the Social Sciences*, P. F. Lazarsfeld, ed. The Free Press, Glencoe, IL, Chap. 4.

[5] Mokken, R. J. (1971). *A Theory and Procedure of Scale Analysis*. Mouton, The Hague, The Netherlands.

[6] Rasch, G. (1960). *Probabilistic Models for Some Intelligence and Attainment Tests*. Danmarks Paedagogiske Institut, Copenhagen, Denmark.

[7] Stouffer, S. A. et al. (1950). *Measurement and Prediction*. Princeton University Press, Princeton, N.J.

(LATENT STRUCTURE ANALYSIS PSYCHOLOGICAL TESTING THEORY RASCH MODEL, THE SOCIOLOGY, STATISTICS IN SOCIOMETRY)

NEIL W. HENRY

TRACKING

Investigators engaged in longitudinal studies of human populations have often found that individual patterns of change over time can be described in relatively simple ways that improve understanding of development and aging and facilitate the prediction of future values of biological variables (*see* LONGITUDINAL DATA ANALYSIS). Some investigators have emphasized the high correlation* between successive measurements of the same individuals [8], others have noted the tendency for individuals to maintain relative or percentile rank over time [5, 8], and still others have noted that individual patterns of growth or aging can be described by simple functions of time [9]. These related phenomena have been described as *tracking*, often without benefit of a precise definition for the term. Recently, several papers in the statistical literature have provided explicit definitions and described statistical methods for investigating tracking in longitudinal data sets. We discuss three different approaches.

One definition is based on the notion that individuals remain at a constant percentile of the population distribution over time.

Definition 1 (McMahan).   A population is said to track with respect to an observable characteristic if, for each individual, the expected value of the deviation from the population mean remains fixed over time [7].

To develop a quantitative analysis of tracking, McMahan assumes that the data of interest arise from a polynomial growth curve* model. Since other definitions of tracking are also related to growth curve analysis, we begin with a brief description of the growth curve model for a balanced design. For a more general formulation, see ref. 6.

Let y_i, $i = 1, 2, \ldots, N$, be a $p \times 1$ vector of observations of the biological variable of interest at p different time points and $\mathbf{X}$ be a $p \times k$ within-individual design matrix*. The two-stage growth curve model is defined by
Stage I:

$$\mathbf{y}_i | \boldsymbol{\beta}_i \sim N(\mathbf{X}\boldsymbol{\beta}_i, \sigma^2 \mathbf{I}),$$

where $N(\cdot, \cdot)$ denotes the multinormal distribution* and $\boldsymbol{\beta}_i$ is a $k \times 1$ vector of individual growth curve parameters;
Stage II:

$$\boldsymbol{\beta}_i \sim N(\boldsymbol{\beta}, \boldsymbol{\Sigma}).$$

The two-stage model implies that $\mathbf{y}_i$ has the marginal distribution

$$\mathbf{y}_i \sim N(\mathbf{X}\boldsymbol{\beta}, \mathbf{X}\boldsymbol{\Sigma}\mathbf{X}' + \sigma^2 \mathbf{I}).$$

The total variance of $\mathbf{y}$ consists of between-individual variation $\mathbf{X}\boldsymbol{\Sigma}\mathbf{X}'$ and error variance $\sigma^2 \mathbf{I}$.

Returning to the issue of tracking, Definition 1 implies that

$$E[y_{ij} | \boldsymbol{\beta}_i] = \mu_j + \kappa_i \delta_j,$$
$$i = 1, \ldots, N, \ j = 1, \ldots, p,$$

where μ_j and δ_j are the population mean and interindividual standard deviation respectively, at time j; κ_i is constant for the ith individual and $E(\kappa_i) = 0$, $\mathrm{var}(\kappa_i) = 1$ in the population. This can be expressed in matrix form as

$$\mathbf{X}\boldsymbol{\beta}_i = \mathbf{X}\boldsymbol{\beta} + \kappa_i \boldsymbol{\delta},$$

where $\boldsymbol{\delta}' = (\delta_1, \ldots, \delta_p)$ and $\delta_i^2 = (\mathbf{X}\boldsymbol{\Sigma}\mathbf{X}')_{ii}$. Hence, tracking is equivalent to a restriction on the interindividual covariance matrix, namely,

$$\mathbf{X}\boldsymbol{\Sigma}\mathbf{X}' = \boldsymbol{\delta}\boldsymbol{\delta}'.$$

This definition leads to a quantitative index of tracking for an arbitrary $\mathbf{X}$ and $\boldsymbol{\Sigma}$. One index,

$$\zeta = \boldsymbol{\delta}'\mathbf{X}\boldsymbol{\Sigma}\mathbf{X}'\boldsymbol{\delta}/(\boldsymbol{\delta}'\boldsymbol{\delta})^2,$$

represents the fraction of the total interindividual variation, $\mathrm{tr}(\mathbf{X}\boldsymbol{\Sigma}\mathbf{X}')$, attributable to tracking. Since ζ will be positive even when $\mathbf{X}\boldsymbol{\Sigma}\mathbf{X}'$ is diagonal (implying mutually independent deviations and usually regarded as representing no tracking), McMahan defined a corrected index

$$\tau = \frac{\boldsymbol{\delta}'(\mathbf{X}\boldsymbol{\Sigma}\mathbf{X}' - \mathbf{D}_\delta^2)\boldsymbol{\delta}}{\boldsymbol{\delta}'(\boldsymbol{\delta}\boldsymbol{\delta}' - \mathbf{D}_\delta^2)\boldsymbol{\delta}}$$

where $\mathbf{D}_\delta$ is a diagonal matrix containing the elements of $\boldsymbol{\delta}$.

McMahan also defined tracking indices for deviations standardized by the interindividual component of population variance. These definitions clarify the interpretation of the two indices. If the tracking indices are defined in terms of $\mathbf{R} = \mathbf{D}_\delta^{-1}\mathbf{X}\mathbf{\Sigma}\mathbf{X}'\mathbf{D}_\delta^{-1}$, the correlation matrix* of the expected deviations, and if $\mathbf{1}$ denotes a vector of 1's, then

$$\zeta_\rho = \mathbf{1}'\mathbf{R}\mathbf{1}/(\mathbf{1}'\mathbf{1})^2 = \mathbf{1}'\mathbf{R}\mathbf{1}/p^2,$$

$$\tau_\rho = [\mathbf{1}'(\mathbf{R} - \mathbf{I})\mathbf{1}]/[\mathbf{1}'(\mathbf{1}\mathbf{1}' - \mathbf{I})\mathbf{1}]$$

$$= [\mathbf{1}'(\mathbf{R} - \mathbf{I})\mathbf{1}]/[p(p-1)].$$

Then ζ_ρ is the average of the elements of $\mathbf{R}$ and τ_ρ is the average of the off-diagonal elements, the interoccasion correlation coefficients. McMahan estimated $\mathbf{\Sigma}$ by maximum likelihood* (Anderson [1]). Substitution of functions of $\hat{\mathbf{\Sigma}}$ into the various expressions for tracking coefficients gives their estimates. McMahan also proposed use of the jackknife* to compute confidence intervals for each index. Although he considered only balanced data, the definitions extend directly to incomplete data sets and $\mathbf{\Sigma}$ can be estimated by either the maximum likelihood or the restricted maximum likelihood criterion, as described by Laird and Ware [6]. For unbalanced data sets, the concept remains clear but the indices are not well defined, since the design matrix $\mathbf{X}$ varies among individuals.

McMahan's definition of tracking is related to principal components* or factor analysis*. There is, however, a fundamental difference in that the vector $\boldsymbol{\delta}$ is fixed rather than computed from the spectral decomposition of $\mathbf{X}\mathbf{\Sigma}\mathbf{X}'$ as principal components analysis would require. For example, the one-factor model can be written as

$$\mathbf{y}_i = \boldsymbol{\mu} + \mathbf{a}f_i + \boldsymbol{\epsilon}_i,$$

where $\boldsymbol{\mu}$ is the vector of population means of the p occasions, $\mathbf{a}$ is a $p \times 1$ vector of unobservable factor loadings, f_i is a scalar factor with a standard normal distribution in the population, and $\boldsymbol{\epsilon}_i$ is the vector of errors or singularities. This formulation generalizes directly to several factors, but there is no natural definition of a tracking index corresponding to the factor analytic representation.

Often, individual patterns of change over time are regular but not as simple as Definition 1 would imply. For example, individuals

may have different rates of development or aging. Hence, Ware and Wu [9] take the view that observations arising from a more complex growth curve model also exhibit a form of tracking. For example, if

$$\mathbf{X} = \begin{pmatrix} 1, \ldots, 1 \\ t_1, \ldots, t_p \end{pmatrix},$$

the expected value of each individual's deviations from the population mean varies linearly with time.

This view of tracking suggests:

Definition 2. A population is said to track with respect to an observable characteristic if, for each individual, the expected values of the serial measurements are given by a polynomial function of time.

This definition suggests a more general index of tracking based on the proportion of total variation explained by a polynomial growth curve of lower order than that required to achieve good fit to the variance–covariance matrix. However, the model implied by Definition 1 is not a growth curve model unless interindividual variation is constant over time.

One motivation for interest in tracking is the desire to predict future values of known risk factors for disease. For example, since high blood pressure is a risk factor for cardiovascular disease in adult life, investigators have sought to determine whether children whose blood pressures are high for their age become adults with high blood pressure level. Ware and Wu [9] noted that, from standard multivariate normal theory, if

$$\mathbf{y}_i = \begin{pmatrix} \mathbf{y}_{1i} \\ \mathbf{y}_{2i} \end{pmatrix} \begin{matrix} p_{i-1} \\ 1 \end{matrix},$$

$$\mathbf{X}_i = \begin{pmatrix} \mathbf{X}_{1i} \\ \mathbf{X}_{2i} \end{pmatrix} \begin{matrix} p_{i-1} \\ 1 \end{matrix},$$

and the observations arise from a two-stage growth curve model, then

$$E[\mathbf{y}_{2i} - \mathbf{X}_{2i}\boldsymbol{\alpha}]$$

$$= (\mathbf{X}_{2i}\mathbf{\Sigma}\mathbf{X}_{1i}')(\mathbf{X}_{1i}\mathbf{\Sigma}\mathbf{X}_{1i}' + \sigma^2\mathbf{I})^{-1}$$

$$\times (\mathbf{y}_{1i} - \mathbf{X}_{1i}\boldsymbol{\alpha})$$

with prediction variance

$$\sigma^2 + \mathbf{X}_{2i}\Sigma\mathbf{X}'_{2i} - (\mathbf{X}_{2i}\Sigma\mathbf{X}'_{1i})$$

$$\times(\mathbf{X}_{1i}\Sigma\mathbf{X}'_{1i} + \sigma^2\mathbf{I})^{-1}(\mathbf{X}_{1i}\Sigma\mathbf{X}'_{2i}).$$

For example, if, as in the intraclass correlation* model, $\mathbf{X}$ is $p_i \times 1$ and a column of 1's, the expected deviation at occasion p_i is the average of previous deviations corrected for regression to the mean*.

Foulkes and Davis [5] introduced a third definition, namely:

Definition 3. A population tracks with respect to an observable characteristic if, for each pair of individuals, the expected values of that characteristic remain in the same rank order over time.

For a time interval $[T_1, T_2]$, they defined an index of tracking

$$\gamma(T_1, T_2) = \Pr\{ f(t, \boldsymbol{\beta}_1) \geqslant f(t, \boldsymbol{\beta}_2)$$

$$\text{or } f(t, \boldsymbol{\beta}_2) \geqslant f(t, \boldsymbol{\beta}_1)$$

$$\text{for all } t \in [T_1, T_2]\},$$

where $f(t, \boldsymbol{\beta}_i)$ $[f(t, \boldsymbol{\beta}_j)]$ is the growth curve of expected values of the ith (jth) randomly chosen individual.

Foulkes and Davis recommend estimation of $\boldsymbol{\beta}_i$ by standard methods and estimate $\gamma(T_1, T_2)$ by

$$\hat{\gamma}(T_1, T_2) = \binom{n}{2}^{-1} \sum_{1<j} \phi(\mathbf{y}_i, \mathbf{y}_j, T_1, T_2),$$

$$\phi(y_i, y_j, T_1, T_2) = \begin{cases} 1, & \text{if} f(t,\hat{\boldsymbol{\beta}}_i) \geqslant f(t,\hat{\boldsymbol{\beta}}_j) \\ & \text{or} f(t,\hat{\boldsymbol{\beta}}_i) \leqslant f(t,\hat{\boldsymbol{\beta}}_j) \\ & \text{for all } t \in [T_1, T_2], \\ 0, & \text{otherwise.} \end{cases}$$

They appeal to the theory of U statistics* to establish asymptotic normality* and estimate the variance of $\hat{\gamma}(T_1, T_2)$. They also emphasize that $\gamma(T_1, T_2)$ depends heavily on the choice of the interval $[T_1, T_2]$.

These three definitions represent three closely related attempts to characterize patterns of growth and aging in populations. McMahan's definition represents a strong and intuitively appealing form of tracking. When it is satisfied, both description and prediction are straightforward. Even when it is not satisfied, however, relatively simple models for serial observations can be useful. For example, Fletcher et al. [4] have shown that individuals lose pulmonary capacity at different rates and that these rates remain constant over time in the absence of changes in smoking habits or health status. Although serial measurements of forced vital capacity do not track in the sense proposed by Mc-Mahan, the relatively simple patterns of temporal change facilitate prediction.

References

[1] Anderson, T. W. (1958). *Multivariate Statistical Analysis*. Wiley, New York.

[2] Berenson, G. S., Foster, T. A., Frank, G. C., Freuchs, R. R., Srinivasan, S. R., Voors, A. W., and Webber, L. S. (1978). *Circulation*, **57**, 603–612.

[3] Clarke, W. R., Schrott, H. G., Leaverton, P. E., Connor, W. E., and Lauer, R. M. (1978). *Circulation*, **58**, 626–634.

[4] Fletcher, C., Peto, R., Tinker, C., and Speizer, F. E. (1976). *The Natural History of Chronic Bronchitis and Emphysema*. Oxford University Press, New York.

[5] Foulkes, M. A. and Davis, C. E. (1981). *Biometrics*, **37**, 439–446.

[6] Laird, N. M. and Ware, J. H. (1982). *Biometrics*, **38**, 963–974.

[7] McMahan, C. A. (1981). *Biometrics*, **37**, 447–455.

[8] Rosner, B., Hennekens, C. H., Kass, E. H., and Miall, W. E. (1977). *Amer. J. Epidemiol.*, **106**, 306–313.

[9] Ware, J. H. and Wu, M. C. (1981). *Biometrics*, **37**, 427–438.

(BIOSTATISTICS
EPIDEMIOLOGICAL STATISTICS
GROWTH CURVES
LONGITUDINAL DATA ANALYSIS)

J. H. WARE

TRAFFIC FLOW PROBLEMS

Road traffic in developed countries may absorb up to 20% of national expenditure and thus be of major economic importance. Areas of concern include, first, the safe and efficient movement of vehicles over the road network, and second, the planning of new facilities. The former requires understanding of the behavior of vehicles and their drivers in the traffic stream and the latter, while primarily an economic problem not discussed here, requires the measurement and forecasting of traffic volumes. Only a few of the more important aspects can be referred to here; for a fuller appreciation of the range of problems, see the bibliography. Problems of road accidents are not discussed; comparisons of accident frequencies between places, time periods, or people require statistical tests which in straightforward cases can be based on the chi-squared distribution*.

DELAY, SPEEDS, AND CONGESTION

Problems of traffic delay and congestion can be divided into those that occur at junctions and those that arise on lengths of road away from junctions. In each case, understanding can be obtained by observing normal traffic, by setting up controlled experiments, and by setting up probability models whose behavior can be explored either by mathematical analysis or, increasingly, by computer simulation* techniques. Best results are usually obtained by combining these approaches.

The optimum design and setting of traffic light signals is of major practical importance and has received much attention. At a single junction, the main problem is to arrange the red–green sequences and their response to the pattern of vehicle arrivals, so as to minimise average delay. Important concepts include the arrival rate (vehicles arriving per unit time on each approach), saturation flow (vehicles per unit time during a green period), and lost time (time for which no traffic is flowing on any approach when the lights change). Detailed studies of such quantities have been made, so far as possible in normal traffic, and the results used to set up probability models. These are usually too complex for explicit mathematical analysis, but

various working rules have been derived to allow calculation of optimum cycle times and green–red split between conflicting flows.

In recent years, studies of signal-controlled junctions have been extended to deal with *area traffic control*, the use of computer-based systems to give integrated control of perhaps hundreds of junctions in an area. Often such control systems are not traffic-responsive, but methods of incorporating such responsiveness are now becoming available. Both computer simulation and real-life experimentation can assist in the designing of area control schemes.

Similar problems arise with junctions not controlled by traffic signals. There is a greater need to use queueing theory* because a central concern is how long a vehicle at the head of a queue will have to wait to find a sufficiently long gap in another traffic stream that it wishes to cross or join. Often it is sufficient to assume random traffic arrivals and constant service time, but very careful calibration of queueing models is required because such assumptions are rarely exactly true. As in the case of traffic signals, the most useful methods of estimating delays and capacities, and of improving junction design, arise from careful combination of observational and theoretical approaches. Studies of this kind are required both for the design of new junction layouts and for their economic evaluation.

Away from junctions, there has been considerable interest in *car-following theory*, the precise way in which one vehicle in a traffic stream positions itself with reference to the vehicle in front of it, especially when the latter may be accelerating or decelerating. This is of concern in relation to the capacity of roads and how speed varies with the amount of traffic, and also in the contexts of safety, especially on high-speed roads, and the control of traffic in tunnels.

An aspect of traffic flow away from junctions that has a strong statistical content is the understanding of following and overtaking behaviour on multilane roads. Various theories have been postulated to represent the processes of catching up to the vehicle in front and changing lanes in order to overtake when the adjacent lane contains a sufficient gap. Some insight can be obtained from simplified theoretical models and some success has been achieved with the incorporation of such behaviour in simulation models of more realistic situations.

At a more empirical level, there is much interest in *speed-flow relations*, which give average speed on a road as a function of the amount of traffic that it is carrying. There are conceptual problems in fluctuating traffic levels and when traffic backs up from one section of road to another. Simple regression models may not be applicable.

MEASUREMENT OF FLOWS

Much highway planning depends on knowledge of traffic flow (vehicles per hour or day) on individual links of the road network, the types of vehicles involved, and the origins and destinations of their journeys. Information can be gathered at the roadside (counting vehicles or stopping them to interview the drivers) or by means of off-road surveys (of random samples of households, for example). Modern analytical techniques, relating traffic volumes to causal socioeconomic variables, together with the use of computers for recording and processing data, tend to favour the more detailed information obtainable from off-road surveys, but both approaches have a continuing place.

With simple roadside traffic counts, whose object is to measure some kind of long term average traffic flow, perhaps over a year as a whole or over peak hours only, there is a problem of sample design. Ideally, a number of short counts would be spread uniformly over the period of interest; the greater the frequency and duration of counts, the greater the accuracy. In practice, it is usually more convenient to make a small number of longer counts and then to apply correction factors to allow for the times of day, days of the week, and seasons of the year at which the counts were taken. Typically, manual counts

may cover up to several 12 or 16 hr days, while counts made with portable automatic counters may be of a week or two duration.

Roadside interview surveys raise similar sampling problems. In addition, there usually needs to be sampling of vehicles from the traffic stream. Careful formulation of procedures is required to prevent an undue tendency to stop vehicles at the head or tail of bunches of traffic.

A recent development is the use of likelihood* methods to estimate patterns of origins and destinations using data from roadside counts only. The practical usefulness of such methods is not yet clear.

Off-road surveys, while providing the most comprehensive source of data, present the most severe statistical problems. A typical survey* might cover the area of a single town. This would involve a detailed survey of a sample of households in the town to give the pattern of personal travel made by residents. This would need to be supplemented by a roadside survey at all points on the periphery of the survey area, primarily to cover nonresident traffic, and surveys of establishments generating commercial vehicle trips. The household survey might cover of the order of 1% of households, chosen from the list of voters, and ask them for details of their travel over a period of up to a week. Such surveys are subject to a number of sources of bias; these include inadequate sample frame*, problems of nonresponse*, and the difficulty of obtaining full and accurate records from participating households. Because of these problems, it is normal to use roadside counts to check and if necessary adjust the results of such surveys.

Further errors arise in making the forecasts for planning new roads. Such errors are of two main kinds. First, there are errors in estimating the response of traffic to altered conditions. For example a new road may reduce flow levels on an existing road, as a result of which it may be less congested; one needs to understand both the extent of this change in congestion level and the extent to which this will influence the choice of route by drivers. Second, there are errors in forecasting to a future year. This requires the forecasting of such factors as economic growth, changes in population, and changes in car ownership. These need to be considered both nationally and in relation to local circumstances.

Transport planners have in recent years developed sophisticated models to represent the numbers of trips made by a household, where the destinations will be, what mode of travel will be used, and what route will be taken. Successful use of such models for forecasting requires detailed understanding of travel behaviour, and this has not always been available.

The importance of forecasting long-term traffic growth is illustrated by the magnitudes involved; in Great Britain, for example, traffic has grown by a factor of about 4 over the past 30 years and is expected to increase by about a further 50% in the next 30 years.

Bibliography

Perhaps the easiest way of obtaining an overall idea of recent work is by reference to the specialist journals. These include:

> *Accident Analysis and Prevention*
> *ITE Journal* (US, monthly, Institute of Transportation Engineers)
> *Traffic Engineering and Control* (Great Britain, monthly)
> *Transportation Research*
> *Transportation Science*

For the United States, issues of *Transportation Research Record* (Transportation Research Board, Washington) cover a wide range, as do the publications in Great Britain of the Transport and Road Research Laboratory (Crowthorne, Berks). Among the books available are:

Baerwald, J. E., (ed.) (1976). *Transportation and Traffic Engineering Handbook*. Prentice-Hall, Englewood Cliffs, NJ.

Greenshields, B. D. and Weida, F. M. (1978). *Statistics with Applications to Highway Traffic Analyses*. Eno Foundation, Westport, CT.

Morlok, E. K. (1978). *Introduction to Transportation Engineering and Planning*. McGraw-Hill Kogakusha, Tokyo, Japan.

Organisation for Economic Co-operation and Development (1982). *Car Ownership and Use*, Report of Research Group AP2. O.E.C.D., Paris, France.

Pignataro, L. J. (1973). *Traffic Engineering Theory and Practice*. Prentice-Hall, Englewood Cliffs, NJ.

Road Research Laboratory (1965). *Research on Road Traffic*. H.M.S.O., London, England.

Theoretical aspects are well covered in the proceedings of a series of international symposia, the latest being

Hurdle, V. F., Hauer, E., and Steuart, G. N., eds. (1983). *Proceedings of the Eighth International Symposium on Transportation and Traffic Theory*. University of Toronto Press, Toronto, Canada.

(QUEUEING THEORY
SURVEY SAMPLING)

J. C. TANNER

TRAFFIC INTENSITY   *See* TRAFFIC FLOW PROBLEMS

TRANSCENDENTAL LOGARITHMIC FUNCTIONS

A transcendental logarithmic function is a particular flexible functional form, i.e., capable of approximating a wide variety of functions. A transcendental logarithmic function may be viewed as a Taylor series* approximation (in logarithms) of a general differentiable function; see, e.g., Christensen et al. [1]. For a recent discussion of various flexible functional forms (including Box–Cox transformations*) see, e.g., Gallant and Golub [2] and the literature referenced therein.

References

[1] Christensen, L. R., Jorgenson, D. W., and Lau, L. J. (1973). *Rev. Econ. Statist.*, **55**, 28–45.

[2] Gallant, A. R. and Golub, G. H. (1984). *J. Econ.*, **26**, 295–321.

(MATHEMATICAL FUNCTIONS,
   APPROXIMATIONS TO)

TRANSECT METHODS

The term "transect methods" refers to a class of techniques which have been developed for use in surveying wildlife* communities. The major goal of such a survey is to estimate the population densities of the inhabitants (animals, trees, birds, etc.). The defining common feature of these methods is that they are *plotless*; instead of counting all individuals in predetermined plots of known areas, we record only those individuals that are easily detected, along with auxiliary measurements. In this way, larger areas are covered more quickly. The auxiliary measurements provide estimates on the size of the area which is effectively surveyed in this way. Although it is assumed implicitly that many individuals will be undetected, these misses will theoretically affect both the total counts and the estimates of effective area equally, thereby giving density estimates which remain accurate.

Once an estimate of population density has been obtained, it can be used to estimate a variety of other indices—population totals, species diversity*, total biomass, habitat breadth, etc. In considering any of these, we may be interested in the specific community as it exists at the time of the survey or the emphasis may be on community structure theory, in which we attempt to estimate features of an *ensemble* of communities from which the specific community surveyed was sampled. It is the former approach that is generally of use in making management decisions, and that on which we will focus. Extrapolation to the latter requires assumptions about the distribution of communities in the ensemble and about the process governing the selection of the one that was observed. Some estimation procedures may have desirable properties in one case but undesirable in the other.

There are several transect field techniques currently in use. Each assumes that a tran-

sect is a straight line randomly located within the survey's target region.

Line Transect* Method: An observer traverses the transect at a uniform speed, recording all objects detected and the shortest ("right-angle") distance from each to the transect. This method is best suited for surveying large homogeneous regions that are easily traversed.

Variable Circular Plot Method: The observer visits stations marked at regular intervals along the transect. Observations are made during a fixed period of time at each station with radial distances being recorded to each detected object. This method is best suited to surveys of heterogeneous regions of difficult terrain.

Line Intercept* Method: Only those objects that intersect the transect line, such as fallen logs or shrub canopies, are recorded, along with the lengths of their intersections. This method has been used widely in vegetational studies (*see* STATISTICS IN FORESTRY).

Strip Transect Method: This resembles the line transect method, but right-angle distance measurements are replaced by indicators of whether detected objects are within a fixed-distance strip around the transect when first detected. This method is appealing where distance estimates are likely to be inaccurate, but it requires near certain detection within the strip.

Flushing Distance Transect Method: This also resembles the line transect method, but the observer records the observer-to-object distance at detection rather than the right-angle distance. This method is appropriate in situations where the observer's presence influences the objects' detectabilities, as with flushed game birds.

Statistical treatment of transect data is based on well-known procedures which adjust sample observations for the probability of occurrence. This probability of observing an individual within the target (surveyed) region is called *detectability*. The methodology does not require that detectability be known completely, but does assume that (a) detections of different individuals are independent events, (b) there exists a known region of perfect detectability, and (c) we know a priori the contours of detectability for the survey method used. The contours can be thought of as lines in the target region on which detectability is constant; thus, contours in a line transect survey run parallel to the transect, whereas they form concentric circles around stations in a variable circular plot survey. Typically, detectability decreases with increasing distance from the observer so that the region in (b) is at zero distance from the transect.

The ideal auxiliary measurement to record in a transect survey is an indicator of the detectability contour upon which an individual is observed. The indicator most often used is the distance from the transect or station. Using Y to denote the contour on which an individual in the target region is placed, we let $A(Y)$ be the area of the target region interior to this contour; e.g., $A(Y) = 2LY$ for a line transect of length L or $A(Y) = \pi Y^2$ for a variable circular plot survey. The function $g(y) = \Pr[\text{detection}|Y = y]$ is the *detectability curve*. We assume that an individual's random location satisfies the uniformity condition $\Pr[Y \leq y] = A(y)/\mathscr{A}$, where $\mathscr{A}$ is the total area of the target region. This may be achieved by assuming that individuals are placed completely at random within the target region or by positioning transects perpendicularly to anticipated density contours.

The data from a transect survey thus consist of the number of detections n and the observed detection distances $Z_1, \ldots, Z_n$; Z denotes the same measurement as Y. The purpose of such notation is to distinguish between the sets of measurements for individuals *present* (Y) and those for individuals *observed* (Z), as the properties of these sets are quite different. The plan of statistical analysis is then twofold: First, the unknown detectability must be estimated, which we do through the use of estimated effective area; then we estimate density by adjusting n for detectability.

If we are surveying a region with uniform density (D) of individuals, it is quite natural to define the *effective area* surveyed, a, as that area making $Da = \mathscr{E}[n]$. This leads to the general definition

$$a = \int_0^\infty g(y) A'(y)\, dy$$

for effective area.

Let $X_j = A(Z_j)$ for $j = 1, \ldots, n$ be the areas within contours of detected individuals. These areas are independent and identically distributed variates having the probability density function (PDF)

$$f_X(x) = g\big(A^{-1}(x)\big)/a, \qquad x \geq 0.$$

To complete the specification of the statistical model, assume that $g(A^{-1}(x)) = h(x/a)$, where $h(\cdot)$ is a detectability curve "kernel" that does not involve a. Thus, the effective area surveyed becomes a simple scale parameter. Inference procedures for transect models have been studied in the following cases: (i) the detectability curve kernel is known; (ii) the kernel is known to belong to a parametric class of kernels; (iii) nothing is assumed about the shape of the kernel, other than $h(0) = 1$. The parametric models of (i) and (ii) are summarized by Ramsey [5]. Nonparametric procedures for (iii) are provided by Burnham and Anderson [1] and Burnham et al. [2]. Here is a brief summary.

(i) The exponential power kernel, where

$$h(u) = \exp\left[-\{\Gamma(1 + 1/\gamma)u\}^\gamma\right]$$

for some $\gamma > 0$, has received the most attention in the literature. With γ known, this yields a single parameter exponential family density for the areas with complete minimal sufficient statistic*

$$T_\gamma = n^{-1} \sum_{j=1}^n X_j^\gamma$$

for the unknown effective area surveyed. The quantity $S_\gamma = nT_\gamma\{\Gamma(1 + 1/\gamma)/a\}^\gamma$ is pivotal*, following a gamma distribution* with unit scale and shape n/γ.

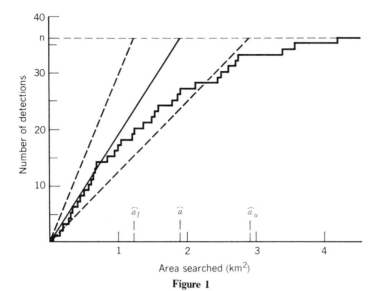

Figure 1

(ii) The exponential power kernel with γ unknown provides a flexible class of distributions. Maximum likelihood* leads to the equation

$$\left\{\log \hat{\gamma} + \log T_{\hat{\gamma}} + \psi(1 + 1/\hat{\gamma})\right\}/\hat{\gamma} = V_{\hat{\gamma}}/T_{\hat{\gamma}},$$
(1)

where $V_{\gamma} = (d/d\gamma)T_{\gamma} = n^{-1}\sum_{j=1}^{n} X_j^{\gamma}\log X_j$ and where $\psi(\cdot)$ is the digamma function*. Equation (1) is solved iteratively for $\hat{\gamma}$. Then

$$\hat{a} = \hat{\gamma}^{1/\hat{\gamma}}\Gamma(1 + 1/\hat{\gamma})T_{\hat{\gamma}}^{1/\hat{\gamma}}.$$

The large sample approximate variance is

$$\mathrm{Var}[\log(\hat{a})]$$
$$= (n\gamma)^{-1}\left[1 - \left\{(1 + 1/\gamma)\right.\right.$$
$$\left.\left.\times\psi'(1 + 1/\gamma)\right\}^{-1}\right]^{-1}.$$

(iii) With minimal assumptions about the shape of the detectability curve, thinking of effective area as $a = 1/f_X(0)$ is preferable to the preceding scale parameter view because resultant estimators emphasize information from detections near the transect. A host of ad hoc nonparametric estimators exist. Robust estimators with good properties are obtained by fitting low order polynomials or trigonometric series to the CDF, the PDF, or the logarithm of the PDF of the area measurements.

Practitioners agree that estimates of variability based on replicate sampling or jackknifing* with a single sample are preferred over estimates based on specific models for the detectability curve.

Once effective area is estimated (by $\hat{a}$, say), density is estimated as $D = n/\hat{a}$. The properties of this estimator may be determined using standard techniques for treating ratios. Conditional on N, the distribution of n is binomial* $(N, a/\mathscr{A})$.

An example of transect data appears in Fig. 1, displaying the cumulative detections curve, a plot of the numbers of detections against the area searched. Typically, the density of detections—the *slope* of the curve—is highest near the transect (area = 0) but declines at higher areas because of reduced detectability of individuals further from the transect. The estimate of the effective area surveyed is visualized in such a picture as the area covered when the slope at the origin is projected to the entire sample size. The estimate, along with its 95% confidence interval*, was determined as in (ii).

The methods presented apply well to line transect, variable circular plot, and (to a lesser extent) strip transect surveys. Treatments of line intercept and flushing distance transect surveys require special models and variations on these methods. The reader is referred to the literature for these.

References

[1] Burnham, K. P. and Anderson, D. R. (1976). Mathematical models for non-parametric inferences from line transect data. *Biometrics*, **32**, 235–237.

[2] Burnham, K. P., Anderson, D. R., and J. L. Laake (1980). Estimation of density from line transect sampling of biological populations. Wildlife Monograph No. 72. *J. Wildlife Management Suppl.*, **44**.

[3] DeVries, P. G. (1979). Line intersect sampling—statistical theory, applications, and suggestions for extended use in ecological inventory. In *Sampling Biological Populations*, R. M. Cormack, G. P. Patil, and D. S. Robson, eds. International Cooperative Publishing House, Fairland, MD, pp. 1–70.

[4] Gates, C. E. (1979). Line transect and related issues. In *Sampling Biological Populations*, R. M. Cormack, G. P. Patil, and D. S. Robson, eds. International Cooperative Publishing House, Fairland, MD, pp. 71–154. (A good survey with excellent bibliography.)

[5] Ramsey, F. L. (1979). Parametric models for line transect surveys. *Biometrika*, **66**, 505–512.

[6] Seber, G. A. F. (1982). *The Estimation of Animal Abundance and Related Parameters*, 2nd ed. Griffin, London, England.

Bibliography

Seber, G. A. F. (1986). *Biometrics*, **42**, 267–292.

(ECOLOGICAL STATISTICS
LINE INTERCEPT SAMPLING
LINE INTERSECT SAMPLING
LINE TRANSECT SAMPLING
STATISTICS IN ANIMAL SCIENCE
STATISTICS IN FORESTRY
WILDLIFE SAMPLING)

FRED L. RAMSEY

TRANSFER FUNCTION MODEL

This is a time series model of the form

$$Y_t = C + \nu(B)X_t + N_t,$$

where C is a constant, Y_t is the output series, X_t is the input series, N_t is the disturbance series, independent of X_t, and $\nu(B) = \sum_{i=0}^{\infty} \nu_i B^i$ is a polynomial in the backshift operator* B (such that $BX_t = X_{t-1}$) (Box and Jenkins [1]).

It is also called a rational distributed lag model. *See also* BOX–JENKINS MODEL and LAG MODELS, DISTRIBUTED. The specification of this model in practice was investigated, among others, by Tsay [2], who also provides numerous references.

References

[1] Box, G. E. P. and Jenkins, G. M. (1976). *Time Series Analysis: Forecasting and Control*. Holden-Day, San Francisco, CA.

[2] Tsay, R. S. (1985). *J. Bus. Econ. Statist.*, **3**, 228–236.

(AUTOREGRESSIVE INTEGRATED
   MOVING AVERAGE (ARIMA)
   MODELS
ECONOMETRICS
TIME SERIES)

TRANSFORMATIONS

INTRODUCTION

Linear transformations arise in many ways in statistics. This article focuses on nonlinear transformations, which may be of parameters, of response variables, or of explanatory variables. The use of transformations in the simplification of distributions and the analysis of data has a long history, the former going back at least to Fisher's z transformation* of the correlation* coefficient [21, especially Sec. 10] and the latter being reviewed by Bartlett [8]. The strong emphasis in the early work on transformation of response variables lay on stabilizing variance so that the formal conditions for the analysis of variance* were more nearly satisfied (*see* EQUALIZATION OF VARIANCE). For a more recent review of transformations and related matters, see Cox [17].

TRANSFORMATION OF PARAMETERS

Full specification of a parametric model calls for a family of probability distributions and normally also for a choice of parametrization. If the purpose is solely the prediction of the expected response at new values of the explanatory variables, the choice of parametrization is a matter of numerical analysis*. Approximate orthogonality and a nearly quadratic form* of the likelihood function* will usually speed convergence of iterative procedures. Systematic methods of achieving these desiderata are not available. Some important general practical comments on parameter transformation are given by Ross [30, 31].

For interpretation it is desirable that the component parameters have a clear and relevant physical meaning and that as many as possible are stable across replicate sets of data. Often dimensionless parameters will be an advantage. Thus in fitting a gamma distribution*, the mean and index (or coefficient of variation*) will often be the most sensible parameters. In multiple regression*, if the models

$$Y_i = \beta_0 + \beta_1 x_{1i} + \cdots + \beta_p x_{pi} + \epsilon_i, \quad (1)$$

$$\log Y_i = \gamma_0 + \gamma_1 \log x_{1i} + \cdots + \gamma_p \log x_{pi} + \epsilon_i' \quad (2)$$

fit about equally well, the second has the advantage that the γ's are dimensionless and so have a more direct qualitative interpretation. In particular, if the variables are lengths, areas, weights, etc., the γ's can be directly related to simple integer exponents which will often have a natural interpretation. An example [11] from the textile industry leads to the establishment of plausible dimensionless quantities.

If the data are not homogeneous, one objective is to confine the variation between sets to as few parameters as possible. Canonical regression analysis can be used to find invariant parameters.

Formal statistical inference is made simpler by such requirements as normal likelihood, stability of variance, or zero asymptotic skewness. Hougaard [24] discusses transformations for single parameter models to achieve these and other criteria. McCullagh and Nelder [28] give log-likelihood plots,

for various parameter transformations, of the Poisson distribution* (p. 131) and the gamma (p. 152). Close validity of asymptotic confidence limits based on the normal distribution is discussed by Sprott [34]. Bates and Watts [9] and Hamilton et al. [22] consider nonlinear least squares* in detail.

Approximate independence of estimation of the components of parameters together with approximate normality are properties which are desirable for statistical as well as numerical analytical reasons. However, approximate normality is unnecessary if confidence regions* are obtained directly from the log-likelihood function. But, in multiparameter problems, approximate normality greatly simplifies summary analysis.

TRANSFORMATION OF RESPONSE VARIABLES

A major problem, in principle, in transforming variables is the conflict between achieving simple forms for the systematic part and for the random part of the model. In normal theory linear problems it is desirable to have: (i) additivity, i.e. simple structure for the systematic part, typically absence of interactions; (ii) constancy of variance; (iii) normality of distribution. For generalized linear* and nonlinear models there are usually direct analogues of (i)–(iii). Very often (i) will be the primary requirement because most scientific interest attaches to the systematic component; see Box and Cox [11].

The simplest situation is where a single homogeneous set of data is to be transformed to a specific distributional form, e.g., normal or exponential. To within the limits set by discreteness, this can always be done exactly by a probability integral transformation* [26, pp. 18–19]. The extent to which more tractable transformations lead to normality is investigated by Efron [19]. Maximum likelihood estimation* within a parametric family often leads essentially to symmetry [18]. One special case is the gamma distribution when $y^{1/3}$ yields the Wilson–Hilferty transformation*. The corresponding multivariate problems are much more difficult except for component by component transformation [2, 3].

With more complex data, one approach is to apply linear model methods to responses transformed to have approximately constant variance. If a theoretical or empirical relationship between var (Y_i) and $E[Y_i]$ is known, asymptotic arguments lead to an approximate variance stabilizing transformation* [26, pp. 98–102], which however is usually not the best transformation to normality [19]. Empirically it is a good idea to plot log variance against log mean for replicated data, leading often to

$$\text{var}(Y_i) \propto \{E[Y_i]\}^{2a},$$

indicating a transformation from Y_i to Y_i^{1-a} ($a \neq 1$) and to log Y_i ($a = 1$). For the Poisson and binomial distributions, proportionality rather than equality between observed and theoretical variances is enough to give some justification for the respective transformations $\sqrt{Y_i}$ and $\sin^{-1}\sqrt{Y_i}$. There is here a relationship with the theory of quasi-likelihood* [27, 36], which develops analogs for general models of the second order assumptions of linear model theory.

Examples of variance stabilizing transformations in the analysis of data are given by Snedecor and Cochran [32, Secs. 11.14–11.18]. These methods, once widely used, have been largely eclipsed due to the recognition that the primary aspect in model choice is usually the form of the systematic part of the relation. The following points are relevant:

(i) For data not too extensive or complex, fitting of nonlinear models is easy using such computer programs as GLIM* [28].

(ii) For balanced or nearly balanced* data, for example, from complex factorial experiments* with several error strata, simple analysis of variance* techniques applied to transformed data remain an attractive method of analysis.

(iii) In some applications, back transformation to express the final conclusions on the original scale is desirable.

(iv) Switzer [35] has given an elegant discussion of transformations of the two sample problem to translation form. This relates the notions of constancy of variance, absence of interaction, and of simplicity of structure.

(v) Some difficulties in establishing covariance stabilizing transformations for multivariate data are described by Song [33].

In general, however, the achievement of simple stable relations in the systematic part of the model is the primary objective of transformations. Box and Cox* [11] restricted attention almost entirely to the parametric family

$$y(\lambda) = (y^\lambda - 1)/\lambda, \qquad \lambda \neq 0,$$
$$y(0) = \log y, \qquad (3)$$

assuming provisionally that, for some unknown value of λ, the transformed responses follow a simple model, for example a normal theory general linear model*. The parameter estimate $\hat{\lambda}$ and associated confidence limits can then be obtained by maximum likelihood. Numerous examples and generalizations are given by Atkinson [7, Chaps. 6–9].

A disadvantage of the transformation (3) is that the scale of the observations depends upon λ, so that the Jacobian* of the transformation must be included in the likelihood. This is equivalent to working with the normalized transformation

$$z(\lambda) = (y^\lambda - 1)/(\lambda \dot{y}^{\lambda-1}), \qquad \lambda \neq 0,$$
$$z(0) = \dot{y} \log y, \qquad (4)$$

where $\dot{y}$ is the geometric mean* of the observations. The log-likelihood of the observations is then proportional to the log of the residual sum of squares of $z(\lambda)$. Because $z(\lambda)$ has the dimension y for all λ, values of the estimated parameters in the linear model do not, as they do for the transformation

(3), depend critically on the unknown value of λ.

TRANSFORMATIONS OF EXPLANATORY VARIABLES

It will usually be best to examine together transformations of response and explanatory variables. However, there may be occasions when transformation of the response is inappropriate because, for example, the untransformed response has a unique physical interpretation or because the error structure is simple on the untransformed scale.

With monotonic regression on a single explanatory variable, an empirical transformation to linearity is always possible. One way of handling several such sets of data is to compare the transformations. If the transformations are restricted to parametric families, more formal procedures are available. For example if several explanatory variables are to be considered for a power transformation, maximum likelihood procedures can be used which are analogous to those for power transformations of the response [12].

Sometimes particular transformations are of special interest, for instance from x to $\log x$. Then the best procedure is usually to fit the comprehensive model

$$Y_i = \beta_0 + \beta_1 x_i + \beta_2 \log x_i + \epsilon_i. \quad (5)$$

Tests of hypotheses about the parameters may indicate consistency with either or neither simple form. Since the vectors $\{x_i\}$ and $\{\log x_i\}$ will often be nearly linearly dependent, interpretation of the individual estimates in (5) should, as in econometric* models, be made very cautiously [23].

If the data require that both β_1 and β_2 are different from zero, a search for a different transformation will usually be indicated. Very occasionally it will be necessary to choose between fitting a linear regression on x only or on only $\log x$, with no other options open; see for example Fisher and McAleer [20]. But more commonly the choice would be between models (1) and (2) in which both response and explanatory variables are transformed.

TRANSFORMATIONS OF PROBABILITIES AND PROPORTIONS

For binomial data the probit*, logit*, and arcsine transformations are almost linear over the central range, with probabilities roughly between 0.2 and 0.8, so that largely equivalent results are to be expected unless extreme probabilities play an important role [16, p. 28]. The analysis is most simply brought within the framework of generalized linear models by writing the logit of the probability of success θ as a function of the linear model, namely

$$\mathrm{logit}(\theta_i) = \log\{\theta_i/(1 - \theta_i)\} = \mathbf{x}_i^T\boldsymbol{\beta}. \quad (6)$$

An advantage of (6) is that it is readily inverted to give the probability as a function of the linear model $\mathbf{x}_i^T\boldsymbol{\beta}$. It depends very much on the context whether the probit or the logit is the more amenable to mathematical analysis.

A general parametric family of invertible transformations which includes (6) is given by Aranda-Ordaz [4]. Like (6), this transformation is symmetrical in that it does not distinguish between "success" and "failure." Aranda-Ordaz also describes an asymmetrical transformation which includes the complementary log log model. These parametric families provide a means of testing consistency with specific models, particularly the logit, and for indicating how critical the choice of transformation is.

For continuous responses which arise as proportions, so that $0 \leqslant E[Y_i] \leqslant 1$, the power transformation (4) for nonnegative responses does not generally yield simple models. Atkinson [7, Chap. 7] studies the transformation derived from (6) on replacement of θ_i by y_i. A rather different problem arises in the analysis of compositional data where there are many responses, for example, the proportions of various minerals in an ore, but few or no explanatory variables.

Aitchison [1] considers several transformation models, including the logit normal.

Outliers*, Diagnostics, and Robustness*

When a transformation is estimated empirically rather than from a priori arguments, individual observations can have an appreciable influence on the estimated transformation. The methods of regression diagnostics* can be used to identify such influential observations*. Although we illustrate these ideas for the parametric family of power transformations (4), the ideas apply to a wide class of parametric transformations of response and explanatory variables. For similar ideas for the link function of a generalized linear model*, see Pregibon [29].

The log-likelihood for the power transformation (4) is identical with that from the normal theory linear model

$$z(\lambda) = \mathbf{X}\beta + \epsilon, \qquad (7)$$

with $z(\lambda)$ given by (4). To determine whether a transformation is required, a likelihood ratio test* can be calculated for the null hypothesis $\lambda = \lambda_0 = 1$, corresponding to no transformation. A disadvantage of the likelihood ratio test is that iterative calculation of the maximum likelihood estimate $\hat{\lambda}$ is required. An alternative, yielding asymptotically equivalent results, is the score test*, which depends only on quantities calculated under the null hypothesis. A second advantage of the score test is that it can yield a graphical indication of the contribution of individual observations to the inference.

A simple approximation to the score test is found [5] by expansion of (7) in a Taylor series to yield the linearized model

$$z(\lambda_0) \simeq \mathbf{X}\beta - (\lambda - \lambda_0)w(\lambda_0) + \epsilon$$
$$= \mathbf{X}\beta + \gamma w(\lambda_0) + \epsilon, \qquad (8)$$

where $w(\lambda_0) = \partial z(\lambda)/\partial\lambda$ evaluated at $\lambda = \lambda_0$. A variable such as $w(\lambda_0)$ was called a constructed variable by Box [10]. The hypothesis $\lambda = \lambda_0$ is equivalent to the hypothesis $\gamma = 0$, that is, that there is no regression on $w(\lambda_0)$ in (8). The form of (8) is identical to that discussed in equations (9)–(11) of REGRESSION DIAGNOSTICS*, where

the form of the t test for γ is derived. The effect of individual observations is indicated by an added variable plot, that is, a scatter plot of the residuals of $z(\lambda_0)$ against those of $w(\lambda_0)$, in both cases after regression on X. The modified Cook statistic [REGRESSION DIAGNOSTICS, eq. (8)] provides information on observations that are influential for the value of the test statistic and hence on the evidence for a transformation.

As an example, consider the analysis of a 3^{4-1} factorial experiment* [25] in which the response is related to the tensile strength of an industrial nylon yarn. When a model with two factor interactions in all except one factor is fitted to the original form of the response, there is appreciable evidence of the need for a transformation: the score test, that is the t test for γ in (8), has the value 2.98, apparently indicating that a transformation is needed. The added variable plot of residual* and residual constructed variables, Fig. 1, seems to support this result, although the contribution of observation 11 may be appreciable. Just how appreciable it is can be judged from the half normal plot* of the modified Cook statistic C_i, Fig. 2. If the evidence for a transformation were spread throughout the data, this plot would, apart from sampling fluctuations, be a straight line. But in Fig. 2 the value of C_{11} is 6.85, with the next highest value near 1. If this observation is deleted, the value of the score statistic is reduced to a nonsignificant 0.57.

A fuller discussion of these procedures as well as of this example is given by Atkinson [6] and by the discussants of that paper. Related methods for highly unbalanced data are described by Cook and Wang [14] and by Atkinson [7, Sec. 12.3]. For our present purpose the important point is that the graphical methods, particularly the half normal plot of the modified Cook statistic, led to the discovery that one observation on its own was providing all the evidence of the need for a transformation. In the absence of a sensitivity analysis, the likelihood ratio test would fail to detect this dependence.

Identification of influential observations is a main purpose of diagnostic techniques.

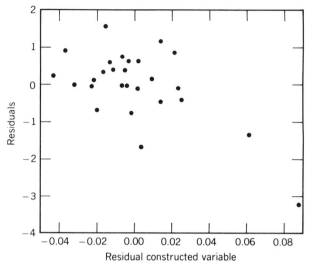

Figure 1 John's 3^{4-1} experiment: added variable plot of the constructed variable $w(1)$ for the hypothesis of no transformation.

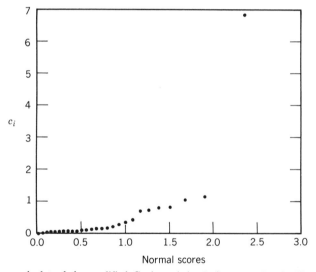

Figure 2 Half normal plot of the modified Cook statistic C_i for regression in Fig. 1; the effect of observation 11 is clearly apparent.

Cook and Weisberg [15] contrast identification with accommodation, which is the purpose of robust statistical methods. Carroll [13] compares several robust estimates of the transformation parameter λ for John's 3^{4-1} experiment. The estimates depend to only a limited extent on the egregious observation 11 and so do not indicate the need for a transformation. Without more detailed knowledge of the investigation that gener-

ated the data, it is impossible to say whether observation 11 should be included. Automatic use of robust methods is unwise.

BIBLIOGRAPHIC NOTE ON REFERENCES

The majority of the references are to specialized points in our entry. However, many general aspects of the use of parametric families of transformations in the analysis of data are considered by Box and Cox [11]. The extension of these procedures to diagnostic analysis is described by Atkinson [7, Chaps. 6–8]. Cox [17] provides both a fuller discussion of many topics only touched on in this entry and a list of references to earlier work on transformations.

References

[1] Aitchison, J. (1982). *J. R. Statist. Soc. B*, **44**, 139–177.

[2] Andrews D. F., Gnanadesikan, R., and Warner, J. L. (1971). *Biometrics*, **27**, 825–840.

[3] Andrews, D. F., Gnanadesikan, R., and Warner, J. L. (1973). In *Multivariate Analysis III*, P. R. Krishnaiah, ed. Academic Press, New York, pp. 95–106.

[4] Aranda-Ordaz, F. J. (1981). *Biometrika*, **68**, 357–363.

[5] Atkinson, A. C. (1973). *J. R. Statist. Soc. B*, **35**, 473–479.

[6] Atkinson, A. C. (1982). *J. R. Statist. Soc. B*, **44**, 1–36.

[7] Atkinson, A. C. (1985). *Plots, Transformations and Regression*. Oxford University Press, Oxford, England.

[8] Bartlett, M. S. (1947). *Biometrics*, **3**, 39–52.

[9] Bates, D. M. and Watts, D. G. (1980). *J. R. Statist. Soc. B*, **42**, 1–25.

[10] Box, G. E. P. (1980). *J. R. Statist. Soc. A*, **143**, 383–430.

[11] Box, G. E. P. and Cox, D. R. (1964). *J. R. Statist. Soc. B*, **26**, 211–252.

[12] Box, G. E. P. and Tidwell, P. W. (1962). *Technometrics*, **4**, 531–550.

[13] Carroll, R. J. (1982). *Appl. Statist.*, **31**, 149–152.

[14] Cook, R. D. and Wang, P. C. (1983). *Technometrics*, **25**, 337–343.

[15] Cook, R. D. and Weisberg, S. (1983). *J. Amer. Statist. Ass.*, **78**, 74–75.

[16] Cox, D. R. (1970). *The Analysis of Binary Data*. Chapman and Hall, London and New York.

[17] Cox, D. R. (1977). *Math. Operat. Statist., Ser. Statist.*, **8**, 3–22.

[18] Draper, N. R. and Cox, D. R. (1969). *J. R. Statist. Soc. B*, **31**, 472–476.

[19] Efron, B. (1982). *Ann. Statist.*, **10**, 323–339.

[20] Fisher, G. R. and McAleer, M. (1981). *J. Econometrics*, **16**, 103–119.

[21] Fisher, R. A. (1915). *Biometrika*, **10**, 507–521.

[22] Hamilton, D. C., Watts, D. G., and Bates, D. M. (1982). *Ann. Statist.*, **10**, 383–393.

[23] Hendry, D. F. and Richard, J.-F. (1983). *Int. Statist. Rev.*, **51**, 111–163.

[24] Hougaard, P. (1982). *J. R. Statist. Soc. B*, **44**, 244–252.

[25] John, J. A. (1978). *Appl. Statist.*, **27**, 111–119.

[26] Kendall, M. G. and Stuart, A. (1977). *The Advanced Theory of Statistics*, Vol. 1, 4th ed. Griffin, London/Macmillan, New York.

[27] McCullagh, P. (1983). *Ann. Statist.*, **11**, 59–67.

[28] McCullagh, P. and Nelder, J. A. (1983). *Generalized Linear Models*. Chapman and Hall, London and New York.

[29] Pregibon, D. (1980). *Appl. Statist.*, **29**, 15–24.

[30] Ross, G. J. S. (1970). *Appl. Statist.*, **19**, 205–221.

[31] Ross, G. J. S. (1975). *Bull. I. S. I.*, **46**(2), 585–593.

[32] Snedecor, G. W. and Cochran, W. G. (1967). *Statistical Methods*, 6th ed. Iowa University Press, Ames, IA.

[33] Song, C. C. (1982). *Ann. Statist.*, **10**, 313–315.

[34] Sprott, D. A. (1973). *Biometrika*, **60**, 457–465.

[35] Switzer, P. (1976). *Biometrika*, **63**, 13–25.

[36] Wedderburn, R. W. M. (1974). *Biometrika*, **61**, 439–447.

(APPROXIMATIONS TO DISTRIBUTIONS
BOX AND COX TRANSFORMATION
BOX–MULLER TRANSFORMATION
CURVE FITTING
EQUALIZATION OF VARIANCE
EQUIVARIANT ESTIMATORS
INFLUENCE FUNCTION
INFLUENTIAL OBSERVATIONS
JOHNSON'S SYSTEM OF DISTRIBUTIONS
NONLINEAR MODELS
REGRESSION DIAGNOSTICS
WILSON–HILFERTY TRANSFORMATION)

A. C. ATKINSON
D. R. COX

TRANSFORMATION VARIABLE *See* EQUIVARIANT ESTIMATORS

TRANSFORMS *See* INTEGRAL TRANSFORMS

TRANSIENT STATE *See* MARKOV PROCESSES

TRANSITION PROBABILITIES *See* MARKOV PROCESSES

TRANSITIVITY PARADOX *See* FALLA-CIES, STATISTICAL

TRAPEZOIDAL DISTRIBUTIONS

These are continuous distributions with probability density functions of the form

$$f_X(x) = \begin{cases} 0, & x < a, \\ (x-a)(b-a)^{-1}h, & a \leqslant x < b, \\ h, & b \leqslant x < c, \\ (d-x)(d-c)^{-1}h, & c \leqslant x < d, \\ 0, & x \geqslant d, \end{cases}$$

where $h = 2(d + c - b - a)^{-1}$. The name *trapezoidal* reflects the shape of a graph of $f_X(x)$ against x. Triangular* and uniform* distributions are special cases.

Use of these distributions in risk* analysis has been advocated (e.g., by Pouliquen [1]).

Reference

[1] Pouliquen, L. Y. (1970). Risk Analysis in Project Appraisal, *World Bank Staff Occasional Papers* **11**. Johns Hopkins University Press, Baltimore, MD.

TRAPEZOIDAL RULE

This is a very simple graduation* formula, obtained by approximating the integral of $f(x)$ by the linear function of x,

$$\int_a^{a+h} f(x)\, dx = \tfrac{1}{2}h\{f(a) + f(a+h)\}.$$

If the second derivative $f^{(2)}(x)$ is continuous, the error is $\frac{1}{12}h^3 f^{(2)}(\xi)$ for some value ξ between a and $(a + 1)$.

(NUMERICAL INTEGRATION
SIMPSON'S RULE
THREE-EIGHTHS RULE
WEDDLE'S RULE)

TRAVELING SALESMAN PROBLEM

The traveling salesman problem derives from the following situation. A salesman is re-quired to visit each city in his territory. He leaves from a home city, visits each of the other cities exactly once, and finally returns to the home city. The problem is to find an itinerary that minimizes the total distance traveled by the salesman. More formally, we can state the problem as follows: Given a network of n nodes (cities) with arcs possessing real-valued lengths (intercity distances), find the shortest-length path (itinerary/tour) in the network that visits each node exactly once, except that the initial (home) node and final node are the same. We let the n integers $1, 2, \ldots, n$ denote the nodes, with the home node being 1. The ordered pair (i, j) denotes the arc joining nodes i and j, and the number C_{ij} is the length of the arc joining node i to node j. As an example, consider a five-city problem. We number the cities 1, 2, 3, 4, and 5, with city 1 being the home city. We know the distances between each city and assume a city can be reached from any other city. The distance table is given in Table 1; note that the problem is asymmetric in that the distance C_{ij} is not necessarily equal to C_{ji}. A tour through the cities $1, 3, 2, 5, 4, 1$, with a distance of 53 is illustrated by the following figure:

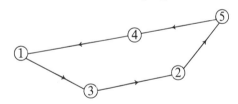

A traveling salesman's tour that passes through each city once is also called a Hamiltonian circuit, named after the mathematician William Hamilton. He first studied such circuits by determining one that passes

Table 1

City	1	2	3	4	5
1	0	17	10	15	17
2	18	0	6	12	20
3	12	5	0	14	19
4	12	11	15	0	7
5	16	21	18	6	0

through the 20 vertices of a dodecahedron, with the only allowable routes being along the edges of the figure. In general, a *Hamiltonian circuit* in a graph is one that passes through each node exactly once. A Hamiltonian circuit of shortest distance is a solution to the traveling salesman problem. In its present form, the traveling salesman problem is due to Merrill Flood, with its origins and history reviewed in refs. 5 and 6.

The traveling salesman problem belongs to the class of combinatorial problems that are termed *NP-complete*. Such problems are considered difficult in that there are no known efficient (polynomial) algorithms for solving them [9]. This does not mean that we cannot solve particular instances of a traveling salesman problem. When n is very small, we could enumerate all possible tours and select the one with the minimum distance. But this is not a practical process for real-world problems as there are $(n - 1)!/2$ tours for a problem whose distances are symmetric, i.e., $C_{ij} = C_{ji}$. For moderate size n (up to 200 cities), the partial enumerative procedures based on the branch and bound* algorithm [8] have proven effective. For example, a 180-city problem was solved in 441 sec. on a UNIVAC 1108 computer [11]. Other successful approaches include the transformation of the traveling salesman problem into an integer-programming* problem and combining the procedures of linear programming, branch and bound, and cutting planes to find the minimum solution to the integer program. Such a procedure was used to solve a 318-city symmetric problem in about 37 min. on an IBM 370/168 computer [4]. For asymmetric problems, a Lagrangian relaxation method has been used to solve problems with 325 cities in an average time of 50 sec. on a CDC 7600 computer [1].

Variations to the traveling salesman problem include the m salesmen problem in which m tours are to be found such that each of the n cities (except the home city) is visited once. This can be transformed into a single salesman problem. A similar problem is the capacitated traveling salesman problem in which each city i requires a given amount d_i of a product. The salesman can only carry an amount k when starting a tour. A solution calls for the construction of many tours, each tour having a total demand less than k, with the sum of the distances of all the tours a minimum. This problem arises in the routing and scheduling of delivery services [3]. Of special interest is the solution of the vehicle routing problem in which the d_i are random variables with known distributions. Other problems related to the traveling salesman problem include the Königsberg bridge problem (*see* GRAPH THEORY), the Chinese postman problem, and the problem of finding a minimal spanning tree in a network [7, 12]. (Also, *see* DENDRITES.)

The survey papers and book [2, 6, 7a, 10] review the theoretical and computational aspects of the traveling salesman problem.

References

[1] Balas, E. and Christofides, N. (1981). *Math. Program.*, **21**, 19–46.

[2] Bellmore, M. and Nemhauser, G. (1968). *Operat. Res.*, **16**, 538–558.

[3] Bodin, L., Golden, B., Assad, A., and Ball, M. (1983). *Comp. Operat. Res.*, **31**, 65–211.

[4] Crowder, H. and Padberg, M. W. (1980). *Manag. Sci.*, **26**, 495–509.

[5] Flood, M. M. (1956). *Operat. Res.*, **4**, 61–75.

[6] Held, M., Hoffman, A. J., Johnson, E. L., and Wolfe, P. (1984). *IBM J. Res. Development*, **28**, 476–486.

[7] Lawler, E. (1976). *Combinatorial Optimization*. Holt, Rinehart, and Winston, New York.

[7a] Lawler, E., Lenstra, J. K., Kan, A. H. G. R., and Shmoya, D. B. (1985). *The Traveling Salesman Problem*. Wiley, New York.

[8] Little, J. D. C., Murty, K., Sweeney, D. W., and Carel, C. (1973). *Operat. Res.*, **21**, 972–989.

[9] Papadimitriou, C. H. and Steiglitz, K. (1982). *Combinatorial Optimization*. Prentice-Hall, Englewood Cliffs, NJ.

[10] Parker, R. G. and Rardin, R. L. (1983). *Naval Res. Logist. Quart.*, **30**, 69–96.

[11] Smith, T. H. C., Srinivasan, V., and Thompson, G. L. (1977). *Ann. Discrete Math. (Amsterdam)*, **1**, 495–506.

[12] Tucker, A. (1984). *Applied Combinatorics*, 2nd ed. Wiley, New York.

(DECISION THEORY
GRAPH THEORY
NETWORK ANALYSIS
OPTIMIZATION)

SAUL I. GASS

TREATMENTS *See* ANALYSIS OF VARI-
ANCE; MAIN EFFECTS

TREE *See* CLASSIFICATION; DENDRITES;
DENDROGRAMS

TREE EXPERIMENTS *See* STATISTICS IN
FORESTRY

TRENCH MATRIX *See* TOEPLITZ
MATRICES

TREND

A great deal of information in physical, biological, and social sciences occurs in the
form of time series where observations are
dependent and the nature of this dependence
is of interest in itself. Time series are generally compiled for consecutive and equal
period, such as weeks, months, quarters, and
years. Traditionally, four types of movement
have been distinguished in the analysis of
time series* namely, the trend, the cycle,*
the seasonal variations* (for subannual data),
and the irregular fluctuations. As a matter of
statistical description, a given series can always be represented by one of these components or a sum of several of them. The four
components are usually interrelated and for
most series, they influence one another.

The trend corresponds to sustained and
systematic variations over a long period of
time. It is associated with the structural
causes of the phenomenon in question, for
example, population growth, technological
progress, new ways of organization, or capital
accumulation. For the majority of socioeconomic time series, the trend is very important because it dominates the total variation of the series.

The identification of trend has always
posed a serious statistical problem. The

problem is not one of mathematical or analytical complexity but of conceptual complexity. This problem exists because the trend
as well as the remaining components of a
time series are latent (nonobservables) variables and, therefore, assumptions must be
made on their behavioral pattern. The trend
is generally thought of as a smooth and slow
movement over a long term. The concept of
"long" in this connection is relative and
what is identified as trend for a given series
span might well be part of a long cycle once
the series is considerably augmented, such as
the Kondratieff economic cycle. Kondratieff
[12] estimated the length of this cycle to be
between 47 and 60 years. Often, a long cycle
is treated as a trend because the length of
the observed time series is shorter than one
complete face of this type of cycle.

To avoid the complexity of the problem
posed by a statistically vague definition,
statisticians have resorted to two simple
solutions: One consists of estimating trend
and cyclical fluctuations together, calling this
combined movement *trend-cycle*; the other
consists of defining the trend in terms of the
series length, denoting it as the longest non-
periodic movement.

TREND MODELS

Within the large class of models identified
for trend, we can distinguish two main categories, deterministic trends and stochastic
trends.

Deterministic trend models are based on
the assumption that the trend of a time
series can be approximated closely by simple
mathematical functions of time over the entire span of the series. The most common
representation of a deterministic trend is by
means of polynomials or of transcendental
functions. The time series from which the
trend is to be identified is assumed to be
generated by a nonstationary process where
the nonstationarity results from a deterministic trend (*see also* TIME SERIES, NONSTA-
TIONARY). A classical model is the regression

or error model (Anderson [1]) where the observed series is treated as the sum of a systematic part or trend and a random part or irregular. This model can be written as

$$Z_t = Y_t + U_t, \qquad (1)$$

where U_t is a purely random process, that is, $U_t \sim$ i.i.d.$(0, \sigma_U^2)$ (independent and identically distributed with expected value 0 and variance σ_U^2.)

In the case of a polynomial trend,

$$Y_t = a_0 + a_1 t + a_2 t^2 + \cdots + a_n t^n, \quad (2)$$

where generally $n \leqslant 3$. The trend is said to be of a deterministic character because it is not affected by random shocks which are assumed to be uncorrelated with the systematic part.

Model (1) can be generalized by assuming that U_t is a second-order linear stationary stochastic process*, that is, its mean and variance are constant and its autocovariance is finite and depends only on the time lag.

Besides polynomials in time, other suitable mathematical functions are used to represent deterministic trends. Three of the most widely applied functions, known as *growth curves**, are the modified exponential, the Gompertz*, and the logistic*.

The modified exponential trend can be written as

$$Y_t = a + bc^t, \quad a \text{ real}, b \neq 0, c > 0, c \neq 1. \tag{3}$$

For $a = 0$, model (3) reduces to the unmodified exponential trend

$$Y_t = bc^t = Y_0 e^{\alpha t}; \qquad b = Y_0, \alpha = \log c. \tag{4}$$

When $b > 0$ and $c > 1$, and so $\alpha > 0$, model (4) represents a trend that increases at a constant relative rate α. For $0 < c < 1$, the trend decreases at the rate α. Models (3) and (4) are solutions of the differential equation

$$dY_t/dt = \alpha(Y - a), \qquad \alpha = \log c, \tag{5}$$

which specifies the simple assumption of noninhibited growth.

Several economic variables during periods of sustained growth or of rapid inflation, as well as population growths measured in relative short periods of time, can be well approximated by trend models (3) and (4). But in the long run, socioeconomic and demographic time series are often subject to obstacles that slow their time path, and if there are no structural changes, their growths tend to a stationary state. Quetelet* made this observation with respect to population growth and Verhulst [17] seems to have been the first to formalize it by deducing the logistic model. Adding to eq. (5) an inhibit factor proportional to $-Y^2$, the result is

$$dY/dt = \alpha Y - \beta Y^2 = \alpha Y(1 - Y/k),$$
$$k = \alpha/\beta, \alpha, \beta > 0, \tag{6}$$

which is a simple null form of the Ricatti differential equation. Solving eq. (6), we obtain the logistic model,

$$Y_t = k(1 + ae^{-\alpha t})^{-1}, \tag{7}$$

where $a > 0$ is a constant of integration.

Model (7) belongs to a family of S-shaped curves generated from the differential equation (see Dagum [6] and INCOME DISTRIBUTION MODELS):

$$dY_t/dt = Y_t \Psi(t) \Phi(Y_t/k), \qquad \Phi(1) = 0. \tag{8}$$

Solving eq. (8) for $\Psi = \log c$ and $\Phi = \log(Y_t/k)$, we obtain the Gompertz curve used to fit mortality table data; that is,

$$Y_t = kb^{c^t}, \qquad b > 0, b \neq 1, 0 < c < 1, \tag{9}$$

where b is a constant of integration.

It should be noted that differencing will remove polynomial trends and suitable mathematical transformations plus differencing will remove trends from nonlinear processes; e.g., for (7) using

$$Z_t = \log[Y_t/(k - Y_t)]$$

and then taking differences gives $\Delta Z_t = \alpha$.

The second major class of trend models is the one that assumes the trend to be a stochastic process, most commonly that the series from which the trend will be identified follows a homogeneous linear nonstationary stochastic process (Yaglom [17]). Processes of this kind are nonstationary, but applying a homogeneous filter, usually the difference filter, we obtain a stationary process in the differences of a finite order. In empirical applications, the nonstationarity is often present in the level and/or slope of the series; hence, the order of the difference is low. An important class of homogeneous linear nonstationary processes are the ARIMA (autoregressive integrated moving averages processes)* which can be written as (Box and Jenkins [2])

$$\phi_p(B)\Delta^d Y_t = \theta_q(B)a_t,$$

$$a_t \sim \text{i.i.d.}(0,\sigma_a^2), \qquad (10)$$

where B is the backshift operator such that $B^n Y_t = Y_{t-n}$; $\phi_p(B)$ and $\theta_q(B)$ are polynomials in B of order p and q, respectively, and satisfy the conditions of stationarity and invertibility; $\Delta^d = (1-B)^d$ is the difference operator of order d and a_t is a purely random process. Model (10) is also known as an *ARIMA process** of order (p,d,q). If $p = 0$, the process follows an *IMA model.*

Two common stochastic trend models are the IMA(0,1,1) and IMA(0,2,2) which take the form, respectively,

$$(1-B)Y_t = (1-\theta B)a_t, \qquad |\theta| < 1,$$
$$a_t \sim \text{i.i.d.}(0,\sigma_a^2) \qquad (11)$$

or, equivalently,

$$Y_t = Y_{t-1} + a_t - \theta a_{t-1}, \qquad (12)$$

and

$$(1-B)^2 Y_t = (1 - \theta_1 B - \theta_2 B^2)a_t, \qquad (13)$$
$$\theta_2 + \theta_1 < 1,\ \theta_2 - \theta_1 < 1,\ -1 < \theta_2 < 1,$$
$$a_t \sim \text{i.i.d.}(0,\sigma_a^2),$$

or, equivalently,

$$Y_t = 2Y_{t-1} - Y_{t-2} + a_t - \theta_1 a_{t-1} - \theta_2 a_{t-2}. \qquad (14)$$

The a's may be regarded as a series of random shocks that drives the trend and θ can be interpreted as measuring the extent to which the random shocks or "innovations" incorporate themselves into the subsequent

history of the trend. For example, in model (11), the smaller the value of θ, the more flexible the trend; the higher the value of θ, the more rigid the trend (less sensitive to new innovations). For $\theta = 1$, model (11) reduces to one type of random walk* model which has been used mainly for economic time series such as stock market price data (Granger and Morgenstern [9]). In such models, as time increases the random variables tend to oscillate about their mean value with an ever increasing amplitude. The use of stochastic models in business and economic series has received considerable attention during recent years (see, for example, Nelson and Plosser [13] and Harvey [10]).

Economists and statisticians are also often interested in the "short" term trend of socioeconomic time series. The short term trend generally includes cyclical fluctuations and is referred to as *trend-cycle.* Most seasonal adjustment methods such as census X-11* (Shiskin et al [15]) and X-11-ARIMA (Dagum [7]) also produce stochastic trendcycle estimates applying a set of linear filters to the original series. Studies on the properties of such filters has been made by Young [18] and by Dagum and Laniel [8]. Other important studies related to trend-cycle estimators have been made by Burman [4], Cleveland and Tiao [5], Box et al. [3], Kenny and Durbin [11]. and Pierce [14].

References

[1] Anderson, T. W. (1971). *The Statistical Analysis of Time Series.* Wiley, New York.

[2] Box, G. E. P. and Jenkins, G. M. (1970). *Time Series Analysis: Forecasting and Control.* Holden-Day, San Francisco, CA.

[3] Box, G. E. P., Hillmer, S. C., and Tiao, G. C. (1978). Analysis and modelling of seasonal time series. In *Seasonal Analysis of Economic Time Series,* A. Zellner, ed. U.S. Bureau of Census, Washington, DC.

[4] Burman, J. P. (1980). Seasonal adjustment by signal extraction. *J. R. Statist. Soc. A,* **143,** 321–337.

[5] Cleveland, W. P. and Tiao, G. C. (1976). Decomposition of seasonal time series: A model for the census X-11 program. *J. Amer. Statist. Ass.,* **71,** 581–587.

[6] Dagum, C. (1985). Analyses of income distribution and inequality by education and sex in Canada. In *Advances in Econometrics*, Vol. IV, R. L. Basmann and G. F. Rhodes, Jr., eds. JAI Press, Greenwich, CN, pp. 167–227.

[7] Dagum, E. B. (1980). The X-11-ARIMA seasonal adjustment method. Statistics Canada, Ottawa, Canada, Catalogue No. 12–564.

[8] Dagum, E. B. and Laniel, N. J. D. (1987). Revisions of trend-cycle estimators of moving average seasonal adjustment methods. *J. Bus. Econ. Statist.*, **5**, 177–189.

[9] Granger, C. W. J. and Morgenstern, O. (1970). *Predictability of Stock Market Prices*. D.C. Heath, Lexington, MA.

[10] Harvey, A. G. (1985). Trends and cycles in macroeconomic time series. *J. Bus. Econ. Statist.*, **3**, 216–227.

[11] Kenny, P. B. and Durbin, J. (1982). Local trend estimation and seasonal adjustment of economic and social time series. *J. R. Statist. Soc. Ser. A*, **145**, 1–41.

[12] Kondratieff, N. (1925). Long economic cycles. *Voprosy Konyuktury*, Vol. 1, No. 1. (English translation: *The Long Wave Cycle*, Richardson and Snyder, New York, 1984.)

[13] Nelson, C. R. and Plosser, C. I. (1982). Trends and random walks in macroeconomic time series: Some evidences and implications. *J. Monetary Econ.*, **10**, 139–162.

[14] Pierce, D. A. (1975). On trend and autocorrelation. *Commun. Statist.*, **4**, 163–175.

[15] Shiskin, J., Young, A. H., and Musgrave, J. C. (1967). The X-11 variant of the census method II seasonal adjustment program. *Technical Paper No. 15*, U.S. Department of Commerce, U.S. Bureau of Census, Washington, DC.

[16] Verhulst, P. F. (1838). Notice sur la loi que la population suit dans son accroissement. *Correspondance Mathématique et Physique*, A. Quételet, ed. Tome X, pp. 113–121.

[17] Yaglom, A. M. (1962). *An introduction to the Theory of Stationary Random Functions*. Prentice-Hall, Englewood Cliffs, NJ.

[18] Young, A. H. (1968). Linear approximations of the census and BLS seasonal adjustment methods. *J. Amer. Statist. Ass.*, **63**, 445–471.

(GRADUATION
SEASONALITY
TIME SERIES
X-11 METHOD)

CAMILO DAGUM
ESTELE BEE DAGUM

TREND-FREE BLOCK DESIGNS

In comparative experiments, subsets of the available experimental units (plots) are treated alike in specified ways and the objective is to compare the effects of treatments through observation or measurement of a response variable. The efficiencies of treatment comparisons are improved in two principal ways: the use of blocking and the use of covariates. Modern experimental design uses these concepts, sometimes in combination, and maintains simplicity of data analysis through design symmetries or balance.

Sometimes response to a plot is affected by its position (spatial or temporal) in the block, and experimental efficiency may be further improved through restriction of the randomization of treatments to plots within blocks. Latin square* and Youden square* designs do this. Plot position may be described by covariates and the analysis of covariance* used on the assumption that a common polynomial trend of specified degree in one or more dimensions exists over the plots within each block. If treatment positions within blocks can be chosen so that treatment comparisons are unaffected by the trend, a *trend-free block design* results. Then data analysis is simplified and the design has desirable optimality properties.

Basic references to trend-free block designs are refs. 3 and 9. *See* ANALYSIS OF VARIANCE; ANALYSIS OF COVARIANCE; BLOCKS, RANDOMIZED COMPLETE; BLOCKS, BALANCED INCOMPLETE; DESIGN OF EXPERIMENTS; and PARTIALLY BALANCED DESIGNS for background information.

MODELS AND DESIGNS

Consider trends in one dimension only; extensions to several dimensions are discussed in ref 3. In usual design notation, suppose that v treatments are applied to plots arranged in blocks of size $k \leq v$. Each plot receives one treatment and each treatment occurs at most once in a block. A polynomial trend of prespecified degree p, common to all blocks, is assumed to exist over the plots in a block and to be a function of the plot position $t = 1, \ldots, k$.

A model representing the experiment, when the effects of treatments, blocks, and trend are additive, is obtained by the addition of trend terms to a general block design model, and written as

$$y_{jt} = \mu + \sum_{i=1}^{v} \delta_{ijt}\tau_i + \beta_j + \sum_{\alpha=1}^{p} \theta_\alpha\phi_\alpha(t) + \epsilon_{jt},$$

(1)

$j = 1, \ldots, b$, $t = 1, \ldots, k$, where y_{jt} is the observation on plot position t of block j; μ, τ_i, and β_j are, respectively, the usual mean, treatment, and block parameters; the ϕ_α, $\alpha = 1, \ldots, p$, are orthogonal polynomials of degree α; $\sum_{\alpha=1}^{p}\theta_\alpha\phi_\alpha(t)$ is the trend effect on plot t, not dependent on the block j, with θ_α being the regression coefficient* for trend component α; and $p < k$. The ϵ_{jt} are identically and independently distributed random errors with zero means. Designation of the treatment applied to plot (j, t) is effected through indicator variables $\delta_{ijt} = 1$ or 0 as treatment i is or is not on plot (j, t), $i = 1, \ldots, v$.

A block design modelled by (1) is said to be *trend-free* relative to the trend in that model if the treatment and block sums of squares may be calculated as if the trend terms were deleted from the model. A necessary and sufficient condition for a design under model (1) to be trend-free is that

$$\sum_{j=1}^{b} \sum_{t=1}^{k} \delta_{ijt}\phi_\alpha(t) = 0 \qquad (2)$$

for all $i = 1, \ldots, v$ and $\alpha = 1, \ldots, p$.

EXAMPLES

A $v \times v$ Latin square with rows regarded as blocks is a complete block design free of polynomial trend effects to degree $v - 1$. A $v \times k$ Youden square design is a balanced incomplete block design free of trend effects to degree $k - 1$.

A simple interpretation of (2) is that the effect $\phi_\alpha(t)$ of each trend component α should sum to zero for the set of plots t assigned to each treatment i. In Table 1, a complete block design free of linear and quadratic trend components, is shown for $v = 7$, $b = 6$. The first two rows show values proportional to values of $\phi_\alpha(t)$ for $\alpha = 1, 2$ and the remaining rows are blocks with treatments designated by letters. The format of Table 2 is similar except that two blocks are shown in each row of that table, the example providing a balanced incomplete block design* free of a linear trend for $v = 5$, $b = 10$, $k = 3$. Data are shown for the plots in Table 2.

ANALYSIS

Trend-free block designs are derived from known block designs through appropriate restrictions on the assignment of treatments to plots within blocks. The designs permit

Table 1   Complete block design $v = 7$, $b = 6$, $p = 2$.

-3	-2	-1	0	1	2	3
5	0	-3	-4	-3	0	5
A	B	C	D	E	F	G
F	A	C	G	D	B	E
G	D	F	A	E	B	C
C	E	G	F	A	B	D
D	B	G	E	A	C	F
E	B	F	C	D	G	A

Table 2   Balanced incomplete block design $v = 5$, $b = 10$, $k = 3$, $p = 1$.

-1	0	1	-1	0	1
D 5.03	E 3.16	A 0.80	A 1.91	B 1.60	C 1.53
E 5.14	B 1.81	D 4.49	C 4.86	D 5.87	A 3.26
A 3.92	B 3.91	D 5.42	E 3.66	B 3.45	C 1.97
C 3.69	E 4.66	A 2.96	D 6.02	B 3.31	C 4.50
C 4.26	D 2.81	E 4.06	A 5.06	B 3.94	E 4.93

the use of standard estimators of treatment and block effects and standard calculations of total, block, and treatment sums of squares and mean squares in the analysis of variance table. Each of the p trend components adds an additional row to the analysis of variance table with one degree of freedom (d.f.). The regression coefficient* θ_α is estimated by W_α/b and the sum of squares is W_α^2/b, where

$$W_\alpha = \sum_{j=1}^{b} \sum_{t=1}^{k} \phi_\alpha(t) y_{jt}, \qquad \alpha = 1, \ldots, p.$$

$$(3)$$

The error sum of squares is obtained by subtraction with the standard number of degrees of freedom reduced by p.

The analysis of variance table for Table 2 is given in Table 3. In the example, $W_1 = [-(5.03 + 5.14 + \cdots + 5.06) + (0.80 + 4.49 + \cdots + 4.93)]/\sqrt{2} = -6.81$, the divisor being required since the values of $\phi_1(t)$ are $(-1\sqrt{2}, 0, 1/\sqrt{2})$, their sum of squares necessarily being unity. The sum of squares for the linear trend is $W_1^2/b = 4.64$. Note that the balanced incomplete block* design, without the trend component, would have had 16 degrees of freedom for the error sum of squares.

It is seen that trend-free block designs retain simplicity in computations. Under the normality assumption for the distributions of the ϵ_{ij} in (1), the usual F-test* for treatment effects may be made.

EXISTENCE

Given a known block design, it may or may not be possible to create the corresponding trend-free block design. Thus existence theorems become important. Condition (2) is an existence condition and the following theorem provides another.

Suppose that each treatment in a block design has r replications and model (1) applies. Define

$$S_\alpha(k, r) = r \sum_{t=1}^{k} t^\alpha/k, \qquad \alpha = 1, \ldots, p.$$

$$(4)$$

Theorem. Given v, b, k, r, and p in model (1), a trend-free block design exists if and only if there exists a $v \times b$ matrix $\mathbf{W}$ with nonnegative elements such that

(i) each column of $\mathbf{W}$ has integers $1, \ldots, k$ as elements along with $(v - k)$ zero elements, and,

(ii) for any $\alpha = 1, \ldots, p$, the sum of the αth powers of elements for any single row of $\mathbf{W}$ is $S_\alpha(k, r)$.

Note that w_{ij}, the typical element of $\mathbf{W}$, is zero if treatment i does not appear in block j and otherwise is the plot position of treatment i in block j. The matrix $\mathbf{W}$ is a representation of the combinatorial structure of a trend-free block design and assists in design construction.

Existence results for trend-free complete block designs when $p = 1$ or 2 are summarized: When $p = 1$, a trend-free complete block design exists if and only if one of the following cases holds: (i) b is even or (ii) both v and b are odd. In case (i), designs for $b = 2$ can be obtained by randomizing the

Table 3   Analysis of Variance for Data in Table 2.

Source of Variation	d.f.	Sum of Squares	Mean Square
Blocks (unadjusted)	9	24.59	2.73
Treatments (adjusted for blocks)	4	15.75	3.94
Linear Trend	1	4.64	4.64
Error	15	8.70	0.58
Total	29	53.68	—

Table 4   Trend-Free Complete Block Designs for $v = 2n + 1$, $b = 3$, $p = 1$.

1	$2 \cdots n$	$n + 1$	$n + 2 \cdots 2n$	$2n + 1$
$2n$	$2n - 2 \cdots 2$	$2n + 1$	$2n - 1 \cdots 3$	1
$2n + 1$	$2n - 1 \cdots 3$	1	$2n \cdots 4$	2

Table 5   Parametric Combinations (v, b) with Trend-Free Complete Block Designs, $v \leqslant 16$, $p = 2$, m a Positive Integer.

v	b	v	b	v	b	v	b
3	$3m$	7	$b \geqslant 6$	11	$b \geqslant 7$	15	$3m, m > 1$
4	$2m, m > 1$	8	$2m, m \neq 1, 3$	12	$6m$	16	$2m, m > 2$
5	5 or $b \geqslant 8$	9	$3m, m > 1$	13	$b \geqslant 7$		
6	$6m$	10	$2m, m > 2$	14	$2m, m > 2$		

treatments to plots in the first block and reversing the treatment order in the second block, and those for $b > 2$ can be obtained by combining two-block designs. In case (ii), designs for $b = 3$ can be constructed systematically as in Table 4, wherein numbers represent treatments and rows are blocks, and those for $b > 3$ can be constructed by combining trend-free designs with two and three blocks. When $p = 2$, a necessary condition (from the Theorem) for the existence of a trend-free design is that both $b(v + 1)/2$ and $b(v + 1)(2v + 1)/6$ be integers. In Table 5, the parameter combinations (v, b), $v \leqslant 16$, for which a trend-free complete block design exists when $p = 2$ are listed.

Additional theorems on the existence of trend-free block designs are given in ref. 9. Existence theorems for trend-free incomplete block designs are difficult, particularly if $p > 1$, although construction is often easy. For example, the design of Table 2 may be changed so that it is trend-free of both a linear and a quadratic trend component. In ref. 9, it is conjectured that, given a connected block design with r replications of each treatment, the corresponding trend-free block design, free of a linear trend component, may be constructed if and only if $r(k + 1)/2$ is an integer. This is a necessary condition, but proof of sufficiency is given only for $k = 2$. Recent unpublished research has led to counterexamples, showing that the condition is not sufficient for larger values of k. Apparently, some additional balance properties are required, even though design construction is usually possible.

REMARKS

Model (1) assumes a trend in one dimension. It can be easily generalized to cases with higher-dimensional trends without additional complexity to the trend-free conditions and the analysis of variance; see ref. 3.

The restricted randomization of the trend-free block design is possible only when the trend in model (1) is known to a prespecified degree prior to the experimentation. If the presence of trend is not known beforehand but detected during or after the experimentation, analysis of covariance with trend terms as covariates can be utilized to remove the confounding between treatments and trend and therefore to increase the sensitivity of treatment comparisons; one such example is discussed in Federer and Schlottfeldt [7].

Designs constructed for a given model may be vulnerable to incorrect specifications of the degree of the trend. To guard against this, a design free of a trend with degree up to $p + 1$ may be considered even when a p-degree trend is assumed in the model. It is apparent that trend-free block designs benefit from correct specification of the trend, and the price for this benefit may be increased vulnerability to bias from incorrect trend specification.

Strong optimality properties, including the usual *A*-, *D*-, and *E*-optimality, are possessed by trend-free block designs under model (1) relative to the corresponding block design with unrestricted randomization of treatments over plots within blocks used with covariance analysis. See OPTIMALITY CRITERIA.

NEARLY TREND-FREE BLOCK DESIGNS

When a trend-free block design does not exist, it may be possible to generate a "nearly trend-free block design" with high efficiency, although some simplicity in calculation is lost. In ref. 10, concepts of nearly trend-free block designs are presented in general and complete block designs with $p = 1$ or 2 are studied in particular. It is found that, when $p = 1$, nearly trend-free block designs can be constructed and they are *A*- and *D*-optimal. When $p = 2$, highly efficient designs can be obtained.

PRIOR CONCEPTS

Before the introduction of trend-free block designs in ref. 3, the concept had been used by others. Cox [4, 5] considered the assignment of treatments to plots ordered in space or time without blocking and with a trend extending over the entire sequence of plots. Box [1] and Box and Hay [2] in similar experimental sequences investigated choices of levels of quantitative factors. Hill [8] combined the designs of Cox and Box to form new designs to study the effects of both qualitative and quantitative factors in the presence of trends. Daniel and Wilcoxon [6] provided methods of sequencing the assignments of factorial treatment combinations to experimental units to achieve better estima-

tion of specified factorial effects, again in the presence of a trend in time or distance.

References

[1] Box, G. E. P. (1952). *Biometrika*, **39**, 49–57.

[2] Box, G. E. P. and Hay, W. A. (1953). *Biometrics*, **9**, 304–319.

[3] Bradley, R. A. and Yeh, C. M. (1980). *Ann. Statist.*, **8**, 883–893.

[4] Cox, D. R. (1951). *Biometrika*, **38**, 321–323.

[5] Cox, D. R. (1952). *J. R. Statist. Soc. B*, **14**, 211–219.

[6] Daniel, C. and Wilcoxon, F. (1966). *Technometrics*, **8**, 259–278.

[7] Federer, W. T. and Schlottfeldt, C. S. (1954). *Biometrics*, **10**, 282–290.

[8] Hill, H. M. (1960). *Technometrics*, **2**, 67–82.

[9] Yeh, C. M. and Bradley, R. A. (1983). *Commun. Statist. Theor. Meth.*, **12**, 1–24.

[10] Yeh, C. M., Bradley, R. A., and Notz, W. I. (1985). *J. Amer. Statist. Ass.*, **80**, 985–992.

(ANALYSIS OF COVARIANCE
ANALYSIS OF VARIANCE (ANOVA)
BLOCKS, BALANCED INCOMPLETE
BLOCKS, RANDOMIZED COMPLETE
CURVE FITTING
DESIGN OF EXPERIMENTS
LATIN SQUARES, LATIN CUBES, LATIN
RECTANGLES, ETC.
OPTIMUM DESIGN OF EXPERIMENTS
PARTIALLY BALANCED DESIGNS
REGRESSION, POLYNOMIAL
YOUDEN SQUARE)

RALPH A. BRADLEY
CHING-MING YEH

TREND IN COUNT DATA, TESTS FOR

Suppose that a nonnegative discrete random variable Y_i is related to a regression* variable x_i with

$$E[Y_i] = w_i f(x_i), \qquad i = 1, \ldots, k,$$

where w_i is a known design variable or a known incidental variable and $f(\cdot)$ is an unknown positive monotone continuous function. Some examples of $f(x)$ are linear, logistic, probit, arcsine, extreme value, and one hit model functions of $a + bx$.

The null hypothesis is

$$H : f(x) = \text{constant} \quad \text{for all } x$$

and the alternative hypothesis, called the *trend alternative*, is

$$K : f(x) < f(y) \quad \text{for } x < y.$$

We will assume $x_1 < x_2 < \cdots < x_k$ and thus $f(x) < f(x_2) < \cdots < f(x_k)$ under the trend alternative. [Sometimes we are interested in testing $f(x_1) > f(x_2) > \cdots > f(x_k)$, which we will call the *reverse trend alternative*.] We will consider the test of trend alternative when $(Y_1, Y_2, \ldots, Y_k)$ are independent binomial* random variables, multinomial* random variables ($\Sigma Y_i = n$), and independent Poisson* random variables.

Example 1. The District of Columbia has one of the highest infant mortality rates in the nation. The infant mortality rates among black populations between 1971 and 1981 were published in *The Washington Post*, Metro Section (Tuesday, November 30, 1982) and are shown in Table 1.

A question is whether there is a decreasing trend in the mortality rate. Let Y_i, w_i, and p_i denote the number of deaths, number of live births, and the probability of mortality for the year i. The probability p_i may be a function of year and, under reverse trend

Table 1   Washington, DC Infant Mortality Data.

Year	No. of Live Births	No. of Deaths	Mortality Rate per 1,000 Live Births
1971	12,131	368	30.3
1972	10,518	296	28.1
1973	9,413	245	26.0
1974	8,737	255	29.2
1975	8,462	254	30.0
1976	8,293	230	27.7
1977	8,515	251	29.5
1978	8,004	229	28.6
1979	8,053	199	24.7
1980	7,884	210	26.6
1981	7,749	196	25.3

alternative p, we have $p_1 > p_2 > \cdots > p_{11}$.

Example 2. Bellet et al. [3] cross-classified 121 Caucasian female patients with primary cutaneous malignant melanoma by years at risk and involvement of nonmelanocytic noncutaneous malignant neoplasms, yielding the data in Table 2 (taken from their Table 2). Because the malignant disease in younger patients tends to be more aggressive, thus possibly causing other malignancies, one might suspect that the younger patients are more prone to the multiple malignancies. The numbers of patients with other primary cancers ($Y_1, \ldots, Y_k$) form a conditional multinomial* random vector with parameters ($n = 11$, $k = 13$, $p_1, \ldots, p_{13}$). If all the patients are equally susceptible, then p_i would be proportional to the expected number E_i in the ith age group. Letting $w_i = E_i/\Sigma E_j$,

Table 2   Multiple Neoplasms Data.

Group No.	Years at Risk	Number of Subjects	The Number of Patients with Other Primary Cancers	
			Expected $(E_i)^a$	Observed (Y_i)
1	15–24[b]	8	0.02951	1
2	25–29	6	0.03774	0
3	30–34	6	0.0618	2
4	35–39	9	0.153	0
5	40–44	11	0.3135	1
6	45–49	9	0.4239	0
7	50–54	10	0.707	0
8	55–59	15	1.50	1
9	60–64	18	2.412	2
10	65–69	8	1.392	0
11	70–74	8	1.752	1
12	75–79	5	1.350	1
13	80 +	8	2.584	2

[a] Obtained by multiplying the number of subjects in the age group with the probability that a healthy subject develops cancer during the time at risk. The probability is age specific and is estimated from the incidence rates for all nonskin cancers extracted from the Third National Cancer Survey. For details, see Bellet et al. [3].
[b] The two age groups, 15–19 and 20–24, pooled.

we write

$$p_i = w_i f(x_i),$$

where x_i denotes the group number of the ith age group and $f(\cdot)$ is an unknown function accounting for the age effect. Then we may write

$$E[Y_i] = nw_i f(x_i)$$

and under the reverse age trend, $f(x_1) > f(x_2) > \cdots > f(x_{13})$.

Example 3. Poirier et al. [14] tested 20 chemical compounds in strain A mice by intraperitoneal injection and investigated their potency of inducing lung adenoma in the test mice. The potency is measured by the average number of adenomatic lesions, and a significant dose–response is one of the indications for the potential carcinogenicity. Most compounds were tested in four groups: zero, low, medium, and high dose groups. Twenty animals were assigned to each dose of each compound and a varying number to various zero dose groups. One of the chemicals tested yielded the data in Table 3.

The number of lung adenoma is considered as a Poisson variate when it is not excessive. If Y_i, w_i, and x_i denote the total number of tumors, the number of animals necropsied, and the dose level of the ith dose group, we have

$$E[Y_i] = w_i f(x_i),$$

where $f(\cdot)$ is an unknown function. Under the trend alternative (i.e., positive dose–response), $f(x)$ is increasing in x.

Table 3   Compound Methyl Iodide.

	Zero Dose	Low Dose	Medium Dose	High Dose	Sum
No. of Animals Necropsied	154	19	20	11	
Total No. of Tumors	34	4	6	6	50

More examples may be found in Lee [13].

For the binomial, multinomial, and Poisson distributions, the trend test statistic may be based on

$$\sum i Y_i \quad \left[\text{or } \sum (k - i + 1) Y_i \right.$$
$$\left. \text{for a reverse trend} \right],$$

if regression variables x_i are equally spaced or ordinal categories. The statistic may by based on

$$\sum x_i Y_i \quad \left(\text{or } - \sum x_i Y_i \text{ for a reverse trend} \right),$$

if x is a quantitative variable and $f(x)$ is a linear function of x or a smooth function of $a + bx$. The test based on $\sum i Y_i$ will be called the *monotone trend test* (MTT) and the test based on $\sum x_i Y_i$ will be called the *linear trend test* (LTT).

BINOMIAL DISTRIBUTION CASE

Let Y_i be an independent binomial random variable with the sample size n_i and the probability of success $f(x_i)$. Then we have

$$E[Y_i] = n_i f(x_i).$$

Let $\bar{x} = \sum n_i x_i / \sum n_i$, $\bar{i} = \sum i n_i / \sum n_i$, $\tilde{p} = \sum Y_i / \sum n_i$, and $\tilde{q} = 1 - \tilde{p}$. The LTT is: Reject the null hypothesis of no trend if $Z_L \geqslant Z_{1-\alpha}$, where

$$Z_L = \frac{\sum x_i Y_i - \tilde{p} \sum n_i x_i}{\left[\tilde{p}\tilde{q} \sum n_i (x_i - \bar{x})^2 \right]^{1/2}} \geqslant z_{1-\alpha},$$

and $z_{1-\alpha}$ is an upper α cutoff point of the standard normal distribution. In the MTT, Z_M replaces Z_L, where

$$Z_M = \left(\sum i Y_i - \tilde{p} \sum i n_i \right) \Big/ \left[\tilde{p}\tilde{q} \sum n_i (i - \bar{i})^2 \right]^{1/2}.$$

Note that Z_L and Z_M are normalized versions of $\sum x_i Y_i$ and $\sum i Y_i$.

Armitage [1] and Cochran [5] proposed the statistic Z_L^2 to determine whether $f(x)$ is a linear function. Namely, if $f(x) = a + bx$, then Z_L^2 determines the significance of H: $b = 0$. The departure from linearity can be determined by the statistic

$$X_{\text{linear}}^2 = X^2 - Z_L^2,$$

where $X^2 = \sum n_i(\tilde{p}_i - \tilde{p})^2/\tilde{p}\tilde{q}$ with $\tilde{p}_i = Y_i/n_i$. The asymptotic null distribution of X^2_{linear} is the chi-squared distribution with $k - 2$ degrees of freedom. The exact null distribution of $\sum x_i Y_i$ conditional on $\sum Y_i$ can be determined from the multivariate hypergeometric* distribution [7, 19].

Departure from the trend alternative can be tested by examining

$$X^2_{\text{monotone}} = X^2 - Z^2_M,$$

where X^2_{monotone} is asymptotically chi-squared with $(k - 2)$ degrees of freedom under the hypothesis of no such departure.

The MTT and LTT are both uniformly most powerful tests [7] and asymptotically efficient, namely the $C(\alpha)$ tests [18] (see OPTIMAL C(α) TESTS).

It is possible that, for some i, $\tilde{p}_i > \tilde{p}_{i+1}$, even if the trend alternative is true. Departures from the trend can be corrected by an "amalgamation process" (also called "isotonic regression*"), thus producing order restricted estimates of p_i's, denoted by $\tilde{p}_1^* \leqslant \tilde{p}_2^* \leqslant \cdots \leqslant \tilde{p}_k^*$ [2]. The trend alternative can be tested by

$$X^{*^2} = \sum n_i(\tilde{p}_i^* - \tilde{p})^2/(\tilde{p}\tilde{q}).$$

Collings et al. [6] compared the power of X^{*^2} and Z^2_L statistics, and concluded that the two tests are of the same power in detecting the trend alternative.

When the independent variable x_i represents a qualitative ordered classification, the common practice is to assign score i to x_i and apply Z^2_L to test the trend alternative [see refs. 1 and 6 (pp. 243–248), for example]. This, in effect, is applying the MTT statistic Z_M to detect the trend alternative.

Cox [8, p. 65] noted that the LTT is "nonparametric" in the sense that "the null hypothesis and the distribution used to obtain a significance level hold very generally." Tarone and Gart [18] note that the LTT is "asymptotically nonparametric" because it is the $C(\alpha)$ test. The same conclusions can be made on the MTT while noting that the MTT is more broadly applicable because x could be qualitative or quantitative. Even if

x is quantitative and thus the LTT is applicable, it is generally recommended to apply both LTT and MTT statistics.

Tarone [16] modified the LTT statistics to incorporate historical control data. When there are several statistics to be combined, Tarone and Gart [18] made the following efficient combination: Let Y_{ij} be $B[n_{ij}, p(x_{ij})]$, $j = 1, \ldots, k$, $i = 1, \ldots, I$, where $x_{ij} = x_{i'j} = x_j$ for all $i \neq i'$ and $x_1 < x_2 < \cdots < x_k$. The efficient combined statistic, when the $p(x)$'s are small, is

$$\sum\left(\sum x_j Y_{ij}\right)\bigg/\left(\tilde{p}_i\tilde{q}_i\right)^{1/2},$$

where $\tilde{p}_i = \sum Y_{ij}/\sum n_{ij}$. When the $p(x)$'s are moderate, the efficient combined statistic is

$$\sum\sum x_j y_{ij}.$$

Wood [20] studied the regression of Y_{ij} when $p(x_{ij})$ is a known function of x_{ij} (most notably a linear function).

Bennett [4] obtained the LTT by taking a rank transformation of the binary data.

The test of the reverse trend can be tested for the data in Example 1 by the MTT. In this case, the LTT and MTT are the same test, because the interval between two adjacent independent variables is the same for all. From the data in Table 1, we obtain $\sum(k - i + 1)y_i = 17795$, $\tilde{p}\sum(k - i + 1)n_i = 17458.64$, and $\tilde{p}\tilde{q}\sum n_i(i - \bar{i})^2 = (167.5)^2$, producing $Z_M = (17795 - 17458.64)/167.5 = 2.0$. The decreasing trend is significant at the p-value* of 0.022.

MULTINOMIAL DISTRIBUTION CASE

Let $\mathbf{Y} = (Y_1, Y_2, \ldots, Y_k)$ be a multinomial random vector $(\sum Y_i = n)$ and $(p_1, p_2, \ldots, p_k)$ be corresponding multinomial probabilities. The k categories have covariates $(x_1, \ldots, x_k)$ and weights $(w_1, w_2, \ldots, w_k)$. The probability p_i is assumed to be related to w_i and x_i by

$$p_i = w_i f(x_i)/n,$$

where $f(\cdot)$ is an unknown monotone function. Let $\pi_i = f(x_i)/n$. Under the trend alternative, we have $\pi_1 < \pi_2 < \cdots < \pi_k$ and $\Sigma w_i \pi_i = 1$.

Under the null hypothesis H, we may assume π_i is 1 for all i and thus $\Sigma w_i = 1$. (We do not lose any generality by this assumption. Suppose that $\pi_i = \pi_0 \neq 1$ under H. Because of the requirement $\Sigma w_i \pi_i = 1$, we have $\pi_0 = 1/\Sigma w_i$ under H, and redefine w_i by $w_i/\Sigma w_j$, thus having $\Sigma w_i = 1$.)

For the data in Example 2, $(1, 0, 2, 0, 1, 0, 0, 1, 2, 0, 1, 1, 2)$ forms an observed value of the random vector **Y**, and the value of w_i is proportional to $E_i/\Sigma E$; under the reverse age trend, $\pi_1 > \pi_2 > \cdots > \pi_k$.

It has been shown that MTT and LTT are *maximin* tests (*see* MINIMAX TESTS), namely maximizing the minimum power over alternative hypotheses where $\pi_{i+1}/\pi_i \geq \lambda^* > 1$ with λ^* not specified. In general, the MTT is more generally applicable [13].

For small to moderate n, the exact null distribution of $\Sigma i Y_i$ and $\Sigma x_i Y_i$ for given $(w_1, w_2, \ldots, w_k)$ can be obtained by exact enumeration. Their exact power for given λ^*, n, and $(w_1, w_2, \ldots, w_k)$ can also be determined. The null mean and variance of $\Sigma i Y_i$ are

$$n \sum i w_i \quad \text{and} \quad n \left\{ \sum i^2 w_i - \left(\sum i w_i \right)^2 \right\}.$$

The following statistic is asymptotically standard normal under H:

$$T_M = \frac{\sum i Y_i - n \sum i w_i}{\left[n \left\{ \sum i^2 w_i - \left(\sum i w_i \right)^2 \right\} \right]^{1/2}}.$$

Similarly, we can standardize the LTT statistic into

$$T_L = \frac{\sum x_i y_i - n \sum x_i w_i}{\left[n \left\{ \sum x_i^2 w_i - \left(\sum x_i w_i \right)^2 \right\} \right]^{1/2}}.$$

MTT and LTT statistics are asymptotically efficient. Departures from the linear and monotone trends can be similarly tested as in the binomial distribution case.

The data in Table 2 show a significant reverse age trend in susceptibility to the second primary cancer ($p = 0.0044$).

POISSON DISTRIBUTION CASE

Let $Y_i, \ldots, Y_k$ be independent Poisson random variables with $E[Y_i] = w_i f(x_i)$, $i = 1, 2, \ldots, k$. The constant w_i is a known design constant. For example, Y_i may be a sum of w_i independent and identically distributed Poisson random variables with expected value $f(x_i)$; or Y_i may be the number of rare events (meeting Poisson postulates) during the w_i-long interval, where the expected number of events per unit interval is $f(x_i)$. The trend alternative in $f(x)$ is to be tested.

For Example 3, Y_i denotes the total number of tumors in the dose group i, w_i the number of animals necropsied, and x_i the dose level. The question is whether there exists a dose–response.

The LTT and MTT statistics for the multinomial distribution can be applied to testing the trend alternative in $f(x)$, because k independent Poisson random variables $(Y_1, Y_2, \ldots, Y_k)$, given their sum $\Sigma Y_i = N$, form a multinomial random vector. Conditional on $\Sigma Y_i = N$, the T_M statistic will detect the monotone trend and the T_L statistic will detect the linear trend. These are conditional maximin tests.

There are alternate approaches to testing the trend alternative. One approach is to estimate the regression coefficient using maximum likelihood* (ML) estimation methods and test the hypothesis on the coefficient (refs. 9 and 10, among others). Another approach is the best asymptotic normal (BAN) estimation method [11]. These methods require that $f(x)$ is linearly related to x, namely $f(x) = a + bx$. Gart's method [10] is exact and is applicable only when the intercept constant is zero, while others rely upon asymptotic distributions of the estimate of b.

The log-likelihood, assuming $f(x) = a + bx$, is

$$L(a, b) = \sum y_i \log(a + bx_i)$$
$$- \sum (a + bx_i) w_i + C,$$

where C is a constant independent of a and b. Denote $\partial L/\partial a$, $\partial L/\partial b$, $\partial^2 L/\partial a \partial b$,

$\partial^2 L/\partial a^2$, and $\partial^2 L/\partial b^2$ by L_a, L_b, L_{ab}, L_{aa}, and L_{bb}. The information matrix and score vector are denoted by

$$\mathbf{I} = -\begin{pmatrix} L_{aa} L_{ab} \\ L_{ab} L_{bb} \end{pmatrix}, \quad \mathbf{S} = \begin{pmatrix} L_a \\ L_b \end{pmatrix}.$$

The ML estimators of a and b are simultaneous solutions of $L_a = 0$ and $L_b = 0$. Since the closed form solutions are not possible, an iterative approach is taken, most popularly the Newton–Raphson* iteration. When $E[Y_i]$ is large, the convergence is usually achieved after one or two iterations.

The BAN estimator [12] is the first-round scoring estimator. Let a_0 and b_0 be the least-squares estimates of a and b in $E[Y_i] = w_i(a + bx_i)$, $i = 1, \ldots, k$. Using a_0 and b_0 as the initial estimate, the first-round scoring estimators of a and b are

$$\begin{pmatrix} a_1 \\ b_1 \end{pmatrix} = \begin{pmatrix} a_0 \\ b_0 \end{pmatrix} + \mathbf{I}_0^{-1} \mathbf{S}_0,$$

where $\mathbf{I}_0$ and $\mathbf{S}_0$ are the information matrix $\mathbf{I}$ and score vector $\mathbf{S}$ evaluated at a_0 and b_0.

If the least-squares* estimates a_0 and b_0 cause $a_0 + b_0 x$ to have a very small positive value or even a negative value for some x, it is apparent that we cannot obtain a_1 and b_1. Such a problem arises for small $E[Y_i]$'s. The singularity of the information matrix could also cause similar difficulties (see Lee [12]). Computational difficulties are usually expected when applying the BAN method (as well as any methods requiring the inversion of the information matrix) to small to moderate sample sizes.

A simulation study of small sample powers of the BAN method and conditional MTT and LTT tests showed that none of them displayed clear superiority to the others [12]. It is not surprising that the three tests are comparable in power because (1) BAN estimation procedure is based on the likelihood function and (2) statistics $\Sigma i Y_i$ and $\Sigma x_i Y_i$ are asymptotically efficient (see ref. 17).

From the viewpoint of computational ease, the LTT and MTT methods are preferred.

They are asymptotically nonparametric and efficient (using the same argument as in ref. 19). Their properties under the binomial distribution are carried over, because the Poisson distribution is a limiting distribution of the binomial random variable with a small probability and large index.

Very often in applications, we do not know functional forms of the relationship between the response variable and the independent variable, but we want to determine whether there is a monotone relationship. A general recommendation for such cases is to use the statistic $\Sigma i Y_i$ [1, p. 378], which is the MTT. The recommendation is theoretically justified in ref. 13.

From the data in Example 3, we obtain

$$\Sigma i Y_i = 84,$$

$$n \Sigma i w_i = 72.55,$$

$$n \left[\Sigma i^2 w_i - \left(\Sigma i w_i \right)^2 \right] = 38.36,$$

where $n = 50$, $w_1 = 154/204$, $w_2 = 19/204$, $w_3 = 20/204$, and $w_4 = 11/204$. From the preceding, we obtain the MTT statistic

$$T_M = 1.85,$$

which indicates no significance at the significance level of 0.05.

Because the sample size is large and the probability is small in Example 1, the data can also be analyzed as if they are Poisson data.

References

[1] Armitage, P. (1955). *Biometrics*, **11**, 375–386. (Nontechnical method paper for testing linear trends in count data.)

[2] Barlow, R. E., Bartholomew, D. J., Bremner, J. M., and Brunk, H. D. (1972). *Statistical Inference Under Order Restrictions*. Wiley, New York. (Theoretical book on isotonic regression problems.)

[3] Bellet, R. E., Vaisman, I., Mastrangelo, M. J., and Lustbader, E. (1977). *Cancer*, **40**, 1974–1981. (Source paper for the data in Example 2.)

[4] Bennett, J. M. (1981). *Biom. J.*, **23**, 719–720. (Short note on the algebraic relationship between rank transformation and regression in count data.)

[5] Cochran, W. G. (1954). *Biometrics*, **10**, 417–451. (Excellent expository paper describing statistical methods for analyzing count data with relatively simple structure.)

[6] Collings, B. J., Margolin, B. M., and Oehlert, G. W. (1981). *Biometrics*, **37**, 775–794. (Isotonic regression in binomial proportions with applications to biological problems and comparative power studies.)

[7] Cox, D. R. (1958). *J. R. Statist. Soc. B*, **20**, 215–242. (Excellent basic paper introducing logistic regression for binary data.)

[8] Cox, D. R. (1970). *The Analysis of Binary Data.* Chapman and Hall, London, England. (Excellent theoretical monograph on how to analyze count data.)

[9] Frome, E. L., Kutner, M. H., and Beauchamp, J. J. (1973). *J. Amer. Statist. Ass.*, **68**, 935–940. (Expository paper on regression analysis of Poisson data.)

[10] Gart, J. J. (1964). *Biometrika*, **51**, 517–521. (Theoretical paper on simple regression analysis of Poisson data with zero intercept.)

[11] Jorgenson, D. W. (1961). *J. Amer. Statist. Ass.*, **56**, 235–245. (BAN method for multiple regression of Poisson data.)

[12] Lee, Y. J. (1980). *ASQC Technical Conference Transactions*, Atlantic City, NJ, pp. 683–691. (Comparison of three methods in simple regression of Poisson data via simulation.)

[13] Lee, Y. J. (1980). *J. Amer. Statist. Ass.*, **75**, 1010–1014. (Theoretical paper on testing trend in count data.)

[14] Poirier, L. A., Stoner, G. D., and Shimkin, M. B. (1975). *Cancer Research*, **35**, 1411–1415. (Source paper for the data in Example 3.)

[15] Snedecor, G. W. and Cochran, W. G. (1972). *Statistical Methods.* Iowa State University Press, Ames, IA. (Excellent intermediate level book on applications and theories of statistics.)

[16] Tarone, R. E. (1982a). *Biometrics*, **38**, 215–220. (Theoretical paper on simple regression in binomial proportions with historical control.)

[17] Tarone, R. E. (1982b). *Biometrics*, **38**, 457–462. (Theoretical paper on simple regression in Poisson means with historical control.)

[18] Tarone, R. E. and Gart, J. J. (1980). *J. Amer. Statist. Ass.*, **75**, 110–116. (Theoretical paper on testing trends in binomial proportions.)

[19] Thomas, D. G., Breslow, N., and Gart, J. J. (1977). *Comput. and Biomed. Res.*, **10**, 373–381. (Method paper on testing linear trend and departure from linear trend in count data.)

[20] Wood, C. L. (1978). *Biometrics*, **34**, 496–504. (Method paper on one-way ANOVA model with covariate in binomial proportions.)

(ISOTONIC REGRESSION
P-VALUES
TREND TESTS)

YOUNG JACK LEE

TREND IN PROPORTIONS, TEST FOR

Simplicity and effectiveness are both features of the test for trend* in proportions proposed by Yates [9], Cochran [3], and Armitage [1]; as a consequence, this test is well-nigh unbeatable for its intended purpose. The scenario envisioned for this trend test* is one in which N independent random variables have been observed, say $Y_1, \ldots, Y_N$. Each observation is binomially distributed, but the binomial* parameters may vary with the index, i.e., for each $i = 1, 2, \ldots, N$ and y a nonnegative integer,

$$\Pr[Y_i = y] = \binom{n_i}{y} p_i^y (1 - p_i)^{n_i - y}.$$

The index i is assumed to reflect an ordering associated with a single factor or covariate, whose N levels distinguish the set of observations. The "factor-level" terminology is common in designed experiments, where the N levels of the factor define N experimental groups, each yielding a binomial response. The test for trend, however, has far broader applicability and can be employed whenever count data can be represented in a $2 \times N$ contingency table* with ordered columns representing ordered levels of an explanatory variable and responses in different columns representing independent observations.

Two different cases need to be distinguished. If the explanatory variable is naturally associated with a quantitative measurement, say dose of a test substance, then the N levels can each be assigned a numerical value, say d_i, $i = 1, \ldots, N$. On the other hand, the levels of the explanatory variable may be ordered, but not quantitative, such as the degree of injury suffered by a burn victim. In the latter case, the recommendation by Yates [9], Cochran [3], and Armitage

[1] is to assign a meaningful score d_i to the ith level; a common assignment of scores when there is little to argue otherwise is the equispaced one, i.e., $(d_{i+1} - d_i)$ equal to a positive constant for all i. A brief discussion of alternative principles for constructing scores, together with references for further study, is given by Snedecor and Cochran [6, p. 246].

The observed proportions are denoted by $\hat{p}_i = Y_i/n_i$, $i = 1, \ldots, N$. The test for trend in these proportions as a function of the explanatory variable under study is based upon the regression coefficient b that results from the weighted linear regression* of the $\hat{p}_i$ on the scores d_i, $i = 1, \ldots, N$. The weights reflect the fact that the variance of $\hat{p}_i$ is inverse to n_i. A formula for b is given by

$$b = \sum_{i=1}^{N} n_i d_i (\hat{p}_i - \hat{p})/s^2, \qquad (1)$$

where $\hat{p} = \Sigma Y_i/\Sigma n_i$, $s^2 = \Sigma n_i (d_i - \bar{d})^2$, and $\bar{d} = \Sigma n_i d_i/\Sigma n_i$. Equivalent formulae used by various authors include

$$b = \sum_{i=1}^{N} n_i (d_i - \bar{d})(\hat{p}_i - \hat{p})/s^2$$

$$= \Sigma n_i (d_i - \bar{d})\hat{p}_i/s^2$$

$$= \Sigma (d_i - \bar{d}) Y_i/s^2. \qquad (2)$$

The variance of b is

$$\mathrm{var}(b) = \sum_{i=1}^{N} n_i p_i (1 - p_i)(d_i - \bar{d})^2/s^4. \qquad (3)$$

If the factor under study is without effect, i.e., if

$$H_0: p_i = p$$

holds, then

$$\mathrm{var}(b) = p(1 - p) \sum_{i=1}^{N} n_i (d_i - \bar{d})^2/s^4$$

$$= p(1 - p)/s^2; \qquad (4)$$

(4), in turn, may be estimated by

$$\widehat{\mathrm{var}}(b) = \hat{p}(1 - \hat{p})/s^2.$$

The test for trend in the proportions $\hat{p}_i$ is

then based upon the test statistic

$$Z = b/\left[\widehat{\mathrm{var}}(b)\right]^{1/2}$$

$$= \sum_{i=1}^{N} (d_i - \bar{d}) Y_i \Big/ \left[\hat{p}(1 - \hat{p})s^2\right]^{1/2}. \qquad (5)$$

If H_0 obtains, this test statistic is asymptotically distributed as a standard normal random variable. When a one-tailed test for increasing trend is desired, large values of Z are referred to a table of the tail area of a standard normal distribution to obtain an observed significance level. A two-tailed test is most easily constructed by referring the value of Z^2 to a table of the tail area of a chi-square distribution* with 1 degree of freedom.

A number of important properties have been established.

1. Cox [5] established that the trend test based upon (5) is uniformly most powerful unbiased for logistic alternatives to H_0.

2. Tarone and Gart [8] established that the trend test is the optimal $C(\alpha)$-test* for any alternative to H_0 in which p_i is specified by a monotone twice differentiable function of d_i, i.e., it is asymptotically locally optimal for any smooth monotone alternative to H_0.

3. Collings et al. [4] extensively compared the trend test to the isotonic* test proposed by Barlow et al. [2, pp. 192–194] and concluded that both tests have exact significance levels under H_0 that are in close agreement with their corresponding nominal levels for fairly small values of n_i; moreover, for testing for monotone trend, the more complicated isotonic test offers little advantage over the trend test, unless there is a substantial downturn in response at high levels of the explanatory variable—a phenomenon that does occur in various biomedical contexts.

4. Tarone [7] extended the use of the trend test to the important case where one level of the explanatory variable repre-

sents a "control" condition for which there exists historical information regarding the associated response.

In view of the results of Tarone and Gart [8], it is an error to consider the trend test in (5) to be a test for linearity in the proportions; the regression is linear, but the response need not be. With the exception of the caveat raised in the preceding item 3, the evidence is persuasive that the test is a highly effective procedure for testing for monotone trend in proportions, with wide applicability in the biological, physical, and social sciences.

References

[1] Armitage, P. (1955). *Biometrics*, **11**, 374–386.
[2] Barlow, R. E., Bartholomew, D. J., Bremner, J. M., and Brunk, H. D. (1972). *Statistical Inference Under Order Restrictions*. Wiley, New York.
[3] Cochran, W. G. (1954). *Biometrics*, **10**, 417–451.
[4] Collings, B. J., Margolin, B. H., and Oehlert, G. W. (1981). *Biometrics*, **37**, 775–794.
[5] Cox, D. R. (1958). *J. R. Statist. Soc., Ser. B*, **20**, 215–242.
[6] Snedecor, G. W. and Cochran, W. G. (1967). *Statistical Methods*. Iowa State Press, Ames, IA.
[7] Tarone, R. E. (1982). *Biometrics*, **38**, 215–220.
[8] Tarone, R. E. and Gart, J. J. (1980). *J. Amer. Statist. Ass.*, **75**, 110–116.
[9] Yates, F. (1948). *Biometrika*, **35**, 176–181.

(CONTINGENCY TABLES
ISOTONIC INFERENCE
ORDER-RESTRICTED INFERENCE
TREND IN COUNT DATA, TESTS OF
TREND TESTS)

BARRY H. MARGOLIN

TREND TESTS

Trend tests are generally described as "*k*-sample tests of the null hypothesis of identical distribution against an alternative of linear order," i.e., if sample i has distribution function F_i, $i = 1, \ldots, k$, then the null hypothesis

$$H_0: F_1 = F_2 = \cdots = F_k$$

is tested against the alternative

$$H_1: F_1 \geqslant F_2 \geqslant \cdots \geqslant F_k \qquad (1)$$

(or its reverse), where at least one of the inequalities is strict. These tests can be thought of as special cases of tests of regression* or correlation* in which association is sought between the observation and its ordered sample index. They are also related to analysis of variance* except that the tests are tailored to be powerful against the subset of alternatives H_1 instead of the more general set $\{ F_i \neq F_j,$ some $i \neq j \}$.

Different tests arise from requiring power* against specific elements or subsets of this rather extensive set of alternatives. Particular attention has been focussed on local alternatives ($\beta \downarrow 0$) in the linear model

$$y_{ij} = \alpha + \beta \mu_i + \epsilon_{ij},$$
$$i = 1, \ldots, k, \ j = 1, \ldots, n_i, \qquad (2)$$

where ϵ_{ij} are mutually independent and identically distributed (i.i.d.) with mean zero and some specified distribution function F and density f, n_i denotes the size of sample i, and $\mu_1 \leqslant \mu_2 \leqslant \cdots \leqslant \mu_k$ with at least one strict inequality, i.e., the alternative is restricted to

$$H_1^R: F_i(\cdot) = F(\cdot - \beta \mu_i),$$
$$\mu_1 \leqslant \mu_2 \leqslant \cdots \leqslant \mu_k. \qquad (3)$$

The *relative spacings*

$$\delta_i = (\mu_{i+1} - \mu_i)/(\mu_k - \mu_i), \ i = 1, \ldots, k-1,$$

are important in this respect, and much work either explicitly or implicitly has assumed them all to be equal.

Tests also differ as to whether or not they are parametric, i.e., they use the actual observed y_{ij} values, or are nonparametric and are based only on functions of the ranks of the $\{ y_{ij} \}$.

In addition, special tests have also been devised for different experimental designs, notably randomised complete block designs*, proportions in $2 \times k$ tables, k-group life-tables*, and $r \times c$ contingency tables*.

Specific tests are now described.

SIMPLE *k*-SAMPLE PROBLEMS

Shifts μ_i Assumed Known

In the parametric case, efficiency against local alternatives in (3) leads to tests of the form [12, 14]

$$T = \sum_i \sum_j (z_i - \bar{z}) g(y_{ij} - \bar{y}), \quad (4)$$

where

$$\bar{z} = N^{-1} \sum_{i=1}^{k} n_i z_i, \quad N = \sum_{i=1}^{k} n_i,$$

and $\bar{y}$ is an estimate of the center of the $\{y_{ij}\}$, usually the sample mean or median. Maximal power is achieved when $g(y) = \phi(y) \equiv (\log f(y))' = f'(y)/f(y)$ and $z_i = \mu_i$. Often the μ_i are unknown and in the absence of further information, $z_i = i$ is taken. In the parametric case, the function $g(\cdot)$ is usually taken to be the identity $g(y) = y$. Other choices of g suffer from the problem of leading to tests that depend upon estimates of the location and scale of the $\{y_{ij}\}$. The inner product form of (4) is noted, especially its resemblance to tests for regression or correlation—the only difference being that the second variable is replaced by z_i, a measure of the hypothesized differences between groups.

More often, a nonparametric version of (4) is considered. To do this, the distribution F must be known or hypothesized. The value chosen may not be the correct one (i.e., F) and, in the following text, we shall denote the conjectured null hypothesis by G. The test is modified by replacing $g(y_{ij} - \bar{y})$ by the expected order statistic* (assuming all y_{ij} have df G)

$$E_G\left[g(y_{ij})|R_{ij}\right],$$

where R_{ij} denotes the rank of y_{ij} in the combined sample. This is difficult to calculate exactly in most cases and can be replaced by the asymptotically equivalent expression

$$g\left(G^{-1}\left(\frac{R_{ij}}{N+1}\right)\right).$$

In this last case, the test statistic has mean zero and variance

$$\mathrm{Var}(T) = \frac{1}{N-1}\left\{\sum_i n_i(z_i - \bar{z})^2\right\}$$

$$\times \left[\sum_i \sum_j \left\{g\left(G^{-1}\{R_{ij}/(N+1)\}\right) - \bar{g}\right\}^2\right],$$

$$(5)$$

where

$$\bar{g} = N^{-1}\sum_i \sum_j g\left(G^{-1}\{R_{ij}/(N+1)\}\right), \quad (6)$$

whilst in the parametric case, the mean is again zero and the permutational variance is as before except that $(y_{ij} - \bar{y})$ replaces $G^{-1}(R_{ij}/(N+1))$ in (5) and (6) and N replaces $(N-1)$ in (5). The unconditional variance is

$$\left\{\sum_i n_i(z_i - \bar{z})^2\right\}$$

$$\times \left[\int g^2(y)f(y)\,dy - \left\{\int g(y)f(y)\,dy\right\}^2\right].$$

Asymptotic normality* can be shown under general conditions assuming either the null hypothesis or contiguous alternatives in both the parametric and nonparametric cases, and the normal approximation is generally quite good if $g(G^{-1}(u))$ is symmetric about some point and $\max_i k(z_i - z_{i-1})/(z_k - z_1)$ is not too large.

When $g(y) = \phi(y) \equiv f'(y)/f(y)$, $G = F$, and $z_i = \mu_i$, the rank test can be shown to be fully efficient [14]. A major reason for preferring rank tests is that their efficiency is generally greater when the specification of (g, G) is incorrect. Asymptotic relative efficiencies* are easily computed for local alternatives of the form (3). The asymptotic relative efficiency of tests with inefficient values of z_i in both parametric and nonparametric cases is reduced by the factor [14]

$$\frac{\{\sum n_i(\mu_i - \bar{\mu})(z_i - \bar{z})\}^2}{\{\sum n_i(\mu_i - \bar{\mu})^2\}\{\sum n_i(z_i - \bar{z})^2\}},$$

where $\bar{\mu} = N^{-1}\sum n_i\mu_i$. In the parametric case, the loss of efficiency for incorrect (g, G) is given by the factor

$$\frac{\left\{\int_0^1 g\left(F^{-1}(u)\right)\phi\left(F^{-1}(u)\right)\,du\right\}^2}{\left\{\int_0^1\left(g\left(F^{-1}(u)\right)-\bar{g}\right)^2\,du\right\}\left\{\int_0^1\phi^2\left(F^{-1}(u)\right)\,du\right\}},$$

where $\bar{g} = \int_0^1 g\left(F^{-1}(u)\right)\,du$.

In the nonparametric case, this factor is

$$\frac{\left\{\int_0^1 g\left(G^{-1}(u)\right)\phi\left(F^{-1}(u)\right)\,du\right\}^2}{\left\{\int_0^1\left(g\left(G^{-1}(u)\right)-\bar{g}\right)^2\,du\right\}\left\{\int_0^1\phi^2\left(F^{-1}(u)\right)\,du\right\}},$$

where now $\bar{g} = \int_0^1 g\left(G^{-1}(u)\right)\,du$. (Note that $\bar{\phi} = 0$.)

Special Cases

1. **$k = 2$.** The classic Chernoff–Savage two-sample rank tests are obtained by taking $z_i = 0$ or 1 according as $i = 1$ or $i = 2$.

2. **Wilcoxon–Type Trend Test*.** $F =$ logistic, $g(F^{-1}(u)) = u$, $z_i = i$. Simplest type of rank test for trend in which overall ranks are correlated against sample number. (See ref. 11 for examples of its use.)

3. **Normal Scores*.** $F = \Phi(\text{normal})$, $g(F^{-1}(u)) = \Phi^{-1}(u)$. Maximal power is obtained for normal errors.

4. **Savage Scores*.** $F(y) = 1 - e^{-y}$, $g(F^{-1}(u)) = 1 - \log(1 - u)$. This gives maximum power by exponential scale models.

5. **Sign Tests*.** Easy to compute sign tests with good efficiency have been developed in ref. 10.

Example 1. Consider the hypothetical data in Table 1, which consist of 14 observations in four ordered groups. If we let z_i denote group membership, then

$$\bar{z} = \tfrac{1}{14}\{3(1) + 2(2) + 4(3) + 5(4)\}$$

$$= \tfrac{39}{14} = 2.79.$$

If we take $g(y) = y$, then (4) gives a parametric test for trend and

$$T = (1 - 2.79)(2.5 + 3.1 + 8.8)$$
$$+ (2 - 2.79)(1.3 + 4.8)$$
$$+ (3 - 2.79)(3.9 + 7.1 + 10.3 + 11.7)$$
$$+ (4 - 2.79)(6.2 + 7.6 + 11.5$$
$$+ 18.1 + 21.3)$$
$$= 55.13,$$

Table 1

	Value y_{ij}	Rank R_{ij}	Exact Normal Score	Approx. Normal Score $\Phi^{-1}(R_{ij}/(N+1))$
Group 1	2.5	2	−1.208	−1.111
$z_i = 1$	3.1	3	−0.901	−0.842
$n_i = 3$	8.8	9	0.267	0.253
Group 2	1.3	1	−1.703	−1.501
$z_i = 2$	4.8	5	−0.456	−0.430
$n_i = 2$				
Group 3	3.9	4	−0.662	−0.623
$z_i = 3$	7.1	7	−0.088	−0.084
$n_i = 4$	10.3	10	0.456	0.430
	11.7	12	0.901	0.842
Group 4	6.2	6	−0.267	−0.253
$z_i = 4$	7.6	8	0.088	0.084
$n_i = 5$	11.5	11	0.662	0.623
	18.1	13	1.208	1.111
	21.3	14	1.703	1.501

$\bar{y} = 8.44$, $\sum y_{ij}^2 = 1436.38$, $s^2 = 31.36$, $N = 14$.

$$\text{Var}(T) \cong \tfrac{1}{14}\big\{3(1 - 2.79)^2 + 2(2 - 2.79)^2$$
$$+ 4(3 - 2.79)^2 + 5(4 - 2.79)^2\big\}$$
$$\times \Big\{ \sum_{i,j}(y_{ij} - \bar{y})^2 \Big\}$$
$$= 574.89.$$

Thus an approximately normal deviate is given by $Z = T/\text{Var}^{1/2}(T) = 55.13/(574.89)^{1/2} = 2.30$. If a trend in only one direction was plausible or of interest a one-sided test is appropriate, giving an approximate significance level of 0.01. More often a two-sided test is appropriate and the significance level should be doubled. The Wilcoxon test is obtained by replacing the y_{ij} by their ranks. This yields

$$T = 40.50, \qquad \text{Var}(T) \cong 321.25,$$
$$Z \cong 2.26, \quad \text{a very similar value.}$$

When exact normal scores are used we obtain $T = 9.24$, $\text{Var}(T) \cong 16.65$, and $Z = 2.26$, whilst approximate normal scores yield

$$T = 8.40, \qquad \text{Var}(T) \cong 13.67, \qquad Z = 2.27,$$

which emphasizes the very high degree of correlation between these last three statistics.

An alternative collection of nonparametric trend test criteria expressed as a linear combination of two-sample Chernoff–Savage test criteria has also been developed [21, 23]. They take the form

$$\sum_{u<v} a_{uv}T_{uv}, \tag{7}$$

where T_{uv} is any two-sample Chernoff–Savage* statistic based on groups u and v; $u, v = 1, \ldots, k$, $u < v$, and a_{uv} are constants. These tests can only be made equivalent to the nonparametric form of (4) when g is the identity, i.e., for Wilcoxon–type tests. It is not clear for what alternatives these tests are fully efficient and the approach based on (4) would appear to have many advantages.

Shifts μ_i Unknown

The means μ_i are rarely known and the choice of $z_i = i$ has the effect of assuming that the differences between neighboring

groups are equal. This is a small subset of H_1^R and may lead to weak tests when the alternative is of the form

$$H_1^E\colon \mu_1 = \mu_2 = \cdots = \mu_{k-1} < \mu_k.$$

Bartholomew [3, 4] and later Chacko [6] have studied the problem when the ϵ_{ij} are normal with the same variance. They determine the likelihood ratio* test for H_0 versus H_1^R. This involves estimating the sample means by maximum likelihood subject to the order constraint $\mu_1 \leqslant \cdots \leqslant \mu_k$, and performing a one-way analysis of variance* on the reduced problem. More specifically, let

$$\bar{y}_i = n_i^{-1}\sum_j y_{ij}$$

and

$$\bar{y}_{[j,k]} = \sum_{i=j}^{k} n_i \bar{y}_i \Big/ \sum_{i=j}^{k} n_i.$$

The constrained MLE estimate of μ_i, denoted $\tilde{\mu}_i$, is given by

$$\tilde{\mu}_i = \max_{1 \leqslant r \leqslant i} \min_{i \leqslant s \leqslant k} \bar{y}_{[r,s]}$$

and this can be computed by the following iterative procedure:

(i) Initially set $\tilde{\mu}_i = \bar{y}_i$.

(ii) If $\tilde{\mu}_i > \tilde{\mu}_{i+1}$ for some i, replace these two groups by a single group with the mean $\bar{y}_{[i,i+1]}$ of the combined group.

(iii) Continue to do this until the remaining k_0 groups satisfy $\tilde{\mu}_1 \leqslant \cdots \leqslant \tilde{\mu}_{k_0}$.

It can be shown that (8) will always be reached regardless of the order in which the collapsing takes place. Let m_i denote the sample size of the ith reduced group. The test then becomes a simple χ^2 test on the reduced data set, i.e.,

$$\bar{\chi}_k^2 = \sum_{i=1}^{k_0} m_i(\tilde{\mu}_i - \bar{y})^2 / \sigma^2,$$

where σ^2 is the variance of an individual observation (which may be estimated; see the following text) and $\bar{y}$ is the grand mean. The bar indicates that this is a reduced problem and will not have a χ^2 distribution under the null hypothesis, but is a mixture of χ^2 distributions with a random number

$(k_0 - 1)$ of degrees of freedom. Under H_0,

$$\Pr\left[\bar{\chi}_k^2 > C\right] = \sum_{j=2}^{k} p_{jk}\Pr\left[\chi_{j-1}^2 > C\right],$$

where $\Pr[\chi_{j-1}^2 > C]$ is the tail probability for the ordinary χ^2 distribution on $(j - 1)$ degrees of freedom (df) and p_{jk} are constants (denoting the probabilities of j groups in the reduced problem) which have not been evaluated in general. For $k \leqslant 4$, they are given in ref. 4. When the sample sizes are all equal (more specifically, the means $\bar{y}_i$ have the same variance), it has been shown that $p_{jk} = |S_k^j|/k!$, where S_k^j is the Stirling number* of the first kind. When σ^2 is unknown, it can be estimated by

$$s^2 = \frac{1}{N}\sum_i\sum_j (y_{ij} - \bar{y})^2$$

and χ_{j-1}^2 must be replaced by N times a beta* variable with parameters $\frac{1}{2}(j - 1)$ and $\frac{1}{2}(N - j)$, i.e.,

$$\Pr\left[\bar{\chi}_k^2 > C\right]$$
$$= \sum_{j=1}^{k_0} P_{jk}\Pr\left[\beta_{(j-1)/2,(N-j)/2} > CN^{-1}\right].$$

A nonparametric version can be obtained by replacing the y_{ij} with ranks or normal scores [16].

Example 2. For the data in Table 1, we have $(n_1,\ldots,n_4) = (3,2,4,5)$, $(\bar{y}_1,\ldots,\bar{y}_4)$ $= (4.80, 3.05, 8.25, 12.94)$. Combining the first two groups gives $y_{[1,2]} = 4.10$ and we arrive at the reduced sample (m_1, m_2, m_3) $= (5, 4, 5)$, $(\mu_1, \mu_2, \mu_3) = (4.10, 8.25, 12.94)$. The sample mean is $\bar{y} = 8.44$ and the variance is estimated by $s^2 = 31.36$. Thus $k_0 = 3$ and $s^2 = 31.36$. Thus $k_0 = 3$ and

$$\bar{\chi}_4^2 = \left\{5(4.10 - 8.44)^2 + 4(8.25 - 8.44)^2\right.$$
$$\left. + 5(12.94 - 8.44)^2\right\}/31.36$$
$$= 195.57/31.36 = 6.236.$$

The coefficients $(p_{24},\ldots, p_{44})$ are $(0.285, 0.235, 0.215)$ (cf. [4]) and the beta significance levels are

$$\Pr\left[\beta_{1/2,6} > 6.236/14\right] = 0.01,$$
$$\Pr\left[\beta_{1,51/2} > 0.445\right] = 0.03,$$
$$\Pr\left[\beta_{3/2,5} > 0.445\right] = 0.11$$

so that

$$\Pr\left[\bar{\chi}_4^2 > 6.236\right] = 0.03.$$

Abelson and Tukey [1] considered the problem from a different point of view. Instead of estimating the μ_i by restricted maximum likelihood*, they sought a contrast $\{c_i\}$ which would maximize the minimum correlation coefficient squared, r^2, between $\{c_i\}$ and all $\{\mu_i\}$ subject to $\mu_1 \leqslant \cdots \leqslant \mu_k$. When the sample sizes n_i are all equal, they obtain the solution

$$c_i = \left[(i - 1)\left(1 - \frac{i-1}{k}\right)\right]^{1/2} - \left[i\left(1 - \frac{i}{k}\right)\right]^{1/2},$$
$$i = 1,\ldots, k.$$

A simple approximation to the optimal $\{c_i\}$ is the so-called linear two–four contrast [1], which is obtained by taking c_i to be linear in i and then increasing the end-group coefficients fourfold and doubling the penultimate group coefficients.

For large k, the maximum r^2 was approximated by $2/\{2 + \log(k - 1)\}$.

Example 3. This method is illustrated by the data in Table 2 which consists of two observations from each of eight ordered categories. When the maximin scores for the groups are used in (4), we obtain

$$T = 21.03, \quad \mathrm{Var}(T) \cong 44.92, \quad Z \cong 3.14,$$

whereas the linear 2–4 scores give

$$T = 362.30, \quad \mathrm{Var}(T) \cong 13590.53,$$
$$Z \cong 3.11.$$

Table 2

Group	Values	Maximin Score	Linear 2–4 score
1	$-1.4, 0.4$	-0.935	-16
2	$6.5, 3.8$	-0.290	-6
3	$4.1, 3.9$	-0.145	-2
4	$6.1, 8.1$	-0.045	-1
5	$2.3, 3.8$	0.045	1
6	$4.6, 0.9$	0.145	2
7	$8.1, 5.3$	0.290	6
8	$9.0, 12.3$	0.935	16

For comparison, use of the group ordering scores $z_i = i$ gives

$$T = 78.00, \quad \text{Var}(T) \cong 960.95, \quad Z = 2.52.$$

General Trend Alternatives

A method for testing against the most general trend alternative (1) is based on an extension of the Smirnov statistic. Conover [8] analyzed the statistic

$$\sup_{\substack{1 \le i \le k-1 \\ t}} \left\{ \hat{F}_{i+1}(t) - \hat{F}_i(t) \right\},$$

where $\hat{F}_j$ is the empirical distribution of sample j, $j = 1, \ldots, k$, and gave distributional results.

RANDOMIZED BLOCK DESIGNS

A large amount of work has been done on trend tests in block designs. Here the basic model is

$$y_{ij} = \beta \mu_i + b_j + \epsilon_{ij},$$
$$i = 1, \ldots, k, \, j = 1, \ldots, n,$$

where b_j are unknown block effects and all other terms are as in (1), except that the sample sizes are all the same ($n_i = n$). To cancel the block effects, two types of nonparametric procedures have been proposed [19].

Within Block Tests (W-Tests)

Here the data are ranked *within* blocks. Let R_i^j be the ranks of y_{ij}, $i = 1, \ldots, k$, within block j and define

$$w = \sum_{j=1}^{n} w_j,$$

where w_j is a simple linear rank statistic

$$w_j = \sum_{i=1}^{k} z_i g(R_i^j).$$

Among Block Tests (A-Levels)

Let

$$A = \sum_{u < v} a_{uv} T_{uv},$$

where a_{uv} are constants and T_{uv} is a simple rank test based on *differences* of paired samples. [This is different from (7).] Specifically, set $\Delta_{uv}^{(j)} = y_{uj} - y_{vj}$ and let $R_{uv}^{(j)}$ be the rank of $|\Delta_{uv}^{(j)}|$ among $|\Delta_{uv}^{(l)}|$, $l = 1, \ldots, n$. Then for $u, v = 1, \ldots, k$, $u \ne v$,

$$T_{uv} = \sum_{j=1}^{n} g(R_{uv}^{(j)}) I(\Delta_{uv}^{(j)}),$$

where

$$I(x) = \begin{cases} 1, & x > 0, \\ 0, & x = 0, \\ -1, & x < 0. \end{cases}$$

TRENDS IN CONTINGENCY TABLES

Trends in Proportions in $2 \times k$ Tables

This material has been summarized by Armitage [2]. If n_i is the total number of observations in column i and p_i is the proportion of these observations which lie in row 1, then the test takes the form

$$\left\{ \sum_i n_i p_i (z_i - \bar{z}) \right\}^2 \bigg/ \left\{ \sum_i n_i (z_i - \bar{z})^2 \right\},$$

which after normalization will asymptotically have a χ^2 distribution on 1 df. Again column scores $z_i = i$ are usually taken.

Trends in Life Tables

In k-sample comparisons of survival, the rank tests (4) need to be modified to accommodate censoring. This is usually done via generalized Savage scores ξ_{ij}, where i indexes group and j an individual within the group. The usual trend test is of the form

$$\sum_i \sum_j (z_i - \bar{z}) \xi_{ij},$$

where $z_i = i$ is commonly assumed. See refs. 9 and 18.

Trends in General $r \times c$ Tables

No natural linear order exists in this case, but an obvious partial order can be defined. Tests based on Kendall's τ have been proposed [17a], as have tests against the alternative that odds ratios defined at each cell are all greater than unity. These tests are generally 1 df tests and are more powerful than the usual $(r - 1) \times (c - 1)$ df test of heterogeneity when the restricted alternative holds. If p_{ij} denotes the cell probabilities $i = 1, \ldots, r$, $j = 1, \ldots, c$, tests that are powerful when the following odds ratios are constant (and greater than unity) have been proposed:

1. (See ref. 20)

$$\theta_{ij} = \left(\sum_{\substack{k \leqslant i \\ l \leqslant j}} p_{kl} \right) \left(\sum_{\substack{k > i \\ l > j}} p_{kl} \right) \Bigg/ \left[\left(\sum_{\substack{k \leqslant i \\ l > j}} p_{kl} \right) \left(\sum_{\substack{k > i \\ l \leqslant j}} p_{kl} \right) \right].$$

2. (See ref. 7)

$$\theta_{ij} = p_{ij} \left(\sum_{\substack{k > i \\ l > j}} p_{kl} \right) \Bigg/ \left[\left(\sum_{k > i} p_{kj} \right) \left(\sum_{l > j} p_{il} \right) \right].$$

3. (See ref. 13)

$$\theta_{ij} = p_{ij} p_{i+1, j+1} \big/ \left(p_{i, j+1} p_{i+1, j} \right).$$

References

[1] Abelson, R. P. and Tukey, J. W. (1963). *Ann. Math. Statist.*, **34**, 1347–1369. (Maximum likelihood method of the section Shifts μ_i Unknown.)

[2] Armitage, P. (1955). *Biometrics*, **11**, 375–386. (Tests for trends in proportions.)

[3] Bartholomew, D. J. (1961). *J. R. Statist. Soc. B*, **23**, 239–272. (Overview paper of the method described in the section Shifts μ_i Unknown.)

[4] Bartholomew, D. J. (1959). *Biometrika*, **46**, 328–335. (Gives coefficients for the method described in the section Shifts μ_i Unknown.)

[5] Berenson, M. L. (1982). *Psychometrika*, **47**, 265–280; Erratum 535–539. (Numerical study of different tests in block design setup.)

[6] Chacko, V. J. (1963). *Ann. Math. Statist.*, **34**, 945–956. (Further mathematical results on method of Bartholomew.)

[7] Clayton, D. and Cuzick, J. (1985). *J. R. Statist. Soc. A*, **148**, 82–117. (Trend tests in $r \times c$ tables.)

[8] Conover, W. J. (1967). *Ann. Math. Statist.*, **38**, 1726–1730. (Extension of Smirnov tests to trend alternatives.)

[9] Cox, D. R. (1972). *J. R. Statist. Soc. B*, **74**, 187–220. (Trends in life tables.)

[10] Cox, D. R. and Stuart, A. (1955). *Biometrika*, **42**, 80–95. (Quick sign tests.)

[11] Cuzick, J. (1985). *Statist. in Medicine*, **4**, 87–90 and 543–547. (Examples of the use of Wilcoxon trend test. Second reference is to ensuing correspondence on general subject of trend tests.)

[12] Gibbons, J. D. (1971). *Non-parametric Statistical Inference*. McGraw-Hill, New York. (Introductory text on all aspects of rank tests.)

[13] Goodman, L. A. (1981). *J. Amer. Statist. Ass.*, **76**, 320–334. (Also *ibid.*, **74**, 537–552). (Trend tests in $r \times c$ tables.)

[14] Hajék, J. and Šidák, Z. (1967). *Theory of Rank Tests*. Academic, New York. (Classic work on rank tests including trend tests.)

[15] Hollander, M. and Wolfe, D. A. (1973). *Nonparametric Statistical Methods*. Wiley, New York. (General applied text on nonparametric methods.)

[16] Johnson, R. A. and Mehrotra, K. G. (1971). *J. Indian Statist. Ass.*, **9**, 9–23. (Nonparametric versions of tests in the section Shifts μ_i Unknown.)

[17] Jonckheere, A. R. (1954). *Biometrika*, **41**, 133–145. (One of the earliest papers on nonparametric trend tests.)

[17a] Kendall, M. G. and Stuart, A. (1979). *The Advanced Theory of Statistics*, 4th ed., **2**, Chapter 33. (Methods for analyzing categorical data.)

[18] Mantel, N. (1963). *J. Amer. Statist. Ass.*, **58**, 690–700. (Trends in life tables.)

[19] Pirie, W. R. (1974). *Ann. Statist.*, **2**, 374–382. (Theoretical paper on trend tests in block designs.)

[20] Plackett, R. L. (1981). *The Analysis of Categorical Data*. Griffin, London, England. (Trend tests in $r \times c$ tables.)

[21] Puri, M. L. (1965). *Commun. Pure Appl. Math.*, **18**, 51–63. [Tests of the form given by (7).]

[22] Puri, M. L. and Sen, P. K. (1971). *Nonparametric Methods in Multivariate Analysis*. Wiley, New York. (Rigorous mathematical treatment of nonparametric procedures.)

[23] Tryon, P. V. and Hettmansperger, T. P. (1973). *Ann. Statist.*, **1**, 1061–1070. [Tests of the form given by (7).]

(DISTRIBUTION-FREE METHODS
NORMAL SCORES
SCORE STATISTICS
TREND IN COUNT DATA, TESTS FOR
TREND IN PROPORTIONS, TEST FOR)

JACK CUZICK

TRIANGLE, PASCAL'S *See* PASCAL'S TRIANGLE

TRIANGLE TEST

The triangle test, sometimes called the *sensory difference test*, is applied to determine whether respondents are able to distinguish between two test products, one of which is a variant of the other. The classic setting involves taste testing of food products, but other problems involve sensory measurements such as color, odor, touch, or sound. Because it is difficult to quantify sensory perceptions, the experimental design is set up in such a way that each person is presented with three product specimens, two of which are the *same* product and one of which is the other product. The identification of each specimen is disguised and the order of presentation is randomized. Each respondent is asked to identify the sample which differs from the other two.

Suppose we have a random sample of n respondents. Let Y denote the number among them who correctly identify the odd sample among the three. If persons are unable to distinguish the two products (the null hypothesis), then Y will have a binomial distribution* with parameters n and $p = \frac{1}{3}$. If persons actually are able to distinguish the two products (the alternative hypothesis), Y will tend to be larger than anticipated under the null hypothesis binomial* distribution. So we will believe that persons can distinguish the two products if $Y \geq k$, where k is chosen so that

$$\alpha = P\left[Y^* \geq k | Y^* \text{ is binomial } n, p = \tfrac{1}{3}\right].$$

Here α denotes the significance level of the test. When n is large, we will believe the products are distinguishable provided

$$Y \geq n/3 + (z_\alpha)(2n/9)^{1/2},$$

where z_α is the upper αth percentile of a standard normal distribution.

Woodward and Schucany [5] show data from an experiment comparing two types of potato chips. Among the $n = 105$ respondents, $Y = 55$ correctly identified the odd brand. With $\alpha = 0.05$, the rule rejects the null hypothesis if

$$Y \geq (105/3) + (1.645)(2 \times 105/9)^{1/2}$$
$$= 42.95.$$

These data are convincing that persons are able to distinguish these two brands of potato chips.

The origins of the triangle test are obscure. Its mathematical properties were developed by Hopkins and Gridgeman [4] and Bradley [1]. Modifications of the test have also been proposed. Bradley and Harmon [2] investigate a model for settings in which the respondent specifies a degree of difference between the odd sample and the other two. Woodward and Schucany [5] treat settings in which the respondent indicates a preference for one of the two products he identifies. Cash [3] investigates settings in which respondents are not forced to select an odd sample when they fail to perceive a difference in the three products.

References

[1] Bradley, R. A. (1963). *Biometrics*, **19**, 385–397. (Compares triangle test to competing experimental designs.)
[2] Bradley, R. A. and Harmon, T. (1964). *Biometrics*, **20**, 608–625. (Modifies test to include the degree of difference between the two products.)
[3] Cash, W. S. (1983). *Biometrics*, **39**, 251–255. (Modifies procedure to include response that person cannot identify the odd product.)
[4] Hopkins, J. W. and Gridgeman, N. T. (1955). *Biometrics*, **11**, 63–68. (Compares the triangle test to other designs.)
[5] Woodward, W. A. and Schucany, W. R. (1977). *Biometrics*, **33**, 31–39. (Modifies procedure where respondent specifies a preference for one of the two products.)

(PAIRED COMPARISONS)

RONALD H. RANDLES

TRIANGULAR COORDINATES

See BARYCENTRIC COORDINATES. Additional references:

Mosteller, F. and Tukey, J. W. (1968). Data analysis including statistics. In *Handbook of Social Psychology*, G. Lindzey and E. Aronson, eds. Addison-Wesley, Reading, MA.

Shelton, W. C. (1972). *Amer. Statist.*, **26**(5), 17–19. (Application to three-dimensional data.)

TRIANGULAR DISTRIBUTION *See* UNIFORM (RECTANGULAR) DISTRIBUTION

TRIANGULAR INEQUALITY *See* DISTANCE FUNCTIONS; CLASSIFICATION

TRIANGULAR PLOT

Consider a contingency table* (as in Table 1) and the corresponding row proportions table (Table 2). Of these two representations, the row proportions are the more relevant. They show for example that characteristics A and B are very similar and that C and E are very different. These similarities and dissimilarities can be plotted in a three-dimen-

Table 1   Contingency Table.

	a	b	c	T
A	2	3	5	10
B	8	10	22	40
C	4	3	13	20
D	21	21	18	60
E	1	18	1	20
T	36	55	59	150

Table 2   Row Proportions Table.

	p_1	p_2	p_3	T
A	0.2	0.3	0.5	1.0
B	0.2	0.25	0.55	1.0
C	0.2	0.15	0.65	1.0
D	0.35	0.35	0.30	1.0
E	0.05	0.90	0.05	1.0

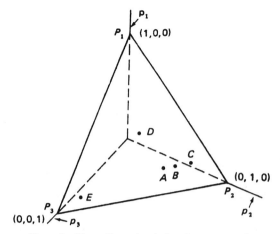

Figure 1   Three-dimensional plot of row proportions.

sional scatter plot* (Fig. 1). However, the points of the scatter plot are not distributed in three-dimensional space but in two dimensions. They all lie in a plane and within the triangle outlined in the figure. This is because for each data point $\sum_{i=1}^{3} p_i = 1$. We thus resort to *barycentric coordinates** to provide a two-dimensional *triangular plot* of the same data (Fig. 2). The triangle is equilateral. The vertices P_1, P_2, and P_3 correspond to any *rows* with, respectively, the extreme proportions $(1, 0, 0)$, $(0, 1, 0)$, and $(0, 0, 1)$ and any one vertex is at unit distance from its opposite side. A particular row such as D with $p_1 = 0.35$, $p_2 = 0.35$, and $p_3 = 0.30$ is plotted as indicated in Fig. 2 in barycentric coordinates. The same is done

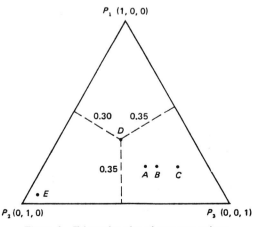

Figure 2   Triangular plot of row proportions.

for points A, B, C, and E. Note the relative location of points on the triangular plot.

(BARYCENTRIC COORDINATES
CONTINGENCY TABLES)

TRIANGULATION

In statistics, this term is a name coined by Mosteller [2] to describe a procedure for obtaining estimates of the magnitude of a characteristic related to some social or physical phenomenon (such as the average number of miles driven per year for American motorists or the number of persons on welfare in a particular state) using different statistical sources and techniques.

See Mosteller [2] and Hoaglin et al. [1] for details.

References

[1] Hoaglin, D. C., Light, R. J., McPeek, B., Mosteller, F., and Stoto, M. A. (1982). *Data for Decisions*. Abt Books, Cambridge, MA.

[2] Mosteller, F. (1977). In *Statistics and Public Policy*, W. Fairley and F. Mosteller, eds. Addison-Wesley, Reading, MA, pp. 163–184.

(REPRESENTATIVE SAMPLE
SURVEY SAMPLING)

TRIDIAGONAL MATRIX

A square matrix $\mathbf{B}$ such that $b_{ij} = 0$ for $|i - j| > 1$, $b_{ij} \neq 0$ for $|i - j| \leqslant 1$. The only nonzero elements are all in a band along the diagonal. A birth and death (Markov) process* has a tridiagonal transition matrix, since the only allowable transitions are to the adjacent states.

Example.

$$\mathbf{B} = \begin{bmatrix} 0.3 & 0.7 & 0 & 0 \\ 0.4 & 0.2 & 0.4 & 0 \\ 0 & 0.1 & 0.8 & 0.1 \\ 0 & 0 & 0.6 & 0.4 \end{bmatrix}$$

is a tridiagonal transition matrix.

TRIEFFICIENCY

The *triefficiency* of an estimator is defined by Beaton and Tukey [1] as the minimum of its efficiencies for random samples from the normal* (Gaussian), one-wild*, and slash* distributions. Further details are available in ref. 2.

References

[1] Beaton, A. E. and Tukey, J. W. (1974). *Technometrics*, **16**, 147–185.

[2] Hoaglin, D. C., and Mosteller, F., and Tukey, J. W. (1983). *Understanding Robust and Exploratory Data Analysis*. Wiley, New York.

(EFFICIENCY)

TRIGAMMA FUNCTION

The derivative of the digamma* or psi function* $[\Psi(x) = d \log \Gamma(x)/dx]$. It is customarily denoted by

$$\Psi'(x) = \frac{d^2 \log \Gamma(x)}{dx^2}.$$

TRIM ARRAY   *See* STRENGTH AND OTHER PROPERTIES OF AN ARRAY

TRIMEAN

The following linear combination of order statistics*

$$0.25 X_{([n/4]+1)} + 0.5 M_e + 0.25 X_{(n-[n/4])}$$

is called the *trimean* (here M_e is the median of the sample of size $X_1, X_2, \ldots, X_n$ from a symmetric distribution and $X_{(i)}$ is the ith order statistic). This robust estimator of the location parameter of a symmetric distribution is similar to Gastwirth's [1] estimator:

$$G = 0.3 X_{([n/3]+1)} + 0.4 M_e + 0.3 X_{(n-[n/3])}.$$

Reference

[1] Gastwirth, J. L. (1971). *J. Amer. Statist. Ass.*, **61**, 929–98.

(ORDER STATISTICS
ROBUST ESTIMATION)

TRIM-LOSS PROBLEM *See* LINEAR
PROGRAMMING

TRIMMED AND WINSORIZED MEANS, TESTS FOR

Modifications of classical normal-theory procedures less affected by outliers* or long-tailed distributions are often desirable.

Let $y_1 \leqslant y_2 \leqslant \cdots \leqslant y_n$ be the ordered observations in a random sample of size n from a single distribution. The *g-times trimmed mean*

$$\bar{y}_{tg} = \frac{1}{h} \sum_{i=g+1}^{n-g} y_i,$$

where $h = n - 2g$, is a location estimate resistant to such effects. Let

$$\bar{y}_{wg} = \frac{1}{n} \left\{ (g+1)y_{g+1} + \sum_{i=g+2}^{n-g-1} y_i \right.$$

$$\left. + (g+1)y_{n-g} \right\},$$

$$\mathrm{SSD}_{wg} = (g+1)(y_{g+1} - \bar{y}_{wg})^2$$

$$+ \sum_{i=g+2}^{n-g-1} (y_i - \bar{y}_{wg})^2$$

$$+ (g+1)(y_{n-g} - \bar{y}_{wg})^2.$$

These are the *g-times Winsorized* mean and sum of squared deviations, respectively, where *Winsorizing* (after Charles P. Winsor) in effect consists of replacing at each end the g most extreme observations with the value of the next most extreme one. *See also* TRIMMING AND WINSORIZATION.

ONE-SAMPLE TESTS

Suppose we have a random sample from a single distribution, ordered as above, and are interested in location. Tukey and McLaugh-

lin [8] studied $\bar{y}_{tg}$ and concluded that a robust estimate of its variability could be based on SSD_{wg}. They proposed a trimmed t test using

$$t_{tg} = \frac{\bar{y}_{tg} - \mu}{\left[\mathrm{SSD}_{wg} / \{ h(h-1) \} \right]^{1/2}},$$

where μ is the hypothesized mean, and recommended that the null distribution of t_{tg} be approximated by Student's t with $h - 1$ *degrees of freedom* (df) (t_{h-1}).

In order that this test (and the one following) be valid, the distribution of the original observations must be symmetric, at least in the null case. This is most often plausible where the observations are differences, on the same experimental units, between two conditions. (Validity of the associated confidence intervals requires symmetry in general.)

Dixon and Tukey [1] proposed a similar test resistant to the effects of outliers or long tails. This Winsorized t test uses

$$t_{wg} = \frac{h-1}{n-1} \frac{\bar{y}_{wg} - \mu}{\left[\mathrm{SSD}_{wg} / \{ n(n-1) \} \right]^{1/2}}.$$

Dixon and Tukey suggested that, provided $g \leqslant n/4$, t_{wg} can be referred to tables of the same t_{h-1} distribution as before. Yuen [11] compared the power* of t_{tg} and t_{wg}, finding little difference.

For either test the t_{h-1} approximation assumes g to be selected independently of the values of the sample at hand. Thus g should be chosen *prior* to examining the data values. If not, ref. 1 (pp. 85–86) illustrates how the (conservative) Bonferroni* adjustment can be applied to the tabulated points of t_{h-1}.

For either test, corresponding $100(1 - \alpha)\%$ confidence intervals for μ can be formed in the usual way, as the set of μ-values acceptable at level α, given the sample.

TWO-SAMPLE TESTS

Suppose we have (separately) ordered random samples from each of two distributions and, assuming equal underlying variances, are interested in difference in location. Yuen and Dixon [12] proposed a trimmed t test using

$$T_{tg} = \frac{\bar{y}_{2tg} - \bar{y}_{1tg.} - \Delta}{\left[\dfrac{SSD_{1wg} + SSD_{2wg}}{h_1 + h_2 - 2}\left(\dfrac{1}{h_1} + \dfrac{1}{h_2}\right)\right]^{1/2}},$$

where Δ is the hypothesized mean difference, $h_i = n_i - 2g_i$, and, on any subscripted quantity, the first subscript refers to the sample upon which it is based. Yuen and Dixon recommended Student's t with $h_1 + h_2 - 2$ degrees of freedom ($t_{h_1 + h_2 - 2}$) as an approximation to the null distribution of T_{tg}, although for $n_i < 7$ this may not be sufficiently accurate. They provided a table of empirical percentage points for $4 \leqslant n_1 = n_2 \leqslant 6$ and $g_1 = g_2 = 1$.

The Yuen–Dixon test was proposed originally for symmetric distributions and has been studied most extensively for them, but limited results for skewed distributions in ref. 5 seem to show that it may have some utility in those cases as well.

There is no requirement that $g_1 = g_2$, and, in fact, when $n_1 \neq n_2$ one might want to make the g_i's as nearly proportional to the n_i's as possible. One thus would be using the same central fraction of each sample in computing $\bar{y}_{1tg}$ and $\bar{y}_{2tg}$. When, however, the major concern is with contamination by relatively rare outliers, one might prefer to choose, say, $g_1 = g_2 = 1$, even when $n_1 \neq n_2$. Results described in ref. 5 suggest the test performs reasonably well for both symmetric and skewed distributions in this situation, tending to err on the conservative side.

Fung (née Yuen) and Rahman [3] proposed a similar test resistant to the effects of outliers or long tails. This Winsorized t test is based on the statistic T_{wg}, which is identical to T_{tg} except that $\bar{y}_{iwg}$ replaces $\bar{y}_{itg}$ in the numerator, $i = 1, 2$. (This statement, rather than the formula for T_{wg} on p. 339 of ref. 3, is correct [2].) As before, for $n_i \geqslant 7$ the null distribution of T_{wg} may be approximated by the $t_{h_1 + h_2 - 2}$ distribution.

Forms of T_{tg} and T_{wg} are available for use when the variances of the two distributions

may not be equal [10, 3]. Denote these respectively by T_{tg}^* and T_{wg}^*. Their use was suggested by Welch's approximation to the null distribution of the classical t when modified to have the unpooled-variances form of the standard error [9]. The starred test statistics differ from the unstarred only in that in each case the denominator is replaced by

$$\sqrt{\frac{s_{1wg}^2}{h_1} + \frac{s_{2wg}^2}{h_2}},$$

where $s_{iwg}^2 = SSD_{iwg}/(h_i - 1)$, $i = 1, 2$. The null distribution of either T_{tg}^* or T_{wg}^* may be approximated as t_f, where

$$\frac{1}{f} = \frac{c^2}{h_1 - 1} + \frac{(1-c)^2}{h_2 - 1},$$

$$c = \frac{s_{1wg}^2/h_1}{s_{1wg}^2/h_1 + s_{2wg}^2/h_2}.$$

It is not fully clear how to test equality of Winsorized variances, but it may not be necessary. With respect to the classical t, there is evidence that where variance heterogeneity is possible, it is better to dispense with a test on variances and use Welch's test* unconditionally [4]. One might expect similar results here.

Fung and Rahman [3] compared the power of T_{tg} and T_{wg} for uncontaminated and contaminated normal distributions. They also compared T_{tg}^* and T_{wg}^* under similar conditions but with unequal underlying variances. They found very little difference between the trimmed and Winsorized tests in either case.

The t approximations to the null distributions of all the preceding statistics assume the g_i to be selected independently of the values of the samples at hand. Thus the g_i should be chosen *prior* to examining the data values. If not, the Bonferroni adjustment can be applied, using the cardinality of the set of candidate pairs (g_1, g_2).

For any of the preceding tests, corresponding $100(1 - \alpha)\%$ confidence intervals for Δ can be formed in the usual way, as the set of Δ values acceptable at level α, given the observations.

CENSORING

An additional application for the two-sample case concerns censoring. If only the largest and/or smallest several observations are censored in either or both samples, the g_i can be chosen to enable the use of any of these procedures (so long as the choice is made independently of the uncensored values actually observed). This application is relevant to the one-sample case as well, except that in dealing with observations which are differences, it seems unlikely that censoring would be triggered by extreme values for the differences.

k-SAMPLE TESTS

Tests using trimmed means have been proposed for the multisample location problem, both with [7] and without [6] the requirement of equal underlying variances.

References

[1] Dixon, W. J. and Tukey, J. W. (1968). *Technometrics*, **10**, 83–98. (Proposes the t_{wg} test and studies its critical values under normality.)

[2] Fung, K. Y. Personal communication.

[3] Fung, K. Y. and Rahman, S. M. (1980). *Commun. Statist. B*, **9**, 337–347. (Propose and evaluate T_{wg}, T_{wg}^* tests and compare them to $T_{tg} T_{tg}^*$ tests, respectively.)

[4] Gans, D. J. (1981). *Commun. Statist. B*, **10**, 163–174. (Investigates use of Welch's test conditionally and unconditionally.)

[5] Gans, D. J. (1984). *J. Statist. Comput. Simul.*, **19**, 1–21. (As part of broader study, obtains robustness results for the T_{tg} test.)

[6] Lee, H. and Fung, K. Y. (1983). *J. Statist. Comput. Simul.*, **18**, 125–143. (Proposes and evaluates k-sample location tests not requiring equal variances.)

[7] Lee, H. and Fung, K. Y. (1985). *Sankhyā B*, **47**, 186–201. (Proposes and evaluates k-sample location tests assuming equal variances.)

[8] Tukey, J. W. and McLaughlin, D. H. (1963). *Sankhyā A*, **25**, 331–352. (Propose the t_{tg} test after outlining the investigation leading to it.)

[9] Welch, B. L. (1949). *Biometrika*, **36**, 293–296. (Gives a formula for df approximation to the null distribution of Student's t with unpooled variances.)

[10] Yuen, K. K. (1974). *Biometrika*, **61**, 165–170. (Proposes and evaluates T_{tg}^* test.)

[11] Yuen, K. K. (1975). *Canad. J. Statist.*, **3**, 71–80. (Compares power of T_{tg}, T_{wg}, and Student's t tests.)

[12] Yuen, K. K. and Dixon, W. J. (1973). *Biometrika*, **60**, 369–374. (Propose and evaluate T_{tg} test.)

(CENSORED DATA
OUTLIERS
ROBUSTIFICATION
TRIMMING AND WINSORIZATION
WELCH TESTS)

DANIEL J. GANS

TRIMMING AND WINSORIZATION

Trimming and *Winsorization* refer, respectively, to the removal and to the modification of the extreme values of a sample. For example, to symmetrically trim a univariate sample of size N, one removes the k smallest and k largest order statistics for some specific $k < N/2$. The sample is symmetrically Winsorized by setting the k smallest order statistics* equal to the $(k + 1)$th order statistic* $X_{k+1:N}$ and setting the k largest order statistics equal to $X_{N-k:N}$. Standard statistics can then be calculated from the trimmed or Winsorized sample. Many classical statistics such as the sample mean and variance are extremely sensitive to outliers; the purpose of the trimming or Winsorization is to reduce this sensitivity.

Methods of outlier* detection can locate discordant data for possible removal. Trimming and Winsorization differ from outlier rejection rules in that a fixed proportion of the sample is deleted or changed regardless of the values of the observations. An exception is adaptive trimming which determines k from a characteristic of the sample such as the kurtosis. The trimmed or Winsorized observations are not necessarily outliers or contaminants and they are not labeled as such. The purpose of their removal or modification is solely to improve statistical efficiency.

Trimming and Winsorization are both used to robustify statistical inferences. Hoaglin et al. [10] provide a comprehensive introduction to robustness; Huber [16] gives a more advanced treatment. *See also* ROBUSTIFICATION.

Winsorize is a term introduced by Dixon [4], who attributes the idea to Charles P. Winsor. Dixon was concerned particularly with the possibility that extreme values are poorly determined or unknown to the statistician.

The trimmed mean has a long history. Huber [14] quotes an anonymous author from 1821, who explained that in certain provinces of France the mean yield of land was estimated by averaging the middle 18 yields from 20 consecutive years. In a stimulating article, Stigler [22] describes early work by mathematicians on the problem of outliers and robust estimation*. Stigler gives an interesting account of a long neglected paper by Daniell [3], who describes the asymptotic properties of the trimmed mean and other weighted averages of order statistics.

UNIVARIATE SAMPLES

For univariate samples the trimmed mean is by far the most widely studied trimmed or Winsorized statistic. The Winsorized variance and, to a lesser extent, the trimmed variance also appear in the literature.

The *α-trimmed mean* for $\alpha = k/N$ is

$$\bar{x}_\alpha = \left(\sum_{i=k+1}^{N-k} X_{i:N} \right) \Big/ (N - 2k),$$

where N is the sample size. If one specifies α such that αN is noninteger, then some cases can be partially trimmed so that the fractions trimmed above and below are each α. Specifically, let $k = [\alpha N] + 1$, where $[\cdot]$ is the greatest integer function. Then

$$\bar{x}_\alpha = \Big\{ (k - \alpha N)(X_{k:N} + X_{N=k+1:N})$$

$$+ \sum_{i=k+1}^{N-k} X_{i:N} \Big\} \Big/ \{ N(1 - 2\alpha) \}.$$

If the sample comes from a continuous distribution F, then $\bar{x}_\alpha$ consistently estimates the population trimmed mean

$$\mu_\alpha = \int_{F^{-1}(\alpha)}^{F^{-1}(1-\alpha)} x \, dF(x) / (1 - 2\alpha).$$

If F is symmetric about μ, then $\mu_\alpha = \mu$ for all α and $\bar{x}_\alpha$ is an unbiased estimator of μ.

Two important characteristics of a robust estimator are the breakdown point and the influence function*, introduced by Hampel ([8, and 9], respectively). The *breakdown point* is the fraction of contaminants that the estimator can tolerate before the bias caused by the contamination becomes arbitrarily large. The breakdown point of $\bar{x}_\alpha$ is α. To understand the meaning of the breakdown point, suppose that a certain fraction $(1 - \gamma)$ of the sample comes from the target population, but that the remaining fraction γ are contaminants. If a contaminant is similar to the observations from the target population, then it may not be trimmed, but on the other hand it will not grossly bias the estimate. If $\gamma \leqslant \alpha$, then all contaminants which are sufficiently far removed from the sample of the target population will be trimmed. Thus a disastrously large bias can only occur if $\gamma > \alpha$.

The *influence function* at (x, F), IF(x, F), is the sensitivity of the estimator to a contaminant with value x when the sample comes from the distribution F. An estimator has bounded-influence if IF is a bounded function. The IF of $\bar{x}_\alpha$ is given by

$$(1 - 2\alpha)\text{IF}(x, F)$$

$$= \begin{cases} F^{-1}(\alpha) - \mu, & x < F^{-1}(\alpha); \\ x - \mu, & F^{-1}(\alpha) < x < F^{-1}(1 - \alpha); \\ F^{-1}(1 - \alpha) - \mu, & x > F^{-1}(1 - \alpha); \end{cases}$$

when F is symmetric about μ, so $\bar{x}_\alpha$ has bounded influence.

Under some reasonable regularity conditions, the large sample variance of an estimator is

$$N^{-1} \int (\text{IF}(x, F))^2 \, dF(x),$$

which for the trimmed mean is the *popula-*

tion Winsorized variance

$$\sigma_\alpha^2 = \left\{ \alpha \left(F^{-1}(\alpha) - \mu \right)^2 \right.$$

$$+ \alpha \left(F^{-1}(1 - \alpha) - \mu \right)^2$$

$$\left. + \int_{F^{-1}(\alpha)}^{F^{-1}(1-\alpha)} (x - \mu)^2 \, dF(x) \right\}$$

$$\times \left\{ N^{-1}(1 - 2\alpha)^{-2} \right\}.$$

To estimate σ_α^2, one uses the *sample Winsorized variance*

$$s_\alpha^2 = \frac{k \left(x_{k:N} - \hat{\mu} \right)^2 + k \left(x_{N-k+1:N} - \hat{\mu} \right)^2 + \sum_{i=k+1}^{N-k} \left(x_{i:N} - \hat{\mu} \right)^2}{N(1 - 2\alpha)^2}$$

where $\hat{\mu}$ equals either $\bar{x}_\alpha$ or the Winsorized sample mean. The advantage of the Winsorized variance as compared to the trimmed variance is that it estimates the sampling variability of the *trimmed* mean.

An interesting feature of the influence function of $\bar{x}_\alpha$ is that IF(x, F) is *not* zero if $x < F^{-1}(\alpha)$ or $x > F^{-1}(1 - \alpha)$, so that the trimmed observations do have influence. This was unexpected and, in fact, the original rationale behind using the Winsorized mean instead of the trimmed mean was that extreme values should have some influence. An intuitive way of seeing that extreme values have influence on $\bar{x}_\alpha$ is to consider the effect of replacing one observation at random by an extremely large value. This contaminant will certainly be trimmed, but the probability that the case it replaces was among the original upper-trimmed observations is only α. Thus with probability $(1 - \alpha)$, $\bar{x}_\alpha$ will increase. The influence function of the Winsorized mean is similar to that of $\bar{x}_\alpha$ but has jumps at $F^{-1}(\alpha)$ and $F^{-1}(1 - \alpha)$. Perhaps the comparison of these influence functions is the reason that $\bar{x}_\alpha$ is now widely preferred to the Winsorized mean.

The trimmed mean has good efficiency when sampling from a normal distribution, where the untrimmed mean is optimal. For $N = 10$, the efficiency of the sample mean relative to $\bar{x}_\alpha$ is only 1.009, 1.048, and 1.148, respectively, for $\alpha = 0.05, 0.10,$ and 0.25

(Andrews et al. [1, exhibit 5-6]). However, the Winsorized and trimmed variances are not particularly efficient for the normal distribution (Sarhan and Greenberg [21] and Dixon [4]).

Several proposals have been made for using the data to choose α. Jaeckel [17] suggests calculating s_α^2 for all α in some fixed interval $\alpha_0 \leqslant \alpha \leqslant \alpha_1$. The α minimizing s_α^2 is then used. Asymptotically, this method chooses the trimmed mean with the smallest variance (see Theorem 1 of Jaeckel [17]).

Hogg [12] suggests choosing α from some fixed finite set, e.g., $\frac{1}{8}, \frac{1}{4}, \frac{3}{8}, \frac{1}{2}$, depending on the value of

$$Q = \left[\bar{U}(\beta_1) - \bar{L}(\beta_1) \right] / \left[\bar{U}(\beta_2) - \bar{L}(\beta_2) \right].$$

Here $\bar{U}(\beta)$ and $\bar{L}(\beta)$ are the averages of the $[N\beta]$ largest (respectively smallest) order statistics and $0 < \beta_1 < \beta_2 \leqslant \frac{1}{2}$. ($\beta_1 = 0.05$ and $\beta_2 = 0.5$ are mentioned by Hogg.) Q measures the "heaviness" of the sample's tails. The larger the value of Q, the larger α should be. Hogg [11] had suggested using the sample kurtosis* to measure the length of the tails, but in ref. 12 he states that Q is a better indicator. Hogg [12] also mentions the use of outer means for small values of Q corresponding to very light-tailed distributions. The α-*outer mean* is the average of the $[\alpha N]$ largest and smallest order statistics, precisely the cases deleted from the trimmed mean. For a further discussion of Hogg's estimators, see Prescott [19].

Tukey and McLaughlin [24] were the first to suggest the ratio of $\bar{x}_\alpha$ to s_α as a "*t* statistic" for testing and interval estimation. Gross [7] used simulation to estimate the 2.5 percentage points of the pivotal statistic $\sqrt{N}(\bar{x}_\alpha - \mu)/s_\alpha$ for $N = 10$ and 20 and $\alpha = 0.1, 0.25, 0.35$ across a spectrum of symmetric distributions. The large sample distribution of $\sqrt{N}(\bar{x}_\alpha - \mu)/s_\alpha$ is standard normal, so that an approximate $(1 - \gamma)$ confidence interval* for μ is

$$\bar{x}_\alpha \pm s_\alpha Z_{\gamma/2} / \sqrt{N},$$

where $Z_{\gamma/2}$ is the $100(\gamma/2)$ upper percentile of the standard normal distribution. The accuracy of this stated coverage probability for small N is suspect. The focus of Gross's work is to improve the accuracy by replacing $Z_{\gamma/2}$ with another constant depending on N, α, and γ (which are known) but not the unknown population distribution. Gross concentrated on the common choice of $\gamma = 0.05$. He found that for fixed N and α, these 2.5 percentage points for the pivot were rather constant across the distributions he studied. By replacing $Z_{0.025}$ by the maximum 2.5 percentage point among these distributions, one obtains a conservative but reasonably efficient 95 percent confidence interval. It is conservative, since the coverage probability is at least 95% for any of the distributions under study. Presumably these distributions are sufficiently representative that the coverage probability will not drop much below 95% for any symmetric distribution. At a specific distribution, a large loss of efficiency could occur if the 2.5 percentage point for that distribution were well below the maximum 2.5 percentage point; this circumstance, however, did not occur in Gross's study.

$\bar{x}_\alpha$ has the same influence function as Huber's [13] location M-estimator*. However, the M-estimator has a breakdown point of 0.5 and in this regard it is superior to $\bar{x}_\alpha$. Trimming has the advantage of simplicity and ease of computation.

MULTIVARIATE LOCATION AND DISPERSION

The simplicity of trimming in univariate situations does not carry over into multivariate analysis. Nonetheless, trimming methods have been proposed for multivariate data.

Gnanadesikan and Kettenring [6] give an ingenious method for estimating bivariate correlations using the identity $\text{cov}(X, Y) = \{\text{var}(X + Y) - \text{var}(X - Y)\}/4$. Let $S_*(X)$ be a robust estimate of the standard deviation of $X_1, \ldots, X_N$, for example, the trimmed or Winsorized standard deviation. Let

$U_i = X_i/S_*(X)$ and $V_i = Y_i/S_*(Y)$ be the standardized coordinates of a bivariate sample (X_i, Y_i), $i = 1, \ldots, N$. They suggest

$$\rho_{XY}^* = \frac{S_*^2(U + V) + S_*^2(U - V)}{S_*^2(U + V) + S_*^2(U - V)}$$

and $\sigma_{XY}^* = \rho_{XY}^* S_*(X) S_*(Y)$ as robust estimates of the correlation and covariance between X and Y.

If a sample of N observations from $\mathbf{X} = (X_1, \ldots, X_p)$ is at hand, one can estimate its mean vector and covariance matrix robustly with univariate location and scale estimates separately for each coordinate and a bivariate covariance estimate for each pair X_i, X_j, $i \neq j$. However, this procedure is not invariant to affine reparametrization and the estimated covariance need not be positive semidefinite.

An alternative is to identify and trim multivariate extreme values and then compute the mean and covariance of the remaining cases. Gnanadesikan and Kettenring [6] suggest trimming the k cases with the largest values of $d_i^2 = (\mathbf{X}_i - \hat{\mu})'\hat{\Sigma}^{-1}(\mathbf{X}_i - \hat{\mu})$, where $\hat{\mu}$ and $\hat{\Sigma}$ are the mean and covariance of the untrimmed observations; $\hat{\mu}$, $\hat{\Sigma}$, and $d_1, \ldots, d_N$ are calculated iteratively, starting with the untrimmed mean and variance. *Peeling** is the removal of extreme points of the convex hull of the data, a process which can be repeated. Titterington [23] discusses a recursion for calculating the ellipsoid of minimal content covering the sample and suggests trimming points on its surface.

Depth trimming is described by Donoho and Huber [5]. Let D_k be the intersection of p-dimensional half-spaces containing at least $N + 1 - k$ observations. A point in D_k but not D_{k+1} has depth k. Let k^* be the maximal depth of the sample. The mean of all points of depth k^* has a particularly high breakdown point, approaching $\frac{1}{3}$ for centrosymmetric distributions. In contrast, the breakdown point of the peeled mean and of multivariate M-estimators* is at most $(p + 1)^{-1}$. See Donoho and Huber [5] for further discussion. Therefore, based on our very limited understanding of the various methods of multivariate trimming, depth trimming seems preferable.

REGRESSION

Ruppert and Carroll [20] investigate two methods of trimming a regression sample before applying least-squares* analysis. The first was suggested by Koenker and Bassett [18] and the second uses residuals from a preliminary estimate.

Let

$$y_i = \mathbf{X}_i'\beta + \epsilon_i$$

be the regression model and suppose ϵ_i has a symmetric distribution F. For $0 < \theta < 1$ define $\rho_\theta(x) = x(\theta - 1)$ if $x < 0$ and $\rho_\theta(x) = x\theta$ if $x \geq 0$. The θth regression quantile* is defined by Koenker and Bassett [18] as any solution to

$$\sum_{i=1}^{N} \rho_\theta(y_i - \mathbf{X}_i'\mathbf{b}_\theta) = \min!$$

If $r_{i\theta} = y_i - \mathbf{x}_i'\mathbf{b}_\theta$ are the residuals* from $\mathbf{b}_\theta$, then for large samples $\{i: r_{i\theta} < 0\}$ is approximately the same set as $\{i: \epsilon_i < F^{-1}(\alpha)\}$. Koenker and Bassett [18] suggest trimming any observation whose residual $r_{i\alpha}$ is negative or whose residual $r_{i(1-\alpha)}$ is positive. Ruppert and Carroll [20] prove that the large-sample covariance matrix of the trimmed least-squares estimator is $(\sum_{i=1}^{N}\mathbf{X}_i\mathbf{X}_i')^{-1}$ times the Winsorized variance of F. The covariance of the least-squares estimator and the large-sample covariance of Huber's [15] M-estimators* are also proportional to $(\sum_{i=1}^{N}\mathbf{X}_i\mathbf{X}_i')^{-1}$.

If trimming is based on residuals from a preliminary estimator, then the trimmed least-squares estimator does not have a covariance matrix with this pattern except if the preliminary estimate is $(\mathbf{b}_\alpha + \mathbf{b}_{1-\alpha})/2$.

Chen and Dixon [2] proposed two trimmed (or Winsorized) regression estimators which are applicable only if each $\mathbf{X}_i$ value is replicated a fixed number of times, say M. The stratified procedure requires $M \geq 3$. For each $\mathbf{x}_i$, let $\bar{y}_i$ be the trimmed (or Winsorized) mean (for some α) of the M

responses at x_i. The stratified estimator is the least-squares* estimator based on $(\bar{y}_1, \mathbf{X}_1), \ldots, (\bar{y}_N, \mathbf{X}_N)$. For all N and $M \geq 3$, its covariance matrix is exactly $(\sum_{i=1}^{N}\mathbf{X}_i\mathbf{X}_i')^{-1}$ times the variance of the trimmed (or Winsorized) mean in a sample of size M. Except when M is unusually large, the possible values of α are limited unless one uses partial trimming as already described.

The pooled estimator of Chen and Dixon is more flexible, but its variance matrix is apparently unknown even asymptotically. For each $\mathbf{X}_j$, let $r_{1j}, \ldots, r_{Mj}$ be the residuals of the corresponding y values from their mean. Let $r_{(1)} \leq \cdots \leq r_{(MN)}$ be the ordered values of the pooled residuals. To trim the sample, one removes all cases corresponding to $r_{(1)}, \ldots, r_{(g)}$ or $r_{(MN-g+1)}, \ldots, r_{(MN)}$ for $g = [\alpha MN]$. To Winsorize the sample, one replaces y_{ij} by $r_{(g+1)}$ if $r_{ij} \leq r_{(g)}$ and replaces y_{ij} by $r_{(MN-g)}$ if $r_{ij} \geq r_{(MN-g+1)}$. The least-squares estimator is then computed from the trimmed or Winsorized sample. Chen and Dixon [2] present a detailed Monte Carlo* study of the stratified and pooled estimators.

ROLE OF TRIMMING AND WINSORIZATION IN STATISTICAL ANALYSIS

The automatic removal or modification of extreme values should be contrasted with outlier tests and diagnostics which highlight observations for further study. Subsequent investigation may suggest that these extreme values should be removed because they are simply "bad" data or, for example, in regression with an aberrant x value, because they are outside the scope of the study. On the other hand, the conclusion may be reached that the extreme values should be unchanged but that the statistical model is inadequate and needs to be modified. It is clear that in many cases outlier detection can lead to more insightful data analysis than automatic trimming or Winsorization.

Nonetheless, trimming and Winsorization have several important roles. They can be used for the rapid processing of data such as in automatic control, where human inspection of the data is impossible. Also, the comparison of a trimmed, Winsorized, or otherwise robustified statistic with the same statistic from the original sample can detect the presence of outliers, and these outliers can be identified using the residuals from a robust estimator. Finally, there are populations that have heavier tails than the Gaussian distribution. Outliers will then occur relatively often but are simply part of the usual sampling variability, and trimming or Winsorization can lead to considerable increase in the efficiency of estimators and tests.

References

[1] Andrews, D. F., Bickel, P. J., Hampel, F. R., Huber, P. J., Rogers, W. H., and Tukey, J. W. (1972). *Robust Estimates of Location*. Princeton University Press, Princeton, N.J.

[2] Chen, E. H. and Dixon, W. J. (1972). *J. Amer. Statist. Ass.*, **67**, 664–671.

[3] Daniell, P. J. (1920). *Amer. J. Math.*, **42**, 222–236.

[4] Dixon, W. J. (1960). *Ann. Math. Statist.*, **31**, 385–391.

[5] Donoho, D. L. and Huber, P. J. (1983). In *A Festschrift for Erich Lehmann*, P. J. Bickel, K. A. Doksum, and J. L. Hodges, eds. Wadsworth, Belmont, CA, pp. 157–184. (Interesting discussion of the breakdown point.)

[6] Gnanadesikan, R. and Kettenring, J. R. (1972). *Biometrics*, **28**, 81–124.

[7] Gross, A. M. (1976). *J. Amer. Statist. Ass.*, **71**, 409–416.

[8] Hampel, F. R. (1971). *Ann. Math. Statist.*, **42**, 1887–1896.

[9] Hampel, F. R. (1974). *J. Amer. Statist. Ass.*, **69**, 383–393. (Good discussion of the influence function and its importance in characterizing the behavior of estimators.)

[10] Hoaglin, D. C., Mosteller, F., and Tukey, J. W. (1983). *Understanding Robust Exploratory Data Analysis*. Wiley, New York. (Good introduction to robust statistics.)

[11] Hogg, R. V. (1967). *J. Amer. Statist. Ass.*, **62**, 1179–1186.

[12] Hogg, R. V. (1974). *J. Amer. Statist. Ass.*, **69**, 909–927.

[13] Huber, P. J. (1964). *Ann. Math. Statist.*, **35**, 73–101.

[14] Huber, P. J. (1972). *Ann. Math. Statist.*, **43**, 1041–1067.

[15] Huber, P. J. (1973). *Ann. Statist.*, **5**, 799–821.

[16] Huber, P. J. (1981). *Robust Statistics*. Wiley, New York. (Introduction to robustness at an advanced mathematical level. Covers many properties of trimmed and Winsorized statistics.)

[17] Jaeckel, L. A. (1971). *Ann. Math. Statist.*, **42**, 1540–1552.

[18] Koenker, R. and Bassett, G., Jr. (1978). *Econometrica*, **46**, 33–50.

[19] Prescott, P. (1978). *J. Amer. Statist. Ass.*, **73**, 133–140.

[20] Ruppert, D. and Carroll, R. J. (1980). *J. Amer. Statist. Ass.*, **75**, 828–838.

[21] Sarhan, A. E. and Greenberg, B. G. (1956). *Ann. Math. Statist.*, **27**, 427–470.

[22] Stigler, S. M. (1973). *J. Amer. Statist. Ass.*, **68**, 872–879.

[23] Titterington, D. M. (1978). *Appl. Statist.*, **27**, 227–234.

[24] Tukey, J. W. and McLaughlin, D. H. (1963). *Sankhyā A*, **25**, 331–352. (Of historical interest.)

(CENSORING
L-STATISTICS
ORDER STATISTICS
OUTLIERS
PEELING
REGRESSION QUANTILES
RESIDUALS
ROBUST ESTIMATION
ROBUSTIFICATION
ROBUST STATISTICS)

DAVID RUPPERT

TRINOMIAL DISTRIBUTION

A multinomial distribution* with three cells. Although the distribution is conveniently expressed as

$$\Pr[N_1 = n_1, N_2 = n_2, N_3 = n_3]$$
$$= \frac{N!}{n_1! n_2! n_3!} p_1^{n_1} p_2^{n_2} p_3^{n_3},$$
$$n_1 + n_2 + n_3 = N, \; p_i \geqslant 0,$$
$$p_1 + p_2 + p_3 = 1,$$

it is really a *bivariate* distribution because of the condition on the n's.

(MULTINOMIAL DISTRIBUTIONS)

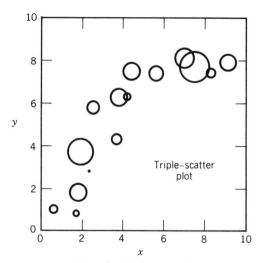

Figure 1 Triple scatter plot.

TRIPLE SCATTER PLOT

A term suggested by Anscombe [1] for graphical representation of three real variables (x_i, y_i, z_i). The x and y values are portrayed by the centers of the circles and form an ordinary Cartesian graph. The value of the third variable is represented by the areas of the circles. Figure 1 is taken from ref. 2 (p. 536).

References

[1] Anscombe, F. J. (1973). *Amer. Statist.*, **27**, 17–21.
[2] Cleveland, W. S. and McGill, R. (1984). *J. Amer. Statist. Ass.*, **79**, 531–554.

(GRAPHICAL REPRESENTATION OF DATA
MULTIVARIATE GRAPHICS
SCATTER PLOTS)

TRIPOTENT MATRIX

A square matrix **A** is *tripotent* if $A^3 = A$. The eigenvalues of any tripotent matrix are -1, 0, and 1. Anderson and Styan [2] have extended Cochran's theorem* for quadratic forms* to tripotent matrices. Anderson and Fang [1] have studied the distribution of quadratic forms with tripotent matrices of normal and elliptically distributed random variables.

References

[1] Anderson, T. W. and Fang, K. T. (1982). Distribution of Quadratic Forms and Cochran's Theorem for Normal and Elliptically Contoured Distributions and their Applications. *Tech. Rep. No. 53*, Dept. Statistics, Stanford University, Stanford, CA.
[2] Anderson, T. W. and Styan, G. P. H. (1982). *Stat. Prob. CRR*, 1–23. [Original version: *Tech. Rep. No. 43*, Dept. Statistics, Stanford University, Stanford, CA (1980).]

(IDEMPOTENT MATRICES
QUADRATIC FORMS)

TROJAN SQUARE

This is really a Graeco-Latin* (GL) or hyper-Graeco-Latin (HGL) square, of dimension $r \times r$, with g "treatment factors" which are, in fact, formed by a subdivided set of rg true factors into g groups of r treatments each. The analysis follows that appropriate to GL or HGL squares. Just as in these cases, it has to be supposed that the combined effect of treatments applied to the same plot is the sum of the appropriate treatment effects—that is, the effect is *additive*, with no interaction*.

Furthermore, the tables of estimated values of effects should be amalgamated into a single table comparing values for all rg treatments.

Darby and Gilbert [1] gave the name "Trojan" to indicate that the design could be employed in a wide variety of situations.

Reference

[1] Darby, L. A. and Gilbert, N. (1958). *Euphytica*, **7**, 183–188.

(ANALYSIS OF VARIANCE
GRAECO-LATIN SQUARES
LATIN SQUARES, LATIN CUBES, LATIN
  RECTANGLES, ETC.)

"TRUE" PRIOR

A term used in Bayesian* analysis to signify the prior distribution that would be used for the whole parameter space (Ω) if it were known exactly. (Usually Ω is really too complex for the statistician to be able to specify the prior distribution completely.)

Detailed discussions of this concept and related philosophical issues are contained in the refs. 1 and 2.

References

[1] Diaconis, P. and Ylvisaker, D. (1983). Quantifying Prior Opinion. *Tech. Rep. No. 207*, Dept. Statistics, Stanford University, Stanford, CA.

[2] Krasker, W. (1984). *Ann. Statist.*, **12**, 751–757.

(BAYESIAN INFERENCE
PRIOR DISTRIBUTION
PRIOR PROBABILITIES)

TRUNCATED DATA

Data are called *truncated* on the right/left if sample values larger/smaller than a fixed constant are not recorded or not observed.

(CENSORED DATA
CENSORING
TRUNCATION, COEFFICIENT OF)

Jean Dickinson Gibbons

TRUNCATED SAMPLING *See* CURTAILED SAMPLING

TRUNCATION, COEFFICIENT OF

The *coefficient of truncation* is a descriptive statistic for a sample of size n that is a function of its median M, first order statistic* $X_{1:n}$, and nth order statistic $X_{n:n}$. The coefficient is defined as

$$\text{TRUN} = \left| \frac{2M - X_{n:n} - X_{1:n}}{X_{n:n} - X_{1:n}} \right|$$

and is used for exploratory data analysis* in Gibbons and Stavig [1]. This coefficient ranges between 0 and 1. The extreme value of 0 is attained when the sample median is equal to the midpoint $(X_{1:n} + X_{n:n})/2$ and the extreme value of 1 is attained when the sample median is equal to either $X_{1:n}$ or $X_{n:n}$. Thus the value is close to 1 when an unusually large proportion of the observations are at one extreme and within a narrow range. This would generally occur for a set of censored* data or truncated* data. A truncated distribution may be an appropriate model for data with a large coefficient of truncation.

Reference

[1] Gibbons, J. D. and Stavig, G. R. (1980). In *Sociological Methodology*, K. F. Schuessler, ed. Jossey-Bass, San Francisco, CA, pp. 545–558.

(CENSORED DATA
CENSORING
TRUNCATED DATA)

Jean Dickinson Gibbons

TRUNCATION METHODS IN PROBABILITY THEORY

For given finite numbers $a < b$, define the function

$$I(x; a, b) = \begin{cases} 1, & \text{if } a < x \leqslant b, \\ 0, & \text{otherwise.} \end{cases}$$

The following two truncations of the random variable X are frequently applied in probability theory:

$$Y(a, b) = XI(X; a, b) + aI(X; -\infty, a) + bI(X; b, +\infty), \qquad (1)$$

$$Z(a, b) = XI(X; a, b). \qquad (2)$$

((1) is also known as Winsorization*.) Both truncations approximate X through bounded random variables, and thus the truncated versions have moments* of all (positive) orders. This leads to a powerful method of proof of limit theorems* of probability theory, called the method of truncation, which consists of two steps. First, one proves a limit theorem for bounded random variables and then shows that the same statement holds for unbounded random variables X by approximating X by either $Y(a,b)$ or $Z(a,b)$, where a and b are appropriate constants.

For illustration, let us look at the following familiar proof of the weak law of large numbers. (*See* CONVERGENCE OF SEQUENCES OF RANDOM VARIABLES.) It states that if $X_1, X_2, \ldots, X_n$ are independent copies of the random variable X whose expectation E is finite, then for arbitrary $\epsilon > 0$,

$$\lim_{n=+\infty} P\left(\left|\frac{X_1 + X_2 + \cdots + X_n}{n} - E\right| \geqslant \epsilon\right)$$
$$= 0. \qquad (3)$$

With the additional assumption of finite variance for X, the proof of (3) is immediate from the Chebyshev inequality*. However, since $V(X)$ is not known to be finite, one can proceed as follows. We first prove a limit similar to (3) for the sequence $Z_i = Z_i(-cn, cn)$, where $c > 0$ is an arbitrary fixed number and Z_i is the truncation (2) of X_i. Clearly, the random variables Z_i are independent and identically distributed. Furthermore, the Z_i are bounded and, thus, $E^* = E[Z_i]$ and the variance $V(Z_i)$ are finite and satisfy the relations

$$\lim E[Z_i] = E, \qquad n \to +\infty, \qquad (4)$$

and

$$V(Z_i) \leqslant E[Z_i^2]$$
$$= \int_{-cn}^{cn} x^2 \, dF(x)$$
$$\leqslant cn \int_{-cn}^{cn} |x| \, dF(x),$$

where $F(x)$ is the distribution function of X; that is,

$$V(Z_i) \leqslant cnE[|X|] = c\mu n. \qquad (5)$$

Therefore, by the Chebyshev inequality,

$$P\left(\left|\frac{Z_1 + Z_2 + \cdots + Z_n}{n} - E^*\right| \geqslant \frac{1}{2}\epsilon\right)$$
$$\leqslant \frac{4c\mu}{\epsilon^2 n}$$

and, thus, since because of (4), $|E - E^*| < \frac{1}{2}\epsilon$ for n sufficiently large,

$$P\left(\left|\frac{Z_1 + Z_2 + \cdots + Z_n}{n} - E\right| \geqslant \epsilon\right)$$
$$\leqslant \frac{4c\mu}{\epsilon^2 n}. \qquad (6)$$

We now try to replace the Z_i by the X_i in (6). For this aim, we observe that

$$P(Z_i \neq X_i) = \int_{|x| > cn} dF(x)$$
$$\leqslant \frac{1}{cn} \int_{|x| > cn} |x| \, dF(x)$$
$$< \frac{c}{n}$$

for n sufficiently large, the last estimate being valid in view of the finiteness of E. We thus have

$$P\left(\sum_{i=1}^{n} X_i \neq \sum_{i=1}^{n} Z_i\right)$$
$$\leqslant \sum_{i=1}^{n} P(X_i \neq Z_i) < c. \qquad (7)$$

Combining (6) and (7), we obtain that for all sufficiently large n,

$$P\left(\left|\frac{X_1 + X_2 + \cdots + X_n}{n} - E\right| \geqslant \epsilon\right)$$
$$\leqslant P\left(\left|\frac{Z_1 + Z_2 + \cdots + Z_n}{n} - E\right| \geqslant \epsilon\right)$$
$$+ P\left(\sum_{i=1}^{n} X_i \neq \sum_{i=1}^{n} Z_i\right)$$
$$\leqslant \frac{4c\mu}{\epsilon^2 n} + c.$$

Because $c > 0$ is arbitrary, this last estimate implies (3).

In other instances, one can use the Borel–Cantelli lemma* to justify the approximation of X by one of its truncations. Indeed, if $X_1, X_2, \ldots$ are random variables and $a_i < b_i$ are real numbers such that

$$\sum_{i=1}^{+\infty} \left[P(X_i \leqslant a_i) + P(X_i > b_i) \right] < +\infty,$$

then the Borel–Cantelli lemma yields that with probability 1, $X_i = Y_i(a_i, b_i) = Z_i(a_i, b_i)$ for all but a finite number of values of i. Thus, with proper choices of a_i and b_i, one immediately has that the normalized sums

$$\left(\sum_{i=1}^{n} X_i \right) \Big/ B_n, \quad \left(\sum_{i=1}^{n} Z_i(a_i, b_i) \right) \Big/ B_n$$

and

$$\left(\sum_{i=1}^{n} Y_i(a_i, b_i) \right) \Big/ B_n$$

all have the same asymptotic properties whenever $B_n > 0$ and $B_n \to +\infty$ with n.

There are limit theorems of probability theory in which truncations are unavoidable. As a matter of fact, theorems such as Kolmogorov's three series theorem*, the two series theorem*, Lindeberg's form of the central limit theorem*, and others give necessary and sufficient conditions for their conclusions in terms of some truncations of the random variables involved. For example, the Lindeberg* form of the central limit theorem* states that if $X_1, X_2, \ldots, X_n$ are independent random variables with $E[X_i] = 0$ and finite variances $V(X_i)$ such that $S_n^2 = V(X_1) + V(X_2) + \cdots + V(X_n) \to +\infty$ with n, then the asymptotic distribution of $(X_1 + X_2 + \cdots + X_n)/S_n$ is standard normal if and only if the truncated variables $Z_i = Z_i(-cS_n, cS_n)$ satisfy the limit relation

$$\lim_{n \to +\infty} \frac{1}{S_n} \sum_{i=1}^{n} E\left[Z_i^2 \right] = 1$$

for arbitrary value of $c > 0$.

Although at (1) and (2) we defined two-sided truncations, these same forms, in fact, are applied when one of the constants a and b becomes infinite, i.e., either $a \to -\infty$ or $b \to +\infty$. One of the best known one-sided truncations is the positive part (or the negative part) of a random variable X which is, of course, $Y(0, +\infty) = Z(0, +\infty)$, the equation being valid with probability 1.

In the theory of order statistics*, conditional distributions can be expressed by unconditional distributions of truncated random variables (see ref. 1, p. 20). Such an observation can be utilized in characterizing distributions by properties of order statistics. For several examples, see ref. [2], particularly the discussions on pp. 21 and 51 and several theorems in Chap. 3.

References

[1] David, H. A. (1981). *Order Statistics*, 2nd ed. Wiley, New York.

[2] Galambos, J. and Kotz. S. (1978). Characterizations of Probability Distributions. *Lect. Notes Math.*, **675**. Springer Verlag, Heidelberg, Germany.

(LIMIT THEOREMS
LINDEBERG CONDITION
TRIMMING AND WINSORIZATION)

JANOS GALAMBOS

TRUNCATION, NONPARAMETRIC ESTIMATION UNDER

Two kinds of incompletely reported data need to be distinguished: *censored* and *truncated* data. In the words of Hald [6, p. 144] who was the first to use both terms systematically, truncation is "sampling an incomplete population" or, as we would prefer to say today, sampling from a conditional distribution. On the other hand censoring occurs "when we are able to sample the complete population but the individual values of observations below (or above) a given value are not specified."

The theory of right-censored data (the situation where the exact value of the random variable is unknown if it exceeds a certain value) is now elaborate, primarily motivated by life testing applications to biostatistics* and reliability*. See Kaplan and Meier [11], Cox [4], and Aalen [1] for pioneering contributions to nonparametric estimation; Kalbfleisch and Prentice [10] or Cox and Oakes [5] for survey monographs.

Table 1   Example of Calculation of the Product-Limit and Nelson–Aalen Estimators Under Left-Truncation

i	t_i	Y_i	$R(Y_i)$	Y_i^{-1}	$1 - \hat{F}(Y_i)$	$\hat{\Lambda}(Y_i)$
1	0.073	0.145	2	0.500	0.500	0.500
2	0.136	0.409	4	0.250	0.375	0.750
3	0.241	0.435	3	0.333	0.250	1.083
4	0.496	0.499	4	0.250	0.188	1.333
5	0.483	0.768	4	0.250	0.141	1.583
6	0.377	0.789	3	0.333	0.094	1.917
7	0.549	0.823	2	0.500	0.047	2.417
8	0.205	0.953	1	1.000	0.000	3.417

Left-truncation is the situation where the individual is only included if its life time exceeds a certain value. Nonparametric estimation in this situation (perhaps combined with right-censoring) was mentioned in passing by Kaplan and Meier [11]. In the simplest situation one considers an underlying random variable X with continuous distribution function F. Let $t_1, \ldots, t_n$ be truncation times (initially assumed deterministic), and consider $Y_1, \ldots, Y_n$ to be independent, with Y_i following the conditional distribution of X given $X > t_i$. At any time x one may calculate the *number at risk*

$$R(x) = \#\{t_i < x\} - \#\{Y_i \geqslant x\}.$$

The *product-limit estimator** of F is then given by

$$1 - \hat{F}(x) = \prod_{Y_i \leqslant x} \left(1 - \frac{1}{R(Y_i)}\right).$$

A similar estimator, the *Nelson–Aalen estimator**, may be derived for the integrated hazard $\Lambda(x) = -\log(1 - F(x))$:

$$\hat{\Lambda}(x) = \sum_{Y_i \leqslant x} \frac{1}{R(Y_i)}.$$

Left-truncation is equivalent to *delayed entry*, and as such the example in Table 1 should be next to self-explanatory. For early examples and hypothesis testing* theory see Hyde [7, 8] and Andersen et al. [3].

For mathematical theory, it is convenient to assume that the truncation times t_i are realizations of independent identically distributed random variables T_i. If $X_1, X_2, \ldots$ are iid random variables, independent of the T_i, and following the distribution F "of interest," then one observes n independent replications of $(T_i, X_i) | X_i > T_i$. Woodroofe [15] gave a comprehensive discussion of the exact estimation theory as well as consistency and asymptotic normality of the product-limit estimator. Woodroofe also surveyed the history and application of this theory within astronomy but did not put it into the context of survival analysis*. This was done by Wang et al. [14] and by Keiding and Gill [12] who went on to show how the exact and asymptotic properties of the estimators may be obtained as corollaries from statistical theory of counting processes and Markov processes by Aalen and Johansen [2] and Johansen [9].

More complicated patterns of truncation will usually admit no explicit solution. Still, a simple algorithm may be given for the nonparametric maximum likelihood* estimator. This was shown in the important paper by Turnbull [13], whose basic idea as far as truncation is concerned was to consider the expected number of "ghosts" corresponding to each observed value. The ghosts of an observation are those fictitious individuals with life times close to the observed one which were never observed because of the truncation; Turnbull showed how an EM-type algorithm may then be specified.

References

[1] Aalen, O. O. (1978). *Ann. Statist.*, **6**, 701–726.

[2] Aalen, O. O. and Johansen, S. (1978). *Scand. J. Statist.*, **5**, 141–150.

[3]   Andersen, P. K., Borgan, Ø., Gill, R. D., and Keiding, N. (1982). *Int. Statist. Rev.*, **50**, 219–258; Correction, **52**, 225.

[4]   Cox, D. R. (1972). *J. R. Statist. Soc. B*, **34**, 187–220.

[5]   Cox, D. R. and Oakes, D. (1984). *Analysis of Survival Data*. Chapman and Hall, London, England.

[6]   Hald, A. (1952). *Statistical Theory with Engineering Applications*. Wiley, New York.

[7]   Hyde, J. (1977). *Biometrika*, **64**, 225–230.

[8]   Hyde, J. (1980). In *Biostatistics Casebook*, R. G. Miller, Jr., B. Efron, B. W. Brown, Jr., and L. E. Moses, eds. Wiley, New York, pp. 31–46.

[9]   Johansen, S. (1978). *Scand. J. Statist.*, **5**, 195–199.

[10] Kalbfleisch, J. D. and Prentice, R. L. (1980). *The Statistical Analysis of Failure Time Data*. Wiley, New York.

[11] Kaplan, E. L. and Meier, P. (1958). *J. Amer. Statist. Ass.*, **53**, 457–481.

[12] Keiding, N. and Gill, R. D. (1987). *Research Rep. No. 87/3*, Statistical Research Unit, University of Copenhagen, Denmark. (Also issued as Rep. No. MS-R8702, Centre for Mathematics and Computer Science, Amsterdam, The Netherlands.

[13] Turnbull, B. W. (1976). *J. Roy. Statist. Soc. B*, **38**, 290–295.

[14] Wang, M.-C., Jewell, N. P., and Tsai, W.-Y. (1986). *Ann. Statist.*, **14**, 1597–1605.

[15] Woodroofe, M. (1985). *Ann. Statist.*, **13**, 163–177; Correction **15**, 883.

(CENSORED DATA
KAPLAN-MEIER ESTIMATOR
LIFE TABLES)

NIELS KEIDING

TSCHUPROW'S COEFFICIENT T *See* ASSOCIATION, MEASURES OF

t-STATISTICS, t-TESTS   *See* STUDENT'S t TESTS; t-DISTRIBUTION

TTT-PLOT; TTT-TRANSFORM   *See* TOTAL TIME ON TEST TRANSFORM

TUKEY–KRAMER INTERVALS   *See* MULTIPLE COMPARISONS

TUKEY'S CONFIDENCE INTERVAL FOR LOCATION

The Tukey procedure constructs a distribution-free* confidence interval* for the median of a symmetric population. (It may also be used to estimate the difference between treatments with paired data, as will be shown later.) Let $Z_1, \ldots, Z_n$ denote a random sample from a population that is symmetric about θ. Form the Walsh averages*

$$\frac{Z_i + Z_j}{2} \text{ for } 1 \leqslant i \leqslant j \leqslant n.$$

There will be $n(n + 1)/2$ of these averages. Order these Walsh averages from smallest to largest. Using a table of the null distribution of the Wilcoxon signed rank* statistic (for example, Table A.4 in Hollander and Wolfe [1]), we find a positive integer k satisfying

$$P_{H_0}[T^+ \geqslant k] = \alpha/2.$$

Here T^+ denotes the Wilcoxon signed rank test statistic for testing $H_0: \theta = 0$, namely,

$$T^+ = \sum_{i=1}^{n} \Psi(Z_i) R_i^+,$$

where $\Psi(t) = 1, 0$ as $t > , \leqslant 0$ and where R_i^+ denotes the rank of $|Z_i|$ among $|Z_1|, \ldots, |Z_n|$, ranking from smallest to largest. Form

$$r = n(n + 1)/2 + 1 - k.$$

The $100(1 - \alpha)\%$ Tukey confidence interval for θ is then

$$[L, U],$$

where L is the rth smallest Walsh average and U is the rth largest Walsh average. When the sample size n is large, we can approximate r via

$$r \approx \frac{n(n + 1)}{4}$$

$$- Z_{(\alpha/2)}\left[\frac{n(n + 1)(2n + 1)}{24}\right]^{1/2},$$

where $Z_{(\alpha/2)}$ is the point on a standard normal distribution with probability $\alpha/2$ above it.

To illustrate this procedure, we examine data reported by Shasby and Hagerman [5]

Table 1   Walsh Average Computations

	166	167	169	170	173	174	189
166	166						
167	166.5	167					
169	167.5	168	169				
170	168	168.5	169.5	170			
173	169.5	170	171	171.5	173		
174	170	170.5	171.5	172	173.5	174	
189	177.5	178	179	179.5	181	181.5	189

measuring the effects of early athletic training of adolescent boys. Seven males ages 12–13 underwent a 12 week conditioning period in distance running followed by a 4 month deconditioning period of relative inactivity. At the end of the deconditioning period, the heart rates (beats per minute) of these boys were measured while performing a 5 minute 6 mile per hour run on a treadmill. The seven values were: 166, 189, 170, 173, 174, 167, and 169. In Table 1 we index the columns and rows with these values arranged in ascending order. The entries within Table 1 comprise the $n(n + 1)/2 = (7)(8)/2 = 28$ Walsh averages. Using Table A.4 in Hollander and Wolfe [1], we see that for $n = 7$,

$$P_{H_0}[T^+ \geqslant 26] = 0.023.$$

Therefore, $r = 28 + 1 - 26 = 3$. The three smallest Walsh averages in Table 1 are 166, 166.5, and 167. The three largest Walsh averages are 181, 181.5, and 189. With $\alpha/2 = 0.023$, we see that $1 - \alpha = 0.954$. Therefore, the 95.4% Tukey confidence interval for the median heart rate of adolescent boys following this type of athletic training extends from the third smallest to the third largest Walsh average, namely [167, 181] beats per minute.

The Tukey procedure yields the confidence interval that corresponds to the Wilcoxon signed rank test. It consists of all the location parameter values that would not be rejected by a two-sided Wilcoxon signed rank test if that value were used as the null hypothesis. Thus the distribution-free property and asymptotic efficiencies of the Tukey

interval are analogous to those of the Wilcoxon signed rank test.

The confidence interval is also applied to paired data in order to estimate the difference between the two treatment populations. Let $(X_1, Y_1), \ldots, (X_n, Y_n)$ denote a random sample of paired replicates from a two treatment experiment. Here X_i denotes the response of the ith subject to treatment 1 and Y_i denotes the response of that same subject to treatment 2. To estimate the difference in the response levels of the two treatments, we form $Z_i = Y_i - X_i$ for $i = 1, \ldots, n$ and construct the Tukey confidence interval for the median of the Z_i responses.

The confidence interval was proposed by Tukey [6] and described by Moses in Walker and Lev [7, Chap. 18]. A graphical description of the procedure is found in Moses [3] and Hollander and Wolfe [1], among others. Noether [4] showed that the use of the closed interval $[L, U]$ makes the probability of including θ at least $1 - \alpha$, whether the symmetric underlying distribution is discrete or continuous. The interval $[L, U]$ corresponds directly to the Wilcoxon signed rank test and includes θ with probability $1 - \alpha$ for continuous symmetric distributions. Lehmann [2] demonstrated the relationship between this interval and the Wilcoxon signed rank test. He also developed asymptotic efficiencies for this interval and its use in providing a consistent estimator of $\int f^2(t) \, dt$.

References

[1]   Hollander, M. and Wolfe, D. A. (1973). *Nonparametric Statistical Methods*. Wiley, New York. (Stat-

istical methods text containing procedure description and tables for its implementation.)

[2] Lehmann, E. L. (1963). *Ann. Math. Statist.*, **34**, 1507–1512. (Develops the asymptotic properties of the confidence interval.)

[3] Moses, L. E. (1965). *Technometrics*, **7**, 257–260. (Graphical description of the procedure is presented in answer to a query.)

[4] Noether, G. E. (1967). *J. Amer. Statist. Ass.*, **62**, 184–188. (Shows the effects of discrete populations on the confidence interval coverage.)

[5] Shasby, G. B. and Hagerman, F. C. (1975). *J. Sports Med.*, **3**, 97–105. (Source of example data.)

[6] Tukey, J. W. (1949). The Simplest Signed Rank Tests. *Tech. Rep. No. 17*, Statist. Res. Group, Princeton University, Princeton, NJ. (Source of method for finding the confidence interval.)

[7] Walker, H. M. and Lev, J. (1953). *Statistical Inference*, 1st ed. Holt, Rinehart, and Winston, New York. (Chapter 18 in this text is written by L. Moses and contains a description of this confidence interval procedure.)

(CONFIDENCE INTERVALS AND REGIONS
DISTRIBUTION-FREE METHODS
WALSH AVERAGES
WILCOXON SIGNED RANK TEST)

RONALD H. RANDLES

TUKEY'S *g*- AND *h*-DISTRIBUTIONS
See G- AND H-DISTRIBUTIONS

TUKEY'S HANGING ROOTOGRAM

Given random sample values $X_1, \ldots, X_n$, a density estimator called a "histogram" is defined as

$$\tilde{f}(x; a, h_n) = \tilde{f}(x)$$

$$= \tfrac{1}{2} \left\{ \hat{F}_n \left(a + (k+1)h_n \right) - \hat{F}_n \left(a + kh_n \right) \right\}$$

for $a + kh_n \leqslant x < a + (k+1)h_n$, $k = 0, \pm 1, \pm 2$, $\hat{F}_n(x) = $ [number of X_i's $\leqslant x$]$/n$, and $X_1, \ldots, X_n$ is a random sample. As a measure of discrepancy between the population from which the sample originat-

ed and a hypothesized density f_0, Tukey proposed the *hanging rootogram*

$$\sqrt{\tilde{f}} - \sqrt{f_0}.$$

See, e.g., Bickel and Doksum [1] for properties of this measure of discrepancy.

Reference

[1] Bickel, P. J. and Doksum, K. A. (1977). *Mathematical Statistics: Basic Ideas and Selected Topics*. Holden-Day, San Francisco, CA.

(DENSITY ESTIMATION
GOODNESS OF FIT
GRAPHICAL REPRESENTATION OF DATA
HISTOGRAM
KERNEL ESTIMATORS
KOLMOGOROV–SMIRNOV TESTS)

TUKEY'S INEQUALITY FOR OPTIMAL WEIGHTS

If $Y_1, Y_2, \ldots, Y_k$ are mutually independent random variables with common expected value η, but with possibly different variances $\sigma_1^2, \sigma_2^2, \ldots, \sigma_k^2$, respectively, then the *weighted average*

$$\overline{Y}_w = \sum_{i=1}^{k} w_i Y_i, \quad \text{with } \sum_{i=1}^{k} w_i = 1, \ w_i \geqslant 0,$$

is an unbiased* estimator of η. The variance of $\overline{Y}_w$ is

$$\text{var}\left(\overline{Y}_w\right) = \sum_{i=1}^{k} w_i^2 \sigma_i^2.$$

This is minimized with respect to the *weights* $w_1, w_2, \ldots, w_k$ and subject to the condition $w_1 + w_2 + \cdots + w_k = 1$ by weights

$$w_j^* = \frac{\sigma_j^{-2}}{\sum_{i=1}^{k} \sigma_i^{-2}}, \qquad j = 1, 2, \ldots, k.$$

The corresponding (optimal) weighted average

$$\overline{Y}_{w^*} = \sum_{i=1}^{k} w_i^* Y_i$$

has variance var $\left(\overline{Y}_{w^*}\right) = \left(\sum_{i=1}^{k} \sigma_i^{-2}\right)^{-1}$.

Tukey [1] has derived the inequality

$$\mathrm{var}(\overline{Y}_w) \leqslant \frac{(1+b)^2}{4b}\mathrm{var}(\overline{Y}_{w*}),$$

where

$$b = \left\{ \max_{i=1,\ldots,k} \left(w_i^*/w_i \right) / \min_{i=1,\ldots,k} \left(w_i^*/w_i \right) \right\}.$$

In particular if $w_1 = w_2 = \cdots = w_k = k^{-1}$, corresponding to $\overline{Y}_w$ being the arithmetic mean $\overline{Y}$, then $b = \max_i(\sigma_i^2)/\min_i(\sigma_i^2)$ and Tukey's inequality shows that

$$\frac{\mathrm{var}(\overline{Y})}{\mathrm{var}(\overline{Y}_{w*})}$$

$$\leqslant \frac{1}{4} \frac{\left\{ \min_{i=1,\ldots,k}\left(\sigma_i^2\right) + \max_{i=1,\ldots,k}\left(\sigma_i^2\right) \right\}^2}{\left\{ \min_{i=1,\ldots,k}\left(\sigma_i^2\right)\max_{i=1,\ldots,k}\left(\sigma_i^2\right) \right\}}.$$

For $b = 4$, $\{ \frac{1}{4}(1+b)^2/b \}^{1/2} = 1.25$, so that if $\max_i(\sigma_i^2)/\min_i(\sigma_i^2) = 4$, the standard error of $\overline{Y}$ cannot exceed that of the optimal weighted average by more than 25%.

Reference

[1] Tukey, J. W. (1948). *Ann. Math. Statist.*, **19**, 91–92.

(ARITHMETIC MEAN
ESTIMATION, POINT
WEIGHT BIAS)

TUKEY'S LINE

A term used in exploratory data analysis* to denote a straight line fitted to two-dimensional (x, y) data by finding the cross-medians of the first one-third and the last one-third of cases (observations) along the horizontal (x axis), drawing a straight line through these cross-medians and then moving this line parallel to itself until one-half of the observations lie below (above) the line. The Tukey line is used as a "first smooth"* to remove linearity from a relationship. Residuals* are then examined to search for nonlinearity.

(DATA ANALYSIS
GRAPHICAL REPRESENTATION OF
    STATISTICAL DATA

OUTLIERS
ROUGH
SMOOTH)

TUKEY'S MEDIAN POLISH

Median polish is a technique invented by J. W. Tukey (see Tukey [1, p. 366]) for extracting row and column effects in a two-way data layout using medians rather than arithmetic means, and therefore possessing the good robustness properties held by other median-like procedures: *see* MEDIAN ESTIMATES and SIGN TESTS. For a good explanation and examples, see Velleman and Hoaglin [2, Chap. 8].

METHOD

Add to the two-way layout a column of row effects and a row of column effects, both initially all zeros, and a single overall effect term; see Example 1. In every row (including the row of column effects), subtract the row median from all entries and add the row median to the row effect. Operate similarly on columns instead of rows, then return to operate on rows, then columns, . . . , etc. The procedure terminates when the two-way layout of residuals has zero median in every row and column, and where the row and column effects each have median zero. Thus, if x_{ij} is the entry in row i, column j, if r_{ij} is the corresponding residual, μ the overall effect, α_i the ith row effect, and β_j the jth column effect, then $x_{ij} = \mu + \alpha_i + \beta_j + r_{ij}$, with $\mathrm{median}_i(\alpha_i) = 0 = \mathrm{median}_j(\beta_j)$, and $\mathrm{median}_i(r_{ij}) = 0 = \mathrm{median}_j(r_{ij})$, all i, j. This decomposition is completely analogous to that in two-way analysis of variance* (ANOVA), using medians instead of means. Note that: (i) The method is nonunique. Starting by operating on columns rather than rows may lead to a different (but qualitatively similar) answer. (ii) Instead of terminating in a finite number of steps, the iterations might converge geometrically to the solution.

Example 1. Two-way table:

$$
\begin{array}{cccc|c}
18 & 11 & 14 & 15 & 0 \\
13 & 10 & 15 & 14 & 0 \\
9 & 17 & 15 & 7 & 0 \\
\end{array}
\begin{array}{c}
\\ \text{row} \\ \text{effects} \\
\end{array}
$$

$$
\begin{array}{cccc|c}
0 & 0 & 0 & 0 & 0 \\
\end{array}
$$
column effects   overall effect

$\xrightarrow{\text{(rows)}}$

$$
\begin{array}{cccc|c}
3.5 & -3.5 & -0.5 & 0.5 & 14.5 \\
-0.5 & -3.5 & 1.5 & 0.5 & 13.5 \\
-3 & 5 & 3 & -5 & 12 \\
\hline
0 & 0 & 0 & 0 & 0 \\
\end{array}
$$

$\xrightarrow{\text{(columns)}}$

$$
\begin{array}{cccc|c}
4 & 0 & -2 & 0 & 1 \\
0 & 0 & 0 & 0 & 0 \\
-2.5 & 8.5 & 1.5 & -5.5 & -1.5 \\
\hline
-0.5 & -3.5 & 1.5 & 5 & 13.5 \\
\end{array}
$$

$\xrightarrow{\text{(rows)}}$

$$
\begin{array}{cccc|c}
4 & 0 & -2 & 0 & 1 \\
0 & 0 & 0 & 0 & 0 \\
-2 & 9 & 2 & -5 & -1.5 - 0.5 = -2 \\
\hline
-0.5 & -3.5 & 1.5 & 0.5 & 13.5 \\
\end{array}
$$

When operating on columns first, the median polish converges toward the same result at a geometric rate. For example, after four iterations the table is

$$
\begin{array}{cccc|c}
4.25 & -0.25 & -1.75 & 0.25 & 0.75 \\
0 & -0.5 & 0 & 0 & 0 \\
-2 & 8.5 & 2 & -5 & -2 \\
\hline
-0.5 & -3 & 1.5 & 0.5 & 13.5 \\
\end{array}
$$

Example 2. Polishing is nonunique. When starting by rows (columns) first, the table

$$
\begin{array}{ccccc}
13 & 11 & 15 & 13 & 14 \\
13 & 12 & 15 & 14 & 19 \\
9 & 14 & 15 & 10 & 15 \\
\end{array}
$$

yields the result(s)

$$
\begin{array}{ccccc|c}
1 & 0 & 1 & 0 & 0 & -1 \\
0 & 0 & 0 & 0 & 4 & 0 \\
-4 & 2 & 0 & -4 & 0 & 0 \\
\hline
-1 & -2 & 1 & 0 & 1 & 14 \\
\end{array} \; ,
$$

$$
\begin{array}{ccccc|c}
0 & -1 & 0 & 0 & -1 & 0 \\
0 & 0 & 0 & 1 & 4 & 0 \\
-4 & 2 & 0 & -3 & 0 & 0 \\
\hline
0 & -1 & 2 & 0 & 2 & 13 \\
\end{array} \; ,
$$

each time in just two iterations.

Remarks.

1. The median polish is a least absolute deviations method of fitting (*see* MEDIAN ESTIMATES AND SIGN TESTS) and so possesses efficiency properties typical of median estimates, namely, modest for normal data but very good for contaminated or long-tailed data.

2. The method is a versatile data-analytic tool. It can be applied to higher order classifications than two-way layouts, and to situations with more than one observation per cell.

References

[1] Tukey, J. W. (1977). *Exploratory Data Analysis.* Addison-Wesley, Reading, MA.

[2] Velleman, P. F. and Hoaglin, D. C. (1981). *The ABC of Exploratory Data Analysis.* Duxbury Press, Belmont, CA.

(MEDIAN ESTIMATES AND SIGN TESTS MEDIAN POLISH)

B. M. BROWN

TUKEY'S QUICK TEST

This is a distribution-free test of the equality of location of two distributions, proposed by Tukey [6] following a suggestion of W. E. Duckworth. It has the special feature that the critical value (for given significance level α) varies only slowly with sample size. As a consequence, the test can be applied without recourse to elaborate (often, to any) tables.

Using random samples of sizes n_1, n_2 from populations Π_1, Π_2 the test statistic T is calculated in the following way.

1. If the greatest and least of the $(n_1 + n_2)$ observed values come from the same population, $T = 0$.
2. Otherwise T is the sum of the number of values from Π_1 greater (less) than any from Π_2 and the number of values from Π_2 less (greater) than any from Π_1.

For example, with observed values (after ordering)

$5, 10, 12, 13, 14, 17$ from Π_1 $(n_1 = 6)$

$3, 4, 7, 9, 11$ from Π_2 $(n_2 = 5)$

we have $T = 2 + 4 = 6$ (corresponding to the observations $3, 4$ from Π_2 and $12, 13, 14, 17$ from Π_1). If an additional observation 18 were to come from Π_2, then we would have $T = 0$. As in an approximation, ties can be counted $\frac{1}{2}$, so that if 12 rather than 18 were the additional observation, we would have $T = 5\frac{1}{2}$.

Large values of T are regarded as being significant for differences in location of Π_1 and Π_2. For a considerable range of values of n_1 and n_2, with n_1/n_2 not greatly different from 1, exact (conservative) minimal values for significance are

for $\alpha = 0.05$,   8;

for $\alpha = 0.01$,   10;

for $\alpha = 0.001$,   14.

Tukey, however, recommended the use of 7, 10, and 13 as rough minimal values.

Exact tables are given in ref. 1 for all n_1, n_2 with $\max|n_1 - n_2| \leq 20$. The power of

the test has been studied by Neave and Granger [3]. A modified test less affected by outliers appears in ref. 2. A graphical confidence interval procedure is described by Sandelius [5]. There is an interesting discussion of the problem by Rosenbaum [4].

References

[1] Gans, D. J. (1981). *Technometrics*, **23**, 193–195.
[2] Neave, H. R. (1966). *J. Amer. Statist. Ass.*, **61**, 949–964.
[3] Neave, H. R. and Granger, C. W. J. (1968). *Technometrics*, **10**, 509–522.
[4] Rosenbaum, S. (1965). *J. Amer. Statist. Ass.*, **60**, 1118–1126.
[5] Sandelius, M. (1968). *Technometrics*, **10**, 193–194.
[6] Tukey, J. W. (1959). *Technometrics*, **1**, 31–48.

(DISTRIBUTION-FREE TESTS)

D. J. GANS

TUKEY'S SIMULTANEOUS COMPARISON PROCEDURE *See* MULTIPLE COMPARISONS

TUKEY'S TEST FOR NONADDITIVITY

In a randomized block (RB) design for t treatments and b blocks (*see* BLOCKS, RANDOMIZED COMPLETE for notation) or in a two-way classification with one observation per treatment combination, the usual model is

$$y_{ij} = \mu + \tau_i + \beta_j + \epsilon_{ij}, \qquad (1)$$

$$i = 1, \ldots, t; \; j = 1, \ldots, b;$$

$$\sum_i \tau_i = \sum_j \beta_j = 0, \; \epsilon_{ij} \sim N(0, \sigma^2).$$

This model is additive; it was first thought that without replication of treatment–block combinations there are no degrees of freedom to test for interaction* γ_{ij} in the model

$$y_{ij} = \mu + \tau_i + \beta_j + \gamma_{ij} + \epsilon_{ij},$$

$$\sum_i \gamma_{ij} = \sum_j \gamma_{ij} = 0, \qquad (2)$$

i.e., the analysis of variance* of (1) assumes

Table 1   Analysis of Variance: RB Design

Source	Degrees of Freedom	Sum of Squares	Mean Square
Treatments	$t - 1$	S_A	$MS_A = S_A/(t - 1)$
Blocks	$b - 1$	S_B	$MS_B = S_B/(b - 1)$
Nonadditivity	1	S_{AB}	$MS_{AB} = S_{AB}$
Error	$(t - 1)(b - 1) - 1$	S_E	$MS_E = S_E/((t - 1)(b - 1) - 1)$
Total	$bt - 1$	S_T	

that $\gamma_{ij} \equiv 0$. In 1949, however, Tukey [9] developed heuristically a test for nonadditivity, deriving a sum of squares with 1 degree freedom. Although Tukey intended to test for the presence of general interaction, his test has especially good power when the interaction is a product of the treatment and block effects [4], i.e., when

$$y_{ij} = \mu + \tau_i + \beta_j + \lambda\tau_i\beta_j + \epsilon_{ij}, \quad (3)$$

and it has therefore been associated latterly with the model (3), as a test of the hypothesis $H: \lambda = 0$. Let

$$S_B = b \sum_{i=1}^{t} (\bar{y}_{i.} - \bar{y}_{..})^2,$$

$$S_A = t \sum_{j=1}^{b} (\bar{y}_{.j} - \bar{y}_{..})^2$$

$$S_T = \sum_i \sum_j (y_{ij} - \bar{y}_{..})^2,$$

$$S_{AB} = \frac{\left[\sum_i\sum_j \left\{ y_{ij}(\bar{y}_{i.} - \bar{y}_{..})(\bar{y}_{.j} - \bar{y}_{..}) \right\}\right]^2}{S_A S_B},$$

$$S_E = S_T - S_A - S_B - S_{AB}.$$

Then the analysis of variance shown in Table 1 yields Tukey's test statistic MS_{AB}/MS_E, which has an F-distribution with $(1, (b - 1) \times (t - 1) - 1)$ degrees of freedom under the hypothesis H. H is rejected at a level α if the value of the statistic exceeds the $100(1 - \alpha)$ percent point of this F-distribution. If H is rejected, treatment effects should be studied within each block separately; Cressie [1] introduced transformations* that may largely remove the additivity and enable the usual tests for treatment and block effects to be carried out with some validity. If H is acceptable, the usual tests for treatment and block effects can be performed on the basis of Table 1.

The key result is derived by Rao [8, Sec. 4e.1], and a numerical example appears in Ostle and Mensing [7, Sec. 11.3].

The first association of Tukey's test statistic with model (3) was by Ward and Dick [10]. Ghosh and Sharma [2] expressed the power of the test in the form of an integral; see also Hegemann and Johnson [3]. Competing test procedures of Johnson and Graybill [3, 4] and Onukogu [5, 6] were developed to test for interaction in the model

$$y_{ij} = \mu + \tau_i + \beta_j + \lambda\alpha_i\gamma_j + \epsilon_{ij},$$
$$\sum_i \alpha_i = \sum_j \gamma_j = 0,$$
$$\sum_i \alpha_i^2 = \sum_j \gamma_j^2 = 1. \quad (4)$$

Although these tests are more powerful than Tukey's in the model (4), Tukey's seems to perform better in model (3) [3, 5] and Onukogu's test statistic reduces to Tukey's in that case. Johnson and Graybill [4] list several references for extensions of Tukey's test to design other than RB and to other interaction functions.

References

[1] Cressie, N. A. C. (1978). *Biometrics*, **34**, 505–513.

[2] Ghosh, M. N. and Sharma, D. (1963). *J. R. Statist. Soc. B*, **25**, 213–219.

[3] Hegemann, D. and Johnson, D. E. (1976). *J. Amer. Statist. Ass.*, **71**, 945–948.

[4] Johnson, D. E. and Graybill, F. A. (1972). *J. Amer. Statist. Ass.*, **67**, 862–868.

[5] Onukogu, I. B. (1980). *Metron*, **38**(3-4), 131–137.

[6] Onukogu, I. B. (1981). *Metron*, **39**(1-2), 229–242.

[7] Ostle, B. and Mensing, R. W. (1975). *Statistics in Research*, 3rd ed. Iowa State University Press, Ames, IA.

[8] Rao, C. R. (1973). *Linear Statistical Inference and Its Applications*, 2nd ed. Wiley, New York.

[9] Tukey, J. W. (1949). *Biometrics*, **5**, 232–242.

[10] Ward, G. C. and Dick, I. D. (1952). *New Zealand J. Sci. Technology*, **33**, 430–436.

(ANALYSIS OF VARIANCE
BLOCKS, RANDOMIZED COMPLETE
INTERACTION
INTERACTION MODELS
TUKEY'S LINE)

CAMPBELL B. READ

TUKEY'S TEST FOR ORDERED ALTERNATIVES

See WILCOXON-TYPE TESTS FOR ORDERED ALTERNATIVES IN RANDOMIZED BLOCKS

TVERSKY'S MODEL OF CHOICE

See LUCE'S CHOICE AXIOM AND GENERALIZATIONS

TWICING

Twicing is the operation of applying the same smoother to a data series twice, as follows. If the relationship between a data series and the output from a smoother is represented additively as

$$\text{data} = \text{smooth}_1 + \text{rough}_1,$$

then twicing is the operation of applying the same smoother to the rough:

$$\text{rough}_1 = \text{smooth}_2 + \text{rough}_2,$$

whence

$$\text{data} = (\text{smooth}_1 + \text{smooth}_2) + \text{rough}_2.$$

Thus, the final smooth from twicing results from smoothing* the rough and adding it back to the original smooth. The purpose of twicing is to restore potentially interesting features in the data that may have been smoothed away and left in rough_1. By smoothing rough_1, such features will return to the overall smooth of the data.

If the data series and its smooth are represented by $\{x_n\}$ and $S\{x_n\}$, respectively, and $I\{x_n\}$ leaves the series unchanged, then the result of twicing may be expressed algebraically as

$$\{x_n\} = S\{x_n\} + (I - S)\{x_n\},$$

$$(I - S)\{x_n\} = S(I - S)\{x_n\} + (I - S)$$
$$\times (I - S)\{x_n\}$$

$$\Rightarrow \quad \{x_n\} = S\{x_n\} - S(I - S)\{x_n\}$$
$$+ (I - S)^2\{x_n\},$$

showing that $S(2I - S)$ is the resultant smoother from the twicing operation.

The term twicing is attributed to Tukey [3, p. 526]. Generally, but by no means exclusively, twicing applies to data values that are indexed by time. A discussion of its effects on a time series* is given in ref. 1, p. 113. Illustrations of analyses using twicing may be found in refs. 2–4.

References

[1] Hamming, R. V. (1977). *Digital Filters*. Prentice-Hall, Englewood Cliffs, NJ.

[2] McNeil, D. R. (1977). *Interactive Data Analysis*. Wiley, New York.

[3] Tukey, J. W. (1977). *Exploratory Data Analysis*. Addison-Wesley, Reading, MA.

[4] Velleman, P. F. and Hoaglin, D. C. (1981). *Applications, Basics, and Computing of Exploratory Data Analysis*. Duxbury, Boston, MA.

(EXPLORATORY DATA ANALYSIS
GRADUATION
SUMMATION [n]
WAVE-CUTTING INDEX
WHITTAKER–HENDERSON
  GRADUATION FORMULAS)

KAREN KAFADAR

TWO-ARMED BANDIT PROBLEM   *See* ONE- AND TWO-ARMED BANDIT PROBLEMS

TWO-BY-TWO (2 × 2) TABLES

Data in the form

	A	Not A	Total
I	a	b	m = a + b
II	c	d	n = c + d
Total	r	s	N = r + s
	= a + c	= b + d	= m + n

can arise in many different ways: (1) A sample of N may be taken from a population in which each unit is classified as A or not A, and as I or II; (2) a total of N experimental units (e.g., "patients") may be assigned by some random procedure to one of two treatments I or II, and the outcomes for each unit classified as A or not A (e.g., "cured" or "not cured"); (3) I and II may refer to distinct populations from which samples of sizes m and n are taken, the sample items being classified as A or not A; (4) a set of N units may be classified as I or II and as A or not A, the number classified as I and the number classified as A being determined in advance. In case (1), N is taken as fixed and there remain three quantities to be experimentally determined. These three quantities correspond to three underlying parameters which may conveniently be taken to be P_I, the probability that an item is classified as I, p_1, the probability that an item is classified as A, given that it is I, and p_2, the probability that an item is classified as A, given that it is II. We define $Q_I = 1 - P_I$, $q_1 = 1 - p_1$, and $q_2 = 1 - p_2$. P_I, p_1, and p_2 can be simply estimated by m/N, a/m, and c/n. In case (2), m and n may or may not be fixed in advance, and correspondingly P_I may or may not be regarded as a parameter, while p_1 and p_2 will be so regarded. If P is a parameter, it will usually be known. In case (3) only p_1 and p_2 will be parameters. In case (4) we usually have a "nonparametric" situation.

We confine our discussion to these four cases, while noting that many other possibilities exist. In a genetical experiment, N may be taken as fixed, while P_I and the column marginal probability $P_I p_1 + Q_I p_2$ may be taken as known, equal to $\frac{1}{4}$, for example, and inverse sampling may have been carried out, with a and c fixed in advance, leaving m and n to be experimentally determined.

When the number of parameters of interest is equal to the number of experimentally determined quantities, no special inferential problems arise, except those common to all cases of discrete and perhaps asymmetric distributions. When a, b, c, d are all large, no serious problems arise, because asymptotic likelihood theory works well. That the "well worn topic" of the 2×2 table has been, and continues to be, a prolific source of controversy since the days of Karl Pearson* is due mainly to the fact that cases are frequent where a, b, c, d, are not all large and where only one parameter is of interest, the others being "nuisance parameters."

This single parameter of interest relates to the "difference" between p_1 and p_2, and the first source of debate is the choice of measure of this "difference." While in special circumstances $p_1 - p_2$ may be a natural choice, and in others the ratio p_1/p_2 may be so, as a general purpose measure there are strong grounds for choosing the *odds ratio** $\theta = p_1 q_2 / (p_2 q_1)$, or some function of it such as its logarithm or Yule's $Q = (\theta - 1)/(\theta + 1)$ (suggested by Yule in 1900). Choice of $p_1 - p_2$ would equate a change from $p_1 = 0.02$ to $p_2 = 0.01$ with a change from $p_1 = 0.49$ to $p_2 = 0.48$, while choice of p_1/p_2 would equate the first change with one from 0.96 to 0.48. When p_1 and p_2 are both small, θ approximates to p_1/p_2, while when p_1 and p_2 are near $\frac{1}{2}$, $\log \theta$ is nearly proportional to $p_1 - p_2$. The transformation to Q makes the parameter range from -1 to $+1$; other advantages of Q and θ were pointed out by Yule.

A further reason for choosing θ as a "measure" of difference is, that it can be estimated with high efficiency on its own without reference to the remaining nuisance parameter needed to specify p_1 and p_2. This remaining parameter may be taken to be $\phi = p_1 p_2 / (q_1 q_2)$, the square of the geometric mean of the odds. Then

$$p_1 = \frac{(\phi\theta)^{1/2}}{1 + (\phi\theta)^{1/2}}, \quad \text{while}$$

$$p_2 = \frac{(\phi/\theta)^{1/2}}{1 + (\phi/\theta)^{1/2}}.$$

If we imagine the experimental result to be learned in three stages

(E_1) in which we learn the values of m and n,

(E_2) in which we learn the values of r and s,

(E_3) in which we learn the value of a (and hence those of $b = m - a$, $c = r - a$, and $d = n - r + a$),

we can see that E_1 may give information about P_I, if this is not yet known, but it tells us nothing about the difference between p_1 and p_2. Then E_2 will tell us much about the geometric mean of the odds $\phi^{1/2}$, but very little, if anything, about the difference measure θ. True, when $m = n = 1$, $(r, s) = (1, 1)$ will suggest that there *is* a difference, while $(r, s) = (2, 0)$ will suggest otherwise, but $(r, s) = (1, 1)$ gives no clue as to which way the difference goes. For larger values of m and n, the values of r and s are even less informative as to the difference. For information about θ, we are then left with E_3. The probability of observing the value a, *given* (m, n) and (r, s) is

$$\Pr[a|\theta] = \frac{\theta^a}{a!b!c!d!}$$

$$\times \left\{ \sum_u \frac{\theta^u}{u!(m-u)!(r-u)!(n-r+u)!} \right\}^{-1},$$

(A)

where the summation is taken over all values of u that make u, $m - u$, $r - u$, and $n - r + u$ nonnegative. Since this depends only upon θ, it provides information about θ irrespective of the nuisance parameter.

Given (m, n), the probability of the column totals (r, s) is

$$\Pr[(r, s)|\theta, \phi]$$

$$= \sum_u \binom{m}{u} p_1^u q_1^{m-u} \binom{n}{r-u} p_2^{r-u} q_2^{n-r+u}$$

$$= \frac{m!n!\phi^{r/2}}{\left\{1 + (\phi\theta)^{1/2}\right\}^m \left\{1 + (\phi/\theta)^{1/2}\right\}^n}$$

$$\times \left\{ \sum_u \frac{\theta^{u-r/2}}{u!(m-u)!(r-u)!(n-r+u)!} \right\}^{-1}.$$

When $m = n$, this expression is unchanged if θ is replaced by $1/\theta$, and it remains nearly independent of θ for plausible values of ϕ—that is, when $\phi^{1/2}$ is reasonably near to r/s—as can be checked numerically in any given case. This confirms the intuitive argument already given showing that practically all the information about θ is provided by the conditional distribution (A).

Much of the literature on 2×2 tables is concerned with approximations to (A), but with modern microcomputers the coefficients of the polynomial involved $(\sum_u H(u)\theta^u)$ are easily determined using the recurrence relation

$$H(u + 1) = \frac{(m - u)(r - u)}{(u + 1)(n - r + u + 1)} H(u),$$

so that approximations are scarcely necessary.

If it be accepted that the parameter of interest is θ, and it is the value of a, given the marginal totals, that provides practically all the information about θ, inference about θ will be based on an observation $X = a$, where X is an observable with probability distribution

$$\Pr[X = a|\theta] = \frac{H(a)\theta^a}{\sum_u H(u)\theta^u}.$$

If it appears that the value of a is too small to be compatible with a hypothetical value $\theta = \theta_0$ (often $= 1$, implying $p_1 = p_2$), the associated P value is

$$\underline{P}(\theta_0, a)$$

$$= \Pr[X = a|\theta_0] + \Pr[X = a - 1|\theta_0]$$

$$+ \cdots + \Pr[X = \max(r - n, 0)|\theta_0]$$

and the hypothesis $\theta = \theta_0$ will be rejected against alternatives $\theta < \theta_0$ at significance level α if $\underline{P}(\theta_0, a) \leq \alpha$. Conversely, if it appears that the value of a is too large to be compatible with $\theta = \theta_0$, the P value is

$$\bar{P}(\theta_0, a) = \Pr[X = a|\theta_0]$$
$$+ \Pr[X = a + 1|\theta_0]$$
$$+ \cdots + \Pr[X = \min(r, m)|\theta_0]$$

and $\theta = \theta_0$ will be rejected against alternatives $\theta > \theta_0$ at level α if $\bar{P}(\theta_0, a) \leq \alpha$. Further if, with P fixed, θ_0 is found such that $\underline{P}(\theta_0, a) = P$, then θ_0 will be an upper confidence bound for θ with confidence coefficient $(1 - P)$. Similarly, $\bar{P}(\theta_0, a) = P$ gives a lower confidence bound.

The likelihood function* for θ, given $X = a$, is $L(\theta) = \theta^a/\Sigma_u H(u)\theta^u$ and unless one of a, b, c, d is zero, in which case the maximum likelihood estimate $\hat{\theta}$ of θ is infinite, taking $2\{\ln L(\hat{\theta}) - \ln L(\theta)\}$ to be distributed as χ^2 with 1 degree of freedom gives very close approximations to confidence intervals for θ. In particular, if $\underline{\theta}$ and $\bar{\theta}$ are the two roots of the equation $\ln L(\hat{\theta}) - \ln L(\theta) = 1.92$, the interval $(\underline{\theta}, \bar{\theta})$ will cover the true value of θ very nearly 95% of the time. Here $\hat{\theta}$ may be taken as equal to $ad/(bc)$.

If the true expected values of (a, b, c, d) were $(a + x, b - x, c - x, d + x)$, then the true value of θ would be

$$(a + x)(d + x)/(b - x)(c - x)$$

and the value of χ^2 with Yates' correction, for x positive, would be

$$\chi_c^2 = (x - \tfrac{1}{2})^2\{1/(a + x) + 1/(b - x)$$
$$+ 1/(c - x) + 1/(d + x)\}$$

and taking x as the positive root of the equation $\chi_c^2 = 3.84$ will give the upper value $\bar{\theta}$ of a 95% confidence interval. Yates' "correction for continuity" reduces the *magnitude* of the deviation by $\tfrac{1}{2}$ to allow for discreteness of the true distribution, so to obtain the lower value $\underline{\theta}$ we replace $(x - \tfrac{1}{2})$ by $(x + \tfrac{1}{2})$ and take the negative root. Then $(\underline{\theta}, \bar{\theta})$ will cover the true value very nearly 95% of the time. This approximation is due to R. A. Fisher*. Both these approximations break down when one of $a, b, c,$ or d is zero. In such cases, only one-sided confidence bounds are possible.

CONTROVERSY

No account of 2×2 tables would be complete without a sketch of the long series of controversies which have attended them.

After Karl Pearson's unsuccessful attempt to persuade Yule* to prefer Pearson's "tetrachoric r" to Yule's Q, the next controversy arose when Yule and Greenwood [13] pointed to a discrepancy between Pearson's χ^2 test for association, based on

$$\chi^2 = \frac{(a - mr/N)^2}{mr/N} + \frac{(b - ms/N)^2}{ms/N}$$
$$+ \frac{(c - nr/N)^2}{nr/N} + \frac{(d - ns/N)^2}{ns/N}$$
$$= \frac{N(ad - bc)^2}{mnrs},$$

and the test based on a normal approximation to the difference of two binomial estimates,

$$z^2 = \frac{\{(a/m) - (c/n)\}^2}{(rs/mnN)} = \frac{N(ad - bc)^2}{mnrs}.$$

According to accepted practice at the time, χ^2 would have been assigned degrees of freedom $k - 1$ (k is the number of cells), i.e., $4 - 1 = 3$, whereas z^2, the square of a standard normal variate, approximately, would have only 1 degree of freedom. Soon afterward, Fisher showed that, in general, the appropriate number of degrees of freedom is $k - p - 1$, where p is the number of nuisance parameters efficiently fitted from the data. In particular, for the 2×2 table, $p = 2$, removing the discrepancy noted by Yule and Greenwood. As the Fisherian concept of precisely defined statistical models, with parameters classifiable as "of interest" or as "nuisances," has gained ground (Stigler [10]) Fisher's treatment of the degrees of freedom problem has become generally accepted, though people remain who point out that the distinction between "of interest" and "nuisance" is not always clear cut, nor is fully efficient estimation always possible.

Further controversy arose when Barnard [2] pointed to the great discrepancy, when model (3) applies, between the $\underline{P}(0, a)$ value of $1/20$ assigned by Fisher's "exact" test to the table with $(a, b, c, d) = (3, 0, 0, 3)$ and the upper bound $\bar{\alpha} = 1/64$ to the frequency of type I error. Barnard [3] elaborated his views with a "CSM" test, designed to reduce the discrepancy, and Pearson [9] pointed out that taking the expression for z to be standard normal would usually give a good approximation to $\bar{\alpha}$. Neither Pearson nor Barnard disputed the appropriateness of $\underline{P}(0, a)$ in the case of model (4). Fisher attacked Barnard's proposals in print [7] and in correspondence he drew Barnard's attention to the need to relate the interpretation of a "level of significance" to the sensitivity of the test employed. By averaging over all possible values of r, the CSM test lumped together data which could detect small changes in θ (r large) with data which were insensitive to changes in θ (r near to or even equal to 0).

The need to relate the interpretation of a level of significance to the sensitivity of a test is clear when failure to attain a 5% level is, as often, interpreted as "little or no evidence" against the null hypothesis. Such an interpretation requires that, with the data available, the test be sensitive enough to detect any important departures from the null hypothesis.

In 1949, Barnard [4] acknowledged acceptance of Fisher's view. But it was not until much later that Barnard [5] (following Pitman) showed that, to minimize the long run frequency of errors in Neyman* and Pearson's sense, the critical significance level applied to insensitive data *must* (other things being equal) be larger than the critical level applied to sensitive data. With the 2×2 table, the sensitivity is strongly dependent on the observed r value, and lumping all r values together is inadmissible.

Before Barnard [3], Wilson [11] had put forward a view differing from Fisher's. He later withdrew it, but later still, in an unpublished note to Barnard, Wilson said he remained in doubt.

Yates [12] (with the discussion and papers therein referenced) gives a recent review of this continuing controversy.

An elegant treatment of the 2×2 table from the point of view of Neyman's theory of hypothesis testing* is presented in E. L. Lehmann [8]. It is there proved, for example, that the uniformly most powerful unbiased test* of H: $p_1 = p_2$ against K: $p_1 \neq p_2$ at significance level (or "size") α rejects H whenever $\underline{P}\ (0, a) < \frac{1}{2}\alpha$, and also, by a randomisation* procedure, when

$$\underline{P}(0, a) - \Pr[X = a|0] < \tfrac{1}{2}\alpha < \underline{P}(0, a).$$

In this latter case, H is rejected with probability

$$\tfrac{1}{2}\alpha - \underline{P}(0, a) + \Pr[X = a|0].$$

H is also rejected when $\overline{P}(0, a) < \frac{1}{2}\alpha$, and when a corresponding randomisation* rule so indicates. The difficulties arising from lumping together data of widely differing sensitivity are very marked with this test, since, with non-zero probability α we are required by it to reject H when $a = c = 0$, whatever the (nonzero) values of m and n.

If the notion is accepted that the significance level of a result must be related to the informativeness of the data, then the equation, often made, between what Neyman calls the "size" of a test and the significance level, must be abandoned in those cases, as here, where the informativeness of the data cannot be known in advance.

A further cause of controversy with 2×2 tables has been the fact that the distribution of X is discrete and often asymmetrical. Difficulties of interpretation, therefore, arise similar to those met with in testing the hypothesis that an observable Y has a binomial distribution with index n and probability parameter $p = \frac{1}{3}$. A prescribed "size" will not usually be attainable, and two-sided tests raise difficulties due to asymmetry. Reference is made to Cox and Hinkley [6]. Fisher's suggestion, to double the P value* to make a "one-sided" test "two-sided" is a simple, justifiable convention. Another possible approach is indicated by Anscombe [1a].

All the previously mentioned difficulties disappear, of course, when prior distributions for all the unknown parameters are available. A discussion of the resulting Bayesian* approach is given by Albert and Gupta [1].

References

[1]   Albert, J. H. and Gupta, A. K. (1983). *J. Amer. Statist. Ass.*, **78**, 708–717.

[1a] Anscombe, F. J. (1981). *Computing in Statistical Science Through APL*, Springer, New York, pp. 288–289.

[2]   Barnard, G. A. (1945). *Nature*, **156**, 177.

[3]   Barnard, G. A. (1947). *Biometrika*, **34**, 123–128.

[4]   Barnard, G. A. (1949). *J. R. Statist. Soc. B*, **11**, 115–149.

[5]   Barnard, G. A. (1982). *Statistics and Probability*. G. Kallianpur, P. R. Krishnaiah, and J. K. Ghosh, eds. North-Holland, Amsterdam, Netherlands.

[6]   Cox, D. R., and Hinkley, D. V. (1974). *Theoretical Statistics*. Chapman and Hall, London, England.

[7]   Fisher, R. A. (1945). *Nature*, **156**, 388.

[8]   Lehmann, E. L. (1959). *Testing Statistical Hypotheses*. Wiley, New York.

[9]   Pearson, E. S. (1947). *Biometrika*, **34**, 139–167.

[10] Stigler, S. M. (1976). *Ann. Statist.*, **4**, 498–499.

[11] Wilson, E. B. (1942). *Proc. Nat. Acad. Sci. Wash.*, **28**, 94–100.

[12] Yates, F. (1984). *J. R. Statist. Soc. A*, **147**, 426–463.

[13] Yule, G. U. and Greenwood, M. (1915). *Proc. Roy. Soc. Medicine* (*Epidemiology*), **8**, 113–190.

(CONTINGENCY TABLES
CONTINUITY CORRECTION
FISHER'S EXACT TEST
HYPOTHESIS TESTING
INFERENCE, STATISTICAL-I, II.
P-VALUES
RANDOMIZED TESTS
SIGNIFICANCE TESTS,
    HISTORY AND LOGIC
TETRACHORIC CORRELATION
    COEFFICIENT)

G. A. BARNARD

TWO-BY-TWO TABLES: MISSING VALUES

If one of the four cell frequencies in a 2×2 table is missing (so that marginal frequencies also are unknown), estimation of the missing value must depend on making some assumptions about the process producing the table. The simplest assumption is that of exact proportionality, so that in the table

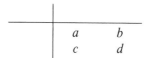

we have $a/b = c/d$. Then, if the frequency d is missing, it would be estimated as bc/a. This estimate is reasonable to use if the assumption of independence between the two factors is reasonable. (See Chandrasekar and Deming [1].)

The assumption of independence is often untenable. Several modifications of the formula bc/a have been suggested, especially with reference to situations where the coverages of two procedures for collecting demographic data are being compared, and the two-way table is formed from the categories "recorded" or "not recorded" for both procedures. (The frequency d corresponds to "not recorded" for both procedures.)

Greenfield [2, 3] proposes the estimator

$$\hat{d}_G = \left(A + \tfrac{1}{4}B^2 \right)^{1/2} - \tfrac{1}{2}B,$$

where

$$A = \frac{bc\left\{ bc - \hat{r}^2(a + b)(a + c) \right\}}{\hat{r}^2(a + b)(a + c) - a^2},$$

$$B = \frac{(b + c)(a + b)(a + c) + 2abc}{\hat{r}^2(a + b)(a + c) - a^2},$$

and

$$\hat{r} = \frac{a - (bc)^{1/2}}{\left\{ (a + b)(a + c) \right\}^{1/2}}.$$

Simplified modified values of A and B, proposed by Greenfield and Tam [4], are

$$A = \frac{bc\left\{ 3(bc)^{1/2} - a \right\}}{(bc)^{1/2} - a},$$

$$B = \frac{(b + c)\left\{ a - (bc)^{1/2} \right\}^2 + 8abc}{\left\{ (bc)^{1/2} + a \right\}\left\{ (bc)^{1/2} - 3a \right\}}.$$

[In ref. 4, the even simpler formula $\hat{d}_{GT} = (bc)^{1/2}$ is proposed for use in some cases.]

Nour [6] suggests the estimator

$$\hat{d}_N = \frac{2abc}{a^2 + bc}.$$

Macarthur [5] modifies this slightly to

$$\hat{d}_M = \frac{2abc + a^2}{a^2 + bc + a}.$$

References

[1] Chandrasekar, C. and Deming, W. E. (1949). *J. Amer. Statist. Ass.*, **44**, 101–115.

[2] Greenfield, C. C. (1975). *J. R. Statist. Soc. A*, **138**, 51–61.

[3] Greenfield, C. C. (1983). *J. R. Statist. Soc. A*, **146**, 273–280.

[4] Greenfield, C. C. and Tam, S. M. (1976). *J. R. Statist. Soc. A*, **139**, 96–103.

[5] Macarthur, E. W. (1983). *J. R. Statist. Soc. A*, **146**, 85–86.

[6] Nour, El-S. (1982). *J. R. Statist. Soc. A*, **145**, 106–116.

(CONTINGENCY TABLES
TETRACHORIC CORRELATION
  COEFFICIENT
TWO-BY-TWO TABLES)

TWO-PHASE SAMPLING *See* SURVEY
SAMPLING

TWO-SAMPLE PROBLEM *See* LOCA-
TION TESTS

TWO-SERIES THEOREM

The two-series theorem is related to the following concept. We say that the series ΣX_n of random variables, where throughout this entry Σ signifies summation over n running from 1 to $+\infty$, is *essentially convergent* if there are constants a_n, $n \geqslant 1$, such that the series $\Sigma(X_n - a_n)$ converges with probability 1. If the random variables X_n, $n \geqslant 1$, are independent, then the classical zero–one law* implies that if ΣX_n is not essentially convergent, then, for all sequences c_n, $n \geqslant 1$, $\Sigma(X_n - c_n)$ diverges with probability 1.

For formulating the two-series theorem, we introduce the following notation. Let M_n

be a median of X_n and let Z_n^c be the truncation* of $X_n - M_n$ at c,

$$Z_n^c = \begin{cases} X_n - M_n, & \text{if } |X_n - M_n| < c, \\ 0, & \text{otherwise,} \end{cases}$$

where $c > 0$ is an arbitrary constant. Now, the two-series theorem states that if $X_1, X_2, \ldots$ are independent random variables, then ΣX_n is essentially convergent if and only if the two series

$$\sum P(X_n - M_n \neq Z_n^c)$$

and

$$\sum V(Z_n^c)$$

converge. In the definition of essential convergence, one can always choose $a_n = M_n + E[Z_n^c]$.

There is a close relation between the two-series theorem and Kolmogorov's three-series theorem* even though they deal with different concepts of convergence of ΣX_n. This is clear from the last part of the two-series theorem in which the centering constants a_n, $n \geqslant 1$, are explicitly given. Other properties show this close relation as well. One such property is the following equivalence theorem: If $X_1, X_2, \ldots$ are independent random variables, then ΣX_n is essentially convergent if and only if ΣY_n converges almost surely, where $Y_n = X_n - X_n'$ with X_n and X_n' being independent and identically distributed random variables. The transformation of X_n to Y_n is known as *symmetrization*. Evidently, the convergence or divergence of ΣY_n is decided by Kolmogorov's three-series theorem*.

Let us restate the equivalence theorem by means of characteristic functions*. Let $\varphi_n(t)$ be the characteristic function of X_n; then the characteristic function of Y_n is $|\varphi_n(t)|^2$. Now, in view of the zero–one law* and the continuity theorem of characteristic functions in the case of independent random variables X_n, ΣX_n is almost surely convergent if and only if the product $\varphi_1(t)\varphi_2(t) \cdots \varphi_n(t)$ converges to $\varphi(t)$, which is a characteristic function. On the other hand, by the equivalence theorem, ΣX_n is essentially convergent if and only if the product $|\varphi_1^2(t)\varphi_2^2(t) \cdots \varphi_n^2(t)|$ converges to $\varphi^*(t)$, which is a characteristic function. One can decide whether $\varphi(t)$ and $\varphi^*(t)$ are characteristic functions by the common rule, established in ref. 1: If $\varphi(t) \neq 0$ [or $\varphi^*(t) \neq 0$] on a set of positive Lebesgue measure, then $\varphi(t)$ [$\varphi^*(t)$] is a characteristic function. Note that $\varphi_n(t)$ is sensitive to an additive constant a_n in $X_n - a_n$, while $|\varphi_n(t)|$ is not. So, one could have "guessed" that the same type of rule should apply to the convergence of ΣX_n in terms of $\varphi_n(t)$ as to its essential convergence in terms of $|\varphi_n(t)|$. But, as was indicated, the proof leading to such rules requires the most powerful, and some of the most beautiful, results of probability theory.

Detailed proofs of most of the results mentioned in this article can be found, e.g., in Sec. 3.12 of ref. 2.

References

[1] Kawata, J. and Udagawa, M. (1949). *Kodai Math. Sem. Rep.*, **3**, 15–22.

[2] Loève, M. (1963). *Probability Theory*, 3rd ed. Van Nostrand, New York.

(CHARACTERISTIC FUNCTIONS
CONVERGENCE OF SEQUENCES
  OF RANDOM VARIABLES
KOLMOGOROV'S THREE-SERIES
  THEOREM
TRUNCATION METHODS IN PROBABILITY
  THEORY)

JANOS GALAMBOS

TWO-SEX PROBLEM *See* MATHEMATICAL THEORY OF POPULATION

TWO-STAGE LEAST SQUARES

The two-stage least squares (2SLS) estimator, independently proposed in refs. 6 and 4, is one of the most popular estimators in econometrics*. It is generally useful in regression* equations where some of the regressor variables may be correlated with the error term.

Consider a regression equation

$$\mathbf{y} = \mathbf{Z}\boldsymbol{\alpha} + \mathbf{u}, \qquad (1)$$

where $\mathbf{y}$ is a T-vector of observable random variables, $\mathbf{Z}$ is a $T \times G$ matrix of observable random variables (some elements may be known constants) of rank G, $\boldsymbol{\alpha}$ is a G-vector of unknown parameters, and $\mathbf{u}$ is a T-vector of unobservable random variables with zero mean. If $\mathbf{Z}$ and $\mathbf{u}$ satisfy the conditions

$$\operatorname*{plim}_{T \to \infty} T^{-1}\mathbf{Z}'\mathbf{u} = 0 \qquad (2)$$

and

$$\operatorname*{plim}_{T \to \infty} T^{-1}\mathbf{Z}'\mathbf{Z} \quad \text{exists and is nonsingular}, \qquad (3)$$

the least squares estimator is consistent.

The 2SLS estimator becomes useful when (2) does not hold. We suppose there exists a $T \times H$ matrix $\mathbf{S}$ (of *instrumental variables**) with rank $H \geqslant G$ such that

$$\operatorname*{plim}_{T \to \infty} T^{-1}\mathbf{S}'\mathbf{u} = \mathbf{0}, \qquad (4)$$

$$\operatorname*{plim}_{T \to \infty} T^{-1}\mathbf{S}'\mathbf{S} \quad \text{exists and is nonsingular}, \qquad (5)$$

and

$$\operatorname*{plim}_{T \to \infty} T^{-1}\mathbf{S}'\mathbf{Z} \quad \text{exists and has rank } G. \quad (6)$$

Then the 2SLS estimator $\hat{\boldsymbol{\alpha}}$ defined by

$$\hat{\boldsymbol{\alpha}} = (\mathbf{Z}'\mathbf{P}\mathbf{Z})^{-1}\mathbf{Z}'\mathbf{P}\mathbf{y}, \qquad (7)$$

where $\mathbf{P} = \mathbf{S}(\mathbf{S}'\mathbf{P}\mathbf{S})^{-1}\mathbf{S}'$, is consistent.

The suggestive name of the estimator is based on the observation that $\mathbf{P}\mathbf{Z}$ is the least squares* predictor of $\mathbf{Z}$ obtained from the regression of $\mathbf{Z}$ on $\mathbf{S}$ and $\hat{\boldsymbol{\alpha}}$ is the least squares estimator in the regression of $\mathbf{y}$ on $\mathbf{P}\mathbf{Z}$.

Model (1) can arise in a variety of ways. For example, suppose that the columns of $\mathbf{Z}$ consist of *exogenous variables* (*see* ECONOMETRICS) and the lagged values of $\mathbf{y}$ and that the elements of $\mathbf{u}$ are correlated. Such a model is called a *distributed-lag model** in econometrics. A proper set of instrumental variables from which the 2SLS estimator can be defined for this model could be a set of the exogenous variables and their various lagged values.

Another specific case of model (1) is defined by

$$y = Y\gamma + X_1\beta + u \tag{8}$$

and

$$Y = X\Pi + V, \tag{9}$$

where Y is a $T \times N$ matrix of observable random variables, X is a $T \times K$ matrix of exogeneous variables (or known constants), X_1 consists of $K_1 \leqslant K - N$ columns of X, V is a $T \times N$ matrix of unobservable random variables with zero mean, and γ, β, and Π are vectors or matrices of unknown parameters.

Equation (8) can be interpreted as a single equation in a *simultaneous equations model*, and eq. (9) can be regarded as the *reduced-form* equation for Y solved from the simultaneous equations model. Together, (8) and (9) may be called the *limited-information simultaneous equations model*. Here, the appropriate 2SLS estimator of $\alpha \equiv (\gamma', \beta')'$ is obtained by putting $Z = (Y, X_1)$ and $S = X$ in the general formula (7).

Suppose that the rows of a $T \times (N + 1)$ matrix (u, V) are independent and identically distributed (i.i.d.) drawings from an $(N + 1)$-variate normal distribution with zero mean and nonsingular covariance matrix Σ. Suppose also that there are no constraints on γ, β, Π, and Σ. Then the maximum likelihood estimator of $\alpha \equiv (\gamma', \beta')'$ is called the *limited-information maximum likelihood* (LIML) *estimator*[*]. If the rows of (u, V) are i.i.d. with zero mean and covariance matrix Σ (they need not be normally distributed) and if $\text{plim}_{T \to \infty} T^{-1}X'X$ exists and is nonsingular, the 2SLS and LIML estimators have the same asymptotic normal distribution. Specifically, we can show

$$\sqrt{T}(\hat{\alpha} - \alpha) \to N(0, \sigma^2 A^{-1}), \tag{10}$$

where σ^2 is the element in the northwest corner of Σ and $A = \text{plim}_{T \to \infty} T^{-1}Z'PZ$.

The exact distributions of 2SLS and LIML have been obtained in simple cases (see a recent survey in ref. 5). It has been shown that LIML does not have moments of any order and 2SLS has moments of order up to and including $K - N - K_1$. The choice between the two estimators is an unsettled issue in econometrics. 2SLS is computationally simpler and its estimates tend to be more stable, whereas LIML has been recently shown to converge to a normal distribution more rapidly (see ref. 3).

Asymptotic normality[*] (10) has been derived under the assumption $E[uu'] = \sigma^2 I$. If $E[uu'] \equiv \Lambda$, where Λ is not necessarily a scalar times an identify matrix, an asymptotically more efficient estimator can be found. An example is the generalized two-stage least squares (G2SLS) estimator (see ref. 7) defined as $(Z'P\hat{\Lambda}^{-1}Z)^{-1}ZP\hat{\Lambda}^{-1}y$, where $\hat{\Lambda}$ is a consistent estimator of Λ. Applying this method to a simultaneous equations model leads to the *three-stage least squares estimator*.

Now, we go back to model (1) and consider generalizing it to a nonlinear function. Suppose the tth equation is given by

$$f(y_t, z_t, \alpha) = u_t, \tag{11}$$

where y_t is an observable scalar random variable, z_t is a vector of observable random variables possibly correlated with u_t, α is a vector of unknown parameters, and u_t is an unobservable scalar random variable with zero mean. Write (11) in vector notation as

$$f = u, \tag{12}$$

where f and u are T-vectors whose tth elements appear in (11). The nonlinear two-stage least squares (NL2S) estimator proposed by ref. 1 minimizes $f'Pf$, where P is defined after (7). It is shown in ref. 1 that, under reasonable assumptions, the estimator is consistent and asymptotically normal. If $\{u_t\}$ are i.i.d. with variance σ^2, its asymptotic covariance matrix is given by

$$\sigma^2 \left[\text{plim}_{T \to \infty} T^{-1}(\partial f'/\partial \alpha)P(\partial f/\partial \alpha') \right]^{-1}.$$

For a further discussion of the subject, see ref. 2.

References

[1] Amemiya, T. (1974). *J. Econometrics*, **2**, 105–110.

[2] Amemiya, T. (1985). *Advanced Econometrics*. Harvard University Press, Cambridge, MA.

[3] Anderson, T. W. (1982). In *Advances in Econometrics*, W. Hildenbrand, ed. Cambridge University Press, Cambridge, MA, pp. 109–122.

[4] Basmann, R. L. (1957). *Econometrica*, **25**, 77–83.

[5] Phillips, P. C. B. (1983). In *Handbook of Econometrics*, Vol. 1, Z. Griliches and M. D. Intriligator, eds. North-Holland, Amsterdam, Netherlands, pp. 449–516.

[6] Theil, H. (1957). *Rev. Int. Statist. Inst.*, **25**, 41–51.

[7] Theil, H. (1961). *Economic Forecasts and Policy*, 2nd ed. North-Holland, Amsterdam, Netherlands.

(ECONOMETRICS
FIX-POINT METHOD
LEAST SQUARES
TIME SERIES)

TAKESHI AMEMIYA

TWOING INDEX

An index of "distinctiveness" or diversity* used in some recursive partitioning* procedures. If a node is split into two nodes, say L and R, in proportions $p_L : p_R$ ($p_L + p_R = 1$) such that the proportion of items from class (j) in $L(R)$ is P_{jL} (P_{jR}), the *twoing index* of the split is

$$\tfrac{1}{4} p_L p_R \left\{ \sum_j |P_{jL} - P_{jR}| \right\}^2 .$$

The greater the index, the more effective the split.

(DIVERSITY INDICES
RECURSIVE PARTITIONING (*see*
    SUPPLEMENT))

TYPE BIAS

A bias attributable to the type of mean used in calculating an index number.

(INDEX NUMBERS)

TYPE I ERROR   *See* HYPOTHESIS TESTING

TYPE II ERROR   *See* HYPOTHESIS TESTING

TYPE I AND TYPE II VARIABLES   *See* DESIGN OF EXPERIMENTS

TYPE III AND TYPE IV ERRORS

In addition to the well-known type I (incorrect rejection of a valid hypothesis) and type II (incorrect nonrejection of a hypothesis, which is not valid), several other types of error in hypothesis testing* have been formulated. These are intended to bring into consideration some of the practical considerations which necessarily have to be ignored in the mathematical formulation of definition of errors of types I and II.

"Type III error" is often used to denote errors arising from choosing the wrong hypotheses to test.

"Type IV error" (a term introduced by Marascuilo and Levin [2–4]) is used to describe incorrect interpretation of a correct rejection of a hypothesis. See, e.g., Betz and Gabriel [1] for further details and an account of some relevant controversy.

References

[1] Betz, M. A. and Gabriel, K. R. (1978). *J. Educ. Statist.*, **3**, 121–143.

[2] Levin, J. R. and Marascuilo, L. A. (1973). *Psychol. Bull.*, **80**, 308–309.

[3] Marascuilo, L. A. and Levin, J. R. (1970). *Amer. Educ. Res. J.*, **7**, 397–421.

[4] Marascuilo, L. A. and Levin, J. R. (1976). *Amer. Educ. Res. J.*, **13**, 61–65.

(HYPOTHESIS TESTING
INFERENCE, STATISTICAL-I, II)

TYPICAL VALUES

Hartigan [3] introduced "typical values" as building blocks for constructing nonparametric confidence intervals*. Let θ be the unknown parameter for which we want a confidence interval*. The random variables $X_1, X_2, \ldots, X_N$ are called *typical values for* θ if each of the intervals between the ordered random variables $-\infty < X_{(1)} < X_{(2)} < \cdots < X_{(N)} < +\infty$ contains θ with equal probability. That is, we require $P\{X_{(i)} < \theta < X_{(i+1)}\} = 1/(N + 1)$ for each $i \in \{0, 1, \ldots, N\}$, where $X_{(0)} \equiv -\infty$ and $X_{(N+1)} \equiv +\infty$. Given such a set of typical values, we may construct a confidence interval for θ by taking the union of adjacent intervals. The confidence interval may be two-sided or one-sided, but the level of confidence is constrained to be an integer multiple of $1/(N + 1)$.

The simplest setting in which typical values are available is as follows: Suppose $Y_1, Y_2, \ldots, Y_n$ are independent and identically distributed (i.i.d.) continuous symmetric random variables. The center of symmetry for their common distribution is θ, a location parameter. For each possible nonempty subset of $\{Y_1, Y_2, \ldots, Y_n\}$, we compute the arithmetic mean of that subset of random variables. (These subsets of $\{Y_1, Y_2, \ldots, Y_n\}$ are sometimes referred to as "subsamples".) The resulting collection of $N = 2^n - 1$ subset-means constitutes a set of typical values for θ. The confidence intervals that we obtain in this situation have three noteworthy advantages. First, they are nonparametric—the user does not have to know the common distribution of the Y_i's. Second, the level of confidence is exact—not an approximation or a conservative lower bound. Last, they are valid for any sample size n, no matter how small. [If we were willing to add the strong parametric assumption that the Y_i's are normally distributed, then exact confidence intervals for θ could be obtained via Student's t-distribution* (of form $\overline{Y} \pm t_{(n-1)}s/n^{1/2}$). In the nonparametric scenario, we could obtain *approximate* confidence intervals for θ via the central limit theorem* (of form $\overline{Y} \pm zs/n^{1/2}$), assuming that Y_i has finite variance; the coverage probability for these intervals is only valid asymptotically as $n \to \infty$.]

Example. The observations $Y_1 = 0.069$, $Y_2 = 0.686$, $Y_3 = 0.299$ are a random sample of size $n = 3$ from a continuous symmetric distribution. There are $N = 7$ subset-means

$$X_1 = 0.069, \qquad X_2 = 0.686,$$
$$X_3 = 0.299, \qquad X_4 = 0.3775,$$
$$X_5 = 0.184, \qquad X_6 = 0.4925,$$
$$X_7 = 0.351\overline{3},$$

which form a set of typical values for θ, the center of symmetry. After ordering we have

$$X_{(1)} = 0.069, \qquad X_{(2)} = 0.184,$$
$$X_{(3)} = 0.299, \qquad X_{(4)} = 0.351\overline{3},$$
$$X_{(5)} = 0.3775, \qquad X_{(6)} = 0.4925,$$
$$X_{(7)} = 0.686.$$

Therefore, the interval $(0.184, 0.4925)$ is a 50% confidence interval for θ; the half-line $(-\infty, 0.686)$ contains θ with confidence level 87.5%. In this example, the true value of θ was actually $\frac{1}{2}$.

These confidence intervals based on all possible subset-means, as previously described, have two practical disadvantages. First, if n is even moderately large, then the number of subset-means which must be computed and ordered becomes prohibitively large. For example, when $n = 20$ we find $2^n - 1 = 1,048,575$. In practice, we need not compute all $2^n - 1$ subset-means, because certain smaller collections of subset-means retain the property of being typical values. One approach is to select at random and without replacement N subsets out of the possible $2^n - 1$ subsets; the N corresponding subset-means are then typical values for θ (Efron [1]). Another approach is to calculate subset-means only for a "balanced" collection of N subsets; Hartigan [3] shows that such a collection of subset-means forms a set of typical values for θ. [The collection of all 2^n subsets of $\{Y_1, Y_2, \ldots, Y_n\}$ is a group G under the binary operation $A \circ B = (A \cup B) \cap (A \cap B)^c$; the unit element of this group is ϕ. A collection C of nonempty subsets of $\{Y_1, Y_2, \ldots, Y_n\}$ is called *balanced* if (C, ϕ) is a subgroup of G.]

A second practical consideration is that of interval length. Although the long-run coverage probabilities for the intervals are exact, the lengths of the actual confidence intervals may vary greatly from one realization of $Y_1, Y_2, \ldots, Y_n$ to the next. Since the typical value confidence interval technique is nonparametric, the lengths of the resulting intervals are generally not optimal for the underlying distribution of Y_i. Forsythe and Hartigan [2] and Maritz [5] investigate the lengths of confidence intervals constructed from typical values.

Typical values are also available in other settings. Hartigan [3] shows that the Y_i's need not be identically distributed; if they are independent, continuous, and share the same center of symmetry θ, then the subset-means are still typical values. Hartigan [3], Efron [1], and Maritz [5] consider calculating for each subset of $\{Y_1, Y_2, \ldots, Y_n\}$ an M-estimate* of θ, rather than the arithmetic mean. Efron [1] shows that subset-medians behave like a crude version of typical values in this situation: $Y_1, Y_2, \ldots, Y_n$ are i.i.d. continuous random variables whose distribution (which is not necessarily symmetric) has median θ. Finally, Hartigan [4] shows that a general statistic computed on N randomly selected subsets of $\{Y_1, Y_2, \ldots, Y_n\}$ yields N random variables which behave asymptotically (as $n \to \infty$) like typical values, provided that the general statistic is central.

References

[1] Efron, B. (1982). *The Jackknife, the Bootstrap and Other Resampling Plans*. SIAM, Philadelphia, PA, Chaps. 9 and 10.

[2] Forsythe, A. and Hartigan, J. (1970). *Biometrika*, **57**, 629–639.

[3] Hartigan, J. (1969). *J. Amer. Statist. Ass.*, **64**, 1303–1317.

[4] Hartigan, J. (1975). *Ann. Statist.*, **3**, 573–580.

[5] Maritz, J. (1979). *Biometrika*, **66**, 163–166.

(CENTRAL STATISTICS
CONFIDENCE INTERVALS AND REGIONS
DISTRIBUTION-FREE METHODS
NONPARAMETRIC CONFIDENCE
    INTERVALS
RESAMPLING TECHNIQUES)

ED CARLSTEIN

TYPOLOGY *See* ARCHAEOLOGY, STATISTICS IN

U

U-FUNCTION (UNIVERSAL PROBABILITY GENERATING FUNCTION)

The ordinary probability generating function* (PGF) for a *discrete* random variable X is a function of the form

$$\phi(z) = \sum p_k z^{a_k},$$

where $\Pr[X = a_k] = p_k$ and z is the argument of the function.

For convolution* (sum) of n mutually independent discrete random variables, the product of the corresponding PGFs is used:

$$\phi(z) = \prod_{1 \le i \le n} \phi_i(z),$$

or, explicitly,

$$\phi(z) = \sum_{\mathbf{k}_s \in K} \prod_{1 \le i \le n} p_{ik_{is}} z^{\sum_{1 \le i \le n} a_{ik_{is}}},$$

where $\mathbf{k}_s = (k_{1s}, \ldots, k_{ns})$, $1 \le k_{is} \le u_i$, u_i is the number of values of the random variable X_i; the a's and u's are finite, and K is the set of distinct vectors $\mathbf{k}_s$, $|K| = \prod_{1 \le i \le n} u_i$. Ushakov [1] has proposed and developed a *universal* PGF (*U-function*) which is useful in applications of reliability theory*. For a single random variable, the U-function is written in the same manner as the "usual" probability generating function

$$\phi_i(z) = \sum_{1 \le k_i < u_i} p_{ik_i} z^{a_{ik_i}},$$

where k_i is the number of distinct a_{ik_i} for a given i. The universal generating function for n random variables is written as

$$\phi^F(z) = \bigcup_{1 \le i \le n}^F \phi_i(z)$$

$$= \bigcup_{1 \le i \le n}^F \sum_{1 \le k_i \le u_i} p_{ik_i} z^{a_{ik_i}},$$

where the transform $\bigcup^F$ differs from the ordinary product of polynomials in that the exponents of z are not added, but an operation F *is performed on them*. This operation is defined by

$$\bigcup_{1 \le i \le n}^F \sum_{1 \le k_i \le u_i} p_{ik_i} z^{a_{ik_i}}$$

$$= \sum_{k_s \in K} \prod_{1 \le i \le n} p_{ik_{is}} z^{F(a_{1k_{ls}}, \ldots, a_{nk_{ns}})}.$$

The U-function possesses all the properties of ordinary PGFs. In the polynomial $\phi^F(z)$

the coefficient of the term z^A is the probability that $F(X_1, \ldots, X_n) = A$.

Example 1. Consider n resistors in series each having random resistance α_{ik}, i.e., $p_{ik} = \Pr[\alpha_{ik} = a_{ik}]$ Then resistance of a chain of n resistors is defined by

$$F_1\left(a_{ik_{is}}; i = \overline{1, n}\right) = \sum_{1 \leqslant i \leqslant n} a_{ik_{is}}.$$

Here the U-function is the ordinary PGF.

Example 2. If the resistors are in parallel, the resistance of the system is determined by

$$F_2\left(a_{ik_{is}}; i = \overline{1, n}\right) = \left(\sum_{1 \leqslant i \leqslant n} a_{ik_{is}}^{-1}\right)^{-1}.$$

The U-function, in this case, is

$$\phi^{F_2}(z) = \sum_{k_s \in K} \prod_{1 \leqslant i \leqslant n} p_{ik_{is}} z^{\left(\sum_{1 \leqslant i \leqslant n} a_{ik_{is}}^{-1}\right)^{-1}}.$$

Example 3. Consider a series connection of n elements each of which is either *working* ($\alpha_i = 1$) or *failed* ($\alpha_i = 0$) ($i = 1, 2, \ldots, n$). If $p_i = \Pr[\alpha_i = 1]$ and $q_i = \Pr[\alpha_i = 0] = 1 - p_i$, we have

$$F_3\left(\mathbf{a}; i = \overline{1, n}\right) = \prod_{1 \leqslant i \leqslant n} a_i = \min_{1 \leqslant i \leqslant n} \{a_i\}.$$

Example 4. If the elements in Example 3 are connected in parallel, then

$$F_4\left(\mathbf{a}; i = \overline{1, n}\right) = 1 - \prod_{1 \leqslant i \leqslant n} (1 - a_i)$$

$$= \max_{1 \leqslant i \leqslant n} \{a_i\}.$$

The expected value of a random variable X for any operation F is given by

$$E[X] = \frac{d}{dz}\phi^F(z)\bigg|_{z=1}.$$

Higher moments can be derived by the standard methods (*see* GENERATING FUNCTION). For bivariate and multivariate extensions, see Ushakov [1]. The U-function allows us to write down and solve complex combinatorial problems using a convenient algorithmic form. Note that the usual PGF

is unsuitable when operations other than summation (such as taking max or min, computing products, etc.) are being studied.

Reference

[1] Ushakov, I. A. (1986). *Tekh. Kibern.*, **25**(6), 40–42 (in Russian). [English translation: *Sov. J. Comput. System Sci.*, **25**(4), 79–82 (1987).]

(CHARACTERISTIC FUNCTIONS
GENERATING FUNCTIONS
RELIABILITY)

ULAM'S METRIC

Ulam's metric on permutations was introduced by S. M. Ulam in 1972. It was originally used in DNA research to measure the distance between two strings of molecules (Ulam [6, 7], Waterman et al. [8]). In a statistical context, Ulam's metric measures the distance between two rankings of the same n items.

The definition of this metric is as follows. Let

$$\pi: \{1, \ldots, n\} \to \{1, \ldots, n\},$$

$$\sigma: \{1, \ldots, n\} \to \{1, \ldots, n\}$$

be two permutations [the interpretation for ranking data being that $\pi(i)$ is the rank assigned to item i]. Ulam's distance between π and σ is

$$U(\pi, \sigma) = n - \begin{pmatrix} \text{the length of the} \\ \text{longest increasing} \\ \text{subsequence (LLIS) in} \\ \sigma\pi^{-1}(1), \ldots, \sigma\pi^{-1}(n) \end{pmatrix}$$

$$= n - \begin{pmatrix} \text{the maximal number} \\ \text{of items ranked} \\ \text{in the same order} \\ \text{by } \pi \text{ and } \sigma \end{pmatrix}.$$

For example, if π, σ are the permutations

$$\pi(1) = 5, \quad \pi(2) = 1, \quad \pi(3) = 2,$$

$$\pi(4) = 4, \quad \pi(5) = 3,$$

$$\sigma(1) = 3, \quad \sigma(2) = 4, \quad \sigma(3) = 1,$$

$$\sigma(4) = 5, \quad \sigma(5) = 2$$

corresponding to the rankings

$$\pi = \text{item 2} \quad \text{item 3} \quad \text{item 5} \quad \text{item 4} \quad \text{item 1}$$

$$\begin{array}{ccccc} \uparrow & & & & \uparrow \\ \text{1st} & \cdot & \cdot & \cdot & \text{5th} \\ \text{place} & & & & \text{place} \\ \downarrow & & & & \downarrow \end{array}$$

$$\sigma = \text{item 3} \quad \text{item 5} \quad \text{item 1} \quad \text{item 2} \quad \text{item 4,}$$

then

$$U(\pi,\sigma) = 5 - \text{LLIS in } \sigma\pi^{-1}(1), \ldots,$$
$$\sigma\pi^{-1}(5)$$
$$= 5 - \text{LLIS in } \sigma(2), \sigma(3), \sigma(5),$$
$$\sigma(4), \sigma(1)$$
$$= 5 - \text{LLIS in } 4, 1, 2, 5, 3$$
$$= 2,$$

because $1, 2, 3$ is the longest increasing subsequence and has length 3. Alternatively, and more directly, $U(\pi, \sigma) = 5 - 3 = 2$, because items $3, 5, 1$ are ranked in the same order by π and σ, and are a maximal collection of items with this property.

Analogous to Kendall's tau* or Spearman's rho*, Ulam's metric can be used as a nonparametric measure of association between the rankings π and σ. Ulam's metric has the special feature that if two rankings π and σ agree on the relative ordering of a large subset of the n items, then the distance $U(\pi, \sigma)$ will be small. For example, if items 1 through $n - 1$ are ranked in the same relative order by π and σ, but if item n is ranked first by π and last by σ, then $U(\pi, \sigma) = 1$. This property is especially suitable for DNA research: Ulam's distance between two sequences of molecules will be quite small whenever the two sequences are almost identical, except that one sequence is shifted by a few spaces relative to the other.

Ulam's metric is more difficult to compute efficiently than either Kendall's tau or Spearman's rho. A clever algorithm, of order n^2, for numerically *computing* Ulam's distance was discovered by Sellers [5].

Under the null hypothesis of no correlation between rankings (i.e., all the $n!$ permutations equally likely), the mean of Ulam's metric is (asymptotically) $n - 2\sqrt{n}$, but the asymptotic variance and asymptotic distribution are as yet unknown. See Logan and Shepp [4].

Further discussion of Ulam's metric from a statistical point of view can be found in Gordon [3], Diaconis [2], and Critchlow [1].

References

[1] Critchlow, D. (1985). Metric methods for Analyzing Partially Ranked Data. *Lect. Notes Statist.*, **34**. Springer-Verlag, New York.

[2] Diaconis, P. (1982). Group Theory in Statistics. *Harvard University Lecture Notes*. Institute of Mathematical Statistics Lecture Notes–Monograph Series, **11**.

[3] Gordon, A. D. (1979). A measure of the agreement between rankings. *Biometrika*, **66**, 7–15.

[4] Logan, B. F. and Shepp, L. (1977). Variational problems for random Young tableau. *Adv. Math.*, **26**, 206–222.

[5] Sellers, P. H. (1974). On the theory and computation of evolutionary distances. *SIAM J. Appl. Math.*, **26**, 787–793.

[6] Ulam, S. M. (1972). Some ideas and prospects in biomathematics. *Ann. Rev. Biophys. Bioeng.*, **1**, 277–292.

[7] Ulam, S. M. (1981). Future applications of mathematics in the natural sciences. *American Mathematical Heritage: Algebra and Applied Mathematics*. Texas Tech. University, Mathematics Series, **13**, 101–114.

[8] Waterman, M. S., Smith, T. F., and Beyer, W. A. (1976). Some biological sequence metrics. *Adv. Math.*, **20**, 367–387.

(DISTANCE FUNCTIONS
LEVY DISTANCE (METRIC)
PROBABILITY SPACES, METRICS AND
  DISTANCES ON
SPEARMAN RANK CORRELATION
  COEFFICIENT
TOTAL VARIATION, DISTANCE OF
WASSERSTEIN'S DISTANCE)

D. CRITCHLOW

ULTRAMETRIC

A distance function* $d(\mathbf{X}_1, \mathbf{X}_2)$ between vectors $\mathbf{X}_1$ and $\mathbf{X}_2$ which, in addition to the axioms

$$d(\mathbf{X}_1, \mathbf{X}_2) = d(\mathbf{X}_2, \mathbf{X}_1) \qquad (1)$$

and

$$d(\mathbf{X}_1, \mathbf{X}_3) \leqslant d(\mathbf{X}_1, \mathbf{X}_2) + d(\mathbf{X}_2, \mathbf{X}_3) \quad (2)$$

also satisfies

$$d(\mathbf{X}_1, \mathbf{X}_2) \leqslant \max[d(\mathbf{X}_1, \mathbf{X}_3), d(\mathbf{X}_2, \mathbf{X}_3)], \quad (3)$$

is called an *ultrametric*.

Benzécri [1] and Johnson [2] describe some applications in taxonomy* and cluster analysis (see also Lebart et al. [3]).

References

[1] Benzécri, J. P. (1965). Problèmes et méthodes de la taxonomie. In *L'Analyse des Données. I. La Taxonomie*, J. P. Benzécri, ed. Dunod, Paris, France.

[2] Johnson, S. C. (1967). *Psychometrika*, **32**, 241–254.

[3] Lebart, L., Morineau, A, and Warwick, K. M. (1984), *Multivariate Descriptive Statistical Analysis*, Wiley, New York.

(CLASSIFICATION
DISTANCE FUNCTIONS
HIERARCHICAL CLUSTER ANALYSIS
PROBABILITY SPACES, METRICS AND
   DISTANCES ON)

ULTRASPHERICAL POLYNOMIALS

For fixed $\nu > -\frac{1}{2}$, let $A_p^{(\nu)}(x)$ be a polynomial of degree p in x chosen so that the system of polynomials $(A_p^{(\nu)}(x): p = 0, 1, \ldots)$ is an orthogonal system with respect to the beta probability measure

$$W^{(\nu)}(x)$$

$$= \begin{cases} 0, & x \leqslant -1, \\ \int_{-1}^{x} (1 - x^2)^{\nu - 1/2} \, dx / B(\tfrac{1}{2}, \nu + \tfrac{1}{2}), \\ & -1 \leqslant x \leqslant +1, \\ 1, & x \geqslant = 1. \end{cases}$$

Aside from a multiplicative constant, this defines the ultraspherical (Gegenbauer) polynomials. If it be required that

$$\int_{-1}^{+1} \left\{ A_p^{(\nu)}(x) \right\}^2 dW^{(\nu)}(x) = A_p^{(\nu)}(1),$$

then $A_0^{(0)}(1) = 1$,

$$A_p^{(\nu)}(1) = \frac{\{(p + \nu)/\nu\}\Gamma(p + 2\nu)}{\{p!\Gamma(2\nu)\}}$$

if $p \neq 0$ and, with $[a]$ the integral part of a and $(p, \nu) \neq (0, 0)$,

$$A_p^{(\nu)}(x) = \frac{(\nu + p)}{\Gamma(\nu + 1)}$$

$$\times \sum_{s=0}^{[p/2]} \frac{(-1)^s \Gamma(\nu + p - s)}{s!(p - 2s)!} (2x)^{p - 2s}.$$

The inverse relationship is

$$x^p = \frac{p!}{2^p} \sum_{s=0}^{p/2} \frac{A_{p-2s}^{(\nu)}(x)}{s!(\nu + 1)(\nu + 2) \cdots (\nu + p - s)}.$$

Let $\mathscr{S}^m = \{\mathbf{u}: \mathbf{u} \in \mathbb{R}^m, \|\mathbf{u}\| = 1\}$ denote the sphere of unit column vectors in $\mathbb{R}^m$, let $\mathscr{A}$ denote the σ-algebra of subsets generated by the class of sets $\{\{\mathbf{u}: \mathbf{u} \in \mathscr{S}^m, \mathbf{u}'\boldsymbol{\xi} \leqslant t\}: -\infty < t < \infty, \boldsymbol{\xi} \in \mathscr{S}^m\}$, and let $U^{(m)}\{\cdot\}$ denote the probability measure on sets in $\mathscr{A}$ for which, with $\nu = \frac{1}{2}m - 1$, $U^{(m)}\{\mathbf{u}: \mathbf{u} \in \mathscr{S}^m, \mathbf{u}'\boldsymbol{\xi} \leqslant t\} = W^{(\nu)}(t)$. In particular, $U^{(m)}\{\cdot\}$ is invariant over orthogonal transformations on sets in $\mathscr{A}$ and $U^{(m)}\{\mathscr{S}^m\} = 1$. A random vector u with probability measure $U^{(m)}\{\cdot\}$ will be described as being *uniformly distributed* on $\mathscr{S}^m$. The fundamentally important result is that if $\mathbf{u}$ is uniformly distributed on $\mathscr{S}^m$, then

$$\mathscr{E}_{\mathbf{u}} A_p^{(\nu)}(\mathbf{u}'\boldsymbol{\xi}) A_q^{(\nu)}(\mathbf{u}'\boldsymbol{\eta})$$

$$= \begin{cases} 0, & p \neq q, \\ A_p^{(\nu)}(\boldsymbol{\xi}'\boldsymbol{\eta}), & p = q. \end{cases}$$

Given a function $g(\mathbf{u}'\boldsymbol{\xi})$, defined for $\mathbf{u} \in \mathscr{S}_m$, we may write

$$g(\mathbf{u}'\boldsymbol{\xi}) = \sum_{p \geqslant 0} C_p^{(\nu)}[g] A_p^{(\nu)}(\mathbf{u}'\boldsymbol{\xi}),$$

where

$$C_p^{(\nu)}[g] = \int_{\mathscr{S}_m} g(\mathbf{U}'\boldsymbol{\xi}) A_p^{(\nu)}(\mathbf{u}'\boldsymbol{\xi}) \, dU^{(m)}(\mathbf{u}).$$

When $m = 2$, since $A_p^{(0)}(\cos \theta) = 2\cos(p\theta)$, these last two equations reduce to the familiar formulae for the Fourier transform* of the function $g(\cos \theta)$ and for the Fourier–Stieltjes coefficients. The ultraspherical polynomials provide, therefore, a generalization, from the circle to the sphere of arbitrary dimension, of the methods of Fourier transforms. As an example of the statistical applications, if $g(\mathbf{u'}\boldsymbol{\xi})$ is a density on $\mathscr{S}_m$, then $c_p^{(\nu)}[g]$ is the expectation of $A_p^{(\nu)}(\mathbf{u'}\boldsymbol{\xi})$ with respect to this density. Thus $g(\cdot)$ could be approximated by replacing $c_p^{(\nu)}[g]$ by its sample equivalent.

A comprehensive discussion of Gegenbauer polynomials is in ref 1; some further statistical applications are in ref. 2.

References

[1] Erdelyi, A., Magnus, W., Oberhettinger, F., and Tricomi, F. G. (1953). *Higher Transcendental Functions*, Vol. 2, McGraw-Hill, New York.

[2] Saw, J. G. (1984). Ultraspherical polynomials and statistics on the *m*-sphere. *J. Multivariate Anal.* **14**, 105–113.

(ORTHOGONAL EXPANSIONS)

J. G. SAW

ULTRASTRUCTURAL RELATIONSHIPS

The term *ultrastructural relationship* (USR) was introduced by Dolby [4] to describe a type of linear relation between observed pairs (X_1, X_2). Cox [3] independently formulated the same model. The pairs occur in n groups of r replicates. For the jth replicate of the ith group, the model may be represented in the form

$$X_{1ij} = \mu_i + \delta_{ij} + \epsilon_{1ij},$$

$$X_{2ij} = \alpha + \beta(\mu_i + \delta_{ij}) + \epsilon_{2ij},$$

where (δ_{ij}), (ϵ_{1ij}), and (ϵ_{2ij}) are independent normal errors with zero expectations and variances σ_δ^2, σ_1^2, and σ_2^2, respectively, and (μ_i) are unknown parameters. Thus the expectations lie on the line $E[X_2] = \alpha + \beta E[X_1]$; relationships that are nonlinear in $E[X]$ have not yet received much attention. This model includes as special cases the functional relationship (FR) when $\sigma_\delta^2 = 0$,

and the structural relationship (SR) when $n = 1$ or all μ's are equal. One could think of an USR as a FR with unobservable points $(\mu_i, \alpha + \beta\mu_i)$ being the centres of the groups, each group forming a SR with identical slope β.

Care should be taken to check that the parameters in the model are identifiable*. The means and variances between and within groups are just sufficient to identify the parameters, but they fail if error pairs (ϵ_1, ϵ_2) have an unknown correlation*. The unreplicated case $r = 1$ is clearly not identifiable without imposing extra assumptions, e.g., that variance ratios are known, as in refs. 3–5.

Estimation of the parameters by maximum likelihood* runs into some trouble. First, the maximum may occur on a boundary corresponding to zero for a variance parameter, so that its derivatives are not all zero. Second, elimination of the (μ_i) between the likelihood equations may seriously bias the estimators of the other parameters, so that as n tends to infinity they may be inconsistent. For the variance parameters, it is often obvious how to "correct" for the lost degrees of freedom. The slope estimator and information matrix can be unreliable for the same reason and require subtler modification; see Morton [7] and Patefield [8] for proposed methods (the latter applies such adjustments to Dolby's original simple numerical example). If n is fixed and r tends to infinity, consistency holds under fairly general conditions, since moments are consistent estimators of their expectations. Gleser [5, 6] discusses these points further.

One may extend the definition of USR to higher dimensions so that there are several linear constraints between the expectations of variables $(X_1, \ldots, X_p)$. Similar problems arising from likelihood estimation occur; a possible method has been proposed by Morton [7], where brief details are given for the case where the μ's are one dimensional. The model is closely related to that discussed in Anderson [1], where the covariance matrix of $(X_1, \ldots, X_p)$ does not have the natural restrictions imposed by the nonnegative variance parameters of the USR model; see Gleser [5]. Anderson [2] shows how these and other models are related.

References

[1] Anderson, T. W. (1951). *Ann. Math. Statist.*, **22**, 327–351. Correction: *Ann. Statist.*, **8**, 1400 (1980).

[2] Anderson, T. W. (1984). *Ann. Statist.*, **12**, 1–45. 1982 Abraham Wald Memorial Lectures. (An excellent survey on estimating linear relationships, including factor analysis* and structural equations* in economics.)

[3] Cox, N. R. (1976). *Biometrika*, **63**, 231–237.

[4] Dolby, G. R. (1976). *Biometrika*, **63**, 39–50.

[5] Gleser, L. J. (1983). *Proc. Bus. Econ. Statist. Sect.*, 47–66. American Statistical Association, Washington, DC.

[6] Gleser, L. J. (1985). *Biometrika*, **72**, 117–124.

[7] Morton, R. (1981). *Biometrika*, **68**, 735–737.

[8] Patefield, W. M. (1978). *Biometrika*, **65**, 535–540.

(CONSISTENCY
FACTOR ANALYSIS
FUNCTIONAL EQUATIONS
MAXIMUM LIKELIHOOD ESTIMATION
REGRESSION (various entries)
STRUCTURAL MODELS)

R. MORTON

UMBRELLA ALTERNATIVES

A term introduced by Mack and Wolfe [1] to describe alternatives of the form

$$H_1: F_1(x) \geqslant \cdots \geqslant F_{l-1}(x) \geqslant F_l(x)$$
$$\leqslant F_{l+1}(x) \leqslant \cdots \leqslant F_k(x)$$

(for all x with at least one strict inequality for at least one x value) for the k-sample problem, when testing whether the CDFs $F_1(x), \ldots, F_k(x)$ are identical.

The motivation for this term is due to the configuration of the corresponding population medians.

The integer l is called the "point" or "peak" of the umbrella. For a detailed discussion, see Mack and Wolfe [1].

Reference

[1] Mack, G. A. and Wolfe, D. A. (1981). *J. Amer. Statist. Ass.*, **76**, 175–181.

(DISTRIBUTION-FREE TESTS
HYPOTHESIS TESTING
LEHMANN ALTERNATIVES
STOCHASTIC ORDERING)

UNACCEPTABLE QUALITY LEVEL (UQL)

This is also known as limiting quality level (LQL), rejectable quality level (RQL), and lot tolerance percent defective (LTPD). It is a percentage, or proportion, of nonconforming units in a lot which the consumer requires that an acceptance inspection scheme will accept with no more than a specified (usually quite low) probability.

Sampling plans* indexed by UQL are intended for use when emphasis is placed on the quality of individual lots.

(ACCEPTANCE SAMPLING
QUALITY CONTROL, STATISTICAL)

UNAVAILABILITY

This is a term used in reliability theory*. It denotes the probability that a component is in failed state at time t, given it was "as good as new" at $t = 0$. It is, of course, simply 1 minus the survival function for the lifetime of the component.

(SURVIVAL ANALYSIS)

UNBALANCEDNESS OF DESIGNS, MEASURES OF

Even in the simple one-way classification*, the equality of the sample sizes ($n_j = n$) has some advantages, not only for fixed effect* models (model I), but also for random effect models (model II). In model I the consequences of unequal variances are not serious if each $n_j = n$. Further, if $n_j = n$, the probability of committing a type II error is a minimum and if the null hypothesis H_0 is rejected, some further tests may be desirable, and some of these are not available unless the sample sizes are equal. In model II, in estimating components of variance*, the equality of the sample sizes ($n_j = n$) is sufficient to obtain optimal estimators according to the ANOVA method, but not so in the so-called unbalanced case (at least two n_j's are different). Often there is only a "small deviation" from the equality of the sample sizes, that is from the so-called balanced case. Therefore, it is justified to ask how to measure unbalancedness. A first step in measuring it was done by Hess [3].

DEFINITION OF A MEASURE OF UNBALANCEDNESS IN A ONE-WAY CLASSIFICATION

Let a design of the one-way classification with m groups be denoted by

$$(1) \qquad D = \{n_1, n_2, \ldots, n_m\},$$

$$n_j \geqslant 1, \; \sum_{j=1}^{m} n_j = N,$$

and

$$(2) \qquad n_1 \leqslant n_2 \leqslant \cdots \leqslant n_m,$$

without loss of generality. If $n_j = n$ for each j, the design D is said to be *balanced*; if at least two n_j's are different, D is said to be *unbalanced*.

Let

$$\bar{n} = \frac{1}{m} \sum_{j=1}^{m} n_j$$

be the mean number of observations in each group. A measure of unbalancedness should be

(i)   a simple function of the n_j's, symmetric in its arguments:

(ii) invariant under k-fold replications of the design

$$\{n_1, \ldots, n_m, n_1, \ldots, n_m, \ldots,$$
$$n_1, \ldots, n_m\}$$

as well as under the case that in each of the m groups the number of observa-

tions is multiplied by a constant k:

$$\{kn_1, \ldots, kn_m\}.$$

The following measure fulfills the given requirements:

$$(3) \qquad \gamma(D) = \frac{m}{\bar{n}\sum(1/n_j)}.$$

Moreover, the measure reflects in a specific way properties of statistical analysis (see Ahrens and Pincus [1]).

It is easy to see that γ is the ratio of the well known harmonic mean* H and the arithmetic mean* A, and because $A \geqslant H$, we have

$$1 \geqslant \frac{H}{A} = \gamma.$$

If one defines the m-vectors $(n_1^{1/2}, \ldots, n_m^{1/2})'$ and $(n_1^{-1/2}, \ldots, n_m^{-1/2})'$, the measure γ will be the squared cosine between both vectors, so that

$$0 < \gamma \leqslant 1.$$

The case $\gamma = 0$ is excluded because $n_j > 0$ for each j. $\gamma = 1$ if the design D is balanced: the closer γ is to 0, the more unbalanced the design.

A FURTHER DEFINITION AND EXTENSIONS

If one defines

$$(4) \qquad \nu(D) = \frac{1}{m\sum(n_j/N)^2},$$

one gets a new measure of unbalancedness which also satisfies conditions (i) and (ii); moreover, it is a function of the well known coefficient of variation*

$$c = \sqrt{\frac{1}{m}\sum(n_j - \bar{n})^2}\Big/\bar{n},$$

namely

$$\nu = \frac{1}{1 + c^2}.$$

If the m-vectors $(\bar{n}, \ldots, \bar{n})'$ and $(n_1, \ldots, n_m)'$ are given, one obtains

$$0 < \nu(D) \leqslant 1,$$

with an interpretation like that of γ in relation to balancedness and unbalancedness.

Extension of γ and ν to a k-fold hierarchical design seems to be obvious if one defines a measure of unbalancedness for the lowest level of the hierarchical design and composes it by means of suitable weights to an "overall measure" of unbalancedness of the whole design. For a two-fold hierarchical design, see Ahrens and Pincus [1].

CLASSIFICATION OF UNBALANCED ONE-WAY LAYOUT DESIGNS

In the following we use only the measure γ. Let m and $N = \Sigma n_j$ be fixed; then the variety of n_j sets for given m and N is very large even for moderately great m and N. For example, if $N = 40$ and $m = 10$, there are 3580 different n_j sets. To each set is associated a real number γ ($0 < \gamma \leqslant 1$). Of course, for a given design (n_j set) one can calculate a number γ, but for practical aims a decomposition of the set of all unbalanced designs for given N and m is more favourable.

The Most Balanced Design

If $m | N$, then $\bar{n}$ is a positive integer number and

$$D_b = \{ \underbrace{\bar{n}, \ldots, \bar{n}}_{m \text{ times}} \}$$

is the "most balanced" design, $\gamma(D_b) = 1$. If $m \nmid N$, then $N = mp + q$, $p, q \in \mathcal{N}$ and the "most balanced" design has the form

(5)

$$D = \{ \underbrace{p, \ldots, p,}_{(m-q) \text{ times}} \underbrace{(p+1), \ldots, (p+1)}_{q \text{ times}} \}.$$

If one asks for all designs (n_j sets) $D = \{n_1, \ldots, n_m\}$ for given N and m, one obtains them by starting with $n_1 = 1$ and using the lexicographic order under the condition that $n_1 \leqslant n_2 \leqslant \cdots \leqslant n_m$.

The process continues with $n_1 = 2$ and so on, where n_1 runs from 1 to $[N/m]$ ($[a]$ is the integer part of a.)

Example 1. When $N = 15$, $m = 4$, $[N/m] = 3$, one gets

$$\{1, 1, 1, 12\} \quad \{2, 2, 2, 9\} \quad \{3, 3, 3, 6\}$$
$$\{1, 1, 2, 11\} \qquad \vdots \qquad \vdots$$
$$\vdots \qquad \vdots \qquad \{3, 4, 4, 4\}$$
$$\vdots \qquad \{2, 4, 4, 5\}$$
$$\{1, 4, 5, 5\}$$

There are classes starting with $n_1 = 1$, $n_1 = 2$, and $n_1 = 3 = [N/m]$. One can fix in each class the so-called "worst" and "best" design, i.e., the "most unbalanced" and the "most balanced" design.

The "Most Unbalanced" and the "Most Balanced" Design for Fixed N and m and Given n_1

Any real design $\{n_1, \ldots, n_m\}$ belongs to a class of designs all starting with a given n_1. To find the "most unbalanced" and the "most balanced" design within this class, we indicate the following true statement: In the class of all designs starting with a fixed n_1, the relation

(6) $$\gamma(\{n_1, \ldots, n_1, N - (m-1)n_1\})$$
$$\leqslant \gamma(\{n_1, \ldots, n_m\})$$

holds. Using this relation for Example 1, the "most unbalanced" designs in the classes starting with $n_1 = 1$, $n_1 = 2$, or $n_1 = 3$, respectively, are $\{1, 1, 1, 12\}$, $\{2, 2, 2, 9\}$, and $\{3, 3, 3, 6\}$, respectively.

Now one can also determine [according to (5) in the class of all designs starting with a fixed n_1] the "most balanced" design

(7) $$D_{n_1}^* = \{ n_1, \underbrace{p, \ldots, p,}_{(m-q-1) \text{ times}}$$
$$\underbrace{(p+1), \ldots, (p+1)}_{q \text{ times}} \},$$

where $N - n_1 = (m-1)p + q$.

Using Example 1, one gets the "most balanced" design starting with $n_1 = 2$ as follows:

$$N - n_1 = 13 = 3 \cdot 4 + 1, \qquad p = 3, q = 1.$$

Therefore,

$$D_2^* = \{2, 4, 4, 5\}.$$

Example 2. Let $D = \{11, 8, 6, 24, 15\}$ be a real design with $m = 5$ groups and $N = 64$ observations. It is equivalent to $D = \{6, 8, 11, 15, 24\} = \{n_1, n_2, \ldots, n_5\}$. One can determine the measure γ of unbalancedness and the "most unbalanced" and "most balanced" designs in the class of all designs with $m = 5$ and $N = 64$ starting with $n_1 = 6$:

$$D_6 = \{6, 6, 6, 6, 40\}, \qquad \gamma(D_6) = 0.565,$$

$$D = \{6, 8, 11, 15, 24\}, \qquad \gamma(D) = 0.795,$$

$$D_6^* = \{6, 14, 14, 15, 15\}, \quad \gamma(D_6^*) = 0.882.$$

FINAL REMARKS

The use of unbalanced data causes difficulties, mostly, in the context of the estimation of components of variance, especially in comparing different estimation procedures which are based on the MSE functions. Studying the dependency of the MSE function on the unbalancedness, infinitely many n_j sets could be used and statements like "very unbalanced" or "almost balanced", and so on are often made (Searle [4]).

A measure of unbalancedness enables quantification of the degree of unbalancedness of a given design and consequently it is possible to include the degree of unbalancedness as a parameter in a statistical analysis or in a simulation study (see Ahrens and Sanchez [2]).

References

[1] Ahrens, H. and Pincus, R. (1981). *Biom J.*, **23**, 227–235.

[2] Ahrens, H. and Sanchez, J. (1982). *Biom. J.*, **24**, 649–661.

[3] Hess, J. L. (1979). *Biometrics*, **35**, 645–649.

[4] Searle, S. R. (1971). *Biometrics*, **27**, 1–76.

(ANALYSIS OF VARIANCE
BALANCING IN EXPERIMENTAL DESIGN
DESIGN OF EXPERIMENTS
FIXED-, RANDOM-, AND MIXED-EFFECTS
    MODELS
GENERAL BALANCE
OPTIMUM DESIGN OF EXPERIMENTS)

H. AHRENS
J. SANCHEZ

UNBIASED TESTS   *See* UNBIASEDNESS

UNBIASEDNESS

UNBIASED POINT ESTIMATION

The bias* (or systematic error) of an estimator $\delta(X)$ of an unknown quantity $g(\theta)$ is the difference between the expected value of $\delta(X)$ and $g(\theta)$:

$$b(\theta) = E_\theta[\delta(X)] - g(\theta). \qquad (1)$$

An estimator with positive bias $b(\theta) > 0$ tends systematically to overestimate. When the bias is constant, $b(\theta) = b$ independent of θ—for example, when a scale is incorrectly calibrated—the bias can be eliminated by replacing $\delta(X)$ by $\delta'(X) = \delta(X) - b$. This simple device is not available when the bias depends on the (unknown) parameter θ, and it may then not be possible to remove the bias completely. A general bias-reducing technique was proposed by Quenouille [28, 29]. This method, as part of the subject of jackknifing* into which it developed, is reviewed, for example, in Efron [9].

An estimator $\delta(X)$ of $g(\theta)$ is said to be *unbiased* if its bias is identically zero, i.e., if

$$E_\theta[\delta(X)] = g(\theta) \quad \text{for all } \theta. \qquad (2)$$

A function $g(\theta)$ is called *U-estimable* if it has an unbiased estimator.

Example 1. Let $X_1, \ldots, X_n$ be independent identically distributed random variables, each with density $f(x - \theta)$ having finite first moment, but otherwise unknown.

(i) Both the sample mean and the sample median are unbiased estimators of θ if $f(x)$ is even (i.e., symmetric about 0).

(ii) If $f(x)$ is not symmetric about 0, the mean continues to be an unbiased estimator of the population mean $\theta = E[X_i]$, but the median typically is not.

Example 2. Under the assumptions of Example **1(ii)**, if the X_i have finite variance, then:

(i) $S^2 = \Sigma(X_i - \bar{X})^2/(n - 1)$ is an unbiased estimator of σ^2 when $n > 1$.

(ii) When $n = 1$, no unbiased estimator of σ^2 exists.

(iii) Part **(ii)** remains true even when f is known to be normal.

Example 3. Let X be the number of successes in n binomial* trials with success probability p. Then:

(i) X/n is an unbiased estimator of p.

(ii) $X(n - X)/\{n(n - 1)\}$ is an unbiased estimator of pq.

(iii) A necessary and sufficient condition for a function $g(p)$ to have an unbiased estimator is that $g(p)$ be a polynomial of degree less than or equal to n. Thus in particular, for any n there exists no unbiased estimator of $1/p$. [*Note:* Parts **(i)** and **(ii)** are consequences of Examples 1 and 2, respectively.]

Unbiasedness is not a requirement in the sense that biased estimators should be ruled out of consideration (see examples at end of section). However, since a large bias typically is undesirable, it seems reasonable to determine the best estimator within the class of unbiased ones (if this is not empty) and then to examine its properties. This somewhat pragmatic view seems close to that of Gauss, who introduced the concept of lack

of systematic error [10] and applied it when it suited him, but in other cases disregarded it.

Several bodies of statistical theory are concerned with obtaining the best among all unbiased estimators or among a class C of such estimators. The following are some of the principal examples.

A. Locally minimum variance unbiased (LMVU) estimators. It is desired to determine the unbiased estimator which has the smallest variance when θ has some given value θ_0. The theory of such estimators is developed by Barankin [2], Stein [31], and Bahadur [1].

B. Uniformly minimum variance unbiased (UMVU) estimators. It sometimes turns out that the LMVU estimator at θ_0 is independent of θ_0 and then *uniformly* minimizes the variance among all unbiased estimators of $g(\theta)$ (is UMVU). An example is provided by the estimator S^2 of σ^2 in Example 2(i) in the two cases (a) that f is completely unknown and (b) that f is known to be normal. In general, a UMVU estimator exists for all U-estimable $g(\theta)$ if (and under mild conditions also only if) the minimal sufficient statistic for the given family of distributions is complete* (Lehmann and Scheffé [21], Bahadur [1]). For a detailed discussion and many applications, see Lehmann [20]. *See also* MINIMUM VARIANCE UNBIASED ESTIMATION.

C. Best linear unbiased estimator (BLUE). Let $Y_1, \ldots, Y_n$ be random variables with expectations $E[Y_i] = \eta_i$ and covariance matrix $\Sigma = \sigma^2 V$, where V is a known positive definite matrix and σ^2 an unknown scalar. Then the BLUE of $\theta = \Sigma c_i \eta_i$ (the c's known) is the linear function $\delta = \Sigma a_i Y_i$ which minimizes $\text{Var}(\Sigma a_i Y_i)$ subject to the condition $E[\Sigma a_i Y_i] = \theta$. Two cases are of particular interest.

(i) When the Y's are independent and have common variance (so that V is the identity matrix), the simplest form of the *Gauss–Markov theorem** states that the BLUE's coincide with the least squares estimators*. For a discussion of this result and some generalizations, see, for example, Kendall and Stuart [14, Chap. 19].

(ii) If $X_1, \ldots, X_n$ are independently distributed with a common density $(1/\sigma)f((x - \theta)/\sigma)$ and the Y's are the order statistics* $X_{(1)} < \cdots < X_{(n)}$, the BLUE (which is a linear function of the $X_{(i)}$ but not of the X_i) depends on f through V. For details of the theory of these estimators, initiated by Lloyd [22] and Downton [7], see, for example, David [6].

BLUE's also play a role in the theory of survey sampling*, where the definition of both "linear" and "unbiased" depends on whether one is dealing with a fixed finite population or a superpopulation model. (For details, see, for example, Cassel et al. [5].)

Although large biases are undesirable, small biases may be quite acceptable. Biased estimators may be employed for a number of reasons.

(i) It is possible that no unbiased estimator exists. This is illustrated in Example **3 (iii)**.

(ii) There are situations in which unbiased estimators exist but all are unsatisfactory. For example, it may happen that all unbiased estimators take on values outside the range of the estimand. This is the case in particular whenever a UMVU estimator exists and takes on values outside this range. (More general conditions are given by Hoeffding [11].) An example is provided by the estimation of ξ^2 on the basis of a sample from $N(\xi, \sigma^2)$ with σ known. The UMVU estimator is $\overline{X}^2 - \sigma^2/n$, which takes on negative values with positive probability. The same difficulty arises in the estimation of variance components*.

A more extreme instance occurs in the estimation of $e^{-3\lambda}$ when X has a Poisson distribution with $E[X] = \lambda$.

The only unbiased estimator in this case is $(-2)^X$, which bears no meaningful relation to the quantity being estimated. Such estimators are discussed in detail in Lehmann [19].

(iii) Even when a UMVU estimator is sensible, a biased estimator may be preferable, for example, if it has uniformly smaller risk. For instance, in Example **2 (i)**, if f is normal and $S'^2 = \Sigma(X_i - \overline{X})^2/(n + 1)$, we have

$$E\left[(S'^2 - \sigma^2)^2\right] < E\left[(S^2 - \sigma^2)^2\right],$$

despite the fact that S^2 is the UMVU estimator of σ^2.

Example 4. Let X_i be independently distributed as $N(\theta_i, 1)$ for $i = 1, \ldots, p$, and consider the estimation of $(\theta_1, \ldots, \theta_p)$ by means of $\delta = (\delta_1, \ldots, \delta_p)$. For $p \geqslant 3$, the (biased) James–Stein estimator* $Y = (Y_1, \ldots, Y_p)$, suggested by James and Stein [12],

$$Y_i = \left(1 - \frac{p - 2}{(n - 1)S^2}\right) X_i$$

with

$$S^2 = \sum (X_i - \overline{X})^2 \Big/ (n - 1)$$

has uniformly smaller risk than the best unbiased estimator $(X_1, \ldots, X_p)$ for many loss functions, including $E[\sum_{i=1}^n (\delta_i - \theta_i)^2]$. For a discussion of such "shrinkage" estimators*, see, for example, Efron [8] and Lehmann [20].

UNBIASED TESTS AND CONFIDENCE SETS

A level α test of a hypothesis $H: F \in \mathscr{F}_0$ against the alternatives $K: F \in \mathscr{F}_1$, where F is the (unknown) distribution of the observations is said to be unbiased if

$$P_F(\text{rejection } H) \geqslant \alpha \quad \text{for all } F \in \mathscr{F}_1, \quad (3)$$

i.e., if the test is at least as likely to reject under any alternative as under any of the distributions of H.

In a number of important situations in which no uniformly most powerful (UMP) test exists, there exists a test which is UMP among all unbiased tests. This is the case in particular for the problems of testing either (i) H: $\theta_1 \leqslant \theta_1^0$ against $\theta_1 > \theta_1^0$ or (ii) H: $\theta_1 = \theta_1^0$ against $\theta_1 \neq \theta_1^0$ in the exponential family* of densities

$$p_\theta(x) = \beta(\theta)e^{\Sigma_{i=1}\theta_i T_i(x)}h(x). \quad (4)$$

For details and other examples, see Lehmann [18].

Example 5. Let $X_1, \ldots, X_n$ be i.i.d. as $N(\xi, \theta^2)$. Then the standard tests of the hypothesis (i) or (ii) with $\theta_1 = \xi$, $\theta_1^0 = 0$ (σ unknown) are UMP unbiased. The same is true for (i) when $\theta_1 = \sigma^2$, θ_1^0 arbitrary (ξ unknown). For testing (ii) with $\theta_1 = \sigma^2$, there exists a UMP unbiased test which accepts when $C_1 \leqslant \Sigma(X_i - \bar{X})^2 \leqslant C_2$ for suitable C_1, C_2. These constants are not those corresponding to either the equal tails or the likelihood ratio test*.

The UMP unbiased character of the tests in all these examples is due primarily to the effect of the unbiasedness condition for alternatives close to the hypothesis. In Example 5(i) with $\theta_1 = \xi$, $\theta_1^0 = 0$, it has been shown that if attention is restricted to alternatives $\xi > \xi^* > 0$, it is possible to improve the power* over that of the t-test (Brown and Sackrowitz [4]).

For problems in which the hypothesis specifies more than one parameter, for example, that the vector $(\theta_1, \ldots, \theta_s)$ of means in a multivariate normal distribution has a specified value $(\theta_1^0, \ldots, \theta_s^0)$, UMP unbiased tests typically will not exist but the standard tests often are unbiased. For some results of this kind, see Perlman [26] and Perlman and Olkin [27] and the literature cited therein.

If the distribution of X depends on parameters θ and ϑ, confidence intervals $\underline{\theta} \leqslant \theta \leqslant \bar{\theta}$ for θ at confidence level $1 - \alpha$ are said to be *unbiased* if

$$P_{\theta,\vartheta}\left[\underline{\theta} \leqslant \theta' \leqslant \bar{\theta}\right] \leqslant 1 - \alpha$$

$$\text{for all } \theta' \neq \theta \text{ and all } \vartheta, \quad (5)$$

i.e., if the probability of covering a false value θ' is always less than or equal to $1 - \alpha$. Results concerning unbiased tests of H: $\theta = \theta_0$ with acceptance region $A(\theta_0)$ yield analogous results concerning unbiased confidence sets $S(x)$ for θ through the relation

$$\theta \in S(x) \quad \text{if and only if} \quad x \in A(\theta). \quad (6)$$

In this way, it is seen, for instance from the results of Example 5, that the confidence intervals $\bar{X} - CS \leqslant \xi \leqslant \bar{X} + CS$ uniformly minimize the probability of covering a false value ξ' among all unbiased confidence intervals with the same confidence level.

The concept of *unbiased test* is due to Neyman and Pearson [24]; that of *unbiased confidence sets* and the theory based on the correspondence (6) to Neyman [23]. A detailed account of this theory is given in Lehmann [18].

A GENERAL CONCEPT OF UNBIASEDNESS

The classical unbiasedness concepts for point estimation, tests, and confidence intervals already discussed can be subsumed under a general decision-theoretic definition of unbiasedness which also has many other applications. A decision procedure $\delta(X)$ is said to be *L-unbiased* with respect to the loss function $L(\theta, d)$ or *risk unbiased* if it satisfies

$$E_\theta L(\theta', \delta(X)) \geqslant E_\theta L(\theta, \delta(X))$$

$$\text{for all } \theta, \theta'. \quad (7)$$

Example 6.

(i) If a real-valued $g(\theta)$ is to be estimated and the loss function is $L(\theta, d) = [d - g(\theta)]^2$, this definition reduces to (2).

(ii) If, instead, the loss function is $L(\theta, d) = |d - g(\theta)|$, (7) reduces to the requirement that $g(\theta)$ be a median of the distribution of $\delta(X)$. An estimator satisfying this condition is said to be *median-unbiased**.

Example 7. Consider the problem of testing H: $\theta \in \omega_0$ against the alternatives K: $\theta \in \omega_1$, denote the decisions to accept or reject by d_0 and d_1, respectively, and let the loss be a for false acceptance, b for false rejection, and 0 if the correct decision is taken. Then (7) reduces to

$$P_\theta(d_1) \leqslant \alpha \quad \text{for } \theta \in \omega_0,$$
$$\geqslant \alpha \quad \text{for } \theta \in \omega_1,$$

where $\alpha = a/(a + b)$, i.e., to the unbiasedness condition for testing already discussed.

Example 8.

(i) For estimation of θ by confidence sets $S(x)$, let the loss be zero if $S(x)$ covers the true value θ and one if it does not. Then (7) reduces to (5).

(ii) For estimation of θ by means of confidence intervals $I(x)$: $\underline{\theta} \leqslant \theta \leqslant \bar{\theta}$, let the loss function be

$$L(\theta, I(x)) = (\theta - \underline{\theta})^2 + (\bar{\theta} - \theta)^2.$$

Then (7) reduces to

$$E_\theta\left[\frac{\underline{\theta} + \bar{\theta}}{2}\right] = \theta, \qquad (8)$$

i.e., to the condition that the midpoint of the interval be an unbiased estimator of θ.

Applications of L-unbiasedness to multiple decision problems can be found in Lehmann [17] and Karlin and Rinott [13].

Risk unbiasedness was introduced by Lehmann [16], who also explores the relationship between unbiasedness and invariance* (or equivariance). Existence problems are investigated in Klebanov [15] and Rojo [30], and the relationship between unbiased and Bayes estimation by Bickel and Blackwell [3] and Noorbaloochi and Meeden [25].

References

[1] Bahadur, R. R. (1957). On unbiased estimates of uniformly minimum variance. *Sankhyā*, **18**, 211–224.

[2] Barankin, E. W. (1949). Locally best unbiased estimates *Ann. Math. Statist.*, **20**, 477–501.

[3] Bickel, P. J. and Blackwell, D. (1967). A note on Bayes estimates. *Ann. Math. Statist.*, **38**, 1907–1911.

[4] Brown, L. D. and Sackrowitz, H. (1984). An alternative to Student's *t*-test for problems with indifference zones. *Ann. Statist.*, **12**, 451–469.

[5] Cassel, C., Särndal, C., and Wretman, J. H. (1977). *Foundations of Inference in Survey Sampling.* Wiley, New York.

[6] David, H. A. (1981). *Order Statistics*, 2nd ed. Wiley, New York.

[7] Downton, F. (1953). A note on ordered least-squares estimation. *Biometrika*, **40**, 457–458.

[8] Efron, B. (1975). Biased versus unbiased estimation. *Adv. Math.*, **16**, 259–277.

[9] Efron, B. (1982). *The Jackknife, the Bootstrap and other Resampling Plans.* SIAM, Philadelphia, PA.

[10] Gauss, C. F. (1821). *Theoria combinationis observationum erroribus minimis obnoxiae.* (An English translation can be found in Gauss' collected works.)

[11] Hoeffding, W. (1983). Unbiased range-preserving estimators. In *Festschrift for Erich Lehmann*, P. J. Bickel, K. A. Doksum, and J. L. Hodges, eds. Wadsworth, Belmont, CA.

[12] James, W. and Stein, C. (1961). Estimation with quadratic loss. In *Proc. Fourth Berkeley Symp. Math. Statist. Prob.*, **1**, 311–319. Univ. of California, Berkeley, CA.

[13] Karlin, S. and Rinott, Y. (1983). Unbiasedness in the sense of Lehmann in *n*-action decision problems. In *Festschrift for Erich Lehmann*, P. J. Bickel, K. A. Doksum, and J. L. Hodges, eds. Wadsworth, Belmont, CA.

[14] Kendall, M. G. and Stuart, A. (1979). *The Advanced Theory of Statistics*, 4th ed., Vol. 2. Hafner, New York.

[15] Klebanov, L. B. (1976). A general definition of unbiasedness. *Theory Prob. Appl.*, **21**, 571–585.

[16] Lehmann, E. L. (1951). A general concept of unbiasedness. *Ann. Math. Statist.*, **22**, 587–592.

[17] Lehmann, E. L. (1957). A theory of some multiple decision problems. I. *Ann. Math. Statist.*, **28**, 1–25.

[18] Lehmann, E. L. (1986). *Testing Statistical Hypotheses.* 2nd edn. Wiley, New York.

[19] Lehmann, E. L. (1981). An interpretation of completeness and Basu's theorem. *J. Amer. Statist. Ass.*, **76**, 335–340.

[20] Lehmann, E. L. (1983). *Theory of Point Estimation*, Wiley, New York.

[21] Lehmann, E. L. and Scheffé, H. (1950). Completeness, similar regions, and unbiased estimation. *Sankhyā*, **10**, 305–340.

[22] Lloyd, E. H. (1952). Least squares estimation of location and scale parameters using order statistics. *Biometrika*, **39**, 88–95.

[23] Neyman, J. (1937). Outline of a theory of statistical estimation based on the classical theory of probability. *Philos. Trans. R. Soc.*, **236**, 333–380.

[24] Neyman, J. and Pearson, E. S. (1936, 1938). Contributions to the theory of testing statistical hypotheses. *Statist. Res. Mem.*, **I**, 1–37; **II**, 25–57.

[25] Noorbaloochi, S. and Meeden, G. (1983). Unbiasedness as the dual of being Bayes, *J. Amer. Statist. Ass.*, **78**, 619–623.

[26] Perlman, M. D. (1980). Unbiasedness of the likelihood ratio tests for equality of several covariance matrices and equality of several multivariate normal populations. *Ann. Statist.*, **8**, 247–263.

[27] Perlman, M. D. and Olkin, I. (1980). Unbiasedness of invariant tests for MANOVA and other multivariate problems. *Ann. Statist.*, **8**, 1326–1341.

[28] Quenouille, M. H. (1949). Approximate tests of correlation in time-series. *J. R. Statist. Soc. B*, **11**, 68–84.

[29] Quenouille, M. H. (1956). Notes on bias in estimation. *Biometrika*, **43**, 353–360.

[30] Rojo, J. (1983). On Lehmann's General Concept of Unbiasedness and Some of Its Applications. Ph.D. thesis, University of California, Berkeley, CA.

[31] Stein, C. (1956). Inadmissibility of the usual estimator for the mean of a multivariate normal distribution. In *Proc. Third Berkeley Symp. Math. Statist. Prob.*, **1**, pp. 197–206. University of California, Berkeley, CA.

(BIAS
CONFIDENCE INTERVALS AND REGIONS
ESTIMABILITY
ESTIMATION, POINT
HYPOTHESIS TESTING
MINIMUM VARIANCE UNBIASED
   ESTIMATION)

E. L. Lehmann

UNCERTAINTY, COEFFICIENT OF

A measure of association* often used in social sciences (see, e.g., ref. 1) for nominal-level variables. It is based on concepts of entropy and mutual information for discrete random variables introduced by C. E. Shannon in 1948 (*see* ENTROPY). One form of this coefficient, when X is an "indepen-

dent" discrete random variable and Y is the "independent" one is

$$\frac{H(Y) - H(Y|X)}{H(Y)},$$

where

$$H(Y) = -\sum p(y_j)\log p(y_j)$$

and

$$H(Y|X) = -\sum_k \sum_i p(y_j|x_k)\log p(y_j|x_k)$$

are entropies. The coefficient measures the proportion by which uncertainty in the dependent variable is reduced by knowledge of the independent one. See, e.g., ref. 1 and Theil [2] for further details.

References

[1] Nie, N. H., Hull, C. H., Jenkins, J. G., Steinbrenner, K., and Bent, D. H. (1975). *SPSS*, 2nd ed. McGraw-Hill, New York.

[2] Theil, H. (1967). *Economics and Information Theory*. Rand McNally, Chicago, IL.

(ASSOCIATION, MEASURES OF
DEPENDENCE, MEASURES AND
   INDICES OF
ENTROPY
KULLBACK INFORMATION)

UNEQUAL PROBABILITY SAMPLING

A term closely related to *probability proportional to size* (pps) *sampling**, but slightly more general. It refers to the situation where the *sample frame** either contains a measure of size already (e.g., populations of districts, turnovers of business units, or numbers of students in schools) or contains data from which such a measure can readily be constructed, usually by a linear summation. For example, if the frame consists of a list of farms which run only sheep and beef cattle, and it is known that each cow or steer is economically equivalent to eight sheep, a reasonable (measure of) size Z_i for the ith farm is

$$Z_i = S_i + 8B_i,$$

where S_i is the number of sheep on the ith farm and B_i is the number of beef cattle on that farm. The optimum probabilities of inclusion in the sample are then usually proportional to a nonlinear function of size, rather than to size itself. The reasons for this conclusion are outlined in the following text.

As shown by Brewer [1], optimum unequal probability sampling can effectively be regarded as the special case of optimum (Neyman) sample allocation for a *stratified design**, where each stratum consists of a single population unit. The optimum inclusion probabilities (i.e,, the Neyman sample fractions) for any stratified design are proportional to the stratum standard deviations of the variable being estimated. When strata are small, the relevant stratum standard deviations can often be regarded as proportional to a function of the average size of units in the stratum. Hence, in the special case where, effectively, each stratum consists of a single population unit, the optimum inclusion probabilities can also be regarded as proportional to a function of size. (Superpopulation models are frequently employed where this relationship is treated as exact. The function of size used in specifying the model is usually a nonlinear one. Quite often it is a fractional power.)

This, then, is the typical situation where the optimum inclusion probabilities in an unequal probability sampling design are proportional not to size but to a function of size. For this reason the expression "unequal probability sampling" is sometimes preferred, as being more general in its reference than "*probability proportional to size sampling*"*, but in all other respects the two may be treated as equivalent.

Reference

[1] Brewer, K. R. W. (1979). *J. Amer. Statist. Ass.*, **74**, 911–915. (Demonstrates relationship between Neyman allocation and optimum unequal probability sampling. Uses a superpopulation model of the type described.)

Bibliography

Brewer, K. R. W. and Hanif, M. (1983). *Sampling with Unequal Probabilities*. Springer-Verlag, New York. (A comprehensive reference on selection procedures and estimators.)

Cassel, C.-M., Särndal, C.-E., and Wretman, J. H. (1977). *Foundations of Survey Sampling*. Wiley, New York. (Contains a discussion of a generalised regression estimator which has certain desirable properties and is specifically designed for use with unequal probability sampling.)

Cochran, W. G. (1977). *Sampling Techniques*, 3rd ed. Wiley, New York. (A very popular textbook on sampling. Good for single stage sampling, but uses a rather cumbersome notation in the multistage situation.)

Hájek, J. (1981). *Sampling from a Finite Population*. Dekker, New York. (An idiosyncratic but refreshingly different approach to the topic.)

Hansen, M. H. and Hurwitz, W. N. (1943). *Ann. Math. Statist.*, **14**, 333–362. (The first use of unequal probability sampling in the literature.)

Horvitz, D. G. and Thompson, D. J. (1952). *J. Amer. Statist. Ass.*, **47**, 663–685. (Introduces the classical Horvitz–Thompson estimator*.)

Kendall, M. G. and Stuart, A. (1976). *The Advanced Theory of Statistics*, Vol. 3. 3rd Edn., Griffin, London. (Chapters 39 and 40 provide a theoretical coverage of sampling survey theory. Chapter 39 in particular is heavily oriented toward unequal probability sampling.)

Kish, L. (1965). *Survey Sampling*. Wiley, New York. (Good on the practical side of unequal probability sampling, particularly in the multistage situation.)

Rao, J. N. K. (1978). In *Contributions to Survey Sampling and Applied Statistics*, H. A. David, ed. Academic, New York, pp. 69–87. (A useful theoretical summary and bibliography.)

Schreuder, H. T., Sedransk, J., Ware, K. D., and Hamilton, D. A. (1971). *Forest Sci.*, **17**, 103–118. (Unequal probability sampling in forestry developed almost independently of other applications until these authors came together.)

Vos, J. W. E. (1974). *Statistica Neerlandica*, **28**, 11–49, 69–107. (A survey of methods in use up to the time of publication.)

(OPTIMUM STRATIFICATION
PROBABILITY PROPORTIONAL TO SIZE
(PPS) SAMPLING
STRATIFIED DESIGNS)

K. R. W. Brewer

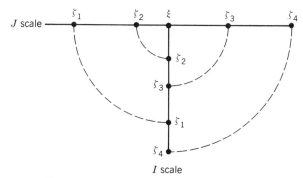

Figure 1   Relationship between *I*-scale and *J*-scale.

UNFOLDING

The unfolding model is based on the concept of a joint space. Consider an experiment in which the preferences of several individuals for each of a set of stimuli are measured. It is assumed that each individual and each stimulus can be represented by points in a Euclidean space called the *joint space*. The points in this space corresponding to the individuals are called *ideal* points. It is assumed that an individual's ideal point corresponds to his ideal stimulus and that the closer to this point a stimulus point lies, the greater his preference for that stimulus will be.

The unfolding model can also be applied more generally to data consisting of distances or dissimilarities between objects in one set and objects in another distinct set. In this context it will be shown that unfolding can be regarded as a special case of multidimensional scaling* (MDS) with blocks of missing information in the matrix of estimated dissimilarities between all objects from both sets. As an example of an unfolding analysis of distance perceptions, a number of people could be asked to estimate the distance from their homes to various city landmarks. It would be possible to use unfolding analysis to produce a map of the city with points for each landmark and for each individual's home.

To see why the model is called the unfolding model, consider a joint space consisting of the points on the real line. This is the original one-dimensional joint space postulated by Coombs (see, for example, ref. 2). Figure 1 shows an ideal point ξ and four stimulus points ζ_1, ζ_2, ζ_3, and ζ_4 lying in the joint space. This space is called the joint scale or *J-scale*. The order of preference for an individual with ideal point ξ would be ζ_2, ζ_3, ζ_1, and last, ζ_4. This ordering is given by the *I-scale*, which can be thought of as the *J*-scale folded at the ideal point ξ, as shown by the dotted lines in the figure. Thus, given a number of observed or inferred *I*-scales, the problem is to unfold these into a common underlying *J*-scale. The preference rankings for an individual are monotonically related to the estimated distances between the individual's ideal point and the stimulus points in the joint space. The ranks can thus be treated as dissimilarity measures and the unfolding problem can be seen as a nonmetric multidimensional scaling problem with missing data. Points representing both individuals and stimuli are to be scaled, but the portions of the dissimilarity matrix representing dissimilarities between individuals and stimulti are missing and only the rectangular portion of the matrix corresponding to dissimilarities between individuals and stimuli is observed. Since the dissimilarities are only assumed to be monotonically related to distances for each individual in the preceding example, it would not matter if some people used miles in estimating distances, some used kilometers, and some gave the time in minutes taken to walk there.

Any program for nonmetric MDS that has the facility to specify a separate monotonic transformation for each individual could in theory be used to perform an unfolding analysis. In practice, however, the large proportion of missing information and the necessity for separate transformations make the problem more difficult and standard nonmetric MDS programs usually produce unsatisfactory results.

Heiser [5] and Evers-Kiebooms and Delbeke [3] have produced procedures especially designed for unfolding analysis based on the methodology of nonmetric MDS. Heiser's procedure overcomes some of the numerical problems by placing restrictions on the monotonic transformations to avoid pathological cases. This procedure is implemented in the SMACOF-III computer program, which is referred to in ref. 5. This reference also gives a very comprehensive survey on the literature of unfolding up to 1981.

If the observed or inferred dissimilarities provide direct estimates of the distances in the joint space rather than merely being monotonically related to estimates of these distances, then the problem reduces to a metric unfolding problem. An algebraic solution for such a problem is given by Schönemann [7]. This solution can prove unstable in the fallible case, but an efficient iterative least squares procedure is given by Greenacre [4]. Carroll [1] gives an interesting

survey of both metric and nonmetric unfolding methods. Probabilistic unfolding models have been formulated for analysing preference data. Models of this kind are given by Zinnes and Griggs [9] and Schönemann and Wang [8]. There are also "quasinonmetric" versions of Schönemann and Wang's model that attempt to avoid the difficulties of the nonmetric approach while retaining much of its flexibility by allowing the nonmetric transformation function to be any member of a general but well-behaved family of functions. Muller [6] gives a model of this type which is an extension of Schönemann and Wang's model and also provides a maximum likelihood estimation* procedure.

An example of an unfolding analysis using this procedure is based on research on human values carried out at the National Institute for Personnel Research of the South African Council for Scientific and Industrial Research. It was hypothesised that the ideal points and the 10 adjectives "accepting" (AC), "competitive" (CO), "caring" (CA), "flexible" (FL), "controlled" (CN), "independent" (IN), "conforming" (CF), "tolerant" (TO), "disciplined " (DI), and "cooperative" (CP) could be represented in a two-dimensional joint space. The hypothesised dimensions were "open–closed" and "individualist–collectivist." Subjects were asked to choose the better human quality

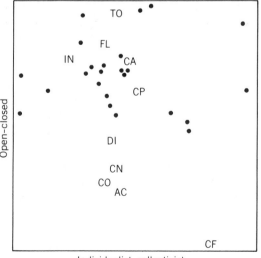

Figure 2 Unfolding solution in two dimensions.

from each of the 45 possible pairs formed from these adjectives. An unfolding analysis in two dimensions yielded the results shown in Figure 2. The vertical axis is the "open–closed" dimension which contrasts "tolerant" and "flexible" with "controlled" and "conforming." The stimulus points do not vary so much on the horizontal axis, but the highest coordinate values are for "conforming" and "cooperative" and the lowest for "independent," so this can tentatively be identified as the "individualist–collectivist" dimension. Most of the ideal points lie on the "open" side of the "open–closed" axis as could be expected from research workers. In interpreting the results of an unfolding analysis, it is important to remember that the distances between an ideal point and the stimulus points are inversely related to relative preference for that individual. Two people with ideal points that are close together will have the same order of preference for the stimuli, but, in absolute terms, one may like all the stimuli while the other dislikes all of them.

To sum up, it seems that, while early experience with unfolding as a practical analysis technique was rather negative, the newer procedures are proving more successful. Some recent computational experience suggests that the restricted nonmetric method and the quasinonmetric method tend to produce similar results. This gives greater confidence in the results of these analyses compared with earlier methods.

References

[1] Carroll, J. D. (1972). In *Multidimensional Scaling: Theory and Applications in the Behavioral Sciences*, Vol. 1, R. N. Shepard, A. K. Romney, and S. B. Nerlove, eds. Seminar Press, New York, pp. 105–155.

[2] Coombs, C. H. (1964). *A Theory of Data*. Wiley, New York.

[3] Evers-Kiebooms, G. and Delbeke, L. (1982). *Psychologica Belgica*, **22**, 99–119.

[4] Greenacre, M. J. (1978). Some Objective Methods of Graphical Display of a Data Matrix. *Special Report*, Dept. of Statistics and Operations Research, University of South Africa, Republic of South Africa.

[5] Heiser, W. J. (1981). Unfolding Analysis of Proximity Data. Doctoral dissertation, University of Leiden, The Netherlands.

[6] Muller, M. W. (1984). Multidimensional Unfolding of Preference Data by Maximum Likelihood. *Special Report PERS/374*, National Institute for Personnel Research, Republic of South Africa.

[7] Schönemann, P. H. (1970). *Psychometrika*, **35**, 349–366.

[8] Schönemann, P. H. and Wang, M. M. (1972). *Psychometrika*, **37**, 275–309.

[9] Zinnes, J. L. and Griggs, R. A. (1974). *Psychometrika*, **39**, 327–350.

(MEASURES OF SIMILARITY, DISSIMILARITY, AND DISTANCE MULTIDIMENSIONAL SCALING STATISTICS IN PSYCHOLOGY)

M. MULLER

UNICLUSTER DESIGN

A unicluster sampling design with exactly k clusters is implemented by choosing in a random manner one subset of the population out of a given nonrandom partition into k subsets. A systematic sampling design* with one single random starting point is an example of a unicluster design; other meaningful examples are yet to be discovered.

In the "general theory of survey sampling*," the unicluster designs crop up now and then, not because they are in any way particularly interesting but mainly because they are annoying exceptions to general theorems. For example, in the nonexistence theorem of Godambe [1] this situation occurs: In a unicluster design, and only in such a design, does there exist a uniform minimum variance (UMV) linear unbiased estimator. (The reason for this existence is very humdrum: In such a design there exists only one unbiased linear estimator and that estimator is of course UMV in the class consisting just of itself.) For more details on this point, see [2]; for another example, see [3].

References

[1] Godambe, V. P. (1955). *J. R. Statist. Soc. B*, **17**, 269–278.

[2] Lanke, J. (1973). *Metrika*, **20**, 196–202.

[3] Lanke, J. and Ramakrishnan, M. K. (1974). *Ann. Statist.*, **2**, 205–215.

(SURVEY SAMPLING
SYSTEMATIC DESIGN)

JAN LANKE

UNIDIMENSIONALITY, TESTS OF

The notion of the latent unidimensionality of a data set or joint distribution has at its roots the ability of a single unobservable quantity to "explain" the dependence structure of the observed data. The search for a unidimensional representation of data by an underlying factor was strongly motivated by initial attempts in group mental testing in the first half of the twentieth century (Spearman [15], Thurstone [17], and others). These early psychometricians were trying to discover if a single "general intelligence" factor could explain performance in a battery of various tests, such as an analogies/reading comprehension/arithmetic combination. (*See* FACTOR ANALYSIS.)

What is meant by "explaining the data by a single variable" can be best described by a common example drawn from elementary statistics: Consider a group of first year law students' test scores on (1) the LSAT (a standardized test used by law schools to determine basic aptitudes of its entering class) and (2) a final exam in a first year basic law course. Correlation between these two measures is commonly in the 0.3–0.6 range (see Freedman et al. [8, Chap. 10]). Now suppose that an additional measure is made available, first year grade point average (GPA), for example. If one looks only at those students having a GPA in a narrow range, the correlations between LSAT and basic law course examination scores will commonly disappear; the only unexplained variation in a restricted group such as this is measurement error. Here, GPA is the variable that allows a unidimensional representa-

tion of the pair (LSAT score, first year score). GPA is observable; in many applications in psychometrics and other fields, however, the explanatory variable is hidden (henceforth called latent). McDonald [14, p. 101] puts it most succinctly: The primary property of a unidimensional data set is that "the partial correlations of the [observed variables] are all zero if the [hidden variable] is partialled out."

To put the preceding concepts into a mathematical context, suppose

$$\mathbf{X} = (X_1, \ldots, X_n)$$

are n observable random variables,

$$\Theta = (\Theta_1, \ldots, \Theta_d)$$

are d latent variables, and

$$f_i(x_i | \Theta)$$

is the conditional density (or mass, if discrete) function of X_i conditioned on Θ. If the latent variables explain the association between the observables, then

$$f(\mathbf{x} | \Theta) = \prod_{i=1}^{n} f_i(x_i | \Theta),$$

i.e., $\mathbf{X}$ is an independent set of variables when Θ is conditioned upon; here $f(\mathbf{x} | \Theta)$ is the joint conditional distribution of the observables. A unidimensional set is one in which Θ is a scalar, so that only one latent variable produces independence:

$$f(\mathbf{x} | \theta) = \prod_{i=1}^{n} f_i(x_i | \theta).$$

Applications of this model can be found in several areas, including genetics, social sciences, and psychometrics, where the existence of an underlying quantity is suspected to cause association (see Holland and Rosenbaum [9]).

1. Genetics. Consider n offspring of a fixed parent pair and let x_i indicate for the ith offspring the presence ($x_i = 1$) or absence ($x_i = 0$) of a characteristic completely determined by the presence (Aa or AA) or absence (aa) of a single dominant allele (A). Let the scalar θ be the conditional probability of the characteristic occuring in a randomly chosen offspring given the parents' unobservable genotype (e.g., father might be Aa, mother aa). Thus,

$$\theta = P[\,X_i = 1 | \text{parents' pair of genotypes}\,].$$

Then the x_i's are independent conditional on θ and satisfy a latent variable model.

2. Systems Reliability* (Engineering). The x_i's indicate the functioning of the ith component of a system, $i = 1, \ldots, n$, where $x_i = 1$ if it is functioning and $x_i = 0$ if it is not. A latent set of variables that could be useful here is a set of various stress forces that affects all of the components. If enough such stresses are modeled, then the x_i's are conditionally independent, satisfy a latent variable model, and further, if independence is present conditioned on only one stress, unidimensionality holds.

3. Latent Trait* Models (Psychometrics). The modeling of causative factors in test question responses is widely used nowadays, especially since the advent of standardized testing. It is often desired to know whether the test measures a single unobservable trait, such as "mathematical ability" or other accepted psychometric construct. The observed variables in the model are scores (x_i) on individual items on a test (say n of them), with 0 for an incorrect response and 1 for a correct response. The replications are N test examinees. Under these models, the item scores are independent once the latent traits are conditioned upon.

Most work using latent variable models has centered on the preceding application, for which detecting test unidimensionality is crucial for two reasons:

1. It is vital that a test that purports to measure a certain ability (latent trait) is not contaminated by other abilities.

2. Many test theorists consider it foundational that a test designed to measure individual differences must measure a unified ability, i.e., is the ability to be tested really a single trait?

Henceforth, then, discussion of statistical tests of unidimensionality are confined to this area.

Many approaches for detecting test unidimensionality have been developed: however, many are ad hoc procedures not based on a model at all, making the very definition of "unidimensionality" unclear. Of the statistical tests of unidimensionality derived from a latent trait model, only a few of the more important procedures will be discussed.

LINEAR FACTOR ANALYTIC APPROACH

(*See* FACTOR ANALYSIS; see also Lord and Novick [11] and Hulin et al. [10].) Initial attempts at assessing dimensionality used a modified linear factor analysis model, which assumes that an observed 0/1 item response is dichotomized by a threshold mechanism operating on an underlying normally distributed item response (useful if the use of factor analysis is to be defended). The tetrachoric correlation* matrix of the item responses is factored (using communality estimates in the diagonal of the observed correlation matrix) and dimensionality indices are calculated from the resultant eigenvalues and factor loadings (cf. Tucker et al. [18] and Drasgow and Lissak [7]). Typically, a dominant eigenvalue or recognizable pattern in the loadings occurs when unidimensionality is present. The indices detect departures from unidimensionality in a variety of testing situations but have unknown distributions, even asymptotic ones, so estimates of the indices' standard errors are not obtainable (asymptotic theory on factor loading estimates has been done by Christofferson [6], but he supplies no test of unidimensionality). An additional foundational problem in these classically oriented analyses is the undesirable detection of spurious "item difficulty" and "examinee guessing" dimensions.

Most approaches to formulating statistical tests of unidimensionality involve working with *item characteristic curves* (henceforth

ICCs)

$$E[X_i|\theta] = P_i(\theta), \qquad i = 1,\ldots,n,$$

where each ICC $P_i(\cdot)$ is assumed to be a nondecreasing nonlinear function of the latent trait θ. The interpretation of $P_i(\theta)$ is the "probability of a correct response to item i, given ability θ."

NONLINEAR FACTOR ANALYSIS (MCDONALD [13])

McDonald states that a set of responses is unidimensional if and only if it fits a nonlinear factor model with one factor. Typically he assumes that (1) each item's ICC is a normal ogive with item-dependent scale and location parameters (called a two-parameter normal ogive family) and (2) the single latent variable is normally distributed. He then implements the nonlinear factor model by calculating the covariance structure of the observed $\mathbf{x}$ as a function of θ, using a Fourier decomposition of $P_i(\theta)$ into orthogonal polynomials (thus, in effect, linearizing the analysis) and fitting this covariance matrix to the observed matrix of the responses by least squares, applying an asymptotic test of fit to the residuals. See McDonald [12, 13] for details.

GOODNESS OF FIT TESTS TO A UNIDIMENSIONALITY MODEL ASSUMING PARAMETRIZED ICCs

Bartholomew [1], Bejar [2], Bock [4], and others have specified various parametric forms for ICCs in devising tests of unidimensionality. For brevity, we refer only to the details of Bock's procedure. Bock basically assumes a one-dimensional normal distribution on the latent variabale and calculates multinomial probabilities of each of the possible $2^{(\text{number of items})}$ response patterns. He also assumes that the number of examinees greatly exceeds this number, so that each response pattern's relative frequency can be accurately estimated. If unidimensionality

holds, these response pattern probabilities have a certain form; a likelihood ratio test of fit to this form, against a general multinomial alternative, is applied. Rejection of unidimensionality occurs if the likelihood ratio statistic

$$G^2 = 2 \sum_{\text{all response patterns } l} r_l \log\left[\frac{r_l}{Nh(l)}\right]$$

is too large. Here r_l is the observed number of occurrences of response patterns l, $h(l)$ is the estimated probability of response pattern l under assumption of unidimensionality, and N is the number of examinees taking the test. The test is based on the theoretical fact that G^2 is approximately $\chi^2(2^n - 3)$ when unidimensionality holds. The calculation of the probabilities $h(l)$ is done by assuming each ICC to be a two-parameter normal ogive function, estimating the item-dependent location and scale parameters by a method called marginal maximum likelihood (Bock and Aitkin [5]), and evaluating numerically

$$h(l) = \int \phi(\theta) \prod_{i=1}^{n} P_i(\theta)\, d\theta,$$

$\phi(\cdot)$ denoting the density function of θ (standard normal).

NONPARAMETRIC ICCs, A CONTINGENCY TABLE* PROCEDURE (HOLLAND AND ROSENBAUM [9])

In this framework, only unidimensionality and monotonicity of the ICCs are assumed, hence justifying the term "nonparametric." A brief description of the procedure is as follows: Consider a pair of items i and i'. Condition on the total score of the remaining items to produce a layered 2×2 contingency table with $n - 2$ layers. Apply a Mantel–Haenzel* weighted combination of odds ratio test (MH test) (Bishop, Fienberg, and Holland [3, pp. 146–148]) to test for *negative partial association*, i.e., $\mathrm{Cov}(X_i, X_{i'}|\Sigma_{k \neq i, i'} X_k) < 0$ for each $i \neq i'$. Note that there are $n(n - 1)/2$ pairings of items possible so that $n(n - 1)/2$ tests are being performed. If a substantial number of these item-pair tests show negative partial association, it is evidence that a large number of items are pairwise disparate, thus indicating that a unidimensional model is not plausible.

The theoretical basis for the approach is that in a unidimensional test, conditional association of X_i and $X_{i'}$ holds, i.e.,

$$\text{Cov}\left(X_i, X_{i'} \mid \sum_{k \neq i,\, i'} X_k = c\right) \geqslant 0$$

$$\text{for all } 1 \leqslant i \neq i' \leqslant n$$
$$\text{and all } 0 < c < n - 2$$

(see Holland and Rosenbaum [9, Def. 3.4 and Theorem 6] for an explanation). A major strength of this approach is that the $n(n-1)/2$ individual tests statistics are approximately normally distributed under the null hypothesis if the number of examinees is large.

NONPARAMETRIC ICCs, A
TEST-SPLITTING APPROACH (STOUT [14])

Stout's procedure also assumes unidimensionality and monotonicity as does that of Holland and Rosenbaum. It involves (1) splitting the test into two subtests [one a "pure" (i.e., unidimensional) test, called the *assessment test*, the other consisting of the remaining items and called the *partitioning test*], then (2) conditioning on the total score of the partitioning test, a statistic based on the responses on the assessment test is calculated. The procedure uses an explanatory factor analysis of the item responses to select the items that have the most similar factor loadings to form the assessment test.

The evidence for multidimensionality occurs with large positive values of the statistic, which is based on conditional covariances of item responses, as is the method of Holland and Rosenbaum. Specifically, the statistic is based on the fact that under a unidimensional model,

$$\text{Cov}\left(X_i, X_{i'} \mid \sum_{k=1}^{m} X_k = c\right) \to 0,$$

as the number of items approaches infinity

and the number of examinees slowly approaches infinity for all $0 < c < m - 1$ and all i, i' on the assessment test, where $\sum_{k=1}^{m} X_k$ is the total score on the partitioning test. Rejection of unidimensionality occurs if, for a long partitioning test, the sum of the preceding covariances is judged sufficiently greater than zero. An additional property of Stout's statistic is that when unidimensionalaity holds, it is asymptotically standard normal as the number of test items approaches infinity and the number of examinees slowly become infinite. It was found necessary to make a correction for finite test length bias. A large scale Monte Carlo* simulation study was conducted showing that the bias-corrected statistic displays good power when unidimensionality fails and displays faithful adherence to the nominal level of significance when unidimensionality holds. Moreover, an improved asymptotic theory for the bias-corrected statistic was obtained with the removal of the requirement that the number of examinees must go "slowly" to infinity.

Although very different in detail, the Holland and Rosenbaum method and the Stout method are foundationally similar. The primary differences in the two methods are twofold. First, Holland and Rosenbaum base their procedure on the fact that the conditional covariances [$\text{Cov}(X_i, X_{i'} \mid \sum_{k=1}^{m} X_k = c)$] are always nonnegative when unidimensionality holds. By contrast, Stout bases his procedure on the fact that all such covariances approach zero asymptotically when unidimensionality holds. Second, Holland and Rosenbaum, in practice, compare pairs of test items (i and i') that are suspected to be dissimilar, causing negative correlations when unidimensionality fails, while Stout looks at a set of similar items (the "pure" assessment test), conditioned on the total score of the partitioning test, which is dissimilar to the assessment test when unidimensionality fails; if unidimensionality holds and hence the partitioning test is also "pure" and similar to the assessment subtest, then effectively one has conditioning on the single latent variable, and thus the covariances of the assessment test items are asymptotically negligible.

CONCLUSION

In summary, while the stand-alone factor analysis approach fails, these other approaches in detection of a unidimensionality latent variable model all show promise. At the time of the writing of this article, there is no consensus of either theoreticians or practitioners about which method or methods are best.

References

[1] Bartholomew, D. J. (1980). Factor analysis for categorical data. *J. R. Statist. Soc. B*, **42**, 293–321.

[2] Bejar, I. I. (1980). A procedure of investigating the unidimensionality of achievement tests based on item parameter estimates. *J. Educ. Meas.*, **17**, 283–296. (A test of unidimensionality of a set of test items using a logistic ICC family.)

[3] Bishop, Y. M., Fienberg, S. E., and Holland, P. W. (1977). *Discrete Multivariate Analysis: Theory and Practice*, MIT Press, Cambridge, MA.

[4] Bock, R. D. (1984). Contributions of empirical Bayes and marginal maximum likelihood methods to the measurement of individual differences. *Proc. 23rd Int. Conf. Psychology*, to appear.

[5] Bock, R. D. and Aitkin, M. (1981). Marginal maximum likelihood estimation of item parameters: Application of an EM algorithm. *Psychometrika*, **46**, 443–459.

[6] Christoffersson, A. (1975). Factor analysis of dichotomous variables. *Psychometrika*, **40**, 5–32.

[7] Drasgow, F. and Lissak, R. (1983). Modified parallel analysis: A procedure for examining the latent dimensionality of dichotomously scored item responses. *J. Appl. Psychol.*, **68**, 363–373.

[8] Freedman, D., Pisani, R., and Purves, R. (1978). *Statistics*. W. W. Norton and Company, New York, Chap. 10.

[9] Holland, P. W. and Rosenbaum, P. R. (1986). Conditional association and unidimensionality in monotone latent variable models. *Ann. Statist.*, **14**, 1523–1543. (A technical treatment by a mathematical statistician to determine if a test is unidimensional, using categorical data analysis.)

[10] Hulin, C. L., Drasgow, F., and Parsons, L. K. (1983). *Item Response Theory*. Dow Jones–Irwin, Homewood, IL. (A text on item response theory;

various ways of assessing the unidimensionality of a set of test items are discussed.)

[11] Lord, F. M. and Novick, M. R. (1968). *Statistical Theories of Mental Test Scores*. Addison-Wesley, Reading, MA. (The first definitive text of test theory; unidimensionality under test theory models is assumed.)

[12] McDonald, R. P. (1974). Difficulty factors in binary data. *Brit. J. Math. Statist. Psychol.*, **27**, 82–99.

[13] McDonald, R. P. (1980). A simple comprehensive model for the analysis of covariance structures: Some remarks on applications. *Brit. J. Math. Statist. Psychol.*, **33**, 161–183.

[14] McDonald, R. P. (1981). The dimensionality of test and items. *Brit. J. Math. Statist. Psychol.*, **34**, 100–117.

[15] Spearman, C. (1904). "General intelligence" objectively determined and measured. *Amer. J. Psychol.*, **15**, 201–293.

[16] Stout, W. (1987). A nonparametric approach for assessing latent trait unidimensionality. *Psychometrika*, **52**. (A technical treatment by a mathematical statistician to determine if a test is unidimensional, using probability theory.)

[17] Thurstone, L. L. (1938). Primary mental abilities. *Psychometric Monographs*, **1**.

[18] Tucker, L. R., Humphreys, L. G., and Roznowski, M. A. (1986). Comparative accuracy of five indices of dimensionality of binary types. *Tech. Rep. No. N00014-84-K-0186*, Office of Naval Research.

(FACTOR ANALYSIS
ITEM RESPONSE THEORY
LATENT STRUCTURE ANALYSIS
LATENT TRAIT THEORY
STATISTICS IN PSYCHOLOGY)

ROBIN SHEALY
WILLIAM STOUT

UNIFORM DISTRIBUTION MODULO 1

Let $x_1, x_2, \ldots$ be a sequence of real numbers and consider their fractional parts $\{x_n\} = x_n - [x_n]$, i.e., the numbers x_n reduced modulo 1. The sequence $x_1, x_2, \ldots$ is said to be *uniformly distributed modulo 1* (u.d. mod 1) if for each $x \in [0, 1]$,

$$\lim_{n \to \infty} \frac{1}{n} \#(k \leqslant n: \{x_k\} < x) = x, \quad (1)$$

in other words, if in the long run each interval contained in $[0, 1]$ gets its fair share of points $\{x_n\}$. Examples of sequences u.d. mod 1 are given by $x_n = n\theta$, or $x_n = p_n\theta$, where θ is an irrational number and p_n is the nth prime number.

Formulated in terms of probability theory, a sequence $x_1, x_2, \ldots$ is u.d. mod 1 if the averages $\mu_n := (1/n)\Sigma_{k \leqslant n}\delta\{x_k\}$ of the point masses at $\{x_k\}$ converge weakly to Lebesgue measure λ on $[0, 1)$ or if the empirical distribution $F_n(x) := (1/n)\Sigma_{k \leqslant n}I_{[0, 1)}(X_k)$ of the constant random variables X_k with value $\{x_k\}$ converges to the uniform distribution over $[0, 1)$.

An often efficient way to determine whether or not a sequence is u.d. mod 1 is provided by *Weyl's criterion*: A sequence $x_1, x_2, \ldots$ is u.d. mod 1 if and only if

$$\lim_{n \to \infty} \frac{1}{n} \sum_{k \leqslant n} e^{2\pi i m x_k} = 0 \qquad (2)$$

for all integers $m \neq 0$. If, for instance $x_n = n\theta$ with θ irrational, then the expression inside the limit is bounded by

$$2\left[n|1 - e^{2\pi i m\theta}|\right]^{-1} \to 0$$

for all integers $m \neq 0$ since $e^{2\pi i m\theta} \neq 1$.

To obtain an intuitive understanding of Weyl's [7] criterion, think of the real axis wrapped around a circle of perimeter 1. Then the numbers $\{x_n\}$ correspond to a point on the circle at an angle $2\pi x_n$. Weyl's criterion says, in effect, that for a sequence $x_1, x_2, \ldots$ to be u.d. mod 1 it is necessary and sufficient that the barycenter of the corresponding points on the circumference of the circle approach the origin as $n \to \infty$; this not just for the points themselves, but also for each of their nonzero integral multiples.

There are criteria similar to (2) for a sequence $x_1, x_2, \ldots$ to be u.d. mod 1, also due to H. Weyl [7], where, on the left side of (2), the exponential function is replaced by either a continuous or a Riemann integrable function f with period 1 and where, accordingly, 0 on the right side is replaced by

$$\int_0^1 f(x)\, dx.$$

A number $0 < \omega < 1$ is called *normal to base b* ($\geqslant 2$, integer) if in its b-adic expansion $\omega = \Sigma_{n=1}^{\infty} d_n(\omega)b^{-n}$, where $d_n(\omega) = 0, 1, \ldots, b - 1$, each block of digits occurs with its proper frequency. If, for instance, $b = 2$, then the block 11101 should occur with asymptotic frequency $1/32$. *Borel's normal number theorem* states that all ω outside a set of Lebesgue measure zero are normal to any given base b. In present day language this is just the strong law of large numbers* applied to the sequence of independent Bernouli trials with probability of success $p = b^{-k}$, where k is the number of digits in the given block. The connection with uniform distribution mod 1 is given by a theorem of Wall [6]: ω is normal to base b if and only if the sequence $x_n = b^n\omega$ is u.d. mod 1.

The central concept of the quantitative theory of uniform distribution is the *discrepancy*. The discrepancy D_N of the finite sequence $x_1, \ldots, x_N$ is given by

$$D_N := \sup_{0 \leqslant \alpha < \beta \leqslant 1} \left| \frac{1}{N} \sum_{k \leqslant N} I_{[\alpha, \beta)}(\{x_k\}) - (\beta - \alpha) \right|.$$

For an infinite sequence $x_1, x_2, \ldots$ the discrepancy D_N means the discrepancy of the first N elements; the sequence is u.d. mod 1 if and only if $\lim_{N \to \infty} D_N = 0$. This follows by an argument used in the proof of the Glivenko–Cantelli theorem* on empirical distribution functions. The discrepancy is a quantitative measure of how uniform the distribution mod 1 of a sequence $x_1, x_2, \ldots$ is over intervals. A very important theorem is the following inequality of Erdös and Turán [1]. For any positive integer M

$$D_N \leqslant \frac{4}{M+1} + \frac{4}{\pi} \sum_{m=1}^{M} \frac{1}{N} \left| \sum_{n=1}^{n} e^{2\pi i m x_n} \right|.$$

The discrepancy of an infinite sequence cannot be too small. According to a theorem of W. M. Schmidt [5], there is an absolute constant $c > 0$ such that $D_N \geqslant c \log N$ infinitely often, no matter how the sequence $x_1, x_2, \ldots$ is chosen. Such results belong to the area of "irregularities of distribution." On the other hand, sequences $x_1, x_2, \ldots$ have been constructed with $D_N = O(\log N)$. Such "low discrepancy sequences" have importance for numerical integration* because of the following inequality of Koksma [3] and its higher dimensional extension due to Hlawka [2]:

$$\left| \frac{1}{N} \sum f(x_n) - \int_0^1 f(t) \, dt \right| \leqslant V(f) D_N$$

for any finite sequence $x_1, \ldots x_N$ contained in $[0, 1)$ and any function f of bounded variation $V(f)$ on $[0, 1)$.

Of particular interest to probabilists is the metric theory of uniform distribution. The following is a typical example. Let $a_1, a_2, \ldots$ be a sequence of real numbers with $a_n - a_{n-1} \geqslant \delta > 0$ for all $n \geqslant 2$. Then for all ω outside a set of Lebesgue measure zero, the sequence $a_1 \omega, a_2 \omega, \ldots$ is u.d. mod 1. Paraphrased in terms of probability theory this result reads as follows. Let X be a random variable with uniform distribution over a finite interval. Then the sequence of random variables $\{a_1 X\}, \{a_2 X\}, \ldots$ satisfies the conclusion of the Glivenko–Cantelli theorem. If the sequence $a_1, a_2, \ldots$ is lacunary, i.e., if $a_n / a_{n-1} \geqslant q > 1$, then $\{a_1 X\}, \{a_2 X\}, \ldots$ even satisfies the bounded Chung–Smirnov law of the iterated logarithm* for empirical distribution functions.

The theory of uniform distribution has been extended in many ways. If the limit in (1) equals $g(x)$, then the sequence $x_1, x_2, \ldots$ is said to have *asymptotic distribution function mod 1 $g(x)$*. The elements of the sequence can be vectors in $\mathbb{R}^s$ or elements of a compact space. Instead of considering the arithmetic mean before taking the limit in (1), general summation methods via Toeplitz matrices* have been studied extensively.

The standard reference for these topics up to 1974 is ref. 4. Since then research efforts have been mostly concentrated around questions on irregularities of distribution, on upper bounds for the discrepancy of number-theoretically interesting sequences, and on questions on the distribution of sequences in

abstract spaces and their connection with abstract harmonic analysis.

References

[1] Erdös, P. and Turán, P. (1948). On a probelm in the theory of uniform distribution, I. *Indag. Math.*, **10**, 370–378.

[2] Hlawka, E. (1961). Funktionen von beschränkter Variation in der Theorie der Gleichverteilung, *Ann. Math. Pure. Appl.*, (4) **54**, 325–333.

[3] Koksma, J. F. (1942/3). Een algemeene stelling uit de theorie der gelijkmatige verdeehling modulo 1. *Mathematica B* (*Zutphen*), **11**, 7–11.

[4] Kuipers, L. and Niederreiter, H., (1974). *Uniform Distribution of Sequences*. Wiley, New York.

[5] Schmidt, W. M. (1972). Irregularities of distribution. VII. *Acta Arith.*, **21**, 45–50.

[6] Wall, D. D. (1949). Normal Numbers, *Ph.D. thesis*, University of California, Berkeley, CA.

[7] Weyl, H. (1916). Über die Gleichverteilung von Zahlen mod Eins. *Math. Ann.*, **77**, 313–352.

(BARYCENTRIC COORDINATES
CONVERGENCE OF SEQUENCES
   OF RANDOM VARIABLES
DIRECTIONAL DISTRIBUTIONS
PROBABILITY THEORY (OUTLINE)
UNIFORM (OR RECTANGULAR)
   DISTRIBUTIONS
WRAPPED DISTRIBUTIONS)

WALTER PHILIPP

UNIFORMITY, TESTS OF

INTRODUCTION

There are many statistical situations where data either occur naturally uniformly distributed or where a transformation will give values which, when a given hypothesis H_0' is true, will be uniformly distributed. A test for H_0' is then transformed into a test of uniformity, namely, of the new hypothesis H_0 that a set of univariate values, say U, comes from a continuous uniform distribution with limits a and b: Often these limits are 0 and 1.

Tests of uniformity for univariate observations form the primary subject of this article.

There is an extensive literature on such tests, and many of them have already been discussed in these volumes (see, for example, ANDERSON–DARLING STATISTIC; EDF STATISTICS; KOLMOGOROV–SMIRNOV TESTS OF FIT; NEYMAN'S AND OTHER SMOOTH GOODNESS-OF-FIT TESTS; NEYMAN'S TEST FOR UNIFORMITY). Further details on the test procedures, including some historical development and tables to make the tests, are given in the review article by Stephens [33]. This review contains many references, and for the most part references given here will be to basic or introductory articles on the tests or to newer work not discussed in Stephens [33].

Suppose $U(a, b)$ denotes the uniform continuous distribution (also called the rectangular distribution) with limits a, b. A set of i.i.d. (identically and independently distributed) random variables from $U(0, 1)$ will be called *uniforms*: When these are placed in ascending order, they become *ordered uniforms*.

An example of a naturally occurring uniform distribution is the distribution of errors when data are "rounded-off," say to the nearest integer; the error between the true value and the recorded value might then, in most situations, be $U(-\frac{1}{2}, \frac{1}{2})$.

Important transformations that give uniform variates are the probability integral transformation* (PIT) and two transformations that change exponential random variables to uniforms. Use of the latter makes it possible to change a test for exponentiality to a test for uniformity, so that they become important in examining renewal processes and in reliability theory, queuing theory*, and survival analysis, where exponentials can be expected to occur. There are also several transformations that take uniforms to a new set of uniforms; they are of interest in possibly increasing the power of a test.

The Probability Integral Transformation (PIT)

Suppose random variable X has a continuous distribution $F(x; \theta)$, where θ denotes a vector of parameters occurring in the specification of the distribution. Then if $U = F(X; \theta)$, the new random variable U has the $U(0, 1)$ distribution. Thus a test of H_0': a set X is a random sample from $F(x; \theta)$, can be converted by the transformation $U_i = F(X_i; \theta)$, $i = 1, \ldots, n$, to the test of

H_0: a set of n values U

is a random sample from $U(0, 1)$.

TESTS OF UNIFORMITY

Many tests have been devised for the test of H_0; We shall group them under several headings. Suppose U_i, $i = 1, \ldots, n$, denotes a sample of U values and let $U_{(1)} < U_{(2)} < \cdots < U_{(n)}$ be the order statistics*.

A Likelihood Ratio Test* Statistics

When the alternative to uniformity is very precisely defined, likelihood ratio methods can be used to give a test statistic that will have optimal properties in detecting that alternative. Thus the mean $\overline{U}$ is the likelihood ratio (LR) statistic for testing $k = 0$ (uniformity) against $k > 0$ when the U density is $f(u) = \{k/(e^k - 1)\}e^{ku}$, $k \geqslant 0, 0 < u < 1$, and Moran's M, discussed in C, is the LR statistic when the U come from JX in a test for exponentiality on X, against gamma* alternatives (see Tests Arising from Tests of Exponentiality).

When the alternative to the uniform distribution is $f(u) = (k + 1)u^k$, $k > -1$, $0 < u < 1$, the LR test statistic for a test that $k = 0$ (uniformity) against $k \neq 0$ is $P = -2\sum_{i=1}^{n}\log_e U_i$. This statistic was suggested by Fisher in connection with combining results from several significance tests; see Practical Applications of Tests.

B EDF Statistics

These are based on the empirical distribution function (EDF). Well-known statistics in this family include the Kolmogorov–Smirnov, Cramér–von Mises, and Anderson–Darling statistics (*see* EDF STATISTICS).

Some other test statistics, closely related to these, have been suggested in connection with tests on the periodogram in time series analysis; they are based on the quantities $V_i = U_{(i)} - i/(n + 1)$, $i = 1, \ldots, n$, that is, on the displacements of the $U_{(i)}$ from their expected values. These statistics include:

$$C^+ = \max_i V_i,$$

$$C^- = \max_i (-V_i),$$

$$C = \max(C^+, C^-),$$

$$K = C^+ + C^-,$$

$$T_1 = \sum_{i=1}^{n} V_i^2/n,$$

$$T_2 = \sum_{i=1}^{n} |V_i|/n.$$

EDF statistics and the preceding statistics are usually used with upper tail values significant, that is, H_0 is rejected for large values. On occasion it might be necessary to use the lower tail to detect values that are too regularly spaced to be a random sample of uniforms.

C Statistics Based on Spacings*

Many test statistics have been based on the spacings D_i between the $U_{(i)}$, defined by $D_i = U_{(i)} - U_{(i-1)}$, $i = 1, \ldots, n + 1$, where $U_{(0)} \equiv 0$ and $U_{(n+1)} \equiv 1$, by definition. Test statistics are often of the form $T = \sum_{i=1}^{n+1} g(D_i)$, where $g(\cdot)$ is a suitable function. Spacings are discussed by Pyke [25, 26]. When the set U comes from the J transformation (to be discussed) in connection with testing exponentiality, spacings are proportional to the values X to be tested and have a natural appeal. Two spacings statistics which have attracted attention are Greenwood's [12] statistic $G = \sum_{i=1}^{n+1} D_i^2$ and Moran's [17, 18] statistic $M = -2\sum_{i=1}^{n+1} \log_e\{(n + 1)D_i\}$. These have been adapted in recent years for use with k spacings, defined by $D_{ki} = U_{(ki)} - U_{(ki-k)}$, $i = 1, \ldots, r$, where k is a fixed integer, and where, for simplicity, suppose $r = (n + 1)/k$

is also an integer. These are the spacings between the observations, taken k at a time. The k spacings do not overlap. Alternatively the statistics may be used with overlapping sets of spacings, defined as kth order gaps, $R_{ki} = U_{(i+k)} - U_{(i)}$, for k a fixed integer and $i = 0, 1, 2, \ldots, n + 1 - k$. These statistics are discussed in Stephens [33, 34]. Greenwood's and Moran's statistics, or their adaptions, will detect different types of alternatives to uniformity. Although the picture is at present far from complete, Greenwood's G appears to be a good omnibus statistic in this class.

D Neyman Tests

Neyman [19] suggested that the alternative to uniformity should be approximated by a density expressed as a series of Legendre polynomials; then likelihood ratio methods were used to devise a test statistic for H_0, based on testing that coefficients in the series, other than the first, were zero. (*See* NEYMAN'S AND OTHER SMOOTH GOODNESS-OF-FIT TESTS and NEYMAN'S TEST FOR UNIFORMITY.) The test statistic N_k takes the form $N_k = \sum_{i=1}^{k} W_i^2$, where each individual W_i is a term (here called a component) calculated from the data. On H_0, these components are asymptotically independently distributed, each with a standard normal distribution, so that N_k has a χ_k^2 distribution. In order to apply the test, the order k, that is, the number of components to use, must be decided. The first two components depend on the mean $\bar{U}$ and the variance $S^2 = \sum_{i=1}^{n}(U_i - 0.5)^2/n$. For finite samples, tables are available for tests based on N_k, for $k \leqslant 4$.

E Graphical Methods

On H_0, the expected value of $U_{(i)}$ is $i/(n + 1)$. A plot of $U_{(i)}$ against i should be close to the line $L: U_{(i)} = i/(n + 1)$. Quesenberry and Hales [27] have given bands around L in which the observations should lie and Schweder and Spjøtvoll [28] discuss graphical procedures in connection with analysing significance levels from several tests (see

Practical Applications of Tests). A test of H_0 may be based on how close the points are to L; since L is specified, this is naturally done using the residuals $U_{(i)} - i/(n + 1)$, which are the v_i of Section B, and this leads to statistics of the C, K, T type in that section. Use of the correlation coefficient between $U_{(i)}$ and i is not sufficient for a test, since a full sample would give a good correlation even if uniform on only a small subset of $(0, 1)$; see Stephens [32] for further discussion.

F Tests of Uniformity with Limits (a, b)

When testing that U has a uniform distribution with known limits (a, b), the transformation $U' = (U - a)/(b - a)$ reduces the test to a test of H_0 on set U'. When (a, b) are not known, a conditional test can be made. Treat $U_{(1)}$ and $U_{(n)}$ as a and b and test H_0 for the $n - 2$ values U' given by the remaining values of U. Alternatively the correlation coefficient R between the $U_{(i)}$ and i can now be used; tables are given in Stephens [32].

G Tests Based on Censored Data

Many of the preceding test statistics have been adapted for right-censored, left-censored, or doubly-censored data; they are discussed in Stephens [31–33]. In most cases it is much harder to adapt the statistics for randomly censored data; Koziol [14] and O'Reilly and Stephens [20] give some procedures for certain kinds of random censoring. A general discussion of tests for censored data is given by Michael and Schucany [16].

Other tests of uniformity include those for the discrete uniform distribution, tests for the circle, and tests for uniformity in higher dimensions.

H Tests for the Discrete Uniform Distribution

A discrete uniform distribution is given by the multinomial distribution with equiprob-able cells. More generally, suppose that the multinomial distribution has k cells, with p_i the probability of an observation in the ith cell. The discrete test of uniformity is the test of

$$H_{0D}: \text{all } p_i = 1/k.$$

The classical test statistic is the Pearson χ^2 (see CHI SQUARE STATISTIC) but a particular problem arises if the expected numbers in the cells are small, so that the usual chi square approximation for χ^2 does not hold. The exact distributions of χ^2 and of the likelihood-ratio statistic are discussed, and tables given, by Good et al. [11], and more recently by Fattorini [8, 9]. A quick test, based on the maximum number of observations in any one cell, has been given by Dijkstra et al. [5], but note that in their numerical comparisons, effective significance level is sometimes considerably higher than nominal, giving unduly enhanced values for power.

A discrete version of Neyman's test was given by Barton [2]. Also, EDF tests have been adapted for the discrete situation; a Kolmogorov–Smirnov test was given by Pettitt and Stephens [23], Watson's U^2 has been adapted by Freedman [10] and the Cramér–von Mises W^2 by Stephens [35]. A problem in applying the discrete test is that often there is no natural ordering for the cells, and different orderings may produce different values of certain test statistics.

I Tests of Uniformity on the Circle

Many of the test statistics so far discussed will take different values with choice of origin if used to test whether observations U recorded on the circumference of a circle are uniform. EDF statistics which are invariant are Watson's U^2 and Kuiper's V; they were originally introduced for this problem. Ajne [1] also developed two other invariant statistics. Another test, based on the resultant of the vectors from the center of the circle to the points on the circumference, is given in J.

J Tests in Higher Dimensions

It will sometimes be of interest to test if points appear uniformly distributed in higher dimensions, for example, over a plane or the surface of a sphere, or inside a volume such as a sphere or a cube. Many techniques can be developed based on the properties of such a set of points, but it is usually difficult to provide distribution theory and percentage points for test statistics. The efficacy of such tests will again be dependent on the alternatives to be expected, and it may be argued that a test in higher dimensions (say p) should not be reduced to one test value, but should have at least p values to compare. On the sphere, tests are required against unimodal alternatives and against alternatives with density concentrated at two opposite poles or on a great circle. A simple test of uniformity of points on a circle or a sphere (usually denoting directions, say of magnetism in rock samples), which has power against the unimodal von Mises alternative, is as follows. Suppose vectors OU_i join the center O to points U_i on a hypersphere of radius 1 and dimension p, and let R be the length of the resultant of the vectors OU_i. The asymptotic distribution of R is given by $pR^2/n = \chi_p^2$. Tables for finite n, for two and three dimensions, are also available (Stephens [33]). A review of tests for the circle and the hypersphere is given by Prentice [24]. Other tests for the sphere are in Stephens [32] and Kielson et al. [13]. Some tests have been compared by Diggle et al. [4]. Tests of uniformity in high dimensions have been proposed by Smith and Jain [30], and a power study for tests of uniformity on the unit hypercube in high dimensions is described by Fattorini [8].

TRANSFORMATIONS FROM UNIFORMS TO UNIFORMS

Suppose U_i, $i = 1, \ldots, n$, is a set of uniforms with order statistics $U_{(i)}$. Let D_i be the spacings between the $U_{(i)}$. Suppose further that $D_{(i)}$, $i = 1, \ldots, n + 1$, denote the *ordered*

spacings and defnie new variables D' by $D' = (n + 2 - i)(D_{(i)} - D_{(i-1)})$, $i = 1, \ldots, n + 1$, with $D_{(0)} \equiv 0$. The set D' is another set of unordered uniform spacings (Sukhatme [36]) and a set of ordered uniforms U'' can clearly be built up by $U''_{(i)} = \sum_{j=1}^{i} D_j$, $i = 1, \ldots, n$. The extension of this definition gives $U''_{(n+1)} \equiv 1$. This transformation will be called the G *transformation* and we write $U'' = GU$; it was discussed by Sukhatme [36]. For some alternatives to uniformity, a test on U'' will be more powerful than that on U, so that use of G has sometimes been advocated as a method of increasing the power of the uniform test; see, for example, Durbin [7] and Seshadri et al. [29]. Another uniforms-to-uniforms transformation is given by $U'''_i = \{U_{(i)}/U_{(i+1)}\}^i$, $i = 1, \ldots, n$, with $U_{(n+1)} \equiv 1$. The U''' will be i.i.d. uniforms provided the U set are i.i.d. uniforms. In some circumstances, use of U''' rather than U might also increase the power of a test. A disadvantage of such transformations is that properties of the final set may be hard to interpret in terms of the original values. There should, therefore, be some reason, based perhaps on useful interpretation or possibly power against a wide range of alternatives, to justify their use.

TESTS ARISING FROM TESTS OF EXPONENTIALITY

Testing for exponentiality is an important area in which two transformations have considerable importance. They transform values X which are to be tested to come from an exponential distribution, to new values U or U' which, on the null hypothesis, should be uniforms. Thus the test on X is transformed to a test for uniformity on U or on U'. The two transformations will be called the J and K transformations. The mathematics of these is discussed in Seshadri et al. [29]; see also corrections in Dufour et al. [6]. The J transformation can be described in the context of analysing a series of events, where it is often used. If the events are occurring randomly in time or, more technically, if they form a realisation of a Poisson process*, the time intervals between successive events should be exponential and independent; that is, the intervals X will be i.i.d. with distribution $\text{Exp}(0, \beta)$, where $\text{Exp}(\alpha, \beta)$ refers to the distribution

$$F(x: \alpha, \beta) = 1 - \exp\{-(x - \alpha)/\beta\},$$
$$x > \alpha, \beta > 0.$$

Here the scale parameter β is unknown. Suppose, therefore, the events occur at times T_i given by $0 < T_1 < T_2 < \cdots < T_n < T$, where T may be either an arbitrary stopping time or may be the time of the $(n + 1)$th event. Then the X_i are given by $X_1 = T_1$, $X_i = T_i - T_{i-1}$, $i = 2, \ldots, n$, and $X_{n+1} = T - T_n$. It may then be shown that the values $U_{(i)}$ defined by $U_{(i)} = T_i/T$, $i = 1, \ldots, n$, will be ordered uniforms. Thus $n + 1$ intervals yield n ordered uniforms. The J transformation is a natural one to use for a series of events, since it is a simple scaling of the actual times as they would appear recorded on a line, and the indexing of the times may be relevant. For example, if the times denote industrial accidents, one might hope that the intervals between times are not all $\text{Exp}(0, \beta)$, but are getting longer as time passes; the U set will then appear to be closer to 0 than a truly uniform sample. Or if the times denote a natural phenomenon, such as the eruption of a geyser, it might be suggested that the intervals are very regular; then the $U_{(i)}$ will appear to be very evenly spaced in the $(0, 1)$ interval. We shall call such U values *super-uniform*. As presented here this transformation operates on the times T rather than on the intervals X; it can be regarded as a transformation of the exponentials X since the T can be created from the X by $T_i = \sum_{j=1}^{i} X_j$, $i = 1, \ldots, n$. This transformation of the X values to the U values will be called the J *transformation*, and we write $U = JX$.

The i.i.d. $\text{Exp}(0, \beta)$ values X can be transformed to another set of ordered uniforms U' by first creating the set U using transformation J as previously described and then transforming the set U by transformation G, discussed in the third section, to give the final set U'. The transformation, called K, may be written $U' = KX$. At first sight this transformation seems somewhat arbitrary, but it has uses in analysing lifetime data arising in reliability* and survival analysis*. In particular, because the G transformation first orders the spacings between the U values and the spacings are proportional to the X values, the transformation can be used with right-censored data such as often occur when lifetimes are analysed. There are also useful interpretations of the patterns of the set U' in terms of the failure rate of the X distribution, when H_0 is rejected for the set U' and so it is concluded that the X are not exponential. Further discussion of applications of the J and K transformations is in Stephens [34].

PRACTICAL APPLICATION OF TESTS

A Test Statistics and Patterns of U Values

In choosing which tests for uniformity to use, the user will naturally be guided by the particular application. For example, for a general univariate test of fit using the PIT, a test should be chosen to give good power against the alternatives it is desired to detect most effectively. Test statistics are effective against different alternatives, although many are highly correlated, and it is not easy to make overall classification in terms of power, but some useful guidelines can be found by observing the patterns of U values that will lead to significant values of different test statistics. For example, EDF statistic D^+ will be large when the U set tends toward 0, and D^- will be large when it tends toward 1; statistics D, W^2, and A^2 will also be large for U values moving toward either 0 or 1. The Fisher statistic P will also detect such movement, being large when the values are close to 0 and small when they are close to 1. In terms of testing that X comes from a known distribution $F(x; \theta)$, a drift of U values toward 0 suggests that the hypothesized distribution has a mean that is larger than the true mean and a drift toward 1 suggests that the suggested mean is too small. (Other parameters may of course be incorrectly specified or the functional form may also be wrong.) Such patterns of U values will of course be detected also by $\overline{U}$ or, usually less effectively, by the median $\tilde{U}$, which has a beta distribution.

The EDF statistics U^2 and V will detect a clustering of U values, either at one point in the interval or in two clusters at each end (corresponding to one cluster when the interval is placed around a circle); such patterns suggest misspecification of the variance in the test of fit.

B   Components

In recent years, EDF statistics W^2, U^2, and A^2 have been partitioned into components analogous to those of Neyman's N_k. The first two components of both Neyman's statistic and of the Anderson–Darling A^2 are functions of $\overline{U}$ and of S^2 defined in Neyman's tests. If either of these is significant, there is an easy interpretation in terms of the U values, and this in turn can be roughly translated to properties of the original X set, if the U were derived from X by the PIT. The first components of W^2 will also tend to detect changes in mean or variance of the U population and, therefore, of the X population. The first component of U^2 combines mean and variance, somewhat like N_2. Thus individual components of test statistics tend to reflect different aspects of the original X population. As test statistics themselves, they may have good power against selected alternatives, sometimes greater than the entire statistic. (For Neyman's statistic, the entire statistic will be N_∞.) Nevertheless as test statistics they must be used with care, since against other, perhaps equally important alternatives, they will have low power. For example, the first component of N_k, that is, N_1, or the first components of W^2 or A^2, will detect changes in mean but not changes in variance. For a consistent test against all alternatives, all components, that is, the entire statistic, must be used. For further references and discus-sion on the use of components, see Best and Rayner [3] and Stephens [31].

These observations can suggest when certain statistics will be better than others for a particular test situation. In general, it appears that EDF statistics give good omnibus tests. In many goodness-of-fit problems involving the PIT, the statistician is particularly concerned to detect outliers; then the EDF statistic A^2 will be effective.

C   Testing for Superuniformity

There may be occasions when a set U, particularly after a transformation such as J, can be superuniform, that is, more evenly spaced than expected from a random sample of uniforms. An interesting example occurs when J is applied to the dates of Kings and Queens of England; this is discussed by Pearson [22], together with several other interesting data sets. Most of the preceding tests, as usually used, will not detect superuniforms, although they can often be easily adapted to do so (usually by using the tail opposite to that normally used).

D   Combining Significance Tests*

The statistic $P = -2\sum_{i=1}^{n} \log_e U_i$ was suggested by Fisher to combine the results of n independent significance tests. Suppose H_{0i} is the null hypothesis for the ith test and let H_{00} be the overall null hypothesis that all H_{0i} are true. Let p_i be the significance level of the test statistic for H_{0i}. When H_{0i} is true, p_i should be $U(0, 1)$; H_{00} is then tested using P, with $U_i = p_i$. This test is easy to apply; on H_{00}, P has a χ^2_{2n} distribution. When some or all of the individual null hypotheses are not true, test procedures will usually give corresponding low values of p_i, so that H_{00} will be rejected for large P. Other closely related statistics with the same distribution on H_{00} as P are $P_2 = -2\sum_{i=1}^{n} \log_e q_i$, $P_3 = -2\sum_{i=1}^{n} \log_e 2r_i$, and $P_4 = -2\sum_{i=1}^{n} \log_e(1 - 2r_i)$, where here $q_i = 1 - p_i$ and r_i is the minimum of p_i and q_i. Pearson [21] suggested the possibility of using these alternative statistics to test H_{00}; on occasion, for example, P_2 can be more sensitive than P. These possibilities are discussed in Stephens [33]. All these statistics provide a test that the p values are uniform, and of course other tests of uniformity could as well be used.

E Nonuniformity

Since transformations are often used to produce a set of uniforms, it might be appropriate to conclude with some cautionary remarks on when uniformity is *not* to be expected. This will be so, for example, when the U set is derived from the PIT and when some parameters, unknown in the distribution, are replaced by estimates. In this situation, even when the estimates are efficient, the U set will be superuniform, giving much lower values of, say, EDF statistics, than if the set were uniform; this remains so even as the sample size grows bigger. Also, spacings from a distribution for X other than the exponential, will, when suitably normalised, behave like exponentials asymptotically and under certain conditions. However, they cannot then be used with, say, the J transformation, to produce uniforms for testing the original distribution for X: The constraints on the spacings are sufficiently strong that the resulting U values tend again to be superuniform and special tables must be produced for test statistics (Lockhart et al. [15]).

References

[1]   Ajne, B. (1968). *Biometrika*, **55**, 343–354.

[2]   Barton, D. E. (1955). *Skand. Aktuar.*, **39**, 1–17.

[3]   Best, D. J. and Rayner, J. C. W. (1985). *Sankhyā*, **47**, 25–35.

[4]   Diggle, P. J., Fisher, N. I., and Lee, A. J. (1985). *Aust. J. Statist.*, **27**, 53–59.

[5]   Dijkstra, J. B., Rietjens, T. J. M., and Steutel, F. W. (1984). *Statistica Neerland.*, **38**, 33–44.

[6]   Dufour, R., Maag, U., and van Eeden, C. (1984). *J. R. Statist Soc. B*, **46**, 238–241.

[7]   Durbin, J. (1961). *Biometrika*, **48**, 41–55. (Discusses how the G transformation can increase power.)

[8]   Fattorini, L. (1984a). *Metron*, **42**, 53–66.

[9]   Fattorini, L. (1984b). *Metron*, **42**, 207–212.

[10] Freedman, L. S. (1981). *Biometrika*, **68**, 708–711.

[11] Good, I. J., Gover, T. N., and Mitchell, G. J. (1970). *J. Amer. Statist. Ass.*, **65**, 267–283.

[12] Greenwood, M. (1946). *J. R. Statist. Soc. A*, **109**, 85–110.

[13] Kielson, J., Petrondas, D., Sumita, U., and Wellner, J. (1983). *J. Statist. Comput. Simul.*, **17**, 195–218.

[14] Koziol, J. A. (1980). *Biometrika*, **67**, 693–696.

[15] Lockhart, R. A., O'Reilly, F. J., and Stephens, M. A. (1986). *J. R. Statist. Soc. B*, **48**, 344–352.

[16] Michael, J. R. and Schucany, W. R. (1986). In *Goodness-of-fit Techniques*, R. B. d'Agostino and M. A. Stephens, eds. Marcel Dekker, New York, Chap. 11. (General review of testing with censored data.)

[17] Moran, P. A. P. (1947). *J. R. Statist. Soc. B*, **9**, 92–98.

[18] Moran, P. A. P. (1951). *J. R. Statist. Soc. B*, **13**, 147–150.

[19] Neyman, J. (1937). *Skand. Aktuar.*, **20**, 149–199.

[20] O'Reilly, F. J. and Stephens, M. A. (1988). *Technometrics*, to appear.

[21] Pearson, E. S. (1938). *Biometrika*, **30**, 134–148.

[22] Pearson, E. S. (1963). *Biometrika*, **50**, 315–325. (Very complete paper on J transformation followed by EDF tests, illustrated on four sets of data.)

[23] Pettitt, A. N. and Stephens, M. A. (1977). *Technometrics*, **19**, 205–210.

[24] Prentice, M. J. (1978). *Ann. Statist.*, **6**, 169–176.

[25] Pyke, R. (1965). *J. R. Statist. Soc. B*, **27**, 395–449. (Very broad review of spacings.)

[26] Pyke, R. (1972). *Proc. Sixth Berkeley Symp. Prob. Math. Statist.*, **1**, 417–427.

[27] Quesenberry, C. P. and Hales, S. (1980). *J. Statist. Comput. Simul.*, **11**, 41–53.

[28] Schweder, T. and Spjøtvoll, E. (1982). *Biometrika*, **69**, 493–502.

[29] Seshadri, V., Csörgő, M. and Stephens, M. A. (1969). *J. R. Statist. Soc. B*, **31**, 499–509. (Discusses G, J, and K transformations.)

[30] Smith, S. P. and Jain, A. K. (1984). *IEEE Trans. Pattern Anal. Mach. Intell.*, **6**, 73–81.

[31] Stephens, M. A. (1986a). In *Goodness-of-fit Techniques*, R. B. d'Agostino and M. A. Stephens, eds. Marcel Dekker, New York, Chap. 4. (Contains EDF statistics for testing uniformity, with many tables.)

[32] Stephens, M. A. (1986b). In *Goodness-of-fit Techniques*, R. B. d'Agostino and M. A. Stephens, eds. Marcel Dekker, New York, Chap. 5. (Contains correlation coefficient tests.)

[33] Stephens, M. A. (1986c). In *Goodness-of-fit Techniques*, R. B. d'Agostino and M. A. Stephens, eds.

Marcel Dekker, New York, Chap. 8. (A general review of tests of uniformity, with references and tables.)

[34] Stephens, M. A. (1986d). In *Goodness-of-fit Techniques*, R. B. d'Agostino and M. A. Stephens, eds. Marcel Dekker, New York, Chap. 10. (A general review of tests of exponentiality, including those based on *J* and *K* transformations.)

[35] Stephens, M. A. (1987). *Tech. Rep.*, Dept. of Mathematics and Statistics, Simon Fraser University, Burnaby, B.C. V5A 1S6, Canada

[36] Sukhatme, P. V. (1937). *Ann. Eugen.* (*London*), **8**, 52–56. (Discussion of *G* transformations.)

(COMBINATION OF DATA
EDF STATISTICS
EXPONENTIAL DISTRIBUTIONS
SPACINGS
VON MISES DISTRIBUTION)

MICHAEL A. STEPHENS

UNIFORMITY TRIALS

This term originally referred to agricultural experiments in which there are no specifically introduced differences (in treatment, variety, etc.). The object of such experiments is to assess "natural" variation (e.g., in soil fertility) and to use this information in planning future experiments and the interpretation (especially in regard to formulation of a statistical model*) of the data obtained from such experiments.

Uniformity trials are also used in nonagricultural situations, though they are often known under different names, for example, control of calibration* experiments.

It is prudent—in the absence of additional evidence—not to rely too precisely on numerical estimates obtained from uniformity trials. Interaction* between changes in conditions (from uniformity to later trials) and natural variation can affect models based on precise numerical values for uniformity trials. The uniformity trial results can, however be used for comparative purposes—for assessing the kinds of natural variations and their (approximate) relative magnitudes—with some confidence.

Reference 1 (Table II, pp. 122–125) contains summaries of results of a number of (agricultural) uniformity trials.

Reference

[1] Neyman, J., Iwaszkiewicz, K., and Kolodziejezyk, S. (1935). *J. R. Statist. Soc. Suppl.*, **2**, 107–154 (discussion 154–180).

(AGRICULTURE, STATISTICS IN
ANALYSIS OF COVARIANCE
ANALYSIS OF VARIANCE
DESIGN OF EXPERIMENTS)

UNIFORMIZATION (OF MARKOV CHAINS)

A widely used technique of equalizing the rates in which a transition occurs from each state in a continuous-time Markov chain*. It is effected by introducing transitions from a state to itself. If the actual rate of leaving state i is ν_i, then a uniform nominal rate of leaving, ν, greater than any ν_i, is used, but only a fraction ν_i/ν of the transitions are real; the remainder are "transitions" from state i to itself. See, for example, ref. 1 for further details.

Reference

[1] Ross, S. M. (1983). *Stochastic Processes*. Wiley, New York.

(MARKOV PROCESSES
STOCHASTIC PROCESSES)

UNIFORM METRIC

Given two cumulative distribution functions $F_1(\cdot)$, $F_2(\cdot)$, the metric defined by

$$\pi(F_1, F_2) = \sup_x |F_1(x) - F_2(x)|$$

is called the *uniform* metric. (Also, *Kolmogorov* metric.)

(DISTANCE FUNCTIONS
KOLMOGOROV–SMIRNOV STATISTICS
PROBABILITY SPACES, METRICS AND
   DISTANCES ON)

UNIFORM (OR RECTANGULAR) DISTRIBUTIONS

DISCRETE RECTANGULAR DISTRIBUTION

Let X_1 be a random variable taking values $0, 1, 2, \ldots, N$ with equal probability $(N + 1)^{-1}$. Then X_1 has a *discrete rectangular* or *uniform distribution* with mean $\frac{1}{2}N$, variance $\frac{1}{12}N(N + 2)$, and third and fourth central moments equal to zero and $\frac{1}{240}N(N + 2)$ $\{3(N + 1)^2 - 7\}$, respectively. The probability generating function* is

$$\frac{(1 - t^{N+1})}{\{(N + 1)(1 - t)\}}.$$

A more general form of the distribution is given by

$$\Pr[X_2 = a + jc] = 1/(N + 1),$$
$$j = 0, 1, 2, \ldots, N. \quad (1)$$

The rth central moment of X_2 is derived by multiplying that of X_1 by c^r; the mean of X_2 is $a + \frac{1}{2}Nc$.

In 1812, Laplace* derived an early central limit* property for discrete rectangular variables [11, Sec. 18 of Chap. IV]. However, (1) is more frequently approximated by a continuous rectangular distribution, for example, in the construction of tables of random numbers (*see* GENERATION OF RANDOM VARIABLES and Johnson and Kotz [7, Sec. 10.2]), and in the estimation of N, which is discussed later.

CONTINUOUS RECTANGULAR DISTRIBUTION

Let X have probability density function (PDF)

$$f(x) = \begin{cases} (b - a)^{-1}, & a < x < b, \\ 0, & \text{otherwise.} \end{cases} \quad (2)$$

Then X has a (continuous) *rectangular* or *uniform distribution*, with cumulative distribution function (CDF)

$$F(x) = \begin{cases} 0, & x \leqslant a, \\ (x - a)/(b - a), & a < x < b, \\ 1, & x \geqslant b. \end{cases}$$

We denote (2) by $U(a, b)$. Other forms include $U(a - h, a + h)$ and the standard form $U(0, 1)$. For $U(a - h, a + h)$, the mean is a, the variance is $h^2/3$, the odd central moments are zero, and the $(2r)$th (even) central moment is $h^{2r}/(2r + 1)$; the mean deviation is $\frac{1}{2}h$. The characteristic function* is $e^{ita}\sin(th)/(th)$.

The uniform distribution has been in use since the eighteenth century (*see* LAWS OF ERROR), the earliest uses being in its discrete form (Simpson [17]). For further properties, see Johnson and Kotz [8, Chap. 25].

Applications

The $U(-h, h)$ distribution is used for the distribution of round-off error* of numerical values to the nearest k decimal places, where $h = \frac{1}{2} \times 10^{-k}$. Uniform distributions feature in deriving Sheppard's corrections for grouping*, in models for traffic flow (Allan [1]), and as approximations to discrete uniform distributions.

Properties

(i) Let X have a continuous CDF, F. Then $F(X)$ has a $U(0, 1)$ distribution. This is the probability integral transformation*.

(ii) Let X have a $U(0, 1)$ distribution. Then $-2 \ln X$ has a chi-square distribution* with 2 degrees of freedom.

(iii) (Rohatgi [16, Sec. 5.3]). Let F be any CDF and let X have a $U(0, 1)$ distribution. Then there exists a function g such that $g(X)$ has CDF F.

(iv) If $X_{1,n} \leqslant X_{2,n} \leqslant \cdots \leqslant X_{n,n}$ are order statistics* in a random sample of size n from a $U(0, 1)$ distribution, then (a) $X_{r,n}$ has a beta distribution* with parameters r and $n - (r - 1)$, $r = 1, \ldots, n$; (b) the spacings* $D_i = X_{i,n} - X_{i-1,n}$ (with $X_{0,n} \equiv 0$) have a joint exchangeable Dirichlet distribution* and are identically distributed with a common beta $(1, n)$ distribution (Rao [15]); (c) the range* $X_{n, n} - X_{1, n}$ has a beta $(n - 1, 2)$ distribution.

Property **(i)** leads to Kolmogorov-Smirnov-type tests of goodnes-of-fit* of data to a specified continuous CDF $F_0(\cdot)$. Under the null hypothesis the Kolmogorov-Smirnov statistics* have distributions depending, not on F_0, but on order statistics from a $U(0, 1)$ parent; *see* EDF STATISTICS and Rohatgi [16, pp. 539–540]. Property **(iii)** has been used in the generation of random variables*, discrete and continuous.

Characterization

X has a $U(0, 1)$ distribution if and only if

1. for X having support $[0, 1]$ and for all $0 \leqslant x < y \leqslant 1$, $\Pr(x < X \leqslant y)$ depends only on $y - x$ (Rohatgi [16, Sec. 5.3]);
or
2. $E[-\ln(1 - X)|X > y] = -\ln(1 - y) + 1, 0 \leqslant y < 1$ (Hamdan [4]);
or
3. for a random sample with order statistics $X_{1, n} \leqslant X_{2, n} \leqslant \cdots \leqslant X_{n, n}$ from the distribution of X, of bounded support and continuous density, the spacings $X_{1, n}$ and $X_{2, n} - X_{1, n}$ are identically distributed (Huang et al. [6]).

For other characterizations see Kotz [10] and CHARACTERIZATIONS OF DISTRIBUTIONS.

INFERENCE

Consider estimation of θ in the family of $U(0, \theta)$ distributions, $\theta > 0$. The largest order statistics $X_{n, n}$ in a sample of size n is complete and sufficient for the family. Let

$$\hat{\theta}_1 = X_{n, n},$$

$$\hat{\theta}_2 = (n + 1) X_{n, n}/n,$$

$$\hat{\theta}_3 = (n + 2) X_{n, n}/(n + 1).$$

Then $\hat{\theta}_1$ is the maximum likelihood estimator (MLE), $\hat{\theta}_2$ the minimum variance unbiased estimator*, and $\hat{\theta}_3$ has minimum mean square error* (MSE) among estimators of θ depending only on $X_{n, n}$. Thus (Rao [15])

$$\text{MSE}(\hat{\theta}_1) = \theta^2/(n + 1)^2$$

$$> \text{MSE}(\hat{\theta}_2) = \theta^2/\{n(n + 2)\}$$

$$> \text{MSE}(\hat{\theta}_3) = 2\theta^2/\{(n + 1)(n + 2)\}.$$

$\hat{\theta}_2$ is also the best linear unbiased estimator (BLUE) of θ among linear combinations of the order statistics or spacings. In a censored sample where only the smallest m order statistics are available, the BLUE of θ is $[(n + 1)/m]X_{m, n}$, with variance

$$\frac{(n - m + 1)\theta^2}{m(n + 2)}.$$

Estimation of N in the discrete rectangular distribution on the integers $1, \ldots, N$ is frequently approached by a continuous rectangular approximation. In the "taxi problem"*, where the numbers of cabs are noted on a street corner until one is flagged down, or serial numbers of captured enemy tanks are recorded, the total number of cabs or tanks is to be estimated. The estimate $[(n + 1)/n]X_{n, n}$ is the sum of the gaps (spacings) between consecutive serial numbers observed, plus the average gap length; see Rao [15] and Noether [13].

A $100(1 - \alpha)\%$ confidence interval for θ is given by $(x_{n, n}, x_{n, n}/(1 - \alpha)^{1/n})$, where $x_{n, n}$ is the observed value of $X_{n, n}$ [18].

The MLEs $\hat{a}$ and $\hat{h}$ of a and h based on a sample from a $U(a - h, a + h)$ family are the midrange* of the sample and $\frac{1}{2}$(range of the sample), respectively. The BLUEs of a and h are $\hat{a}$ and $[(n + 1)/(n - 1)]\hat{h}$, with variances $2h^2/[(n + 1)(n + 2)]$ and $2h^2/[(n - 1)(n + 2)]$, respectively; $\hat{a}$ and $\hat{h}$ are uncorrelated but not independent. See Johnson and Kotz [8, Chap. 25], where estimators for censored samples are also discussed.

For a discussion of tests of uniformity, *see* NEYMAN'S TEST FOR UNIFORMITY. A comparative study of the power of various tests was made by Miller and Quesenberry [12] and by Quesenberry and Miller [14]. *See also* DIRECTIONAL DATA ANALYSIS.

RELATED DISTRIBUTIONS

The mean of two independent $U(a - h, a + h)$ variables has a *triangular distribution*, with PDF

$$f(x) = [h - |x - a|]/h^2,$$
$$a - h \leqslant x \leqslant a + h,$$

symmetrical about $x = a$. The mean is a and the variance is $h^2/6$. The rth central moments vanish if r is odd and equal $2h^r/[(r + 1)(r + 2)]$ if r is even. The properties of a general asymmetric triangular distribution with standard form

$$f(x) = \begin{cases} 2x/H, & 0 \leqslant x \leqslant H, \\ 2(1 - x)/(1 - H), & H \leqslant x \leqslant 1, \end{cases}$$

having mean $(1 + H)/3$, variance $(1 - H + H^2)/18$, and median $\sqrt{\frac{1}{2}\max(H, 1 - H)}$, are discussed by Ayyangar [2] and by Johnson and Kotz [8, Chap. 25]. The CDF of the sum S_n of n independent $U(0, 1)$ random variables is given by

$$\Pr(S_n \leqslant x) = \sum_{j=0}^{n} (-1)^j \binom{n}{j}(x - j)^n/n!,$$

$$0 \leqslant x \leqslant n.$$

The sum of three independent $U(-1, 1)$ variables has been used [5] to approximate the standard normal distribution. The PDF is

$$q(x) = \begin{cases} (3 - x^2)/8, & |x| \leqslant 1, \\ (3 - |x|)^2/16, & 1 \leqslant |x| \leqslant 3, \\ 0, & |x| \geqslant 3. \end{cases}$$

BIVARIATE UNIFORM DISTRIBUTIONS

If (X, Y) has a continuous joint distribution with $U(0, 1)$ marginal CDFs, then (X, Y) is said to have a *bivariate uniform* (BVU) *distribution*. Kimeldorf and Sampson [9] point out that X and Y in this framework are independent if and only if their joint PDF is constant.

Let $H(x, y)$ be any continuous joint CDF having marginal CDFs $F(x)$ and $G(y)$. Then the *uniform representation* of H is given by the BVU distribution with joint CDF

$$U_H(u, v) = H(F^{-1}(u), G^{-1}(v)),$$

$$0 \leqslant u \leqslant 1, 0 \leqslant v \leqslant 1.$$

The form of U_H enables us to determine if X and Y are independent, or the form of dependence* between X and Y; these are the same for all BV distributions in certain equivalence classes. The authors give uniform representations of Plackett* and Morgenstern distributions (*see* FARLIE–GUMBEL–MORGENSTERN DISTRIBUTIONS), of the Marshall–Olkin* BV exponential distribution, of BV Cauchy*, and of Pareto distributions*. Barnett [3] gives contour plots of these forms of dependence and for the uniform representation of the BV normal distribution.

References

[1] Allan, R. R. (1966). *Proc. Third. Conf. Aust. Road Res. Board*, **3**, 276–316.

[2] Ayyangar, A. A. K. (1941). *Math. Student*, **9**, 85–87.

[3] Barnett, V. (1980). *Commun. Statist. A*, **9**, 453–461; correction, **10**, 1457 (1981).

[4] Hamdan, M. A. (1972). *Technometrics*, **14**, 497–499.

[5] Hoyt, J. P. (1968). *Amer. Statist.*, **22**, 25–26.

[6] Huang, J. S., Arnold, B. C. and Ghosh, M. (1979). *Sankhyā B*, **41**, 109–115.

[7] Johnson, N. L. and Kotz, S. (1969). *Distributions in Statistics: Discrete Distributions*. Wiley, New York.

[8] Johnson, N. L. and Kotz, S. (1970). *Distribution in Statistics: Continuous Univariate Distributions* 2. Wiley, New York.

[9] Kimeldorf, G. and Sampson, A. (1975). *Commun. Statist. A*, **4**, 617–627.

[10] Kotz, S. (1974). *Int. Statist. Rev.*, **42**, 39–65.

[11] Laplace, P. S. (1812). *Théorie Analytique des Probabilités*. Paris. (See Volume 7 of *Oeuvres Completes de Laplace*, published 1878 and 1912 by Gauthier-Villars, Paris, France.)

[12] Miller, R. L. and Quesenberry, C. P. (1979). *Commun. Statist. B*, **8**, 271–290.

[13] Noether, G. (1971). *Introduction to Statistics. A Fresh Approach.* Houghton Mifflin, Boston, MA. (A lucid nonmathematical discussion of the "gap estimate" in the taxi problem; see Chap. 1.)

[14] Quesenberry, C. P. and Miller, R. L. (1977). *J. Statist. Comput. Simul.*, **5**, 169–191.

[15] Rao, J. S. (1981). *Metrika*, **28**, 257–262.

[16] Rohatgi, V. K. (1976). *An Introduction to Probability Theory and Mathematical Statistics.* Wiley, New York.

[17] Simpson, T. (1756). *Philos. Trans. R. Soc.*, **49**, 82–83.

[18] Wilks, S. S. (1962). *Mathematical Statistics.* Wiley, New York.

(DIRECTIONAL DATA ANALYSIS
GENERATION OF RANDOM VARIABLES
NEYMAN'S TEST FOR UNIFORMITY
PROBABILITY INTEGRAL
    TRANSFORMATION)

CAMPBELL B. READ

UNIMODAL REGRESSION

In many applications the a priori information about the regression function is limited to the general shape, such as unimodality of the curve. A regression* function $E(Y|X = x) = f(x)$ is *unimodal* if it has one and only one local maximum. The equally important case of one and only one local minimum can be treated in the same way, with obvious modifications; thus, only the case of one local maximum will be described. To be specific, there exists an $X = x_m$ such that

$$f(x_i) \leqslant f(x_j) \leqslant f(x_m) \quad \text{if } x_i < x_j < x_m$$

and

$$f(x_k) \leqslant f(x_l) \leqslant f(x_m) \quad \text{if } x_k > x_l > x_m.$$

The regression is *strictly unimodal* if all the preceding inequalities are strict.

An example of a case where unimodal regression and especially the estimate of the maximum is of interest is the development of the concentration of a drug with time. The concentration in a living body of a one-time administrated drug will generally first increase because of the uptake and then, after a maximum, decrease as the drug is broken down or secreted. For further details about this and other applications, see Frisén [4].

As the restriction on the regression is of ordinal kind, it can be regarded as a member of the class of isotonic regression*. (See Barlow et al. [1] for a comprehensive treatment of this class.) However, there is not a simple order restriction as in monotonic regression. Unimodal regression consists of an up-phase, where $f(x)$ is monotonically increasing with x, and a down-phase, where the regression is monotonically decreasing with x.

If a unique partition in monotonic phases were known, then it would be possible to use the least-squares* monotonic regression method for each phase. In the present case the turning point is unknown. For each possible partition, the least-squares solution can be obtained by the fitting of monotonic regression to each phase. The sum Q of the squared deviations between observed and fitted values can be calculated for each of these cases. The least-squares estimators $\hat{Y}_i$ of $E(Y|X = x_i)$ for the case of an unknown partition give the solution corresponding to the least Q. It is, in fact, sufficient to investigate only partitions just before each observed local maximum because partitions after these will give the same solution and because for all other partitions, the Q can be improved. An algorithm that gives the least-squares solution in the case of an unknown turning point was described in Frisén [3] and the FORTRAN program is available from the author. A SAS procedure based on this program and a manual (Pehrsson and Frisén [5]) is available from the Gothenburg Computer Central, Box 19070, S-40012 Göteborg, Sweden.

The computational procedure is illustrated by a small numerical example given in Table 1. A regression with one maximum is fitted to eight observed points (x_i, y_i), $i = 1, \ldots, 8$, all with the same weight. $\hat{y}_i(1–2, 3–8)$ are the least-squares estimates obtained by ordinary monotonic regression (see isotonic regression) under the condition that the regression

Table 1

i	1	2	3	4	5	6	7	8
x_i	1	2	3	3.5	4	4.5	7	8
y_i	1	2.5	4	3	2	5	1.5	0
$\hat{y}_i$ (1–2, 3–8)	1	2.5	4	3.$\dot{3}$	3.$\dot{3}$	3.$\dot{3}$	1.5	0
$\hat{y}_i$ (1–5, 6–8)	1	2.5	3	3	3	5	1.5	0

is monotonically increasing for the points 1–2 and monotonically decreasing for the points 3–8. $\hat{y}_i(1$–$5, 6$–$8)$ are the corresponding regression estimates for the partition in points 1–5 and 6–8. The sum of squared deviations is 2 for the partition 1–2, 3–8 and 42/9 for the partition 1–5, 6–8. Thus $\hat{y}_i(1$–$2, 3$–$8)$ give the least-squares solution.

As is seen in this example, the largest observed value is not necessarily the estimated maximum. There are no parameters of some assumed function to estimate, but for each observed value of X, a least-squares* estimate of the expected value of Y will be obtained. The solution is a compromise between a jumping connection between observed values, which would not utilize all the available information, and a perfectly smooth but possibly misleading mathematical function. Some statistical properties of the estimation method and some modifications are found in Frisén [4] and Dahlbom [2].

References

[1] Barlow, R. E., Bartholomew, D. J., Bremner, J. M., and Brunk, H. D. (1972). *Statistical Inference under Order Restrictions*. Wiley, Chichester, England and New York.

[2] Dahlbom, U. (1986). Some Properties of Estimates of Unimodal Regression. *Research Report 1986:6*, Dept. of Statistics, University of Göteborg, Göteborg, Sweden.

[3] Frisén, M. (1980). U-shaped regression. *Compstat. Proc. Comput. Statist.*, **4**, 304–307.

[4] Frisén, M. (1986). Unimodal regression. *The Statistician*, London, **35**, 479–485.

[5] Pehrsson, N-G. and Frisén, M. (1983). The UREGR procedure. Manuals for SAS-Procedures. Gothenburg Computer Central, Göteborg, Sweden.

(ISOTONIC INFERENCE
REGRESSION (various entries))

M. FRISÉN

UNIMODALITY

This is the property of having a single mode* or modal interval. For a univariate distribution with cumulative distribution function* $F(x)$, it is defined by requiring the existence of at least one value M such that for all

$$x_1 < x_2 < M < x_3 < x_4,$$

we have

$$F\left(\tfrac{1}{2}(x_1 + x_2)\right) \leqslant \tfrac{1}{2}\{F(x_1) + F(x_2)\}$$

and

$$F\left(\tfrac{1}{2}(x_3 + x_4)\right) \geqslant \tfrac{1}{2}\{F(x_3) + F(x_4)\}.$$

The set of numbers $\mu'_1, \mu'_2, \ldots, \mu'_{2r}$ can be the first $2r$ moments* of a unimodal distribution if and only if the determinants

$$\begin{vmatrix} 1 & 2\mu'_1 & \cdots & (s+1)\mu'_s \\ 2 & 3\mu'_2 & \cdots & (s+2)\mu'_{s+1} \\ \vdots & \vdots & & \vdots \\ (s+1)\mu'_s & (s+2)\mu'_{s+1} & \cdots & (2s+1)\mu'_{2s} \end{vmatrix}$$

are nonnegative for $s = 1, 2, \ldots, r$.

For any unimodal distribution with finite first four moments the (central-) moment ratios* $\beta_1 = \mu_3^2/\mu_2^3$, $\beta_2 = \mu_4/\mu_2^2$ must be such that the value of β_2, for given β_1, exceeds that given by the parametric equations

$$\beta_1 = \frac{108\theta^4}{(1-\theta)(1+3\theta)^3},$$

$$5\beta_1 - 9 = \frac{72\theta^2(\theta-1)}{(1-\theta)(1+3\theta)^2} \qquad (0 \leqslant \theta \leqslant 1)$$

(Johnson and Rogers [1]).

The restriction that a distribution is unimodal allows considerable improvement on bounds for probabilities of Chebyshev type (see Royden [2]).

Note that a discrete distribution*, for which $F(x)$ is a step function*, cannot be unimodal according to the definition we have used (except in the degenerate case when $\Pr[X = \xi] = 1$ for some ξ), because if

$$F(x) = P_i(> 0) \quad \text{for } \xi_i \leqslant x < \xi_i + h$$

$$= P_{i+1}(> P_i) \text{ for } \xi_{i+h} \leqslant x < \xi_i + 2h,$$

then for $0 < g < \frac{1}{2}h$,

$$F\left(\tfrac{1}{2}(\xi_i + g + \xi_i + 2h - g)\right)$$

$$= F(\xi_i + h) = P_{i+1} > \tfrac{1}{2}(P_i + P_{i+1})$$

$$= \tfrac{1}{2}\{F(\xi_i + g) + F(\xi_i + 2h - g)\},$$

but

$$F\left(\tfrac{1}{2}(\overline{\xi_i + g} + \overline{\xi_i + h + g})\right)$$

$$= F(\xi_i + \tfrac{1}{2}h + g) = P_i < \tfrac{1}{2}(P_i + P_{i+1})$$

$$= \tfrac{1}{2}\{F(\xi_i + g) + F(\xi_i + h + g)\}.$$

It is customary to define the term "*discrete unimodality*" for a distribution for which

$$\Pr[X = \xi_i] = p_i$$

with $\dots \xi_{i-1} < \xi_i < \xi_{i+1} \left(\sum_i p_i = 1\right)$,

by the requirement that there is a number j such that

$$\dots p_{j-2} \leqslant p_{j-1} \leqslant p_j \geqslant p_{j+1} \geqslant p_{j+2} \dots.$$

References

[1] Johnson, N. L. and Rogers, C. A. (1951). *Ann. Math. Statist.*, **22**, 433–439.

[2] Royden, H. L. (1953). *Ann. Math. Statist.*, **24**, 361–376.

Bibliography

Dharmadhikari, S. and Joag-dev, K. (1987). *Unimodality, Convexity, and Applications*. Academic, New York.

(CHEBYSHEV INEQUALITIES
MULTIVARIATE UNIMODALITY
THREE-SIGMA RULE)

UNINFORMATIVENESS OF A LIKELIHOOD FUNCTION

A survey* population consisting of N labelled individuals can be denoted by $\mathscr{P} = \{i: i = 1, \dots, N\}$; the individuals i may be farms or households and the like. Let $\mathbf{y} = (y_1, \dots, y_i, \dots, y_N)$ be the vector of variates under study, e.g., y_i may be the produce of the farm i. Since the "census" is not available, a sample s ($s \subset \mathscr{P}$) is drawn using a suitable sampling design p. If $\mathscr{S} = \{s: s \subset \mathscr{P}\}$, $p: \mathscr{S} \to [0,1]$, $\Sigma_{\mathscr{S}} p(s) = 1$. After the sample s is drawn, the variate values y_i: $i \in s$ are ascertained through a survey. Here the data $\mathbf{Y}_s$ consist, in addition to the variate values y_i: $i \in s$, of the "labels"* s, $\mathbf{Y}_s = (s, y_i: i \in s)$. Now the probability distribution on the space $\{\mathbf{Y}_s\}$ of all the possible data $\mathbf{Y}_s$ is determined (apart from the fixed sampling design p) by the population vector $\mathbf{y}' = (y_1', \dots, y_i', \dots, y_N')$ say, as

$$\text{Prob}_{\mathbf{y}'}(\mathbf{Y}_s) = \begin{cases} p(s) & \text{if } y_i = y_i' \text{ for } i \in s, \\ 0 & \text{otherwise} \end{cases}$$

$$(1)$$

Thus for *given* data $\mathbf{Y}_s$, the *likelihood function* for different population vectors $\mathbf{y}'$ is given by

$$\text{Prob}_{\mathbf{y}'}(\mathbf{Y}_s). \tag{2}$$

There are two peculiar features of the likelihood function of $\mathbf{y}'$ defined by (2). (i) Since the likelihood function is defined only up to a constant multiple, it remains the same if in (1) and (2), $p(s)$ is replaced by a constant, say 1. That is, the likelihood* function is independent of the sampling design p used to select the sample s. (ii) The likelihood function in (2) is independent of the coordinate y_i' of $\mathbf{y}'$ if $i \notin s$. That is, the likelihood function is uninformative of the $\mathbf{y}$ values associated with the individuals i of the population $\mathscr{P}$, which are not included in the sample s.

The feature (ii) is sometimes described as *uninformativeness* of the likelihood function given by (2). Now in the general statistical theory, the likelihood function plays a central role in inference. Hence the *uninformative-*

ness of the likelihood function given by (2) poses a fundamental problem for inference. Another component of the problem is as follows. It seems intuitive that the sampling design p employed to draw the sample s should be related in some sense to the inference. Surely in some situations one would expect stratified random sampling to yield an estimate with smaller variance than the one given by simple random sampling (*see* OPTIMUM STRATIFICATION). Yet as seen in (i) the likelihood function in (2) is independent of the sampling design.

The problem of uninformative likelihood function that is independent of the sampling design was first formulated in Godambe [2]. Various approaches to deal with the problem are suggested by Ericson [1], Godambe [2, 3], Royall [4], and Thompson [5].

References

[1] Ericson, W. A. (1969). *J. R. Statist. Soc. B*, **31**, 195–224.

[2] Godambe, V. P. (1966). *J. R. Statist. Soc. B*, **28**, 310–328.

[3] Godambe, V. P. (1982). *J. Amer. Statist. Ass.*, **77**, 393–406.

[4] Royall, R. M. (1976). *Biometrika*, **63**, 605–614.

[5] Thompson, M. E. (1984). *J. Statist. Plann. Inf.*, **10**, 323–334.

(LABELS
LIKELIHOOD
SAMPLING DESIGN)

V. P. GODAMBE

UNION–INTERSECTION PRINCIPLE

INTRODUCTION

The union–intersection principle is a principle suggested by Roy [6] for designing tests, particularly for multivariate* problems. It has the advantage over other principles (such as the likelihood ratio* principle) that the union–intersection test often has relatively simple associated simultaneous confidence intervals*.

Let $\boldsymbol{\theta}$ be an unknown (typically vector-valued) parameter and consider the problem of testing the null hypothesis that $\boldsymbol{\theta} \in B$. Suppose that there exist sets C_s, such that $\boldsymbol{\theta} \in B$ if and only if $\boldsymbol{\theta} \in C_s$ for all s, and that there exist sensible size α^* tests ψ_s for testing that $\boldsymbol{\theta} \in C_s$. Let ψ be the test that rejects $\boldsymbol{\theta} \in B$ if and only if there exists at least one s such that ψ_s rejects the hypothesis that $\boldsymbol{\theta} \in C_s$. This test ψ is called a *union–intersection* test for this problem. Its motivation is fairly obvious. We accept the hypothesis $\boldsymbol{\theta} \in B = \bigcap C_s$ if and only if we accept that $\boldsymbol{\theta} \in C_s$ for all s.

Before looking at some multivariate examples of union–intersection tests, we make some elementary comments about them. The first is that the size α of ψ is not the same as the common size α^* of the tests ψ_s. The second comment is that the union–intersection principle is not uniquely defined. There may be several collections of sets C_s, such that $B = \bigcap C_s$ (see the next section). For this reason, it is difficult to prove any general properties of these tests.

In many problems of multivariate analysis*, the union–intersection principle is applied in the following way. Suppose we observe $\mathbf{X}_1, \ldots, \mathbf{X}_n$ independent, $\mathbf{X}_i \sim N(\boldsymbol{\mu}_i, \Sigma)$ and we are interested in testing that $(\boldsymbol{\mu}_1, \ldots, \boldsymbol{\mu}_n, \Sigma)$ is in some set B. For all $\mathbf{s} \in R^p$, $\mathbf{s} \neq \mathbf{0}$, let $Y_i^s = \mathbf{s}'\mathbf{X}_i \sim N_1(v_i^s, (\sigma^s)^2)$, where $v_i^s = \mathbf{s}'\boldsymbol{\mu}_i$, $(\sigma^s)^2 = \mathbf{s}'\Sigma\mathbf{s}$. Let C_s be sets such that $(\boldsymbol{\mu}_1, \ldots, \boldsymbol{\mu}_n, \Sigma) \in B$ if and only if $(v_1^s, \ldots, v_n^s, (\sigma^s)^2) \in C_s$. The problem in which we observe $Y_1^s, \ldots, Y_n^s$ and are testing that $(v_1^s, \ldots, v_n^s, (\sigma^s)^2) \in C_s$ is a univariate problem, and there are often sensible size α^* tests for that problem. The union–intersection test associated with those univariate tests is one that rejects if at least one of the univariate tests is rejected.

EXAMPLES

Consider the model in which we observe $\mathbf{X}_1, \ldots, \mathbf{X}_n$ independent, $\mathbf{X}_i \sim N_p(\boldsymbol{\mu}, \Sigma)$, Σ

> 0. Suppose we want to test that $\mu = \mathbf{0}$. Let $\mathbf{s} \in R^p$, $\mathbf{s} \neq \mathbf{0}$, and let $Y_i^s = \mathbf{s}'\mathbf{X}_i$, $\nu^s = \mathbf{s}'\mu$, $(\sigma^s)^2 = \mathbf{s}'\Sigma\mathbf{s}$. Then $\mu = \mathbf{0}$ if and only if $\nu^s = 0$ for all $\mathbf{s} \neq \mathbf{0}$. Let $\overline{Y}^s = (n)^{-1}\Sigma Y_i^s$ and $(T^s)^2 = (n-1)^{-1}\Sigma(Y_i^s - \overline{Y}^s)^2$ be the sample mean and the sample variance of the Y_i^s. The test we would use to test that $\nu^s = 0$ is just the one sample t-test in which we reject if

$$F_s = (t_s)^2 = n(\overline{Y}^s)^2 / (T^s)^2 > F_{1, n-1}^{\alpha^*}.$$

The union–intersection test rejects if at least one of the F-tests rejects, i.e. if and only if $\sup F_s$ is too large. (Note that the critical value $F_{1, n-1}^{\alpha^*}$ does not depend on $\mathbf{s}$.) Let $\overline{\mathbf{X}} = n^{-1}\Sigma\mathbf{X}_i$ and

$$\mathbf{T} = (n-1)^{-1}\Sigma(\mathbf{X}_i - \overline{\mathbf{X}})(\mathbf{X}_i - \overline{\mathbf{X}})'$$

be the sample mean vector and sample covariance matrix of the $\mathbf{X}_i$. Then $\overline{Y}^s = \mathbf{s}'\overline{\mathbf{X}}$, $(T^s)^2 = \mathbf{s}'\mathbf{T}\mathbf{s}$ and

$$V = \sup F_s = \sup n(\mathbf{s}'\overline{\mathbf{X}})^2 / \mathbf{s}'\mathbf{T}\mathbf{s}$$

$$= n\overline{\mathbf{X}}'\mathbf{T}^{-1}\overline{\mathbf{X}}.$$

Therefore, the union–intersection test rejects if Hotelling's T^2 is too large. This test is the same as the likelihood ratio test for this problem. The critical value for this test comes from the fact that

$$c(n, p)V \sim F_{p, n-p}(n\mu'\Sigma^{-1}\mu),$$

$$c(n, p) = (n-p)/\{p(n-1)\}.$$

We now find simultaneous confidence intervals* associated with this test. Let $F^* = c(n, p)n(\overline{\mathbf{X}} - \mu)'\mathbf{T}^{-1}(\overline{\mathbf{X}} - \mu) = \sup c(n, p)n(\mathbf{s}'(\overline{\mathbf{X}} - \mu))^2 / \mathbf{s}'\mathbf{T}\mathbf{s}$. Then $F^* \sim F_{p, n-p}$ also. Therefore,

$$P\Big(\mathbf{s}'\mu \in \mathbf{s}'\overline{\mathbf{X}} \pm \big(c(n, p)nF_{p, n-p}^{\alpha}\mathbf{s}'\mathbf{T}\mathbf{s}\big)^{1/2}$$

$$\text{for all } \mathbf{s}\Big)$$

$$= P\Big(c(n, p)n(\mathbf{s}'(\overline{\mathbf{X}} - \mu))^2 / \mathbf{s}'\mathbf{T}\mathbf{s}$$

$$\leq F_{p, n-p}^{\alpha} \text{ for all } \mathbf{s} \neq \mathbf{0}\Big)$$

$$= P\big(F^* \leq F_{p, n-p}^{\alpha}\big) = 1 - \alpha.$$

This result establishes a set of $(1 - \alpha)$ simultaneous confidence intervals for the set of all $\nu^s = \mathbf{s}'\mu$. Note that these simultaneous confidence intervals are compatible with the union–intersection test in that the hypothe-

sis $\mu = \mathbf{0}$ is rejected with that test if and only if at least one of the simultaneous confidence intervals does not contain 0. We think of those choices for $\mathbf{s}$ such that 0 is not in the associated simultaneous confidence interval as the linear combinations which are causing the hypothesis $\mu = \mathbf{0}$ to be rejected.

We note that the preceding test is not the only union–intersection test that could be derived for this problem. Let us restrict attention to those vectors $\mathbf{s}$ which have one 1 and the rest 0's. The $\mu = \mathbf{0}$ if and only if $\nu_s = 0$ for all these vectors $\mathbf{s}$. Now, $\sup F_s$ over this set is just $\max n(\overline{X}_i)^2 / T_{ii}$, where $\overline{X}_i$ is the ith component of $\overline{X}$ and T_{ii} is the ith diagonal element of T. This test is not the same as Hotelling's T^2 test already derived. This example illustrates that the union–intersection test depends on the class of sets C_s such that $B = \cap C_s$.

Now, consider the problem of testing that $\Sigma = \mathbf{I}$. This problem is the same as testing that $(\sigma^s)^2 = \mathbf{s}'\mathbf{s}$ for all $\mathbf{s}$. For the univariate problem, the usual test rejects if $U_s = (T^s)^2 / \mathbf{s}'\mathbf{s} > d$ or $U_s < c$. Therefore, the union–intersection test rejects if $r_1 = \sup U_s > d$ or $r_p = \inf U_s < c$. However, r_1 and r_p are the largest and smallest eigenvalues of $\mathbf{T}$. Tables of critical values for c and d so that the test has size 0.05 are given, for example, in Pearson and Hartley [5]. Note that this test is not the same as the likelihood ratio test for this problem.

We now find compatible simultaneous confidence intervals for the set of all $(\sigma^s)^2 = \mathbf{s}'\Sigma\mathbf{s}$. Let $\mathbf{T}^* = \Sigma^{-1/2}\mathbf{T}\Sigma^{-1/2}$ and let r_1^* and r_p^* be the largest and smallest eigenvalues* of $\mathbf{T}^*$. The distribution of $\mathbf{T}^*$ is the same as the null distribution of $\mathbf{T}$, and therefore the distribution of (r_1^*, r_p^*) is the same as the null distribution of (r_1, r_p). Let c and d be chosen so that the union–intersection test has size α. Using the fact that $r_1^* = \sup \mathbf{s}'\mathbf{T}\mathbf{s}/\mathbf{s}'\Sigma\mathbf{s}$ and $r_p^* = \inf \mathbf{s}'\mathbf{T}\mathbf{s}/\mathbf{s}'\Sigma\mathbf{s}$, we see that

$$1 - \alpha = P\big(c \leq r_p^* \leq r_1^* \leq d\big)$$

$$= P\big(c \leq \mathbf{s}'\mathbf{T}\mathbf{s}/\mathbf{s}'\Sigma\mathbf{s} \leq d \text{ for all } \mathbf{s} \neq \mathbf{0}\big)$$

$$= P\big(\mathbf{s}'\mathbf{T}\mathbf{s}/d \leq \mathbf{s}'\Sigma\mathbf{s} \leq \mathbf{s}'\mathbf{T}\mathbf{s}/c \text{ for all } \mathbf{s}\big)$$

and we have a set of $(1 - \alpha)$ simultaneous confidence intervals. Note that these simultaneous confidence intervals are compatible with the union–intersectiontest in that the hypothesis $\mathbf{\Sigma} = \mathbf{I}$ is rejected with this test if and only ifthere is at least one $\mathbf{s}$ such that the

interval for $(\sigma^s)^2 = \mathbf{s}'\mathbf{\Sigma}\mathbf{s}$ does not contain $\mathbf{s}'\mathbf{s}$. We think of such choices for $\mathbf{s}$ as the ones that are causing the null hypothesis to be rejected.

We now consider one more example which is somewhat different from the other two. Let

$$\mathbf{Y}_i = \begin{pmatrix} \mathbf{Y}_{11} \\ \mathbf{Y}_{12} \end{pmatrix}, \qquad \mathbf{T} = \begin{pmatrix} \mathbf{T}_{11} & \mathbf{T}_{12} \\ \mathbf{T}_{12} & \mathbf{T}_{12} \end{pmatrix},$$

$$\mu = \begin{pmatrix} \mu_1 \\ \mu_2 \end{pmatrix}, \qquad \mathbf{\Sigma} = \begin{pmatrix} \mathbf{\Sigma}_{11} \mathbf{\Sigma}_{12} \\ \mathbf{\Sigma}_{21} \mathbf{\Sigma}_{22} \end{pmatrix},$$

where μ_1 and $\mathbf{Y}_{i1}$ are $q \times 1$ and $\mathbf{\Sigma}_{11}$ and $\mathbf{T}_{11}$ are $q \times q$. Consider testing that $\mathbf{\Sigma}_{12} = \mathbf{0}$. To use the union–intersection principle for this problem, let $\mathbf{a} \in R^q, \mathbf{b} \in R^{p-q}, \mathbf{a} \neq \mathbf{0}, \mathbf{b} \neq \mathbf{0}$, $U_i^{\mathbf{a}} = \mathbf{a}'\mathbf{Y}_{i1}$, $V_i^{\mathbf{b}} = \mathbf{b}'\mathbf{Y}_{i2}$. Then the $(U_i^{\mathbf{a}}, V_i^{\mathbf{b}})'$ are a sample from a bivariate normal distribution* with correlation coefficient $\rho^{\mathbf{a},\mathbf{b}} = \mathbf{a}'\mathbf{\Sigma}_{12}\mathbf{b}/(\mathbf{a}'\mathbf{\Sigma}_{11}\mathbf{a}\mathbf{b}'\mathbf{\Sigma}_{22}\mathbf{b})^{1/2}$ and $\mathbf{\Sigma}_{12} = \mathbf{0}$ if and only if $\rho^{\mathbf{a},\mathbf{b}} = 0$ for all $\mathbf{a}$ and $\mathbf{b}$. Let $r^{\mathbf{a},\mathbf{b}}$ be the sample correlation coefficient computed from the $U_i^{\mathbf{a}}$ and $V_i^{\mathbf{b}}$. The usual test for the bivariate problem rejects if $(r^{\mathbf{a},\mathbf{b}})^2$ is too large. Hence the union–intersection test rejects if $W = \sup(r^{\mathbf{a},\mathbf{b}})^2$ is too large. However, $(r^{\mathbf{a},\mathbf{b}})^2 = (\mathbf{a}'\mathbf{T}_{12}\mathbf{b})^2/(\mathbf{a}'\mathbf{T}_{11}\mathbf{a}\mathbf{b}'\mathbf{T}_{22}\mathbf{b})$ and W is the largest eigenvalue of $\mathbf{T}_{11}^{-1}\mathbf{T}_{12}\mathbf{T}_{22}^{-1}\mathbf{T}_{21}$, i.e., the largest sample canonical correlation coefficient*. Therefore, the union–intersection test for this problem rejects if the largest sample canonical correlation coefficient is too large. A table of critical values for this test is also available, for example, in Pearson and Hartley [5]. There do not seem to be any fairly simple simultaneous confidence intervals associated with this test, primarily because there is no natural pivotal quantity whose distribution is the same as the null distribution of W.

FURTHER COMMENTS

The union–intersection principle has been applied to many other multivariate prob-

lems, most importantly to multivariate analysis of variance, where it leads to Roy's largest root test and the associated simultaneous confidence intervals. It has also been applied to multivariate nonparametric testing problems (see Chinchilli and Sen [2] and their earlier papers). In most situations, it leads to sensible tests (invariant, unbiased, and admissible), although any optimality properties must be established for each problem. No general optimaltiy results are known. In many situations it leads to reasonable associated simultaneous confidence intervals.

The union–intersection principle was developed in Roy [6]. The associated simultaneous confidence intervals were derived in Roy and Bose [8]. Among testbooks in multivariate analysis, Roy [7] has the most detailed discussion of the union–intersection principle. Kshirsagar [3] and Arnold [1] use the union–intersection principle, among others, to derive tests. Arnold [1] finds the associated simultaneous confidence intervals. The work of Morrison [4] is a more applied book which also uses the union–intersection principle to find tests and gives the associated simultaneous confidence intervals.

References

[1] Arnold, S. F. (1981). *The Theory of Linear Models and Multivariate Analysis*. Wiley, New York.

[2] Chinchilli, V. M. and Sen, P. K. (1981). *Sankhyā B*, **43**, 152–171.

[3] Kshirsagar, A. (1972). *Multivariate Analysis*. Dekker, New York.

[4] Morrison, D. (1976). *Multivariate Statistical Models*. McGraw-Hill, New York.

[5] Pearson, E. S. and Hartley, H. O. (1972). *Biometrika Tables for Statisticians*, Vol. 2, Cambridge University Press, Cambridge, England.

[6] Roy, S. N. (1953). *Ann. Math. Statist.*, **24**, 220–238.

[7] Roy, S. N. (1957). *Some Aspects of Multivariate Analysis*. Wiley, New York.

[8] Roy, S. N. and Bose, R. C. (1953). *Ann. Math. Statist.*, **24**, 513–536.

(HOTELLING'S T^2
HYPOTHESIS TESTING
LIKELIHOOD RATIO PRINCIPLE

(MULTIVARIATE ANALYSIS
ROY'S CHARACTERISTIC ROOT
STATISTIC)

STEVEN F. ARNOLD

UNION OF SETS

The union of two sets A and B is the set composed of elements belonging to A or to B (or to both). It is conventionally denoted $A \cup B$.

(SYMMETRIC DIFFERENCE
VENN DIAGRAM)

UNIQUENESS THEOREM (SHANNON)
See ENTROPY

UNITARY MATRIX

This is a generalization of the concept of orthogonal matrix*, admitting complex numbers. A square matrix U is *unitary* if

$$U\overline{U}' = \overline{U}'U = I,$$

where $\overline{U}$ is the complex conjugate of U (and $\overline{U}'$ is the transpose of $\overline{U}$). An example of a unitary matrix is

$$\begin{pmatrix} -\frac{1}{2} & 2+i \\ \frac{1}{10}(2-i) & 1 \end{pmatrix}.$$

The determinant of a unitary matrix is either 1 or -1. The product of two unitary matrices is a unitary matrix. Unitary matrices are used in the theory of multivariate analysis* and multivariate distributions.

(TRIPOTENT MATRIX)

UNITED NATIONS, STATISTICAL OFFICE OF THE

This office is part of the Department of International Economics and Social Affairs in the Secretariat of the United Nations (New York, NY 10017). It collects and pub-lishes statistics showing the main economic and social characteristics of individual countries, of regions, and of the world as a whole. Its recurrent publications are: *Statistical Yearbook*, *Monthly Bulletin of Statistics*, *Statistical Pocketbook* (*World Statistics in Brief*), *Demographic Yearbook*, *Population and Vital Statistics Report*, *Compendium of Human Settlements Statistics*, *Compendium of Social Statistics*, *Yearbook of International Trade Statistics*, *World Trade Annual* and *Supplement*, *Commodity Trade Statistics*, *Yearbook of World Energy Statistics*, *Yearbook of Industrial Statistics*, *Yearbook of Construction Statistics*, and *Yearbook of National Accounts Statistics*. The Office also publishes statistical classifications, guidelines, and methodological and technical studies such as *Handbook of Household Survey* (1984) and *Handbook of Vital Statistics Systems and Methods*, Vol. 1 (1985), a *Directory of International Statistics*, a *Directory of Environmental Statistics*, sample survey reports, bibliographies, and reports of training centers.

UNIVERSAL DOMINATION *See* STOCHASTIC DOMINATION AND UNIVERSAL DOMINATION

UNLIKELIHOOD

The *unlikelihood* of a particular simple (specific) hypothesis, or probability model, when several possible hypotheses are under consideration, is a weighted linear combination (with nonnegative weights) of all other *likelihoods*. Thus, if the observed data x have density (likelihood) $f_i(x)$ according to hypothesis i, then $u_j = \sum_{i \neq j} w_{ij} f_i(x)$ is the *unlikelihood* of hypothesis j. A method of inference or decision, called *minimum unlikelihood*, is one in which hypothesis j is chosen as correct only when u_j is minimal; it reduces to *maximum likelihood** if w_{ij} is constant.

These concepts were introduced by Lindley [3], who argued that in such multiple-decision situations, decision procedures should be characterized and partially ordered by their error probabilities p_{ij}—the probability of choosing hypothesis $j \neq i$ when i obtains—rather than the risk functions of decision theory*. He showed that all reasonable procedures are minimum unlikelihood, for some suitable weights w_{ij}.

If the weights are interpreted as the product of a prior probability and a loss for incurring that particular kind of error, then u_j is proportional to the posterior risk of choosing hypothesis j, and minimum unlikelihood is then a Bayes procedure—even though Lindley's objective was not to minimize average risk. However, Lindley's problem may be viewed from a standard decision theory perspective, by artificially enlarging the parameter space; see ref. 1.

That the method of minimum unlikelihood provides solutions to multiple-decision problems in which all error probabilities are to be controlled (bounded) was further developed in ref. 1. Unlikelihood, as a basis for a sequential stopping rule*, was considered in ref. 2.

References

[1] Hall, W. J. (1958). *Ann. Math. Statist.*, **29**, 1079–1094.

[2] Hall, W. J. (1980). In *Asymptotic Theory of Statistical Tests and Estimation*, I. M. Chakravarti, ed. University of North Carolina Press, Chapel Hill, NC, pp. 325–350.

[3] Lindley, D. V. (1953). *J. R. Statist. Soc. B*, **15**, 30–76.

(BAYESIAN INFERENCE
DECISION THEORY
MAXIMUM LIKELIHOOD ESTIMATION
MULTIPLE DECISION PROCEDURES)

W. J. HALL

UNRELIABILITY

A term sometimes used for the complement of reliability*. Thus if probability of survival to time τ is used as an index of reliability,

the corresponding index of unreliability would be the probability of failure in the time τ.

(MANN–GRUBBS METHOD
RELIABILITY)

UNRELIABILITY FUNCTION

This is an alternative name for the "failure function" (CDF*) of a distribution of lifetime (T):

$$\Pr[T \leq t] = F(t).$$

It is not very commonly used.

(CUMULATIVE DISTRIBUTION FUNCTION
LIFETIME DISTRIBUTIONS
RELIABILITY
SURVIVAL ANALYSIS)

UNTILTING

A common statistical model involves the fitting of a straight line to n pairs of points (y, x), where y and x are naturally paired by being measured on the same experimental unit. The model, usually referred to as the simple linear regression* model (SLR), is $\hat{y} = b_0 + b_1 x$, where b_0 and b_1 represent the intercept and slope of the line and $\hat{y}$ is the fit at the value x.

The data analysis for this model typically begins with the plotting of the n paired observations (y_i, x_i), $i = 1, \ldots, n$, in the form of a scatter plot*. In order to assess the appropriateness of the SLR model to the n pairs of points (*see* EYE ESTIMATES), the residuals* which represent the difference between the observed and fitted values, are computed and analyzed. Thus, the residual for the ith point r_i is computed as $r_i = y_i - \hat{y}_i = y_i - b_0 - b_1 x_i$ for $i = 1, \ldots, n$.

A variety of methods have been proposed to calculate the values of b_0 and b_1. An eye-ball fit to the data may be obtained by simply using a ruler to draw a line that best describes the scatter of points. The most

Table 1

| Data | | Residuals | |
x	y	$r = y - (24.4 + 1.4x_i)$	$r' = r - (3.7 + 0.32x)$
2.0	23.0	−4.20	−1.14
2.6	26.0	−2.04	0.83
4.0	30.0	0.00	2.42
4.7	27.5	−3.48	−1.28
5.4	31.5	−0.46	1.51
6.3	33.0	−0.22	1.46
7.0	34.5	0.30	1.76
8.3	34.0	−2.02	−0.98
9.0	37.0	0.00	0.82
10.0	37.0	−1.40	−0.90

commonly used technique to obtain b_0 and b_1 is the method of least squares*. Another method is the three-group resistant line*, a technique that allows for the fitting of a straight line to data without the line being unduly influenced by one or more unusual observations. The least squares method, while having the benefit of ease of computation, is nonresistant to such anomalies. The notion of the resistant line is illustrated in detail in Tukey [3] and Hoaglin et al. [1].

A nonzero value of b_1, or slope term, indicates a dependent relationship (though not necessarily a causal relationship; see Mosteller and Tukey [2]) between y and x. Tukey [3] refers to the slope in such a scatter plot as the *tilt*.

The residuals r_i that result from subtracting the fit from the observed data will now have the property that a scatter plot of the n observations (r_i, x_i) should exhibit less tilt than the original plot of (y_i, x_i). This remaining tilt, or nonzero slope, may be further reduced and eventually eliminated by continuing the process of fitting a straight line to the plot of r_i versus x_i and obtaining new residuals. The process is repeated until the x and the residuals show no apparent tilt. Tukey has labeled this technique *untilting*.

The final values of b_0 and b_1 are obtained by adding together the fits at each stage of the untilting process. Thus,

$$\text{final } b_0 = \text{sum of } b_0\text{'s,}$$
$$\text{final } b_1 = \text{sum of } b_1\text{'s,}$$

where the sum is over all stages required to untilt the relationship between x and y.

To illustrate the untilting process, consider the fictitious (x, y) data presented in Table 1 and displayed in Fig. 1. An initial fit, obtained by eye, of $\hat{y}_i = 24.4 + 1.4x_i$ (see the line in Fig. 1) is used to obtain the residuals $r_i = y_i - \hat{y}_i$. A scatter plot of (r_i, x_i) (Fig. 2) reveals a tilt and, therefore, a second line, again obtained by eye, $\hat{r}_i = -3.7 + 0.32x_i$ is fitted. The residuals r_i', after this

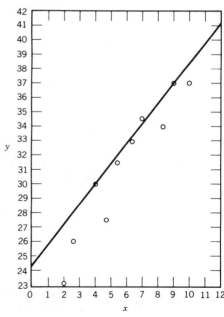

Figure 1 Scatter plot of the (y, x) data and the line $\hat{y} = 24.4 + 1.4x$.

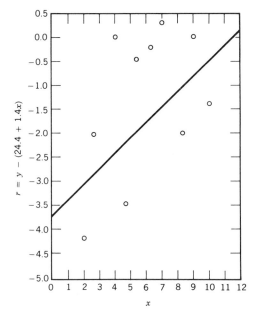

Figure 2 Scatter plot of the (r, x) data and the line $\hat{r} = -3.7 + 0.32x$.

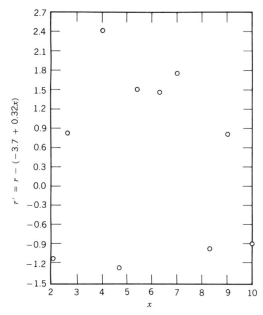

Figure 3 Scatter plot of the (r', x) data.

second stage, are displayed in Fig. 3. Since no apparent tilt remains, the untilting process is complete and the final fit is calculated as

$$b_0 = 24.4 + (-3.7) = 20.7,$$
$$b_1 = 1.4 + 0.32 = 1.72.$$

Untilted analysis refers to the process of untilting and the subsequent analysis of the residuals. The untilted analysis may reveal the need for the original data to be expressed in a different scale. Transformations* of y or x or both y and x are sometimes necessary to linearize the relationship between x and y and facilitate the residual analysis.

References

[1] Hoaglin, D. C., Mosteller, F., and Tukey, J. W. (1983). *Understanding Robust and Exploratory Data Analysis*. Wiley, New York. (Presents some of the statistical theory and motivation for EDA and robust procedures.)

[2] Mosteller, F. and Tukey, J. W. (1977). *Data Analysis and Regression—A Second Course in Statistics*. Addison-Wesley, Reading, MA. (Includes EDA plus a good introduction to regression analysis.)

[3] Tukey, J. W. (1977). *Exploratory Data Analysis*. Addison-Wesley, Reading, MA. (Excellent reference by the author who developed the subject.)

(EXPLORATORY DATA ANALYSIS REGRESSION (various entries) TRANSFORMATIONS)

JEFFREY B. BIRCH

UP-AND-DOWN METHOD *See* STAIRCASE METHOD

UPGMA CLUSTERING METHOD

This is also called *group average method* or *average linkage method*. It is a hierarchal clustering method wherein the distance between two clusters is the average of the distances between all possible pairs of individuals, one from one group and one from the other. Using an obvious notation, the distance between clusters $I \equiv \{x_{1, I}, \ldots, x_{n_I, I}\}$ and $J \equiv \{x_{1, J}, \ldots, x_{n_J, J}\}$ is

$$d(I, J) = (n_I n_J)^{-1} \sum_i \sum_j d(x_{i, I}, x_{j, J}).$$

At each step, the two (or more) clusters closest to each other are merged.

(HIERARCHAL CLUSTER ANALYSIS)

UPPER-EASING INEQUALITY *See* MARTINGALES

UPTIME

A term used in reliability* theory to denote the total time during which a system is in acceptable operating condition.

URN MODELS

INTRODUCTION

Urn models are useful in illustrating many of the basic ideas and problems in probability theory*. Most introductory textbooks in probability theory are replete with examples and exercises dealing with balls drawn at random from randomly selected urns. When introducing the basic concept of a *random experiment*, Parzen [27] writes in his well-known textbook: "More generally, if one believes that the experiment of drawing a ball from an urn will, in a long series of trials, yield a white ball in some definite proportion (which one may not know) of the trials of the experiment, then one has asserted (i) that the drawing of a ball from such an urn is a random phenomenon and (ii) that the drawing of a white ball is a random event."

Because urn models can be easily visualized and are very flexible, they are adaptable to a wide range of situations. Numerous results in discrete probability theory can be derived from simple urn models. In place of urns, it is of course possible to resort to experiments based on games of chance*, such as coin tossing or dice rolling. However, urn models do have certain advantages over things like lotteries, dice, decks of cards, and chessboards in that the latter are commonly associated with certain numbers such as 6, 52, 64. Moreover, properties of playing cards and dice are not familiar to all people. Alfréd Rényi [31] mentions a Hungarian colleague of his who taught statistics in Ethiopia and who had great difficulty in acquainting his students with basic probability concepts because games of chance were forbidden and virtually unknown in Ethiopia.

The urn filled with balls of different colors, or with tickets bearing some ciphers or letters, remains the most common conceptual model for producing random results. This model is continuously used in statistics and probability courses as a didactic tool, and in our analyses as a means of translating practical problems into mathematical ones. Pólya [30], whose name is intimately associated with the development of urn models, observes "Any problem in probability appears comparable to a suitable problem about bags containing balls, and any random mass-phenomenon appears similar in certain essential aspects to successive drawings of balls from a system of suitably combined bags." Discussing instruction in the new school subject of stochastics, Heitele [17] supports Pólya's view, remarking that "in principle, it is possible to assign urn models to the greater part of chance experiments, at least those with a countable sample space." He then goes on to say that an intuitive grasp of basic concepts for later analytic knowledge is more important in "stochastics" than anywhere else in view of the elusive character of many probability concepts and the large number of paradoxes inherent therein. Finally, the book by Johnson and Kotz [20], exclusively devoted to urn models and their applications, bears eloquent witness to the continued importance of urn models in probability and statistics.

URN MODELS, EXAMPLES

An urn model is constructed by assuming a set of urns containing balls of different colors. In specific cases, we consider sequences of experiments in which balls selected from the urns are possibly replaced into the urns according to certain rules. These rules may require that we add balls or remove them from certain urns at various stages of the experiment. These rules may also call for certain balls to change color during the experiment. When a ball is drawn at random from an urn containing N balls, we normally assume that each ball has the same probability of being chosen. Freudenthal [13] regards the urn model as the expression of three postulates: (i) the constancy of the probability distributions, (ii) the random character of the choice, and (iii) the independence of successive choices, whenever the drawn balls are put back into the urn. This is ensured by the solidity of the vessel, preventing any consciously selective choice.

When considering urn experiments, we are usually interested in random variable(s) determined by

(i)   the composition of the urns at the nth draw, e.g., the distribution of the number of balls of various colors in the urn(s);

(ii) the outcome of the first n draws, for example, the number of balls of a particular color obtained in the n first draws;

(iii) the waiting time until one or more specified conditions are satisfied.

The random variables just mentioned give rise to and include quite a remarkable range of important discrete probability distributions, and also some less common ones. By considering limiting forms when certain parameters tend to infinity, the field of applications can be extended to a number of continuous distributions. A unified approach to limit theorems* for urn models is presented by Holst [18].

It may prove useful at this point to consider some examples of urn models. The examples we give illustrate problems of respectable antiquity.

Example 1. Bernoulli's model. Balls are drawn at random from an urn containing red and black balls in a fixed proportion, say

Np red and $N(1 - p)$ black balls. Each ball drawn is replaced and the urn is thoroughly shaken before the next drawing is made. In this way, the results of the drawings become mutually independent. The number of red balls obtained in n such drawings is a random variable, say X, whose distribution is given by the well-known *binomial* formula*

$$\Pr[X = x] = \binom{n}{x} p^x (1 - p)^{n-x},$$

$$x = 0, 1, \cdots, n, \ n = 1, 2, \ldots,$$

representing one of the basic distributions of statistical theory.

Example 2. Bayes' Theorem*. This theorem is often and perhaps best illustrated by an urn model (see, e.g., Parzen [27]). The model used to illustrate *Bayes' theorem* usually consists of several urns $U_1, U_2, \ldots, U_n$ together with a preassigned distribution of m balls of k colors into the urns, so that there are $m(i, j) \geqslant 0$ balls of color j in the urn U_i. The numbers $m(i, j)$ are known in advance, but the urns are identical in appearance, so initially it is assumed that the identity of the urns is indeterminate. One randomly selects an urn and then randomly draws a ball from this urn. The experiment is possibly repeated (with or without replacement of the balls) a certain number of times and the outcome is recorded. Given this outcome, Bayes' formula is then invoked to calculate the probability for the identification of the urns.

Example 3. The Law of Succession. This example is also of ancient vintage and illustrative in the spirit of Laplace. The formulation given here is essentially that of Feller [12]. Imagine a set of $N + 1$ urns, each containing N black and white balls; the urn number k contains k black and $N - k$ white balls, $k = 0, 1, \ldots, N$. An urn is chosen at random and n random drawings are made from the urn, the ball being replaced each time. Now suppose that all n balls turn out to be black (event A). We seek the conditional probability that the next drawing will also yield a black ball (event B). If the first choice falls on urn number k, then the probability of extracting in succession n black balls is obviously $(k/N)^n$. Hence, the probability of event A is

$$\Pr[A] = \frac{1^n + 2^n + \cdots + N^n}{N^n(N+1)}.$$

The event AB means that $n + 1$ draws yield $n + 1$ black balls and, therefore,

$$\Pr[AB] = \Pr[B]\Pr[A|B]$$

$$= \frac{1^{n+1} + 2^{n+1} + \cdots + N^{n+1}}{N^{n+1}(N+1)}.$$

The required probability is $\Pr[A|B] = \Pr[AB]/\Pr[B]$. For large N, we obtain approximately

$$\Pr[A|B] \approx \frac{n+1}{n+2}.$$

The formula can be interpreted roughly as follows: If all compositions of an urn are equally likely and if n trials yielded black balls, the probability of a black ball at the next trial is $(n+1)/(n+2)$. This is the so-called *law of succession* of Laplace* (1812). Laplace himself illustrated the use of the formula by computing the probability that the sun will rise tomorrow, given that it has risen daily for 5000 years or $n = 1826{,}213$ days. It is said that Laplace was ready to bet 1826,214 to 1 in favor of regular habits of the sun. Before the ascendence of modern probability theory, the notion of equal probabilities was often used as synonymous for "no advance knowledge." An historical study would be necessary to render justice to Laplace and to understand his intentions, as Feller points out when presenting this example. (*See also* LAPLACE'S LAW OF SUCCESSION.)

Example 4. Lexis' Model. An urn model in the form of an arrangement of urns and balls together with a sampling rule may also be used to set up models of *structures of variation*. An example of historical importance is due to Lexis* and quoted by Johnson and Kotz [20]. Lexis considered a set of urns containing different proportions of red and black balls. If we plan to draw n balls

from the urns, we may then either

1. choose an urn at random and then draw a ball n times, replacing the ball each time, or
2. choose as nearly equal a number of balls as possible from each urn.

Considering the extreme case when we have two urns, one containing only red balls and one containing only black balls, we see that the first method leads to maximal variation in the number of red balls chosen (this must be either 0 or n), whereas the second method leads to more stable results (number of red balls constant $= n/2$, if n is even). Case 1 is termed *supernormal dispersion* and case 2, *subnormal dispersion*, as compared with the results of simple random sampling with replacement from an urn containing red and black balls in proportions p and $1 - p$, respectively (cf. Example 1).

URN MODELS FOR DEPENDENCE AND CONTAGION

Stochastic Processes*

We now turn to urn experiments devised to model *dependence structures*. These models give rise to simple discrete *stochastic processes* in which each draw is the natural time unit.

Mathematical models of systems that vary in time in a random manner are known as stochastic processes. Numerical observations made as the process continues indicate its evolution. Historically, however, the term stochastic process has been reserved for families, usually infinite, of *dependent* variables. A basic problem is therefore to devise suitable simple forms of dependency; that is, to discover new types of stochastic processes that are useful, or mathematically elegant, or which can conform otherwise to the investigator's criterion of importance.

The fundamental ingredient that made it possible to go beyond the assumption of independence, which dominated probability and statistics until the early part of the century, into the realm of dependence is the essential concept of *conditioning* (*see* CONDITIONAL PROBABILITY AND EXPECTATION). In modern probability theory, models are frequently introduced by specifying appropriate conditional probabilities or properties; this is the case, for instance, with certain urn models, *Markov chains*, and *martingales**. Of the host of possibilities for constructing families of dependent stochastic variables, only a few forms of dependence have been systematically studied and found acceptable for practical work.

To illustrate the concepts we have just discussed, let us again consider a few examples of urn models.

Example 5. A Markov Chain (Kemeny and Snell [21]).

An urn contains two unpainted balls. At a sequence of times a ball is chosen at random, painted either red or black, and then put back into the urn. If the ball drawn was unpainted, the choice of color is made at random. If it is colored, its color is changed. A *state* is defined by the three numbers x, y, z, where x is the number of unpainted balls, y is the number of red balls, while, finally, z is the number of black balls. The *transition matrix* of probabilities for changing states is given by

	(011)	(020)	(002)	(200)	(110)	(101)
(011)	0	$\frac{1}{2}$	$\frac{1}{2}$	0	0	0
(020)	1	0	0	0	0	0
(002)	1	0	0	0	0	0
(200)	0	0	0	0	$\frac{1}{2}$	$\frac{1}{2}$
(110)	$\frac{1}{4}$	$\frac{1}{4}$	0	0	0	$\frac{1}{2}$
(101)	$\frac{1}{4}$	0	$\frac{1}{4}$	0	$\frac{1}{2}$	0

We have here a sequence of experiments possessing the *Markov property*: Given the present state of the system, the past states have no influence on the future development of the process. If we have information that the process starts in some particular state, then we have enough information to determine all relevant probabilities. What we have just described in urn model terms is a special case of a *Markov chain* (*see* MARKOV PROCESSES).

Example 6. The Ehrenfest Model.

A classical mathematical model of diffusion through a membrane is the famous Ehrenfest model, which may be viewed as a *random walk** on a finite set of states with the boundary states reflecting. The random walk is restricted to $i = -N, -N + 1, \ldots, 0, 1, \ldots, N - 1, N$, with transition probabilities given by

$$P_{ij} = \begin{cases} (N - i)/2N & \text{if } j = i + 1, \\ (N + i)/2N & \text{if } j = i - 1, \\ 0 & \text{otherwise.} \end{cases}$$

The physical interpretation of the model is as follows. Imagine two containers containing a total of $2N$ balls. Suppose the first container, labeled A, holds k balls and the second, B, $2N - k$ balls. A ball is selected at random (all selections equally likely) from among the totality of the $2N$ balls and moved to the other container. Each selection generates a transition of the process. Clearly, the balls fluctuate between the two containers with a drift from the one with the larger concentration of balls. To reduce this to an urn model it suffices to call the balls in container A red, the others black. Then at each drawing the ball drawn is replaced by a ball of the opposite color. It is clear that in this case the process can continue as long as we please (if there are no red balls, a black ball is drawn automatically and replaced by a red one).

In the Ehrenfest model of heat exchange between two isolated bodies, the temperature of the bodies is symbolized by the number of balls in the containers and the heat exchange is random as in the kinetic theory of gases. In his famous book on mathematical biology, Lotka [23] uses a similar urn model to illustrate the phenomenon of *reversibility*.

URN MODELS FOR CONTAGION

General Remarks

There are several ways in which probability models representing some form of *contagion* can be constructed; that is, models in which the occurrence of an event has the effect of changing the probability of additional such events. Fortunately, many cases of particular practical importance can be subsumed under a class of relatively simple methods of construction.

To indicate the approach and to give a touch of concreteness to the discussion, let us consider an industrial plant liable to accidents. The occurrence of an accident might be pictured as the result of a superhuman game of chance (cf. Feller [12, p. 109]). Fate has in store an urn containing black and white balls. At regular intervals a ball is drawn at random, a black ball signifying an accident. Now if the chance of an accident remains constant in time, the composition of the urn is always the same. But it is conceivable that each accident has an *aftereffect*, in that it either increases or decreases the chance of new accidents. This corresponds to an urn whose composition changes according to certain rules that depend upon the outcome of the preceding drawings. Obviously, there is considerable latitude in the choice of these rules. The following is general enough to illustrate the possibilities (Feller [12, p. 109]): "An urn contains a white balls and b black balls. A ball is drawn at random. It is replaced and, moreover, c balls of the color drawn and d balls of the opposite color are added. A new random drawing is made from the urn containing $a + b + c + d$ balls, and this procedure is repeated." We note that the urn model specifies certain *conditional probabilities*, from which basic probabilities can be calculated. However, despite the model's apparent simplicity, explicit probabilities are not readily available except in the most important and best known special cases, to which we will return presently.

Although a number of stochastic processes have been developed primarily for the study of diffusion phenomena such as the spread of contagious diseases, they seem equally applicable to the study of socially "contagious" phenomena. In his classic text, the sociologist Coleman [5] writes: "Probability

models are often described in terms of particular ways of drawing balls from an urn...It goes without saying that the model becomes a theory of a social process whenever the social process parallels the methods of drawing balls from the urn which generated the model. Models of medical and social contagion are often generated by such urn schemes...."

An historical note

*Mixtures** are probability distributions of a compound kind, usually derived from distributions dependent on parameters, by regarding these parameters as themselves having probability distributions. Mixtures have played an important role in the construction of so-called *contagious distributions**. An early example was given by Greenwood and Yule [15], who, in an analysis of accident data, derived the classical *negative binomial distribution** as a mixture distribution. Other contagious distributions in the form of mixtures were derived by Neyman [26], later referred to as *Neyman type* A, B, and C distributions*.

The idea of using urn schemes to model contagion phenomena seems to be due to Pólya. The development of these models stems largely from a paper by Eggenberger and Pólya [10] from 1923 and the distributions resulting from these models are known as *Pólya* or *Pólya–Eggenberger distributions* (there are differences in usage among different authors). In their first, seminal paper, Pólya and Eggenberger use the expression *Chancenvermehrung durch Erfolg* (increase in chance due to success) for what Feller terms aftereffect. Some further analysis is given in a later paper (in French) by Pólya [29], in which he classifies dependence structures and suggests the term "contagion". Despite some confusion caused by this suggestive name, it has become customary in statistical literature to speak of contagion instead of aftereffect, the word preferred by Feller.

The Pólya–Eggenberger urn scheme served as a prototype for many models discussed in the literature. To quote only one example here, Friedman [14], in a paper entitled "A simple urn model," extends the Pólya–Eggenberger model; among five special cases studied by him are the Ehrenfest model (see Example 6) and a safety campaign model.

THE PÓLYA–EGGENBERGER DISTRIBUTION

Genesis and First Properties

The common feature of the Pólya family of urn models is that sampling is done with some sort of replacement, but it is not simple replacement of the ball last drawn. The way in which the replacement is modified determines the particular subclass of distributions arising. In particular, the kind of replacement depends upon the results of the sampling, i.e., upon the results of the preceding drawings (corresponding to Pólya's concept of *influence globale*).

The genesis of what we will call the Pólya–Eggenberger distribution is conveniently described as follows. Initially, there are a red balls and b black balls in the urn. One ball is drawn at random and then replaced together with c balls of the same color. This procedure is repeated n times. Here, after each drawing, the number of balls of the color drawn increases, whereas the balls of opposite color remain unchanged in number. In effect the drawing of either color renders more probable the repetition of the same color at the next drawing, and we have a rough model of phenomena such as contagious diseases, where each occurrence increases the probability of further occurrences.

The analytic simplicity of the Pólya–Eggenberger model is due to the following symmetry property. Any sequence of n drawings resulting in x red balls and $n - x$ black balls has the same probability as the event of extracting first x red balls and then $n - x$ black balls. Since there are $\binom{n}{x}$ outcomes with x red balls, we can write down the probability distribution of the random variable X, the number of red balls drawn, namely

$$\Pr[X = x]$$
$$= \binom{n}{x} \frac{\prod_{i=0}^{x-1}(a + ic) \prod_{j=0}^{n-x-1}(b + jc)}{\prod_{i=0}^{n-1}(a + b + ic)}, \quad (1)$$

$x = 0, 1, \ldots, n$. This is the probability function of the Pólya–Eggenberger distribution with parameters a, b, c, and n.

If we set $\alpha = a/c$ and $\beta = b/c$, (1) can be rewritten more compactly as

$$\Pr[X = x]$$
$$= p(x) = \binom{n}{x} \alpha^{[x]} \beta^{[n-x]} / (\alpha + \beta)^{[n]}$$
$$(2)$$

where $\alpha^{[x]} = \alpha(\alpha + 1) \cdots (\alpha + x - 1)$. In this notation the rth *factorial moment** of the variable X is

$$E(X^{(r)}) = \mu_{(r)} = n^{(r)} \alpha^{[r]} / (\alpha + \beta)^{[r]},$$
$$r = 1, 2, \ldots,$$

where $X^{(r)} = X(X - 1) \cdots (X - r + 1)$. In particular the first two central moments—the mean and the variance—are

$$\mu = n\alpha/(\alpha + \beta) = na/(a + b)$$

and

$$\sigma^2 = \frac{n\alpha\beta(\alpha + \beta + n)}{(\alpha + \beta)^2(\alpha + \beta + 1)}$$
$$= \frac{nab(a + b + nc)}{(a + b)^2(a + b + c)},$$

respectively. Letting $p = a/(a + b)$, $q = 1 - p$, and $\gamma = c/(a + b)$, leads to convenient reparametrization of (1) and (2), namely

$$\Pr[X = x] = \binom{-p/\gamma}{x}\binom{-q/\gamma}{n - x} \Big/ \binom{-1/\gamma}{n},$$
$$(3)$$

where

$$\binom{-a}{x} = \frac{(-a)(-a - 1) \cdots (-a - x + 1)}{x!}$$
$$= (-1)^x \frac{a(a + 1) \cdots (a + x - 1)}{x!}.$$

Expressed in these parameters, the variance can be written

$$\sigma^2 = npq(1 + n\gamma)/(1 + \gamma). \quad (4)$$

The reader will recognize the factor npq in (4) as the variance of the binomial distribution* with parameters n and p.

The probability generating function* can be compactly written in terms of the Gaussian hypergeometric function (see Kemp and Kemp [22]):

$$G(z) = \text{const.}_2F_1(-n, -p/\gamma;$$
$$-q/\gamma, -n + 1; z). \quad (5)$$

Some Interrelations

First we note that the Pólya–Eggenberger distribution (1) with parameters $a, b, 0$, and n is the same as the binomial distribution with parameters n and $p = a/(a + b)$. Similarly, the Pólya–Eggenberger distribution with parameters $a, b, c = -1$ and n is the same as the hypergeometric distribution* with parameters $a + b$, a, and n, while $a, b, c = 1$ and n yield the negative hypergeometric distribution*. Moreover, if in (1) the parameters satisfy $a = b = c$, then the distribution reduces to the uniform distribution* over $(0, 1, \ldots, n)$. These relationships are apparent from the genesis of the Pólya–Eggenberger distribution as already described.

The Pólya–Eggenberger distribution can be derived as a mixture distribution as follows. Let the random variable X have a binomial distribution with parameters n and p, where p in turn is a random variable having a beta distribution with parameters α and β. The unconditional distribution of the random variable X is the Pólya–Eggenberger distribution with parameters n, α, β, i.e., the discrete distribution (2) (Boswell and Patil [3]).

In his classic paper, Pólya [29] classifies limiting cases as follows. Let $\rho = a/(a + b)$ describe the initial composition of the urn, i.e., the proportion of red balls at the start. Further let the parameter $\gamma = c/(a + b)$ denote the contagion factor. In easily understood French, Pólya distinguishes between

the following cases:

(a) *événements usuels.* ρ constant as n tends to infinity;

(b) *événements rares.* ρ tends to zero but such that $n\rho \to \lambda$ constant, as n tends to infinity;

(1) *événements independents.* $\gamma = 0$;

(2) *contagion faible.* $\gamma \to 0$ such that $n\gamma \to$ a constant as n tends to infinity;

(3) *contagion forte.* $\gamma > 0$, constant, i.e., $n\gamma \to \infty$.

Let us look a little more closely at Pólya's case: rare events and weak contagion, i.e., in the discrete distribution (3) we let $n \to \infty$, $p \to 0$, $\gamma \to 0$, so that $np \to \lambda$ and $n\gamma \to 1/\theta$. We then obtain the following limiting form of the Pólya–Eggenberger distribution:

$$\Pr[X = x]$$
$$= \binom{\lambda\theta + x - 1}{x}\left(\frac{\theta}{\theta + 1}\right)^{\lambda\theta}\left(\frac{1}{\theta + 1}\right)^x,$$
$$x = 0, 1, \ldots . \quad (6)$$

The reader will recognize (6) as the negative binomial distribution* with parameters θ and $\lambda\theta$. In a similar manner, the Poisson distribution* is obtained as a limiting form of the Pólya–Eggenberger distribution when in Pólya's classification scheme we have rare independent events.

Related or Modified Distributions

Consider an urn containing a white balls and b black balls. The balls are drawn at random, each ball replaced along with c additional balls of the same color before the next ball is drawn. The number of black balls drawn until we have obtained exactly k white balls has the inverse Pólya–Eggenberger distribution with parameters a, b, c, and k and probability function

$$P(x) = \binom{k + x - 1}{x}$$
$$\times \frac{\displaystyle\prod_{i=0}^{k-1}(a + ic)\prod_{j=0}^{x-1}(b + jc)}{\displaystyle\prod_{i=0}^{k+x-1}(a + b + ic)}. \quad (7)$$

Reparametrizing as before, we can write

$$P(x) = \left(\begin{array}{c} -q/\gamma \\ x \end{array} \right) \left(\begin{array}{c} 1/\gamma - 1 \\ -k - x \end{array} \right) \bigg/ \left(\begin{array}{c} p/\gamma - 1 \\ -k \end{array} \right),$$

(8)

which is the inverse Pólya–Eggenberger distribution with parameters p, γ, and k. The inverse Pólya–Eggenberger distribution (8) with parameters a, b, 1, and k is the same as the generalized hypergeometric distribution* type IV with integer-valued parameters $-b$, $a + b - 1$, and $-k$.

Applications

Eggenberger [9] applied the Pólya–Eggenberger urn model to data taken from Swiss smallpox records. He had at his disposal the number of smallpox deaths for each month of the years 1877–1900, a total of about 1600. He classified the months with respect to the number of deaths. Eggenberger's problem was to describe the empirical frequency distribution of the months by a theoretical probability distribution. He first tried the Poisson distribution with a poor fit, explained by Eggenberger as resulting from the extremely epidemic character of smallpox. He then proceeded to consider the model of contagion launched by himself and Pólya [10]. The limit passage characterized by *événements rares* and *contagion faible* was suggested by the phenomenon considered, and Eggenberger was thus led to apply the limiting form (6) of the Pólya–Eggenberger distribution, i.e., a special case of the negative binomial distribution.

A further application was reported by Newbold [25], who gave a summary of the statistical side of some investigations of accident occurrences. Among the questions she discussed were:

1. Does any definite tendency exist under uniform conditions of risk for certain people to sustain more accidents than others?
2. If such a tendency exists, to what extent can it be modified by the occurrence of accidents?

Newbold applied the Pólya–Eggenberger distribution in the limiting form (6), which she called the *Greenwood–Yule curve*. She estimated the two parameters by the method of moments, obtaining fairly good agreement.

In Eggenberger's and Newbold's applications we find two different ways of interpreting model parameters. Eggenberger considers all his units, i.e., all months, as originally equally exposed to smallpox, and explains the heterogeneity of his records as being the result of an influence between smallpox cases. Newbold, on the other hand, explains the heterogeneity as being the result of different personal liabilities. Thus, while Eggenberger considers the parameter c (or γ) in (1) or (3) to be a measure of dependence, or contagion, Newbold considers the same parameter to be a measure of the variance of the risk distribution. The following example may serve to illustrate Newbold's approach.

Example 7. Urn Model for Stratification. The proneness to accidents may vary from person to person or from profession to profession, and we imagine that each person (or profession) has its own urn. The experiment of observing a randomly chosen person's accidents during n time periods has the following counterpart in terms of an urn model: There is a given set of urns with varying composition. We choose an urn with a certain probability and then select n balls from the urn at random.

Suppose there are just two urns I and II, containing red balls in the proportions ρ_1 and ρ_2, respectively. The choice of urn is determined by the probabilities P_{I} and $P_{\mathrm{II}} = 1 - P_{\mathrm{I}}$. Let us now compute the conditional probability of a second red ball, given that the first was red. We find

$$\Pr\left[\begin{array}{c|c} \text{second} & \text{first} \\ \text{red} & \text{red} \end{array} \right]$$
$$= \left(P_{\mathrm{I}}\rho_1^2 + P_{\mathrm{II}}\rho_2^2 \right) / \left(P_{\mathrm{I}}\rho_1 + P_{\mathrm{II}}\rho_2 \right).$$

For example, suppose we have $P_{\mathrm{I}} = 0.20$, $\rho_1 = 0.5$, and $\rho_2 = 0.05$. Then

$$\Pr\left[\begin{array}{c|c} \text{second} & \text{first} \\ \text{red} & \text{red} \end{array} \right] = \frac{13}{35},$$

Table 1   Observed Frequencies of the Number of Claims During a 10-year Period (policy years 3–7 and 8–12) and the Expected Frequencies According to the Pólya Distribution (6). (From ref. 24, Table IXA.)

Number of Claims	Observed Frequencies	Expected Frequencies
0	324	347.6
1	299	270.8
2	209	204.9
3	151	153.4
4	121	114.3
5	68	84.9
6	59	63.1
7	49	46.6
8	44	34.4
9	27	25.5
10	12	18.9
11	15	13.9
12	11	10.2
13	8	7.6
14	4	5.5
15	4	4.1
$\geqslant 16$	12	11.3
	1417	1417.0

$\chi^2 = 14.6$, 14 d.f., p-value: $0.30 < p < 0.50$

while the probability for red on any draw is constant, equal to 0.14.

The occurrence of the event "a red ball" changes—in our case increases—the probability of observing further such events. Yet there is no contagion involved and the composition of the urns remains the same throughout the experiment. In technical terms: We sample here from a mixture distribution or a stratified population. In our application, the occurrence of an accident indicates that the person comes from a high risk stratum.

A further early application of the present family of distributions is given in Lundberg's [24] book, first published in 1940, which treated the application of random processes to sickness and accidental statistics. The data shown in Table 1 are taken from the records of *Sjukförsäkrings AB Eir*, a company granting benefits to workers disabled through

accident or illness. The data comprised all "male" policies taken out during the years 1918–1922 and that were in effect for at least 12 years. The distribution fitted in Table 1 is the limiting form (6), called Pólya's distribution by Lundberg and, in Pólya's terminology, corresponding to the case of "rare events" and "weak contagion." The observed frequencies were computed using the method of moments.

A Note on Usage

Hald [16], Patil and Joshi [28], Bosch [2], Coleman [5], and Feller [12], among others, call the distribution (1) the *Pólya distribution*, while Johnson and Kotz [20], in their book on urn models, refer to it as the *Pólya–Eggenberger distribution*. We adhere to the latter convention in this article. As noted by Johnson and Kotz, Eastern European authors rightly term the distribution (1) the *Markov* or *Markov–Pólya distribution*, since Markov discussed this and similar distributions as early as 1917, i.e., several years before Pólya and Eggenberger published their paper. As an historical curiosity, it may be noted that they do refer to Markov's textbook *Wahrscheinlichkeitsrechnung* from 1912, which contains nothing resembling the scheme suggested by Pólya and Eggenberger. Later, however, in his 1931 article summarizing a series of lectures in Paris, Pólya [29] quotes Markov's 1917 article for a discussion of distributions of the present type. Very probably, someone had called his attention to this reference.

URN MODELS AND LAGRANGIAN DISTRIBUTIONS

As we have seen, urn models have been used to analyze a number of problems concerning such physical phenomena as contagious events, random walks*, ballot results (Engelberg [11]), and games of chance. Recently, urn models have been invoked to explain and motivate what are known as Lagrange-type distributions, which goes to show that the urn approach is still a potential source for theory construction (*see* LAGRANGE AND RELATED PROBABILITY DISTRIBUTIONS). In the urn models we have discussed so far, the individual's (sampler's) decision or strategy does not play any role. The introduction of some factor based on decision or strategy into the probability models may seem desirable for explaining many patterns of variation in data, especially those that are concerned with living beings. Consul [6] and later Consul and Mittal [8] study urn models in which the probability of success in a game depends upon the decision of the player. Let us take a closer look at Consul's model.

Example 8. Consul's Urn Scheme. Two urns I and II are given, containing a white balls (I) and b black balls (II), respectively. Let N and θ be two further known positive integers. A person is allowed to choose an integer k such that $0 \leqslant k \leqslant N$, before making one draw from urn I and N draws from urn II. The rules of the game are as follows:

(i) First $k\theta$ black balls are added to urn I, while $k\theta$ white and $(N - k)\theta$ black balls will be added to urn II before any further draws are made.

(ii) If the person gets a black ball from urn I in the first draw, the game is lost. If, on the other hand, he gets a white ball from urn I, he is then allowed to make N further draws from urn II with replacement. He wins a prize if he gets exactly k white balls in these N draws.

The joint probability of drawing a white ball from urn I and then exactly k white balls in N repeated trials from urn II is clearly given by

$$p(k) = \frac{a}{a + k\theta} \binom{N}{k} \left(\frac{a + k\theta}{a + b + N\theta} \right)^k$$

$$\times \left(\frac{b + (N - k)\theta}{a + b + N\theta} \right)^{N-k},$$

$$k = 0, 1, \ldots, N. \qquad (8)$$

Under what conditions does (8) represent an honest probability distribution? In other words, do the terms defined by (8) sum to unity? The answer is obtained directly from Abel's generalization of the binomial theorem, as presented by Riordan [32]. To prove this result, Consul [6] quotes a theorem by Jensen, which in turn derives from a Lagrange expansion*. The resulting discrete probability distribution is called the *quasibinomial distribution**. If θ is set equal to zero, (8) reduces to the common binomial distribution. If we reparametrize $a/(a + b + N\theta) = p$ and $\theta/(a + b + N) = \gamma$, (8) reduces to the somewhat simpler form

$$P(k) = \binom{N}{k} p(p + \gamma k)^{k-1}(1 - p - k\gamma)^{N-k},$$

$$k = 0, 1, \ldots, N. \qquad (9)$$

Consul's urn scheme has been extended by Janardan and Schaeffer [19], among others. They develop an urn model in the form of a generalized Markov–Pólya scheme for voting in small groups, where the outcome is dependent upon the strategy of a campaign leader. Their starting point is a model by Coleman [5], resulting in the Pólya–Eggenberger distribution (1). The assumptions behind Coleman's model are tantamount to assuming that the process starts with a probability p (constant), of voting for candidate A, and then letting this probability change during the course of the election day or campaign, so that after $u + v$ people have cast their votes, of which u are for candidate A, the probability that the next vote will be for candidate A is a linear function of u for given $u + v$, namely:

$$\Pr\left[\begin{array}{c|c} \text{next vote} & u \text{ out of} \\ \text{for } A & u + v \text{ for } A \end{array}\right]$$

$$= \frac{p + ur}{1 + (u + v)r}.$$

Janardan and Schaeffer [19] extend this model by letting the number of votes cast for A also depend on the outcome of a previous election in the group, in which k persons voted for A. This gives a conditional probability:

$$\Pr\left[\begin{array}{c|c} \text{next vote} & u \text{ out of } u + v \text{ and} \\ \text{to } A & k \text{ at previous election} \end{array}\right]$$

$$= \frac{p + ks + ur}{1 + Ns + (u + v)r}.$$

Using Lagrange and power series expansions, Janardan and Schaeffer give a rigorous proof that the resulting unconditional expression does indeed define a probability distribution, which they name the generalized Markov–Pólya distribution.

Typically there is interdependence between the voters in the same sociological unit, be it precinct, ward, county, or *Wahlkreis*, which arises through social influence and a common cultural background. Although the notion of sequential trials that is seen to lie behind Coleman's model and its generalization appears to be incompatible with the notion of mutual influence, this can be thought of as being reflected in the contagion factor, i.e., the *c* additional replacements after each draw.

By modifying slightly the preceding urn schemes and by considering limiting cases, a remarkable range of discrete probability distributions is derived. Thus, to exemplify, by letting *N* tend to infinity and *p* to zero in (9) so that $Np \to M$, we obtain the discrete distribution (noting that $N\gamma \to \theta$)

$$\Pr[X = k] = M(M + k\theta)^{k-1} e^{-M - k\gamma}/k!,$$

$$k = 0, 1, 2, \dots. \qquad (10)$$

This is the so-called *generalized Poisson distribution* or *Lagrangian Poisson distribution*. It has important applications in queuing theory* as the *first busy period distribution* in a certain type of queue system (see, e.g., Consul [7]).

CONCLUDING EXAMPLE

All of our examples up to now have dealt with probability concepts. However, urn models are also useful for illustrating purely statistical principles. This article on urn models may appropriately be concluded with the following example, due to Basu [1].

Example 9. Sufficiency in Survey Sampling. Balls are drawn one at a time with replacement from an urn containing *N* identical balls, *Np* of which are red, the rest being black. Here, after *n* draws the sample observation consists of a sequence of red or black balls. The number *r* of red balls in the sample then constitutes a complete, sufficient statistic* for the parameter *p*. Hence *r/n* is the uniformly minimum variance unbiased estimator* of the parameter *p*. Now suppose that the *N* balls are distinguishable from one another, or suppose we put distinguishing marks on them before they are replaced into the urn. Consider now the sample observation in which for each of the *n* balls that are drawn we make a note of its color and of whether this particular ball was drawn before or not. Here the sample can be represented as a sequence of *n* reds and blacks with cross marks at *v* places indicating at which drawings we had distinct balls. The sample observation is thus more detailed than in the first case. If ρ is the number of distinct red balls, then the statistic (ρ, v) is sufficient and by the Rao–Blackwell theorem*, ρ/v (whose conditional expectation is *r/n*) is a better estimator than *r/n* of the parameter *p* in the sense of having smaller variance.

Basu's [1] seminal paper was among the first in a series of attempts to apply modern inference theory to estimation problems in survey sampling*. For further details see the well-known textbook by Cassel et al. [4].

Probability theory* is a well-established part of mathematics and, in principle, words like "urn," "choice," "event," and so forth can be dispensed with in favor of precise mathematical terms. However, these words and the ideas they convey still add intuitive significance to the subject and suggest analytical methods and new investigations, besides making the subject more accessible to workers from applied fields. Urn sampling experiments provide a picturesque background for various applications of statistics and probability theory, as we have seen in the preceding sections. We use the descriptive language of urns and balls, but the same urn model often admits a great variety of different interpretations.

A good deal of the strength of the classical urn models is due to the fact that they are often strongly suggested by the analysis of real-life situations. However, in the interest of a balanced perspective, we must guard against the danger of transferring the logical necessity of the suggestive urn scheme, or parts of it, to parts of the theory. Differently stated, this suggests that the assumptions of the urn model may impose unnecessary limitations on the theory. An urn model formulation may serve useful purposes in the early development of the theoretical ideas. In the course of time, however, references to the particular model tend to become superfluous, as one comes to realize the immediate applicability of the mathematical theory. In the important field of stochastic processes, for example, one might legitimately object that we do not require the sometimes misleading language appropriate to urn models; correct application of the proper mathematical models is all that is necessary.

References

[1]   Basu, D. (1958). *Sankhyā B*, **20**, 287–294.

[2]   Bosch, A. J. (1963). *Statist. Neerland.*, **17**, 201–213.

[3]   Boswell, M. T. and Patil, G. P. (1970). In *Random Counts in Scientific Work*, Vol. I, pp. 3–22.

[4]   Cassel, C. M., Särndal, C. E. and Wretman, J. H. (1977). *Foundations of Inference in Survey Sampling*. Wiley, New York.

[5]   Coleman, J. S. (1964). *Introduction to Mathematical Sociology*. Free Press of Glencoe, Collier-Macmillan, London, England.

[6]   Consul, P. C. (1974). *Sankhyā B*, **36**, 301–309.

[7]   Consul, P. C. (1975). *Commun. Statist.*, **4**, 555–563.

[8]   Consul, P. C. and Mittal, S. P. (1975). *Biom. J.*, **17**, 67–75.

[9]   Eggenberger, F. (1924). *Mitteilungen der Vereinigung schweizerischer Versicherungs-Mathematiker*, Bern, Switzerland.

[10]   Eggenberger, F. and Pólya, G. (1923). *Z. Angew. Math. Mech.*, **3**, 279–289.

[11]   Engelberg, O. (1965). *Z. Wahrsch. verw. Gebiete*, **3**, 271–275.

[12]   Feller, W. (1957). *An Introduction to Probability Theory and its Applications*, Vol. I, 2nd ed. Wiley, New York.

[13]   Freudenthal, H. (1960). *Synthèse*, **12**, 202–212.

[14]   Friedman, B. (1949). *Commun. Pure Appl. Math.*, **2**, 59–70.

[15]   Greenwood, M. and Yule, G. U. (1920). *J. R. Statist. Soc.*, **83**, 255–279.

[16]   Hald, A. (1960). *Technometrics*, **2**, 275–340.

[17]   Heitele, D. (1975). *Educational Studies in Mathematics*, Vol. 6. Reidel, Dordrecht, Netherlands.

[18]   Holst, L. (1979). *J. Appl. Prob.*, **16**, 154–162.

[19]   Janardan, K. G. and Schaeffer, D. J. (1977). *Biom. J.*, **20**, 87–106.

[20]   Johnson, N. L. and Kotz, S. (1977). *Urn Models and Their Application*. Wiley, New York.

[21]   Kemeny, J. G. and Snell, L. (1960). *Finite Markov Chains*. Van Nostrand, NY.

[22]   Kemp, A. W. and Kemp, C. D. (1975). In *Statistical Distributions in Scientific Work*, Vol. I, G. P. Patil et al. eds. Reidel, Dordrecht, Netherlands, pp. 31–40.

[23]   Lotka, A. J. (1956). *Elements of Mathematical Biology*. Dover, New York.

[24]   Lundberg, O. (1964). *On Random Processes and their Application to Sickness and Accident Statistics*, 2nd ed. Almqvist and Wiksell, Uppsala, Sweden.

[25]   Newbold, E. (1927). *J. R. Statist. Soc.*, **90**, 487–547.

[26]   Neyman, J. (1939). *Ann. Math. Statist.*, **10**, 35–57.

[27]   Parzen, E. (1960). *Modern Probability and Its Applications*. Wiley, New York.

[28]   Patil, G. G. and Joshi, S. W. (1968). *A Dictionary and Bibliography of Discrete Distributions*. Oliver and Boyd, Edinburgh, Scotland.

[29]   Pólya, G. (1931). *Ann. Inst. H. Poincaré*, **1**, 117–161.

[30]   Pólya, G. (1954). *Patterns of Plausible Inference*. Princeton University Press, Princeton, NJ.

[31]   Rényi, A. (1970). In *The Teaching of Probability and Statistics*, L. Råde, ed., pp. 273–282.

[32]   Riordan, J. (1968). *Combinatorial Identities*. Wiley, New York.

Bibliography

Bartholomew, D. J. (1982). *Stochastic Models for Social Processes*, 3rd ed. Wiley, New York.

Consul, P. C. and L. R. Shenton (1975). In *Statistical Distributions in Scientific Work*, G. P. Patil, ed. Reidel, Dordrecht, The Netherlands, pp. 41–57.

Holst, L. (1979). A unified approach to limit theorems for urn models, *J. Appl. Prob.*, **16**, 154–162.

Johnson, N. L. and S. Kotz (1969). *Discrete Distributions*. Wiley, New York.

Kemp, A. (1968). *Sankhyā A*, **30**, 401–410.

Kendall, M. G. and W. R. Buckland (1982). *Dictionary of Statistical Terms*, 4th ed. Oliver and Boyd, Edinburgh, Scotland.

Markoff (Markov), A. A. (1912). *Wahrscheinlichkeitsrechnung*. Teubner, Leipzig, Germany.

Pólya, G. (1954). *Patterns of Plausible Inference*. Princeton University Press, Princeton, NJ.

Pearson, E. S., ed. (1978). *The History of Statistics in the 17th and 18th Centuries. Lectures by Karl Pearson*. Griffin, London, England.

Shenton, L. R. (1981). *Bull. Math. Biol.*, **43**, 327–340.

Shenton, L. R. (1983). *Bull. Math. Biol.*, **45**, 1–9. (Shenton's reinforcement–depletion urn model is an interesting extension of the Pólya–Eggenberger urn scheme.)

Todhunter, I. (1863). *A History of the Mathematical Theory of Probability* (from the time of Pascal to that of Laplace). Chelsea, New York.

Wenocur, R. S. (1981). Waiting times and return periods related to order statistics: an application of urn models. In *Statistical Distributions in Scientific Work*, C. Taillie et al., eds. Reidel, Dordrecht, Netherlands, pp. 419–433.

(CONTAGIOUS DISTRIBUTIONS
GENERALIZED HYPERGEOMETRIC
   DISTRIBUTIONS
HYPERGEOMETRIC DISTRIBUTIONS
LAGRANGE AND RELATED PROBABILITY
   DISTRIBUTIONS
MARKOV PROCESSES
NEGATIVE BINOMIAL DISTRIBUTIONS
OCCUPANCY PROBLEMS)

SVEN BERG

U-SHAPED HAZARD FUNCTION *See* BATHTUB CURVE

[Additional bibliography:

Glaser, R. E. (1980). *J. Amer. Statist. Ass.*, **71**, 480–487.

Hjorth, J. (1980). *Technometrics*, **22**, 99–108.]

U-STATISTICS

One way to generalize the sample mean is to form the average, over the sample, of a given function $h(x_1, \ldots, x_m)$ of several arguments. The class of statistics that may be formed in such a way, for various choices of the "kernel" h, was introduced and studied in a fundamental paper by Hoeffding [32]. It turns out that many statistics of interest may be represented exactly as such "*U*-statistics" for suitable choices of kernel; moreover, most other statistics of interest may be approximated by *U*-statistics, for the purpose of theoretical analysis of their large-sample behavior. Consequently, an understanding of *U*-statistics plays a central role in the study of statistics of arbitrary type.

Our formulation will include multisample *U*-statistics, but for simplicity the remainder of our treatment will be confined to the one-sample case. Regrettably, many interesting topics and useful references will be omitted from the present brief introduction, which must be limited to the most central ideas and results. For additional details, extensions and topics, the reader is referred to sections on *U*-statistics in the books by Puri and Sen [46], Randles and Wolfe [48], Serfling [55], Sen [53], and Denker [20], and to a forthcoming comprehensive monograph by Janssen et al. [38].

FORMULATION, EXAMPLES, AND APPLICATIONS

The One-Sample Case

Let $X_1, \ldots, X_n$ be independent observations on a (possibly multidimensional) distribution F. (Indeed, the observations X_i may be taken as random elements of an arbitrary space.) For any function $h(x_1, \ldots, x_m)$ and any sample size $n \geq m$, the *U*-statistic corresponding to the kernel h is obtained by averaging h over the observations

$$U_n = U(X_1, \ldots, X_n)$$
$$= n_{(m)}^{-1} \sum h(X_{i_1}, \ldots, X_{i_m}),$$

where the summation is taken over all $n_{(m)} = n(n-1) \cdots (n-m+1)$ m-tuples $(i_1, \ldots, i_m)$ of distinct elements from $\{1, \ldots, n\}$. If h is not symmetric in its arguments, it may be replaced by the symmetric kernel

$$(m!)^{-1} \sum_p h(x_{i_1}, \ldots, x_{i_m}),$$

where $\sum_p$ denotes summation over the $m!$ permutations $(i_1, \ldots, i_m)$ of $(1, \ldots, m)$, without changing the value of U_n. Therefore, for convenience and without loss of generality, we assume that h is symmetric, in which case we may write U_n [with probability (wp1)] in the form

$$U_n = \binom{n}{m}^{-1} \sum_c h(X_{i_1}, \ldots, X_{i_m}),$$

where $\sum_c$ denotes summation over the $\binom{n}{m}$ combinations of distinct elements $\{i_1, \ldots, i_m\}$ from $\{1, \ldots, n\}$. We see that U_n is an unbiased estimator of the parameter

$$\theta(F) = E_F[h(X_1, \ldots, X_m)]$$

$$= \int \cdots \int h(x_1, \ldots, x_m) \, dF(x_1)$$

$$\cdots dF(x_m).$$

A U-statistic may be represented (wp1) as the result of conditioning the kernel on the *order statistic** $\mathbf{X}_{(n)} = (X_{n1}, \ldots, X_{nn})$ associated with $X_1, \ldots, X_n$, i.e.,

$$U_n = E[h(X_1, \ldots, X_m)|\mathbf{X}_{(n)}]. \quad (1)$$

(Here $X_{n1} \leqslant X_{n2} \leqslant \cdots$ denote the ordered values of the X_i's.) In the case of a statistical model for which the order statistic is complete sufficient* for the "parameter" F, we see from (1) that the U-statistic provides the *minimum variance unbiased* estimator* of $\theta(F)$.

Often the kernel h is selected to proved an estimator or test statistic. For example, $h(x) = x$ yields $U_n = \bar{X}$ (the sample mean), $h(x_1, x_2) = \frac{1}{2}(x_1 - x_2)^2$ yields $U_n = s^2$ (the sample variance), and $h(x_1, x_2) = \mathbf{1}\{x_1 + x_2 \leqslant 0\}$ yields a version of the Wilcoxon one-sample statistic. For observations $\mathbf{X}_i$ on a bivariate distribution function (df) F, the kernel $h(\mathbf{x}_1, \mathbf{x}_2) = \mathbf{1}\{(x_{22} - x_{21})(x_{12} - x_{11}) > 0\}$ yields a U-statistic U_n for which $2U_n - 1$ equals Kendall's sample correlation coefficient (τ)*. Many other statistics of practical interest may be expressed exactly as U-statistics.

Of equal or greater importance, however, is the theoretical role played by U-statistics, as the basic terms in decompositions of statistics of more general form into sums of terms of simpler form. In such decompositions, the leading terms then serve as approximations to the given statistics. See, for example, Serfling [55, Chapter 6] regarding the decomposition of statistics given as "von Mises differentiable statistical functions," or Rubin and Vitale [51] and Vitale [64] regarding decomposition of square-integrable symmetric statistics. In such a fashion, U-statistic theory plays a fundamental role in the analysis of statistics of rather general and arbitrary type.

A current interest, in connection with the study of *random fields**, is the formulation of U-statistics on multidimensionally indexed arrays of random variables $\{X_i, \ \mathbf{i} \leqslant \mathbf{n}\}$, where $\mathbf{n}$ is of the form $\mathbf{n} = (n_1, \ldots, n_r)$. The special case of the sample mean has been treated in work by Wichura [67] and Smythe [62], for example; the general case is developed in Christofides [18], Chapter 4. Another current trend is the study of bootstrapped U-statistics (see Bickel and Freedman [9], Athreya et al. [4], and Helmers [28]).

Multisample U-Statistics

Extension to the case of several samples is straightforward. Consider c independent collections of independent observations $\{X_1^{(1)}, \ldots, X_{n_1}^{(1)}\}, \ldots, \{X_1^{(c)}, \ldots, X_{n_c}^{(c)}\}$ taken from distributions $F^{(1)}, \ldots, F^{(c)}$, respectively. Let $\theta = \theta(F^{(1)}, \ldots, F^{(c)})$ denote a parametric function for which there is an unbiased estimator, i.e.,

$$\theta = E\Big[h\big(X_1^{(1)}, \ldots, X_{m_1}^{(1)};$$

$$\cdots; X_1^{(c)}, \ldots, X_{m_c}^{(c)}\big)\Big],$$

where, without loss of generality, h is assumed to be symmetric within each of its c blocks of arguments. Corresponding to the kernel h and assuming $n_1 \geqslant m_1, \ldots, n_c \geqslant m_c$, the U-statistic for estimation of θ is defined as

$$U_n = \left[\prod_{j=1}^{c} \binom{n_j}{m_j} \right]^{-1} \sum h \left(X_{i_{11}}^{(1)}, \ldots, X_{i_{1m_1}}^{(1)}; \right.$$
$$\left. \cdots; X_{i_{c1}}^{(c)}, \ldots, X_{i_{cm_c}}^{(c)} \right).$$

Here $\sum$ denotes summation over all combinations $\{i_{j1}, \ldots, i_{jm_j}\}$ of distinct elements from $\{1, \ldots, n_j\}$, $1 \leqslant j \leqslant c$, and $\mathbf{n} = (n_1, \ldots, n_c)$.

As an example, consider the *Wilcoxon two-sample statistic**. Given independent observations $\{X_1, \ldots, X_{n_1}\}$ and $\{Y_1, \ldots, Y_{n_2}\}$ from continuous df's F and G, respectively, an unbiased estimator for

$$\theta(F, G) = \int F \, dG = P(X \leqslant Y)$$

is

$$U_{(n_1, n_2)} = \frac{1}{n_1 n_2} \sum_{i=1}^{n_1} \sum_{j=1}^{n_2} \mathbf{1}(X_i \leqslant Y_j).$$

The extension of Hoeffding's treatment of one-sample U-statistics to the multisample case was initiated by Lehmann [41]. For a more recent treatment, including a variety of examples, see Randles and Wolfe [48, Secs. 3.4–3.6].

USEFUL REPRESENTATIONS

Here we briefly indicate three quite useful says of representing U-statistics in terms of structures that have been treated extensively by probability theory, thus enabling the use of this theory to develop properties of U-statistics.

An Average of Averages of Independent and Identically Distributed (I.I.D.) Random Variables (R.V.'s)

Because its summands are not all mutually independent, the U-statistic is *not* an average of i.i.d. random variables. However, by grouping the summands appropriately [Hoeffding [34]], one may write U_n as an average of $n!$ terms, each of which is an average of $[n/m]$ i.i.d. terms. (Here $[\cdot]$ denotes "greatest integer part.") For example, one such average is given by

$$[n/m]^{-1} \{ h(X_1, \ldots, X_m) + h(X_{m+1}, \ldots, X_{2m}) + \cdots \}.$$

These $n!$ averages are identically distributed, but mutually dependent.

Any one of these averages may be regarded as a computational approximation to the U-statistic, providing an also unbiased estimator of θ which has greater variance than U_n but may be computed in $O(n)$ instead of $O(n^m)$ steps. This could be relevant in conducting large sample Monte Carlo studies to determine the values of a parameter of the form $\theta(F) = E_F[h]$ for some specified h and various choices of F.

A Reverse Martingale

It was noted by Berk [5] that U-statistics are *reverse martingales**. Provided that $E_F[|h|] < \infty$, we have (wp1)

$$E \left[U_n | \mathbf{X}_{(n+1)}; X_{n+2}, X_{n+3}, \ldots \right] = U_{n+1},$$
$$\text{all } n \geqslant m. \quad (2)$$

In particular, the ordinary sample mean $\overline{X}$ is a reverse martingale.

A Linear Combination of Forward Martingales

One can define (Hoeffding [33]) for each $d = 1, \ldots, m$, a *forward martingale** sequence $\{S_{dk}, d \leqslant k \leqslant n\}$ satisfying (wp1) $E_F[S_{dn}|X_1, \ldots, X_k] = S_{dk}$, $d \leqslant k \leqslant n$, for which

$$U_n - \theta = \sum_{d=1}^{m} \binom{m}{d} \binom{n}{d}^{-1} S_{dn}. \quad (3)$$

Moreover, the terms $S_{1n}, \ldots, S_{mn}$ appearing in (3) are orthogonal.

In particular, for $m = 1$ and $h(x) = x$, the term S_{1n} is given by $\sum_{i=1}^{n}(X_i - \theta)$ and (3) reduces to $\overline{X} - \theta = n^{-1}\sum_{i=1}^{n}(X_i - \theta)$. For $m = 2$ and $h(x_1, x_2) = \frac{1}{2}(x_1 - x_2)^2$, U_n is the sample variance $s^2 =$

$(n - 1)^{-1}\sum_1^n (X_i - \bar{X})^2$ and we have

$$S_{1n} = \sum_1^n g(X_i),$$

$$S_{2n} = \sum_{1 \leqslant i < j \leqslant n} g(X_i, X_j),$$

where $g(x) = \frac{1}{2}[(x - \mu)^2 - \sigma^2]$ and $g(x_1, x_2) = -\mu^2 + x_1\mu + x_2\mu - x_1 x_2$.

In general, the term $\binom{n}{d}^{-1} S_{dn}$ appearing in (3) is itself a U-statistic, based on a kernel $g_d(x_1, \ldots, x_d)$. We shall make further comments on this aspect later, in terms of some concepts and quantities yet to be introduced.

VARIANCE AND OTHER MOMENTS

Exact variance formulas for U-statistics are given, along with some asymptotic results for these and other moments.

The Variance of a U-Statistic

Associated with the kernel h of a one-sample U-statistic, and assuming $E_F[h^2] < \infty$, we introduce auxiliary functions $h_m = h$ and, for $1 \leqslant d \leqslant m - 1$,

$h_d(x_1, \ldots, x_d)$

$$= E_F[h(x_1, \ldots, x_d, X_{d+1}, \ldots, X_m)],$$

and define parameters $\zeta_0 = 0$ and, for $1 \leqslant d \leqslant m$,

$$\zeta_d = \mathrm{Var}_F\{h_d(X_1, \ldots, X_d)\}.$$

Then for the variance of U_n we have

$\mathrm{Var}_F\{U_n\}$

$$= \binom{n}{m}^{-1} \sum_{d=1}^m \binom{m}{d}\binom{n-m}{m-d}\zeta_d \quad (4a)$$

$$= \frac{m^2\zeta_1}{n} + O(n^{-2}), \qquad n \to \infty. \quad (4b)$$

It can be seen that the first term of (4b) is the variance of the first term of (3). Indeed, (3) is the basis of an alternative formula to (4a) for the variance of U_n.

In particular, for U_n the sample variance considered earlier, these definitions and for-

mulas lead to

$$\mathrm{Var}_F\{s^2\} = \binom{n}{2}^{-1}[2(n - 2)\zeta_1 + \zeta_2]$$

$$= \frac{\mu_4 - \sigma^4}{n} + O(n^{-2}).$$

Other Moments

Generalizations of (4a) to central moments of order other than 2 are difficult even to state, but, fortunately, simple asymptotic results are easily obtained and quite useful in practice. For real $r > 1$ and assuming $E_F[|h|^r] < \infty$, we have

$$E[|U_n - \theta|^r] = O(n^{-c(r)}), \qquad n \to \infty,$$

where

$$c(r) = r - 1, \qquad 1 < r < 2,$$
$$= \tfrac{1}{2}r, \qquad r \geqslant 2.$$

RELATED V-STATISTICS

The parameter $\theta(F)$ estimated by a U-statistic with kernel h may be regarded as a *functional* in the variable F and, accordingly, in the spirit of von Mises [65], may be estimated alternatively by evaluating this functional at a suitable estimator of F. This notion yields the "V-statistic*"

$$V_n = \theta(F_n) = n^{-m}\sum h(X_{i_1}, \ldots, X_{i_m}),$$

where the sum is over all m-tuples $(i_1, \ldots, i_m)$ from $\{1, \ldots, n\}$ and F_n denotes the usual empirical df. Under moment restrictions of order ν (any positive integer ν), we have

$$E[|U_n - V_n|^\nu] = O(n^{-\nu}), \qquad n \to \infty,$$

showing the close connection between the two statistics. In particular this reflects a general equivalence between the use of U-statistics and the use of V-statistics as the basic components in the theory of differentiable statistical functions. Also, many of the results for U-statistics to be mentioned in the sequel have analogues (which we shall not bother to state) for V-statistics.

PROJECTION APPROXIMATION

Applying a technique introduced in Hoeffding [32] and later popularized in general form by Hájek [26], we may associated with a *U*-statistic a special i.i.d.-average $\hat{U}_n$, called the *projection*, defined by the formula

$$\hat{U}_n - \theta = \frac{m}{n} \sum_{i=1}^{n} \left[h_1(X_i) - \theta \right].$$

In fact, this is the first term in the expansion (3).

For many purposes, the projection provides an adequate and very attractive approximation to the *U*-statistic, as is suggested by the relation

$$E\left[(U_n - \hat{U}_n)^2 \right] = O(n^{-2}), \qquad (5)$$

which can be obtained from (4b) by using the fact that $U_n - \hat{U}_n$ is itself a *U*-statistic with a kernel H satisfying $E[H] = 0$ and $\zeta_1^{(H)} = 0$. It also follows from (3) by showing that the terms other than the first are *U*-statistics with kernels for which the corresponding ζ_1's are all zero.

ALMOST SURE BEHAVIOR

It is a desirable regularity property of statistical estimators that they converge to the parameter of interest as the sample size becomes large. We are pleased to report that *U*-statistics indeed possess this feature. Here we characterize this behavior in terms of the strong law of large numbers* (SLLN) and the law of the iterated logarithm* (LIL).

The SLLN

The classical SLLN for i.i.d. averages extends to *U*-statistics. In the one-sample case, we have: If $E_F[|h|] < \infty$, then wp1 $U_n \to \theta$ as $n \to \infty$. This was proved by Hoeffding [33] using the forward martingale representation and later noted by Berk [5] to follow from the reverse martingale SLLN. A similar conclusion for the *c*-sample case has been established by Sen [52] under the more stringent moment assumption $E_F[|h|(\log^+|h|)^{c-1}] < \infty$. (It is an open question whether this

assumption may be relaxed to simply first moment finite.)

The rate of convergence in the SLLN for *U*-statistics is the same as in the case of i.i.d. averages: Under the condition $E_F[|h|^\nu] < \infty$, where $\nu \geqslant 2$, we have

$$P\left(\sup_{k \geqslant n} |U_k - \theta| > \epsilon \right) = o(n^{1-\nu}), \qquad (6)$$

for any $\epsilon > 0$ (Grams and Serfling [25] and Janssen [35]). Also, extension to the multisample case has been established (Christofides [18]) without requiring more stringent moment assumptions.

In some applications, for example involving cross-validation in density estimation, it is of interest to establish the SLLN for *U*-statistics of a given form uniformly over kernels h belonging to a given class; for results in this direction, see Nolan and Pollard [45].

The LIL

The classical LIL also was shown (Serfling [54]) to extend to *U*-statistics: If $E_F[h^2] < \infty$ and $\zeta_1 > 0$, then wp1

$$\overline{\lim_{n \to \infty}} \frac{n^{1/2}(U_n - \theta)}{\left(2m^2 \zeta_1 \log \log n \right)^{1/2}} = 1.$$

Analogues for the case $\zeta_1 = 0 < \zeta_2$ have also been derived (Dehling et al. [19]).

ASYMPTOTIC DISTRIBUTION THEORY

The most far-reaching property of the class of *U*-statistics is the *asymptotic normality* which holds under typical conditions. More generally, in the case $\zeta_0 = \cdots = \zeta_{d-1} = 0 < \zeta_d$, the normalized random variable $n^{d/2}(U_n - \theta)$ has a nondegenerate limit distribution, which happens to be normal in the case $d = 1$ and otherwise is much more complicated in form (a weighted sum of independent χ^2 random variables in the case $d = 2$, for example). Rubin and Vitale [51] exhibited the general form of the limit law as that of a linear combination of products of Hermite polynomials of independent $N(0, 1)$

random variables. An alternative representation of the limit law in terms of multiple Wiener integrals has been developed by Dynkin and Mandelbaum [22] and Mandelbaum and Taqqu [43]; this is related to earlier work of Filippova [23]. It should be noted that these general forms for the limit law of a U-statistic are given in fact for general (square integrable) symmetric statistics and thus include the case of V-statistics, for example.

For simplicity, we will confine attention primarily to the case of asymptotic normality but we include a number of related considerations.

The Central Limit Theorem* (CLT)

The following was established by Hoeffding [32]. It may be derived from (5) in conjunction with the classical CLT applied to the projection.

Central Limit Theorem.   If $E_F[h^2] < \infty$ and $\zeta_1 > 0$, then $n^{1/2}(U_n - \theta) \to_d N(0, m^2\zeta_1)$, as $n \to \infty$.

For example, for the sample variance we suppose that $\zeta_1 = (\mu_4 - \sigma^4)/4$ is positive and finite, and conclude

$$n^{1/2}(s^2 - \sigma^2) \to_d N(0, \mu_4 - \sigma^4).$$

The preceding CLT remains true for the *Studentized* U-statistic, i.e., the statistic $n^{1/2}s_n^{-1}(U_n - \theta)$ converges in distribution to standard normal, where $n^{-1}s_n^2$ is the *jack-knife** estimator of the variance of U_n (see Arvesen [3]).

As may be expected, a vector of several U-statistics based on the same sample is asymptotically multivariate normal* under typical restrictions. Also, c-sample U-statistics follow a similar asymptotic distribution theory and, for example, for the Wilcoxon two-sample statistic we have as approximate distribution

$$N\left(\frac{1}{2}, \frac{1}{12}\left(\frac{1}{n_1} + \frac{1}{n_2}\right)\right)$$

as $\min(n_1, n_2) \to \infty$.

The rate of convergence in the CLT for U-statistics has also been investigated, initially by Grams and Serfling [25], including the multisample case, and subsequently with sharpening in the one-sample case by Chan and Wierman [17], Callaert and Janssen [14], Helmers and van Zwet [30], and Korolyuk and Borovskikh [39]. In the one-sample case, under the assumptions of the CLT and the further assumption that $\eta = E_F[|h_1(X)|^3] < \infty$, we thus have, as an extension of the classical Berry–Esséen theorem*,

$$\sup_x |P(\sigma_n^{-1}(U_n - \theta) \leqslant x) - \Phi(x)|$$

$$\leqslant C\eta(m^2\zeta_1)^{-3/2}n^{-1/2}, \qquad (7)$$

where C is an absolute constant, Φ denotes the standard normal df, and σ_n denotes the standard deviation of U_n (which may be replaced by the standard deviation of $\hat{U}_n$ if desired). The nonuniform generalization of (7) has been obtained by Zhao and Chen [19]; generalization to randomly indexed U-statistics has been developed by Aerts and Callaert [1]; extension to Studentized U-statistics is due to Callaert and Veraverbeke [16], Zhao [68], and Helmers [27, 28]. A variety of further extensions, for example to particular non-i.i.d. cases, may be found in the literature; these serve to illustrate the general robustness of the preceding CLT and associated rates of convergence.

Under more restrictive conditions on h and F, one can establish an *Edgeworth expansion** for U_n, whereby in (7) Φ is replaced by a more intricate function involving the third and fourth cumulants of U_n and the right-hand side becomes $o(n^{-1})$. At the cost of such additional complexity, a more refined approximation to the distribution of U_n is thus obtained. For developments of this type, see Bickel [8], Götze [24], Callaert et al. [15], Bickel et al. [10], and Helmers [28].

Another line of generalization concerns weighted and "incomplete" statistics, whereby each term $h(X_{i_1}, \ldots, X_{i_m})$ in U_n becomes weighted by a coefficient $w(i_1, \ldots, i_m)$ depending only on the indices $i_1, \ldots, i_m$. For various results and applications, see Blom [11], Brown and Kildea [12], Sievers [59],

Shapiro and Hubert [57], Brown and Silverman [13], Weber [66], Lee [40], Berman and Eagleson [7], and Herndorff [31].

In connection with some cases of incomplete *U*-statistics, the relevant limit distribution is Poisson* rather than normal. This also can arise, although not in every instance, when the *U*-statistic is based on a kernel $h_n(\cdot)$ depending upon n (but still having fixed order); see Rao Jammalamadaka and Janson [49] for recent results for such choices of kernel and for further references on the case of Poisson limits.

In a number of problems there arises a *U*-statistic whose kernel h is defined in terms of some unknown parameters. Extended asymptotic distribution theory for the corresponding *U*-statistic with these parameters estimated is developed by Randles [47] and de Wet and Randles [21], following earlier work by Sukhatme [63]. Making use of such results, the CLT for *U*-statistics based on *trimmed samples* has been obtained by Janssen et al. [37].

PROBABILITY INEQUALITIES AND RELATED RESULTS

The foregoing convergence theory may be augmented with bounds or estimates for the probability that a *U*-statistic exceeds its mean by a "moderate" or "large" deviation. As an example, for the case that $a \leqslant h(\cdot) \leqslant b$, from Hoeffding [34] we have the exponential-rate bound

$$P(U_n - \theta \geqslant t) \leqslant e^{-2[n/m]t^2/(b-a)^2}, \quad (8)$$

for $t > 0$ and $n \geqslant m$. Further results, including cases with t in (8) replaced by a sequence $t_n = O(n^{-1/2}(\log n)^{1/2})$, have been developed by Hoeffding [34], Rubin and Sethuraman [50], and Berk [6].

ASSOCIATED STOCHASTIC PROCESSES

Two basic types of stochastic process have been associated with the terms $h(X_{i_1},$ $\dots, X_{i_m})$ involved in a *U*-statistic defined in connection with given data $\{X_i\}$.

The Sum Process

Analogous to the partial-sum stochastic process defined in association with sequences $\{X_i\}$ of random variables, we can define relevant "sum-processes" for *U*-statistics. Two versions have been introduced, each of which is in fact a sequence of processes. In one case the nth process in the sequence is a random function based on $U_1, \dots, U_n$ and summarizes the past history of $\{U_i, i \leqslant n\}$. In the other sequence, the nth random function is based on $U_n, U_{n+1}, \dots$, and summarizes the future history of $\{U_i, i \geqslant n\}$. In each case the nth random function, suitably normalized, converges in distribution to a Wiener process* (see Miller and Sen [44] and Loynes [42], respectively, for initial results, and also Mandelbaum and Taqqu [43] and Christofides [18] for further development and additional discussion and references on these "invariance principles").

The Empirical Process*

Analogous to the classical empirical df F_n associated with a sequence $\{X_i, i \leqslant n\}$, it is useful to define in addition, in connection with any given kernel $h(x_1, \dots, x_m)$, the empirical df H_n of the terms $h(X_{i_1}, \dots, X_{i_m})$. For results on the behavior of the related empirical process and its role in various applications, see Silverman [60, 61], Serfling [56], Janssen et al. [36], Aerts et al. [2], Helmers et al. [29], and Shorack and Wellner [58].

References

[1] Aerts, M. and Callaert, H. (1986). *Sequential Anal.*, **5**, 19–35.

[2] Aerts, M., Janssen, P., and Mason, D. M. (1985). Glivenko–Cantelli convergence for weighted empirical and quantile processes of *U*-statistic structure. *Proc. IVth Vilnius Conference*, to appear.

[3] Arvesen, J. N. (1969). *Ann. Math. Statist.*, **40**, 2076–2100.

[4]   Athreya, K. B., Ghosh, M., Low, L. Y., and Sen, P. K. (1984). *J. Statist. Plann. Inf.*, **9**, 185–194.

[5]   Berk. R. H. (1966). *Ann. Math. Statist.*, **37**, 51–58.

[6]   Berk. R. H. (1970). *Ann. Math. Statist.*, **41**, 894–907.

[7]   Berman, M. and Eagleson, G. K. (1983). *J. Appl. Prob.*, **20**, 47–60.

[8]   Bickel, P. J. (1974). *Ann. Statist.*, **2**, 1–20.

[9]   Bickel, P. J. and Freedman, D. A. (1981). *Ann. Statist.*, **9**, 1196–1217.

[10] Bickel, P. J., Götze, F., and van Zwet, W. R. (1986). *Ann. Statist.*, **14**, 1463–1484.

[11] Blom, G. (1976). *Biometrika*, **63**, 573–580.

[12] Brown, B. M. and Kildea, D. G. (1978). *Ann. Statist.*, **6**, 828–835.

[13] Brown, T. C. and Silverman, B. (1979). *J. Appl. Prob.*, **16**, 428–432.

[14] Callaert, H. and Janssen, P. (1978). *Ann. Statist.*, **6**, 417–421.

[15] Callaert, H., Janssen, P., and Veraverbeke, N. (1980). *Ann. Statist.*, **8**, 299–312.

[16] Callaert, H. and Veraverbeke, N. (1981). *Ann. Statist.*, **9**, 194–200.

[17] Chan, Y. K. and Wierman, J. (1977). *Ann. Statist.*, **5**, 136–139.

[18] Christofides, T. C. (1987). Maximal Probability Inequalities for Multidimensionally Indexed Semimartingales and Convergence Theory of *U*-Statistics, Ph.D. dissertation, Department of Mathematical Sciences, Johns Hopkins University, Baltimore.

[19] Dehling, H., Denker, M., and Philipp, W. (1984). *Z. Wahrsch. verw. Gebiete*, **67**, 139–167.

[20] Denker, M. (1985). *Asymptotic Distribution Theory in Nonparametric Statistics*. Vieweg, Wiesbaden.

[21] de Wet, T. and Randles, R. H. (1987). *Ann. Statist.*, **15**, 398–412.

[22] Dynkin, E. B. and Mandelbaum, A. (1983). *Ann. Statist.*, **11**, 739–745.

[23] Filippova, A. A. (1962). *Theor. Prob. Appl.*, **7**, 24–57.

[24] Götze, F. (1979). *Z. Wahrsch. verw. Gebiete*, **50**, 333–355.

[25] Grams, W. F. and Serfling, R. J. (1973). *Ann. Statist.*, **1**, 153–160.

[26] Hájek, J. (1968). *Ann. Math. Statist.*, **39**, 325–346.

[27] Helmers, R. (1985). *Canad. J. Statist.*, **13**, 79–82.

[28] Helmers, R. (1987). On the Edgeworth Expansion and the Bootstrap Approximation for a Studentized *U*-Statistic. *Tech. Rep.*, Centre for Mathematics and Computer Science, Amsterdam.

[29] Helmers, R., Janssen, P., and Serfling, R. (1985). Glivenko–Cantelli Properties of Some Generalized Empirical df's and Strong Convergence of Generalized *L*-Statistics. *Tech. Rep.*, Department of Mathematical Sciences, Johns Hopkins University, Baltimore.

[30] Helmers, R. and van Zwet, W. R. (1982). The Berry–Esséen bound for *U*-statistics. In *Statistical Decision Theory and Related Topics, III*, Vol. 1, S. S. Gupta and J. O. Berger. Academic Press, New York, pp. 497–512.

[31] Herndorff, N. (1986). *Metrika*, **33**, 179–188.

[32] Hoeffding, W. (1948). *Ann. Math. Statist.*, **19**, 293–325.

[33] Hoeffding, W. (1961). The strong law of large numbers for *U*-statistics. Univ. of North Carolina Institute of Statistics Mimeo Series No. 302.

[34] Hoeffding, W. (1963). *J. Amer. Statist. Ass.*, **58**, 13–30.

[35] Janssen, P. (1981). *Metrika*, **28**, 35–46.

[36] Janssen, P., Serfling, R., and Veraverbeke, N. (1984). *Ann. Statist.*, **12**, 1369–1379.

[37] Janssen, P., Serfling, R., and Veraverbeke, N. (1987). *J. Statist. Plann. Inf.*, **16**, 63–74.

[38] Janssen, P., Serfling, R., and Veraverbeke, N. (1988). *Theory and Applications of U-Statistics*, to appear.

[39] Korolyuk, V. S. and Borovskikh, Yu. V. (1984). *Asymptotic Analysis of Distributions of Statistics* (in Russian). Naukova Dumka, Kiev, USSR.

[40] Lee, A. J. (1982). *Aust. J. Statist.*, **24**, 275–282.

[41] Lehmann, E. L. (1951). *Ann. Math. Statist.*, **22**, 165–179.

[42] Loynes, R. M. (1970). *Proc. Amer. Math. Soc.*, **25**, 56–64.

[43] Mandelbaum, A. and Taqqu, M. S. (1984). *Ann. Statist.*, **12**, 483–496.

[44] Miller, R. G., Jr. and Sen, P. K. (1972). *Ann. Math. Statist.*, **43**, 31–41.

[45] Nolan, D. and Pollard, D. (1987). *Ann. Statist.*, **15**, 780–799.

[46] Puri, M. L. and Sen, P. K. (1971). *Nonparametric Methods in Multivariate Analysis*. Wiley, New York.

[47] Randles, R. H. (1982). *Ann. Statist.*, **10**, 462–474.

[48] Randles, R. H. and Wolfe, D. A. (1979). *Introduction to the Theory of Nonparametric Statistics*. Wiley, New York.

[49] Rao Jammalamadaka, S. and Janson, S. (1986). *Ann. Prob.*, **14**, 1347–1358.

[50] Rubin, H. and Sethuraman, J. (1965). *Sankhyā*, **27A**, 325–346.

[51] Rubin, H. and Vitale, R. A. (1980). *Ann. Statist.*, **8**, 165–170.

[52] Sen, P. K. (1977). *Ann. Prob.*, **5**, 287–290.

[53] Sen, P. K. (1981). *Sequential Nonparametrics: Invariance Principles and Statistical Inference*. Wiley, New York.

[54] Serfling, R. J. (1971). *Ann. Math. Statist.*, **42**, 1974.

[55] Serfling, R. J. (1980). *Approximation Theorems of Mathematical Statistics*. Wiley, New York.

[56] Serfling, R. J. (1984). *Ann. Statist.*, **12**, 76–86.

[57] Shapiro, C. P. and Hubert, L. (1979). *Ann. Statist.*, **7**, 788–794.

[58] Shorack, G. R. and Wellner, J. A. (1986). *Empirical Processes with Applications to Statistics*. Wiley, New York.

[59] Sievers, G. L. (1978). *J. Amer. Statist. Ass.*, **73**, 628–631.

[60] Silverman, B. W. (1976). *Adv. Appl. Prob.*, **8**, 806–819.

[61] Silverman, B. W. (1983). *Ann. Statist.*, **11**, 745–751.

[62] Smythe, R. T. (1973). *Ann. Prob.*, **2**, 906–917.

[63] Sukhatme, B. V. (1958). *Ann. Math. Statist.*, **29**, 60–78.

[64] Vitale, R. A. (1984). An expansion for symmetric statistics and the Efron–Stein inequality. In *Inequalities in Statistics and Probability*, Y. L. Tong, ed. IMS Lecture Notes–Monograph Series, pp. 112–114.

[65] von Mises, R. (1947). *Ann. Math. Statist.*, **18**, 309–348.

[66] Weber, N. C. (1981). *Scand. J. Statist.*, **8**, 120–123.

[67] Wichura, M. J. (1969). *Ann. Math. Statist.*, **40**, 681–687.

[68] Zhao, L. (1983). *Science Exploration*, **3**, 45–52.

[69] Zhao, L. and Chen, X. (1983). *Scientia Sinica (Ser. A)*, **26**, 795–810.

(DISTRIBUTION-FREE METHODS
EXCHANGEABILITY
KENDALL'S TAU
MANN–WHITNEY–WILCOXON STATISTIC)

Robert J. Serfling

USUAL PROVIDER CONTINUITY INDEX

This is an index designed to measure "continuity-of-care" in relation to patients' choice of medical advisor. For an individual who sees j different doctors in a period of n visits, the index is [3]

$$I_v = \{\max(n_1, \ldots, n_j) - 1\}/(n-1),$$

where n_i is the number of times the individual visits the ith doctor.

If the patient always visits the same doctor, then $j = 1$, $n_1 = n$, and $I_v = 1$. Generally $0 < I_v \leq 1$.

For a group of individuals, the arithmetic average of I_v values is used as an index for the whole group. As an alternative, the I_v's may be weighted by the number of visits (n).

Other, competing indices include:

(i)   the *continuity-of-care index* [2]

$$I_c = \{n(n-1)\}^{-1} \sum_{i=1}^{j} n_i(n_i - 1)$$

$$= \{n(n-1)\}^{-1} \left(\sum_{i=1}^{j} n_i^2 - n \right)$$

(since $\sum_{i=1}^{j} n_i = n$),

(ii)   the *K index* [3]

$$I_K = (n - j)/(n - 1),$$

and

(iii)   the *sequential continuity index* [4], which is $I_s = (n-1)^{-1}\{$number of visits immediately preceded by a visit to the same doctor$\}$.

All of these indices can take values in the range 0 to 1 inclusive. A value of 1 corresponds to perfect continuity ($j = 1$, $n_1 = n$).

Ejlertsson and Berg [3] have compared these indices. They note that

$$I_c \leq I_v \leq I_K \quad \text{and} \quad I_s \leq I_K.$$

The index I_c can be written as

$$\{n(n-1)\}^{-1}$$
$$\times \left\{ \sum_{i=1}^{j} (n_i - nj^{-1})^2 + n(nj^{-1} - 1) \right\}.$$

The second term represents the effect of numbers of different doctors visited; the first reflects fidelity to one doctor among the j visited.

References

[1] Bice, T. E. and Boxerman, S. B. (1977) *Med. Care*, **15**, 347–349.

[2] Breslau, N. and Haug, M. R. (1976) *J. Hlth. Soc. Behav.*, **17**, 339–352.

[3] Ejlertsson, G. and Berg, S. (1984) *Med. Care*, **22**, 231–239.

[4] Steinwachs, D. M. (1979) *Med. Care*, **17**, 551–565.

(BIOSTATISTICS)

UTHOFF-TYPE TESTS FOR HEAVY TAILS

Uthoff [3, 4] devised most powerful scale and location invariant tests (MPIT) for testing normal versus uniform* and normal versus Laplace* (double exponential) distributions. His work was extended by Smith [2] and Franck [1] to normal versus symmetric stable distributions* with characteristic functions

$$\phi(t) = \exp(i\beta t - |\delta t|^{\theta}), 1 \leqslant \theta < 2.$$

(Note that $\theta = 1, 2$ correspond to normal and Cauchy* distributions, respectively.)

The test statistics are relatively complicated in form but they have quite good power as compared with standard tests of normality*.

References

[1] Franck, W. E. (1981), *J. Amer. Statist. Ass.*, **76**, 1002–1005.

[2] Smith, V. K. (1975). *J. Amer. Statist. Ass.*, **70**, 662–665.

[3] Uthoff, V. A. (1970). *J. Amer. Statist. Ass.*, **65**, 1597–1600.

[4] Uthoff, V. A. (1973). *Ann. Statist.*, **1**, 170–174.

(DEPARTURE FROM NORMALITY, TESTS FOR
STABLE DISTRIBUTIONS)

UTILITY THEORY

Utility theory is the study of quantitative representations of people's preferences and choices. In statistics, the theory of subjective expected utility developed by Ramsey [35], de Finetti [13], and Savage [36] offers a well reasoned and mathematically elegant basis for analyses of decision making under uncertainty (*see* DECISION THEORY). Economists rely on theories of ordinal utility for investigations in consumer preferences and equilibrium theory [4, 10], and on the expected utility theory of von Neumann and Morgenstern [39] for analyses of conflict situations (*see* GAME THEORY) and choice among risky prospects. Probabilistic theories of choice and utility developed by mathematical psychologists [31, 32, 38] attempt to model coherent patterns of variability and vagueness in people's judgments and choices. Because probabilistic utility is discussed under LUCE'S CHOICE AXIOM AND GENERALIZATIONS and RANDOM UTILITY MODELS, it will not be included in the present entry.

Numerous utility theories have arisen from different structural formulations and from different assumptions about preferences within these formulations. Most of these share two things in common. First, they presume the existence of a nonempty set X of objects for comparative preference or choice. Examples for X that suggest the variety within utility theory are election candidates, pageant contestants, multidimensional commodity bundles, restaurant menus, job offers, medical diagnoses, fertilizer treatments, information-eliciting experiments, investment proposals, insurance policies, gambling* strategies, levels of nuclear armament, and, in general, courses of action with uncertain consequences.

Second, most nonrandom utility theories presume that the person whose preferences are at issue has an asymmetric binary relation $\succ$ of preference on X. For x and y in X,

$x \succ y$ indicates that the person prefers x to y.

Asymmetry says that if $x \succ y$, then it is false that $y \succ x$. If neither $x \succ y$ nor $y \succ x$, it is common to write $x \sim y$ and say that the person is indifferent between x and y: A choice between x and y by coin flip will be quite satisfactory. Because $\succ$ is asymmetric, its induced indifference relation $\sim$ is reflexive ($x \sim x$) and symmetric (if $x \sim y$, then $y \sim x$). The union of $\succ$ and $\sim$ will be written as $\succsim$, so that $x \succsim y$ when the person prefers x to y or is indifferent between them.

The simplest imaginable utility representation for $(X, \succ)$ is the real valued function ϕ on the set $X \times X$ of all ordered pairs (x, y) defined by

$$\phi(x, y) = 1, \quad \text{if } x \succ y,$$
$$\phi(x, y) = -1, \quad \text{if } y \succ x,$$
$$\phi(x, y) = 0, \quad \text{if } x \sim y.$$

This is the indicator function for $\succ$: It represents preferences numerically in the sense that, for all x and y in X,

$$x \succ y \Leftrightarrow \phi(x, y) > 0.$$

Although this representation is nearly trivial and presumes only the asymmetry of $\succ$ for its validity, it has its uses. For example, in the theory of voting, if ϕ_i represents the preferences of voter i, then $\Sigma \phi_i(x, y) > 0$ exactly when more voters prefer x to y than prefer y to x.

ORDERED PREFERENCES

Most utility theories assume that $\succ$ is transitive as well as asymmetric:

if $x \succ y$ and $y \succ z$, then $x \succ z$.

Transitivity imparts an ordering to preferences, thereby opening up the possibility of ordered utility representations that assign a numerical value $u(x)$ to each x separately rather than to each ordered pair (x, y).

It is not unreasonable in some situations to assume that $\succ$ is transitive without also assuming that $\sim$ is transitive [15]. Nontransitive indifference sometimes arises from an inability to distinguish small differences that together add up to a noticeable difference. Thus, you may be indifferent between i and $i + 1$ grains of sugar in your coffee for $i = 1, 2, \ldots, 999$, yet prefer 1 grain to 1000 grains. For another example that introduces probability into the picture, suppose $\succ$ is transitive on a set X of monetary gambles and the person is indifferent between an even-chance gamble for \$0 or \$100 and \$38 as a sure thing, indifferent between the gamble and \$37.50 as a sure thing, yet prefers \$38 to \$37.50. Then $\sim$ is not transitive. When $\succ$

but not $\sim$ is assumed to be transitive, so that preferences are only partially ordered, and when X is countable, there is [16] a real-valued function u on X such that, for all x and y in X,

$$x \succ y \Rightarrow u(x) > u(y).$$

A great variety of u functions may satisfy this representation. When $x \sim y$ and $x \neq y$, some of these could have $u(x) > u(y)$ and others could have $u(y) > u(x)$ if $\sim$ is not transitive. In any event, if $\sim$ is nontransitive, then no u can have $u(x) = u(y)$ for all instances of $x \sim y$.

The latter problem disappears if both $\succ$ and $\sim$ are transitive. When this is true, $\sim$ is an equivalence relation (reflexive, symmetric, transitive) that partitions X into indifference classes that are totally ordered by $\succ$, and $\succ$ or $\succeq$ is often referred to as a *weak order* or *complete preorder*. The collection of indifference classes in this case, which may be smooth curves or surfaces when X is an n-dimensional Euclidean space and suitable continuity conditions are satisfied, is often referred to as an *indifference map*.

If $\succ$ is a weak order and there are only countably many indifference classes in X, then there is a real-valued function u on X such that, for all x and y in X,

$$x \succ y \Leftrightarrow u(x) > u(y). \tag{1}$$

Consequently, all x within an indifference class have the same utility $u(x)$.

If X contains uncountably many indifference classes, then (1) could fail, owing to the possibility that there is an insufficient number of real numbers to order the classes. For example, if X is the real plane and $\succ$ is the lexicographic order $\succ_L$ defined by

$$(x_1, x_2) \succ_L (y_1, y_2)$$
$$\Leftrightarrow [x_1 > y_1 \text{ or } (x_1 = y_1, x_2 > y_2)],$$

then $\succ$ is a weak order, but no real-valued function u satisfies (1). In this case, $\succ$ can be represented numerically using two real-valued functions ordered lexicographically, namely, $u_1(x_1, x_2) = x_1$ and $u_2(x_1, x_2) = x_2$ with $x \succ y \Leftrightarrow (u_1(x), u_2(x)) \succ_L$

$(u_1(y), u_2(y))$. A survey of lexicographic utility is given in ref. 17.

Following the work of Cantor [7] in set theory, it was recognized that (1) holds when $\succ$ is a weak order if and only if there is a countable subset Y of X such that, for all x and z in X, if $x \succ z$, then $x \succeq y \succeq z$ for some y in Y. This unintuitive condition is not widely used since important structural contexts allow the formulation of continuity axioms for $\succ$ that are sufficient for (1) in the presence of weak order. For example, if X is the real plane, if $\succ$ is a weak order with preference increasing in each dimension, and if for any $x \succ y \succ z$ there is a convex combination of x and z that is indifferent to y, then (1) can be satisfied. See refs. 12 and 16 for additional information.

Representation (1) characterizes "ordinal utility" since any other real valued function v on X that preserves the ordering of $\succ$ also qualifies as a utility function. The term "cardinal utility" is reserved for functions u that satisfy (1) *and* other conditions such that v also satisfies these conditions if and only if there are real numbers $a > 0$ and b such that, for all x in X,

$$v(x) = au(x) + b. \qquad (2)$$

One route to a cardinal utility representation is through axioms for ordered preference *differences* [16, Chap. 6]. If you can weak order all positive differences between monetary amounts—your difference in preference between $50 and $10 exceeds your difference in preference between $210 and $160, and so forth—then a u that satisfies (1) and preserves your ordering of preference differences—$u(50) - u(10) > u(210) - u(160)$, and so forth—may be a cardinal utility function.

A different route to cardinal utility that does not directly involve comparisons of preference differences is discussed in the next section. A third approach to cardinal utility [11, 16, 29] that also uses only simple preference comparisons arises in additive utility theory when X is the product of other sets, say

$$X = X_1 \times X_2 \times \cdots \times X_n,$$

so that each x is an n-tuple $x = (x_1, \ldots, x_n)$. The index i for X_i could identify a specific commodity, an evaluative feature of a product, or a specific time interval. If $\succ$ on X is a weak order that satisfies appropriate continuity assumptions and independence axioms among the X_i, then there will be real-valued functions u_i on X_i $(i = 1, \ldots, n)$ such that (1) holds when u is defined by

$$u(x_1, x_2, \ldots, x_n)$$
$$= u_1(x_1) + u_2(x_2) + \cdots + u_n(x_n). \qquad (3)$$

Moreover, u will be uniquely specified up to the transformations of (2), given that it satisfies (1) and (3), and the u_i in (3) will have similar uniqueness properties.

EXPECTED UTILITY

During the early part of the eighteenth century, Gabriel Cramer and Daniel Bernoulli [6] sketched a theory to explain why maximization of expected profit or wealth was often violated by the choices of prudent individuals among risky monetary options. For example, you may prefer $10,000 as a sure thing to an even-chance lottery for $0 or $22,000 even though the lottery has greater expected value ($11,000). Or consider the famous St. Petersburg paradox*, in which a person can buy for $100 a wager that returns 2^n if the first "head" in a sequence of flips of a fair coin occurs on flip n. Although the wager has an infinite expected return ($\frac{1}{2}(2) + \frac{1}{4}(4) + \frac{1}{8}(8) + \cdots$), few people would buy it.

More than two centuries after the pioneering work of Cramer and Bernoulli*, von Neumann* and Morgenstern [39] crystallized their underlying logic in a set of axioms for a preference relation on a set of probability measures and showed that these axioms imply the existence of a utility function u that satisfies (1) and the additional condition that the utility of a probability measure—or the risky prospect to which it refers—equals the mathematical expectation of the utilities of its possible outcomes.

The abstract theory of von Neumann and Morgenstern refers neither to outcomes, monetary or otherwise, nor to mathematical expectation, nor even to risky prospects or probability measures [16, 24]. However, probability measures will be used here because of their obvious ties to decision making. The axioms to be cited follow ref. 25.

Let P be a convex set of probability measures defined on an algebra of subsets of a given set. Convexity says that $\lambda p + (1 - \lambda)q$ is in P whenever λ is in $[0, 1]$ and p and q are in P. The abstract theory uses three axioms for $(P, \succ)$. The first is weak order. The second is a continuity condition that says that if $p \succ q \succ r$, then $\alpha p + (1 - \alpha)r \succ q$ and $q \succ \beta p + (1 - \beta)r$ for some α and β strictly between 0 and 1. The third axiom is an independence or additivity condition that asserts, for all p, q, and r in P and λ in $(0, 1)$, that

$$p \succ q \Rightarrow \lambda p + (1 - \lambda)r \succ \lambda q + (1 - \lambda)r. \tag{4}$$

These axioms are necessary and sufficient for the existence of u on P which satisfies (1) $[p \succ q \Leftrightarrow u(p) > u(q)]$ and is *linear*, i.e.,

$$u(\lambda p + (1 - \lambda)q)$$
$$= \lambda u(p) + (1 - \lambda)u(q), \tag{5}$$

whenever p and q are in P and λ is in $[0, 1]$. Moreover, such a u is unique up to the transformations of (2) and is therefore a "cardinal utility" function.

This is the essence of the von Neumann–Morgenstern utility theory. If we suppose that X is the set (outcomes, pure strategies, etc.) on which the measures in P are defined, and that P contains every one-point measure with $u(x)$ defined as $u(p)$ when $p(x) = 1$, then the linearity property (5) implies the expected-utility form

$$u(p) = \sum_X p(x)u(x)$$

for each measure in P that has finite support. Additional axioms are needed to obtain $u(p) = \int u(x)\, dp(x)$ for all p in P when P contains more general measures. In particular, it is necessary to assume that if

$p(Y) = 1$ for a subset Y of X, then $(y \succ q$ for all y in $Y) \Rightarrow p \succsim q$ and $(q \succ y$ for all y in $Y) \Rightarrow q \succsim p$. Apart from technical details, this appealing dominance principle in conjunction with the three basic axioms yields the general integral form for expected utility. Numerous other extensions and generalizations of the von Neumann–Morgenstern theory are discussed in refs. 19, Part I, and 28.

Although exception has been taken to each of the three basic axioms as a reasonable description, if not as a normative principle, of choice behavior between risky prospects, the independence axiom (4) has come under especially severe criticism. The evidence [2, 26, 27, 33, 37] shows that many people systematically violate (4) in several ways when X is taken as the traditional set of increments to wealth. A famous example from Allais [1] illustrates one of these ways. Suppose p, q, r, and s are lotteries with monetary prizes that obtain with the following probabilities:

$$p(\$500,000) = 1;$$
$$q(\$2500,000) = 0.10,$$
$$q(\$500,000) = 0.89,$$
$$q(\$0) = 0.01;$$
$$r(\$2500,000) = 0.10,$$
$$r(\$0) = 0.90;$$
$$s(\$500,000) = 0.11,$$
$$s(\$0) = 0.89.$$

Many people prefer p to q because of the sure-thing aspect of p, and also prefer r to s because r has a much larger prize with nearly the same probability as the prize in s. However, $p \succ q$ and $r \succ s$ are inconsistent with the expected utility model, or with (4) and its converse: $p \succ q$ gives

$$u(\$500,000) > (0.10)u(\$2500,000)$$
$$+ (0.89)u(\$500,000)$$
$$+ (0.01)u(\$0),$$

which reduces to

$$(0.11)u(\$500,000) > (0.10)u(\$2500,000)$$
$$+ (0.01)u(\$0),$$

whereas a similar reduction for $r \succ s$ gives the opposite inequality.

Remarks on recent work on utility theories that accommodate this and other violations of the von Neumann–Morgenstern axioms appear in the final section.

RISK ATTITUDES

Risk attitudes constitute a particularly fruitful offshoot of basic expected utility theory. This term applies to curvature properties of u on X when X is an interval of monetary amounts, u increases on X, and u on P with $u(p) = \int u(x)\, dp(x)$ satisfies $u(p) > u(q) \Leftrightarrow p \succ q$. It is only tangentially related to the conventional notion of risk as the possibility of something bad happening (*see* RISK MEASUREMENT, FOUNDATIONS OF).

The theory of risk attitudes was systematically developed by Pratt [34] and Arrow [3] although its roots go back to Bernoulli [6]. When u'' denotes the second derivative of u on X, u is said to be *risk averse* on an interval if $u'' < 0$ on the interval, *risk neutral* if $u'' = 0$, and *risk seeking* or *risk loving* if $u'' > 0$. Let $E(p)$ be the actuarial expected value of p, and let $C(p)$ be the certainty equivalent of p, with $C(p)$ the monetary amount x for which $x \sim p$. Then, for nondegenerate measures p, $C(p) < E(p)$ indicates risk aversion, $C(p) = E(p)$ indicates risk neutrality, and $C(p) > E(p)$ indicates risk seeking.

Risk aversion is evident in Bernoulli's logarithmic utility function for wealth, it is fundamental to several notions of stochastic dominance and preference for variance-reducing contractions [40], and it has been a predominant theme in economic analysis [3]. Additional notions of increasing and decreasing risk aversion, which refer to how $-u''(x)/u'(x)$ varies in x, are sometimes useful in differentiating between behaviors for economic phenomena.

The generality of risk aversion was challenged by an explanation [23] of the simultaneous behaviors of insurance buying and gambling* by low-income consumers, which requires that u be risking seeking over some interval. More recent studies [22, 26, 37] indicate that many people who are risk-averse in the gains region are also risk-seeking in the loss region.

SUBJECTIVE EXPECTED UTILITY

Although there are more than a dozen theories of subjective expected utility [18], the earliest complete theory (Savage [36]) remains among the most attractive for its insightfulness and elegance. Savage's personalistic theory of preference between decisions in the face of uncertainty was inspired by Ramsey's [35] earlier sketch of a similar theory, the von Neumann–Morgenstern axiomatization of expected utility, and de Finetti's [13] pioneering work on the theory of personal probability (*see* AXIOMS OF PROBABILITY). A complete summary of Savage's theory, including proofs, is given in Ref. 16, Chap. 14. An outline follows.

We begin with a set S of *states* of the world that describe alternative realizations of the environment about which the person is uncertain, and a set X of *consequences* that describe what happens if the person follows a specific course of action and a specific state in S obtains, or is the true state. An *event* is any subset of S; event A obtains if it contains the true state. The complement of A is denoted by A^c.

A Savage *act* is a function $f: S \to X$ that assigns a consequence $f(s)$ to each state s in S. Act f is *constant* if $f(s) = f(t)$ for all s and t in S, and it is *simple* if it is constant on each event in some finite partition of S. The part of act f defined on event A is denoted f_A. Savage's set of acts $F = \{f, g, \dots\}$ is a large subset of X^S that contains the simple acts.

Seven axioms are used for $(F, \succ)$. One is weak order. Two others are independence axioms:

$$[f_A = f'_A,\ g_A = g'_A,\ f_{A^c} = g_{A^c},\ f'_{A^c} = g'_{A^c}]$$
$$\Rightarrow [f \succ g \Leftrightarrow f' \succ g']; \qquad (6)$$

$$[x \succ y,\ f_A \equiv x,\ f_{A^c} \equiv y,\ g_B \equiv x,\ g_{B^c} \equiv y,$$
$$\text{and likewise for } x', y', f', g']$$
$$\Rightarrow [f \succ g \Leftrightarrow f' \succ g']. \qquad (7)$$

The first of these, which follows the spirit of (4), says that $\succ$ is independent of states (in A^c) that have identical consequences for the two acts. For the second, define $A \succ^* B$ (A is more probable than B) to mean that $f \succ g$ when $x \succ y$ (for constant acts), f yields the preferred x if A obtains and y otherwise, and g yields the preferred x if B obtains and y otherwise. Then (7) is used to ensure that $\succ^*$ is an unambiguous weak order on the set of events.

Savage's remaining four axioms involve two innocuous technical conditions, a continuity axiom based on finite partitions of S and a dominance principle similar to the one that extends expected utility for simple measures in P to more general measures. The seven axioms imply that there is a real-valued function u on X and a finitely additive probability measure ρ on the set of events such that, for all acts f and g, and for all events A and B,

$$f \succ g \Leftrightarrow \int_S u(f(s))\,d\rho(s)$$

$$> \int_S u(g(s))\,d\rho(s),$$

$$A \succ^* B \Leftrightarrow \rho(A) > \rho(B),$$

$$0 < \lambda < 1 \Rightarrow \rho(C)$$

$$= \lambda\rho(B) \quad \text{for some } C \subseteq B,$$

with ρ uniquely determined and u unique up to the transformations of (2). In addition, u is bounded. Thus, if preferences satisfy the axioms, the person's judgments reveal an underlying subjective probability measure ρ on events that combines with a utility function u on consequences by mathematical expectation to preserve the order of $\succ$ on acts.

Many of the other axiomatizations of subjective expected utility [18] obtain a similar representation that avoids debatable structural assumptions used by Savage, such as the availability of constant acts and the infinite divisibility of S expressed by $0 < \lambda < 1 \Rightarrow \rho(C) = \lambda\rho(B)$ for some $C \subseteq B$. Other extensions and generalizations are discussed in ref. 19, Part II.

Like (4) for the von Neumann–Morgenstern theory, independence condition (6) has been challenged as a reasonable principle of choice. One of Ellsberg's [14] examples makes the point. An urn contains 30 red balls and 60 others that are black and yellow in unknown proportion. Four acts based on a ball to be chosen at random are:

f: win \$1000 if red is chosen,

    nothing otherwise;

g: win \$1000 if black is chosen,

    nothing otherwise;

f': win \$1000 if red or yellow is chosen,

    nothing otherwise;

g': win \$1000 if black or yellow is chosen,

    nothing otherwise.

Many people prefer f to g and g' to f', which suggests a preference for greater specificity or lower ambiguity. For example, $f \succ g$ because 30 balls are known to be red, but anywhere from 0–60 could be black. Similarly, $g' \succ f'$ because 60 balls are known to be black or yellow, whereas anywhere from 30–90 could be red or yellow. But $f \succ g$ and $g' \succ f'$ violate (6) when $A = \{$red, black$\}$ and $A^c = \{$yellow$\}$. Viewed another way, $f \succ g$ indicates that "red" is believed to be more probable than "black," while $g' \succ f'$ indicates that "not red" is believed to be more probable than "not black."

NEW DIRECTIONS

For many years the so-called paradoxes of Allais and Ellsberg were regarded by most decision theorists as aberrations that revealed faulty reasoning or tricks played on naïve subjects by experimenters, or else were deemed too rare for serious consideration. This has been changing in the past few years, and investigations of utility models and axiomatizations that make allowance for judgments that are inconsistent with the expected utility paradigms are gaining prominence. The primary thrust of this research is the avoidance of independence conditions like (4) and (6).

Approaches that generalize the von Neumann–Morgenstern paradigm include Kahneman and Tversky's prospect theory [26, 27], Chew and MacCrimmon's ratio model [9], Machina's generalized utility analysis [33], Fishburn's skew-symmetric bilinear theory [20], and Bell's regret model [5]. A related model based on a notion of regret in Savage's states-of-the-world context is proposed by Loomes and Sugden [30].

Efficient axiomatizations have been provided for the Chew–MacCrimmon [8, 21] and Fishburn utility representations. For the first of these, a relaxation of (4) implies that $(P, \succ)$ can be represented by *two* linear functions u and w on P with w nonnegative, as

$$p \succ q \Leftrightarrow u(p)w(q) > u(q)w(p).$$

This reduces to the von Neumann–Morgenstern model if w is constant. Fishburn's axioms [20] relax both transitivity and (4) to obtain

$$p \succ q \Leftrightarrow \phi(p, q) > 0,$$

where ϕ on $P \times P$ is skew symmetric $[\phi(q, p) = -\phi(p, q)]$ and linear separately in each argument. The preceding *uw* model is implied by the ϕ model when transitivity is restored.

References

[1] Allais, M. (1953). *Econometrica*, **21**, 503–546.

[2] Allais, M. and Hagen, O., eds. (1979). *Expected Utility Hypotheses and the Allais Paradox*. Reidel, Dordrecht, The Netherlands. (Exposition of Allais's approach. Many other contributions: MacCrimmon and Larsson especially valuable.)

[3] Arrow, K. J. (1965). *Aspects of the Theory of Risk Bearing*. Yrjö Jahssonin Säätiö, Helsinki, Finland.

[4] Arrow, K. J. and Hahn, F. H. (1971). *General Competitive Analysis*. Holden-Day, San Francisco, CA.

[5] Bell, D. E. (1982). *Operat. Res.*, **30**, 961–981.

[6] Bernoulli, D. (1738). *Comm. Acad. Sci. Imper. Petropolitanae*, **5**, 175–192 [transl. *Econometrica*, **22**, 23–36 (1954)].

[7] Cantor, G. (1895; 1897). *Math. Annalen*, **46**, 481–512; **49**, 207–246.

[8] Chew, S. H. (1983). *Econometrica*, **51**, 1065–1092.

[9] Chew, S. H. and MacCrimmon, K. R. (1979). Alpha-Nu Choice Theory. *Working Paper No. 669*, University of British Columbia, Vancouver, Canada.

[10] Debreu, G. (1959). *Theory of Value*. Wiley, New York.

[11] Debreu, G. (1960). In *Mathematical Methods in the Social Sciences*, *1959*, K. J. Arrow, S. Karlin, and P. Suppes, eds. Stanford University Press, Stanford, CA, pp. 16–26.

[12] Debreu, G. (1964). *Int. Econ. Rev.*, **5**, 285–293.

[13] de Finetti, B. (1937). *Ann. Inst. H. Poincaré*, **7**, 1–68 [transl. in *Studies in Subjective Probability*, Kyburg and Smokler, eds. Wiley, New York (1964)].

[14] Ellsberg, D. (1961). *Quart. J. Econ.*, **75**, 643–669.

[15] Fishburn, P. C. (1970a). *Operat. Res.*, **18**, 207–228. (Survey of intransitive indifference.)

[16] Fishburn, P. C. (1970b). *Utility Theory for Decision Making*. Wiley, New York. [Reprinted Krieger, Huntington, NY (1979).]

[17] Fishburn, P. C. (1974). *Management Sci.*, **20**, 1442–1471. (Extensive survey of lexicographic orders and related material.)

[18] Fishburn, P. C. (1981). *Theory and Decision*, **13**, 139–199.

[19] Fishburn, P. C. (1982a). *The Foundations of Expected Utility*. Reidel, Dordrecht, The Netherlands. (Unified treatment of previous research by author on expected utility.)

[20] Fishburn, P. C. (1982b). *J. Math. Psychol.*, **26**, 31–67.

[21] Fishburn, P. C. (1983). *J. Econ. Theory*, **31**, 293–317.

[22] Fishburn, P. C. and Kochenberger, G. A. (1979). *Decision Sci.*, **10**, 503–518.

[23] Friedman, M. and Savage, L. J. (1948). *J. Polit. Econ.*, **56**, 279–304.

[24] Herstein, I. N. and Milnor, J. (1953). *Econometrica*, **21**, 291–297.

[25] Jensen, N. E. (1967). *Swedish J. Econ.*, **69**, 163–183.

[26] Kahneman, D. and Tversky, A. (1979). *Econometrica*, **47**, 263–291.

[27] Kahneman, D. and Tversky, A. (1982). *Sci. American*, **246**, 160–173. (Informative, valuable, readable exposition of risky choice behavior.)

[28] Keeney, R. L. and Raiffa, H. (1976). *Decisions with Multiple Objectives*. Wiley, New York. (Extensive theoretical and practical discussion of multiattribute decision making.)

[29] Krantz, D. H., Luce, R. D., Suppes, P., and Tversky, A. (1971). *Foundations of Measurement*, Vol. I. Academic, New York. (Rigorous treatment of measurement theory under weak orderings.)

[30] Loomes, G. and Sugden, R. (1982). *Econ. J.*, **92**, 805–824.

[31] Luce, R. D. (1959). *Individual Choice Behavior.* Wiley, New York.

[32] Luce, R. D. and Suppes, P. (1965). In *Handbook of Mathematical Psychology*, Vol. III, R. D. Luce, R. R. Bush, and E. Galanter, eds. Wiley, New York, pp. 249–410. (Dated but excellent survey on utility and subjective probability.)

[33] Machina, M. J. (1982). *Econometrica*, **50**, 277–323.

[34] Pratt, J. W. (1964). *Econometrica*, **32**, 122–136.

[35] Ramsey, F. P. (1931). In *The Foundations of Mathematics and Other Logical Essays*. Routledge and Kegan Paul, London, England, 156–198. [Reprinted in *Studies in Subjective Probability*, Kyburg and Smokler, eds. Wiley, New York (1964).]

[36] Savage L. J. (1954). *The Foundations of Statistics.* Wiley, New York. [Second revised edition, Dover, New York (1972). Immensely influential work on Bayesian decision theory.]

[37] Schoemaker, P. J. H. (1982). *J. Econ. Lit.*, **20**, 529–563. (Extensive recent survey of expected utility.)

[38] Tversky, A. (1972). *Psychol. Rev.*, **79**, 281–299.

[39] von Neumann, J. and Morgenstern, O. (1944). *Theory of Games and Economic Behavior.* Princeton University Press, Princeton, NJ. [Monumental treatise on game theory. Proof of expected utility theorem first appears in the second edition (1947).]

[40] Whitman, G. A. and Findlay, M. C., eds. (1978). *Stochastic Dominance.* Heath, Lexington, MA. (Useful introduction to theory and applications of stochastic dominance.)

(AXIOMS OF PROBABILITY
DECISION THEORY
GAME THEORY
LUCE'S CHOICE AXIOM AND
    GENERALIZATIONS
RANDOM UTILITY MODELS
RISK MEASUREMENT, FOUNDATIONS OF)

PETER C. FISHBURN

UTILIZATION ANALYSIS

In recent years there has been an increased use of statistics and statistical methodology in legal applications (*see* LAW, STATISTICS IN). This has been particularly true in the area of equal opportunity analysis in which quantification of concepts plays an important role because of the frequency with which large amounts of relevant but unstructured data are encountered. One such concept is that of *utilization analysis*, which refers to analysis of the number of individuals in specific classes whose employment rights are legally protected. Examples of such classes include females, blacks, Hispanics, and minorities in general. The issue of interest here arises because after passage of the United States' Civil Rights Act of 1964, Title VII of which dealt with equal employment opportunity, the United States Department of Labor implemented regulations requiring government contractors of sufficient size to develop Affirmative Action programs. It specifically calls upon them to perform utilization analyses for minorities and women. For the purpose of this discussion, we are concerned with the definition of underutilization provided in Section 60-2.11 (b): " 'Underutilization' is defined as having fewer minorities or women in a particular job classification then would reasonably be expected by their availability."

One reason that the analysis of underutilization is of special interest to statisticians is that the development of its treatment involves matters that appear clear from a statistical point of view but the partial resolution of which has required years of debate, explanation, and ultimately judicial process. Even after extensive consideration, there remain basic unresolved legal issues.

Once these regulations went into effect, it became the responsibility of government contractors to implement them as written. It, therefore, became necessary to interpret the components of this definition.

The definition of underutilization is complex and requires other than legal expertise to be interpreted in an accurate and useful manner. Specifically, there are three aspects of the definition that themselves require classification and interpretation. These are (i) job classification, (ii) availability, and (iii) "fewer... than would reasonably be expected."

Although these concepts may be considered independently, the actual implementation of any reasonable procedure finds them to be related. In the following, both their definitions and their interrelationships are considered.

JOB CLASSIFICATION

The regulations themselves provide the basic definition: "job classification herein meaning one or a group of jobs having similar content, wage rates, and opportunities." The measurement of, and judgments concerning, such similarities are clearly a subjective matter and no further guidelines are provided. In practice, two characteristics that frequently affect such decision making are the existence of information which will permit reasonable estimation of availability rates, the proportions by class of individuals with the necessary knowledge, skills, and/or abilities to successfully perform the jobs (as well as the financial incentive to do so), and the numbers of incumbents in the positions.

If a job classification is "too small," it may turn out that, with a reasonable availability rate, there is no circumstance under which underutilization would be declared. On the other hand, if a job classification is made larger by adding job titles, then the underutilization decision applies to the totality of titles included and remedial actions may not properly be demanded on a more limited set.

Job classification may be determined by applying the preceding definition in conjunction with specific information about the individual jobs. It is common for there to be more than one acceptable alternative, but classification definition is in no sense arbitrary.

AVAILABILITY

The guidelines provide substantial direction in terms of obtaining or, as statisticians would prefer, estimating availability rates. They state: "(1) In determining whether minorities are being underutilized in any job classification the contractor will consider at least all of the following factors: (i) The

minority population of the labor area surrounding the facility; (ii) The size of the minority unemployment force in the labor area surrounding the facility; (iii) The percentage of the minority workforce as compared with the total workforce in the immediate labor area; (iv) The general availability of minorities having requisite skills in the immediate labor area; (v) The availability of minorities having requisite skills in the area in which the contractor can reasonably recruit; (vi) The availability of promotable and transferable minorities within the contractor's organization; (vii) The existence of training institutions capable of training persons in the requisite skills; and (viii) The degree of training which the contractor is reasonably able to undertake as a means of making all job classes available to minorities."

A parallel equivalent set of factors is also provided for females.

There have been attempts to further quantify this concept by several government monitoring agencies. The most common approach, known generally as the *eight factor rule*, consists of determination of a weighted average of numbers corresponding to the items for which consideration is required. Unfortunately, many such approaches either mix incommensurate quantities or add unrealistic requirements. In fact, the relevance of some of the factors themselves has been questioned, e.g., unemployment data which do not relate to specific job classifications.

Quantification of availability may reasonably include weighting estimated availability rates from several sources, but in order to be interpretable this must be done in a rational way. For many job classifications, it appears to be sufficient to consider only an appropriate external labor force rate. The other factors may be considered in selecting that rate, thereby satisfying the regulations which do not themselves require a weighted average of the eight factors to be computed. Clearly, there is room for legal argument as well as statistical. Most commonly, at least two factors are taken into account—representing internal and external availability. To the extent that mutually exclusive sources are available, additional factors are generally considered to be justified but with the additional complications of determining the weights to be used. To the extent that the appropriate personnel history data are available, they may be used toward this end.

Determination of an appropriate estimated availability rate may be accomplished by use of data from the Census Bureau*, Bureau of Labor Statistics*, Labor Department, Education Department, Equal Employment Opportunity Commission, or such other sources as trade and professional associations. The most commonly used of these is Census data because of both its detail and large sample size. The 1980 Census is a primary source of data for the construction of availability estimates for the next decade.

From a practical point of view, a major change in data availability took place when, in the spring of 1983, the Bureau of the Census released their EEO special file permitting consideration of the detailed occupation information collected in the 1980 Decennial Census. Although they are distributing this data only in the form of magnetic tape or microfiche, tabulations are available from the private sector.

These commercially available reports can include the same level of detail for each of females, Hispanics, blacks, Asian Americans, American Indians, other minorities, nonwhites, and whites as for the total population. Further, that level of detail can include all 514 lines of detailed occupation at the state, SMSA, county, or city (population 50,000 or more) level.

It is of interest that the Bureau of the Census (BOC) data tapes are organized by BOC job groups and not by the commonly used EEO-1 categories of the Equal Employment Opportunity Commission (EEOC). A crosswalk has, however, been agreed upon by the BOC and the EEOC which permits reorganization of the data on an EEO-1 category basis. Since individual codes are reassigned to determine the correspondence, it is, in a practical sense, virtually impossible to adjust the raw Census data to an EEO-1 category basis without the use of a computer.

Finally, it should be noted that regardless of data source, it may be possible to estimate availability as a proportion without estimating corresponding counts. For example, external availability rates for managing civil engineers may be estimated by the corresponding rates among all civil engineers—thereby assuming that possession of the required management skills is independent of race and sex. Alternatively, one might consider using rates based on all managers assuming that for them possession of civil engineering skills is independent of race and sex. This type of subjectivity cannot be eliminated but may be made more acceptable by combining multiple estimates or making use of conservative worst-case results.

"FEWER... THAN WOULD REASONABLY BE EXPECTED"

The final component of the analysis, and from the statistician's point of view the most interesting, is selection of an appropriate statistical methodology to determine whether there are "fewer... than would reasonably be expected" of the group under consideration. The emphases for the purpose of this discussion are on the words "expectation" and "reasonably."

The term "expectation" has an intuitive meaning to most people that coincides with the statistician's well defined term "expected value." This is particularly true when dealing with proportions and percentages as we are here. When asked how many times a fair coin is expected to come up heads if tossed 100 times, virtually every serious response is 50. One would expect, therefore, little difficulty with using the two terms interchangeably. Such turns out to be the case. In practice, lawyers and statisticians, regardless of their other positions, appear to have no problem with this concept.

This brings us to the last and most crucial aspect of utilization analysis—quantification of the term "reasonable." In February 1974, the Office of Federal Contract Compliance issued *Technical Guidance Memorandum No. 1* on Revised Order 4 in an attempt to resolve this matter. They concluded that underutilization existed whenever an observed number was less than its corresponding expected value. Ignoring small sample situations, this would occur, on the average, approximately half the time. A company with 100 underutilization analyses (job classifications) could expect to be underutilized in approximately 50 cases for each class analyzed. The proposed interpretation meant that in order to avoid declarations of underutilization, all contractors had to be "at least average" in all job classifications for all groups considered. Although this could be termed an unreasonable expectation, various attempts at justification were made. Generally, they came down to a government position that it would not hurt a company to make such declarations. There were contractors and attorneys who did not agree. The corporate attorneys did not want to face their companies' own declarations of underutilization in future discrimination suits nor did they want to run the risk of being charged with (reverse) discrimination as a result of implementing a technically incorrect analysis.

As a formal matter, some small leeway was provided to avoid such problems by not requiring an underutilization declaration when observed and expected numbers differed by only a fraction of a person. For example, when a company had two mechanical engineers, of whom none were black, and an estimated availability rate for black mechanical engineers of 2%, it would not have to declare itself underutilized in the employment of black mechanical engineers.

The government was, in fact, ignoring the word "reasonably" rather than defining it. Their interpretation could reasonably have been argued to be correct without that word, but with it some latitude appeared necessary.

At this point the statistician as an analyst became involved. Under the assumptions of independent identically distributed (common availability rate) hires, the sex and (dichotomized) race composition of the incumbents of a job classification may be considered as the results of Bernoulli processes. This permits computation of probabilities by use of the binomial distribution*. The statistician is able to compute the probability of an event at least as extreme as the one observed, i.e., the occurrence of as few or fewer of the group under consideration in the job classification under consideration, as was actually observed.

The independence assumption may reasonably be questioned under real world considerations such as successive applicants being friends or relatives, but it also appears reasonable to assume that a lack of independence strong enough to have a substantial effect on the computed probabilities would manifest itself in recognizable situations. In such cases, adjustments may be made through defining the accessions to be considered. An example of this is the absorption by one firm of the employees of a second firm from a geographic area different in characteristics from that of the first—such as a large metropolitan firm taking over a rural firm. As with any other modeling scheme, the statistician must be alert to situations that depart sufficiently from the assumptions so as to make the model inappropriate.

In the preceding mechanical engineer example, the probability of there being as few or fewer black mechanical engineers as was observed (zero) is the probability of there being exactly zero of them. Using the values assumed, this may be computed to be 0.98 × 0.98 or 0.96. It appears somewhat unreasonable to consider an event with probability so close to certainty as unreasonable. In fact, even had there been 32 mechanical engineers, the chance of seeing no blacks would be 0.52.

As with other applications of statistics in the courts, determination of the level at which a result may be called "unreasonable" or a probability may be called "too low" or a result may be called "statistically significant" is properly up to the court. The statistician may compute the probability but the court is the arbiter of its meaning.

The government's concept of "reasonably" was rejected in a Federal Court involving two corporations using the binomial approach (*The Firestone Synthetic Rubber & Latex Co. and Koppers Company v. F. Ray Marshall, et al.*; United States District Court for the Eastern District of Texas, Beaumont Division, CA No. B-80-499).

UNDERREPRESENTATION

The EEOC has implemented a somewhat different use of the term underutilization. This agency, responsible for monitoring the workforces of federal agencies, has defined workforce utilization analysis to be a two step analysis. The first step is development of a workforce profile and the second step is an assessment of underrepresentation.

Although not referred to as part of utilization analysis, a workforce profile is also required by Revised Order 4 from private sector employees. Underrepresentation in EEOC terms has undergone a succession of changes in definition. Currently it is a comparison versus the corresponding civilian labor force for nonprofessional positions and versus the differentiated (corresponding) portion of the civilian labor force for professional positions. The major differences between underutilization and underrepresentation are that the former is a dichotomy, i.e., underutilization for a specific job group and class either exists or does not and that determination of whether it exists must allow for reasonable deviation from expectation, whereas underrepresentation is measured by an index (agency rate ratio to appropriate civilian labor force rate) and makes no allowance for chance deviation.

Bibliography

Baldus, D. C. and Cole, J. W. L. (1980). *Statistical Proof of Discrimination*. McGraw-Hill, New York. (A general reference for legal and statistical aspects of discrimination analysis.)

Glazer, N. (1975). *Affirmative Discrimination: Ethnic Inequality and Public Policy*. Basic Books, New York. (A discussion of issues related to equal opportunity and affirmative action in the United States.)

U.S. Equal Employment Opportunity Commission (1979). *Affirmative Action Guidelines: Technical Amendments to the Procedural Regulations*. Federal Register 44(14), 4422–4430.

(LAW, STATISTICS IN)

CHARLES R. MANN

V

VACANCY

Consider the spatial pattern formed by placing sets at random into k-dimensional Euclidean space $\mathbb{R}^k$ (e.g., random spheres into Euclidean space $\mathbb{R}^3$). *See* COVERAGE and TARGET COVERAGE for examples of possible mechanisms. If $\mathcal{R}$ is a subset of $\mathbb{R}^k$, then the *vacancy* $V(\mathcal{R})$ within $\mathcal{R}$ is just the content of that part of $\mathcal{R}$ that is not covered by any sets. For example, if $\mathbb{R}^k$ is Euclidean space $\mathbb{R}^3$, then $V(\mathcal{R})$ equals the volume of the uncovered part of $\mathcal{R}$. If the sets represent air bubbles in foam plastic, then $V(\mathcal{R})$ is the amount of plastic, as distinct from air, within $\mathcal{R}$.

Write $\|\mathcal{S}\|$ for the k-dimensional content of a set $\mathcal{S} \subseteq \mathbb{R}^k$. The ratio $p(\mathcal{R}) = E[V(\mathcal{R})]/\|\mathcal{R}\|$ of expected vacancy to total content, is called the *porosity* of the region $\mathcal{R}$. (See Serra [6, pp. 487–488].) It equals the expected fraction of $\mathcal{R}$ that is uncovered. Porosity has greatest physical significance when the coverage process is first-order stationary, that is, when the chance that the point $\mathbf{x}$ is covered does not depend on $\mathbf{x}$. In that circumstance, $p(\mathcal{R})$ does not depend on $\mathcal{R}$.

Porosity and expected vacancy are fundamental characteristics of a random medium, and porosity is scale-invariant. Each can be estimated from lower-dimensional sections. (*See* STEREOLOGY.) For example, suppose the coverage process in question is in $\mathbb{R}^3$ and first-order stationary. If the coverage process represents a mineral in an ore body, then a drill core is a one-dimensional (i.e., linear) section. If the process represents cells in an organism, then a tissue section on a microscope slide is a two-dimensional section. Each section generates a coverage process in its own right, and each has the same porosity as the original process in $\mathbb{R}^3$.

A formal mathematical definition of vacancy may be given as follows. Let $\mathcal{X}$ denote the union in $\mathbb{R}^k$ of the sets comprising the coverage process. Given $\mathbf{x} \in \mathbb{R}^k$, define

$$I[\mathbf{x} \text{ not covered}] = \begin{cases} 1, & \text{if } \mathbf{x} \notin \mathcal{X}, \\ 0, & \text{otherwise.} \end{cases}$$

Then the vacancy within $\mathcal{R}$ is

$$V(\mathcal{R}) = \|\tilde{\mathcal{X}} \cap \mathcal{R}\|$$

$$= \int_{\mathbf{x} \in \mathcal{R}} I[\mathbf{x} \text{ not covered}] \, d\mathbf{x}, \quad (1)$$

where $\tilde{\mathcal{X}}$ denotes the complement in $\mathbb{R}^k$ of $\mathcal{X}$. Of course, $V(\mathcal{R})$ is a random variable, and quantities such as $E[V(\mathcal{R})]$ and $\mathrm{var}[V(\mathcal{R})]$ should be interpreted in the usual manner for random variables.

Formula (1) leads to simple expressions for moments* of vacancy. In particular, the expected vacancy within $\mathcal{R}$ is

$$E[V(\mathcal{R})] = \int_{\mathbf{x} \in \mathcal{R}} \Pr[\mathbf{x} \text{ not covered}] \, d\mathbf{x}.$$

If the coverage process is first-order stationary, then $p \equiv \Pr[\mathbf{x} \text{ not covered}]$ does not depend on $\mathbf{x}$, implying that $E[V(\mathcal{R})] = p\|\mathcal{R}\|$. In this case, p is identical to porosity. If the coverage process is also second-order stationary, such that

$$q(\mathbf{y}) \equiv \Pr[\mathbf{x} \text{ not covered}, \mathbf{x} + \mathbf{y} \text{ not covered}]$$

does not depend on $\mathbf{x}$, then the variance of vacancy equals

$$\text{var}[V(\mathcal{R})] = \int\int_{(\mathbf{x},\mathbf{y}) \in \mathcal{R}^2} q(\mathbf{x} - \mathbf{y}) \, d\mathbf{x} \, d\mathbf{y}$$
$$- \left(p\|\mathcal{R}\|\right)^2.$$

We shall consider the example of vacancy in a Boolean model. The latter is generated by centring independent and identically distributed random sets at points of a homogeneous Poisson process* in $\mathbb{R}^k$, and is both first- and second-order stationary. If the sets are distributed as the random set $\mathcal{S}$, then $p = \exp(-\alpha\lambda)$, where $\alpha \equiv E[\|\mathcal{S}\|]$ is the mean content of $\mathcal{S}$ and λ is the intensity of the Poisson process. Furthermore, $q(\mathbf{x}) = \exp\{-\lambda E[\|\mathcal{S} \cup (\mathbf{x} + \mathcal{S})\|]\}$, where $\mathbf{x} + \mathcal{S} \equiv \{\mathbf{x} + \mathbf{y} : \mathbf{y} \in \mathcal{S}\}$. Therefore $E[V(\mathcal{R})] = \|\mathcal{R}\|e^{-\alpha\lambda}$ and

$$\text{var}[V(\mathcal{R})]$$
$$= e^{-2\alpha\lambda} \int\int_{(\mathbf{x},\mathbf{y}) \in \mathcal{R}^2} \Big(\exp\big\{\lambda E\big[\|\mathcal{S}$$
$$\cap (\mathbf{x} - \mathbf{y} + \mathcal{S})\|\big]\big\} - 1\Big) \, d\mathbf{x} \, d\mathbf{y}.$$

See [2, Chap. 3] and [5, p. 61 ff] for derivations of these formulas, including those for p and $q(\mathbf{x})$.

Porosity and vacancy are related to the *point variogram* (*see also* VARIOGRAM) of a second-order stationary coverage process. The point variogram is defined by

$$\gamma(\mathbf{y}) \equiv \Pr[\mathbf{x} \text{ covered}, \mathbf{x} + \mathbf{y} \text{ not covered}]$$
$$= p - q(\mathbf{y}).$$

See Serra [6, p. 280] for details and discussion.

In most cases of practical interest, the events "$\mathcal{R}$ completely covered" and "$V(\mathcal{R}) = 0$" differ only by an event of probability 0. This means that the problem of determining the probability that a given region $\mathcal{R}$ is completely covered (for examples, *see* COVERAGE) can be regarded as an offshoot of that of determining the distribution of vacancy. Only for Poisson-related models (e.g., Boolean models), and usually only in the case of dimension $k = 1$, can either of these problems be solved with any degree of exactness. In other cases, both problems admit solutions that are asymptotic approximations for large regions $\mathcal{R}$ or for extreme values of the parameters; see, e.g., [1–4].

The first two moments of vacancy lead to a useful upper bound for the probability of coverage:

$$\Pr[V = 0] \leqslant \text{Var}[V]/E[V^2].$$

See [2, Sec. 3.2] for a derivation.

References

[1] Baddeley, A. (1980). *Adv. Appl. Prob.*, **12**, 447–461. (Proves a central limit theorem for vacancy in the case of stationary, non-Boolean coverage processes.)

[2] Hall, P. (1988). *Introduction to the Theory of Coverage Processes*. Wiley, New York. (Chapter 3 is devoted to properties of vacancy, with emphasis on vacancy in Boolean models.)

[3] Hüsler, J. (1982). *J. Appl. Prob.*, **19**, 578–587. (Proves limit theorems for vacancy in the case of arcs on a circle.)

[4] Mase, S. (1982). *J. Appl. Prob.*, **19**, 111–126. (Proves a central limit theorem for vacancy in the case of stationary, non-Boolean coverage processes.)

[5] Matheron, G. (1975). *Random Sets and Integral Geometry*. Wiley, New York. [Describes an axiomatic theory for processes of random sets, such as Boolean models (p. 61). Formulas for moments of vacancy in Boolean models are simple consequences of Matheron's expression (3-3-1), p. 62.]

[6] Serra, J. (1982). *Image Analysis and Mathematical Morphology*. Academic, New York. (Point variogram and porosity are defined on pp. 280 and 487, respectively. Part IV discusses models for coverage processes, and their application and analysis.)

(COVERAGE
SERRA'S CALCULUS
STEREOLOGY
TARGET COVERAGE)

Peter Hall

VALAND'S (MANTEL AND VALAND'S) NONPARAMETRIC MANOVA TECHNIQUE

Let $\Omega = \{\omega_1, \ldots, \omega_N\}$ designate a finite population of N objects, let $\mathbf{x}'_I = (x_{1I}, \ldots, x_{rI})$ denote r response measurements involving object ω_I, $I = 1, \ldots, N$, let R_{kI} be the tie-adjusted rank-order statistic* corresponding to x_{kI} relative to the ordered values among the kth of the r response measurements, $1 \leqslant R_{kI} \leqslant N$, let $\Delta_{I,J}$ represent a symmetric distance function* associated with objects ω_I and ω_J; let $S_1, \ldots, S_g$ designate an exhaustive partitioning of the N objects comprising Ω into g disjoint groups; and let $n_i \geqslant 2$ be the number of objects in group S_i, $i = 1, \ldots, g$. Mantel and Valand's statistic [1] is given by

$$\delta = \left[\sum_{j=1}^{g} \binom{n_j}{2} \right]^{-1} \sum_{i=1}^{g} \binom{n_i}{2}$$
$$\times \sum_{I<J} \Delta_{I,J} \psi_i(\omega_I) \psi_i(\omega_J),$$

where

$$\Delta_{I,J} = \sum_{k=1}^{r} |R_{kI} - R_{kJ}|,$$

$\Sigma_{I<J}$ is the sum over all I and J such that $1 \leqslant I < J \leqslant N$, and $\psi_i(\omega_I)$ is 1 if ω_I belongs to S_i and 0 otherwise. The null hypothesis of this technique implies that each of the $N!/(\prod_{i=1}^{g} n_i!)$ possible allocation combinations of the N objects to the g groups occur with equal chance. If δ_0 is an observed value of δ, then the P-value* is the probability that $\delta \leqslant \delta_0$ under the null hypothesis.

Mantel and Valand's nonparametric MANOVA technique is a special case of Multiresponse Permutation Procedures* (MRPP) [2, 3]. Contrary to a statement by

Mantel and Valand [1], Mielke [2] showed that the asymptotic distribution of δ is not a normal distribution when $n_i = N/g$, $i = 1, \ldots, g$. If δ is replaced by the more efficient statistic (when the n_i's are unequal) given by

$$\delta' = \frac{2}{N} \sum_{i=1}^{g} (n_i - 1)^{-1}$$
$$\times \sum_{I<J} \Delta_{I,J} \psi_i(\omega_I) \psi_i(\omega_J),$$

then the asymptotic distribution of δ' is never a normal distribution [3]. While Mantel and Valand's choice for $\Delta_{I,J}$ is a metric distance function which differs, for $r \geqslant 2$, from the more intuitive metric distance function corresponding to Euclidean distance, this choice provides a monumental improvement over many of the most commonly used rank tests, which are based on a nonmetric distance function corresponding to squared Euclidean distance [3, 4].

The following example demonstrates a distinct advantage of Mantel and Valand's test over the Mann–Whitney*–Wilcoxon (squared-distance) test. Consider univariate data ($r = 1$) confined to two groups ($g = 2$) where 40 objects ($N = 40$) are allocated equally to both groups ($n_1 = n_2 = 20$). Suppose the rank-order statistics of the 20 response measurements in the first group (S_1) are $\{5, 6, 7, 8, 9, 10, 11, 12, 13, 14, 15, 16, 17, 18, 19, 20, 37, 38, 39, 40\}$. Similarly, the rank-order* statistics of the 20 response measurements in the second group (S_2) are $\{1, 2, 3, 4, 21, 22, 23, 24, 25, 26, 27, 28, 29, 30, 31, 32, 33, 34, 35, 36\}$. The P-value associated with Mantel and Valand's test is 0.0075, whereas the corresponding P-value associated with the Wilcoxon–Mann–Whitney test is 0.1298. The substantial difference in P-value results is a consequence of Mantel and Valand's (the Wilcoxon–Mann–Whitney) test being a median- (mean-)based technique that does (does not) satisfy the congruence principle [4]. To further clarify this difference in P-value results, note that the medians (means) of S_1 and S_2 are 14.5 and 26.5 (17.7 and 23.3), respectively.

References

[1] Mantel, N. and Valand, R. S. (1970). *Biometrics*, **26**, 547–588. (The present technique is introduced.)

[2] Mielke, P. W. (1978). *Biometrics*, **34**, 277–282. (Shows that the normal approximation for δ usually yields erroneous inferences for this technique.)

[3] Mielke, P. W. (1984). In *Handbook of Statistics*, Vol. 4, P. R. Krishnaiah and P. K. Sen, eds. North-Holland, Amsterdam, pp. 813–830. (Describes various properties of MRPP that pertain to this technique.)

[4] Mielke, P. W. (1985). *J. Atmospheric Sci.*, **42**, 1209–1212. (Describes the congruence principle and provides examples that demonstrate inferential travesties associated with statistical tests based on squared Euclidean distance.)

(DISTRIBUTION-FREE METHODS
MANN–WHITNEY–WILCOXON STATISTIC
MULTIRESPONSE PERMUTATION
   PROCEDURES
PERMUTATION TESTS
RANK-ORDER TESTS
RANK TESTS)

PAUL W. MIELKE, JR.

Editorial Note

In 1971, R. S. Valand changed his name to Ran S. Sharma.

VALIDITY

In the theory and applications of mental testing, this term is used to denote the correlation of a test score with some criterion. It is, in effect, a measure of the predictive value of a test. Compare with reliability* of a test. A detailed, "classical" discussion of this concept can be found in Gulliksen [1].

The *coefficient of validity* is a (product-moment) correlation coefficient*. The term is used, especially in psychology, when the two variables X and Y, between which the correlation is calculated, are supposed to measure the "same thing." For example, X might be a test score and Y the value of the quantity the test is supposed to measure. The correlation coefficient between X and Y is then called the "validity coefficient of X."

Reference

[1] Gulliksen, H. (1950). *Theory of Mental Tests*. Wiley, New York.

(PSYCHOLOGICAL TESTING THEORY
STATISTICS IN PSYCHOLOGY)

VALUE-ITERATION METHOD *See* MARKOV PROCESSES

VANDERMONDE CONVOLUTION

An elementary combinatorial identity

$$\binom{n_1 + n_2}{k} = \sum_i \binom{n_1}{i}\binom{n_2}{k - i},$$

where the limits of summation are

$$\max(0, k - n_2) \leqslant i \leqslant \min(k, n_1).$$

It is useful in analyses involving hypergeometric* distributions, and, in particular, in the theory of sampling from finite populations*.

[The identity reflects the fact that k objects can be selected from a set of $(n_1 + n_2)$ objects by splitting into two subsets of sizes n_1, n_2 and choosing i from the first subset and $(k - i)$ from the second.]

(COMBINATORICS)

VANDERMONDE MATRIX

This is an $n \times n$ matrix of the form

$$\begin{bmatrix} 1 & x_1 & x_1^2 & \cdots & x_1^{n-1} \\ 1 & x_2 & x_2^2 & \cdots & x_2^{n-1} \\ \vdots & \vdots & \vdots & & \vdots \\ 1 & x_n & x_n^2 & \cdots & x_n^{n-1} \end{bmatrix}$$

The determinant of this matrix is $\prod_{i < j}(x_j - x_i)$. These matrices are used in the design of experiments*, in particular, in the theory of polynomial models. See, e.g., Raktoe et al. [2] for further details; uses of the matrix in the derivation of multivariate distributions* are described in [1].

References

[1] Krishnaiah, P. R., ed. (1980). In *Handbook of Statistics*, Vol. 1. North-Holland, Amsterdam, pp. 745–971.

[2] Raktoe, B. L., Hedayat, A., and Federer W. T. (1981). *Factorial Designs*. Wiley, New York.

(RESPONSE SURFACES)

VAN DER WAERDEN TEST *See* NOR-MAL SCORES TESTS

VAN MONTFORT–OTTEN TEST

In extreme-value theory it is often of interest to test whether the distribution from which a set of data has been drawn is of extreme-value type I rather than types II or III. This is equivalent to testing whether the shape parameter is 0 in the generalized extreme-value distribution* defined by the cumulative distribution function

$$F(x; \xi, \alpha, k)$$

$$= \begin{cases} \exp\left[-\{1 - k(x - \xi)/\alpha\}^{1/k}\right], \\ \qquad\qquad\qquad\qquad k \neq 0, \\ \exp\left[-\exp\{-(x - \xi)/\alpha\}\right], \\ \qquad\qquad\qquad\qquad k = 0. \end{cases}$$

Van Montfort and Otten [4] have derived such a test based on the spacings* $x_{(i)} - x_{(i-1)}$, $i = 2, \ldots, n$, of an ordered random sample $x_{(1)} \leqslant x_{(2)} \leqslant \cdots \leqslant x_{(n)}$. Let

$$l_i = \frac{x_{(i)} - x_{(i-1)}}{m_i - m_{i-1}},$$

where $m_i = -\log[-\log\{i/(n + 1)\}]$ is, apart from a linear transformation, the large-sample expectation of $x_{(i)}$ when $k = 0$, and let $W_i = l_i/(l_2 + l_3 + \cdots + l_n)$, $i = 2, \ldots, n$. The test statistic is

$$A = n^{1/2} \sum_{i=2}^{n} W_i d_i,$$

where

$$d_i = \left(\Delta_i - \bar{\Delta}\right)/\sigma_\Delta,$$

$$\Delta_i = \log\left[-\log\{(i - \tfrac{1}{2})/(n + 1)\}\right],$$

$$\bar{\Delta} = (n - 1)^{-1} \sum_{i=2}^{n} \Delta_i,$$

$$\sigma_\Delta = \left\{(n - 1)^{-1} \sum_{i=2}^{n} \left(\Delta_i - \bar{\Delta}\right)^2\right\}^{1/2}.$$

The hypothesis that $k = 0$ is rejected if the calculated value of A is too far from 0. Significantly negative values of A suggest that $k < 0$ (type II extreme-value distribution) and significantly positive values suggest that $k > 0$ (type III extreme-value distribution). Otten and van Montfort [3] have given critical values for the test. Hosking [1] found that an adequate approximation to the critical values is obtained by treating the statistic $A^* = A + 0.2$ as having a standard normal distribution when $k = 0$. Hosking [1] recommends the use of the test based on A^* as an alternative to the likelihood-ratio test* for samples of size 100 or less: The A^*-test is quick to compute and has good power, particularly for discriminating between the cases $k = 0$ and $k < 0$.

Example. A series of 35 annual maximum floods of the river Nidd at Hunsingore, Yorkshire, England, is given in NERC [2, p. 235]. The ordered data and the calculations needed for the van Montfort–Otten test are presented in Table 1. The final *P*-value* of 0.251 indicates that the hypothesis that the underlying flood frequency distribution is extreme-value type I should not be rejected.

References

[1] Hosking, J. R. M. (1984). *Biometrika*, **71**, 367–374.

[2] NERC (1975). *Flood Studies Report*, Vol. 1. Natural Environment Research Council, London.

[3] Otten, A. and van Montfort, M. A. J. (1978). *J. Hydrol.*, **37**, 195–199.

[4] van Montfort, M. A. J. and Otten, A. (1978). *Math. Operat. Statist. Ser. Statist.*, **9**, 91–104.

Table 1. Calculations for the van Montfort–Otten Test Applied to the Annual Maximum Floods of the River Nidd

i	$x_{(i)}$	l_i	W_i	d_i
1	65.08			
2	65.60	2.41	0.0015	1.5968
3	75.06	62.58	0.0385	1.4299
4	76.22	9.42	0.0058	1.3014
5	78.55	21.75	0.0134	1.1927
6	81.27	28.06	0.0173	1.0960
7	86.73	60.69	0.0373	1.0073
8	87.76	12.10	0.0074	0.9242
9	88.89	13.85	0.0085	0.8449
10	90.28	17.58	0.0108	0.7685
11	91.80	19.65	0.0121	0.6942
12	91.80	0.00	0.0000	0.6211
13	92.82	13.48	0.0083	0.5488
14	95.47	35.08	0.0216	0.4768
15	100.40	64.99	0.0400	0.4048
16	111.54	145.47	0.0895	0.3322
17	111.74	2.57	0.0016	0.2588
18	115.52	47.70	0.0293	0.1841
19	131.82	200.70	0.1234	0.1077
20	138.72	82.47	0.0507	0.0292
21	148.63	114.36	0.0703	−0.0518
22	149.30	7.42	0.0046	−0.1360
23	151.79	26.32	0.0162	−0.2240
24	153.04	12.52	0.0077	−0.3165
25	158.01	46.83	0.0288	−0.4144
26	162.99	43.76	0.0269	−0.5190
27	172.92	80.55	0.0495	−0.6316
28	179.12	45.87	0.0282	−0.7544
29	181.59	16.42	0.0101	−0.8899
30	189.02	43.56	0.0268	−1.0421
31	213.70	124.48	0.0766	−1.2166
32	226.48	53.54	0.0329	−1.4228
33	251.96	84.14	0.0517	−1.6768
34	261.82	23.46	0.0144	−2.0115
35	305.75	62.08	0.0382	−2.5120

$A = n^{1/2} \Sigma W_i d_i = -1.348$

$A^* = A + 0.2 = -1.148$

For a standard normal random variate Z,
$P(|Z| > 1.148) = 0.251$

(EXTREME-VALUE DISTRIBUTION
HYPOTHESIS TESTING
 SPACINGS)

J. R. M. HOSKING

VAN VALEN'S TEST

Van Valen's test [6] is designed to detect a difference in the amount of variation in two or more multivariate samples, when mean values are unknown and are not necessarily equal. The standard test in this situation is the multivariate Bartlett test*. However, the Bartlett test requires the assumption that the samples come from multivariate normal distributions*, and is not robust* to departures from this assumption. There is always the possibility that a significant result is due to nonnormality rather than to unequal covariance matrices (Seber [5, p. 449]).

Van Valen's test uses the same principle as Levene's robust test for homogeneity of variances*. Let x_{ijk} denote the observed data value for variable X_k for the ith individual in the jth sample, $k = 1, 2, \ldots, p$; $i = 1, 2, \ldots, n_j$; $j = 1, 2, \ldots, m$, and let M_{jk} denote the mean value for X_k in the jth sample. The test involves transforming the data values into $y_{ijk} = |x_{ijk} - M_{jk}|$, so that samples displaying a great deal of variation will have high-mean y-values and samples with little variation will have low-mean y-values. The question of whether samples display significantly different amounts of variation then becomes a question of whether the transformed values have significantly different means. If $\mathbf{y}_j = (y_{j1}, y_{j2}, \ldots, y_{jp})'$ denotes the vector of means for the transformed values in sample j, then the null hypothesis is that the expected values of $\mathbf{y}_1, \mathbf{y}_2, \ldots, \mathbf{y}_m$ are equal, whereas the alternative hypothesis is that they are not equal.

TWO-SAMPLE TESTS

The mean vectors $\mathbf{y}_1$ and $\mathbf{y}_2$ can be compared for two samples using Hotelling's T^2-test*. However, Van Valen suggested instead that the "distance" of the ith individual in sample j from the "centre" of the sample can be calculated as

$$d_{ij} = \sqrt{\left\{ \sum_{k=1}^{p} y_{ijk}^2 \right\}}, \qquad (1)$$

and the sample means of the d-values can be compared using a two-sample t-test*. If one sample is more variable than another, then the mean d-value will be higher in the more variable sample. To ensure that all variables are about equally important in the test, they should be standardized* before calculating the d-values. One way to achieve this is to make each variable have a variance of unity for both samples lumped together. Levene's test has been found to be more robust if deviations from medians are used instead of deviations from means. It may therefore be better to define M_{jk} as the sample median rather than the mean for the transformation from x to y values.

It is implicit in the calculations for Van Valen's test that when differences in variation occur, one sample is expected to be more variable than the other for all variables so that the differences are accumulated in the d-values. If this is not the case, then the test is liable to have low power. This reservation does not apply if Hotelling's T^2-test is used to compare the sample mean vectors of absolute deviation from means or medians.

SEVERAL-SAMPLE TESTS

With more than two samples, the generalization of two-sample tests is straightforward. Once deviations from sample means or medians have been calculated, the mean vectors can be compared using multivariate analysis of variance*, or d-values can be calculated using (1) and they can be used as the data for a one-factor analysis of variance*. Either way, a significant result indicates that the amount of variation is not constant for the populations sampled.

PROPERTIES OF THE TEST

Little is known about the properties of Van Valen's test. It is an ad hoc procedure that relies for its validity on the known general robustness of the t-test and analysis of variance, in much the same way as Levene's test does.

Power* and robustness have been studied in the context of stabilizing selection on biological populations, where a sample taken after selection is expected to display less variation than a sample taken before selection, because of the high death rate for extreme individuals. For this type of situation, simulations carried out by Manly [2, 3] indicate that:

(a) The multivariate Bartlett test on two covariance matrices is not particularly powerful and is severely affected by nonnormal distributions.

(b) When the correlations between variables are close to 0, Van Valen's test has reasonable power in comparison with a likelihood-ratio test* aimed specifically at detecting stabilizing selection. However, when correlations between variables are 0.9, Van Valen's test has only moderate power in comparison with this likelihood-ratio test.

(c) Van Valen's test is affected very little by nonnormality and generally has approximated the correct size.

(d) Hotelling's T^2-test* on absolute deviations from sample means has very low power to detect stabilizing selection.

Example. As an example of the use of the test, consider the data indicated in Table 1 for five body measurements on female sparrows, originally collected by Bumpus [1]. There is one sample of 21 sparrows that survived a severe storm and a second sample of 28 sparrows that died as a result of the storm. According to Darwin's theory of natural selection, it is reasonable to suppose that the storm eliminates a number of sparrows with unusual body measurements because they were not as "fit" as average sparrows, and hence that the population of survivors is less variable than the population of nonsurvivors.

The transformed data and d-values for Van Valen's test are indicated in Table 2. For example, the first value in the sample of

Table 1 Body Measurements of Female Sparrows

Bird	X_1	X_2	X_3	X_4	X_5
			Survivors		
1	156	245	31.6	18.5	20.5
2	154	240	30.4	17.9	19.6
3	153	240	31.0	18.4	20.6
⋮					
21	159	236	31.5	18.0	21.5
Mean	157.4	241.0	31.43	18.50	20.81
SD	3.32	4.18	0.73	0.42	0.76

Bird	X_1	X_2	X_3	X_4	X_5
			Nonsurvivors		
1	155	240	31.4	18.0	20.7
2	156	240	31.5	18.2	20.6
3	160	242	32.6	18.8	21.7
⋮					
28	164	248	32.3	18.8	20.9
Mean	158.4	241.6	31.48	18.45	20.84
SD	3.88	5.71	0.85	0.66	1.15

	X_1	X_2	X_3	X_4	X_5
			Pooled samples		
Mean	158.0	241.3	31.46	18.47	20.83
SD	3.62	5.16	0.81	0.58	1.01

X_1 = total length, X_2 = alar extent, X_3 = length of beak and head, X_4 = length of humerus, and X_5 = length of keel of sternum, all in millimeters. The full data are given by Manly [4, p. 2].

survivors for X_1 is 0.28, which was obtained as follows:

(a) The data were standardized to have a 0 mean and a variance of 1 for this variable for all 49 birds. This was achieved by coding X_1 to $(X_1 - 158.0)/3.62$. For the first survivor the coded value is -0.55.

(b) The median of the transformed data values (-0.27) was found for the survivors.

(c) The absolute deviation from the sample median for the first survivor was then calculated as $|-0.55 - (-0.27)| = 0.28$, as recorded.

Table 2 Absolute Deviations from Sample Medians for the Sparrow Data and d-Values Calculated Using Equation (1)

Bird	X_1	X_2	X_3	X_4	X_5	d
			Survivors			
1	0.28	1.00	0.25	0.00	0.10	1.07
2	0.83	0.00	1.27	1.07	1.02	2.12
3	1.11	0.00	0.51	0.18	0.00	1.23
⋮						
21	0.55	0.80	0.13	0.90	0.92	1.61
					Mean	1.76
					SD	0.64

Bird	X_1	X_2	X_3	X_4	X_5	d
			Nonsurvivors			
1	1.11	0.40	0.13	0.90	0.00	1.48
2	0.83	0.40	0.00	0.54	0.10	1.07
3	0.28	0.00	1.40	0.54	1.02	1.83
⋮						
28	1.38	1.20	1.02	0.54	0.20	2.17
					Mean	2.27
					SD	1.06

The last column in Table 2 shows the d-values found by squaring and adding the values in the previous columns. A t-test to compare the two mean d-values then gives a test statistic of -1.92 with 47 degrees of freedom using the formula

$$t = (\bar{d}_1 - \bar{d}_2)\Big/\left\{s\sqrt{(1/n_1 + 1/n_2)}\right\},$$

where $\bar{d}_j$ is the mean d-value in sample j and s is the usual pooled estimate of the within-sample standard deviation. This t-value is significantly low at the 5% level, indicating less variation for survivors than for nonsurvivors. For the same data the multivariate Bartlett test gives a test statistic of $M = 11.89$, which is not significant at the 5% level when compared with a chi-squared distribution with 15 degrees of freedom.

References

[1] Bumpus, H. C. (1898). *Biological Lectures*. Marine Biology Laboratory, Woods Hole, MA, pp. 209–226.

[2] Manly, B. F. J. (1985). *Evolutionary Theory*, **7**, 205–217.

[3] Manly, B. F. J. (1986). In *Pacific Statistical Congress*, I. S. Francis, B. F. J. Manly, and F. C. Lam, eds. North-Holland, Amsterdam, pp. 339–344.

[4] Manly, B. F. J. (1986). *Multivariate Statistical Methods: A Primer*. Chapman and Hall, London. (A short introductory test on multivariate analysis that includes examples of the use of Van Valen's test.)

[5] Seber, G. A. F. (1984). *Multivariate Observations*. Wiley, New York. (A comprehensive text on multivariate analysis that discusses the limitations of the multivariate Bartlett test and alternative procedures for testing for equal covariance matrices.)

[6] Van Valen, L. (1978). *Evolution Theory*, **4**, 33–43. (Erratum in *Evolution Theory*, **4**, 202.)

(ANALYSIS OF VARIANCE
HOTELLING'S T^2-TEST
LIKELIHOOD-RATIO TEST
LEVENE'S ROBUST TEST OF
 HOMOGENEITY OF VARIANCES
MULTIVARIATE ANALYSIS OF VARIANCE
MULTIVARIATE BARTLETT TEST
MULTIVARIATE NORMAL DISTRIBUTION
POWER
ROBUSTIFICATION AND ROBUST
 STATISTICS
STUDENT'S t-TESTS)

BRYAN F. J. MANLY

VAN ZWET TAIL ORDERING

Given two cumulative distribution functions $F(x)$ and $G(x)$, we say that F has *lighter tails* than G (and G has *heavier tails* than F) if the function $G^{-1}(F(x))$ is convex for $x \geqslant 0$. This relationship is represented by $F \lessdot G$.

The definition is due to Van Zwet [3] and has been used by Hettmansperger [1] for comparison of score statistics*. (Some other definitions of tail ordering are discussed in [2].)

References

[1] Hettmansperger, T. P. (1984). *Statistical Inference Based on Ranks*. Wiley, New York.

[2] Hettmansperger, T. P. and Keenan, M. A. (1975). In *Statistical Distributions in Scientific Work*, Vol.

6, C. Taillie, G. P. Patil, and B. A. Baldessari, eds. Reidel, Dordrecht and Boston, pp. 161–172.

[3] Van Zwet, W. R. (1970). Convex Transformations of Random Variables. *Math. Centre Tracts*, **7**. Math. Centre, Amsterdam.

(HEAVY-TAILED DISTRIBUTIONS
ORDERING OF DISTRIBUTIONS BY
 DISPERSION
ORDERING OF DISTRIBUTIONS, PARTIAL
STOCHASTIC ORDERING
TAIL ORDERING)

VARIABILITY, HARMONIC

For a positive random variable X, Brown [1] introduced the concept of the *harmonic measure of variability*

$$c^2 = 1 - \left\{ E[X] E[X^{-1}] \right\}^{-1}.$$

Provided the expected values exist, $0 \leqslant c^2 < 1$.

The measure was introduced as an index of accuracy likely to be achieved when using the method of statistical differentials* (based on Taylor expansions*) in obtaining approximations to moments of functions of random variables.

Brown [1] showed, inter alia, that if $g(x)$ is a strictly monotonic function of x with $g(0)$ finite, then

$$0 \leqslant E[g(X)] - g(E[X]) \leqslant c^2 g(0)$$

and

$$\mathrm{var}\left(g(X) \right) \leqslant c^2 g(0).$$

Reference

Brown, M. (1985). *Ann. Statist.*, **13**, 1239–1243.

(VARIANCE, UPPER BOUNDS)

VARIABLES, METHOD OF

A term sometimes used in acceptance sampling* to refer to inspection of a product by measurement of some continuous type(s) of characteristic (such as weight, volume, tensile strength, etc.) as opposite to observation of an attribute* [such as passing a go–no go gauge, having surface blemish(es), etc.].

(ACCEPTANCE SAMPLING
CONTROL CHARTS
QUALITY CONTROL, STATISTICAL
TOLERANCE LIMITS)

VARIABLES PLANS *See* SAMPLING PLANS

VARIANCE

The square of the standard deviation*. For a population it is the second central moment*

$$\text{var}(X) = \mu_2(X) = E\big[\{X - E[X]\}^2\big].$$

(MEAN SQUARED ERROR
RANGES
STANDARD DEVIATION
STANDARD ERROR
VARIANCE, SAMPLE)

VARIANCE COMPARISONS BASED ON PERMUTATION THEORY

Let $X_1, \ldots, X_m$ and $Y_1, \ldots, Y_n$ be random variables (RVs) representing independent samples for which both $(X - \mu)/\sigma$ and $(Y - \nu)/\tau$ have continuous CDF G having finite fourth central moment μ_4. The parameters μ, ν, σ, τ are unknown and $\Delta \equiv \tau/\sigma$ is the parameter of interest. The classical F-test*, appropriate for normal G, assumes $F \equiv s_Y^2/s_X^2$ has an $F_{n-1, m-1}$ distribution; here s_X^2 and s_Y^2 are the sample variances.

This is an extremely nonrobust test. In fact, it is even asymptotically wrong, since $[mn/(m + n)]^{1/2}(F - \Delta^2)/(2^{1/2}\Delta^2)$ is asymptotically $N(0, 1 + \gamma_2/2)$, where the kurtosis $\gamma_2 \equiv (\mu_4/\sigma^4) - 3$ is 0 for normal G but is typically nonzero otherwise. Dividing the normalized form of F above by

$(1 + \hat{\gamma}_2/2)^{1/2}$, though asymptotically correct, is still highly unsatisfactory, even in moderate samples.

Now the $m + n$ unobservable RVs $U_i \equiv \Delta(X_i - \mu)$, $1 \le i \le m$, and $U_{m+i} \equiv (Y_i - \nu)$, $1 \le i \le n$, are independent and identically distributed (iid). The permutation distribution of the unobservable RV $F^* \equiv (m/n)\sum_{i=m+1}^{m+n} U_i^2/\sum_{i=1}^{m} U_i^2$ takes on $\binom{m+n}{m}$ different values with equally likely probability. This permutation distribution is shown in [2] to be approximately an F-distribution with d^*n and d^*m degrees of freedom, where

$$d^* \equiv 1/\{1 + (b_2^* - 3)/2\}$$

and

$$b_2^* \equiv (m + n + 2) \sum_1^{m+n} U_i^4 \Big/ \Big(\sum_1^{m+n} U_i^2\Big)^2.$$

When μ and ν are known this can be used to approximate the distribution of F^* under the null hypothesis that $\Delta = \Delta_0$.

Even if μ and ν are unknown, it is shown in [2] that a.s. $F^* = F/\Delta^2 + o((m + n)^{-1/2+\delta})$ and $b_2^* = b_2(\Delta) + o((m + n)^{-1/2+\delta})$ for any $\delta > 0$ provided $\mu_4 < \infty$; here

$$b_2(\Delta) \equiv (m + n)$$

$$\cdot \frac{\Delta^4\sum_{i=1}^{m}\big(X_i - \overline{X}\big)^4 + \sum_{i=1}^{n}\big(Y_i - \overline{Y}\big)^4}{\Big\{\Delta^2\sum_{i=1}^{m}\big(X_i - \overline{X}\big)^2 + \sum_{i=1}^{n}\big(Y_i - \overline{Y}\big)^2\Big\}^2}.$$

This is sufficient to imply that the following approximate permutation F-test (the *APF-test*) is asymptotically correct provided $\mu_4 < \infty$.

In summary, the APF-test rejects the hypothesis $H: \Delta \le \Delta_0$ in favor of $K: \Delta > \Delta_0$ in case

$$F/\Delta_0^2 > F_{d(n-1), d(m-1)}^{(\alpha)};$$

here $F = S_Y^2/S_X^2$ is the classical variance ratio and $1/d \equiv 1 + \hat{\gamma}_2(\Delta_0)/2 \equiv 1 + \{b_2(\Delta_0) - 3\}/2$. That is, the classical quantity F/Δ_0^2 is referred to the percentage point of an F-distribution in which the classical normal theory degrees of freedom $n - 1$ and $m - 1$ are now altered by the random factor d that depends on the data.

Computing the interval of Δ's for which the natural two-sided version of this test fails to reject provides an approximate confidence interval for Δ.

Monte Carlo results reported in [1] and [2] show that this test performs well in case G is a normal, uniform or double exponential CDF. It is shown in [1] that the power of this test is approximately equal to that of one of its chief competitors, the jackknifed version of the classical F-test.

References

[1]   Miller, R. (1968). *Ann. Math. Statist.*, **39**, 567–582.

[2]   Shorack, G. (1969). *J. Amer. Statist. Ass.*, **64**, 999–1013.

(F-TEST
JACKKNIFE
LEVENE'S ROBUST TEST OF
    HOMOGENEITY OF VARIANCES
RANDOMIZATION TESTS
VARIANCE-RATIO)

GALEN R. SHORACK

VARIANCE COMPONENTS

Although there is a vast amount of literature dealing with variance components, the basic concept is simple. Suppose an observable random variable Y, with variance σ_Y^2, is the sum $A + E$ of two nonobservable independent variables A and E. Let σ_A^2 denote the variance of A and σ_E^2 denote the variance of E. Then, due to the independence of A and E, $\sigma_Y^2 = \sigma_A^2 + \sigma_E^2$. σ_A^2 and σ_E^2 are called *variance components* of σ_Y^2.

In a variety of practical problem areas, it is important to make statistical inference statements relative to variance components. As early as 1861, the astronomer Airy [1] concerned himself with telescopic observations of the same phenomenon b_i times for the ith night, for $i = 1, 2, \ldots, a$ nights. Airy

desired information pertinent to the nightly variation σ_A^2 and the within-night variation σ_E^2. The essence of Airy's problem is also the essence of problems familiar to many investigators. For illustrative purposes, we consider an animal-science setting involving litters of pigs. Let Y_{ij} be the weight gain in a fixed time interval for the jth animal in the ith litter.

It is helpful to think of the observations Y_{ij} as variables created by adding realizations of variables associated with sampling from two hypothetical populations; one is a population of litter effects and the other a population of nonlitter-related pig effects. Crucial to the estimation procedure is the fact that one realization of an A_i variable is added to several (in this case b) different values of E_{ij} variables. If one E value was added to each A value, then the effects would be confounded* and the components σ_A^2 and σ_E^2 would be nonestimable. To describe this situation statistically, we assume the random-effects model $Y_{ij} = \mu + A_i + E_{ij}$, where all variables A_i, $i = 1, \ldots, a$, and E_{ij}, $j = 1, \ldots, b$, are mutually independent, $\mathscr{E}[A_i] = 0$, $\mathscr{E}[E_{ij}] = 0$, $\mathrm{Var}(A_i) = \sigma_A^2$, and $\mathrm{Var}(E_{ij}) = \sigma_E^2$. In this setting A_i is the random effect produced by the ith litter and σ_A^2 is the variance component attributable to variation in litters. $E_{ij} = Y_{ij} - \mu - A_i$ is the residual effect or the random individual within-litter pig effect and σ_E^2 is the variance component due to variation other than litters. *See also* FIXED-, RANDOM-, AND MIXED-EFFECTS MODELS AND MODELS I, II, AND III. The problem is to make inference statements relative to σ_A^2 and σ_E^2. We first focus on point estimates of the components.

To quote Searle [47], who has studied variance components problems in detail, "The basic principle for estimating variance components has been and to a large extent still is, that of equating quadratic functions of the observations to their expected values. Obvious candidates for such functions are those of the analysis of variance table."

Suppose $a = 6$ litters are chosen at random from a very large number of available litters. Consider the data given in Table 1.

Table 1   Weight Gains for 30 Pigs

Litter	1	2	3	4	5	6
	4.17	4.83	5.95	4.43	3.86	4.98
	5.21	5.70	6.50	4.85	4.12	5.58
Weight Gains	4.60	5.91	6.34	3.92	4.70	6.10
	4.35	4.92	6.09	4.25	3.90	5.75
	4.54	5.72	5.87	4.64	4.41	5.29

The one-way classification analysis of variance* (AOV) is shown in Table 2.

The estimation procedure is to set numerical values of the MSs equal to expected mean squares and solve the resulting equations

$$\hat{\sigma}_E^2 + 5\hat{\sigma}_A^2 = 2.936,$$

$$\hat{\sigma}_E^2 = 0.153.$$

We find $\hat{\sigma}_A^2 = 0.557$ and thus conclude that the component due to litters is quite large relative to the within-litters component.

Unbiasedness* is among the desirable properties that AOV point estimators of variance components enjoy. Among their undesirable properties is the fact that estimates can be negative and indeed often are. Furthermore, no closed form exists for the distribution of estimators such as $\hat{\sigma}_A^2 = (\text{MSA} - \text{MSE})/b$. The support for the distribution is all real numbers. AOV estimators are linear combinations of chi-square* related variables.

If the inference mode is testing instead of point estimation, the analysis of variance along with normality assumptions give mean squares, which are distributed independently as multiples of central chi-square variables. In particular, for the balanced one-way*

situation, which we have thus far focused upon,

$$\frac{\text{SSA}}{\sigma_E^2 + b\sigma_A^2} \sim \chi^2(a-1) \text{ independent of}$$

$$\frac{\text{SSE}}{\sigma_E^2} \sim \chi^2[a(b-1)].$$

Thus under the hypothesis $H_0 : \sigma_A^2 = 0$, the ratio $\text{MSA}/\text{MSE} \sim F\{(a-1), a(b-1)\}$ and serves as a test statistic for testing H_0 vs. $\sigma_A^2 > 0$. For the pig weight-gain data, the F-ratio is $F = 2.936/0.153 = 19.25$.

If the inference mode is that of a confidence interval*, then the procedure varies, for a multitude of suggestions have been presented. The reason for this is that for most components only approximate interval statements can be derived. An exception is the exact interval statement

$$P\left[\frac{\text{SSE}}{\chi_{1-\alpha/2}^2} < \sigma_E^2 < \frac{\text{SSE}}{\chi_{\alpha/2}^2}\right] = 1 - \alpha.$$

This statement is an immediate consequence of the fact that under normality, SSE/σ_E^2 is distributed as a central chi-square* variable.

There exist exact intervals for certain functions of variance components, which are important in studies conducted by geneticists and animal breeders. An exact interval can be obtained for $\sigma_E^2/(\sigma_A^2 + \sigma_E^2)$ from the fact that

$$\frac{\sigma_E^2}{\sigma_A^2 + \sigma_E^2} \frac{\text{MSA}}{\text{MSE}} \sim F\{a-1, a(b-1)\}.$$

Manipulation of the statement for $\sigma_E^2/(\sigma_A^2 + \sigma_E^2)$ leads directly to probability interval statements for σ_A^2/σ_E^2 and $\sigma_A^2/(\sigma_A^2 + \sigma_E^2)$, the

Table 2   AOV for Weight-Gain Data

Source	Degrees of Freedom df	Sum of Squares SS	Mean Square MS	Expected Value of MS EMS
Total corrected	$ab - 1 = 29$	SST = 18.34		
Among litters	$a - 1 = 5$	SSA = 14.68	MSA = 2.936	$\sigma_E^2 + 5\sigma_A^2$
Within litters	$a(b - 1) = 24$	SSE = 3.66	MSE = 0.153	σ_E^2

Table 3 A Two-Fold Nested Data AOV

Source	df	MS	EMS
Among A's (houses)	$a-1$	$\text{MSA} = \dfrac{\text{SSA}}{a-1}$	$\sigma_E^2 + c\sigma_B^2 + bc\sigma_A^2$
B's in A's (benches in houses)	$a(b-1)$	$\dfrac{\text{SSB} - \text{SSA}}{a(b-1)}$	$\sigma_E^2 + c\sigma_B^2$
C's in B's (plants on benches)	$ab(c-1)$	$\dfrac{\text{SSC} - \text{SSB}}{ab(c-1)}$	σ_E^2

last function being known as the *intraclass correlation.*

The concepts presented for the balanced one-way classification situation can be extended in many ways. Consider three variance components that arise in conjunction with a random-effects model for balanced two-way factorial data. Let factor A be studied at randomly chosen levels and factor B at b levels. Denote the observation for the ith level of A and the jth level of B by Y_{ij} and suppose $Y_{ij} = \mu + A_i + B_j + E_{ij}$. The parameters of interest are the variances σ_A^2, σ_B^2, and σ_E^2 for the respective populations from which A_i, B_j, and E_{ij} were selected. Provided we have a full $a \times b$ factorial experiment* (no missing observations), the AOV procedure is to let

$$\text{MSA} = \hat{\sigma}_E^2 + b\hat{\sigma}_A^2, \quad \text{MSB} = \hat{\sigma}_E^2 + a\hat{\sigma}_B^2$$

and $\text{MSE} = \hat{\sigma}_E^2$, from which we obtain

$$\hat{\sigma}_A^2 = \frac{1}{b}[\text{MSA} - \text{MSE}],$$

$$\hat{\sigma}_B^2 = \frac{1}{a}[\text{MSB} - \text{MSE}],$$

and $\hat{\sigma}_E^2 = \text{MSE}$.

Under normality assumptions, the F-ratios MSA/MSE and MSB/MSE serve as test statistics for the hypotheses $\sigma_A^2 = 0$ and $\sigma_B^2 = 0$.

For two- or more-fold balanced nested data, the AOV procedure is similar and will be illustrated by considering data from a greenhouse experiment involving a houses, with b benches in each house, and c plants

on the jth bench in the ith greenhouse. The random-effects nested model for such an experimental situation can be written $Y_{ijk} = \mu + A_i + B_{ij} + E_{ijk}$ with variance components σ_A^2, σ_B^2, and σ_E^2. Table 3 is an appropriate AOV table.

Equating MSs to EMSs and solving, we obtain point estimates

$$\hat{\sigma}_A^2 = \frac{1}{bc}[\text{MSA} - \text{MSBs in } A\text{'s}],$$

$$\hat{\sigma}_B^2 = \frac{1}{c}[\text{MSBs in } A\text{'s} - \text{MSCs in } B\text{'s}],$$

$$\hat{\sigma}_C^2 = \text{MSCs in } B\text{'s}.$$

Table 3 provides MSs for F-ratio test statistics to do hypothesis testing relative to the variance components, under normal assumptions, according to the following scheme.

For testing	Use	Distributed under H_0 as
$\sigma_B^2 = 0$	$\dfrac{\text{MSBs in } A\text{'s}}{\text{MSCs in } B\text{'s}}$	$F\{a(b-1), ab(c-1)\}$
$\sigma_A^2 = 0$	$\dfrac{\text{MSA}}{\text{MSBs in } A\text{'s}}$	$F\{a-1, a(b-1)\}$
$\sigma_B^2 = \sigma_A^2 = 0$	$\dfrac{\text{MSA}}{\text{MSCs in } B\text{'s}}$	$F\{a-1, ab(c-1)\}$

Extensions of the basic ideas are not always straightforward. There are complicating factors when:

(1) the data are unbalanced,
(2) the model is a mixed effects model, and when
(3) interactions* are present.

Table 4   An AOV for Two-Way Unbalanced Data

Source	df	Reductions
Levels of B	$b - 1$	$R(\beta|\mu)$
A adjusted for B	$a - 1$	$R(A|\mu, \beta)$
Interaction adjusted	$s - a - b + 1$	$R(A \times B|\mu, A, \beta)$
Error	$N - s$	SSE

One complicating factor is that of determining expected mean squares. Another is choosing among the often many possible quadratic forms* to equate with their expectations. Henderson [26] presented three methods, which have been widely used. Method I uses quadratic forms (not necessarily sums of squares) analogous to those already discussed for the balanced data cases. Method I unfortunately gives biased estimates of variance components in mixed-model situations. Although this is well-known, the ease of the method has made it popular and for mixed-model data two approaches are commonly employed. The first is to ignore the fixed effects and the second is to treat the fixed effects as random effects. Both approaches have been investigated and deemed unsatisfactory.

Henderson's method II was designed to overcome the bias problem inherent in method I when applied to mixed models. The idea is to correct the data for the mixed effects and then use method I on the corrected data. The correction process produces correlated error terms, which then lead to a biased estimate of σ_E^2, but this bias can also be corrected.

Henderson's method of correcting the data for fixed effects is but one of several ways that can and have been applied. This lack of uniqueness detracts from the usefulness of the method. There are other difficulties. If there are interactions between any of the fixed effects with a random effect, then one cannot transform to a model where method I can be directly applied. This is a serious limitation.

Henderson's method III (the fitting constants method) is often employable when methods I and II would be unsatisfactory.

The method can be used for any situation. We will illustrate its use in an unbalanced two-way cross-classification situation with interaction. Suppose there are N observations. Let the model be $Y_{ijk} = \mu + \beta_j + A_i + (A\beta)_{ij} + E_{ijk}$. Factor B is fixed but factor A is random as are interaction and error. Suppose s cells contain observations. We desire to estimate the variance components σ_A^2, σ_{AB}^2, and σ_E^2. Seeing there are three components, the method consists of setting three reductions in sums of squares equal to their expectations. The method is not unique. The problem remains unsolved as to which quadratic forms should be used. One choice is to write an AOV for "A after B" (see Table 4).

$\hat{\sigma}_E^2$ is always estimated from SSE/$(n - s)$ but other choices are

$$R(A \times B|\mu, A, \beta) = C_1\hat{\sigma}_{A\beta}^2 + (s - a - b - 1)\hat{\sigma}_E^2$$

and

$$R(A, A \times B|\mu, \beta) = C_2\hat{\sigma}_{A\beta}^2 + C_3\hat{\sigma}_A^2 + (s - b)\hat{\sigma}_E^2.$$

Until 1967, Henderson's methods were the only methods used with unbalanced data. Since that year many alternative approaches have been presented. Among these are:

ML: Maximum likelihood*.

REML: Restricted maximum likelihood*, which is an adaptation of ML where the idea is to maximize the likelihood of that part of the sufficient statistics, which is location invariant.

MINQUE*:   Minimum norm quadratic unbiased estimation.

MINQUE0:   A variation of MINQUE.

I-MINQUE: Iterative MINQUE, which is essentially the same as REML.

MIVQUE: Minimum variance quadratic unbiased estimation.

Several comparative studies have been done but the intractability of the problem, even in the normal assumption case, has limited the findings. There seem to be few optimal properties in the unbalanced cases. In balanced cases variance-component estimators obtained by the analysis-of-variance method are minimum variance quadratic unbiased. Furthermore, when normality is assumed the estimators are minimum variance unbiased*.

Two noteworthy survey papers should be mentioned. Crump [11] presented the status of the subject as of 1951 and 20 years later Searle [46], in an invited paper, summarized the situation to 1971.

References

[1] Airy, G. B. (1861). *On the Algebraical and Numerical Theory of Errors of Observations and the Combination of Observations.* MacMillan, London, England.

[2] Crump, S. L. (1951). *Biometrics*, **7**, 1–16.

[3] Henderson, C. R. (1953). *Biometrics*, **9**, 226–252.

[4] Searle, S. R. (1971). *Biometrics*, **27**, 1–76.

[5] Searle, S. R. (1977). Variance Components Estimation: A Thumbnail Review. *Biometrics Unit Mimeo Series BU-612-M*, Cornell University, Ithaca, NY.

Bibliography

Anderson, R. L. (1965). *Technometrics*, **7**, 75–76.

Anderson, R. L. and Bancroft, T. A. (1952). *Statistical Theory in Research.* McGraw-Hill, New York.

Blischke, W. R. (1966). *Biometrics*, **22**, 553–565.

Blischke, W. R. (1968). *Biometrics*, **24**, 527–540.

Broemeling, L. D. (1969a). *J. Amer. Statist. Ass.*, **64**, 660–664.

Broemeling, L. D. (1969b). *Biometrics*, **25**, 424–427.

Bulmer, M. G. (1957). *Biometrika*, **44**, 159–167.

Bush, N. and Anderson, R. L. (1963). *Technometrics*, **5**, 421–440.

Crump, S. L. (1946). *Biometrics Bull.*, **2**, 7–11.

Cunningham, E. P. (1969). *Biometrika*, **56**, 683–684.

Cunningham, E. P. and Henderson, C. R. (1968). *Biometrics*, **24**, 13–25. Correction, **25**, 777–778.

Eisenhart, C. (1947). *Biometrics*, **3**, 1–21.

Gaylor, D. W. and Hartwell, T. D. (1969). *Biometrics*, **25**, 427–430.

Graybill, F. A. (1954). *Ann. Math. Statist.*, **25**, 367–372.

Graybill, F. A. (1961). *An Introduction to Linear Statistical Models*, Vol. I. McGraw-Hill, New York.

Graybill, F. A. (1976). *Theory and Application of the Linear Model*, Duxbury, Belmont, CA.

Graybill, F. A. (1979). *J. Amer. Statist. Ass.*, **74**, 368–374.

Graybill, F. A. and Hultquist, R. A. (1961). *Ann. Math. Statist.*, **32**, 261–269.

Graybill, F. A. and Wortham, A. W. (1956). *J. Amer. Statist. Ass.*, **51**, 266–268.

Hartley, H. O. (1967). *Biometrics*, **23**, 105–114. Correction, **23**, 853.

Hartley, H. O. and Rao, J. N. K. (1967). *Biometrika*, **54**, 93–108.

Hartley, H. O. and Searle, S. R. (1969). *Biometrics*, **25**, 573–576.

Harville, D. A. (1977). *J. Amer. Statist. Ass.*, **72**, 320–338.

Herbach, L. H. (1959). *Ann. Math. Statist.*, **30**, 939–959.

Kapadia, C. H. and Weeks, D. L. (1963). *Biometrika*, **50**, 327–336.

Kaplan, J. S. (1983). *J. Amer. Statist. Ass.*, **78**, 476–477.

Klotz, J. H., Milton, R. C., and Zacks, S. (1969). *J. Amer. Statist. Ass.*, **64**, 1383–1402.

LaMotte, L. R. (1973). *Biometrics*, **29**, 311–330.

Low, L. Y. (1964). *Biometrika*, **51**, 491–494.

Low, L. Y. (1969). *J. Amer. Statist. Ass.*, **64**, 1014–1030.

Oktaba, W. (1968). *Biom. Zeit.*, **10**, 97–108.

Patterson, H. D. and Thompson, R. (1971). *Biometrika*, **58**, 545–554.

Pukelsheim, F. (1976). *J. Multivariate Anal.*, **6**, 626–629.

Rao, C. R. (1971a). *J. Multivariate Anal.*, **1**, 257–275.

Rao, C. R. (1971b). *J. Multivariate Anal.*, **1**, 445–456.

Rao, C. R. (1973). *Linear Statistical Inference and Its Applications*, 2nd ed. Wiley, New York.

Rao, J. N. K. (1968). *Biometrics*, **24**, 963–978.

Scheffé, H. (1956). *Ann. Math. Statist.*, **27**, 23–36.

Scheffé, H. (1959). *The Analysis of Variance.* Wiley, New York, Chaps. 7 and 8.

Searle, S. R. (1961). *Ann. Math. Statist.*, **32**, 1161–1166.

Searle, S. R. (1968). *Biometrics*, **24**, 749–788.

Searle, S. R. (1971). *Linear Models.* Wiley, New York.

Seeger, P. (1970). *Technometrics*, **12**, 207–218.

Smith, D. W. and Hocking, R. R. (1978). *Commun. Statist. A*, **7**, 1253–1266.

Smith, D. W. and Murray, L. W. (1984). *J. Amer. Statist. Ass.*, **79**, 145–151.

Tukey, J. W. (1956). *Ann. Math. Statist.*, **27**, 722–736.

Tukey, J. W. (1957a). *Ann. Math. Statist.*, **28**, 43–56.

Tukey, J. W. (1957b). *Ann. Math. Statist.*, **28**, 378–386.

Wang, Y. Y. (1967). *Biometrika*, **54**, 301–305.

Weeks, D. L. and Graybill, F. A. (1961). *Sankhyā A*, **23**, 261–268.

Weeks, D. L. and Graybill, F. A. (1962). *Sankhyā A*, **24**, 339–354.

Williams, J. S. (1962). *Biometrika*, **49**, 278–281.

Winsor, C. P. and Clarke, G. L. (1940). *Sears Foundation J. Marine Res.*, **3**, 1.

(ANALYSIS OF VARIANCE
FACTORIAL EXPERIMENTS
FIXED-, RANDOM-, AND MIXED-EFFECTS
  MODELS
MEASUREMENT ERROR
MINQE
MODELS I, II, AND III
ONE-WAY ANALYSIS OF VARIANCE)

ROBERT HULTQUIST

VARIANCE DILATION

Let X and Y be two random variables with distribution functions F and G, respectively. We say that G is a *dilation* of F (denoted by $F \prec G$ or $X \prec Y$) if there exist three random variables X', Y', and Z' defined on some common probability space such that $X =^d X'$, $Y =^d Y'$, and $X' = E[Y'|Z']$ almost surely (*see* GEOMETRY IN STATISTICS: CONVEXITY for more details). As will be shown, if $X \prec Y$, then, roughly speaking, Y is "more variable" than X.

It can be shown (see, e.g., the references in Shaked [4]) that $F \prec G$ if and only if

$$\int_{-\infty}^{x} [G(y) - F(y)] \, dy \geq 0, \qquad x > -\infty,$$

and

$$\int_{-\infty}^{\infty} [G(y) - F(y)] \, dy = 0, \qquad (1)$$

provided the integrals exist. Also $F \prec G$ if and only if

$$E[\psi(X)] \leq E[\psi(Y)], \qquad (2)$$

for all convex functions ψ for which the expectations exist. Condition (1) is useful for identifying random variables X and Y, which are ordered by dilation. Condition (2) is

useful for obtaining inequalities from the dilation relation.

From (2) it follows that if $X \prec Y$, then $E[X] = E[Y]$. Taking $\psi(X) = X^2$, it is seen from (2) that if $X \prec Y$, then $E[X^2] \leq E[Y^2]$. Since $E[X] = E[Y]$ it follows that if $X \prec Y$, then

$$\text{Var}(X) \leq \text{Var}(Y), \qquad (3)$$

provided the variances exist.

Suppose X and Y have (discrete or continuous) densities f and g, respectively. Denote by $S^-(f - g)$ the number of sign changes of the difference $f(x) - g(x)$ as x varies over $(-\infty, \infty)$ (in the continuous case) or over $\{\ldots, -1, 0, 1, \ldots\}$ (in the discrete case), where the x's such that $f(x) = g(x)$ are discarded. If $E[X] = E[Y]$ and if

$$S^-(f - g) = 2$$

and the sign sequence is $-, +, -,$
$$\qquad (4)$$

then $X \prec Y$. Condition (4) corresponds to the fact that g has heavier tails than f. It follows that (4) is a sufficient condition for the inequality $\text{Var}(X) \leq \text{Var}(Y)$.

Consider a one-parameter exponential family* of distributions with (continuous or discrete) densities, having parameter ρ, of the form

$$\tilde{f}(x : \rho) \propto \exp\{\psi(\rho)x + \chi(\rho)\}, \qquad (5)$$

where ψ and χ are some fixed functions. Let g be some mixture of the densities in (5), that is, let

$$g(x) \propto \int_{-\infty}^{\infty} \exp\{\psi(\rho)x + \chi(\rho)\} \, dH(\rho),$$

where H is some distribution on $(-\infty, \infty)$. Denote $\bar{\rho} \equiv \int_{-\infty}^{\infty} \rho \, dH(\rho)$. Let f be of the form (5) with parameter $\bar{\rho}$, that is,

$$f(x) \equiv \tilde{f}(x; \bar{\rho}).$$

Then (Shaked [4]) $E[X] = E[Y]$ and $S^-(f - g) = 2$ and the sign sequence is $-, +, -$. Thus g has heavier tails than f and as a consequence it is seen that mixing dilates the variance (see, e.g., Shaked [4] and Titterington et al. [7, p. 51]).

This observation was extended to some two-parameter exponential families in

Shaked [5]. It was put in a general framework of stochastic parametric convexity in Schweder [3] and in Shaked and Shanthikumar [6].

Another variance dilation result has been obtained by Hoeffding [1]. Let $p_1, \ldots, p_n$ be parameters of the independent Bernoulli random variables $X_1, \ldots, X_n$, respectively. That is,

$$\Pr\{ X_i = 1 \} = p_i,$$
$$\Pr\{ X_i = 0 \} = 1 - p_i, \quad i = 1, \ldots, n.$$

Denote $\bar{p} = (1/n) \sum_{i=1}^{n} p_i$ and let $Y_1, \ldots, Y_n$ be independent Bernoulli random variables with parameter $\bar{p}$, that is, $\Pr\{ Y_i = 1 \} = \bar{p}$, $\Pr\{ Y_i = 0 \} = 1 - \bar{p}$, $i = 1, 2, \ldots, n$. Then

$$\operatorname{Var}\left(\sum_{i=1}^{n} X_i \right) \leqslant \operatorname{Var}\left(\sum_{i=1}^{n} Y_i \right).$$

Extensions of this result have been obtained by several authors. See Marshall and Olkin [2, pp. 359–360] for details and references.

References

[1] Hoeffding, W. (1956). *Ann. Math. Statist.*, **27**, 713–721.

[2] Marshall, A. W. and Olkin, I. (1979). *Inequalities: Theory of Majorization and Its Applications*. Academic, New York.

[3] Schweder, T. (1982). *Scand. J. Statist.*, **9**, 165–169.

[4] Shaked, M. (1980). *J. R. Statist. Soc. B*, **42**, 192–198.

[5] Shaked, M. (1985). *Sankhyā A*, **47**, 117–127.

[6] Shaked, M. and Shanthikumar, J. G. (1988). *Adv. Appl. Prob.*, (to appear).

[7] Titterington, D. M., Smith, A. F. M., and Makov, V. E. (1985). *Statistical Analysis of Finite Mixture Distributions*. Wiley, New York.

(ORDERING DISTRIBUTIONS BY
  DISPERSION
STOCHASTIC ORDERING)

MOSHE SHAKED

VARIANCE FUNCTION

A mathematical function

$$V(Y|\mathbf{x})$$

giving the variance of a random variable Y as a function of the values x of a vector variable. The function is used in response surface* analysis and regression* analysis. It is sometimes termed the *array variance*.

VARIANCE INFLATION FACTOR

This is a term used in multiple linear regression* analysis. It is the ratio of the variance of the least-squares estimator of the partial regression coefficient* $(\beta_{0i.1,2,\ldots,i-1,i+1,\ldots,p})$ of X_0 on X_i—where X_i is one of p predictor ("independent") variables $X_1, \ldots, X_p$—to the variance of the least-squares estimator of β_{0i}, the regression coefficient of X_0 on X_i alone, based on a random sample from the general linear model*

$$\mathbf{X}_0 = \mathbf{X}\ \boldsymbol{\beta} + \mathbf{Z} \qquad p < n,$$
$$\,_{n \times 1} \quad _{n \times p}\ _{p \times 1} \quad _{n \times 1}$$

where $Z_1, \ldots, Z_n$ are mutually uncorrelated, and with expected value 0 and variance σ^2.

The variance inflation factor (VIF) depends on the configuration of the sample values of the predictor variables. It is, in fact, equal to $(1 - R_{i.1,2,\ldots,i-1,i+1,\ldots,p}^2)^{-1}$, where R^2 is the multiple correlation* coefficient of X_i on the remaining $(p - 1)$ predictor variables. (Some authors call R^2 the *square* of the multiple correlation coefficient or the multiple *determination* coefficient.) Clearly, large values of R^2 are undesirable, because they lead to large variances for the estimators of partial regression* coefficients.

The VIFs can be used as indicators of multicollinearity*, as suggested by Chatterjee and Price [2]. Further discussion of this aspect of their use is given in Belsley et al. [1] and Neter et al. [3].

References

[1] Belsley, D. A., Kuh, E., and Welsch, R. E. (1980). *Regression Diagnostics: Identifying Influential Data and Sources of Collinearity*. Wiley, New York.

[2] Chatterjee, S. and Price, B. (1977). *Regression Analysis by Example*. Wiley, New York.

[3] Neter, J., Wasserman, W., and Kutner, M. H. (1983). *Applied Linear Regression Models*. Irwin, Homewood, IL.

(COEFFICIENT OF DETERMINATION
MULTICOLLINEARITY
MULTIPLE CORRELATION
MULTIPLE LINEAR REGRESSION
PARTIAL REGRESSION
RIDGE REGRESSION)

VARIANCE, INNOVATION

If $\{X_t\}$ is a discrete-time, real-valued, and purely nondeterministic stationary process*, then it may be represented as

$$X_t = \sum_{j=0}^{\infty} \Phi_j a_{t-j}, \quad \text{with } \Phi_0 = 1,$$

where $\{a_r\}$ is a sequence of uncorrelated variables each with mean 0 and variance σ^2. This variance—*the innovation variance*—is related to the spectral density $f(\lambda)$ of $\{X_t\}$ by the relation

$$\sigma^2 = \exp\left[\frac{1}{2\pi} \int_{-\pi}^{\pi} \log\{2\pi f(\lambda)\} \, d\lambda\right].$$

In practical applications, σ^2 is usually unknown. Davis and Jones [2] introduced an estimator of σ^2 defined by

$$\hat{\sigma}^2 = \exp\left[\frac{1}{n} \sum_{j=1}^{n} \log\{2\pi I(\lambda_j)\} + \gamma\right],$$

where $\gamma \approx 0.57722$ is the Euler constant*, based on the periodogram* ordinates

$$I(\lambda_j) = \frac{1}{2\pi N} \left|\sum_{t=1}^{n} X_t \exp(-i\lambda_j t)\right|^2,$$

$$\lambda_j = \frac{2\pi j}{N}, \quad j = 1, \ldots, n = [(N-1)/2],$$

where $[x]$ is the greatest integer not exceeding x, and $X_1, \ldots, X_N$ are the observations generated by $\{X_t\}$ and $i = \sqrt{-1}$.

This estimator has been studied by Bhansali [1], Janacek [4], and Hannan and Nicholls [3] among others. Simulation studies by Hannan and Nicholls [3] indicate that the bias of $\hat{\sigma}^2$ may be large in finite samples.

Pukkila and Nyquist [5] proposed an estimator of σ^2 based on a "tapered" time series. The time series is multiplied by a weight function v_t, $t = 1, \ldots, N$, the so-called *taper*, to obtain a new time series $Y_t = v_t X_t$, $t = 1, \ldots, N$. The tapered time series is used to construct a modified periodogram with ordinates

$$I^T(\lambda_j)$$

$$= \frac{1}{2\pi N_1} \left|\sum_{t=1}^{N} v_t(X_t - \bar{y}) \exp(-i\lambda_j t)\right|^2,$$

where

$$\bar{y} = \sum_{t=1}^{N} \frac{Y_t}{N_2} = \sum_{t=1}^{N} \frac{v_t X_t}{N_2},$$

$$\text{with} \quad N_1 = \sum_{t=1}^{N} v_t^2, \quad N_2 = \sum_{t=1}^{N} v_t,$$

and the estimator of Pukkila and Nyquist is given by

$$\hat{\sigma}_T^2 = \exp\left[\frac{1}{N} \sum_{j=1}^{N} \log\{2\pi I^T(\lambda_j)\} + \gamma\right],$$

where

$$v_t = \begin{cases} (t/N)/c, & 0 < t/N \leq c, \\ 1, & c \leq t/N \leq 1-c, \\ \{1 - (t-1)/N\}, & 1-c \leq t/N \leq 1, \\ 0, & \text{otherwise,} \end{cases}$$

with $c = 6.25/N$ ($N \geq 50$). This estimator seems to have a large bias-reducing, as well as variance-reducing effect—especially for the case of estimating the innovation variance of a stationary univariate time series in the frequency domain.

References

[1] Bhansali, R. J. (1974). *J. R. Statist. Soc. B*, **36**, 61–73.

[2] Davis, H. T. and Jones, R. H. (1968). *J. Amer. Statist. Ass.*, **63**, 141–149.

[3] Hannan, E. J. and Nicholls, D. F. (1977). *J. Amer. Statist. Ass.*, **72**, 834–840.

[4] Janacek, G. (1975). *Biometrika*, **62**, 175–180.

[5] Pukkila, T. and Nyquist, H. (1985). *Biometrika*, **72**, 317–323.

VARIANCE, INTERVIEWER

Interviewer variance is the portion of error in a survey* or census* statistic that arises from different results being obtained by different interviewers or enumerators. It is an error component that is thus limited to telephone or face-to-face data collections or others that utilize interviewers in the data-collection process. It is generally included within a model of the survey that assumes that interviewer selection, training, supervision, and evaluation procedures are fixed properties of a data-collection* design, but that the specific persons chosen as interviewers might vary over replications (trials) of the data collection. Interviewer variance is the variability in the values of statistics over such replications of the data collection, due to sample persons being interviewed by different people. Omitted from interviewer variance are errors shared by all potential interviewers and thus not producing variation over replications that use a different interviewing corps.

Most work on interviewer variance focuses on variability in answers obtained from respondents (sometimes called response error), but most empirical estimates also include variability in nonresponse* and noncoverage errors. Hence interviewer variance, when estimated empirically, might be more appropriately thought of as component of nonsampling error variance associated with interviewers.

An influential work in interviewer variance is the Hansen et al. [6] model, which decomposes the variance of a sample mean $V(\bar{y}_t)$ into three components

$$V(\bar{y}_t) = E\left[(\bar{y} - \bar{Y})^2 + (\bar{y}_t - \bar{y})^2 \right.$$
$$\left. + 2(\bar{y}_t - \bar{y})(\bar{y} - \bar{Y})\right],$$

where $\bar{y}_t$ is the mean of the sample on the tth conceptual trial of the data collection, $\bar{y}$ is the expected value of the sample mean

over trials, and, $\bar{Y}$ is the expected value of the sample mean over all possible samples using the same sample design.

The previous expression has three components. The first term is the sampling variance of the mean, reflecting the fact that different samples of persons will produce different mean values. The second is a measurement variance or response variance component. The third term is a covariance between response and sampling deviations. The measurement variance component reflects variation over trials, including variation due to the use of different interviewers for the data collection. The difference between the expected value of a respondent's answers over trials and an answer on a particular trial is termed a "response deviation" and defined as $d_{ij} = y_{it} - y_t$, for the ith respondent, where y_{it} is the answer by that respondent on the tth trial. If one lets σ_d^2 signify the variance over trials of response deviations, then

$$E\left[(\bar{y}_t - \bar{y})^2\right] = \frac{1}{n}\left[\sigma_d^2 + (n-1)\rho\sigma_d^2\right],$$

where n is the number of respondents subject to the design feature producing the correlation and the covariance across respondents in their response deviations on the same trial is $\sigma_d^2\rho$. In most formulations of the model focusing on interviewers, the covariance is posited to arise within interviewer assignments (and n is the interviewer's workload). That is, respondents of the same interviewer are seen to have correlated response deviations because of shared influences on their answer from the interviewer's behavior. Following these observations, the term $\rho\sigma_d^2$ is sometimes called the "correlated response variance" term and σ_d^2, the "simple response variance." In many empirical studies of interviewer variance, no other source of correlation among response deviations of different sample persons is acknowledged; in most, those sources are assumed to be of negligible import.

Estimation of variance components* associated with interviewers can be accomplished through interpenetration, replication, or a combination of the two methods. Interpenetrated designs* (see Mahalanobis [11] assign to each interviewer a random subset of all cases in the sample. Since in expectation and in the absence of interviewer effects, all interviewer assignments would yield the same values on survey statistics, deviations from those expectations are used to measure interviewer variance. Reinterview designs are used in the same way, assigning respondents at random to two different interviewers for two interviews. With interpenetration alone, the correlation of response deviations due to interviewers cannot be estimated in an unbiased fashion. However, if one assumes no covariance between sampling and response deviations, an estimator for $\sigma_d^2 \rho$ can be constructed. Fellegi [4] demonstrated how the combination of interpenetration and reinterviewing would permit estimation of the simple response variance, the correlation due to interviewers, and the covariance between sampling and response deviations, with small bias terms.

There are many models and estimators for interviewer variance components. Unfortunately, there is no standardized terminology or notation in the field, and readers must carefully examine definitions of model terms and the nature of expectations to perceive differences among them. For example, Kish [8] uses a slightly different approach than Hansen et al. [6], with a linear response error model and one-way analysis of variance* to estimate an intraclass correlation* associated with interviewers. The intraclass correlation estimated from this approach is not the same as the ρ above. O'Muircheartaigh [12] and Lessler [10] comment on these and other differences among the models. Many of the differences concern treatment of covariances between various sampling and nonsampling errors* and assumptions about the sources of variability in the measurement process.

Despite the large number of data-collection efforts using interviewers, there are relatively few studies of interviewer variance. The majority of the studies are imbedded in censuses (e.g., Hanson and Marks [7], Fellegi [4], and Krotki and Hill [9]). There are a few

personal-interview sample surveys with such estimates (e.g., Bailey et al. [2]), but these are rare because of the increased travel charges associated with interpenetrated designs using this mode. With the growing use of centralized telephone surveys* in the United States, there is increased interest in estimation of interviewer variance because of reduced marginal costs of interpenetrated designs (e.g., Stokes [13] and Groves and Magilavy [5]).

Most of the developments of estimators of interviewer variance components are based on assumptions of interpenetration of simple random samples, with equal workloads assigned to each interviewer, and no response. The complications of estimating correlated response variance in the presence of a complex sample design is discussed by Bailey et al. [2] and Biemer and Stokes [3]. The complications of variable workload sizes is addressed by Kish [8]. Stokes [13] discusses the estimation of interviewer effects when interpenetration is restricted by telephone interviewers working different shifts. There are few explicit treatments of the variance of the interviewer variance estimates, but they are discussed in Bailey et al. [2] and Groves and Magilavy [5]. Anderson and Aitken [1] discuss the complication of estimating interviewer effects on responses to dichotomous variables.

References

[1] Anderson, D. A. and Aitkin, M. (1985). *J. R. Statist. Soc. B*, **47**, 203–210.

[2] Bailey, L., Moore, T. F., and Bailar, B. (1978). *J. Amer. Statist. Ass.*, **70**, 23–30.

[3] Biemer, P. P. and Stokes, S. L. (1985). *J. Amer. Statist. Ass.*, **80**, 158–166.

[4] Fellegi, I. P. (1964). *J. Amer. Statist. Ass.*, **59**, 1016–1041.

[5] Groves, R. M. and Magilavy, L. J. (1986). *Public Opinion Quart.*, Summer, 1986.

[6] Hansen, M. H., Hurwitz, W. N., and Bershad, M. A. (1961). *Bull. Int. Statist. Inst.*, **38**, 359–374.

[7] Hanson, R. H. and Marks, E. S. (1958). *J. Amer. Statist. Ass.*, **53**, 635–655.

[8] Kish, L. (1962). *J. Amer. Statist. Ass.*, **57**, 92–115.

[9] Krotki, K. P. and Hill, C. J. (1978). *Survey Methodology*, **4**(2), 87–99.

[10] Lessler, J. T. (1985). In *Surveys of Subjective Phenomena*, C. Turner and E. Martin, eds. pp. 405–440.

[11] Mahalanobis, P. C. (1946). *J. R. Statist. Soc.*, **109**, 327–378.

[12] O'Muircheartaigh, C. A. (1977). In *The Analysis of Survey Data*, Vol. 2, C. A. O'Muircheartaigh and C. Payne, eds. Wiley, New York, pp. 193–239.

[13] Stokes, S. L. (1986). *Proceedings of the Second Annual Research Conference*. U.S. Bureau of the Census, Washington, D.C.

(VARIANCE COMPONENTS
RANDOMIZED RESPONSE
SURVEY SAMPLING
TELEPHONE SURVEYS, COMPUTER
ASSISTED)

ROBERT M. GROVES

VARIANCE-RATIO

The immediate meaning of this term is the ratio of one variance to another—possibly, by extension, the ratios of a set of variances to each other.

Usually, the term refers to the ratio of two *estimates* of population variance based on the sample variances. Thus if $(X_{11}, \ldots, X_{1n_1}), (X_{21}, \ldots, X_{2n_2})$ denote values for random samples of size n_1, n_2 from two populations Π_1, Π_2, respectively, the variance-ratio is

$$\frac{(n_1 - 1)^{-1}\sum_{i=1}^{n_1}(X_{1i} - \overline{X}_1)^2}{(n_2 - 1)^{-1}\sum_{i=1}^{n_2}(X_{2i} - \overline{X}_2)^2},$$

$$\text{where } \overline{X}_j = n_j^{-1}\sum_{i=1}^{n_j} X_{ji}, \quad j = 1, 2.$$

If the population distributions are normal with variances σ_1^2 σ_2^2, then the above statistic is distributed as

$$\frac{\sigma_1^2}{\sigma_2^2} \; (F \text{ with } n_1 - 1,$$

$$n_2 - 1 \text{ degrees of freedom}).$$

For this reason, F is sometimes called the "variance-ratio distribution."

The term "variance-ratio" is also applied to mean square ratios calculated for analy-sis-of-variance* tables.

(ANALYSIS OF VARIANCE
F-DISTRIBUTION)

VARIANCE-RATIO TRANSFORMATION

This refers to the transformation

$$z = \tfrac{1}{2}\log_e F$$

applied to a variable having an F_{ν_1, ν_2} distribution*. It is also known as Fisher's *z*-transformation*, but is to be distinguished from Fisher's *z'*-transformation* of sample correlation coefficients from a bivariate normal population*.

(ANALYSIS OF VARIANCE
FISHER'S *z*-TRANSFORMATION)

VARIANCE, SAMPLE

**THE SAMPLE VARIANCE AS A
DESCRIPTIVE STATISTIC**

When a sample $X_1, X_2, \ldots, X_n$ of n values is given, from a distribution $F(x)$, two important descriptive statistics are the sample mean $\overline{X}$ and the sample variance S^2 defined by

$$\overline{X} = \sum_i X_i/n$$

and

$$S^2 = \sum_i (X_i - \overline{X})^2 / (n - 1).$$

Let D be defined as $\sum_i (X_i - \overline{X})^2$; then $S^2 = D/(n - 1)$.

The standard deviation of the sample is S, the square root of the sample variance.

The sample mean is a measure of location, that is, it describes roughly where the observations are centered if they are recorded on the X-axis; however, in the analysis of data, a measure of dispersion—how spread out the observations are—is needed also, and the sample standard deviation is such a measure. The standard deviation provides a measure of spread that has the same units as X, and so S can also be displayed on the line. When the parent population is normal, with mean μ and standard deviation σ, roughly 95% of the observations can be expected to lie between $c_L = \mu - 2\sigma$ and $c_U = \mu + 2\sigma$, so $2S$ is sometimes marked off on either side of $\bar{X}$ to give boundaries $\hat{c}_L = \bar{X} - 2S$ and $\hat{c}_U = \bar{X} + 2S$. Although these values are only estimates of c_L and c_U, they are used as a simple graphical device to suggest where most of the observations should lie if the population were normal.

Other measures of dispersion have been suggested (for example, the average absolute deviation from the mean), and it has sometimes been argued that S should be preferred, simply on the grounds that S^2 can be easily manipulated algebraically. This is certainly true, but the more pressing claim to preference must surely be the fact that the sample variance estimates the population variance σ^2 and this parameter, together with the population mean μ, are the most important parameters entering into the sampling behaviour of many sample statistics. For example, the central limit theorem* shows that the asymptotic behaviour of a sample mean depends, under regularity conditions, only on μ and σ of the parent population. As previously defined, S^2 is in fact an unbiased estimator of σ^2. In earlier days, n rather than $(n - 1)$ was sometimes preferred for the divisor of D, and there has been a recent tendency to return to this practice on the grounds that the sample variance is an average and division by n, although introducing bias, is more appealing to a nonmathematical statistician.

CONFIDENCE INTERVALS BASED ON A SAMPLE FROM A NORMAL DISTRIBUTION

It was mentioned already that the sample mean $\bar{X}$ has properties that depend on the mean and variance of the parent population.

When this population is normal, these properties (and those of S^2) lead to precise techniques to give confidence intervals for these parameters. They are based on the following important results concerning the statistics $\bar{X}$ and D (or equivalently, S, or S^2):

(1) $\bar{X}$ and D are independently distributed.

(2) $Z = \sqrt{n}(\bar{X} - \mu)/\sigma$ has a $N(0, 1)$ distribution, where $N(\mu, \sigma^2)$ refers to a normal distribution with mean μ and variance σ^2.

(3) $W = D/\sigma^2$ has a chi-square distribution* with $(n - 1)$ degrees of freedom (df).

The first result gives a characterization of the normal distribution: The sample mean $\bar{X}$ and variance S^2 are independent only if the sample is from a normal parent population. The second result provides a well known method for deriving a confidence interval for μ and σ is known. The first two results combined lead to the result that $\sqrt{n}(\bar{X} - \mu)/S$ has a Student-t distribution* with $(n - 1)$ df, and this leads to a confidence interval for μ using the sample standard deviation when σ is unknown.

Statistics such as D, namely $\sum_i (X_i - d)^2$, where d is an estimate of the mean of X, arise throughout statistical work; many years ago, E. J. G. Pitman coined the useful term *squariance* for such a sum of squares. A squariance, derived from normal variates and divided by an appropriate population variance, is often, as in (3), distributed as a chi-square variable; this occurs in the analysis of variance*, for example. Because of this, squariances are often used for testing and for deriving confidence intervals. Thus, (3) may be used to give the confidence interval for σ^2 at level $100(1 - \alpha)\%$:

$$D/c_U < \sigma^2 < D/c_L, \qquad \text{(A)}$$

where c_U and c_L are, respectively, the upper and lower percentage points of χ^2_{n-1}, each at level $\alpha/2$.

The ratio of two independent sample variances from normal populations is used as a test statistic for equality (or more generally a given ratio) of two population variances, or to provide a confidence interval for such a ratio using the F distribution. Independent squariances, each believed to come from normal populations with common variance σ^2, can be added and the sum divided by the sum of the degrees of freedom, to give a pooled estimate of σ^2; this procedure is used frequently in the analysis of variance, since independent squariances arise in the decomposition of the total squariance. The pooling procedure is based on the fact that the sum of independent χ^2 variates is also χ^2 distributed. Squariances are also used in Bartlett's* test for equality of k population variances.

THE SAMPLE VARIANCE FOR NONNORMAL POPULATIONS

Samples from a nonnormal population give squariances $D = \Sigma_i (X_i - \overline{X})^2$, which, when divided by population variances σ^2, do not have χ^2 distributions; the true distribution depends on the parent population and is in almost all instances very difficult to find. However, the moment of S^2/σ^2 can be calculated; Church [2] gave the first four moments, which depend on the first eight population moments because of the squares that arise in S^2. The moments can also be found using k statistics, which are described in Kendall and Stuart [4, Vol. 1, Chap. 12]. When the population moments are known, so that those of S^2/σ^2 can be found, they can be used to approximate the distribution of S^2/σ^2 by curve fitting methods. For example, using expansions of Edgeworth type but based on the χ^2 density (Roy and Tiku [7]) or using Pearson curves or a generalised χ^2 density (Solomon and Stephens [8]). Pearson curves give good results over a wide range of parent populations, although in extreme cases the distribution of sample variance could have more than one mode, and Pearson curves cannot have this form. These approximations to the distribution of S^2/σ^2 depend on knowing the higher moments of the parent population and are mostly useful to study the behaviour of S^2; it is not easy to adapt (A) to give a confidence interval for σ^2 when only sample data are available.

Suppose the parent population has skewness* $\beta_1 = \mu_3^2/\mu_2^3$ and kurtosis* $\beta_2 = \mu_4/\mu_2^2$, where μ_i is the ith population moment about the mean, and let $\gamma_2 = \beta_2 - 3$. The value of γ_2 measures the deviation of the kurtosis* from the normal value $\gamma_2 = 0$. Box [1] showed that to a good approximation, $D/\{\sigma^2(1 + \gamma_2/n)\}$ could be regarded as χ_{n-1}^2 distributed; thus the kurtosis, rather than the skewness, appears to have the most influence on the distribution of S^2 and on the error in confidence level for σ^2 that would arise if (A) were used. Nevertheless, a very extensive study (without benefit of modern computers) by Le Roux [5] and followup work by Pearson and Please [6] and by Juritz and Stephens [3] indicate that β_1 and possibly higher moments can also play a role. Juritz and Stephens [3] give graphs to show the true confidence level attained using (A), say $L\%$, plotted against the parent β_2; several curves appear, indicating the influence of β_1 also.

Box's result suggests that a crude approximation to a confidence interval for σ^2 might be the interval $T/c_U < \sigma^2 < T/c_L$, where c_U, c_L are points from χ_{n-1}^2 as before and where $T = D/(1 + g_2/n)$; here $g_2 = b_2 - 3$, where $b_2 = m_4/m_2^2$ and where m_i is the ith sample moment about the mean. Statistic g_2 is a biased estimator of γ_2 and is notoriously susceptible to sampling fluctuations. Also the preceding studies suggest that somehow $b_1 = m_3^2/m_2^3$ should be incorporated into the calculation of the interval.

Modern computing methods, notably Monte Carlo sampling* and curve-fitting* techniques, were extensively used in the abovementioned studies on S^2, and such methods are increasingly used to study data from nonnormal populations. The bootstrap* is another computer-intensive method used to estimate the sampling variance of a statistic, say t, derived from a single data set. The data set is successively resampled to give many values of t, and this sample of t values is used to find the sample variance of t in the usual way. Properties of the bootstrap are still being investigated in many contexts. Variances and squariances are fundamental in statistical work and knowledge of their sampling properties is important; these properties must certainly influence the sensitivity of many analyses of data in ways that are as yet not always clear.

References

[1] Box, G. E. P. (1953). *Biometrika*, **40**, 318–335.

[2] Church, A. E. R. (1925). *Biometrika*, **17**, 79–83.

[3] Juritz, J. M. and Stephens, M. A. (1981). Effect of Non-normality on Confidence Level for the Variance. *Tech. Rep.*, Dept. of Mathematics, Simon Fraser University, Burnaby, BC, Canada.

[4] Kendall, M. G. and Stuart, A. (1958). *The Advanced Theory of Statistics*. Griffin, London.

[5] Le Roux, J. N. (1931). *Biometrika*, **23**, 134–190.

[6] Pearson, E. S. and Please, N. W. (1975). *Biometrika*, **62**, 223–241.

[7] Roy, J. and Tiku, M. L. (1962). *Sankhyā*, **24**, 181–184.

[8] Solomon, H. and Stephens, M. A. (1983). *Canad. J. Statist.*, **11**, 149–154.

(BOOTSTRAPPING
RESAMPLING PROCEDURES)

H. SOLOMON
M. A. STEPHENS

VARIANCE-STABILIZING TRANSFORMATIONS *See* EQUALIZATION OF VARIANCES

VARIANCE, UPPER BOUNDS

As a measure of maximum dispersion from the mean, upper bounds on variance have applications in many areas of statistics, including variance estimation, robustness* studies, nonparametric theory, stochastic processes*, and census* sampling.

BASIC INEQUALITIES

We begin by listing some basic variance upper bounds. The inequalities will be presented roughly in order of increasing knowledge about the underlying distribution. In what follows, continuity assumptions, used here for the sake of simplicity, may be relaxed somewhat. See the references for details.

1. If X is any random variable (RV) on $[a, b]$, then

$$\text{Var}(X) \leqslant (b - a)^2/4. \tag{1}$$

The bound will be achieved if and only if X has a Bernoulli distribution*, placing one-half of the probability mass at each of a and b.

2. Jacobson [16] has shown that if X has a continuous unimodal* probability density on $[a, b]$, with at least one mode in (a, b), then

$$\text{Var}(X) \leqslant (b - a)^2/9. \tag{2}$$

3. Let X be a unimodal random variable with support $[a, b]$, mean μ, and piecewise continuous density f. Several authors have derived sufficient conditions for the bound

$$\text{Var}(X) \leqslant (b - a)^2/12 \tag{3}$$

to hold. Gray and Odell [14] have shown that if X is symmetric unimodal with support $[a, b]$, then (3) holds.

Define the mean value of f on the interval (u, v) as

$$M(f; u, v) = \frac{1}{v - u} \int_u^v f(x)\, dx. \tag{4}$$

Let $m = (a + b)/2$. Jacobson [16] has shown that if $\lim_{\delta \to 0} f(m - \delta) \geqslant M(f; a, m)$ and $\lim_{\delta \to 0} f(m + \delta) \leqslant M(f; m, b)$, then (3) holds.

Suppose f has a unique interior mode. Then there exist numbers $u, v \in (a, b)$ such that $f(u) = f(v) = 1/(b - a)$. Let $m = (a + b)/2$. Seaman et al. [32] have shown that if $u \leqslant m \leqslant v$, and $\Pr[a \leqslant X \leqslant z] = z$, for some $z \in (2m - v, 2m - u)$, then (3) holds. It can be shown that if $\mu = m$, and $f(x) \leqslant$

$1/(b - a)$ for $x = a$ and $x = b$, then (3) holds.

4. Let X be a continuous RV on $[a, b]$ with mean μ. Then

$$\text{Var}(X) \leqslant (b - \mu)(\mu - a). \quad (5)$$

Note that if X is on $[0, 1]$, then

$$\text{Var}(X) \leqslant \mu(1 - \mu). \quad (6)$$

Suppose X is unimodal on $[0, 1]$ with unique mode $m \in (0, 1)$. Then it can be shown (see [32]) that

$$\text{Var}(X) \leqslant \frac{2\mu(1 + m) - m}{3} - \mu^2. \quad (7)$$

It is interesting to note that, in this case, Johnson and Rogers [17] have shown that a *lower* bound for the variance is $(\mu - m)^2/3$.

5. Let X be a RV with support $\mathcal{X}$ and mean μ. Partition $\mathcal{X}$ into sets A and R so that $\mathcal{X} = A \cup R$ and $A \cap R = \varnothing$. Let $\alpha = \Pr[X \in R]$, $0 < \alpha < 1$. Denote the expectation and variance of X with respect to distributions confined to $\mathcal{X}$, A or R by $E[S]$ and $\text{Var}(S)$, respectively, $S = \mathcal{X}$, A or R. Then Rayner [26] has shown that

$$\sup_a B_a(A) \leqslant \text{Var}(\mathcal{X}) \leqslant \inf_a B_a(R), \quad (8)$$

where $B_a(S) = \text{Var}(S) + (E[S] - a)^2 - (\mu - a)^2$. Rayner has applied this result to obtain variance bounds for discrete random variables [see (12)] and to study certain properties of the power functions* of hypothesis tests* for members of exponential families*.

6. Bounds for RVs with discrete distributions have been discussed by Muilwijk [23], Moors and Muilwijk [22], and Rayner [26]. Suppose X has a discrete probability distribution on the set $\{x_1, x_2, \ldots, x_n\}$, where $x_1 < x_2 < \cdots < x_n$ and $\Pr[X_i = x_i] = p_i$. If $\mu = \sum_{i=1}^{n} p_i x_i$, then

$$\text{Var}(X) \leqslant (x_n - \mu)(\mu - x_1). \quad (9)$$

This bound is due to Muilwijk [23]. Note the similarity to the inequality in (5). Moors and Muilwijk [22] have obtained a slightly tighter bound assuming more is known about the underlying distribution. Let X be a discrete RV defined as above with $\Pr[X = x_i] = r_i/s$,

$i = 1, \ldots, n$, where r_i and s are positive integers such that $\sum_{i=1}^{n} r_i = s$. Let B denote the fractional part of $s(\mu - x_1)/(x_n - x_1)$. Then

$$\text{Var}(X) \leqslant (x_n - \mu)(\mu - x_1)$$
$$- (B - B^2)(x_n - x_1)^2/s. \quad (10)$$

Rayner [26] has used (8) to obtain bounds for the variance of a discrete RV. Let X be a RV such that $\Pr[X = x_i] = p_i$ as defined previously. For $R = \{x, y\}$, $x < y$, define $t = \Pr[X = x]/\Pr[X = x \text{ or } y]$ and $d = y - x$. Then

$$\text{Var}(X) \leqslant \begin{cases} (x_n - \mu)(\mu - x_1) - c, \\ \quad \text{if } R = \{x_1, x_2\}, \\ (x_n - \mu)(\mu - x_1) + c, \\ \quad \text{if } R = \{x_{n-1}, x_n\}, \end{cases} \quad (11)$$

where $c = d\{x_n - \mu - t(x_n - x_1)\}$. The first bound in (11) is tighter than (9) if and only if $t < (x_n - \mu)/(x_n - x_1)$. The second bound in (11) is tighter than (9) if and only if $t > (x_n - \mu)/(x_n - x_1)$.

7. Let X be a RV with a standard normal distribution*. Suppose g is an absolutely continuous* function of X and $g(X)$ has finite variance. Chernoff's inequality* [8] states that

$$\text{Var}[g(X)] \leqslant E[\{g'(X)\}^2], \quad (12)$$

with equality if and only if g is linear in X. Chernoff has applied this bound to a problem in information theory* concerned with probability bounds on decoding error. Chernoff has proved this result as an isoperimetric problem in the calculus of variations [9]. This inequality has also been studied by Chen [7] and Borovkov and Utev [3]. It has been generalized by several authors. Cacoullos and Papathanasiou [5, 6] have established an improved bound for the case of $E[X] \neq 0$. Cacoullos [4] and Cacoullos and Papathanasiou [5, 6] have established similar bounds for various nonnormal distributions as well as some lower bounds. Klaassen [18] has derived an upper bound similar to Chernoff's but for arbitrary random variables. We now present Klaassen's result. Let X be a real-valued random variable with density f with respect to a σ-finite measure μ. Let $\pi: R^2 \to R$ be a measurable function such that $\pi(x, \cdot): R \to R$ does not change sign for almost all real x. Furthermore, let $g: R \to R$ be a measurable function such that $G: R \to R$ is well defined by

$$G(x) = \int \pi(x, y)g(y)\,d\mu(y) + c,$$

for some constant c. Finally, let $h: R \to R$ be a nonnegative measurable function such that $H: R \to R$ is well defined by

$$H(x) = \int \pi(x, y)h(y)\,d\mu(y).$$

If $\mu(\{x \in R: g(x) \neq 0, f(x)h(x) = 0\}) = 0$, then

$$\mathrm{Var}[G(X)] \leqslant$$

$$E\left[\frac{g^2(X)}{f(X)h(X)} \int \pi(z, X)H(z)f(z)\,d\mu(z)\right].$$
$$(13)$$

VARIANCE INEQUALITIES AND DISPERSION ORDERINGS

Certain dispersion orderings* imply variance orderings. One common dispersion ordering declares a distribution function G to be more dispersed than F, denoted $F \prec_{\mathrm{disp}} G$, if and only if

$$F^{-1}(v) - F^{-1}(u) \leqslant G^{-1}(v) - G^{-1}(u),$$

for any u, v such that $0 < u < v < 1$. Shaked [35] has shown that if $F \prec_{\mathrm{disp}} G$, then $\mathrm{Var}(X|F) \leqslant \mathrm{Var}(X|G)$.

Shaked [34] and Whitt [37] have discussed the following dispersion ordering. Let $S(h)$ be the number of sign changes of the function h. A distribution function G is said to be more dispersed than F, written $F \prec * G$, if and only if $S(G - H) = 1$ with sign sequence $+, -$. If $F \prec * G$ and $E[X|F] = E[X|G]$, then $\mathrm{Var}(X|F) \leqslant \mathrm{Var}(X|G)$.

INEQUALITIES FOR SPECIAL APPLICATIONS

Upper bounds on variance have been applied in a variety of areas. We will now briefly survey the literature on such inequalities.

Variance bounds manifest themselves in many ways in the study of the properties of inferential procedures. Rayner's work [26] applying nested variance bounds to the study of power functions has already been mentioned.

Variance bounds are useful in the establishment of asymptotic normality* of various statistics. For example, in his development of the asymptotic theory of linear rank statistics*, Hájek [15] has derived an upper bound on the variance of such statistics. A discussion of this bound and its use in the study of the large sample theory* of linear rank statistics can also be found in Puri and Sen [24]. Let $X_1, X_2, \ldots, X_N$ be independent random variables with continuous distribution functions $F_1, F_2, \ldots, F_N$, respectively. Let $a(1), \ldots, a(N)$ be a set of scores and $c(1), \ldots, c(N)$ a set of regression coefficients*. Define the linear rank statistic $L_N = \sum_{i=1}^{N} c(i)a(R_i)$, where R_i denotes the rank of X_i among $X_1, X_2, \ldots, X_N$. Hájek has proved that

$$\mathrm{Var}(L) \leqslant 21\left\{ \max_{1 \leqslant i \leqslant N} \left[c(i) - \bar{c}\right]^2 \right.$$

$$\left. \times \sum_{i=1}^{N} \left[a(i) - \bar{a}\right]^2 \right\}, \quad (13)$$

where

$$\bar{c} = \frac{1}{N}\sum_{i=1}^{N} c(i) \quad \text{and}$$

$$\bar{a} = \frac{1}{N}\sum_{i=1}^{N} a(i).$$

Hájek used this bound in establishing asymptotic normality of rank statistics for square-integrable score functions*.

In studying variances associated with inference procedures, the upper bounds themselves are not always the principle subject of interest. In some cases, finding the distributions that achieve the bound is the primary goal. Gray and Odell [14] have considered such least-favorable distributions in relation to the problem of maximizing functions of random variables, where the function, the underlying random variable, or both, are not completely known.

Another example arises in the quantitative measure of robustness*. One can use the maximum asymptotic variance of a function of the bias of an estimator as such a measure of robustness. Let $X_1, \ldots, X_n$ be a random sample from a distribution $F(x - \theta)$, where F is a member of a class $\mathscr{F}$ of approximately normal symmetric distributions. Let $\hat{\theta}_n$ be an M-estimator* of the unknown location parameter θ. A measure of quantitative robustness of the M-estimator is $\sup\{V_n(F) : F \in \mathscr{F}\}$, where V_n is the asymptotic variance of $n^{1/2}(\hat{\theta}_n - \theta)$, under suitable regularity conditions. Collins [10] has obtained a necessary and sufficient condition for a member of $\mathscr{F}$ to maximize V_n. Collins and Portnoy [11] have extended this result and have shown that the maximum asymptotic variance is obtained using contaminating distributions, which are convex combinations of at most two pairs of symmetric point masses.

Several authors have presented upper and lower bounds on the variance of the estimator of success probability in inverse sampling*. These include Mikulski and Smith [21], Sathe [30], Sahai and Buhrman [29], Korwar et al. [19], and Sahai [27, 28]. D. van Dantzig [36], Birnbaum [2], and McK. Johnson [20] have obtained upper and lower bounds on the variance of the Mann–Whitney–Wilcoxon U-statistic*. For symmetric distributions with shift alternatives, McK. Johnson [20] has obtained upper and lower bounds on the variance of U-statistics.

Other areas of application include stochastic processes* [12], queuing theory* [25], risk analysis* [13, 31], variance estimation [33], and census surveys* [1].

References

[1]   Biemer, P. (1982). *Amer. Statist. Ass. Proc. Surv. Res. Sect.*, **3**, 318–322. (Estimating upper and lower bounds of census nonsampling variance.)

[2]   Birnbaum, Z. W. (1956). *Proc. Third Berkeley Symp. Math. Statist. Prob.*, **1**, 13–17. (On use of the Mann–Whitney statistic.)

[3]   Borovkov, A. A. and Utev, S. A. (1984). *Theory Prob. Appl.*, **28**, 219–228. (On an inequality and a related characterization of the normal distribution.)

[4]   Cacoullos, T. (1982). *Ann. Prob.*, **10**, 799–809. (Upper and lower bounds for the variance of functions of random variables.)

[5]   Cacoullos, T. and Papathanasiou, V. (1985). *Statist. Prob. Lett.* **3**, 175–184. (Upper bounds for the variance of functions of random variables.)

[6]   Cacoullos, T. and Papathanasiou, V. (1986). *Statist. Prob. Lett.*, **4**, 21–23. (Bounds for the variance of functions of random variables by orthogonal polynomials and Bhattacharya bounds.)

[7]   Chen, L. H. Y. (1982). *J. Multivariate Anal.*, **12**, 306–315. (An inequality for the multivariate normal distribution.)

[8]   Chernoff, H. (1980). *Ann. Statist.*, **8**, 1179–1197. (Identification of an element of a large population in the presence of noise.)

[9]   Chernoff, H. (1981). *Ann Statist.*, **9**, 533–535. (A note on an inequality involving the normal distribution.)

[10] Collins, J. R. (1977). *Ann. Statist.*, **5**, 646–657. (Upper bounds on asymptotic variances of M-estimators of location.)

[11] Collins, J. R. and Portnoy, S. L. (1981). *Ann. Statist.*, **9**, 569–577. (Maximizing the variance of M-estimators, using the generalized method of moment spaces.)

[12] Daley, D. J. (1978). *Stoch. Processes Appl.*, **7**, 255–264. (Bounds for the variance of certain stationary point processes.)

[13] Goldstein, M. (1974). *J. Appl. Prob.*, **11**, 409–412. (Some inequalities on variances.)

[14] Gray, H. L. and Odell, P. L. (1967). *SIAM Rev.*, **9**, 715–720. (On least-favorable density functions.)

[15] Hájek, J. (1968). *Ann. Math. Statist.*, **39**, 325–346. (Asymptotic normality of simple linear rank statistics under alternatives.)

[16] Jacobson, H. I. (1969). *Ann. Math. Statist.*, **40**, 1746–1752. (The maximum variance of restricted unimodal distributions.)

[17] Johnson, N. L. and Rogers, C. A. (1951). *Ann. Math. Statist.* **22**, 433–439. (Inequalities on moments of unimodal distributions.)

[18] Klaassen, C. A. J. (1984). *Ann. Prob.*, **13**, 966–974. (On an inequality of Chernoff.)

[19] Korwar, R. M., Prasad, G., and Sahai, A. (1983). *Commun. Statist. A*, **12**, 1807–1812. (A generalized improvement procedure for variance bounds for MVUE estimators in inverse sampling.)

[20] McK. Johnson, B. (1975). *Ann. Statist.*, 3, 955–958. (Bounds on the variance of the *U*-statistic for symmetric distributions with shift alternatives.)

[21] Mikulski, P. W. and Smith, P. J. (1976). *Biometrika*, **63**, 216–217. (A variance bound for unbiased estimation in inverse sampling.)

[22] Moors, J. J. A. and Muilwijk, J. (1971). *Sankhyā B*, **33**, 385–388. (An inequality for the variance of a discrete random variable.)

[23] Muilwijk, J. (1966). *Sankhyā B*, **28**, 183 (Note on a theorem of M. N. Murthy and V. K. Sethi.)

[24] Puri, M. L. and Sen, P. K. (1985). *Nonparametric Methods in General Linear Models*. Wiley, New York.

[25] Ramalhoto, M. F. (1984) *Adv. Appl. Prob.*, **16**, 929–932. (Bounds for the busy period of the M/G/∞ queue.)

[26] Rayner, J. C. W. (1975). *Indian J. Statist.*, **37**, 135–138. (Variance bounds.)

[27] Sahai, A. (1980). *J. Statist. Plann. Inf.* **4**, 213–216. (Improved variance bounds for unbiased estimation in inverse sampling.)

[28] Sahai, A. (1983). *Statistica*, **43**, 621–624. (On a systematic sharpening of variance bounds for MVUEs in inverse binomial sampling.)

[29] Sahai, A. and Buhrman, J. M. (1979). *Statist. Neerlandica*, **33**, 213–215. (Bounds for the variance of an inverse binomial estimator.)

[30] Sathe, Y. S. (1977). *Biometrika*, **64**, 425–426. (Sharper variance bounds for unbiased estimation in inverse sampling.)

[31] Seaman, J. W. and Odell, P. L. (1985). *Adv. Appl. Prob.*, **17**, 679–681. (On Goldstein's variance bound.)

[32] Seaman, J. W., Young, D. M., and Turner, D. W. (1987). *The Math Scientist*, **12**.

[33] Seaman, J. W., Odell, P. L., and Young, D. M. (1987). *Indust. Math.*, **37**, 65–75.

[34] Shaked, M. (1980). *J. R. Statist. Soc. B*, **42**, 192–198. (Mixtures from exponential families.)

[35] Shaked, M. (1982). *J. Appl. Prob.*, **19**, 310–320. (Dispersive ordering of distributions.)

[36] van Dantzig, D. (1951). *Kon. Ned. Akad. Wet. Proc.*, *Ser. A*, **54**, 1–8. (On the consistency and the power of Wilcoxon's two-sample test.)

[37] Whitt, W. (1985). *J. Appl. Prob.*, **22**, 619–633. (Uniform conditional variability ordering of probability distributions.)

(CRAMÉR–RAO INEQUALITY
INEQUALITIES FOR EXPECTED SAMPLE
  SIZES
INEQUALITIES ON DISTRIBUTIONS:
  BIVARIATE AND MULTIVARIATE
INEQUALITIES ON DISTRIBUTIONS:
  UNIVARIATE
INFORMATION AND CODING THEORY
ORDERING DISTRIBUTIONS BY
  DISPERSION
ORDERING OF DISTRIBUTIONS, PARTIAL
U-STATISTICS
VARIABILITY, HARMONIC
VARIANCE)

J. W. SEAMAN, JR.
P. L. ODELL

VARIANT

In quality-control* literature, the adjective "variant" refers to an item or an event that is classified differently from the majority of others of its kind or type. A unit of product that contains at least one variant characteristic or attribute is called a *variant* unit. (This term is less specific than *defective* or *nonconforming* unit.)

(QUALITY CONTROL, STATISTICAL)

VARIATE DIFFERENCE METHOD

The variate difference method is a procedure to separate the stochastic and permanent components of time series* by successive differencing. The method utilizes the well-known theorem that the Kth difference of a Kth-order polynomial is constant and its variance therefore vanishes. The scope of this method is twofold. First, it yields an estimate for the lowest order of differencing, which sufficiently eliminates the influence of the permanent component of a time series on its variance. This estimate can be used to apply smoothing formulas on time series, which are more flexible than the usual constant coefficient time polynomials. Second, the variate difference method provides an estimate of the variance of the stochastic component of a time series. This estimate can be utilized for all errors-in-variables methods and tests on multicollinearity*.

The variate difference method has been independently developed by Anderson [1, 2] and Gosset* ("Student") [5]. Large-sample tests have been provided by Anderson [2], Zaycoff [17], and Tintner [12]. Strecker [10, 11] applied the variate difference method for the case of a multiplicative, instead of an additive, stochastic component (*see* QUO-TIENT METHOD). The exact distributions of the empirical variances of a time series and the tests for constancy after differenc-ing enough times have been established by Tintner [14] and Rao and Tintner [8, 9]. The problem of multiple choice has been raised by Anderson [3] and Hogg [7]. This problem refers to the high correlation of the series of tests, which arises when we are starting with first differences and go on to see whether further differencing will reduce the properly adjusted variance of our time series* signifi-cantly. New results concerning this problem and a wide variety of applications of the variate difference method are discussed in Tintner et al. [15].

The variate method cannot be applied to time series that show pronounced seasonal peak patterns or other types of zig-zag move-ments (Wald [16]). For time series generated by stochastic difference equations, the variate difference method is not appropriate, either (Tintner [12], Haavelmo [6], and Tintner [13]). Special problems arise in the case of autocorrelated stochastic components. Tintner [12] suggested a method of proper selections of subsets of the observations at the sacrifice of a considerable loss of infor-mation.

METHOD

Let x_t be an observation of a time series, $t = 1, 2, \ldots, N$. We assume that x_t consists of two additive and independent compo-nents

$$x_t = m_t + u_t, \qquad E[m_s u_t] = 0,$$

$$s, t = 1, 2, \ldots, N. \qquad (1)$$

m_t is the systematic component, which is assumed to be a locally smooth function of time that allows a sufficient approximation through a polynomial of finite and not too high order K, $K = k_0 - 1$, in the neighbour-hood of t. This excludes peak seasonal varia-tions and observations from stochastic dif-ference equations for which the variate difference method is not an appropriate pro-cedure to separate the stochastic and sys-tematic component from each other. The advantage of the variate difference method is that it avoids unnecessary assumptions about the functional pattern of the systematic com-ponent over the whole observation period, which are typical for all analytical "trend regression" methods. For the stochastic com-ponent u_t, we assume mean 0, constant vari-ance, and no autocorrelation;

$$E[u_t] = 0, \qquad E[u_t^2] = \sigma^2,$$

$$E[u_t u_s] = 0, \quad \text{if } t \neq s. \qquad (2)$$

If the systematic component m_t is repre-sented locally by a smooth function of time, its contribution to the variance of the total time series x_t can be eliminated by taking successive differences of finite order. Con-sider finite differences of the observed time series

$$\Delta x_t = x_{t+1} - x_t, \qquad \Delta^k x_t = \Delta\left(\Delta^{k-1} x_t\right),$$

$$\Delta^k x_t = \Delta^k m_t + \Delta^k u_t. \qquad (3)$$

For the stochastic component u_t, we have

$$E\left[\left(\Delta^k u_t\right)^2\right] = \sigma^2 \,_{2k}C_k,$$

$$_{2k}C_k = \binom{2k}{k} \qquad (4)$$

and

$$\sigma_k^2 = E\left[\left(\Delta^k u_t\right)\right]^2 / _{2k}C_k = \sigma^2,$$

$$\text{for all } k \geqslant 0. \quad (5)$$

If we have eliminated the systematic or per-manent component m_t by taking enough differences, $k = k_0$, the properly adjusted theoretical variance of the time series x_t will fall together with the variance of the stochas-tic component and remain constant for all differences of higher order $k_0, k_0 + 1, k_0 + 2, \ldots$.

What we are looking for is a test, which tells us how many times we have to difference the time series x_t in order to eliminate its smooth systematic component m_t. The above argument suggests testing whether further differencing of our time series reduces the empirical variance (adjusted properly) significantly.

Define the *empirical reduced variance* of the observations in the following way:

$$V_0' = s_0^2 = \sum_{t=1}^{N} \frac{(x_t - \bar{x})^2}{N - 1},$$

$$\bar{x} = \sum_{t=1}^{N} \frac{x_t}{N},$$

$$V_k' = s_k^2 = \sum_{t=1}^{N-k} \frac{(\Delta^k x_t)^2}{(N - k)._{2k}C_k},$$

$$k = 1, 2, \ldots . \quad (6)$$

If the systematic component m_t is eliminated or at least greatly reduced by differences of order k_0, then we have approximately

$$V_{k_0}' \approx V_{k_0+1}' \approx \cdots \approx V_{k_0+n}'. \quad (7)$$

According to Anderson [2] and Zaycoff [17], the test statistic

$$R_k' = \frac{(V_k' - V_{k+1}')}{V_k'} H_{kN},$$

$$H_{kN} \quad (8)$$

$$= \sqrt{\frac{\dfrac{(N - k)(N - k - 1)}{N}}{b_k'.Nb_k'' - \dfrac{(N - k)(N - k - 1)}{N}}}$$

is approximately normally distributed for large samples with mean 0 and variance 1, assuming that u_t is normally distributed. Tables for H_{kN}, b_k', and b_K'' for alternative values of k and N are to be found in Tintner [12], Tintner et al. [15] and Strecker [11].

In order to apply the variate difference method, we have to calculate the R_k'. If we find a value of k_0, so that R_{k_0-1}' is significant but R_{k_0}' is not, we can assume that the systematic component m_t is eliminated sufficiently after taking $k = k_0$ differences ($|R_{k=k_0}'| < 3$) and that V_{k_0}' is an estimate for the variance of the stochastic component u_t.

These results can be used to apply moving average* formulas in order to yield an estimate of the "smooth" (systematic) component m_t of the time series x_t. Let $k_0 = 2n$ for k_0 even and $k_0 = 2n + 1$ for k_0 odd. A smoothing formula that minimizes

$$E\left[(u_t + f_t)^2\right]$$

is

$$f_t = b_1 \Delta^{2n}(u_{t-n}) + b_2 \Delta^{2n+2}(u_{t-n-1}) + \cdots,$$

where b_i are constants, to be determined by least squares. Hence

$$m_t \approx m_t' = x_t + f_t,$$

and we get the smoothing formula

$$m_t' = \sum_{s=1}^{m} g_{mn}(s)(x_{t+s} + x_{t-s}) + g_{mn}(0)x_t. \quad (9)$$

Values of $g_{mn}(s)$ are tabulated by Tintner [12] and coincide with least-squares regression weights, which result from fitting a polynomial of degree n to $2m + 1$ ($m \geq n$), consecutive observations. n depends on the degree of the polynomial and m represents the desired accuracy. This moving average formula m_t' coincides with Sheppard's smoothing formula and can be taken as an estimate of the systematic component m_t.

SMALL-SAMPLE RESULTS

Small-sample tests have been established by Tintner et al. [15], following the methods proposed by Anderson [4]. In order to derive exact tests, the assumption of a circular distribution has to be made;

$$x_{N+t} = x_{-N+t} = x_t. \quad (10)$$

Tables of the critical values at 1% and 5% error probability for alternative sample sizes are given in Tintner et al. [15] for the variance ratios V_1'/V_0', V_2'/V_1', and V_3'/V_2'. In empirical applications the noncircular empirical variances are taken as an approximation for the theoretically circular variances and the V_k' are calculated in the same way as in (6).

The estimation of the degree of the polynomial, which represents the systematic component of a time series, involves the problem of multiple choice. If we proceed, as suggested above, by successively testing whether first, second, ..., k_0-th differences will result in significant reductions of the properly adjusted empirical variances V'_{k_0}, we face the problem that the tests are not independent and additionally involve unknown parameters. Following a method suggested by Anderson [3] and Hogg [7], Rao and Tintner [8, 9] and Tintner et al. [15] developed a procedure that takes into account the problem of multiple choice. Instead of starting with a test of the significance of the variance reduction after the first difference has been taken, and then going on until a k_0 has been found, so that V_{k_0+1}/V_{k_0} is not significantly different from 1 any more, the order of testing the degree of the underlying polynomial is reversed. First, we have to specify the highest possible order q (a priori), and then we have to calculate V'_{q+1}/V'_q, V'_q/V'_{q-1}, and so on, until a k_0 appears, so that V'_{k_0+1}/V'_{k_0} is *not* significant, but V'_{k_0}/V'_{k_0-1} *is* significant. Tintner et al. [15] give tables for the lower critical values at 1% and 5% error probability for alternative sample sizes between 20 and 100.

Tintner et al. [15] discuss applications of the variate difference method on problems of multicollinearity, weighted regression, and diffusion processes*.

References

[1] Anderson, O. (1914). Nochmals über "The elimination of spurious correlation due to position in time or space." *Biometrika*, **10**, 269–279.

[2] Anderson, O. (1929). Die Korrelationsrechnung in der Konjunkturforschung, Kurt Schroeder-Verlag Bonn (reprinted in O. Anderson (1963). In *Ausgewählte Schriften*, Vol. 1, H. Kellerer, W. Mahr, G. Schneider, and H. Strecker, eds. (1963), J. C. B. Mohr (Siebeck), Tübingen, Federal Republic of Germany, pp. 164–301.

[3] Anderson, T. W. (1962). The choice of the degree of a polynomial regression as a multiple choice problem. *Ann. Math. Statist.*, **33**, 255–265.

[4] Anderson, T. W. (1971). *Time Series Analysis*. Wiley, New York.

[5] Gosset, W. S. ("Student") (1914). The elimination of spurious correlation due to position in time or space. *Biometrika*, **10**, 179–180.

[6] Haavelmo, T. (1941). A note on the variate difference method. *Econometrica*, 9, 74–79.

[7] Hogg, R. V. (1961). On the resolution of statistical hypotheses. *J. Amer. Statist. Ass.*, **56**, 978–989.

[8] Rao, J. N. K. and Tintner, G. (1962). The distribution of the ratio of the variances of variate differences in the circular case. *Sankhyā A*, **24**, 385–394.

[9] Rao, J. N. K. and Tintner, G. (1963). On the variate difference method. *Austral. J. Statist.*, **5**, 105–116.

[10] Strecker, H. (1949). Die Quotientenmethode, eine Variante der "Variate Difference" Methode. *Mitteilungsbl. für Math. Statist.*, **1**, 115–130.

[11] Strecker, H. (1970, 1971). Ein Beitrag zur Analyse von Zeitreihen: Die Quotientenmethode, eine Variante der "Variate Difference" Methode. *Metrika*, **16**, 130–187; **17**, 257–259. (Refs. 10 and 11 are reprinted in *Survey Methods of Statistics–Selected Papers of H. Strecker*, M. Beckman and R. Wiegert, eds. (1987). Vandenhoeck and Ruprecht, Göttingen, Federal Republic of Germany, pp. 248–325.)

[12] Tintner, G. (1940). *The Variate Difference Method*. Principia, Bloomington, IN.

[13] Tintner, G. (1941). The variate difference method: A reply. *Econometrica*, 9, 163–164.

[14] Tintner, G. (1955). The distribution of the variances of variate differences in the circular case. *Metron*, **17**(3), 43–52.

[15] Tintner, G., Rao, J. N. K., and Strecker, H. (1978). *New Results in the Variate Difference Method*. Vandenhoeck and Ruprecht, Göttingen, Federal Republic of Germany.

[16] Wald, A. (1936). *Berechnung und Ausschaltung von Saisonschwankungen*. Springer-Verlag, Vienna, Austria.

[17] Zaycoff, R. (1937). Über die Ausschaltung der zufälligen Komponente nach der "Variate Difference" Methode. Publications of the Statistical Institute for Economic Research, State University of Sofia, Bulgaria, No. 1, 75–120.

Note

Professor Gerhard Tintner died on November 13, 1983, before he could finish this article. The co-authors tried to finish this task by following most closely G. Tintner's writings on this subject, which cover his complete scientific career.

(AUTOREGRESSIVE-INTEGRATED
MOVING AVERAGE (ARIMA) MODELS

CURVE FITTING
FINITE DIFFERENCES, CALCULUS OF
QUOTIENT METHOD
REGRESSION, POLYNOMIAL
TIME SERIES
TREND)

HEINRICH STRECKER
GERHARD TINTNER
ANDREAS WÖRGÖTTER
GABRIELE WÖRGÖTTER

VARIATION-DIMINISHING PROPERTY

See TOTAL POSITIVITY

VARIATION RATIO

A simple measure of dispersion often used in behavioral sciences. It is applicable to grouped data*, and especially to nominal* scales. It is defined as follows:

$$VR = 1 - f_{modal}/N,$$

where f_{modal} is the number of "cases" in the estimated modal category and N the total number of "cases." Thus it measures the degree to which the cases are concentrated in the modal category.

VR is insensitive to the distribution of cases among nonmodal categories and is dependent on the categorization procedure. However, in the case of nominal scales, lack of ordering of categories prevents construction of more refined measures. See, e.g., Blalock [1] for more details.

Reference

[1] Blalock, H. M., Jr. (1979). *Social Statistics*, 2nd ed., McGraw-Hill, New York.

(GROUPED DATA
NOMINAL DATA)

VARIATION-REDUCTION PROPERTY

See TOTAL POSITIVITY

VARIMAX METHOD

In the factorisation $\Omega = AA' + \Phi$, (*see* FACTOR ANALYSIS) the $p \times m$ matrix $\mathbf{A}$ contains the unknown factor loadings, the parameters a_{ij}. They measure the (linear) relationship between the ith (observable) stochastic variable x_i and the jth (nonobservable) stochastic variable y_j (also called the "common factor") in the factor model $\mathbf{x} = \mathbf{Ay} + \epsilon$. The factorisation is not unique, as the loading matrix $\mathbf{A}$ can be postmultiplied by an arbitrary orthogonal matrix* $\mathbf{T}$, that is, $\mathbf{AT(AT)'} = \mathbf{ATT'A'} = \mathbf{AA'}$.

It is clear that the postmultiplication of $\mathbf{A}$ is equivalent to a rotation of the vector of factors $\mathbf{y}$, namely, $\mathbf{x} = (\mathbf{AT})(\mathbf{T'y}) + \epsilon$. This nonuniqueness property can be exploited to simplify the structure of an estimated $\mathbf{A}$. The vector of factors $\mathbf{y}$ is rotated in such a way that the estimated loadings show columnwise more contrast. The aim is to obtain a relatively large number of near-vanishing elements, a limited number of very large elements, and very few elements of intermediate size. As a result the conclusion of an analysis can be simplified, and sometimes be more meaningful.

A well-known family of procedures to obtain a simple(r) structure is the "varimax." It is derived from *Kaiser's varimax*, which will be discussed shortly. For its presentation, see [3, 4]. For an illustration, see [6, Chap. 8.7].

Varimax uses as a criterion for the simplicity of the structure the expression

$$\sum_{j=1}^{m} \left[\left(\sum_{i=1}^{p} a_{ij}^4 \right) - \frac{1}{p} \left(\sum_{i=1}^{p} a_{ij}^2 \right)^2 \right], \quad (1)$$

where the a_{ij} are estimated factor loadings. Expression (1) is proportional to the within-group variance of the *squared* estimates, there being m groups (factors).

Using matrix algebra, one can express the simplicity criterion as

$$tr(\mathbf{A} \odot \mathbf{A})' \mathbf{N} (\mathbf{A} \odot \mathbf{A}), \quad (2)$$

where $\odot$ denotes the Hadamard (or Schur) product, the matrix $\mathbf{N}$ corrects quantities for their group averages, and tr stands for matrix trace*.

The varimax procedure consists in the determination of an orthogonal matrix $\mathbf{T}$ that *maximizes* the value of the criterion $\text{tr}(\mathbf{B} \otimes \mathbf{B})'\mathbf{N}(\mathbf{B} \otimes \mathbf{B})$, where $\mathbf{B} = \mathbf{AT}$ and $\mathbf{A}$ is a given estimate of the loadings matrix. Maximization can be achieved by taking the critical points of the Lagrangian function

$$\varphi(\mathbf{T}) = \tfrac{1}{4}\text{tr}\,\mathbf{C}'\mathbf{NC} - \text{tr}\,\mathbf{G}(\mathbf{TT}' - \mathbf{I}), \quad (3)$$

where $\mathbf{G}$ is a (symmetric) Lagrange multiplier matrix, $\mathbf{C} = \mathbf{B} \ominus \mathbf{B}$, $\mathbf{B} = \mathbf{AT}$. Usually some sort of normalization is administered to the loadings matrix. For further mathematics, see [2, Chap. 18.4; 5, Chap. 6.3; 7, 8]. The varimax family has various members. Instead of working with the variance of the squared estimates, one can use the variance of some positive even power of the estimated factor loadings. For further details, see [2, Chap. 18]. The varimax procedure, as it was described, uses an arbitrary estimate of the loadings matrix as a starting point. It has been argued that a better approach might be to use more basic information such as the data matrix or the correlation matrix. See [2, Chap 19]. Several authors have pointed out that there are situations in which the varimax, as discussed previously, fails. They describe such situations and offer remedies. Examples are "weighted varimax," proposed by [2], and "predictive varimax," proposed by [9].

References

[1]   Cureton, E. E. and Mulaik, S. A. (1975). *Psychometrika*, **40**, 183–195.

[2]   Horst, P. (1965). *Factor Analysis of Data Matrices*. Holt, Rinehart, and Winston, New York.

[3]   Kaiser, H. F. (1958). *Psychometrika*, **23**, 187–200.

[4]   Kaiser, H. F. (1959). *J. Educ. Psychol. Meas.*, **19**, 413–420.

[5]   Lawley, D. N. and Maxwell, A. E. (1971). *Factor Analysis as a Statistical Method*. Butterworth, London.

[6]   Morrison, D. F. (1976). *Multivariate Statistical Methods*. McGraw-Hill, New York.

[7]   Neudecker, H. (1981). *Psychometrika*, **46**, 343–345.

[8]   Sherin, R. J. (1966). *Psychometrika*, **31**, 535–538.

[9]   Williams, W. T. and Campbell, S. J. (1978). *Biometrie-Praximetrie*, **18**, 75–81.

(FACTOR ANALYSIS
ROTATION TECHNIQUES)

H. Neudecker

VARIOGRAM

The variogram is a function that characterizes the second-order dependence properties of a stochastic process* defined on $\mathbb{R}^m$, the m-dimensional Euclidean space. Its most important use is for spatial prediction, or *kriging**, usually in two or three dimensions. Let $\{Z(\mathbf{s}) : \mathbf{s} \in D \subset \mathbb{R}^m\}$ be a real-valued stochastic process defined on a domain D of $\mathbb{R}^m$, and suppose

$$\text{var}(Z(\mathbf{s} + \mathbf{h}) - Z(\mathbf{s})) = 2\gamma(\mathbf{h}),$$
$$\text{for all } \mathbf{s}, \mathbf{s} + \mathbf{h} \in D. \quad (1)$$

The quantity $2\gamma(\cdot)$, which is a function only of the *difference* between the spatial locations $\mathbf{s}$ and $\mathbf{s} + \mathbf{h}$, has been called the *variogram* by Matheron [9], although earlier appearances in the scientific literature can be found. It has been called a *structure function* by Kolmogorov [6] (physics) and Gandin [3] (meteorology), and a *mean squared difference* by Jowett [5] (time series). Nevertheless, it has been Matheron's (mining) terminology that has persisted.

Clearly, $\gamma(-\mathbf{h}) = \gamma(\mathbf{h})$ and $\gamma(\mathbf{0}) = 0$. If $\gamma(\mathbf{h}) \to c_0 > 0$ as $\mathbf{h} \to \mathbf{0}$, then c_0 is called the *nugget effect*, made up of both measurement error* and micro-scale variation [1]. Furthermore, Matheron [10] shows that the variogram must satisfy the conditional negative semidefiniteness condition; namely, $\sum_{i=1}^{r}\sum_{j=1}^{r} a_i a_j 2\gamma(\mathbf{s}_i - \mathbf{s}_j) \leqslant 0$, for any finite number of spatial locations $\{\mathbf{s}_i : i = 1, \ldots, r\}$ and real numbers $\{a_i : i = 1, \ldots, r\}$ satisfying $\sum_{i=1}^{r} a_i = 0$. When $2\gamma(\mathbf{h}) = 2\gamma(\|\mathbf{h}\|)$, the variogram is said to be *isotropic*. Various parametric models for the variogram are presented in Journel and Huijbregts [4].

Replacing (1) by the stronger assumption

$$\text{cov}(Z(\mathbf{s} + \mathbf{h}), Z(\mathbf{s})) = C(\mathbf{h}),$$
$$\text{for all } \mathbf{s}, \mathbf{s} + \mathbf{h} \in D, \quad (2)$$

and specifying the mean function to be constant

$$E[Z(\mathbf{s})] = m, \quad \text{for all } \mathbf{s} \in D, \qquad (3)$$

defines the class of *second-order* (or wide-sense) *stationary* processes* in D, with covariance function $C(\cdot)$. Time-series* analysts usually assume (2) and work with the quantity $\rho(\cdot) \equiv C(\cdot)/C(0)$. Conditions (1) and (3) define the class of *intrinsically stationary* processes, which is now seen to contain the class of second-order stationary processes.

Assuming only (2),

$$\gamma(\mathbf{h}) = C(\mathbf{0}) - C(\mathbf{h});$$

that is, the *semivariogram* (one-half of the variogram) is related very simply to the covariance function. Furthermore, as $\|\mathbf{h}\| \to \infty$ it is often seen that $2\gamma(\mathbf{h}) \to 2C(\mathbf{0})$, a quantity called the *sill*. And any vector $\mathbf{r}_0$ for which $2\gamma(\mathbf{r}_0) = 2C(\mathbf{0})$, but $2\gamma(\mathbf{r}_0(1 - \epsilon)) < 2C(\mathbf{0})$ for any $\epsilon > 0$, is called the *range* in the direction $\mathbf{r}_0/\|\mathbf{r}_0\|$.

An example of a process for which $2\gamma(\cdot)$ exists but $C(\cdot)$ does not, is a one-dimensional standard Wiener process $\{W(t) : t \geqslant 0\}$, where $2\gamma(h) = |h|$, $-\infty < h < \infty$, but $\text{cov}(W(t), W(u)) = \min(t, u)$. Thus the class of intrinsically stationary processes strictly contains the class of second-order stationary processes.

Estimation of the variogram has received little attention. Suppose $\{Z(\mathbf{s}_i) : i = 1, \ldots, n\}$ are observations on the intrinsically stationary process [i.e., the process satisfies (1) and (3)], taken at the n spatial locations $\{\mathbf{s}_i : i = 1, \ldots, n\}$. The obvious nonparametric estimator is unbiased:

$$2\hat{\gamma}(\mathbf{h}) = \sum_{N(\mathbf{h})} \left(Z(\mathbf{s}_i) - Z(\mathbf{s}_j) \right)^2 \Big/ |N(\mathbf{h})|,$$

$$\mathbf{h} \in \mathbb{R}^m, \qquad (4)$$

where the average in (4) is taken over $N(\mathbf{h}) = \{(\mathbf{s}_i, \mathbf{s}_j) : \mathbf{s}_i - \mathbf{s}_j = \pm \mathbf{h}\}$, and $|N(\mathbf{h})|$ is the number of distinct elements in $N(\mathbf{h})$. Note that in $\mathbb{R}^1$, $2\hat{\gamma}(1)$ is used by von Neumann et al. [11] to estimate the constant variance of *independent* data, whose means vary. Cressie and Hawkins [2] propose an estimator based on a Gaussian process, but *robust* to small departures from Gaussianity,

$$2\bar{\gamma}(\mathbf{h})$$

$$= \frac{\left\{ \sum_{N(\mathbf{h})} |Z(\mathbf{s}_i) - Z(\mathbf{s}_j)|^{1/2} \Big/ |N(\mathbf{h})| \right\}^4}{0.457 + 0.494/|N(\mathbf{h})|},$$

$$\mathbf{h} \in \mathbb{R}^m. \qquad (5)$$

In Matheron's *Geostatistics* and Gandin's *Objective Analysis*, the variogram is then used to define coefficients in an optimal linear predictor. Suppose it is desired to predict $Z_B \equiv \int_B Z(\mathbf{s})\, d\mathbf{s} / \int_B d\mathbf{s}$, using a linear function of the data,

$$\hat{Z}_B = \sum_{i=1}^{n} \lambda_i Z(\mathbf{s}_i), \qquad (6)$$

which is unbiased and minimizes $E[(Z_B - \hat{Z}_B)^2]$. Then the optimal λ satisfies an $(n + 1)$-dimensional linear equation that depends on $2\gamma(\mathbf{s}_i - \mathbf{s}_j)$, $i, j = 0, 1, \ldots, n$, any $\mathbf{s}_0 \in B$. Matheron [9] calls this procedure *kriging* after Krige [8], whereas Gandin [3] calls it *optimal interpolation*. In fact, this optimal linear predictor (6) can be found earlier in the literature; see Kolmogorov [7] and Wiener [12]. Cressie [1] investigates various kriging options when the process satisfies (1) but does not necessarily satisfy the constant-mean assumption (3). The effect on kriging, of using an estimated and fitted variogram rather than the true variogram, is not well understood.

References

[1] Cressie, N. (1986). *J. Amer. Statist. Ass.*, **81**, 625–634.

[2] Cressie, N. and Hawkins, D. M. (1980). *J. Internat. Ass. Math. Geol.*, **12**, 115–125.

[3] Gandin, L. S. (1963). *Objective Analysis of Meteorological Fields*. GIMIZ, Leningrad (Israel Program for Scientific Translations, Jerusalem, 1965).

[4] Journel, A. G. and Huijbregts, C. J. (1978). *Mining Geostatistics*. Academic, New York.

[5] Jowett, G. H. (1952). *Appl. Statist.*, **1**, 50–59.

[6] Kolmogorov, A. N. (1941). *Dok. Akad. Nauk SSSR*, **30**, 229–303. Reprinted in *Turbulence: Classic Papers on Statistical Theory*, S. K. Friedlander and L. Topping, eds. Wiley, New York (1961).

[7] Kolmogorov, A. N. (1941). *Izv. Akad. Nauk SSSR Ser. Mat.*, **5**, 3–14.

[8] Krige, D. G. (1951). *J. Chem. Metallurgical and Mining Soc. S. Africa*, **52**, 119–139.

[9] Matheron, G. (1963). *Econ. Geol.*, **58**, 1246–1266.

[10] Matheron, G. (1971). The Theory of Regionalized Variables and Its Applications. *Cahiers du Centre de Morphologie Mathématique No. 5*, Fontaine-bleau, France.

[11] von Neumann, J., Kent, R. H., Bellinson, H. R., and Hart, B. I. (1941). *Ann. Math. Statist.*, **12**, 153–162.

[12] Wiener, N. (1949). *Extrapolation, Interpolation and Smoothing of Stationary Time Series.* MIT Press, Cambridge, MA.

Acknowledgment

The preparation of this entry was supported by National Science Foundation Grants DMS-8503693 and SES-8401460.

(KRIGING
STOCHASTIC PROCESSES)

NOEL CRESSIE

VARTIA INDEX   *See* LOG-CHANGE INDEX NUMBERS

VASICEK ENTROPY-BASED TEST OF NORMALITY

This test uses the sample entropy* as a criterion. It was suggested by Vasicek [3].

Let $X_{(1)} \leqslant X_{(2)} \leqslant \cdots \leqslant X_{(n)}$ be an ordered random sample. Then for $m < \frac{1}{2}n$, the Vasicek test statistic is

$$K_{m,n} = \frac{n}{2mS} \left\{ \prod_{i=1}^{n} \left(X_{(i+m)} - X_{(i-m)} \right) \right\}^{1/2},$$

where $S^2 = n^{-1}\sum_{i=1}^{n}(X_{(i)} - \overline{X}^2)$, $\overline{X} = n^{-1}\sum_{i=1}^{n}X_{(i)}$, and

$$X_{(i)} = X_{(1)}, \quad \text{for } i < 1,$$
$$X_{(i)}L = X_{(n)}, \quad \text{for } i > n.$$

The critical region* (leading to rejection of the hypothesis that X has a normal distribution in the population from which the random sample is taken) is of the form $K_{m,n} > K_{m,n}(\alpha)$, where α denotes the significance level* of the test.

If the population distribution is normal, then $K_{m,n}$ converges in probability to $\sqrt{(2\pi e)}$ as $m, n \to \infty$ with $m/n \to \alpha$. If the population distribution were not normal, with probability density function $f(x)$ and finite variance σ^2, then $K_{m,n}$ would converge in probability to

$$\sigma^{-1}\exp\{H(f)\} \qquad \left(< \sqrt{(2\pi)} \right),$$

where

$$H(f) = \int_{-\infty}^{\infty} f(x) \log f(x) \, dx$$

is the entropy of the distribution.

This test is consistent* with respect to *all* alternatives not possessing a singular continuous component. Vasicek found that (for $n = 20$, $\alpha = 0.05$) the test appeared to be more powerful than tests based on the Shapiro–Wilk* W_1 and on EDF statistics* (such as the Kolmogorov–Smirnov* D, the Cramér–von Mises* W^2, Kuiper's* V, Watson's U^2, and the Anderson–Darling* A^2) with respect to exponential*, beta* (2,1), and uniform* alternatives, but inferior to all these tests with respect to Cauchy* alternatives.

Hui [2] has developed a multivariate extension of the test. Dudewicz and Van der Meulen [1] have developed a test of uniformity* based on a similar approach.

References

[1] Dudewicz, E. J. and Van der Meulen, E. C. (1981). *J. Amer. Statist. Ass.*, **76**, 967–974.

[2] Hui, T. (1983). On Tests of Multivariate Normality. *Dissertation Abstracts International*, **45**(2), 606–608. University Microfilms International, Ann Arbor, MI.

[3] Vasicek, O. (1976). *J. R. Statist. Soc. B*, **38**, 54–59.

(DEPARTURE FROM NORMALITY, TESTS
   FOR
EDF STATISTICS
ENTROPY
GOODNESS OF FIT
KOLMOGOROV–SMIRNOV STATISTICS
SHAPIRO–WILK TEST)

VEC OPERATOR

INTRODUCTION

The vec operator stacks a matrix column by column. If $A = (a_{ij})$ is a matrix with m rows and n columns, then vec A is the mn-vector $(a_{11}, a_{21}, \ldots, a_{m1}, \ldots, a_{1n}, \ldots, a_{mn})'$. The idea of stacking the elements of a matrix goes back at least to Sylvester [13]. The notation "vec" was introduced in [5]. The use of the vec operator in multivariate statistics, econometrics, psychometrics, etc., often occurs in conjunction with the Kronecker product*, the trace operator, and the commutation matrix* (for reviews, see, e.g., [1, 3, 6]). The basic results are (i) vec $ABC = (C' \otimes A)$vec B, with $\otimes$ the (right) Kronecker product (probably due to Roth [12]), and (ii) tr $ABCD = (\text{vec} A')'(D' \otimes B)(\text{vec} C)$.

A useful operator is the commutation matrix $P_{m,n}$ of order $mn \times mn$, which can be defined implicitly as $P_{m,n}$vec $A' = $ vec A uniformly for any A $(m \times n)$. As vec $A' = (a_{11}, a_{12}, \ldots, a_{1n}, \ldots, a_{m1}, \ldots, a_{mn})$, P changes the running order of a vector of double-indexed variables. Useful properties of P are

(i) $P_{m,n}P_{n,m} = I_{mn}$,

(ii) $P_{m,p}(A \otimes B)P_{q,n} = B \otimes A$ for $A: m \times n$ and $B: p \times q$, and

(iii) $\text{vec}(A \otimes B) = (I_n \otimes P_{m,q} \otimes I_p)(\text{vec} A \otimes \text{vec} B)$ (see [11]).

APPLICATIONS

A wide variety of applications show the usefulness of the vec operator. Some applications are:

(i) *Solving linear equations.* Let A, B, C (all $m \times m$) be known. Then the solution of $AX + XB = C$ is vec $X = (I_m \otimes A + B' \otimes I_m)^{-1}$ vec C, if the inverse exists.

(ii) *Variance of a matrix.* If X is a matrix of random variables with expectation M, the variance of X can be defined as

$$D(X) = E\left[(\text{vec} X)(\text{vec} X)'\right] - (\text{vec} M)(\text{vec} M)'.$$

If each row is independently $N(\mu, \Sigma)$, then $D(X) = \Sigma \otimes I$.

(iii) *Matrix derivatives*.* The matrix derivative $\partial Y / \partial X$ is not unambiguous as to the ordering of the elements. One possible ordering is as in ∂ vec $Y / \partial (\text{vec} X)'$, which preserves the chain rule (see [8]).

EXTENSIONS

The redundancy in stacking a symmetric matrix can be avoided by using the operator vech or v (e.g., different authors [2, 7] use different symbols), which leaves out supra-diagonal elements. An elaborate algebra has been developed (e.g., [7]) to handle vech and to bridge the gap between expressions in vec A and vech A. The same holds for operators that stack only the diagonal elements or the infradiagonal elements (see [9]). Useful ways of stacking multidimensional matrices are given in [4].

References

[1] Balestra, P. (1976). *La Dérivation Matricielle.* Sirey, Paris.

[2] Henderson, H. V. and Searle, S. R. (1979). *Canad. J. Statist.*, **7**, 65–81.

[3] Henderson, H. V. and Searle, S. R. (1981). *Linear and Multilinear Algebra*, **9**, 271–288.

[4] Kapteyn, A., Neudecker, H., and Wansbeek, T. J. (1986). *Psychometrika*, **51**, 269–275.

[5] Koopmans, T. C., Rubin, H., and Leipnik, R. B. (1950). *Statistical Inference in Dynamic Economic Models.* Wiley, New York.

[6] Magnus, J. R. and Neudecker, H. (1979). *Ann. Statist.*, **7**, 381–394.

[7] Magnus, J. R. and Neudecker, H. (1980). *SIAM J. Algebraic Discrete Math.*, **1**, 422–449.

[8] Magnus, J. R. and Neudecker, H. (1985). *J. Math. Psychol.*, **29**, 474–492.

[9] Neudecker, H. (1983). *Linear and Multilinear Algebra*, **14**, 271–295.

[10] Neudecker, H. (1985). *Linear Algebra Appl.*, **70**, 257–262.

[11] Neudecker, H. and Wansbeek, T. J. (1983). *Canad. J. Statist.*, **11**, 221–231.

[12] Roth, W. E. (1934). *Bull. Amer. Math. Soc.*, **40**, 461–468.

[13] Sylvester, J. (1984). *C. R. Acad. Sci. Paris*, **99**, 117–118, 409–412, 432–436.

(KRONECKER PRODUCT OF MATRICES
MATRIX DERIVATIVES
TRACE)

TOM WANSBEEK

VECTORIAL DATA *See* DIRECTIONAL
DATA

VEHICLE SAFETY, STATISTICS IN

In a typical year in the United States, there are 6 million motor vehicle crashes resulting in 45,000 fatalities and over 3 million injuries [9]. Automated data bases are maintained by the federal government and the individual states, serving multiple purposes: record keeping on individual drivers or vehicles; summary statistics on the number of crashes or casualties; detailed descriptive analyses of collision or injury mechanisms and their relative frequency; and evaluation of the effectiveness of safety devices installed in vehicles—such as seat belts, air bags, penetration resistant windshields, or dual master brake cylinders.

State agencies maintain census files of crashes investigated by police. The files often do not have detailed vehicle and injury information needed for descriptive analyses or evaluations. The Federal government and the motor vehicle manufacturers have been sponsoring in-depth investigations of samples of crashes since the 1950s. Only since the mid 1970s, however, have the crashes been selected by probability sampling. Since severe crashes are rare, stratified sampling with unequal proportions is used to enlarge the number of severe crashes in the sample (e.g., perform detailed investigations of all fatal crashes and smaller proportions of crashes at lower severities). Cluster sampling (of counties or police jurisdictions) is needed because travel is a major expense. The National Accident Sampling System is a stratified two stage cluster sample of the nation's crashes [1, 3].

The most complex statistical use of accident data is the evaluation of the effectiveness of vehicle safety equipment. It has rarely been possible to perform controlled experiments where two fleets are identical except for the presence or absence of one safety device. That is because typical safety devices such as high penetration resistant windshields are installed in every care of a certain model year (1966 in this case) and thereafter, but in few if any cars before that model year. The appropriate accident or injury rate (facial lacerations due to windshield contact) in the "after" cars (model year 1966 and later) is tested to see if it is lower than in the "before" cars. The problem is that there are other factors—some real and some spurious—that make the "after" cars have lower injury rates than the "before" cars. Statistical techniques separate the effect of the safety device from these extraneous effects. Since the dependent and primary independent variables are usually categorical (lacerated versus not lacerated, type of windshield), categorical data analysis techniques are needed. The most important extraneous factors that cause more recent cars to have lower accident and injury rates than older cars are:

1. Other safety devices installed a few years before or after the one being evaluated.

2. Older cars are generally driven in a more unsafe manner than new ones and tend to be in more severe accidents. Evidence for this effect is sometimes provided by "control" variables on the data file, such as crash speed, vehicle weight, or driver age, whose values for older cars are typically the ones associated with higher injury risk.

3. Owners of older cars often do not bother to report minor accidents. That results in spuriously high reported injury rates.

4. Better highways and other factors unrelated to vehicle equipment have caused a secular trend toward lower accident and injury rates. As a result, the cars of earlier model years had relatively high injury rates even when they were young.

Researchers gradually became aware of these biases *circa* 1970 and began to develop remedies [6]. One approach to sorting out the effect of a safety device from nuisance factors is to use multidimensional contingency table analyses, such as those developed by Koch, Kullback, Bishop, and Fienberg. Koch and his students applied GENCAT to vehicle safety statistics in the early 1970s [11]. Early versions of this approach could only handle a few control variables and were unsuitable for data gathered by stratified sampling with unequal proportions among the strata. In a 1981 evaluation of the effect of energy absorbing steering assemblies on the injury risk of drivers, Kahane developed a method especially suited to categorical data obtained from accident files [4]. The method resembles stepwise regression* in that control variables are added one at a time to the model, based on how much of the bias in the effectiveness estimate they are responsible for. Mantel–Haenszel* estimators are used to estimate the effect of the main independent variable (type of steering assembly) on the dependent variable (injury severity) after standardizing (poststratifying) the data for various combinations of control variables (speed, vehicle weight, driver age, etc.), but only after the celled data are first "smoothed" by multidimensional contingency table* analysis. In the first step the procedure is applied to each of the three-dimensional tables formed by the dependent variable, the principal independent variable, and one of the control variables. The control variable which results in the greatest change from the original effectiveness estimate is selected. In the second step the procedure is applied to the four-way tables formed by the dependent variable, the principal independent variable, the control variable selected in the preceding step, and one of the remaining control variables, and so on. A jackknife* technique is used to get an empirical measure of sampling error with this complex effectiveness estimator and these data from a stratified sample.

This procedure only works well with the in-depth accident files, because the states' accident censuses lack detailed potential control variables. With state files, one approach is simply to gather enough accident cases for an adequate sample of cars of the very first model year with the safety device versus the last one before it was installed. Then, as a check, extend the analysis to ±2 model years, ±3 model years, etc. It is also possible to use multiple regression if there are accident data from many different calendar years: In the earliest years, the unimproved cars are still young while in the latest, the improved cars are already old. The accident rate is modeled as a function of pre/postimprovement, vehicle age, and calendar year. Thus the vehicle age and calendar year biases (items 2, 3 and 4 just listed) are removed [5].

Controlled experiments are possible when crashes are staged in the laboratory or proving ground, with anthropomorphic dummies as surrogates for vehicle occupants. By the 1970s it became clear that a principal statistical task is to relate a categorical dependent variable (injured versus uninjured) to continuous intermediate variables (stress measurements on the dummy). A variety of statistical techniques have been used to model these relationships: logistic regression on aggregate data [7], logistic regression* on disaggregate data, using maximum likelihood* techniques [8], probit analysis [2], and the Weibull distribution* [10].

References

[1] Edmonds, H. J., Hanson, R. H., Morganstein, D. R., and Waksberg, J. (1979). *National Accident Sampling System: Sample Design*, *Phases 2 and 3*, National Highway Traffic Safety Administration, Washington, DC.

[2] Eppinger, R. H., Marcus, J. H., and Morgan, R. M. (1984). *Development of Dummy and Injury Index for NHTSA's Thoracic Side Impact Protection Research Program*. Society of Automotive Engineers, Warrendale, PA.

[3] Kahane, C. J. (1976). *National Accident Sampling System: Selection of Primary Sampling Units*. National Highway Traffic Safety Administration, Washington, DC.

[4] Kahane, C. J. (1981). *An Evaluation of Federal Motor Vehicle Safety Standards for Passenger Car Steering Assemblies*. National Highway Traffic Safety Administration, Washington, DC.

[5] Kahane, C. J. (1983). *An Evaluation of Side Marker Lamps for Cars, Trucks and Buses*. National Highway Traffic Safety Administration, Washington, DC.

[6] Kahane, C. J. (1984). *The National Highway Traffic Safety Administration's Evaluations of Federal Motor Vehicle Safety Standards*. Society of Automotive Engineers, Warrendale, PA.

[7] Kahane, C. J. (1986). *An Evaluation of Child Passenger Safety: The Effectiveness and Benefits of Safety Seats*. National Highway Traffic Safety Administration, Washington, DC.

[8] Kroell, C. K., Allen, S. D., Warner, C. Y., and Perl, T. R. (1986). Interrelationship of Velocity and Chest Compression in Blunt Thoracic Impact to Swine II. In *Thirtieth Stapp Car Crash Conference*. Society of Automotive Engineers, Warrendale, PA.

[9] *National Accident Sampling System, 1983—A Report on Traffic Accidents and Injuries in the U.S.* (1985). National Highway Traffic Safety Administration, Washington, DC.

[10] Ran, A., Koch, M., and Mellander, H. (1984). Fitting Injury versus Exposure Data into a Risk Function. In *1984 International IRCOBI Conference on the Biomechanics of Impacts*. Research Institute for Road Vehicles, Delft, Netherlands.

[11] Reinfurt, D. W., Silva, C. Z., and Hochberg, Y. (1976). *A Statistical Analysis of Seat Belts Effectiveness in 1973–75 Model Cars Involved in Towaway Crashes*. National Highway Traffic Safety Administration, Washington, DC.

(CONTINGENCY TABLES
TRAFFIC FLOW, STATISTICS OF)

CHARLES J. KAHANE

VENN DIAGRAM

Venn diagrams [named after the British logician J. Venn (1834–1923)] are used to illustrate some properties of the *algebra of sets**. They are also called *Ballantine diagrams*.

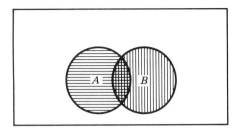

In the above diagram the points interior to the *rectangle* constitute the "universal set*." Arbitrary sets within the universal set (*subsets* of the universal set) are represented by points interior to circles within the rectangle. In the diagram, set A is indicated by horizontal hatching, set B by vertical hatching. The intersection $A \cap B$ appears as the cross-hatched area.

VERSHIK'S DISTRIBUTION

These constitute a class of multivariate distributions defined by Vershik [7]. The class is defined by the property that all linear functions of the random variable having the same variance have the same distribution. Formally, denoting the random variables by $\mathbf{Y} = (Y_1, \dots, Y_p)'$, then all linear functions $\mathbf{a'Y}$ with the same value of var($\mathbf{a'Y}$) have the same distribution. Clearly, apart from possible multiplicative factors, this is also the marginal distribution of each of the p random variables.

This class is a (relatively minor) generalization of the class of spherical distributions* (see also refs. 2, 5, and 6). For joint distributions with finite second moments it is identical with the family of elliptic distributions (e.g., Chu [1] and Kelker [4]), having the property that the joint density function ($f_{\mathbf{Y}}(\mathbf{y})$) depends only on the value of a positive definite quadratic form in $\mathbf{y}$. (*See* ISOTROPIC DISTRIBUTIONS).

For further details see ref. 3.

References

[1] Chu. K. C. (1973). *IEEE Trans. Auto. Control.* **AT-18**, 499–505.

[2] Dempster, A. P. (1969). *Elements of Continuums Multivariate Analysis.* Addison-Wesley, Reading, MA.

[3] Devlin, S. J., Gnanadesikan, R., and Kettenring, J. R. (1976). In *Essays in Probability and Statistics (Ogawa Volume).* Shinko Tsusho, Tokyo, Japan.

[4] Kelker, D. (1970). *Sankhyā A*, **32**, 419–430.

[5] Kingman, J. F. C. (1972). *Biometrika*, **59**, 492–494.

[6] Lord, R. D. (1954). *Biometrika*, **41**, 44–55.

[7] Vershik, A. M. (1964). *Theor. Prob. Appl.*, **9**, 353–356.

(EXCHANGEABILITY
ISOTROPIC DISTRIBUTIONS
SPHERICAL DISTRIBUTIONS)

VERSIERA   *See* WITCH OF AGNESI

VESTNIK STATISTIKI

Vestnik Statistiki (The Messenger of Statistics) commenced publication in Moscow in 1919 as the organ of the Central Statistical Office (Tsentralnoe Statisticheskoe Upravlenie) of the USSR, and ceased publication in 1929 due to ideological and political pressures on statistics prevailing in the USSR at that time ([2, Sec. 4]). During 1919–1929, it represented the central repository of scientific Soviet statistical thinking, publishing (*inter alia*) methodological developments in mathematical statistics by such researchers as B. S. Iastremsky, N. S. Chetverikov, E. E. Slutsky*, V. I. Romanovsky*, A. A. Chuprov*, E. J. Gumbel, A. Ia. Khinchin, and N. V. Smirnov. Chetverikov was active in the *Vestnik* from the outset in his role as Director of the Section for Statistical Methodology at the Central Statistical Office, and a leader of its Circle for Mathematical Statistics and Probability Theory. Iastremsky became one of the three editors in about 1927. The journal by then had begun to take a politically doctrinaire

ideological line under the influence of another editor, M. N. Smit. During 1919–1929, there were 35 issues of the journal; a table giving a listing of issues by year, number, and volume is given in [2], since the labeling on the issues themselves is incomplete.

The functions of the *Vestnik* in respect of official statistics were taken over in 1930 by the journal *Planovoe Khozaistvo (Planned Economy)* until the *Vestnik* resumed publication in 1949. Detailed subject, table, and author indexes for the 50 years, 1919–1968, are given in [1], although specification is by year, issue, and page, but not volume number.

References

[1] Onoprienko, G. K., Onoprienko, A. N., and Gelfand, V. S. (1971). *Bibliograficheskii Ukazatel Statei i Materialov po Statistike i Uchetu. Zhurnal Vestnik Statistiki za 50 let (1919–1968).* Statistika, Moscow, USSR.

[2] Seneta, E. (1985). A sketch of the history of survey sampling in Russia. *J. R. Statist. Soc. A*, **148**, 118–125.

E. SENETA

VINOGRAD THEOREM

A theorem in matrix algebra, often used in the theory of multivariate analysis* for problems involving decomposition of "matrices." It states (Vinograd [1]):

Let $\mathbf{A}$ and $\mathbf{B}$ be two real matrices of dimensions $r \times s$ and $r \times t$, respectively, with $s \leqslant t$. Let $\mathbf{A}'$ ($\mathbf{B}'$) be the transpose of $\mathbf{A}$ ($\mathbf{B}$). The equality $\mathbf{AA}' = \mathbf{BB}'$ holds if and only if there exists a $s \times t$ orthogonal matrix $\mathbf{T}$ (i.e., $\mathbf{TT}' = \mathbf{I}_s$ where $\mathbf{I}_s$ is the $s \times s$ identity matrix) such that $\mathbf{AT} = \mathbf{B}$.

Reference

[1] Vinograd, B. (1950). *Proc. Amer. Math. Soc.*, **1**, 159–161.

(MATRICES IN STATISTICS
MULTIVARIATE ANALYSIS)

VINTAGE METHOD   *See* HEDONIC INDEX
NUMBERS

VIRTUAL WAITING-TIME PROCESS
See TAKÁCS PROCESS

VITAL STATISTICS

Vital statistics may be defined as the end
product of the collection, compilation, and
presentation of a select group of statistical
data associated with the occurrence of
"vital" events, defined as those related to
certain life-and-death processes. As a minimum, they usually include live births, deaths,
fetal deaths, marriages, and divorces. Some
countries also include adoptions, legitima-
tions, annulments, and legal separations in
the definition of vital events, and abortions
as a separate component of fetal deaths.
Today, in most developed countries and in
many developing countries, an individual
document, or a book-type record, is prepared for each vital event soon after occurrence and recorded or registered with some
governmental agency. These documents contain the raw material from which vital statistics for a specified geographic area are
derived.

The most important items, from a vital
statistics point of view, usually found on the
live birth and fetal death documents are date
and place of occurrence, place of residence
of mother, her marital status, ages and ethnic
group of parents, sex and birth order of
child, gestation, birth weight, and name of
attendant at birth. The death record may
include date and place of occurrence, the
decedent's place of residence, age, sex, ethnic
group, and occupation, cause of death, and
certifier. Marriage* and divorce records may
include date and place of occurrence, age
and place of residence of bride and groom or
divorcees, and previous marital status of
bride and groom. These data are compiled

and often analyzed for a specified geographic area, and presented as vital statistics
for the area.

Clear definitions of the components of
vital statistics are necessary for accurate statistics. The World Health Organization and
the United Nations made detailed recommendations in the early 1950s toward the
standardization of definitions and they have
been adopted by most countries that collect
vital statistics data. Briefly stated, a *live birth*
is the complete expulsion or extraction from
its mother of a product of conception, irrespective of birth weight or duration of pregnancy, which shows some evidence of life
after separation. Evidence of life is established by the presence of breathing,
heartbeat, or movement of any voluntary
muscle. The definition of *fetal death* is similar to live birth except that there is no evidence of life after separation. *Death* is the
permanent disappearance of all evidence of
life at any time after live birth. *Marriage* is
the legal union of persons of opposite sex,
and *divorce* is the legal dissolution of a
marriage.

Responsibility for reporting vital events
varies from country to country. In most
countries, the parents or the attendant at
delivery are responsible for reporting a live
birth or fetal death and the information relating to it; the nearest relative, funeral director, or attending physician is responsible
for reporting a death; the officiant or bride
and groom are responsible for reporting a
marriage; and the court or parties to a divorce are responsible for reporting a divorce.
For more detailed definitions of these events
and registration responsibilities, see Shryock
and Siegel [8].

HISTORICAL DEVELOPMENT

Registration of vital events has had a long
history, which began with the ecclesiastical
recording of burials, baptisms, and weddings
in Europe and in the Orient dating back to
the Middle Ages and before. This was fol-

lowed by governmental or civil registration of births, deaths, and marriages in several countries of the Western Hemisphere in the sixteenth and seventeenth centuries primarily for legal and sometimes military purposes (*see also* BIOSTATISTICS). According to the *United Nations Handbook of Vital Statistics Methods* [10], secularization of registration of vital events continued with the adoption of the Napoleonic Code in France in 1804. The civil section of that code strongly influenced the development of vital registration systems throughout Western Europe, Latin America, and parts of the Middle East. It placed responsibility for recording births, deaths, and marriages on the state and set forth provisions for determining who would report events and what items would be included in the records. Registration of vital events in the Orient is believed to have had its origin in China, independent of the European influence, and spread to Japan and gradually to other countries in Asia.

Aside from registration of vital events for legal and military reasons, the utility of vital statistics for other uses lay dormant for a long time. John Graunt, an English statistician, is generally credited with being the first to use registration records for statistical purposes. In 1662, he published his *Natural and Political Observations* [3] based upon his creative work with the London Bills of Mortality. The Bills of Mortality were established by ordinance in 1532 and required parish priests in London to record the burials on a weekly basis upon reports of death from "ancient matrons." In 1538, the law was extended to include weekly entries of weddings and baptisms by Anglican priests. Graunt devised methods of estimating mortality rates by age, when age was not part of the death record, and developed methods of estimating population size from birth rates based on burials and baptisms. Sir William Petty also made important contributions to the field of vital statistics at about the same time through extensive studies with the London Bills of Mortality, including the development of life-table concepts. In 1837, the Births, Marriages, and

Deaths Registration Act became effective in England and a central records office was established. Dr. William Farr (1807–1883), another English statistician and physician, made history in that office as an early pioneer in the creation of a national system of vital statistics. In 1839, he was appointed to the General Register Office as Compiler of Abstracts, where he spent the next 40 years in developing the vital statistics system of England and Wales. He is credited with initiating the first regular publication of vital statistics by a government office.

In the United States, governmental efforts toward a vital statistics registration system on a national basis began with the creation of the Death Registration Area in 1880 with two states and a few cities as members, and the Birth Registration Area in 1915 with ten States and the District of Columbia initially meeting the registration requirements for admission. The admission requirements were adoption of a recommended model law and standard certificates, plus evidence of 90% completeness in registration of births or deaths. In 1933, all of the then 48 contiguous states had met the requirements for registration of births and deaths. The Marriage and Divorce Registration Areas, begun in 1957 and 1958, respectively, are still developing. The civil vital registration system in the United States is, therefore, of recent origin.

Although registration of vital events is probably the most reliable method of obtaining vital statistics, it is by no means the only method used for this purpose. Censuses and surveys are widely used, particularly in developing countries, to provide vital statistics information on a periodic basis. Both of these methods have limitations related to the fact that they provide vital statistics data only at irregular intervals and generally only for large geographic areas. They also have problems in obtaining complete and accurate reporting of vital events because of memory loss on the part of respondents with respect to time and place of occurrence, dissolution of households following the death of certain members, deliberate concealment of the fact of death, and other reasons. Nevertheless, historically, censuses and surveys have played a major role in the acquisition of vital statistics, especially during the period prior to the development of civil registration in a country. See World Fertility Survey [12], Chandrasekar and Deming [2], and Marks et al. [6] for methodological aspects of sample surveys in vital statistics and some applied results.

Sample registration areas as a means of obtaining estimates of vital statistics have been used in several countries, including India, Pakistan, Turkey, and Peru. One variant of this method involves the selection of a probability sample of registration areas in the country, implementation of intensive registration procedures in those selected areas, and making inferences from the sample to a broader universe. Some others are experimental in nature designed only to yield estimates for the specific area and to improve registration procedures for broader use at a later date. For further details on the historical development of vital statistics, see Greenwood [4], United Nations [10, 11], Shryock and Siegel [8], Benjamin [1], Powell [7], and International Institute for Vital Registration and Statistics [5].

USES OF VITAL STATISTICS

The uses of vital statistics are numerous and varied, particularly when population data obtained from periodic population censuses are available as a supplement. This is the common situation since census data are usually available wherever one finds satisfactory vital statistics information. Some of the more important uses of vital statistics will be discussed here. Other uses may be found in standard demography* textbooks, such as Spiegelman [9] and Benjamin [1].

Fertility*

Fertility is measured by the number of births occurring in a given population over a specified period of time. Characteristics of the births and of the parents are useful in the measurement of population change and in predictions of live births for the future. Variables of prime importance are ages of parents, marital status, parity of mother, order of birth of the child, interval since previous birth or since marriage, sex of child, education, and ethnic group of parents. These variables are generally available from vital statistics in terms of number of events cross-tabulated by these characteristics to describe the population of births in a geographic area over a specified time period for comparison with other areas and other time periods.

Fertility measures take on an added dimension when vital statistics on births are combined with population data to produce certain rates and ratios for comparative purposes over time and place. Notable among these are the crude and age-specific birth rates, general and age-specific fertility rates, marital fertility rates, legitimate and illegitimate rates and ratios, age–sex adjusted birth rates, total fertility rate, gross and net reproduction rates, and a host of others including birth probabilities.

Mortality*

The demographic analysis of mortality is dependent on vital statistics describing the event of death and the characteristics of the decedent. Of particular importance in mortality analysis is age, sex, and ethnic group of the decedent. Cause of death is fundamental in epidemiologic and public health studies in combination with other information available from vital statistics and population data. Crude and age-specific death rates, cause-specific rates and ratios, and adjusted death rates of many varieties are just a few of the many indices used in mortality research.

Age-specific mortality from vital statistics is the primary ingredient of the life table*, a statistical technique commonly used by demographers, actuaries, and others interested in longevity, probability of survival from one age to another, expectation of life, and other related applications. The most common type of life table is generated from age-specific mortality rates as of a particular period of time, usually from one to three years. In recent years, however, the life-table method has been applied to many areas outside the field of mortality as, for example, in public health, contraceptive evaluation, and economics, and the list of applications continues to grow. The foundation of the method rests on vital statistics, however, and its history can be traced back to the early pioneers mentioned above.

Marriage and Divorce

The utility of marriage and divorce statistics is not as well defined and developed as that of births and deaths. The latter events directly affect population growth, whereas the former do so only indirectly. Nevertheless, there is considerable interest in the vital statistics of marriages and divorces, particularly among sociologists, for their economic and social implications as well as their impact on fertility and mortality. Indices derived from marriage and divorce registration data in combination with population data include crude and general marriage rates, age–sex-specific and order-specific rates, marriage probabilities, nuptiality tables using life-table principles, adjusted rates, and others. A similar set of statistical indices based on divorce statistics is available and commonly used for specialized purposes.

Other Uses

Included in this category are vital statistics useful in themselves without reference to a population base. In the initiation, execution, and evaluation of public-health programs, it is useful to know, for example, the number of infant deaths, number of deaths by cause of death, number of births with congenital malformations, and number of births to teenagers. This information is valuable in assessing the magnitude of potential health problems in the community and in determining the need for intervention programs. Some

methods of population estimates and projections are dependent upon the numbers of births and deaths occurring in selected geographic areas during a particular time period. An example is the use of the basic demographic equation

$$P_t = P_0 + B - D + M,$$

where P_t is the estimated population at the end of the period; P_0 is the population at the beginning of the period; and B, D, and M represent births, deaths, and net migration*, respectively, during the period. Several variations of this equation are used in the study of population change.

PUBLICATION OF VITAL STATISTICS

Publication practices of countries collecting vital statistics vary considerably with regard to the form of publication and, indeed, whether or not formal publication is carried out at all. For a number of years the United Nations has attempted to obtain available vital statistics from member countries and publish them on a periodic basis. Two of the most widely used of these publications are the *Demographic Yearbook* and the *Statistical Yearbook*, which have been published annually since 1948. These publications are not confined exclusively to vital statistics but contain other types of demographic and statistical data for each country. The U.S. Bureau of the Census* also publishes international data on topics related to demography, which in some cases include vital statistics gleaned from various in-country sources.

Vital statistics data for the United States are published annually by the National Center for Health Statistics of the Department of Health and Human Services, in the *Vital Statistics of the United States* series: Volume I, *Natality*; Volume II, *Mortality*; and Volume III, *Marriage and Divorce*. Although the titles and agency responsible have changed several times over the years, volumes of mortality statistics are available from 1900 and natality volumes from 1915. Prior to 1933, data were presented only for those states belonging to the national registration areas for births and deaths, discussed above.

References

[1] Benjamin, B. (1959). *Elements of Vital Statistics*. Quadrangle Books, Chicago, IL.

[2] Chandrasekar, C. and Deming, W. E. (1949). *J. Amer. Statist. Ass.*, **44**, 101–115.

[3] Graunt, J. (1665). *Natural and Political Observations Mentioned in a Following Index, and Made Upon the Bills of Mortality*, 4th ed. Oxford (printed by William Hall for John Martyn and James Allestry, Printers to the Royal Society), London, England.

[4] Greenwood, M. (1948). *Medical Statistics from Graunt to Farr*. Cambridge University Press, Cambridge, England.

[5] International Institute for Vital Registration and Statistics (1984). *Improving Civil Registration*, F. E. Linder and I. M. Moriyama, eds. Bethesda, MD.

[6] Marks, E. S., Seltzer, W., and Krotki, K. J. (1974). *Population Growth Estimation: A Handbook of Vital Statistics Measurement*. The Population Council, New York.

[7] Powell, N. P. (1975). The Conventional Vital Registration System. *Scientific Report Series No. 20*, International Program of Laboratories for Population Statistics, University of North Carolina at Chapel Hill, NC.

[8] Shryock, H. S. and Siegel, J. S. (1971). *The Methods and Materials of Demography*, Vols. I and Vol. II. U.S. Department of Commerce, Bureau of the Census, Washington, D.C.

[9] Spiegelman, M. (1968). *Introduction to Demography*. Harvard University Press, Cambridge, MA.

[10] United Nations (1955). *Handbook of Vital Statistics Methods, Series F, No. 7*, United Nations Statistical Office, New York.

[11] United Nations (1973). Principles and Recommendations for a Vital Statistics System. *Statistical Papers, Series M, No. 19, Rev. 1*, United Nations, New York.

[12] World Fertility Survey (1984). *World Fertility Survey: Major Findings and Implications*. Alden Press Oxford, London and Northampton, England.

(ACTUARIAL STATISTICS, LIFE
BIOSTATISTICS
DEMOGRAPHY
FERTILITY
LIFE TABLES
MARRIAGE
MIGRATION
RATES, STANDARDIZED)

J. R. ABERNATHY

V-MASK   *See* CUMULATIVE SUM CONTROL CHARTS

V_N-TEST   *See* GOODNESS OF FIT

VOLTERRA DERIVATIVE   *See* STATISTICAL FUNCTIONALS

VOLTERRA EXPANSIONS

These are models of the form

$$Y_t = \mu + \sum_{u=-\infty}^{\infty} a_u \epsilon_{t-u} + \sum_{u,v=-\infty}^{\infty} a_{uv} \epsilon_{t-u} \epsilon_{t-v}$$

$$+ \sum_{u,v,w=-\infty}^{\infty} a_{uvw} \epsilon_{t-u} \epsilon_{t-v} \epsilon_{t-w} + \cdots,$$

where $\epsilon_t : -\infty < t < \infty$ is a strictly stationary process*. [The $(n+1)$st term involves a n-fold summation.] The successive terms on the right-hand side are usually referred to as the linear, quadratic, cubic,... terms. They are the most general form of nonlinear *stationary* time-series models. Unfortunately, they involve a large number of parameters. Keenan [1] discusses tests of linearity (vs. second-order Volterra expansion) in a stationary time series.

Reference

[1] Keenan, D. McR. (1985). *Biometrika*, **72**, 39–44.

(STOCHASTIC PROCESSES
TIME SERIES)

VON MISES, RICHARD MARTIN EDLER

Born: April 19, 1883, in Lemberg (Lvov), Russia.

Died: July 14, 1953, in Boston, Massachusetts.

Contributed to: theoretical and practical statistical analysis, theory of probability, geometry, aerodynamics, hydrodynamics, (calculus).

Von Mises, one of the greatest mathematicians in the field of applied mathematics in the twentieth century, was born in Lemberg (Lvov), the son of Arthur von Mises, who held a doctoral degree from the Institute of Technology in Zürich and who was a prominent railroad engineer in the civil service. Among his ancestors are engineers, physicians, bankers, philologists, and bibliophiles.

Von Mises studied mathematics, physics, and mechanical engineering at the Vienna Technical University during 1901–1906. After finishing his studies, he became an assistant to the young professor of theoretical mechanics and hydrodynamics, Georg Hamel, at the Technical University in Brünn (Brno). In 1908, von Mises graduated in Vienna with an inaugural dissertation entitled "Ermittlung der Schwungmassen im Schubkurbelgetriebe," obtaining the *venia legendi*; one year later he was offered a chair as associate professor of applied mathematics at the University of Strasbourg. At the beginning of World War I, he joined the Flying Corps of the Austro-Hungarian Army and acquired a pilot's license, but after a short time he was recalled from service in the field in order to act as a technical instructor and teacher in flight theory to German and Austrian officers. These lectures were published in several printings under the title *Fluglehre*. In 1919, he was a professor of stress analysis, hydro- and aerodynamics at the Technical University in Dresden and then went to the University to Berlin, where he was appointed as professor of applied mathematics and the first director of the newly founded Institutes of Applied Mathematics. In the following years von Mises was engaged in developing practical and theoretical research in his institute and introduced a modern curriculum for the students, including lectures, seminars, and practical exercises.

When, in 1933, the night of fascism descended in Germany, von Mises had to quit his successful position in Berlin. He accepted an invitation to organize mathematical education at the University of Istanbul. Six years later he emigrated to the United States, became a lecturer in the School of Engineering at Harvard University, and, soon after, was appointed Gordon McKay Professor of Aerodynamics and Applied Mathematics. In the first years the situation was not easy for von Mises. He represented the traditions of European applied mathematics, which was influenced by the ideas of the École Polytechnique with its high theoretical level. Von Mises tried to transplant these European traditions into the United States, where there was more emphasis on practical know-how. But he soon gathered interested pupils and organized an active scientific life. In 1921, von Mises founded the journal *Zeitschrift für Angewandte Mathematik und Mechanik* and he later edited *Advances in Applied Mechanics*. von Mises died after a long illness in Boston in 1953. In recognition of his many-sided life-work he was made a member of several academies in Europe and America.

Von Mises enriched several fields of applied mathematics. A feature of his work was a striving for clarity and understanding, and explanation of observations based strictly on the principles of mechanics. His work reflects the scholarship for which von Mises stood, an inseparable blend of teaching and research that it is difficult to overestimate. There is no doubt, that von Mises has had a great influence on the development of applied mathematics in the first half of the twentieth century.

Shortly before his death, von Mises divided the topics of his researches into eight themes: practical analysis, differential and integral calculus, mechanics, hydrodynamics, aerodynamics, constructive geometry, theory of probability and statistics, and philosophy. He also obtained important results in plasticity (yield condition of plasticity), in the theory of approximation and development of numerical methods for the solution of differential equations, equation systems, and calculation of integrals. von Mises extended the notion of "vector" in his *Motorrechnung* and created new mathematical tools for mechanics. During his period of greatest activity in Berlin, he organized a classic two-volume treatise, *Die Differential- und Integralgleichungen der Mechanik und Physik*, with a galaxy of outstanding coauthors, including Ph. Frank, who edited the second part of this work, which deals with physical problems. Von Mises himself wrote important contributions on the mathematical theory of fluids, hydrodynamics, boundary-value problems of second order, tensor analysis, and other topics in the first volume, with great pedagogical dexterity and experience.

Perhaps his greatest achievement was a new approach to the foundations of the theory of probability.

The basic notion upon which von Mises' probability theory is established is the concept of "Kollektiv" (collective), a theoretical model of the empirical sequences or populations. It consists of an infinite sequence of events, experiments, or observations, each of them leading to a definite numerical result, provided that the sequence of results fulfills two conditions: existence of limiting frequencies and invariance of this limit for any appropriately selected subsequence of phenomena as a specific expression of randomness. Then the limiting value of relative frequency is called the " 'probability of the attribute in question within the collective involved.' The set of all limiting frequencies within one collective is called its distribution." Hence von Mises' theory of probability is a frequency theory. By combining four fundamental operations, called *place selection*, *mixing*, *partition*, and *combination*, it is possible to settle all problems in probability theory. In von Mises' sense, it is not the task of this theory to determine the numerical values of probabilities of certain events; it is a mathematical theory like the mathematical theory of electricity or mechanics, based on experience but operating by means of the methods of analysis of real variables and theory of sets. Indeed, von Mises' frequency theory can be formulated in terms of measure theory* for definable cases, but, as a consequence of holding fast to conceptual verification of the collectives, he did not seem to be able to construct a theory of stochastic processes*. Also it is impossible to discover new theorems and propositions with von Mises' concept, the merit of which is still today the immediate connection between the theory of probability and observed regularities of random mass events.

There were severe objections against the notion of the "collective," the existence of a limiting value of relative frequency in a collective, the selection of any subsequences having probabilities, and especially against the "irregularity" condition [1]. After long discussions it was possible, with the modern mathematical methods of the theory of algorithms and some computational complexity, to show the connections between the existence of limit values of relative frequency in a sequence and its random character [3, 18, 19]. In particular, it was demonstrated that a form of convergence (apparent convergence) can be defined, which is relevant to von Mises' definition of probability, with simultaneous realization of irregularity [4, 6, 9, 17, 19].

Von Mises extended his probability approach to diverse problems in statistics. In 1931, he wrote a paper "Über einige Abschätzungen von Erwartungswerten" [11], in which he calculated not only the approximate values, but also the limits of their failures. He dealt with the limits of a distribution function [14] and laws of large numbers for statistical functions [12] and developed a generalization of the classical limit theorem [13].

The von Mises distribution* on the circle is defined by the density

$$M_2(\theta k) = M_2(\theta) = \frac{1}{2\pi I_0(k)} \exp\{k \cos\theta\},$$

where $-\pi \leqslant \theta < \pi$, $k > 0$, is a concentration parameter. The sequence $\{E \cos n\theta\}_{n=0}^{\infty}$ gives the Fourier coefficients of the von Mises distribution given by $I_t(k) = I(t; k)$, the modified Bessel function* of the first kind. The distribution is infinitely divisible* for all values of the parameter in all dimensions [5, 8]. The Cramér–von Mises test* [16] tests the significance of the difference between the empirical distribution of a stochastic variable in a sample and a hypothetical distribution function [2]. Von Mises did not believe that statistical explanations in physics are of only transient utility.

In his philosophical thought, he described himself as a "positivist" [15], but more precisely he should be called an "inconsequent materialist," since he assumed the existence of a world outside of the consciousness of a certain mind.

References

[1] Chinčin, A. J. (1961). Častotaja teorija R. Mizesa i sovremennye idei teorii verojatnostej. *Voprosy Filos.*, **1**, 92–102; **2** 77–89.

[2] Darling, D. A. (1957). The Kolmogorov–Smirnov, Cramér–von Mises tests. *Ann. Math. Statist.*, **28**, 823–838.

[3] Feller. W. (1939). Über die Existenz von sogen. Kollektiven. *Fund. Math.*, **32**, 87–96.

[4] Fine, T. L. (1973). *Theories of Probability. An Examination of Foundations.* Wiley, New York.

[5] Kent, J. P. (1977). The infinite divisibility of the von Mises–Fisher distribution for all values of the parameter in all dimensions. *Proc. Lond. Math. Soc.* (*3*), **35**, 359–384.

[6] Kolmogorov, A. N. (1963). On tables of random numbers. *Sankhyā A*, **24**, 369–376.

[7] Kolmogorov, A. N. (1965). Tri podchoda k opredeleniju ponjatija količestvo informacii. *Problemy Peredači Informacii*, **1**, 3–11.

[8] Lewis, T. (1975). Probability functions which are proportional to characteristic functions and the infinite divisibility of the von Mises distribution. In *Perspectives in Probability and Statistics: The M. S. Bartlett Volume.* Academic, New York, pp. 19–28.

[9] Loveland, D. W. (1966). A new interpretation of the von Mises' concept of random sequences. *Z. Math. Logik Grundl. Math.*, **12**, 279–294.

[10] Martin-Löf, P. (1966). The definition of random sequences. *Inform. Control*, **6**, 602–619.

[11] Mises, R. von (1931). Über einige Abschätzungen von Erwartungswerten. *J. Reine Angew. Math.*, **165**, 184–193.

[12] Mises, R. von (1936). Das Gesetz der großen Zahlen für statistische Funktionen. *Monatsh. Math. Phys.*, **4**, 105–128.

[13] Mises, R. von (1938). Generalisation des théorèmes de limites classiques. *Coll. Théor. Prob., Genève, Act. Sci. Ind.* N. 737-57-66.

[14] Mises, R. von (1939a). Limits of a distribution function if two expected values are given. *Ann. Math. Statist.*, **10**, 99–104.

[15] Mises, R. von (1939b). *Kleines Lehrbuch des Positivismus.* The Hague, Netherlands.

[16] Mises, R. von (1947). On the asymptotic distribution of differentiable statistical functions. *Ann. Math. Statist.*, **18**, 309–348.

[17] Schnorr, C. P. (1971). *Zufälligkeit und Wahrscheinlichkeit. Eine Algorithmische Begründung der Wahrscheinlichkeitstheorie.* Springer, Berlin.

[18] Ville, J. A. (1936). Sur la notion de collectif. *C. R. Acad. Sci. Paris*, **203**, 26–27.

[19] Wald, A. (1937). Die Widerspruchsfreiheit des Kollektivbegriffs der Wahrscheinlichkeitsrechnung. *Ergebn. Math. Koll.*, **8**, 38–72.

Bibliography

Mises, R. von (1952). *Wahrscheinlichkeit, Statistik und Wahrheit*, 3rd ed. Vienna, Austria. [English translation: *Probability, Statistics and Truth.* New York (1957).]

Selected Papers of Richard von Mises, 2 vols. American Mathematical Society, Providence, RI.

In regard to von Mises' life there are only some short published reports:

Basch. A. (1953). Richard v. Mises zum 70. Geburtstag. *Österreich. Ingen.-Archiv.*, **VII**(2), 73–76.

Birkhoff, G. (1983). Richard von Mises' years at Harvard. *Z. Angew. Math. Mech.*, **43**, 283–284.

Collatz, L. (1983). Richard von Mises als numerischer Mathematiker. *Z. Angew. Math. Mech.*, **43**, 278–280.

Cramér, H. (1953). Richard von Mises' work in probability and statistics. *Ann. Math. Statist.*, **24**, 657–662.

Rehbock, F. (1954). Richard von Mises. *Physikal. Blätter*, **10**, 31.

Sauer, R. (1953). Richard von Mises. 19. 4. 1883–14. 7. 1953. *Jahrbuch Bayr. Akad. Wiss.*, 27–35.

Temple, G. (1953). Prof. R. von Mises. *Nature*, **172**, 333.

A biographical sketch, based on new archival material is contained in:

Bernhardt, H. (1979). Zum Leben und Wirken des Mathematikers Richard von Mises. *NTM*, **16**(2), 40–49.

Bernhardt, H. (1980). Zur Geschichte des Institutes für angewandte Mathematik an der Berliner Universität 1920–1933. *NTM*, **17**(1), 23–41.

A full treatment is given in:

Bernhardt, H. (1984). Richard von Mises und sein Beitrag zur Grundlegung der Wahrscheinlichkeitsrechnung im 20. dissertation, Jahrhundert, Humboldt Universität, Berlin, unpublished, pp. 255.

(FOUNDATIONS OF PROBABILITY)

HANNELORE BERNHARDT

VON MISES DISTRIBUTION *See* DIRECTIONAL DISTRIBUTIONS

VON MISES EXPANSIONS

A statistic $T(X_1, \ldots, X_n)$, based on the random sample of observations $X_1, \ldots, X_n$, may sometimes be expanded in a series of terms of decreasing probabilistic importance. If the statistic is symmetric in the observations, then it may be considered as a functional of the empirical distribution function* (EDF) F_n so that $T(X_1, \ldots, X_n) = T(F_n)$. As a simple example, the sample mean $\bar{X}$ may be written as $T_0(F_n) = \int x \, dF_n(x)$. The functional T may sometimes be extended to some convex class of distribution functions $\mathscr{F}$. The definition of the sample mean extends in this way to the class of distribution functions possessing a mean. One may write $T_0(F) = \int x \, dF$ for a distribution function F. (*See also* STATISTICAL FUNCTIONALS.)

In [6] von Mises introduced modified Volterra functional derivatives and used them to obtain expansions of statistics $T(F_n)$.

If for all $F, G \in \mathscr{F}$, one can write
$$\frac{d^r}{dt^r} T(F + t(G - F)) \bigg|_{t=0}$$

$$= \int \cdots \int T^{(r)}\{F; y_1, \ldots, y_r\} \prod_{i=1}^{r} d(G - F)(y_i), \tag{1}$$

where $T^{(r)}\{F; y_1, \ldots, y_r\}$ is symmetric in the y_i's and does not depend on G, then $T^{(r)}$ is called an *rth derivative* of $T(F)$. It can be made unique by insisting that for $i = 1, 2, \ldots, r$,

$$T^{(r)}\{F; y_1, \ldots, y_{i-1}, x, y_{i+1}, \ldots, y_r\} \, dF(x)$$
$$= 0. \tag{2}$$

One may then write
$$\frac{d^r}{dt^r} T(F + t(G - F)) \bigg|_{t=0}$$

$$= \int \cdots \int T^{(r)}\{F; y_1, \ldots, y_r\} \prod_{i=1}^{r} dG(y_i). \tag{3}$$

As is discussed in [7, p. 578], one may evaluate the higher derivatives step by step, along the lines of

$$\left[T^{(r)}(F; y_1, \ldots, y_r)\right]^{(1)}\{F; y_{r+1}\}$$

$$= T^{(r+1)}\{F; y_1, \ldots, y_{r+1}\}$$

$$- \sum_{i=1}^{r} T^{(r)}\{F; y_1, \ldots, y_{i-1}, y_{i+1}, \ldots, y_{r+1}\}. \tag{4}$$

For instance, if

$$T_1(F) = \int \{y - x \, dF(x)\}^2 \, dF(y) = \sigma_F^2,$$

the variance of F, then taking $r = 0$, $T_1^{(0)}(F) = T_1(F)$, and by direct evaluation of the derivative

$$\frac{d}{dt} T_1(F + t(G - F)),$$

$$T_1^{(1)}(F; y_1) = -\sigma_F^2 + (y_1 - \mu_F)^2,$$

where μ_F is the mean of F. Then, again by direct differentiation, and the rule (4) given previously

$$\left[-\sigma_F^2 + (y_1 - \mu_F)^2 \right]^{(1)} \{ F; y_2 \}$$

$$= \sigma_F^2 - (y_2 - \mu_F)^2 + 2(y_2 - \mu_F)(\mu_F - y_1)$$

$$= T_1^{(2)} \{ F; y_1, y_2 \} - \left[-\sigma_F^2 + (y_2 - \mu_F)^2 \right],$$

so

$$T_1^{(2)} \{ F; y_1, y_2 \} = -2(y_2 - \mu_F)(y_1 - \mu_F).$$

All the higher derivatives are identically 0.

A formal expansion of $T(F + t(G - F))$ in powers of t gives

$$T(G) = T(F) + \sum_{r=1}^{\infty} \frac{t^r}{r!}$$

$$\times \int \cdots \int T^{(r)} \{ F; y_1, \ldots, y_r \}$$

$$\times \prod_{i=1}^{r} dG(y_i),$$

and putting $t = 1$ and $G = F_n$, one obtains $T(X_1, \ldots, X_n) = T(F)$

$$+ \sum_{r=1}^{\infty} \sum_{i_1=1}^{n} \cdots \sum_{i_r=1}^{n} T^{(r)} \{ F; X_{i_1}, \ldots, X_{i_r} \} \Big/ (n^r r!).$$

(5)

Expansion (5) is the *von Mises expansion* of T. For example,

$$T_1(F_n) = \sum_{1}^{n} (X_i - \overline{X})^2$$

$$= \sigma_F^2 + \frac{1}{n} \sum_{i_1=1}^{n} \left[-\sigma_F^2 + (X_{i_1} - \mu_F)^2 \right]$$

$$+ \frac{1}{2n^2} \sum_{i_1=1}^{n} \sum_{i_2=1}^{n} \left[-2(X_{i_1} - \mu_F) \right.$$

$$\left. \times (X_{i_2} - \mu_F) \right].$$

The derivatives introduced in [6] are not the only possible choices. Reeds [4] contains an extended discussion of Gâteaux and Fréchet derivatives (*see* STATISTICAL FUNCTIONALS), as well as compact differentiation.

Von Mises expansions have been used in two ways. Following the lead given in [6] are several papers giving rigorous derivations of asymptotic distributions for statistics possessing expansions, at least to a low order.

There is some difficulty in giving general theorems, which cover enough cases of special interest. A good general account of these developments is given in [5]. The other main use for von Mises expansions has been in heuristically obtaining results on moments for jackknifed estimators. A pioneering paper [2] by Hinkley has been augmented by the papers of Frangos [1] and Knott and Frangos [3]. A complete summary of results that can be applied to jackknifing is given in [7], but no such applications are discussed in that paper. The results on variances to $O(n^{-2})$ are that if T is jackknifed with groups of m observations, the variance of the jackknife estimator is

$$\sigma_{11}/n + \sigma_{22}/\{2n(n-m)\}, \qquad (6)$$

and the mean value of the jackknife* estimate of variance for T is

$$\sigma_{11}/n + \sigma_{12}/\{n(n-m)\}$$

$$+ \sigma_{13}(n-m-1)/\{n(n-m)^2\}$$

$$+ \sigma_{22}(2n-3m-1)/\{2n(n-m)^2\}.$$

(7)

In (6) and (7) $\sigma_{11} = \mathrm{Var}[T^{(1)}\{F; X_1\}]$; $\sigma_{12} = \mathrm{Cov}[T^{(1)}\{F; X_1\} T^{(2)}\{F; X_1, X_1\}]$; $\sigma_{13} = \mathrm{Cov}[T^{(1)}\{F; X_1\}; T^{(3)}\{F; X_1, X_2, X_2\}]$; $\sigma_{22} = \mathrm{Var}[T^{(2)}\{F; X_1, X_2\}]$.

References

[1] Frangos, C. C. (1980). *Biometrika*, **67**, 715–718.

[2] Hinkley, D. V. (1978). *Biometrika*, **65**, 13–21.

[3] Knott, M. and Frangos, C. C. (1983). *Biometrika*, **70**, 501–504.

[4] Reeds, J. A. (1976). On the Definition of von Mises Functionals. Ph.D. dissertation, Harvard University, Cambridge, MA.

[5] Serfling, R. J. (1980). *Approximation Theorems of Mathematical Statistics*. Wiley, New York.

[6] von Mises, R. (1947). *Ann. Math. Statist.*, **18**, 309–348.

[7] Withers, C. S. (1983). *Ann. Statist.*, **11**, 577–587.

(ASYMPTOTIC EXPANSION
STATISTICAL DIFFERENTIALS, METHODS

OF
STATISTICAL FUNCTIONALS
VOLTERRA EXPANSIONS)

M. KNOTT

VON MISES–FISHER MATRIX DISTRI-
BUTION *See* MATRIX-VALUED DISTRIBU-
TIONS

VON NEUMANN STATISTIC *See* DUR-
BIN–WATSON TEST; MEAN SQUARE SUCCES-
SIVE DIFFERENCES

VOTING MODELS *See* PROBABILISTIC
VOTING MODELS

VOTING PARADOX

Voting is a mechanism that aggregates individual preference orderings on alternatives
into a collective social ordering. The trouble
with majority voting is that the method may
cause intransitivity. This problem was known
to the French philosopher Condorcet and
discussed in his famous 1785 *Essai*.
Condorcet had a pessimistic outlook on the
prospects of a transitive outcome from a
voting procedure in which simple majority
rule is applied.

The *paradox of voting* is a phenomenon
that may arise whenever a group of individuals endeavors to choose among a set of at
least three alternatives by simple majority
rule. The paradox is said to occur if in
pairwise voting there does not emerge a maximal element or dominant choice—a
Condorcet winner. The term has been used
alternatively to refer to the case in which one
alternative is favored by the voters, but there
is intransitivity among some other alternatives. Obviously, the two definitions are not
equivalent. In the case of three alternatives
and strong preference orderings, however,
the absence of a Condorcet winner and the
presence of circularity are indeed equivalent
conditions.

The usual example of the paradox of voting is posed in the following terms. Consider
a situation in which there are three persons
(1, 2, and 3) and three alternatives (A, B,
and C) and social choice is expressed in
pairwise comparisons. Conditions that yield
the paradox are:

1. $A \succ B \succ C$ ($\succ$ reads "is preferred to");
2. $C \succ A \succ B$; and
3. $B \succ C \succ A$.

The reader will notice that we have a Latin
square* arrangement in that each alternative
occurs once in each of the three possible
ordering positions. Inspection shows that
majority voting in this case leads to the
anomalous outcome: $A \succ B \succ C \succ A$, there
being a 2–1 majority in each comparison.

Economists are usually interested in the
transitivity property of the decision rule,
whereas political scientists focus attention
on the presence of a Condorcet winner. In a
more general context and under reasonable
conditions on collective choice rules, Arrow
[1] proved that there exists no mechanism
(rule, method, etc.) that integrates *every* pattern of individual orderings into a transitive
collective ordering. This is known as *Arrow's
impossibility theorem*. However, the election
of a Condorcet winner is appealing for a
number of reasons, among them its adherence to the principle of selection according to the will of the majority. Furthermore,
to select a Condorcet winner may promote
social stability in that no other alternative
can displace a Condorcet winner by a direct
vote between the two.

When the paradox of voting occurs, commonly used procedures for selecting one
candidate from a group of candidates are
susceptible to instability and strategic manipulation. Therefore it is important to have
some idea of how likely the paradox is in
various types of situations. Several aspects of
this problem have been of interest, two of
which are mentioned here. First, one has
tried to place restrictions on individual preference profiles necessary and/or sufficient

for the resulting social ordering to be transitive. Thus, e.g., single-peaked preferences disallow patterns such as the Latin square type in the previous example (cf. Sen [8]). The second approach is one of probabilistic analysis: finding the probability that certain types of anomalies occur, given that voters are allowed to choose among all possible preference profiles according to some probability distribution. Since empirical data are not readily available, resulting probability estimates are necessarily of an a priori nature. A notable exception, however, is the work of Riker [7], who cites several instances that suggest that the paradox has actually occurred in the U.S. Congress.

A more formal presentation of the problem is as follows. We have a voting body of N individuals and a set of p alternatives. For the sake of simplicity, we will assume that N is odd, thereby avoiding tied votes. A pattern of individual preferences is a vector of frequencies: $\mathbf{x} = (x_1, \ldots, x_M)$, where x_i is the number of voters who have the ith preference ordering, M the total number of different preference orderings, and $\Sigma x_i = N$. For a triplet of alternatives facing the voting body, there are six strong (13 weak) conceivable orderings. In the same notation as before, the six strong orderings are enumerated

1. $A \succ B \succ C$,   4. $B \succ C \succ A$,
2. $A \succ C \succ B$,   5. $C \succ A \succ B$,
3. $B \succ A \succ C$,   6. $C \succ B \succ A$.

The social ordering induced by majority rule is now

$$\xi \succ \eta \Leftrightarrow d(\xi, \eta) > 0, \qquad (1)$$

where

$$d(A, B) = x_1 + x_2 + x_5 - x_3 - x_4 - x_6,$$
$$d(A, C) = x_1 + x_2 + x_3 - x_4 - x_5 - x_6,$$
$$d(B, C) = x_1 + x_3 + x_4 - x_2 - x_5 - x_6,$$
$$\text{etc.} \quad (2)$$

The voting paradox is now said to occur for a preference pattern $\mathbf{x} = (x_1, \ldots, x_6)$, such that the rule d, defined by (1) and (2), leads to an intransitive outcome. Analogous defi-

nitions and notation may be used in the case of four or more alternatives. However, as mentioned at the outset, there are then two different definitions of the paradox.

Given that members of the voting body are assumed to choose among the preference orderings according to some probability distribution, probabilities for the occurrence of the paradox of voting can now be calculated. A vector of probabilities $\mathbf{p} = (p_1, \ldots, p_6)$, assigning voters to preference profiles, is called a *culture* (Garman and Kamien [4]). An impartial culture, in this terminology, means taking equal probabilities $p_1 = p_2 = \cdots = p_6 = 1/6$, for the assignment of voters to preference profiles.

The probability for the occurrence of the voting paradox is obtained if we sum multinomial* probabilities

$$P(\mathbf{x}) = \frac{N!}{x_1! \cdots x_6!} \left(\frac{1}{6}\right)^N,$$

over the region determined by the two sets of inequalities

$$d(A, B) > 0, \qquad d(B, C) > 0,$$
$$d(C, A) > 0,$$

and

$$d(A, C) > 0, \qquad d(C, B) > 0,$$
$$d(B, A) > 0,$$

with $d(\xi, \eta)$ defined by (2). If we let the number of voters N tend to infinity and approximate the multinomial probabilities by a multivariate normal distribution, then we obtain a result known as *Guilbaud's number*

$$1 - \frac{3}{\pi} \cos^{-1} \sqrt{3} = 0.0877.$$

This is the probability for the occurrence of the paradox in an "infinite" voting body faced with three alternatives.

Paradox probabilities have been tabulated for different numbers of voters and alternatives, e.g., by Niemi and Weisberg [6] and Garman and Kamien [4]. Not surprisingly, published tables show that the likelihood of the paradox increases rapidly as the number of alternatives voted upon increases.

The equally likely assumption underlying the concept of an impartial culture and leading to multinomial probabilities is tantamount to assuming that all 6^N combinations of voters and preference orderings are equally likely. An alternative randomness assumption is to assume that there are $\binom{5+N}{N}$ distinguishable combinations, all equally probable. In this case, called "anonymous preference profiles" by Gehrlein and Fishburn [5], a simple formula for the probability of a Condorcet winner is obtained (cf. also Berg and Bjurulf [2]):

$$\Pr[\text{Condorcet winner}|N \text{ voters}]$$

$$= \frac{15}{16} \frac{(N+3)^2}{(N+2)(N+4)}, \qquad N \text{ odd.}$$

Hence for an assembly of "infinite size," we get a slightly lower estimate of the paradox probability than that given by Guilbaud's number, namely,

$$1/16 = 0.0625.$$

Finally, we note that in statistical mechanics (*see* STOCHASTIC MECHANICS), in discussions of random distributions of particles into compartments, the two randomness assumptions used here are known under the names Maxwell–Boltzmann and Bose–Einstein respectively (see Feller [3, Secs. I.6 and II.5] and FERMI–DIRAC STATISTICS).

References

[1] Arrow, K. J. (1951). *Social Choice and Individual Values*. Wiley, New York.

[2] Berg, S. and Bjurulf, B. (1983). *Public Choice*, **40**, 307–316.

[3] Feller, W. (1957). *An Introduction to Probability Theory and Its Applications*, Vol. 1, 2nd ed. Wiley, New York.

[4] Garman, M. B. and Kamien, M. I. (1968). *Behav. Sci.*, **13**, 306–316.

[5] Gehrlein, W. V. and Fishburn, P. C. (1976). *Public Choice*, **26**, 1–18.

[6] Niemi, R. M. and Weisberg, H. (1968). *Behav. Sci.*, **13**, 317–323.

[7] Riker, W. H. (1965). In *Mathematical Applications in Political Science*, J. Claunch, ed. Southern Methodist University, Dallas, TX, pp. 41–60.

[8] Sen, A. K. (1966). *Econometrica*, **34**, 491–499.

Bibliography

For a general background to the problem, the following two books can be recommended:

Black, D. (1958). *The Theory of Committees and Elections*. Cambridge University Press, Cambridge, England.

Brams, S. J. (1976). *Paradoxes in Politics*. The Free Press, New York.

The complete title of Condorcet's essay is: *Essai sur l'Application de l'Analyse à la Probabilité des Decisions Rendus à la Pluralité des Voix* (Paris, 1785).

Condorcet's life and work are described in the recent biography:

Baker, K. M. (1975). *Condorcet. From Natural Philosophy to Social Mathematics*. The University of Chicago Press, London and Chicago.

(BALLOT PROBLEMS
ELECTION FORECASTING
POLITICAL SCIENCE, STATISTICS IN)

SVEN BERG

V-STATISTICS

THE METHOD DESCRIBED

The *V*-statistics are used in nonparametric (distribution-free*) procedures to test the equality of location* parameters of a continuous population $F(x)$ and a hypothetical continuous population $G(x)$. The data consist of a sample $X_1, \ldots, X_n$ with the empirical distribution function* (EDF) $F_n(x)$ defined by (number of $X_i \leqslant x)/n$; it is the proportion of values less than or equal to x. Let $d_j = F_n(q_j) - G(q_j)$ be the deviations at the quantiles $q_j = G^{-1}(j/k)$, $j = 1, \ldots, k-1$. The statistic is

$$V(n, k) = 2n \sum_{j=1}^{k-1} d_j.$$

If ρ_j be the ranks of the quantiles q_j in the set $\{x_i, q_j | i = 1, \ldots, n; j = 1, \ldots, k-1\}$,

Table 1 Critical values $V_{1-\alpha}$ (smallest value of c for which $\Pr[V(n, k = 8) > c] \leqslant \alpha$)

	One-Sided Significance Level α			
n	0.05	0.025	0.01	0.005
2	12	14	—	—
3	15	17	19	21
4	18	20	22	24
5	19	21	25	27
6	20	24	28	30
7	21	25	29	33
8	24	28	32	34
9	25	29	33	37
10	26	30	36	38
11	27	31	37	41
12	28	32	38	42
13	29	35	41	43
14	30	36	42	46
15	31	37	43	47
16	32	38	44	48
17	33	39	45	51
18	34	40	46	52
19	35	41	49	53
20	36	42	50	54

we have

$$V(n, k) = (n + k)(k - 1) - 2\sum_{j=1}^{k-1} \rho_j.$$

The V-statistic is sensitive to alternatives $F(x) = G(x + \Delta)$ and is similar to the Wilcoxon–Mann–Whitney* two-sample statistics after replacement of the second sample by quantiles of the hypothetical distribution.

The mean and variance of $V(n, k)$ under the null hypothesis $H_0: \Delta = 0$ are

$$E[V(n, k)] = 0,$$

$$\mathrm{Var}[V(n, k)] = \frac{n}{3}(k^2 - 1).$$

Table 1 shows the critical values for the significance levels 0.05, 0.025, 0.01, and 0.005, $n = 2, \ldots, 20$ and $k = 8$. An extended table of critical values is given by Rey and Riedwyl [6].

For large values of n, the statistic

$$z = \{V(n, k) + 1\}/\{\tfrac{1}{3}n(k^2 - 1)\}^{1/2},$$

with continuity correction*, has an asymptotic standard normal distribution and may be used to test H_0. The asymptotic relative efficiency* (ARE) of $V(n, k)$ to the t-test* in the one-sample case is $3/\pi$ as $k \to \infty$. For $k = 6$ the ARE is greater than 0.9. For practical purposes we use $6 \leqslant k \leqslant 20$.

NUMERICAL EXAMPLE

The following data represent 20 measurements of mortar compressive strengths: 611, 619, 620, 621, 622, 634, 635, 638, 638, 639, 640, 641, 643, 648, 651, 656, 656, 666, 677, 686. We test the two-sided hypothesis that the sample is drawn from a known normal distribution with mean 650 and standard deviation 20. For this example we group the data into $k = 8$ classes. This leads to the quantiles: 627.0, 636.5, 643.6, 650.0, 656.4, 663.5, 673.0.

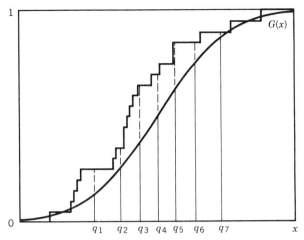

Figure 1   The deviations for computation of $V(20, 8)$.

Figure 1 shows how the deviations d_j arise from the EDF and the hypothetical distribution function at the quantiles* q_j.

For the computation of $V(n, k)$ we need the ranks ρ_j of the quantiles within the set of the 20 measurements. We get: 6, 9, 16, 18, 22, 23, 25. With $n = 20$ and $k = 8$ the test statistic can now easily be computed as

$$V(20, 8) = -42.$$

$|V| = 42$ is equal to $V_{0.975} = 42$ and is therefore just significant at the two-sided significance level of 5%. Using the normal approximation, we get $Z = 2.001$, which is also significant.

FURTHER READING

The $V(n, k)$-statistic was proposed by Riedwyl [7] and studied further by Carnal and Riedwyl [1]. Generalizations for grouped data were given by Maag et al. [3]. Rey [5] discusses the power* and robustness* against extreme observations for small sample size. It is also shown that in a normal distribution case with known standard deviation, $V(n, k)$ is more powerful than Student's t for small n. Holzherr [2] studied the behaviour of the test for a normally distributed variable with the variance estimated from the sample.

RELATED STATISTICS

The V-statistics are related to distribution-free* and goodness-of-fit* tests.

$V(n, 2)$ is equivalent to the sign test*. The maximum of d_j, $j = 1, \ldots, k - 1$, is asymptotically equivalent to the Kolmogorov–Smirnov test, and the sum of squares of d_j is related in the same way to the Cramér–von Mises test. It would be possible to simplify the calculations and the tabulation of critical values for small n by looking at a smaller number of deviations between the empirical and theoretical distribution function of the classical tests as we do in the replacement of $V(n, n)$ by $V(n, k)$, $k < n$. The Mann–Whitney U-test or Wilcoxon rank-sum test of $N = m + n$ observations can equivalently be defined by the sum of N deviations between the two EDFs at the sample points instead of the sum of the ranks of one sample. See Hüsler and Riedwyl [3].

In the one-sample location problem the V-statistic is the nonparametric alternative to the parametric normal Z-test, when the hypothetical distribution is completely known. The V-statistics could be generalized by looking at arbitrary grouping intervals [3, 5] or extending to the k-independent sample problem.

References

[1] Carnal, H. and Riedwyl, H. (1972). *Biometrika*, **59**, 465–467.

[2] Holzherr, E. (1975). Ph.D. dissertation, University of Berne, Switzerland.

[3] Hüsler, J. and Riedwyl, H. (1988). *Biom. J.* (to appear).

[4] Maag, U. R., Streit, F., and Drouilly, P. A. (1973). *J. Amer. Statist. Ass.*, **68**, 462–465.

[5] Rey, G. (1979). *Biom. J.*, **21**, 259–276.

[6] Rey, G. and Riedwyl, H. (1977). *J. Statist. Comp. Simul.*, **6**, 75–81.

[7] Riedwyl, H. (1967). *J. Amer. Statist. Ass.*, **62**, 390–398.

(DISTRIBUTION-FREE METHODS
EDF STATISTICS
MANN-WHITNEY-WILCOXON STATISTIC
PERMUTATION TESTS
RANK ORDER STATISTICS
U-STATISTICS)

H. RIEDWYL

W

W_N^2-**TEST** *See* GOODNESS OF FIT

WAGR TEST

This is another name for the sequential *t*-test (*see* SEQUENTIAL ANALYSIS). The initials stand for Wald, Anscombe, Girshick, and Rushton, four statisticians who were instrumental in the formulation of the test.

(SEQUENTIAL T^2-TEST)

WAITING TIMES *See* TAKÁCS PROCESS

WAKEBY DISTRIBUTIONS

These distributions were defined by H. A. Thomas in 1976. He gave them the name *Wakeby distributions* because he conceived them at his summer home overlooking Wakeby Pond in Cape Cod, Massachusetts. According to Hosking [1], the widespread and successful use of Wakeby distributions in hydrology* is due to certain general properties of the distributions.

(i) By suitable choice of parameter values, it is possible to mimic the extreme-value*, log-normal*, and log-gamma distributions.

(ii) There are five parameters, more than most of the common systems of distributions. This allows for a wider variety of shapes.

(iii) The distributions have finite lower bounds, which is physically reasonable for many hydrological observations.

(iv) Some Wakeby distributions have heavy upper tails and can give rise to occasional high outliers*, a phenomenon often observed in hydrology. (See Makalas et al. [4].)

(v) The form of the distribution (see below) is well suited to simulation*. (See also Houghton [2].)

The distributions are defined in terms of the quantile* function (inverse CDF) $x(F)$, where

$$\Pr[X \leqslant x(F)] = F.$$

For Wakeby distributions,

$$x(F) = \xi + \theta\beta^{-1}\{1 - (1 - F)^{\beta}\}$$
$$- \gamma\delta^{-1}\{1 - (1 - F)^{\delta}\}. \quad (1)$$

513

For $\beta = 0$ $(\delta = 0)$ the second (third) term on the right-hand side of (1) is taken as $\log(1 - F)$. The five parameters $\xi, \theta, \beta, \gamma, \delta$ are constrained by the conditions:

(a) $\gamma \geqslant 0$, $\theta + \gamma \geqslant 0$.
(b) Either $\beta + \delta > 0$ or $\beta + \gamma = \delta = 0$.
(c) If $\theta = 0$, then $\beta = 0$.
(d) If $\gamma = 0$, then $\delta = 0$.

For $\delta > 0$ and $\gamma > 0$, as $x \to \infty$,

$$F(x) \sim 1 - (\delta x / \gamma)^{-1/\delta},$$

$$f(x) = F'(x) \sim \gamma^{1/\delta}(\delta x)^{-1-1/\delta}$$

The upper tail of the distribution behaves like that of a Pareto distribution*.

The moments of all orders exist provided $\delta \leqslant 0$. If δ is positive, $E[X^r]$ exists for $0 \leqslant r < \delta^{-1}$. When the rth moment exists,

$$E\left[\left\{x - (\xi + \theta\beta^{-1} - \gamma\delta^{-1})\right\}^r\right]$$

$$= \sum_{j=0}^{r} (-1)^{r-j} \binom{r}{j} (\theta/\beta)^i (\gamma/\delta)^{r-j}$$

$$\times \{1 - r\delta + j(\beta + \delta)\}^{-1}.$$

The probability moment

$$\alpha_r = E\left[X\{1 - F(X)\}^r\right]$$

has the simple expression

$$(r + 1)^{-1}\left\{\xi + \theta(r + \beta + 1)^{-1} \right.$$
$$\left. + \gamma(r - \delta + 1)^{-1}\right\}, \qquad r \geqslant 0.$$

(Note that $\alpha_0 = \xi + \theta(\beta + 1)^{-1} + \gamma(-\delta + 1)^{-1} = E[X]$.)

Given an ordered random sample of n values $x_1 \leqslant \cdots \leqslant x_n$, the value of α_r is unbiasedly estimated by

$$a_r = n^{-1} \sum_{j=1}^{n} \binom{n-j}{r} x_j \Big/ \binom{n-1}{r}.$$

See Landwehr et al. [3], who recommend the biased, but more stable, estimator

$$\tilde{a}_r = n^{-1} \sum_{j=1}^{n} \left\{1 - n^{-1}(j - 0.35)\right\}^r x_j.$$

The estimators a_r and $\tilde{a}_r$ are asymptotically equivalent. Hosking [1] has shown that

as $n \to \infty$ the joint distribution of $\{n^{1/2}(a_r - \alpha_r),$ $r = 0, 1, \ldots, m - 1\}$—and so also that of $\{n^{1/2}(\tilde{a}_r - \alpha_r), r = 0, 1, \ldots, m - 1\}$ —is multinormal with zero mean.

References

[1] Hosking, J. R. M. (1986). The Wakeby Distribution. *IBM Research Report RC 12302*, IBM, Yorktown Heights, NY.

[2] Houghton, J. C. (1978). *Water Resour. Res.*, **14**, 1105–1109.

[3] Landwehr, J. M., Matalas, N. C., and Wallis, J. R. (1979). *Water Resour. Res.*, **15**, 1055–1064, 1361–1379.

[4] Matalas, N. C., Slack, J. R., and Wallis, J. R. (1975). *Water Resour. Res.*, **11**, 815–826.

(EXTREME-VALUE DISTRIBUTIONS
HYDROLOGY, STOCHASTIC
LAMBDA DISTRIBUTIONS
PARETO DISTRIBUTION)

WALD, ABRAHAM

Born: October 31, 1902, in Cluj, Hungary (now Romania).
Died: December 13, 1950, in Travancore, India.
Contributed to: decision theory, sequential analysis, geometry, econometrics.

Abraham Wald was born in Cluj on October 31, 1902. At that time, Cluj belonged to Hungary, but after World War I it belonged to Romania. Menger [2] states that when he first met Wald in Vienna in 1927, Wald spoke "fluent German, but with an unmistakable Hungarian accent." A short time after Wald emigrated to the United States, he spoke fluent English, with an accent.

Wald would not attend school on Saturday, the Jewish sabbath, and as a result he did not attend primary or secondary school, but was educated at home by his family. On the basis of this education, he was admitted to and graduated from the local university. He entered the University of Vienna in 1927 and received his Ph.D. in mathematics in

1931. Wald's first research interest was in geometry, and he published 21 papers in that area between 1931 and 1937. Two later papers, published in 1943 and 1944, were on a statistical generalization of metric spaces, and are described in [4]. A discussion of Wald's research in geometry by Menger [2] describes it as deep, beautiful, and of fundamental importance.

During the 1930s, economic and political conditions in Vienna made it impossible for Wald to obtain an academic position there. To support himself, he obtained a position as tutor in mathematics to Karl Schlesinger, a prominent Viennese banker and economist. As a result of this association, Wald became interested in economics and econometrics*, and published 10 papers on those subjects, plus a monograph [6] on seasonal movements in time series*. Morgenstern [3] describes this monograph as developing techniques superior to all others. Wald's first exposure to statistical ideas was a result of his research in econometrics. This research in economics and econometrics is described by Morgenstern [3] and Tintner [5]. Once again, his contributions to these areas are characterized as of fundamental importance.

Austria was seized by the Nazis early in 1938, and Wald came to the United States in the summer of that year, as a fellow of the Cowles Commission. This invitation from the Cowles Commission probably saved Wald's life, for almost all of the members of his family in Europe were murdered by the Nazis. In the fall of 1938, Wald became a fellow of the Carnegie Corporation and started to study statistics at Columbia University with Harold Hotelling*. Wald stayed at Columbia as a fellow of the Carnegie Corporation until 1941, lecturing during the academic year 1939–1940. In 1941, he joined the Columbia faculty and remained there for the rest of his life. During the war years, he was also a member of the Statistics Research Group at Columbia, doing war-related research. The techniques he developed for estimating aircraft vulnerability are still used and are described in a reprint [16] published in 1980. *See also* MILITARY STATISTICS.

Wald's first papers in statistics were published in 1939, and one of them [7] is certainly one of his most important contributions to statistical theory. J. Wolfowitz, who became a close friend and collaborator of Wald soon after Wald arrived in New York, in [22] describes this 1939 paper by Wald as probably Wald's most important single paper. In this paper, Wald points out that the two major problems of statistical theory at that time, testing hypotheses and estimation, can both be regarded as simple special cases of a more general problem—known nowadays as a "statistical decision problem." This generalization seems quite natural once it is pointed out, and the wonder is that nobody had thought of it before. Perhaps it needed a talented person with a fresh outlook to see it.

Wald does much more than merely point out the generalization in this paper. He defines loss functions, risk functions*, a priori distributions, Bayes decision rules, admissible decision* rules, and minimax decision rules, and proves that a minimax decision rule has a constant risk under certain regularity conditions. It is interesting that in this paper Wald states that the reason for introducing a priori distributions on the unknown parameters is that it is useful in deducing certain theorems on admissibility* and in the construction of minimax decision rules: That is, he is not considering the unknown parameters to be random variables. This paper did not receive much attention at first, but many other papers, by Wald and others, have extended statistical decision theory*. Wald's 1950 book [13] contains most of the results developed up to that year. This book is accessible only to those with a strong background in mathematics. The 1952 paper [14] summarizes the basic ideas and is more easily read.

Wald's other great contribution to statistical theory is the construction of optimal statistical procedures when sequential sampling is permitted. (Sequential sampling is any sampling scheme in which the total number of observations taken is a random variable.) Unlike statistical decision theory, the concept of sequential sampling is not due to Wald and is not included in the 1939 paper on decision theory. Just who first thought of sequential sampling is apparently not known. In [11] it is stated that Captain G. L. Schuyler of the U.S. Navy made some comments that alerted M. Friedman and W. Allen Wallis to the possible advantages of sequential sampling, and Friedman and Wallis proposed the problem to Wald in March 1943. Wald's great contribution to sequential analysis was in finding optimal sequential procedures. He started by considering the problem of testing a simple hypothesis against a simple alternative using sequential sampling and conjectured that the subsequently famous sequential probability ratio test is optimal for this problem, in the sense that among all test procedures with preassigned upper bounds on the probabilities of making the wrong decision, the sequential probability ratio test minimizes the expected sample size under both the hypothesis and the alternative. Early in his investigation, he was able to show that this conjecture is at least approximately true, but it was not until his 1948 paper with Wolfowitz [21] that a proof that it is exactly true was given. This proof was based on the ideas that Wald had developed in his work on statistical decision theory and thus united his two major contributions to statistical theory. The 1947 book [11] describes the results on sequential analysis known up to that time in an elementary manner and is accessible to anyone with a knowledge of elementary probability theory. The 1950 book [13] incorporates sequential sampling into statistical decision theory.

Besides statistical decision theory and sequential analysis, Wald made many other fundamental contributions to statistical theory, some of which will be described briefly. In [9] he derived the large-sample distribution of the likelihood ratio test* under alternatives to the hypothesis being tested and proved the asymptotic optimality of the test. In collaboration with Wolfowitz, he made fundamental contributions to nonparametric statistical inference in [17-20]. He wrote a pioneering paper on the optimal design of experiments [8], a field which became very active a few years later. In [10] Wald generalized a theorem of von Neumann on zero-sum two-person games. In collaboration with Mann [1], he developed statistical inference for stochastic difference equations. [12] contains a new proof of consistency of maximum likelihood* estimators.

Wald was an excellent teacher, always precise and clear. He was a master at deriving complicated results in amazingly simple ways. His influence on the teaching of statistics extended far beyond the students who actually attended his classes, because, with his permission, the Columbia students reproduced the notes they took in his classes. These reproduced notes were only supposed to be circulated to other Columbia students, but they had a much wider circulation than that.

In late 1950, Wald was in India, at the invitation of the Indian Government, lecturing on statistics. He was accompanied by his wife. On December 13, 1950, both were killed in a planecrash.

A fuller account of Wald's research and a list of 104 of his publications are contained in [15]. This list is complete except for the 1980 reprint [16].

References

[1]   Mann, H. B. and Wald, A. (1943). *Econometrica*, **11**, 173-220.

[2]   Menger, K. (1952). *Ann. Math. Statist.*, **23**, 14-20.

[3]   Morgenstern, O. (1951). *Econometrica*, **19**, 361-367.

[4]   Schweizer, B. and Sklar, A. (1983). *Probabilistic Metric Spaces*. North-Holland, New York.

[5]   Tintner, G. (1952). *Ann. Math. Statist.*, **23**, 21-28.

[6]   Wald, A. (1936). *Berechnung und Ausschaltung von Saisonschwankungen*. Springer, Vienna.

[7]   Wald, A. (1939). *Ann. Math. Statist.*, **10**, 299-326.

[8]   Wald, A. (1943). *Ann. Math. Statist.*, **14**, 134-140.

[9]   Wald, A. (1943). *Trans. Amer. Math. Soc.*, **54**, 426-482.

[10] Wald, A. (1945). *Ann. Math.*, **46**, 281-286.

[11] Wald, A. (1947). *Sequential Analysis*. Wiley, New York.

[12] Wald, A. (1949). *Ann. Math. Statist.*, **20**, 595-601.

[13] Wald, A. (1950). *Statistical Decision Functions.* Wiley, New York.

[14] Wald, A. (1952). *Proc. Intern. Congress of Mathematicians.* Harvard University Press, Cambridge, MA.

[15] Wald, A. (1955). *Selected Papers in Statistics and Probability.* McGraw-Hill, New York.

[16] Wald, A. (1980). *A Method of Estimating Plane Vulnerability Based on Damage of Survivors.* Center for Naval Analyses, Washington, D.C.

[17] Wald, A. and Wolfowitz, J. (1939). *Ann. Math. Statist.*, **10**, 105–118.

[18] Wald, A. and Wolfowitz, J. (1940). *Ann. Math. Statist.*, **11**, 147–162.

[19] Wald, A. and Wolfowitz, J. (1943). *Ann. Math. Statist.*, **14**, 378–388.

[20] Wald, A. and Wolfowitz, J. (1944). *Ann. Math. Statist.*, **15**, 358–372.

[21] Wald, A. and Wolfowitz, J. (1948). *Ann. Math. Statist.*, **19**, 326–339.

[22] Wolfowitz, J. (1952). *Ann. Math. Statist.*, **23**, 1–13.

(DECISION THEORY
GAME THEORY
SEQUENTIAL ANALYSIS
WALD'S DECISION THEORY)

L. WEISS

WALD(–BARTLETT) SLOPE ESTIMATION

Consider a situation in which variables X_i and Y_i are related by the equation

$$Y_i = \alpha + \beta X_i, \qquad i = 1, \ldots, N,$$

with both variables subject to random measurement errors* ϵ_i and η_i, respectively. It can be assumed that $E[\epsilon_i] = E[\eta_i] = 0$. It is known that if simple least-squares* procedures are used to estimate β, measurement errors in x_i will attenuate the estimate. Wald [11] derived an estimator of β calculated from measured values $x_i = X_i + \epsilon_i$ and $y_i = Y_i + \eta_i$ that is consistent *provided that the following assumptions have been met*:

(i)   The random variables $\epsilon_i, \ldots, \epsilon_N$ each have the same distribution and they are uncorrelated, that is, $E[\epsilon_i \epsilon_j] = 0$ for $i \neq j$. The variance of ϵ_i is finite.

(ii) The random variables $\eta_i, \ldots, \eta_N$ each have the same distribution and they are uncorrelated, that is, $E[\eta_i \eta_j] = 0$ for $i \neq j$. The variance of η_i is finite.

(iii) The random variables ϵ_i and η_j, $i = 1, 2, \ldots, N$; $j = 1, 2, \ldots, N$, are uncorrelated, that is, $E[\epsilon_i \eta_j] = 0$.

(iv) A single linear relation holds between the true values X_i and Y_i, that is to say, $Y_i = \alpha + \beta X_i$, $i = 1, 2, \ldots, N$.

(v)   The limit inferior of $\{(X_1 + \cdots + X_m) - (X_{m+1} + \cdots + X_N)\}N^{-1}$ ($N = 2, 4, \ldots, \infty$), is positive, where the total number of cases N is even and $m = N/2$.

Wald's estimator b is formed as follows:

$$b = \frac{\bar{y}_1 - \bar{y}_2}{\bar{x}_1 - \bar{x}_2}$$

$$= \frac{(y_1 + y_2 + \cdots + y_m) - (y_{m+1} + \cdots + y_N)}{(x_1 + x_2 + \cdots + x_m) - (x_{m+1} + \cdots + x_N)},$$

where there are two groups G_1 and G_2 with means $\bar{x}_1$ and $\bar{y}_1$ and $\bar{x}_2$ and $\bar{y}_2$, respectively. The elements are ordered in increasing magnitude on X. G_1 contains half of the observations $(1, 2, \ldots, m)$ and G_2 contains the remainder $(m + 1, m + 2, \ldots, N)$.

In addition, it is assumed that the grouping is unrelated to the error terms (which essentially implies that measurement errors in x_i are negligible). We then face the practical problem of finding a method of grouping the scores is independent of the error terms ϵ_i and η_i.

Bartlett [1] modified Wald's procedure and suggested dividing the distribution ordered on x_i into *thirds*. The smallest *third* becomes G_1 and the largest G_2. Additional related modifications into *quarters* (the smallest *quarter* being G_1 and the largest G_2) and an iteration that requires calculations of two slope estimates were developed by Nair and Banerjee [8] and Lindley [5], respectively.

Carter and Blalock [2] investigated in some detail robustness of Wald-type estimators, given varying degrees of departure from the assumption that the grouping procedure is independent of measurement errors.

They conclude, inter alia, that for normally distributed parent values the quarter-grouping method provides less biased slope estimates than least squares*. However, for skewed parent distributions, the grouping techniques produce estimates that are even more biased than least-squares estimates.

Related investigations were carried out by Madansky [6] in connection with optimum cutoff points for the groups in Bartlett's procedure. It should be noted that some confusion exists in the literature concerning these procedures especially in economics and econometrics* (see, e.g., Pakes [9] for a detailed, although not completely accurate, discussion of this problem). See also Moran [7] (for a more general treatment), Gupta and Aman [3], Theil and Van Ijzeren [10], and Lancaster [4] (for the case of heteroscedastic data).

References

[1] Bartlett, M. S. (1949). *Biometrics*, **5**, 207–212.

[2] Carter, L. F. and Blalock, H. M. (1970). *Appl. Statist.*, **19**, 34–41.

[3] Gupta, Y. P. and Aman, U. (1970). *Statist. Neerlandica*, **24**, 109–123.

[4] Lancaster, T. (1968). *J. Amer. Statist. Ass.*, **63**, 182–191.

[5] Lindley, D. V. (1947). *J. R. Statist. Soc. Suppl.*, **9**, 218–225.

[6] Madansky, A. (1959). *J. Amer. Statist. Ass.*, **54**, 173–205.

[7] Moran, P. A. P. (1971). *J. Multivariate Anal.*, **1**, 232–255.

[8] Nair, K. R. and Banerjee, K. S. (1943). *Sankhyā*, **6**, 331.

[9] Pakes, A. (1982). *Int. Econ. Rev.*, **23**, 491–497.

[10] Thiel, H. and Van Ijzeren, J. (1956). *Rev. Inst. Int. Statist.*, **24**, 17–26.

[11] Wald, A. (1940). *Ann. Math. Statist.*, **11**, 284–300.

(LINEAR REGRESSION)

WALD DISTRIBUTION *See* INVERSE GAUSSIAN DISTRIBUTION; SEQUENTIAL ANALYSIS

WALD STATISTICS; WALD TESTS *See* WALD'S *W*-STATISTICS

WALD TEST OF MARGINAL SYMMETRY *See* MARGINAL SYMMETRY

WALD–WOLFOWITZ TWO-SAMPLE TEST *See* RUNS

WALD'S DECISION THEORY

The centerpiece of the work of Abraham Wald* in mathematical statistics was his creation of a unified mathematical basis for statistical decision making. This sketch of his formulation and results should be read in conjunction with DECISION THEORY (DT). It is, however, helpful to note that Wald worked from a "frequentist" perspective, rather than from the more "Bayesian"* point of view taken in DT.

FEATURES OF THE THEORY

Wald sought a flexible and general framework within which a unified theory could be developed. There were a number of important features of his approach. First, the traditional problems of hypothesis testing*, point and interval estimation* are all regarded as particular cases of a general multi-decision problem. Second, the theory allows for multistage or sequential experimentation. Third, the question of design of experiments* is regarded as part of the general decision problem. Fourth, the general theory is most conveniently cast in terms of randomized decision* rules; though a considerable effort is made to determine when nonrandomized rules will suffice.

ELEMENTS OF THE THEORY

The statistician has (potentially) available observations on a sequence $X =$

$(X_1, X_2, \ldots)$ of (not necessarily independent) random variables having joint distribution F. At the outset, the statistician is assumed to know nothing except that F belongs to a family $\{F_\theta : \theta \in \Theta\}$. There is a space D of decisions d, one of which the statistician has (eventually) to make. If F_θ is the actual distribution governing the data, and if decision d is taken, then a loss $L(\theta, d)$ is incurred. The total loss is the sum of the loss due to the decision made and the cost of the observations.

Experimentation can, in general, be conducted sequentially. A *statistical decision function*, in its most general form, is a rule, which at the mth stage ($m = 0, 1, 2, \ldots$, the zeroth stage being at the outset of the experiment before any observations have been taken) tells the statistician whether or not to take further observations (at the zeroth stage, whether to take any observations), on which random variables to take observations (if at all), and which decision to take (if no further observations are to be taken). At the mth stage, the decision function $\delta_m(d|x_1, \ldots, x_m)$ is a function of the preceding observations, and is, in general, a probability distribution over the available possibilities. For such a *randomized* decision rule, the actual decision is made by an independent chance mechanism governed by the distribution $\delta_m(d|x_1, \ldots, x_m)$.

A nonrandomized decision rule restricts the values of $\delta_m(d|x_1, \ldots, x_m)$ to 0 and 1, and the (statistically objectionable) use of the independent chance experiment is avoided. The rule δ can then be more simply described in terms of a stopping rule $\tau = \tau(x_1, \ldots, x_m) \in \{1, 2, \ldots\}$ and terminal decision rules $d_m(x_1, \ldots, x_m) \in D$, which are defined whenever $\tau = m$.

Wald assesses the merit of a candidate decision rule δ by averaging the loss plus cost incurred by the prescriptions $\{\delta_m\}$ with respect to the distribution F_θ of the data X, obtaining a *risk function* $r(\theta, \delta)$. In the special case of a fixed sample size experiment with n cost-free observations $x = (x_1, \ldots, x_n)$, the risk of a nonrandomized decision rule $d_n(x)$ would be given explicitly

by

$$r(\theta, d_n) = E_\theta L(\theta, d_n(X))$$
$$= \int L(\theta, d_n(x)) F_\theta(dx).$$

Comparison of two decision rules is made on the basis of their risk functions. A decision rule δ_1 is (strictly) better than a decision rule δ_2 if $r(\theta, \delta_1) \leqslant r(\theta, \delta_2)$ for all $\theta \in \Theta$ (with strict inequality holding for at least one θ). The rule δ_1 is called *admissible* if there is no uniformly better decision rule (*see* ADMISSIBILITY).

Wald's risk function depends on the decision rule δ and the parameter θ. It should be distinguished from the (Bayes) risk function $R(\xi, \delta)$ discussed below and in DT, in which a further average is taken with respect to a (prior) probability $\xi(d\theta)$ on θ.

BAYES AND MINIMAX* DECISION RULES

Wald's theory studies these rules both for their intrinsic interest and for their role in the construction and characterization of complete classes of decision rules.

If $\xi(d\theta)$ is a (prior) probability distribution on Θ, the Bayes or integrated risk of a decision rule is given by $R(\xi, \delta) = \int_\Theta r(\theta, \delta) \xi(d\theta)$, and if

$$R(\xi, \delta^*) = \inf_\delta R(\xi, \delta), \qquad (1)$$

then δ^* is termed a *Bayes decision rule* with respect to ξ. More generally, δ^* is called a *wide-sense* Bayes rule if property (1) is approximately true. That is, if there exist prior distributions $\{\xi_i\}$, $i = 1, 2, \ldots$, such that

$$\lim_{i \to \infty} \left[R(\xi_i, \delta^*) - \inf_\delta R(\xi_i, \delta) \right] = 0.$$

A *minimax* decision rule is one for which $\sup_{\theta \in \Theta} r(\theta, \delta)$ is a minimum. A prior distribution ξ_0 is *least favorable** (from the point of view of the statistician) if the minimum average risk relative to ξ_0 is larger than for any other prior distribution: $\inf_\delta R(\xi_0, \delta) \geqslant \inf_\delta R(\xi, \delta)$ for all ξ.

In the extreme generality of Wald's formulation, it is important to know whether decision functions satisfying the above criteria exist. A (weak) notion of convergence is introduced for decision rules so that the class of possible rules forms a compact space. Under appropriate restrictions on loss and cost functions, Wald then shows that:

(a) Minimax rules exist, as do Bayes rules, for any choice of ξ.

(b) Minimax rules are wide-sense Bayes.

(c) If Θ is compact in the appropriate topology, then a least-favorable distribution ξ_0 exists, and a minimax rule is always Bayes with respect to ξ_0.

Wald's decision theory is intimately related to game theory*, in particular, to two-person zero-sum games in which Nature takes the role of player 1 and the statistician is player 2. Wald notes that parts of the theory of minimax solutions (for example, the existence of a minimax "value" for a game) is of no particular intrinsic interest in the statistical setting. Wald's attention to minimax theory derives from two sources: its role in deriving the complete class results and the idea that a minimax solution may be reasonable in a decision problem in which a prior distribution does not exist or is unknown to the experimenter.

COMPLETE CLASS THEOREMS

A class of decision rules is called *complete* if for any rule δ not in C, there exists a strictly better rule δ^* belonging to C. C is called *essentially complete* if δ^* is only required to be better than δ, and not strictly so. In principle, if C is complete or at least essentially complete, then one can ignore all decision rules not belonging to C. In this way the choice of a decision rule is reduced to the choice of a member of C. Thus, in Wald's theory, the discovery and description of complete classes occupies a central position. (Of course, even if a minimal complete class can be identified in a given problem, a further criterion needs to be employed to choose a particular rule, or group of rules, for use.)

The basic complete class theorems assert, under boundedness conditions on loss, cost, and risk functions, that the class of Bayes rules is essentially complete. If the parameter space is compact, then Bayes rules actually form a complete class. In general, however, a complete class is obtained by extending the class to include wide-sense Bayes rules.

Wald notes that the ideas of prior distributions and Bayes rules are used as mathematical tools in describing complete classes. The theory does not require the statistician to adopt a Bayesian approach and to postulate the existence of a specific prior distribution. The obituary article by Wolfowitz [15, pp. 4 ff] contains a useful perspective on Wald's attitude to the theory generally, to the use of Bayes solutions in particular, and on contemporary reaction to Wald's results.

As an example of a success of decision theory within Wald's lifetime, one can cite his proof (jointly with Wolfowitz [14]) of the optimum character of the sequential probability ratio test (SPRT). Introduced as a basic element of his work on sequential analysis*, the conjectured optimality of the SPRT was only proved after the framework of decision theory was extended to cover sequential experimentation and manipulations of Bayes solutions exploited.

In general, however, the results of Wald's theory are not cast in a form ready for application to specific questions. Explicit description of complete classes and determination of minimax rules is often technically challenging and has been the object of much subsequent research. (See, for example, [3, 6], or the survey in [1].) Happily, contemporary availability of powerful and cheap computing may provide a convenient alternative to explicit analytic solutions or approximations in the analysis of competing decision rules.

FURTHER DEVELOPMENTS

Wald's work in statistical decision theory began with his 1939 paper [11] and continued intensively in the 1940s. It was presented in detail in his 1950 book [12], and would have undoubtedly continued but for his early death. A brief and accessible account of his ideas is given in the address [13], on which, together with [15], the present entry is partly based. Ferguson [4] has given an historical account of the development of the decision model, going back to Daniel Bernoulli.

Since 1950, Wald's theory has been extended and refined by many workers, including L. Brown, J. Kiefer, L. LeCam, E. Lehmann, and C. Stein. This work is too extensive to discuss here; some sources are listed in the references to DT and in the References and Bibliography that follow. Major themes have included the development of necessary and sufficient conditions for admissibility, the inadmissibility of maximum likelihood procedures in multiparameter contexts, the role of invariance in decision problems with natural group structures, decision theoretic study of confidence sets and conditional confidence procedures, model selection, and properties of sequential procedures.

The conceptual framework established by Wald (and its antecedents in Neyman–Pearson* theory and game theory) has been influential in a number of fields of mathematical statistics, including asymptotic theory [9] simultaneous or "Stein" estimation [1, 10], robust estimation* [5], optimal experimental design* [7], multiple comparisons*, and robust Bayesian analysis [2].

References

[1]   Berger, J. O. (1988). *Multivariate Estimation—A Synthesis of Bayesian and Frequentist Approaches.* SIAM, Philadelphia, PA. (To appear.)

[2]   Berger, J. O. (1980, 1985). Books on *Statistical Decision Theory* (see listing under Bibliography).

[3]   Brown, L. D. (1971). Admissible estimators, recurrent diffusions and insoluble boundary value problems. *Ann. Math. Statist.*, **42**, 855–903.

[4]   Ferguson, T. S. (1976). Development of the decision model. in *On the History of Statistics and Probability*, D. B. Owen, ed. Dekker, New York, Basel (Switzerland).

[5]   Huber, P. J. (1964). Robust estimation of a location parameter. *Ann. Math. Statist.*, **35**, 73–101.

[6]   Johnstone, I. M. (1986). Admissible estimation, Dirichlet principles and recurrence of birth–death chains on Z_+^D. *Prob. Theor. Rel. Fields*, **71**, 231–271.

[7]   Kiefer, J. (1974). General equivalence theory for optimum designs (approximate theory). *Ann. Statist.*, **2**, 849–879.

[8]   LeCam, L. (1955). An extension of Wald's theory of statistical decision functions. *Ann. Math. Statist.*, **26**, 69–81.

[9]   LeCam, L. (1986). *Asymptotic Methods in Statistical Decision Theory.* Springer, New York.

[10] Stein, C. (1956). Inadmissibility of the usual estimator of the mean of a multivariate normal distribution. *Proc. Third Berkeley Symp. Math. Statist. Prob.* **1**, 197–206. University of California Press, Berkeley, CA.

[11] Wald, A. (1939). Contributions to the theory of statistical estimation and testing hypotheses. *Ann. Math. Statist.*, **10**, 299–326.

[12] Wald, A. (1950). *Statistical Decision Functions.* Wiley, New York.

[13] Wald, A. (1952). Basic ideas of a general theory of statistical decision rules. *Proc. Intern. Congress of Mathematicians*, **1**, 231–243.

[14] Wald, A. and Wolfowitz, J. (1948). Optimum character of the sequential probability ratio test. *Ann. Math. Statist.*, **19**, 326–339.

[15] Wolfowitz, J. (1952). Abraham Wald, 1902–1950. *Ann. Math. Statist.*, **23**, 1–13.

Bibliography

(This list supplements that given in DECISION THEORY.)

Berger, J. O. (1980). *Statistical Decision Theory: Methods and Concepts*, Springer, New York.

Berger, J. O. (1985). *Statistical Decision Theory and Bayesian Analysis.* Springer, New York. (Two editions of an introductory graduate-level text presenting both frequentist and Bayesian concepts. There is some difference of perspective between the two editions.)

Brown, L. D. (1986). Foundations of Exponential Families. *IMS Lecture Notes—Monographs Series.* Volume 9.

Chentsov, N. N. (1972). *Statistical Decision Rules and Optimal Inference* (translation). American Mathematical Society, Providence, RI. (An advanced monograph that develops Wald's theory in the context of modern differential geometry.)

Diaconis, P. and Stein, C. (1983). *Lecture Notes on Statistical Decision Theory.* (Unpublished, Stanford University, Stanford, CA.) (Graduate-level course taught many times by Stein, one of the principal developers of Wald's theory.)

Ferguson T. (1967). *Mathematical Statistics: A Decision Theoretic Approach*. Academic, New York. (A widely used introductory graduate-level textbook presenting mathematical statistics from the point of view of Wald's theory.)

Lehmann, E. L. (1959, 1986). *Testing Statistical Hypotheses*, 1st and 2nd ed. Wiley, New York.

Lehmann, E. L. (1981). *Theory of Point Estimation*. Wiley, New York. (Popular introductory graduate-level textbook containing introductions to mathematical statistics using the concepts of the Neyman–Pearson–Wald school.)

(ADMISSIBILITY
BAYESIAN INFERENCE
DECISION THEORY
DESIGN OF EXPERIMENTS
ESTIMATION, POINT
GAME THEORY
HYPOTHESIS TESTING
LEAST FAVORABLE DISTRIBUTIONS
MINIMAX DECISION RULES
MINIMAX ESTIMATION
MINIMAX TESTS
SEQUENTIAL ANALYSIS)

IAIN JOHNSTONE

WALD'S EQUATION

Let $X_1, X_2, \ldots$ be a sequence of random variables and N a *stopping number**; then N is positive integer-valued and provides a stopping rule so that sampling ceases after N variables in the sequence have been observed. A stopping variable by definition requires that the event "$N = n$" depends upon $X_1, X_2, \ldots, X_n$ only, for $n = 1, 2, \ldots$, i.e., that this event belongs to the σ-field generated by $X_1, \ldots, X_n$. Suppose that $E[X_i] = \mu$ and that $E[N] < \infty$. Then Wald's equation states that

$$E[X_1 + X_2 + \cdots + X_N] = \mu E[N]. \quad (1)$$

This does not always hold, but Wald* [4, p. 53] was motivated to derive it so that he could establish important properties of the sequential probability ratio test in sequential analysis*. Various sets of conditions under which (1) holds have been obtained. The

most useful of these, in addition to those already mentioned, are as follows.

(i) $\{X_i\}$ is an independent and identically distributed (i.i.d.) sequence of variables and N is independent of $\{X_i\}$.

(ii) $X_1, X_2, \ldots$ are mutually independent; $E[|X_i|] \leqslant A < \infty$ for all i and some finite A. The event "$N \geqslant i$" depends only on $X_1, X_2, \ldots, X_{i-1}$ [3, 5]. (In sequential analysis, the X_i are frequently the log-likelihood ratios of observations $\{Y_i\}$ in hypothesis testing and are independent; *see* RANDOM SUM DISTRIBUTIONS and SEQUENTIAL ANALYSIS.)

(iii) [1, Theorem 2.3]. $\{X_n\}$ is a martingale* sequence, where X_n is $\mathscr{F}_n$-measurable and $\mathscr{F}_m \subset \mathscr{F}_n \subset \mathscr{F}$ in a probability space $(\Omega, \mathscr{F}, P)$ for all $m < n$, $1 \leqslant n < \infty$. $E[X_1] = 0$ and N is a stopping variable. $E[X_N]$ exists and

$$\liminf_{n \to \infty} \int_{[N > n]} |x_n| \, dP = 0. \quad (2)$$

Under these conditions $E[X_N] = 0$, or more generally,

$$E[X_N | \mathscr{F}_n] = X_n \quad \text{on } [N \geqslant n],$$
$$n = 1, 2, \ldots.$$

This result is the most general. In order for $E[X_N] < \infty$ and (2) to hold, it suffices either that N is bounded almost surely (a.s.) that $|X_n| < c$ a.s. on $[N > n]$, for some c and n, or that $E[\sum_{n=1}^N \sigma_n^2] < \infty$ [where $\sigma_n^2 = \mathrm{var}(X_n)$], or that

$$E\left[\sum_{n=1}^N E[|X_n - X_{n-1}|] | \mathscr{F}_{n-1}) \right] < \infty.$$

Chow et al. [1] and Woodroofe [5] refer to (1) as Wald's lemma, but the latter term is commonly applied to another result; *see* FUNDAMENTAL IDENTITY OF SEQUENTIAL ANALYSIS.

Example. The sequence $\{X_n\}$ is mutually independent, defined by

$$\mathrm{Pr}(X_n = -2^n) = \mathrm{Pr}(X_n = 2^n) = 0.50,$$
$$n = 1, 2, \ldots,$$

so that $E[X_n] = 0$ for all n. Let N be the first n for which $\sum_{i=1}^{n} X_i \leqslant -2$. Then $E[\sum_{i=1}^{N} X_i] \leqslant -2$ a.s. and $E[N] < \infty$, but Wald's equation (1) does not hold. This indicates that the condition $E[|X_i|] \leqslant A < \infty$ in (ii) is necessary.

An extension of Wald's equation to variances is frequently linked to (1). Suppose that condition (ii) holds, and, additionally, that var$(X_n) = \sigma^2 < \infty$ for all n, that

$$E\left[(X_n - \mu)^2 | N \geqslant i\right] \leqslant B < \infty,$$

for all $n < i$, and that $E[N^2] < \infty$. Then [3], if for all n, $X_n' = X_n - \mu$ and $S_n' = \sum_{i=1}^{n} X_i'$,

$$\text{var}(S_N') = E\left[S_N'^2\right] = \sigma^2 E[N]. \quad (3)$$

See also Chow et al. [1, p. 24]. Equation (3) thus holds for i.i.d. variables $X_1', X_2', \ldots$ having mean 0, but no such result holds for an i.i.d. sequence $X_1, X_2, \ldots$ having mean $\mu \neq 0$, for which

$$E\left[(S_N - N\mu)^2\right] = \sigma^2 E[N] < \infty,$$

but where var(S_N) may even be infinite. Chow and Teicher [2] discuss this apparent paradox.

References

[1] Chow, Y. S., Robbins, H., and Siegmund, D. (1971). *Great Expectations: The Theory of Optimal Stopping*. Houghton Mifflin, Boston.

[2] Chow, Y. S. and Teicher, H. (1966). *Ann. Math. Statist.*, **37**, 388–392.

[3] Johnson, N. L. (1959). *Ann. Math. Statist.*, **30**, 1245–1247. Correction (1961), ibid., **32**, 1344.

[4] Wald, A. (1947). *Sequential Analysis*. Wiley, New York.

[5] Woodroofe, M. (1975). *Probability with Applications*. McGraw-Hill, New York, pp. 338–339.

(OPTIMAL STOPPING RULES
RANDOM SUM DISTRIBUTIONS
SEQUENTIAL ANALYSIS
STOPPING NUMBERS AND STOPPING
  TIMES
WALD'S IDENTITY)

CAMPBELL B. READ

WALD'S IDENTITY—APPLICATIONS

Wald's fundamental identity of sequential analysis* is one relating the generating function* of a random walk* in one dimension to statistical properties of the absorption time and absorption point, in the presence of one or two absorbing barriers. It was derived by Wald in this development of sequential analysis (*see* SEQUENTIAL ANALYSIS) and plays a central role in that field [14–16] (*see* POWER; AVERAGE SAMPLE NUMBER). A precise statement of the identity is given in FUNDAMENTAL IDENTITY OF SEQUENTIAL ANALYSIS and we use the notation introduced there. Let the individual steps of the random walk be identically distributed random variables with the common generating function $M(t)$, let S_n be the position of the random walk at the nth step, and let N be the first step at which the condition

$$b < S_n < a, \qquad \text{where } b < 0, \, a > 0, \quad (1)$$

is violated. Wald's identity states that

$$E\left[e^{tS_N}\{M(t)\}^{-N}\right] = 1. \quad (2)$$

This identity can be used quite trivially to relate moments of S_N to moments of N by differentiating (2) with respect to t and setting $t = 0$. If μ is the (finite) mean of a single step of the random walk and σ^2 its (finite) variance, application of this procedure leads to the identities

$$E[S_N] = \mu E[N],$$

$$E\left[(S_N - N\mu)^2\right] = \sigma^2 E[N]. \quad (3)$$

Higher moments are readily generated by this means. The results obtainable using Wald's identity can be applied in a variety of fields, and the key results in sequential analysis are valid for more general random walks. The results cited in FUNDAMENTAL IDENTITY OF SEQUENTIAL ANALYSIS allow one to generate an approximation to the probability of absorption at a particular end of a line. A similar technique leads to an approximate value for the generating function of the stopping time [3].

APPLICATIONS AND GENERALIZATIONS

The largest number of applications has come from the area of sequential analysis. Wald's identity can be used to furnish an approximation for both the average sample number (ASN) and operating characteristic* (OC) for a sequential design, and it was for this purpose that it was devised. In these applications it is customary to make the approximation that $S_N = a$ or b in order to use (3). Many authors have tried to improve the degree of approximation by removing or modifying the assumptions made on S_N. Kemp [5] has produced more accurate approximations for the ASN and OC, which are relatively simple and appear to work well. Tallis and Vagholkar [13] have also produced an effectively more precise approximation to the same quantities. Both of these papers apply their results to normally distributed steps, finding a marked improvement over results furnished by the application of Wald's identity.

A second area of generalization is that of sums of dependent random variables. Miller [3, 10] appears to have been the first to study this problem, and Kemperman has related material in [6]. Arjas [1, 2] motivated by a study of the semi-Markov* queue has also derived a generalized Wald's identity. A quite general version of Wald's identity for dependent random variables has been given by Franken and Lisek [4]. Applications of Wald's identity have appeared in many contexts. The theory of queues* and dams* has seen several useful applications [11], and for this purpose Wald's identity has been generalized to processes that evolve in continuous time [3]. Other applications have been made to models for neutron firing [7] and reliability [12].

References

[1] Arjas, E. (1972). *Adv. Appl. Prob.*, **4**, 258–270. (A general derivation of Wald's identity for semi-Markov processes.)

[2] Arjas, E. (1972). *Adv. Appl. Prob.*, **4**, 271–284. (Application of the generalized Wald's identity to the semi-Markov queue.)

[3] Cox, D. R. and Miller, H. D. (1965). *The Theory of Stochastic Processes*. Wiley, New York. (A clear account of Wald's identity using a method developed by Miller of proving it.)

[4] Franken, P. and Lisek, B. (1982). *Zeit. Wahrscheinlichkeitsth. Verwand. Geb.*, **60**, 143–150.

[5] Kemp, K. W. (1958). *J. R. Statist. Soc. B*, **20**, 379–386.

[6] Kemperman, J. H. B. (1961). *The Passage Problem for a Stationary Markov Chain*. University of Chicago Press, Chicago. (Overlaps some of the cited work by H. D. Miller on the derivation of Wald's identity.)

[7] Kryukov, V. I. (1976). *Adv. Appl. Prob.*, **8**, 257–277.

[8] Miller, H. D. (1961). *Ann. Math. Statist.*, **32**, 549–560.

[9] Miller, H. D. (1962). *Proc. Camb. Philos. Soc.*, **58**, 268–285. (A derivation of the fundamental identity used to prove generalized versions of Wald's identity.)

[10] Miller, H. D. (1962). *Proc. Camb. Philos. Soc.*, **58**, 286–298. (Absorption probabilities for sums of random variables defined on a Markov chain.)

[11] Phatarfod, M. R. (1982). *Stoch. Proc. Appl.*, **13**, 279–282.

[12] Serfozo, R. F. (1973). *Manag. Sci.*, **20**, 1314–1315.

[13] Tallis, G. M. and Vagholkar, M. K. (1965). *J. R. Statist. Soc. B*, **27**, 74–81.

[14] Wald, A. (1944). *Ann. Math. Statist.*, **15**, 283–296. (The original statement of Wald's identity.)

[15] Wald, A. (1946). *Ann. Math. Statist.*, **17**, 493–497.

[16] Wetherill, G. B. and Glazebrook, K. D. (1986). *Sequential Methods in Statistics*, 3rd. ed. Chapman and Hall, London.

(AVERAGE SAMPLE NUMBER
CHARACTERISTIC FUNCTIONS
DAM THEORY
FUNDAMENTAL IDENTITY OF
  SEQUENTIAL ANALYSIS
GENERATING FUNCTIONS
PASSAGE TIMES
QUEUEING THEORY
RANDOM WALKS
SEQUENTIAL ANALYSIS)

G. WEISS

WALD'S LEMMA *See* FUNDAMENTAL IDENTITY OF SEQUENTIAL ANALYSIS; WALD'S EQUATION

WALD'S *W*-STATISTICS

Let $\Pi_i : N_p(\boldsymbol{\xi}_i, \boldsymbol{\Sigma})$, $i = 1, 2$, be two *p*-variate $[\mathbf{X} = (X_1, \ldots, X_p)]$ multinormal* populations with common variance–covariance matrix $\boldsymbol{\Sigma}$. It is required to assign an individual to Π_1 or Π_2 on the basis of observed values $\mathbf{x}$. When all parameters are known, an optimal classification rule (minimizing the probability of incorrect decision) is "assign to Π_1 if

$$u_0 = \left\{ \mathbf{x} - \tfrac{1}{2}(\boldsymbol{\xi}_1 + \boldsymbol{\xi}_2) \right\}' \boldsymbol{\Sigma}^{-1}(\boldsymbol{\xi}_1 - \boldsymbol{\xi}_2) > K,"$$

$$(1)$$

where K depends on prior probabilities of Π_1 and Π_2. This is also the minimum distance rule*. Note that $u_0 = \mathbf{x}' \boldsymbol{\Sigma}^{-1}(\boldsymbol{\xi}_1 - \boldsymbol{\xi}_2)$ + constant; the first term on the right-hand side is *Fisher's linear discriminant function** [3].

When $\boldsymbol{\xi}_1$, $\boldsymbol{\xi}_2$ and $\boldsymbol{\Sigma}$ are not known, but are estimated from random samples of sizes n_1, n_2 from Π_1, Π_2, respectively, by the sample arithmetic means $\overline{\mathbf{X}}_1, \overline{\mathbf{X}}_2$ and the pooled sample variance–covariance matrix [with divisor $(n_1 + n_2 - 2)$] $\mathbf{S}$, Wald [12] proposed using the classification statistic

$$W = \left\{ \mathbf{x} - \tfrac{1}{2}(\overline{\mathbf{X}}_1 + \overline{\mathbf{X}}_2) \right\}' \mathbf{S}^{-1}(\overline{\mathbf{X}}_1 - \overline{\mathbf{X}}_2),$$

$$(2)$$

obtained by "plugging in" the sample estimators in place of the (unknown) parameters.

Wald showed that the limiting distribution of W as $n_1, n_2 \to \infty$ is the same as that of u_0. Specifically, for samples from Π_1 the distribution of W tends to normal with mean $\tfrac{1}{2}\Delta^2$ and variance Δ^2; for samples from Π_2 it tends to normal with expected value $- \tfrac{1}{2}\Delta^2$ and variance Δ^2, where

$$\Delta^2 = (\boldsymbol{\xi}_1 - \boldsymbol{\xi}_2)' \boldsymbol{\Sigma}^{-1}(\boldsymbol{\xi}_1 - \boldsymbol{\xi}_2).$$

(This is the *Mahalanobis distance**.)

More generally, if $\mathbf{T}$ is an estimator of a vector parameter $\boldsymbol{\theta}$, and it is desired to test the hypothesis that $\mathbf{g}(\boldsymbol{\theta}) = \mathbf{0}$ for some vector-valued twice-differentiable function $\mathbf{g}$,

then quadratic forms*

$$W = [\mathbf{g}(\mathbf{T})]' \mathbf{D}^{-1} \mathbf{g}(\mathbf{T}) \qquad (3)$$

may be used as test statistics, rejecting the hypothesis for large values of W, where $\mathbf{D}$ is an estimator of the covariance matrix of the vector $\mathbf{g}(\mathbf{T}) - \mathbf{g}(\boldsymbol{\theta})$. Test statistics of this kind are called *Wald statistics*, after their introduction by Wald [11, 12]. He introduced them with $\mathbf{T}$ as a maximum likelihood* estimator (MLE) of $\boldsymbol{\theta}$ based on independent observations from a family of distributions indexed by $\boldsymbol{\theta}$, satisfying some regularity conditions. Conditions more general than those of Wald under which W has an asymptotic chi-square or noncentral chi-square distribution* are derived in Stroud [10].

When $\mathbf{T}$ is a MLE of $\boldsymbol{\theta}$, it is natural to compare the large-sample performance of W with that of the likelihood-ratio* (LR) statistic $-2 \log \Lambda$ and of Rao's score statistic* S, since all three have the same asymptotic chi-square distribution; for definitions, assumptions, and basic properties, see Rao [9, Sec. 6e]. Now let θ be scalar and $g(\theta) \equiv \theta$. Then [3, 4] the test based on S is more powerful in the neighborhood of $\theta = 0$ than either Wald's test or the LR test, under certain conditions; furthermore, Wald's test under these conditions is the least powerful of the three. However, the three tests perform almost equally near $\theta = 0$ for families with small statistical curvature*.

In testing a restricted model in which the restriction on $\boldsymbol{\theta}$ is nonlinear in the parameters against an unrestricted model, a transformation to a form that is algebraically equivalent under the null hypothesis could alter the numerical value of the Wald statistic in finite samples, so that a set of data could conceivably lead to contrary conclusions based on one such form of W or another; the numerical values of $-2 \log \Lambda$ and of S in these circumstances, however, would be unchanged [6]. This shortcoming of W does not arise, for example, in testing general linear restrictions on coefficients in the multivariate general linear model*. The exact finite sample distribution of W in the latter case has been obtained by Phillips [8].

In multidimensional contingency-table* analysis, Wald statistics can be produced via weighted least-squares* techniques if $g(\theta)$ is expressible in terms of linear functions of some unknown vector parameters ϕ; see refs. 2 and 7. *See also* MARGINAL SYMMETRY, and ref. 1 for applications in tests of marginal symmetry and quasisymmetry*.

More general conditions than those of Wald under which W has an asymptotic central or noncentral chi-square distribution are derived in Stroud [10].

References

[1] Bhapkar, V. P. (1979). *Biometrics*, **35**, 426.

[2] Bhapkar, V. P. and Koch, G. G. (1968). *Technometrics*, **10**, 107–123.

[3] Chandra, T. K. and Joshi, S. N. (1983). *Sankhyā A*, **45**, 226–246.

[4] Chandra, T. K. and Mukerjee, R. (1985). *Sankhyā A*, **47**, 271–284.

[5] Fisher, R. A. (1936). *Ann. Eugen. (Lond.)*, **7**, 179–188.

[6] Gregory, A. W. and Veall, M. R. (1985). *Econometrica*, **53**, 1465–1468.

[7] Grizzle, J. E., Starmer, C. F., and Koch, G. G. (1969). *Biometrics*, **25**, 489–504.

[8] Phillips, P. C. B. (1986). *Econometrica*, **54**, 881–895.

[9] Rao, C. R. (1973). *Linear Statistical Inference and Its Applications*, 2nd ed. Wiley, New York. [The first edition (1965) includes a conjecture (omitted here), that the test based on the score statistic is locally more powerful than the LR or Wald test. Rao's conjecture was largely justified in ref. 3.]

[10] Stroud, T. W. F. (1971). *Ann. Math. Statist.*, **42**, 1412–1424. (This paper contains a very readable background discussion of Wald's original test procedure and its asymptotic optimality properties.)

[11] Wald, A. (1943). *Trans. Amer. Math. Soc.*, **54**, 426–482.

[12] Wald, A. (1945). *Ann. Math. Statist.*, **16**, 117–186.

(CHI-SQUARE TESTS
DISCRIMINANT ANALYSIS
DISTANCE FUNCTIONS
LIKELIHOOD-RATIO TESTS
MULTIVARIATE ANALYSIS
SCORE STATISTICS

WELCH'S v-CRITERION, MULTIVARIATE EXTENSION)

WALLIS AND MOORE PHASE-FREQUENCY TEST

This test is aimed at detecting departure from randomness* in a sequence of values $X_1, X_2, \ldots, X_n$ (such as a time series*). It is based on the signs of the successive differences $X_1 - X_0, X_2 - X_1, \ldots, X_n - X_{n-1}$. Wallis and Moore [1] term a sequence of like signs a *phase*; their test is based on the total number of phases (whether of $+$ or $-$ sign). Omitting the first and last phases in the sequence, the number of phases H is determined. If the sequence is indeed random (and the X's are identically distributed), then

$$E[H] = \tfrac{1}{3}(2n - 7)$$

and

$$\mathrm{var}(H) = \tfrac{1}{90}(16n - 29).$$

For $n \geqslant 10$, a reasonably good test can be based on the assumption that H is normally distributed, if a continuity correction* is used. The correction is not needed if $n \geqslant 25$. Exact tables (based on a chi-squared distribution*) are given in [1].

Reference

[1] Wallis, W. A. and Moore, G. H. (1941). *J. Amer. Statist. Ass.*, **36**, 401–409.

(RUN LENGTHS, TESTS OF (*See Supplement*))

WALLIS' FORMULA

$$\frac{1}{2}\pi = \frac{2}{1} \cdot \frac{2}{3} \cdot \frac{4}{3} \cdots \frac{2k}{2k-1} \cdot \frac{2k}{2k+1} \cdots$$

$$= \prod_{j=1}^{\infty} \left\{ 4j^2 (4j^2 - 1)^{-1} \right\}.$$

The formula is named after John Wallis [1] (1616–1703).

References

[1] Wallis, J. (1655). *Arithmetica Infinitorum*, Oxford, England.

WALSH AVERAGES

Hodges and Lehmann [2] discussed a general method for deriving a point estimate of location from a test statistic. Given a sample $X_1, \ldots, X_n$ from a continuous, symmetric population, the Hodges–Lehmann estimator, derived from the Wilcoxon signed rank statistic*, is the median of the $n(n + 1)/2$ pairwise averages $(X_i + X_j)/2, 1 \leqslant i \leqslant j \leqslant n$. In fact, the Wilcoxon statistic, which is the sum of ranks of the positive observations when ranked among all the absolute values, is precisely equal to the number of positive pairwise averages. Walsh [5, 6] proposed various tests and confidence intervals based on the pairwise averages, and Tukey [4] established the connection between the Wilcoxon signed rank statistic and the pairwise averages and dubbed them the *Walsh averages*.

The Walsh averages are fundamental in describing statistical inference based on general linear rank tests in the one-sample location model. Let $a(1) \leqslant \cdots \leqslant a(n)$ be a nonconstant sequence of scores and define

$$S(\theta) = \sum_{i=1}^{n} a[R_i(\theta)] \mathrm{sign}(X_i - \theta),$$

where $R_i(\theta)$ is the rank of $|X_i - \theta|$ among $|X_1 - \theta|, \ldots, |X_n - \theta|$, and $\mathrm{sign}(z) = 1, 0, -1$ as $z > 0, = 0, < 0$, respectively. Bauer [1] showed that $S(\theta)$ is a decreasing step function, which has steps only at some or all of the Walsh averages. Let $X_{(1)} \leqslant \cdots \leqslant X_{(n)}$ denote the ordered sample, then, for $i \leqslant j$, the step size at $(X_{(i)} + X_{(j)})/2$ is $2[a(j - i) - a(j - i + 1)]$, where $a(0) \equiv 0$. Hence the value of $S(\theta)$ at a value just to the right of a Walsh average is $\sum_1^n a(i)$, less the accumulated jumps.

The Hodges–Lehmann estimate of θ, derived from $S(\theta)$, is the solution $\theta = \hat{\theta}$ of $S(\theta) = 0$, where equality is interpreted to

mean the Walsh average where $S(\theta)$ steps across 0 or the average of the two Walsh averages that determine the interval of zeros. If $P_\theta(S(\theta) < k) = \alpha/2$, then the acceptance region of the test determines a $(1 - \alpha)100\%$ nonparametric confidence interval* for θ. This interval has appropriately chosen Walsh averages as its end points.

To illustrate the calculations, we consider the first four used Rolls Royce prices in the *New York Times*, September 17, 1982, Section B. The prices in thousands of dollars are 25, 42, 50, 63, and we will suppose this is a sample from a symmetric population of prices. We will use the absolute normal scores* statistic with

$$a(i) = \Phi^{-1}[(i + n + 1)/\{2(n + 1)\}]$$

and

$$\Phi^{-1}(u) \doteq 4.91[u^{0.14} - (1 - u)^{0.14}],$$

where Φ^{-1} is the inverse of the standard normal distribution function. The approximation, based on Tukey's λ-distribution, is sufficient for most practical purposes and is easy to use on a calculator. (See Joiner and Rosenblatt [3].) The absolute normal scores for $n = 4$ are 0.25, 0.52, 0.84, and 1.28.

The table shows the Walsh average (and its rank), the jump in $S(\theta)$, and the value of $S(\theta)$ just to the right of the Walsh average. Note that max $S(\theta) = \Sigma a_i = 2.89$.

	25	42	50	63
25	25(1) −0.50 2.39			
42	33.5(2) −0.54 1.85	42(4) −0.50 0.71		
50	37.5(3) −0.64 1.21	46(6) −0.54 −0.71	50(7) −0.50 −1.21	
63	44(5) −0.88 −0.17	52.5(8) −0.64 −1.85	56.5(9) −0.54 −2.39	63(10) −0.50 −2.89

Hence the Hodges–Lehmann estimate of θ is the Walsh average, 44. If we had used the Wilcoxon signed rank statistic, the estimate is the median of the Walsh averages, $(44 + 46)/2 = 45$. Finally, if $a(i) \equiv 1$ so we have the sign statistics, then the estimate is the sample median, $(42 + 50)/2 = 46$, itself a Walsh average.

References

[1]   Bauer, D. F. (1972). *J. Amer. Statist. Ass.*, **67**, 687–690.

[2]   Hodges, J. L., Jr. and Lehmann, E. L. (1963). *Ann. Math. Statist.*, **34**, 598–611.

[3]   Joiner, B. L. and Rosenblatt, J. R. (1971). *J. Amer. Statist. Ass.*, **66**, 394–399.

[4]   Tukey, J. W. (1949). *Memorandum Report No. 17*, Statistical Research Group, Princeton University, Princeton, NJ.

[5]   Walsh, J. E. (1949a). *Ann. Math. Statist.*, **20**, 64–81.

[6]   Walsh, J. E. (1949b). *J. Amer. Statist. Ass.*, **44**, 342–355.

(HODGES-LEHMANN ESTIMATORS
LOCATION PARAMETER, ESTIMATION OF
L-STATISTICS
NONPARAMETRIC CONFIDENCE
    INTERVALS
ORDER STATISTICS
WILCOXON SIGNED RANK TEST)

T. P. HETTMANSPERGER

WALSH FUNCTIONS *See* WALSH-FOU-
RIER TRANSFORMS

WALSH–FOURIER TRANSFORMS

Walsh functions* $\{W(n, x),\ n = 0, 1, 2, \ldots,\ 0 \leqslant x < 1\}$ are defined as follows:

(i) $W(0, x) = 1,\ 0 \leqslant x < 1$.

(ii)   Let n be dyadic, i.e. $n = \sum_{i=0}^{\infty} x_i 2^i$, where $x_i = 0$ or $x_i = 1$; and $x_i = 0$ for $i > m$. Then

$$W(n, x) = \prod_{i=1}^{n} \{ r_{m_i}(x) \},$$

where $m_1, \ldots, m_r$ correspond to the coefficients $x_{m_i} = 1$. These functions were intro-

duced by Walsh [9] and studied extensively by Fine [2–4].

If f is any function of period 1, Lebesgue integrable on $[0, 1]$, then it can be expanded in a Walsh–Fourier series

$$f(x) \sim \sum_{n=0}^{\infty} a_n W(n, x),$$

with coefficients $a_n = \int_0^1 f(x) W(n, x)\, dx$. Generalized Walsh functions are defined for nonnegative real arguments. See, e.g., Chrestenson [1] and Selfridge [8] for details. For applications of Walsh–Fourier transforms in time-series* analysis, see, e.g., Morettin [5] and the references therein. The natural class of stationary processes*, which is analyzed by Walsh functions, is the class of dyadic stationary processes, i.e., stationary processes unchanged under dyadic shifts.

For further properties of Walsh functions, see Révész [6] and (with special reference to the law of iterated logarithm*) Révész and Wschebor [7].

References

[1]   Chrestenson, N. E. (1955). *Pacific J. Math.*, **5**, 17–31.

[2]   Fine, N. J. (1949). *Trans. Amer. Math. Soc.*, **65**, 372–414.

[3]   Fine, N. J. (1950). *Trans. Amer. Math. Soc.*, **69**, 66–77.

[4]   Fine, N. J. (1957). *Trans. Amer. Math. Soc.*, **86**, 246–255.

[5]   Morettin, P. A. (1981). *Proc. 43rd Session of the ISI*, **49**, Book 3, 1211–1230. (Detailed list of references.)

[6]   Révész, P. (1968). *The Laws of Large Numbers*, Academic, New York.

[7]   Révész, P. and Wschebor, M. (1965). *Publ. Math. Inst. Hung. Acad. Sci.*, **9A**, 543–554.

[8]   Selfridge, R. G. (1955). *Pacific J. Math.*, **5**, 451–480.

[9]   Walsh, J. L. (1923). *Amer. J. Math.*, **45**, 5–24.

(RADEMACHER FUNCTIONS
WALSH FUNCTIONS)

WALSH INDEX *See* INDEX NUMBERS

WANDERING-QUARTER SAMPLING

A practical method of sampling inter-item distances of *spatial* distributions of hidden or unmapped (point) items, proposed by Catana [1].

Wandering-quarter sampling is a variant of *T*-square sampling*, which in turn is a variant of nearest neighbor sampling*. In each of these sampling schemes, sampling sites $O_1, O_2, \ldots, O_n$ are chosen randomly in the region of interest, for example, by choosing exact coordinates on a map of the region. Each sampling site is then visited, and the location of the nearest item to that sampling site is then found, say at P_i for sampling site O_i.

In wandering-quarter sampling, the nearest item to that at P_i is then found, subject to the condition that the direction from the new item to P_i lies within 45° of the direction from P_i to O_i. From this new item, say at R_i, a new search for its nearest item can again be conducted (once more subject to the condition that the direction from the new nearest item lies within 45° of the direction from P_i to O_i), and so on. See Fig. 1. In Catana's original version of wandering-quarter sampling, *Catana sampling*, the original search direction for the location of the original nearest point at P_i, is itself restricted to lie within a prespecified direction.

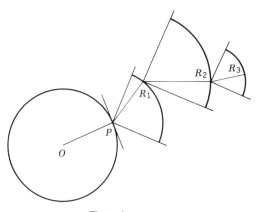

Figure 1

Reference

[1] Catana, A. J. (1963). *Ecology*, **44**, 349–360.

Bibliography

Diggle, P. J. (1983). *Statistical Analysis of Spatial Point Processes*. Academic, London.

Ripley, B. D. (1981). *Spatial Statistics*. Wiley, New York.

(SPATIAL PROCESSES
SPATIAL SAMPLING)

W. G. S. HINES

WARD'S CLUSTERING ALGORITHM

Ward's clustering algorithm is a popular procedure within the set of algorithms called agglomerative hierarchical methods. These methods apply a routing strategy to reproduce a hierarchical or treelike structure among *n* objects. Starting with *n* clusters, where each object is a cluster, an agglomerative hierarchical method proceeds in a stagewise manner to reduce the number of clusters one at a time until all *n* objects are in one cluster. *See also* HIERARCHICAL CLUSTER ANALYSIS.

WARD'S ALGORITHM

The following steps describe the usual implementation of an (agglomerative) hierarchical cluster analysis:

1. Define a triangular matrix that shows a measure of similarity or proximity between each pair of *n* objects (*see* PROXIMITY DATA). This matrix has $n(n-1)/2$ entries and is often constructed by computing proximity measures such as Euclidean distances or correlations based on an original $n \times m$ data matrix of *n* objects and *m* attributes or variables.

2. Search the proximity matrix for the most similar pair of clusters and join these two clusters. The proximity value between the two merged clusters is called the criterion or objective-function value for stage k, z_k.

3. Update the proximity matrix by recomputing proximity values between the new cluster and all other clusters. The new proximity matrix has one less row (or column) than the preceding proximity matrix.

4. Repeat steps **2** and **3** until all objects reside in one cluster. The result is a treelike structure that shows which two clusters were merged at each stage k, $k = 1, \ldots, n - 1$, and the corresponding criterion values z_k for each stage, where stage k corresponds to $n - k$ clusters.

Differences among hierarchical methods primarily center around two procedural steps: the definition of the "most similar" pair of clusters (step **2**) and the method of updating similarity measures from one stage to the next (step **3**).

Lance and Williams [4] developed a generalized transformation model that elegantly defines the measure of proximity in step **3** for six popular hierarchical models. Later, this transformation model was extended by Wishart [12] to include Ward's method, which is sometimes called the *minimum variance* or *Ward's error sum of squares* method.

Interestingly, the original article by Ward [11] described a generalized hierarchical method similar to, but no less general than, the four-step description above. In particular, Ward specified that the "loss" from joining two groups (i.e., the criterion value in step **2**) is best expressed by whatever objective function makes sense to the investigator and then described various objective functions that he used in his research for the Air Force [11, p. 237]. Indeed, in a subsequent communication, Ward's own preference for naming his model was the MAXOF (MAXimize an Objective Function) clustering model.

In his numerical example, Ward used the "sum of squared deviations about the group mean" or "error sum of squares," which in multidimensional Euclidean space is defined for cluster c as

$$\text{ESS}_c = \sum_{j=1}^{m} \sum_{i=1}^{n_c} \left(x_{cij} - \bar{x}_{cj} \right)^2, \quad (1)$$

where m is the number of attributes; n_c is the number of objects in cluster c; x_{cij} is the measure (raw, standardized, etc.) of attribute j on object i within cluster c; and $\bar{x}_{cj}$ is the mean of the jth attribute in cluster c. The overall error sum of squares objective function in stage k is then given by

$$\text{ESS}_k = \sum_{c=1}^{n-k} \text{ESS}_c, \quad (2)$$

and the "loss" or increase in ESS based on the fusion of two clusters in stage k is given by

$$z_k = \text{ESS}_k - \text{ESS}_{k-1}, \quad (3)$$

which defines an "error sum of squares" or "minimum variance" criterion for step **3**.

Subsequently, Wishart [12, p. 167] showed that the criterion in (3) is equivalent to one-half of the squared Euclidean distance between two joined single-object clusters and proved that the use of a squared Euclidean distance proximity matrix is functionally equivalent to Ward's ESS example and implementable through the transformation function first described by Lance and Williams.

Thus Ward's early choice of an ESS example, Wishart's link to Euclidean distance, and the Lance and Williams transformation function, the attractive but not necessarily valid conceptualization of clusters as swarms in Euclidean space, the "closeness" of clusters based on the proportionality between increase in the ESS and the squared Euclidean distance separating merged-cluster centroids [1, p. 143], and subsequent implementations by commercial computer packages all came together to transform Ward's perfectly general algorithm into an algorithm with an exclusive, distance-based, minimum variance focus.

See Ward [11, p. 241] and Anderberg [1, p. 43] for numerical examples of Ward's method based on the ESS criterion. Anderberg [1] and Everitt [3] are excellent sources for descriptions of the various hierarchical algorithms.

STOPPING RULES

Users of hierarchical clustering algorithms often wish to determine the "best" number of clusters. Mojena [7] and Mojena and Wishart [8] proposed and evaluated three statistical rules for this task, based on the behavior of the criterion vector $\mathbf{z}$ as a monotonically increasing function. These rules predict a "significant" increase from z_k to z_{k+1}, if any, which implies an undesirable fusion, and the stage with $n - k$ clusters as "best." Ward's method, together with a simple upper-tail rule, gave consistently good results across Monte Carlo* data sets that conceptualized clusters as compact swarms in Euclidean space. Morey et al. [9] further confirmed successful results with this rule, along with an alternative rule based on an adaptation of Cattell's scree test. Binder [2] has proposed a Bayesian approach to estimating the best number of clusters, but its usefulness is restricted to small problems. See [10] for other procedures.

SOFTWARE PACKAGES

There are many sources of computer software for implementing cluster analyses, but two commercially available packages stand out, both of which include Ward's method: SAS and CLUSTAN (*see* STATISTICAL SOFTWARE).

SAS [10] is a comprehensive system for data analysis that is widely used in universities and other research-oriented groups. It implements seven clustering procedures, including four common hierarchical methods. It uses the $n \times m$ multivariate data matrix as input, has procedures for printing dendrograms*, and prints reports that include a criterion for estimating the best number of clusters.

CLUSTAN [13] is by far the most comprehensive clustering package available. It includes 28 clustering procedures, 10 of which are hierarchical. Input data options include multivariate data matrices and user-defined similarity matrices. Users have a choice of 40 proximity measures, depending on data types (numeric or binary) and user needs. The package also includes relocation routines for improving an initial clustering, various forms of graphical output, stopping-rule procedures from Mojena [7], the ability to read data files previously created through the SPSS software system, and a conversational pre-processor.

EVALUATION

Ward's method figures prominently in the literature that addresses the evaluation of clustering algorithms. The effectiveness of Ward's method as a clustering procedure can be viewed from various perspectives. First, does it give an optimal solution with respect to minimum error sum of squares? Second, just how good is it in recovering cluster structure, i.e., in identifying both the correct number of clusters and the correct membership of objects? Finally, how does it compare to other clustering procedures?

Ward's method is a heuristic rather than an optimization algorithm. As such, it does not ensure that the resulting clustering yields an overall minimum variance solution. Indeed, it would be very surprising if it were to yield an optimal or even near-optimal solution except for trivial data sets. Optimal solutions to the ESS clustering problem have been generated by dynamic and 0–1 integer programming* formulations, but these severely limit the size (n) of the problem due to storage and computational constraints. An attractive strategy suggested by Wishart [13] and implemented by Morey et al. [9] is to generate an initial solution by Ward's method and then systematically reassign objects by using relocation techniques.

The evaluation literature primarily reports on the recovery performances of clustering techniques and on their comparisons. The literature tends to favor Ward's method, although results are mixed. It would now appear that some tentative conclusions are emerging based on work by Morey et al. [9] and Milligan et al. [6], and on the thorough review by Milligan [5]:

1. Ward's method performs quite well across a variety of data sets that include Monte Carlo mixtures, ultrametric* data, and real data; however, performance can vary widely depending on the selection of clustering parameters such as proximity measures, and on certain data-set characteristics such as cluster size and cluster overlap.

2. The ESS focus of Ward's method dictates the use of squared Euclidean distance as a measure of proximity; yet this measure of association may not be warranted for all studies, as it mixes together object associations due to shape, scatter, and height. If only shape is of interest, then correlation-type measures are more appropriate. In this case, an algorithm such as the group average method based on a correlation criterion can give more legitimate clustering results.

3. The extent of cluster overlap affects the performance of various algorithms. Ward's method appears to give the best recovery as overlap increases, but the group average method seems to outperform Ward's method with nonoverlapping structures.

4. Ward's method tends to fuse small clusters and appears to favor the creation of clusters having roughly the same number of observations. The group average method is as good or better when clusters are of unequal size.

5. Ward's method and other hierarchical algorithms are not very robust with respect to various types of error perturbations, such as outliers*.

6. Ward's method is sensitive to profile elevation, with a tendency to give distinct, but not necessarily valid, clusters along the principal component of a multivariate distribution. If the elevation component is pervasive, then the solution can be valid, as in the alcohol abuse study by Morey et al. [9].

References

[1] Anderberg, M. R. (1973). *Cluster Analysis for Applications*. Academic, New York.

[2] Binder, D. A. (1981). *Biometrika*, **68**, 275–285.

[3] Everitt, B. S. (1980). *Cluster Analysis*, 2nd ed. Heinemann, London, England.

[4] Lance, G. N. and Williams, W. T. (1967). *Computer J.*, **9**, 373–380.

[5] Milligan, G. W. (1981). *Multivariate Behav. Res.*, **16**, 379–407.

[6] Milligan, G. W., Soon, S. C., and Sokol, L. M. (1983). *IEEE Trans. Pattern Analysis Machine Intelligence*, **PAMI-5**, 40–47.

[7] Mojena, R. (1977). *Computer J.*, **20**, 359–363.

[8] Mojena, R. and Wishart, D. (1980). *COMPSTAT 1980 Proc.* Physica-Verlag, Vienna, Austria, pp. 426–432.

[9] Morey, L. C., Blashfield, R. K., and Skinner, H. A. (1983). *Multivariate Behav. Res.*, **18**, 309–329.

[10] *SAS USER'S GUIDE: Statistics* (1982). SAS Institute, Cary, NC.

[11] Ward, J. H. (1963). *J. Amer. Statist. Ass.*, **58**, 236–244.

[12] Wishart, D. (1969). *Biometrics*, **22**, 165–170.

[13] Wishart, D. (1982). *CLUSTAN User Manual*. Program Library Unit, Edinburgh University, Edinburgh, Scotland.

(DENDROGRAMS
HIERARCHICAL CLUSTER ANALYSIS
PROXIMITY DATA
RECURSIVE PARTITIONING (*See Supplement*))

Richard Mojena

WARING–HERDAN MODELS

This is a term used in linguistic literature, for models of vocabulary distribution based on Waring distributions. See Herdan [1] and a survey by Holmes [2] for details.

References

[1] Herdan, G. (1964). *Quantitative Linguistics*. Butterworth, London.

[2] Holmes, D. I. (1985). *J. R. Statist. Soc. A*, **148**, 328–341.

(FACTORIAL SERIES DISTRIBUTIONS
LINGUISTICS, STATISTICS IN
LITERATURE AND STATISTICS)

WARING'S DISTRIBUTION *See* FACTORIAL SERIES DISTRIBUTIONS; YULE DISTRIBUTION

WARING LIMITS   *See* CONTROL CHARTS

WARNING LINES

These are lines on a control chart* indicating a mild degree of departure from a desired level of control. More or less, conventionally, a deviation likely to be exceeded (in absolute value) with probability of about 5%, if production is in the desired state of control, is used as the basis for warning lines. In an $\overline{X}$ chart*, if the target average is μ and the population standard deviation (assumed known) is σ, the warning lines for a sample of size n are at $\mu \pm 2\sigma/\sqrt{n}$. "Action" lines, on the other hand, are customarily at $\mu \pm 3\sigma/\sqrt{n}$. Of course, in practical use, "warning" and "action" can take various forms.

(CONTROL CHARTS
QUALITY CONTROL, STATISTICAL)

WASSERSTEIN DISTANCE

For a metric space X with metric ρ, the Wasserstein distance $W(P, Q)$ between probability measures P and Q is defined as

$$W(P, Q) = \inf E[\rho(\xi, \eta)],$$

where the inf is taken over all possible pairs of random variables ξ and η with distributions P and Q, respectively (Wasserstein [9]). For the case of one-dimensional Euclidean space $X = R^1$, with usual Euclidean metric

$$W(P, Q) = \int_{-\infty}^{\infty} |F(x) - G(x)| \, dx,$$

where F and G are distribution functions from P and Q, respectively.

Rüschendorf [6] presents a general definition of the form

$$D(P, Q) = \int_0^1 |F^{-1}(u) - G^{-1}(u)| \, du.$$

This form of distance was studied as early as 1956 by Dall'Aglio [1], and in 1958 by Kantorovič and Rubinstein [4]. Calculations of Wasserstein distances are given in Vallander [8].

For the relation between Prohorov's distance and Wasserstein distance, see Dobrushin [2] and Strassen as cited by Rüschendorf [6]. Using the concept of multivariate quantile transformation, Rüschendorf [6] obtained explicit formulas for the Wasserstein distance between multivariate distributions in certain cases. These may be compared with the Fréchet distance between multinormal distributions* [3, 5].

A detailed discussion of Wasserstein distance and its applications in the theory of empirical processes* is given in Shorack and Wellner [7].

References

[1] Dall'Aglio, G. (1956). *Annali Scuola Normale Sup. di Pisa*, **10**, 35–74.

[2] Dobrushin, R. L. (1970). *Theory Prob. Appl.*, **15**, 458–486.

[3] Dowson, D. C. and Landau, B. V. (1982). *J. Multivariate Anal.*, **12**, 450–455.

[4] Kantorovič, L. and Rubinstein, G. (1958). *Vestn. Leningrad Univ. Mat.*, **13**(7), 52–59.

[5] Olkin, I. and Pukelsheim, F. (1982). *Linear Algebra Appl.*, **48**, 257–263.

[6] Rüschendorf, L. (1985). *Zeit. Wahrscheinlichkeitsth. verwand. Geb.*, **70**, 117–130.

[7] Shorack, G. B. and Wellner, J. A. (1986). *Empirical Processes with Applications to Statistics*. Wiley, New York.

[8] Vallander, S. S. (1973). *Theory Prob. Appl.*, **18**, 784–786.

[9] Wasserstein, L. N. (1969). *Prob. Peredachi Infor-matsii*, **5**(3), 64–73 (in Russian).

(PROBABILITY SPACES, METRICS AND
    DISTANCES ON)

WATSON'S DISTRIBUTION

This is the distribution of the ratio

$$R = \sum_{j=1}^{k} \lambda_j W_j \bigg/ \sum_{j=1}^{k} \mu_j W_j,$$

where $\mu_j > 0$ for all $j, \lambda_1/\mu_1 > \lambda_2/\mu_2 > \cdots > \lambda_n/\mu_n$, and $W_1, W_2, \ldots, W_k$ are mutually independent random variables, each having the standard exponential distribution* with PDF

$$f_W(w) = e^{-w}, \qquad 0 < w.$$

It is a special case of the distribution of the ratio of quadratic forms in normal variables (Box [1]). Watson [3] obtained the (equivalent) formulas

$\Pr[R \geqslant r]$

$$= \sum_{i=1}^{m} \left[\frac{(\lambda_i - r\mu_i)^{k-1}}{\Pi_{j=1, j \neq i}^{k} \left\{ (\lambda_i - \lambda_j) - r(\mu_i - \mu_j) \right\}} \right],$$

$$\text{for } \frac{\lambda_m}{\mu_m} \geqslant r \geqslant \frac{\lambda_{m+1}}{\mu_{m+1}},$$

$\Pr[R \leqslant r]$

$$= \sum_{i=m}^{k} \left[\frac{(r\mu_i - \lambda_i)^{k-1}}{\Pi_{j=1, j \neq i}^{k} \left\{ (\mu_i - \mu_j)r - (\lambda_i - \lambda_j) \right\}} \right],$$

$$\text{for } \frac{\lambda_{m-1}}{\mu_{m-1}} \geqslant r \geqslant \frac{\lambda_m}{\mu_m}.$$

For computations, one would use the formula with the smaller number of terms. The mean and variance of the distribution have been obtained by Dent and Broffitt [2].

References

[1] Box, G. E. P. (1954). *Ann. Math. Statist.*, **25**, 290–302.

[2] Dent, W. T. and Broffitt, J. D. (1974). *J. R. Statist. Soc. B*, **36**, 91–98.

[3] Watson, G. S. (1955). *Austral. J. Phys.*, **8**, 402–407.

(DURBIN–WATSON TEST
QUADRATIC FORMS
SERIAL CORRELATIONS)

WATSON'S U^2

The Watson U^2-statistic [23] for testing goodness of fit* is a statistic derived from the empirical distribution function (EDF)* of a given random sample. It measures the discrepancy between the EDF and the hypothesized distribution, which for the present is assumed to be continuous; the distribution may be completely specified, or may contain unknown parameters. Suppose $F_n(x)$ is the EDF of a sample of x-values of size n, and $F(x; \theta)$ is the hypothesized distribution, with θ, the vector of parameters, fully known; then

$$U^2 = n \int_{-\infty}^{\infty} \left[F_n(x) - F(x; \theta) \right.$$

$$\left. - \int_{-\infty}^{\infty} \left\{ F_n(x) - F(x; \theta) \right\} dF(x; \theta) \right]^2 dF(x; \theta).$$

$$(1)$$

For practical computations, U^2 can be calculated as follows. Let $x_{(1)} < x_{(2)} < \cdots < x_{(n)}$ be the order statistics* of the sample and let $z_{(i)} = F(x_{(i)}; \theta)$, $i = 1, \ldots, n$; the $z_{(i)}$ will also be in ascending order and will lie between 0 and 1. Then

$$U^2 = \sum_{i=1}^{n} \left\{ z_{(i)} - (2i - 1)/(2n) \right\}^2$$

$$- n(\bar{z} - 0.5)^2 + 1/(12n). \qquad (2)$$

U^2 is a modification of the well-known Cramér–von Mises statistic*, now usually called W^2; it was introduced by Watson for use with observations P_i recorded on the circumference of a circle (*see* DIRECTIONAL DATA). For such observations, x is the arc length on the circumference, from an origin O to P_i. The value of W^2 depends on the choice of origin O, but U^2 is invariant. Thus different statisticians, presented with test values on a circle and choosing different origins, will find the same value of the test statistic U^2, whereas values of W^2 will differ. In fact, U^2 is the minimum value of W^2

as the origin is varied. Kuiper's statistic V is the corresponding invariant modification of the Kolmogorov–Smirnov* D.

DISTRIBUTION THEORY AND TABLES

Suppose the null hypothesis is H_0: a sample of x-values, $x_1, x_2, \ldots, x_n$ comes from a continuous distribution $F(x; \theta)$, with θ specified. This situation will be called case 0. The asymptotic distribution of U^2 on H_0 was given by Watson [23]; as for other Cramér–von Mises statistics, the asymptotic distribution is that of a sum of weighted independent chi-square variables, each with d.f. = 1. For U^2 the weights occur in equal pairs, so that the distribution is a sum of weighted exponentials, and percentage points are easily calculated. An interesting result is that the asymptotic distributions of $\pi^2 U^2$ and of nD^2 are the same; this seems difficult to explain on intuitive grounds. Also, the asymptotic distribution of U^2 is a convolution of the asymptotic distribution of $W^2/4$ with itself.

For small samples, Stephens [12, 13] gave exact distribution theory for U^2 on H_0; this is based on the fact that the values $z_i = F(x_i; \theta)$ will be uniformly distributed between limits 0 and 1, written $U(0, 1)$. Moments of U^2 were also found and percentage points were calculated by fitting Pearson curves. The distribution of U^2 was also discussed by Pearson and Stephens [9] and by Tiku [22]. Subsequently, Stephens [15] found a technique to make U^2 available for case 0 tests using only the asymptotic points.

Large values of U^2 lead to rejection of H_0. The value of U^2 is found from (2), then modified to U^* given by

$$U^* = \left(U^2 - \frac{0.1}{n} + \frac{0.1}{n^2} \right)\left(1.0 + \frac{0.8}{n} \right),$$

and U^* is referred to the upper-tail asymptotic percentage points. These are 0.152, 0.187, 0.222, and 0.268 for $\alpha = 0.10$, 0.05, 0.025, and 0.01, respectively. In a later work, Pettitt and Stephens [10] have given asymptotic points for right-censored data of either type 1 or type 2. For n finite, percentage points for type 1 censored data* converge so rapidly to the asymptotic points that no further tables are necessary; for type 2 censored data Stephens [21] has given points found by Monte Carlo methods.

POWER FOR CASE 0

Although U^2 was designed for values on a circle, it can be a useful statistic also for values on a line. U^2 will detect if the z-values, which should be $U(0, 1)$, are tending to lie in a single cluster somewhere along the line, or in two clusters at 0 and 1 (which, wrapped around a circle of circumference 1, would make a single cluster). In contrast, EDF statistics* W^2 and A^2 will be significant if z-values have moved strongly toward 0 or toward 1. In practical terms, this means that if vector θ contains location parameter α and scale parameter β, U^2 will detect if β has been wrongly specified in $F(x; \theta)$, and W^2 and A^2 will detect if α has been wrongly specified. Asymptotic power of U^2 has been discussed by Stephens [16], using the technique of decomposing the statistic into orthogonal components, by a Fourier series expansion of $\sqrt{n}\{F_n(z) - z\}$, where $F_n(z)$ is the EDF of the $z_{(i)}$ above. Monte Carlo comparisons of tests for uniformity, including U^2, have been given by Stephens [17], Locke and Spurrier [3], and Miller and Quesenberry [8].

TESTS WITH UNKNOWN PARAMETERS

EDF statistics, including U^2, may be adapted to test for $F(x; \theta)$ with unknown components in the parametric vector θ. These components must then be estimated from the given sample; let $\hat{\theta}$ be the vector with estimated components where necessary, and let $\hat{U}^2$ be the definition of U^2 as in (1) but with $F(x; \hat{\theta})$ replacing $F(x; \theta)$. Similarly, $\hat{z}_i = F(x; \hat{\theta})$ replaces z_i in the computational formula (2), to give $\hat{U}^2$. Asymptotic distribution theory for $\hat{U}^2$, when the estimators of unknown components of θ are maximum likelihood or other asymptotically efficient estimators, has been given by Stephens [17]–[21] for tests for the normal, exponential, extreme-value, logistic, and Cauchy distributions; for some of these distributions, modifications of the case 0 type above have been devised so that the tests can be made with a complete sample of size n, using only the asymptotic percentage points. Pettitt and Stephens [11] and Lockhart and Stephens [4, 5] have given percentage points for tests for the gamma distribution* and the von Mises distribution*. A test for the Weibull distribution*, when only the scale and/or shape parameters are unknown, can be reduced to a test for extreme-value distribution*; Lockhart and Stephens [6] have given points to make the test, when, in addition, the location (origin) of the Weibull distribution is unknown. It would be difficult to develop the asymptotic theory in analytic form for censored data, and points for tests with such data are not available. Most of the tables mentioned above are collected together in Stephens [21].

When statistics such as W^2, A^2, and U^2 are used with estimated parameters, the variation in ability to detect different types of alternative tends to disappear. Power studies indicate that in many situations (except perhaps for the Cauchy distribution), U^2 appears marginally less powerful than W^2 or A^2 when used for points on a line; again, of course, only U^2 should be used for tests for the von Mises distribution, which is a distribution on a circle. For further discussion, see Stephens [21].

TEST FOR DISCRETE DISTRIBUTION

Freedman [2], has adapted the U^2-test for a discrete distribution. The null distribution now depends on the number of cells n and on the probabilities p_i of falling into the cells; Freedman shows how the distribution may be approximated when these parameters are specified. The discrete version of U^2 is suitable for data such as measurements of angles, which have been grouped; also for counted data, recorded, say, monthly, and displayed in 12 cells around a circle denoting one year. Freedman [2] gives an example.

MULTISAMPLE TESTS

Watson [24] developed $U^2_{M,N}$ for testing that two samples of sizes N and M have the same (unspecified) continuous distribution and showed that the null asymptotic distribution is the same as for the one-sample statistic. Burr [1] gave percentage points and Stephens [14] gave moments and percentage points for $U^2_{M,N}$. Maag [7] extended Watson's work to a k-sample U^2-test. These multisample tests are essentially based on the ranks of the observations in their own samples and in the combined sample; although not many studies have been made, these tests appear to be comparable in power to other rank tests*.

References

[1] Burr, E. J. (1963). *Ann. Statist.*, **34**, 1091–1098.

[2] Freedman, L. S. (1981). *Biometrika*, **68**, 708–711.

[3] Locke, C. and Spurrier, J. D. (1978). *Commun. Statist. A.* **7**, 241–258.

[4] Lockhart, R. A. and Stephens, M. A. (1985a). *Biometrika*, **72**, 647–653.

[5] Lockhart, R. A. and Stephens, M. A. (1985b). Goodness-of-Fit Tests for the Gamma Distribution. *Tech. Rep.*, Dept. of Mathematics and Statistics, Simon Fraser University, British Columbia, Canada.

[6] Lockhart, R. A. and Stephens, M. A. (1986). Tests for the Weibull Distribution Based on the Empirical Distribution Function. *Tech. Rep.*, Dept. of Mathematics and Statistics, Simon Fraser University, British Columbia, Canada.

[7] Maag, U. R. (1966). *Biometrika*, **53**, 579–583.

[8] Miller, F. L. and Quesenberry, C. P. (1979). *Commun. Statist. B*, **8**, 271–290.

[9] Pearson, E. S. and Stephens, M. A. (1962). *Biometrika*, **49**, 397–402.

[10] Pettitt, A. N. and Stephens, M. A. (1976). *Biometrika*, **63**, 291–298.

[11] Pettitt, A. N. and Stephens, M. A. (1983). EDF Statistics for Testing for the Gamma Distribution. *Tech. Rep.*, Dept. of Statistics, Stanford University, Stanford, CA.

[12] Stephens, M. A. (1963). *Biometrika*, **50**, 303–313.

[13] Stephens, M. A. (1964). *Biometrika*, **51**, 393–397.

[14] Stephens, M. A. (1965). *Biometrika*, **52**, 661–663.

[15] Stephens, M. A. (1970). *J. R. Statist. Soc. B*, **32**, 115–122.

[16] Stephens, M. A. (1974a). *Ann. Inst. H. Poincaré Sect. B*, **10**, 37–54.

[17] Stephens, M. A. (1974b). *J. Amer. Statist. Ass.*, **69**, 730–737.

[18] Stephens, M. A. (1976). *Ann. Statist.*, **4**, 357–369.

[19] Stephens, M. A. (1977). *Biometrika*, **64**, 583–588.

[20] Stephens, M. A. (1979). *Biometrika*, **66**, 591–595.

[21] Stephens, M. A. (1986). In *Goodness-of-fit Techniques*, R. B. D'Agostino and M. A. Stephens, eds. Dekker, New York, Chap. 4.

[22] Tiku, M. L. (1965). *Biometrika*, **52**, 630–633.

[23] Watson, G. S. (1961). *Biometrika*, **48**, 109–114.

[24] Watson, G. S. (1962). *Biometrika*, **49**, 57–63.

(EDF STATISTICS
KOLMOGOROV–SMIRNOV STATISTICS
RÉNYI-TYPE DISTRIBUTIONS)

M. A. Stephens

WAVE-CUTTING INDEX

This is an index reflecting the effectiveness of a linear m-point graduation* formula

$$u_x^* = a_n u_{x-n} + \cdots + a_1 u_{x-1} + a_0 u_x$$
$$+ a_1 u_{x+1} + \cdots + a_n u_{x+n},$$
$$m = 2n + 1,$$

or

$$u_x^* = a_n u_{x-n+1/2} + \cdots + a_1 u_{x-1/2}$$
$$+ a_1 u_{x+1/2} + \cdots + a_n u_{x+n-1/2},$$
$$m = 2n,$$

in reducing the effect of local variation (e.g., of wild values near u_x).

The index is

$$a_0 + 2a_1 + 2a_2, \quad \text{for } m = 2n + 1,$$
$$2a_1 + 2a_2 + a_3, \quad \text{for } m = 2n.$$

For all graduation formulas satisfying the basic condition that $u_x^* = \theta$ if $u_y = \theta$ for all y, the sum of the coefficients is 1, so the index represents the "proportion" of the graduated value contributed by observed values near x.

The *lower* the value of the index, the *greater* the wave-cutting power. For Spencer's 21-point formula*, for example, the value of the wave-cutting index is 0.766; for *Hardy's wave-cutting formula*,

$$u_x^* = \{[5][3]\{[3] + [5] - [7]\}/65\}u_x$$

(where $[\cdot]$ is the summation operator $[2r + 1]u_x = \sum_{j=-r}^{r} u_{x+j}$), the index value is 0.415. For further examples, see [1, p. 287].

Reference

[1] Benjamin, B. and Pollard, J. H. (1980). *The Analysis of Mortality and Other Actuarial Statistics*, 2nd ed. Heinemann, London, England.

(GRADUATION
SUMMATION [n]
TWICING)

WEAK CONVERGENCE: STATISTICAL APPLICATIONS

Weak convergence, also known as convergence in law, is a way of describing how a sequence of probability distributions Q_n converges to a limit distribution Q as n increases. Formally, Q_n converges weakly to Q if $Q_n(A)$ converges to $Q(A)$ for every Borel set A whose boundary has Q-probability 0. For distributions in a Euclidean space, weak convergence amounts to pointwise convergence of the cumulative distribution functions, at every continuity point of the limit cumulative distribution function (see Billingsley [3] and CONVERGENCE OF SEQUENCES OF RANDOM VARIABLES). The importance of weak convergence stems from the central limit theorem*, which gives conditions under which the distribution of a normalized average converges weakly to a normal, Poisson, or other limit distribution.

Statistical applications of the theory of weak convergence are widespread. Suppose $\mathbf{x}_n = (X_1, \ldots, X_n)$ is a random sample with distribution $P_{\theta, n}$, which belongs to a specified family of distributions $\{P_{\phi, n} : \phi \in \Theta_n\}$.

The parameter value θ is unknown. The parameter space Θ_n can be finite or infinite dimensional and may vary with sample size n. Such a probability model amounts to a mathematically tractable way of generating hypothetical samples, which are intended to mimic important features of the data. The behavior of a statistical procedure $T_n = T_n(X_n)$ under the model $\{P_{\phi, n} : \phi \in \Theta_n\}$ can often be approximated by studying the limit distribution of T_n. Weak convergence results for *triangular arrays* (see Gnedenko and Kolmogorov [5]) are important tools for this purpose. Several illustrations follow. Since θ may be infinite dimensional, the discussion below also extends to time-series* analysis and nonparametrics. Sequential analysis* also depends heavily on weak convergence theory.

ESTIMATION

Consider the situation where the parameter space Θ_n is a fixed subset of the real line, $\hat{\theta}_n = \hat{\theta}_n(x_n)$ is an estimate of θ, and the distribution $H_n(\theta)$ of the centered estimate $n^{1/2}(\hat{\theta}_n - \theta)$ converges weakly to a $N(0, \sigma^2(\theta))$ distribution. This pointwise limit can be deceptive. In an example Hodges (*see* HODGES SUPEREFFICIENCY) the components of x_n are independent identically distributed $N(\theta, 1)$ random variables. The estimate $\hat{\theta}_n = \overline{X}_n$, the sample mean, if $|\overline{X}_n| > n^{-1/4}$; otherwise $\hat{\theta}_n = c\overline{X}_n$, with $0 < |c| < 1$. The asymptotic variance $\sigma^2(\theta)$ is then 1 if $\theta \neq 0$, but is c^2 if $\theta = 0$. At first glance, this result appears to contradict the minimax* property of the sample mean as an estimate of θ. However, if $\theta_n = n^{-1/2}h$, where h is fixed, then $H_n(\theta_n)$ converges weakly to the $N((c - 1)h, c^2)$ distribution. The bias and mean squared error of the Hodges estimate may thus be substantial for θ near 0; and the weak convergence of its distribution $H_n(\theta)$ to the normal limit is not uniform over any compact neighborhood of $\theta = 0$.

The key idea here—the need to study weak convergence of $\{H_n(\theta_n); n \geq 1\}$ for all sequences $\{\theta_n\}$ that converge to θ at an interesting rate—was developed by Stein, Rubin, Le Cam, Hájek, and others (see Chernoff [4] and Ibragimov and Has'minskii [8]). It applies equally well to infinite-dimensional parameters. The concepts of locally asymptotically minimax estimate and of least-dispersed regular estimate are two important outcomes, which provide asymptotic justifications for maximum likelihood* and related estimates. Robustness* of the estimate $\hat{\theta}_n$ can be studied by embedding the model $\{P_{\phi, n} : \phi \in \Theta_n\}$ into a larger model, usually nonparametric, and by then considering triangular array convergence of the distribution or risk of $\hat{\theta}_n$ within this supermodel (see Rieder [11]).

CONFIDENCE SETS AND THE BOOTSTRAP*

Suppose the asymptotic variance $\sigma^2(\theta)$ of the estimate $\hat{\theta}_n$ is a continuous function of θ and suppose $H_n(\theta_n)$ converges weakly to the $N(0, \sigma^2(\theta))$ distribution for every sequence $\{\theta_n = \theta + n^{-1/2}h\}$. An asymptotic confidence interval for θ is

$$A_n = \left\{ \theta : n^{1/2}|\hat{\theta}_n - \theta| \right.$$
$$\left. \leqslant \sigma(\hat{\theta}_n)\Phi^{-1}(1 - \alpha/2) \right\},$$

where Φ is the standard normal CDF. The coverage probability of A_n converges to $1 - \alpha$ in a locally uniform way:

$$\lim_{n \to \infty} \sup\left\{ \left|P_{n, \phi}[\phi \in A_n] - (1 - \alpha)\right| : \right.$$
$$\left. n^{1/2}|\phi - \theta| \leqslant c \right\} = 0,$$

for every θ and every finite positive c.

Let $K_n(\theta)$ denote the distribution of $n^{1/2}(\hat{\theta}_n - \theta)/\sigma(\hat{\theta}_n)$. A natural estimate of $K_n(\theta)$ is $K_n(\hat{\theta}_n)$, which Efron has termed a *bootstrap estimate* [*see* RESAMPLING PROCEDURES and BOOTSTRAPPING-II (in *Supplement*)]. Given the sample x_n, suppose x_n^* is an artificial sample drawn from the fitted model $P_{\hat{\theta}_n, n}$. Let $\hat{\theta}_n^*$ denote the value of the estimator $\hat{\theta}_n$, recomputed from the bootstrap sample x_n^*. Then $K_n(\hat{\theta}_n)$ is the conditional distribution of $n^{1/2}|\hat{\theta}_n^* - \hat{\theta}_n|/\sigma(\hat{\theta}_n^*)$, given

x_n. This interpretation of the bootstrap distribution leads to Monte Carlo algorithms for approximating it numerically. A bootstrap analog to confidence interval A_n is

$$B_n = \left\{ \theta : n^{1/2} \left| \hat{\theta}_n - \theta \right| \leqslant \sigma\left(\hat{\theta}_n\right) d_n \right\},$$

where d_n is a $(1 - \alpha)$th quantile* of the bootstrap distribution $K_n(\hat{\theta}_n)$. The coverage probability of B_n is typically $1 - \alpha + O(n^{-3/2})$, locally uniformly in θ, whereas that of A_n is only $1 - \alpha + O(n^{-1})$ (see Hall [7]). The bootstrap approach to confidence sets makes no overt use of asymptotic theory, but often relies on triangular array weak convergence results for its justification (see Beran [1]). The bootstrap approach is particularly valuable when the limit law is intractable.

POWER FUNCTIONS

Suppose θ has the partition $\theta = (\xi, \eta)$. A test $\psi_n = \psi_n(x_n)$ for the null hypothesis $\xi = \xi_0$ has power function $\beta_n(\xi, \eta) = E(\psi_n | P_{\theta, n})$. Typically, $\lim_{n \to \infty} \beta_n(\xi, \eta)$ is α when $\xi = \xi_0$ and is 0, α, or 1 when $\xi \neq \xi_0$. Calculating $\lim_{n \to \infty} \beta_n(\xi_0 + n^{-1}h, \eta)$ as a function of h—the asymptotic power function—often yields a better approximation to the power function of ψ_n. Introduced by Neyman [9], this triangular array approach was developed by Wald [12] and others. Contiguity arguments together with the central limit theorem often provide a way to derive the asymptotic power function (see Hájek and Šidák [6] and CONTIGUITY). The concept of a locally asymptotically maximin test, which stems from this work, provides asymptotic justification for likelihood ratio* and related tests. Robustness of a test's level and power can be assessed by studying triangular array convergence of its power function within a supermodel (see Rieder [10]). The power function of a test of a composite hypothesis can be estimated directly from the sample by bootstrap methods (see Beran [2]). Since ξ, η can be infinite-dimensional parameters, the discussion here includes goodness-of-fit* tests.

References

[1] Beran, R. (1984). *Jb. Dt. Math. Ver.*, **86**, 14–30. (Surveys bootstrap asymptotics through 1982.)

[2] Beran, R. (1986). *Ann. Statist.*, **14**, 151–173. (Describes bootstrap tests and bootstrap estimates of power functions.)

[3] Billingsley, P. (1968). *Convergence of Probability Measures*. Wiley, New York. (An influential account of weak convergence theory.)

[4] Chernoff, H. (1956). *Ann. Math. Statist.*, **27**, 1–22. (A fine survey of asymptotic theory in the early 1950s.)

[5] Gnedenko, B. V. and Kolmogorov, A. N. (1954). *Limit Distributions for Sums of Independent Random Variables*. Addison-Wesley, Reading, MA. (A definitive account of the topic.)

[6] Hájek, J. and Šidák, Z. (1967). *Theory of Rank Tests*. Academic, New York. (A fine mathematical account of rank tests, contiguity, and asymptotic power.)

[7] Hall, P. (1985). Unpublished preprint. (Analyzes the error in coverage probability of bootstrap confidence intervals.)

[8] Ibragimov, I. A. and Has'minskii, R. Z. (1981). *Statistical Estimation: Asymptotic Theory*. Springer, New York. (A modern, highly mathematical treatment.)

[9] Neyman, J. (1937). *Skand. Aktuarietidskr.*, **20**, 149–199. (Introduces asymptotic power in studying smoothed tests.)

[10] Rieder, H. (1978). *Ann. Statist.*, **6**, 1080–1094. (One possible shrinking neighborhood analysis of robust tests.)

[11] Rieder, H. (1983). *Trans. Ninth Prague Conference on Information Theory, Statistical Decision Functions, Random Processes*. Academia, Prague, pp. 77–89. (Surveys shrinking neighborhood theories of robust estimation.)

[12] Wald, A. (1943). *Trans. Amer. Math. Soc.*, **54**, 426–482. (A definitive early study of the likelihood ratio test and its asymptotic optimality.)

(BOOTSTRAPPING-II (*in Supplement*)
CONTIGUITY
CONVERGENCE OF SEQUENCES OF
   RANDOM VARIABLES
HODGES SUPEREFFICIENCY
LARGE SAMPLE THEORY
LIMIT THEOREM, CENTRAL
NEYMAN'S AND OTHER SMOOTH
   GOODNESS-OF-FIT TESTS
RESAMPLING PROCEDURES)

R. J. BERAN

WEAK STATIONARITY

A stochastic process* $\{ X_t, t \in E \}$ is *weakly stationary* if its mean and variance are the same for all t, and if the covariance between X_{t_1} and X_{t_2} depends only on the absolute difference $|t_1 - t_2|$. Formally,

$$EX_t = \mu,$$

$$EX_{t_1}X_{t_2} - EX_{t_1}EX_{t_2} = r\big(|t_1 - t_2|\big),$$

for all t, t_1, t_2 in E. The space E can be, for example, $(-\infty, +\infty)$ or $\{0, 1, 2, \dots\}$.

The process X_t is *strictly stationary* if the probability structure is invariant under time shifts, i.e., if the distribution of $(X_{t_1+t}, \dots, X_{t_n+t})$, $n \geq 1$, does not depend on t. Strict stationarity implies weak stationarity (provided X_t has a finite variance). The converse, in general, is false. For Gaussian processes*, however, strict and weak stationarity are equivalent.

The Ornstein–Uhlenbeck process* is an example of a stationary process $\{ X_t, -\infty < t < +\infty \}$ that is Gaussian (and Markovian). As a second illustration, consider the following discrete-time process:

$$X_0 = a\epsilon_0,$$
$$X_t = \rho X_{t-1} + \epsilon_t, \qquad t = 1, 2, \dots,$$

where $\epsilon_t, t = 0, 1, 2, \dots$, are uncorrelated random variables with mean 0 and variance 1. The process $\{ X_t, t = 0, 1, 2, \dots \}$ is AR (1) (autoregressive of order 1). It is not weakly stationary if $|\rho| \geq 1$. If $|\rho| < 1$ and if $a = (1 + \rho^2)^{-1/2}$, then X_t is weakly stationary. It is strictly stationary if the ϵ_t are also Gaussian.

In time-series* analysis, one focuses on the covariance structure of the underlying process and not on its finite-dimensional distributions. When weak stationarity holds, one has

$$r(t) = \int_{-\infty}^{+\infty} e^{it\lambda}\, dF(\lambda),$$

where $F(\lambda)$ is a *spectral distribution*, and therefore the time series can be analyzed in the spectral domain (*see* SPECTRAL ANALYSIS). In fact, if $F(\lambda)$ is absolutely continuous, i.e., if $dF(\lambda) = f(\lambda)\, d\lambda$, then the covariance function $r(t)$ is simply the Fourier transform (*see* INTEGRAL TRANSFORMS) of the *spectral density* $f(\lambda)$,

$$r(t) = \int_{\infty}^{+\infty} e^{it\lambda} f(\lambda)\, d\lambda.$$

(SPECTRAL ANALYSIS
STATIONARY PROCESSES
STOCHASTIC PROCESSES
TIME SERIES)

MURAD S. TAQQU

WEAKLY ERGODIC CHAIN *See* MARKOV PROCESSES

WEAR PROCESSES

Wear processes refer to a class of reliability* models in which the failure of an item is modeled as resulting from an accumulation of wear over time with failure occurring when this wear has exceeded some threshold (either fixed or random). This model has two special cases of sufficient interest to merit separate entries. These are shock models* and cumulative damage models*. We shall only briefly comment on these; the remainder of our effort here will be in those aspects of wear processes that have not been covered in these entries.

Shock models deal with a macroscopic analysis of the failure process. An item is presumed to be subject to shocks that occur over time. Various assumptions can be made on this counting process. The item is generally assumed to have an inherent ability to survive a random number of these shocks. A cumulative damage model deals with a more microscopic perspective on failure. The shocks are assumed to cause random damages, which accumulate in some fashion, usually additively. Failure occurs when the accumulated damage exceeds some threshold (fixed or random). These two classes of models can be seen to be equivalent in a mathematical sense. In any cumulative damage model, there is some random number of shocks before the accumulated damage exceeds the threshold. Conversely, one can view a shock model as a cumulative damage model in which the damages are degenerate random variables taking the value 1 with probability 1 and the threshold is the random number of shocks before failure.

In the cumulative damage model, one may think of the accumulated damage as being a stochastic process* $\{W(t): t \geq 0\}$ in which $W(t) = \sum_{i=1}^{N(t)} Z_i$, where the collection $\{Z_i\}$ represents the random damages caused by the shocks and where the process $\{N(t): t \geq 0\}$ is the counting process governing the shocks or times at which these random damages occur. The sample paths of such a cumulative damage process thus are typically step functions.

We shall now turn our attention to processes in which the sample paths of the process $\{W(t): t \geq 0\}$ describing the accumulated wear by time t are more general. Namely, we consider the case in which wear accumulates continuously over time, instead of, or in addition to, wear which accumulates in the form of random damages at random points in time. We also permit the possibility of repair (i.e., a decrease in the accumulated wear).

In an early paper on shock models, cumulative damage models, and wear processes, Esary et al. [6] consider a wear process that begins at 0, has nonnegative increments, is a Markov process*, and satisfies a technical condition that, for fixed time, the greater the wear the more prone the item is to further wear and that for equal amounts of wear, an older device is more susceptible to wear than a younger device. There are a number of stochastic processes that satisfy these conditions, for example, processes starting at the origin and having nonnegative, stationary, and independent increments. These include the Poisson process*, compound Poisson process, and the infinitesimal renewal process*. For such wear processes in which the threshold is deterministic, the distribution of the time until failure belongs to the class of increasing failure-rate average (IFRA) distributions.

In a sequence of two papers, Adbel-Hameed [1, 2] studied the gamma wear process, in which the wear sustained by time t has a gamma distribution* with probability

density function

$$f(x, t) = \frac{\lambda e^{-\lambda x}(\lambda x)^{t-1}}{\Gamma(t)}, \qquad t \geq 0.$$

When the threshold is random and $\bar{G}(x)$ represents the probability that the device survives x units of wear, the reliability function of the device can be represented as

$$\bar{F}(t) = \lambda \int_0^\infty \frac{e^{-\lambda x}(\lambda x)^{t-1}}{\Gamma(t)} \bar{G}(x) \, dx.$$

In [1], Abdel-Hameed shows that the life-distribution properties of $\bar{G}(x)$, such as increasing failure rate (IFR), increasing failure-rate average (IFRA), and new better than used (*see* HAZARD-RATE AND OTHER CLASSIFICATIONS OF DISTRIBUTIONS and RELIABILITY, PROBABILISTIC) are inherited as corresponding properties of the reliability function $\bar{F}(t)$. In [2], he studies optimal replacement policies for such devices.

Çinlar [4] studies shock and wear models using Markov additive processes. He investigates the failure-time distribution under random-threshold models and multiplicative killing-type failure mechanisms. One result of interest is that when the deterioration process is assumed to be a gamma process, whose shape parameter varies as a function of a Brownian motion* process, the lifetime distribution is Weibull*. Another paper by Çinlar [5] is expository in nature; some general models for deterioration processes and lifetime distributions are discussed. Deterioration processes are modeled by continuous Markov, continuous semi-Markov, right-continuous Markov, Markov additive, and semi-Markov processes*.

Gottlieb [7] has investigated sufficient conditions on a wear process for the life distribution to be IFR. He describes various classes of stochastic processes for which these sufficient conditions are satisfied, and he investigates conditions that lead to an IFRA result. In a 1984 paper, Gottlieb and Levikson [8] investigate optimal replacement policies for a wear process resulting from a self-repairing shock model in which random damages accumulate additively. However, between shocks, the cumulative damage process decreases. Conditions under which a control limit policy is optimal are given.

Pieper and Tiedge [10] apply stochastic processes, including Wiener processes with drift and related multiplicative processes to modeling wear, in the engineering setting. Various distributions arising from such models with either a fixed or a random threshold are discussed including the Birnbaum–Saunders and inverse Gaussian* distributions. Related inference questions are also considered.

Processes in which the time until the accumulated wear reaches a fixed threshold has a new better-than-used distribution are called NBU *processes*. They have been studied by Marshall and Shaked [9] and by Shanthikumar [11]. If wear results by independent, stochastically increasing random damages accumulating additively, if the times between damages are NBU and independent of the damages, and if the cumulative damage decreases deterministically in a certain technical fashion between damages, then the wear process is a NBU process [9]. A second result of Marshall and Shaked assumes that the times between damages are exponential and that there is a deterministic increase in wear between damages. A NBU process also results from this model. Shanthikumar has generalized this in various ways including allowing the times between damages to be DFR and allowing the damage and the times between damages to be dependent. Work on various multivariate extensions of this has been carried out.

Other research has been conducted in which the damage process remains constant between damages. One model, which does deserve mention, is that of Assaf et al. [3]. In their cumulative damage model, unlike others, damages may be negative. Under suitable conditions, the first-passage time to any threshold has a Pólya type 2* (PF$_2$) density.

References

[1] Adbel-Hameed, M. S. (1975). *IEEE Trans. Rel.*, **R-24**, 152–154.

[2] Adbel-Hameed, M. S. (1977). In *The Theory and Applications of Reliability*, Vol. 1, C. P. Tsokos and I. N. Shimi, eds. Academic, New York, pp. 397–412.

[3] Assaf, D., Shaked, M., and Shanthikumar, J. G. (1985). *J. Appl. Prob.*, **22**, 185–196.

[4] Çinlar, E. (1977). In *The Theory and Applications of Reliability*, Vol. 1, C. P. Tsokos and I. N. Shimi, eds. Academic, New York, pp. 193–214.

[5] Çinlar, E. (1984). In *Reliability Theory and Models.*, M. S. Abdel-Hameed, E. Çinlar, and J. Quinn, eds. Academic, New York, pp. 3–41.

[6] Esary, J. D., Marshall, A. W., and Proschan, F. (1973). *Ann. Prob.*, **1**, 627–649.

[7] Gottlieb, G. (1980). *J. Appl. Prob.*, **17**, 745–752.

[8] Gottlieb, G. and Levikson, B. (1984). *J. Appl. Prob.*, **21**, 108–119.

[9] Marshall, A. W. and Shaked, M. (1983). *Adv. Appl. Prob.*, **15**, 601–615.

[10] Pieper, V. and Tiedge, J. (1983). *Math. Operationsforschung, Ser. Statist.*, **14**, 485–502.

[11] Shanthikumar, J. G. (1984). *Adv. Appl. Prob.*, **16**, 667–686.

(CUMULATIVE DAMAGE MODELS
RELIABILITY, PROBABILISTIC
SHOCK MODELS
SURVIVAL ANALYSIS)

WILLIAM S. GRIFFITH

WEATHER MODIFICATION—I

Weather-modification experimentation has been largely statistical; the study of the effectiveness of cloud seeding has mostly dealt with statistical aggregates. After some 40 years, the results are still equivocal: A few significant findings are balanced by a long series of nonsignificant ones. Why is the issue still moot? What has gone wrong with the process of experimentation and analysis that such an accumulation of data has not provided definitive information on the effectiveness of seeding? Is there a lesson here for statistics?

THE PHYSICAL BASIS [7]

The physical basis for weather modification is the introduction of chemical nucleating agents into clouds that contain super-cooled (below freezing level) water. Under laboratory conditions this was shown to produce ice nuclei and droplets, and it was therefore expected to induce precipitation in actual clouds, whether by forming new nuclei, by augmenting existing drops, or by enhancing cloud buoyancy. Actual seeding with dry ice indeed produced visible changes in individual clouds. Verification of effects on precipitation was more difficult, since the weather is notoriously variable. Some physicists had hoped to isolate meteorological phenomena that act with greater regularity than rainfall or hail on the ground (see METEOROLOGY, STATISTICS IN). But it was well understood that a large stochastic component would remain in any weather data. The occasional enthusiast will produce impressive series of photographs showing some cloud that changed shape after being seeded. But as Samuel Johnson said as early as 1759 [11], "Might not some other cause produce this occurrence? The Nile does not always rise on the same day [9]." Clearly, weather experimentation needs a statistical design in which randomization* of treatment must play a central role. Clearly also, the results must be evaluated statistically and tested for significance.

COMMERCIAL OPERATIONS AND EARLY EVALUATION

An early statistical experiment was Langmuir's 1950 week-on–week-off cloud seeding from Alamagordo; it was largely inconclusive [14]. Private enterprise did not wait for verification, and a budding cloud-seeding industry developed in the 1950s, especially in the United States. At a relatively low price per acre, farmers and growers bought future rainfall, to be induced mostly by ground burners that generated smoke containing silver iodide. Rainfall was often observed following seeding and acclaimed as evidence of success; lack of rainfall on other occasions raised doubts. The evidence was equivocal. As Fisher* [9] remarked, "At-

tempts to seed clouds remind me of the burnt offerings of old; on the ground there was an impressive display of fire and smoke which ascended to the heavens: whether it had any effect on Heaven was, however, more doubtful."

In an attempt to assess cloud seeding, a special advisory committee sponsored a statistical analysis of past operations [11] and came to cautiously optimistic conclusions. Its report was forcefully criticized in the statistics literature, principally on the grounds that the available evidence had come from commercial operations without randomized allocation of treatments. The criticism pointed out the likelihood of bias and insisted that valid inferences could only be drawn from properly randomized experiments [3, 6].

The analysis of data from commercial seeding had compared data from operations, with historical records of precipitation on the same areas and on surrounding areas. Influential statisticians argued that these analyses were of no possible value, because of the demonstrated possibility of bias. Their attitude discouraged any further analysis of commercial seeding, even of operations that had been carefully controlled and well documented. And yet evaluation of data from such sources is in principle no different from observational studies* in other areas, such as medicine [4]. Though it is generally agreed that randomized experimentation is preferable, this total dismissal of observational studies in weather modification is surprising; it may have been an overreaction to the sanguine claims of commercially motivated cloud-seeding operators.

RANDOMIZED EXPERIMENTS

Insistence on randomization* won the day and was incorporated in the cloud-seeding experiments that were started in the 1960s and 1970s in many countries. Most of these were designed with the help of statisticians, and involved randomly allocating some units to be seeded and other units to serve as controls. Many of them also used data from nearby, upwind, areas as concomitant variables and employed various designs to reduce error variability. The requirements of rigorous design led to careful definition of units and variables [10, 13].

Meteorologists would have liked to experiment with physically meaningful units such as clouds, storms, systems, or fronts, but the state of the art did not allow these to be precisely defined in real time. What, indeed, is a cloud? It may be reasonably well defined at any one moment, but a seeding effect takes time and so a cloud unit has to be defined over time. Clouds, however, grow or dissipate, split or merge within short time spans. In many weather situations there seem to be no individual clouds at all: rather, a more or less continuous mass of moisture over hundreds of miles. Since clouds were difficult to define visually, there were attempts to define them operationally in terms of radar echoes of ice nuclei—which surely are relevant to the precipitation-forming mechanism. But the definition remained elusive, especially if it had to be done in real time so that experimenters could proceed to seed the particular cloud that is identified on the radar screen, and then have response data collected specifically for that cloud.

If a cloud was difficult to define operationally, what of smaller units such as towers, or larger units such as "storms" or "systems?" Efforts have been made to define these in terms that can be used in real time and analyzed objectively. But the units remain vague in definition and impossible to treat or observe independently. Thus, the seeding material from one tower may well be carried over to nearby towers, and the dynamic effects on one cloud's build up may draw moisture from other clouds nearby.

Since experimentation with such "natural" units was found to be impractical, units were usually defined in time and space, i.e., a given area over a number of hours, a day, a year, etc. Correspondingly, the response, instead of being some property of a cloud, was defined in terms of precipitation on the given area in that period of time. Thus the experimental unit might have been an afternoon from 2 to 8 P.M. and the response precipitation in central Illinois. Units of this kind were suitable for a statistical experiment as they permitted a randomized decision to seed or not to seed (or to seed in different ways if that was the object of the experiment) to be made ahead of time and the observations to be collected in a predefined objective manner.

Interaction between meteorologists and statisticians led to a series of well-designed experiments, which were monitored carefully. The statisticians insisted on adherence to rigid experimental protocol, designed to avoid possible biases and to permit testing the significance of results. Experimenters were not always happy with the conservative attitude, but would rather have incorporated their new ideas into the experiment as it continued. A small illustration of the difference in the professions' outlooks will illustrate this. The target for seeding in a well-known experiment had been defined a priori but, several months into the experiment, pilots and meteorologists wished to reduce the target area. They claimed that parts of it were difficult to fly over and they cited the evidence of their flight paths. The statistician suspected they might have been flying where they found suitable clouds, so that redefining, ex post facto, the area to fit the treatment would have introduced a bias. The statistician was formally right, but the others did not appreciate his rigid attitude as they thought it would reduce the experiment's sensitivity.

The results of these experiments on inducing precipitation over fixed time–space units were tested for significance of seeding effects. Most of the results were inconclusive, but a few indicated significant effects. The validity of these tests was generally accepted, though there was some acrimonious controversy on whether randomization had been properly carried out and whether some data might have been doctored. It was curious to see the statisticians who had advocated randomization so fervently now questioning its practical application [5].

EXPLORATORY ANALYSES
AND MULTIPLICITY

About the time of the analysis of these experiments, John Tukey's ideas and terminology of "exploratory" vs. "confirmatory" data analysis* came into vogue. The experiments had been designed to test precise hypotheses of null seeding effects against hopeful alternatives of positive effects on rainfall and snow, or negative effects on hail. Statisticians had rightly been concerned to insist on "confirmatory" analyses of this kind. In that period, it will be recalled, the profession was very squeamish about any other type of analysis at all, and statisticians may sometimes have discouraged any study of experimental results beyond tests of significance*. Now, finally, statistical practice began to be liberated from the exclusive dominance of testing for significance.

It is perhaps not surprising that this liberation was accompanied by confusion regarding the proper role of confirmation and exploration, and that some experimenters seized upon the kind of statistical ideas that most strongly supported their own expectations and hopes. Where new hypotheses were suggested by exploration of data, they were embellished by tests of significance carried out meaninglessly on the same data. From the initial priggish insistence that the only proper statistics were the single tests of significance, the stage now moved to promiscuous testing of every new proposition.

It is more disturbing that many statisticians were also confused between the two modes of analysis, as is evident by their readiness to compute multiple significance levels, or P-values*, from the data of any single experiment. They attached P-values to subsets of experimental data, to redefined responses, to readjustments for newly proposed concomitant variables, and to a variety of transformations and alternative tests. An instance is known of eminent statisticians testing as many as 180 subhypotheses of an experiment and then publishing the most striking of the many P-values. These are prime examples of the problem of multiplicity [1]. The statisticians' own confusion cannot have contributed to a better understanding of weather-modification experimentation.

There also was a reverse form of multiplicity. Where experiments resulted in significant seeding effects, some meteorologists proceeded to reanalyze the data meticulously by checking a variety of concomitant variables. Some of these reanalyses obviously suggested that the "significant" effect might have resulted from an unfortunate randomization. In other words, they suggested that a "type I error*" had occurred in that a more favorable sample had been selected for seeding and a less favorable one for control. The multiplicity of these reanalyses was generally as unproductive as the multiplicity of analyses [11].

This stage of analysis, subanalysis, and reanalysis was accompanied by often intemperate debate on the interpretation of the results [20]. Some statisticians saw fit to stretch their legitimate role of skeptical watchdogs of unbiased scientific procedure and adopted an inquisitorial attitude in casting doubt on the honesty or competence of experimenters and data analysts. They also preferred to test their own hypotheses, rather than ones proposed by subject matter specialists. At the same time, some meteorologists chafed at the constraints of a rigorous experimental protocol, which did not allow the use of physically meaningful experimental units and had no mechanism for evolutionary adaptation to the new insights into cloud physics and improvements in technology. Since most experiments did not produce the hoped-for results, many meteorologists despaired of the "statistical" approach and wished to revert to simpler exploration of clouds and seeding.

RECENT EXPERIMENTS

A few experiments on time–space weather modification continued, mostly outside the United States, but by the middle 1970s it was generally agreed that more elaborate experiments were required. Some such projects were organized and included collection of radar data and in-cloud observations by means of specially instrumented aircraft. It was hoped that this would show how seeding affected the formation of precipitation. The execution of such experiments and the collection of adequate samples of data did not turn out to be as straightforward as had been hoped for, and the results obtained were largely inconclusive.

With the paucity of positive findings and the difficulty of funding, weather-modification experimentation has ground to a halt in the middle 1980s. Any future studies are likely to be dependent on new insights into the physics of precipitation. The experiments that will then be designed are likely to be much more complex, both in terms of the physics and the statistics [20].

The disappointing results of this 40-year effort of field trials with cloud seeding must probably be ascribed to the unexpectedly complex nature of precipitation formation, rather than to shortcomings of the investigators and their studies. And yet, one may ask what the statisticians' contribution to this research effort has been and where it might have failed.

THE CONTRIBUTION OF STATISTICIANS

On the whole, there is little doubt that statisticians have made a crucial contribution to weather experimentation by providing good designs with randomized controls and allowing valid testing of significance. No less important was their insistence on workable definitions of units and measurements and on clear experimental protocol. Their insistence on rigid adherence to protocol and to predetermined methods of evaluation, as well as their proverbial professional skepticism, have not always endeared them to experimenters who lived with the challenge of "taming the weather." Carrying out operations and observations from bouncing airplanes, with the exhilaration of seeming to control the clouds,

did not always foster patience with earthbound skeptics and their tables of random numbers. And yet the cooperation between the two professions, with their very different outlooks, was successful on the whole.

The contribution made by formal mathematical statistics, on the other hand, has been rather minor. Much effort has been spent by statisticians in devising "optimal" analyses of cloud-seeding experiments. Using the classical approach of mathematical statistics, they based their derivations on various distributional assumptions about the response variable and its relation to the concomitants, on assuming that the units were independent of each other, and that the effect of seeding was fixed from unit to unit, either additively to natural precipitation or multiplicatively. They then derived optimal tests mathematically. At times, they relaxed some of the distributional assumptions and proposed nonparametric* tests instead. At other times they studied multivariate* tests and a variety of other techniques and compared them with each other to select the "best." They never seemed to deal with the well-known dependence of precipitation on successive units, presumably because it would have made the mathematics intractable.

It is doubtful whether this effort did much to improve the evaluation of weather-modification experiments. As a rule, confirmatory analyses by different statistical techniques gave essentially the same results. And power studies did not indicate that the "optimal" method was much better than most other methods. The underlying assumptions were rarely tested but much debated. An ironic example is the discussion of whether seeding effects would be additive or multiplicative, a discussion based on neither physical considerations nor experimental data, but entirely on the statisticians' predilections.

It is curious that all this effort and controversy took place while a valid approach was available, in a source as well known as Fisher's *The Design of Experiments* [8], and studied extensively by Kempthorne and his students. Testing by rerandomizing the treatment allocation did not require tenuous assumptions about distributions, false assumptions about independence, or idiosyncratic assumptions about the form of the effects. Isolated early uses of such tests [12] did not convince most statisticians and meteorologists to adopt them until a 1975 report by Brillinger et al. [2] reintroduced them with a novel terminology. Then, rather belatedly, they became fashionable. One cannot but wonder why it took so long.

Some insight on the role of statistics in the assessment of experimental evidence may be obtained from a review article by a senior British meteorologist [17] who evaluated the results of three randomized weather experiments. In reviewing tabulated data and tests, he used the quoted *P*-values* incidentally to give an idea of strength of evidence. He did not use the common statistics textbook recipe of choosing a single statistic to provide a unique decision about each experiment's outcome. Instead, he assessed each experiment's results in terms of how well its various subanalyses supported each other and how much meteorological sense he could make of them. How very different from the statisticians' meta-analyses, which pooled the *P*-values of a collection of experiments! Statistical theory does not seem to reflect the way experimental scientists view statistical evidence.

References

[1] Braham, R. R., Jr. (1979). *J. Amer. Statist. Ass.*, **74**, 57–68. (With discussion, especially J. A. Flueck, K. R. Gabriel, W. Kruskal, and P. W. Mielke.)

[2] Brillinger, D. R., Jones, L. V., and Tukey, J. W. (1978). *The Management of Weather Resources: Report of the Statistical Task Force to the Weather Modification Advisory Board*, Vol. 2. U.S. Government Printing Office, Washington, D.C.

[3] Brownlee, K. A. (1960). *J. Amer. Statist. Ass.*, **55**, 446–453.

[4] Changnon, S. A., Hsu, C-F., and Gabriel, K. R. (1981). *J. Weather Modification*, **13**, 195–199.

[5] *Commun. Statist. A*, **8**, 953–1153. (Special 1979 Issues on Weather Modification, including contributions by K. R. Gabriel, P. W. Mielke, and E. J. Smith.)

[6] Cohen, N. M. (1987). Personal communication.

[7] Dennis, A. S. (1980). *Weather Modification by Cloud Seeding*. Academic, New York.

[8] Fisher, R. A. (1935). *The Design of Experiments*. Oliver and Boyd, London.

[9] Fisher, R. A. (1962). Personal communication.

[10] Flueck, J. A. and Mielke, P. W. (1978). *Amer. Meteor. Soc. Monograph*, **38**, 225–235.

[11] Gabriel, K. R. (1980). *Bull. Amer. Meteorol. Soc.*, **62**, 62–69.

[12] Gabriel, K. R. and Feder, P. (1969). *Technometrics*, **11**, 149–160.

[13] Gabriel, K. R. and Petrondas, D. (1983). *J. Climate Appl. Meteorol.*, **22**, 626–631.

[14] Howell, W. (1978). *J. Appl. Meteorol.*, **17**, 1753–1757.

[15] Johnson, S. (1959). *The History of Rasselas, Prince of Abyssinia*. (Reissued in 1962 by Appleton-Century-Crofts, New York.)

[16] LeCam. L. and Neyman, J. (1967). *Proc. Fifth Berkeley Symp. Math. Statist. Prob.*, **5**. University of California Press, Berkeley, CA (Especially contributions by J. Neyman and E. M. Scott.)

[17] Mason, J. (1980). *Meteorol. Mag.* **109**, 335–344.

[18] Silverman, B. A. (1986). "Weather modification as a water management-tool," *Preprint, Tenth Conf. Water. Manage.*, American Meteorological Society, Boston, MA.

[19] Thom, H. C. S. (1957). In *Final Report of the Advisory Committee on Weather Control*, Vol. 2. U.S. Government Printing Office, Washington, D.C. pp. 25–50.

[20] Wegman, E. J. and DePriest, D. J. eds. (1980). *Statistical Analysis of Weather Modification Experiments*. Dekker, New York. (Including contributions by R. A. Bradley, O. Kempthorne, and J. Neyman.)

(DATA ANALYSIS
GEOGRAPHY, STATISTICS IN
HYDROLOGY, STOCHASTIC
HYPOTHESIS TESTING
METEOROLOGY, STATISTICS IN
MULTIPLE COMPARISONS
OBSERVATIONAL STUDIES
P-VALUES
RANDOMIZATION-I, II
SIGNIFICANCE TESTS, HISTORY AND
   LOGIC
WEATHER MODIFICATION-II (*See Supplement*))

K. R. GABRIEL

WEBER FUNCTION

This is a solution of the nonhomogeneous Bessel equation

$$z^2 y'' + zy' + (z^2 - \nu^2) y$$
$$= -(z + \nu)\pi^{-1} - (z - \nu)\pi^{-1}\cos\nu\pi,$$
$$\nu > 0.$$

It can be expressed as

$$E_\nu(z) = \frac{1}{\pi}\int_0^\pi \sin(\nu\phi - z\sin\phi)\,d\phi.$$

This is also known as the *Bessel function* $Y_\nu(z)$ of the second kind of order ν (*see* BESSEL FUNCTIONS). Usually the term "Weber function" is reserved for the case when $\nu = n$ is an integer.

In this case an explicit expression for this function is

$$Y_n(x) = \lim_{p \to n} \frac{\cos(p\pi)J_p(x) - J_{-p}(x)}{\sin(p\pi)}$$

$$= \frac{2}{\pi} - \left[\ln\left(\frac{x}{2}\right)\right]J_n(x)$$

$$- \frac{1}{\pi}\sum_{r=0}^{n-1}\frac{(n-r-1)!}{r!}\left(\frac{2}{x}\right)^{n-2r}$$

$$- \frac{1}{\pi}\sum_{r=0}^{\infty}\frac{(-1)^r}{r!(n+r)!}\left(\frac{x}{2}\right)^{n+2r}$$

$$\times \left[\Phi(r+n) + \Phi(r)\right], \quad n \geq 1,$$

with

$$\Phi(x) = \sum_{s=1}^r s^{-1}, \quad \Phi(0) = 0,$$

and

$$J_n(x) = \sum_{\nu=0}^{\infty}\frac{(-1)^\nu}{\nu!\Gamma(n+\nu+1)}\left(\frac{x}{2}\right)^{n+2\nu}$$

is the Bessel function of the first kind of order n.

Bibliography

Weber, H. F. (1879). *Zurich Vierteljahresschrift*, **24**, 33–76.

(BESSEL FUNCTIONS)

WEDDLE'S RULE

This is the following quadrature* formula, using values of the integrand at seven equally spaced values of the variable,

$$\int_a^{a+6h} f(x)\,dx$$

$$\doteq h\big[f(a) + f(a+2h) + f(a+4h)$$
$$+ f(a+6h) + 5\{f(a+h)$$
$$+ f(a+5h)\} + 6f(a+3h)\big].$$

It gives the exact value of the integral if $f(x)$ is polynomial of degree 5 or less. If $f(x)$ is a polynomial of degree 6, the error is only

$$\frac{h}{140}\delta_h^6 f(a+3h)$$

$$= \frac{h}{140}\big[\{f(a) + f(a+6h)\}$$
$$- 6\{f(a+h) + f(a+5h)\}$$
$$+ 15\{f(a+2h) + f(a+4h)\}$$
$$- 20f(a+3h)\big],$$

which is likely to be very small. (δ_h denotes "central difference"—*see* FINITE DIFFERENCES, CALCULUS OF.) The remainder is

$$-2.6 \times 10^{-8}(6h)^7 f^{(6)}(\xi_1)$$
$$-6.4 \times 10^{-10}(6h)^9 f^{(8)}(\xi_2)$$
$$= -0.00728 h^7 f^{(6)}(\xi_1)$$
$$-0.00645 h^9 f^{(8)}(\xi_2),$$

for some ξ_1, ξ_2 between a and $a + 6h$.

The formula

$$\int_a^{a+6h} f(x)\,dx \doteq \frac{h}{140}\big[41\{f(a) + f(a+6h)\}$$

$$+ 216\{f(a+h) + f(a+5h)\}$$
$$+ 17\{f(a+2h) + f(a+4h)\}$$
$$+ 272f(a+3h)\big]$$

gives the exact value of the integrals if $f(x)$ is a polynomial of degree 6 or less, though the coefficients are not so simple as for Weddle's rule.

Bibliography

Chataravarti, I. M., Laha, R. G., and Roy, J. (1967). *Handbook of Methods of Applied Statistics*, Vol. 1, Wiley, New York, pp. 38–41.

Milne-Thomson, L. M. (1933). *Calculus of Finite Differences*. Macmillan, London, p. 172.

Weddle, T. (1854). *Cambridge and Dublin Math. J.*, **9** (*Cambridge Math. J.*, **13**), 79–80.

(NUMERICAL INTEGRATION
SHOVELTON'S FORMULA
SIMPSON'S RULE
THREE-EIGHTHS RULE
TRAPEZOIDAL RULE)

WEDGE ESTIMATOR

This is an estimator of the partial regression* coefficients β in the multiple linear regression* model

$$\mathbf{Y} = \mathbf{X}\beta + \epsilon$$

(*see* MULTIPLE REGRESSION), obtained by modifying the ridge regression* estimator. It is

$$\mathbf{b}_W = (\mathbf{Z}'\mathbf{Z})^{-1}\mathbf{Z}\mathbf{Y},$$

where $\mathbf{Z} = \mathbf{X} + k\mathbf{X}(\mathbf{X}'\mathbf{X})^{-1}$ with k arbitrary. (The corresponding ridge regression estimator is

$$\mathbf{b}_R = (\mathbf{X}'\mathbf{X} + k\mathbf{I})^{-1}\mathbf{X}'\mathbf{Y}.)$$

The estimator $\mathbf{b}_W$ was developed in Riddell and von Hohenbalkan [1].

Reference

[1] Riddell, W. C. and von Hohenbalkan, B. (1978). Unpublished manuscript, Dept. of Economics, University of Alberta, Edmonton, Canada.

(RIDGE REGRESSION)

WEIBULL DISTRIBUTION

DEFINITIONS AND HISTORICAL REMARKS

The Weibull distribution was named for Professor Waloddi Weibull of Sweden who suggested it as a distribution for a variety of applications [75, 76]. The agreement he demonstrated between his observations and those predicted with the fitted Weibulls was very impressive. He considered the problems of yield strength of a Bofors steel, fiber strength of Indian cotton, length of syrtoideas, fatigue life of an St-37 steel, statures of adult males born in the British Isles, and breadth of beans of *Phaseolus vulgaris*. The distribution was used as early as 1933 by Rosin and Rammler [69] in describing the "laws governing the fineness of powdered coal."

The Weibull includes the exponential distribution* as a special case and is sometimes thought of as a generalization of the exponential distribution. The exponential model was developed and applied rather extensively in the 1950s and the Weibull began to be seriously considered as a competing model in the 1960s [6, 29, 47, 73], especially in problems in which the time to failure was the response of interest [27, 32, 47, 59].

A random variable W follows the three-parameter Weibull distribution [denoted by $W \sim \text{WE3}(a, b, c)$] if its cumulative distribution is given by

$$F_W(w) = 1 - \exp\{-[(w - a)/b]^c\},$$
$$a < w < \infty,$$

where $b, c > 0$. The parameters a and b are location and scale parameters and c is a shape parameter. The density function is given by

$$f_W(w) = \frac{c}{b}\left(\frac{w - a}{b}\right)^{c-1}$$
$$\times \exp\{-[(w - a)/b]^c\},$$
$$\text{for } a < w < \infty.$$

The reliability function is given by

$$R_W(w) = \exp\{-[(w - a)/b]^c\},$$
$$\text{for } a < w < \infty.$$

The hazard function, also known as the failure rate, is given by

$$h(w) = c(w - a)^{c-1}/b^c, \quad \text{for } w > a.$$

For values of c less than 1, the hazard function is a decreasing function, for $c = 1$ it is constant, and for $c > 1$ it is an increasing function of w. The value of c, the shape parameter, is thus an important parameter and often has a characteristic or predictable value depending upon the fundamental nature of the problem being studied.

It is fairly common to assume the value of the location parameter (a) is known, often 0. Setting $a = 0$ or letting $X = W - a$ leads to the common two-parameter Weibull distribution [denoted by $X \sim$ WE2(b, c)].

RELATION TO EXTREME-VALUE DISTRIBUTIONS

The Weibull distribution is one of the extreme-value distributions* (a type 3) derived by Fisher and Tippett [22]. It is also related to their type 1 extreme-value distributions*. A random variable Y has a type 1 distribution for minima if its cumulative distribution is given by

$$F_Y(y) = 1 - \exp\{-\exp[(y - \alpha)/\beta]\},$$
$$-\infty < y < \infty,$$

where $\beta > 0$ and α is unrestricted. α is a location parameter and β is a scale parameter and we denote this distribution by EVS(α, β). It is easily seen that if $X \sim$ WE2(b, c) and if we let $Y = \ln X$, then $Y \sim$ EVS($\ln b, 1/c$). It thus follows that the two models are in effect the same model and any statistical procedure developed for one applies to the other. In particular, all general results for location and scale distributions may be called upon when developing procedures for the two-parameter Weibull model [15, 23, 50, 51].

A random variable Z is said to have a type 1 distribution for maxima if

$$F_Z(z) = \exp[-\exp[(z - \alpha')/\beta']],$$
$$-\infty < z < \infty,$$

where $\beta' > 0$ and α' is unrestricted. We denote this distribution by EVL(α', β') and note that α' and β' are location and scale parameters. It is easily seen that if $Y \sim$ EVS(α, β) and we let $Z = -Y$, we have $Z \sim$ EVL($-\alpha, \beta$). Thus we also have that if $W \sim$ WE2(b, c) and $Z = -\ln W$, then $Z \sim$ EVL($-\ln b, 1/c$). It follows then that all results for type 1 for maximum distributions also have corresponding results for Weibull distributions (see, e.g., [5] and [50]).

PROPERTIES

(a) **Moments.**

$$E\left[(W - a)^r\right] = b^r\Gamma(1 + r/c) \text{ for } r > 0.$$

(b) If $W \sim$ WE3(a, b, c), then $[(W - a)/b]^c$ is a standard exponential random variable.

(c) If $X_1, X_2, \ldots, X_n$ are independent and identically distributed (i.i.d.) random variables, then $X_m = \min(X_1, \ldots, X_n)$ has a Weibull distribution if and only if the X_i's are Weibull random variables [16]. If $X_i \sim$ WE3(a, b, c), then $X_m \sim$ WE3($a, bn^{-1/c}, c$).

(d) Let T be the time to first occurrence of a nonhomogeneous Poisson process* with intensity $V(t) = ct^{c-1}/b^c$. Then $T \sim$ WE2(b, c).

(e) Moments of order statistics* are available in the literature. Means, variances, and covariances of Weibull order statistics are given by Weibull [77] for $c^{-1} = 0.1(0.1)1.0$ and $n = 5(5)20$, and also by Govindarajulu and Joshi [23] for $c = 1$, 2, 2.5, 3(1)10 and $n = 2(1)10$. Harter [27] gives means for $c = 0.5(0.5), 4.0(1)8$ and $n = 1(1)40$.

Means, variances, and covariances of log-Weibull and thus extreme-value order statistics are provided by Mann [47] for $n = 2(1)25$ (see also [50]). White [78] also gives means for $n = 1(1)100$.

STATISTICAL INFERENCE FOR THE TWO-PARAMETER WEIBULL

Early users [60, 76] of the two-parameter Weibull model were mainly engineers; they used graphical methods for estimation of the parameters. Such methods served them well and also provided a subjective method for judging the adequacy of the model.

As noted earlier the log transform of a Weibull random variable has a type 1 extreme-value distribution, which is a location and scale parameter distribution. Thus the methods of estimation available for such families can be applied to log-Weibull data. Lieblein and Zelen [45] and White [78] worked in this way in developing best linear unbiased estimates. Later, Mann [48, 49, 51] suggested the use of best linear invariant estimators (BLIEs) and gave tables of weights to be used for samples of size 25 or less. All of these methods are actually developed for estimating the parameters in a type 1 extreme-value distribution; for example, Mann's estimators are invariant under location and scale transformation of the log-Weibull data, and it is in the log-Weibull data that the estimators are linear.

Approximate best linear estimators are considered by McCool [58], Hassanein [31], Chan and Kabir [12], Mann and Fertig [56], and Downton [15]. In a notable paper, Johns and Lieberman [32] used Monte Carlo simulation to obtain lower confidence bounds on reliability based upon estimators that are asymptotic approximations to the BLIEs.

Simple linear estimators that have high relative efficiency, especially for heavily censored samples, have been developed for complete and censored samples. See, for example, Bain [5] and Engelhardt and Bain [19].

Maximum likelihood* estimation of the parameters has been considered by Harter and Moore [30], Cohen [14], and Leone et al. [44], among others. The maximum likelihood estimators (MLEs) based upon a complete sample of size n are given by the solution to the equations

$$\frac{1}{\hat{c}} - \frac{\sum X_i^{\hat{c}} \ln X_i}{\sum X_i^{\hat{c}}} + \frac{\sum \ln X_i}{n} = 0,$$

$$\hat{b} = \left(\frac{\sum X_i^{\hat{c}}}{n} \right)^{1/\hat{c}}.$$

The existence and uniqueness of the solution to these equations was shown by Pike [65] and McCool [59]. They are easily solved by using Newton's method to solve first for $\hat{c}$ and then using this to obtain $\hat{b}$.

For the two-parameter Weibull the MLEs are asymptotically efficient and they are asymptotically normally distributed. Let $\mathbf{U} = (\sqrt{n}\,(\hat{b} - b), \sqrt{n}\,(\hat{c} - c))$ so that asymptotically $\mathbf{U} \sim \mathrm{MVN}(\mathbf{0}, \mathbf{V})$, where

$$\mathbf{V} = \begin{bmatrix} 1.109b^2/c^2 & 0.257b \\ 0.257b & 0.608c^2 \end{bmatrix}.$$

Confidence intervals for the parameters based upon maximum likelihood* estimates (MLEs) are given by Thoman et al. [73] and by McCool [60–62]. These make use of the fact that the pivotal quantities $\hat{b}/b$ and $\hat{c}(\ln \hat{b} - \ln b)$ have densities that do not depend upon the values of b and c. This property is true for complete samples and also for samples with censoring on the order statistics. The necessary tables for certain censoring patterns are given by Billman et al. [11].

Confidence intervals for the reliability at some time of interest, based upon a maximum likelihood estimate of reliability, are given by Thoman et al. [74] for complete samples and by Billman et al. [11] for certain censored samples. These were developed by Monte Carlo* methods, using the fact that the distribution of the MLE of reliability depends only upon the true value of the reliability. Tables of these and other methods for obtaining confidence intervals in the Weibull setting may be found in Bain [5] and Mann and Fertig [56]. Kingston and Patel [36] consider interval estimation of the largest reliability of K populations and simultaneous confidence intervals for K reliabilities.

Sample sizes needed to choose the better of two Weibull populations are given by Rademaker and Antle [67]. Qureishi [66] and Thoman and Bain [72] considered the problem of choosing the better of two Weibull populations assuming a common shape parameter. Kingston and Patel [35] consider the problem of choosing m or less populations from K populations, which contain the best l populations in terms of reliability with prescribed probability of correct selection.

Lawless [39, 40] considers inference procedures based upon MLEs conditional upon the ancillary* statistics. The resulting confidence intervals are just slightly shorter than the simpler ones based only upon the MLEs.

Approximations to the tables necessary for statistical inference based upon MLEs are given by Bain and Engelhardt [9]. They show that approximately

$$(0.822)nc^2/\hat{c}^2 \sim \chi^2(0.822(n-1)),$$

$$\sqrt{n-1}\,\hat{c}(\ln\hat{b} - \ln b)/1.053 \sim t(n-1),$$

where χ^2 and t denote the usual chi-squared* and Student's t-distributions*. These lead to convenient tests of hypotheses and confidence intervals. Bain and Engelhardt [9] also give approximate tolerance limits and confidence limits on reliability.

Prediction limits based upon MLEs have been studied by Antle and Rademaker [3], Lawless [38], and Mann [52]. Bayesian prediction procedures are considered by Evans and Nigin [21] and Ashour and Raswan [4].

Many other simplified estimators based upon two or more observations have been proposed. A number of these are reviewed by Johnson and Kotz [33] and Mann and Fertig [56]. Bain and Antle [6] review some least-squares-type estimators. They have the advantage of being applicable with various censoring or missing observations; however, their efficiency is not as good as maximum likelihood estimation.

The discussion in this section assumes that the two unknown parameters are b and c. Another two-parameter Weibull of interest has a and b unknown. This particular model, a location- and scale-parameter distribution, has received little attention. A characteristic value of c may indeed be known in many situations. If this is the case it was shown by Rockette et al. [68] and by Peto and Lee [64] that the maximum likelihood estimates for a and b always exist and are unique. Weibull probability paper may be obtained from TEAM, Box 25, Tamworth, NH 03886.

STATISTICAL INFERENCES FOR THE THREE-PARAMETER WEIBULL

As noted earlier, if either a or c is assumed known the maximum likelihood estimates of the remaining parameters always exist and are unique. When all three are unknown, the likelihood function is unbounded as $a \to \min X_i$. However, if one assumes $c \geqslant 1$, then the likelihood function is bounded and often has a maximum. However, one should not rush into solving the likelihood equations because if they have one solution, they will for sure have at least two solutions [68].

A picture that is useful for the likelihood function is obtained by fixing the location parameter a at several values near and to the left of $\min X_i$ and solving for the respective $\hat{b}$ and $\hat{c}$ and the maximized likelihood as a function of a. A plot of this maximized likelihood vs. a will show that as a moves left from X_i, the maximized likelihood will decrease and then either

(i) continue decreasing, or
(ii) have a minimum very near $\min X_i$ and then a relative maximum farther to the left.

If the maximized likelihood has no relative maximum to the left of $\min X_i$, then one has $\hat{a} = \min X_i$, $\hat{c} = 1$, and $\hat{b} = \Sigma(X_i - \hat{a})/n$ as the maximum likelihood estimates in the space $c \geqslant 1$.

If the maximized likelihood as a function of a has a relative maximum to the left of $\min X_i$, then its value there should be compared with its value at the corner point described above ($c = 1$, etc.) and the point at which the likelihood is larger should be chosen as the maximum likelihood estimate.

The above process is not at all difficult to employ because with a known the likelihood equations for b and c are easily solved. It is possible that there may indeed be more than two solutions to the likelihood equations when all three parameters are unknown. However, this does not seem to occur in practice [68].

Lemon [43] studies the three-parameter Weibull as does Cohen [14], Zanakis [80], and Wyckoff et al. [79]. The last paper gives confidence limits for c and a and b unknown. Also see Harter and Dubey [28], Bain and Thoman [10], Bain and Antle [7], and Mann and Fertig [55], and the discussions on this distribution in MANN-FERTIG STATISTIC and MAXIMUM LIKELIHOOD ESTIMATION.

COMPETING MODELS

Models that generally compete with the Weibull are the exponential*, lognormal*, and the gamma*. For testing H_0: one-parameter exponential vs. H_1: WE2(b, c), one should simply test $H_0 : c = 1$ vs. $H_1 : c \neq 1$ by use of the maximum likelihood estimate of c and the tables given by Thoman et al. [73].

For testing H_0: two-parameter exponential vs. H_1: WE3(a, b, c), one should use either the method in Antle et al. [2] or Engelhardt and Bain [18].

For choosing between a Weibull and a lognormal model, the ratio of maximized likelihoods [17] provides a very good test. The Weibull and the lognormal are very difficult to distinguish and sample sizes of 50 or more are needed in order to have a large probability of correct selection.

Mann and Fertig [55] give a test for $H_0 : a = a_0$ in the three-parameter Weibull setting. This provides a useful test of whether the three-parameter Weibull is needed.

Bain and Engelhardt [8] consider discrimination between the Weibull and gamma distributions* on the basis of probability of correct selection. Chandra et al. [13] provide percentage points for goodness-of-fit* tests for the extreme-value or Weibull distributions with unknown parameters based on the Kolmogorov–Smirnov* or Kuiper* statistics.

Some results for mixed or compound Weibull distributions are provided by McNolty et al. [63] and Harris and Singpurwalla [26]. Multivariate Weibull distributions are considered by Lee [42] and Johnson and Kotz [34].

Klein and Basu [37] consider accelerated life-testing* under competing causes of failure and Shaffer [71] studies confidence bands for minimum fatigue life under K stress levels.

References

[1] Antle, C. E. (1972). Choice of Model for Reliability Studies and Related Topics. *Tech. Reps. ARL 72-0108* and *AD 751340*, ARL, Wright-Patterson AFB, OH.

[2] Antle, C. E., Klimko, L. A., Rockette, H. E., and Rademaker, A. W. (1975). Upper bounds for the power of invariant tests for the exponential distribution with Weibull alternatives. *Technometrics*, **17**, 357–360.

[3] Antle, C. E. and Rademaker, A. (1972). Confidence intervals for the maximum of a set of future observations from the extreme value distribution. *Biometrika*, **59**, 475–477.

[4] Ashour, S. K. and Rashwan, D. R. (1981). Bayesian predictions for compound Weibull model. *Commun. Statist. A*, **10**, 1613–1624.

[5] Bain, L. J. (1978). *Statistical Analysis of Reliability and Life-Testing Models*. Dekker, New York.

[6] Bain, L. J. and Antle, C. E. (1967). Estimation of parameters in the Weibull distribution. *Technometrics*, **9**, 621–627.

[7] Bain, L. J. and Antle, C. E. (1970). Inferential Procedures for the Weibull and Generalized Gamma Distributions. *Tech. Rep. ARL 70-0266*, ARL, Air Force Systems Command, USAF, Wright-Patterson AFB, OH.

[8] Bain, L. J. and Engelhardt, M. (1980a). Probability of correct selection of Weibull versus gamma based on likelihood ratio. *Commun. Statist. A*, **9**, 375–381.

[9] Bain, L. J. and Engelhardt, M. E. (1980b). Simple approximate distributional results for confidence and tolerance limits for the Weibull distribution based on maximum likelihood estimators. *Technometrics*, **23**, 15–20.

[10] Bain, L. J. and Thoman, D. R. (1968). Some tests of hypotheses concerning the three-parameter Weibull distribution. *J. Amer. Statist. Ass.*, **63**, 853–860.

[11] Billman, B. R., Antle, C. E., and Bain, L. J. (1972). Statistical inference from censored Weibull samples. *Technometrics*, **14**, 831–840.

[12] Chan, L. K. and Kabir, A. B. M. L. (1969). Optimum quantities for the linear estimation of the parameters of the extreme-value distribution

in complete and censored samples. *Naval Res. Logist. Quart.*, **16**, 381–404.

[13] Chandra, M., Singpurwalla, N. D., and Stephens, M. A. (1981). Kolmogorov statistics for tests of fit for the extreme-value and Weibull distributions. *J. Amer. Statist. Ass.*, **76**, 729–731.

[14] Cohen, A. C. (1974). Multi-censored sampling in three-parameter Weibull distribution. *Technometrics*, **17**, 347–352.

[15] Downton, F. (1966). Linear estimates of parameters in the extreme-value distribution, *Technometrics*, **8**, 3–17.

[16] Dubey, S. D. (1966). Characterization theorems for several distributions and their applications. *J. Indust. Math.*, **16**, 1–22.

[17] Dumonceaux, R. H. and Antle, C. E. (1973). Discrimination between the lognormal and the Weibull distribution. *Technometrics*, **15**, 923–926.

[18] Engelhardt, M. and Bain, L. J. (1975). Tests of two-parameter exponentiality against three-parameter Weibull alternatives. *Technometrics*, **17**, 353–356.

[19] Engelhardt, M. and Bain, L. J. (1977). Simplified statistical procedures for the Weibull or extreme-value distribution. *Technometrics*, **19**, 323–331.

[20] Engelhardt, M. and Bain, L. J. (1979). Prediction limits and two-sample problems with complete or censored Weibull data. *Technometrics*, **21**, 233–237.

[21] Evans, I. G. and Nigin, A. M. (1980). Bayesian prediction for two-parameter Weibull lifetime models. *Commun. Statist. A*, **9**, 659–672.

[22] Fisher, R. A. and Tippett, L. M. C. (1928). Limiting forms of the frequency distribution of the largest or smallest member of a sample. *Proc. Camb. Philos. Soc.*, **24**, 180–190.

[23] Govindarajulu, Z. and Joshi, M. (1968). Best linear unbiased estimation of location and scale parameters of Weibull distribution using ordered observations. *Statist. Appl. Res., JUSE*, **15**, 1–14.

[24] Gross, A. J. and Clark, V. A. (1975). *Survival Distributions: Reliability Applications in the Biomedical Sciences*. Wiley, New York.

[25] Hager, H. W., Bain, L. J., and Antle, C. E. (1971). Reliability estimation for the generalized gamma distribution and robustness of the Weibull model. *Technometrics*, **13**, 547–557.

[26] Harris, C. M. and Singpurwalla, N. (1968). Life distributions derived from stochastic hazard functions. *IEEE Trans. Rel.*, **R-17**, 70–79.

[27] Harter, H. L. (1969). *Order Statistics and Their Use in Testing and Estimation: Estimator Based on Order Statistics from Various Populations*, Vol. 2. U.S. Government Printing Office, Washington, D.C.

[28] Harter, H. L. and Dubey, S. D. (1967). Theory and Tables for Tests of Hypotheses Concerning the Mean and the Variances of a Weibull Population. *Tech. Rep. ARL 67-0059*, Wright-Patterson AFB, OH.

[29] Harter, H. L. and Moore, A. H. (1967). Asymptotic variances and covariances of maximum-likelihood estimators, from censored samples, of the parameters of Weibull and gamma populations. *Ann. Math. Statist.*, **38**, 557–570.

[30] Hartr, H. L. and Moore, A. H. (1968). Maximum-likelihood estimation, from doubly censored samples, of the parameters of the first asymptotic distribution of extreme values. *J. Amer. Statist. Ass.*, **63**, 889–901.

[31] Hassanein, K. M. (1972). Simultaneous estimation of the parameters of the extreme value distribution by sample quantiles. *Technometrics*, **14**, 63–70.

[32] Johns, M. V. and Lickerman, G. J. (1966). An exact asymptotically efficient confidence bound for reliability in the case of the Weibull distribution. *Technometrics*, **8**, 135–175.

[33] Johnson, N. L. and Kotz, S. (1970). *Distributions in Statistics: Continuous Univariate Distributions—1*. Wiley, New York.

[34] Johnson, N. L. and Kotz, S. (1975). A vector multivariate hazard rate. *J. Multivariate Anal.*, **5**, 53–66.

[35] Kingston, J. V. and Patel, J. K. (1980). A restricted subset selection procedure for Weibull populations. *Commun. Statist. A*, **9**, 1371–1383.

[36] Kingston, J. V. and Patel, J. K. (1981). Interval estimation of the largest reliability of K Weibull populations. *Commun. Statist. A*, **10**, 2279–2298.

[37] Klein, J. P. and Basu, A. P. (1981). Weibull accelerated life tests when there are competing causes of failure. *Commun. Statist. A*, **10**, 2073–2100.

[38] Lawless, J. F. (1973). On the estimation of safe life when the underlying life distribution is Weibull. *Technometrics*, **15**, 857–865.

[39] Lawless, J. F. (1975). Construction of tolerance bounds for the extreme-value and Weibull distributions. *Technometrics*, **17**, 255–261.

[40] Lawless, J. F. (1978). Confidence interval estimation for the Weibull and extreme value distributions. *Technometrics*, **20**, 355–364.

[41] Lawless, J. F. and Mann, N. R. (1976). Tests for homogeneity of extreme-value scale parameters, *Commun. Statist. A*, **5**, 389–405.

[42] Lee, L. (1979). Multivariate distributions having Weibull properties. *J. Multivariate Anal.*, **9**, 267–277.

[43] Lemon, G. H. (1975). Maximum likelihood estimation for the three parameter Weibull distribution based on censored samples. *Technometrics*, **17**, 247–254.

[44] Leone, F. C., Rutenberg, Y. H., and Topp, C. W. (1960). Order Statistics and Estimators for the Weibull Population. *Tech. Reps. AFOSR TN 60-489* and *AD 237042*, Air Force Office of Scientific Research, Washington, D.C.

[45] Lieblein, J. and Zelen, M. (1956). Statistical investigations of the fatigue life of deep groove ball bearings. Research Paper 2719. *J. Res. Natl. Bur. Stand.*, **57**, 273–316.

[46] Littell, R. C., McClave, J. T., and Offen, W. W. (1979). Goodness-of-fit tests for the two-parameter Weibull distribution. *Commun. Statist. B*, **8**, 257–269.

[47] Mann, N. (1965). *Point and Interval Estimates for Reliability Parameters when Failure Times have the Two-Parameter Weibull Distribution*. Ph.D. dissertation, University of California at Los Angeles, Los Angeles, CA.

[48] Mann, N. (1967a). Results on Location and Scale Parameter Estimation with Application to the Extreme-Value Distribution. *Tech. Rep. ARL 67-0023*, Office of Aerospace Research, USAF, Wright-Patterson AFB, OH.

[49] Mann, N. (1967b). Tables for obtaining the best linear invariant estimates of parameters of the Weibull distribution. *Technometrics*, **9**, 629–645.

[50] Mann, N. (1968). Results on Statistical Estimation and Hypothesis Testing with Application to the Weibull and Extreme-Value Distributions. *Tech. Reps. ARL 68-0068* and *AD 672979*, ARL, Wright-Patterson AFB, OH.

[51] Mann, N. (1971). Best linear invariant estimation for Weibull parameters under progressive censoring. *Technometrics*, **13**, 521–533.

[52] Mann, N. R. (1976). Warranty periods for production lots based on fatigue-test data. *Eng. Fracture Mech.*, **8**, 123–130.

[53] Mann, N. (1977). An *F*-approximation for two-parameter Weibull and log-normal tolerance bounds based on possibly censored data. *Naval Res. Logist. Quart.* **9**, 187–196.

[54] Mann, N. and Fertig, K. W. (1973). Tables for obtaining confidence bounds and tolerance bounds based on best linear invariant estimates of parameters of the extreme-value distribution. *Technometrics*, **15**, 87–101.

[55] Mann, N. R. and Fertig, K. W. (1975). A goodness-of-fit test for the two-parameter vs. three parameter Weibull, confidence bounds for threshold. *Technometrics*, **17**, 237–246.

[56] Mann, N. and Fertig, K. W. (1977). Efficient unbiased quantile estimators for moderate-size complete samples from extreme-value and Weibull distributions, confidence bounds and tolerance and prediction intervals. *Technometrics*, **19**, 87–93.

[57] Mann, N. R., Schafer, R. E., and Singpurwalla, N. D. (1974). *Methods for Statistical Analysis of Reliability and Life Data*, Wiley, New York.

[58] McCool, J. I. (1965). The construction of good linear unbiased estimates from the best linear estimates for a smaller sample size. *Technometrics*, **7**, 543–552.

[59] McCool, J. I. (1970). Inferences on Weibull percentiles and shape parameter from maximum likelihood estimates. *IEEE Trans. Rel.*, **R-19**, 2–9.

[60] McCool, J. I. (1974). Inferential Techniques for Weibull Populations. *Tech. Reps. 74-0180 and AD A-009645*, ARL, Wright-Patterson AFB, OH.

[61] McCool, J. I. (1975). Inferential Techniques for Weibull Populations II. *Tech. Rep. ARL 76-0233*, ARL, Wright-Patterson AFB, OH.

[62] McCool, J. I. (1979). Analysis of single classification experiments based on censored samples from the two-parameter Weibull distribution. *J. Statist. Plann. Inf.*, **3**, 39–68.

[63] McNolty, F., Doyle, J., and Hansen, E. (1980). Properties of the mixed exponential failure process. *Technometrics*, **22**, 555–566.

[64] Peto, R. and Lee, P. N. (1973). Weibull distributions for continuous carcinogenesis experiments. *Biometrics*, **29**, 457–470.

[65] Pike, M. (1966). A suggested method of analysis of a certain class of experiments in carcinogenesis. *Biometrics*, **22**, 142–161.

[66] Qureishi, A. S. (1964). The discrimination between two Weibull processes. *Technometrics*, **6**, 57–75.

[67] Rademaker, A. W. and Antle, C. E. (1975). Sample size for selecting the better of two populations. *IEEE Trans. Rel.*, **R-24**, 17–20.

[68] Rockette, H., Antle, C. E., and Klimko, L. A. (1974). Maximum likelihood estimation with the Weibull model. *J. Amer. Statist. Ass.*, **69**, 246–249.

[69] Rosen, P. and Rammler, B. (1933). The laws governing the fineness of powdered coal. *J. Inst. Fuels*, **6**, 29–36.

[70] Schafer, R. E. and Sheffield, T. S. (1976). On procedures for comparing two Weibull populations. *Technometrics*, **18**, 231–235.

[71] Shafer, R. B. (1974). Confidence bands for minimum fatigue life. *Technometrics*, **16**, 113–123.

[72] Thoman, D. R. and Bain, L. J. (1969). Two sample tests in the Weibull distribution. *Technometrics*, **11**, 805–815.

[73] Thoman, D. R., Bain, L. J., and Antle, C. E. (1969). Inferences on the parameters of the Weibull distribution. *Technometrics*, **11**, 445–460.

[74] Thoman, D. R., Bain, L. J., and Antle, C. E. (1970). Reliability and tolerance limits in the Weibull distribution. *Technometrics*, **12**, 363–371.

[75] Weibull, W. (1939). A statistical theory of the strength of materials. *Ing. Vetenskaps Akad. Handl.*, **151**, 1–45.

[76] Weibull, W. (1951). A statistical distribution function of wide applicability. *J. Appl. Mech.*, **18**, 293–297.

[77] Weibull, W. (1967). Estimation of Distribution Parameters by a Combination of the Best Linear Order Statistic Method and Maximum Likelihood. *Tech. Rep. AFML 67-105*, Air Force Materials Laboratory, Wright-Patterson AFB, OH.

[78] White, J. S. (1967). The Moments of log-Weibull Order Statistics. *General Motors Research Publication GMR-717*, General Motors Corporation, Warren, Michigan.

[79] Wyckoff, J., Bain, L. J., and Engelhardt, M. E. (1980). Some complete and censored sampling results for the three-parameter Weibull distribution, *J. Statist. Comp. Simul.*, **11**, 139–152.

[80] Zanakis, S. H. (1979). A simulation study of some simple estimators for the three-parameter Weibull distribution. *J. Statist. Comp. Simul.* **9**, 101–116.

Editorial Note

The *double Weibull* distribution, with PDF

$$f_X(x) = \tfrac{1}{2}c|x|^{c-1}\exp(-|x|^c), \qquad c > 0,$$

which is found by reflecting the standard Weibull about the origin, is studied by N. Balakrishnan and S. Kocherlakota in *Sankhyā B*, **47** (1985), 161–178.

(EXTREME-VALUE DISTRIBUTIONS
HAZARD RATE AND OTHER
   CLASSIFICATIONS OF DISTRIBUTIONS
LIFE TESTING
MANN-FERTIG STATISTIC
MAXIMUM LIKELIHOOD ESTIMATION
MULTIVARIATE WEIBULL
   DISTRIBUTIONS
WEIBULL PROCESSES)

CHARLES E. ANTLE
LEE J. BAIN

WEIBULL–EXPONENTIAL DISTRIBUTION

Zacks [1] introduced a three-parameter Weibull–exponential distribution having a CDF

$$F(x; \lambda, \alpha, \tau)$$

$$= 1 - \exp\left\{-\lambda x - \left(\lambda(x-\tau)_+\right)^{\alpha}\right\},$$

$$x \geqslant 0,$$

where $Y_+ = \max(0, Y)$, $\lambda > 0$ (scale parameter), $\alpha \geqslant 1$ (shape parameter), and $\tau \geqslant 0$ (change-point parameter).

This family possesses the following nondecreasing failure-rate function (*see* HAZARD RATE AND OTHER CLASSIFICATIONS OF DISTRIBUTIONS):

$$h(t; \lambda, \alpha, \tau)$$

$$= \begin{cases} \lambda, & \text{if } 0 \leqslant t < \tau, \\ \lambda + \lambda^{\alpha}\alpha(t-\tau)^{\alpha-1}, & \text{if } \tau < t, \end{cases}$$

which is a superposition (for $t \geqslant \tau$) of a Weibull* hazard rate on a constant hazard. As $\tau \to \infty$, the distribution approaches an exponential distribution* with parameter λ. The family is suitable for modeling systems that, after a certain length of time, enter a wear-out phase in which components of the system have an increasing hazard rate. Zacks [1] also develops an adaptive Bayesian* estimator for τ (λ and α being assumed to be known), when τ has a prior CDF of the form

$$\xi(\tau) = \left\{ p + (1-p)[1 - e^{-\Psi(\tau-\tau_0)}] \right\},$$

$$\text{for } \tau \geqslant \tau_0,$$

where $0 < p < 1$, $0 < \Psi < \infty$, and τ_0 is a time point chosen so that the true change point* τ exceeds it with prior probability* 1.

Reference

[1] Zacks, S. (1984). *Operat. Res.*, **32**, 741–749.

(ADAPTIVE ESTIMATION)

WEIBULL–POISSON PROCESS

This is a nonhomogeneous Poisson* process with intensity function

$$\lambda(t) = (\beta/\theta)(t/\theta)^{\beta-1}, \qquad t > 0; \beta, \theta > 0.$$

It has been studied by Bain and Engelhardt [1, 5], Bain et al. [2], Crow [3, 4], Finkelstein [6], Lee and Lee [7], and Saw [8], among others. There has been special interest in constructing tests of hypotheses on the value of β, with θ regarded as a nuisance parameter* [3]. (If $\beta = 1$, we have a simple homogeneous Poisson process.)

Powers of such tests have been studied in [3].

References

[1] Bain, L. J. and Engelhardt, M. E. (1980). *Technometrics*, **22**, 421–426.

[2] Bain, L. J., Engelhardt, M. E., and Wright, F. M. (1985). *J. Amer. Statist. Ass.*, **80**, 419–422.

[3] Crow, L. H. (1974). In *Reliability and Biometry*, F. Proschan and R. J. Serfling, eds. SIAM, Philadelphia, pp. 379–410.

[4] Crow, L. H. (1982). *Technometrics*, **24**, 67–72.

[5] Engelhardt, M. E. and Bain, L. J. (1978). *Technometrics*, **20**, 167–169.

[6] Finkelstein, J. (1976). *Technometrics*, **18**, 115–117.

[7] Lee, L. and Lee, S. K. (1978). *Technometrics*, **20**, 41–45.

[8] Saw, J. G. (1975). *Commun. Statist.*, **4**, 777–782.

(POISSON PROCESSES
WEIBULL DISTRIBUTION
WEIBULL PROCESSES)

WEIBULL PROCESSES

THE WEIBULL PROCESS MODEL

A *Weibull process* is a useful model for phenomena that are changing over time. Essentially, all of the work on Weibull processes has been motivated by applications in which occurrences are failures of a repairable system, and it will be discussed in this framework, although it could be applied more generally. Improvement of a system, or reliability* growth, may occur if the system is in a developmental program. On the other hand, if only minimal repairs are made each time a failure occurs, the system will be deteriorating with time. As noted by Ascher [1], care should be taken to distinguish between a Weibull process, which models a repairable system, and a Weibull distribution, which models a nonrepairable system. Other terms, such as "Duane model" are sometimes used.

It was discovered by Duane [6], while examining data on the reliability growth of various repairable systems, that the number of system failures vs. operating time appeared to be approximately linear when plotting the logarithms of these quantities. Subsequently, Crow [4] proposed a stochastic analog in which the number of system failures is assumed to occur according to a nonhomogeneous Poisson process*, $\{ N(t) : t \geqslant 0 \}$, with mean value function of the form $m(t) = E[N(t)] = \lambda t^{\beta}$, and intensity function of the form $\nu(t) = dm(t)/dt = \lambda \beta t^{\beta - 1}$. This is consistent with the empirical work of Duane since $\ln m(t) = \beta \ln t + \ln \lambda$ is a linear function of $\ln t$. Another common parameterization is $m(t) = (t/\theta)^{\beta}$. $\beta < 1$ corresponds to improvement of the system, and $\beta > 1$ corresponds to deterioration.

An alternative characterization of a Weibull process is given by the sequence of successive failure times $T_1, T_2, \ldots$, where T_n represents the time until the nth failure. The primary reason for the terminology "Weibull process" is the fact that the time to first failure T_1 has the Weibull distribution* with hazard rate function* $\nu(t) = (\beta/\theta)(t/\theta)^{\beta - 1}$. It is also true that the conditional failure time T_n, given $T_1 = t_1, \ldots, T_{n-1} = t_{n-1}$, follows a Weibull distribution, which is truncated below the point t_{n-1}.

In order to estimate or test hypotheses about the parameters of a Weibull process, it is necessary to adopt a method of gathering data. The most common way to obtain data is "failure truncation," in which the process is observed for a fixed number n of failures. This leads, in a natural way, to an ordered set of data $0 < t_1 < t_2 < \cdots < t_n$. Another way to obtain data is "time truncation," in which the process is observed for a fixed length of time t. In this case, the data have one of the following forms: Either (1) $N(t) = 0$, or (2) $N(t) = n > 0$ and $0 < t_1 < t_2 < \cdots < t_n < t$. Notice that, with time truncation, the observed number of occurrences is part of the data set. It will be advantageous to consider the cases of failure and time truncation separately.

FAILURE TRUNCATION

Estimation and inference procedures, in this case, are discussed by numerous authors (see [2, 4, 5, 7, 8, 12]). Suppose $T_1, T_2, \ldots, T_n$ are the first n successive failure times of a Weibull process. The joint probability density function (PDF) is

$$f(t_1, t_2, \ldots, t_n) = (\beta/\theta)^n \prod_{i=1}^{n} (t_i/\theta)^{\beta-1}$$
$$\times \exp\left[-(t_n/\theta)^{\beta}\right],$$

where β and θ are, respectively, shape and scale parameters.

The joint maximum likelihood* estimators (MLEs) are

$$\hat{\beta} = n \bigg/ \sum_{i=1}^{n-1} \ln(T_n/T_i)$$

and

$$\hat{\theta} = T_n/n^{1/\hat{\beta}}.$$

The variable $Z = 2n\beta/\hat{\beta}$ is a pivotal quantity* (i.e., its distribution is free of unknown parameters), and it has the chi-square distribution* with $2(n-1)$ degrees of freedom. Thus confidence limits can be derived. Lower $1 - \epsilon$ confidence limits will be given, although upper limits and two-sided confidence intervals can be obtained in a similar manner. A lower $1 - \epsilon$ confidence limit for β is given by

$$\beta_L = \left[\hat{\beta}/(2n)\right]\chi^2_{2(n-1), \epsilon},$$

where $\chi^2_{2(n-1), \epsilon}$ is the 100ϵ percentile of $\chi^2_{2(n-1)}$.

The variable $W = (\hat{\theta}/\theta)^{\beta}$ is also a pivotal quantity and its cumulative distribution function (CDF) is given by

$$F_W(w) = \int_0^{\infty} H\left(2(nw)^{z/2n}; 2n\right)$$
$$\times h(z; 2(n-1))\, dz,$$

where $H(z; r)$ and $h(z; r)$ represent the CDF and PDF of χ^2_r. A lower $1 - \epsilon$ confidence limit for θ is given by

$$\theta_L = \hat{\theta}/w_{1-\epsilon}^{1/\hat{\beta}},$$

where $w_{1-\epsilon}$ is the $100(1 - \epsilon)$ percentile of

W. Tabulated percentiles of W, obtained by Monte Carlo simulation*, are given in [8]. A normal approximation is discussed in [2].

It is also possible to construct confidence limits on the current system reliability following truncation of the process. Suppose a system is in a developmental program until a time at which changes in the system cease. If the changes cease at the time of the nth failure $T_n = t_n$, and if it is assumed that the intensity $\nu(t_n)$ remains constant thereafter, then the subsequent times between failures of the system will be independent and exponentially distributed with failure rate $\nu(t_n)$. The "current system reliability" for some specified length of time t_0 would be $R(t_0) = \exp[-\nu(t_n)t_0]$. Confidence limits can be based on the pivotal quantity $Q = \nu(T_n)/\hat{\nu}(T_n)$, where $\hat{\nu}(T_n) = n\hat{\beta}/T_n$. The CDF of Q is

$$F_Q(q) = \int_0^{\infty} H\left(4n^2 q/z; 2n\right) h(z; 2(n-1))\, dz.$$

A lower $1 - \epsilon$ confidence limit for $\nu(T_n)$ is

$$\nu_L = \hat{\nu}(T_n)q_{\epsilon},$$

where q_{ϵ} is the 100ϵ percentile of Q, and a lower $1 - \epsilon$ confidence limit for $R(t_0)$ is

$$R_L = \exp\left[-\hat{\nu}(T_n)q_{1-\epsilon}t_0\right].$$

This is also related to the instantaneous mean time between failure (MTBF) $M(t) = 1/\nu(t)$. The corresponding lower $1 - \epsilon$ confidence limit is

$$M_L = 1/\left\{\hat{\nu}(T_n)q_{1-\epsilon}\right\}.$$

Tabulated values ρ_1 and ρ_2 are given in [5] such that $\rho_1\hat{M}(T_n) < M(T_n) < \rho_2\hat{M}(T_n)$ is a $1 - \epsilon$ confidence interval for $M(T_n)$, where $\hat{M}(T_n) = T_n/(n\hat{\beta})$. A normal approximation is also discussed.

It is also possible to construct prediction limits for future failure times. A lower $1 - \epsilon$ prediction limit for the $(n + k)$th future failure time T_{n+k} is a statistic $T_L = T_L(n, k, 1 - \epsilon)$ such that

$$P[T_L < T_{n+k}] = 1 - \epsilon.$$

Such a statistic can be based on the pivotal quantity $Y = (n - 1)\hat{\beta}\ln(T_{n+k}/T_n)$. The resulting prediction limit is

$$T_L = T_n \exp[y_\epsilon / \{(n-1)\hat{\beta}\}],$$

where y_ϵ is the 100ϵ percentile of Y. The CDF of Y is

$$F_Y(y) = \sum_{i=1}^{k} m_i \{ 1 - [1 + (n+i-1)$$
$$\times y / \{n(n-1)\}]^{-(n-1)} \},$$

with

$$m_i = \frac{(-1)^{i-1}(n+k-1)!}{(n-1)!(k-i)!(i-1)!(n+i-1)!}.$$

For the case $k = 1$, an explicit form is obtained,

$$T_L = T_n \exp\{[(1-\epsilon)^{-1/(n-1)} - 1]/\hat{\beta}\}.$$

A convenient approximation for the general case is

$$T_L = T_n \exp[rf_{r,2(n-1),\epsilon} / \{2(n-1)c\hat{\beta}\}],$$

where $f_{r,2(n-1),\epsilon}$ is the 100ϵ percentile of the F-distribution with $r = 2(\sum_{i=n}^{n+k-1} 1/i)^2 / (\sum_{i=n}^{n+k-1} 1/i^2)$ and $2(n-1)$ degrees of freedom, and

$$c = \left(\sum_{i=n}^{n+k-1} 1/i\right) \bigg/ \left\{ n \left(\sum_{i=n}^{n+k-1} 1/i^2\right) \right\}.$$

TIME TRUNCATION

Estimation and inference procedures, in this case, are discussed in [2, 4, and 5]. Suppose $N(t) = n$ is the number of failures in the time interval $(0, t]$. If $n = 0$, then a limited amount of statistical analysis is possible on $m(t) = E[N(t)]$. In particular, the MLE is $\hat{m}(t) = 0$, and an upper $1 - \epsilon$ confidence limit on $m(t)$ is given by $\frac{1}{2}\chi^2_{2,1-\epsilon}$. Suppose $n > 0$. The joint PDF of the successive failure times $T_1, T_2, \ldots, T_n$ and $N(t)$ is

$$f(t_1, t_2, \ldots, t_n, n)$$
$$= (\beta/\theta)^n \prod_{i=1}^{n} (t_i/\theta)^{\beta-1} \exp[-(t/\theta)^\beta],$$

for $0 < t_1 < t_2 < \cdots < t_n < t$.

The joint MLEs are

$$\hat{\beta} = n \bigg/ \sum_{i=1}^{n} \ln(t/T_i)$$

and

$$\hat{\theta} = t/n^{1/\hat{\beta}}.$$

Since $N(t)$ is sufficient for θ, when β is fixed, the conditional distribution of $\hat{\beta}$, given $N(t) = n$, is free of θ. Furthermore, since the conditional distribution of $Z = 2n\beta/\hat{\beta}$, given $N(t) = n$, is chi-square with $2n$ degrees of freedom, a conditional lower $1 - \epsilon$ confidence limit for β is given by

$$\beta_L = [\hat{\beta}/(2n)]\chi^2_{2n,\epsilon}.$$

Due to the time truncation, θ is no longer a scale parameter, and the pivotal property does not hold. An approximate confidence limit for θ is given in [2]. Suppose $w_{n+1,1-\epsilon}$ is the $100(1 - \epsilon)$ percentile of W for a fixed number of failures $n + 1$, as discussed in the case of failure truncation. If we now let n represent the observed number of failures in the time truncation case, an approximate lower $1 - \epsilon$ confidence limit is given by

$$\theta_L = \hat{\theta}\{n[(n+1)w_{n+1,1-\epsilon}]^{-n/(n+1)}\}^{1/\hat{\beta}}.$$

In the present case, the current system reliability would be given by $R(t_0) = \exp[-\nu(t)t_0]$. It should be noted that these results will be applicable only when the entire time interval $(0, t]$ has been observed.

In order to construct lower confidence limits for $R(t_0)$, it is necessary to reparameterize the model in terms of $\nu(t)$ and β. Since it is necessary to have $n > 0$ to draw inferences on any parameter except $m(t)$, it is desirable to condition on $N(t) > 0$. The resulting conditional PDF is

$$f_c(t_1, t_2, \ldots, t_n, n)$$
$$= \frac{[\nu(t)]^n}{\exp\{(t/\beta)\nu(t)\} - 1} \prod_{i=1}^{n} (t_i/t)^{\beta-1},$$

for $0 < t_1 < t_2 < \cdots < t_n < t$ and $n \geqslant 1$. In this form, $Y = N(t)/\hat{\beta}$ is sufficient* for β, when $\nu(t)$ is fixed, so the conditional distribution of $N(t)$ given $Y = y$ is free of β.

The conditional PDF of $N(t)$ given $Y = y$ is

$$f_{N(t)|y}(n; \nu(t))$$

$$= \frac{\{t\nu(t)y\}^n}{n!(n-1)!} \left\{ \sum_{k=1}^{\infty} \frac{\{t\nu(t)y\}^k}{k!(k-1)!} \right\}^{-1},$$

for $n = 1, 2, \ldots$.

A conservative lower $1 - \epsilon$ confidence limit for $\nu(t)$ is given by the largest solution $\nu_L = \nu_1$ of the inequality

$$\sum_{k=n}^{\infty} f_{N(t)|y}(k; \nu_1) \leq \epsilon.$$

A conservative lower $1 - \epsilon$ confidence limit for $R(t_0)$ is given by

$$R_L = \exp(-\nu_2 t_0),$$

where ν_2 is the smallest solution of the inequality

$$\sum_{k=1}^{n} f_{N(t)|y}(k; \nu_2) \leq \epsilon.$$

The corresponding conservative lower $1 - \epsilon$ confidence limit for instantaneous MTBF $M(t)$ is

$$M_L = 1/\nu_2.$$

Tabulated values π_1 and π_2 are given in [5] such that $\pi_1 \hat{M}(t) < M(t) < \pi_2 \hat{M}(t)$ is a $1 - \epsilon$ confidence interval for $M(t)$, where $\hat{M}(t) = t/(n\hat{\beta})$. Normal approximations are also discussed in [2] and [5].

OTHER TOPICS

Multisample procedures are considered by Crow [5], who gives a test of equality of shape parameters for several independent systems. Related work is given by Lee [11], who provides a method for comparing rate functions of several independent processes.

Goodness of fit* is also considered by Crow [5], who proposes a test based on the Cramér–von Mises statistic*. Lee [10] also considers goodness of fit by a different approach. A more general model is developed, which includes the Weibull process and log-linear rate models as special cases. Tests are derived, in this more general framework, for

testing adequacy of the simpler Weibull and log-linear rate models.

Sequential* probability ratio tests for the shape parameter of one or more independent Weibull processes are provided by Bain and Engelhardt [3]. The resulting tests are expressed in terms of the MLEs of the shape parameters in the usual failure truncation situation.

A modification of the Weibull process is proposed by Finkelstein [9]. The modified model yields finite and nonzero instantaneous MTBFs at the start and end of development testing.

References

[1]   Ascher, H. (1979). *IEEE Trans. Rel.*, **R-28**, 119.

[2]   Bain, L. J. and Engelhardt, M. E. (1980). *Technometrics*, **22**, 421–426.

[3]   Bain, L. J. and Engelhardt, M. E. (1982). *IEEE Trans. Rel.*, **R-31**, 79.

[4]   Crow, L. H. (1974). In *Reliability and Biometry, Statistical Analysis of Life Lengths*, F. Proschan and R. J. Serfling, eds., SIAM, Philadelphia, pp. 379–410.

[5]   Crow, L. H. (1982). *Technometrics*, **24**, 67–72.

[6]   Duane, J. T. (1964). *IEEE Trans. Aerospace*, **2**, 563–566.

[7]   Engelhardt, M. and Bain, L. J. (1978). *Technometrics*, **20**, 167–169.

[8]   Finkelstein, J. M. (1976). *Technometrics*, **18**, 115–117.

[9]   Finkelstein, J. M. (1979). *IEEE Trans. Rel.*, **R-28**, 111–113.

[10] Lee, L. (1980). *Technometrics*, **22**, 195–200.

[11] Lee, L. (1980). *Technometrics*, **22**, 427–430.

[12] Lee, L. and Lee, S. K. (1978). *Technometrics*, **20**, 41–45.

Bibliography

Ascher, H. and Feingold, H. (1984). *Repairable Systems Reliability*. Dekker, New York. (There is a good discussion of the confusion that exists between the Weibull process and the Weibull distribution.)

Bain, L. J. (1978). *Statistical Analysis of Reliability and Life-Testing Models*. Dekker, New York. (There is a good summary of statistical methods for the Weibull process.)

Kempthorne, O. and Folks, L. (1971). *Probability, Statistics, and Data Analysis*. The Iowa State University Press, Ames, IA. (There is a derivation that the first failure time of a Weibull process is Weibull distributed.)

Parzen, E. (1962). *Stochastic Processes*. Holden Day, San Francisco, CA. (There is a good discussion of nonhomogeneous Poisson processes and the connection between the counting process and occurrence times.)

(RELIABILITY, PROBABILISTIC STOCHASTIC PROCESSES WEIBULL DISTRIBUTION)

MAX ENGELHARDT

WEIERSTRASS APPROXIMATION THEOREM *See* MATHEMATICAL FUNCTIONS, APPROXIMATIONS TO

WEIGHING DESIGNS

INTRODUCTION

The problem of finding the suitable combinations in which several light objects have to be weighed in a scale to determine their individual weights with maximum possible precision is known as the *weighing problem*, and the suitable combinations as the *weighing designs*. The weighing problem has its origin in a casual illustration furnished by Yates [48]. This illustration subsequently attracted the notice of Hotelling [20] who recast Yates' example in an alternative setting, viewing Yates' illustration as a problem in weighing. Thus came into being Hotelling's (or the Yates–Hotelling) weighing problem.

Besides being useful in routine weighing operations to determine the weights of light objects, the results of research in this area may be useful in chemical, physical, biological, economic, and other sciences. Weighing designs should have applications in any problem of measurements, where the measure of a combination is expressible as a linear combination of the separate measures with numerically equal coefficients.

Dr. Sloane (see [6]) once wrote to the present author as follows: "You might be interested in the enclosed, which gives an application of weighing designs (only we didn't call them by that name!) in measuring frequency spectra."

The following lines from Youden [49, p. 118] may also be of interest in this context: "I confess, with some embarrassment that for the last three investigations over the past years at the NBS (National Bureau of Standards*), I suggested individual and special programs and overlooked the general character and wide applicability of these weighing designs."

Viewing the potential of the subject against the above perspective, a precise formulation of the problem and some of its developments are provided in the following pages in the form of a brief outline.

STATISTICAL FORMULATION OF THE WEIGHING PROBLEM

The weighing problem may be formulated in the least-squares* setup as follows:

The results of N weighing operations to determine the individual weights of p light objects fit into the general linear hypothesis model $\mathbf{Y} = \mathbf{X}\boldsymbol{\beta} + \boldsymbol{\epsilon}$, where $\mathbf{Y}$ is an $N \times 1$ random observed vector of the records of weighings, $\mathbf{X} = (x_{ij})$, $i = 1, 2, \ldots, N$; $J = 1, 2, \ldots, p$, is an $N \times p$ matrix of known quantities, with $x_{ij} = +1$, -1, or 0, if in the ith weighing operation the jth object is placed, respectively, in the left pan, in the right pan, or in none: $\boldsymbol{\beta}$ is a $p \times 1$ vector ($p \leqslant N$) representing the weights of the objects; $\boldsymbol{\epsilon}$ is an $N \times 1$ unobserved random vector of errors such that $E[\boldsymbol{\epsilon}] = \mathbf{0}$, and $E[\boldsymbol{\epsilon}\boldsymbol{\epsilon}'] = \sigma^2\mathbf{I}$.

Consistent with the signs that the elements x_{ij} can take, y_i, $i = 1, 2, \ldots, N$, is taken as positive or negative, depending on whether the balancing weight is placed in the right pan or the left.

As is well known, the least-squares estimates of the weights will be given by $\hat{\boldsymbol{\beta}} = (\mathbf{X}'\mathbf{X})^{-1}\mathbf{X}'\mathbf{Y}$, where $\mathbf{X}'$ is the transpose of the full rank design matrix $\mathbf{X}$, with $\text{cov}(\hat{\boldsymbol{\beta}}) = \sigma^2(\mathbf{X}'\mathbf{X})^{-1} = \sigma^2(c_{ij})$. c_{ii} is the variance factor of the ith estimated weight. When $\mathbf{X}$ is a square matrix of full rank, $\hat{\boldsymbol{\beta}} = (\mathbf{X}'\mathbf{X})^{-1}\mathbf{X}'\mathbf{Y}$

$= \mathbf{X}^{-1}\mathbf{Y}$. If, in particular, $\mathbf{X}$ is orthogonal, $\hat{\beta} = \mathbf{X}^{-1}\mathbf{Y} = (\mathbf{X}'\mathbf{Y})/N$. Also, if we solve for β in $\mathbf{Y} = \mathbf{X}\beta$, when $\mathbf{X}$ is square and of full rank, $\beta = \mathbf{X}^{-1}\mathbf{Y}$.

Two types of problems would arise in practice, one with reference to the spring balance (one-pan balance) and the other with reference to the chemical balance (two-pan balance). In the spring balance, the elements x_{ij} are restricted to the values of $+1$ or 0, whereas in the chemical balance, these elements are either $+1$, -1, or 0.

EFFICIENCY OF WEIGHING DESIGNS

In weighing designs, we search for the elements x_{ij} in such a way that each c_{ii} be the minimum possible. But, since all the estimates might not have equal variance, a weighing design would be called the most efficient, if (i) the average variance were the minimum (A-optimal), or (ii) the $\det|c_{ij}|$ were the minimum (equivalently, $\det|\mathbf{X}'\mathbf{X}|$ were the maximum) (D-optimal), or (iii) the minimum eigenvalue of $\mathbf{X}'\mathbf{X}$ were the maximum (E-optimal). In some cases, any two of these criteria may lead to equivalent results. (For details, see [6].)

DEVELOPMENTS OF THE SUBJECT

In response to Hotelling's call for further mathematical research, the following authors came out with their contributions: Kishen [25], Mood [29], Banerjee (at least 19 papers, summarized in [6]), Rao [39], Kempthorne [24], Raghavarao (at least five papers, summarized in [36]), Zacks [50], Beckman (at least two papers, summarized in [6]), Sihota and Banerjee [40], Hazra and Banerjee [19], Hazra (see [6]), Dey [14, 15], Kulshrestha and Dey [27], Lese and Banerjee [28], Moriguti [30], Sloane and others (several papers, two referred to in [6]), Bhaskararao [10], Youden [49], Swamy [44–46], Bose and Cameron [11], Chakravarti and Suryanarayana [12], Cheng [13], Federer et al. [16],

Galil and Kiefer (at least 2 papers [17, 18]), Jacroux and Wong [22], Kounias and Farmakis [26], Mukherjee and Saha Ray [32], Mukherjee and Huda [31], Raghavarao and Federer [38], Suryanarayana [43], Sinha [41], and others.

The work of the above authors includes the construction of efficient spring balance and chemical balance weighing designs of different dimensions, the treatment of designs under autocorrelation of measurement errors*, determination of total weight, orthogonal designs*, singular weighing designs, fractional weighing designs (randomized and nonrandomized), repeated spring balance designs, biased estimation in weighing designs, etc. The papers written prior to 1975 are mostly summarized in [6].

EFFICIENT WEIGHING DESIGNS

Hadamard matrices provide the most efficient chemical balance designs. (For construction of Hadamard matrices*, see Plackett and Burman [33] and Raghavarao [36].) When, for the number of objects to be weighed, a suitable Hadamard matrix does not exist, we would look for other efficient design matrices of appropriate dimensions. Some research has also been carried out to cover such situations (see the references cited in [6] and also the section entitled Present Directions of Research).

Banerjee [1] has pointed out that the arrangements of a balanced incomplete block design* (BIBD) provide efficient (and in some cases, most efficient) spring balance designs. (Incidentally, a BIBD is an arrangement of v distinct objects into b blocks such that each block contains exactly k distinct objects, each object occurs in exactly r different blocks, and every pair of distinct objects occurs together in exactly λ blocks. In weighing designs, v takes the place of p, the number of objects to be weighed, and b that of the number of weighing operations.) A BIBD does not, in general, provide the estimates as orthogonal linear functions of the observations. But, when such a BIBD is used in a slightly adjusted form, the estimates turn out to be mutually orthogonal (see [2]). Incidentally, Yates' original scheme provides the most efficient spring balance design for the problem under consideration with mutually orthogonal estimates. There exists only one more spring balance design of the same dimension with the same maximum possible efficiency as pointed out by Banerjee [5].

SINGULAR WEIGHING DESIGNS

Despite the best of intentions, a design matrix **X** may not be of full rank, resulting in what may be called a *singular weighing design*. It is well known in least-squares theory that in such a situation, it may not be possible to provide unique unbiased estimates of the individual weights. But, it may be possible to provide unique unbiased estimates of certain linear functions (perhaps, total weight) of the weights. What linear functions will be estimable will depend upon the structure of the design matrix **X** (*see* ESTIMABILITY). When a singular weighing design is encountered, one way to make the design work may be to augment the design matrix suitably by the inclusion of additional weighing operations (i.e., by the addition of rows to the design matrix) to make up for the deficiency in rank. Raghavarao [35] gave a start in this direction. Subsequently, Banerjee [3, 7] and later, Hazra and Banerjee [19] generalized the procedure of augmentation in matrices, bringing, in the sequel, some additional results.

FRACTIONAL WEIGHING DESIGNS

A singular weighing design might also result when, instead of the *full* design matrix **X**, fewer rows of **X** (a fraction of the design) were used for reasons of economy, time, or other considerations. Zacks [50] visualized the possible use of such fractional weighing designs under *randomized procedures*, which would provide an unbiased estimate of *any linear function* of the weights, i.e., of any $\lambda'\beta$ with the minimum possible variance. The probability vector of the *randomization procedure* would, of course, depend upon the linear functional λ (For details, see Zacks [50].)

Some analogous results were later developed by Banerjee [4] with reference to such fractional weighing designs under *nonrandomized procedures*.

CONNECTION WITH FACTORIAL EXPERIMENTS*

Kempthorne [24] discussed the weighing problem from the viewpoint of factorial experiments and indicated how fractional replicates of factorial experiments could provide the most efficient weighing designs.

Connection of weighing designs with factorial experiments could also be traced by way of Hadamard matrices*, which, on one hand, provide the most efficient chemical balance designs, and, on the other hand, saturate orthogonal main effect plans.

The relationship of weighing designs with factorial experiments* could also be traced on a more explicit basis (see Raghavarao [37]).

SPECIFIC APPLICATIONS

Bose and Cameron [1] introduced balanced weighing designs in the context of tournaments and testing experiments. Later, Suryanarayana [43] and Chakravarti and Suryanarayana [12] generalized these ideas to partially balanced weighing designs.

Raghavarao and Federer [38] have found an application of spring balance weighing designs to elicit information on sensitive questions in sample surveys. Also, spring balance weighing designs have been used by Federer et al., [16] to determine the proportion of legume, weed, and grass contents of hay in crop experiments.

Sloane et al. ([42], and other papers) have applied the principles of weighing designs to problems in optics, with a continued research interest in this direction.

PRESENT DIRECTIONS OF RESEARCH

Sinha and Saha [41] have studied the problem of constructing optimal weighing designs with a string property (a design is said to have a *string property*, if the design matrix has a row of ones).

Swamy [45, 46] has constructed some efficient chemical balance weighing designs, given a specific number of weighing operations, as alternatives to repeated designs suggested by Dey [14, 15]. She has also studied [44] the problem of estimating the total weight, using both spring balance and chemical balance weighing designs under certain restrictions.

Mukherjee and Saha Ray [32] and Mukherjee and Huda [31] have developed some asymptotically D-optimal designs under certain restrictions.

Construction of optimum weighing designs in different situations, especially when $n \equiv 2 \pmod 4$ or $n \equiv 3 \pmod 4$, is still being actively pursued, as is reflected in the published papers of Galil and Kiefer [17, 18], Jacroux and Wong [22], Kounias and Farmakis [26], Cheng [13], and possibly others.

References

[1] Banerjee, K. S. (1948). *Ann. Math. Statist.*, **19**, 394–399. (Weighing designs and balanced incomplete blocks.)

[2] Banerjee, K. S. (1950). *Biometrika*, **37**, 50–58. (How balanced incomplete block designs may be made to furnish orthogonal estimates in weighing designs.)

[3] Banerjee, K. S. (1966). *Ann. Math. Statist.*, **37**, 1021–1032. (Singularity in Hotelling's weighing design and a generalized inverse.) (A correction note appears in *Ann. Math. Statist.*, **40**, 719)

[4] Banerjee, K. S. (1966). *Ann. Math. Statist.*, **37**, 1836–1841.

[5] Banerjee, K. S. (1974). *Commun. Statist.*, **3**, 185–190.

[6] Banerjee, K. S. (1975). *Weighing Designs*. Dekker, New York. (This monograph includes a summary of Banerjee's 19 papers as well as a list of other authors' contributions.)

[7] Banerjee, K. S. (1972). *J. Amer. Statist. Ass.*, **67**, 211–212. (Singular weighing designs and a reflexive generalized inverse.)

[8] Beckman, R. J. (1969). Randomized Spring Balance Weighing Designs. Ph.D. dissertation, Kansas State University, Manhattan, KS.

[9] Beckman, R. J. (1972). *Commun. Statist.*, **1**, 561–565.

[10] Bhaskar Rao, M. (1966). *Ann. Math. Statist.*, **37**, 1371–1381. (Weighing designs when n is odd.)

[11] Bose, R. C. and Cameron, J. M. (1968). *J. Res. Natl. Bur. Stand.* **B69**, 323–332.

[12] Chakravarti, I. M. and Suryanarayana, K. V. (1972). *J. Comb. Theory*, **A13**, 426–431.

[13] Cheng, C. S. (1980). *Ann. Statist*, **8**, 436–446. (Optimality of some weighing and 2^n fractional factorial designs.)

[14] Dey, A. (1969). *Ann. Inst. Statist. Math., Tokyo*, **21**, 343–346.

[15] Dey, A. (1972). *J. Indian Soc. Agric. Statist.*, **24**, 119–126.

[16] Federer, W. T., Hedayat, A., Lowe, C. C., and Raghavarao, D. (1976). *Agron. J.*, **68**, 914–918.

[17] Galil, Z. and Kiefer, J. (1980). *Ann. Statist.* **8**, 1293–1306. [D-optimum weighing designs for $n \equiv 3 \pmod 4$.]

[18] Galil, Z. and Kiefer, J. (1982). *Ann. Statist.*, **10**, 502–510.

[19] Hazra, P. K. and Banerjee, K. S. (1973). *J. Amer. Statist. Ass.*, **68**, 392–393.

[20] Hotelling, H. (1964). *Ann. Math. Statist.*, **15**, 297–306. (Some improvements in weighing and other experimental techniques.)

[21] Jacroux, M. and Notz, W. (1983). *Ann. Statist.*, **11**, 970–978.

[22] Jacroux, M. and Wong, C. S. (1983). *J. Statist. Plann. Inf.*, **8**, 231–240. (On the optimality of chemical balance weighing designs.)

[23] Kageyama, S. and Saha, G. M. (1983). *Ann. Inst. Statist. Math., Tokyo*, **35**, 447–452.

[24] Kempthorne, O. (1949). *Ann. Math. Statist.*, **19**, 238–248.

[25] Kishen, K. (1945). *Ann. Math. Statist.*, **16**, 294–300.

[26] Kounias, S. and Farmakis, N. (1984). *J. Statist. Plann. Inf.*, **10**, 177–187. [A construction of D-optimal weighing designs when $n \equiv 3 \pmod 4$.]

[27] Kulshrestha, A. C. and Dey, A. (1970). *Austral. J. Statist.*, **12**, 166–168.

[28] Lese, N. G., Jr. and Banerjee, K. S. (1972). Orthogonal estimates in weighing designs. *Proc. 18th Conf. on Designs in Army Research and Testing.* (Originally, Lese's PhD. dissertation, University of Delaware, Newark, DE.)

[29] Mood, A. M. (1946). *Ann. Math. Statist.*, **17**, 432–446. (On Hotelling's weighing problem.)

[30] Moriguti, S. (1954). *Rep. Statist. Appl. Res. JUSE*, **3**, 1–24. (Optimality of orthogonal designs.)

[31] Mukherjee, R. and Huda, S. (1985). *Commun. Statist. A*, **14**, 669–677. (*D*-optimal statistical designs with restricted string property.)

[32] Mukherjee, R. and Saha Ray, R. (1983). *Tech. Rep. No. 19/83*; Indian Statistical Institute, Calcutta, India.

[33] Plackett, R. L. and Burman, J. P. (1946). *Biometrika*, **33**, 305–325.

[34] Raghavarao, D. (1959). *Ann. Math. Statist.*, **30**, 295–303. (Some optimum weighing designs.)

[35] Raghavarao, D. (1964). *Ann. Math. Statist.*, **35**, 673–680. (Singular weighing designs.)

[36] Raghavarao, D. (1971). *Constructions of Combinatorial Problems in Design of Experiments*, Wiley, New York.

[37] Raghavarao, D. (1975). *Gujarat Statist. Rev.*, **2**, 1–16. (Review article.)

[38] Raghavarao, D. and Federer, W. T. (1979). *J. R. Statist. Soc. B*, **41**, 40–45. (Block total response as an alternative to the randomized response method in surveys.)

[39] Rao, C. R. (1946). *Sankhyā*, **7**, 440.

[40] Sihota, S. S., and Banerjee, K. S. (1974). *Sankhyā B*, **36**, 55–64. (Biased estimation in weighing designs.)

[41] Sinha, B. K. and Saha, R. (1983). *J. Statist. Plann. Inf.*, **8**, 365–374.

[42] Sloane, N. J. A., Fine, T., Phillips, P. G., and Harwit, M. (1969). *Appl. Opt.*, **8**, 2103–2106. (Codes for multiplex spectrometry.)

[43] Suryanarayana, K. V. (1971). *Ann. Math. Statist.*, **42**, 1316–1321.

[44] Swamy, M. N. (1980). *Commun. Statist. A*, **9**, 1185–1190. (Optimum spring balance weighing design for estimating the total weight.)

[45] Swamy, M. N. (1981). *J. Indian Agric. Statist.*, **33**, 23–28.

[46] Swamy, M. N. (1981). *J. Indian Statist. Ass.*, **19**, 177–181.

[47] Swamy, M. N. (1982). *Commun. Statist. A*, **11**, 769–785.

[48] Yates, F. (1935). *J. R. Statist. Soc. Suppl.*, **3**, 181–247. (Complex experiments.)

[49] Youden, N. J. (1962). *Technometrics*, **4**, 111–123. (Systematic errors in physical constants.)

[50] Zacks, S. (1966). *Ann. Math. Statist.*, **37**, 1382–1395. (Randomized fractional weighing designs.)

(DESIGN OF EXPERIMENTS
FACTORIAL EXPERIMENTS
FRACTIONAL FACTORIAL DESIGNS
OPTIMUM DESIGN OF EXPERIMENTS
(*See Supplement*))

K. S. Banerjee

WEIGHT BIAS

A bias attributable to the weighting used in calculating an index number.

(INDEX NUMBERS)

WEIGHT OF AN ARRAY *See* STRENGTH AND OTHER PROPERTIES OF AN ARRAY

WEIGHT OF EVIDENCE *See* STATISTICAL EVIDENCE

WEIGHTED DISTRIBUTIONS

INTRODUCTION

The concept of weighted distributions can be traced to the study of the effect of methods of ascertainment upon estimation of frequencies by Fisher [7]. The initial idea of length-biased sampling appears in Cox [4]. In extending the basic ideas of Fisher, Rao [26] saw the need for a unifying concept and identified various sampling situations that can be modeled by what he called weighted distributions. Within the biomedical context of cell kinetics and the early detection of disease, Zelen [36] introduced weighted distributions to represent what he broadly perceived as length-biased sampling*.

Patil [16] has discovered weighted distributions as stochastic models in the equilibrium study of populations subject to harvesting and predation. See the bibliography by Patil et al. [21] for a comprehensive survey of the literature.

UNIVARIATE WEIGHTED DISTRIBUTIONS

Suppose X is a nonnegative observable random variable (RV) with PDF $f(x; \theta)$, where θ is a parameter. Suppose a realization x of X under $f(x; \theta)$ enters the investigator's record with probability proportional to $w(x, \beta)$, so that

$$\Pr(\text{Recording}|X = x) = w(x, \beta).$$

Here the recording (weight) function $w(x, \beta)$ is a nonnegative function with parameter β representing the recording (sighting) mechanism. Clearly, the recorded x is not an observation on X, but on the RV X^w, say, having PDF

$$f^w(x; \theta, \beta) = \frac{w(x, \beta)f(x; \theta)}{\omega}, \quad (1)$$

where $\omega = E[w(X, \beta)]$ is the normalizing factor, making the total probability equal to unity. The RV X^w is called the *weighted version* of X, and its distribution in relation to that of X is called the *weighted distribution* with weight function w. An important weighted distribution corresponds to $w(x, \beta) = x$, in which case, $X^* = X^w$ is called the *size-biased version* of X. The distribution of X^* is called the *size-biased distribution* of PDF

$$f^*(x; \theta) = \frac{xf(x; \theta)}{\mu}, \quad (2)$$

where $\mu = E[X]$. The PDF f^* is called the length-biased or size-biased version of f, and the corresponding observational mechanism is called length- or size-biased sampling.

The concept of weighted distributions has been used during the last 25 years as a useful tool in the selection of appropriate models for observed data, especially when samples are drawn without a sampling frame that enables random samples to be drawn.

In many situations the model given in (1) is appropriate, and the statistical problems that arise are the determination of a suitable weight function $w(x, \beta)$ and drawing inference on θ.

The following examples from Patil and Rao [18] illustrate a few situations generating weighted distributions.

(i) Truncation. The distribution of a random variable truncated to a set T is a weighted distribution with weight function $w(x) = 1$ for $x \in T$ and 0 elsewhere.

(ii) Missing Data. If the sampling mechanism results in a proportion $1 - w(x)$, $0 \leqslant w(x) \leqslant 1$, of the observations having the value x being omitted, the PDF to use for the analysis of the observed data is with the weight function $w(x)$.

(iii) Damaged Observations. Consider a damage model*, where an observation $X = x$ is reduced to y by a destructive process with PDF $d(y|x)$. See Rao [26]. Then the probability that the observation $X = x$ is undamaged is $d(x|x)$, and the distribution of the undamaged observation is the weighted distribution with $w(x) = d(x|x)$. For example, under the binomial survival model, $d(x|x) = \theta^x$, $0 < \theta < 1$. An investigator recording only undamaged observations will need to work with a corresponding weighted distribution.

(iv) Analysis of Family Data. This is an example of size-biased sampling. The discussion is based on Rao [26].

Consider the data in Table 1, which relates to brothers and sisters in families of 104 boys who were admitted to a postgraduate course at the Indian Statistical Institute.

Table 1 Family Data

Family size	1	2	3	4	5	6	7	8	9	10	11	12	13	15	Total
Number of families	1	6	6	13	12	7	14	11	12	8	6	5	2	1	104
Brothers	1	8	12	34	34	29	59	50	54	46	32	31	16	8	414
Sisters	0	4	6	18	26	13	39	38	54	34	34	29	10	7	312

Let us assume that in families of given size n, the probability of a family with x boys coming into the record is proportional to x. Also, suppose that the number of boys in a family follows a binomial distribution with probability parameter π. Then

$$f(x; \pi) = \binom{n}{x} \pi^x (1 - \pi)^{n-x},$$

$$w(x) = x, \qquad E[w(X)] = \omega = n\pi,$$

$$f^w(x; \pi) = \binom{n-1}{x-1} \pi^{x-1}(1 - \pi)^{n-x},$$

$$E[X^w/n] = \pi + (1 - \pi)/n > \pi,$$

$$E[(X^w - 1)/(n - 1)] = \pi.$$

If k boys representing families of size $n_1, n_2, \ldots, n_k$ report $x_1, x_2, \ldots, x_k$ boys, an unbiased estimate of π is

$$\tilde{\pi} = \left(\sum x_i - k\right)/\left(\sum n_i - k\right)$$

$$= (414 - 104)/(726 - 104) \doteq 1/2,$$

whereas if one wrongly treats x_i's as observations on k randomly drawn families with at least one boy, i.e., as arising from a truncated binomial, then the estimate of π will have a serious upward bias.

WEIGHT FUNCTIONS AND PROPERTIES

The following forms of weight functions $w(x)$ have appeared in the scientific and statistical literature:

1. $w(x) = x^\beta$ for $\beta = 1, 2, 3, 1/2, 2/3, 0 < \beta < 1$.
2. $w(x) = \left(\frac{x}{2}\right)^\beta$ for $\beta = 1, 1/2, 0 < \beta < 1$.
3. $w(x) = (x)^{(\beta)} = x(x - 1) \cdots (x - \beta + 1)$.
4. $w(x) = e^{\beta x}, \theta^x$.
5. $w(x) = x + 1, \alpha x + \beta$.
6. $w(x) = 1 - (1 - \beta)^x$ for $0 < \beta < 1$.
7. $w(x) = (\alpha x + \beta)/(\delta x + \gamma)$.
8. $w(x) = \Pr(Y \leqslant x)$ for some RV Y.
9. $w(x) = \Pr(Y > x)$ for some RV Y.
10. $w(x) = r(x)$, where $r(x)$ is the probability of "survival" of observation x.

An important case arises where T is a subset of the real line and $r(x)$ is defined by $r(x) = 1$ if $x \in T$ and $r(x) = 0$ if $x \notin T$. The resulting weighted distribution is said to be a *truncated version* of the original distribution. If, for an original discrete distribution for example, $r(0) = 0$ and $r(x) = 1$ for $x \neq 0$, the resulting distribution is called a *zero-truncated distribution*.

The weight functions **1–10** are monotone increasing or decreasing functions of x. The following results provide useful comparisons of X^w with X.

Result 3.1 (Patil et al. [20]). The weighted version X^w is stochastically greater or smaller than the original RV X according as the weight function $w(x)$ is monotone increasing or decreasing in x. As a consequence, the expected value of the weighted version X^w is greater or smaller than the expected value of the original RV X according as the weight function $w(x)$ is monotone increasing or decreasing in x.

Result 3.2 (Zelen [36] and Patil and Rao [19]). The expected value of the size-biased version X^* is $E[X^*] = \mu[1 + \sigma^2/\mu^2]$, where $E[X] = \mu$ and $V(X) = \sigma^2$. Furthermore, the harmonic mean of X^* is equal to the mean of the original RV X when it is positive, i.e., $E[1/X^*] = 1/\mu$. Another way of expressing these results is that $E[X^*]E[1/X^*] = 1 + \sigma^2/\mu^2$.

Result 3.3 (Patil and Ord [17]). Let the RV X have PDF $f(x; \theta)$ and have size bias with weight function x^β. Then a necessary and sufficient condition for $f^w(x; \theta, \beta) = f(x; \eta)$, where $\eta = \eta(\theta, \beta)$, is that $f(x; \theta) = x^\theta a(x)/m(\theta)$. In this case, $f^w(x; \theta, \beta) = f(x; \theta + \beta)$. This result holds under certain mild regularity conditions.

Result 3.4 (Mahfoud and Patil [13]). Consider X^w to be a RV X subject to a size bias with the weight function x^β. Then X has a lognormal distribution* if and only if $V(\log X) \equiv V(\log X^w)$ for $\beta > 0$, where V stands for variance. Thus the invariance of the logarithmic variance under size bias of order β characterizes the log-normal distribution.

Example. In sedimentology and various other fields, the distribution of particles is usually analyzed by mass rather than frequency. Sieve analysis is a good example, which provides data consisting of sizes of sieves and corresponding masses of all particles retained by those sieves. It is interesting to note that the mass–size density is nothing but the weighted version of size bias of order 3 of the PDF of the particle size. It can be verified that, if X is lognormal with parameters μ and σ^2, then X^w with $\beta = 3$ is lognormal with parameters $\mu + 3\sigma^2$ and σ^2. This property has been empirically noticed and utilized for inference in sedimentology literature. See, for example, Krumbein and Pettijohn [12] and Herdan [11].

BIVARIATE WEIGHTED DISTRIBUTIONS

Let (X, Y) be a pair of nonnegative RVs with a joint PDF $f(x, y)$ and let $w(x, y)$ be a nonnegative weight function such that $E[w(X, Y)]$ exists. The weighted version of $f(x, y)$ is

$$f^w(x, y) = \frac{w(x, y)f(x, y)}{E[w(X, Y)]}.$$

The corresponding weighted version of (X, Y) is denoted by $(X, Y)^w$. The marginal and conditional distributions of $(X, Y)^w$ are

$$f^w(x) = \frac{E[w(x, Y)|x]f(x)}{E[w(X, Y)]},$$

$$f^w(y|x) = \frac{w(x, y)f(y|x)}{E[w(x, Y)|x]}.$$

Clearly, both are weighted versions of the corresponding marginal and conditional distributions of (X, Y).

Special cases of weight functions of practical interest are

1. $w(x, y) = x^\alpha$;
2. $w(x, y) = w(y)$;
3. $w(x, y) = x + y$;
4. $w(x, y) = x^\alpha y^\beta$;
5. $w(x, y) = \max(x, y)$; and
6. $w(x, y) = \min(x, y)$.

The following results are of some interest.

Result 4.1 (Patil and Rao [19]). Let (X, Y) be a pair of nonnegative RVs with PDF $f(x, y)$. Let $w(x, y) = w(y)$, as is the case in sample surveys involving sampling with probability proportional to size*. Then the random variables X and X^w are related by

$$f^w(x) = \frac{E[w(Y)|x]f(x)}{E[w(Y)]}.$$

Note that X^w is a weighted version of X, and the regression of $w(Y)$ on X serves as the WF.

Result 4.2 (Mahfoud and Patil [14]). Let (X, Y) be a pair of nonnegative independent RVs with joint PDF $f(x, y) = f_X(x)f_Y(y)$ and let $w(x, y) = \max(x, y)$. Then the RVs $(X, Y)^w$ are dependent. Furthermore, the regression of Y^w on X^w by $E[Y^w|X^w = x]$ is a decreasing function of x.

Example. Let (X, Y) be statistically independent lifetimes of two components in parallel forming a kit with lifetime $w = \max(x, y)$. Consider a renewal system of kits. If at time t, one records the lifetimes of the two components of the kit in action, their joint PDF is the weighted version of their natural PDF with weight function $w = \max(x, y)$. The result shows that whereas there is underlying independence in the true lifetimes of the two components/organs, the data so obtained will not reveal that; actually the data will show negative dependence.

POSTERIOR AND WEIGHTED DISTRIBUTIONS

There is a Bayesian analog to the theory of weighted distributions. (See Mahfoud and Patil [14] and Patil et al. [20].)

Result 5.1 (Mahfoud and Patil, [13]). Consider the usual Bayesian inference* in conjunction with (X, θ) having joint PDF $f(x, \theta) = f(x|\theta)f(\theta) = f(\theta|x)f(x)$. The posterior PDF $f(\theta|x) = f(x|\theta)f(\theta)/f(x) = l(\theta|x)f(\theta)/E[l(\theta|X)]$ is a weighted version of the prior* PDF $f(\theta)$. The weight function is the likelihood* function of θ for the observed x.

Result 5.2 (Patil et al. [20]). Consider the usual Bayesian inference in conjunction with (X, θ) with PDF $f(x, \theta) = f(x|\theta)f(\theta) = f(\theta|x)f(x)$. Let $w(x, \theta) = w(x)$ be the weight function for the distribution of $X|\theta$, so that the PDF of $X^w|\theta$ is

$$w(x)f(x|\theta)/\omega(\theta),$$

where

$$\omega(\theta) = E[w(X)|\theta].$$

Then the original and the weighted posteriors are related by

$$f(\theta|x) = \frac{\omega(\theta)f^w(\theta|x)}{E[\omega(\theta)|X^w = x]}.$$

Furthermore, the weighted posterior* RV $(\theta^w|X^w = x)$ is stochastically greater or smaller than the original posterior RV $\theta|X = x$ according as $\omega(\theta)$ is a monotonically decreasing or increasing function of θ.

Examples. Table 2 provides a convenient format.

WEIGHTED DISTRIBUTIONS IN STOCHASTIC POPULATION DYNAMICS

Consider the stochastic differential equation*

$$\frac{1}{x}\frac{dx}{dt} = r(x, t)$$

$$= g(x) + \gamma(t),$$

where $x(t)$ is the population size at time t, $r(x, t)$ is the per capita growth rate of population of size x at time t, $g(x)$ is the biological part of the per capita growth rate dependent on population size x, but independent of time t, and $\gamma(t)$ is the environmental part of the per capita growth rate dependent on time t, but independent of the population size x. Let $\gamma(t)$ be a white-noise process with environmental unpredictability parameter σ^2.

The population size $x(t)$ is then a stochastic integral, and when it exists, its equilibrium PDF

$$f(x) = \frac{\text{const}}{V(x)}\exp\left[2\int\frac{M(x)}{V(x)}\,dx\right]$$

$$= \exp[a\log x + b(x) + c], \quad \text{say,}$$

is a member of the log-exponential family, where $M(x) = xg(x)$ and $V(x) = \sigma^2 x^2$.

Furthermore, if the population is subjected to exploitation (harvesting, predation, etc.) with per capita exploitation rate $h(x)$, the equilibrium PDF $f_h(x)$ of the exploited population size interestingly simplifies to

$$f_h(x) = \frac{w(x)f(x)}{E[w(X)]},$$

Table 2   Posterior and Weighted Distributions

$X|\Theta = \theta$	Θ	$\Theta|X = x$	$w(x)$	$\Theta|X^w = x$
Poisson(θ)	Gamma(k, λ)	Gamma($k + x, 1/(\lambda + 1)$)	x	Gamma($k + x - 1, 1/(\lambda + 1)$)
Binomial(n, θ)	Beta(a, b)	Beta($x + a, n - x + b$)	x	Beta($x - 1 + a, n - x + b$)
Neg-Bin(k, θ)	Beta(a, b)	Beta($k + a, x + b$)	x	Beta($k + 1 + a, x - a + b$)
Exponential(θ)	Gamma(k, λ)	Gamma($k + 1, \lambda/(\lambda x + 1)$)	x	Gamma($k + 2, \lambda/(\lambda x + 1)$)

where $f(x)$ is the natural population equilibrium PDF and

$$w(x) = \exp\left[-2\int \frac{xh(x)}{\sigma^2 x^2}\, dx\right].$$

For further discussion, see Patil [16] and Dennis and Patil [5].

APPLICATIONS OF WEIGHTED DISTRIBUTIONS

A vast number of situations arise in which weighted distributions find their application. For lack of space, we indicate rather than discuss some of these applications.

1. Cell cycle analysis and pulse labeling, Zelen [36].
2. Efficacy of early screening for disease and scheduling of examinations, Zelen [35, 36].
3. Cardiac transplantation, Temkin [33].
4. Estimation of antigen frequencies, Simon [29].
5. Ascertainment studies in genetics, Rao [26, 27] and Stene [30].
6. Renewal theory and reliability, Cox [4] and Zelen [36].
7. Nonrenewable natural resource exploration, Barouch et al. [1].
8. Traffic research, Brown [2].
9. Word association analysis, Haight and Jones [9].
10. Marketing and resource utilization, Morrison [15].
11. Analysis of spatial pattern, Pielou [23].
12. Species abundance and diversity, Engen [6].
13. Transect sampling, Cook and Martin [3], Patil and Rao [19], and Quinn [25].
14. Forest products research, Warren [34].
15. Income inequality and species inequitability, Hart [10] and Taillie [31].
16. Canonical hypothesis in ecology, Preston [24] and Patil and Taillie [22].
17. Particle-size statistics, Gy [8] and Krumbein and Pettijohn [12].
18. Mass–size distributions, Herdan [11] and Schultz [28].
19. Quality of Swiss cheese, Tallis [32].

References

[1] Barouch, E., Chow, S., Kaufman, G. M., and Wright, T. H. (1985). *Stud. Appl. Math.*, **73**, 239–260.

[2] Brown, M. (1972). *Adv. Appl. Prob.*, **4**, 177–192.

[3] Cook, R. D. and Martin, F. B. (1974). *J. Amer. Statist. Ass.*, **69**, 345–349.

[4] Cox, D. R. (1962). *Renewal Theory*. Methuen, New York.

[5] Dennis, B. and Patil, G. P. (1984). *Math. Biosci.*, **68**, 187–212.

[6] Engen, S. (1978). *Stochastic Abundance Models.* Chapman and Hall, London.

[7] Fisher, R. A. (1984). *Ann. Eugen.*, **6**, 13–25.

[8] Gy, P. M. (1982). *Sampling of Particulate Materials.* Elsevier, New York.

[9] Haight, F. A. and Jones, R. B. (1974). *J. Math. Psych.*, **11**, 237–244.

[10] Hart, P. E. (1975). *J. R. Statist. Soc. A*, **138**, 423–434.

[11] Herdan, G. (1960). *Small Particle Statistics*. Elsevier, New York.

[12] Krumbein, W. C. and Pettijohn, F. J. (1938). *Manual of Sedimentory Petrography*. Appleton-Century-Croft, New York.

[13] Mahfoud, M. and Patil, G. P. (1981). In *Statistics in Theory and Practice: Essays in Honor of Bertil Matern*, B. Ranneby, ed. Swedish University Agricultural Science, Umeå, Sweden, pp. 173–187.

[14] Mahfoud, M. and Patil, G. P. (1982). In *Statistics and Probability: Essays in Honor of C. R. Rao*, G. Kallianpur et al., eds. North-Holland, Amsterdam, pp. 479–492.

[15] Morrison, D. G. (1973). *Amer. Statist.*, **27**, 226–227.

[16] Patil, G. P. (1984). In *Proceedings of the Indian Statistical Institute Golden Jubilee International Conference on Statistics: Applications and New Directions*, J. R. Ghosh and J. Roy, eds. Statistical Publishing Society, Calcutta, India, pp. 478–503.

[17] Patil, G. P. and Ord, J. K. (1975). *Sankhyā*, **38**, 48–61.

[18] Patil, G. P. and Rao, C. R. (1977). Weighted distributions and a survey of their applications. In *Applications of Statistics*, P. R. Krishnaiah, ed. North-Holland, Amsterdam, pp. 383–405.

[19] Patil, G. P. and Rao, C. R. (1978). *Biometrics*, **34**, 179–184.

[20] Patil, G. P., Rao, C. R., and Ratnaparkhi, M. V. (1986). *Comm. Statist.-Theor. Meth.*, **15**, 907–918.

[21] Patil, G. P., Rao, C. R., and Zelen, M. (1986). *A computerized bibliography of weighted distributions and related weighted methods for statistical analysis and interpretations of encountered data, observational studies, representativeness issues, and resulting inferences.* Center for Statistical Ecology and Environmental Statistics, Pennsylvania State University, University Park, PA.

[22] Patil, G. P. and Taillie, C. (1979). *Bull. Int. Statist. Inst.*, **44**, 1–23.

[23] Pielou, E. C. (1977). *Mathematical Ecology.* Wiley, New York.

[24] Preston, F. W. (1962). *Ecol.*, **43**, 185–215, 410–432.

[25] Quinn, T. J. (1979). In *Contemporary Quantitative Ecology and Related Econometrics*, G. P. Patil and M. Rosenzweig, eds. International Co-operative, Fairland, MD, pp. 473–491.

[26] Rao, C. R. (1965). In *Classical and Contagious Discrete Distributions*, G. P. Patil, ed. Statistical Publishing Society, Calcutta, India and Pergamon Press, New York, pp. 320–332.

[27] Rao, C. R. (1985). In *Celebration of Statistics: The ISI Centenary Volume*, A. C. Atkinson and S. E. Fienberg, eds., International Statistical Institute, The Hague, Netherlands, pp. 543–569.

[28] Schultz, D. M. (1975). In *Statistical Distributions in Scientific Work*, G. P. Patil, S. Kotz, and J. K. Ord, eds. Reidel, Dordrecht, Netherlands and Boston, MA, pp. 275–288.

[29] Simon, R. (1980). *Amer. J. Epidemiology*, **111**, 444–452.

[30] Stene, J. (1981). In *Statistical Distributions in Scientific Work*, Vol. 6, C. Taillie, G. P. Patil, and B. Baldessari, eds. Reidel, Dordrecht, Netherlands and Boston, MA, pp. 233–264.

[31] Taillie, C. (1979). In *Ecological Diversity in Theory and Practice*, J. F. Grassle et al., eds. International Co-operative, Fairland, MD, pp. 51–62.

[32] Tallis, G. M. (1970). *Biometrics*, **26**, 87–104.

[33] Temkin, N. (1976). Interactive Information and Distributional Length Biased Survival Models. Ph.D. dissertation, University of New York at Buffalo, University Microfilms, Ann Arbor, Michigan.

[34] Warren, W. G. (1975). In *Statistical Distributions in Scientific Work*, G. P. Patil, S. Kotz, and J. K. Ord, eds. Reidel, Dordrecht, Netherlands and Boston, MA, pp. 369–384.

[35] Zelen, M. (1971). *Bull. Int. Statist. Inst., Proc.*, **38**, Session I, pp. 649–661.

[36] Zelen, M. (1974). In *Reliability and Biometry, Statistical Analysis of Life Lengths*, F. Proschan and R. J. Serfling, eds. SIAM, Philadelphia, pp. 701–726.

(DAMAGE MODELS
TRUNCATED DISTRIBUTION)

G. P. PATIL
C. R. RAO
MARVIN ZELEN

WEIGHTED EMPIRICAL PROCESSES: GENESIS AND APPLICATIONS

Weighted empirical processes (WEPs) arise typically in the context of rank* and robust* statistics (both tests and estimates) in linear models. One version of the WEP may be regarded as a direct extension of the classical empirical (distributional) process* to the regression* setup, a second version corresponds to a stochastic process* with regression parameters taking the role of time parameters, whereas a general class of the WEP may be defined in terms of a two-dimensional time-parameter process involving the distributional process as well as the regression parameters. We shall introduce these WEPs in this order.

For a given set $c_{n1}, \ldots, c_{nn}$ of known (regression) constants and independent random variables $X_1, \ldots, X_n$ having distributions $F_1, \ldots, F_n$, respectively (not necessarily all the same), the regression WEP is defined by

$$W_n(x) = \sum_{i=1}^{n} c_{ni}^* \{ I(X_i \leqslant x) - F_i(x) \},$$

$$-\infty < x < \infty, \quad (1)$$

where $I(A)$ stands for the indicator function of the set A and

$$c_{ni}^* = c_{ni} \bigg/ \left\{ \sum_{j=1}^{n} c_{nj}^2 \right\}^{1/2}, \quad \text{for } i = 1, \ldots, n. \quad (2)$$

Often, by means of a transformation $X_i \to Y_i = H_n(X_i)$, $1 \leqslant i \leqslant n$, where $H_n(x)$ is nonnegative and monotone (in x) with

$H_n(-\infty) = 0$ and $H_n(\infty) = 1$, we may reduce (1) to

$$W_n^0 = \sum_{i=1}^{n} c_{ni}^* \{ I(Y_{ni} \leqslant t) - G_{ni}(t) \},$$

$$t \in [0, 1], \quad (3)$$

where G_{ni} is the distribution of Y_{ni} [i.e., $G_{ni}(t) = F_i(H_n^{-1}(t))$, $t \in [0, 1]$], for $i = 1, \ldots, n$. $\{ W_n^0 \}$ is termed the *reduced weighted empirical process* (RWEP).

In the particular case of $c_{ni} = 1$, for every $i = 1, \ldots, n$, the RWEP relates to the usual empirical distributional process (in the comparatively more general setup of possibly nonidentically distributed random variables), whereas the classical empirical distributional process corresponds to the case where the G_{ni} are all the same [namely, $G_{ni}(t) = t$, $\forall\, t \in [0, 1]$, $i = 1, \ldots, n$]. A detailed account of the basic (small- as well as large-sample) properties of these empirical processes is given in PROCESSES, EMPIRICAL. Because of some important statistical uses of the WEP and RWEP, a more application-oriented treatise of them is considered here.

Whereas the early adaptation of the classical empirical process can be traced to the study of the asymptotic distribution theory of the one-sample *Kolmogorov–Smirnov statistics**, the WEP may similarly be identified with the two-sample goodness-of-fit* problem. Suppose that $X_1, \ldots, X_{n_1}$ have the common distribution F and $X_{n_1+1}, \ldots, X_n$, $n = n_1 + n_2$, have the common distribution G, and we want to test for the equality of F and G. For this problem, the classical two-sample Kolmogorov–Smirnov statistic corresponds to (1) with $c_1 = \cdots = c_{n_1} = 1/n_1$ and $c_{n_1+1} = \cdots = c_n = -1/n_2$.

In general, in the one-sample models, in various nonparametric (or robust) statistical inference problems, estimators and/or test statistics are expressed as functionals of the classical empirical processes, and in a similar manner, in the multisample models, these can be expressed in terms of the WEP or RWEPs. This representation greatly simplifies the treatment of the related asymptotic distribution theory, and constitutes by

far the most important statistical use of the WER and RWEPs.

This important application was considered in a more general regression setup by Hájek [6], although the earlier works of Dwass [4] and Chernoff and Savage [3] (in the particular case of two-sample models) bear close scrutiny. Consider the simple regression model: $X_i = \beta c_i + e_i$, where the c_i are known constants, β is the (unknown) regression parameter, and the errors e_i are independent with a common distribution F. Hájek [6] used the WEP in the following manner to test the null hypothesis $H_0 : \beta = 0$ against $\beta \neq 0$. Corresponding to known regression constants $c_1, \ldots, c_n$, let us take $c_{ni} = c_i - \bar{c}_n$, $i = 1, \ldots, n$, where $\bar{c}_n = n^{-1}\sum_{j=1}^{n} c_j$, and we define the c_{ni}^* as in (2). Also, let $Z_{n:1}, \ldots, Z_{n:n}$ be the ordered values corresponding to the $X_1, \ldots, X_n$, and define the vector of anti-ranks $\mathbf{D} = (D_1, \ldots, D_n)'$ by letting $Z_{n:k} = X_{D_k}$ for $k = 1, \ldots, n$ (ties are neglected with probability 1 by assuming the distribution F to be continuous). Then Hájek's extensions of the Kolmogorov–Smirnov statistics* to the regression setup are

$$K_n^+ = \max_{1 \leqslant k \leqslant n} \left\{ \sum_{j=1}^{k} c_{nD_j}^* \right\} \quad \text{and}$$

$$K_n = \max_{1 \leqslant k \leqslant n} \left| \sum_{j=1}^{k} c_{nD_j}^* \right|. \quad (4)$$

If in (1), we let $F_1 = \cdots = F_n = F$, then it is easy to show that

$$K_n^+ = \max \{ W_n(Z_{n:k}); k \leqslant n \}$$

$$= \sup \{ W_n(x) : x \in (-\infty, \infty) \}, \quad (5)$$

and a similar representation holds for K_n.

Thus the asymptotic theory for the WEP in (1) [viz., weak convergence*, law of iterated logarithm*, etc.] can be incorporated to derive parallel results for the Kolmogorov–Smirnov statistics for regression alternatives. Similar results have also been worked out for the Rényi as well as the Cramér–von Mises statistics*, and the WEP plays the basic role in this context (viz., Hájek and Šidák [9], Chaps. 5 and 6).

The statistics in (4) and (5) have important applications in time-sequential* analysis arising typically in clinical trials* and other life-testing* experimentations, where the observations are gathered sequentially over time. In the context of interim analysis in clinical trials, it is common to monitor the accumulating data set with a view to achieving an early termination on sound statistical grounds. In this progressive censoring* setup, at the successive failure points $(Z_{n:k})$, one may look at the $W_n(Z_{n:k})$, and hence the problem reduces to that of drawing a conclusion based on K_n or K_n^+ (or more generally some weighted versions of them). An extension of (4) to the case of a multiple regression model where the $\mathbf{c}_i$ are p-vectors (so is $\boldsymbol{\beta}$), for some $p \geqslant 1$, has been considered by Sinha and Sen [37, 38]. For the simple as well as multiple regression model, Sinha and Sen [35–38] have incorporated the WEP in a slightly more general setup and considered appropriate time-sequential* tests arising typically in the progressive censoring schemes.

The WEP (or RWEP) have been very appropriately incorporated in the study of the asymptotic distribution theory of rank-order statistics (and derived R-estimators) in linear models. As has been mentioned earlier, the two-sample rank statistics may be expressed in terms of the WEP, and taking this lead, Hoeffding [10] suggested a simple way of incorporating the WEP in the study of the asymptotic distribution theory of linear rank statistics in a regression setup; for some further work in this direction, we may refer to Koul and Staudte [26] and Ghosh and Sen [5]. The earlier works of Hájek [7] and Pyke and Shorack [29, 30] deserve mention in this context. We shall discuss this in more detail later on.

Let us consider next the second type of WEP. In the context of robust (M-) estimation* in linear models, the following type of WEP arises. For given $c_{n1}, \ldots, c_{nn}$ and $X_1, \ldots, X_n$, one may consider a suitable score function $\psi = \{\psi(x), -\infty < x < \infty\}$, and corresponding to another set $\{d_{n1}, \ldots, d_{nn}\}$

of real constants, define a process

$$W_n(b) = \sum_{i=1}^{n} c_{ni}^* \psi(X_i - bd_{ni}),$$

$$b \in B = [-K, K], \quad \text{for some } K \in (0, \infty).$$
(6)

For example, taking $d_{ni} = c_{ni}$, $i = 1, \ldots, n$, and equating $W_n(b)$ to 0 (in a meaningful way), one gets the usual M-estimator of the regression coefficient. The weak convergence of $W_n = \{W_n(b); b \in B\}$ to some (drifted) Gaussian function provides the key to the study of asymptotic distribution theory of M-estimators* of regression parameters. Actually, in (6), one may easily use for c_{ni} and d_{ni} some p-vectors (so also for b), and this will lead to a p-variate WEP with a p-dimensional time parameter $\mathbf{b} \in B^p$. Such general WEPs have been studied systematically by Jurečková and Sen [16–18], where applications to some problems in sequential analysis* have also been stressed.

Let us consider the third type of WEPs, which are more general than the others referred to earlier and which play a very significant role in the asymptotic distribution theory of rank and other nonparametric statistics. Typically, a linear rank statistic* is defined as $L_n = L_n(\mathbf{X}_n) = \sum_{i=1}^{n} c_{ni}^* a_n(R_{ni})$, where $a_n(1), \ldots, a_n(n)$ are suitable scores*, the c_{ni}^* are defined as before, and R_{ni} is the rank of X_i among $X_1, \ldots, X_n$ for $i = 1, \ldots, n$. Suppose that in the above definition, we replace $\mathbf{X}_n$ by $\mathbf{X}_n - b\mathbf{d}_n$, where $\mathbf{d}_n = (d_{n1}, \ldots, d_{nn})'$ is a vector of known constants and b is real; the resulting ranks are then denoted by $R_{ni}(b)$.

For real x and b, we define

$$S_n(x, b) = \sum_{i=1}^{n} c_{ni}^* I(X_i \leqslant x + bd_{ni}),$$

$$-\infty < x < \infty, \; -\infty < b < \infty. \quad (7)$$

Also, the usual empirical distribution based on $\mathbf{X}_n - b\mathbf{d}_n$ is defined as $H_n(x, b) = n^{-1}\sum_{i=1}^{n} I(X_i \leqslant x + bd_{ni})$, $-\infty < x < \infty$, $-\infty < b < \infty$. If the scores $a_n(k)$ are generated by a score-generating function ϕ, such that $a_n(k) = \phi_n(k/n)$, $k = 1, \ldots, n$, $\phi_n(\cdot)$

$\to \phi(\cdot)$, then we have

$$L_n(\mathbf{X}_n - b\mathbf{d}_n)$$

$$= \int_{-\infty}^{\infty} \phi_n(H_n(x, b)) \, dS_n(x, b),$$

$$\text{for } b \in (-\infty, \infty). \quad (8)$$

This representation enables us to use the two-dimensional time-parameter WEP in (7) in the study of asymptotic distribution theory of rank statistics and more general rank processes. In particular, if in (7), we take $b = 0$, we obtain the stochastic component of $W_n(x)$ in (1), hence $W_n(x)$ in (1) can be used [along with the usual expansion of (8) around the true distributions when $b = 0$] to express the normalized form of $L_n(\mathbf{X}_n)$ in terms of WEPs, and this has been worked out in detail by Hoeffding [10], Pyke and Shorack [29, 30], and others.

More generally, we may consider a rank process defined by

$$\{ L_n(\mathbf{X}_n - b\mathbf{d}_n) - L_n(\mathbf{X}_n) + b\gamma\Delta_n;$$

$$b \in B = [-K, K] \}, \quad (9)$$

where K is a finite positive number, Δ_n depends on the c_{ni}^* and d_{ni}, and γ is a functional of the underlying F (and ϕ). By using (8), we are able to express this rank process in terms of the general WEP in (7), so that (7) provides a very convenient way of studying the asymptotic behavior of such processes. Actually, on any compact B, (9) converges to a null process, in distribution, as n increases; in the literature, this is known as the Jurečková [11] *linearity of linear rank statistics* in the regression parameter. Parallel results on signed WEP and related signed rank statistics (or processes) were obtained by Koul [19–21], van Eeden [39], and Jurečkova [12], among others. The relations in (7) and (8) also prompted Sen and Ghosh [32] and Ghosh and Sen [5] to incorporate directly the WEP in the study of the almost-sure convergence of rank statistics and to improve the Jurečková result to an almost-sure convergence result too.

To stress further the role of WEP, we also mention its adaptation in rank estimation

theory. If the scores $a_n(k)$ are monotone increasing (in k) and c_{ni}^* and d_{ni} are concordant, then $L_n(\mathbf{X}_n - b\mathbf{d}_n)$ is monotone nonincreasing in b, and this fact was tacitly used by Adichie [1] in providing an R-estimator* of the regression coefficient by equating $L_n(\mathbf{X}_n - b\mathbf{d}_n)$ to 0. Equation (9) may be used to express the normalized form of this R-estimator in terms of $L_n(\mathbf{X}_n - \beta\mathbf{d}_n)$, and deeper results in this direction (based on the WEP) are due to Jurečková [14, 15] and others.

In this context (as well as in others), it has been observed that the asymptotic behavior of the WEP for local (contiguous) alternatives provides a basic framework for parallel results on various nonparametric tests and estimates. This approach has been elegantly developed by the Czechoslovakian school under the pioneering guidance of the late professor J. Hájek; a detailed account of some of these developments is given in Sen [31, Chaps. 4–6]; see also the recent book by Shorack and Wellner [34]. The general WEP in (7) has also been extended to the case of vector c_{ni}^*, d_{ni} (and b); for some nice use of these WEPs, we may refer to Jurečková [13] and Sen and Puri [33], among others.

In linear models, it is not uncommon to encounter nuisance scale or other parameters. In this context, (9) provides a convenient way of eliminating these nuisance parameters by using their R-estimators and the related aligned rank statistics. Hájek [8] pointed out the rationality of this approach (and the inaccuracy of earlier ones). Koul and Sen [25] have shown that this alignment principle works out well for the multiparameter linear models when the error distributions are symmetric. The WEPs play a basic role in this study too.

Other notable uses of the WEP include the following:

(i)   Linear combinations of order statistics* for the regression model (viz., Bickel [2]), where the ordering of the observations is (essentially) done by assigning weights based on the regression constants.

(ii)   Minimum distance* type estimator of the regression parameter (viz., Koul and DeWet [24], Koul [23], and Millar [27]), where a quadratic functional of the WEP is directly involved in the minimization process.

(iii)   Goodness of fit* in linear models in the presence of nuisance parameters (viz., Koul [22] and Pierce and Kopecky [28]), where estimators of these nuisance parameters are incorporated in the definition of residuals, and WEPs are then constructed for these residuals. These are often called *weighted residual empirical processes* (WREP).

Faced with this genesis background and variety of applications of the WEP, RWEP, and WREP, several workers have studied the asymptotic theory of these processes under increasing degrees of generalizations. Some detailed accounts of this theory are given in Shorack and Wellner [34], where other references are also cited.

References

[1]   Adichie, J. N. (1967). *Ann. Math. Statist.*, **38**, 894–904.

[2]   Bickel, P. J. (1973). *Ann. Statist.*, **1**, 597–617.

[3]   Chernoff, H. and Savage, I. R. (1958). *Ann. Math. Statist.*, **29**, 972–994.

[4]   Dwass, M. (1957). *Ann. Math. Statist.*, **28**, 424–431.

[5]   Ghosh, M. and Sen, P. K. (1972). *Sankhyā A*, **34**, 33–52.

[6]   Hájek, J. (1965). In *Bernoulli–Bayes–Laplace Seminar, Berkeley*. University of California Press, Berkeley, CA, pp. 45–60.

[7]   Hájek, J. (1968). *Ann. Math. Statist.*, **39**, 325–346.

[8]   Hájek, J. (1970). In *Nonparametric Techniques in Statistical Inference*, M. L. Puri, ed. Cambridge University Press, pp. 1–17.

[9]   Hájek, J. and Šidák, Z. (1967). *Theory of Rank Tests*. Academic, New York.

[10]   Hoeffding, W. (1970). In *Nonparametric Techniques in Statistical Inference*, M. L. Puri, ed. Cambridge University Press, London, England, pp. 18–19.

[11]   Jurečková, J. (1969). *Ann. Math. Statist.*, **40**, 1889–1900.

[12]   Jurečková, J. (1971a). *Sankhyā A*, **33**, 1–18.

[13]   Jurečková, J. (1971b). *Ann. Math. Statist.*, **42**, 1328–1338.

[14]   Jurečková, J. (1973). *Ann. Statist.*, **1**, 1046–1060.

[15]   Jurečková, J. (1977). *Ann. Statist.*, **5**, 664–672.

[16]   Jurečková, J. and Sen, P. K. (1981a). *Sankhyā A*, **43**, 190–210.

[17]   Jurečková, J. and Sen, P. K. (1981b). *J. Statist. Plann. Inf.*, **5**, 253–266.

[18]   Jurečková, J. and Sen, P. K. (1984). *Statist. Dec. Suppl.*, 31–46.

[19]   Koul, H. L. (1969). *Ann. Math. Statist.*, **40**, 1950–1979.

[20]   Koul, H. L. (1970). *Ann. Math. Statist.*, **41**, 1768–1773.

[21]   Koul, H. L. (1971). *Ann. Math. Statist.*, **42**, 466–476.

[22]   Koul, H. L. (1982). *Colloq. Math. Soc. Janos Bolyai*, **32**, 537–565.

[23]   Koul, H. L. (1985). *Sankhyā A*, **47**, 57–84.

[24]   Koul, H. L. and DeWet, T. (1983). *Ann. Statist.*, **11**, 921–932.

[25]   Koul, H. L. and Sen, P. K. (1985). *Statist. Prob. Lett.*, **3**, 111–115.

[26]   Koul, H. L. and Staudte, R. G., Jr. (1972). *Ann. Math. Statist.*, **43**, 832–841.

[27]   Millar, P. W. (1981). *Zeit. Wahrscheinlichkeitsth. Verwand. Geb*, **55**, 73–89.

[28]   Pierce, D. A. and Kopecky, K. J. (1979). *Biometrika*, **66**, 1–5.

[29]   Pyke, R. and Shorack, G. (1968a). *Ann. Math. Statist.*, **39**, 755–771.

[30]   Pyke, R. and Shorack, G. (1968b). *Ann. Math. Statist.*, **39**, 1675–1685.

[31]   Sen, P. K. (1981). *Sequential Nonparametrics*. Wiley, New York.

[32]   Sen, P. K. and Ghosh, M. (1971). *Ann. Math. Statist.*, **42**, 189–203.

[33]   Sen, P. K. and Puri, M. L. (1977). *Zeit. Wahrscheinlichkeitsth. Verwand. Geb.*, **39**, 175–186.

[34]   Shorack, G. and Wellner, J. A. (1986). *Empirical Processes with Applications to Statistics*. Wiley, New York.

[35]   Sinha, A. N. and Sen, P. K. (1979a). *Commun. Statist. A*, **8**, 871–898.

[36]   Sinha, A. N. and Sen, P. K. (1979b). *Calcutta Statist. Ass. Bull.*, **28**, 57–82.

[37]   Sinha, A. N. and Sen, P. K. (1982). *Sankhyā B*, **44**, 1–18.

[38]   Sinha, A. N. and Sen, P. K. (1984). *Indian Statist. Inst. Golden Jubilee* Vol., 237–253.

[39]   van Eeden, C. (1972). *Ann. Math. Statist.*, **43**, 791–802.

P. K. Sen

WEIGHTED LEAST SQUARES

Suppose that a multiple linear regression*
model is given by

$$y_i = \beta_1 x_{1i} + \cdots + \beta_k x_{ki} + \epsilon_i,$$
$$i = 1, \ldots, n, \qquad (1)$$

with responses $y_1, \ldots, y_n$, regression coefficients $\beta_1, \ldots, \beta_k$, independent variables $x_{1i}, \ldots, x_{ki}$, $i = 1, \ldots, n$, and errors $\epsilon_1, \ldots, \epsilon_n$. Ordinary least squares* (OLS) estimates $\hat{\beta}_1, \ldots, \hat{\beta}_k$ of $\beta_1, \ldots, \beta_k$ are chosen to minimize

$$\sum_{i=1}^{n} \{ y_i - \beta_1 x_{1i} - \cdots - \beta_k x_{ki} \}^2.$$

This estimation procedure presupposes that the errors are uncorrelated with equal variances. If $\mathrm{var}(\epsilon_i) = \sigma^2/w_i$, $i = 1, \ldots, n$, for known weights $w_1, \ldots, w_n$, then weighted least-squares (WLS) estimates of $\beta_1, \ldots, \beta_k$ are obtained by minimizing

$$\sum_{i=1}^{n} w_i (y_i - \beta_1 x_{1i} - \cdots - \beta_k x_{ki})^2. \quad (2)$$

This can be done by constructing pseudo-variates*

$$y_i^* = \sqrt{w_i}\, y_i, \qquad x_{ji}^* = \sqrt{w_i}\, x_{ji},$$

where $i = 1, \ldots, n$ and $j = 1, \ldots, k$, and calculating OLS estimates via the model

$$y_i^* = \beta_1 x_{1i}^* + \cdots + \beta_k x_{ki}^* + \epsilon_i^*,$$

where $\epsilon_1^*, \ldots, \epsilon_n^*$ now have equal variances. See Box et al. [4] for an elementary discussion and examples.

Suppose that the ϵ_i's are also correlated with positive-definite covariance matrix $\sigma^2 \mathbf{V}$, where $\mathbf{V}$ is known, and write eq. (1) in matrix notation as

$$\mathbf{Y} = \mathbf{X}\boldsymbol{\beta} + \boldsymbol{\epsilon}, \qquad (3)$$

where now $\mathbf{X}$ is an $n \times k$ matrix of rank k and $E[\boldsymbol{\epsilon}] = \mathbf{0}$. Since a nonsingular matrix $\mathbf{P}$ exists satisfying $\mathbf{V} = \mathbf{PP}'$, the transformation

$$\mathbf{Y}^* = \mathbf{P}^{-1}\mathbf{Y} = \mathbf{X}^*\mathbf{B} + \boldsymbol{\epsilon}^*$$

leads to uncorrelated identically distributed errors $\epsilon_1^*, \ldots, \epsilon_n^*$ with common variance σ^2. The sum of squares to be minimized is then

$$\boldsymbol{\epsilon}^{*\prime}\boldsymbol{\epsilon}^* = (\mathbf{Y}^* - \mathbf{X}^*\boldsymbol{\beta})'(\mathbf{Y}^* - \mathbf{X}^*\boldsymbol{\beta}) \quad (4a)$$
$$= (\mathbf{Y} - \mathbf{X}\boldsymbol{\beta})'\mathbf{V}^{-1}(\mathbf{Y} - \mathbf{X}\boldsymbol{\beta}) \quad (4b)$$
$$= (\mathbf{Y} - E[\mathbf{Y}])'\mathbf{V}^{-1}(\mathbf{Y} - E[\mathbf{Y}]). \quad (4c)$$

The OLS solution to minimizing (4a) is the WLS solution to minimizing (4b) and leads to WLS estimators

$$\hat{\boldsymbol{\beta}}^* = (\mathbf{X}'\mathbf{V}^{-1}\mathbf{X})^{-1}\mathbf{X}'\mathbf{V}^{-1}\mathbf{Y}. \quad (5)$$

If the space spanned by the columns of $\mathbf{V}^{-1}\mathbf{X}$ coincides with that spanned by the columns of $\mathbf{X}$, then the WLS estimates $\hat{\boldsymbol{\beta}}$ coincide with the OLS estimates $\hat{\boldsymbol{\beta}}$; see Seber [11, Sec. 3.6] for details.

As with OLS, WLS estimators constructed with suitable weights in the classical general linear model* lead to best linear unbiased estimators of the parameters $\boldsymbol{\beta}$. This includes models with heterogeneous variances leading to minimization of (2), grouped data, replicates with unequal group sizes, and multivariate regression estimators as in (5).

In practice, the covariance matrix $\sigma^2 \mathbf{V}$ is usually unknown. When the responses are independent, $\sigma^2 \mathbf{V}$ is diagonal, with $\mathrm{var}(Y_i) = \sigma_i^2$, say. Replacing $\sigma_1^2, \ldots, \sigma_n^2$ by sample variances $s_1^2, \ldots, s_n^2$ in the preceding leads to estimated WLS estimators $\hat{\boldsymbol{\beta}}^{**}$, which have the same large-sample properties as $\hat{\boldsymbol{\beta}}^*$ (Schmidt [10, p. 71]). Following a Monte Carlo study, Kleinjen et al. [7] gave recommendations for estimating $\boldsymbol{\beta}$ with small to moderate samples in experimental design situations.

In logistic (or logit) regression, k binomial populations with parameters $p_1, \ldots, p_k$, are analyzed via the model

$$p_i = \frac{\exp(\alpha + \boldsymbol{\beta}'\mathbf{x}_i)}{1 + \exp(\alpha + \boldsymbol{\beta}'\mathbf{x}_i)}, \qquad i = 1, \ldots, k,$$

$$(6)$$

with explanatory variables $\mathbf{x}_i = (x_1, \ldots, x_q)$, say, so that the logit

$$\log\{ p_i/(1 - p_i)\} = \alpha + \boldsymbol{\beta}'\mathbf{x}_i. \quad (7)$$

If n_i independent observations are made on the ith group, resulting in Y_i occurrences of the event of interest (so that p_i is estimated by $\hat{p}_i = Y_i/n_i$), the corresponding observed logit is

$$l_i = \log[\hat{p}_i/(1 - \hat{p}_i)]. \qquad (8)$$

[If any $y_i = 0$, replace it by 0.5; if any $y_i = n_i$, let $\hat{p}_i = n_i/(n_i + 1)$, in order to avoid infinite values of l_i.] Then WLS estimates of $\alpha, \beta_1, \ldots, \beta_q$ are obtained by minimizing

$$\sum_{i=1}^{k} (l_i - \alpha - \boldsymbol{\beta}'\mathbf{x}_i)^2/v_i, \qquad (9)$$

where

$$v_i = (1/y_i + 1/(n_i - y_i))$$

estimates the variance of l_i, $i = 1, \ldots, k$. Under the model (6) the minimized weighted sum of squares in (9) is Berkson's minimum logit chi-square statistic, used to test the goodness of fit* of the model if the n_i's are not too small. An excellent elementary presentation, with an example using National Football League field goal kicking data, is given by Morris and Rolph [9, Sec. 8.2]. For a detailed discussion, *see* LOGISTIC REGRESSION and Bishop et al. [3, p. 355].

WLS estimation in logit regression illustrates an approach to estimating parameters in log-linear models arising from multidimensional contingency tables* (*see also* CONTINGENCY TABLES and Bishop et al. [3, pp. 352–357]). The general WLS approach in this setup was developed by Grizzle et al. [6].

With discrete or grouped data* $\mathbf{Y}$, minimization of quadratic forms (4c), where $E(\mathbf{Y})$ is a function of a vector of parameters $\boldsymbol{\theta}$, frequently leads to WLS estimates $\hat{\boldsymbol{\theta}}$ of $\boldsymbol{\theta}$ that are best asymptotically normal, and the substitution of $\hat{\boldsymbol{\theta}}$ for $\boldsymbol{\theta}$ in the quadratic form yields a statistic with an asymptotic chi-square distribution. In (4c) $\mathbf{V}$ may be replaced by a matrix $\mathbf{M}$, which converges in probability to $\mathbf{V}$ or to $\sigma^2\mathbf{V}$ (such as the covariance matrix of the responses), or minimization of (4c) may be subject to certain constraints, or the responses may be transformed. In such cases WLS leads to minimum chi-square* estimates; Pearson's chi-square and minimum logit chi-square are two such instances. Berkson [2] contended that "the basic principle of estimation is minimum chi-square, not maximum likelihood." Berkson's contention is not a defense of WLS estimation, but MINIMUM CHI-SQUARE indicates the key role of WLS in many of the minimizing procedures. *See also* CHI-SQUARE TESTS; GROUPED DATA; LOGISTIC REGRESSION; MOST PROBABLE NUMBER; POISSON REGRESSION.

Where data in the form of time series* are used to predict future observations, WLS forecasts based on exponential smoothing can be obtained. The procedure is to discount past observations by giving greater weight to recent observations. Under certain conditions such discounted least-squares estimates lead to minimum mean square error forecasts [1, Chap. 2, 3, 7]; *see* PREDICTION AND FORECASTING.

In some circumstances the weights (or the matrix of weights) involved in computing WLS estimates are themselves functions of the fitted values. In such cases, initial estimates of the parameters are used to calculate appropriate weights and in turn improved estimates, the process being repeated iteratively until some desired degree of accuracy has been reached. Thus the estimation problem for certain generalized linear models* reduces to iteratively reweighted least squares*; see McCullagh and Nelder [8, Sec. 2.5] for a comprehensive and detailed discussion. In robust regression* points further away from a regression line or surface receive less weight than points closer to the line; *see* ITERATIVELY REWEIGHTED LEAST SQUARES for an overview and references.

Other fields in which WLS techniques figure are listed among the related entries; see also the *Current Index to Statistics**. Rather than comprising a unified methodology on its own, WLS is tailored to the requirements of the specific problem at hand, as are the computer programs (in SAS and BMDF for the classical linear model, GLIM* for generalized linear models, EM algorithms for certain iterative procedures, etc.) that derive the estimates.

There is some difference in nomenclature in the literature. Searle (*see* GENERAL LINEAR MODEL) and Seber [11] call the minimization of (4b)—and analogous estimation problems in which the rank of **V** is less than full—the *generalized least-squares* (GLS) problem, and Seber further restricts the use of the term "weighted least squares" to the case in which **V** is a diagonal matrix. Among sources covering classical estimation problems with normally distributed errors, Draper and Smith [5, Sec. 2.11] use the WLS terminology to cover all of these cases, as we have done here. The same is true of writers dealing with models having nonnormally distributed errors. In one class of generalized linear model problems, however, the GLS terminology seems to be more appropriate, namely, that in which the covariance matrix **V** in (4c) is not a matrix of constants but (for example) a matrix whose elements are known functions of $E(\mathbf{Y})$. For further discussions, *see* QUASI-LIKELIHOOD FUNCTIONS and McCullagh and Nelder [8, Chap. 8].

References

[1] Abraham, B. and Ledolter, J. (1983). *Statistical Methods for Forecasting*. Wiley, New York.

[2] Berkson, J. (1980). *Ann. Statist.*, **8**, 457–487. (Includes discussion by B. Efron, J. K. Ghosh, L. LeCam, J. Pfanzagl, and C. R. Rao.)

[3] Bishop, Y. M. M., Fienberg, S. E., and Holland, P. W. (1975). *Discrete Multivariate Analysis*: *Theory and Practice*. MIT Press, Cambridge, MA.

[4] Box, G. E. P., Hunter, W. G., and Hunter, J. S. (1978). *Statistics for Experimenters*. Wiley, New York.

[5] Draper, N. R. and Smith, H. (1981). *Applied Regression Analysis*, 2nd ed. Wiley, New York.

[6] Grizzle, J. E., Starmer, C. F., and Koch, G. G. (1969). *Biometrics*, **25**, 489–504.

[7] Kleinjen, J., Brent, R., and Brouwers, R. (1981). *Commun. Statist. B*, **10**, 303–313.

[8] McCullagh, P. and Nelder, J. A. (1983). *Generalized Linear Models*. Chapman and Hall, London.

[9] Morris, C. N. and Rolph, J. E. (1981). *Introduction to Data Analysis and Statistical Inference*. Prentice-Hall, Englewood Cliffs, NJ.

[10] Schmidt, P. (1976). *Econometrics*. Dekker, New York.

[11] Seber, G. A. F. (1977). *Linear Regression Analysis*. Wiley, New York.

(CHI-SQUARE TESTS
GENERALIZED LINEAR MODELS
GENERAL LINEAR MODEL
GROUPED DATA
ITERATIVELY REWEIGHTED LEAST
    SQUARES
LEAST SQUARES
LOGISTIC REGRESSION
MINIMUM CHI-SQUARE
MOST PROBABLE NUMBER
ORDINAL DATA
POISSON REGRESSION
PREDICTION AND FORECASTING
PROBABILITY PLOTTING
PSEUDO-VARIATES
QUASI-LIKELIHOOD FUNCTIONS
RESISTANT TECHNIQUES
ROBUST REGRESSION
SCHWEPPE-TYPE ESTIMATORS
WALD'S *W*-STATISTICS)

CAMPBELL B. READ

WEIGHTED LEAST-SQUARES RANK ESTIMATORS

Suppose $X_1, \ldots, X_n$ represents a random sample from an absolutely continuous* distribution. Suppose, further, that the density function $f(x - \theta)$ is symmetric about θ, so that the location parameter θ is the *median* (and mean if it exists).

Let $a(1) \leqslant \cdots \leqslant a(n)$ be a sequence of nonconstant scores and consider the rank* statistic

$$S_n(t) = \sum_{i=1}^{n} a[R_i(t)]\text{sign}(X_i - t), \quad (1)$$

where $R_i(t)$ is the rank of $|X_i - t|$ among $|X_1 - t|, \ldots, |X_n - t|$. When $a(i) \equiv 1$, $a(i) = i$, or $a(i) = \Phi^{-1}[(i + n + 1)/\{2(n + 1)\}]$ where Φ is the standard normal CDF, we have the sign statistic, the Wilcoxon signed rank statistic*, or the van der Waerden one-sample normal scores* statistic, respectively.

Note that $S_n(t)$ is a decreasing step function and $ES_n(\theta) = 0$. Hence a natural estimator of θ is $\hat{\theta}$ defined as a solution $t = \hat{\theta}$ of the estimating equation

$$S_n(t) = 0. \quad (2)$$

Since $S_n(t)$ is a step function, we take as our solution either the point at which $S_n(t)$ steps across zero or the midpoint of the interval of values for which $S_n(t)$ is identically zero.

When $a(i) = 1$, $\hat{\theta}$ is the median of the sample, and when $a(i) = i$, $\hat{\theta}$ is the median of the $n(n + 1)/2$ pairwise Walsh averages* $(X_i + X_j)/2$, $1 \leqslant i \leqslant j \leqslant n$. The estimator, derived from a nonparametric rank statistic, was first proposed by Hodges and Lehmann [2] (*see* HODGES-LEHMANN ESTIMATORS). For other scores, such as the one-sample normal scores, an explicit representation of the estimator is not available. Hence numerical methods are needed to solve (2). Even in the case $a(i) = i$, numerical methods can be useful since it may be time consuming to find the median of the $n(n + 1)/2$ averages.

A simple iterative procedure can be based on the method of *weighted* least squares*. We rewrite (1) in the following form:

$$S_n(t) = \sum_{i=1}^{n} w_i(t)(X_i - t), \quad (3)$$

where

$$w_i(t) = \begin{cases} a[R_i(t)]/|X_i - t|, \\ \qquad\qquad \text{if } |X_i - t| > 0, \\ 0, \qquad \text{otherwise.} \end{cases}$$

Now, using $\hat{\theta}_0$ to denote an initial estimate, generally the sample mean or median, a one-step solution of (3) is given by

$$\hat{\theta}_1 = \frac{\sum w_i(\hat{\theta}_0) X_i}{\sum w_i(\hat{\theta}_0)} = \hat{\theta}_0 + \frac{S_n(\hat{\theta}_0)}{\sum w_i(\hat{\theta}_0)}. \quad (4)$$

The estimator $\hat{\theta}_k$ is defined by performing k iterations and is called the k-step weighted least-squares rank estimator. This approach extends quite easily to estimation in the general linear model*.

The properties of $\hat{\theta}_k$ are related to the properties of the initial estimate $\hat{\theta}_0$ and to the properties of $\hat{\theta}$, the true solution to (2). As k increases, $\hat{\theta}_k$ is more like $\hat{\theta}$ and the effects of $\hat{\theta}$ become negligible. Under mild regularity conditions, $\hat{\theta}_k$ has an approximately normal distribution with mean θ and variance σ_k^2 for large n. The formula for σ_k^2 is rather complicated; see Theorem 2 of Cheng and Hettmansperger [1]. However, in many circumstances, σ_k^2 is essentially the same as the asymptotic variance of $\hat{\theta}$ after a few iterations. Thus $\hat{\theta}_k$ and $\hat{\theta}$ have essentially the same efficiency properties. Furthermore, after 4 or 5 iterations, even starting with the nonrobust $\hat{\theta}_0 = \bar{X}$, the breakdown values and the *influence curve** for $\hat{\theta}_k$ are very similar to those for $\hat{\theta}$. Hence, for most practical purposes, after a few iterations $\hat{\theta}_k$ may be used in place of $\hat{\theta}$.

As k increases convergence of $\hat{\theta}_k$ to $\hat{\theta}$ can be guaranteed by incorporating an interval-halving procedure with the iteration of (4). Generally, however, we can treat $\hat{\theta}_k$ as an estimator in its own right with properties very similar to those of $\hat{\theta}$. Simply iterating (4) a few times yields numerical values quite close to $\hat{\theta}$.

As a simple example we consider Example 12.4 of Noether [5, p. 130]. Nine examination scores are given in order as follows: 62, 70, 74, 75, 77, 80, 83, 85, and 88. If we take $a(i) = i$, then $\hat{\theta} = 77.5$, the median of the 45 pairwise averages. A Minitab program [6] was written to iterate (4) using an extreme starting value of $\hat{\theta}_0 = 50$. We find for $k = 1, 2, \ldots, 5$, $\hat{\theta}_k = 79.39$, 78.08, 77.71, 77.36, and 77.46, respectively. Hence, even with a poor starting value, $\hat{\theta}_k$ was close to $\hat{\theta}$ after only a few iterations.

Using an appropriate linear approximation to (1), Newton–Raphson*-types of iterations can be performed to define another k-step estimator. This approach requires the estimation of the slope of the linear approximation and is a bit more complicated than a weighted least-squares method. However, good numerical results are also achieved with this method. For a discussion of these linearized rank estimators in the linear model see Kraft and van Eeden [3] and McKean and Hettmansperger [4].

References

[1] Cheng, K. S. and Hettmansperger, T. P. (1983). *Commun. Statist. A*, **12**, 1069–1086. (This reference contains the details of the weighted least-squares rank estimates for location as well as linear models.)

[2] Hodges, J. L., Jr. and Lehmann, E. L. (1963). *Ann. Math. Statist.*, **34**, 598–611.

[3] Kraft, C. H. and van Eeden, C. (1972). *Ann. Math. Statist.*, **43**, 42–57.

[4] McKean, J. W. and Hettmansperger, T. P. (1978). *Biometrika*, **65**, 571–579.

[5] Noether, G. E. (1976). *Introduction to Statistics, A Nonparametric Approach*, 2nd ed. Houghton Mifflin, Boston.

[6] Ryan, T. A., Jr., Joiner, B. L., and Ryan, B. F. (1982). *Minitab Reference Manual*. Minitab Project, Inc., University Park, PA.

(INFLUENCE FUNCTIONS
RANK ORDER STATISTICS
SCORE STATISTICS
WEIGHTED LEAST SQUARES)

THOMAS P. HETTMANSPERGER

WEIGHTED MEAN *See* TUKEY'S INEQUALITY FOR OPTIMAL WEIGHTS

WEIGHTED NORMAL PLOTS

Weighted normal plots, proposed by Dempster and Ryan [2], provide a model-checking technique that is particularly sensitive to nonnormality of the random effects in a simple one-way* comparisons model. Suppose $Y_1, \ldots, Y_n$ are independent observations from the model

$$Y_i = \mu_i + \epsilon_i, \qquad (1)$$

where the pairs (μ_i, ϵ_i) are independent and normally distributed with means μ and 0, and variances σ^2 and σ_i^2, respectively. The outcomes $Y_1, \ldots, Y_n$ might be summary statistics from grouped data* of unequal sample sizes or with different sampling variances in each group, as illustrated later with an example.

Overall goodness of fit* may be assessed by a normal probability plot* of the standardized variables

$$Z_i = (Y_i - \mu) / (\sigma^2 + \sigma_i^2)^{1/2}.$$

This approach involves plotting Z_i against $\Phi^{-1}[F_n(Z_i)]$, where $\Phi(\cdot)$ is the standard normal cumulative distribution function (CDF) and $F_n(\cdot)$ is an empirical CDF, which, ignoring end-point adjustments for now, can be written as

$$F_n(x) = \sum_{i=1}^{n} I(x - Z_i)/n, \qquad (2)$$

where $I(x) = 1$ for $x \geqslant 0$ and 0 otherwise. Endpoint adjustments are ignored for the present.

For the purpose of checking the normality of the random effects in model (1), a weighted normal plot may be more efficient. The idea is to weight each Z_i according to the amount of information* contained about the random effects μ_i. For instance, a variable Z_i should be downweighted if the error variance σ_i^2 is relatively large. Suitable weights are given by $W_i = (\sigma^2 + \sigma_i^2)^{-1}$ and may be incorporated by replacing $F_n(\cdot)$ in (2) by

$$F_n^*(x) = \sum_{i=1}^{n} I(x - Z_i)W_i \Big/ \sum_{i=1}^{n} W_i. \qquad (3)$$

In practice, the weighted empirical CDF (3) must be adjusted at the end points, otherwise the function $\Phi^{-1}[F_n^*(\cdot)]$ cannot be calculated at the largest Z_i. The following formula performs well:

$$F_n^*(x) = (5/8)W_{(1)}/S,$$

for $x \leqslant Z_{(1)}$, and

$$F_n^*(x) = \left[(9/8)W_{(1)} + W_{(2)} + \cdots\right.$$
$$\left. + (1/2)W_{(i)}\right]/S,$$

for $x \in (Z_{(i-1)}, Z_{(i)}]$, where $Z_{(i)}$ is the ith order statistic*, and

$$S = \left[(9/8)W_{(1)} + \cdots + W_{(i)}\right.$$
$$\left. + \cdots + (9/8)W_{(n)}\right].$$

In the case of equal weights, $F_n^*(\cdot)$ yields the widely used plotting positions recommended by Blom [1]:

$$F_n(x) = (i - 3/8)/(n + 1/4),$$
$$x \in (Z_{(i-1)}, Z_{(i)}].$$

Dempster and Ryan [2] show that weighted and unweighted plots behave similarly, except that the pointwise variation of the weighted plot increases by the factor $(1 + v/m^2)$, where m and v are the mean and variance of the weights. However, the extra variability of the weighted plot is countered by increased sensitivity to detect violations of the normality assumption on the random effects in model (1).

Weighted normal plots are illustrated here by an example involving a comparison of stomach surgery rates in 21 regions of Vermont (Miao [3]). For each of the 21

regions, a logistic model was used to provide estimates $\hat{\lambda}_i$ of age-adjusted surgery rates in each region. To smooth these estimators and to model the interregional variation, model (1) was applied using the estimators $\hat{\lambda}_i$ as Y_i and the estimated sampling variance of each $\hat{\lambda}_i$ as σ_i^2. The corresponding weighted normal plot, displayed in Fig. 1, suggests that model (1) fitted quite well. For this example, the factor $(1 + v/m^2)$ was 1.12.

References

[1] Blom, G. (1958). *Statistical Estimates and Transformed Beta Variables*. Wiley, New York.

[2] Dempster, A. P. and Ryan, L. M. (1985). *J. Amer. Statist. Ass.*, **80**, 845–850.

[3] Miao, L. L. (1977). An Empirical Bayes Approach to Analysis of Interarea Variation: A study of Vermont Surgical Mortality Rates, Ph.D. Thesis, Statistics Department, Harvard University, Cambridge, MA.

(HALF-NORMAL PLOTS
PROBABILITY PLOTS)

LOUISE M. RYAN

WEIGHTED QUASILINEAR MEAN

Fishburn [1] has defined the *quasilinear mean* of a random variable X with CDF $F_X(x)$, relative to a function $f(x)$, as

$$f^{-1}(E[f(X)]) = f^{-1}\left(\int_{-\infty}^{\infty} f(x)\,dF(x)\right).$$

He also defines the *weighted quasilinear mean* relative to $f(x)$, with weight function $g(x)$, as

$$f^{-1}(E[f(x)g(x)]/E[g(x)])$$
$$= f^{-1}\left(\frac{\int_{-\infty}^{\infty} f(x)g(x)\,dF(x)}{\int_{-\infty}^{\infty} g(x)\,dF(x)}\right).$$

These concepts have applications in utility theory*.

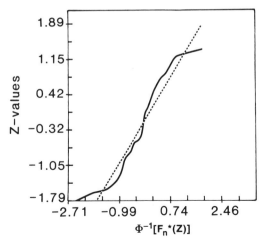

Figure 1 Weighted normal plot of stomach surgery rates in Vermont.

Reference

[1] Fishburn, P. C. (1986). *Econometrica*, **54**, 1197–1205.

(GEOMETRIC MEAN
UTILITY THEORY)

WEIGHTED SYMMETRY

A random variable X is said to be *symmetric* about the value θ if

$$P[X > \theta + t] = P[X < \theta - t],$$

for every $t > 0$. Assuming a density (or probability function) exists, it must therefore satisfy

$$f(\theta + t) = f(\theta - t),$$

for every $t > 0$. Thus symmetry around θ means the density is the same t units above θ as it is t units below θ. Weighted symmetry extends this characterization to cover cases in which the probability above θ is $\lambda (> 0)$ times the probability below θ. Therefore

$$f(\theta + t) = \lambda f(\theta - t),$$

for every $t > 0$, that is, the density (probability function) t units above θ is the same as λ times the density (probability function) t units below θ. Thus the density above and below θ has the same shape but potentially differs in total probability content. More generally, X is *weighted symmetric* around θ if for some $\lambda > 0$,

$$P(X > \theta + t) = \lambda P(X < \theta - t),$$

for every $t > 0$. See Wolfe [2] for tests using weighted symmetry. Ref. 1 contains an interesting application.

References

[1] Parent, E. A. (1965). *Tech. Rep. No. 80*, Dept. of Statistics, Stanford University. (Uses weighted symmetry to perform a sequential signed rank test.)

[2] Wolfe, D. A. (1974). *J. Amer. Statist. Ass.*, **69**, 819–822. (Characterizes weighted symmetry and uses these characterizations to perform tests of hypotheses.)

Ronald H. Randles

WEILER'S INDEX OF GOODNESS OF FIT

Weiler [3] developed an "index of discrepancy," which measures the deviation of a distribution of frequencies from a hypothetical one.

Suppose a random event can have s possible outcomes $A_1, \ldots, A_s$ with unknown probabilities $p_1, \ldots, p_s$, respectively ($\Sigma_{j=1}^s p_j = 1$). To test the hypothesis

$$H_0: p_j = \pi_j, \qquad j = 1, \ldots, s,$$

using a random sample of size n with A_j observed n_j times ($j = 1, \ldots, s$; $\Sigma_{j=1}^s n_j = n$), we can use the *sample* version of Weiler's index, which is

$$\hat{\phi}^2 = \frac{\pi_0}{1 - \pi_0} \cdot \frac{X^2}{n},$$

where $\pi_0 = \min(\pi_1, \ldots, \pi_s)$ and

$$X^2 = \sum_{j=1}^s (n\pi_j)^{-1} (n_j - n\pi_j)^2$$

is the familiar chi-squared* statistic.

The sample version is a (positively biased) estimator of the *population* Weiler index

$$\phi^2 = \frac{\pi_0}{1 - \pi_0} \sum_{j=1}^s \frac{(p_j - \pi_j)^2}{\pi_j}.$$

The modified sample version

$$\hat{\phi}_1^2 = \frac{\pi_0}{1 - \pi_0} \left(X^2 + 1 - \sum_{j=1}^s \frac{n_j}{n\pi_j} \right) \frac{1}{n - 1}$$

is an unbiased estimator of ϕ^2, but it can take negative values.

In the special case $\pi_1 = \pi_2 = \cdots = \pi_s = s^{-1} (= \pi_0)$, we have

$$\phi^2 = \{ n(s - 1) \}^{-1} X^2,$$

which is closely related to Simpson's diversity index D (*see* DIVERSITY INDICES). [The index $I = s(s - 1)^{-1}D$ was introduced independently by Mueller and Schuessler [2].]

Agresti and Agresti [1] give a survey of these and related indices.

References

[1] Agresti, A. and Agresti, B. F. (1978). In *Sociological Methodology*, K. F. Schuessler, ed. Jossey-Bass, San Francisco, pp. 204–237.

[2] Mueller, J. H. and Schuessler, K. F. (1961). *Statistical Reasoning in Sociology*. Houghton Mifflin, Boston.

[3] Weiler, H. (1966). *Technometrics*, **8**, 327–334.

(CONTINGENCY TABLES
DIVERSITY INDICES
GOODNESS OF FIT
MULTINOMIAL DISTRIBUTION
SOCIOLOGY, STATISTICS IN
SURPRISE INDEX)

WEISS TEST OF INDEPENDENCE (OF VARIABLES)

A familiar example of testing independence is a contingency table*, in which we test the independence of two or more criteria of classification, each containing a finite number of classes. The classes in the table are sometimes formed by grouping continuous random variables. The boundaries of these classes must be nonrandom for the usual statistical analysis to be valid. But in testing the independence of a set of continuous random variables, whose joint distribution is unknown, using nonrandom boundaries (that is, boundaries set before observations are taken) can result in a situation where it is impossible to carry out a meaningful analysis. For example, every observed point might fall in just one of the cells of the contingency table, and then no light would be shed on whether or not the variables are independent. In the tests described below, we use order statistics* as boundaries. Also, the number of cells may increase as the sample size increases. In deciding how many cells there should be, there are two conflicting considerations. The larger the number of cells, the smaller is the amount of information lost due to the grouping. But if the number of cells is too large, the joint distribution of the cell frequencies will no longer

approach a normal distribution as the number of observations increases, and the analysis becomes too complicated to be of any practical use.

We use the following notation. n is a positive integer, and we observe n independent and identically distributed s-tuples $(X_1(i), \ldots, X_s(i); \ i = 1, \ldots, n)$ of continuous random variables. The joint distribution of the s components is unknown, and the problem is to test the hypothesis that the s components are mutually independent. K_n and L_n are positive integers, with

$$n/(K_n + 2) < L_n,$$

$$\lim_{n \to \infty} \{L_n - n/(K_n + 2)\} = \lim_{n \to \infty} K_n^{3s}/n = 0.$$

(Note that these assumptions allow K_n to approach infinity as n increases, but do not require it.) Let $Y_j(1) < \cdots < Y_j(n)$ denote the ordered values of $X_j(1), \ldots, X_j(n)$ and define $Y_j(0)$ as $-\infty$ and $Y_j((K_n + 2)L_n)$ as $+\infty$. If $h_1, \ldots, h_s$ are integers, each between 1 and $K_n + 2$ inclusive, we define $N_n(h_1, \ldots, h_s)$ as the number of values of i for which the s-tuple $(X_1(i), \ldots, X_s(i))$ satisfies the conditions $(Y_j((h_j - 1)L_n) < X_j(i) < Y_j(h_j L_n))$ for $j = 1, \ldots, s$. The $(K_n + 2)^s$ quantities $(N_n(h_1, \ldots, h_s))$ are analogous to the cell frequencies in an s-dimensional contingency table, and they will be used to construct tests of independence.

Let $F_j(x)$ denote the marginal cumulative distribution function for $X_j(i)$. The values of the $(K_n + 2)^s$ quantities $(N_n(h_1, \ldots, h_s))$ would not be affected if $X_j(i)$ were replaced by $F_j(X_j(i))$ for $j = 1, \ldots, s$ and $i = 1, \ldots, n$. Of course, $F_j(x)$ is unknown, so we cannot actually replace $X_j(i)$ by $F_j(X_j(i))$, but for the analysis below it will be useful to imagine that this has been done. Then we can assume that the marginal distribution of $X_j(i)$ is the uniform distribution over $(0, 1)$, and we can write the joint probability density function of the s components of $(X_1(i), \ldots, X_s(i))$ as $1 + r(x_1, \ldots, x_s)$ over the s-dimensional unit cube C_s, and 0 outside C_s, where

$$\eta = \int \cdots \int_{C_s} r(x_1, \ldots, x_s)\, dx_1 \cdots dx_s = 0.$$

Then the hypothesis of independence is equivalent to the hypothesis that $r(x_1, \ldots, x_s) = 0$ for all $(x_1, \ldots, x_s)$ in C_s.

A statistic analogous to the usual statistic used for testing independence in a contingency table is

$$n^{-1}\left[(K_n + 2)^s \sum_{h_s=1}^{K_n+2} \cdots \right.$$

$$\left. \times \sum_{h_1=1}^{K_n+2} \left\{ N_n(h_1, \ldots, h_s) - n(K_n + 2)^{-s} \right\}^2 \right],$$

which we denote by W_n. Assuming that K_n approaches infinity as n increases, subject to the restrictions on K_n given above, the distribution of W_n is approximately a noncentral chi-square distribution* with $(K_n + 1)^s$ degrees of freedom and noncentrality parameter

$$\lambda = n \int \cdots \int_{C_s} r^2(x_1, \ldots, x_s)\, dx_1 \cdots dx_s.$$

If the desired level of significance is α, the hypothesis of independence is rejected if W_n is greater than the $(1 - \alpha)$th quantile of the central chi-square distribution* with $(K_n + 1)^s$ degrees of freedom. This test is consistent against any alternative for which $\lambda = 0$.

We can find much better tests than the one based on W_n, if we are willing to make some mild assumptions about the joint distribution of $(X_1(i), \ldots, X_s(i))$. Since the test based on W_n is consistent against any fixed alternative, to describe the improvement we use the familiar device of letting the alternative approach the hypothesis being tested as the sample size n increases. Specifically, we assume that the joint probability density function of the s components of $(X_1(i), \ldots, X_s(i))$ is $1 + n^{-1/2}r(x_1, \ldots, x_s)$ over C_s, with $\eta = 0$ and $\lambda < \infty$. The test based on W_n is useless against such alternatives: The power of the test approaches the level of significance as n increases.

We expand $r(x_1, \ldots, x_s)$ as a Fourier cosine series over C_s,

$$\sum_{j_1=0}^{\infty} \cdots \sum_{j_s=0}^{\infty} A(j_1, \ldots, j_s) 2^{D(j_1, \ldots, j_s)}$$

$$\times \prod_{i=1}^{s} \cos(j_i \Pi x_i),$$

where $D(j_1, \ldots, j_s)$ is equal to one-half the number of positive integers among $j_1, \ldots, j_s$. Since $\eta = 0$, it follows that $A(0, \ldots, 0) = 0$. Then the hypothesis of independence is equivalent to the hypothesis that $A(j_1, \ldots, j_s) = 0$ for every s-tuple $(j_1, \ldots, j_s)$ of nonnegative integers containing at least one positive integer. We can estimate $A(j_1, \ldots, j_s)$ by the following function of the quantities $(N_n(h_1, \ldots, h_s))$:

$$n^{1/2} 2^{D(j_1, \ldots, j_s)}$$

$$\times \sum_{i_s=1}^{K_n+2} \cdots \sum_{i_1=1}^{K_n+2} \left\{ N_n(i_1, \ldots, i_s) - n/(K_n + 2)^s \right\}$$

$$\times \prod_{b=1}^{s} \left(\cos \pi j_b \left(i_b - \frac{1}{2} \right) n^{-1} L_n \right).$$

Denote this estimator by $A_n^*(j_1, \ldots, j_s)$.

It is shown in [4] that if $(j_1^{(i)}, \ldots, j_s^{(i)};\ i = 1, \ldots, m)$ are m different sets of nonnegative integers, each set containing at least one positive integer, then the joint distribution of $(A_n^*(j_1^{(i)}, \ldots, j_s^{(i)});\ i = 1, \ldots, m)$ approaches the distribution of m independent normal random variables, each with unit variance and with respective means $(A(j_1^{(i)}, \ldots, j_s^{(i)});\ i = 1, \ldots, m)$, as n increases. This fact can be used to construct a variety of tests of the hypothesis of independence. For example, from the theory of Fourier series*, if $r(x_1, \ldots, x_s)$ has partial derivatives that are bounded in absolute value, then $A(j_1, \ldots, j_s)$ will be small in absolute value if at least one of the s values $j_1, \ldots, j_s$ is large. Suppose there is a positive integer T such that we are willing to assume that $A(j_1, \ldots, j_s)$ is close to 0 if any of the values $j_1, \ldots, j_s$ is above T. Define

$$V(T) = \sum_{j_1=0}^{T} \cdots \sum_{j_s=0}^{T} \left(A_n^*(j_1, \ldots, j_s) \right)^2,$$

where $A_n^*(0, \ldots, 0)$ is defined to be 0. If n is large, the distribution of $V(T)$ is approximately noncentral chi-square* with $(T + 1)^s - 1$ degrees of freedom and noncentrality parameter

$$\sum_{j_1=0}^{T} \cdots \sum_{j_s=0}^{T} \left(A(j_1, \ldots, j_s) \right)^2.$$

If the desired level of significance is α, the

hypothesis of independence is rejected if $V(T)$ is greater than the $(1 - \alpha)$th quantile of the central chi-square distribution with $(T + 1)^s - 1$ degrees of freedom.

The test based on $V(T)$ is an all-purpose test, in the sense that it has reasonably good power against the very wide class of alternatives for which the noncentrality parameter ζ is a respectable distance above 0. Suppose we are particularly concerned about an alternative to the hypothesis for which $r(x_1, \ldots, x_s)$ has known Fourier coefficients $(\bar{A}(j_1, \ldots, j_s); \ j_i = 0, \ldots, T)$. Define

$$R(T) = \sum_{j_1 = 0}^{T} \cdots \sum_{j_s = 0}^{T} \bar{A}(j_1, \ldots, j_s)$$
$$\times A_n^*(j_1, \ldots, j_s).$$

The optimal test against this particular alternative is to reject the hypothesis if $R(T)$ is greater than

$$\sqrt{\sum_{j_1 = 0}^{T} \cdots \sum_{j_s = 0}^{T} \left(\bar{A}(j_1, \ldots, j_s) \right)^2 \, G^{-1}(1 - \alpha)},$$

where G denotes the standard normal cumulative distribution function.

We can combine the special-purpose test based on $R(T)$ and the all-purpose test based on $V(T)$ into a single test, as follows: Reject the hypothesis if either $V(T) > c_1$ or $R(T) > c_2$, where c_1 and c_2 are chosen to give the desired level of significance. By varying the relative values of c_1 and c_2, we can increase the power of the test against the special alternative at the expense of power against other alternatives, or vice versa. Details are given in [4].

Historical Note. There are many tests of independence of continuous random variables. Some of the better known ones are Hoeffding's test* for $s = 2$ [3], a general purpose test for general s based on the empirical distribution function* [2], and rank tests* for $s = 2$ which are optimal against specific alternatives [1].

References

[1] Bell, C. B. and Doksum, K. A. (1967). *Ann. Math. Statist.*, **38**, 429–446.

[2] Blum, J., Kiefer, J., and Rosenblatt, M. (1961). *Ann. Math. Statist.*, **32**, 485–498.

[3] Hoeffding, W. (1948). *Ann. Math. Statist.*, **19**, 546–557.

[4] Weiss, L. (1985). *Naval Res. Logist. Quart.*, **32**, 337–346.

(CONTINGENCY TABLES
DEPENDENCE, TESTS FOR
FOURIER COEFFICIENTS
HOEFFDING'S INDEPENDENCE TEST
HYPOTHESIS TESTING)

L. WEISS

WEISS-TYPE ESTIMATORS OF SHAPE PARAMETERS

The random variable X is said to have a three-parameter Weibull distribution* if its cumulative distribution function is $1 - \exp\{ -[(x - \alpha)/\beta]^{1/\gamma} \}$ if $x \geq \alpha$ and 0 if $x < \alpha$. Here β and γ are positive. α is known as the *location parameter*, β as the *scale parameter*, and γ as the *shape parameter*.

Suppose $X_1, \ldots, X_n$ are independent and identically distributed random variables, the common distribution being a Weibull distribution with all three parameters unknown. The problem of estimating the parameters is difficult, largely because the form of a good estimator of α changes greatly as γ crosses from below 1 to above 1. A consistent estimator of γ can be constructed as follows. Let t_1, t_2 be fixed values with $0 < t_1 < t_2 < 1$, and let $[A]$ denote the largest integer not greater than A. Let $X_1' < X_2' < \cdots < X_n'$ denote the ordered values of $X_1, X_2, \ldots, X_n$. Then

$$\frac{\log\left(X_{[nt_2]}' - X_1' \right) - \log\left(X_{[nt_1]}' - X_1' \right)}{\log \log\left(1/(1 - t_2) \right) - \log\log\left(1/(1 - t_1) \right)}$$

is a consistent estimator of γ.

Weiss [1] investigated the following problem. Suppose $X_1, \ldots, X_n$ are independent and identically distributed with common probability density function $f(x)$, given as follows:

$$f(x) = 0, \quad \text{if } x < \alpha,$$
$$f(x) = c(x - \alpha)^{1/\gamma - 1}[1 + r(x - \alpha)],$$
$$\text{for } x > \alpha,$$

where c, α, γ are unknown parameters with $c > 0$ and $\gamma > 0$; $r(y)$ is an unknown function except that we know that $|r(y)| \le Ky^A$ for all y in some interval $[0, \Delta]$; where K, A, and Δ are unknown positive values. That is, in a neighborhood of α, this density function is like a Weibull density, which itself satisfies these conditions. A wide class of densities, which also satisfies the conditions, consists of densities truncated* on the left at α. That is, $f(x) = g(x)/\int_\alpha^\infty g(x)\,dx$ if $x > \alpha$, and $f(x) = 0$ if $x < \alpha$, where $g(x)$ is a density satisfying mild conditions.

Let $U_1, U_2, \ldots$ denote independent random variables, each with CDF $1 - e^{-u}$ for $u \ge 0$. Let $\{k(n);\ n = 1, 2, \ldots\}$ be any sequence of positive integers such that $\lim_{n \to \infty} k(n) = \infty$ and $\lim_{n \to \infty} k(n)/n^\delta = 0$ for every $\delta > 0$. [For example, $k(n) = [\log n]$.] Then Weiss showed that for all asymptotic (as $n \to \infty$) probability calculations, we can represent X_i' as

$$\alpha + \left[\frac{1}{c\gamma n}(U_1 + \cdots + U_i)\right]^\gamma,$$

$$\text{for } i = 1, \ldots, k(n).$$

Based on this, Weiss showed that

$$\frac{\log\left(X_{k(n)}' - X_1'\right) - \log\left(X_{[k(n)/2]}' - X_1'\right)}{\log 2}$$

is a consistent estimator of γ. Weissman [2] investigated the consistent estimators of γ of the form

$$\frac{\log\left(X_{k(n)}' - X_1'\right) - \log\left(X_{m(n)}' - X_1'\right)}{\log\left(k(n)/m(n)\right)},$$

for all integers $m(n) < k(n)$, and showed that the value of $m(n)$ that minimizes the asymptotic variance of the estimator is $m(n) = [0.2032 k(n)]$. This choice gives an asymptotic variance about 25% lower than $m(n) = [\frac{1}{2}k(n)]$, this latter being the one used in the estimator constructed by Weiss.

The estimators constructed by Weiss and Weissman use only three order statistics. No doubt they could be improved by using more, but the computations become very difficult. Even in the simpler case when we are sampling from a Weibull distribution, the maximum likelihood* estimators are so complicated that they have not yet been satisfactorily analyzed.

References

[1] Weiss, L. (1971). *Naval Res. Logist. Quart.*, **18**, 111–114.

[2] Weissman, I. (1981). *Naval Res. Logist. Quart.*, **28**, 603–605.

(CONSISTENT
ORDER STATISTICS
WEIBULL DISTRIBUTION)

L. WEISS

WELCH TESTS

In the one-way* layout to compare the means of k normally distributed populatons, it may not be valid in some cases to assume homogeneous variances. Hence the ANOVA* F-test* is not applicable, and the Welch [19] test was proposed to fill this void. An important special case ($k = 2$) is the famous Behrens–Fisher* problem. This special case was solved by Welch [18] several years earlier than the general case. His solution for $k = 2$ was refined and tabled by Aspin [1, 2] and has become known as the Aspin–Welch test (AWT). Further tables were later provided by Trickett et al. [17]. Competing solutons to the Behrens–Fisher problem have been suggested by Fisher [8], Lee and Gurland [11] (denoted LG), Cochran [6], and Welch himself [18; 2, Appendix]. All these tests depend on normality, and Yuen [21] and Tiku and Singh [16] attempt more robust solutions. Some competing procedures for general k are due to Brown and Forsythe [5], James [9], and Bishop and Dudewicz [3]. The Welch and Brown-Forsythe tests have been extended by Roth [13] to the case where the k populations have a natural ordering (e.g., different dosages of the same drug) and a trend test* is desired to detect differences in the means that are monotone as a function of this ordering.

Another (unrelated) Welch [20] test was designed in mixed or random effects models to provide confidence intervals for variance components*, whose estimators are often distributed as linear combinations of chi-squared variates. Basically, Welch provides correction terms to the confidence limits obtained via the Satterthwaite [14, 15] approximation*, which is based on a single chi-squared variate. These corrections were long among the most widely advocated methods (see, e.g., Mendenhall [12, pp. 352–354]), with no perceived major drawbacks except tedious computations. However, Boardman's [4] simulations showed that the Welch corrections are actually detrimental to achieving the nominal confidence coefficients. Hence they have fallen justifiably into disfavor and will not be discussed further.

We now explore the details and properties of the Welch tests described in the first paragraph, and we begin with the AWT. The test statistic is

$$t' = \frac{\bar{x}_1 - \bar{x}_2}{\sqrt{s_1^2/n_1 + s_2^2/n_2}}, \qquad (1)$$

and Welch [18], for a slightly more general problem, derives the percentage points of t' as a power series in $1/f_i = 1/(n_i - 1)$ for $i = 1, 2$. The P fractile $(0 < P < 1)$ of t' is explicitly given to order $(1/f_i)^2$ by

$$\alpha \left[1 + \frac{(1 + \alpha)^2}{4} \frac{\Sigma\left(s_i^4/n_i^2 f_i\right)}{\left(\Sigma s_i^2/n_i\right)^2} \right.$$

$$+ \frac{(3 + 5\alpha^2 + \alpha^4)}{3} \frac{\Sigma\left(s_i^6/n_i^3 f_i^2\right)}{\left(\Sigma s_i^2/n_i\right)^3}$$

$$\left. - \frac{(15 + 32\alpha^2 + 9\alpha^4)}{32} \frac{\Sigma\left(s_i^4/n_i^2 f_i\right)^2}{\left(\Sigma s_i^2/n_i\right)^4} \right], \qquad (2)$$

where $\alpha = \Phi^{-1}(P)$ and Φ is the standard normal CDF. Note that the constant term in (2) reflects simply the normal approximation to t'.

Welch also suggests a method that refers to ordinary t-tables. This is done by equating the first two moments of t' to those of a t-distribution* with f degrees of freedom. The solution for f is

$$\frac{1}{f} = \frac{c^2}{f_1} + \frac{(1 - c)^2}{f_2}, \qquad (3)$$

where $c = (s_1^2/n_1)/(s_1^2/n_1 + s_2^2/n_2)$. Welch [18] originally suggested replacing f_i in (3) by $f_i + 2$, $i = 1, 2$, and blank, but he later repudiated this suggestion [2, Appendix]. He showed that critical values based on (3) agree with the correct ones based on (2) to order $(1/f_i)$, but they differ in the $(1/f_i)^2$ term.

The AWT is more powerful (i.e., has lower critical values) to varying degrees, and hence gives narrower confidence intervals, in general, than the essentially Bayesian* Behrens–Fisher solution, the Cochran method, or the Welch test based on (3); the latter two are widely used due to the computational simplicity of referring to ordinary t-tables, (3) being far more accurate. This accuracy is evaluated from the tables of LG [11], whose general method for this whole class of size and power calculations revealed that the AWT operates closest by far to the nominal level from among a set of seven competing tests. LG then proposed their own test, which is almost identical to the AWT in both size and power, and recommended it on the grounds of greater simplicity. However, it requires five constants that depend on the sample sizes and the nominal level, which are provided only for $5 \leqslant n_1 \leqslant n_2 \leqslant 10$ at the 0.05 level.

The Welch [19] test for general k compares the statistic

$$W^* = \frac{\Sigma w_j\left(\bar{x}_j - \hat{\mu}\right)^2/(k - 1)}{1 + [2(k - 2)/(k^2 - 1)]\Sigma h_j}, \qquad (4)$$

to the $F(k - 1, f)$ distribution, where

$$w_j = n_j/s_j^2, \quad \hat{\mu} = \Sigma w_j x_j/W, \quad W = \Sigma w_j,$$

$$h_j = (1 - w_j/W)^2/(n_j - 1),$$

$$f = (k^2 - 1)/(3\Sigma h_j).$$

It and the Brown–Forsythe [5] test both reduce to the Welch test based on (3) when

$k = 2$. The derivation of W^*, like that of the AWT, stems from a power series in $(1/f_i)$. Welch shows that W^* agrees to order $1/f_i$, but not to order $(1/f_i)^2$, with the James [9] test, which is based on a chi-squared (not F) approximation. Brown and Forsythe [5] demonstrate via simulations that, in general, their procedure and W^* both outperform the James test; furthermore, W^* tends to be better than their procedure when extreme means are associated with small variances, and vice versa. Importantly, both procedures lose little power in the equal variance case relative to the "optimal" ANOVA F-test, which is hence NOT recommended for the one-way layout. Dykstra and Werter [7] refine the James test and claim from their simulations that this refinement is on balance superior to the other tests; however, their numerical tables seem to support this conclusion only mildly when $k = 6$ and not at all when $k = 4$. In any case, the Welch test is quite competitive. Incidentally, Johansen [10] rederives the Welch test as a special case of a more general result on residuals* from a weighted linear regression*.

Roth's [13] extension of W^* to the Welch trend test (WT) for ordered populations is basically obtained by first amalgamating the population means using isotonic regression* for simple order with weights $w_j = n_j/s_j^2$. Conditionally on the results of the amalgamation process, the statistic W^* (when applied to the amalgamated populations) is multiplied by an appropriate constant so that its conditional distribution is similar to that of $\bar{E}^2$, which is the trend analog of the ANOVA F-test. Roth also developed the Brown–Forsythe trend test (BFT), and his simulations showed that WT is generally (but by no means uniformly) the better of the two, tending to have larger type I error rates but compensating for this with gains in power too great to be explained merely by the differences in level. Conditions under which WT is superior to BFT (and vice versa) are analogous to the above-mentioned findings of Brown and Forsythe in the nontrend situation. Analogously as well, the $\bar{E}^2$-test does not seem to gain much power (and

hence is not recommended) even when the variances are equal, unless the sample sizes are as small as 2 or 3.

References

[1] Aspin, A. A. (1948). *Biometrika*, **35**, 88–96. (Refines the AWT and tables some critical values.)

[2] Aspin, A. A. (1949). *Biometrika*, **36**, 290–296. [Contains an appendix by Welch commenting on Aspin's work and proposing the test based on (3).]

[3] Bishop, T. A. and Dudewicz, E. J. (1978). *Technometrics*, **20**, 419–430.

[4] Boardman, T. J. (1974). *Biometrics*, **30**, 251–262. (Simulations and references involving 12 procedures for confidence intervals for variance components.)

[5] Brown, M. B. and Forsythe, A. B. (1974). *Technometrics*, **16**, 385–389.

[6] Cochran, W. G. (1964). *Biometrics*, **20**, 191–195. (Points out important drawbacks to his own procedure.)

[7] Dykstra, J. B. and Werter, P. S. P. J. (1981). *Commun. Statist. B*, **10**, 557–569.

[8] Fisher, R. A. (1941). *Ann. Eugen., Lond.*, **11**, 141–172. [The original fiducial (essentially Bayesian) solution to the Behrens–Fisher problem.]

[9] James, G. S. (1951). *Biometrika*, **38**, 324–329.

[10] Johansen, S. (1980). *Biometrika*, **67**, 85–92.

[11] Lee, A. F. S. and Gurland, J. (1975). *J. Amer. Statist. Ass.*, **70**, 933–941. [Extensive bibliography on the entire subject. Also outlines general method for computing size and power of Welch-type tests (numerically tabled for several tests), and proposes a new test for $k = 2$.]

[12] Mendenhall, W. (1968). *Introduction to Linear Models and the Design and Analysis of Experiments*. Wadsworth, Belmont, CA.

[13] Roth, A. J. (1983). *J. Amer. Statist. Ass.*, **78**, 972–980. (Defines WT and BFT with detailed numerical examples.)

[14] Satterthwaite, F. E. (1941). *Psychometrika*, **6**, 309–316.

[15] Satterthwaite, F. E. (1946). *Biometrics Bull.*, **2**, 110–114.

[16] Tiku, M. L. and Singh, M. (1981). *Commun. Statist. A*, **10**, 2057–2071. (Robust statistic for $k = 2$ that outperforms Yuen's.)

[17] Trickett, W. H., Welch, B. L., and James, G. S. (1956). *Biometrika*, **43**, 203–205. (More tables of critical values for the AWT.)

[18] Welch, B. L. (1947). *Biometrika*, **34**, 28–35. [Proposes both the AWT and the test based on (3) with f_i replaced by $f_i + 2$.]

[19] Welch, B. L. (1951). *Biometrika*, **38**, 330–336. (Welch's k-sample test.)

[20] Welch, B. L. (1956). *J. Amer. Statist. Ass.*, **51**, 132–148.

[21] Yuen, K. K. (1974). *Biometrika*, **61**, 165–170. [Obtains level of Welch test based on (3) for many nonnormal distributions and proposes a more robust statistic.]

Editorial Note

In more recent papers, Aucamp (1986) (*J. Statist. Comp. Simul.*, **24**, 33–46) proposes the critical region

$$|t'| > z_{1-\alpha/2} \left[1 + 2\hat{C}^2 f_1^{-1} + 2(1 - \hat{C})^2 f_2^{-1} \right]^{1/2},$$

with $\Phi(z_{1-\alpha/2}) = 1 - \alpha/2$ and

$$\hat{C} = \frac{s_1^2}{n_1} \left[\frac{s_1^2}{n_1} + \frac{s_2^2}{n_2} \right]^{-1},$$

and Matuszewski and Sotres (1986) (*Comp. Statist. Data Anal.*, **3**, 241–249) propose rejection of the null hypothesis if the 80% confidence intervals for the two individual means do not overlap—giving a significance level of approximately 5%.

(BEHRENS–FISHER PROBLEM
ISOTONIC INFERENCE
TREND
TREND TESTS)

ARTHUR J. ROTH

WELCH'S v-CRITERION (MULTIVARIATE EXTENSION)

Suppose we have $n = \sum_{t=1}^{k} n_t$ p-variate observations, classified into k groups. Denoting the random variable corresponding to the jth observation in the tth group by the $p \times 1$ column vector $\mathbf{X}_{tj}$, the assumed model is

$$\mathbf{X}_{tj} = \boldsymbol{\mu} + \boldsymbol{\alpha}_t + \mathbf{E}_{tj} \cdot,$$

$(t = 1, \ldots, k; \ j = 1, \ldots, n_t)$, where $\boldsymbol{\mu}$ and $\boldsymbol{\alpha}_t$ are each constant $p \times 1$ vectors, such that

$$\sum_{t=1}^{k} n_t \boldsymbol{\alpha}_t = \mathbf{0},$$

and $\mathbf{E}_{tj}$ is a random $p \times 1$ error vector. The joint distribution of the p-variates of $\mathbf{E}_{tj}$ does not depend on j and the expected value of each is 0. The common distribution

is often assumed to be multinormal*, and the $\mathbf{E}_{tj}$'s are assumed mutually independent.

In multivariate analysis of variance* (MANOVA), the null hypothesis (of no difference among group means)

$$H_0 : \boldsymbol{\alpha}_1 = \boldsymbol{\alpha}_2 = \cdots = \boldsymbol{\alpha}_k = \mathbf{0} \quad (1)$$

is tested against the class of alternative hypotheses that at least one of the equalities in (1) is violated.

For a general model with $p > 1$ and $k > 2$ and with the variance–covariance matrices of the distributions of the $\mathbf{E}_{tj}$'s unknown, James [1] proposed the test statistic

$$T_v^2 = \sum_{t=1}^{k} (\overline{\mathbf{X}}_t - \overline{\mathbf{X}})' \mathbf{W}_t (\overline{\mathbf{X}}_t - \overline{\mathbf{X}}),$$

where

$$\overline{\mathbf{X}}_t = n_t^{-1} \sum_{j=1}^{n_t} \mathbf{X}_{tj},$$

$$\mathbf{S}_t = (n_t - 1)^{-1} \sum_{j=1}^{n_t} (\mathbf{X}_{tj} - \overline{\mathbf{X}}_t)(\mathbf{X}_{tj} - \overline{\mathbf{X}}_t)',$$

$$\mathbf{W}_t = (n_t^{-1} \mathbf{S}_t)^{-1}, \qquad \mathbf{W} = \sum_{t=1}^{k} \mathbf{W}_t,$$

$$\overline{\mathbf{X}} = \mathbf{W}^{-1} \sum_{t=1}^{k} \mathbf{W}_t \overline{\mathbf{X}}_t.$$

See related entries and James [1] for further details.

Reference

[1] James, G. S. (1954). *Biometrika*, **41**, 19–43.

(BEHRENS–FISHER PROBLEM
MULTIVARIATE ANALYSIS OF VARIANCE
WALD'S W-STATISTICS
WELCH TESTS)

WELL-CALIBRATED FORECASTS

Consider a forecaster who at the beginning of each period n in a sequential* process $n = 1, 2, \ldots$ must specify the probability that some particular event A_n will occur during that period. It is assumed that when the forecaster specifies the probability of A_n, he or she is aware of the values of various variables which may be relevant to the occurrence of A_n. In particular, the knowledge of which of the previous events $A_1, \ldots, A_{n-1}$ actually occurred will usually be part of the forecaster's information at the beginning of period n.

For example, the forecaster might be a meteorologist, who at the beginning of each day must specify the probability that it will rain during that day at a particular location, or an economist, who at the beginning of each weekly or quarterly period must specify the probability that a particular interest rate or stock market average will rise during the period. In another context, the forecaster might be a medical diagnostician, who must specify the probability that a patient has a particular disorder on the basis of an examination and who subsequently learns the patient's true condition. *See* METEOROLOGY, STATISTICS IN and MEDICAL DIAGNOSIS, STATISTICS IN.

In this article the discussion will be presented in the context of a weather forecaster, who day after day must specify the probability x that there will be at least a certain amount of rain at some given location during a specified time interval of the day. We refer to the occurrence of this well-specified event simply as "rain." The probability x specified by the forecaster on any particular day is called the prediction for that day. To begin, we make the assumption that x is restricted to a given finite set of values $0 = x_0 < x_1 < \cdots < x_k = 1$, and let $X = \{x_0, \ldots, x_k\}$ denote the set of possible predictions. The forecaster's predictions might, in general, be based on mathematical or stochastic models of the weather, computer simulations, subjective judgments, or a combination of these methods.

The notion of the *calibration** of a forecaster's predictions pertains to a comparison of the predicted probabilities with the actual outcomes of the events being predicted. Consider the forecaster's predictions over a large number of days, and for each value $x_j \in X$, let $\rho(x_j)$ denote the proportion of

days on which it actually rained among all those days for which the prediction was x_j. The function $\rho(x_j)$ is called the forecaster's *calibration curve*. Loosely speaking, the forecaster is said to be *well calibrated* if $\rho(x_j)$ is approximately equal to x_j for each $x_j \in X$. In other words, the forecaster is well calibrated if among all those days for which the predicted probability of rain was x_j, the proportion of days on which it actually rained was approximately x_j.

In evaluating a forecaster's performance, it is usually regarded as desirable that he or she be well calibrated. However, as the following argument indicates, it is typically not difficult for the forecaster to make sure that this criterion will be satisfied. Suppose, for example, that the forecaster calculates a calibration curve $\rho(x_j)$ after a certain number of days, and notes that $\rho(x_j) < x_j$ for some particular value of x_j. Then, to increase the value of $\rho(x_j)$, the forecaster need only wait until there is a day when he or she is almost certain that it will rain, and to state the value $x = x_j$ as the prediction for that day. Similarly, if $\rho(x_j) > x_j$ for some value of x_j, then the forecaster can decrease the value of $\rho(x_j)$ by making the prediction $x = x_j$ on days when he or she is almost certain that it will not rain. In this way, by stating predictions other than actual subjective probabilities of rain, the forecaster can usually manipulate $\rho(x_j)$ so that it is approximately equal to x_j for all values of $x_j \in X$. These considerations throw doubt on the usefulness, and even on the meaning, of the forecaster's predictions when the property of being well calibrated is regarded as a primary criterion.

Even if a forecaster's "honest" subjective probabilities make him or her well calibrated, the predictions may not be very useful. Consider, for example, the extreme case of a forecaster who knows nothing about meteorology and simply states the same prediction $x = \mu$ every day, where μ is the long-run relative frequency of rain. Then the forecaster will be well calibrated, but the predictions will obviously be useless to anyone who already knows the value of μ. At the other extreme is a forecaster whose prediction each day is either $x = 0$ or $x = 1$, and who is always correct. That forecaster is also well calibrated and the predictions convey perfect information about whether or not it will rain each day. Methods of comparing and evaluating forecasters based on concepts of *refinement* and *scoring rules* are presented by DeGroot and Fienberg [3, 4] and DeGroot and Eriksson [2]. Empirical studies of calibration are described by Murphy and Winkler [9] and Lichtenstein et al. [6].

For any fixed finite number of days, the property of being well calibrated can only be stated, as we have done, in terms of approximate equality between $\rho(x_j)$ and x_j. Consider now an infinite sequence of days, and let X^* denote the subset of X containing the values of x_j for which the prediction $x = x_j$ is made on an infinite number of days in the sequence. For each value of $x_j \in X^*$, let $\rho(x_j)$ denote the limiting relative frequency of rain among all those days for which the prediction was $x = x_j$. In this context, a forecaster is said to be well calibrated if $\rho(x_j) = x_j$ for each value of $x_j \in X^*$.

A more stringent definition of being well calibrated over an infinite sequence of days, but one that is more general in the sense that it does not require that the set X of possible predictions be finite, has been given by Dawid [1]. His approach is to select an infinite subsequence of days and to compare the limiting relative frequency of rain over this subsequence with the limit of the average of the predicted probabilities for the subsequence. More precisely, let ρ denote the limiting relative frequency of rain and let

$$\pi = \lim_{k \to \infty} \frac{1}{k} \sum_{i=1}^{k} \pi_{n_i},$$

where π_{n_i} is the prediction made on the ith day of the subsequence. Then the forecaster is said to be well calibrated if $\rho = \pi$ for every subsequence that is selected in accordance with certain conditions. The selection of a subsequence is constrained only by the requirement that the decision to include or not include any particular day in the subsequence must be based on the information that is available to the forecaster at the time that the prediction for that day must be made. Thus, the selection of days can be made sequentially, but the decision to include a particular day cannot be based on whether or not it actually rained on that day or on any other future observations. Under these conditions, Dawid shows that if the forecaster specifies a joint probability distribution of all of the variables that will be observed over the entire infinite sequence of days, and if the prediction made by the forecaster each day is the conditional probability of rain given all of the past data, as calculated from this joint distribution, then the forecaster will be well calibrated with probability 1 (where this probability is again calculated under the forecaster's joint distribution).

Another approach to the concept of calibration—one that is completely subjective—takes into account both the forecaster and a decision maker, who will learn (and presumably use) the forecaster's prediction. In this context, a forecaster is said to be *well calibrated for a particular decision maker* if, after learning the prediction x on any given day, the decision maker's subjective probability of rain on that day also becomes x (Lindley [7]). This approach has the advantage that it is based on individual days rather than on limiting frequencies over some hypothetical infinite sequence of "similar" days. However, it introduces a dependence between the forecaster and the decision maker that may impede its applicability (French [5]).

The usefulness of concepts of calibration in problems of combining expert opinion is discussed in Morris [8] and Schervish [10].

References

[1] Dawid, A. P. (1982). *J. Amer. Statist. Ass.*, **77**, 605–613.

[2] DeGroot, M. H. and Eriksson, E. A. (1985). In *Bayesian Statistics 2*, J. M. Bernardo, M. H. DeGroot, D. V. Lindley, and A. F. M. Smith, eds. Elsevier Science Publishers (North-Holland), Amsterdam, pp. 99–118.

[3]   DeGroot, M. H. and Fienberg, S. E. (1982). In *Statistical Decision Theory and Related Topics III*, S. S. Gupta and J. O. Berger, eds. Academic, New York, pp. 291–314.

[4]   DeGroot, M. H. and Fienberg, S. E. (1983). *The Statistician*, **32**, 12–22.

[5]   French, S. (1983). *Tech. Rep.* 114, Department of Decision Theory, University of Manchester, Manchester, England (unpublished).

[6]   Lichtenstein, S., Fischhoff, B., and Phillips, L. D. (1977). In *Decision Making and Change in Human Affairs*, H. Jungerman and G. de Zeeuw, eds. Reidel, Dordrecht, Netherlands, pp. 275–324.

[7]   Lindley, D. V. (1982). *J. R. Statist. Soc. A*, **145**, 117–126.

[8]   Morris, P. A. (1977). *Manag. Sci.*, **29**, 24–32.

[9]   Murphy, A. H. and Winkler, R. L. (1977). *J. R. Statist. Soc. Ser. C*, **26**, 41–47.

[10] Schervish, M. J. (1984). *Tech. Rep. 294*, Department of Statistics, Carnegie-Mellon University, Pittsburgh, PA (unpublished).

(BAYESIAN INFERENCE
CALIBRATION
DECISION THEORY
FORECASTING
PREDICTION)

MORRIS H. DEGROOT

WESTENBERG TEST OF DISPERSION

Although there is no single measure of the dispersion of a continuous probability distribution over a population of purely ordinal data* (not necessarily numerical), the dispersions of two such populations can be compared if their elements X and Y, respectively, can be put in a single order. We can then take, as a criterion of equality of dispersion, equality of the probabilities that X and Y lie between the quartiles* of the pooled populations. In order to test the null hypothesis H_0 that these probabilities are equal, given random samples (of sizes m and n) from the two populations, we construct a 2×2 frequency table (*see* TWO-BY-TWO TABLES). The columns relate to the categories $\tilde{q}_1 \leqslant Z < \tilde{q}_3$ and ($Z < \tilde{q}_1$ or $Z \geqslant \tilde{q}_3$), where $\tilde{q}_1$ and $\tilde{q}_3$ are the quartiles of the pooled data; the first and second rows relate to the

X's and Y's, respectively. We reject H_0 if the proportions in the two rows are sufficiently different, using the Fisher exact test* or, as an approximation, a χ^2 test* for significance.

Although the Westenberg test was designed primarily for purely ordinal data, it can be applied usefully to data measured on an interval scale, i.e., a scale such that differences can be ordered. Dispersion can then be measured by the interquartile range* (IQR), defined as $q_3 - q_1$. It can readily be seen that if the quartiles of one of the distributions lie between those of the other, H_0 is equivalent to equality of IQRs. This applies in particular if the distributions have a common median ($X = Y = 0$) and differ only in scale, so that X and Y/θ have the same distribution; H_0 is then the hypothesis $\theta = 1$.

If, however, the distributions differ in location in the sense that the quartiles of neither lie between those of the other, the test can reject H_0 even if the IQRs are equal. This unsatisfactory feature of the Westenberg test can sometimes be overcome by first changing one of the variates, say Y, using $Y + \Delta$ instead. This device is, however, not available for data that are no more than ordinal.

The test can readily be extended to more than two populations. Given a random sample from each, a frequency table is set up, with the two column categories defined as before, and with one row for each sample (*see* CATEGORICAL DATA). A χ^2 test for significance may be used.

The test was introduced by Westenberg [3] in 1948; Bradley [1] gives a much clearer exposition and a useful discussion. It is the simplest of a group of quasirank-sum tests applicable to nonnumerical ordinal data; the Ansari–Bradley test* and the Siegel–Tukey test* belong to this group. For any of these tests the pooled sample data have first to be ordered in the usual way. Quasiranks* are then assigned to them so that the lowest quasiranks go to data at the extremes of the natural order and the highest go to data at the centre. A sum of quasiranks (for the X's say) is used as test statistic. For the Westenberg test the quasiranks are all 0 or 1.

EXAMPLE

The tabulated X's and Y's are ordered samples from two populations; it may be assumed that they are measured on the same interval scale. We wish to test the null hypothesis H_0 that the dispersions of the two populations are equal against the alternative hypothesis that the Y population is more widely dispersed than the X population.

Westenberg

X	Y	Y'
60.0	75.0	62.8
47.6	67.9	55.7
47.0	67.0	54.8
47.0	60.1	47.9
45.8	58.5	46.3
45.2	57.5	45.3
43.3	50.7	38.5
43.1	47.5	35.3
40.4	44.7	32.5
38.2	43.1	30.9

Ansari-Bradley

X	Rank	Y''	Rank
60.0	2	62.5	1
47.6	$5\frac{1}{2}$	55.4	3
47.0	$7\frac{1}{2}$	54.5	4
47.0	$7\frac{1}{2}$	47.6	$5\frac{1}{2}$
45.8	10	46.0	9
45.2	10	45.0	9
43.3	8	38.2	$4\frac{1}{2}$
43.1	7	35.0	3
40.4	6	32.2	2
38.2	$4\frac{1}{2}$	30.6	1

If we use the Westenberg test we must first replace the variate Y by $Y' = Y - 12.2$ in order to bring the midpoints of the interquartile ranges of the X's and the Y's into coincidence. It can then be seen that the number of X's between the quartiles of the pooled data is 8. Accordingly the relevant frequency table is

8	2	10
2	8	10
10	10	20

Reference to a table of critical values for the

Fisher exact test* shows that H_0 can be rejected at significance level 0.025 but not 0.01.

If, for comparison, we use the Ansari–Bradley test*, we must first change the variate Y to $Y'' = Y - 12.5$ in order to bring the medians of the X's and the Y's into coincidence. Quasiranks are then assigned as shown in the table; average ranks are given to ties. The sum W of the quasiranks of the X's is 68. Reference to a table of upper-tail probabilities for the null distribution of W (e.g., Table A6 in Hollander and Wolfe [2]) shows that the corresponding upper-tail probability is 0.0282. Accordingly H_0 can be rejected at significance level 0.05 but not 0.025.

References

[1] Bradley, J. V. (1968). *Distribution-Free Statistical Tests*. Prentice-Hall, Englewood Cliffs, N.J.

[2] Hollander, M. and Wolfe, D. A. (1973). *Nonparametric Statistical Methods*. Wiley, New York.

[3] Westenberg, J. (1948). *Proc. Kon. Nederl. Akad. Wetensch.*, **51**, 252–261.

(ANSARI–BRADLEY W-STATISTICS
CAPON TEST
CATEGORICAL DATA
DISTRIBUTION-FREE METHODS
F-TESTS
HOLLANDER EXTREME TEST
KLOTZ TEST
MOOD'S DISPERSION TEST
ORDINAL DATA
RANK TESTS
SCALE TESTS
SIEGEL–TUKEY TEST
TWO-BY-TWO TABLES)

F. C. POWELL

WEYL INEQUALITY

If $\mathbf{A}$ and $\mathbf{B}$ are symmetric $n \times n$ matrices with eigenvalues* (characteristic roots) $\lambda_1(\mathbf{A}) \geqslant \lambda_2(\mathbf{A}) \geqslant \cdots \geqslant \lambda_n(\mathbf{A})$ and $\lambda_1(\mathbf{B}) \geqslant \lambda_2(\mathbf{B}) \geqslant \cdots \geqslant \lambda_n(\mathbf{B})$, respectively, then

$$\lambda_i(\mathbf{A}) + \lambda_j(\mathbf{B}) \geqslant \lambda_{i+j-1}(\mathbf{A} + \mathbf{B}).$$

This is Weyl's inequality (Bellman [1]). Applications in the theory of multivariate analysis* are described in Seber [2].

References

[1] Bellman, R. (1960). *Introduction to Matrix Analysis*. McGraw-Hill, New York.

[2] Seber, G. A. F. (1984). *Multivariate Analysis*. Wiley, New York.

WHEELER AND WATSON'S TEST

Imagine samples from two unimodal distributions on a circle to be marked on the unit circle as n_1 O's and n_2 X's. The sum of the vectors to the O's and the sum of the vectors to the X's indicate the mean directions of the two distributions. To test whether these are the same without making a distributional assumption, J. L. Hodges suggested respacing the sample points $2\pi/(n_1 + n_2)$ apart and computing R_1^2, the squared length of the sum of the vectors in sample 1. On the null hypothesis that the two samples come from the same population, R_1^2 may be referred to the randomization distribution obtained by picking n_1 points at random from the $n_1 + n_2$ equally spaced points. Wheeler and Watson [1] give the mean and variance of R_1^2 in this distribution and the approximation $\exp\{-a/E[R_1^2]\}$ to the tail probabilities $\Pr[R_1^2 \geqslant a]$, which overestimates slightly. In any specific case, it is easy to do a simulation to get a precise estimate of $\Pr[R_1^2 \geqslant a]$.

Reference

[1] Wheeler, S. and Watson, G. S. (1964). A distribution-free two-sample test on a circle. *Biometrika*, **51**, 256–257.

(DIRECTIONAL DATA
DIRECTIONAL DISTRIBUTIONS)

G. S. WATSON

WHERRY CLEAN-UP METHOD

A procedure for correcting erroneous initial communality* estimates in factor analysis*,
suggested by Wherry [3]. The procedure entails searching for similarities between a row of residuals and a set of corresponding loadings for a given factor (or a mirror image thereof). When such a pattern is observed, the loading of that variable on the factor is changed and the amount of change is used as a multiplier of the loadings of other factors on that factor to reduce the size of the residuals. Wherry's method can be used as a shortcut for the Dwyer [1] extension model, which, in turn, is closely related to the minimum residual factor (MINRES*) method of Harman and Jones [2].

Further details are given in ref. 4.

References

[1] Dwyer, P. S. (1937). *Psychometrika*, **2**, 173–178.

[2] Harman, H. H. and Jones, W. H. (1966). *Psychometrika*, **34**, 351–368.

[3] Wherry, R. J. (1949). *Psychometrika*, **14**, 231–241.

[4] Wherry, R. J. (1984). *Contributions to Correlation Analysis*. Academic, New York.

(COMPONENT ANALYSIS
FACTOR ANALYSIS
STATISTICS IN PSYCHOLOGY)

WHITE NOISE *See* NOISE

WHITE TESTS OF MISSPECIFICATION

White tests of misspecification* are statistical tests designed to detect model misspecifications which invalidate the usual maximum likelihood* inference procedures. The theory of these tests is discussed in detail in White [6, 8]. Any test of the hypotheses H_0 or H_0' formulated here can be regarded as performing a White test (e.g., tests based on the Durbin–Watson statistic*, certain Lagrange multiplier statistics* for autocorrelation or heteroscedasticity*, or the standard skewness and kurtosis statistics). The conclusion to be drawn from rejection of these hypotheses is always that one has statistical evidence of a model misspecification which invalidates the usual maximum likelihood methods for inference, and which implies that the supposed maximum likelihood estimator is in fact inefficient and possibly inconsistent. This possible inconsistency can be investigated analytically or by other specification tests, such as the Hausman tests* (Hausman [2]) or Newey's conditional moment tests (Newey [5]; White [8]).

Let the joint likelihood function posited for a sample of data $\mathbf{X}^n = (X_1, \ldots, X_n)$ be denoted $f_n(\mathbf{X}^n, \boldsymbol{\theta})$, where $\boldsymbol{\theta}$ takes values in a parameter space $\Theta \subset \mathbb{R}^p$, and let $\hat{\boldsymbol{\theta}}_{\mathrm{ML}}$ be the maximum likelihood* estimator.

When the model is correctly specified, $\hat{\boldsymbol{\theta}}_{\mathrm{ML}}$ is generally a consistent estimator for the true parameters $\boldsymbol{\theta}^0$. This is generally not true otherwise, but see Gourieroux et al. [1], Levine [4], and White [7].

Two other consequences of correct specification are also important. Let the conditional likelihood of X_t given $\mathbf{X}^{t-1}$ be denoted $f_{t|t-1}(X_t|\mathbf{X}^{t-1}; \boldsymbol{\theta})$. The first consequence is that

$$E\left(\nabla \ln f_{t|t-1}^0 | \mathbf{X}^{t-1}\right) = \mathbf{0},$$

where we write the "scores" $\nabla \ln f_{t|t-1}^0 = \nabla \ln f_{t|t-1}(X_t|\mathbf{X}^{t-1}; \boldsymbol{\theta}^0)$ and $\nabla \equiv \partial/\partial \boldsymbol{\theta}$, so that

$$\mathbf{I}_n^0 = \mathbf{K}_n^0, \tag{1}$$

with

$$\mathbf{I}_n^0 = \mathbf{var}\left[n^{-1/2}\nabla \ln f_n(\mathbf{X}^n, \boldsymbol{\theta}^0)\right],$$

$$\mathbf{K}_n^0 = n^{-1}\sum_{t=1}^n E\left(\nabla \ln f_{t|t-1}^0 \nabla \ln f_{t|t-1}^{0'}\right).$$

The second is that

$$E\left(\nabla^2 \ln f_{t|t-1}^0 + \nabla \ln f_{t|t-1}^0 \nabla \ln f_{t|t-1}^{0'}\right) = \mathbf{0},$$

implying that

$$\mathbf{H}_n^0 = -\mathbf{K}_n^0, \tag{2}$$

$$\mathbf{H}_n^0 = E\left[n^{-1}\nabla^2 \ln f_n(\mathbf{X}^n, \boldsymbol{\theta}^0)\right].$$

Equations (1) and (2) are called *information matrix equalities*. Together, they justify the usual estimators for the asymptotic covariance matrix of $\hat{\boldsymbol{\theta}}_{\mathrm{ML}}$. Although the usual estimators are consistent when the model is correctly specified, they are not necessarily consistent otherwise. Thus, misspecification can adversely affect inference by invalidat-

ing the usual asymptotic covariance estimator for $\hat{\boldsymbol{\theta}}_{\mathrm{ML}}$.

The "White tests" detect such misspecifications using the information matrix equalities (1) and (2). For this reason they are also called *information matrix tests* for misspecification. Tests based on equation (1) are called *dynamic information matrix tests*.

Appropriate hypotheses based on (2) can be stated as

$$H_0: \mathbf{A}_0 \mathrm{vech}\left[\mathbf{H}_n^0 + \mathbf{K}_n^0\right] = \mathbf{0}$$

vs.

$$H_a: \mathbf{A}_0 \mathrm{vech}\left[\mathbf{H}_n^0 + \mathbf{K}_n^0\right] \neq \mathbf{0},$$

where $\mathbf{A}_0$ is any given $q \times p(p+1)/2$ matrix of real numbers, $q \leqslant p(p+1)/2$. Let the caret denote evaluation of a function at $\hat{\boldsymbol{\theta}}_{\mathrm{ML}}$. A statistic proposed by White [8] for testing H_0 vs. H_a is computed as nR^2 from the artificial regression in which the constant unity is regressed on the variables $\nabla \ln \hat{f}_{t|t-1}$ and

$$\mathbf{A}_0 \mathrm{vech}\left[\nabla^2 \ln \hat{f}_{t|t-1} + \nabla \ln \hat{f}_{t|t-1}' \nabla \ln \hat{f}_{t|t-1}\right],$$
$$t = 1, \ldots, n,$$

where R^2 is the (constant unadjusted) squared multiple correlation coefficient*. The statistic nR^2 has the χ_q^2 distribution approximately in large samples under H_0. The test is sensitive to misspecifications analogous to neglected heteroscedasticity* in the classical linear regression* framework. See White [6] for further discussion and examples.

The dynamic information matrix tests are formal tests of the hypothesis

$$H_0': \mathbf{A}_0 \mathrm{vec}\left[E\left(\nabla \ln f_{t|t-1}^{0'}\left[\nabla \ln f_{t-1|t-2}^0, \ldots, \right.\right.\right.$$
$$\left.\left.\left. \nabla \ln f_{t-\lambda|t-\lambda-1}^0\right]\right)\right] = \mathbf{0}$$

vs.

$$H_a': H_0' \text{ is false,}$$

where $\mathbf{A}_0$ is a given $q \times p^2\lambda$ matrix of real numbers, $q \leqslant p^2\lambda$, vec is the operator that stacks the elements of a $p \times r$ matrix into a $pr \times 1$ column vector, and λ is a given finite integer. This null hypothesis essentially allows one to look for autocorrelation in the scores of any order up to λ.

White [8] gives general conditions under which an asymptotic chi-square statistic can be computed as $(n - \lambda)R^2$ from the artificial regression in which the constant unity is regressed on explanatory variables $\nabla \ln \hat{f}_{t|t-1}$, and

$$\mathbf{A}_0 \text{vec} \big[\nabla \ln \hat{f}_{t|t-1}, \big(\nabla \ln \hat{f}_{t-1|t-2}, \dots,$$
$$\nabla \ln \hat{f}_{t-\lambda|t-\lambda-1} \big) \big], \; t = \lambda + 1, \dots, n.$$

As before, R^2 is the (constant unadjusted) squared multiple correlation coefficient. Under H_0', $(n - \lambda)R^2$ has the χ_q^2 distribution approximately in large samples. The test is sensitive to misspecifications analogous to neglected autocorrelation in the classical linear regression framework. See White [8] for additional discussion and a variety of examples.

The statistics just given are in fact merely convenient choices for testing H_0 or H_0'. Any other statistic which tests these hypotheses can form the basis for a White test (e.g., those mentioned at the outset). These alternative statistics may perform better than those just given in particular samples.

References

[1] Gourieroux, C., Monfort, A., and Trognon, A. (1984). *Econometrica*, **52**, 681–700 (Discusses the behavior of the maximum likelihood estimator when the distribution generating the data differs from the assumed likelihood function.)

[2] Hausman, J. (1978). *Econometrica*, **46**, 1251–1272. (Proposes a test for model misspecification based on the difference between two consistent estimators, one efficient and the other inefficient.)

[3] Lancaster, T. (1984). *Econometrica*, **52**, 1051–1055. (Discusses a procedure for simplifying computation of the White test.)

[4] Levine, D. (1983). *J. Econometrics*, **23**, 337–342. (Gives conditions under which dynamic misspecification does not lead to inconsistency of the maximum likelihood estimator.)

[5] Newey, W. (1985). *Econometrica*, **53**, 1047–1070. (Proposes a general framework for specification testing.)

[6] White, H. (1982). *Econometrica*, **50**, 1–26. (Discusses maximum likelihood estimation of misspecified models and proposes the information matrix test.)

[7] White, H. (1984). *Discussion Paper 84-3*, Department of Economics, University of California, San Diego, CA. (Discusses maximum likelihood estimation when the conditional mean is correctly specified, but other misspecification may be present.)

[8] White, H. (1985). Paper presented to Fifth World Congress, Econometric Society, Cambridge, Massachusetts (to appear). (Extends results of Newey [5] to the case of dependent observations and proposes the dynamic information matrix tests.)

(ECONOMETRICS
HAUSMAN'S SPECIFICATION TEST
MISSPECIFICATION, TESTS FOR)

HALBERT WHITE

WHITTAKER–HENDERSON GRADUATION FORMULAS

HOW THE PROBLEM OF GRADUATION ARISES

Man has always had a great desire to study the processes of nature. In the course of his study, he sometimes has the good fortune to discover a "law of nature," such as the law of gravity or the law of radioactive decay. These laws can be represented by mathematical formulas which adequately explain the processes observed. These formulas are smooth continuous functions.

The problem of graduation* arises when man is studying a process which he believes should behave in a smooth and continuous manner, but for which he has not been able to discover any "law" or formula that it follows. He must, therefore, rely upon empirical evidence to represent the process. If the process is based upon probabilities, then actual observations of the process will exhibit irregularities or chance fluctuations, often referred to as errors of observation.

A series of observed values u_x'', then, may be thought of as having two components. The first is a smooth series u_x that represents the underlying "law" that the process under study follows. The second component is a superimposed irregular series e_x that accounts for the irregularities appearing in the observed series. If we write $u''_x = u_x + e_x$ and we let G be a graduation operator, then our goal is to have $G(u''_x) = G(u_x) + G(e_x) = u_x$. We must realize, however, that the graduation will not be wholly successful in distinguishing between random errors and points of inflection in the process being studied.

FORMULATION OF THE GRADUATION OPERATOR

The goals of the graduation operator can be restated by describing the desired qualities of the result of the graduation. We choose qualities that can be quantified so that our analysis will lead to mathematical formulas. The two desired qualities are smoothness and fit. By *smoothness* we mean that we desire the graduate values to be smooth; by fit we mean that we desire the graduated values be close to the ungraduated values. The two qualities are basically inconsistent in the sense that there is a limit to how far one quality can be improved without making the other quality worse.

A test of smoothness that has given good results is a summation of the squares of some order of the finite differences* of the graduated series u_x. Normally, second or third differences are used; in this article we will use second differences. Therefore, the test of smoothness can be written as $S = \Sigma(\Delta^2 u_x)^2$, where the smaller the sum, the smoother the graduated series. ($\Delta u_x = u_{x+1} - u_x$ and $\Delta^2 u_x = u_{x+2} - 2u_{x+1} + u_x$.)

A test statistic that has given good results is a summation of the weighted squares of the deviation of the graduated values from the ungraduated values. Therefore, the statistic can be written as $F = \Sigma W_x(u''_x - u_x)^2$, where again the smaller the sum, the closer the fit.

Whittaker—Henderson Type B graduation formulas minimize the function

$$F + kS = \sum W_x(u''_x - u_x)^2 + k\sum(\Delta^2 u_x)^2,$$

where u''_x is an observed or ungraduated value for integral values $x = 1$ to $x = n$, u_x is the corresponding graduated value, W_x is the weight assigned to the value u''_x, often an exposure or sample size, and k is a constant that determines the relative importance we are placing on smoothness as compared to fit, referred to as the *coefficient of smoothness*.

The case when $W_x = 1$ for all x refers to a Whittaker–Henderson Type A graduation formula. Although this type of formula generally does not give results as good as the Type B formula, it involves easier calculations and, therefore, was more widely used before the advent of computers.

The preceding equation can be written and solved most easily in matrix notation.

Let $\mathbf{u}''$ denote the column vector of n rows whose values are the ungraduated values $u''_1, u''_2, \ldots, u''_n$, $\mathbf{u}$ denote the corresponding vector of graduated values, $\mathbf{W}$ denote the $n \times n$ diagonal matrix whose diagonal elements are the weights $w_1, w_2, \ldots, w_n$ (a diagonal matrix is a matrix in which all elements not on the diagonal are zero), and $\mathbf{K}$ denote the $(n - 2) \times n$ matrix such that the $n - 2$ elements of the vector $\mathbf{Ku}$ are the $n - 2$ second differences of the elements of u. As an example, if $n = 5$, then

$$\mathbf{K} = \begin{bmatrix} 1 & -2 & 1 & 0 & 0 \\ 0 & 1 & -2 & 1 & 0 \\ 0 & 0 & 1 & -2 & 1 \end{bmatrix}.$$

If $\mathbf{x}$ is a column vector of n rows whose elements are $x_1, x_2, \ldots, x_n$, and if $\mathbf{x}^T$ is its transpose, then $\mathbf{x}^T\mathbf{x} = \sum_{i=1}^n x_i^2$; also, $\mathbf{x}^T\mathbf{wx} = \sum_{i=1}^n w_i x_i^2$. Then,

$$F + kS = (\mathbf{u} - \mathbf{u}'')^T \mathbf{W}(\mathbf{u} - \mathbf{u}'') + k\mathbf{u}^T\mathbf{K}^T\mathbf{Ku}.$$

Let us define the $n \times n$ matrix $\mathbf{A}$, $\mathbf{A} = \mathbf{W} + k\mathbf{K}^T\mathbf{K}$. It can be proven that $\mathbf{F} + k\mathbf{S}$ assumes its minimum value when $\mathbf{Au} = \mathbf{Wu}''$, which represents a system of n equations in the unknowns $u_1, u_2, \ldots, u_n$.

NUMERICAL SOLUTION TO THE PROBLEM

The solution of the equation $\mathbf{Au} = \mathbf{Wu}''$ is most easily solved by a factorization of the matrix **A** into triangular matrices, so that $\mathbf{A} = \mathbf{L}\mathbf{L}^T$, where **L** is lower triangular and $\mathbf{L}^T$ is upper triangular. This factorization is possible because **A** is a positive definite matrix. A matrix **A** is positive definite if it is symmetrical and if $\mathbf{x}^T\mathbf{A}\mathbf{x}$ is positive for any nonzero vector **x**. This method of solution is called the Choleski factorization method (*see* LINEAR ALGEBRA, COMPUTATIONAL), or sometimes the square-root method. The elements of **L** are given by

$$l_{11} = \sqrt{a_{11}},$$

$$l_{ij} = \frac{1}{l_{jj}}\left(a_{ij} - \sum_{c=1}^{j-1} l_{ic}l_{jc}\right),$$

$$i = 2, 3, \ldots, n,$$
$$j = 1, 2, \ldots, i-1,$$

$$l_{ii} = \sqrt{a_{ii} - \sum_{c=1}^{i-1} l_{ic}^2}, \qquad i = 2, 3, \ldots, n.$$

This transforms the problem of solving the equation $\mathbf{A}\mathbf{u} = \mathbf{W}\mathbf{u}''$ into the problem of solving two triangular systems of equations $\mathbf{L}\mathbf{u}' = \mathbf{W}\mathbf{u}''$ and $\mathbf{L}^T\mathbf{u} = \mathbf{u}'$, where $\mathbf{u}' = \mathbf{L}^T\mathbf{u}$ and $\mathbf{A}\mathbf{u} = \mathbf{L}\mathbf{L}^T\mathbf{u} = \mathbf{W}\mathbf{u}''$.

The triangular systems of equations are, of course, easily solved. The elements of $\mathbf{u}'$ are given by

$$u_1' = w_{11}u_1''/l_{11},$$

$$u_2' = \left(w_{22}u_2'' - l_{21}u_1'\right)/l_{22}, \ldots,$$

$$u_i' = \left(w_{ii}u_i'' - \sum_{c=1}^{i-1} l_{ic}u_c'\right)\Big/l_{ii}, \ldots,$$

until $i = n$. The equation $\mathbf{L}^T\mathbf{u} = \mathbf{u}'$ can be solved in a similar manner for **u**, except by working from bottom to top, thus solving for u_n first.

NUMERICAL EXAMPLE

The following chart shows a numerical example of the Whittaker–Henderson graduation formula. The vectors $\mathbf{u}''$, **W**, $\mathbf{W}\mathbf{u}''$, $\mathbf{u}'$, and **u** are shown in the table, while the matrices **A** and **L** are shown immediately after. In this example, the coefficient of smoothness k has been chosen as 200. This value is roughly 10 times the value of the weights, a relationship that has been shown to provide good results. The graduator usually will experiment with several coefficients of smoothness, trying to find the result that best smooths unwanted fluctuations while best preserving the underlying pattern of the data. A coefficient of smoothness of 0 will just reproduce the original ungraduated data, while a coefficient that approaches infinity will (in the case where second differences are used as the measure of smoothness) approach the least-squares* line that fits the data. It should be noted that, regardless of the coefficient of smoothness, $\sum_{i=1}^{n} w_{ii}u_i'' = \sum_{i=1}^{n} w_{ii}u_i$ and $\sum_{i=1}^{n} iw_{ii}u_i'' = \sum_{i=1}^{n} iw_{ii}u_i$. Thus, the weighted average value of the graduated series will equal the weighted average value of the ungraduated series and their first moments around their mean are equal.

x	u''	W	Wu''	u'	u
1	21	9	189	13.073	19.7
2	24	13	312	42.829	24.6
3	33	17	561	84.472	29.5
4	32	20	640	114.611	34.5
5	36	24	864	138.808	39.9
6	48	19	912	159.504	45.8
7	55	16	880	239.054	51.8
8	56	10	560	282.609	57.7

$$\mathbf{A} = \begin{bmatrix} 209 & -400 & 200 & 0 & 0 & 0 & 0 & 0 \\ -400 & 1013 & -800 & 200 & 0 & 0 & 0 & 0 \\ 200 & -800 & 1217 & -800 & 200 & 0 & 0 & 0 \\ 0 & 200 & -800 & 1220 & -800 & 200 & 0 & 0 \\ 0 & 0 & 200 & -800 & 1224 & -800 & 200 & 0 \\ 0 & 0 & 0 & 200 & -800 & 1219 & -800 & 200 \\ 0 & 0 & 0 & 0 & 200 & -800 & 1016 & -400 \\ 0 & 0 & 0 & 0 & 0 & 200 & -400 & 210 \end{bmatrix},$$

$$\mathbf{L} = \begin{bmatrix} 14.457 & 0 & 0 & 0 & 0 & 0 & 0 & 0 \\ -27.669 & 15.731 & 0 & 0 & 0 & 0 & 0 & 0 \\ 13.834 & -26.523 & 17.948 & 0 & 0 & 0 & 0 & 0 \\ 0 & 12.714 & -25.785 & 19.837 & 0 & 0 & 0 & 0 \\ 0 & 0 & 11.143 & -25.844 & 20.782 & 0 & 0 & 0 \\ 0 & 0 & 0 & 10.082 & -25.957 & 21.062 & 0 & 0 \\ 0 & 0 & 0 & 0 & 9.624 & -26.123 & 15.523 & 0 \\ 0 & 0 & 0 & 0 & 0 & 9.496 & -9.788 & 4.902 \end{bmatrix}$$

GENERALIZATION TO MORE THAN ONE DIMENSION

The problem of graduating a basic set of data that is a function of more than one variable sometimes arises. The basic theory to handle this case has already beeen discussed. Formulation of the two-dimensional problem will show how the generalization proceeds. Let $\mathbf{U}''$ denote an $m \times n$ array of ungraduated values, $\mathbf{W}'$ denote the $m \times n$ array of corresponding weights, $\underset{v}{\Delta}^2$ denote the vertical second difference operator such that

$$\underset{v}{\Delta}^2 U_{ij} = U_{(i+2)j} - 2U_{(i+1)j} + U_{ij},$$

$\underset{h}{\Delta}^2$ denote the horizontal second difference operator such that

$$\underset{h}{\Delta}^2 U_{ij} = U_{i(j+2)} - 2U_{i(j+1)} + U_{ij},$$

v denote the vertical coefficient of smoothness, and h denote the horizontal coefficient of smoothness. Thus, the expression to be minimized is given by

$$\sum_{i=1}^{m} \sum_{j=1}^{n} W'_{ij} (U_{ij} - U''_{ij})^2$$

$$+ v \sum_{i=1}^{m-2} \sum_{j=1}^{n} \left(\underset{v}{\Delta}^2 U_{ij} \right)^2 + h \sum_{i=1}^{m} \sum_{j=1}^{n-2} \left(\underset{v}{\Delta}^2 U_{ij} \right)^2.$$

This can be done using the same formulas used in the one-dimensional case by creating a one-dimensional array $\mathbf{u}''$ obtained by arranging the rows of the ungraduated values of $\mathbf{U}''$ vertically into one column vector of length mn. The vector $\mathbf{u}$ will denote the corresponding graduated values. Then create the $mn \times mn$ matrix $\mathbf{V}$ such that the elements of the vector $\mathbf{Vu}$ are the $n(m-2)$ vertical second differences of the elements of u and $2n$ zeros, while $\mathbf{H}$ is an $mn \times mn$

matrix such that the elements of the vector $\mathbf{Hu}$ are the $m(n-2)$ horizontal second differences of the elements u and $2m$ zeros. Finally, create the $mn \times mn$ diagonal matrix $\mathbf{W}$ by arranging the rows of the array $\mathbf{W}'$ one right after the other along the diagonal. Then if we let $\mathbf{A} = \mathbf{W} + v\mathbf{u}^T \mathbf{V}^T \mathbf{Vu} + h\mathbf{u}^T \mathbf{H}^T \mathbf{Hu}$, the system of equations to be solved is $\mathbf{Au} = \mathbf{Wu}''$, which can be solved just as in the one-dimensional case.

References

[1] Chan, F. Y., Chan, L. K., and Mean, E. R. (1982). *Scand. Actuar. J.*, **1**, 57–61.

[2] Chan, F. Y., Chan, L. K., and Yu, M. H. (1984). *Trans. Soc. Actuar.*, **36**, 183–204.

[3] Greville, T. N. E. (1974). *Part 5 Study Notes: Graduation*. Society of Actuaries, Chicago, IL.

[4] Knorr, F. E. (1984). *Trans. Soc. Actuar.*, **36**, 213–240.

[5] London, D. (1986). *Graduation: The Revision of Estimates*. ACTEX Publications, Abingdon and Winsted, CT.

[6] Lowrie, W. B. (1982). *Trans. Soc. Actuar.*, **34**, 329–350.

[7] McKay, S. F. and Wilkin, J. C. (1977). Derivation of a Two-Dimensional Whittaker–Henderson Type B Graduation Formula. Published as an appendix to F. Bayo and J. C. Wilkin, *Experience of Disabled Worker Benefits under OASDI 1965–74*. Actuarial Study Number 74, DHEW Publication Number (SSA) 77-11521, Government Printing Office, Washington, DC.

[8] Miller, M. D. (1946). *Elements of Graduation*. The Actuarial Society of America and American Institute of Actuaries.

(FINITE DIFFERENCES, CALCULUS OF GOODNESS OF FIT
GRADUATION
LAWLIKE RELATIONSHIPS
LINEAR ALGEBRA, COMPUTATIONAL
SMOOTHNESS PRIORS)

JOHN C. WILKIN

WHITTAKER-TYPE DISTRIBUTIONS

Apart from the inverse Gaussian distribution*, no other stable distributions* with characteristic exponent $\alpha \in (0, 1)$ are known to have a density that can be expressed in terms of elementary functions.

However, series expansions of such densities are available for rational values of α. In the special case $\alpha = 2/3$ the series can be expressed in terms of Whittaker functions (see Zolotarev [4], Pollard [3] and Bar-Lev and Enis [1] for further details). Explicitly, a stable distribution with characteristic exponent $\rho = 2/3$ has density

$$g_{2/3}(x) = \frac{1}{2(3\pi)^{1/2}} \frac{1}{x} e^{-2/(27x^2)}$$
$$\times W_{-1/2, -1/6}\left(\frac{-4}{27x^2}\right), \quad (1)$$

where $W_{\lambda, \mu}(z)$ is a solution of the Whittaker equation

$$W'' + \left[\frac{\frac{1}{4} - \mu^2}{z^2} + \frac{\lambda}{z} - \frac{1}{4}\right]W = 0,$$

given explicitly by

$$W_{\lambda, \mu}(z) = \frac{\Gamma(-2\mu)}{\Gamma\left(\frac{1}{2} - \lambda - \mu\right)} M_{\lambda, \mu}(z)$$
$$+ \frac{\Gamma(2\mu)}{\Gamma\left(\frac{1}{2} - \lambda + \mu\right)} M_{\lambda, -\mu}(z),$$

where

$$M_{\lambda, \mu}(z) = z^{\mu + 1/2} e^{-z/2}$$
$$\times F_1\left(\mu - \lambda + \tfrac{1}{2}; 2\mu + 1; z\right)$$

[$F_1(\cdot; \cdot; \cdot)$ being a confluent hypergeometric function]. For $\lambda = 0$, equation (1) reduces to a Bessel equation. Whittaker-type distributions are natural exponential families* (see, e.g., Morris [2]). They possess power variance functions of the form $V(\mu) = \alpha\mu^\gamma$ generated by stable distributions; Bar-Lev and Enis [1] present recent applications.

References

[1] Bar-Lev, S. K. and Enis, P. (1986). *Ann. Statist.*, **14**, 1507–1522.

[2] Morris, C. (1982, 1983). *Ann. Statist.*, **10**, 65–80; **11**, 515–529.

[3] Pollard, H. (1946). *Bull. Amer. Math Soc.*, **52**, 908–910.

[4] Zolotarev, V. M. (1961). *Selected Trans. Math. Statist. Probab.*, **1**, 163–167.

(NATURAL EXPONENTIAL FAMILY STABLE DISTRIBUTIONS)

WHITTEMORE'S COLLAPSIBILITY

An important consideration in the analysis of cross-classified discrete data is whether a multidimensional contingency table* can be collapsed (that is, summed over variables) into a table of lower dimension without losing information about the relationships among the remaining variables. A table of lower dimension is desirable because it is simpler, has higher counts per cell, and can yield more efficient parameter estimates than the usual estimators based on the full table [7].

Collapsing some tables, however, can give misleading results, as in the $2 \times 2 \times 2$ example of Table 1. At each location the proportion of females responding yes exceeds the proportion of males responding yes, but in the table collapsed over location this is reversed. The reversal, known as the amalgamation (or Simpson's) paradox* [3, 8, 9, 10, 12], is the result of the proportionally higher numbers of women at location 2, where yes responses are less common.

Collapsibility of a multidimensional table of cell probabilities can be defined in terms of the parameters of the loglinear model for

Table 1

	Location 1		Location 2		Both Locations	
	Male	Female	Male	Female	Male	Female
Yes	4	3	1	3	5	6
No	6	3	9	18	15	21

cell expectations. Let m_{ijk} be the expected count in the ijk cell of a three-dimensional table and let m_{ij} be the expected count in the ij cell of the table collapsed over variable 3. ANOVA-like models for the three-dimensional table and the two-dimensional collapsed table are

$$\log m_{ijk} = u + u_{1(i)} + u_{2(j)} + u_{3(k)} + u_{12(ij)}$$
$$+ u_{13(ik)} + u_{23(jk)} + u_{123(ijk)}$$

and

$$\log m_{ij} = u + u_{1,3(i)} + u_{2,3(j)} + u_{12,3(ij)},$$

with the constraints that u terms sum to zero over each subscript. Models for several types of independence can be obtained from these models by setting combinations of terms equal to zero [2]. For example, a model for conditional independence of variables 1 and 2 given 3 is obtained by setting $u_{123(ijk)}$ and $u_{12(ij)}$ to zero. Bishop et al. [2] define a three-dimensional table to be *collapsible over variable 3* with respect to the interaction of variables 1 and 2 if $u_{12(ij)} = u_{12,3(ij)}$ for all i, j, that is, if the interaction of variables 1 and 2 is unchanged by collapsing over variable 3.

Whittemore [11] notes the need for a stronger notion of collapsibility, citing Table 2, which is collapsible over each variable with respect to the interaction of the other two, yet some important information about the structure of the table is lost by collapsing. Although $u_{12(ij)} = u_{12,3(ij)}$ in Table 2, the $u_{12(ij)}$ term itself is not an acceptable measure of the relationship between variables 1 and 2 when $u_{123(ijk)} \neq 0$, much in the same way that main effects in a two-way ANOVA are not necessarily meaningful in the presence of an interaction*. Whittemore

defines *strict* collapsibility, adding to collapsibility the condition that $u_{123(ijk)} = 0$ for all i, j, k. [In Table 1, $u_{123(ijk)} = 0$; in Table 2, $u_{123(ijk)} = (-1)^{i+j+k}\ln(16)/8$.]

Results are available that characterize strict collapsibility in a three-dimensional table. Simpson [9], using an equivalent formulation in terms of cross-product ratios, shows that a $2 \times 2 \times 2$ table is strictly collapsible over variable 3 with respect to the interaction of variables 1 and 2 if and only if the following conditions hold:

(i) $u_{123(ijk)} = 0$ for all i, j, k, and

(ii) $u_{13(ik)} = 0$ for all i, k or $u_{23(jk)} = 0$ for all j, k.

That is, variable 3 must be conditionally independent of at least one of the remaining variables, given the other. Whittemore [11] demonstrates that these conditions are also equivalent to strict collapsibility in an $I \times J \times K$ table, when $K = 2$, but only sufficient for strict collapsibility when K exceeds 2. Ducharme and LePage [7], using cross-product ratios, show that the two conditions are equivalent to the strict collapsibility of all $I \times J \times K^*$ subtables, $K^* \leq K$, obtainable by collapsing over some of the categories of variable 3. Davis [6] defines strict *partial* collapsibility of the full table onto an $I \times J \times K^*$ subtable along the same lines as Whittemore's strict collapsibility, and shows that the preceding two conditions are equivalent to strict partial collapsibility onto all subtables having $K^* = K - 1$.

Definition of collapsibility of an n-dimensional table is a straightforward generalization of the three-dimensional case. An n-dimensional table of probabilities, cross-classified by the variables in the set F, is collapsible over $B \subseteq F$ with respect to $A \subseteq F \setminus B$ if the interaction u_A from the original table equals the corresponding interaction $u_{A,B}$ from the collapsed table [2]. The table is strictly collapsible if, in addition, u_Z is zero for all Z such that $A \subseteq Z \subseteq F$ and $Z \cap B$ is not empty [11]. Generalizations of the three-dimensional results as well as nec-

Table 2

	Location 1		Location 2		Both Locations	
	Male	Female	Male	Female	Male	Female
Yes	6	3	3	6	9	9
No	3	6	6	3	9	9

essary and sufficient conditions for both collapsibility and strict collapsibility, given in terms of the cell probabilities, are in ref. 11.

Asmussen and Edwards [1] give several equivalent definitions of collapsibility of a loglinear *model* (compared to Whittemore's collapsibility of a table of cell probabilities with respect to a particular interaction) and give necessary and sufficient conditions in terms of the interaction graphs of Darroch et al. [4]. Davis [5] extends Whittemore's definition and shows that their notion of collapsibility can be characterized in terms of Whittemore's.

References

[1] Asmussen, S. and Edwards, D. (1983). *Biometrika*, **70**, 567–578. (Gives definitions of collapsibility of models and conditions for collapsibility in terms of interaction graphs.)

[2] Bishop, Y. M., Fienberg, S. E., and Holland, P. W. (1975). *Discrete Multivariate Analysis: Theory and Practice*. MIT Press, Cambridge, MA. (An excellent reference on categorical data; erroneously identifies conditional independence as necessary for collapsibility.)

[3] Blyth, C. R. (1972). *J. Amer. Statist. Ass.*, **67**, 364–366. (Examines Simpson's paradox and its relationship to Savage's sure-thing principle.)

[4] Darroch, J. N., Lauritzen, S. L., and Speed, T. P. (1980). *Ann. Statist.*, **8**, 522–539. (Uses undirected graphs to connect ideas of Markov fields and conditional independence in log linear models.)

[5] Davis, L. J. (1986). *Commun. Statist. Theory Meth.*, **15**, 2541–2554. (Extends the work of Whittemore and connects it to the work of Asmussen and Edwards.)

[6] Davis, L. J. (1987). *Statist. Prob. Lett.*, **5**, 129–134. (Defines notions of partial collapsibility and gives equivalent conditions.)

[7] Ducharme, G. R. and LePage, Y. (1986). *J. R. Statist. Soc. B*, **48**, 197–205. (Defines stronger and weaker forms of collapsibility than strict collapsibility and gives tests in terms of odds ratios.)

[8] Good, I. J. and Mittal, Y. (1987). *Ann. Statist.*, **15**, 694–711. (Examines the amalgamation paradox with respect to several measures of association for $2 \times 2 \times 2$ tables and suggests sampling designs to avoid it.)

[9] Simpson, E. H. (1951). *J. R. Statist. Soc. B*, **13**, 238–241. (Corrected the then prevailing notion that $2 \times 2 \times 2$ tables can be collapsed if there is no second order interaction.)

[10] Shapiro, S. H. (1982). *Amer. Statist.*, **36**, 43–46. (Graphs odds ratios in two-dimensional plots to explain collapsibility and Simpson's paradox.)

[11] Whittemore, A. S. (1978). *J. R. Statist. Soc. B*, **40**, 328–340. (Defines strict collapsibility for n-dimensional tables.)

[12] Yule, G. U. (1903). *Biometrika*, **2**, 121–134. (An early reference to the "paradox" Simpson noted in 1951.)

(CATEGORICAL DATA
CONTINGENCY TABLES
MULTIDIMENSIONAL CONTINGENCY
  TABLES
SIMPSON'S PARADOX
TWO-BY-TWO TABLES)

PHILLIP L. CHAPMAN
PAUL W. MIELKE, JR.

WHITTINGHILL–POTTHOFF TESTS

See POTTHOFF–WHITTINGHILL TESTS

WHITTLE EQUIVALENCE THEOREM

The Whittle theorem [2] is a generalisation of an earlier theorem concerning D-optimality due to Kiefer [1] (*see* OPTIMUM DESIGN OF EXPERIMENTS in the *Supplement*).

Consider the regression* model

$$y(\mathbf{x}) = \boldsymbol{\theta}' f(\mathbf{x}) + \epsilon,$$

where $\mathbf{x}$ is the independent variable, taking values in a *design space* $\mathscr{X}$, $\boldsymbol{\theta}$ is a k-dimensional vector of regression coefficients*, and ϵ is a residual of zero mean and constant variance, values of ϵ for distinct observations being statistically independent. If one takes N observations with values $\mathbf{x}_j$ ($j = 1, 2, \ldots, N$) for $\mathbf{x}$, then the *design matrix*

$$\mathbf{M}(\xi) = \frac{1}{N} \sum_{j=1}^{N} f(\mathbf{x}_j) f(\mathbf{x}_j)'$$

$$= \int f(\mathbf{x}) f(\mathbf{x})' \xi(d\mathbf{x})$$

characterises the precision of the minimum variance linear unbiased estimate $\hat{\boldsymbol{\theta}}$ of $\boldsymbol{\theta}$ derived from the experiment, in that

$$\text{cov}(\hat{\boldsymbol{\theta}}) = \mathbf{M}(\xi)^{-1}.$$

Here ξ is a measure on the design space, and one wishes to choose ξ, the *design measure*, to maximise $\mathbf{M}(\xi)$, in some sense.

Suppose that as criterion one chooses some scalar function of $\mathbf{M}(\xi)$, which can then be regarded as a function $\phi(\xi)$ of the measure ξ. A design maximising the criterion function ϕ is termed ϕ-*optimal*. A ϕ-optimal design will exist if $f(\mathbf{x})$ is continuous and $\mathcal{X}$ compact.

Define the directional derivative

$$\Phi(\xi, \eta)$$
$$= \lim_{\alpha \downarrow 0} \alpha^{-1} \left[\phi\{(1 - \alpha)\xi + \alpha\eta\} - \phi(\xi) \right],$$

where η is an alternative design measure. Define also

$$\overline{D}(\xi) = \sup_{\eta} \Phi(\xi, \eta),$$

the maximal rate of increase of $\phi(\cdot)$ at ξ. Finaly, let δ_x be a measure concentrated entirely on the value x.

The Whittle Equivalence Theorem

(a) If ϕ is concave, then a ϕ-optimal design measure ξ^* can be equivalently characterised by any of the three conditions
  (i) ξ^* maximises ϕ.
  (ii) ξ^* minimises $\overline{D}(\xi)$.
  (iii) $\overline{D}(\xi^*) = 0$.

(b) The point (ξ^*, ξ^*) is a saddle point of Φ in that

$$\Phi(\xi^*, \eta) \leqslant 0 = \Phi(\xi^*, \xi^*) \leqslant \Phi(\xi, \xi^*).$$

(c) If ϕ is also differentiable, then the support of ξ^* is contained in the set of x for which $\Phi(\xi^*, \delta_x) = 0$, in that $\Phi(\xi^*, \delta_x) = 0$ almost everywhere in ξ^* measure.

The classic special case is that of D-optimality, for which $\phi(\xi) = |\mathbf{M}(\xi)|$ and $\overline{D}(\xi) = \sup_x f(\mathbf{x})' \mathbf{M}(\xi)^{-1} f(\mathbf{x}) - k$. $\overline{D}(\xi)$ is related to the maximal mean square error of a prediction from the fitted regression line.

Another case is that of G-optimality, when one wishes to choose ξ to minimise the variance of the estimate $\hat{\psi} = \mathbf{c}'\hat{\boldsymbol{\theta}}$ of $\psi = \mathbf{c}'\boldsymbol{\theta}$.

One can then take $\phi(\xi) = -\mathbf{c}'\mathbf{M}(\xi)^{-1}\mathbf{c}$ and criteria (b), (c) of the theorem amount to

$$\sup_x \left| \mathbf{c}'\mathbf{M}(\xi)^{-1} f(\mathbf{x}) \right|^2 \geqslant \mathbf{c}'\mathbf{M}(\xi)^{-1}\mathbf{c}$$

with equality for a G-optimal design. That is,

$$\sup_x \left| \text{cov}(\hat{\psi}, y(\mathbf{x})) \right|^2 \geqslant \text{var}(\hat{\psi}).$$

References

[1] Kiefer, J. (1962). *Ann. Math. Statist.*, **37**, 792–796.
[2] Whittle, P. (1973). *J. R. Statist. Soc. B*, **35**, 123–130.

Bibliography

Fedorov, V. (1972) *Theory of Optimal Experiments.* Academic, London. (English transl.)

Silvey, S. D. (1980) *Optimal Design.* Chapman and Hall, London, England.

(DESIGN OF EXPERIMENTS
OPTIMUM DESIGN OF EXPERIMENTS
  (*Supplement*)
REGRESSION (various entries))

P. WHITTLE

WHOLESALE PRICE INDEX   *See* PRO-DUCERS' PRICE INDEX

WIENER CHAIN

An intuitively appealing discrete analog of the Wiener process (*see* BROWNIAN MOTION) suggested by Kreith [1].

It is defined as a stationary Markov process* $W_0, W_1, W_2, \ldots$ with state space consisting of all integers for which

(i) $W_0 = 0$;
(ii) $(W_n + n)\sigma^{-1}$ has a binomial distribution with parameters $(2n, \frac{1}{2})$—a ("$n\sigma^2/2$-centered binomial";
(iii) if $n_1 < n_2 < n_3 < \cdots$ then $W_{n_2} - W_{n_1}, W_{n_3} - W_{n_2}, \cdots$ are mutually independent.

If $W_0, W_1, W_2, \ldots$ is a Wiener chain then $W_{m+n} - W_m$ has a $n\sigma^2/2$-centered binomial distribution, for all m.

Kreith [1] gives details of motivation and background, and discusses a stochastic difference equation yielding a continuous version, corresponding to a Wiener process.

Reference

[1] Kreith, K. (1985). *Amer. Math. Monthly*, **92**, 281–284.

(BROWNIAN MOTION
MARKOV PROCESS
WIENER PROCESS)

WIENER MEASURE

Early studies of the behavior of a particle performing the random type motion first described by the naturalist Robert Brown date back to the first years of the century (Bachelier, Einstein, Smoluchowski; references to authors mentioned in the text but not in the list *in fine* can be found in the bibliography of Knight's monograph [3]). The first mathematically satisfactory description of the resulting stochastic process* was given by Wiener (1923). Assuming that this Brownian motion* $\{B(t), t \geqslant 0\}$, where $B(t)$ represents displacement at time t in a chosen direction, has continuous trajectories (with probability 1), one is led to consider the space C of continuous functions $\omega(t)$, $t \geqslant 0$, and seek to determine for C a probability measure W which assigns to sets of continuous functions the very probability that a trajectory of B be one of the functions in the set. Basic are cylinder sets of the form $S(t_1, \ldots, t_n; A_1, \ldots, A_n) = \{\omega: \omega(t_i) \in A_i, i = 1, \ldots, n\}$, where $0 = t_0 < t_1 < \cdots < t_n$ and the A_i are Borel sets of the line.

To such a set W one must assign the weight

$$\int_{A_1} dx_1 \cdots \int_{A_n} dx_n \prod_{i=1}^{n} p(t_i - t_{i-1}; x_i - x_{i-1}),$$

where $p(t; x) = (2\pi t)^{-1/2} \exp\{-x^2/(2t)\}$ is the normal density with mean zero and variance t, at the argument x. This expresses the fact that B passes through the successive "gates" $A_1, \ldots, A_n$ by independent normal increments of variances $t_i - t_{i-1}$, as postulated for standard Brownian motion. Wiener showed both that B has continuous paths and that there exists exactly one such probability measure W defined on the σ-algebra $\mathscr{C}$ generated by the cylinder sets. This W is called *Wiener measure*.

The fundamental importance of this work only became widely appreciated after Kolmogorov, a decade later, set forth the general notion of a probability space (*see* AXIOMS OF PROBABILITY).

The probability space $(C, \mathscr{C}, W)$ on which B is defined as the coordinate process $B(t, \omega) = \omega(t)$, $\omega \in C$, constitutes the prototype for the canonical description of a stochastic process*, one in which the space of trajectories of the process is probabilized in accordance with the law of that process, which is then realized as a coordinate process. Considering C rather than the subspace of those functions ω with $\omega(0) = 0$ permits assigning to Brownian motion arbitrary initial value $B(0) = x$. This is done by using instead of W the probability W^x defined by $W^x(A) = W(A + x)$ with $A + x = \{\omega \in C: \omega - x \in A\}$ and $\omega - x$ the function $(\omega - x)(t) = \omega(t) - x$.

The effective evaluation of a Wiener integral

$$\int_C f(\omega) W(d\omega)$$

is in general difficult. Some classes of integrands f have been studied by Cameron and Martin (1944–1945), a typical example being for $g(\omega) = \int_0^T \omega^2(t)\, dt$,

$$\int_C \exp\{-\lambda g(\omega)\} W(d\omega), \qquad \lambda > 0. \quad (1)$$

The Laplace transform of the functional g of the Brownian path is thus determined.

Brownian motion is the prime example of a *Gaussian process*. A Gaussian measure on C is the law of a Gaussian process with continuous trajectories, so Wiener measure constitutes the basic example.

Brownian motion over a finite time interval $[0, T]$ has the representation $B(t) = \sum_1^\infty Z_n h_n(t)$, where the Z_n are independent standard normal variables and the functions h_n are obtained from an arbitrarily chosen complete orthonormal sequence over $[0, T]$. Using this, Shepp [5] has characterized Gaussian measures on C, equivalent to W (having the same null sets), and computed certain integrals generalizing (1). The random Fourier series for $B(t)$, given when $T = 1$ by

$$h_n(t) = 2^{1/2}\sin\left(n - \tfrac{1}{2}\right)\pi t \big/ \left[\left(n - \tfrac{1}{2}\right)\pi\right],$$

is well suited because it transforms $g(\omega) = \int_0^1 (\omega(t) - \mu(t))^2\, dt$, for square-integrable μ, into a weighted series of noncentral chi-square* variables from which (1) is easy to compute. Rothman and Woodroofe [4] have used this for a one-sample test of symmetry relative to 0.

Wiener measure is invariant with respect to many 1-1 maps of C. A simple one is $\omega(t) \to t\omega(1/t)$, $t > 0$. Grintsyavicius [2] describes classes of such maps and deduces some Wiener integrals.

Diffusions generalize Brownian motion in another direction, being Markov processes with continuous paths. Laws of diffusions are induced on C, from W, by suitable time-change maps, well explained by Freedman [1] (*see also* DIFFUSION PROCESSES).

References

[1] Freedman, D. (1983). *Brownian Motion and Diffusion*. Springer, New York. (A good reference to start with.)

[2] Grintsyavicius, A. K. (1982). *Litov. Math. Sbornik*, **22**(3), 55–66 (English transl.). (Research paper on transformations preserving W.)

[3] Knight, F. (1981). *Essentials of Brownian Motion and Diffusion*. Amer. Math. Soc., Providence, RI. (A reference monograph.)

[4] Rothman, E. D. and Woodroofe, M. (1972). *Ann. Math. Statist.*, **43**, 2035–2038.

[5] Shepp, L. A. (1966). *Ann. Math. Statist.*, **37**, 321–354.

Bibliography

Norbert Wiener special issue (1966). *Bull. Amer. Math. Soc.*, **72**(No. 1), Part II. (Review and discussion of Wiener's work by leading specialists.)

(AXIOMS OF PROBABILITY
BROWNIAN MOTION
GAUSSIAN PROCESSES
DIFFUSION PROCESSES
STOCHASTIC PROCESSES)

J. P. IMHOF

WIENER, NORBERT

Born: November 26, 1894, Columbia, Missouri.

Died: March 19, 1964, Stockholm, Sweden.

Contributed to: cybernetics, stochastic processes, mathematical physics, communication theory.

Norbert Wiener was born in 1894 in the United States. His father, descended from a family of rabbinical scholars, had migrated from Russia and, without a university education, became a professor of Slavic languages at Harvard. Under his father's influence Norbert Wiener became a child prodigy, entering Tufts College in Boston at 11 and graduating with a Ph.D. from Harvard at 18. His early interest was in natural science, particularly biology, rather than mathematics, but failure in the laboratory led his father to suggest philosophy and his Ph.D. thesis was on mathematical logic.

In 1913 he went to Cambridge, England, and was influenced by Bertrand Russell and G. H. Hardy; Wiener claimed the latter was the "master in my mathematical training." Russell pointed out to Wiener that a mathematical logician might well learn something of mathematics. He published his first mathematical work soon after and his first substantial paper [10] in 1914. After a period as an instructor in philosophy at Harvard and in mathematics at Maine he became an instructor in mathematics at Massachusetts Institute of Technology.

Motivated partly by Einstein's work on Brownian motion he developed the idea of representing each Brownian path as a point in a function space on which a probability measure is defined (*see* BROWNIAN MOTION and WIENER MEASURE). He showed that almost all paths were continuous but not differentiable. His ideas were presented fully [11] in 1923. They have had an influence on modern probability of the most profound kind, as can be seen from the following.

If $X(t)$ denotes a one-dimensional Wiener process, then the increment $X(t + \delta) - X(t)$ is independent of $X(s)$, $s \leqslant t$, and has a distribution independent of t. As a consequence of this (and the continuity of sample paths) $X(t)$ is Gaussian. If Y_k is a sequence of sums of independent identically distributed (i.i.d.) random variables with zero mean and finite variance, then the sequence has the same probabilistic structure as $X_k = X(T_1 + T_2 + \cdots + T_k)$ for suitable random variables, T_k, that are functions of the X_j, $j \leqslant k$, but i.i.d. (*see* FUNCTIONAL LIMIT THEOREMS: INVARIANCE PRINCIPLES). Thus much of the limit theory of probability can be studied via $X(t)$. Moreover $X(t)$ itself represents chaotic behaviour, as the previous comment about differentiability shows. Thus $dX(t)$ can intuitively be thought of as the driving term (or input) to a stochastic differential equation* (or system) out of which comes a solution with the relatively organised behaviour expected of physical phenomena. In this sense, the phenomena might be regarded as having been explained.

In 1926 he began his work on generalized harmonic analysis, which was fully developed in ref. 13. Beginning from the notion of a measurable function $f(t)$, he introduced the autocorrelations

$$\gamma(t) = \lim_{T \to \infty} \frac{1}{2T} \int_{-T}^{T} f(s + t)f(s) \, ds,$$

assumed to exist. From these he constructed what would now be called the spectral distribution function (*see* SPECTRAL ANALYSIS). Moreover he showed how the mean square

$\gamma(0)$ could be represented as a linear superposition of contributions from every oscillatory frequency, each contribution being the squared modulus of the contribution of that frequency to $f(t)$ itself. Wiener's characteristically constructive method is here demonstrated and can be contrasted with the axiomatic approach commencing from a probability space and a Hilbert space of square integrable functions, over that space, generated by the action of a unitary group of translations on an individual function. (See ref. 1, pp. 636–637, for some history of this.) Out of his work on generalized harmonic analysis grew his work on Tauberian* theory [12], one aspect of which is his celebrated theorem that if $f(\omega)$, $\omega \in [-\pi, \pi]$, has an absolutely convergent Fourier series and $f(\omega)$ is never zero, then $\{f(\omega)\}^{-1}$ also has an absolutely convergent Fourier series. The Wiener–Lévy theorem [4, p. 280] is the natural generalisation of this. His work from 1926 to 1930 was gathered together in his book [14].

In 1933, with Paley named as co-author, he produced ref. 7, Paley being dead at the time of writing. Apart from the (so-called) Paley–Wiener theory (see ref. 4, pp. 173–177, for a discussion), this contained also mathematical results relating to the Wiener–Hopf equation (see below) necessary for Wiener's later work on linear prediction (*see* PREDICTION AND FILTERING, LINEAR). For the latter Wiener sought to discover a weight function $K(s)$, that, for a stationary process* $f(t)$ with finite variance, minimised

$$\lim_{T \to \infty} \frac{1}{2T} \int_{-T}^{T} \left| f(t + a) \right.$$
$$\left. - \int_{0}^{\infty} f(t - s) \, dK(s) \right|^2 dt,$$

i.e., that minimised the mean square error in predicting $f(t + a)$ from $f(s)$, $s \leqslant t$. He reduced the solution of this to that of the Wiener–Hopf equation

$$\int_{0}^{\infty} \gamma(t - s) \, dK(s) = \gamma(t), \qquad t \geqslant 0,$$

which had previously arisen in connection

with the distribution of stellar atmosphere temperature. Wiener solved this by the methods mentioned in connection with ref. 7. This work was published in ref. 15, publication having been delayed due to restrictions because of a supposed need for military secrecy. At much the same time Kolmogoroff [5, 6] had been working on the same problem. (See ref. 15, p. 59, for a discussion by Wiener of the question of priority.) Though perhaps less constructive, Kolmogoroff's work, which began from the Wold decomposition [20] of a discrete time stationary process into a purely nondeterministic part and a perfectly predictable part, was in some ways more general. Wiener commenced from a direct representation of $f(t + 1)$ in terms of $f(s)$, $s \leqslant t$ (i.e., an autoregressive representation), whereas Kolmogoroff commenced from the representation of $f(t)$ in terms of the prediction error or innovation sequence (i.e., moving average* representation). Kolmogoroff made use of fundamental results due to Szegö [9], that, for example, express the variance of the innovations as the geometric mean of the spectral density. *See also* WIENER-KOLMOGOROV PREDICTION (*Supplement*).

Wiener became well known to the general scientific public for his basically philosophical work on cybernetics [16]. This was concerned with the analogy between man as a self-regulating system, receiving sensory data and pursuing certain objectives, and mechanical or electrical servomechanisms. He made some attempt with Siegel [8] to use his differential space as a basis for a theory of quantum systems. With Masani [18, 19] he extended prediction theory to the multivariate case. In ref. 17 he sought to describe a general class of random processes obtained from a Brownian motion as a sum of homogeneous multilinear functionals of that process, of various degrees. This has led to much further research. (See ref. 2.)

The breadth of Wiener's work means that his name occurs throughout mathematics and probability. An example over and above those already mentioned arises in connection with a test for recurrence in a random walk*.

This is turn relates to potential theory and to diffusion. (See ref. 3, p. 257.) Norbert Wiener died in 1964.

References

[1] Doob, J. L., (1953). *Stochastic Processes*. Wiley, New York.

[2] Hida, T. (1970). *Stationary Stochastic Processes*. Princeton University Press, Princeton, NJ.

[3] Itô, K. and McKean, H. P., (1965). *Diffusion Processes and their Sample Paths*. Springer-Verlag, Berlin.

[4] Katznelson, Y. (1968). *An Introduction to Harmonic Analysis*. Wiley, New York.

[5] Kolmogoroff, A. N. (1939). Sur l'interpolation et extrapolation des suites stationnaires, *C.R. Acad. Sci.*, *Paris*, **208**, 2043–2045.

[6] Kolmogoroff, A. N. (1941). Interpolation and extrapolation of stationary random sequences, *Izv. Akad. Nauk SSSR*, *Ser. Math.* **5**, 3–14.

[7] Paley, R. E. A. C. and Wiener, N. (1934). *Fourier Transforms in the Complex Domain*. Amer. Math. Soc., Providence, RI.

[8] Siegel, A. and Wiener, N. (1955). The differential space of quantum systems, *Nuovo Cimento (10)*, 2, 982–1003.

[9] Szegö, G. (1920). Beiträge zur theorie der Toeplitzschen formen. *Math. Z.* **6**, 167–202.

[10] Wiener, N. (1914). A simplification of the logic of relations. *Proc. Camb. Philos. Soc.*, **27**, 387–390.

[11] Wiener, N. (1923a). Differential space, *J. Math. Phys.*, **2**, 131–174.

[12] Wiener, N. (1923b). Tauberian theorems, *Ann. Math.*, **33**, 1–100.

[13] Wiener, N. (1930). Generalized harmonic analysis, *Acta Math.*, **55**, 117–258.

[14] Wiener, N. (1933). *The Fourier Integral and Certain of its Applications*. C.U.P., New York.

[15] Wiener, N. (1949). *Extrapolation, Interpolation and Smoothing of Stationary Time Series*. The MIT Press, Cambridge, Mass. and Wiley, New York.

[16] Wiener, N. (1950). *The Human Use of Human Beings*. Houghton Mifflin, Boston.

[17] Wiener, N. (1958). *Nonlinear Problems in Random Theory*. The MIT Press, Cambridge, Mass. and Wiley, New York.

[18] Wiener, N. and Masani, P. (1957). The prediction theory of multivariate stochastic processes. I. The regularity condition, *Acta Math.*, **98**, 111–150.

[19] Wiener, N. and Masani, P. (1958). The prediction theory of multivariate stochastic processes, II. The linear predictor. *Acta Math.*, **99**, 93–137.

[20] Wold, H. (1938). *A Study in the Analysis of Stationary Time Series.* Almqvist and Wiksell, Uppsala, Sweden.

Bibliographical Notes

Wiener wrote a most interesting two volume autobiography that serves to give some impression of the greatness of his mind. The volumes are:

Wiener, N. (1953). *Ex-prodigy: My Childhood and Youth.* Simon and Schuster, New York.

Wiener, N. (1956). *I Am a Mathematician. The Later Life of a Prodigy.* Doubleday, Garden City, NY.

In 1966 the American Mathematical Society devoted a special edition of its *Bulletin* (Vol. 72, No. 1, Part II) to Wiener. This contains articles by distinguished scholars about Wiener and his work, including a very interesting article about the man himself by N. Levinson and a complete bibliography of his writing.

E. J. HANNAN

WIENER PROCESS   *See* BROWNIAN MOTION

WIENER–HOPF EQUATION *See* PREDICTION AND FILTERING, LINEAR; WIENER, NORBERT

WIJSMAN'S REPRESENTATION THEOREM

Let X be a random variable taking its values in a space $\mathscr{X}$, and suppose that the statistical problem is invariant under a group G of transformations of $\mathscr{X}$ (*see* INVARIANCE CONCEPTS IN STATISTICS). Let $T = t(X)$ be a maximal invariant under G and suppose one wants to restrict inference to procedures that depend only on T. In general, one would have to derive the distribution P^T of T given any distribution P of X. However, for cer-

tain questions (especially in testing problems) one may get by with a density ratio $p_2^T(t)/p_1^T(t)$, given any two distributions P_1 and P_2 of X.

For instance, suppose that $X = (X_1, \ldots, X_n)$ is a sample from a normal population with mean μ and variance σ^2, both unknown, and that one wants to test one value of $\mu/\sigma = \gamma$ against another; say $\gamma = \gamma_1$ versus $\gamma = \gamma_2$. The problem is invariant under the transformations $X_i \to cX_i$, $i = 1, \ldots, n$, with any $c > 0$. There are many choices of a maximal invariant* $T = t(X)$; for instance,

$$T = \left(\operatorname{sgn} X_1, X_2/X_1, \ldots, X_n/X_1 \right)$$

(after deleting from $\mathscr{X}$ the points with $x_1 = 0$). For any choice of T, the distribution of T depends only on γ, so that for tests depending on T the two hypotheses become simple. A most powerful test based on T is then obtained via the Neyman–Pearson lemma* with the help of $p_2^T(t)/p_1^T(t)$, where p_i^T is the density of T under γ_i. An application of the general representation (1) given below yields

$$p_2^T(t)/p_1^T(t)$$

$$= \int_0^\infty p_2(c\mathbf{x}) c^{n-1}\, dc \Big/ \int_0^\infty p_1(c\mathbf{x}) c^{n-1}\, dc$$

in which $\mathbf{x} = (x_1, \ldots, x_n)$ and for $i = 1, 2$, $p_i(x)$ is the density of X for any μ_i, σ_i such that $\mu_i/\sigma_i = \gamma_i$. The value of this ratio of integrals depends on x only through $t(x)$. One of the advantages of this representation is that it does not depend on any particular choice of a maximal invariant.

In the general situation it will be assumed throughout that G is locally compact so that it admits a left invariant measure (Haar* measure), say μ_G; i.e., for any measurable subset A of G we have $\mu_G(gA) = \mu_G(A)$ for every $g \in G$. (In the preceding example, G is the set of positive numbers c under multiplication and the invariant measure is dc/c or any constant multiple of it.) Furthermore, it is assumed that on $\mathscr{X}$ there is a relatively invariant measure λ with multiplier χ; i.e., $\lambda(gB) = \chi(g)\lambda(B)$ for every measurable subset B of $\mathscr{X}$, in which χ is a positive continuous function on G. (In the example λ is Lebesgue measure on Euclidean n space and $\chi(c) = c^n$.)

Suppose that P_1 and P_2 are two distributions on $\mathscr{X}$ such that $P_i(dx) = p_i(x)\lambda(dx)$, $i = 1, 2$. Then under additional conditions that will be stated below we have

$$\left(p_2^T/p_1^T \right)(t(x))$$

$$= \frac{\int p_2(gx)\chi(g)\mu_G(dg)}{\int p_1(gx)\chi(g)\mu_G(dg)}, \quad (1)$$

in which the right-hand side depends on x only through $t(x)$. The idea of obtaining the density ratio of T as a ratio of two integrals over G was first advanced by Stein [7]. An early application can be found for instance in ref. 5. The possibility of the representation (1) is especially important in problems where G is not completely specified but only known to belong to a certain family. For an application, see ref. 9.

The result (1) was first proved by Wijsman [8] for $\mathscr{X}$ Euclidean, λ Lebesgue measure on $\mathscr{X}$, and G a group of linear transformations*. This was extended in ref. 10 to affine transformations. It also had to be assumed that $\mathscr{X}$ under the action of G is a *Cartan G-space* [6]. This is defined as a pair $(\mathscr{X}, G)$ such that every $x \in \mathscr{X}$ has a neighborhood V with the property that the set of group elements $g \in G$ for which $V \cap gV \neq \emptyset$ has compact closure. This guarantees a sufficiently regular action of G on $\mathscr{X}$. It is in particular satisfied if G consists of translations, orthogonal transformations, and transformations of the form $S \to CSC'$ in which S is $p \times p$ positive definite (usually a covariance matrix) and C runs through a subgroup of nonsingular matrices. This has been found to hold in all interesting multivariate applications.

The validity of (1) has also been proved by Andersson [1] under the assumption that the action of G is *proper*. [Assuming $\mathscr{X}$ and G locally compact and the action continuous, the action is called proper if the inverse image of a compact set under the mapping $G \times \mathscr{X} \to \mathscr{X} \times \mathscr{X}$ given by $(g, x) \to (x, gx)$

is compact.] A comparison of the assumption of Cartan G-space versus proper action can be found in ref. 11. Equation (1) has also been proved under weaker assumptions by Bondar [2] and by Farrell [3, 4].

References

[1] Andersson, S. A. (1982). *Ann. Statist.*, **10**, 955–961.

[2] Bondar, J. V. (1976). *Ann. Statist.*, **4**, 866–877.

[3] Farrell, R. H. (1976). *Lecture Notes in Math.*, **520**. Springer, Berlin.

[4] Farrell, R. H. (1985). *Multivariate Calculation. Use of the Continuous Groups.* Springer, Berlin.

[5] Giri, N. C. (1964). *Ann. Math. Statist.*, **35**, 181–189.

[6] Palais, R. S. (1961). *Ann. Math.*, **73**, 295–323.

[7] Stein, C. (1959). Notes on Multivariate Analysis. Stanford University (unpublished), Stanford, CA.

[8] Wijsman, R. A. (1967). *Proc. Fifth Berkeley Symp. Math. Statist. Prob.*, Vol. 1. University of California Press, Berkeley, CA, pp. 389–400.

[9] Wijsman, R. A. (1967). *Ann. Math. Statist.*, **38**, 8–24.

[10] Wijsman, R. A. (1972). *Proc. Sixth Berkeley Symp. Math. Statist. Prob.*, Vol. 1. University of California Press, Berkeley, CA, pp. 109–128.

[11] Wijsman, R. A. (1985). *Ann. Statist.*, **13**, 395–402.

(HAAR DISTRIBUTIONS
INVARIANCE CONCEPTS IN STATISTICS
TRANSFORMATIONS)

Robert A. Wijsman

WILCOXON, FRANK

Born: September 2, 1892, County Cork, Ireland.

Died: November 18, 1965, Tallahassee, Florida.

Contributed to: (statistics) rank tests, multiple comparisons, sequential ranking, factorial design; (chemistry) fungicidal action, synthesis of plant growth substances, insecticides research (pyrethrins, Parathion, Malathion).

Frank Wilcoxon was born in Glengarriffe Castle, near Cork, Ireland, to wealthy American parents on September 2, 1892. His father was a poet, outdoorsman, and hunter. Wilcoxon spent his boyhood at Catskill, New York and developed his lasting love of nature and water there. Adolescence seems to have been difficult, with a runaway period during which he was briefly a merchant seaman in New York harbor, a gas pumping station attendant in an isolated area of West Virginia, and a tree surgeon. As this period ended, he was enrolled in Pennsylvania Military College and, although the school's system did not agree with his ideas of personal freedom, he received the B.S. degree in 1917.

Wilcoxon entered Rutgers University in 1920 after a World War I position with the Atlas Powder Company at Houghton, Michigan and received the M.S. degree in chemistry in 1921. He continued his education at Cornell University and received the Ph.D. degree in physical chemistry in 1924. At Cornell, he met Frederica Facius and they were married on May 27, 1926. Frank and Freddie later became well known and loved in the statistical community, particularly through their regular participation in the Gordon Research Conference on Statistics in Chemistry and Chemical Engineering.

From 1924 to 1950, Frank Wilcoxon was engaged in research related to chemistry. In 1925 he went to the Boyce Thompson Institute for Plant Research under a Crop Protection Institute fellowship sponsored by the Nichols Copper Company and worked on colloid copper fungicides until 1927. He then (1928–1929) worked with the sponsoring company in Maspeth in Queens. In 1929 he returned to the Boyce Thompson Institute and remained there until 1941, working on the chemistry and mode of action of fungicides and insecticides. With a leave of absence from 1941 to 1943, he designed and directed the Control Laboratory of the Ravenna Ordnance Plant operated by the Atlas Powder Company. Wilcoxon joined the American Cyanamid Company in 1943 and continued with that company until his retirement in 1957, first with the Stamford Research Laboratories as head of a group developing insecticides and fungicides and then as head of the statistics group of the Lederle Division in Pearl River, New York. He served as a consultant to various organizations, including the Boyce Thompson Institute, until 1960, when he joined the faculty of the new Department of Statistics at Florida State University. There he was active in research and teaching and contributed to the development of the department until his death in 1965.

Frank Wilcoxon made many contributions to chemistry and biochemistry with some 40 publications in the field. His first paper [13] on acidimetry and alkalimetry was published in 1923 in *Industrial and Engineering Chemistry* with van der Meulen. A series of papers with S. E. A. McCallan was written on the mode of action of sulphur and copper fungicides. A series of papers with Albert Hartzell dealt with extracts of pyrethrum flowers and resulted in a mercury reduction method for the determination of pyrethrin I. Also, in his work with the Boyce Thompson Institute, he synthesized a number of plant growth substances, including alpha-naphthaleneacetic acid, which were studied with P. W. Zimmerman and led to a series of papers on the action of these growth substances on plants. In his work with the American Cyanamid Company he led the research group that studied the insecticide Parathion and which developed the less toxic Malathion.

Wilcoxon's interest in statistics began in 1925 with a study of R. A. Fisher's* book, *Statistical Methods for Research Workers*. This study was done in a small reading group in which his colleague at the Boyce Thompson Institute, W. J. Youden*, participated also. Wilcoxon became increasingly interested in the application of statistics in experimentation and this is apparent in a number of his papers on spore germination tests. His first publication in a statistics journal [14] dealt with uses of statistics in plant pathology. His research contributions in statistics were to range over rank tests*, multiple comparisons*, sequential ranking* (*see*

RANK ORDER STATISTICS), factorial experiments*, and bioassay*. Throughout his research, he sought statistical methods that were numerically simple and easily understood and applied.

Wilcoxon introduced his two rank tests, the rank-sum test for the two-sample problem and the signed-rank test for the paired-samples problem, in his 1945 paper [15]. This paper and a contemporary one by Mann and Whitney [10] led the way to the extensive development of nonparametric statistics, including Wilcoxon's own further contributions. (See WILCOXON SIGNED-RANK TEST; MANN–WHITNEY–WILCOXON STATISTIC; LINEAR RANK TESTS; WILCOXON-TYPE TESTS FOR ORDERED ALTERNATIVES IN RANDOMIZED BLOCKS.) There is no doubt that this paper was Wilcoxon's most important contribution. Although there were independent proposals of the two-sample statistic (see Kruskal [9] for historical notes), the paper became a major inspiration for the development of nonparametric methods. In addition, the statistical methodology introduced has had a major impact in applied statistics, particularly for applications in the social sciences, becoming one of the most popular of statistical tools.

Research on Wilcoxon test theory continued. He, himself, provided additional tables [16] in 1947 and again, with Katti and Wilcox [20], in 1963. Serfling [12] studied the properties of the two-sample rank-sum test in a setting where dependence is allowed within samples. Hollander et al. [8] studied the properties of the rank-sum test in a model where the assumption of independence between samples is relaxed. Hettmansperger [7] provides a modern treatment of nonparametric inference based on ranks.

Wilcoxon was interested in extensions of his basic rank procedures to new situations. He was an initiator of research in nonparametric sequential methods. Since nonparametric techniques were so successful in fixed sample-size situations, he felt that their good properties would naturally carry over to the sequential setting. This idea led to a number of sequential rank procedures developed under the leadership of Wilcoxon and Bradley [2, 3, 17, 18]. Other researchers continued the development of nonparametric sequential methods after this early work by Bradley and Wilcoxon and a comprehensive development of the theory is given by Sen [11]. Wilcoxon was interested also in the problem of testing whether two p-variate populations, $p \geqslant 2$, are equivalent and two of his proposals for this problem motivated refs. 1 and 4.

In experiments in the natural sciences and in the behavioral sciences, typically there are more than just one or two conditions (treatments, etc.) under investigation. Often multiple hypotheses need to be tested. In such settings, it is important to control the overall or experimentwise error rate. Wilcoxon recognized this and had a strong interest in multiple comparisons. In particular, his 1964 revision of the booklet [19] (joint with Roberta A. Wilcox) *Some Rapid Approximate Statistical Procedures*, features multiple comparison procedures based on Wilcoxon rank sums for the one-way layout and multiple comparison procedures based on Friedman rank sums for the two-way layout. The booklet played a significant role in the (now) widespread use of nonparametric multiple comparison procedures. In the early 1960s Wilcoxon also suggested and largely directed a dissertation on multiple comparisons by Peter Dunn-Rankin, part of which was published as a joint paper [6].

Other areas of Wilcoxon research seem less well known. From his research on insecticides and fungicides, he, with J. T. Litchfield, Jr. and K. Nolen, developed an interest in and a series of papers on a simplified method of evaluating dose–effect experiments. Daniel and Wilcoxon [5] devised fractional factorial designs* robust against linear and quadratic trends, anticipating to some degree the concept of trend-free block designs.

While Frank Wilcoxon was not an academician for much of his career, he was a teacher and a student throughout his life. His enthusiasm for statistics and his encouragement of others led many to more intensive study of the subject. It was typical of the man that, prior to visits to Russia in 1934 and 1935, he undertook a study of the language and retained a remarkably proficient reading knowledge throughout his life.

Frank Wilcoxon was recognized by his associates. He was a Fellow of the American Statistical Association and of the American Association for the Advancement of Science. He was a leader in the development of the Gordon Research Conference on Statistics in Chemistry and Chemical Engineering and a past Chairman of that Conference.

Karas and Savage list the publications of Wilcoxon. This and other material are listed in the bibliography.

References

[1] Bradley, R. A. (1967). *Proc. Fifth Berkeley Symp.*, Vol. 1, L. LeCam and J. Neyman, eds. Univ. of Calif. Press, Berkeley, CA, pp. 593–605.

[2] Bradley, R. A., Martin, D. C., and Wilcoxon, F. (1965). *Technometrics*, **7**, 463–483.

[3] Bradley, R. A., Merchant, S. D., and Wilcoxon, F. (1966). *Technometrics*, **8**, 615–623.

[4] Bradley, R. A., Patel, K. M., and Wackerly, D. D. (1971). *Biometrics*, **27**, 515–530.

[5] Daniel, C. and Wilcoxon, F. (1966). *Technometrics*, **8**, 259–278.

[6] Dunn-Rankin, P. and Wilcoxon, F. (1966). *Psychometrika*, **31**, 573–580.

[7] Hettmansperger, T. P. (1984). *Statistical Inference Based on Ranks*. Wiley, New York.

[8] Hollander, M., Pledger, G., and Lin, P. (1974). *Ann. Statist.*, **2**, 177–181.

[9] Kruskal, W. H. (1957). *J. Amer. Statist. Ass.*, **52**, 356–360.

[10] Mann, H. and Whitney, D. R. (1947). *Ann. Math. Statist.*, **18**, 50–60.

[11] Sen, P. K. (1981). *Sequential Nonparametrics*. Wiley, New York.

[12] Serfling, R. J. (1968). *Ann. Math. Statist.*, **39**, 1202–1209.

[13] van der Meulen, P. A. and Wilcoxon, F. (1923). *Ind. and Eng. Chem.*, **15**, 62–63.

[14] Wilcoxon, F. (1945). *Biometrics Bull.*, **1**, 41–45.

[15] Wilcoxon, F. (1945). *Biometrics Bull.*, **1**, 80–83.

[16] Wilcoxon, F. (1947). *Biometrics*, **3**, 119–122.

[17] Wilcoxon, F. and Bradley, R. A. (1964). *Biometrics*, **20**, 892–895.

[18] Wilcoxon, F., Rhodes, L. J., and Bradley, R. A. (1963). *Biometrics*, **19**, 58–84.

[19] Wilcoxon, F. and Wilcox, R. A. (1964). *Some Rapid Approximate Statistical Procedures*. Stamford Research Laboratories, Pearl River, New York. (Revision of a 1947, revised 1949, booklet by Wilcoxon.)

[20] Wilcoxon, F., Katti, S. K., and Wilcox, R. A. (1970). *Selected Tables in Mathematical Statistics*, Vol. 1, H. L. Harter and D. B. Owen, eds. Markham, Chicago, IL, pp. 171–259. (Originally prepared and distributed in 1963, revised 1968, by Lederle Laboratories, Pearl River, New York and the Department of Statistics, Florida State University, Tallahassee, Florida.)

Bibliography

Anon. (1965). *New York Times*, Nov. 19, p. 39, col. 2.

Bradley, R. A. (1966). *Biometrics*, **22**, 192–194.

Bradley, R. A. (1966). *Amer. Statist.*, **20**, 32–33.

Bradley, R. A. and Hollander, M. (1978). *International Encyclopedia of Statistics*, Vol. 2, W. H. Kruskal and J. M. Tanur, eds. The Free Press, New York, pp. 1245–1250.

Dunnett, C. W. (1966). *Technometrics*, **8**, 195–196.

Karas, J. and Savage, I. R. (1967). *Biometrics*, **23**, 1–11.

McCallan, S. E. A. (1966). Boyce Thompson Institute for Plant Research, *Contributions*, **23**, 143–145.

(BIOASSAY, STATISTICAL METHODS IN
FACTORIAL EXPERIMENTS
FISHER, RONALD AYLMER
LINEAR RANK TESTS
MANN–WHITNEY–WILCOXON STATISTIC
MULTIPLE COMPARISONS
RANK ORDER STATISTICS
RANK TESTS
SEQUENTIAL RANK ESTIMATORS
SIGNED RANK STATISTICS
TREND FREE BLOCK DESIGNS
WILCOXON SIGNED RANK TEST
WILCOXON-TYPE SCALE TESTS
WILCOXON-TYPE TESTS FOR ORDERED
   ALTERNATIVES IN RANDOMIZED
BLOCKS)

RALPH A. BRADLEY
MYLES HOLLANDER

WILCOXON RANK-SUM TEST *See* MANN-WHITNEY-WILCOXON STATISTIC

WILCOXON SCORES

These are used in the construction of linear rank statistics* estimating location. The scores are the coefficients $a_{n1}, a_{n2}, \ldots, a_{nm}$ of the order statistics* $X_1' \leqslant X_2' \leqslant \cdots \leqslant X_n'$ in the statistic

$$\sum_{i=1}^{n} a_{ni} X_i'.$$

For Wilcoxon scores

$$a_{ni} = i.$$

Reference

[1] Randles, R. H. and Wolfe, D. A. (1979). *Introduction to the Theory of Nonparametric Statistics.* Wiley, New York.

(*L*-STATISTICS
NORMAL SCORES
SAVAGE SCORES)

WILCOXON SIGNED RANK TEST

Wilcoxon [33] proposed a simple yet powerful test for the location of a symmetric population. (It is also applicable to paired data, as will be illustrated later.) Suppose $Z_1, \ldots, Z_n$ denotes a random sample from a population that is continuous and symmetric around θ. We want to test whether or not the population is located at some specific number θ_0, i.e., the null hypothesis is H_0: $\theta = \theta_0$. The first step is to subtract θ_0 from every Z_i, forming

$$Z_1 - \theta_0, Z_2 - \theta_0, \ldots, Z_n - \theta_0.$$

Now find the absolute values of these differences and rank the absolute values from smallest to largest. (Assign the rank of 1 to the smallest, rank 2 to the second smallest, ..., rank n to the largest.) The test statistic T^+ is the sum of the ranks which correspond to positive values of $Z_i - \theta_0$. To test H_0: $\theta = \theta_0$ against H_a: $\theta > \theta_0$ ($\theta < \theta_0$ or

$\theta \neq \theta_0$), we reject H_0 whenever

$$T^+ \geqslant k_\alpha \ \left[T^+ \leqslant n(n+1)/2 - k_\alpha \ \text{or} \right.$$

$$\left. |T^+ - n(n+1)/4| \geqslant k_{\alpha/2} - n(n+1)/4 \right].$$

Here k_α represents the value satisfying

$$\alpha = P\left[T^+ \geqslant k_\alpha \text{ when } H_0 \text{ is true} \right].$$

Values of k_α may be found, for example, in Table A.4 of Hollander and Wolfe [14]. When the sample size n is large, k_α may be aproximated by means of

$$k_\alpha \doteqdot \frac{n(n+1)}{4}$$

$$+ Z_\alpha \left[\frac{n(n+1)(2n+1)}{24} \right]^{1/2},$$

where Z_α is the point on a standard normal distribution with probability α above it.

This test procedure applies to paired data also. Let $(X_1, Y_1), \ldots, (X_n, Y_n)$ denote a random sample of paired observations from some bivariate population. For instance, X_i might denote a measurement on the ith person before treatment and Y_i, the same type of measurement made after treatment. Analysis of these settings often focuses on the magnitude of the treatment effect which can be measured by forming

$$Z_1 = Y_1 - X_1, Z_2 = Y_2 - X_2, \ldots,$$

$$Z_n = Y_n - X_n.$$

The Z's are analyzed in the fashion described earlier. The assumption of symmetry on the distribution of Z_i is very appropriate in these cases, since it is often assumed that, aside from the treatment effect θ, the two measurements would be interchangeable. That is, the pair $(X_i, Y_i - \theta)$ would have the same distribution as $(Y_i - \theta, X_i)$. Typically, in these problems the null hypothesis corresponds to no treatment effect, that is, $\theta_0 = 0$.

We illustrate the Wilcoxon signed rank test on data collected by Faria and Elliott [6] while investigating the existence of a 23-day physical biorhythm pattern. Eight female gymnasts had their physical performance recorded during periods which were predicted to be ones of high and low physical performances according to the individual athlete's

Table 1   VO₂ max of Athletes During High and Low Biorhythm Periods

Person	Y_i High Period	X_i Low Period	$Z_i = Y_i - X_i$ Difference	Abs. Value of Diff.	Rank of Abs. Value	Sign
A	50.89	44.12	6.77	6.77	7	+
B	40.37	42.21	−1.84	1.84	3	−
C	43.60	35.77	7.83	7.83	8	+
D	40.75	43.09	−2.34	2.34	5	−
E	41.23	41.83	−0.60	0.60	1	−
F	41.38	42.69	−1.31	1.31	2	−
G	53.01	48.84	4.17	4.17	6	+
H	45.04	42.87	2.17	2.17	4	+

biorhythm sine curve pattern. The study was double blind in that neither the athlete nor the person recording performance knew the state of the athlete's biorhythm pattern. The data in Table 1 show the maximum oxygen uptake (VO₂ max) for each athlete while on a treadmill test (averaged over two such measurements) during both high and low biorhythm periods.

In this example the null hypothesis is "no biorhythm pattern differences between high and low periods, $H_0: \theta = 0$", and the alternative would be increased oxygen uptake during high biorhythm periods, i.e., $H_a: \theta > 0$. With a significance level of $\alpha = 0.098$, Table A.4 in Hollander and Wolfe [14] yields $k_\alpha = 28$. Thus, we reject H_0, if $T^+ \geqslant 28$. Table 1 shows that the sum of the ranks corresponding to positive differences is

$$T^+ = 7 + 8 + 6 + 4 = 25.$$

Therefore the data are not sufficiently convincing that VO₂ max levels during high biorhythm periods exceed those during low periods.

The Wilcoxon signed rank test is distribution-free* in that its null distribution is appropriate under a very weak set of assumptions about the Z_i population, namely that the population is symmetric and continuous. Extensive tables for the null distribution are given by Wilcoxon et al. [34]. Other large sample approximations for the null distribution are discussed by Fellingham and Stoker [7] and Bickel [3]. The one-sided test is lo-

cally most powerful for detecting a shift in a logistic distribution among tests based on the signs and the ranks of the absolute values. (See Fraser [8].) It is also asymptotically locally efficient in the same setting among all α-level tests. (See Hájek and Sidak [10], p. 279.) Its tolerance to bad data is also good. (See Hettmansperger [11], p. 124.)

The test has excellent efficiencies, as derived by Pitman [19]. Hodges and Lehmann [12] show that for continuous symmetric distributions, its Pitman efficiency* relative to the one-sample t-test* has a lower bound of 0.864. This efficiency for some common underlying distributions takes on the value 1.0 for a uniform distribution, $3/\pi$ for a normal, $\pi^2/9$ for a logistic, 1.5 for a double exponential, and $+\infty$ for a Cauchy. The test's Bahadur efficiency* was established by Klotz [16]. Small sample efficiencies were investigated by Klotz [16] and Arnold [1]. Examples of its performance in Monte Carlo* studies may be found in Randles and Wolfe [23] and Nath and Duran [18]. Handling ties and zeros is studied by Pratt [20] and Cureton [5]. Also see Putter [21]. The performance under serial correlation* is investigated by Gastwirth and Rubin [9].

Extensions of this method to other settings have been investigated by many authors. The confidence interval corresponding to the Wilcoxon signed rank test was proposed by Tukey [30] and described by Moses in Chapter 18 of Walker and Lev [31]. Also, see Sen [26] and Lehmann [17] for further properties and extensions of this interval. Hodges and Lehmann [13] develop the point estimator corresponding to the test. Quade [22] and Salama and Quade [24] extend the method to randomized block* experiments. Two-stage and sequential versions have also been proposed. (See Weed and Bradley [32], Spurrier and Hewett [29], and Sen [27].) Multivariate extensions were proposed by Bickel [2], Sen and Puri [28], Killeen and Hettmansperger [15], and Brown and Hettmansperger [4]. Extensions to censored data* have been investigated by Woolson and Lachenbruch [35] and Schemper [25].

References

[1] Arnold, H. (1965). *Ann. Math. Statist.*, **36**, 1767–1778. (Small sample power of signed rank test.)

[2] Bickel, P. J. (1965). *Ann. Math. Statist.*, **36**, 160–173. (Marginal signed rank statistics used to construct a multivariate test.)

[3] Bickel, P. J. (1974). *Ann. Statist.*, **2**, 1–20. (Uses Edgeworth expansion to approximate the null distribution of rank statistics.)

[4] Brown, B. M. and Hettmansperger, T. P. (1985). *Tech. Report*, Dept. of Statist., Penn. State Univ., University Park, PA. (Use bivariate quantiles to develop an affine invariant analogue to the signed rank test for the bivariate location problem.)

[5] Cureton, E. E. (1967). *J. Amer. Statist. Ass.*, **62**, 1068–1069. (Large sample normal approximation for the signed rank test in the presence of zeros and ties.)

[6] Faria, I. E. and Elliott, T. L. (1980). *J. Sports Med. Phys. Fitness*, **20**, 81–85. (Source of the data for the example.)

[7] Fellingham, S. A. and Stoker, D. J. (1964). *J. Amer. Statist. Ass.*, **59**, 899–905. (Approximates the null distribution of the signed rank test.)

[8] Fraser, D. A. S. (1957). *Ann. Math. Statist.*, **28**, 1040–1043. (Shows the locally most powerful property of the signed rank test among tests based on the signs and ranks of the absolute values.)

[9] Gastwirth, J. L. and Rubin, H. (1971). *J. Amer. Statist. Ass.*, **66**, 816–820. (Effects of serial correlation among data on the signed rank test.)

[10] Hájek, J. and Sidak, Z. (1967). *Theory of Rank Tests*. Academic, New York.

[11] Hettmansperger, T. P. (1984). *Statistical Inference Based on Ranks*. Wiley, New York. (Advanced text mixing applications and theory of rank tests and estimators.)

[12] Hodges, J. L., Jr. and Lehmann, E. L. (1956). *Ann. Math. Statist.*, **27**, 324–35. (Discusses properties of the Pitman efficiency of the signed rank test compared to the one-sample *t*-test.)

[13] Hodges, J. L., Jr. and Lehmann, E. L. (1963). *Ann. Math. Statist.*, **34**, 598–611. (Source of point estimators of θ corresponding to signed rank test.)

[14] Hollander, M. and Wolfe, D. A. (1973). *Nonparametric Statistical Methods*. Wiley, New York. (Introductory text containing description and tables for the Wilcoxon signed rank test, as well as references to related work.)

[15] Killeen, T. J. and Hettmansperger, T. P. (1972). *Ann. Math. Statist.*, **43**, 1507–1516. (Proposes a multivariate signed rank test and compares tests via Bahadur efficiency.)

[16] Klotz, J. (1963). *Ann. Math. Statist.*, **34**, 624–632. (Small sample power and efficiency of the signed rank test.)

[17] Lehmann, E. L. (1963). *Ann. Math. Statist.*, **34**, 1507–1512. (Develops the asymptotic properties of the confidence interval corresponding to the signed rank test.)

[18] Nath, R. and Duran, B. S. (1984). *J. Statist. Comp. Simul.*, **20**, 235–260. (*t* tests compared to T^+ in a Monte Carlo.)

[19] Pitman, E. J. G. (1948). *Notes on Nonparametric Statistical Inference*, Columbia Univ., New York. (Finds the local asymptotic efficiency of the Wilcoxon signed rank test relative to the test based on the sample mean.)

[20] Pratt, J. W. (1959). *J. Amer. Statist. Ass.*, **54**, 655–667. (Discusses methods for handling zeros in the signed rank test.)

[21] Putter, J. (1955). *Ann. Math. Statist.*, **26**, 368–386. (Asymptotic comparison of the randomized and nonrandomized treatments of ties, in the contents of the sign and two-sample rank sum tests.)

[22] Quade, D. (1979). *J. Amer. Statist. Ass.*, **74**, 680–683. (Uses interblock information in the fashion of the signed rank test but in a many population setting.)

[23] Randles, R. H. and Wolfe, D. A. (1979). *Introduction to the Theory of Nonparametric Statistics*. Wiley, New York. (Develops the theory behind rank tests and estimates.)

[24] Salama, I. A. and Quade, D. (1981). *Commun. Statist. A*, **10**, 385–399. (Tests for ordered alternatives in complete blocks with interblock information included.)

[25] Schemper, M. (1984). *Commun. Statist. A*, **13**, 681–684. (Generalizes the signed rank test to interval and censored data.)

[26] Sen, P. K. (1963). *Biometrics*, **19**, 532–52. (Applies the estimation structure corresponding to the signed rank test to the estimation of relative potency in dilution assays.)

[27] Sen, P. K. (1981). *Sequential Nonparametrics*. Wiley, New York. (Describes sequential rank tests, estimators, and corresponding theory.)

[28] Sen, P. K. and Puri, M. L. (1967). *Ann. Math. Statist.*, **38**, 1216–1228. (Develops theory for a class of permutationally distribution-free rank order statistics for the multivariate one-sample location problem.)

[29] Spurrier, J. D. and Hewett, J. E. (1976). *J. Amer. Statist. Ass.*, **71**, 982–987. (Two stage signed rank test proposed.)

[30] Tukey, J. W. (1949). *Memo Rep. 17*, Statistical Research Group, Princeton Univ., Princeton, NJ. (Source of confidence interval for θ from the signed rank test.)

[31] Walker, H. M. and Lev, J. (1953). *Statistical Inference*, 1st ed. Holt, Rinehart and Winston, New York. (Chapter 18, written by L. Moses, includes a description of the confidence interval corresponding to the signed rank test.)

[32] Weed, H. D., Jr. and Bradley, R. A. (1971). *J. Amer. Statist. Ass.*, **66**, 321–326. (Sequential grouped signed rank tests are proposed.)

[33] Wilcoxon, F. (1945). *Biometrics*, **1**, 80–83. (Signed rank and rank sum tests first introduced.)

[34] Wilcoxon, F., Katti, S. K., and Wilcox, R. A. (1970). *Selected Tables in Mathematical Statistics*, Vol. I, Harter, H. L. and Owen, D. B. eds. Markham, Chicago, IL, pp. 171–259. (Null distribution critical values for $n \leq 50$.)

[35] Woolson, R. F. and Lachenbruch, P. A. (1980). *Biometrika*, **67**, 597–600. (Rank tests for location are discussed for censored matched pairs.)

(DISTRIBUTION-FREE METHODS
RANK ORDER STATISTICS
SIGNED RANK STATISTICS
TUKEY CONFIDENCE INTERVAL FOR
LOCATION)

RONALD H. RANDLES

WILCOXON-TYPE SCALE TESTS

The Wilcoxon (or Mann–Whitney*) test is well known as a distribution-free* test of location for two mutually independent random samples. However, the basic idea of this test can also be used to produce tests which are sensitive to scale* differences; these are frequently called Wilcoxon-type scale tests. The test applied to absolute values of deviations from central tendency is an exact ana-

log of the Wilcoxon test and the Sukhatme [10] test separates the deviations into two groups according to sign and then applies the Wilcoxon test to each group separately. A distinct advantage of these scale tests over most others is that they have a corresponding confidence interval procedure to estimate the ratio of scale parameters (see Noether [9]). These test and estimation procedures are rarely covered in books on nonparametric methods; Gibbons [4, 5] and Noether [8] are the exceptions.

NOTATION AND ASSUMPTIONS

Let $X_1, X_2, \ldots, X_m$ and $Y_1, Y_2, \ldots, Y_n$ denote mutually independent random samples from continuous distributions F_x and F_y with known medians M_x and M_y but unknown scale parameters σ_x and σ_y, respectively. Assume the scale distribution model that $(X - M_x)/\lambda$ and $(Y - M_y)$ are identically distributed with $\lambda = \sigma_x/\sigma_y$. The tests covered here are for $H_0: \lambda = \lambda_0$ where λ_0 can be any positive constant but is frequently equal to 1. The corresponding confidence interval procedures give estimates of λ.

TEST PROCEDURES

Both test procedures use the transformed variables $V_i = (X_i - M_x)/\lambda_0$ and $W_j = (Y_j - M_y)$.

For the first Wilcoxon-type test, we pool the $m + n$ absolute values $|V_1|, |V_2|, \ldots, |V_m|$ and $|W_1|, |W_2|, \ldots, |W_n|$, array them from smallest to largest, and assign ranks $1, 2, \ldots, m + n$ according to their relative magnitudes. The test statistic is T_V, the sum of the ranks assigned to the m absolute values of V's.

If we assume that X and Y have symmetric distributions, the test based on T_V is consistent against scale alternatives. A large value of T_V supports the alternative $\lambda > \lambda_0$ and a small value supports the alternative $\lambda < \lambda_0$.

Under H_0, the distribution of T_V is the same as the Wilcoxon rank sum test for location and exact tables are widely available. The most extensive tables are in Wilcoxon et al. [14], but most nonparametric textbooks also have tables. For larger samples we can use the fact that T_V is asymptotically normal with mean $m(m + n + 1)/2$ and variance $mn(m + n + 1)/12$ under H_0. A continuity correction of 0.5 can be incorporated in the test statistic. The asymptotic relative efficiency of this test relative to the F test is $6/\pi^2$ for normal distributions.

For the Sukhatme [10] test, the positive V's and W's are separated from the negative V's and W's. The positives are arrayed and ranked, and the absolute values of the negatives are arrayed and ranked. Let T_+ be the sum of the ranks of the V's in the positive group and T_- the sum of the ranks of the V's in the negative group. Then the test statistic is $T_V = T_+ + T_-$. A large value of T_V supports the alternative $\lambda > \lambda_0$ and a small value supports $\lambda < \lambda_0$. For large samples, the normal approximation can be used with mean $m(m + n + 2)/4$ and variance $mn(m + n + 7)/48$.

For small samples, the distribution of T_V depends on the numbers of positive and negative deviations, but Laubscher and Odeh [7] have generated the null distribution of a statistic which is a linear function of T_V and is in fact the original form of the Sukhatme [10] statistic. This statistic is

$$T = \sum_{i=1}^{m} \sum_{j=1}^{n} D_{ij},$$

where

$$D_{ij} = \begin{cases} 1 & \text{if } W_j < V_i < 0 \text{ or } 0 < V_i < W_j, \\ 0 & \text{otherwise.} \end{cases}$$

This form shows clearly that T is a modified form of the Mann–Whitney* statistic while T_V corresponds to the Wilcoxon statistic. Sukhatme [10] showed that T is consistent against the scale alternative without any assumption of symmetry.

For the statistic T, large values support the alternative $\lambda < \lambda_0$ and small values support $\lambda > \lambda_0$. Laubscher and Odeh [7] give tables of the critical values for $2 \leqslant m, n \leqslant 10$ and the corresponding complete null distributions are available from these authors. For large samples we can use the normal approximation with mean $mn/4$ and variance $mn(m + n + 7)/48$. Laubscher and Odeh [7] show that this normal approximation is adequate for m and n larger than 10 if a continuity correction of 0.5 is used. The asymptotic relative efficiency* of this test relative to the F-test* is $6/\pi^2$ for normal distributions.

TEST PROCEDURES FOR MEDIANS UNKNOWN

If the medians M_x and M_y are unknown, a natural approach would be to substitute the respective sample medians in calculating V and W and then proceed as before.

Fligner and Killeen [3] showed that the first Wilcoxon-type test with sample medians substituted remains distribution-free, has the same consistency properties, and the same asymptotic relative efficiency as when the medians are assumed known.

Sukhatme [11] showed that his test statistic $\tilde{T}$ computed from sample medians is asymptotically distribution-free for bounded and symmetric distributions and found the test highly efficient for exponential distributions. Laubscher and Odeh [7] used a simulation study to assess the accuracy of using their tables of T with the statistic $\tilde{T}$ and found this procedure quite conservative for the normal*, uniform*, and Cauchy* distributions.

CONFIDENCE INTERVAL PROCEDURES

A confidence interval estimate of λ that assumes symmetry and corresponds to the first Wilcoxon-type test is constructed by forming the mn ratios $|X_i - M_x|/|Y_j - M_y|$. These ratios are then arranged from smallest to largest and the confidence interval endpoints are the kth smallest and kth largest ratios. For m and n small, k is the rank of a left-tail critical value at level $\alpha/2$ of the Wilcoxon test statistic found in a table that gives the complete null distribution. The confidence coefficient is then $1 - \alpha$. For larger samples with $m, n \geqslant 12$, we can estimate k by

$$k = 0.5 + \frac{mn}{2} - Z_{\alpha/2} \sqrt{\frac{mn(m + n + 1)}{12}},$$

where $Z_{\alpha/2}$ satisfies $\Phi(Z_{\alpha/2}) = 1 - \alpha/2$. If k is not an integer it should always be rounded downward for a conservative result.

The confidence interval estimate of λ that corresponds to the Sukhatme test is constructed by forming the M ratios $(X_i - M_x)/(Y_j - M_y)$ which are positive because both the numerator and denominator have the same sign. Of course, $M \leqslant mn$ and approaches $(m/2)(n/2)$. These positive ratios are then arrayed and the confidence interval endpoints are the kth smallest and kth largest of the M. The correct value of k can be found from the tables in Laubscher and Odeh [7]. For large samples the value of k can be approximated by

$$k = 0.5 + \frac{mn}{4} - Z_{\alpha/2} \sqrt{\frac{mn(m + n + 7)}{48}}.$$

A third type of confidence interval estimate of λ which is also a Wilcoxon-type procedure can be used if the X and Y variables can take on only positive values. Here σ_x and σ_y should be regarded as measures of variability from zero as opposed to spread about a central value. The procedure is to form the mn ratios X_i/Y_j and array them. The confidence interval endpoints are the kth smallest and kth largest ratio where k is the same as in the first Wilcoxon-type confidence procedure.

MODIFICATIONS AND EXTENSIONS

Sukhatme [12] proposed another modification of his statistic. More recently, Deshpande and Kusum [1] proposed a generalization of Sukhatme's test for distributions

with an equal quantile which is not necessarily the median.

The Kruskal–Wallis test* is the generalization of the Wilcoxon test for more than two samples. It could be used to detect scale differences if the data are transformed to absolute values of deviations from their respective medians. The performance of this procedure is compared with some other nonparametric c sample scale tests in Tsai et al. [13].

An excellent comprehensive survey of nonparametric tests for scale is given in Duran [2].

References

[1] Deshpande, J. V. and Kusum, K. (1984). *Austral. J. Statist.*, **26**, 16–24.

[2] Duran, B. S. (1976). *Commun. Statist. A*, **5**, 1287–1312. (Comprehensive survey of nonparametric tests for scale.)

[3] Fligner, M. A. and Killeen, T. J. (1976). *J. Amer. Statist. Ass.*, **71**, 210–213.

[4] Gibbons, J. D. (1985a). *Nonparametric Methods for Quantitative Analysis.* American Sciences Press, Columbus, OH. (Scale tests and confidence interval procedures are covered in Chap. 5.)

[5] Gibbons, J. D. (1985b). *Nonparametric Statistical Inference.* Dekker, New York. (Scale tests and confidence interval procedures are covered in Chap. 10.)

[6] Kochar, S. C. and Gupta, R. P. (1986) *Commun. Statist. A*, **15**, 231–239.

[7] Laubscher, N. F. and Odeh, R. E. (1976). *Commun. Statist. A*, **5** 1393–1407.

[8] Noether, G. E. (1967). *Elements of Nonparametric Statistics.* Wiley, New York.

[9] Noether, G. E. (1972). *Amer. Statist.*, **26**, 39–41.

[10] Sukhatme, B. V. (1957). *Ann. Math. Statist.*, **28**, 188–194.

[11] Sukhatme, B. V. (1958a). *Ann. Math. Statist.*, **29**, 60–78.

[12] Sukhatme, B. V. (1958b). *Biometrika*, **45**, 544–548.

[13] Tsai, W. S., Duran, B. S., and Lewis, T. O. (1975). *J. Amer. Statist. Ass.*, **70**, 791–796.

[14] Wilcoxon, F., Katti, S. K., and Wilcox, R. A. (1972). *Selected Tables in Mathematical Statistics*, Vol. I. American Mathematical Society, Providence, RI, pp. 171–259.

(ANSARI-BRADLEY W STATISTICS
CAPON TEST
DISTRIBUTION-FREE TESTS

KLOTZ TEST
MANN–WHITNEY-WILCOXON STATISTIC
NONPARAMETRIC METHODS
RANK TESTS
SCALE TESTS
WILCOXON SCORES
WILCOXON SIGNED RANK TEST
WILCOXON-TYPE TESTS FOR ORDERED
  ALTERNATIVES IN RANDOMIZED
  BLOCKS)

JEAN DICKINSON GIBBONS

WILCOXON-TYPE TESTS FOR ORDERED ALTERNATIVES IN RANDOMIZED BLOCKS

INTRODUCTION

Consider the typical two-way layout consisting of n blocks, k treatments, and one observation per cell. The data can be expressed as

$$X_{ij} = \beta_i + \theta_j + e_{ij},$$
$$i = 1, \ldots, n; \ j = 1, \ldots, k, \quad (1)$$

where β_i denotes the ith block effect, θ_j, the jth treatment effect, and the $\{e_{ij}\}$ are independent and identically distributed (i.i.d.) random variables according to the *unknown* continuous distribution F. In this entry, we review some competing nonparametric tests of

$$H_0: \theta_1 = \theta_2 = \cdots = \theta_k \quad (2)$$

versus

$$H_a: \theta_1 \leqslant \theta_2 \leqslant \cdots \leqslant \theta_k, \quad (3)$$

where at least one of the inequalities is strict. Procedures discussed include (i) Page's* [15] and Jonckheere's* [10] tests based on average rank correlation coefficients, (ii) Hollander's [8] and Doksum's [4] tests based on generalizations of Wilcoxon's paired-replicates signed-rank statistic*, and (iii) tests based on weighted rankings due to Tukey [23], Moses [14], and Quade [19].

Examples of treatments which exhibit a natural ordering and where there may be prior reasons to suggest a deviation from H_0 will be in the direction H_a (or some other specific direction such as $\theta_k \leqslant \theta_{k-1} \leqslant \cdots \leqslant \theta_1$) include amounts of practice, levels of fertilizer, intensities of a stimulus, angles of knee flexion for conditioning, and doses of radiation. If there is a priori evidence that a deviation from H_0 is in the direction $\theta_{j_1} \leqslant \theta_{j_2} \leqslant \cdots \leqslant \theta_{j_k}$, for some permutation $(j_1, j_2, \ldots j_k)$ of $(1, 2, \ldots, k)$, the tests described in the next section can be applied by simply relabeling the treatments so that the X values corresponding to treatment j_1 are called the X_{i1}'s, the X values corresponding to treatment j_2 are called the X_{i2}'s, and so forth.

SOME NONPARAMETRIC METHODS BASED ON RANKS

Tests based on average rank correlation coefficients

Let r_{ij}, $i = 1, \ldots, n$, $j = 1, \ldots, k$, be the rank of X_{ij} in the joint ranking of $\{X_{i\alpha}\}_{\alpha=1}^k$. Let $R_j = \sum_{i=1}^n r_{ij}$, $j = 1, \ldots, k$. Then Page's [15] statistic can be written as

$$L = \sum_{j=1}^k j R_j. \quad (4)$$

H_0 is rejected in favor of H_a for large values of L. L is distribution-free* under H_0 and Page gives upper percentiles of the null distribution of L in the $\alpha = 0.001$, 0.01, and 0.05 regions for $k = 3$, $n = 2(1)20$ and $k = 4(1)8$, $n = 2(1)12$. These tables are reproduced as Table A.16 of Hollander and Wolfe [9]. The large sample approximation ($n \to \infty$) treats

$$L^* = \frac{L - E_0(L)}{\{\text{Var}_0(L)\}^{1/2}}$$
$$= \frac{L - nk(k+1)^2/4}{\{nk^2(k+1)^2(k-1)/144\}^{1/2}} \quad (5)$$

as an approximate $N(0, 1)$ random variable under H_0.

The test based on L is equivalent to one based on $\rho = \sum_{i=1}^{n} \rho_i$ where

$$\rho_i = 1 - \left[\left\{ 6 \sum_{j=1}^{k} (r_{ij} - j)^2 \right\} \Big/ \{ k^3 - k \} \right] \tag{6}$$

is Spearman's rank order correlation coefficient* between postulated order and observation order in the ith block. The relationship between L and ρ is

$$L = (k^3 - k)\rho/12 + nk(k + 1)^2/4.$$

For the model

$$X_{ij} = \beta_i + (j - 1)\theta + e_{ij}, \tag{7}$$

where the $\{ e_{ij} \}$ are i.i.d. according to F, Hollander [8] derived the Pitman asymptotic relative efficiency (ARE) (*see* PITMAN EFFICIENCY) of L with respect to the likelihood ratio* statistic t [for testing $\theta = 0$ when the common distribution of the e's is $N(0, \sigma^2)$] for the alternatives

$$H_n: P_n(X_{ij} \leqslant x)$$
$$= F(x - \beta_i - (j - 1)cn^{-1/2}).$$

The ARE is

$$E(L, t)$$
$$= k \left\{ (k + 1)^{-1} \right\} \left[12\sigma^2 \left\{ \int f^2(x)\, dx \right\}^2 \right],$$

where $\sigma^2 = \text{Var}(F)$ and f is the density corresponding to F. Some values of $E(L, t)$ when F is normal are given in Table 4.1 of [8].

Page's statistic is based on *within*-block rankings. Another distribution-free rank test based on within-block rankings is Jonckheere's test [10] which utilizes Kendall's rank correlation coefficient (*see* KENDALL'S TAU). Jonckheere's test is similar in spirit to Page's test and could be considered the forerunner of the latter. Jonckheere rejects H_0 in favor of H_a for large values of $\tau = \sum_{i=1}^{n} \tau_i$, where τ_i is Kendall's between postulated order and observation order in the ith block. For the H_n alternatives, Hollander [8] showed that

the Pitman ARE of Page's test with respect to Jonckheere's test is, for all F,

$$E(\rho, \tau) = \frac{k(2k + 5)}{2(k + 1)^2}.$$

$E(\rho, \tau) = 1$ when $k = 2$ since in this case both procedures are equivalent to the paired-replicates sign test. $E(\rho, \tau)$ increases until it reaches its maximum value of 1.042 at $k = 5$ and then decreases to its limiting value ($k \to \infty$) of 1. Thus, asymptotic relative efficiencies for the H_n alternatives favor ρ over τ.

Tests based on generalized signed-rank statistics

Hodges and Lehmann [6] pointed out that procedures for the two-way layout which only use within-block comparisons sacrifice useful information—information that can be gained from between-block comparisons. Hollander [8] and Doksum [4] generalized Wilcoxon's signed-rank statistic for paired-replicates data to provide test statistics that do make use of between-block information. Let $Y_{uv}^{(i)} = |X_{iu} - X_{iv}|$ and $R_{uv}^{(i)} = $ rank of $Y_{uv}^{(i)}$ in the joint ranking from least to greatest of $\{ Y_{uv}^{(i)} \}_{i=1}^{n}$. Let $T_{uv} = \sum_{i=1}^{n} R_{uv}^{(i)} \psi_{uv}^{(i)}$, where $\psi_{uv}^{(i)} = 1$ if $X_{iu} < X_{iv}$, $= 0$ otherwise, be the Wilcoxon signed-rank statistic* between treatments u and v. Hollander's statistic is

$$Y = \sum_{u < v}^{k} T_{uv}. \tag{8}$$

H_0 is rejected in favor of H_a when Y is significantly large. Y is *not* distribution-free under H_0. In particular, the null variance of Y depends on the underlying F. The null mean of Y is distribution-free and is $E_0(Y) = k(k - 1)n(n + 1)/8$. The null variance is

$$\text{Var}_0(Y) = \frac{n(n + 1)(2n + 1)k(k - 1)}{144}$$
$$\times \{ 3 + 2(k - 2)\rho_0^n(F) \}, \tag{9}$$

where

$$\rho_0^n(F) = \{(n+1)(2n+1)\}^{-1}$$
$$\times \left[1 + 7n - 6n^2 + 24(n-1) \right.$$
$$\left. \times \{\lambda(F)(n-2) + 2\mu(F)\} \right] \quad (10)$$

$$\mu(F) = P(X_1 < X_2; X_1 < X_3 + X_4 - X_5), \quad (11)$$

$$\lambda(F) = P(X_1 < X_2 + X_3 - X_4;$$
$$X_1 < X_5 + X_6 - X_7), \quad (12)$$

and $X_1, X_2, \ldots, X_7$ are i.i.d. according to the continuous distribution F. Consistent estimators $\hat{\mu}(F)$ of $\mu(F)$ and $\hat{\lambda}(F)$ of $\lambda(F)$ are available (cf. refs. 12, 8, 13). An asymptotically distribution-free test is obtained by replacing $\mu(F)$, $\lambda(F)$ by $\hat{\mu}(F)$, $\hat{\lambda}(F)$ in (11) and (12) to obtain $\widehat{\text{Var}}_0(Y)$, then treating $Y' = \{Y - E_0[Y]\}/\{\widehat{\text{Var}}_0(Y)\}^{1/2}$ as an asymptotic $(n \to \infty)$ $N(0,1)$ random variable under H_0. The Pitman ARE of Y with respect to t, for the H_n alternatives, is

$$E(Y, t) = \frac{24(k+1)\sigma^2 \left[\int g^2(x)\, dx \right]^2}{3 + 2(k-2)\rho^*(F)}, \quad (13)$$

where G is the distribution of $X_1 - X_2$ when X_1, X_2 are i.i.d. according to F, g is the density of G, $\sigma^2 = \text{Var}(F)$, and $\rho^*(F) = 12\lambda(F) - 3$. $E(Y, t) > 0.864$ for all F. When F is normal, $E(Y, t) = 0.963$ for $k = 3$ and $\to 0.989$ as $k \to \infty$. These values compare favorably with the corresponding values of Page's test $(0.716, 0.955)$. See Table 4.1 of ref. 8 for more values.

A much simpler procedure than estimating $\mu(F)$ and $\lambda(F)$ is to use upper bounds for $\mu(F)$ and $\lambda(F)$ (see refs. 8 and 9, Section 7.6). Lehmann [12] showed that for all continuous F, $\lambda(F) \leqslant 7/24$ and Hollander [8] showed that for all continuous F, $\mu(F) \leqslant \{(2^{1/2} + 6)/24\} = 0.3089$. For all distri-

butions for which they have been calculated, both $\mu(F)$ and $\lambda(F)$ are quite close to their upper bounds (see Table 1).

Thus a slightly conservative test is to treat

$$Y_U = \frac{Y - \{k(k-1)n(n+1)/8\}}{V_U} \quad (14)$$

as an approximate $N(0,1)$ random variable under H_0, where

$$V_U^2 = \frac{n(n+1)(2n+1)k(k-1)}{144}$$
$$\times \{3 + 2(k-2)\rho_U^n\} \quad (15)$$

and

$$\rho_U^n = \frac{n^2 + 2n(2^{1/2} - 1) + 3 - 2(2^{1/2})}{(n+1)(2n+1)}. \quad (16)$$

ρ_U^n is obtained by replacing $\mu(F)$ and $\lambda(F)$ by their upper bounds in (10).

Lehmann [12] obtained the value of $\lambda(F)$ for F uniform*, normal*, and Cauchy*. Hollander [7] obtained the values of $\lambda(F)$ for F exponential* and the values of $\mu(F)$ for F uniform, normal, and exponential. Mann and Pirie [13] obtained the value of $\lambda(F)$ for F logistic and the values of $\mu(F)$ for F logistic* and Cauchy. Incidentally, the known lower bounds for $\lambda(F)$ and $\mu(F)$ are not nearly as good as the upper bounds. Lehmann showed that for all continuous F, $\lambda(F) \geqslant 1/4$ and Mann and Pirie improved the lower bound to $\lambda(F) \geqslant 5/18$ for all continuous F. Mann and Pirie also showed that for all continuous F, $\mu(F) \geqslant 1/4$.

Doksum [4] proposed a test that is very similar to the test based on Y. Doksum uses the random variables $U_{uv} = T_{uv} - \sum_{i=1}^{n} \psi_{uv}^{(i)}$, and considers the statistic $D = \sum_{u<v}^{k} (U_u - U_v)$, where $U_u = \sum_{j=1}^{k} U_{uj}/k$. An asymptotically equivalent statistic is $D' = \sum_{u<v}^{k} (T_u - T_v)$. The asymptotic null variance of D also

Table 1 Values of $\lambda(F)$ and $\mu(F)$ for Various Distributions

F	Uniform	Normal	Logistic	Exponential	Cauchy
$\lambda(F)$	0.2909	0.2902	0.2898	0.2894	0.2879
$\mu(F)$	0.3083	0.3075	0.3064	0.3056	0.3043

**Table 2   Approximate Significance Levels
of Hollander's and Doksum's Conservative
Tests for Nominal $\alpha = 0.05$**

n	3	5	10	
k	10	5	3	F
Y	0.0492	0.0494	0.0497	Normal
	0.0476	0.0482	0.0491	Cauchy
D	0.0488	0.0487	0.0491	Normal
	0.0471	0.0467	0.0478	Cauchy

depends on $\lambda(F)$ and an asymptotically distribution-free test can be defined by using a consistent estimator of $\lambda(F)$. Alternatively, one could use the upper bound for $\lambda(F)$ and form a slightly conservative procedure. Doksum shows $E(D, Y) \geqslant 1$ for all F and k. We note that $\lim_{k \to \infty} E(D, Y) = 1$ for all F. How close $E(D, Y)$ is to 1 depends on k and on how close $\lambda(F)$ is to Lehmann's upper bound of $7/24$. Since all calculated values of $\lambda(F)$ are very close to $7/24$, only a very slight increase in Pitman efficiency* is obtained by using Doksum's test instead of Hollander's test. Some values of the maximum increase are

F	Normal	Rectangular	Exponential
$\max_k E(D, Y)$	1.00356	1.00193	1.00575.

In each case, the maximum occurs at $k = 5$.

Mann and Pirie [13] investigate the closeness of the levels (to their nominal values) of the conservative test based on Y and the associated conservative test based on D. Table 2 gives approximate significance levels of both tests in cases where F is normal and Cauchy. The nominal α-value is 0.05 and the entries are based on the normal approximation to the null distributions of the test statistics.

Tests based on weighted rankings

Moses' [14] statistic is based on a generalization of an idea due to Tukey [24]. Let $S_i = \text{Range}\{X_{i1}, \ldots, X_{ik}\}$, $i = 1, \ldots, n$, and let Q_i be the rank of S_i among $S_1, \ldots, S_n$. Then Moses' statistic is

$$M = \sum_{i=1}^{n} \rho_i Q_i, \qquad (17)$$

where ρ_i is the Spearman rank order correlation coefficient between treatment order and observation order in the ith block. The M statistic makes use of between-block comparisons and is distribution-free under H_0. (For $k = 2$, M is equivalent to the Wilcoxon signed-rank statistic). Salama and Quade [20] give critical values of $M(k^3 - k)/6$ for the $\alpha = 0.0005, 0.001, 0.0025, 0.005, 0.01, 0.025, 0.05$ and 0.10 regions, for $k = 3$ and $n = 2(1)15$. A large sample approximation ($n \to \infty$) treats

$$M^* = \frac{M}{\left[n(n+1)(2n+1)/\{6(k-1)\} \right]^{1/2}} \qquad (18)$$

as an approximate $N(0, 1)$ random variable under H_0. Salama and Quade [20] give Monte Carlo comparisons between M and Doksum's [4] D, on the basis of expected significance levels.

Tukey [23] suggested a statistic which totalled block ranks only for those blocks for which the treatment observations are in the exact postulated order. As k increases, much information is sacrificed by such a statistic. That is why Moses generalized Tukey's idea to weighted rankings.

Quade [14] used the idea of weighted rankings to produce a class of distribution-free statistics having an asymptotic χ_{k-1}^2 distribution under H_0. The tests in Quade's class can be considered competitors of Friedman's [5] test* of H_0 versus the general alternatives that not all θ's are equal. Pitman ARE results (for $k > 2$) for Quade's tests are given by Tardif [22].

Other tests and references

Pirie and Hollander [17] proposed a distribution-free normal scores* test based on within-block rankings. Their statistic is $W = \sum_{i=1}^{n}\sum_{j=1}^{k} jD_{r_{ij}}^k$, where D_j^k is the expected value of the jth order statistic of a random sample of size k from a $N(0, 1)$ distribution. (These expected values are called normal scores. Teichroew [23] gives the values of D_j^k for $k \leqslant 20$.) H_0 is rejected in favor of H_a for large values of W. The W test is intended as a distribution-free competitor of Page's test based on L. For $k = 2$ and $k = 3$, the two tests are equivalent. For $k \geqslant 4$, for many distributions the test based on W exhibits better efficiency properties than Page's test. In particular, for $k \geqslant 4$ and normal, uniform, and exponential populations, for the H_n alternatives the Pitman ARE of W with respect to L is greater than 1, and for uniform and exponential populations, the limiting $(k \to \infty)$ value of this efficiency is ∞. Pirie and Hollander give null distribution tables of W in the $\alpha = 0.01, 0.05,$ and 0.10 regions for $k = 4$, $n = 1(1)10$ and $k = 5$, $n = 1(1)5$. A large sample $(n \to \infty)$ normal approximation is also provided.

Puri and Sen [18] generalized Hollander's Y test to a Chernoff–Savage [3] class of tests which includes Y as a special case. Let

$$X_{i,uv}^* = X_{iv} - X_{iu},$$

$$u < v = 1, \ldots, k, \ i = 1, \ldots, n,$$

and set

$$S_{uv} = n^{-1} \sum_{\alpha=1}^{n} E_{n\alpha} Z_{uv,\alpha}^{(n)},$$

$$u < v = 1, \ldots, k,$$

where $Z_{uv,\alpha}^{(n)}$ is 1 or 0 according as the αth smallest observation among $|X_{i,uv}^*|$, $i = 1, \ldots, n$, is from a positive or negative X^*, and $E_{n\alpha}$ is the expected value of the αth order statistic of a sample of size n from a distribution with CDF $\psi^*(x) = \psi(x) - \psi(-x)$, $x \geqslant 0$, $= 0$, $x < 0$. ψ must satisfy certain regularity conditions; see Assumptions I–III of Puri and Sen. Puri and Sen propose to reject H_0 in favor of H_a for large values of $V = \sum_{u < v}^{k} S_{uv}$. When ψ is uniform $[-1, 1]$, V reduces to Hollander's Y statistic. When ψ is the standard normal distribution, V is a normal scores statistic denoted as V_Φ, which is not distribution-free under H_0. To form an asymptotically distribution-free test based on V_Φ, Puri and Sen provide a consistent estimator of its null asymptotic variance. It has excellent efficiency properties. For the H_n alternatives, Puri and Sen show that $E(V_\Phi, t) \geqslant 1$ for all F.

Kepner and Robinson [11] generalized the paired-replicates Wilcoxon signed-rank statistic to produce new distribution-free tests of H_0 versus H_a for $k = 3$ and $k = 4$ treatments. They did not develop procedures for cases where $k > 4$. For $k = 3$, let $Y_{i1} = X_{i2} - X_{i1}$ and $Y_{i2} = 2X_{i3} - X_{i1} - X_{i2}$. Then pool the $2n$ Y_{ij}'s and compute

$$T_3 = \sum_{i=1}^{n} \sum_{j=1}^{2} \psi_{ij}^* R_{ij}^+,$$

where $\psi_{ij}^* = \text{sign}(Y_{ij})$ and R_{ij}^+ is the rank of $|Y_{ij}|$ in the pooled collection of the $2n$ absolute values. H_0 is rejected in favor of H_a for large values of T_3. For $k = 4$, let $Y_{i1} = X_{i3} - X_{i1}$, $Y_{i2} = X_{i4} - X_{i2}$, and $Y_{i3} = X_{i4} - X_{i3} + X_{i2} - X_{i1}$. Pool the $3n$ Y_{ij}'s and compute

$$T_4 = \sum_{i=1}^{n} \sum_{j=1}^{3} \psi_{ij}^* R_{ij}^+,$$

where $\psi_{ij}^* = \text{sign}(Y_{ij})$ and R_{ij}^+ is the rank of $|Y_{ij}|$ in the pooled collection of the $3n$ absolute values. H_0 is rejected in favor of H_a for large values of T_4. Under H_0, Kepner and Robinson show that the distribution of T_j, $j = 3, 4$, is the same as that of the Wilcoxon signed rank* statistic for a random sample of size $m = n(j - 1)$. That is, under H_0, T_j has the same distribution as $\sum_{i=1}^{m} \psi_i^* R_i^+$, where $Z_1, \ldots, Z_m$ are independent, each Z_i has a distribution that is symmetric about 0 (the Z's do not have to be identically distributed), $\psi_i^* = \text{sign}(Z_i)$, and R_i^+ is the rank of $|Z_i|$ among $|Z_1|, \ldots, |Z_m|$.

For sequences of alternatives that include the H_n alternatives, Kepner and Robinson compute Pitman AREs of T_3, T_4 with respect to Page's L, Hollander's Y, and the likelihood ratio t test for model (7). Though their calculations (see Table 1 of Kepner–Robinson [11]) show Y to have favorable Pitman AREs compared to L, t, and T_j, $j = 3, 4$, Kepner and Robinson advocate T_3 and T_4 on the basis of ease of computation and the distribution-free property. However, the conservative test based on Y is easy to implement (see the example in the next section), and since it is nearly distribution-free, it should be preferred on the basis of its efficiency properties.

Pitman AREs for shift alternatives such as H_n, where k is fixed and $n \to \infty$, favor between-blocks rank statistics such as Hollander's Y and Doksum's D over within-blocks rank statistics such as Page's L and the Pirie–Hollander normal scores statistic. Pirie [16] computes Pitman AREs for shift alternatives where n is fixed and $k \to \infty$. Those calculations tend to favor the within-blocks rank tests over the between-blocks rank tests. The "k fixed, $n \to \infty$" efficiencies are interpretable for situations where n is large relative to k (the more frequently encountered cases) whereas the "n fixed, $k \to \infty$" calculations provide guidance for the choice of tests when k is large relative to n.

Skillings and Wolfe [21] investigate a class of distribution-free within-blocks rank tests for ordered alternatives that can be applied to the more general randomized block design with $n_{ij} \geqslant 1$ observations in the (i, j) cell.

AN EXAMPLE

Brown and Hollander [2] describe a study by Angel et al. [1] on transfer of information from the manual control system to the oculomotor control system. We quote from their description.

Whenever the hand is moved under visual guidance, several motor systems are employed.

1. The manual control system directs movement of the hand.
2. The oculomotor system, which moves the eyes, contains specialized subsystems for acquiring and tracking visual targets. (a) The saccadic system specializes in moving the eyes from one position to another very rapidly. (b) The smooth pursuit system matches the eye velocity to velocity of the visual target.

When a subject tries to keep his eyes on a continuously moving target, the smooth

pursuit tends to make the eyes follow the target, but occasionally the target "escapes." The saccadic system then produces a sudden jump of the eyes, bringing the target back to the center of the field. The more accurate the smooth pursuit, the fewer saccadic corrections needed. The number of "saccades" per unit time may thus be taken as a rough measure of the inaccuracy of smooth pursuit tracking.

Other workers have shown that visual tracking is more accurate (i.e., fewer saccades occur) when the target is moved by the subject rather than by an external agent. This is taken as evidence that the motor commands directed to the subject's hand are somehow made available to the oculomotor system. Eye movements can thus anticipate hand movements without the delay involved in sensory feedback. If one assumes that information about hand movement is conveyed to the oculomotor system, the question arises: How long does this information persist in usable form? How long does the brain store information about motor commands directed to the hand? The following experiment was designed to answer questions of this nature.

Eye position was recorded while the subject tried to keep his eyes on a visual target. On each experimental run, the subject moved the target actively by means of a joy stick. Target motion was either synchronous with that of the hand or delayed by a fixed amount (0.18, 0.36, 0.72, or 1.44 sec). On each control run, target motion was a tape-recorded copy of the pattern generated by the subject during the previous test run. (Regardless of the delay between joy stick and target motion, the frequency of saccades was generally smaller during the test run than during the matched control run. The results suggest that information about hand movement is available to the oculomotor system for at least 1 sec following manual performance.)

The experiment just described was repeated with 10 normal volunteer subjects. The experimental conditions (different amounts of lag between hand and target movement) were presented in random order. After each run, the subject watched a tape-recorded

Table 3   Saccades During Test / Saccades During Playback

	Lags				
	1 (0 sec)	2 (0.18 sec)	3 (0.36 sec)	4 (0.72 sec)	5 (1.44 sec)
1	0.600	0.880	0.587	1.207	0.909
2	0.610	0.526	0.603	1.333	0.813
3	0.592	0.527	0.976	1.090	0.996
4	0.988	0.841	0.843	1.010	1.112
5	0.611	0.536	0.654	0.910	0.936
6	0.981	0.830	1.135	0.838	0.876
7	0.698	0.520	0.673	0.892	0.857
8	0.619	0.493	0.734	0.948	0.862
9	0.779	0.759	0.764	0.824	8.851
10	0.507	0.712	0.787	0.868	0.880

copy of the target movement generated on the test run. Each run lasted 100 sec. The number of saccades (rapid, jumping eye movements) was determined for each test run and compared with the matched control run.

The data we analyze here, given in Table 3, are defined as $X = \{$saccades during test$\}/\{$saccades during playback$\}$. Do the X's tend to increase as the lags increase? We can address this question using a test of H_0 versus H_a.

We use the data of Table 3 to illustrate the application of Page's L, Hollander's Y, and

Moses' M. Table 4 gives the within-block ranks.

We find, from (4) and (5),

$$L = R_1 + 2R_2 + 3R_3 + 4R_4 + 5R_5 = 516,$$

$$L^* = (516 - 450)/(250)^{1/2} = 4.17.$$

To compute Y, given by (8), we need to calculate the 10 signed-rank statistics $T_{12}, T_{13}, \ldots, T_{45}$. We illustrate the calculation of T_{12} in Table 5.

Similarly, we obtain $T_{13} = 38$, $T_{14} = 52$, $T_{15} = 53$, $T_{23} = 47$, $T_{24} = 55$, $T_{25} = 55$, $T_{34} = 47$, $T_{35} = 48$, and $T_{45} = 19$. Then from (8)

Table 4   Within-block ranks

	Lags				
	1 (0 sec)	2 (0.18 sec)	3 (0.36 sec)	4 (0.72 sec)	5 (1.44 sec)
1	2	3	1	5	4
2	3	1	2	5	4
3	2	1	3	5	4
4	3	1	2	4	5
5	2	1	3	4	5
6	4	1	5	2	3
7	3	1	2	5	4
8	2	1	3	5	4
9	3	1	2	4	5
10	1	2	3	4	5
	$R_1 = 25$	$R_2 = 13$	$R_3 = 26$	$R_4 = 43$	$R_5 = 43$

Table 5   Calculation of T_{12}

i	$X_{i2} - X_{i1}$	$\lvert X_{i2} - X_{i1} \rvert$	$R_{12}^{(i)}$	$\psi_{12}^{(i)}$	$R_{12}^{(i)} \cdot \psi_{12}^{(i)}$
1	0.280	0.280	10	1	10
2	−0.084	0.084	4	0	0
3	−0.065	0.065	2	0	0
4	−0.147	0.147	6	0	0
5	−0.075	0.075	3	0	0
6	−0.151	0.151	7	0	0
7	−0.118	0.178	8	0	0
8	−0.126	0.126	5	0	0
9	−0.020	0.020	1	0	0
10	0.205	0.205	9	1	9
					$T_{12} = 19$

and (14)–(16), we obtain $Y = 433$:

$$\rho_U^{10} = \frac{100 + 20(2^{1/2} - 1) + 3 - 2(2)^{1/2}}{11(21)}$$

$$= 0.470,$$

$$V_U^2 = \frac{10(11)(21)(5)(4)}{144}\{3 + 6(0.470)\}$$

$$= 1867.3,$$

$$Y_U = \frac{433 - 275}{\{1867.3\}^{1/2}} = 3.66.$$

Table 6 assists in the calculation of M. From (18), we obtain $M^* = 38/(96.25)^{1/2} = 3.87$. Thus the normal deviates for the tests based on L, Y, and M are all significantly large ($P \leqslant 0.0001$) and each test indicates strong evidence that there is an increasing trend in the X's as the lags increase.

Table 6   Calculation of M

i	ρ_i	S_i	Q_i	$Q_i \rho_i$
1	0.6	0.620	9	5.4
2	0.6	0.807	10	6.0
3	0.8	0.563	8	6.4
4	0.7	0.271	2	1.4
5	0.9	0.400	6	5.4
6	−0.1	0.305	3	−0.3
7	0.6	0.372	4	2.4
8	0.8	0.455	7	5.6
9	0.7	0.092	1	0.7
10	1.0	0.373	5	5.0
				$M = 38.0$

Another question of interest is: At roughly what time lag do the X values tend to be close to 1, indicating the hand movement information is no longer available to the brain? Angel et al. [1] estimated the lag at which the effect of target control disappears by constructing confidence intervals for the population median at each lag. The midpoint of the 98% confidence interval (based on the Wilcoxon signed-rank test) corresponding to the lag 0.72 sec. is 1.005. It is reasonable to conclude from this, and examination of the confidence intervals for the other lags, that the effect of target control disappears at a lag close to 0.72 sec. See ref. 1 for more details.

References

[1] Angel, R. W., Hollander, M., and Wesley, M. (1973). *Perception and Psychophysics*, **14**, 506–510. (Technical paper; describes an experiment relating to motor memory. The saccades data are given in this paper.)

[2] Brown, B. W., Jr. and Hollander, M. (1977). *Statistics: A Biomedical Introduction*. Wiley, New York. (General reference; contains a concise description of the motor memory experiment of ref. 1.)

[3] Chernoff, H. and Savage, I. R. (1958). *Ann Math. Statist.*, **29**, 972–994. (Technical paper; proves asymptotic normality of a large class of rank statistics and shows that the two-sample normal scores test has Pitman ARE $\geqslant 1$ with respect to the t-test.)

[4] Doksum, K. A. (1967). *Ann. Math. Statist.*, **38**, 878–883. (Technical paper; generalizes the paired-replicates signed-rank test to obtain an asymptotically distribution-free test for ordered alternatives for k treatments.)

[5] Friedman, M. (1937). *J. Amer. Statist. Ass.*, **32**, 675–701. (Technical paper; introduces a distribution-free rank test which arises naturally if the usual two-way layout F-statistic is computed for the ranks instead of the actual observations.)

[6] Hodges, J. L., Jr. and Lehmann, E. L. (1962). *Ann. Math. Statist.*, **32**, 482–497. (Technical paper; gives rank methods for combination of independent experiments in analysis of variance. Shows how additional information can be gained by using intrablock comparisons in addition to interblock comparisons.)

[7] Hollander, M. (1966). *Ann. Math. Statist.*, **37**, 735–738. (Technical paper; provides an asymptotically distribution-free "treatments vs. control"

multiple comparison procedure based on signed-rank statistics.)

[8] Hollander, M. (1967). *Ann. Math. Statist.*, **38**, 867–877. (Technical paper; generalizes the paired-replicates signed-rank test to obtain an asymptotically distribution-free test for ordered alternatives for k treatments.)

[9] Hollander, M. and Wolfe, D. A. (1973). *Nonparametric Statistical Methods*. Wiley, New York. (General reference.)

[10] Jonckheere, A. R. (1954). *Brit. J. Statist. Psych.*, **7**, 93–100. (Technical paper; generalizes the paired-replicates sign test to a distribution-free test for ordered alternatives for k treatments.)

[11] Kepner, J. L. and Robinson, D. H. (1984). *J. Amer. Statist. Ass.*, **79**, 212–217. (Technical paper; generalizes the Wilcoxon paired-replicates signed-rank test to distribution-free tests for ordered alternatives for the cases $k = 3$ and $k = 4$.)

[12] Lehmann, E. L. (1964). *Ann. Math. Statist.*, **35**, 726–734. [Technical paper; provides asymptotically nonparametric methods, including tests and confidence procedures for linear models. Gives bounds on the parameter $\lambda(F)$.]

[13] Mann, B. L. and Pirie, W. R. (1982). *Commun. Statist. A*, **11**, 1107–1117. [Technical paper; improves Lehmann's lower bound on $\lambda(F)$. Studies asymptotic levels of the conservative versions of Hollander's and Doksum's tests.]

[14] Moses, L. E. (1963). Unpublished notes. (The test based on M is introduced.)

[15] Page, E. B. (1963). *J. Amer. Statist. Ass.*, **58**, 216–230. (Technical paper; generalizes the paired-replicates sign test to a distribution-free test for ordered alternatives for k treatments.)

[16] Pirie, W. R. (1974). *Ann. Statist.*, **2**, 374–382; correction, **3**, 796. (Technical paper; computes Pitman AREs of within-blocks rank tests to between-blocks rank tests for models where (i) k is fixed and $n \to \infty$ and (ii) n is fixed and $k \to \infty$.)

[17] Pirie, W. R. and Hollander, M. (1972). *J. Amer. Statist. Ass.*, **67**, 855–857. (Technical paper; introduces a within-blocks distribution-free normal scores tests for ordered alternatives for k treatments.)

[18] Puri, M. L. and Sen, P. K. (1968). *Ann. Math. Statist.*, **39**, 967–972. (Technical paper; generalizes Hollander's ordered alternatives test to a Chernoff–Savage class of tests which includes a test based on normal scores.)

[19] Quade, D. (1979). *J. Amer. Statist. Ass.*, **74**, 680–683. (Technical paper; introduces a class of weighted-rankings test statistics for k treatments.)

[20] Salama, I. A. and Quade, D. (1981). *Commun. Statist.-Theor. Meth.*, **10**, 385–399. (Technical paper; considers Moses' weighted rankings test based on Spearman correlation, and an analogous test based on Kendall's correlation.)

[21] Skillings, J. H. and Wolfe, D. A. (1978). *J. Amer. Statist. Ass.*, **73**, 427–431. [Technical paper; considers a class of distribution-free within-blocks rank tests that are applicable to the randomized block design with $n_{ij} \geqslant 1$ observations in the (i, j) cell.]

[22] Tardif, S. (1987). *J. Amer. Statist. Ass.*, **82**, 637–644. (Technical paper; obtains efficiency results for Quade–Tukey-type tests based on weighted rankings.)

[23] Teichroew, D. (1956). *Ann. Math. Statist.*, **27**, 410–426. (Technical paper; contains tables of expected values of order statistics and products of order statistics for samples of size 20 and less from the normal distribution.)

[24] Tukey, J. W. (1957). *Ann. Math. Statist.*, **28**, 987–992. (Technical paper; introduces some distribution-free rank tests which may be considered the precursors to Moses' test and Quade's tests.)

Acknowledgments

This research was supported by the Air Force Office of Scientific Research AFSC, USAF, under Grant AFOSR-82-K-0007 to the Florida State University. I thank Edsel Peña for checking the calculations for the saccades data.

(DISTRIBUTION-FREE METHODS
JONCKHEERE TEST FOR ORDERED
   ALTERNATIVES
PAGE TEST FOR ORDERED
   ALTERNATIVES
RANK TESTS
SIGNED-RANK STATISTICS)

Myles Hollander

WILD SHOT, WILD VALUE *See* Outliers

WILDLIFE SAMPLING

INTRODUCTION

A major objective in wildlife management is the determination of population size or density of a species in a given area and to monitor changes. Animals are harder to count than plants because animals often hide from us or they move so fast that the same individual may be counted repeatedly or not at all. Some of the important methods for estimating animal numbers or their densities based on (i) direct counts of units as used in quadrat* or transect* sampling and (ii) indirect counts and indices such as capture–recapture*, change-in-ratio, catch-effort methods, and indices based on track and roadside counts, etc.

QUADRATS AND TRANSECTS

Estimates of total counts of animals are often based on sample plots which are rectangular or square in shape called *quadrats*. Quadrat sampling is the preferred technique for most big game surveys. A known number of quadrats are randomly chosen and the population units within each quadrat are counted. Quadrat size and shape will depend upon the habitat, abundance, and mobility of the species.

When a population is randomly distributed, the number of quadrats to be sampled is given by $s = S/(1 + Nc^2)$, where S is the total number of quadrats, N the population size, and C is the coefficient of variation* (C.V.). When animals congregate in shelters according to a negative binomial distribution*, the number (s_1) is given by $s_1 = (s/N + 1/k)/c^2$, where $k > 0$. For a given N and c, $s_1 > s$.

When population density varies over areas, stratified sampling with quadrats selected at random from each area is recommended. Stratified quadrat sampling with optimum allocation of sampling effort was used in aerial surveys of Alaska caribou [31] which reduced variance by more than half over that of simple random sampling. Where it is difficult to count all animals in all the sampled quadrats, two-phase sampling may be adopted. To estimate the number of beavers in a tract, we count only the number of beaver lodges per quadrat from a preliminary sample of quadrats; also, the ratio of the number of beavers to total beaver lodges can be estimated from a subsample of the quadrats. The two estimates can be com-

bined to estimate the number of beavers in the tract. For a specified cost, double-sampling also proved more efficient [30] than single sampling for estimation of age-composition of commerical fish by species. Techniques have been developed for estimating total catch using ratio-estimates based on post-stratification of sampled trips by market categories and subsampling clusters of a given weight for each category.

Improvements in quadrat sampling in aerial surveys were proposed in refs. 5 and 6 by developing models for estimating the magnitude of visibility.

There are basically three transect methods: the line intercept*, the strip, and the line transect*. The line intercept is generally used for estimating population total (N_1 say) of inanimate objects, e.g., den sites of animals. If w_i is the width of a den site parallel to the baseline W intersecting a random transect of length L and if m is the number of den sites intersecting n randomly selected transects, unbiased estimates of N_1 and its error are given by

$$\hat{N}_1 = \frac{1}{n} \sum_{i}^{m} \frac{1}{p_i},$$

$$v(\hat{N}_1) = \frac{1}{n^2} \sum_{i=1}^{m} \frac{(1 - p_i)}{p_i^2}, \tag{1}$$

where $p_i = w_i/W$. This method is also known as *length biased* sampling [8]. An excellent review is given in ref. 10. A further treatment of the subject is in LINE INTERCEPT SAMPLING.

In the strip-transect method all animals within a strip of fixed width are counted. Parallel lines, one strip width ($2W$, say) apart determine the population of strips. All the n animals observed within the sampled strips are counted. The estimate of total number of animals (N_2, say) is given by

$$N_2 = \frac{A \cdot n}{2LW}, \tag{2}$$

where L and A represent the length and population area of the strip. Estimation of marine mammals using strip transects is discussed in ref. 13. When the terrain is difficult and the population area is large, strip sampling will be more appropriate than quadrats. When population areas are irregular in shape, probability proportional to size* (PPS) selection of strips with replacement was adopted in making aerial censuses of wildlife populations in East Africa [17].

The line-transect method is generally used to measure the rare or fast moving terrestrial mammals and birds that are difficult to locate in a specified area. The method utilizes data on all subjects seen as these are flushed on either side of a transect line of length L across a transect of area A for estimating the "effective width" W of the strip covered by the observer as he moves along the transect line. Different methods based on various models proposed for estimating the population are of the form $N = AD$, where D, the population density, is estimated by

$$\hat{D} = \frac{n}{L}\left(\frac{1}{\hat{W}}\right).$$

Assumptions underlying the models are given in LINE TRANSECT SAMPLING. References 15 and 27 deal with parametric estimators and with consequences for departures from the assumptions, particularly for moving animals; nonparametric estimators are dealt with in refs. 3 and 12. Line-transect methods based on elliptic flushing curves are discussed in ref. 2. Refs. 11 and 22 deal with their relative efficiencies. Substantial work has been done on moving populations using radiotelemetry [27].

CAPTURE–RECAPTURE*

A number of M animals from a population are caught, marked, and released. On a second occasion, a sample of n animals is captured. If m be the number of marked animals in the sample, then a biased estimate of its size (N_3) and its variance for a closed population are

$$\hat{N}_3 = \frac{n}{m}M,$$

$$v\left(\hat{N}_3\right) = \frac{\hat{N}_3^2\left(\hat{N}_3 - M\right)\left(\hat{N}_3 - n\right)}{Mn\left(\hat{N}_3 - 1\right)}. \quad (3)$$

This was first given by Petersen [23] using tagged plaice. Since $\text{C.V.}(\hat{N}_3) \simeq m^{-1/2}$, we should have sufficient recaptures in the second sample for $\hat{N}_3$ to be efficient. For scarce populations, inverse sampling [27] is adopted, in which the second sampling proceeds until a specified m is recovered; for a given M, m can be chosen beforehand to give the desired C.V. CAPTURE-RECAPTURE METHODS provides a recent review for "closed" and "open" populations when marking is done over time. Reference 24 provides a design robust to unequal probability of capture and ref. 25 relates the capture probability to auxiliary variables. Reference 21 provides confidence limits for the parameter values by the Jolly–Seber method. A rationale of the procedures in capture-recapture methods, with illustrations from terrestrial invertebrates and small mammals, is provided in ref. 1.

CHANGE-IN-RATIO

The method provides estimates of the population size by removing individual animals if the change-in-ratio of some attribute of the animal, e.g., sex or age composition is known. The method was first noted in ref. 18 for estimating deer and other wildlife populations. Consider a population of animals and assume that a differential in the number of males and females occurs before and after hunting. Then the maximum likelihood* estimator (MLE) of total (N_t) and the number of males (M_t) of the population at times $t = 1, 2$ based on certain assumptions are given by

$$\hat{N}_t = \frac{R_m - Rp_t}{p_1 - p_2}, \quad (4)$$

$$\hat{M}_t = p_t\hat{N}_t,$$

where $p_t = m_t/n_t$, n_t is the sample size at the beginning ($t = 1$) and end ($t = 2$) of the "harvest period," m_t is the number of males in n_t, and $R = R_m + R_f$, where R_m and R_f are the number of males and females caught between $t = 1, 2$. The number of females F_t is estimated by subtraction, $F_t = N_t - M_t$.

The formulas (4) assume (i) a closed population, (ii) all animals have the same probability of capture in the tth sample, and (iii) R_m and R_f are known exactly. A detailed discussion of the method when these assumptions are violated is given in ref. 27, Section 9.1.2.

CATCH-EFFORT

In the catch-effort method one unit of sampling is assumed to catch a fixed proportion of the population. If N is the initial population size, n_t is the size of the sample removed during the tth time period, e_t is the effort applied in the tth time period, k_t and E_t are, respectively, the cumulative catch and effort through time period $(t - 1)$, and $C_t = n_t/e_t$ is the catch per unit effort in the tth time period, we have, from refs. 9 and 20,

$$E[C_t/k_t] = K(N - k_t), \qquad (5)$$

where K is the catchability coefficient. The values of C_t plotted against k_t will be a straight line with intercept KN and slope K, whence N can be estimated.

If (5) holds, it can be shown [9] that

$$E[\log C_t] = \log(KN) - K(\log e)E_t. \qquad (6)$$

Estimates of K and N can also be obtained if the points $(\log C_t, E_t)$ lie on a straight line. Both (5) and (6) have been widely used in fishery work.

INDICES

Indices are estimates of animal populations derived from counts of animal signs, breeding birds, nests, etc. The results do not give estimates of absolute populations, but do indicate trends in populations over years and habitats. Stratified random sample surveys are being annually conducted in the U.S. and Canada for detecting and measuring changes in abundance of nongame breeding birds at the height of the breeding season [14, 26]. Data on counts of birds heard or seen at predetermined stops on predetermined routes are collected by volunteer observers who make roadside counts of birds heard or seen according to a specified sampling scheme. An estimate of change and its standard error for a particular species between two successive years is given by

$$\hat{R} = (\bar{y} - \bar{x})/\bar{x},$$

$$se(\hat{R}) = \left(\frac{\bar{y}}{\bar{x}}\right)\left[\frac{\operatorname{Var} \bar{y}}{y^2} + \frac{\operatorname{Var} \bar{x}}{x^2} - \frac{2\operatorname{cov}(\bar{x}, \bar{y})}{\overline{xy}}\right]^{1/2}, \qquad (7)$$

where $\bar{x}$ and $\bar{y}$ are, respectively, mean number of birds per route based on the first and second year.

MEASUREMENT ERRORS*

A very important source of variation arises from errors in the measurement of a quantity. For example, aerial surveys may be biased due to errors in sighting, recognition, and reliable counting of animals by observers. Mail surveys may be subject to high nonsampling errors [28, 29] due to nonresponse bias, prestige bias (e.g., hunter exaggerating his kill), memory bias, or other factors. Caughley [3] showed that in aerial censusing of large mammals the observer missed a significant number of animals on the transect and that the number missed increased with increasing width of transect, cruising speed, and altitude. He recommended measurement and correction of the bias using regression* models. Air counts of ducks are also subject to visibility bias. Among prairie breeding species about two-thirds are missed [16]. Besides, observers may be biased for the most colourful species, e.g., mallards and canvasbacks as against pintails, which are conspicuous. Also, they may fail to determine the edge of the transect accurately. Consequently, each year apart from the operational transects, air–ground comparison transects were selected purposively between operational transects and surveyed by the same aerial crew during the same time to provide a correction factor for the estimate based on a larger sample of air counts.

Aerial photographs were used (scale 1 : 10,000) to correct for bias in aerial counts for estimating snow goose populations in Hudson Bay [19]. A subsample of the sampled areas was rephotographed using a scale of 1 : 3000. The 1 : 10,000 photographs yielded wide coverage but resulted in a biased count of total geese owing to difficulty in differentiating between geese and topographical features. The 1 : 3000 photographs yielded more accurate counts as well as discernibility between the white and blue phases of the species. Population estimates of both phases were obtained using regression techniques [32].

References

[1] Blower, J. G., Cook, L. M., and Bishop, J. A. (1981). *Estimating the Size of Animal Populations*, Allen and Unwin, London.

[2] Burnham, K. P. (1979). *Biometrics*, **35**, 587–596.

[3] Burnham, K. P. and Anderson, D. R. (1976). *Biometrics*, **32**, 325–336.

[4] Caughley, G. (1974). *J. Wildl. Manag.*, **38**, 921–933.

[5] Cook, R. D. and Jacobsen, J. O. (1979). *Biometrics*, **35**, 735–742.

[6] Cook, R. D. and Martin, F. B. (1974). *J. Amer. Statist. Ass.*, **69**, 345–359.

[7] Cormack, R. M. (1968). *Ocean. Marine Biol.*, **6**, 455–506.

[8] Cox, D. R. (1969). In *New Developments in Survey Sampling*, N. L. Johnson and H. Smith, eds.Wiley, New York, pp. 120–140.

[9] DeLury, D. B. (1947). *Biometrics*, **3**, 145–167.

[10] DeVries, P. G. (1979a). In *Sampling Biological Populations*, R. M. Cormack, G. P. Patil, and D. S. Robson, eds. Satellite Program in Statistical Ecology. International Cooperative Publishing House, Fairland, MD, pp. 1–70.

[11] DeVries, P. G. (1979b). *Biometrics*, **35**, 743–748.

[12] Eberhardt, L. L. (1978). *J. Wildl. Manag.*, **42**, 1–31.

[13] Eberhardt, L. L., Chapman, D. G., and Gilbert, J. R. (1979). *Wildl. Monogr. No. 63*.

[14] Erskine, A. J. (1973). *Can. Wildl. Ser. Prog. Note*, **32**, 1–15.

[15] Gates, C. E. (1969). *Biometrics*, **25**, 317–328.

[16] Hanson, R. C. and Hawkins, A. S. (1975). *Naturalist*, **25**, 8–11.

[17] Jolly, G. M. (1969). *East Afr. Agric. For. J.*, **34**, 46–49.

[18] Kelker, G. H. (1940). *Proc. Utah Acad. Sci., Arts and Lett.*, **17**, 65–69.

[19] Kerbes, R. H. (1975). *Can. Wildl. Serv. Rep. Series*, **35**.

[20] Leslie, P. H. and Davis, D. H. S. (1939). *J. Animal. Ecol.*, **8**, 94–113.

[21] Manly, B. F. J. (1984). *Biometrics*, **40**, 749–758.

[22] Otten, A. and DeVries, P. G. (1984). *Biometrics*, **40**, 1145–1150.

[23] Petersen, C. E. J. (1896). *Rep. Dan. Biol. Statist.*, **6**, 1–48.

[24] Pollock, K. H. and Otto, M. C. (1983). *Biometrics*, **39**, 1035–1049.

[25] Pollock, K. H., Hines, J. E., and Nichols, J. D. (1984). *Biometrics*, **40**, 329–340.

[26] Robbins, C. S. and Vanvelzen, W. T. (1967). *Spec. Sci. Rep. Wildl. No. 102*, U.S. Forestry and Wildlife Service.

[27] Seber, G. A. F. (1980). *The Estimation of Animal Abundance and Related Parameters*, Vol. 2. Griffin, London, England.

[27a] Seber, G. A. F. (1986). *Biometrics*, **42**, 267–292. (Reviews techniques developed beyond those in ref. 27. Contains six sections with 339 references including wildlife (61%), fisheries (21%), and insects and plants (4%) as major groups.)

[28] Sen, A. R. (1972). *J. Wildl. Manag.*, **36**, 951–954.

[29] Sen, A. R. (1973). *J. Wildl. Manag.*, **37**, 485–491.

[30] Sen, A. R. (1986). *Fishery Bull.*, **84**, 409–421.

[31] Siniff, D. B. and Skoog, R. O. (1964). *J. Wildl. Manag.*, **28**, 391–401.

[32] Smith, G. E. J. (1975). Appendix to Kerbes, R. H. (1975). *Can. Wildl. Service. Rep. Series*, **35**.

(CAPTURE-RECAPTURE METHODS
ECOLOGICAL STATISTICS
FISHERIES RESEARCH, STATISTICS IN
STATISTICS IN ANIMAL SCIENCE
STRATIFICATION
SURVEY SAMPLING
TRANSECT METHODS)

A. R. SEN

WILK AND KEMPTHORNE FORMULAS

For ANOVA* of data from cross-classified experiment designs with unequal but proportionate class frequencies, Wilk and Kempthorne [3] have developed general formulas for the expected values of mean squares (EMSs) in the ANOVA table. We will give appropriate formulas for an $a \times b$ two-way cross-classification.

If the number of observations for the factor level combination (i, j) is n_{ij}, then proportionate class frequencies require

$$n_{ij} = nu_i v_j, \quad i = 1, \ldots, a, \ j = 1, \ldots, b.$$

In this case the Wilk and Kempthorne formulas, as presented by Snedecor and Cochran [2] for a variance components* model, are: For main effect* factor A,

$$\text{EMS} = \sigma^2 + \frac{nuv(1 - u^*)}{a - 1}$$
$$\times \left\{ (v^* - B^{-1})\sigma_{AB}^2 + \sigma_A^2 \right\};$$

for main effect of factor B,

$$\text{EMS} = \sigma^2 + \frac{nuv(1 - v^*)}{b - 1}$$
$$\times \left\{ (u^* - A^{-1})\sigma_{AB}^2 + \sigma_B^2 \right\};$$

for main interaction* $A \times B$,

$$\text{EMS} = \sigma^2 + \frac{nuv(1 - u^*)(1 - v^*)}{(a - 1)(b - 1)} \sigma_{AB}^2,$$

where

$$u = \sum_{i=1}^{a} u_i, \quad u^* = \left(\sum_{i=1}^{a} u_i^2 \right) \bigg/ u^2,$$

$$v = \sum_{j=1}^{b} v_j, \quad v^* = \left(\sum_{j=1}^{b} v_j^2 \right) \bigg/ v^2,$$

$\sigma_A^2, \sigma_B^2, \sigma_{AB}^2$ are the variances of the (random) terms representing main effects of A and B and the $A \times B$ interactions in the model, respectively, and σ^2 is the (common) variance of the residual terms.

Detailed numerical examples and discussions are available, for example in Bancroft [1].

References

[1] Bancroft, T. A. (1968). *Topics in Intermediate Statistical Methods*, Vol. 1. Iowa State University Press, Ames, IA.

[2] Snedecor, G. W. and Cochran, W. G. (1967). *Statistical Methods*, 7th ed. Iowa State University Press, Ames, IA.

[3] Wilk, M. B. and Kempthorne, O. (1955). *J. Amer. Statist. Ass.*, **50**, 1144–1167.

(ANALYSIS OF VARIANCE
FACTORIAL EXPERIMENTS
INTERACTION
MAIN EFFECT
VARIANCE COMPONENTS)

WILKS' INNER AND OUTER TOLERANCE INTERVALS

These are intervals, based on a random sample from a continuous population, formed by a pair of order statistics* $(X_r', X_s') (r < s)$ such that *for inner tolerance intervals*,

$$\Pr\left[\xi_{p_1} < X_r' < X_s' < \xi_{p_2} \right] \geq \gamma;$$

for outer tolerance intervals,

$$\Pr\left[X_r' < \xi_{p_1} < \xi_{p_2} < X_s' \right] \geq \gamma,$$

where $p_2 - p_1 = p_0 \, (> 0)$ is the preassigned "tolerance" and ξ_p is the $100p\%$ quantile* of the population. These intervals were originally called "confidence" intervals by Wilks [2], but "tolerance intervals" is now a more appropriate term.

Guenther [1] describes the aims of construction of these inner (outer) intervals as follows:

> The tolerance interval is to capture no more (less) than a specified proportion $(1 - 2p_0)$ of the distribution with high probability, but not so much less (more) than that, while attempting to keep the tail areas from being too small (large).

Methods of determining r and s so as to require the least possible size of sample are described in ref. 1.

References

[1] Guenther, W. C. (1985). *J. Qual. Technol.*, **17**, 40–43.

[2] Wilks, S. S. (1962). *Mathematical Statistics*. Wiley, New York.

(DISTRIBUTION-FREE TOLERANCE
  LIMITS
NONPARAMETRIC TOLERANCE LIMITS
TOLERANCE REGIONS, STATISTICAL)

WILKS' LAMBDA CRITERION *See*
LAMBDA CRITERION, WILKS'S

WILLCOX, WALTER FRANCIS

>*Born:* March 22, 1861 in Reading,
>Massachusetts.
>
>*Died:* October 30, 1964 in Ithaca,
>New York.
>
>*Contributed to:* American Statisti-
>cal Association, history of statistics,
>vital statistics.

In the course of his remarkably long life,
Walter Francis Willcox was influential in the
development and structuring of statistics, and
especially vital statistics*, as an independent
subject in the general field of scientific
method. He taught "an elementary course in
statistical methods with special treatment of
vital and moral statistics" at Cornell University in 1892–1893, soon after joining the
faculty in 1891. He served at Cornell for 40
years, retiring in 1931.

However, his work was by no means confined to university teaching. He was active in
working on the 12th United States Census
(1900) and 1899 to 1901, and initiated and
supervised a large volume of *Supplementary
Analysis and Derivative Tables.* Willcox was
in steady demand as a consultant on collection and analysis of demographic statistics,
especially in regard to international migration. A topic attracting much of his attention
was the basis of apportionment of seats in
the House of Representatives of the U.S.
Congress. Much interesting detail is available in ref. 1, which was produced in connection with Willcox's 100th birthday, and
in the obituary [2], which contains personal
reminiscences.

Willcox was President of the American
Statistical Association* in 1912 and of the
International Statistical Institute* in 1947.
He was very active in the work of both of
these organizations for many years, and
compiled the first extensive list [3] of definitions of statistics, from G. Achenwall in
1749 to R. A. Fisher* in 1934.

In his later years he interested himself in
the history of statistics—in particular of demographic statistics—and produced a number of insightful analyses in this field.

References

[1] Leonard, W. R. (1961). *Amer. Statist.*, **15**(1), 16–19.

[2] Rice, S. A. (1964). *Amer. Statist.*, **18**(5), 25–26.

[3] Willcox, W. F. (1935). *Int. Statist. Inst. Review*, **3**, 388–399.

(DEMOGRAPHY
KIAER, A. N. (*See Supplement*)
MIGRATION
VITAL STATISTICS)

WILLIAMS' TEST (OF TREND)

A problem in isotonic inference* that has
received considerable attention arises when
there are $k + 1$ independent samples from
normal distributions with equal variances
and means known a priori to be monotonically ordered. For example, in an animal
bioassay we believe that $\mu_0 \leqslant \cdots \leqslant \mu_k$,
where μ_0 is the mean of a response of inter-
est in a control group, and $\mu_1, \ldots, \mu_k$ are the
means corresponding to increasing exposures
of the test agent. Williams [5] suggests a
sequential testing procedure to find the
largest index j such that $\mu_0 = \mu_j$. The null
hypotheses $\mu_0 = \mu_k, \mu_0 = \mu_{k-1}, \ldots, \mu_0 = \mu_1$
are tested sequentially until an hypothesis is
not rejected, or all hypotheses have been
tested. Knowledge of the ordering of the
means is utilized by basing the tests on
$\bar{x}_0, \tilde{\mu}_1, \ldots, \tilde{\mu}_k$, where $\bar{x}_0$ is the sample mean
for the control group and $\tilde{\mu}_i$ is the maximum
likelihood estimate (MLE) of μ_i, $i = 1, \ldots, k$
(see discussion of the amalgamation process
under ISOTONIC INFERENCE or in ref. 1).

For a monotonically nondecreasing order of the means the test of $\mu_0 = \mu_k$ is rejected for large values of Williams' statistic

$$\bar{t}_k = (\tilde{\mu}_k - \bar{x}_0)(2s^2/r)^{-1/2}, \qquad (1)$$

where r is the common sample size and s^2 is an unbiased estimate of the common error variance of the sampled distributions. Replacing $\tilde{\mu}_k$ in (1) by the sample mean $\bar{x}_k$ would yield the familiar Student's t-statistic. Hence the notation $\bar{t}_k$. A table of critical values for $\bar{t}_k$ is given in ref. 5 for the case where all sample-sizes are equal and for $\alpha = 0.01$, $\alpha = 0.05$, $k = 1, \ldots, 10$.

In ref. 6, Williams extends his procedure to allow for a control group that is larger than the common sample size of the treatment groups. He also discusses the optimal relative ratio of the control to treatment sample size, provides tables of critical values for increased control replication, and gives approximate critical values for moderate differences between the number of replications within treatment groups.

If the test statistic $\bar{t}_k$ given in (1) exceeds the tabled value $\bar{t}_{k,\alpha}$, then it is suggested [5] that to test the next hypothesis, $\mu_0 = \mu_{k-1}$, the subscript k in (1) should be replaced by $k - 1$. The size of the latter test conditional on having rejected $\mu_0 = \mu_k$, however, exceeds the nominal size α. It is incorrectly stated [5] as equal to α, but the correction is noted in ref. 7. How much larger the test size is than the nominal size α at any of the tests subsequent to the first one has apparently not been established. It is also noted in ref. 7 that the proof at the beginning of Section 6 in ref. 6 that $P_i < \alpha$ is incorrect as stated, but the result remains valid.

A result in ref. 4 applies to the nominal type I error rate. If $\mu_0 \leqslant \cdots \leqslant \mu_k$, then the null hypotheses sequentially tested, H_k: $\mu_0 = \mu_k$, H_{k-1}: $\mu_0 = \mu_{k-1}, \ldots, H_1$: $\mu_0 = \mu_1$, are nested with $H_k \subset \cdots \subset H_1$. If each sequential test that is conducted is of size α, then the probability that no type I error is made is at least $1 - \alpha$. This is because for no type I error to occur the first true null hypothesis in the sequential order must be tested and not rejected. But this event stops the sequential testing.

Due to the monotonic ordering of the means, a test of $\mu_0 = \mu_k$ is equivalent to a test that all the means are equal ($\mu_0 = \cdots = \mu_k$) against the alternative of at least one strict inequality in the ordering. Power comparisons that include Williams' test and others for treating this problem may be found in refs. 2, 3, and 5. The results do not suggest that a single test is preferable in all circumstances. Williams' test, or the modified Williams test [3, 6] wherein $\bar{x}_0$ is replaced by its MLE $\tilde{\mu}_0$ in (1), appears to perform particularly well when k is small ($k \leqslant 2$) and in cases where the means are equal except at the smallest and/or largest index values.

As an example, suppose that mice are exposed to an ether compound for 18 weeks in a laboratory. Four groups of 16 each are exposed to concentrations of 0, 30, 100, and 300 parts per million (ppm). One indicator of a potential detrimental health effect from the exposure is a reduction in body weight. The mean body weights at the end of 18 weeks are 544, 514, 532, and 488 at 0, 30, 100, and 300 ppm, respectively. The estimated variance of the normal distributions is $s^2 = 1426$ with 60 degrees of freedom.

In this case we have a priori knowledge that the means should be nonincreasing as the exposure increases, $\mu_0 \geqslant \mu_1 \geqslant \mu_2 \geqslant \mu_3$. The MLEs of the means are 544, 523, 523, and 488 and $\bar{x}_0 = 544$. The value of $(2s^2/r)^{-1/2}$ is 0.075, so $\bar{t}_3 = (488 - 544)0.075 = -4.2 < -\bar{t}_{3,0.05} = -1.77$ and $\bar{t}_2 = (523 - 544)0.075 = -1.58 > -\bar{t}_{2,0.05} = -1.75$. Thus we conclude that there is a significant reduction in body weight at 300 ppm, but not at 30 or 100 ppm.

References

[1] Barlow, R. E., Bartholomew, D. J., Bremner, J. M., and Brunk, D. (1972). *Statistical Inference under Order Restrictions.* Wiley, New York.

[2] Chase, G. R. (1974). *Biometrika*, **61**, 569–578.

[3] Marcus, R. (1976). *Biometrika*, **63**, 177–183.

[4]   Marcus, R., Peritz, E., and Gabriel, K. R. (1976). *Biometrika*, **63**, 655–660.

[5]   Williams, D. A. (1971). *Biometrics*, **27**, 103–117.

[6]   Williams, D. A. (1972). *Biometrics*, **28**, 519–531.

[7]   Williams, D. A. (1986). Correction note, included in correspondence from Williams to the author.

Bibliography

Abelson, R. P. and Tukey, J. W. (1963). *Ann. Math. Statist.*, **34**, 1347–1369. (Utilizes prior information on spacing of means for a test of equality against an ordered alternative.)

House, D. E. (1986). *Biometrics*, **42**, 187–190. (Provides a nonparametric version of Williams' test for a randomized block design.)

Shirley, E. (1977). *Biometrics*, **33**, 386–389. (Derives a nonparametric equivalent of Williams' test where observations are replaced by their ranks.)

Shorack, C. R. (1967). *Ann. Math. Statist.*, **38**, 1740–1752. (Normal and nonparametric theory for tests in the one-way analysis of variance against an ordered alternative.)

Williams, D. A. (1977). *Biometrika*, **64**, 9–14. (Determines a limiting distribution of the estimated maximum and range of a set of monotonically ordered normal means when all means are in fact equal.)

Williams, D. A. (1986). *Biometrics*, **42**, 183–186. (Suggests a modification to Shirley's test to improve its power.)

(ISOTONIC INFERENCE
MONOTONE RELATIONSHIPS
ORDER RESTRICTED INFERENCES
TREND TESTS)

KENNETH G. BROWN

WINCKLER–GAUSS INEQUALITIES (FOR ABSOLUTE MOMENTS)

These are inequalities on values of absolute moments of any distribution. The rth absolute moment* of a random variable about a (a fixed number) is

$$_a\nu_r = E[|X - a|^r].$$

The Winckler–Gauss inequalities are

$$\{(n + 1)_a\nu_n\}^{1/n} \leqslant \{(r + 1)_a\nu_r\}^{1/r}$$

or, more generally

$$\left\{(n + 1)^{[k]}{}_a\nu_n\right\}^{1/n} \leqslant \left\{(r + 1)^{[k]}{}_a\nu_r\right\}^{1/r},$$

where $g^{[b]} = g(g + 1) \cdots (g + b - 1)$ for all $0 < n < r$. They were first stated by Gauss [2] for a special case. The general case was stated by Winckler [3], who gave an incorrect proof, and later corrected by Beesack [1].

References

[1]   Beesack, P. A. (1984). *J. Math. Anal. Appl.*, **98**, 435–457.

[2]   Gauss, C. F. (1821). Theoria Combinationis Observationem Erroribus Minimis Obnoxiae. In *Werke*. Dieterische Universitäts Drucksache.

[3]   Winckler, A. (1866). *Sitzungsber. Math.-Natur. Kl. Kōn. Akad. Wiss., Wien, Zweite Abt.*, **53**, 6–41.

(BERNSTEIN'S INEQUALITY
CAMP–MEIDELL INEQUALITY
CHEBYSHEV'S INEQUALITY
MARKOV INEQUALITY
WINCKLER–VON MISES TYPE
    INEQUALITIES)

WINCKLER–VON MISES TYPE INEQUALITIES

In 1866, Winckler [3] derived a large number of inequalities relating to cumulative distribution functions (CDFs) of distributions of random variables (X) possessing a continuous CDF, which is unimodal and symmetrical. Similar inequalities, under less restrictive conditions, were obtained by von Mises* [2] in 1938. More recently, these inequalities have been refined by Beesack [1].

Typical inequalities of this kind are of form: "For any real a, and under certain conditions on the CDF,

$$\Pr[|X - a| \geqslant x] \leqslant \left(\frac{r}{r + 1}\right)^r \frac{_a\nu_r}{x^r}$$

$$\text{if } x \geqslant \frac{r}{r + 1}\{(r + 1)_a\nu_r\}^{1/r}$$

and

$$\Pr\left[\,|X - a| \geqslant x\,\right] \leqslant 1 - \frac{x}{\left\{(r+1)_a\nu_r\right\}^{1/r}}$$

$$\text{if } x < \frac{r}{r+1}\left\{(r+1)_a\nu_r\right\}^{1/r},$$

where $_a\nu_r = E[|X - a|^r]$ is the rth absolute moment* of X about a."

If $x = r(r+1)^{-1}\{(r+1)_a\nu_r\}^{1/r}$, both upper bounds are equal to $(r+1)^{-1}$. See Beesack [1] for further details.

References

[1] Beesack, P. A. (1984). *J. Math. Anal. Appl.*, **98**, 435–457.

[2] von Mises, R. (1938). *Bull. Sci. Math.* (2), **62**, 68–71.

[3] Winckler, A. (1866). *Sitzungsber. Math.- Natur. Kl. Kon. Akad. Wiss. Wien, Zweite Abt.*, **53**, 6–41.

(INEQUALITIES ON DISTRIBUTIONS: BIVARIATE AND MULTIVARIATE PROBABILITY INEQUALITIES FOR RANDOM VARIABLES)

WINDOW PLOT

The window plot is a graphical device used to display the average level of the response ($\bar{y}$) and its uncertainty (e.g., $\pm 95\%$ confidence limits, $\pm$(least significant internal)/2). The plot is a rectangular box with a bar in the interior representing the average ($\bar{y}$). The upper and lower edges of the box represent, respectively, the upper and lower limits of the uncertainty of the average. Since the vertical dimension of the box is usually longer than the horizontal dimension, the

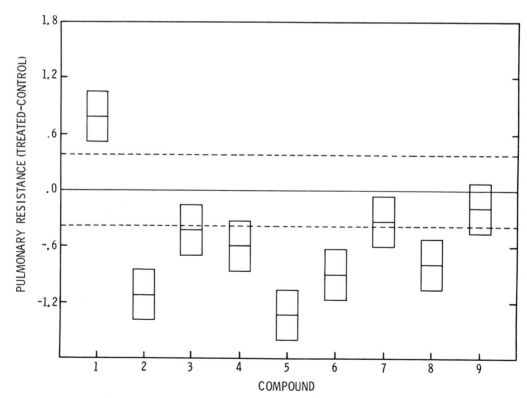

COMPOUND EFFECT ON PULMONARY RESISTANCE

Figure 1 Window plot showing the average effect of several compounds on the pulmonary resistance of dogs (Andrews, et al. [1]).

resulting box looks like a window; hence the name, *window plot*.

The window plot is most useful when it is of interest to display 3–5 or 7 averages. The interval plot, which is similar to the window plot except it has no horizontal dimension, is more effective for the comparison of a large number of averages (e.g., more than 7) because less area is required to display each average. A unique feature of both the window and interval plots is that, if the uncertainty limits used are ± (least significant interval)/2, nonoverlapping boxes or intervals indicate that the associated averages are significantly different at the assigned probability level.

The window plot is also useful in making different kinds of graphical comparions. Figure 1 is useful when it is of interest to compare k treatments, each of which has its own control. The response in this instance is the difference between the control average and treatment average. Two comparisons are of interest. Are the average differences significantly different from zero, and are there any significant differences among the average treatment differences? In Fig. 1 we see that compounds 7 and 9 have no significant effect, because their means lie within the uncertainty limits about zero (i.e., 95% confidence limits = $0 \pm tS_d$, S_d = the standard error of the average difference). It is also clear that all compounds except the first have a negative effect on the response. A comparison of the means and associated uncertainty indicates that there are also differences among the effects of compounds 2–9.

Figure 2 illustrates the comparison-with-a-standard plot. Means whose uncertainty intervals are outside the dashed lines are significantly different from the standard. In

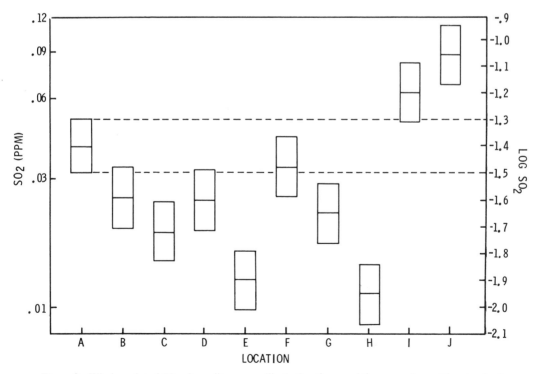

Figure 2 Window plot of SO_2 air quality means, illustrating the use of the comparison-with-a-standard plot. The data were transformed to logarithms before analysis (Andrews, et al. [1]).

Fig. 2 we see that all locations except B, D, F, and I are significantly different from location A. The uncertainty intervals shown in Fig. 2 are honest significant intervals since it was also of interest to compare the SO_2 concentrations of all locations to each other.

The window plot can be used to display any statistic (e.g., median, regression coefficient*) and a measure of its uncertainty. Further details of this approach to the graphical display of data can be found in Andrews et al. [1] and Snee [2].

References

[1] Andrews, H. P., Snee, R. D., and Sarner, M. H. (1980). Graphical display of means. *Amer. Statist.*, **34**, 195–199.

[2] Snee, R. D. (1981). Graphical display and assessment of means. *Biometrics*, **37**, 835–836.

(EXPLORATORY DATA ANALYSIS
GRAPHICAL REPRESENTATION OF DATA
NOTCHED BOX-AND-WHISKER PLOT)

R. D. SNEE

WINDOW WIDTH

Nonparametric curve estimators, for example, in the contexts of regression, density, and spectral density estimation*, may all be thought of as local weighted averages. An interpretation of these estimators is that they produce a curve that may be considered as a series of views of the data through a moving window. This idea was first developed in the setting of time series* analysis; see Blackman and Tukey [1].

The width of the window, which controls the number of points in the local average, is crucial to the performance of the estimator. This is graphically demonstrated in Fig. 1, which considers the special case of scatterplot* smoothing, i.e., nonparametric regression. Figure 1a shows an underlying regression function*, the solid curve, together with some simulated additive Gaussian noise*, represented by the stars.

The dashed curve in Fig. 1b is a moving average* estimate of the regression function. In particular, each point on the dashed curve is found by taking a weighted average of the stars, where the weights are chosen to be proportional to the height of a suitable translation of the "window function," whose graph is shown at the bottom of Fig. 1b. Note that this estimate tends to oscillate rather wildly, because the window width is too narrow, so there are not enough points in each average to provide a stable estimate of the mean.

A means of overcoming this difficulty is to expand the width of the window. The effect of this is shown in Fig. 1c, where the estimate looks much better. The other side of the window width selection problem is demonstrated in Fig. 1d, where the width is too big. Observe that points whose means are quite different are entered into the average, so the resulting estimate has a tendency to smooth away features of the underlying curve, reflected in this example by the peak being too low.

Since the choice of window width is so crucial in this sense, it is often separated out as a parameter of the estimator. This example provides a graphical demonstration of the statement "its (the window width's) choice is one of critical importance" made in KERNEL ESTIMATORS. For a deeper, but easily accessible, treatment of the window selection problem, in the specific context of density estimation, see Silverman [13].

The best known specific examples of settings where window estimators are used include:

(a) Kernel density estimation*, where

$$\hat{f}_h(x) = n^{-1}h^{-1} \sum_{i=1}^{n} K\left(\frac{x - X_i}{h}\right),$$

is used for estimating a probability density function $f(x)$ based on a sample $X_1, \ldots, X_n$ from f, where K is typically a symmetric probability density, as proposed by Rosenblatt [11] and Parzen [8].

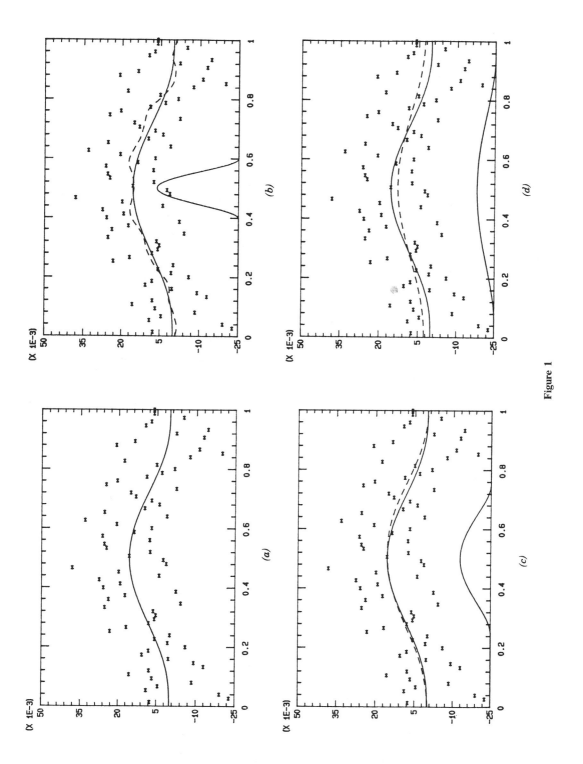

Figure 1

639

(b) Nonparametric regression estimation, where

$$\hat{m}_h(x) = \frac{\sum_{i=1}^{n} K\left(\dfrac{x - X_i}{h}\right) Y_i}{\sum_{i=1}^{n} K\left(\dfrac{x - X_i}{h}\right)},$$

is used for estimating a regression function $m(x)$, based on a sample $(X_1, Y_1), \ldots, (X_2, Y_2)$ with $E[Y_i | X_i] = m(X_i)$, as proposed by Nadaraya [7] and Watson [15]. This is the estimation setting in Fig. 1.

(c) Spectral density or power spectrum estimation, where

$$\hat{f}_h(x) =$$
$$n^{-1} h^{-1} \sum_{k=1}^{M} K\left(\frac{\omega - \omega_k}{h}\right) TS^2(\omega_k),$$

is used for estimating the power spectrum $f(\omega)$ of a stationary time series, by smoothing the periodogram* $TS^2(\omega)$ over the frequencies $\omega_1, \ldots, \omega_M$, as proposed by Daniell [4] (the notation used here is that of PERIODOGRAM ANALYSIS). See Bloomfield [2], Brillinger [3], and Priestley [10] for further discussion of this estimator and for other commonly used notation in this context.

For an extensive list of related nonparametric curve estimators, see Prakasa Rao [9].

Note that, in each case, the parameter h controls the width of the window through which the data enter into the estimator at the point x, or in other words, controls the amount of local averaging that is done. The parameter h in this context is typically called the "smoothing parameter," the "bandwidth" or the "window width"; see Silverman [13, p. 15] for example, although there are a number of other terms used, such as "window size" in Hall [6] and "window" in Stone [14].

The nonparametric curve estimation literature contains several other related uses of the term "window." In time series analysis, it is less typical to separate out the parameter h, so the analog of the entire function

$h^{-1} K(\cdot / h)$, has been given the names "spectral window," "lag window" (see Priestley [10, p. 436]) or "data window" (see Brillinger [3]). In the nonparametric regression setting, this same function has been called the "window function" by Watson [15]. Slightly different usages may be found in density estimation; for example, in Rosenblatt [12] the kernel function $K(\cdot)$ has been called the "window function," and in Fryer [5] the entire estimator $\hat{f}(x)$ was called a "window estimator."

References

[1] Blackman, R. B. and Tukey, J. W. (1959). *The Measurement of Power Spectra from the Point of View of Communications Engineering*. Dover, New York.

[2] Bloomfield, P. (1976). *Fourier Analysis of Time Series: An Introduction*. Wiley, New York.

[3] Brillinger, D. R. (1975). *Time Series: Data Analysis and Theory*. Holt, Rinehart and Winston, New York.

[4] Daniell, P. J. (1946). *J. R. Statist. Soc. (Suppl.)*, **8**, 88–90.

[5] Fryer, M. J. (1977). *J. Inst. Math. Appl.*, **20**, 335–354.

[6] Hall, P. (1983). *Ann. Statist.*, **11**, 1156–1174.

[7] Nadaraya, E. A. (1964). *Theor. Prob. Appl.*, **9**, 141–142.

[8] Parzen, E. (1962). *Ann. Math. Statist.*, **33**, 1065–1076.

[9] Prakasa Rao, B. L. S. (1983). *Nonparametric Functional Estimation*. Academic, New York.

[10] Priestley, M. B. (1981). *Spectral Analysis and Time Series*. Academic, London, England.

[11] Rosenblatt, M. (1956). *Ann. Math. Statist.*, **27**, 832–837.

[12] Rosenblatt, M. (1971). *Ann. Math. Statist.*, **42**, 1815–1842.

[13] Silverman, B. W. (1986). *Density Estimation for Statistics and Data Analysis*. Chapman and Hall, London, England.

[14] Stone, C. J. (1984). *Ann. Statist.*, **12**, 1285–1297.

[15] Watson, G. S. (1964). *Sankhyā A*, **26**, 359–372.

(DENSITY ESTIMATION
KERNEL ESTIMATORS
MOVING AVERAGES
PERIODOGRAM ANALYSIS
SPECTRAL DENSITY)

J. S. Marron

WINGS

A term used in exploratory data analysis* to denote observed values between the extremes (greatest and least) and the "hinges" (the upper and lower quartiles*).

(FIVE-NUMBER SUMMARIES)

WINSORIZATION *See* TRIMMING AND WINSORIZATION

WISHART DISTRIBUTION

INTRODUCTION

The Wishart distribution is a p-dimensional generalization of the χ^2 distribution* (more precisely, of *the distribution of* σ^2 *times* χ^2) and plays the same role for multivariate normal problems as the χ^2 distribution does for univariate normal problems (*see* MULTINORMAL DISTRIBUTIONS). Let $\mathbf{X}_1, \ldots, \mathbf{X}_n$ be independent, $\mathbf{X}_i \sim N_p(\boldsymbol{\mu}_i, \boldsymbol{\Sigma})$, $\boldsymbol{\Sigma} > 0$, i.e., $\mathbf{X}_i$ has a p-dimensional multivariate normal distribution with mean vector $\boldsymbol{\mu}_i$ and covariance matrix $\boldsymbol{\Sigma}$, and let $\mathbf{W} = \Sigma \mathbf{X}_i \mathbf{X}_i'$, $\boldsymbol{\delta} = \Sigma \boldsymbol{\mu}_i \boldsymbol{\mu}_i'$. We say that $\mathbf{W}$ has a p-dimensional *Wishart distribution* with n degrees of freedom on the covariance matrix $\boldsymbol{\Sigma}$ with noncentrality parameter $\boldsymbol{\delta}$ and write $\mathbf{W} \sim W_p(n, \boldsymbol{\Sigma}, \boldsymbol{\delta})$. Note that $\mathbf{W}$ is a $p \times p$ matrix and that $\mathbf{W} \geqslant 0$. From this definition, it is apparent that

$$E\mathbf{W} = n\boldsymbol{\Sigma} + \boldsymbol{\delta},$$

$$\mathbf{AWA}' \sim W_q(n, \mathbf{A\Sigma A}', \mathbf{A\delta A}'), \quad (1)$$

where $\mathbf{A}$ is $q \times p$ of rank q. If $\boldsymbol{\delta} = 0$, we say that $\mathbf{W}$ has a *central Wishart distribution* and write $\mathbf{W} \sim W_p(n, \boldsymbol{\Sigma})$. If $\boldsymbol{\delta} \neq 0$, we say that $\mathbf{W}$ has a *noncentral Wishart distribution*. The rank of $\mathbf{W}$ is the minimum of n and p. Therefore, $\mathbf{W} > 0$ (and hence invertible) if and only if $n \geqslant p$. If $n \geqslant p$, we say that $\mathbf{W}$ has a *nonsingular Wishart distribution* and if $n < p$, we say $\mathbf{W}$ has a *singular Wishart distribution*.

The characteristic function* of the Wishart distribution is

$$\phi(\mathbf{T}) = E \exp(i \operatorname{tr}(\mathbf{TW}))$$
$$= |\mathbf{I} - 2i\mathbf{T\Sigma}|^{-n/2} \exp\left(-\tfrac{1}{2} \operatorname{tr} \mathbf{D}\right)$$

with $\mathbf{D} = \boldsymbol{\Sigma}^{-1}\boldsymbol{\delta}\left[\mathbf{I} - (\mathbf{I} - 2i\mathbf{T\Sigma})^{-1}\right]$, (2)

where $\mathbf{T}$ is a symmetric $p \times p$ matrix such that $\boldsymbol{\Sigma}^{-1} - 2\mathbf{T} > 0$.

The singular Wishart distribution does not have a density function. The density function of the nonsingular Wishart distribution is

$$K|\boldsymbol{\Sigma}|^{-n/2}|\mathbf{w}|^{(n-p-1)/2}$$
$$\times \exp\left[-\tfrac{1}{2}\operatorname{tr}\left\{\boldsymbol{\Sigma}^{-1}(\mathbf{w} + \boldsymbol{\delta})\right\}\right]$$
$$\times {}_0F_1\left(\tfrac{1}{2}n; \tfrac{1}{4}\boldsymbol{\Sigma}^{-1}\boldsymbol{\delta}\boldsymbol{\Sigma}^{-1}\mathbf{w}\right), \quad (3)$$

$\mathbf{w} > 0$, where

$$K^{-1} = 2^{-np/2}\pi^{p(p-1)/4}\prod_{i=1}^{p}\Gamma((n+1-i)/2)$$

and ${}_0F_1(\tfrac{1}{2}n, \tfrac{1}{4}\boldsymbol{\Sigma}^{-1}\boldsymbol{\delta}\boldsymbol{\Sigma}^{-1}\mathbf{w})$ is a hypergeometric function of matrix argument* (e.g., see Muirhead [11], pp. 258–262). If $\boldsymbol{\delta} = 0$, then ${}_0F_1(\tfrac{1}{2}n, 0) = 0$, so that the central Wishart density is given by

$$K|\boldsymbol{\Sigma}|^{-n/2}|\mathbf{w}|^{(n-p-1)/2}\exp\left\{-\tfrac{1}{2}\operatorname{tr}(\boldsymbol{\Sigma}^{-1}\mathbf{w})\right\},$$
$$\mathbf{w} > 0. \quad (4)$$

In the next section we present some basic results about the Wishart distribution. In later sections we indicate how the Wishart distribution occurs in two multivariate models, the multivariate one-sample model and the one-way multivariate analysis of variance* model.

BASIC FACTS

We first present some facts about the Wishart distribution which are straightforward generalizations of well-known facts about the χ^2 distribution:

A. Let the $\mathbf{W}_i$s be mutually independent, $\mathbf{W}_i \sim W_p(n_i, \boldsymbol{\Sigma}, \boldsymbol{\delta}_i)$. Then $\Sigma\mathbf{W}_i \sim W_p(\Sigma n_i, \boldsymbol{\Sigma}, \Sigma\boldsymbol{\delta}_i)$.

B. Suppose that $n\mathbf{V}_n \sim W_p(n, \boldsymbol{\Sigma})$. Then $\mathbf{V}_n$ converges in probability to $\boldsymbol{\Sigma}$.

C. Let $\mathbf{X}' = (\mathbf{X}_1, \ldots, \mathbf{X}_n)$ where $\mathbf{X}_i$ are independent, $\mathbf{X}_i \sim N_p(\boldsymbol{\mu}_i, \boldsymbol{\Sigma})$. Let $\boldsymbol{\mu} = E\mathbf{X}$. Let $\mathbf{A}$ and $\mathbf{B}$ be symmetric $n \times n$ matrices and let $\mathbf{C}$ be $q \times n$.
1. $\mathbf{X}'\mathbf{A}\mathbf{X} \sim W_p(k, \boldsymbol{\Sigma}, \boldsymbol{\mu}'\mathbf{A}\boldsymbol{\mu})$ if and only if $\mathbf{A}^2 = \mathbf{A}$, $k = \text{rank}(\mathbf{A})$.
2. $\mathbf{X}'\mathbf{A}\mathbf{X}$ and $\mathbf{X}'\mathbf{B}\mathbf{X}$ are independent if and only if $\mathbf{A}\mathbf{B} = \mathbf{0}$. $\mathbf{X}'\mathbf{A}\mathbf{X}$ and $\mathbf{C}\mathbf{X}$ are independent if and only if $\mathbf{C}\mathbf{A} = \mathbf{0}$.

(Note that $\mathbf{X}$ is an $n \times p$ matrix and that the independent replication is represented by the rows of $\mathbf{X}$, as in univariate models.) Result C1 implies that $\mathbf{X}'\mathbf{A}\mathbf{X}$ has a Wishart distribution in the multivariate case if and only if $\mathbf{X}'\mathbf{A}\mathbf{X}/\sigma^2$ has a χ^2 distribution in the univariate case and that the degrees of freedom are the same for the multivariate case as for the univariate case. Similarly C2 implies that two quadratic forms* are independent in the multivariate case if and only if they are independent in the univariate case. Using these facts, for example, we could immediately generalize Cochran's theorem to the multivariate case.

We have seen that the Wishart distribution is a generalization of the χ^2 distribution. We now state other relationships between these distributions. Let $\mathbf{a} \in R^p$, $\mathbf{a} \neq \mathbf{0}$.

D. $\mathbf{a}'\mathbf{W}\mathbf{a} / \mathbf{a}'\boldsymbol{\Sigma}\mathbf{a} \sim \chi_n^2(\mathbf{a}'\boldsymbol{\delta}\mathbf{a}/\mathbf{a}'\boldsymbol{\Sigma}\mathbf{a})$.

E. If $\boldsymbol{\delta} = \mathbf{0}$, $n \geq p$, then $\mathbf{a}'\boldsymbol{\Sigma}^{-1}\mathbf{a}/\mathbf{a}'\mathbf{W}^{-1}\mathbf{a} \sim \chi_{n-p+1}^2$.

F. $\text{tr}(\boldsymbol{\Sigma}^{-1}\mathbf{W}) \sim \chi_{np}^2(\boldsymbol{\Sigma}^{-1}\boldsymbol{\delta})$.

G. If $\boldsymbol{\delta} = \mathbf{0}$, $n \geq p$, then $|\mathbf{W}|/|\boldsymbol{\Sigma}| = |\boldsymbol{\Sigma}^{-1}\mathbf{W}| \sim \prod_1^p U_i$, where the U_i are independent, $U_i \sim \chi_{n-i+1}^2$.

We now discuss one of the most important properties of the Wishart distribution. Let $\mathbf{X}$ and $\mathbf{W}$ be independent, $\mathbf{X} \sim N_p(\boldsymbol{\mu}, \boldsymbol{\Sigma})$, $\mathbf{W} \sim W_p(n, \boldsymbol{\Sigma})$, $n \geq p$. Let $F = (n - p - 1)p^{-1}\mathbf{X}'\mathbf{W}^{-1}\mathbf{X}$. Then

$$F \sim F_{p, n-p+1}(\boldsymbol{\mu}'\boldsymbol{\Sigma}^{-1}\boldsymbol{\mu}). \qquad (5)$$

(Note that when $p = 1$, F is the square of an obvious t random variable.) $np(n - p - 1)^{-1}F = T^2$ is called Hotelling's T^{2*}, whose distribution can be determined from (5).

Our next topic is the Bartlett decomposition of the Wishart distribution. Let $\mathbf{W} \sim W_p(n, \mathbf{I})$, $n \geq p$. Let $\mathbf{W} = \mathbf{T}'\mathbf{T}$, where $\mathbf{T}$ is upper triangular with positive diagonal elements. Let t_{ij} be the (i, j) component of $\mathbf{T}$. Then the t_{ij} are independent, $t_{ii}^2 \sim \chi_n^2$, $t_{ij} \sim N(0, 1)$.

The last result in this section is one which is useful in determining the distribution of sample partial correlation* coefficients (*see Supplement*). Partition $\mathbf{W}$ and $\boldsymbol{\Sigma}$ as

$$\mathbf{W} = \begin{pmatrix} \mathbf{W}_{11} & \mathbf{W}_{12} \\ \mathbf{W}_{21} & \mathbf{W}_{22} \end{pmatrix},$$

$$\boldsymbol{\Sigma} = \begin{pmatrix} \boldsymbol{\Sigma}_{11} & \boldsymbol{\Sigma}_{12} \\ \boldsymbol{\Sigma}_{21} & \boldsymbol{\Sigma}_{22} \end{pmatrix},$$

with $\mathbf{W}_{11}$ and $\boldsymbol{\Sigma}_{11}$ $s \times s$. Let $\mathbf{W}_{11.2} = \mathbf{W}_{11} - \mathbf{W}_{12}\mathbf{W}_{22}^{-1}\mathbf{W}_{21}$ and $\boldsymbol{\Sigma}_{11.2} = \boldsymbol{\Sigma}_{11} - \boldsymbol{\Sigma}_{12}\boldsymbol{\Sigma}_{22}^{-1}\boldsymbol{\Sigma}_{21}$. Then

H. $\mathbf{W}_{11} \sim W_s(n, \boldsymbol{\Sigma}_{11})$, $\mathbf{W}_{22} \sim W_{p-s}(n, \boldsymbol{\Sigma}_{22})$. If $\boldsymbol{\Sigma}_{12} = \mathbf{0}$, then $\mathbf{W}_{11}$ and $\mathbf{W}_{22}$ are independent.

I. $\mathbf{W}_{11.2}$ is independent of $\mathbf{W}_{12}$ and $\mathbf{W}_{22}$, and $\mathbf{W}_{11.2} \sim W_s(n - p + s, \boldsymbol{\Sigma}_{11.2})$.

THE MULTIVARIATE ONE-SAMPLE MODEL

We now look at the multivariate one-sample model as an example of how the results in the previous section are applied. In this model, we observe $\mathbf{X}_1, \ldots, \mathbf{X}_n$ independent, $\mathbf{X}_i \sim N_p(\boldsymbol{\mu}, \boldsymbol{\Sigma})$, where $\boldsymbol{\mu}$ and $\boldsymbol{\Sigma} > 0$ are unknown parameters. Let

$$\overline{\mathbf{X}} = n^{-1}\sum\mathbf{X}_i,$$

$$\mathbf{S} = (n - 1)^{-1}\sum(\mathbf{X}_i - \overline{\mathbf{X}})(\mathbf{X}_i - \overline{\mathbf{X}})'.$$

$\overline{\mathbf{X}}$ is called the *sample mean vector* and $\mathbf{S}$ is called the *sample covariance matrix*. Note that $\mathbf{S}$ is a $p \times p$ matrix whose kth diagonal element is just the sample variance computed from the kth components of the $\mathbf{X}_i$ and whose (j, k)th off-diagonal element is just the sample covariance between the jth and kth components of the $\mathbf{X}_i$.

Result **C2** in the last section implies that $\overline{\mathbf{X}}$ and **S** are independent, $(n - 1)\mathbf{S} \sim W_p(n - 1, \Sigma)$. By well-known results about the normal distribution, $\overline{\mathbf{X}} \sim N_p(\mu, n^{-1}\Sigma)$. Therefore, $\overline{\mathbf{X}}$ and **S** are unbiased estimators of μ and Σ [see (1)] and are consistent (see B). $(\overline{\mathbf{X}}, \mathbf{S})$ is a complete sufficient statistic* for this model, so that $\overline{\mathbf{X}}$ and **S** are minimum variance unbiased estimators*. We also note that $\mathbf{S} > 0$ as long as $n > p$, which we henceforth assume.

Now, consider testing that $\mu = \mathbf{0}$. Using the fact that $\overline{\mathbf{X}}$ and **S** are independent, $n^{1/2}\overline{\mathbf{X}} \sim N_p(n^{1/2}\mu, \Sigma)$ and $(n - 1)\mathbf{S} \sim W_p(n - 1, \Sigma)$, we see from (5) that

$$F = c(n, p)\overline{\mathbf{X}}'\mathbf{S}^{-1}\overline{\mathbf{X}} \sim F_{p, n-p}\left(n\mu'\Sigma^{-1}\mu\right),$$

where $c(n, p) = (n - p)np^{-1}(n - 1)^{-1}$. A sensible test is to reject if $F > F_{p, n-p}^{\alpha}$. [Note that when $p = 1$, $F = (n^{1/2}\overline{X}/S)^2 = t^2$ so that this test is a natural generalization of the univariate one sample t-test.] This is called the one-sample Hotelling's T^2 test and is uniformly most powerful invariant, the likelihood ratio* test, unbiased, and admissible. It can also be derived from the union–intersection principle*.

We now show how (5) can be used to generate simultaneous confidence intervals* for the set of all $\mathbf{t}'\mu$, $\mathbf{t} \in R^p$. Let

$$F^* = c(n, p)(\overline{\mathbf{X}} - \mu)'\mathbf{S}^{-1}(\overline{\mathbf{X}} - \mu)$$
$$= c(n, p)\sup_{\mathbf{t}}\left[\{\mathbf{t}'(\overline{\mathbf{X}} - \mu)\}^2/\mathbf{t}'\mathbf{S}\mathbf{t}\right].$$

By (5), $F^* \sim F_{p, n-p}$. Therefore,

$$P\left(\mathbf{t}'\mu \in \mathbf{t}'\overline{\mathbf{X}} \pm \left(\mathbf{t}'\mathbf{S}\mathbf{t}F_{p, n-p}^{\alpha}/c(n, p)\right)^{1/2}\right.$$

$$\text{for all } \mathbf{t})$$

$$= P\left(\left(\mathbf{t}'(\overline{\mathbf{X}} - \mu)\right)^2/\mathbf{t}'\mathbf{S}\mathbf{t}\right.$$

$$\leqslant F_{p, n-p}^{\alpha}/c(n, p) \text{ for all } \mathbf{t} \neq \mathbf{0}\right)$$

$$= P\left(F^* \leqslant F_{p, n-p}^{\alpha}\right) = 1 - \alpha.$$

Any procedures for drawing inferences about the covariance matrix Σ would be based on the sample covariance matrix, and, therefore, properties of these procedures would be based on the nonsingular central Wishart distribution whose density is given in (4). In particular, various sample correlation coefficients (simple, multiple, partial,

canonical) are computed from **S** [or equivalently from $(n - 1)\mathbf{S}$] and hence their joint and marginal densities can be determined from the density in (4). Let $r_n(\rho)$ be the distribution of the sample (simple) correlation coefficient* computed from a sample of size n with (true) correlation coefficient ρ. Let r^* be a sample partial correlation coefficient computed from a sample of size n, conditionally on q variables, with (true) partial correlation coefficient ρ^*. Result **I** with $s = 2$ implies that $r^* \sim r_{n-q}(\rho^*)$, so that any distribution theory developed for sample correlation coefficients can be immediately extended to sample partial correlation coefficients*.

In addition, principal component analysis* and factor analysis* also use the sample covariance matrix **S**, so that their properties are also determined from the Wishart distribution. The determinant $|\Sigma|$ is called the *generalized variance*. It is often used as a real-valued measure of the variability in the sample. $|\mathbf{S}|$ is called the *sample generalized variance*. Its distribution is given in **G**.

MULTIVARIATE ANALYSIS OF VARIANCE

In the one-way multivariate analysis of variance* (MANOVA) model, we observe $\mathbf{X}_{ij}$ independent, $\mathbf{X}_{ij} \sim N(\mu + \alpha_i, \Sigma)$, where μ and α_i are unknown p-dimensional vectors such that $\sum n_i\alpha_i = 0$, and $\Sigma > 0$ is an unknown $p \times p$ matrix. We want to test that the $\alpha_i = 0$. Let

$$N = \sum_i n_i, \overline{\mathbf{X}}_{i\cdot} = n_i^{-1}\sum_i \mathbf{X}_{ij},$$

$$\mathbf{X}_{\cdot\cdot} = N^{-1}\sum_i\sum_j \mathbf{X}_{ij}$$

and

$$\mathbf{T}_1 = \sum_i n_i(\overline{\mathbf{X}}_{i\cdot} - \overline{\mathbf{X}}_{\cdot\cdot})(\overline{\mathbf{X}}_{i\cdot} - \overline{\mathbf{X}}_{\cdot\cdot})',$$

$$\mathbf{T}_2 = \sum_i\sum_j(\mathbf{X}_{ij} - \overline{\mathbf{X}}_{i\cdot})(\mathbf{X}_{ij} - \overline{\mathbf{X}}_{i\cdot})',$$

$$\delta = \sum_i n_i\alpha_i\alpha_i'.$$

We are testing that $\delta = 0$. The procedures that are used for this problem are based on $\mathbf{R}' = (r_1, \ldots, r_b)$, where $r_1 \geqslant r_2 \geqslant \cdots \geqslant r_b$ are the nonzero eigenvalues of $\mathbf{T}_2^{-1}\mathbf{T}_1$. ($b$ is the minimum of $k - 1$ and p.) From **C**, we see that $\mathbf{T}_1$ and $\mathbf{T}_2$ are independent,

$$\mathbf{T}_1 \sim W_p(k - 1, \Sigma, \delta),$$
$$\mathbf{T}_2 \sim W_p(N - k, \Sigma).$$

We must assume that $N - k \geqslant p$, so that $\mathbf{T}_2$ is invertible. Typically, we do not assume that $k - 1 \geqslant p$, so that $\mathbf{T}_1$ may have a singular Wishart distribution. $\mathbf{T}_2$ has a central Wishart distribution, but $\mathbf{T}_1$ has a possibly noncentral Wishart distribution. We are testing that the distribution is a central Wishart distribution. The joint distribution of $\mathbf{R}$ is rather complicated, depending on the nonsingular central Wishart distribution of $\mathbf{T}_2$ and the possibly singular, possibly noncentral Wishart distribution of $\mathbf{T}_1$. Invariance considerations can be used to show that the distribution depends only on the eigenvalues* of $\Sigma^{-1}\delta$. In particular, the null distribution does not depend on any unknown parameters, so that the null distribution of any test statistic based on $\mathbf{R}$ would be completely specified.

FURTHER COMMENTS

In the first two sections, we presented the definition of the Wishart distribution together with some of its properties. In the third section, we looked at the multivariate one-sample model and saw that the sample covariance matrix has a Wishart distribution. For this model, the only relevant Wishart distribution is the nonsingular central Wishart distribution. However, in the one-way MANOVA model, we see that we need the singular Wishart distribution even for finding null distributions of statistics, and need the noncentral Wishart distribution for dealing with power functions (e.g., to show unbiasedness* or admissibility* of tests).

Muirhead [11] presents a detailed treatment of both the central and noncentral Wishart distributions. It is the best book to read on this subject. Other textbooks which present the basic facts about the central Wishart distribution (together with some simple facts about the noncentral case) include Anderson [2], Arnold [3], Eaton [5], Giri [7], and Kshirsagar [9].

We now cite some historical papers. Fisher [6] derives the central Wishart density when $p = 2$. Wishart [13] finds the density function for general p in the central case. Anderson [1] derives the noncentral Wishart density when the noncentrality matrix has rank 1 or 2. Weibull [12] extends this result to the case of rank 3. Herz [8], James [9], and Constantine [4] derive different forms of the noncentral Wishart density for general rank.

References

[1]   Anderson, T. W. (1946). The non-central Wishart distribution and certain problems of multivariate statistics. *Ann. Math. Statist.* **17**, 409–431.

[2]   Anderson, T. W. (1958). *An Introduction to Multivariate Analysis.* Wiley, New York.

[3]   Arnold, S. F. (1981). *The Theory of Linear Models and Multivariate Analysis.* Wiley, New York.

[4]   Constantine, A. G. (1963). Some non-central distribution problems in multivariate analysis. *Ann. Math. Statist.*, **34**, 1270–1285.

[5]   Eaton, M. L. (1983). *Multivariate Statistics, A Vector Space Approach.* Wiley, New York.

[6]   Fisher, R. A., (1915). Frequency distribution of the values of the correlation coefficient in samples from an infinitely large sample. *Biometrika*, **10**, 507–521.

[7]   Giri, N. C. (1977). *Multivariate Statistical Inference.* Academic, New York.

[8]   Herz, C. S. (1955). Bessel functions of matrix argument. *Ann. Math.*, **61**, 474–523.

[9]   James, A. T. (1961). The distribution of noncentral means with known covariance. *Ann. Math. Statist.*, **32**, 874–882.

[10]   Kshirsagar, A. M. (1972). *Multivariate Analysis.* Dekker, New York.

[11]   Muirhead, R. J. (1982). *Aspects of Multivariate Statistical Theory.* Wiley, New York.

[12]   Weibull, M. (1953). The distribution of t- and F-statistics and of correlation and regression coefficients in stratified samples from normal populations with different means. *Skand. Aktuar. (Suppl.)*, **36**, 1–106.

[13]   Wishart, J. (1928). The generalized product moment distribution in samples from a normal multivariate population. *Biometrika*, **20**, 32–52.

(CHI-SQUARE DISTRIBUTION
GENERALIZED VARIANCE
HOTELLING'S T^2
LAMBDA CRITERION, WILKS'S
MULTINORMAL DISTRIBUTIONS
MULTIVARIATE ANALYSIS)

STEVEN F. ARNOLD

WITCH OF AGNESI

This is the name given to a curve with parametric equations

$$x = a \cot \phi; \qquad y = a \sin^2\phi.$$

This results in a curve with y proportional to $(a^2 + x^2)^{-1}$. It is generated by a point P moving in the way indicated in Fig. 1.

The curve is, in fact, in the form of a Cauchy distribution* PDF (x representing the random variable and y the PDF). The area under the curve is

$$a^3 \int_{-\infty}^{\infty} (x^2 + a^2)^{-1} dx$$

$$= \int_{-\pi/2}^{\pi/2} (\cot^2\phi + 1)^{-1}(\sin^2\phi)^{-1} d\phi$$

$$= \int_{-\pi/2}^{\pi/2} d\phi = \pi$$

so the Y scale needs to be multiplied by π^{-1} to produce a PDF. The curve seems to have first appeared in the works of Fermat* in the middle of the seventeenth century. Stigler [1]

gives a fascinating historical discussion of the relations between the curve and the Cauchy distribution.

The name of the curve derives from Marie Gaetena Agnesi (1718–1799), who discussed the curve in 1748. She called the curve *La Versiera* (The Witch). Stigler [1] noted that this name had been used previously (in 1718), and discussed its etymology in some detail.

Reference

[1]   Stigler, S. M. (1974). *Biometrika*, **61**, 375–380.

(CAUCHY DISTRIBUTION
POISSON, SIMÉON-DENIS)

WITT DESIGNS

Witt designs are balanced incomplete block* designs with super balance. They are determined by three parameters, t, k, and v. They have also been called *t designs* with *unit t-set balance*. In a Witt design there are v varieties arranged in blocks of k elements such that every t-set occurs exactly once. It is easy to see that these are special balanced incomplete block designs with

$$b = \binom{v}{t} \bigg/ \binom{k}{t},$$

$$r = \binom{v-1}{t-1} \bigg/ \binom{k-1}{t-1},$$

$$\lambda = \binom{v-2}{t-2} \bigg/ \binom{k-2}{t-2}.$$

Not only are these designs balanced on pairs,

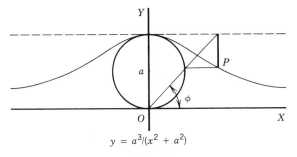

$$y = a^3/(x^2 + a^2)$$

Figure 1

they are balanced on i-sets ($i = 2, 3, \ldots, t$) and the number of occurrences of each i-set is given by

$$\binom{v - i}{t - i} \bigg/ \binom{k - i}{t - i}.$$

The smallest practical design with $t > 2$ is the $(3, 4, 14)$ design generated by cycling $(\infty 124)$ and (3567), modulo 7. As a balanced incomplete block design, this design has parameters $(8, 14, 7, 4, 3)$; every pair occurs three times, and every triple occurs once. Sprott has discussed the features of the particularly interesting Witt design on 24 symbols, block size 8, and $t = 5$; it has parameters $(24, 759, 253, 8, 77)$.

To date, the extra balance features of Witt designs have not been exploited statistically.

Bibliography

Witt, E. (1938). Ueber Steinersche Systems. *Abhandlungen aus dem Mathematischen Seminar der Hansisischen Universitaet*, **12**, 256–264.

Sprott, D. A. (1955). Balanced incomplete block designs and tactical configurations. *Ann. Math. Statist.*, **26**, 752–758.

(BLOCKS, BALANCED INCOMPLETE
CYCLIC DESIGNS
DESIGN OF EXPERIMENTS
t DESIGNS)

RALPH G. STANTON

WOLFOWITZ INEQUALITY

A generalization of the Cramér–Rao inequality* to any *sequential* unbiased estimator $\phi(X)$ of a parametric function $g(\theta)$, which states that

$$\mathrm{Var}_\theta(\phi(X)) \geqslant (g'(\theta))^2 / (E_\theta[N] I(\theta))$$

for every $\theta \in H$ (the space of the values of the parameter). This unequality is valid under similar regularity conditions as its fixed-size sample analog. Here N is the (random) sample size and $I(\theta)$ is the amount of information*

$$I(\theta) = E_\theta\left[\left(\frac{\partial \log f(X, \theta)}{\partial \theta}\right)^2\right]$$

corresponding to density $f_X(\cdot)$.

(In this entry, the subscript θ means "given θ".)

Reference

[1] Wolfowitz, J. (1947). *Ann. Math. Statist.*, **18**, 215–230.

(CRAMÉR–RAO INEQUALITY
SEQUENTIAL ANALYSIS
SEQUENTIAL ESTIMATION)

WOLFOWITZ, JACOB

Born: March 19, 1910 in Warsaw, Poland.

Died: July 16, 1981 in Tampa, Florida.

Contributed to: statistical inference, sequential analysis, inventory theory, queuing theory, information theory, decision theory.

Jacob Wolfowitz was born in Warsaw, Poland on March 19, 1910, and came to the United States with his family in 1920. He received the baccalaureate from the College of the City of New York in 1931. Positions were scarce in 1931, a year of severe economic depression, and he supported himself as a high school teacher while studying for the doctorate in mathematics at New York University. He received the Ph.D. degree in 1942.

Wolfowitz had met Abraham Wald* in the autumn of 1938, when Wald came to Columbia University to study statistics with Harold Hotelling*. Wald and Wolfowitz quickly became close friends and collaborators, their first joint paper [18] appearing in 1939. During the period of United States involvement in World War II, they worked together on war-related research at the Statistics Research Group of Columbia University (*see* MILITARY STATISTICS). In 1945

Wolfowitz became Associate Professor at the University of North Carolina at Chapel Hill. In 1946 he joined the faculty of Columbia University, leaving in 1951 to join the Department of Mathematics at Cornell University. In 1970 he became Professor of Mathematics at the University of Illinois in Urbana. After retiring from the University of Illinois in 1978, he became Distinguished Professor of Mathematics at the University of South Florida in Tampa, a position he held until his death following a heart attack, on July 16, 1981. He had held visiting professorships at the University of California at Los Angeles, at the Universities of Paris and Heidelberg, and at the Technion-Israel Institute of Technology in Haifa.

Wolfowitz's research is remarkable for its combination of breadth and depth. He made important contributions to all of the major areas of mathematical statistics, and also to inventory theory*, queuing theory*, and information theory*. Several of his papers make contributions to several different areas simultaneously.

Wolfowitz's earliest research interest was nonparametric inference. His first two published papers, written jointly with Wald, were on nonparametric inference: ref. 18 constructs a confidence band* for an unknown continuous cumulative distribution function based on a random sample from the distribution; ref. 19 proposes and analyzes the celebrated two-sample test based on runs*. Wolfowitz wrote several other papers on the theory and application of runs, ref. 27 containing an application to quality control*. The term "nonparametric" was originated by Wolfowitz in ref. 26. His interest in nonparametric inference did not end with these early papers. In ref. 7, with Dvoretzky and Kiefer, he proved that the empirical cumulative distribution function (see EDF STATISTICS) is an asymptotically minimax* estimator of the population cumulative distribution function for a variety of reasonable loss functions. In refs. 13 and 15 Kiefer and Wolfowitz extended these results to the problem of estimating joint cumulative distribution functions.

Wolfowitz's research on the minimum distance* method is an application of techniques developed in nonparametric inference to parametric inference. The method estimates the unknown parameters by those values of the parameters that minimize a distance between the empirical cumulative distribution function and the parametric family of cumulative distribution functions. This method gives consistent estimators in some very complicated problems. The papers in refs. 32–35 and the joint paper [9] with Kac and Kiefer cover the development, analysis, and applications of the minimum distance method.

Starting with ref. 28, which discusses the sequential estimation* of a Bernoulli parameter, Wolfowitz made many important contributions to sequential analysis*. In ref. 29 he developed a Cramér–Rao* type of lower bound for the variance of an estimator based on sequential sampling, under certain regularity conditions (see WOLFOWITZ INEQUALITY). In ref. 21 he and Wald studied the structure of Bayes decision rules when sequential sampling is used. One of the papers [20] he was proudest of was written with Wald and proves the optimum character of the Wald sequential probability ratio test. In ref. 8 Dvoretzky, Wald, and Wolfowitz showed that randomization can be eliminated in sequential decision problems under certain conditions. In ref. 22 Wald and Wolfowitz showed that under mild conditions, in sequential decision problems if randomization is used after each observation, we get the same class of risk functions as when randomization is used only once, to choose a nonrandomized decision rule at the start of the process. Dvoretzky, Kiefer and Wolfowitz [4, 5] solved sequential decision problems when observation is continuous over time. In ref. 31 Wolfowitz showed that the optimal sequential estimator of a normal mean when the variance is known is essentially a fixed sample size estimator. In ref. 24 Weiss and Wolfowitz constructed an asymptotically efficient sequential equivalent of Student's t-test*, and in ref. 23 these authors used an adaptive sequential scheme to construct optimal fixed length nonparametric estimators of translation parameters.

Kiefer and Wolfowitz [10] modified the Robbins–Monro stochastic approximation* procedure to estimate the point at which an unknown regression function achieves its maximum (*see* KIEFER-WOLFOWITZ PROCEDURE).

In a regression* model, a particular choice of the values of the independent variables is called a "design." An optimal design is a design which enables the user of the model to estimate given functions of the unknown regression coefficients as efficiently as possible. Kiefer and Wolfowitz [14, 16, 17] made important contributions to the theory underlying the construction of optimal designs (*see* KIEFER-WOLFOWITZ EQUIVALENCE THEOREM).

The inventory* problem is the problem of deciding how much inventory to hold during each of a sequence of time periods, when there are penalties for holding either too much or too little inventory and demand for the product is random. Dvoretzky, Kiefer and Wolfowitz [2, 3, 6] made pioneering contributions to this subject and really started the subject known nowadays as "dynamic programming"; this is the theory of which sequence of nonsampling decisions is optimal, when a decision must be made in each time period in a sequence of time periods. In ref. 6 the authors showed that under certain circumstances the well-known (s, S) policy is optimal: this policy is to order enough to make the total inventory equal to S as soon as the stock on hand goes below s.

Wolfowitz's research on maximum likelihood* estimators started with ref. 30 and led to the development, in collaboration with Weiss, of maximum probability estimators*. For large samples, these estimators have the highest probability of being close to the true unknown parameters, among a wide class of estimators they often coincide with maximum likelihood estimators, but exist in cases where the latter do not. The monograph in ref. 25 describes most of the results in this area.

Kiefer and Wolfowitz [11, 12] made fundamental contributions to the theory of queues with many servers, by showing the existence of limiting distributions of waiting times and queue lengths as time approaches infinity.

Starting in 1957, Wolfowitz [36] devoted a rapidly increasing proportion of his time to what he called coding theorems of information theory*, describing how rapidly information can be sent when random errors occur in the transmission and the probability of correct decipherment must be at least equal to a preassigned value. This problem can be considered as a generalization of statistical decision theory*, in the following sense. In statistical decision theory, there is an unknown parameter with a given set of possible values, and based on observed random variables whose distribution depends on the parameter, we must guess the value of the parameter. In coding theory, we have the additional choice of the set of possible values of the parameter: Each value in the set we choose becomes one of the entries in our codebook, the codebook being simply a list of the words which we are allowed to transmit over the channel. We want to choose as many words as possible, but in such a way as to achieve the desired lower bound on correct decipherment. Wolfowitz proved both direct coding theorems, which state that the codebook can contain at least a certain number of words, and converse theorems which state that the codebook cannot contain more than a certain number of words. His work in this area represents deep generalizations of the theory which existed before he started his research. Most of his results are contained in a monograph [39].

In addition to the mathematical theory, Wolfowitz was interested in practical and philosophical issues. Reference 37 contains a criticism of a set of axioms used to support the Bayesian* approach to statistical decision theory. Reference 38 contains an interesting criticism of the theory of testing hypotheses* for not having practical application.

Wolfowitz was a renowned teacher and lecturer, unsurpassed in his ability to clarify the intuition underlying the most complicated results. He was selected as Rietz Lecturer and as Wald Lecturer by the Institute of Mathematical Statistics, and as Shannon Lecturer by the Institute of Electrical and Electronic Engineers. His list of other academic honors is a long one: an honorary doctorate from the Technion; election to the U.S. National Academy of Sciences and to the American Academy of Arts and Sciences; election as a Fellow of the International Statistics Institute, the Econometric Society, the American Statistical Association, and the Institute of Mathematical Statistics; a term as President of the Institute of Mathematical Statistics; Visiting Professorships at several universities; and selection as a Guggenheim Fellow.

Wolfowitz's reading was not confined to mathematical subjects. He read detective stories for relaxation and kept up with political and social conditions in all of the large nations of the world and many of the smaller ones. He was a man of strong opinions, with a particular detestation of tyranny. He took a leading part in organizing protests against Soviet repression of minorities and dissidents, and was able to aid several victims of such repression.

A fuller account of Wolfowitz's research can be found in ref. 1. A complete list of his 120 publications is given in ref. 40.

References

[1] Augustin, U., Kiefer, J., and Weiss, L. (1980). In *Jacob Wolfowitz: Selected Papers*, J. Kiefer, ed. Springer, New York, pp. ix–xxi.

[2] Dvoretzky, A., Kiefer, J., and Wolfowitz, J. (1952a). *Econometrica*, **20**, 187–222.

[3] Dvoretzky, A., Kiefer, J., and Wolfowitz, J. (1952b). *Econometrica*, **20**, 450–466.

[4] Dvoretzky, A., Kiefer, J., and Wolfowitz, J. (1953a). *Ann. Math. Statist.*, **24**, 254–264.

[5] Dvoretzky, A., Kiefer, J., and Wolfowitz, J. (1953b). *Ann. Math. Statist.*, **24**, 403–415.

[6] Dvoretzky, A., Kiefer, J., and Wolfowitz, J. (1953c). *Econometrica*, **21**, 586–596.

[7] Dvoretzky, A., Kiefer, J., and Wolfowitz, J. (1956). *Ann. Math. Statist.*, **27**, 642–669.

[8] Dvoretzky, A., Wald, A., and Wolfowitz, J. (1951). *Ann. Math. Statist.*, **22**, 1–21.

[9] Kac, M., Kiefer, J., and Wolfowitz, J. (1955). *Ann. Math. Statist.*, **26**, 189–211.

[10] Kiefer, J. and Wolfowitz, J. (1952). *Ann. Math. Statist.*, **23**, 462–466.

[11] Kiefer, J. and Wolfowitz, J. (1955). *Trans. Amer. Math. Soc.*, **78**, 1–18.

[12] Kiefer, J. and Wolfowitz, J. (1956). *Ann. Math. Statist.*, **27**, 147–161.

[13] Kiefer, J. and Wolfowitz, J. (1958). *Trans. Amer. Math. Soc.*, **87**, 173–186.

[14] Kiefer, J. and Wolfowitz, J. (1959a). *Ann. Math. Statist.*, **30**, 271–294.

[15] Kiefer, J. and Wolfowitz, J. (1959b). *Ann. Math. Statist.*, **30**, 463–489.

[16] Kiefer, J. and Wolfowitz, J. (1960). *Canad. J. Math.*, **12**, 363–366.

[17] Kiefer, J. and Wolfowitz, J. (1965). *Ann. Math. Statist.*, **36**, 1627–1655.

[18] Wald, A. and Wolfowitz, J. (1939). *Ann. Math. Statist.*, **10**, 105–118.

[19] Wald, A. and Wolfowitz, J. (1940). *Ann. Math. Statist.*, **11**, 147–162.

[20] Wald, A. and Wolfowitz, J. (1948). *Ann. Math. Statist.*, **19**, 326–339.

[21] Wald, A. and Wolfowitz, J. (1950). *Ann. Math. Statist.*, **21**, 82–99.

[22] Wald, A. and Wolfowitz, J. (1951). *Ann. Math.*, **53**, 581–586.

[23] Weiss, L. and Wolfowitz, J. (1972a). *Z. Wahrsch. verw. Geb.*, **24**, 203–209.

[24] Weiss, L. and Wolfowitz, J. (1972b). *J. R. Statist. Soc. B*, **34**, 456–460.

[25] Weiss, L. and Wolfowitz, J. (1974). *Maximum Probability Estimators and Related Topics*. Springer, New York.

[26] Wolfowitz, J. (1942). *Ann. Math. Statist.*, **13**, 247–279.

[27] Wolfowitz, J. (1943). *Ann. Math. Statist.*, **14**, 280–288.

[28] Wolfowitz, J. (1946). *Ann. Math. Statist.*, **17**, 489–493.

[29] Wolfowitz, J. (1947). *Ann. Math. Statist.*, **18**, 215–230.

[30] Wolfowitz, J. (1949). *Ann. Math. Statist.*, **20**, 601–602.

[31] Wolfowitz, J. (1950). *Ann. Math. Statist.*, **21**, 218–230.

[32] Wolfowitz, J. (1952). *Skand. Aktuar.*, **35**, 132–151.

[33] Wolfowitz, J. (1953). *Ann. Inst. Statist. Math.*, **5**, 9–23.

[34] Wolfowitz, J. (1954). *Ann. Math. Statist.*, **25**, 203–217.

[35] Wolfowitz, J. (1957a). *Ann. Math. Statist.*, **28**, 75–88.

[36] Wolfowitz, J. (1957b). *Illinois J. Math.*, **1**, 591–606.

[37] Wolfowitz, J. (1962). *Econometrica*, **30**, 470–479.

[38] Wolfowitz, J. (1967). *The New York Statistician*, **18**, 1–3.

[39] Wolfowitz, J. (1978). *Coding Theorems of Information Theory*, 3rd ed. Springer, New York.

[40] Wolfowitz, J. (1980). *Selected Papers*. Springer, New York.

(ESTIMATION, POINT
INFORMATION THEORY AND CODING
  THEORY
SEQUENTIAL ANALYSIS
WALD, ABRAHAM)

L. WEISS

WOODBURY MODEL

Woodbury [5] has considered a generalized Bernoulli experiment in which the probability of a success at a given trial depends on the number of successes in previous trials. Let p_r denote the probability of a success in a trial, given that $r - 1$ among the preceding trials resulted in successes. If $\Pr(r|n)$ denotes the probability of r successes in n trials the following recursive relation is valid:

$$\Pr(r|n) = p_r \Pr(r - 1|n - 1)$$
$$+ (1 - p_r)\Pr(r|n - 1),$$
$$r = 0, 1, \ldots, n, \; n \geqslant 1,$$

where $\Pr(-1|n - 1) = 0$ and $\Pr(r|0) = \delta_{0r}$ [$\delta_{ij} = 0$ $(i \neq j)$ and $\delta_{ii} = 1$]. If all p_r's are different, then

$$\Pr(r|n) = \sum_{j=0}^{t} \left\{ (1 - p_j)^r \middle/ \prod_{i \neq j} (p_i - p_j) \right\}.$$

Some special cases have been studied by Rutherford [3] and Chaddha [2]. Suzuki [4] presents an historical sketch of this model and studies in some detail the particular case when $p_r \leqslant p$ for $r(\leqslant m)$ numbers in $(0, 1)$. Further generalizations to the case where the probability of success depends also on the number of trials ($p_{r,n}$, say) are discussed by Suzuki [4] and Alzaid et al. [1].

References

[1] Alzaid, A. A., Rao, C. R., and Shanbhag, D. N. (1986). *Commun. Statist.-Theor. Meth.*, **15**, 643–656.

[2] Chaddha, R. L. (1965). In *Classical and Contagious Discrete Distributions*, G. P. Patil, ed. Stat. Public. Soc. Pergamon Press, Calcutta, pp. 273–290.

[3] Rutherford, R. S. G. (1954). *Ann. Math. Statist.*, **25**, 703–713.

[4] Suzuki, G. (1980). *Ann. Inst. Statist. Math.*, *Tokyo*, **32**, Part A, 143–159.

[5] Woodbury, M. A. (1949). *Ann. Math. Statist.*, **20**, 311–313.

(BERNOULLI DISTRIBUTION
BINOMIAL DISTRIBUTION
LEXIAN DISTRIBUTION
LEXIS, WILHELM)

WOOLHOUSE'S FORMULA

This is a summation formula obtained by equating two Euler–Maclaurin expansions* for the same integral, using differently spaced values of the variable. The formula is

$$\sum_{j=0}^{mn} u_j = n \sum_{j=0}^{m} u_{nj} - 1 - 2(n - 1)(u_0 + u_{mn})$$
$$- \frac{1}{12}(n^2 - 1)(Du_{mn} - Du_0)$$
$$+ \frac{1}{720}(n^4 - 1)(D^3 u_{mn} - D^3 u_0)$$
$$- \cdots,$$

where

$$D^r u_x = \left. \frac{d^r u_x}{dx^r} \right|_{x=k}, \qquad k = 0, mn.$$

The general term in the series is

$$(-1)^{r+1} B_r \frac{n^{2r} - 1}{2^r} \left(D^{2r-1} u_{mn} - D^{2r-1} u_0 \right),$$

where B_r is the rth Bernoulli number*.

By means of this formula summation of $(mn + 1)$ u's can be effected by summation of $(m + 1)$ u's, with some correction terms depending on the derivatives of u_x at the extremes of the range of summation. If the mathematical form (and so the derivatives) of u_x is not known, then Lubbock's formula* which uses only differences, and not derivatives, of u_x can be used.

Some statistical applications of Woolhouse's formula are discussed by Sverdrup [1].

Reference

[1] Sverdrup, E. (1967). *Laws and Choice Variations: Basic Concepts of Statistical Inference*. North Holland, Amsterdam, The Netherlands.

(FINITE DIFFERENCES, CALCULUS OF
LUBBOCK'S FORMULA
SUMMATION [n])

WORKING–HOTELLING–
SCHEFFÉ CONFIDENCE BANDS

Working–Hotelling–Scheffé (WHS) confidence bands (see Fig. 1) determine a confidence region* about a sample regression line or surface to make inferences about the unknown regression line or surface. The bands were developed by Working and Hotelling [11] for simple linear regression* and by Scheffé [8] for the general regression model. It is a special case of the Scheffé method of multiple comparisons* (*see also* SCHEFFÉ'S SIMULTANEOUS COMPARISON PROCEDURE). The procedure will be described first for simple linear regression and then for the general regression model.

The simple linear regression model is given by

$$Y = \beta_0 + \beta_1 x + E,$$

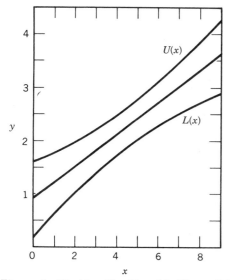

Figure 1 Working–Hotelling–Scheffé confidence bands.

where Y denotes the dependent variable, x denotes the independent variable, β_0 and β_1 denote unknown regression coefficients*, and E denotes random error. The data consist of $n > 2$ observations (x_i, y_i), $i = 1, \ldots, n$ where at least two of the x_i's are unequal. The WHS procedure assumes that the n random errors are independent, identically distributed normal random variables with zero mean and unknown variance σ^2. Denote the sample means of the x_i's and y_i's by $\bar{x}$ and $\bar{y}$, respectively. The sample regression line is given by

$$\hat{Y} = \hat{\beta}_0 + \hat{\beta}_1 x,$$

for $-\infty < x < \infty$, where

$$\hat{\beta}_1 = \frac{\sum_i (x_i - \bar{x})(y_i - \bar{y})}{\sum_i (x_i - \bar{x})^2}$$

and

$$\hat{\beta}_0 = \bar{y} - \hat{\beta}_1 \bar{x}.$$

An unbiased estimator of σ^2 is given by

$$\hat{\sigma}^2 = \sum_i (Y_i - \hat{\beta}_0 - \hat{\beta}_1 x_i)^2 / (n - 2).$$

The WHS confidence bands are given by the functions

$$U(x) = (\hat{\beta}_0 + \hat{\beta}_1 x) + (2 F_{2, n-2, 1-\alpha})^{1/2}$$

$$\times \hat{\sigma} \left[\frac{1}{n} + \frac{(x - \bar{x})^2}{\sum_i (x_i - \bar{x})^2} \right]^{1/2}$$

and

$$L(x) = (\hat{\beta}_0 + \hat{\beta}_1 x) - (2 F_{2, n-2, 1-\alpha})^{1/2}$$

$$\times \hat{\sigma} \left[\frac{1}{n} + \frac{(x - \bar{x})^2}{\sum_i (x_i - \bar{x})^2} \right]^{1/2}$$

defined on $x \in (-\infty, \infty)$, where $F_{a, b, 1-\alpha}$ denotes the $(1 - \alpha)100$ percentile of the F-distribution with a and b degrees of freedom. One is $(1 - \alpha)100\%$ confident that the unknown regression line falls between the two bands, that is, for all x,

$$L(x) < \beta_0 + \beta_1 x < U(x).$$

The width of the WHS bands $U(x) - L(x)$ increases as x moves from $\bar{x}$ in either direction.

The following example illustrates the procedure:

x	0	1	2	3	4	5	6	7	8	9
y	1.0	1.1	1.5	2.0	1.8	2.8	2.6	3.5	3.0	3.6

For these data $n = 10$, $\bar{x} = 4.5$, $\bar{y} = 2.29$, $\Sigma_i(x_i - \bar{x})^2 = 82.5$, $\hat{\beta}_0 = 0.94$, $\hat{\beta}_1 = 0.30$, $\hat{\sigma}^2 = 0.0805$, and $F_{2,8,0.99} = 8.65$. The 99% WHS bands are

$$(0.94 + 0.30x)$$
$$\pm 1.1801\left[0.1 + (x - 4.5)^2/82.5\right]^{1/2}.$$

The general regression model, with p independent variables, is given by

$$Y = \beta_1 x_1 + \dots + \beta_p x_p + E.$$

The data consists of $n > p$ observations $(x_{i1}, \dots, x_{ip}, y_i)$, $i = 1, \dots, n$. Let

$$\mathbf{y} = \begin{bmatrix} y_1 \\ y_2 \\ \vdots \\ y_n \end{bmatrix}, \quad \mathbf{X} = \begin{bmatrix} x_{11} & x_{12} & \cdots & x_{1p} \\ x_{21} & x_{22} & \cdots & x_{2p} \\ \vdots & \vdots & & \vdots \\ x_{n1} & x_{n2} & \cdots & x_{np} \end{bmatrix}.$$

It is assumed that the n random error terms are independent, identically distributed normal random variables with zero means and an unknown variance σ^2 and that $\text{rank}(\mathbf{X}) = p$. The sample regression surface is given by $\hat{Y} = \mathbf{x}'\hat{\beta}$ for $\mathbf{x}' = (x_1, \dots, x_p) \in \mathbb{R}^p$, where $\hat{\beta} = (\mathbf{X'X})^{-1}\mathbf{X'y}$. The unbiased estimator of σ^2 is given by

$$\hat{\sigma}^2 = \mathbf{y}'\left(\mathbf{I} - \mathbf{X}(\mathbf{X'X})^{-1}\mathbf{X'}\right)\mathbf{y}/(n - p),$$

where $\mathbf{I}$ is the identity matrix. The WHS confidence bands are given by the functions

$$U(\mathbf{x}) = \mathbf{x}'\hat{\beta} + \left(pF_{p, n-p, 1-\alpha}\right)^{1/2}$$
$$\times \hat{\sigma}\left[\mathbf{x}'(\mathbf{X'X})^{-1}\mathbf{x}\right]^{1/2}$$

and

$$L(\mathbf{x}) = \mathbf{x}'\hat{\beta} - \left(pF_{p, n-p, 1-\alpha}\right)^{1/2}$$
$$\times \hat{\sigma}\left[\mathbf{x}'(\mathbf{X'X})^{-1}\mathbf{x}\right]^{1/2},$$

for $\mathbf{x} \in \mathbb{R}^p$. One is $(1 - \alpha)100\%$ confident that $L(\mathbf{x}) < \mathbf{x}'\beta < U(\mathbf{x})$ for all $\mathbf{x} \in \mathbb{R}^p$.

The WHS confidence bands form a special case of Scheffé's simultaneous comparison procedure*, where one is simultaneously estimating all linear combinations of the components of β. The confidence region defined by the bands is equivalent to the set of all regression surfaces corresponding to the vectors β_0 such that one cannot reject the null hypothesis $\beta = \beta_0$ using the usual size α general linear model F-test.

If an intercept term is desired, then $x_1 \equiv 1$, and one is $(1 - \alpha)100\%$ confident that $L(\mathbf{x}) < \mathbf{x}'\beta < U(\mathbf{x})$ for all $\mathbf{x}' = (1, x_2, \dots, x_p)$, where $-\infty < x_i < \infty$ for $i = 2, \dots, p$. The probability point $(pF_{p, n-p, 1-\alpha})^{1/2}$ is not changed by the restriction on x_1.

The WHS confidence bands are quite versatile. The probability point $(pF_{p, n-p, 1-\alpha})^{1/2}$ depends only on p, n, and α. Special tables are not needed. Some writers view the property that the band width is proportional to the standard error of $\mathbf{x}'\hat{\beta}$ for each $\mathbf{x}$ as an advantage. Bohrer [2] has shown that under mild conditions such bands have minimum average width over ellipsoidal sets in $\mathbb{R}^p$, among all bands with the same confidence coefficient.

It is often the case that one wishes to estimate $E(Y|\mathbf{x})$ for only a subset A of $\mathbb{R}^p$. This can occur if an independent variable is known to be nonnegative, to be in a finite interval, or to be one of a finite set of values. It can also occur if, due to functional relationships among the independent variables, it is impossible to achieve all $\mathbf{x} \in \mathbb{R}^p$. For example, if $x_1 = t$ and $x_2 = t^2$, then all achievable (x_1, x_2) values lie on the parabola $x_2 = x_1^2$ in $\mathbb{R}^2$. When the WHS confidence bands are used to estimate $E(Y|\mathbf{x})$ for $\mathbf{x} \in A$, they are generally conservative, that is, the true level of confidence for simultaneously estimating $E(Y|\mathbf{x})$ for $\mathbf{x} \in A$ is greater than $(1 - \alpha)100\%$.

The bands can be modified to yield exact $(1 - \alpha)100\%$ confidence bands for $E(Y|\mathbf{x})$ for $\mathbf{x} \in A$ by replacing the probability point $(pF_{p, n-p, 1-\alpha})^{1/2}$ by the number c, such that

$$\Pr\left[\frac{\sup_{\mathbf{x} \in A}|\mathbf{x}'(\hat{\boldsymbol{\beta}} - \boldsymbol{\beta})|}{\left\{\hat{\sigma}^2 \mathbf{x}'(\mathbf{X}'\mathbf{X})^{-1}\mathbf{x}\right\}^{1/2}} \leqslant c\right] = 1 - \alpha.$$

The modified bands are never wider than the WHS bands and are generally narrower. The value of c, and hence the amount of improvement over the WHS bands, depends upon A and $\mathbf{X}$. Calculation of c is difficult, and tables are available for only a limited set of situations. Halperin and Gurian [5] consider the case of simple linear regression with the independent variable x restricted to an interval centered at $\bar{x}$. The amount of improvement ranges from negligible for broad intervals to more than 20% for very narrow intervals. Uusipaikka [10] considers the case of simple linear regression with the independent variable restricted to a finite union of points or intervals. Wynn and Bloomfield [12] consider the quadratic regression model

$$Y = \beta_1 + \beta_2 x + \beta_3 x^2 + E,$$

with no restriction on x. In this case the modified bands are approximately 5% narrower than the WHS bands. Bohrer [1] considers multiple regression with nonnegative independent variables. In this case the amount of improvement increases with p. For $p = 4$, an 8% improvement is achieved. Casella and Strawderman [4] consider multiple regression with $\mathbf{X}'\mathbf{X} = \mathbf{I}$. The restriction

$$\sum_{i=1}^{r} x_i^2 \geqslant q^2 \sum_{i=r+1}^{p} x_i^2$$

is used, where q is a positive constant.

The WHS bands are hyperbolic, and hence not generally of uniform width for all values of $\mathbf{x}$. This lack of uniformity is considered a disadvantage by some writers since the bands are difficult to graph or visualize. Spurrier [9] gives an example where the WHS bands have uniform width. Several authors have presented uniform-width competitors to the WHS bands. Uniform-width bands require the restriction of $\mathbf{x}$ to a subset of $\mathbb{R}^p$ such that the standard error of $\mathbf{x}'\hat{\boldsymbol{\beta}}$ is bounded;

Miller [6, 7] references several articles. Bowden [3] presents a unified theory that yields the WHS bands and some of the uniform-width bands as special cases.

References

[1] Bohrer, R. (1967). *J. R. Statist. Soc. B*, **29**, 110–114.

[2] Bohrer, R. (1973). *Ann. Statist.*, **1**, 766–772.

[3] Bowden, D. C. (1970). *J. Amer. Statist. Ass.*, **65**, 413–421.

[4] Casella, G. and Strawderman, W. E. (1980). *J. Amer. Statist. Ass.*, **75**, 862–868.

[5] Halperin, M. and Gurian, J. (1968). *J. Amer. Statist. Ass.*, **63**, 1020–2027.

[6] Miller, R. G., Jr. (1966). *Simultaneous Statistical Inference.* McGraw-Hill, New York. (Excellent general reference for pre-1966 results.)

[7] Miller, R. G., Jr. (1977). *J. Amer. Statist. Ass.*, **72**, 779–788. (Excellent reference for 1966–1976 results.)

[8] Scheffé, H. (1959). *The Analysis of Variance.* Wiley, New York. (Theoretical text that derives the technique for the general case.)

[9] Spurrier, J. D. (1983). *Commun. Statist. A*, **12**, 969–973.

[10] Uusipaikka, E. (1983). *J. Amer. Statist. Ass.*, **78**, 638–644.

[11] Working, H. and Hotelling, H. (1929). *J. Amer. Statist. Ass. Suppl.* (*Proc.*), **24**, 73–85. (First paper in the area.)

[12] Wynn, H. P. and Bloomfield, P. (1971). *J. R. Statist. Soc. B*, **33**, 202–217.

(CONFIDENCE INTERVALS AND REGIONS
LINEAR REGRESSION
MULTIPLE COMPARISONS
MULTIPLE LINEAR REGRESSION
MULTIVARIATE MULTIPLE
  COMPARISONS
REGRESSION COEFFICIENTS
SCHEFFÉ'S SIMULTANEOUS COMPARISON
  PROCEDURE)

JOHN D. SPURRIER

WRAPPED DISTRIBUTIONS

Suppose a random variable X is measured on a scale with scale interval α, so that the possible measured values are $\{N\alpha\}$, where N is an integer (positive, negative, or zero). Then the value of N scaled to X is $n_\alpha(X)$, where

$$\alpha n_\alpha(X) - \tfrac{1}{2}\alpha < X \leqslant \alpha n_\alpha(X) + \tfrac{1}{2}\alpha.$$

The deviation $U_\alpha = X - n_\alpha(X)\alpha$ has PDF

$$f_{U_\alpha}(u) = \sum_{n=-\infty}^{\infty} f_X(n\alpha - u)$$

$(-\tfrac{1}{2}\alpha \leqslant u \leqslant \tfrac{1}{2}\alpha)$, where $f(\cdot)$ is the PDF of X.

Distributions with PDFs of this type are called *wrapped* (or *wrapped-up*) *distributions*. They appear naturally in models of directional data* (see [1]). Stadje [2] gives a discussion of wrapped distributions in the context of measurement error.

References

[1] Mardia, K. V. (1972). *Statistics of Directional Data*. Academic, New York.

[2] Stadje, W. (1984). *Metrika*, **31**, 303–317.

(CIRCULAR NORMAL DISTRIBUTION
DIRECTIONAL DISTRIBUTIONS
MEASUREMENT ERROR
WRAPPED-UP CAUCHY DISTRIBUTION)

WRAPPED-UP CAUCHY DISTRIBUTION

This is a distribution obtained by wrapping a Cauchy distribution* around a circle and adding up the probability densities (PDFs) coinciding at each point.

With mean angle θ and mean vector length ρ, the PDF of a variable T with this distribution is

$$\frac{1}{2\pi} \frac{1 - \rho^2}{1 + \rho^2 - 2\rho\cos(t - \theta)}, \qquad 0 \leqslant t < 2\pi.$$

The distribution is unimodal and symmetric.

Further details are available in [1].

Reference

[1] Batschelet, E. (1981). *Circular Statistics in Biology*. Academic, New York.

(DIRECTIONAL DISTRIBUTIONS
WRAPPED-UP NORMAL DISTRIBUTIONS)

WRIGHT'S REJECTION PROCEDURE

An early test procedure for rejecting outliers*. It was suggested by Wright [1] and rejects an observation that deviates from the mean by more than three times the standard deviation* (equivalently approximately 4.5 times the probable error*). Note that this criterion is independent of the sample size n.

Reference

[1] Wright, T. W. (1884). *A Treatise on the Adjustment of Observations by the Method of Least Squares*. Van Nostrand, New York.

(CHAUVENET'S CRITERION
OUTLIERS
PEIRCE'S CRITERION
STONE'S REJECTION CRITERION)

WU–HAUSMAN SPECIFICATION TEST

Details of this test are given in HAUSMAN SPECIFICATION TEST. The initial suggestions, due to Wu, are in [1] and [2].

References

[1] Wu, D.-M. (1973). *Econometrica*, **41**, 733–750.

[2] Wu, D.-M. (1974). *Econometrica*, **42**, 529–546.

X

X-11 METHOD

The X-11 *method* refers to a computer program for seasonal adjustment of quarterly or monthly economic time series* maintained by the U.S. Bureau of the Census* and to the methodology employed by the program. The program is the eleventh and last in a sequence of programs developed in the late 1950s and early 1960s at the Census Bureau under the direction of Julius Shiskin. *Technical Paper No.* 15 [12] remains the authoritative document on the contents of the program and contains references on the statistical methodology. The X-11 method is widely used by government agencies and private businesses in the United States and many foreign countries.

The paradigm used in formulating the seasonal adjustment problem is that of unobservable components. Let M_t represent the observed monthly series in Fig. 1 of demand deposits at commercial bnks, with t indexing the months. The series is viewed as containing *seasonal variation* S_t, *trend* and *business cycle variation* P_t, and *random* or *irregular variation* I_t. Whereas the additive representation $M_t = P_t + S_t + I_t$ may be used, one frequently observes seasonal movements proportional to the level of the series, as in Fig. 1. In such cases the alternative representation $M_t = P_t S_t I_t$ is adopted, and the S_t are called *seasonal factors*. The seasonally adjusted series M_t^a is M_t with S_t removed. For the additive representation $M_t^a = M_t - S_t$, whereas for the multiplicative formulation $M_t^a = M_t/S_t$. The quality of the seasonal adjustment process rests on the validity of the component representation and the accuracy of the estimate of S_t.

The techniques for estimating S_t contained in X-11 make use of smoothing or graduation* formulas developed in the early 1900s [9, 14]. These provide several sets of moving average weights or filters, which are applied to the data. The series is first detrended to center it about 0 (additive) or 1 (multiplicative) and then averaged by month (all Januarys, etc.) to get the seasonal deviation from center. This process is repeated with moving averages* of different widths and with downweighting of selected observations based on the size of irregular estimates. The seasonal factor estimate for demand deposits is shown in Fig. 2. The use of local symmetric averages of each month to estimate S_t, rather than a uniform average over all occurrences of the month, allows for

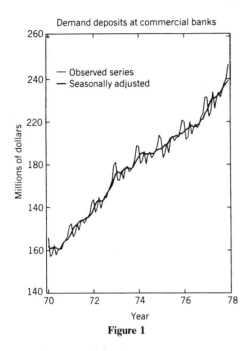

Figure 1

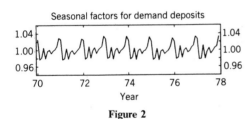

Figure 2

prove adjustments near the end of the series and to obtain seasonal factor projections [5]. The SABL computer program [4] adopts a philosophy similar to X-11, but uses robust versions of moving averages rather than the X-11 outlier procedure. The proceedings of a 1976 conference on seasonal adjustment [16] gives an excellent overview of issues in seasonal adjustment and how they relate to X-11 and the other methodologies. Alternative strategies for seasonal adjustment based on statistical time series models are described there and elsewhere [1, 2, 6–8].

References

[1] Burman, J. P. (1980). *J. R. Statist. Soc. A*, **143**, 321–337.

[2] Cleveland, W. P. and Dempster, A. P. (1980). *Proc. Bus. Econ. Statist. Sect. Amer. Statist. Ass.*, 30–36.

[3] Cleveland, W. P. and Tiao, G. C. (1976). *J. Amer. Statist. Ass.*, **71**, 581–587.

[4] Cleveland, W. S., Dunn, D. M., and Terpenning, I. J. (1976). In *Seasonal Analysis of Economic Time Series*, A. Zellner, ed. U.S. Department of Commerce, Bureau of the Census, Washington, D.C., pp. 201–231.

[5] Dagum, E. B. (1980). *The X-11 ARIMA Seasonal Adjustment Method*. Statistics Canada, Ottawa.

[6] Gersch, W. and Kitagawa, G. (1983). *J. Bus. Econ. Statist.*, **1**, 253–264.

[7] Havenner, A. and Swamy, P. A. V. B. (1981). *J. Econometrics*, **15**, 177–209.

[8] Hillmer, S. C. and Tiao G. C. (1982). *J. Amer. Statist. Ass.*, **77**, 63–70.

[9] Macaulay, F. R. (1931). *The Smoothing of Time Series*. National Bureau of Economic Research.

[10] Nerlove, M. (1965). *J. Amer. Statist. Ass.*, **60**, 442–491.

[11] Rosenblatt, H. M. (1968). *J. Amer. Statist. Ass.*, **63**, 472–501.

[12] Shiskin, J., Young, A., and Musgrave, J. C. (1965). The X-11 Variant of Census Method II Seasonal Adjustment Program. *Technical Paper* 15, U.S. Department of Commerce, Bureau of the Census, Washington, D.C.

[13] Wallis, K. F. (1974). *J. Amer. Statist. Ass.*, **69**, 18–31.

[14] Whittaker, E. and Robinson, G. (1944). *The Calculus of Observations*. Blackie and Son, London, England.

[15] Young, A. (1968). *J. Amer. Statist. Ass.*, **63**, 445–471.

evolving seasonal patterns. Asymmetric versions of these averages are used near the ends of the series.

Some series, e.g., retail sales, reflect a definite daily pattern related to the day of the week. When the daily values are summed to get a monthly value, the effect of having five of some weekdays and four of others is observable. An optional regression procedure is contained in X-11 to handle such series.

Although X-11 is a nonlinear procedure because of its treatment of outliers*, linear approximations excluding this feature have been published and studied [3, 13, 15] in addition to studies of observed results [10, 11]. A recent modification, known as X-11 ARIMA*, uses forecasting* models to im-

[16] Zellner, A., ed. (1976). *Seasonal Analysis of Economic Time Series*. U.S. Department of Commerce, Bureau of the Census, Washington, D.C.

(AUTOREGRESSIVE
  INTEGRATED–MOVING AVERAGE
  (ARIMA) MODELS
FORECASTING
GRADUATION
MOVING AVERAGES
SEASONALITY
TIME SERIES)

WILLIAM P. CLEVELAND

X-BAR CHART *See* CONTROL CHARTS

X–Y PLOTTER

Synonymous with *data plotter*—a unit providing a visual display in the form of a graph on paper.

(GRAPHICAL REPRESENTATION OF
  DATA)

Y

YANSON (JAHNSON), YULII EDUARDOVICH

Born: November 5, 1835(o.s.), in Kiev, Russian Empire.

Died: January 31, 1892(o.s.), St. Petersburg, Russian Empire.

Contributed to: official statistics, economics, demography.

Y. E. Yanson's initial tertiary training was in the historico-philological faculty at Kiev University. In 1861, he was appointed to an academic position in agricultural statistics and political economy and, after several such posts, first taught at St. Petersburg University in 1865. Yanson's role was within the historical development of Russian statistical presentation to important socioeconomic issues, particularly problems of agricultural economics. Sometimes considered a father of the discipline of statistics in the Russian Empire, he is best known for two works: (1) *Comparative Statistics of Russian and Western-European States*, a 2-volume work, the first volume of which appeared in 1878; (2) *The Theory of Statistics*, which appeared in five editions between 1885 and 1913 [5]. Of the latter work, Yanson's contemporary A. I. Chuprov, the father of A. A. Chuprov* and himself a leading figure in the same areas as Yanson, said [1]: "For statistical methodology we have nothing superior to Yanson's book; in regard to the description of statistical establishments and the applications of statistics, it seems there is little comparable to be found in Western-European literature." This book was used as a text for the course of statistics at St. Petersburg University (where Yanson was full professor from 1873), and was studied diligently by Lenin prior to his examination in 1891 as external student by a commission of which Yanson was a member [4]. Lenin was to refer later to Yanson's statistical data in his writings and may be regarded as having gained his statistical technology [3] from this book.

Yanson was elected to the *International Statistical Institute** in 1885 and became corresponding member of the Russian Academy of Science in 1892. He was active in social reform in the manner of the liberal intelligentsia of his milieu, and practically, apart from peasant economics, in the careful planning of censuses and epidemiological investigations. A photograph may be found in [4]. There is a good obituary by A. I. Chuprov [1]; and the encyclopedia entry [2] gives extensive information.

References

[1] Chuprov, A. I. (1893). Yulii Eduardovich Yanson. Obituary—1893 (in Russian). In *Rechi i Stati*, Vol. 1. Sabashnikov, Moscow (1909), pp. 518–525.

[2] *Entsyklopedicheskii Slovar* (1904), Vol. XLI^A. Brokhaus and Efron, St. Petersburg. (The Yanson, Yu. E. entry extends over pp. 681–684.)

[3] Il'in, V. (1908). *Razvitie Kapitalizma v Rossii*, 2nd ed. "Pallada," St. Petersburg. [In English as: Lenin, V. I. (1964). *The Development of Capitalism in Russia*. Progress Publishers, Moscow. V. Il'in was a pseudonym used by Vladimir Il'ich Ulianov, later to become known as Lenin.]

[4] Sipovska, I. V. and Suslov, I. P., eds. (1972). *Istoriia Prepodavaniia i Razvitiia Statistiki v Peterburgskom-Leningradskom Universitete (1819–1971)*. Leningrad University, Leningrad, pp. 22–36.

[5] Yanson, Yu. E. (1887). *Teoriia Statistiki*, 2nd ed. Schröder, St. Petersburg (5th ed., 1913).

E. SENETA

YARNOLD'S CRITERION

A criterion proposed by Yarnold [2] for using the chi-squared distribution* with $k - 1$ degrees of freedom as an approximation to the distribution of

$$X^2 = \sum_{j=1}^{k} \left(N_j - np_j \right)^2 \left(np_j \right)^{-1},$$

where $N_1, N_2, \ldots, N_k$ (with $\sum_{j=1}^{k} N_j = n$) have a multinomial distribution* with parameters $n; p_1, \ldots, p_k$.

According to this criterion, the approximation can be expected to be adequate if

$$np_j \geqslant 5 \times (\text{number of } h\text{'s for which } np_h < 5).$$

Eaton [1] devised a systematic procedure for selecting, in a few steps, the minimum sample size consistent with Yarnold's criterion.

References

[1] Eaton, P. W. (1978). *Amer. Statist.*, **32**, 102–103.

[2] Yarnold, J. K. (1970). *J. Amer. Statist. Ass.*, **65**, 864–886.

(APPROXIMATIONS TO DISTRIBUTIONS
CHI-SQUARE TESTS: NUMERICAL
    EXAMPLES)

YASTREMSKIĬ, BORIS SERGEYEVICH

Born: May 9, 1877, in Dergach, near Kharkov, Ukraine.

Died: November 28, 1962, in Moscow, USSR.

Contributed to: time series, applied statistics.

Son of the well-known Russian revolutionary S. V. Yastremskiĭ, B. S. Yastremskiĭ was the leader of the dogmatic, strictly Marxist–materialistic approach to statistical sciences in the USSR and had a substantial influence on the development of statistics in that country.

He published a total of 97 papers on both the theoretical and practical problems of statistics.

He started his career in 1913 by criticizing the theory of stability of statistical series, which was developed by W. Lexis*, and continued by criticizing the "idealistic treatment" of the law of large numbers* and the "law of averages." He also wrote extensively against Pearson's system of distributions* and the concept of spurious correlation*. He co-authored (with A. Ya. Boyarskiĭ and others) two "Marxist" textbooks on statistics in 1931 and 1936. A summary of his statistical ideas is contained in his last book *Mathematical Statistics*, published in Moscow in 1956. Further details are given in an article [1] commemorating his 90th birthday.

Reference

[1] Boyarskiĭ, A. Ya. and Kil'disher, G. (1967). *Vestnik Statist.*, **5**, 35–40.

YATES' ALGORITHM

This (also called *Yates' technique*) is a system introduced by Yates [3] for facilitating the calculation by hand of estimates of main effects* and interactions* from data of factorial experiments*. The algorithm was first proposed for 2^k experiments (k factors at 2 two levels each). Here we will first describe the analysis for a 2^4 experiment with factors A, B, C, D; and then indicate how the method can be extended to include factors with more than two levels.

We will denote the level of a factor by a subscript attached to the corresponding lower-case level. Thus a_0, a_1 denote the lower and higher levels of A, respectively; $a_0 b_0 c_1 d_0$, for example, denotes the factor level combination in which A, B, and D are at lower, and C at higher level. The experiment gives rise to $2^4 = 16$ observed values, one for each of the 16 possible factor level combinations.

The calculations are set out in the form of a *Yates table* (see Table 1). The rows of the table correspond to the factor level combinations in *standard order*. Starting from $a_0 b_0 c_0 d_0$, this is achieved by increasing the level of each factor in turn (the order of

factors is immaterial), in combination with all preceding combinations of levels of the other factors. Thus following $a_0 b_0 c_0 d_0$, we have $a_1 b_0 c_0 d_0$, and then $a_0 b_1 c_0 d_0$, and so on. See the first column of Table 1. The second column gives the observed values for the 16 factor level combinations.

The calculations are simply addition or subtraction of values in successive rows. The top half of each of the next four columns is obtained by summing successive pairs of values in the preceding column. The lower half is formed by subtracting the upper item of each pair from the lower one. The procedure is repeated four times. (For a 2^k experiment it would be repeated k times and there would be k columns, in place of I–IV.)

Column IV contains estimators of the main effects and interactions (or some multiple thereof) as indicated in the final column.

The corresponding "sum of squares" (with 1 degree of freedom) is

$$(\text{amount in final column})^2/4$$

[generally $(\text{amount in final column})^2/k$].

Table 1   Application of Yates' Algorithm to a 2^4 Experiment

Factor Level Combination	Observed Value	Summation/Subtraction Operation				Effect or Interaction
		I	II	III	IV	
$a_0 b_0 c_0 d_0$	4	10	25	44	120	Total
$a_1 b_0 c_0 d_0$	6	15	19	76	6	A
$a_0 b_1 c_0 d_0$	7	7	41	3	16	B
$a_1 b_1 c_0 d_0$	8	12	35	3	16	$A \times B$
$a_0 b_0 c_1 d_0$	3	19	3	10	-12	C
$a_1 b_0 c_1 d_0$	4	22	0	6	-4	$A \times C$
$a_0 b_1 c_1 d_0$	6	16	2	-3	0	$B \times C$
$a_1 b_1 c_1 d_0$	5	19	1	-1	-6	$A \times B \times C$
$a_0 b_0 c_0 d_1$	9	2	5	-6	32	D
$a_1 b_0 c_0 d_1$	9	1	5	-6	0	$A \times D$
$a_0 b_1 c_0 d_1$	10	1	3	-3	-4	$B \times D$
$a_1 b_1 c_0 d_1$	12	-1	3	-1	2	$A \times B \times D$
$a_0 b_0 c_1 d_1$	7	0	-1	0	0	$C \times D$
$a_1 b_0 c_1 d_1$	9	2	-2	0	2	$A \times C \times D$
$a_0 b_1 c_1 d_1$	10	2	2	-1	0	$B \times C \times D$
$a_1 b_1 c_1 d_1$	9	-1	-3	-5	4	$A \times B \times C \times D$

EXTENSION TO FACTORS WITH MORE THAN TWO LEVELS

For illustrative purposes, we will consider results from a 2×3^2 experiment (factor A at two levels; factor B and C each at those levels). Extension to general numbers of levels is straightforward.

For factor A we have the single comparison $a_1 - a_0$, but for each of B and C there are two orthogonal* comparisons possible (2 "degrees of freedom"). Provided the conditions, of orthogonality to each other and to the sum ($b_0 + b_1 + b_2$, or $c_0 + c_1 + c_2$, as the case may be), hold, the 2 degrees of freedom may be split up arbitrarily. For our illustration we will take the "linear" components $b_2 - b_0, c_2 - c_0$ and the "quadratic" components $b_2 - 2b_1 + b_0, c_2 - 2c_1 + c_0$ (which would be appropriate if the levels of B and C were each equally spaced and we were using a polynomial regression* model).

As before, we start with the factor level combination $a_0 b_0 c_0$ and introduce the higher levels of each factor in turn, as for 2^k experiments (cf. Table 1). This leads to the first column of Table 2.

We now have *three* columns I–III of linear operations, corresponding to *three* factors. The first column (corresponding to A, which has two levels) is obtained exactly the same way as in Table 1. That is, we form sums (corresponding to $a_0 + a_1$) and then differences (corresponding to $a_1 - a_0$) of successive rows in the preceding column. But, in proceeding from column I to column II (corresponding to B, which has three levels), we divide the 18 rows into six sets of three. We first form sums (corresponding to $b_0 + b_1 + b_2$), then differences between the last and first item in each set of three rows (corresponding to $b_2 - b_0$); and then the sum of the first and last, minus twice the middle item (corresponding to $b_2 - 2b_1 +$

Table 2   Yates' Algorithm for a 2×3^2 Experiment

Factor Level Combination	Observed Value	Linear Function Operations			Main Effect or Interaction
		I(A)	II(B)	III(C)	
$a_0 b_0 c_0$	2	40	30	94	Total
$a_1 b_0 c_0$	2	9	20	6	A
$a_0 b_1 c_0$	4	17	44	28	BL
$a_1 b_1 c_0$	5	3	2	0	$A \times BL$
$a_0 b_2 c_0$	8	6	0	4	BQ
$a_1 b_2 c_0$	9	11	4	0	$A \times BQ$
$a_0 b_0 c_1$	1	11	13	14	CL
$a_1 b_0 c_1$	2	15	8	2	$A \times CL$
$a_0 b_1 c_1$	3	18	7	-6	$BL \times CL$
$a_1 b_1 c_1$	3	0	1	0	$A \times BL \times CL$
$a_0 b_2 c_1$	6	1	-2	-4	$BQ \times CL$
$a_1 b_2 c_1$	5	1	1	2	$A \times BQ \times CL$
$a_0 b_0 c_2$	5	1	3	34	CQ
$a_1 b_0 c_2$	6	0	2	6	$A \times CQ$
$a_0 b_1 c_2$	7	-1	-1	4	$BL \times CQ$
$a_1 b_1 c_2$	8	1	-1	6	$A \times BL \times CQ$
$a_0 b_2 c_2$	8	1	0	-2	$BQ \times CQ$
$a_1 b_2 c_2$	10	2	1	0	$A \times BQ \times CQ$

b_0). This operation is repeated in going from column II to column III (corresponding to the three level factor C).

The final column shows the main effects and interactions estimated in column III. The letters L and Q are used to indicate "linear" and "quadratic" components, respectively. The $BL \times CQ$ estimator, for example, is, in fact, the linear combination obtained by formal multiplication from

$$(a_0 + a_1)(b_2 - b_0)(c_2 - 2c_1 + c_0).$$

The divisor for the square of this quantity, in order to obtain the sum of squares for the analysis of variance table, is the sum of squares of the coefficients, which is conveniently calculated as

$$(1^2 + 1^2)(1^2 + 1^2)(1^2 + 2^2 + 1^2)$$
$$= 2 \times 2 \times 6 = 24.$$

For the $A \times BQ \times CQ$ estimator

$$(a_1 - a_0)(b_2 - 2b_1 + b_0)(c_2 - 2c_1 + c_0),$$

the divisor would be

$$(1^2 + 1^2)(1^2 + 2^2 + 1^2)(1^2 + 2^2 + 1^2)$$
$$= 2 \times 6 \times 6 = 72.$$

Yates' technique can be extended to analysis of fractional factorial* experiments and to experiments involving confounding*. For details, see [1, p. 215] and [2, p. 800].

References

[1] Hicks, C. R. (1973). *Fundamental Concepts in Design of Experiments*, 2nd ed. Holt, Rinehart, and Winston, New York.

[2] Johnson, N. L. and Leone, F. C. (1977). *Statistics and Experimental Design: In Engineering and Physical Sciences*, 2nd ed., Vol. 2. Wiley, New York.

[3] Yates, F. (1937). *The Design and Analysis of Factorial Experiments*. Imperial Bureau of Soil Science, Harpenden, England.

(ANALYSIS OF VARIANCE
CONFOUNDING
FACTORIAL EXPERIMENTS
FRACTIONAL FACTORIAL DESIGNS
INTERACTION
MAIN EFFECTS)

YATES' CORRECTION FOR CONTINUITY   *See* CONTINUITY CORRECTIONS

YATES–DURBIN RULE

This is a rule for calculating estimates of sampling error for multistage designs of any degree of complexity. The original version was given by Yates [2] for the case of equal probability of selection. It was subsequently generalized by Durbin [1] to read:

> The estimate of variance in multi-stage sampling is the sum of two parts. The first part is equal to the estimate of variance calculated for the assumption that the first-stage values have been measured without error. The second part is calculated as if the first-stage units selected were fixed strata, the contribution from each first-stage unit being multiplied by the probability of that unit's inclusion in the sample.

Compare with Horvitz and Thompson's* estimator, originally given for the case of single-stage sampling, and with the alternative estimator given by Yates and Grundy [3]. For an illustration of this principle and further discussion, see, e.g., Durbin [1].

References

[1] Durbin, J. (1953). *J. R. Statist. Soc. B*, **15**, 262–269.

[2] Yates, F. (1949). *Sampling Methods for Censuses and Surveys*. Griffin, London, England.

[3] Yates, F. and Grundy, P. (1953). *J. R. Statist. Soc. B*, **15**, 253–261.

(STRATIFIED DESIGNS
STRATIFIED MULTISTAGE SAMPLING
SURVEY SAMPLING)

YOUDEN MISCLASSIFICATION INDEX

Suppose that each individual in an epidemiological* study can be classified as (truly) positive or negative with respect to a particular diagnosis. (This "true diagnosis" might be based on more reliable, but more expensive, methods than those used in a routine test.)

For each individual in a large population suppose that

$$\Pr[\text{test is positive} | \text{true positive}] = 1 - \beta,$$

$$\Pr[\text{test is negative} | \text{true negative}] = 1 - \alpha.$$

A "good" test should have small values for the error probabilities α and β (disregarding cost considerations).

Youden [1] proposed the index

$$J = 1 - (\alpha + \beta),$$

as a measure of the "goodness" of the test.

If the test has no diagnostic value, $\alpha = 1 - \beta$ and $J = 0$.

If the test is always correct, $\alpha = \beta = 0$ and $J = 1$.

Negative values of J (between -1 and 0) can arise if the test results are negatively associated with the true diagnosis.

Reference

[1] Youden, W. J. (1950). *Cancer*, **3**, 32–35.

(CLASSIFICATION
MEDICAL DIAGNOSIS, STATISTICS IN
SENSITIVITY AND SPECIFICITY
TWO-BY-TWO (2 × 2) TABLES)

YOUDEN SQUARE

These experimental designs are not really "squares" at all.

A Youden square is a balanced incomplete blocks* design with the additional property that it can be arranged in columns (corresponding to blocks) in such a way that each treatment appears once in each row. An example with seven treatments and seven blocks each of three plots is shown in Fig. 1.

Blocks Rows	(1)	(2)	(3)	(4)	(5)	(6)	(7)
(i)	A	B	C	D	E	F	G
(ii)	B	C	D	E	F	G	A
(iii)	D	E	F	G	A	B	C

Figure 1 A Youden square.

A	B	C	D
B	C	D	A
C	D	A	B
D	A	B	C

Latin square

A	B	C	D
B	C	D	A
C	D	A	B

Youden square

$(b = t = 4; \ k = r = 3; \ \lambda = 2)$

Figure 2 Construction of a Youden square.

This is a balanced incomplete blocks* design with parameters $b = 7$, $t = 7$, $k = 3$, $r = 3$, and $\lambda = 1$. Note that in a Youden square the number of blocks b must equal the number of treatments t, and the number of plots per block k must equal the number of replications r of each treatment.

The design does not form a "square" but the name "Youden square" (apart from referring to their proponent Youden [3]) may be based on the fact that the design may be regarded as an incomplete Latin square*. (This name was given to them by F. Yates [2].) Indeed, if the last (or any one) row of a $m \times m$ Latin square is deleted, the result is a Youden square with $b = t = m$; $k = r = m - 1$; $\lambda = m - 2$. Figure 2 shows such a construction, with $m = 4$.

However, omission of two or more rows from a Latin square does not, in general, produce a Youden square.

The standard analysis of variance* for data from an experiment designed as a Youden square is the same as for a balanced incomplete randomized block design of the same dimensions, except that a "Between Rows" sum of squares is computed and subtracted from the standard Residual sum of squares. The degrees of freedom are correspondingly reduced by $(r - 1)$, giving Residual degrees of freedom

$$= bk - 1 - (b - 1) - (t - 1) - (r - 1)$$

$$= bk - b - t - r + 2 = (t - 1)(r - 2),$$

since $b = t$ and $k = r$.

On the assumption of the parametric model

observed value

$$= \text{constant} + (\text{Block effect})$$

$$+ (\text{Row effect}) + (\text{Treatment effect})$$

$$+ (\text{Residual}),$$

with independent residuals having constant variance σ^2, the Residual mean square will have expected value t^2; and if normal variation can be assumed, the Rows sum of squares will be independent of the Residual and the Blocks (or Blocks adjusted) and Treatments (or Treatments adjusted) sums of squares.

There is a detailed descripted of construction and analysis of Youden squares in Natrella [1].

References

[1] Natrella, M. G. (1963). *Experimental Statistics.* U.S. Natl. Bur. Stand. Handbk. **9**, Government Printing Office, Washington D.C.

[2] Yates, F. (1936). *J. Agric. Sci.*, **26**, 301–315.

[3] Youden, W. J. (1937). *Contrib. Boyce Thompson Inst.*, **9**, 41–48.

(ANALYSIS OF VARIANCE
BLOCKS, BALANCED INCOMPLETE
LATIN SQUARES, LATIN CUBES, LATIN
RECTANGLES, ETC.)

YUKICH METRIC

Yukich [1] proposed the following metric on the k-dimensional space R^k. The distance between probability measures $P(\cdot)$ and $Q(\cdot)$, with respect to probability density $g(\cdot)$, is

$$d_g(P_1Q)$$
$$= \sup\{|g(x - y)\,d\{P(y) - Q(y)\}|:$$
$$x, y \in R^k\}.$$

See Yukich [1] for details, and for potential applications.

Reference

[1] Yukich, J. E. (1985). *Math. Proc. Camb. Philos. Soc.*, **98**, 533–540.

(PROBABILITY SPACES, METRICS AND
DISTANCES ON)

YULE DISTRIBUTION (*see Supplement*)

YULE, GEORGE UDNY

Born: February 18, 1871, in Beech Hill, near Haddington, Scotland.
Died: June 26, 1951, in Cambridge, England.
Contributed to: correlation theory, distribution theory, stochastic processes, statistics of literary vocabulary.

George Udny Yule was a member of a Scottish family steeped in literary and administrative traditions. After schooling at Winchester, he proceeded at the age of 16 to study engineering at University College, London. His first published work was on research into electric waves under Heinrich Hertz during his sojourn at Bonn 1892. He wrote six papers on electromagnetic theory, but, after taking up a demonstratorship offered him in 1893 by Karl Pearson* (then a professor of applied mathematics at University College), he turned his attention to statistical problems, starting with his 1895 paper, "On the Correlation of Total Pauperism with Proportion of Outrelief." There is a fascinating discussion of this paper in Selvin [3], describing how Yule introduced correlation* coefficients in studying two-way tables in the earlier volumes of the monumental work of Booth [2].

For Yule, Pearson was an inspiring teacher, and Yule made fundamental contributions in 1897 and 1899 to the theory of statistics of regression and correlation. In 1899, Yule left University College for a post at the City and Guilds of London Institute. Between 1902 and 1909, he also gave the Newmarch lectures in statistics at University College. These lectures formed the basis of his famous *Introduction to the Theory of Statistics*, which, during his lifetime, ran to 14 editions [the 11th (1937), 12th (1940), 13th (1945), and 14th (1950) being joint with M.G. Kendall]. He continued to publish papers on association and correlation (1900, 1901, 1903), and was awarded the Guy Medal in Gold in 1911 by the Royal Statistical

Society*. His theoretical works were accompanied by contributions to various economic and sociological subjects (1906, 1907, 1910).

In 1912, Yule was appointed a lecturer in statistics at Cambridge University (later raised to the status of a readership). The years 1920–1930 were the most productive in his career. During this period he introduced the correlogram* and laid the foundations of the theory of autoregressive* series. He was president of the Royal Statistical Society from 1924–1926. In one of his last publications in the *Journal of the Royal Statistical Society**, in late 1933, he showed that German authorities had overestimated the number of Jews entering Germany from Poland and Galicia during and after World War I by a factor of about 5.

In 1931, he retired from his post at Cambridge University. However, he "felt young enough to learn to fly. Accordingly, he went through the intricacies of training, got a pilot's license, and brought a plane. Unfortunately, a heart attack cut short his flying and, to a considerable degree, his scholarly work" (Bates and Neyman [1]). The death of Karl Pearson in 1936 affected him deeply. However, according to M. G. Kendall, "the publication of the revised *Introduction* in 1937 gave him a new lease of life." In the later years, his main work was related to frequency of occurrence of words (particularly nouns) in various texts (*see* ZIPF'S LAW). This research found expression in his last book (1944) on *The Statistics of Literary Vocabulary*.

A great deal of Yule's contributions to statistics resides in the stimulus he gave to students, in discussion with his colleagues, and advice he generously tendered to all who consulted him. His work on correlation and regression is now so standard that only history buffs would consult the original sources; he invented the correlogram and the autoregressive series; he also paved the way for Fisher's derivation of the distributions of partial correlations*. The terms *Yule process* and *Yule distribution* (*see Supplement*) are now firmly established in the literature.

References

[1] Bates, G. E. and Neyman, J. (1952). *Univ. California Publ. Statist.*, **1**, 215–254.

[2] Booth, C. E., ed. (1889–1893). *Life and Labour of the People in London*. Macmillan, London, England (first 4 of 17 volumes).

[3] Selvin, H. C. (1976). *Archives Europ. J. Sociol.*, **17**, 39–51.

Works by G. U. Yule

(1895) *Econ. J.*, **5**, 477–489.
(1897) *Proc. R. Soc. Lond.*, **60**, 477–489.
(1899) *J. R. Statist. Soc.*, **62**, 249–286.
(1900) *Philos. Trans. R. Soc. Lond.*, **197A**, 91–133.
(1901) *Philos. Trans. R. Soc. Lond.*, **197A**, 91–133.
(1903) *Biometrika*, **2**, 121–134.
(1906) *J. R. Statist. Soc.*, **69**, 88–132.
(1907) *J. R. Statist. Soc.*, **70**, 52–87.
(1910) *J. R. Statist. Soc.*, **73**, 26–38.
(1933) *J. R. Statist. Soc.*, **96**, 478–480.
(1944) *The Statistics of Literary Vocabulary*. Cambridge University Press, London, England.

A list of Yule's publications is included in the obituary:

Kendall, M. G. (1952). *J. R. Statist. Soc. A*, **115**, 156–161.

A list of publications is also given in:

Stuart, A. and Kendall, M. G., eds. (1971). *Statistical Papers of George Udny Yule*. Hafner, New York.

YULE SCHEME

A linear autoregressive* scheme of the form

$$u_t = \alpha_1 u_{t-1} + \alpha_2 u_{t-2} + \epsilon_t,$$

where α_1, α_2 are parameters and the ϵ_t's are mutually independent random variables with zero expected value and common variance.

Bibliography

Yule, G. U. (1921). *J. R. Statist. Soc. A*, **84**, 497–526.
Yule, G. U. (1927). *Philos. Trans. R. Soc. Lond.*, **226**, 267–298. (Section III, pp. 280 et seq.)

(AUTOREGRESSIVE-INTEGRATED
   MOVING AVERAGE (ARIMA) MODELS
AUTOREGRESSIVE-MOVING AVERAGE
   (ARMA) MODELS
TIME SERIES)

YULE'S Q

Let $P(A_i B_j) = P_{ij}$, $i, j = 1, 2$, be the population probability of falling into the cell defined by the ith row and jth column of a 2×2 contingency table*, with fixed column totals.

Yule's Q is defined as

$$Q = \frac{P_{11} P_{22} - P_{12} P_{21}}{P_{11} P_{22} + P_{12} P_{21}}.$$

This measure of association* is widely used in social sciences. It can be expressed as

$$Q = (\alpha - 1)/(\alpha + 1),$$

where

$$\alpha = (P_{11}/P_{21})/(P_{12}/P_{22})$$

is the odds ratio*.

The range of possible values of Q is from -1.0 to 1.0 with 0 corresponding to independence. Yule's Q is estimated by

$$\hat{Q} = \frac{f_{11} f_{22} - f_{12} f_{21}}{f_{11} f_{22} + f_{12} f_{21}},$$

where f_{ij}, $i, j = 1, 2$, is the observed frequency corresponding to P_{ij}. The variance of $\hat{Q}$ (under the assumption that $f_{ij} > 0$ for all $i, j = 1, 2$) is estimated by

$$\hat{\sigma}_{(\hat{Q})}^2 = \frac{1}{4}(1 - \hat{Q})^2 \sum_{i=1}^{2} \sum_{j=1}^{2} f_{ij}^{-1}.$$

(ASSOCIATION, MEASURES OF
CONTINGENCY TABLES
TETRACHORIC CORRELATION
YULE'S Y)

YULE'S Y

Yule's Y (also known as *coefficient of colligation*) is defined in a 2×2 contingency table as

$$Y = \frac{\sqrt{P_{11} P_{22}} - \sqrt{P_{12} P_{21}}}{\sqrt{P_{11} P_{22}} + \sqrt{P_{12} P_{21}}} = \frac{\sqrt{\alpha} - 1}{\sqrt{\alpha} + 1}$$

(*see* YULE'S Q for notation). This measure of association has the same properties as Yule's Q, but the absolute value of Y is less than the absolute value of Q unless the "categories" are independent or completely associated. It is estimated by replacing each P_{ij} by f_{ij}. The variance of this estimator of Y (under the assumption that $f_{ij} > 0$, $i, j = 1, 2$) is estimated by

$$\hat{\sigma}_{(\hat{Y})}^2 = \frac{1}{16}(1 - \hat{Y})^2 \sum_{i=1}^{2} \sum_{j=1}^{2} f_{ij}^{-1}.$$

For interpretation of Y and further details, see, e.g., Bishop et al. [1] and Reynolds [2].

References

[1] Bishop, Y. M., Fienberg, S. E., and Holland, P. W. (1975). *Discrete Multivariate Analysis*. MIT Press, Cambridge, MA.

[2] Reynolds, H. T. (1975). *The Analysis of Cross-Classifications*. Free Press–Macmillan, New York.

(ASSOCIATION, MEASURES OF
CONTINGENCY TABLES
DEPENDENCE, MEASURES OF
TETRACHORIC CORRELATION
TWO-BY-TWO (2×2) TABLES
YULE'S Q)

Z

ŽALUDOVÁ'S APPLICATIONS OF LORD'S TEST STATISTIC

Lord [1] proposed replacing the sample standard deviation in a t-test* statistic by the range, or the mean of ranges, from a random sample (or set of subsamples). Assuming normal variation the statistic is essentially distributed as

$$(U + \delta)/\overline{W},$$

where U is a unit normal variable and $\overline{W}$ is independent of U and distributed as the arithmetic mean of m mutually independent variables, each distributed as the range of n mutually independent normal variables.

The noncentral ($\delta \neq 0$) distribution is needed to calculate the power of Lord's t-test [2]. Žaludová [3] showed how this distribution can be used to construct tests of hypotheses on the proportions of normal populations exceeding specified values and also to calculate tolerance intervals* and confidence intervals*. She also reported calculations of relevant tables and gave some graphical representations of the tabular values.

References

[1] Lord, E. (1947). *Biometrika*, **34**, 41–67. Correction, **39**, 442.

[2] Lord, E. (1950). *Biometrika*, **37**, 64–77.

[3] Žaludová, A. H. (1960). *Bull. Int. Inst. Statist.*, **37**(2), 307–310.

(LORD'S TEST STATISTICS
NONCENTRAL t-DISTRIBUTION
PSEUDO-t)

ZAREMBA TEST STATISTIC

This is a modification of the Mann–Whitney–Wilcoxon statistic*, designed to make it applicable to testing a broader class of null hypotheses.

Given two independent random samples of m values $Y_1, Y_2, \ldots, Y_m$ of variables with CDF F and n values $Z_1, Z_2, \ldots, Z_n$ variables with CDF G, the Mann–Whitney–Wilcoxon statistic is defined as

$$W_{YZ} = \sum_{i=1}^{m} \sum_{j=1}^{n} U_{ij},$$

where

$$U_{ij} = \begin{cases} 1, & \text{if } Y_i < Z_j,\ i = 1, \ldots, m; \\ & j = 1, \ldots, n, \\ 0, & \text{otherwise.} \end{cases}$$

If the hypothesis

$$H_0 : F(x) = G(x), \quad \text{for all } x$$

is valid, then

$$E[W_{YZ}] = mn/2,$$

$$\text{var}(W_{YZ}) = mn(m + n + 1)/12,$$

and the distribution of W_{YZ} does not depend on the common CDF. This statistic is used in distribution-free* tests of H_0 against alternative sets

$$H_1 : \Pr[Y < Z] > \tfrac{1}{2}$$

or

$$H_2 : \Pr[Y \geqslant Z] \neq \tfrac{1}{2}.$$

Zaremba [2] sought to test the hypothesis

$$H_0^* : \Pr[Y < Z] = \tfrac{1}{2},$$

against H_1 or H_2. The class of hypotheses H_0^* includes, but is not limited to, H_0. For example, if both F and G are normal with the same expected value, but with different standard deviations, then H_0^* is valid but H_0 is not.

The distribution of W_{YX} under H_0^* is not distribution-free if H_0 is not valid. In fact,

$$\text{var}(W_{YZ})$$
$$= mn\{\alpha + (n - 1)\beta + (m - 1)\gamma\},$$

where $\alpha = \text{var}(U_{ij}) = \tfrac{1}{4}$, $\beta = \text{cov}(U_{ij}, U_{ik})$, $\gamma = \text{cov}(U_{ij}, U_{hj})$, $j \neq k$, $i \neq h$. Zaremba uses the statistic

$$(W_{YZ} - mn/2) / \{\widehat{\text{var}(W_{YZ})}\}^{1/2},$$

with

$$\widehat{\text{var}(W_{YZ})}$$
$$= mn\{0.25 + (n - 1)\hat{\beta} + (m - 1)\hat{\gamma}\},$$

$$\hat{\beta} = \frac{2}{mn(n - 1)} \sum_{i=1}^{m} \sum_{j=1}^{n-1} \sum_{k=j+1}^{n} U_{ij}U_{ih}$$
$$- \left[\frac{W_{XY}}{mn}\right]^2,$$

and

$$\hat{\gamma} = \frac{2}{mn(n - 1)} \sum_{j=1}^{n} \sum_{i=1}^{m-1} \sum_{h=i+1}^{m} U_{ij}U_{hj}$$
$$- \left[\frac{W_{XY}}{mn}\right]^2.$$

(Since $E[W_{YZ}] = mn/2$ under H_0^*, as well as under H_0, it might be reasonable to replace "$-[W_{XY}/mn]^2$" by "$-\tfrac{1}{4}$", but Zaremba did not do this.)

Ferretti and Friedman [1] found that Zaremba's test does indeed have a more stable actual significance level than does W_{YZ}, when H_0^* is valid but H_0 is not; also there is little difference between the powers of the two tests.

References

[1] Ferretti, N. E. and Friedman, S. M. (1985). *Mat. Aplic. Comp.* (*Brasil*), **4**, 157–172.

[2] Zaremba, S. K. (1962). *Monatsh. Mat.*, **66**, 359–370.

(MANN–WHITNEY–WILCOXON STATISTIC)

ZEITSCHRIFT FÜR WAHRSCHEINLICHKEITS-THEORIE UND VERWANDTE GEBIETE

This journal (ZfW, in English: *Journal for Probability Theory and Related Fields*), founded in 1962, is published by Springer-Verlag (Berlin, Heidelberg, New York, Tokyo). It is independent of any society or association, but grants reduced rates to members of the Bernoulli Society*. Its first managing editor was L. Schmetterer, whom K. Krickeberg succeeded in 1971. In April 1985, the journal changed its name to *Probability Theory and Related Fields*, and H. Rost became managing editor. The Editorial Office is at Institut für Angewandte Mathematik der Universität, Im Neuenheimer Feld 294, D-6900 Heidelberg, W. Germany.

The purpose of ZfW is the publication of first-rate original research articles, complete with full proofs, which are of relevance to the development of the areas covered. A few survey articles have also appeared, e.g., on robust statistics. The major part of the work is theoretical, but applied papers are printed as well if they have probabilistic interest, e.g., in control theory, biology, and theoretical physics.

As indicated by its name, ZfW is intended to cover the entire range of subjects connected with probability theory. The core of the material has always belonged to probability theory proper, especially stochastic processes. Classical limit theory now plays a minor rôle, although probabilistic problems involving "fine" analysis will continue to be appreciated, and interesting papers on special laws continue to appear. Modern areas in rapid development are strongly represented, e.g., stochastic calculus on manifolds, random geometry, processes in random environments and spatial processes, and methods of nonstandard analysis.

The relative weight of the "related fields" in ZfW has been changing with time, e.g., in the beginning, operations research was stressed, but is now absent. There are at present few articles in general functional analysis or classical ergodic theory but many on infinite particle systems. Papers on probabilistic methods in number theory are still welcome but rare.

It is part of the editorial policy to encourage the submission of basic work in modern mathematical statistics. This accounts now for over a quarter of all articles. Some of the probabilistic papers published in ZfW are relevant to statistics as well.

In order to handle submissions from all these areas and to take the changing emphasis on the various fields into account, the original editorial board, which included among others K. L. Chung, B. de Finetti, J. Hájek, D. G. Kendall, L. LeCam, J. Neveu, J. Neyman, A. Rényi, and F. Spitzer, has been steadily modified and expanded. It now comprises 32 people: 21 from Europe including the European part of the USSR, 8 from North America, and 3 from East Asia. Among these, 13 have a strong or exclusive interest in statistics. Editors are not appointed for a fixed period but are replaced depending on the development of the field and their own wish. In this way, some members of the original board are still active in the present one, alongside of quite young people.

At the present time (1987), there are three volumes of about 600 pages a year, each one consisting of four issues. Papers may be submitted to any editor or to the Editorial Office. Every paper is reviewed by at least two referees selected by one or two editors. The rejection rate is slightly over one-half, and about three-quarters of the remaining papers are accepted only after revision. Due to rigorous refereeing and frequent and extensive revisions, the average time between submission and notification of acceptance of the final version is fairly long [median 13 months and mean 14.1 months for the 39 papers in Vol. 66 (1984)], whereas the average period from then on until publication is very short (mean 4.6 months for Vol. 66).

Articles may be submitted in English, French, or German, but almost all papers appearing now are written in English.

The contents of Vol. 68, 1985 follow:

Quantum central limit theorems for strongly mixing random variables, by L. Accardi and A. Bach.

Infinitely subadditive capacities as upper envelopes of measures, by B. Anger and J. Lembcke.

Summability methods and almost sure convergence, by N. H. Bingham and M. Maejima.

The survival of branching annihilating random walk, by M. Bramson and L. Gray.

On the length of the longest excursion, by E. Csáki, P. Erdös, and P. Révész.

Inequalities for moments of secant length, by P. J. Davy.

Bayesian estimation in the symmetric location problem, by H. Doss.

The central limit theorem for summability methods of i.i.d. random variables, by P. Embrechts and M. Maejima.

On solutions of one-dimensional stochastic differential equations without drift, by H. J. Engelbert and W. Schmidt.

Sur la convergence étroite des mesures Gaussiennes, by X. Fernique.

Minimax-robust prediction of discrete time series, by J. Franke.

Laws of the iterated logarithm for symmetric stable processes, by P. S. Griffin.

A variational characterization of one-dimensional countable state Gibbs random fields, by B. M. Gurevich.

Mean stochastic comparison of diffusions, by B. Hájek.

Some limit results for lag sums of independent, non-i.i.d. random variables, by D. L. Hanson and R. P. Russo.

On the asymptotic equivalence of L_p metrics for convergence to normality, by C. C. Heyde and T. Nakata.

Evaluating inclusion functionals for random convex hulls, by N. P. Jewell and J. P. Romano.

On independent statistical decision problems and products of diffusions, by I. Johnstone and S. Lalley.

On the asymptotic behaviour of solutions of stochastic differential equations, by G. Keller, G. Kersting, and U. Rösler.

On the limiting behavior of normed sums of independent random variables, by M. J. Klass and R. J. Tomkins.

Random walks with internal degrees of freedom. II. First-hitting probabilities, by A. Krámli and D. Szász.

Finite nearest particle systems, by T. M. Liggett.

Some properties of a special class of self-similar processes, by J. H. Lou.

Non-parametric applications of an infinite dimensional convolution theorem, by P. W. Millar.

A diffusion process in a singular mean-drift-field, by M. Nagasawa and H. Tanaka.

Stochastic calculus for continuous additive functionals of zero energy, by S. Nakao.

A limit theorem for discounted sums, by E. Omey.

Change-of-variance sensitivities in regression analysis, by E. Ronchetti and P. Rousseeuw.

(r, p)-capacity on the Wiener space and properties of Brownian motion, by M. Takeda.

Sur l'intégrabilité des vecteurs gaussiens, by M. Talagrand.

On the Chacon–Jamison theorem, by J. B. Walsh.

Harmonizable stable processes on groups: Spectral, ergodic and interpolation properties, by A. Weron.

A general law of iterated logarithm, by R. Wittmann.

Equilibrium time evolutions and BBGKY hierarchy equations of large classical systems, by H. Zessin.

The mean breadth of a random polytope in a convex body, by H. Ziezold.

K. Krickeberg

ZELEN'S INEQUALITIES *See Supplement*

ZELEN'S RANDOMIZED CONSTANT DESIGNS

These form a class of randomized clinical trials* (RCTs), which obviate the necessity, that arises in classical RCTs, of approaching eligible patients and requesting their consent to either a standard therapy (S) or a new treatment (T).

In their simplest form [2, 3], the patients are randomly assigned to two groups G_1 and G_2. All patients in G_1 receive S; in G_2 those patients who agree to receive T do so, whereas those who do not, receive S. Zelen's contention that all patients must be included if the analysis is to provide a valid comparison between S and T has been examined by McHugh [1], who has shown that the Zelen design yields an estimate of treatment effect that is free of selection bias*.

References

[1] McHugh, R. (1984). *Statist. Med.*, **3**, 215–218.
[2] Zelen, M. (1979). *New England J. Med.*, **300**, 1242–1245.
[3] Zelen, M. (1981). *Surgical Clinics of North America*, pp. 1425–1432.

(CLINICAL TRIALS
DESIGN OF EXPERIMENTS
RANDOMIZATION
SELECTION BIAS)

ZELLNER ESTIMATOR

The term "Zellner estimator (ZE)" is often used for one of the two generalized least-squares estimators (GLSEs) Zellner [13] proposed in what he called a model of "seemingly unrelated regression* (SUR)" equations:

$$y_j = X_j\beta_j + u_j,$$

$$E[u_j] = 0, \quad \text{and} \quad E[u_j u_i'] = \sigma_{ji} I_T, \quad (1)$$

$j, i = 1, \ldots, M$. Here $y_j: T \times 1$, X_j is a $T \times K_j$ fixed matrix of rank K_j, $\beta_j: K_j \times 1$, $u_j: T \times 1$, and σ_{ji} denotes the cross-covariance of the tth elements of u_j and u_i or the cross-covariance of the jth equation and the ith equation in the tth observations. Model (1) is expressed as a multivariate regression model with prior information on the structure of the coefficient matrix:

$$Y = \tilde{X}B + U,$$

with $B = \text{diag}\{\beta_1, \ldots, \beta_M\} : \left(\sum_{j=1}^M K_j\right) \times M,$

$$(2)$$

where $Y = [y_1, \ldots, y_M]$, $\tilde{X} = [X_1, \ldots, X_M]$, $U = [u_1, \ldots, u_M]$, and $\text{diag}\{a_1, \ldots, a_M\}$ denotes the block diagonal matrix with diago-

nal blocks a_j's. It is also expressed as

$$y = X\beta + u,$$

with $X = \text{diag}\{X_1, \ldots, X_M\} : TM \times M,$ (3)

where $y = (y_1', \ldots, y_M')'$, $\beta = (\beta_1', \ldots, \beta_M')'$, and $u = (u_1', \ldots, u_M')'$. In (3), the covariance matrix of u is $\Sigma \otimes I_T$ with $\Sigma = (\sigma_{ji}) : M \times M$ (where $\otimes$ denotes Kronecker product*) and hence a GLSE is given by

$$\hat{\beta}(\hat{\Sigma}) = \left(X'[\hat{\Sigma} \otimes I]^{-1}X\right)^{-1}X'[\hat{\Sigma} \otimes I]^{-1}y,$$
$$(4)$$

where $\hat{\Sigma}$ is some estimator of Σ.

For $\hat{\Sigma}$, Zellner [13] proposed the following two estimators:

$$\tilde{\Sigma} = \left(\tilde{\sigma}_{ji}\right),$$

with $\tilde{\sigma}_{ji} = (y_j - X_j b_j)'(y_j - X_i b_i)/a_{ji}(T)$

$$(5)$$

and

$$S = (Y - \tilde{X}\hat{B})'(Y - \tilde{X}\hat{B})/(T - l)$$
$$= Y'[I - \tilde{X}(\tilde{X}'\tilde{X})^+\tilde{X}']Y/(T - l), \quad (6)$$

where $b_j = (X_j'X_j)^{-1}X_j'y_j$ is the ordinary LSE (OLSE) of β_j, $a_{ji}(T)$ is a normalizing constant such as T or $[(T - K_j)(T - K_i)]^{1/2}$, $\hat{B} = (\tilde{X}'\tilde{X})^+\tilde{X}'Y$, and $l = \text{rank}(\tilde{X})$. Here $(\tilde{X}'\tilde{X})^+$ denotes the Penrose generalized inverse* of $\tilde{X}'\tilde{X}$. Clearly, $\tilde{\Sigma}$ is the estimator of Σ based on the OLS residuals of the M equations, whereas S is the one based on the multivariate residual matrix $Y - \tilde{X}\hat{B}$ when the LS method is applied to the model (2) with the prior information on the structure of B ignored.

To distinguish the two ZEs, the GLSE (4) with $\hat{\Sigma} = \tilde{\Sigma}$ in (5) is sometimes called the RZE (Zellner estimator with the restricted sample covariance matrix) and the GLSE (4) with $\hat{\Sigma} = S$ in (6) the UZE (Zellner estimator with the unrestricted sample covariance matrix). Even when $M = 2$ and u is normal, the explicit form of the maximum likelihood* estimator is hard to derive.

When $M = 2$, the finite sample efficiencies of the RZE and the UZE relative to the OLSE (4) with $\hat{\Sigma} = I$ were studied by Zellner [14], Kmenta and Gilbert [7], Revankar [9], and Mehta and Swamy [8], and the ZEs were shown to be more efficient than the OLSE except when σ_{12} is small and/or T is small. Kariya [4] derived a locally best invariant test for $\sigma_{12} = 0$. Also when $M = 2$, Revankar [10] made a comparison between the RZE and UZE and found that there are certain cases in which the UZE is better than the RZE although the UZE ignores the information on **B**. Generally, the RZE is considered at least more natural than the UZE though the properties of the RZE are more difficult to study. The unbiasedness of the ZEs and the existence of the moments are shown by Kakwani [2] and Kariya and Toyooka [6], and an upper bound for the covariance matrix of the UZE is derived by Kariya [3]. On the other hand, Srivastava [11] pointed out the equivalence between the asymptotic covariance matrices up to $O(T^{-1})$ of the UZE and the RZE, and Kariya and Maekawa [5] derived the valid asymptotic distribution of the UZE. The survey articles by Srivastava and Dwivedi [12] and Judge et al. [1] serve as general references for the literature in the 1960s and 1970s.

References

[1] Judge, G. G., Griffiths, W. E., Hill, R. C. and Lee, T. C. (1980). *The Theory and Practice of Econometrics*. Wiley, New York. pp. 243–296.

[2] Kakwani, N. C. (1967). *J. Amer. Statist. Ass.*, **62**, 141–142.

[3] Kariya, T. (1981). *J. Amer. Statist. Ass.*, **76**, 975–979.

[4] Kariya, T. (1981). *Ann. Statist.*, **9**, 381–390.

[5] Kariya, T. and Maekawa, K. (1982). *Ann. Inst. Statist. Math., Tokyo*, **34**, 281–297.

[6] Kariya, T. and Toyooka, Y. (1985). *Multivariate Analysis VI*, Elsevier, New York. pp. 345–354.

[7] Kmenta, J. and Gilbert, R. F. (1968). *J. Amer. Statist. Ass.*, **63**, 1180–1200.

[8] Mehta, J. S. and Swamy, P. A. V. B. (1976). *J. Amer. Statist. Ass.*, **71**, 634–639.

[9] Revankar, N. S. (1974). *J. Amer. Statist. Ass.*, **69**, 187–190.

[10] Revankar, N. S. (1976). *J. Amer. Statist. Ass.*, **71**, 183–188.

[11] Srivastava, V. K. (1970). *Ann. Inst. Statist. Math., Tokyo*, **22**, 483–493.

[12] Srivastava, V. K. and Dwivedi, T. D. (1979). *J. Econometrics*, **10**, 15–32.

[13] Zellner, A. (1962). *J. Amer. Statist. Ass.*, **57**, 348–368.

[14] Zellner, A. (1963). *J. Amer. Statist. Ass.*, **58**, 977–992. Corrigenda (1972), **67**, 255.

(GENERAL LINEAR MODEL
SEEMINGLY UNRELATED REGRESSIONS)

TAKEAKI KARIYA

ZERO CONTROL METHOD   *See* EDITING STATISTICAL DATA

ZERO DEGREES OF FREEDOM

INTRODUCTION

We usually think of "degrees of freedom*" as a positive integer representing the number of independent pieces of information in a given situation. We might therefore expect the case of "zero degrees of freedom" to represent the uninteresting case in which there is no information. Whereas this is typically true for central distributions, there exist noncentral distributions with zero degrees of freedom that have rich, interesting, and useful structures. We will discuss the chi-squared*, t^*, and F^* distributions. In particular, the noncentral chi-squared distribution* with zero degrees of freedom (a mixture of exact zero values with positive continuous variation) is useful in modeling data containing exact zeroes, in queueing*, and as an asymptotic distribution in time-series* analysis and geometric probability*.

DEFINITIONS AND PROPERTIES

The central chi-squared distribution with zero degrees of freedom χ_0^2 is a degenerate distribution that always takes the value zero. The noncentral chi-squared distribution with zero degrees of freedom $\chi_0^2(\lambda)$ is a mixture* of a discrete distribution degenerate at zero with a continuous positive distribution. As the noncentrality parameter λ approaches zero, the discrete component occurs with high probability and the continuous component (which occurs rarely) is approximately exponentially distributed. As the noncentrality parameter becomes large, the discrete component occurs rarely and the continuous component tends towards a Gaussian distribution.

A random variable $X \sim \chi_0^2(\lambda)$ may be represented as a Poisson mixture of central chi-squared variates with even numbers of degrees of freedom according to the following two-stage procedure. First, choose K from a Poisson distribution* with mean $\lambda/2$, and then choose $X \sim \chi_{2K}^2$. Whenever $K = 0$, we will have $X = 0$; otherwise X will be a positive continuous random variable. The $\chi_0^2(\lambda)$ distribution has the following properties:

Cumulative distribution function:

$$\Pr[X \leqslant x] = 1 - e^{-(\lambda+x)/2} \sum_{k=1}^{\infty} \frac{(\lambda/2)^k}{k!}$$
$$\times \sum_{j=0}^{k-1} \frac{(x/2)^j}{j!}, \qquad x > 0,$$

$$\Pr[X \leqslant 0] = e^{-\lambda/2},$$

and

$$\Pr[X \leqslant x] = 0, \qquad x < 0.$$

Characteristic function:

$$\exp[it\lambda/(1 - 2it)].$$

Moments:

$$E[X^m] = 2^m m! \sum_{k=1}^{m} \left\{ \binom{m-1}{k-1} \left(\frac{\lambda}{2}\right)^k \middle/ k! \right\}.$$

Further details and graphs may be found in Siegel [1].

Properties of t and F distributions with zero degrees of freedom follow easily from the chi-squared case. The central cases are trivial: the t_0 distribution is undefined because it always involves division by zero. The $F(0, n_2)$ distribution is degenerately zero if $n_2 > 0$, whereas $F(n_1, 0)$ is always undefined due to division by zero. The noncentral cases are more interesting: The $t_0(\lambda)$ distribution, a standard Gaussian divided by an independent $\chi_0^2(\lambda)$, takes on real values whenever the denominator is nonzero, i.e., with probability $1 - \exp(-\lambda/2)$. Properties of the singly and doubly noncentral F distribution follow similarly by considering the noncentral chi-squared variates from which it is formed, paying close attention to zeroes in the denominator.

APPLICATIONS

The $\chi_0^2(\lambda)$ distribution is a natural choice for model building in situations where a Poisson number of basic events each contributes an independent exponentially distributed quantity to a total. For example, in queueing theory if customers arrive according to a Poisson process* and if the service times are exponential, then the total service time required for all customers arriving during a fixed period would follow this distribution. Fitting the $\chi_0^2(\lambda)$ distribution to data by maximum likelihood* has been explored by Siegel [4].

The $\chi_0^2(\lambda)$ distribution has also been found useful as an asymptotic limiting distribution in testing for periodicity in a time series* (Siegel [3] and Siegel and Beirlant [5]) when tests are considered which are more powerful than Fisher's test in the case of strong periodicity at multiple periods. The $\chi_0^2(\lambda)$ distribution is also the limiting distribution of the amount of a circle left uncovered (the "vacancy" in geometrical probability) by randomly placed arcs on the circumference of a circle (Siegel [2]).

References

[1] Siegel, A. F. (1979a). The noncentral chi-squared distribution with zero degrees of freedom and testing for uniformity. *Biometrika*, **66**, 381–386. (Definitions, properties, and graphs of the $\chi_0^2(\lambda)$ distribution.)

[2] Siegel, A. F. (1979b). Asymptotic coverage distributions on the circle. *Ann. Prob.* **7**, 651–661. (The $\chi_0^2(\lambda)$ distribution as the asymptotic limit of the random amount of a circle uncovered by randomly placed arcs.)

[3] Siegel, A. F. (1980). Testing for periodicity in a time series. *J. Amer. Statist. Ass.*, **75**, 345–348.

(Extends Fisher's test for periodicity in a time series to make it more powerful in detecting multiple periodicity; the asymptotic distribution of the test statistic follows the $\chi_0^2(\lambda)$ distribution.)

[4] Siegel, A. F. (1985). Modelling data containing exact zeroes using zero degrees of freedom. *J. R. Statist. Soc. B*, **47**, 267–271. [Establishes existence and uniqueness of maximum likelihood estimators for the $\chi_0^2(\lambda)$ distribution. An example fitting the model to snowfall data is given in which data values are either exactly zero (when it didn't snow) or positive (when it did snow).]

[5] Siegel, A. F. and Beirlant, J. (1987). Periodicity testing based on spacing. In *Contributions to the Theory and Applications of Statistics, A Volume in Honor of Herbert Solomon*, A. E. Gefand, ed. Academic, New York, pp. 179–196. [Provides further asymptotic results and tables for using the $\chi_0^2(\lambda)$ distribution in testing for periodicity in a long time series.]

(NONCENTRAL DISTRIBUTIONS
(various entries))

ANDREW F. SIEGEL

ZERO FACTORIAL

The symbol $n!$, where n is a positive integer, stands for

$$n \times (n - 1) \times (n - 2) \times \cdots \times 2 \times 1.$$

This definition does not apply if $n = 0$, but conventionally the value of $0!$ is taken to be 1. This is formally consistent, for example, with the combinatorial (binomial* coefficient) formula

$$\binom{n}{r} = \frac{n!}{r!(n-r)!},$$

since

$$\binom{n}{0} = \frac{n!}{0!n!}$$

is equal to 1.

The rth ascending and descending factorials of n are

$$n^{[r]} = n(n + 1) \cdots (n + r - 1)$$

and

$$n^{(r)} = n(n - 1) \cdots (n - r + 1),$$

respectively. By convention, the "zeroth" ascending and descending factorials are each equal to 1:

$$n^{[0]} = 1 = n^{(0)}.$$

This is analogous to the formula $x^0 = 1$ $(x \neq 0)$ for zero powers.

(COMBINATORICS)

ZERO-TRUNCATED DISTRIBUTION
See DECAPITATED DISTRIBUTION

ZETA DISTRIBUTION   *See Supplement*

ZIPF'S LAW

Consider a set of data values, ordered as

$$x_{(1)} \geqslant x_{(2)} \geqslant \cdots \geqslant x_{(n)},$$

in the reverse of the conventional arrangement, having the largest value ranked first, and so on. We may think of r as the rank and $x_{(r)}$ as the size of the rth data value in the ordered set. Zipf [15] noticed that the relationship

$$rx_{(r)} = \text{constant} \qquad (1)$$

seemed to hold for various kinds of objects, including cities in the United States by population, books by number of pages, words in an essay by their frequency of occurrence, and biological genera by number of species. The *rank–size relation* (1) is known as Zipf's law; its graph is a rectangular hyperbola.

Let x be the size of an object and $f(x)$ its relative frequency of occurrence, where $\int_0^\infty f(x)\,dx = 1$. If n is the number of objects in the data set or collection and $N(x)$ the number of objects with size greater than x, then

$$N(X) = \int_x^\infty nf(u)\,du$$

$$= \text{rank of an object of size } x.$$

Under Zipf's law (1), $xN(x) = \text{constant}$ or $N(x) = K/x$. Hence

$$f(x) = -n^{-1}N'(x) = K'/x^2, \qquad (2)$$

where $K' = K/n$. Equation (2) is the *size–frequency relation* corresponding to (1). Zipf attempted to explain the origins of (1) in the nature of human behavior, through the so-called principle of least effort.

Some writers [4, 8] have criticized (1) and (2) adversely, for one or more of three reasons.

1. Zipf's explanation [15] in terms of human behavior is of doubtful relevance to many of its manifestations and gives no clue to any underlying statistical process.
2. The value of the constant K' in (2) depends on the number n of objects in the study.
3. A statistical rationale for the phenomena observed by Zipf leads to a family of distributions—(4), discussed next—which only includes (2) as a special case.

This last criticism is not entirely fair, because Zipf proposed a generalization of the rank–size relation, namely,

$$r^q x_{(r)} = \text{constant}, \qquad q > 0, \qquad (3)$$

which leads to the size–frequency form with discrete density function

$$f(r) = Ar^{-(1+a)}, \qquad r = 1, 2, \ldots, \qquad (4)$$

where $a > 0$, and

$$A = \zeta(1 + a) = \sum_{r=1}^{\infty} r^{-(1+a)}$$

is the zeta function; $\zeta(u)$ is tabulated in Abramowitz and Stegun [1, p. 811] for $u = 1(1)42$. Equation (4) defines the *discrete Pareto* or *zeta distribution*[7], which includes (2) as the case $a = 1$. Fox and Lasker [4] fitted parameter values to (4) by maximum likelihood estimation* (see Seal [11] for the frequency of surnames in nine districts near London in England), and found that the data gave acceptable fits to the model, with estimates of a lying between 1.76 and 2.88.

The nomenclature is no longer well defined. Hill [5] uses the term "Zipf's law" to denote Zipf's generalization (3) and the family (4) derived from it. He derives (3) from an urn model with two important features. The first is a twofold classification of cities (genera) into regions (families) in a country, and then of cities (genera) within regions (families); the second is a Bose–Einstein scheme of allocating the population (species) to cities (genera) within regions (families); *see* FERMI–DIRAC STATISTICS. The allocation schemes are independent between regions (families). In the city–region context, let $L(R_i)$ be the size of the R_ith largest city in the ith region. Then under some regularity conditions, the ordered values of $L(R_i)$ across regions should yield a Zipf rank–size curve (3), asymptotically. The same should hold if we select a city at random from each region. Finally, a rank–size curve of all cities in the country should be of form (3), approximately. For further discussion, see Hill [5]; the derivation of convergence in probability to the size–frequency form (4) is given in ref. 6, and necessary and sufficient conditions for the same in ref. 13.

These results address the *strong form* of Zipf's law; in this, the proportion of cities, etc., with population r (or grouped around r) has a probability distribution given by (4). In the *weak form* of the law, it is only the expected value of this proportion that has the form (4). Chen [2] discusses general urn models leading to the weak form and gives further references. Rouault [9] shows that sequences of values taken by certain finite Markov chains that exclude passage at any stage to the same state (*see* MARKOV PROCESSES) may follow the distribution (4) for $0 < a < 1$.

Simon [12] describes a stochastic process that leads to a stationary distribution with discrete density function

$$g(r) \propto B(r, a + 1),$$

where $B(\cdot, \cdot)$ is the beta function. For large r,

$$g(r) \sim \Gamma(a + 1) r^{-(a+1)},$$

so that the Yule distribution* (5) approximates (4) in the tails. If $a = 1$, (4) gives

$$g(r) = [r(r + 1)]^{-1}, \quad \sim 1/r^2 \text{ for large } r.$$

Yule [14] also explained the distribution of genera by numbers of species by means of (5) as a limiting distribution.

For applications of Zipf's law to bibliographic data bases see ref. 3 and the references listed therein, and for applications to prediction in geological studies see ref. 10. Rapoport [8] gives an interesting discussion of early papers anticipating Zipf's law.

References

[1]   Abramowitz, M. and Stegun, I. A. (1964). *Handbook of Mathematical Functions*: *Appl. Math. Series No.* 55, National Bureau of Standards, Washington, D.C.

[2]   Chen, W.-C. (1980). *J. Appl. Prob.*, **17**, 611–622.

[3]   Fedorowicz, J. (1982). *J. Amer. Soc. Inf. Sci.*, **33**, 285–293.

[4]   Fox, W. R. and Lasker, G. W. (1983). *Int. Statist. Rev.*, **51**, 81–87.

[5]   Hill, B. M. (1974). *J. Amer. Statist. Ass.*, **69**, 1017–1026.

[6]   Hill, B. M. and Woodroofe, M. (1975). *J. Amer. Statist. Ass.*, **70**, 212–219.

[7]   Johnson, N. L. and Kotz, S. (1969). *Distributions in Statistics: Discrete Distributions*, Vol. 1. Wiley, New York, p. 240.

[8]   Rapoport, A. (1978). *International Encyclopedia of Statistics*, W. H. Kruskal and J. M. Tanur, eds. Free Press, New York, pp. 847–854.

[9]   Rouault, A. (1978). *Ann. Inst. H. Poincaré, B*, **14**, 169–188.

[10]   Rowlands, N. J. and Sampey, D. (1977). *J. Int. Ass. Math. Geol.* **9**, 383–392.

[11]   Seal, H. L. (1952). *J. Inst. Actuaries*, **78**, 115–121.

[12]   Simon, H. A. (1955). *Biometrika*, **42**, 425–440.

[13]   Woodroofe, M. and Hill, B. (1975). *J. Appl. Prob.*, **12**, 425–434.

[14]   Yule, G. U. (1944). *The Statistical Study of Literary Vocabulary*. Cambridge University Press, Cambridge, England.

[15]   Zipf, G. K. (1949). *Human Behavior and the Principle of Least Effort*. Addison-Wesley, Reading, MA.

(FACTORIAL SERIES DISTRIBUTIONS
URN MODELS
YULE DISTRIBUTION)

CAMPBELL B. READ

ZONAL POLYNOMIALS

INTRODUCTION AND GROUP-THEORETIC DEFINITION

Many distributions and moments in multivariate analysis* based on the multivariate normal distribution can be expressed as power series in symmetric functions* of m variables. Often, but not always, these power series may be written in terms of hypergeometric functions of matrix argument. James [13] gives a survey of such distributions for random matrices and their latent roots. The power series involved in multivariate distribution theory can be expanded in terms of one of the many types of symmetric polynomials. For a given basis of the symmetric polynomials, the individual homogeneous polynomials of degree k are usually indexed by partitions

$$\kappa = (k_1, k_2, \ldots, k_m),$$
$$k_1 \geq k_2 \geq \cdots \geq k_m \geq 0,$$
$$\sum_{i=1}^{m} k_i = k,$$

of k into not more than m parts. One particular class of homogeneous symmetric polynomials, namely the class of *zonal polynomials*, yields an enormous simplification of the coefficients in these power series. These polynomials are derived from the group representation theory of Gl(m, R), the general linear group of $m \times m$ real nonsingular matrices, and their study was initiated independently by Hua [9] and James [10]. The general theory of zonal polynomials was developed in a series of papers by James [10–16] and Constantine [3].

We now turn to the group-theoretic definition of zonal polynomials. (An alternate but closely related definition that leads more directly to an algorithm for calculation will be given in the next section). Let V_k be the vector space of homogeneous polynomials $\phi(X)$ of degree k in the $m(m + 1)/2$ different elements of the symmetric $m \times m$ matrix **X**. Corresponding to any congruence transformation

$$\mathbf{X} \to \mathbf{L}\mathbf{X}\mathbf{L}', \qquad \mathbf{L} \in \text{Gl}(m, R),$$

we can define a linear transformation of the space V_k by

$$\phi \to T(\mathbf{L})\phi : (T(\mathbf{L})\phi)(\mathbf{X}) = \phi(\mathbf{L}^{-1}\mathbf{X}\mathbf{L}^{-1'}).$$

This transformation defines a representation of the real linear group $\text{Gl}(m, R)$ in the vector space V_k—i.e., the mapping $\mathbf{L} \to T(\mathbf{L})$ is a homomorphism from $\text{Gl}(m, R)$ to the group of linear transformations of V_k. A subspace $V' \subset V_k$ is *invariant* if

$$T(\mathbf{L})V' \subset V',$$

for all $\mathbf{L} \in \text{Gl}(m, R)$. If, in addition, V' contains no proper invariant subspaces, it is called an *irreducible* invariant subspace. It can be shown that the space V_k decomposes into a direct sum of irreducible invariant subspaces V_κ,

$$V_k = \bigoplus_\kappa V_\kappa,$$

where $\kappa = (k_1, k_2, \ldots, k_m)$, $k_1 \geq k_2 \geq \cdots \geq k_m \geq 0$, runs over all partitions of k into not more than m parts. The polynomial $(\text{tr }\mathbf{X})^k \in V_k$ then has a unique decomposition

$$(\text{tr }\mathbf{X})^k = \sum_\kappa C_\kappa(\mathbf{X}), \qquad (1)$$

into polynomials $C_\kappa(\mathbf{X}) \in V_\kappa$, belonging to the respective invariant subspaces. The polynomial $C_\kappa(\mathbf{X})$ is the *zonal polynomial* corresponding to the partition κ; it is a symmetric homogeneous polynomial of degree k in the latent roots of $\mathbf{X}$. When $m = 1$, (1) becomes $x^k = C_{(k)}(x)$ so that the zonal polynomials of a matrix are analogous to powers of a single variable. Equation (1) holds for all m, with $C_\kappa(\mathbf{X}) \equiv 0$ if the partition κ has more than m parts.

For detailed discussions of the group-theoretic construction of zonal polynomials, the reader is referred to Farrell [6], Kates [17], and the papers of James referenced earlier, particularly James [12, 13]. Another approach to zonal polynomials has been given by Saw [27] and an essentially combinatoric approach has been given by Takemura [32];

a useful survey paper has been written by Subrahmaniam [28].

CALCULATION OF ZONAL POLYNOMIALS

No general formula for zonal polynomials is known and methods for calculating them have been given by James [11, 13, 15], Saw [27], and Kates [17]. The discussion in this section is based on the papers by James [14, 15]; see also Muirhead [24].

Let k be a positive integer and order the partitions of k lexicographically—i.e., if $\kappa = (k_1, \ldots, k_m)$ and $\lambda = (l_1, \ldots, l_m)$ are two partitions of κ, then $\kappa > \lambda$ if $k_i > l_i$ for the first index i for which the parts are unequal. If κ and λ are two partitions of k with $\kappa > \lambda$ and $y_1, \ldots, y_m$ are m variables, the monomial $y_1^{k_1}, \ldots, y_m^{k_m}$ is said to be of *higher weight* than the monomial $y_1^{l_1}, \ldots, y_m^{l_m}$. A definition of zonal polynomials, which leads to a general algorithm for their calculation follows.

Definition.

Let $\mathbf{Y}$ be an $m \times m$ symmetric matrix with latent roots $y_1, \ldots, y_m$ and let $\kappa = (k_1, \ldots, k_m)$ be a partition of k into not more than m parts. The zonal polynomial of $\mathbf{Y}$ corresponding to κ, $C_\kappa(\mathbf{Y})$, is a symmetric homogeneous polynomial of degree k in the latent roots $y_1, \ldots, y_m$ satisfying the following three conditions:

(i) The term of highest weight in $C_\kappa(\mathbf{Y})$ is $y_1^{k_1}, \ldots, y_m^{k_m}$; i.e., $C_\kappa(\mathbf{Y}) = d_\kappa y_1^{k_1}, \ldots, y_m^{k_m} +$ terms of lower weight, where d_κ is a constant.

(ii) $C_\kappa(\mathbf{Y})$ is an eigenfunction of the differential operator $\Delta_\mathbf{Y}$ given by

$$\Delta_\mathbf{Y} = \sum_{i=1}^m y_i^2 \frac{\partial^2}{\partial y_i^2} + \sum_{i=1}^m \sum_{\substack{j=1 \\ j \neq i}}^m \frac{y_i^2}{y_i - y_j} \frac{\partial}{\partial y_i},$$

i.e., $\Delta_\mathbf{Y} C_\kappa(\mathbf{Y}) = \alpha_\kappa C_\kappa(\mathbf{Y})$, where α_κ is a constant.

(iii) As κ runs over all partitions of k the zonal polynomials have unit coeffi-

cients in the expansion of $(\text{tr } \mathbf{Y})^k$; i.e.,
$(\text{tr } \mathbf{Y})^k = (y_1 + \cdots + y_m)^k = \Sigma_\kappa C_\kappa(\mathbf{Y})$.

This definition of zonal polynomials is intimately related to the definition in the first section. Because of its group-theoretic nature, it is known that $C_\kappa(\mathbf{Y})$ must be an eigenfunction of a differential operator called the *Laplace–Beltrami operator*. The differential operator Δ_Y in **(ii)** is derived from this operator.

Using conditions **(i)** and **(ii)**, it can be readily shown that the constant (or eigenvalue) α_κ in **(iii)** is $\alpha_\kappa = \rho_\kappa + k(m - 1)$, where

$$\rho_\kappa = \sum_{i=1}^m k_i(k_i - i), \qquad (2)$$

so that $C_\kappa(\mathbf{Y})$ satisfies the second-order partial differential equation

$$\Delta_Y C_\kappa(\mathbf{Y}) = \left[\rho_\kappa + k(m - 1) \right] C_\kappa(\mathbf{Y}). \quad (3)$$

This forms the basis of an algorithm developed by James [14] for calculating the coefficients of the terms in $C_\kappa(\mathbf{Y})$. Basically what happens is that condition **(i)**, along with the condition that $C_\kappa(\mathbf{Y})$ is a symmetric homogeneous polynomial of degree k, establishes what types of terms appear in $C_\kappa(\mathbf{Y})$. The differential equation (3) for $C_\kappa(\mathbf{Y})$ then gives recurrence relations between the coefficients of these terms, which determine $C_\kappa(\mathbf{Y})$ uniquely up to some normalizing constant. The normalization is provided by condition **(iii)**.

Zonal polynomials can be conveniently expressed in terms of the monomial symmetric functions*. If $\kappa = (k_1, \ldots, k_m)$, the monomial symmetric function of $y_1, \ldots, y_m$ corresponding to κ is defined as

$$M_\kappa(\mathbf{Y}) = \sum y_{i_1}^{k_1} y_{i_2}^{k_2}, \ldots, y_{i_p}^{k_p},$$

where p is the number of nonzero parts in the partition κ and the summation is over the distinct permutations $(i_1, \ldots, i_p)$ of p different integers from the integers $1, \ldots, m$. Condition **(i)** and the fact that $C_\kappa(\mathbf{Y})$ is symmetric and homogeneous of degree k show that $C_\kappa(\mathbf{Y})$ can be expressed in terms

of the monomial symmetric functions as

$$C_\kappa(\mathbf{Y}) = \sum_{\lambda \leqslant \kappa} c_{\kappa, \lambda} M_\lambda(\mathbf{Y}), \qquad (4)$$

where the $c_{\kappa, \lambda}$ are constants and the summation is over all partitions λ of k with $\lambda \leq \kappa$, i.e., λ is below or equal to κ in the lexicographic ordering. Substituting (4) in the partial differential equation (3) and equating coefficients of like monomial symmetric functions on both sides leads to a recurrence relation for the coefficients, namely,

$$c_{\kappa, \lambda} = \sum_{\lambda < \mu \leq \kappa} \frac{\left[(l_i + t) - (l_j - t) \right]}{\rho_\kappa - \rho_\lambda} c_{\kappa, \mu}, \qquad (5)$$

where $\lambda = (l_1, \ldots, l_m)$ and $\mu = (l_1, \ldots, l_i + t, \ldots, l_j - t, \ldots, l_m)$ for $t = 1, \ldots, l_j$, such that when the parts of the partition μ are arranged in descending order, μ is above λ and below or equal to κ in the lexicographical ordering. The summation in (5) is over all such μ, including possibly nondescending ones, and any empty sum is taken to be zero. This recurrence relation determines $C_\kappa(Y)$ uniquely once the coefficient of the term of highest weight is given. From condition **(iii)** it follows that for $\kappa = (k)$, the coefficient of the term of highest weight in $C_{(k)}(\mathbf{Y})$ is unity; i.e., $c_{(k),(k)} = 1$. This determines all the other coefficients $c_{(k), \lambda}$ in the expansion (4) of $C_{(k)}(\mathbf{Y})$ in terms of monomial symmetric functions. These determine, in turn, the coefficient of the term of highest weight in $C_{(k-1, 1)}(\mathbf{Y})$, and once this is known, the recurrence relation gives all the other coefficients, and so on.

The zonal polynomials of orders $k = 2, 3, 4$ are given in Table 1. They have been tabulated to $k = 12$ by Parkhurst and James [25]. For larger values of k, tabulation is prohibitive in terms of space; indeed for $k = 12$, there are 77 zonal polynomials corresponding to the 77 partitions of 12. The algorithm described above forms the basis of a subroutine for calculating zonal polynomials due to McLaren [21]. An alternative method for calculation is due to Kates [17], who expressed them as sums of products of mo-

Table 1.   Coefficients of Monomial Symmetric Functions $M_\lambda(Y)$ in the Zonal Polynomial $C_\kappa(Y)$

		$k = 2$	
		λ	
		(2)	(1,1)
κ	(2)	1	2/3
	(1,1)	0	4/3

		$k = 3$		
		λ		
		(3)	(2,1)	(1,1,1)
	(3)	1	3/5	2/5
κ	(2,1)	0	12/5	18/5
	(1,1,1)	0	0	2

		$k = 4$				
		λ				
		(4)	(3,1)	(2,2)	(2,1,1)	(1,1,1,1)
	(4)	1	4/7	18/35	12/35	8/35
	(3,1)	0	24/7	16/7	88/21	32/7
κ	(2,2)	0	0	16/5	32/15	16/5
	(2,1,1)	0	0	0	16/3	64/5
	(1,1,1,1)	0	0	0	0	16/5

ments of independent normal random variables.

As stated earlier, there is no known general formula for zonal polynomials. Expressions are known for some special cases; see James [13, 14]. One of these special cases is when $Y = I_m$, in which case

$$C_\kappa(I_m) = 2^{2k} k! \left(\tfrac{1}{2} m\right)_\kappa$$

$$\times \frac{\prod_{i<j}^{p}(2k_i - 2k_j - i + j)}{\prod_{i=1}^{p}(2k_i + p - i)!},$$

where p is the number of nonzero parts of the partition κ and

$$(a)_\kappa = \prod_{i=1}^{m}\left(a - \tfrac{1}{2}(i - 1)\right)_{k_i}, \qquad (6)$$

with

$$(a)_k = a(a + 1)\ldots(a + k - 1),$$

$$(a)_0 = 1.$$

Zonal polynomials have so far been defined only for symmetric matrices. The definition can be extended; if Y is symmetric and X is positive definite, then the latent roots of XY are the same as those of $X^{1/2}YX^{1/2}$ and we define $C_\kappa(XY)$ as

$$C_\kappa(XY) = C_\kappa(X^{1/2}YX^{1/2}).$$

Zonal polynomials of Hermitian matrices may also be defined and occur in distributions of random matrices and their latent roots derived from samples from complex multivariate normal distributions. Useful references for such work are James [13] and Krishnaiah [20].

SOME INTEGRALS INVOLVING ZONAL POLYNOMIALS

The evaluation of various integrals involving zonal polynomials is needed in multivariate distribution theory. Some of the more important of these are now listed.

One of the most fundamental properties is due to James [13] and involves averaging over the group $O(m)$ of orthogonal $m \times m$ matrices. The result is

$$\int_{O(m)} C_\kappa(\mathbf{XHYH}')(d\mathbf{H}) = \frac{C_\kappa(\mathbf{X})C_\kappa(\mathbf{Y})}{C_\kappa(\mathbf{I}_m)},$$

$$(7)$$

where $(d\mathbf{H})$ is the invariant Haar* measure on $O(m)$ normalized so that the volume of $O(m)$ is unity.

Three more important integrals are due to Constantine [3], [4] and Khatri [18]. Writing etr($\mathbf{A}$) for exp(tr $\mathbf{A}$), these are

$$\int_{\mathbf{X}>0} \text{etr}(-\mathbf{XZ})(\det \mathbf{X})^{a-(m+1)/2}C_\kappa(\mathbf{XY})(d\mathbf{X})$$

$$= (a)_\kappa \Gamma_m(a)(\det \mathbf{Z})^{-a} C_\kappa(\mathbf{YZ}^{-1}), \quad (8)$$

$$\int_{\mathbf{X}>0} \text{etr}(-\mathbf{XZ})(\det \mathbf{X})^{a-(m+1)/2}C_\kappa(\mathbf{X}^{-1}\mathbf{Y})(d\mathbf{X})$$

$$= \frac{(-1)^k \Gamma_m(a)}{\left(-a + \frac{1}{2}(m+1)\right)_\kappa}(\det \mathbf{Z})^{-a} C_\kappa(\mathbf{YZ}),$$

$$(9)$$

and

$$\int_{0<\mathbf{X}<\mathbf{I}_m} (\det \mathbf{X})^{a-(m+1)/2}$$

$$\times \det(\mathbf{I}_m - \mathbf{X})^{b-(m+1)/2}C_\kappa(\mathbf{XY})(d\mathbf{X})$$

$$= \frac{(a)_\kappa}{(a+b)_\kappa}\frac{\Gamma_m(a)\Gamma_m(b)}{\Gamma_m(a+b)}C_\kappa(\mathbf{Y}). \quad (10)$$

In these integrals

$$\Gamma_m(a) = \pi^{m(m-1)/4}\prod_{j=1}^{m}\Gamma\left(a - \tfrac{1}{2}(i-1)\right)$$

$$(11)$$

(sometimes called the "multivariate gamma function"), and $(a)_\kappa$ is given by (6). In (8) and (9), $\mathbf{Z}$ is a complex symmetric $m \times m$ matrix with Re($\mathbf{Z}$) > 0, $\mathbf{Y}$ is a symmetric $m \times m$ matrix and the integrations are over the space of all positive definite $m \times m$ matrices $\mathbf{X}$; (8) is valid for Re(a) > $\frac{1}{2}(m - 1)$ and (9) is valid for Re(a) > $k_1 + \frac{1}{2}(m - 1)$, where $\kappa = (k_1, k_2, \ldots, k_m)$. In (10), $\mathbf{Y}$ is a symmetric $m \times m$ matrix, the integration

is over all positive definite matrices $\mathbf{X}$ with $\mathbf{I} - \mathbf{X}$ positive definite, and Re(a) > $\frac{1}{2}(m - 1)$ and Re(b) > $\frac{1}{2}(m - 1)$.

The integrals (8) and (10) show that a zonal polynomial has a reproductive property under expectations taken with respect to the Wishart and matrix-variate beta distributions*. Specifically, if $\mathbf{A}$ has the $W_m(n, \Sigma)$ distribution (Wishart* with n degrees of freedom and covariance matrix Σ) with $n > m - 1$ and B is an arbitrary symmetric (nonrandom) $m \times m$ matrix, then (8) shows that

$$E[C_\kappa(\mathbf{AB})] = 2^k\left(\tfrac{1}{2}n\right)_\kappa C_\kappa(\mathbf{B}\Sigma).$$

If $\mathbf{A}$ has a beta (a, b) distribution with density function proportional to

$$(\det \mathbf{A})^{a-(m+1)/2}\det(\mathbf{I}_m - \mathbf{A})^{b-(m+1)/2},$$

$$0 < \mathbf{A} < \mathbf{I}_m,$$

and $\mathbf{B}$ is a fixed $m \times m$ symmetric matrix, (9) shows that

$$E[C_\kappa(\mathbf{AB})] = \frac{(a)_\kappa}{(a+b)_\kappa}C_\kappa(\mathbf{B}).$$

Many other integrals involving zonal polynomials are known. For an excellent survey of these the interested reader is referred to Subrahmaniam [28] and the extensive list of references there.

SOME OTHER PROPERTIES OF ZONAL POLYNOMIALS

In this section, we list some miscellaneous properties of zonal polynomials.

(i) If $\mathbf{Y}$ is a positive definite $m \times m$ matrix with latent roots $y_1, \ldots, y_m$, then $C_\kappa(\mathbf{Y}) > 0$ and $C_\kappa(\mathbf{Y})$ is increasing in each y_i, $i = 1, \ldots, m$ (James [14] and Chattopadhyay and Pillai [2]).

(ii) If $\mathbf{Y}$ is a positive definite $m \times m$ matrix, then

$$C_\kappa(\mathbf{Y})C_\kappa(\mathbf{Y}^{-1}) \geq C_\kappa(\mathbf{I})^2.$$

(iii) A generalization of the binomial expansion, due to Constantine [4], is

$$\frac{C_\kappa(\mathbf{I}_m + \mathbf{Y})}{C_\kappa(\mathbf{I}_m)} = \sum_{s=0}^{k}\sum_{\sigma}\binom{\kappa}{\sigma}\frac{C_\sigma(\mathbf{Y})}{C_\sigma(\mathbf{I}_m)},$$

where the inner summation is over all partitions σ of the integer s. This defines the generalized binomial coefficients $\binom{\kappa}{\sigma}$, which have been tabulated to $k = 4$ by Constantine [4] and to $k = 8$ by Pillai and Jouris [26]. For further results on these coefficients, see Bingham [1] and Muirhead [23].

(iv) It is sometimes useful to express a product of two zonal polynomials in terms of other zonal polynomials, namely,

$$C_\sigma(\mathbf{Y})C_\tau(\mathbf{Y}) = \sum_k g_{\sigma\tau}^\kappa C_\kappa(\mathbf{Y}),$$

where σ is a partition of s, τ is a partition of t, κ runs over all partitions of $k = s + t$, and the $g_{\sigma\tau}^\kappa$ are constants, some of which have been tabulated by Khatri and Pillai [19].

(v) A second-order partial differential equation satisfied by $C_\kappa(\mathbf{Y})$ has already been given in (3). A fourth-order partial differential equation has also been established by Sugiura [30].

(vi) A number of sums involving zonal polynomials have been evaluated by Sugiura and Fujikoshi [31], Fujikoshi [7], Muirhead [22], and Sugiura [29]. We now list some of these, in which $\mathrm{etr}(\mathbf{Y}) \equiv \exp \mathrm{tr}\, \mathbf{Y}$, $s_i = \mathrm{tr}\, \mathbf{Y}^i$, and the constants ρ_κ and π_κ are given by

$$\rho_\kappa = \sum_{i=1}^m k_i(k_i - i),$$

and

$$\pi_\kappa = \sum_{i=1}^m k_i(4k_i^2 - 6ik_i + 3i^2),$$

$$\sum_{k=0}^\infty \sum_\kappa \frac{C_\kappa(\mathbf{Y})}{(k-l)!} = s_1^l \mathrm{etr}(\mathbf{Y}),$$

$$\sum_{k=0}^\infty \sum_\kappa \frac{\rho_\kappa C_\kappa(\mathbf{Y})}{k!} = s_2 \mathrm{etr}(\mathbf{Y}),$$

$$\sum_{k=0}^\infty \sum_\kappa \frac{\rho_\kappa^2 C_\kappa(\mathbf{Y})}{k!}$$

$$= \left(s_2^2 + 4s_3 + s_1^2 + s_2\right)\mathrm{etr}(\mathbf{Y}),$$

$$\sum_{k=0}^\infty \sum_\kappa \frac{k\rho_\kappa C_\kappa(\mathbf{Y})}{k!} = (s_1 s_2 + 2s_2)\mathrm{etr}(\mathbf{Y}),$$

$$\sum_{k=0}^\infty \sum_\kappa \frac{\pi_\kappa C_\kappa(\mathbf{Y})}{k!}$$

$$= \left(4s_3 + 3s_1^2 + 3s_2 + s_1\right)\mathrm{etr}(\mathbf{Y}),$$

$$\sum_\kappa \rho_\kappa C_\kappa(\mathbf{Y}) = k(k-1)s_2 s_1^{k-2},$$

$$\sum_\kappa \rho_\kappa^2 C_\kappa(\mathbf{Y}) = k(k-1)\left[\left(s_2 + s_1^2\right)s_1^{k-2}\right.$$

$$\left. + 4(k-2)s_3 s_1^{k-3} + (k-2)(k-3)s_2^2 s_1^{k-4}\right],$$

$$\sum_\kappa \pi_\kappa C_\kappa(\mathbf{Y})$$

$$= k\left[s_1^k + 3(k-1)\left(s_2 + s_1^2\right)s_1^{k-2}\right.$$

$$\left. + 4(k-1)(k-2)(k-3)s_3 s_1^{k-3}\right].$$

(vii) The effects of a number of differential operators on $C_\kappa(\mathbf{Y})$ have been studied, primarily to establish partial differential equations for hypergeometric functions of matrix argument. For work in this direction, see Muirhead [22], Constantine and Muirhead [5], James [14, 16], Fujikoshi [8], and the survey paper of Subrahmaniam [28].

References

[1] Bingham, C. (1974). An identity involving partitional generalized binomial coefficients. *J. Multivariate Anal.*, **4**, 210–223.

[2] Chattopadhyay, A. K. and Pillai, K. C. S. (1970). On the Maximization of a Matrix Function over the Group of Orthogonal Matrices. *Mimeograph Series No. 248*, Statistics Dept., Purdue University, Lafayette, IN.

[3] Constantine, A. G. (1963). Some noncentral distribution problems in multivariate analysis. *Ann. Math. Statist.*, **34**, 1270–1285.

[4] Constantine, A. G. (1966). The distribution of Hotelling's generalized T_0^2. *Ann. Math. Statist.*, **37**, 215–225.

[5] Constantine, A. G. and Muirhead, R. J. (1972). Partial differential equations for hypergeometric functions of two argument matrices. *J. Multivariate Anal.*, **3**, 332–338.

[6] Farrell, R. H. (1976), *Techniques of Multivariate Calculation*. Springer, New York.

[7] Fujikoshi, Y. (1970). Asymptotic expansions of the distributions of test statistics in multivariate

analysis. *J. Sci. Hiroshima Univ. Ser. A-I*, **34**, 73–144.

[8] Fujikoshi, Y. (1975). Partial differential equations for hypergeometric functions $_3F_2$ of matrix argument. *Canad. J. Statist.*, **3**, 153–163.

[9] Hua, L. K. (1959). *Harmonic Analysis of Functions of Several Complex Variables in the Classical Domains.* Moscow (in Russian).

[10] James, A. T. (1960). The distributions of the latent roots of the covariance matrix. *Ann. Math. Statist.*, **31**, 151–158.

[11] James, A. T. (1961a). The distributions of noncentral means with known covariance. *Ann. Math. Statist.*, **32**, 874–882.

[12] James, A. T. (1961b). Zonal polynomials of the real positive definite symmetric matrices. *Ann. Math.*, **74**, 456–469.

[13] James, A. T. (1964). Distributions of matrix variates and latent roots derived from normal samples. *Ann. Math. Statist.*, **35**, 475–501.

[14] James, A. T. (1968). Calculation of zonal polynomial coefficients by use of the Laplace–Beltrami operator. *Ann. Math. Statist.*, **39**, 1711–1718.

[15] James, A. T. (1973). The variance information manifold and the functions on it. In *Multivariate Analysis III*, P. R. Krishnaiah, ed. Academic, New York, pp. 157–159.

[16] James, A. T. (1976). Special functions of matrix and single argument in statistics. In *Theory and Applications of Special Functions*, R. A. Askey, ed. Academic, New York, pp. 497–520.

[17] Kates, L. K. (1980). Zonal Polynomials. Ph.D. Thesis, Princeton University, Princeton, NJ.

[18] Khatri, C. G. (1966). On certain distribution problems based on positive definite quadratic functions in normal vectors. *Ann. Math. Statist.*, **37**, 468–479.

[19] Khatri, C. G. and Pillai, K. C. S. (1968). On the noncentral distributions of two test criteria in multivariate analysis of variance. *Ann. Math. Statist.*, **39**, 215–226.

[20] Krishnaiah, P. R. (1976). Some recent developments on complex multivariate distributions. *J. Multivariate Anal.*, **6**, 1–30.

[21] McLaren, M. L. (1976). Coefficients of the zonal polynomials. *Appl. Statist.*, **25**, 82–87.

[22] Muirhead, R. J. (1970). Partial differential equations for hypergeometric functions of matrix argument. *Ann. Math. Statist.*, **41**, 991–1001.

[23] Muirhead, R. J. (1974). On the calculation of generalized binomial coefficients. *J. Multivariate Anal.*, **4**, 341–346.

[24] Muirhead, R. J. (1982). *Aspects of Multivariate Statistical Theory.* Wiley, New York.

[25] Parkhurst, A. M. and James, A. T. (1974). Zonal polynomials of order 1 through 12. In *Selected Tables in Mathematical Statistics*, H. L. Harter and D. B. Owen, eds. American Mathematical Society, Providence, RI, pp. 199–388.

[26] Pillai, K. C. S. and Jouris, G. M. (1969). On the moments of elementary symmetric functions of the roots of two matrices. *Ann. Inst. Statist. Math.*, **21**, 309–320.

[27] Saw, J. G. (1977). Zonal polynomials: An alternative approach. *J. Multivariate Anal.*, **7**, 461–467.

[28] Subrahmaniam, K. (1976). Recent trends in multivariate normal distribution theory: On the zonal polynomials and other functions of matrix argument. *Sankhyā A*, **38**, 221–258.

[29] Sugiura, N. (1971). Note on some formulas for weighted sums of zonal polynomials. *Ann. Math. Statist.*, **42**, 768–772.

[30] Sugiura, N. (1973). Derivatives of the characteristic root of a symmetric or a Hermitian matrix with two applications in multivariate analysis. *Commun. Statist.*, **1**, 393–417.

[31] Sugiura, N. and Fujikoshi, Y. (1969). Asymptotic expansions of the non-null distributions of the likelihood ratio criteria for multivariate linear hypothesis and independence. *Ann. Math. Statist.*, **40**, 942–952.

[32] Takemura, A. (1984). *Zonal Polynomials, Lecture Notes–Monograph Series 4*, Institute of Mathematical Statistics, Hayward, CA.

(LATENT ROOT DISTRIBUTIONS
MATRIX-VALUED DISTRIBUTIONS
MATRIX-VARIATE BETA DISTRIBUTION
MULTIVARIATE ANALYSIS
SYMMETRIC FUNCTIONS
WISHART DISTRIBUTION)

ROBB J. MUIRHEAD

ZONE CHART *See* STRATA CHART

Z-TEST OF NORMALITY

Lin and Mudholkar [1] proposed a test of normality based on a random sample of size n, giving sample values $x_1, x_2, \ldots, x_n$ of a variable X.

The statistics

$$y_i = n^{-1}\left[\sum_{\alpha \neq i} x_\alpha^2 - \left(\sum_{\alpha \neq i} x_\alpha\right)^2 / (n-1)\right]^{1/3},$$

$i = 1, \ldots, n$, are calculated, and the product

moment correlation r between x_i and y_i computed, using the n pairs of values (x_i, y_i). Then if X has a normal distribution,

$$Z = \frac{1}{2} \log \left(\frac{1+r}{1-r} \right)$$

has approximately a normal distribution with mean zero and variance $3n^{-1}$ (see FISHER'S z-TRANSFORMATION). A better approximation to the upper $100\alpha\%$ point of the distribution of Z is

$$Z_{1-\alpha} \doteq u_{1-\alpha} + \frac{1}{24} \left(u_{1-\alpha}^3 - 3u_{1-\alpha} \right) \gamma_{2n},$$

where $\sigma_n^2 = 3n^{-1} - 7.32n^{-2} + 53.005n^{-3}$, $\gamma_n = -11.70n^{-1} + 55.06n^{-2}$, and $u_{1-\alpha}$ is the upper $100\alpha\%$ point of the *standard* normal distribution. This approximation is adequate for n as small as 5. Large values of $|Z|$ are regarded as significant.

The power of the test is comparable (although less) than that of the Shapiro–Wilk* test. The Z-test was constructed primarily for detecting asymmetric alternatives to normality but it compares quite well with tests for departure from normality* against long-tailed symmetrical alternatives.

Reference

[1] Lin, C. C. and Mudholkar, G. S. (1980). *Biometrika*, **67**, 455–461.

(DEPARTURES FROM NORMALITY, TESTS FOR)

Z-TRANSFORMATION *See* FISHER'S Z-DISTRIBUTION

Z'-TRANSFORMATION *See* CORRELATION; FISHER'S Z-DISTRIBUTION

ZYSKIND–MARTIN MODELS

Consider general linear models* of the form

$$\mathbf{Y} = \mathbf{X}\boldsymbol{\beta} + \mathbf{Z}.$$

$\mathbf{Y}$ is a $n \times 1$ vector of observed random variables; $\mathbf{X}$ is a $n \times p$ matrix of rank r, $r \leqslant p \leqslant n$, of concomitant variables* (with known values); $\boldsymbol{\beta}$ is a $p \times 1$ vector of regression coefficients* (parameters with unknown values); and $\mathbf{Z}$ is a $n \times 1$ vector of random residuals with

$$E[\mathbf{Z}] = \mathbf{0} \quad \text{and} \quad \text{Var}(\mathbf{Z}) = E[\mathbf{Z}\mathbf{Z}'] = \sigma^2 \mathbf{V},$$

where $\mathbf{V}$ is known.

Zyskind–Martin models [4] have the additional property that the space spanned by columns of $\mathbf{X}$ is contained in the space spanned by columns of $\mathbf{V}$. These models have the property that the best linear unbiased estimator of *any* estimable linear function $\mathbf{a}'\boldsymbol{\beta}$ of the regression coefficients is the same linear function $\mathbf{a}'\boldsymbol{\beta}$ of the solution $\tilde{\boldsymbol{\beta}}$ of the general normal equations*

$$\mathbf{X}'\mathbf{V}^*\mathbf{X}\tilde{\boldsymbol{\beta}} = \mathbf{X}'\mathbf{V}^*\mathbf{Y},$$

where $\mathbf{V}^*$ is *any* conditional inverse of $\mathbf{V}$.

The name *Zyskind–Martin* models was coined by Oktaba [1], who studied some special cases. Estimation of $\tilde{\boldsymbol{\beta}}$ in such models with missing values is discussed in Oktaba et al. [2]. There are some interesting historical remarks in [3].

References

[1] Oktaba, W. (1984). *Biomed. J.*, **26**, 415–424.

[2] Oktaba, W., Kornacki, A., and Wawrzosek, J. (1985). *Biomed. J.*, **27**, 733–740.

[3] Shelton, J. T. and Lowe, V. W. (1986). *Amer. Statist.*, **40**, 78–79.

[4] Zyskind, G. and Martin, F. B. (1969). *SIAM J. Appl. Math.*, **17**, 1190–1202.

(ESTIMABILITY
GENERAL LINEAR MODEL
GENERALIZED INVERSE)

CUMULATIVE INDEX, VOLUMES 1–9